Energy and protein metabolism and nutrition

EAAP – European Federation of Animal Science

UNIVERSITÀ
DEGLI STUDI
DI MILANO

University of Milan

DIPARTIMENTO DI SCIENZE ANIMALI

Department of Animal Science

The European Association for Animal Production wishes to express its appreciation to the *Ministero per le Politiche Agricole e Forestali* and the *Associazione Italiana Allevatori* for their valuable support of its activities

Energy and protein metabolism and nutrition

3rd EAAP International Symposium on Energy and Protein Metabolism and Nutrition
Parma, Italy
6-10 September 2010

EAAP publication No. 127

edited by:

G. Matteo Crovetto

Wageningen Academic
Publishers

ISBN: 978-90-8686-153-8
e-ISBN: 978-90-8686-709-7
DOI: 10.3920/978-90-8686-709-7

ISSN 0071-2477

First published, 2010

Wageningen Academic Publishers
P.O. Box 220
6700 AE Wageningen
The Netherlands
www.WageningenAcademic.com
copyright@WageningenAcademic.com

The 3rd EAAP International Symposium on Energy and Protein Metabolism and Nutrition was organized in Parma (Italy) in the period 6-10th September 2010, by the Department of Animal Science of the Università degli Studi di Milano (Italy) with the logistic support of M.V. Congressi s.p.a., Parma (Italy).

The scientific programme of the Symposium was organized by the Scientific Secretariat of ISEP 2010:
- G. Matteo Crovetto
- Luca Malagutti
- Gianluca Galassi
- Stefania Colombini
- Luca Rapetti

In interaction with the International Scientific Committee:
- Jan Erik Lindberg (President, Sweden)
- Michael Kreuzer (Secretary, Switzerland)
- G. Matteo Crovetto (Italy)
- Hélène Lapierre (Canada)
- Cornelia Metges (Germany)
- James W. Oltjen (USA)
- Isabelle Ortigues Marty (France)
- Masaaki Toyomizu (Japan)

and liaising with the National Scientific Committee:
- Mauro Antongiovanni (Firenze)
- Lucia Bailoni (Padova)
- Alberto Battezzati (Milano)
- Giuseppe Bertoni (Piacenza
- Antonello Cannas (Sassari)
- G. Matteo Crovetto (Milano)
- Andrea Formigoni (Bologna)
- Vincenzo Piccolo (Napoli)
- Gianfranco Piva (Piacenza)
- Afro Quarantelli (Parma)
- Luca Rapetti (Milano)
- Giovanni Savoini (Milano)
- Mauro Spanghero (Udine)

The organizers of the symposium are particularly indebted to the chairpersons of the sessions of ISEP 2010, namely:

1. Energy/protein "omics" and mitochondrial functions
   Anne Collin (France)
2. Stable isotopes and other techniques in metabolism and nutrition
   Cornelia Metges (Germany)
3. Energy/protein tissue metabolism
   Isabelle Ortigues Marty (France)
4. Systemic and local regulation mechanisms
   Hélène Lapierre (Canada)
5. Energy/protein and their interaction on productive functions: ruminants
   Michael Kreuzer (Switzerland)
6. Energy/protein and their interaction on productive functions: monogastrics
   Niels B. Kristensen (Denmark)
7. Energy/protein nutrition and environmental sustainability
   Adrian van Vuuren (The Netherlands)
8. Energy/protein metabolism and nutrition, and health in farm and companion animals
   Giuseppe Bertoni (Italy)
9. Evaluation and modeling of feed value and requirements: ruminants
   James W. Oltjen (USA)
10. Evaluation and modeling of feed value and requirements: monogastrics
    Jaap van Milgen (France)
11. Influence of energy/protein metabolism and nutrition on product quality
    Gianfranco Piva (Italy)
12. *In vitro* techniques and feed nutritive value
    Sergio Calsamiglia (Spain)

Table of contents

Part 2. Stable isotopes and other techniques in metabolism and nutrition

Part 7. Energy/protein nutrition and environmental sustainability

Part 11. Influence of energy/protein metabolism and nutrition on product quality

Preface

Continuing the strategy followed by the previous International Symposium on Energy and Protein Metabolism and Nutrition (ISEP) held in Vichy-France, 2007, the 3rd ISEP focuses on the effort of combining fundamental research with applied research and practical applications. The symposium, as has traditionally been the habit for each ISEP meeting and for the previous separate Energy and Protein symposia for many decades, is based on a workshop spirit, with interaction and collaboration among scientists from different countries and an exchange of expertise and knowledge. Despite the present worldwide economic crisis, the participation of scientists in the symposium is quite satisfactory: 222 papers from 32 countries have been accepted and published. Twenty-six of them are main papers from invited speakers and deserved much more space in terms of number of pages.

Besides many papers dealing with practical aspects of energy and protein nutrition and feeding, several papers presented in the symposium discuss topics related to tissue metabolism and regulation mechanisms. The framework of the 3rd ISEP is therefore very comprehensive and deals with many aspects of animal nutrition and metabolism.

Energy and protein metabolism and nutrition cannot be addressed separately and a better and deeper understanding of nutrient metabolism and nutrition can be achieved only by integrating the outcomes of scientists researching different aspects of this topic. Nutrigenomics is an example of the possibilities of improvement in knowledge and applications that can arise from the synergism between different lines of research. The present situation all over the world is critical economically and in terms of food security, food safety and of environmental impact. Fundamental research is certainly needed but it must also be translated into practical outcomes for the direct benefit of producers and consumers.

Placing a link between energy/protein metabolism and nutrition and food quality is one of the goals of the present symposium and one session is specifically dedicated to this important topic. A second session combines nutrition with environment, considering the attention that pollution and environmental sustainability luckily receive nowadays in an increasing number of countries all over the world. A third session deals with the new possibilities provided by the rapidly growing knowledge on nutrigenomics. Molecular genetics has recently boosted knowledge on many nutritional factors capable of positively influencing animal productions.

Another field of increasing interest for public opinion nowadays is animal welfare. Quite some research deals with the interaction between nutrition and animal health on one side, and the possibility to replace *in vivo* with *in vitro* experiments on the other side. The databases available today are impressive and the possibility to share and get information worldwide makes it easier to develop models to predict both animal requirements and feed nutritive values. Indeed modeling is extensively dealt with in the symposium and deserves two sessions, one dedicated to ruminants and the other to monogastrics.

On behalf of the Scientific Secretariat I want to acknowledge the generous and precious help of the National and International Scientific Committees and of the chairpersons of this symposium. Let me express my gratitude to my colleagues of the scientific secretariat who helped me in many ways. I feel honoured that this scientifically important symposium is held in Italy for the first time. Organizing such an important scientific event is really a challenge and luckily I did not know exactly what I was going to face… As it often happens I learnt a lot and today I would surely do it better, but… that's life! A final thanks to all the contributing authors, to the invited speakers who further increase the qualitative level of the event, and to the participants of the symposium: I feel confident that they will certainly be able to match the scientific interest with the righteous desire, for many of them, of at least a quick, first contact with the Italian culture under its more extensive and comprehensive meaning.

G. Matteo Crovetto

Part 1. Energy/protein omics and mitochondrial functions

Integrating transcriptomic regulation into models of nutrient metabolism in agricultural animals

John P. McNamara
Department of Animal Sciences, Washington State University, Pullman, WA 99164-6351, USA

Abstract

The use of transcriptomic and other technologies has become a mainstay of biological research in the last few years, and this is good. However, there is grave danger that as we become enamored of a technique, we can lose the overall purpose of our research. Although transcriptomic arrays can give much data that is as yet not fully understood, it can also provide at a low cost, tremendous insight into control of metabolism. As animal scientists, our role is to help provide high quality protein and other food for humans. To do this, we must continue to help our farmers be as efficient as possible in their respective environments. The variation among animals, even in one herd and fed the same is still tremendous. Why? How can we use the study of transcriptional regulation to improve efficiency of animal production? We can do it in an ordered systems biology approach that focuses on why and how cells regulate energy and N use, and study this within practical situations applicable on farms. Using existing metabolic models we can design experiments specifically to integrate new data from transcriptional arrays into models that describe nutrient use in farm animals. This approach can focus our research to make faster and large advances in efficiency, and show directly how this can be applied on the farms.

Introduction

With this paper I hope to provide some background and philosophy on systems biology in agricultural research and application, including the use of transcriptomic methodologies; to provide a few specific examples of systems biology work to improve knowledge of energy and protein metabolism; and most importantly, to stimulate a major increase in coordinated, systems based research and analysis in the animal agriculture community worldwide.

Many readers of this paper have extensively more technical expertise than I in the methods used for gene transcription analysis. There are many examples of excellent work worldwide. But what continues to be expressly lacking in a number of efforts is a clear objective based in systems biology and systems research. Such an objective might include critical pre-study reasoning as to *what* should be studied, *why* it should be studied, *why* the system indeed works the way it does (two clearly different questions); and *how* the work, when completed, will contribute to our deep understanding of the system. In such cases the system under study might be defined as the liver, or the dairy cow, or the national poultry flock. Such a limit in itself often underlies a shallow understanding of the real system being studied. Quite often, and for a variety of significant reasons, much energy and protein research, including that with inclusion of transcriptomic methodologies, continues to be 'more of the same', piecemeal studies with a limited objective and little deep reasoning, and an almost complete lack of bio-mathematical reasoning as to how the work will provide quantifiable improvement in understanding of the animal system.

I will borrow heavily on the work of others and acknowledge it lavishly. This is because it is this scientist's opinion that the excellent work of many scientists in the past, and several active today, in providing a framework and purpose for animal science systems research has been forgotten or ignored in the expediency of securing research funding (which is almost never sufficient to truly do a meaningful study); of training graduate students in a limited time frame with limited resources; and the need to publish early and often. Certainly many students graduating today, for no fault of their own, do not have the deep history necessary to understand much of the context of research done today and to critically judge its worth in true expansion of understanding. Many students might be able to provide a definition of 'systems research' or 'systems biology'; but how many would know that in fact this is not a new approach, but dates back decades? Many students could tell you that Watson and Crick were credited with the discovery of the double helix structure; but how many could form a cogent explanation of the initial reasoning and impetus for work in what has come to be called 'molecular biology'?

I hope the history and modern examples I supply encourage people to think about it, pick up the phone, send a message, to begin coordinated national and international efforts to adapt a true systems approach to improving animal agriculture. Such an approach does not mean individuals can no longer work on their 'pet theories' or subjects of personal interest. Rather, it provides more opportunities for such works that would be completely embedded in a focused approach to describe in entirety the 'system' of a dairy cow, or of a beef cow herd, or of a national effort to increase quality protein intake at a reasonable cost to the environment and the consumer. When one considers all the elements that go into the 'system' of animal agriculture locally or worldwide, really considers them, the older ones among us might reflect on all the potential knowledge lost in the piecemeal approach up to now. But hopefully, the younger ones can see the great opportunities ahead to use advanced technologies to solve what in effect is actually a quite simple problem: how to feed everyone in the world a highly nutritious diet including animal products? Our call and challenge as animal scientists is no less than this--how do we ensure a safe and sustainable high quality protein food supply worldwide for future generations?

Systems biology

The new technologies and massive amounts of information stemming from the genomic studies of the last decade have spawned renewed calls for a 'systems approach', which from one perspective can be translated as 'OK, so now that we have all this information from the genome projects, and most people working on those projects never knew or forgot why they were started, what do we do now?' The field of genomics and transcriptomics has already provided a wealth of data, and some knowledge, but many would argue that the former heavily outweighs the latter. For purposes of this paper, I will refer to transcriptomics in a technical way, that of measuring transcripts of mRNA in large arrays of several hundred or thousands of genes (we will not go into any actual technical descriptions, those can be found in many places). There can be other definitions, but this is one germane to our purposes. Given that, where do transcriptomics fit in the system of research in control of animal production?

I think a series of quotes from Cornish-Bowden (2005) helps put 'systems biology' in perspective:

> 'The idea of systems biology is not new: as long ago as 1968, the mathematician and engineer Mihajlo Mesarovic regretted that 'in spite of considerable interest and efforts, the application of systems theory in biology has not quite lived up to expectation'. But what of systems biology today? Does it now look more likely to lead to the expected benefits?'
>
> 'In the 1950s the geneticist and biochemist Henrik Kacser was already urging biologists to take systems seriously: "The problem is … the investigation of systems, i.e. components related or organized in a specific way. The properties of a system are in fact 'more' than (or different from) the sum of the properties of its components, a fact often overlooked in zealous attempts to demonstrate 'additivity' of certain phenomenon. It is with these 'systemic properties' that we shall be mainly concerned..." '
>
> 'In attempting to define systems biology, Olaf Wolkenhauer (University of Rostock, Germany) emphasized the need for a shift in focus away from molecular characterization towards understanding functional activity.'

Dr. Cornish-Bowden used these quotes to help make the point that systems research is not new, and in many cases the true functionality of animal systems has been forgotten or never learned. In animal agriculture, for good reason, we have been focused on 'production' and 'output' in easily measured empirical systems such as body weight, milk production and egg laying. Although much excellent basic biology has been done, very little has been done on the actual control-level of 'the system' namely, gene transcription, mRNA synthesis and enzyme synthesis and degradation. He also makes a very strong point that transcriptomics or any other such activity outside of a clearly stated systems approach misses most of the point. Data are still gathered, but much potential knowledge is missed. It is my opinion however that as long as the data are there (in various repositories), they can be 'mined' by those that have 'matching information' on other levels of the system, to come up with a richer understanding of the links between genetic and environmental control.

Those of us that have done 'basic research' have tended to look at enzyme activity, pathway activity, hormone and receptor concentrations and the like. That is fine, and we have learned from that, but in many cases we did not consider the underlying controls (transcription rates, enzyme synthesis rates) or did not fully relate the cellular information to the animal production level in any systematic mathematical formalism. The former is difficult to do, expensive, and in many ways not necessary to our purposes in animal agriculture. The latter is easy to do, inexpensive and in fact an absolute requirement for our purposes: what are the true biological controls, at the level at which 'control' is exerted, that drive animal production. A description of metabolic control theory and control coefficients is beyond the purpose of this article, but readers are at least encouraged to read some of Kacser, Carson and Cobelli and Cornish-Bowden to understand this (Cornish-Bowden, 2005). I will go into more detail with references on multiple regressions to study the relationship of basic metabolic control, transcriptomics and animal production below.

Purpose of animal agricultural research

With some brief philosophy and background on transcriptomics and systems biology, let us back up to define the critical 'Why' as asked above.

From Baldwin (1995):

> 'There is general agreement among most informed authors that products of animal agriculture will continue to contribute to the world food supply. However, the key challenge of ascertaining how much animals should contribute has not been resolved.'
> 'Our inability to undertake quantitative evaluations of impacts of competing human nutrition strategies on human food availability is due in large part to the fact that current plant and animal production models are normally restricted to single species and have not been interfaced.'
> 'This is a long-term goal that will require the availability of advanced dynamic, mechanistic models of ruminant digestion and metabolism…'

These quotes may seem like obvious statements, but it is interesting the answers one gets to the question of 'Why do we do research in animal agriculture?' either from scientists of long standing to starting graduate students. It might be argued that if this point had been made more strongly and more often in every research and teaching program, a different approach to teaching and research may have evolved.

There already exist two solid and validated frameworks for models of nutrient flux that can provide a basis for a transcriptional control model in cattle and swine. The first is the 40 years of modeling work of Baldwin and colleagues (Baldwin *et al.*, 1987a,b,c; Baldwin, 1995), which has lead to tremendous improvement in understanding of the mechanistic connections between diet and animal performance. The model in question is titled 'Molly' and the full history and detail can be found in the previous reference. The objective of this model was simple: 'To develop a dynamic, mechanistic model of digestion and metabolism in lactating dairy cows suitable for evaluation of hypotheses regarding underlying energetic relationships and patterns of nutrient use.'

In 1968, Dr. Baldwin published an article titled: 'Estimation of theoretical calorific relationships as a teaching technique: a review.' (Baldwin, 1968). In it, he described the aggregate biochemical pathways that in fact were the components of the net energy system of feeding cattle, work which was just wrapping up after about 100 years of effort across the world by many scientists (Lofgreen and Garrett, 1968, NRC, 1968). This connection between the mechanisms of nutrient flux and practical, empirical cattle feeding led to 40 years of work on developing biomathematical models of nutrient use, and 'spun off' many other related efforts. It is this writer's opinion if that all students from then on had read those two papers in one class with a thought to 'where can we really improve our knowledge', we would have a far deeper understanding of agricultural animal systems today.

Stemming from that work came the model of nutrient use in the sow, 'Susie', developed by Pettigrew and colleagues (1992) and since developed and presently being extended to reproduction (McNamara, 2005). This effort began more than 20 years ago, and in 1992, Jim Pettigrew and colleagues gave a start to the first model of nutrition and reproduction in pigs, and a direct quote from that paper is in order (as I cannot say it any better!):

> 'The mechanisms connecting the diet to reproductive performance are presently unknown but may include variations in voluntary feed intake, digestion, absorption, metabolism of absorbed nutrients, and endocrine effects. Clear understanding and manipulation of this connection to optimize long-term sow herd performance requires ability to track, systematically and quantitatively, dietary effects through the various processes to reproductive performance. The project consists of the development of a mathematical model of one component of the connection, the metabolism of absorbed energy-containing nutrients, including amino acids, related to long-term feeding strategies in the lactation phase of the reproductive cycle of sows' (Pettigrew *et al.*, 1992).

These models describe pathway biochemistry, as aggregated pathways in a simple and scientifically correct fashion. There is not an attempt to model every reaction, but to model at the level of biological control most pertinent to the modeling objective. For a thorough discussion of the purposes and practices of metabolic models, see Baldwin (1995). A transcriptomic approach here can have great value in identifying the potential mechanisms involved in control of productive functions and ruling out those not. We will finish with two examples of using systems biology to understand the underlying patterns and control of efficiency and in integrating transcriptomic data into systems models.

Use of a systems approach to study efficiency, including use of transcriptomic technologies

Respiration calorimetry was the bread and butter of energetic and early N efficiency research. By the 1960's, the 'story had been told' in terms of using respiration calorimetry and comparative slaughter to understand underlying biology. Our knowledge had moved past that, and the descriptions of efficiency of DE, ME and NE and N use allowed us to switch our focus to the underlying control mechanisms. For example, in 1968, Baldwin calculated the energy efficiency of milk synthesis at any given composition from the biochemical pathway stoichiometry, almost simultaneously with empirical calculations from respiration calorimetry and energy and N balance studies. Not surprisingly, there was agreement that, assuming the proper amounts and balance of precursors were available, the efficiency of milk synthesis in the mammary gland was about 83% (Baldwin, 1968). That was a constant percentage (variable with variation in milk composition) regardless of the amount of milk. It was thus clear that any increases in efficiency must come from either (1) increasing total milk energy secretion at similar maintenance costs (dilution of maintenance) or (2) improving the efficiency of metabolic functions in organs such as the digestive system, liver, muscle and adipose tissue. It was also known that the efficiency of storing body fat from carbohydrate was about 40%, and from fat 90%. The efficiency of muscle growth was only about 25 to 35%, depending on the stage of maturity and the balance of amino acids provided. This was of course, due to the *normal and required* cost of muscle protein turnover (as Cornish-Bowden noted years later; 2005). Thus if improvements in efficiency were to be made, we needed to understand the underlying functions of the organs of the body.

Recently we used the systems modeling approach to ask the question of 'What patterns of metabolic flux exist in dairy cattle of varying genetic merit and intakes?' Also 'Related to that flux, which genes are changing transcription in the adipose tissue?' This was in direct, if delayed, response to a challenge laid out years earlier by Baldwin (1980): '... when considerable biological variation exists, opportunities for improvement are embedded within the variation…' and: '…observed efficiencies considerably below theoretical are also observed. This raises two important questions: (1) Could we learn to identify animals that are capable of attaining maximum efficiencies and based on genetic selection improve the average efficiency of animal production? (2) If we knew exactly what types of unfortunate metabolic decisions that the less efficient animals were making, could we manipulate the metabolism of those animals such that their efficiencies would approach those of the best animals?' Given that these comments were made in 1980, in retrospect it is clear that many scientists have since then done exactly that (SNP's, QTL's, etc.) but many have not taken on the task of integrating the gene with the metabolism.

Thus, in order to do just that, data were collected from several studies done at WSU, with 1st to 4th parity cows, from 28 d *pre partum* to 120 DIM and included total food intake, nutrient composition of intake, milk and component output, body fat and protein, and transcript levels for several key metabolic control proteins and enzymes expressed in adipose tissue. These cows were all on similar (if not the same diets), from the same herd, spread over several years. The Molly model (Baldwin, 1995; McNamara and Baldwin, 2000); was used to simulate the metabolism of each cow (n=126 from 3 studies) from 0 to 120 DIM. Input variables included daily feed intake and chemical composition, initial body weight, fat and protein content. Outputs included all milk components, and pathway fluxes for lipid and glucose in mammary, body and visceral energy and protein, and changes in body fat and protein. Simulations were then continued until day 305 to predict potential overall efficiency. Body fat, body and visceral protein all varied (P<0.05) in their daily flux, with genetic merit (predicted transmitting ability for milk) and total net energy absorbed being the greatest contributors to variance. Means (ranges) for all cows were 112 (89 to 139) Mcal/d for intake energy, 32.3 (19.9, 41.9) for maintenance; -0.51 (-1.74, -0.015) for change in body energy; and 0.843 (0.826, 0.862) for net energy efficiency (milk energy/ (energy absorbed – maintenance E)). The model predicted response to dietary energy, dietary fiber and dietary protein content within one standard deviation of the observed (P<0.05).

The interesting finding was that variations in maintenance functions (tissue metabolism) affected overall efficiency while mammary efficiency approached the theoretical maxima, as Baldwin predicted 40 years ago (Table 1). Even within a herd of cows quite similar genetically, there was a range of milk productions and feed intakes (as expected) but in fact the variation in metabolic pathways in the adipose, muscle and liver were even more striking). Even within a herd of similar cows on the same diet, use of energy for metabolic functions can vary 100% between animals. Why? There remains significant undefined variation in metabolism that defines the summative energy efficiencies. Studying energy efficiency with a goal of making all cows more efficient must be done in the context of understanding the system where it is controlled; at the pathway level in individual organs.

Table 1. Energetic efficiencies of dairy cattle in early lactation as simulated in Molly from actual data.

Efficiency measure	Milk energy, %, GEI	Milk energy, %, ABSE	Milk energy, % ABSE + BE	Mammary efficiency[1]
Top 20%	26%	43%	44%	84%
Average	23%	38%	38%	84%
Lower 20%	21%	34%	34%	85%
SD	2%	3%	3%	1%

[1] Last measure is milk energy production divided by mammary energy uptake. This is the thermodynamic maximal value.

Similarly to energy use, N use varied as well. Nitrogen intake was 0.66 (0.52, 0.81) kg/d; milk N, 0.21 kg/d (0.16, 0.27), change in body N, -0.016 (-0.06, -0.004), N in urea was 0.31 (0.26, 0.37) and N balance was -0.018 (-0.032, -0.008). Animals varied in non-mammary E and N use, and the model identified ($P<0.05$) differences in E and N in the 20% top versus 20% lowest efficient cows that start a quantitative metabolic control map of efficiency.

So what does this mean in the system of the cow? We can increase efficiency, but we cannot change the laws of thermodynamics. More seriously, we cannot increase efficiency through means that would disrupt the normal cell system. Thus we must pinpoint the critical control mechanisms that vary metabolic rates in the liver, gut tissues, muscle and fat; and ask the questions: Can these efficiencies be changed? And more importantly 'Can they be changed without altering the basic system to the detriment of the animal?' The answer is, of course, yes, because we can identify those animals that are the most efficient utilizers of nutrients and identify their control points. We cannot do this in a piecemeal fashion. Genomics, SNP, and QTL studies alone cannot do this, we must use that information in the context of the animals in their environments, including nutritional environment. This control, connecting of course the SNP and QTL information, to the outputs, includes the control by gene transcription. Approaches to study this have been called nutrigenetics and nutrigenomics.

Nutrigenomics is generally defined as the effect of dietary nutrients on gene transcription: 'Nutrigenomics aims to determine the influence of common dietary ingredients on the genome, and attempts to relate different phenotypes to differences in the cellular and/or genetic response of the biological system'(Mutch *et al.*, 2005). 'Nutrigenetics, on the other hand, aims to understand how the genetic makeup of an individual coordinates their response to diet, and thus considers genetic polymorphisms' (Mutch *et al.*, 2005). The practical application here is to identify the gene variants that relate to differential response to nutrients (Al-Hasani and Joost, 2005).

So finally, transcriptional regulation and transcriptomics

Within the same studies referred to above, we took several samples of adipose tissue from 28 d *pre partum* out to 56, 90 or 180 DIM, depending on the study. Some of this has been published (Sumner and McNamara, 2007) and some presented in various abstracts and talks; this is certainly still a work in progress. Our original interest was in the control of lipolysis. Hormone Sensitive Lipase (HSL) is controlled by phosphorylation, as is the translocation of its colipase Perilipin. Thus, physiological control by the sympathetic nervous system through norepinephrine is a major control point, as is, indirectly, the action of insulin to reduce response to beta-adrenergic stimulation. We had been able to learn a lot about the control of this enzyme, which is of course responsible for providing large amounts of fatty acids for milk fat synthesis as well as energy use by tissues. Over the last 25 years we discovered and confirmed in many ways that activity of HSL and rates of lipolysis are highly related to genetic merit and actual milk production of the cow. Rates of food intake do affect lipolysis, but his is likely through indirect control of anabolic signals. Control of anabolic pathways (lipogenesis), however, is highly related to the actual intakes, as well as affected to a lesser degree by milk production.

Thus, we asked the obvious question: is transcription a major control link in the larger genetic pattern of efficiency, and/or are other metabolic control systems in place. More specifically we asked 'Is the increase in activity of HSL and rates of lipolysis a function of transcription, and if so, by how much?' (As compared to physiological control through enzyme activation). Simultaneously we asked the same question about control of transcription to determine rates of anabolic pathways. It was already known in other species, and more recently in the cow) that in fact anabolic pathways are highly controlled by transcription. For example a drop in glucose absorption signals a cascade of reduction in transcription for control proteins and enzymes that oxidize glucose or convert it to fatty acids.

In 2007, we reported for the first time the level of transcripts for HSL, Perilipin and the B1, B2, and B3 adrenergic receptors. In that study of approximately 20 animals, all of these transcripts increased in amount during lactation, with a peak around 90 DIM, which is when milk production was highest. This indicated a role for increased transcription in control of overall lipolytic activity, but the pattern was more subtle. The increase did not peak until lactation also did, suggesting that this is not an 'early response' to the negative energy balance and increased milk production of early lactation. Rather, this seems to be a secondary response over time. This does not mean it is not important, as a systems approach would show. Just because it was not 'The first physiological response' does not mean it is not quantitatively important. This point of view has not always been shared by those who study physiological control, as often, if a change in control proteins or systems is not the 'first', the 'fastest' or the 'largest', it is not exciting enough to study. Yet in the system, all control is relevant.

When we asked the question of proportional control though multiple regressions, we began to learn more about the system relating transcriptional control with lipolysis. When we regressed the expression of the B2-adrenergic receptor on BW, BCS and empty body fat, we could define only about 10% of the variation. When we focused the regression comparing B2AR transcript on the maximally stimulated rate of lipolysis, again, only about 10% of the variation could be defined. This likely suggests that in fact, only about 10% of the control of lipolysis during lactation can be attributed to an increase in message for this receptor. Given all the other levels of control on lipolysis, and that in fact amount of adrenergic receptor is controlled in a loop of increased stimulation, reduced receptor activity, and attenuation of response (a 'governor', if you will to avoid rapid mobilization).

The other systems analysis we did was to ask how the message for HSL related to lipolytic rate measured *in vivo*. In theory, there may be a tighter relationship as this is the enzyme that catalyzes the reaction directly. However, we could again only find about 12 to 17% of the variation in stimulated lipolysis explained by an increase in HSL mRNA, and, perhaps surprisingly at the time, there was no relationship between HSL message and basal (non-stimulated lipolysis). Thus the inference is that the majority of the control of HSL activity is in fact physiological through activation of the sympathetic nervous system and increased phosphorylation. This is what we demonstrated years earlier in the rat model (McNamara and Murray, 2001). Since this study, in fact, it has been suggested based on some good data, that basal lipolysis is in fact catalyzed by a different enzyme, adipose tissue triacylglycerol lipase (ATGL, Montserrat *et al.,* 2008). This continues to demonstrate the need for a systematic approach to define the quantitative contribution of all control in the system. It also demonstrates the importance of lipolysis to survival, as the amount of control on this very simple reaction is immense and redundant.

From this same study we then conducted an analysis of the gene transcriptome in bovine adipose tissue during the transition from pregnancy to lactation (Sumner *et al.* 2008a, b). We obtained adipose tissue by biopsy at 30 d *pre partum* and 14 d *post partum* and extracted the RNA. This was hybridized to the Affymetrix Genechip® Bovine Genome Array. Animals averaged 29.8 (SEM = 1.3 kg/d of milk for the first 60 DIM (range 18.6 to 44.8 kg/d). They lost 42.6 kg of BW (SEM 8.4, range +9.1 to -113.6) and 0.38 BCS units (SEM 0.10, range 0 to -1.0) from 0 to 14 DIM. This is a normal range for dairy cattle, housed and fed alike and gives a glimpse of the yet unknown effects of genetic variance in a similar population.

Anabolic pathway genes decreased ($P<0.05$), including (mean (% change), (SEM)): steroyl response element binding protein, -25.1, (6.2); glucose transport 1, -57.3 (14.1); thyroid hormone receptor spot 14, -30.8 (7.4); lipoprotein lipase, -48.4 (7.7) and AcCoA carboxylase, -60.6 (13.0). The regression of transcript change on milk production was 0.18 for AcCoA carb and 0.26 for ATP-CL ($P<0.05$). Lipolytic control elements increased, with much variation among animals, including Ca channel subunit 338% (203); B2AR 52.0 (8.8); PKC receptor 10.1 (2.6) and HSL mRNA 23.0 (17.9). The regression of transcript change on milk production was 0.30 and 0.25 for B2AR and HSL mRNA. These latter regressions explain somewhat more of the variation than the ones for HSL and B2AR in lipolysis, which is intriguing. These results lead us to conduct further more in depth studies to integrate transcriptional control into the metabolic model.

We have since conducted additional studies on transcriptional control of metabolism and efficiency in the cow. We identified 1st and 2nd parity animals based on their sire genetic merit and/or previous production, and then fed them to requirements or to 90% of requirements for energy by pair feeding. We then sampled adipose tissue at intervals in late pregnancy and early lactation to measure lipogenesis, lipolysis and transcript amounts through use of the Affymetrix Bovine Gene array, as well as RT-PCR for some of the genes. Hopefully, then we can see more specifically how the transcript, rates of metabolism, body composition, energy intake and milk energy secretion fully relate in the system. To date we do not have the full regressions completed, but some interesting patterns certainly are emerging (Sumner *et al.*, 2010).

A total of 48 cows were grouped by their sire PTAM: High Genetic (PTAM = 870 kg), or Low Genetic (PTAM = 378), and half of each group was fed either to requirements (NE) or to 90% of energy requirements (LE). Other components were fed to requirements. Feed intake from 21 to 1 d *pre partum* was 13.6 (NE) and 12.7 kg (LE) DMI/d (SE = 1.5). From 1 to 56 DIM it was 21.2 and 17.4 kg/d (SE = 1.4). Milk production was 36.1 and 33. 3 kg/d for HG and LG cows from 27 to 56 DIM ($P<0.05$). Adipose tissue biopsies at -21, -7, 7, 28 and 56 days around parturition were used to measure lipolysis, lipogenesis and gene expression. Rates of lipogenesis were lower during lactation and lower in LE cows while lipolysis rates were higher for both conditions ($P<0.05$). The mRNA expression of the beta-2 adrenergic receptor, hormone sensitive lipase and the co-lipase, perilipin, was several-fold higher ($P<0.05$) in animals on restricted energy. The mRNA for caveolin-1 and caveolin-2 decreased 20 to 40% ($P<0.05$) in lactation consistent with the increase in lipolysis and HSL message. The gene expression array showed coordinated decreases in genes regulation lipogenesis (TRPSP14, -26%; AcCoCarb, -76%; LPL, -57%; ATP-Citrate Lyase, -22% as examples) and no change or moderate increases in those controlling lipolysis (Table 2).

Further we were able to run regressions of gene expression on milk production. For the genes listed, in parentheses are the regression coefficients for gene expression versus milk production in the first month of lactation: GLUT1 (0.34); IGFBP3 (0.67); THRP 12 (0.38); LPL (0.18); leptin (0.31). All of these genes controlling anabolic reactions were negatively related with milk production. These regression coefficients give us some mathematical insight into how much control might be exerted on the anabolic pathways by gene expression. There was little relation between milk production and lipolytic control genes, again suggesting that most control on lipolysis is physiological.

So, here now we can use a systems biology approach, based in sound biology, and use the model to ask deeper questions about control of the system. Changes in gene expression alter the maximal

Table 2. Genes coding for metabolic control in adipose tissue of lactating dairy cattle.

Gene	GenBank	-7	7	28	d 28 / d -7	*P*-value
Genes coding for anabolism						
LPL	BG688620	4045	2229	1552	-57%	0.00
FABP5	NM_174315.2	4378	4265	3075	-22%	0.88
GLUT4	NM_174604.1	49	37	36	-26%	0.53
THIHP	CK848521	2679	1148	625	-71%	0.01
ATP CL	CB433477	471	401	351	-22%	0.00
AcCoACarb	NM_174224.2	162	55	39	-73%	0.00
AcCoACarb	BE751005	100	31	21	-76%	0.00
Genes coding for catabolism						
HSL, mRNA	CK769629	39	31	53	-4%	0.28
HSL	BM967863	77	60	80	19%	0.14
b2AR	NM_174231.1	155	138	92	-32%	0.02
CAV1mRNA		1111	905	977	-14%	0.06
cav 1		3549	2813	2714	-23%	0.90
CAV2a		625	481	395	-33%	0.65
CAV 2		834	480	423	-47%	0.04
CAV 2		156	77	74	-52%	0.05

Samples biopsied at times around calving as indicated. Results are signals from the Bovine Affymetrix Gene Array, normalized to an average signal strength of 125.

velocity of lipogenesis and lipolysis. These changes measured in the cows can be used to alter the maximal velocity parameters of lipogenesis and lipolysis in body fat, in direct proportion to the relative change in transcript level, based on the principle that mRNA abundance directly relates to enzyme concentration, but is independent of post-translational modification. We can then add the control by post-translational physiology (already in Molly through anabolic and catabolic hormone control. Integration of these control elements into metabolic models provides the opening to more fully explore the relationships of genotype, phenotype and nutritional environment on the efficiency of dairy cattle. There are some successful examples, requirements and illustrations of this approach in the reference list (McNamara and Pettigrew, 2002a,b; McNamara, 2005, 2006; McNamara and Boyd, 1998; McNamara *et al.*, 1989).

The acceptance of integrative biology is critical

A major barrier to improvement of models remains lack of an accurate description of the phenotype of the animal being modeled, expressed as, for example, gene transcription control, enzyme activity, hormone and receptor kinetics and intracellular signalling. An additional barrier continues to be the thought processes of scientists who are not trained in more complex regulation and theories and are uncomfortable with the ideas or skeptical of the value of integrative biology. The genome projects themselves are starting to change those attitudes, especially in younger scientists, because the central nature of gene transcription in metabolic regulation is better understood now than before, and because the sheer mass of information generated in genomic and transcriptomic work dictates mathematical methods and approaches to bring clarity from the data.

One underlying concept to such integrative work is that the amount and activity of all enzymes and hormones are genetically regulated, from immediate gene transcription and translation, to heritability of variations in hormone and enzyme synthesis and secretion. Some examples may be found in

(Bosch *et al.*, 1999; Daniel *et al.*, 1999; Friedman and Halaas, 1998; Girard *et al.*, 1997). However, some have small heritabilities, or are expressed constitutively, are members of redundant control systems, and are thus not relevant to metabolic control (Cornish-Bowden *et al.,* 2007).

The flood of information from the various genome works, and the ability to generate large volumes of transcriptome data from animal studies has renewed calls for more integration of knowledge, including using bio-mathematical approaches. A model or a modeling approach to research may also be defined as an ordered way of describing knowledge of some real complex system. Such models have been useful in practical systems to describe, for example, drug metabolism, biochemical pathways and nutrient requirements. A quantitative description of metabolic transactions is critical to improve understanding and improvement of nutrient requirements, health and longevity. Models of increasing complexity, ever grounded in validated research data, will continue to improve our quantitative understanding. It is this author's experience that information from genomic research can only be understood with the means of complex model systems, a philosophy shared by others (Dawson, 2004).

We have a long way to go. We need a re-invigorated, multi-investigator, multi-disciplinary, integrated approach to solve the present and future problems of reproduction, and specific to the role of nutrigenetics and nutrigenomics for improved reproduction, this research effort will require construction and testing of mechanistic bio-mathematical models. Finally, we need to train students, scientists and professionals in the importance of using integrative biology and bio-mathematical models to identify, solve and prevent reproductive problems.

References

Al-Hasani, H. and H.-G. Joost, 2005. Nutrition/diet-induced changes in gene expression in white adipose tissue. Best Practice & Research Clin. Endocrinol. & Metab. 19, 589-603.

Baldwin, R.L., 1968. Estimation of theoretical calorific relationships as a teaching technique: A review. 1968. J. Dairy Sci. 51, 104.

Baldwin, R.L., 1995. Modelling ruminant digestion and metabolism. Chapman & Hall, New York, USA: 469-518.

Baldwin, R.L., J. France, and M. Gill, 1987a. Metabolism of the lactating cow. I. Animal elements of a mechanistic model. Journal of Dairy Research 54, 74-105.

Baldwin, R.L., J. France, D.E. Beever, M. Gill and J.H.M. Thornley, 1987c. Metabolism of the lactating cow. III. Properties of mechanistic models suitable for evaluation of energetic relationships and factors involved in the partition of nutrients. J. Dairy Res. 54, 133-145.

Bionaz, M. and J.J. Loor, 2008. ACSL1, AGPAT6, FABP3, LPIN1, and SLC27A6 are the most abundant isoforms in bovine mammary tissue and their expression is affected by stage of lactation. J. Nutr. 138, 1019-1024.

Carson E.R., C. Cobelli and L. Finkelstein, 1981. Modeling and identification of metabolic systems. Am J Physiol. 240, R120-129.

Chagas, L.M., J.J. Bass, D. Blache, C.R. Burke, J.K. Kay, D.R. Lindsay, M.C. Lucy, G.B. Martin, S. Meier, F.M. Rhodes, J.R. Roche, W.W. Thatcher and R. Webb, 2007. New perspectives on the roles of nutrition and metabolic priorities in the subfertility of high-producing dairy cows. J. Dairy Sci. 90, 4022-4032.

Cornish-Bowden, A., 2005. Making systems biology work in the 21st century. Genome Biology 6, 317.

Cornish-Bowden, A., M.L. Cárdenas, J.C. Letelier and J. Soto-Andrade, 2007. Beyond reductionism: metabolic circularity as a guiding vision for a real biology of systems. Proteomics 6, 839-845.

Dawson, K.A., 2006. Nutrigenomics: feeding the genes for improved fertility. Animal Repro. Sci. 96, 312-322.

Garcia, M.D., J.J. Michal, C.T. Gaskins, J.J. Reeves, T.L. Ott, Y. Liu and Z. Jiang, 2006. Significant association of the calpastatin gene with fertility and longevity in dairy cattle. Anim. Gen. 37, 293-307.

Lofgreen, G.P. and W.N. Garrett, 1968. A system for expressing net energy requirements and feed values for growing and finishing beef cattle. J. Anim. Sci. 27, 793-806.

Loor, J.J., H.M. Dan, R.E. Evarts, R. Oliveira, C.A. Green, N.A. Janovick Gruretzky, C.A. Green, S.L. Rodriguez-Zas, H.A. Lewin and J.K. Drackley, 2005. Temporal gene expression profiling of liver from periparturient dairy cows reveals complex adaptive mechanisms in hepatic function. Physiol. Genomics 23, 217-226.

Loor, J.J., H.M. Dann, N.A. Janovick Gruretzky, R.E. Everts, R. Oliveira, C.A. Green, N.B. Litherland, S.L. Rodriguez-Zas, H.A. Lewin and J.K. Drackley, 2006. Plane of nutrition alters hepatic gene expression and function in dairy cows as assessed by longitudinal transcript and metabolic profiling. Physiol. Genomics 27, 29-41

McNamara, J.P., 2005. Research, Improvement and application of mechanistic, biochemical, dynamic models: from Genetics to Kinetics. Chapter 6 In: Mathematical modelling in nutrition and Toxicology. Proceedings of the 8th Internatinoal Conference on Mathematical Modeling in Nutrition and Health Science. Georgia Center for Continuing Education: 87-110.

McNamara, J.P., 2006. Modelling metabolism in the lactating sow: mechanistic modelling at the metabolic level. In: Pig and Poultry Modeling, Symposium, South Africa, April, 2005.

McNamara, J.P. and D.E. Boyd, 1998. Hormones as Quantitative Controllers. In: Kyriazakis, I. (ed.), A Quantitative Biology of the Pig. CAB International, London, UK: 199-225.

McNamara, J.P. and C. Murray, 2001. Sympathetic nervous system activity in adipose tissues during pregnancy and lactation of the rat. J. Dairy Sci. 84, 1382-1389.

McNamara, J.P. and J.E. Pettigrew, 2002a. Protein and energy intake in lactating sows. 1: Effects on milk production and body composition. J. Anim. Science 80, 2442-2451.

McNamara, J.P. and J.E. Pettigrew, 2002b. Protein and energy intake in lactating sows. 2. Challenging parameters of a model of metabolism. J. Anim. Sci. 80, 2452-2460.

McNamara, J.P., J.E. Pettigrew, R.L. Baldwin, B. Walker, W.H. Close and J.W. Oltjen, 1991. Information needed for mathematical modelling of energy use by animals. In: Energy Metabolism in Farm Animals, EAAP Pub. No. 58, Gruppe Emahrung, Zurich Switzerland: 468-472,.

National Research Council, 2001. Nutrient requirements of dairy cattle. 9th Ed. Natl. Acad. Press., Washington, DC.

Pettigrew J.E., M. Gill, J. France and W.H. Close, 1992a. A mathematical integration of energy and amino acid metabolism. Journal of Animal Science 70, 3742-3761.

Pettigrew J.E., M. Gill, J. France and W.H. Close, 1992b. Evaluation of a mathematical model of lactating sow metabolism. Journal of Animal Science 70, 3762-3773.

Montserrat P., H. Hackl, T. Rainer Burkard, A. Prokesch, C Papak, M. Scheideler, G. Hämmerle, R. Zechner, Z. Trajanoski and J.G. Strauss, 2008. Differential transcriptional modulation of biological processes in adipocyte triglyceride lipase and hormone-sensitive lipase-deficient mice Genomics 92, 26-32.

Roche, J.F., 2006. The effect of nutritional management of the dairy cow on reproductive efficiency. Anim. Reprod. Sci. 96, 282-296.

Sumner, J.M., C. Schachtschneider, J.L. Vierck and J.P. McNamara, 2010. Changes in the transcriptome of adipose tissue of the dairy heifer during late pregnancy and lactation as measured by gene array analysis. J. Dairy Sci. 93 (Suppl 1.), in press.

Sumner, J.M and J.P. McNamara, 2007. Expression of lipolytic genes in the adipose tissue of pregnant and lactating Holstein dairy cattle. J. Dairy Sci. 90, 5237-5246.

Mitochondrial energetics and ROS production in chicken skeletal muscle: application of modular kinetic analysis

M. Toyomizu[1], M. Kikusato[1], T. Amo[2] and M.A.K. Azad[1]
[1]Laboratory of Animal Nutrition, Division of Life Sciences, Graduate School of Agricultural Science, Tohoku University, Tsutsumidori-Amamiyamachi 1-1, Aoba-ku, Sendai 981-8555, Japan
[2]Department of Applied Chemistry, National Defense Academy, Hashirimizu 1-10-20, Yokosuka 239-8686, Japan

Abstract

Mitochondrial ATP production or heat production can be a major factor in the conversion efficiency of food/feed energy. An inappropriate control of mitochondrial reactive oxygen species (ROS) production will exert a negative effect on development of tissue and organs, and hence animal growth. To elucidate how alterations in mitochondrial functions specify their ATP production and oxidative damage, modular kinetic analysis was applied in 2 typical mitochondrial studies concerning the differences (a) in oxidative phosphorylation between meat- and laying-type chickens, and (b) in ROS production between heat-exposed chickens and its control group. Skeletal muscle mitochondria from meat-type chickens were found to have a higher oxidative phosphorylation efficiency than laying-type chickens. In the heat-stress conditions, 'substrate oxidation' by the electron transport chain was enhanced, reinforcing mitochondrial superoxide production, probably via an elevation in the membrane potential ($\Delta\Psi$) in avian muscle mitochondria. Thus this method is of value tool in consideration for improving of growth performance.

Introduction

Most of the ATP produced by the cell is generated by mitochondria via oxidative phosphorylation. During substrate oxidation, protons are pumped out of the mitochondrial matrix, creating an electrochemical proton gradient. This proton motive force drives ATP synthesis, although substrate oxidation is not fully coupled to ATP synthesis since some of the gradient is dissipated by protons leaking back to the matrix through natural and/or anion carrier-mediated proton conductance pathways across the mitochondrial inner membrane. Non-productive proton leak pathways are physiologically important since they account for a surprisingly high proportion of cellular and organismal respiration rate, estimated at up to 25% (Rolfe *et al.*, 1999; Brand, 2005). Skulachev (1999) noted some of the main physiological functions of mitochondria as follows: (1) energy conservation, i.e. ATP production, (2) metabolic energy dissipation as heat, (3) substrate production and decomposition, and (4) reactive oxygen species (ROS) generation. ATP production, energy dissipation and control of reactive oxygen species are directly associated with the efficiency of animal production. Thus, ATP production or heat production can be a major determinant of energy conversion efficiency of food/feed energy to ATP for metabolic and mechanical works, depending on the magnitude of uncoupling of oxidative phosphorylation. Inappropriate ROS production that has a negative effect on the development of tissue and organs as well as animal growth can be controlled by the membrane potential ($\Delta\Psi$). To elucidate how alterations in mitochondrial functions would affect their ATP production and oxidative damage, we focused on 2 typical cases related to the differences: (a) in mitochondrial oxidative phosphorylation between meat- and laying-type chickens, and (b) in ROS production between heat-exposed chickens and its control group. By applying modular kinetic analysis to our data set for the mitochondrial studies on skeletal muscle of both cases of chickens, we show its value as a tool in the field of applied animal science.

What is modular kinetic analysis?

Kinetics of proton leak, phosphorylation system and substrate oxidation

To evaluate oxidative phosphorylation in skeletal muscle mitochondria of heat-stressed birds, modular kinetic analysis was used (Hagopian *et al.*, 2005; Amo and Brand, 2007). Mitochondrial oxidative phosphorylation can be divided conceptually into 3 modules connected by $\Delta\Psi$ as the intermediate, i.e. the reactions that: (1) produce $\Delta\Psi$, which are called 'substrate oxidation'; (2) consume $\Delta\Psi$ and synthesize ATP, called 'phosphorylation system'; and (3) consume $\Delta\Psi$ without ATP synthesis, referred to as 'proton leak' (Brand, 1990). To measure kinetic response of proton leak to $\Delta\Psi$, State 4 respiration (non-phosphorylation) of mitochondria was titrated with malonate (to inhibit complex II) in the presence of oligomycin (to prevent any residual ATP synthesis). As with proton leak kinetics, State 4 respiration was titrated with the uncoupler FCCP to measure the kinetic response of substrate oxidation to $\Delta\Psi$. State 3 (maximal rate of ATP synthesis) was obtained using an ADP-regenerating system (100 μM ATP, 20 mM glucose and 6 U/ml hexokinase). Titration of State 3 respiration with malonate (up to 1.5 mM) was done to measure the kinetics of the $\Delta\Psi$-consumers (the sum of phosphorylation system and proton leak).

Coupling efficiency of mitochondrial oxidative phosphorylation

Coupling efficiency of oxidative phosphorylation was calculated from mitochondrial modular kinetic curves. Coupling efficiency is the percentage of mitochondrial respiration rate at a given $\Delta\Psi$ used for ATP synthesis (Amo and Brand, 2007; Affourtit and Brand, 2009). This is calculated by subtracting the oxygen consumption rate of the proton leak from the total oxygen consumption rate (=substrate oxidation) at any chosen $\Delta\Psi$, and expressed as the percentage of the total rate.

Results and discussion

Differences in mitochondrial oxidative phosphorylation between meat-type and laying-type chickens

Meat- and laying-type chickens have been extensively selected for meat and egg production, respectively. As a result, meat-type chickens show very rapid growth, but the reasons they grow faster than laying-type chickens remain unclear. Decreased behavioral activities (Saito *et al.*, 2004) and rates of protein degradation (Saunderson and Leslie, 1988) may contribute to the very rapid growth in meat-type chickens. Although meat- and laying-type chickens tend to spend a similar proportion of time feeding, food intake was twice as high in meat-type chickens than in laying-type (Hocking *et al.*, 1997). Therefore, when food intake was normalized by metabolic body weight, the laying-type chickens ate significantly more than the meat-type chickens, indicating the lower metabolic efficiency or higher energy expenditure in laying-type chickens (Swennen *et al.*, 2007). Indeed, the basal metabolic rate (BMR) of meat-type chicks is lower than that of laying-type chicks from hatching to 500 g body mass (Kuenzel and Kuenzel, 1977). Thus, we hypothesized that the differences of feed efficiency between bird strains might be partly related to their efficiency of mitochondrial oxidative phosphorylation. Therefore, we focused on the physiological functioning of mitochondria isolated from energy-consuming organs, skeletal muscle, applying modular kinetic analysis (Brand, 1998; Amo and Brand, 2007) to determine its efficiency in both chicken types.

Meat-type (Cobb) and laying-type (Julia) male chicks were used. Body weights at 21 and 41 days were individually recorded and feed consumption was also measured to calculate feed efficiency. Meat- and laying-type chicks were killed by decapitation, and breast muscles were rapidly excised (n=6 birds per group).

The meat-type chicks ate 2.7 times more than the laying-type during the measurement period (Table 1), converting food to their body mass about twice as efficiently (0.71±0.01 versus 0.37±0.02). Results of modular kinetic analysis for the oxidative phosphorylation [predicted 'substrate oxidation', 'proton leak' and $\Delta\Psi$-consumers (sum of 'phosphorylation system' and 'proton leak')] for skeletal muscle are shown in Figure 1A. The 'rate of the phosphorylation' module of meat-type chickens was higher in the laying-type (Figure 1B). Mitochondria of meat-type chickens had a lower rate than those of laying-type chickens with regard to the proton leak module (Figure 1C). The coupling efficiency of oxidative phosphorylation was analyzed from the kinetic curves in Figure 1A. The classic measure of the efficiency is the respiratory control ratio (RCR), i.e. respiration with maximal rate of ATP synthesis (State 3) divided by respiration with no ATP synthesis (State 4). The RCR of meat-type chicken mitochondria was slightly, but not significantly, higher than that of laying-type chicken mitochondria (Figure 1D). A more precise measure of coupling efficiency is given by the percentage of respiration coupled to ATP synthesis, calculated by subtracting the rate of proton leak from the total respiration in State 3's membrane potential, and expressing the difference as a percentage of the total rate. Skeletal muscle mitochondria from meat-type chickens had a much higher respiration rate at State 3 (Figure 1B) and a lower proton leak (Figure 1C) than the laying-type ones, resulting in higher coupling efficiency at the membrane potential in State 3 (Figure 1E, left two bars). The coupling efficiency was also estimated across the whole range of rates of ATP synthesis (phosphorylating activity) and values of $\Delta\Psi$ from State 3 to State 4 by postulating that a line of substrate oxidation kinetics was a straight line through the State 3 and State 4 points (Figure 1A) (Brand, 1998). The 2 estimated lines for substrate oxidation apparently are separate (Figure 1A), and the coupling efficiencies at 150, 155 and 160 mV were significantly different (Figure 1E). Thus, meat-type chickens have higher phosphorylating activity and lower proton leak of skeletal muscle mitochondria than laying-type chickens, resulting in higher efficiency of oxidative phosphorylation. Considering that skeletal muscle is a high energy-consuming tissue, the higher coupling efficiency of skeletal muscle mitochondria probably plays an important role in the rapid growth of meat-type chickens.

Mitochondrial uncoupling represents a cellular inefficiency, but it also reduces oxidative stress by attenuating mitochondrial ROS production. Therefore, meat-type chickens that express low uncoupling levels (Figure 1E) will suffer oxidative stress. Indeed, meat-type chickens were much more susceptible to acute and chronic heat-stress, inducing mitochondrial superoxide production and muscular oxidative damage (Mujahid *et al.*, 2005; M.A.K. Azad, unpublished results).

Table 1. Body-weight gain, feed intake and feed efficiency.

Breed	Meat-type (n=6)		Laying-type (n=6)	
	(Age 21-41 day)	(/day)	(Age 21-41 day)	(/day)
Body weight gain (g)	1,189±25	85±2	222±5	15.9±0.4
Feed intake (g)	1,668±11	119±1	614±23	43.9±1.6
Feed efficiency[1]	0.71±0.01		0.37±0.02	

[1] Body weight gain (g) / feed intake (g).

Differences in mitochondrial oxidative phosphorylation and ROS production between heat-exposed chickens and control group

Heat stress is an environmental factor responsible for stimulating superoxide production. Having already shown that mitochondrial superoxide generation can be significantly enhanced in skeletal muscle of birds given acute heat stress (Mujahid *et al.*, 2005), causing oxidative damage to mitochondrial proteins and lipids (Mujahid *et al.*, 2007), we considered whether the mechanisms underlying acute

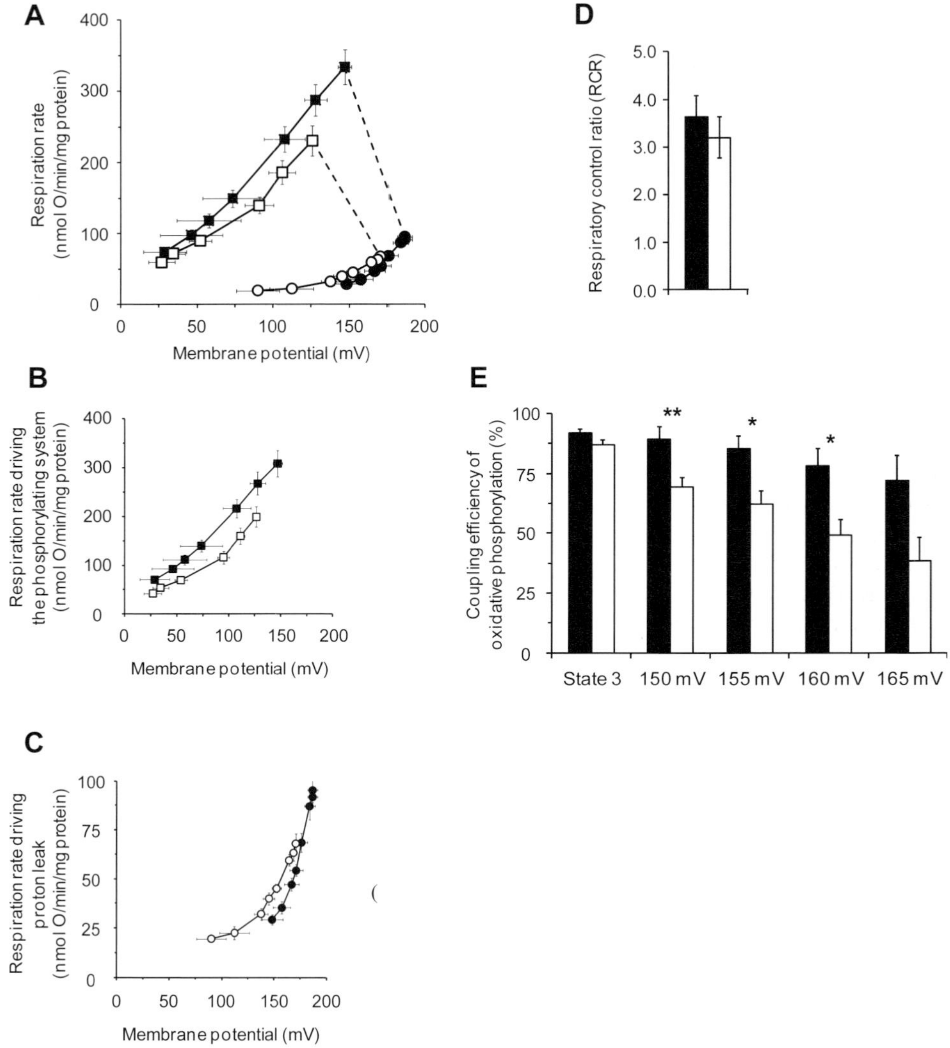

*Figure 1. (A) Oxidative phosphorylation characteristics of skeletal muscle mitochondria. Closed symbols, meat-type chickens; open symbols, laying-type chickens. Results are means ± SEM for 6 chickens per group. Modular kinetic analysis of oxidative phosphorylation in mitochondria isolated from skeletal muscle of meat-type chickens and laying-type chickens. The kinetic responses to membrane potential, $\Delta\Psi$, of respiration driving (C) proton leak ($\Delta\Psi$ titrated with complex II inhibitor malonate, starting in State 4); and (B) the phosphorylating system, calculated by subtracting respiration driving proton leak from respiration driving the $\Delta\Psi$-consumers ($\Delta\Psi$ titrated with malonate starting in State 3; not shown) at each $\Delta\Psi$. (D) Respiratory control ratio (State 3 respiration rate/State 4 respiration rate). (E) Coupling efficiency at the $\Delta\Psi$ of State 3. * = $P<0.05$ and ** = $P<0.01$ for meat- versus laying-type chickens.*

heat stress-induced overproduction of mitochondrial superoxide in chicken skeletal muscle may be involved in enhancing $\Delta\Psi$. Indeed, the $\Delta\Psi$ of muscle mitochondria from heat-stressed bird proved to be elevated, although the proton leak, one factor affecting $\Delta\Psi$, was unchanged (Mujahid *et al.*, 2009). We could therefore postulate that 'substrate oxidation', another factor influencing $\Delta\Psi$, might be responsible for the elevated $\Delta\Psi$. To clarify this matter, we have employed modular kinetic analysis.

After 14 days, the birds were randomly divided into 2 groups (n=12 birds per group). Following a 7-day adaptation period, one of the 2 groups was exposed to 34 °C for 12 h, while the other group was maintained at 24 °C (humidity 55±5%). Birds were provided access to water and the diet *ad libitum* during treatment. After exposure, birds in each group were killed by decapitation, and breast muscles were rapidly excised (n=8 birds per group).

Mitochondrial H_2O_2 production at State 4 in the heat-stressed groups was significantly higher (~2.3 fold) than in the control group (Figure 2), in agreement with previous findings (Mujahid *et al.*, 2005, 2009).

Results of a modular kinetic analysis for the oxidative phosphorylation ('substrate oxidation', 'proton leak' and 'phosphorylation system') for both control and heat-stressed groups (Figure 3) show that oxygen consumption rate due to 'substrate oxidation' in heat-stressed group was higher than in the control group (Figure 3D). There was little difference in proton leak kinetics at any given $\Delta\Psi$ between the groups, but the furthest point (State 4) to the right in the kinetic curve for the heat-stressed group tended towards higher $\Delta\Psi$ and significantly higher oxygen consumption rates than their controls (Figure 3C). Phosphorylation kinetics (Figure 3B) were calculated by subtracting oxygen consumption rate driven by proton leak from oxygen consumption rate driven by the $\Delta\Psi$-consumers at each $\Delta\Psi$ shown in Figure 3A. Phosphorylation kinetics for the heat-stressed group had a slower rate of oxygen consumption at lower $\Delta\Psi$ values than their controls, but the furthest point (State 3) to the right for the heat-stressed group had a significantly higher $\Delta\Psi$ and oxygen consumption rates than their controls.

Korshunov *et al.* (1997) found that mitochondrial ROS production is exquisitely sensitive to $\Delta\Psi$. Brand's group also reported a positive correlation between $\Delta\Psi$ and ROS production in *Drosophila* mitochondria (Miwa *et al.*, 2003). Therefore, one can postulate that the overproduction of mitochondrial ROS in heat-stressed birds may be caused by changes in $\Delta\Psi$ at State 4. This postulation was supported by our observations that simultaneous increases of ROS production and $\Delta\Psi$ at State 4 occurred in muscle mitochondria in the heat-stressed birds. The novel finding is that 'substrate oxidation' by the electron transport chain is enhanced under heat stress conditions, and that these mitochondria have increased $\Delta\Psi$ and ROS production at State 4.

Why does $\Delta\Psi$ increase under heat stress conditions? Rolfe *et al.* (1994) proposed that 'substrate oxidation' as a $\Delta\Psi$-producer, as well as 'proton leak' as a $\Delta\Psi$-consumer, plays a role in controlling the magnitude of $\Delta\Psi$ at State 4. We found that the 'proton leak' was similar in the control and heat-stressed birds, but 'substrate oxidation' was increased in the muscle mitochondria from the latter,

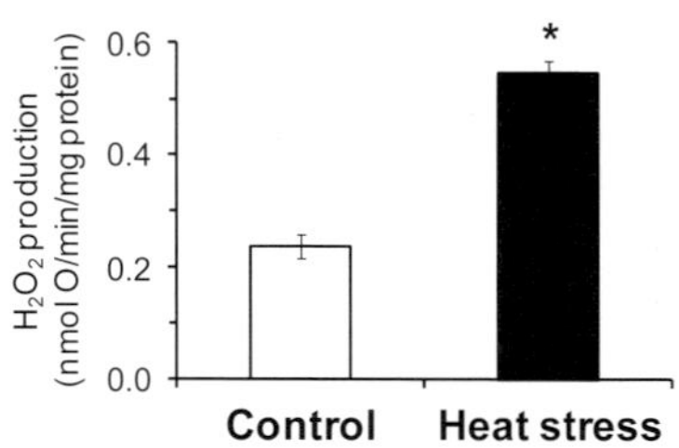

Figure 2. Hydrogen peroxide (H_2O_2) production at State 4 of skeletal muscle mitochondria from control (open bar) and heat-stressed (filled bar) groups. Succinate (4 mM) was used to energize the mitochondria (0.35 mg protein/ml) incubated at 38 °C in a standard assay medium [115 mM KCl, 10 mM KH_2PO_4, 3 mM HEPES, 1 mM EGTA, 2 mM $MgCl_2$ and 0.3% (wt/vol) BSA, 0.1 µM nigericin, pH 7.2]. H_2O_2 production was measured fluorometrically in the absence of rotenone. For details, see General Methods. Values are means ± SE of date from 4 replicates using 4 mitochondrial samples isolated from 2 individual pooled muscle in each group. * = P<*0.05 compared to control groups.*

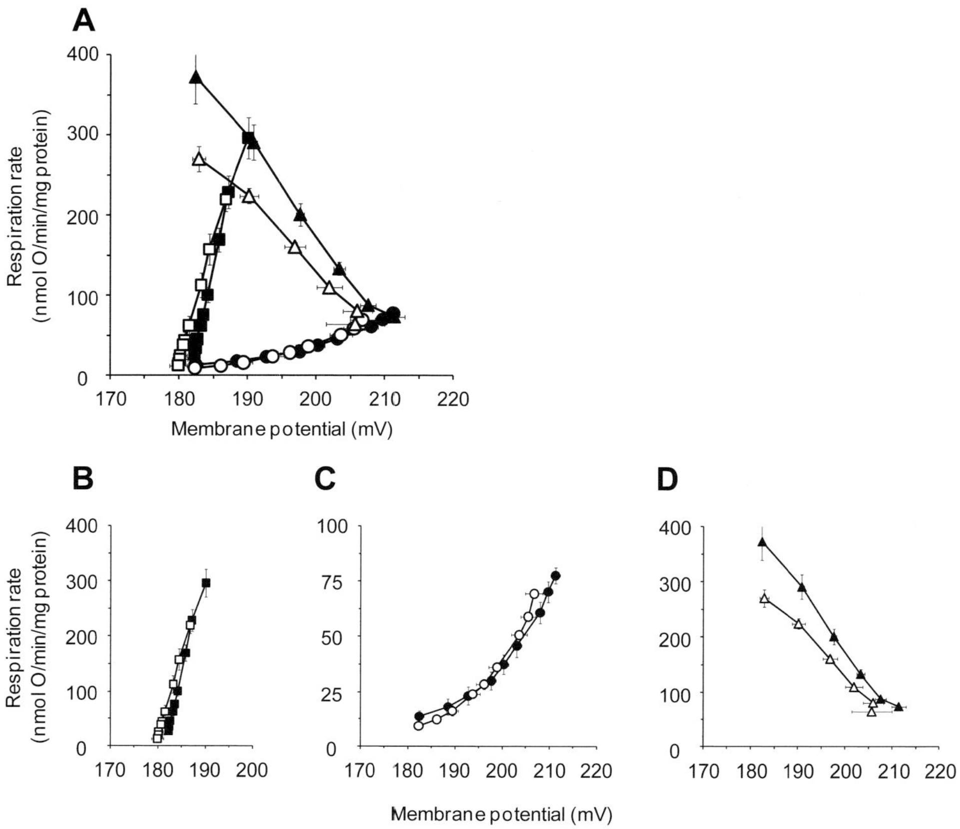

Figure 3. (A) Kinetic responses of the 3 modules of oxidative phosphorylation [(B) phosphorylation system, squares; (C) proton leak, circles; (D) substrate oxidation, triangles] to ΔΨ, in skeletal muscle mitochondria from control (open symbols) and heat-stressed (filled symbols) groups. Succinate (4 mM) was used to energize the mitochondria (0.35 mg protein/ml) incubated at 38 °C in the standard assay medium with 5 μM rotenone. Values are means ± SE of date from 4 replicates using 4 mitochondrial samples isolated from 2 individual pooled muscle in each group.

suggesting that increase in 'substrate oxidation' are primarily responsible for the elevated ΔΨ at State 4 under the heat stress conditions. Therefore, overproduction of mitochondrial ROS was induced by the 'substrate oxidation' stimulating ΔΨ for the heat-stressed birds. Thus, modular kinetic analysis helps to find the underlying mechanism of overproduction of mitochondrial ROS from skeletal muscle in acute heat-stressed chickens.

Conclusion

Mitochondria are the major organelle involved in ATP and reactive oxygen species (ROS) production of cells. The mechanisms involved in their production and efficiencies are complex, inasmuch as mitochondrial bioenergetic system consists of 'substrate oxidation', 'phosphorylation system', and 'proton leak'. However, once modular kinetic analysis are performed for mitochondrial study on oxidative phosphorylation, more information about the underlying mechanism and 'cause and effect relationship' in the complex dynamics can be obtained, e.g.:

a. Skeletal muscle mitochondria from meat-type chickens have a higher oxidative phosphorylation efficiency than laying-type chickens. Skeletal muscle is a very energy consuming tissue, and therefore these differences could play an important role in the high feed efficiency and the rapid growth of meat-type chickens.

b. Under heat stress conditions, 'substrate oxidation' by the electron transport chain is enhanced, reinforcing mitochondrial superoxide production, probably via an elevation in the membrane potential in avian muscle mitochondria, suggesting that acute heat stress enhances 'substrate oxidation' of skeletal muscle mitochondria, leading to greater ROS production.

General methods

Animals and preparation of mitochondria

For both experiments, male chicks (1-day old) were obtained from a commercial hatchery. They were provided access to water and commercial diet *ad libitum* during the treatments. After treatment, birds in each group were selected and killed by decapitation. This method of killing was used in preference to overdose by general anesthetics, which are known to uncouple oxidative phosphorylation (Rottenberg, 1983). *Pectoralis superficialis* muscles were rapidly excised, and a sample of each muscle was placed in ice-cold isolation medium (100 mM KCl, 50 mM Tris/HCl, and 2 mM EGTA, pH 7.4) for mitochondrial isolation (see below). The Animal Care and Use Committee of the Graduate School of Agricultural Science, Tohoku University approved all procedures, and efforts were made to minimize pain or discomfort of the animals.

Isolation of skeletal muscle mitochondria

Muscle mitochondria were isolated by homogenization, protein digestion, and differential centrifugation at 4 °C, as previously described (Cadenas *et al.*, 2002; Mujahid *et al.*, 2009). Muscle was trimmed of fat and connective tissue, blotted dry, weighed, and placed in isolation medium on ice. Tissue was shredded and minced with sharp scissors, rinsed with isolation medium 3 times, stirred for 5 min in protein digestion medium [100 mM KCl, 50 mM Tris/HCl, 2 mM EGTA, 1 mM ATP, 5 mM $MgCl_2$, 0.5% (wt/vol) bovine serum albumin (BSA), and 11.8 units of protease per gram of tissue (Sigma, substilisin type VIII), pH 7.4], and gently homogenized using a Polytron tissue homogenizer. The homogenate was stirred for 6 min before being mixed with the equivalent medium without protease to stop protease activity. The homogenate was rehomogenized in a Potter-Elvehjem homogenizer (6 passages) and centrifuged at 500 *g* for 10 min. The supernatant was filtered through muslin and centrifuged at 10,400 *g* for 10 min. Mitochondrial pellets were resuspended in isolation medium and centrifuged at 10,400 *g* for 10 min, followed by 3,800 *g* for 10 min, and were resuspended in isolation medium. Mitochondrial protein concentration was determined by the bicinchoninic acid (BCA) assay, with BSA as the standard (Brown *et al.*, 1989). All mitochondria were used for the module kinetic analysis on the day of the isolation.

Mitochondrial H_2O_2 production

Mitochondrial H_2O_2 generation rates were determined fluorometrically by measurement of oxidation of 10-acetyl-3,7-dihydroxyphenoxazine (amplex red) coupled to the enzymatic reduction of H_2O_2 by horseradish peroxidase. The rate of H_2O_2 generation after addition of 4 mM succinate was spectrofluorimetrically determined by the change in fluorescence at excitation and emission wavelengths of 544 and 590 nm, respectively.

References

Affourtit, C. and M.D. Brand, 2009. Measuring mitochondrial bioenergetics in INS-1E insulinoma cells. Methods Enzymol. 457, 405-424.

Amo, T., and M.D. Brand, 2007. Were inefficient mitochondrial haplogroups selected during migrations of modern humans? A test using modular kinetic analysis of coupling in mitochondria from cybrid cell lines. Biochem. J. 404, 345-351.

Brand, M.D., 1990. The proton leak across the mitochondrial inner membrane. Biochim. Biophys. Acta 1018, 128-133.

Brand, M.D., 1998. Top-down elasticity analysis and its application to energy metabolism in isolated mitochondria and intact cells. Mol. Cell. Biochem. 184, 13-20.

Brand, M.D., 2005. The efficiency and plasticity of mitochondrial energy transduction. Biochem. Soc. Trans. 33, 897-904.

Brown, R.E., K.L. Jarvis, and K.J. Hyland, 1989. Protein measurement using bicinchoninic acid: elimination of interfering substances. Anal. Biochem. 180, 136-139.

Cadenas, S., K.S. Echtay, J.A. Harper, M.B. Jekabsons, J.A. Buckingham, E. Grau, A. Abuin, H. Chapman, J.C. Clapham, and M.D. Brand, 2002. The basal proton conductance of skeletal muscle mitochondria from transgenic mice overexpressing or lacking uncoupling protein-3. J. Biol. Chem. 277, 2773-2778.

Hagopian, K., M.E. Harper, J.J. Ram, S.J. Humble, R. Weindruch, and J.J. Ramsey, 2005. Long-term calorie restriction reduces proton leak and hydrogen peroxide production in liver mitochondria. Am. J. Physiol. Endocrinol. Metab. 288, E674-684.

Hocking, P.M., B.O. Hughes, and S. Keer-Keer, 1997. Comparison of food intake, rate of consumption, pecking activity and behaviour in layer and broiler breeder males. Br. Poult. Sci. 38, 237-240.

Korshunov, S.S., V.P. Skulachev, and A.A. Starkov, 1997. High protonic potential actuates a mechanism of production of reactive oxygen species in mitochondria. FEBS Lett. 416, 15-18.

Kuenzel, W.J., and N.T. Kuenzel, 1977. Basal metabolic rate in growing chicks *Gallus domesticus*. Poult. Sci. 56, 619-627.

Miwa, S., J. St-Pierre, L. Partridge, and M.D. Brand, 2003. Superoxide and hydrogen peroxide production by *Drosophila* mitochondria. Free Radic. Biol. Med. 35, 938-948.

Mujahid, A., Y. Akiba, and M. Toyomizu, 2009. Olive oil-supplemented diet alleviates acute heat stress-induced mitochondrial ROS production in chicken skeletal muscle. Am. J. Physiol. Regul. Integr. Comp. Physiol. 297, R690-698.

Mujahid, A., N.R. Pumford, W. Bottje, K. Nakagawa, T. Miyazawa, Y. Akiba, and M. Toyomizu, 2007. Mitochondrial oxidative damage in chicken skeletal muscle induced by acute heat stress. J. Poult. Sci. 44, 439-445.

Mujahid, A., Y. Yoshiki, Y. Akiba, and M. Toyomizu, 2005. Superoxide radical production in chicken skeletal muscle induced by acute heat stress. Poult. Sci. 84, 307-314.

Rolfe, D.F.S., A.J. Hulbert, and M.D. Brand, 1994. Characteristics of mitochondrial proton leak and control of oxidative phosphorylation in the major oxygen-consuming tissues of the rat. Biochim. Biophys. Acta 1188, 405-416.

Rolfe, D.F.S., J.M. Newman, J.A. Buckingham, M.G. Clark, and M.D. Brand, 1999. Contribution of mitochondrial proton leak to respiration rate in working skeletal muscle and liver and to SMR. Am. J. Physiol. 276, C692-699.

Rottenberg, H., 1983. Uncoupling of oxidative phosphorylation in rat liver mitochondria by general anesthetics. Proc. Natl. Acad. Sci. USA 80, 3313-3317.

Saito, S., T. Takagi, T. Koutoku, E.S. Saito, H. Hirakawa, S. Tomonaga, T. Tachibana, D.M. Denbow, and M. Furuse, 2004. Differences in catecholamine metabolism and behaviour in neonatal broiler and layer chicks. Br. Poult. Sci. 45, 158-162.

Saunderson, C.L., and S. Leslie, 1988. Muscle growth and protein degradation during early development in chicks of fast and slow growing strains. Comp. Biochem. Physiol. A Comp. Physiol. 89, 333-337.

Skulachev, V.P., 1999. Mitochondrial physiology and pathology; concepts of programmed death of organelles, cells and organisms. Mol. Aspects Med. 20, 139-184.

Swennen, Q., E. Delezie, A. Collin, E. Decuypere, and J. Buyse, 2007. Further investigations on the role of diet-induced thermogenesis in the regulation of feed intake in chickens: comparison of age-matched broiler versus layer cockerels. Poult. Sci. 86, 895-903.

Insights into sarcopenia: interrelation between protein profiles and mitochondrial functionality in rat ageing skeletal muscle

A. Lombardi[1]*, E. Silvestri*[2]*, D. Glinni*[2]*, F. Goglia*[2] *and M. Moreno*[2]
[1]*Dipartimento delle Scienze Biologiche Sezione di Fisiologia ed Igiene, Università degli Studi di Napoli Federico II, Via Mezzocannone8, 80134 Naples, Italy*
[2]*Dipartimetno di Scienze Biologiche ed Ambientali, Università degli Studi del Sannio, Via Port'Arsa 11, Benevento, Italy*

Abstract

Ageing skeletal muscle undergoes a physiological decline that can be considered the consequence of complex multi-factorial processes that are still not completely understood. Several lines of evidence indicate that changes in mitochondria that occur during ageing are key factors in the loss of functionality observed in ageing muscle. In the present manuscript, we summarize recent data on age-related changes in the proteomic profiles of rat skeletal muscle tissue and mitochondria, and the associated changes in the supramolecular organization of the mitochondrial respiratory chain complex, in relation to mitochondrial functionality and the efficiency of oxidative phosphorylation. In addition, since uncoupling protein-3 (UCP3) declines with age and since recent data have indicated its ability to export lipid hydroperoxide from the mitochondrial matrix while at the same time mediating lipid hydroperoxide-induced mitochondrial uncoupling, we will discuss possible roles for UCP3 in both age-related metabolic adaptations of skeletal muscle to oxidative metabolism and oxidative damage to vital mitochondrial components.

Introduction

Among all the tissues of the body, skeletal muscle, by virtue of accounting for about 40% of body mass, plays important roles in energy and metabolic homeostasis, in addition to its roles in coordinated movements and postural control. During ageing, the progressive decline that occurs in the mass and strength function of skeletal muscle is referred to as sarcopenia. Sarcopenia can be considered the consequence of multiple complex processes, the underlying mechanisms of which are not completely understood. In addition to progressive denervation, the age-related changes in skeletal muscle include alterations in excitation-contraction coupling at the level of the triad junction, alterations in contractile fibres and their surrounding connective-tissue anatomy, and increased cellular stress, as well as hormonal and metabolic disturbances (Melton *et al.*, 2000, Greenlund and Nair, 2003). Moreover, pathophysiological cycles of denervation - reinnervation ultimately lead to a loss of motor units or to incomplete recruitment of those that remain. This affects the overall number, size and type of fibres, which in turn leads to weaker contractile properties (Luff, 1998).

Mitochondria play a variety of essential roles in cellular metabolism: e.g. production of ATP through oxidative phosphorylation, regulation of calcium homeostasis in cell-signalling pathways, and the initiation of the signal cascade leading to apoptosis (Green and Reed, 1998). In addition, mitochondria have been implicated in the ageing process as well as in many age-related diseases (Figueiredo *et al.*, 2008).

The free radical theory of ageing, proposed by Harman (1956), asserts that reactive oxygen species (ROS), generated both endogenously and externally, can give rise to the cellular damage that underlies the ageing process. The major source of cellular ROS is mitochondria, which have thus been proposed to play a significant role in the ageing process. These ROS-producing organelles are also the primary target site of oxidative stress (Stadtman, 2002). Oxidative damage to mitochondria amplifies the effect, ROS causing damage to mitochondrial DNA, proteins, and membrane lipids, a self-perpetuating and destructive cycle in which increased ROS production leads to incremental damage and to further ROS production (Harper *et al.,* 2004; Pak *et al.*, 2003).

Over time, a decline in mitochondrial ATP production will reduce the principal cellular source of energy, and this in turn will lead to cellular dysfunction if the oxidative damage exceeds the capacity of the cell to handle removal of both ROS and oxidatively damaged macromolecules. Indeed, an increase in the incidence of oxidative damage to mitochondrial mtDNA has been observed in various tissues with age, and there is strong support for ROS playing a predominant role in these processes (Gokey *et al.*, 2003). The accumulation of mtDNA modifications and mutations that occurs with age may interfere with the synthesis of proteins and with the enzymatic pathways responsible both for the electron transfer along the respiratory electron transport chain and for the synthesis of ATP (Gokey *et al.*, 2003).

Oxidative stress- induced mitochondrial dysfunction may play an important role in the age-related decrease in skeletal muscle mass and function as this tissue, being a metabolically very active tissue and very rich in mitochondria, would be expected to be particularly vulnerable to mitochondrial oxidative stress (Figueiredo *et al.*, 2008). In this context it has been recently shown that increased superoxide *in vivo* accelerates age-associated muscle atrophy through mitochondrial dysfunction and neuromuscular junction degeneration (Jang *et al.*, 2009) Although a reduced activity of the respiratory chain and increased release of reactive oxygen species from mitochondria are well documented during ageing, there is little information as to the mechanism(s) responsible for the changes in mitochondrial function.

In this paper, we review some of our recent data on the proteomic profiles of ageing skeletal muscle in the rats, with particular attention being paid to the mitochondrial protein profile and to the supramolecular organization of the mitochondrial respiratory chain. Proteomic profiles will then be related to mitochondrial functionality and to the efficiency of oxidative phosphorylation. In addition, in relation to the etiology of sarcopenia, we discuss possible interrelated roles among ROS production, mitochondrial efficiency, and the decline in the level of mitochondrial uncoupling protein-3 (UCP3).

Total and mitochondrial protein profiles in the skeletal muscle of aged rats

By using two–dimensional gel electrophoresis (2D-E) followed by MALDI-Tof analysis, we recently compared the protein expression profile in gastrocnemius muscle between young (3 month) and old (24 month) rats (Lombardi *et al.,* 2009). A pattern predominantly of down-regulation in protein expression levels was observed with age, with changes being detected in proteins involved in energy metabolism, cellular signalling, the stress response, cytoskeleton structure, and contraction. Perturbation of the myofibrillar network was evident as alterations in contractile proteins and also in their major regulators. Indeed, in gastrocnemius muscle from the old rats we observed a down-regulation of myosin regulatory light chain2, and an upregulation of myosin light chain 1. Tropomyosin alpha chain isoform and troponin T, two proteins known to bestow calcium sensitivity on actomyosin ATPase activity, were both down-regulated, suggesting an age-related alteration in fibre-shortening velocity. In this context, a study by Capitanio *et al.* (2009) showed significant changes in myosin heavy chain isoforms (MyHC) during aging (when comparing gastrocnemius muscles between 8-month old rats vs. 22-month-old ones), with increases in MyHC-I slow twitch oxidative fibres and MyHC-IIA fast oxidative/glycolytic fibers and a decrement in MyHC-IIB fast glycolytic fibres. The above data suggest a transition from a fast to a slow phenotype, an inference supported by proteomics studies performed by others (O'Connell *et al.,* 2007, 2009; Piec *et al.*, 2005). Beta-tubulin, a component of intermediate filaments, microtubules, and the microfilament cytoskeleton was upregulated in ageing gastrocnemius samples (Lombardi *et al.*, 2009). This evidence, taken together with data obtained by others showing increases in desmin (Capitanio *et al.*, 2009) and gelsolin (Piec *et al.*, 2005), suggests a mechanism affecting the cytoskeleton that compensates for perturbations in myofibrillar structure, so avoiding widespread damage to the myofibres. This could also result from the denervation and reinnervation cycles that occur in ageing skeletal muscle and which may generate muscle atrophy, regeneration cycles, and rearrangement of motor units.

In ageing, the shift in structural fibers towards the slow phenotype indicates an adaptative change towards the use of fatty acid rather than glucose as metabolic fuel. Indeed, glycolytic enzymes such as triosephosphate isomerase, glyoxalase I, and beta enolase were all down-regulated, and pyruvate dehydrogenase E1, which is localized at mitochondrial level, was also decreased in gastrocnemius samples from aged rats (Lombardi *et al.*, 2009).

Other features indicating a perturbation of energy metabolism were down-regulations of creatine kinase, pyruvate kinase, and the NADH-shuttle glycerol 3-phosphate dehydrogenase (Lombardi *et al.*, 2009). Moreover, key mitochondrial enzymes involved in oxidative phoshorylation and the Krebs cycle were also affected by ageing, as described following.

Is well known that in aged tissue mitochondria produce an excessive amount of reactive oxygen species (ROS). We recently reported that the response of ageing skeletal muscle to an elevated ROS production consists of upregulations of specific proteins able to scavenge both ROS and the cytotoxic products resulting from ROS-induced lipid peroxidation, such as reactive aldehyde. Indeed, the levels of the cytoplasmic Cu/Zn superoxide dismutase (SOD) and ferritin heavy chain (H ferritin) isoform, two enzymes involved in ROS scavenging, were significantly increased in muscle from old rats (Lombardi *et al.*, 2009). The ageing gastrocnemius also exhibits activation of a compensatory mechanism that scavenges reactive aldehyde products by increasing the levels of glutathione transferase and mitochondrial aldehyde dehydrogenase. In addition the glyoxalase I protein level was reduced in skeletal muscle from aged rats. This enzyme, together with glyoxalase II, belongs to a widespread enzyme system that catalyzes the glutathione-dependent conversion of methylglyoxal (a compound exerting cytotoxic actions by interacting with subcellular proteins and nucleic acid) to D-lactate, suggesting an accumulation of cytotoxic methylglyoxal within the ageing myocyte that, together with ROS, could underlie the ageing process in skeletal muscle (Lombardi *et al.*, 2009).

Evidence of age-associated protein misfolding has been provided by the finding of upregulations of molecular chaperones (including HSP 27 and disulfide isomerase ER60) (Lombardi *et al.*, 2009).

Although the various proteomic studies on age-related changes in skeletal muscle, including ours, may differ in the list of individual proteins affected (O'Connell *et al.*, 2007; Capitanio *et al.*, 2009, Piec *et al.*, 2005), as a whole the studies performed so far agree in demonstrating that ageing affects metabolic pathways including those involved in energy metabolism, cellular signalling, the stress response, the cytoskeleton, and contraction (Figure 1).

Since mitochondria play a significant role in the ageing process, an evaluation of the age-related alterations in the mitochondrial profile would be expected to increase our understanding of the mechanisms underlying ageing. As above cited, mitochondria are severely effected by ageing, and it is generally believed that dysfunctions of the mitochondria trigger key steps in the ageing process (Dencher *et al.*, 2007). These dysfunctions affect oxidative phosphorylation system and even more the supramolecular organization of respiratory complexes. The oxidative phosphorylation system forms the basis for mitochondrial ATP production. In most organisms it is composed of the ATP synthase complex (complex V) and four oxidoreductase complexes constituting the electron transport chain (ETC): NADH dehydrogenase (complex I), succinate dehydrogenase (complex II), cytochrome *c* reductase (complex III), and cytochrome *c* oxidase (complex IV). All these complexes reside within the inner mitochondrial membrane. Complexes I and II transfer electrons from NADH or FADH2 to ubiquinone, while complex III transfers electrons from ubiquinol to oxygen. Three of the four oxidoreductase complexes couple electron transport with translocation of protons from

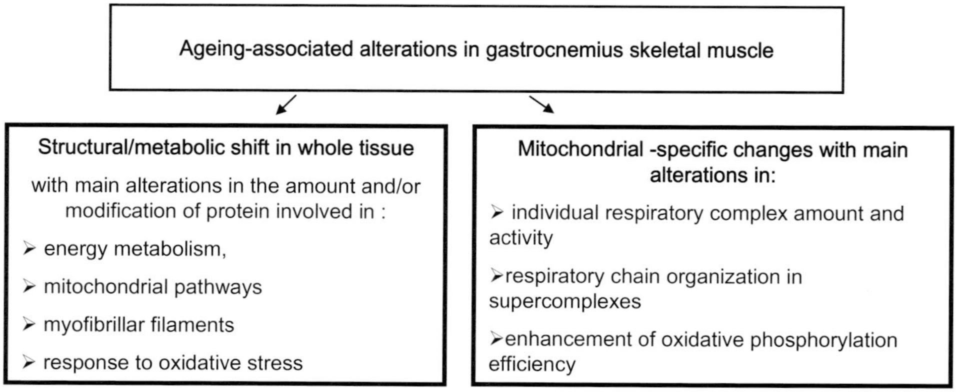

Figure 1. Schema representing the age-related alterations observed in gastrocnemius skeletal muscle at the tissue and mitochondrial levels, as deduced from proteomic and functional studies.

the mitochondrial matrix to the intermembrane space. The generated proton gradient can be used by complex V to catalyze the formation of ATP by the phosphorylation of ADP. These complexes, when isolated in their native state, are functionally active.

By using 2D-E analysis on total gastrocnemius muscle, we found that key mitochondrial enzymes, involved in the Krebs cycle and oxidative phosphorylation were reduced in ageing muscle (namely, isocitrate dehydrogenase, cytochrome c- oxidase, and ATP synthase beta subunits). On the other hand an upregulation of aldehyde dehydrogenase occurred (Lombardi *et al.*, 2009). Using difference gel electrophoresis (DIGE) (an approach that facilitates a sensitive quantitative assessment of changes in soluble proteins), O'Connell *et al.* (2009) compared the skeletal muscle mitochondria-enriched fraction isolated from 26-month-old rats with that from 3-month-old rats. Their main finding was that the expressions of key mitochondrial proteins localized to the mitochondrial inner membrane, the matrix, the inner membrane space, and the outer membrane were increased in the mitochondria from the aged rats. The following proteins showed such an increase: NADH dehydrogenase, the mitochondrial inner membrane protein mitofilin, peroxiredoxin isoform PRX-III, ATPase synthase, succinate dehydrogenase, mitochondrial fission protein Fis1, succinate-coenzyme A ligase, acyl-coenzyme A dehydrogenase, porin isoform VDAC2, ubiquinol-cytochrome c reductase core I protein, and prohibitin (O' Connell *et al.*, 2009). They therefore suggested an enhancement of mitochondrial functionality, in agreement with the idea of an age-related shift to a slower contractile phenotype and more aerobic-oxidative metabolism.

The use of 2D gels of highly purified mitochondria greatly facilitates the analysis and detection of less-abundant proteins (since mitochondrial proteins are distributed across wide ranges of both pH and molecular mass on 2D gels, leading to a more accurate resolution with only a few protein-spot overlaps). However, it allows the analysis only of the soluble subunits of membrane protein complexes, while hydrophobic subunits or aggregates do not focalize in the first dimension and are not represented on the final 2D map. Actually, mitochondrial functionality is strictly related to the abundance and activity of individual mitochondrial complexes, and even more to their supramolecular organization. In fact, current evidence suggests that the assembly of the electron transport chain complexes into respirasomes provides structural support for mitochondrial oxidative phosphorylation by facilitating electron channelling, and perhaps by preventing electron leak and superoxide production (Schagger *et al.*, 2000, 2004). Thus, a reduced amount of a single complex subunit or of an entire complex could not affect mitochondrial functionality.

Blue Native-Polyacrylamide Gel Electrophoresis (BN-PAGE) allows the conservation of protein-protein interaction and activity in the native state, as well as the supramolecular organization of respiratory complexes, although the results depend on the detergent used to solubilize the mitochondrial membranes (Krause, 2006). BN-PAGE of dodecylmaltoside-solubilized mitochondria resolves respiratory complexes individually, and following Coomassie blue staining, the levels of the individual respiratory complexes can be detected. By performing BN-PAGE, we showed that mitochondria from the gastrocnemius muscles of old rats contained significantly lower amounts of complexes I, III, and V, when compared with young ones. On the other hand, we observed an increased amount of complex II and no change in complex IV. In-gel histochemical colorimetric reactions of native complexes allow the evaluation of their catalytic activity. In the gastrocnemius mitochondria of old animals, a lower activity of each of the complexes was detected, suggesting that the reduced activity levels of complexes I, III and V resulted from their presence in smaller amounts (Lombardi *et al.*, 2009). This was not the case for complex II or complex IV, since despite their levels being higher or unchanged, respectively, a reduction in the activity of each was detected. This discrepancy suggests the eventual involvement of ROS attack either directly on the complex or mediated by cardiolipin oxidation, as reported by Paradies *et al.* (2000). Our data, on age-related reduced activities of mitochondrial electron transport chain complexes, obtained in rat skeletal muscle are in agreement with that obtained on mice skeletal muscle by Mansouri *et al.* (2006).

To elucidate whether the ageing process also alters the functional/structural organization of the respiratory chain in terms of the assembly of supercomplexes, we performed BN-PAGE on rat gastrocnemius mitochondria solubilized with the mild detergent digitonin since this extensively retains inner mitochondrial membrane supercomplexes (Krause, 2006). When we compared BN-PAGE of digitonin-solubilised mitochondria with that of dodecylmaltoside solubilised mitochondria, we found that with the former method, monomeric complex I and dimeric complex III were significantly reduced, in both young and old mitochondria. However, they were found assembled in supercomplexes ranging from 1,500 to 2,100 kDA (Lombardi *et al.*, 2009). It should be noted that two major supercomplexes (band *a* and band *b*) and two minor ones (band *c* and band *d*) can be identified. Band *a* contains monomeric complex I and dimeric complex III, band *b* contains monomeric complex I, dimeric complex III, and monomeric complex IV, band *c* contains monomeric complex I, dimeric complex III, and dimeric complex IV, and band *d* contains monomeric complexes I, dimeric complex III, and trimeric complex IV (Schagger *et al.*, 2004).

The supercomplex profile of the mitochondria from old rats was significantly modified in comparison to that obtained from young rats, the heavier supercomplexes (bands *c*, *d*) being better represented in samples from old rats and the lighter complexes being less represented (band *a*). This suggests that, in aged muscle, although reduced levels and activities of respiratory complexes can be detected, there is a significant increase in the supramolecular assembly of respiratory chain complexes into respirasomes. Possibly, this could be a compensatory, ageing-related mechanism in muscle that is functionally directed towards substrate channelling and catalytic enhancement advantaging. If so, skeletal muscle mitochondria would be able to operate even in the presence of a reduced electron supply to the respiratory chain and in the presence of reduced individual complex activity. In addition, assembly into supercomplexes may suggest a compensatory mechanism that reduces the age-related enhancement of mitochondrial oxidative stress. In fact, assembly of complexes into respirasomes has been shown to reduce the level of reactive intermediates such as ubisemiquinones (Schagger and Pfeiffer, 2000).

Age-related changes in skeletal muscle mitochondrial functionality

To gain further insight into the age-related adaptation of skeletal muscle to a more aerobic-oxidative metabolism, studying mitochondrial bioenergetic parameters and mitochondrial oxidative phosphorylation efficiency is of crucial importance.

Upon comparing mitochondria from young and old rats, we did not detect any significant difference between them in their capacity to oxidise substrate, irrespective of the substrate used. Indeed, in skeletal muscle mitochondria from old rats, there was never an inhibition of state 3 respiration (coupled respiration in which the synthesis and export of ATP is at the maximal rate), whether it was detected using FAD-linked substrate (i.e. succinate + rotenone) or NADH-linked substrate (glutamate + malate and pyruvate + malate) (Lombardi *et al.,* 2009; Kerner *et al.*, 2001). In addition, we did not observe a difference (old vs. young) in the maximal capacity of mitochondria to oxidize lipid whether palmitoyl-carnitine or palmitoyl CoA+ carnitine was used as substrate.

It should be taken into account that the state 3 respiration rate is controlled by: (1) the activity of the reactions involved in the oxidation of the substrates and in the production of membrane potential, including the activities of electron transport chain complexes and (2) the activity of reactions that use the membrane potential for the synthesis and the export of ATP, including ATP synthase activity. The absence of an age-related change in state 3 respiration, despite a reduction of the activity of each individual complex of the electron transport chain as well as of the ATP synthase complex, support the hypothesis that supramolecular assembly of respiratory chain complexes into respirasomes (described in the previous chapter) can compensate for the complex being present at the lower levels and activities in old mitochondria.

On the other hand, mitochondria from old rats showed a significant decrease in state 4 respiration (uncoupled respiration in which there is no synthesis of ATP and mitochondria respire principally to balance the flux of protons from the mitochondrial inner membrane space to the matrix). This, in association with an unchanged coupled respiration, reflects an enhancement of oxidative phosphorylation efficiency in mitochondria from old animals. Such an enhancement was confirmed by an experiment designed to detect age-related alterations in the coupling efficiency of skeletal muscle mitochondria (Lombardi *et al.*, 2009). Indeed, when the synthesis of ATP was inhibited, skeletal muscle mitochondria isolated from 3-month-old rats had to respire more than those isolated from 24-month-old ones in order to maintain the same membrane potential. These data indicated that when the driving force acting on protons was the same, their flux across the mitochondrial inner membrane (proton leak) was greater in young mitochondria than in old ones because of a greater proton conductance. This was found to be about 60% higher in young mitochondria than in old ones. What the above indicates is that in young mitochondria, a greater portion of the energy contained in the proton motive force is dissipated as heat and not used for the synthesis of ATP. Several endogenous factors may influence mitochondrial proton leak with hormones playing an important role. Indeed the endocrine system undergoes a lost of functionality with the ageing organism and several tissues become resistant to hormones (Silvestri *et al.*, 2008). Among the hormonal factors able to affect mitochondrial efficiency thyroid hormone (T3) plays a significant role. In this context we reported a progressive enhancement in skeletal muscle proton leak during hypothyroidism-euthyroidism-hyperthyroidism transition (Lanni *et al.*, 1999). However, during aging a decline in thyroid function occurs and a reduction in thyroid hormone levels has been reported (Silvestri *et al.*, 2008); thus the reduction in proton leak observed in skeletal muscle mitochondria isolated from old rats could presumably be the results of a such decline. Age-related increase in mitochondrial efficiency may be responsible for a fattening of older animals as compared to young ones when they are fed the same amount of food (Harper *et al.*, 2008). Such an occurrence should be taken into account when considering aged animals including humans and livestock. Despite studies in humans support a role for mitochondrial efficiency in the disturbance of metabolic homeostasis, evidence for livestock is lacking.

The age-related interactions between mitochondrial functioning and efficiency of respirasomes, feed efficiency and body composition remain to be investigated in farm animals. It should be also taken into account that despite the age-related decline in mitochondrial proton leak is a process that has positive effect in terms of the efficiency of oxidative phosphorylation, it could increase the mitochondrial oxidative stress (such as ROS attack on ETC complexes). This aspect will be discussed in more detail in the following section.

As a whole, the data described above reveal that gastrocnemius mitochondria from old rats, in order to produce a similar amount of ATP, need to oxidize a smaller amount of substrate than those from young rats. In addition, in ageing skeletal muscle mitochondria actuate compensatory mechanisms (super-assembly of complexes and reduced proton leak) that render them more efficient despite lower levels and/or activities of respiratory complexes, thus allowing a metabolic shift toward oxidative metabolism that is coordinated with the structural shift.

The above data on the ageing-associated alteration in gastrocnemius skeletal muscle are summarized in Figure 1.

ROS production, proton leak, and uncoupling protein 3: an interrelated role in sarcopenia?

Mitochondrial ROS production is strongly and positively influenced by proton motive force. Skulachev's group (Korshunov *et al.*, 1998) were the first to suggest that 'mild uncoupling', by reducing the mitochondrial proton motive force, could prevent excess ROS production. The flux of protons from the inner membrane space to the matrix not associated with ATP synthesis (proton leak) is a process that dissipates the proton motive force, with the energy of the gradient being dissipated as heat. Although the physiological role of the proton leak is not entirely clear (Brand and Esteves, 2005), it has been proposed that this process is involved in the regulation of mitochondrial ROS production (Brand and Esteves, 2005; Ecthay, 2007). This could suggest that proton leak might influence ageing related processes by modulating ROS production, and indeed it has been proposed that mitochondrial uncoupling may lead to an increase in lifespan (Brand, 2000; Speakman *et al.*, 2004). The evidence obtained by Speakman *et al.* (2004), of an increased proton leak in skeletal muscle isolated from long-lived mice, seem to support this idea, but evidences to date suggesting a consistent role for uncoupling in life span are few.(reviewed in Mookerjee *et al.*, 2010) In a different study Amara *et al.* (2007) reported that mild mitochondria uncoupling impacts cellular ageing in human muscle *in vivo*.

Actually, two types of proton leak may be present at the mitochondrial level: namely, basal proton leak and inducible proton leak. The basal type is present in mitochondria within every tissue, and may be related both to the lipid environment of the membrane and to specific proteins such as adenine nucleotide translocase (ANT) (Brand *et al.,* 2005b; Shabalina *et al.*, 2006). The inducible type seems to be tightly regulated and is mediated by specific proteins. In skeletal muscle mitochondria, uncoupling protein 3 (UCP3) plays a significant role in the latter process (Brand *et al.,* 2005).

Although the mechanism by which UCP3 catalyzes the proton leak is unclear, there is a broad consensus that UCP3 activates it only in the presence of specific cofactors, with FFA and ROS playing crucial roles (Brand, 2005). It has been shown that the presence of UCP3 is associated with a lower ability of mitochondria to produce ROS, a process that seems to be related to the ability of UCP3 to lower the mitochondrial proton motive force. Other data that support a role for UCP3 in mitochondrial oxidative stress include the following: UCP3-KO mice showed an enhancement of *in vivo* aconitase inhibition by ROS (Vidal-Puig *et al.,* 2000) and an enhancement of both mitochondrial ROS production (Nabben *et al.*, 2008) and H_2O_2 release (Gong *et al.*, 2000). On the other hand, in L6 muscle cells overexpressing UCP3, the production of H_2O_2 was reduced, and an inhibition of UCP3 in rat skeletal muscle by GDP increased H_2O_2 release by mitochondria (Talbot *et al.*, 2004, Lombardi *et al.*, 2008)

The release of superoxide into the matrix is a process that would activate UCP3 uncoupling activity; the mechanism proposed for this process takes into account the formation of membrane lipid peroxidation products such as 4-hydroxynonenal (HNE) (Echtay *et al.*, 2003). By activating UCP3, HNE would provide a negative feedback loop that would limit the mitochondrial production of ROS and the damage they induce (Echtay *et al.*, 2003). An involvement of fatty acid peroxidation in the activation of UCP3 was also suggested by Goglia and Skulachev (2003), who hypothesized the participation of UCP3 in the export of lipid hydroperoxides (LOOH) outside the matrix. Upon actuation of this process, the inner leaflet of the mitochondrial inner membrane would be cleared of LOOH, which that might otherwise trigger a cascade leading to oxidative damage to the mitochondrial DNA and enzymes, and to other mitochondrial matrix-localized components of vital importance (Goglia and Skulachev, 2003). In this context, by using mitochondria isolated from UCP3-null mice and their wild type littermates, we provided evidences that UCP3 is involved in the extrusion of LOOH from the matrix (Lombardi *et al.,* 2010). Indeed we reported that mitochondria are able to release LOOH and that UCP3-null mitochondria displayed a limited capacity to release LOOH compared to their wild-type controls. In addition, we also reported that O_2-released into the matrix and (presumably) the consequent formation of LOOH at the level of the matrix side of the mitochondrial inner membrane are crucial factors in UCP3-mediated LOOH release. The above supports the hypothesis of Goglia and Skulachev (2003).

In addition, by evaluating proton-leak kinetic in both UCP3-null mitochondria and wild type littermates (while modulating mitochondrial endogenous superoxide production, the levels of free fatty acid and lipid hydroperoxide formation) we obtained data indicating that (1) an interrelated role between O_2- and polyunsaturated fatty acid exists in the induction of UCP3 mediated uncoupling and (2) lipid peroxidation is a key event in the activation of such an uncoupling (Lombardi *et al.*, 2010). The above data demonstrate that UCP3 is involved both in mediating the translocation of LOOH across the mitochondrial inner membrane and in LOOH-dependent mitochondrial uncoupling. In effect UCP3-catalyzed extrusion of LOOH from the matrix would reduce lipotoxicity and consequent damage to the matrix component, and at the same time the UCP3-mediated proton leak would lead to a reduced ROS formation.

The above predicts that a lack of UCP3 would result in increased lipid peroxidation (because of higher ROS production), accumulation of LOOH within the matrix, and damage to mitochondrial DNA, RNA, and proteins. Direct evidence for this comes from the observation that mice lacking UCP3 show increased levels of oxidative damage to proteins, lipids, and mtDNA (Brand *et al.*, 2002). Some of these data were confirmed by Asami and colleagues (2008), who showed an increase in lipid peroxidation in UCP-null mice compared to control and UCP3-overexpressing mice, although they did not find any difference in carbonyls between these groups of mice. In skeletal muscle, an age-dependent decline in the mitochondrial UCP3 protein content has been observed. Kerner *et al.* (2001) reported that UCP3 protein was significantly less abundant in skeletal muscle mitochondria from 24-month-old rats than in those from 8-month-old ones. This decline was associated with an inhibition of uncoupled respiration (Kerner *et al.*, 2001). In this context, as described in the previous chapter, we also reported an age-related decline in mitochondrial uncoupled respiration, and in addition we reported that the proton leak in gastrocnemius mitochondria was significant smaller in those isolated from 24-mounth-old rats than in those from 3-month-old rats (Lombardi *et al.*, 2009). It is important to underline that we detected this alteration in an experimental condition in which UCP3 may be activated (i.e. when mitochondria produce high levels of ROS and in the presence of FFA), thereby supporting the relationship between age-related declines in UCP3 levels and proton leak inhibition.

Evidence supporting the idea that the age-related reduction in UCP3 levels is associated with a rise in ROS production was published by Nabben *et al.* (2008), who showed that the increase in mitochondrial ROS generation associated with ageing is blunted in UCP3 over-expressing mice.

Thus, there are experimental evidences showing: (1) a role for UCP3 in mediating both the extrusion of LOOH from the matrix and an LOOH- induced proton leak (Lombardi *et al.,*. 2010), (2) age-related decline in UCP3 levels (Kerner *et al.*, 2001) and (3) the involvement of a reduction in UCP3 levels in the age-related increase in ROS formation (Nabben *et al.*, 2008). On that basis, it is tempting to hypothesize that the decline in UCP3 in ageing mitochondria might contribute to the etiology of mitochondrial dysfunction in elderly individuals, which is characterized by increased levels of oxidative damage to lipids and proteins, and alterations in mitochondrial function. At the same time, an age-dependent decline in UCP3 levels, associated with a reduced proton leak and increased efficiency of mitochondrial oxidative phosphorylation, could underlie skeletal muscle's adaptations to oxidative metabolism (Figure 2).

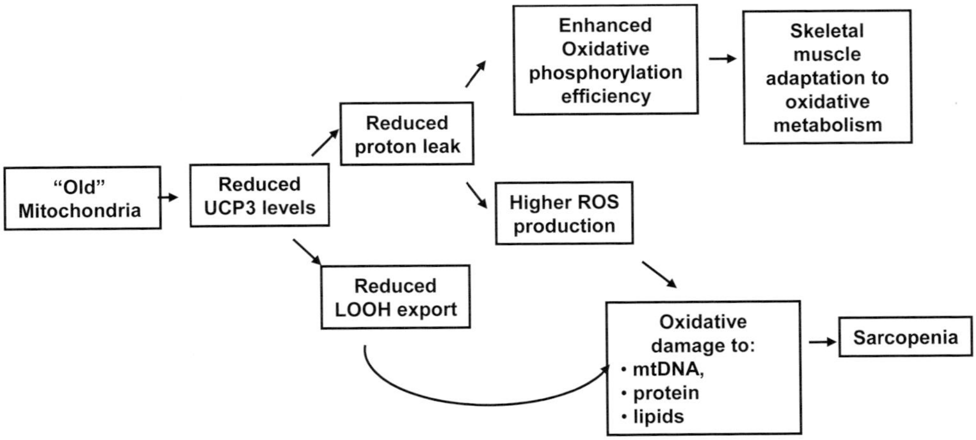

Figure 2. Schematic representation of the possible role of a decline in skeletal muscle UCP3 age-associated metabolic adaptations and sarcopenia.

References

Amara C.E., E.G. Shankland, S.A Jubrias, D.J. Marcinek, M.J. Kushmerick and K.E Conley, 2007. Mild mitochondrial uncoupling impacts cellular aging in human muscles *in vivo*. Proc Natl Acad Sci USA. 104, 1057-62.

Asami, D.K., R.B. McDonald, K. Hagopian, B.A. Horwitz, D. Warman, A. Hsiao, C. Warden and J.J. Ramsey, 2008. Effect of aging, caloric restriction, and uncoupling protein 3 (UCP3) on mitochondrial proton leak in mice. Exp Gerontol. 43, 1069-76.

Brand, M.D. and T. Esteves, 2005. Physiological functions of the mitochondrial uncoupling proteins UCP2 and UCP3 Cell Metab 2, 85-93.

Brand, M.D., J.K. Pakay, A. Oclo, J. Kokoszka, D.C. Wallace, P.S. Brooke and E.J. Cornwall, 2005b. The basal proton conductance of mitochondria depends on adenine nucleotide translocase content. Biochem. J. 392, 353-362.

Brand, M.D., 2000. Uncoupling to survive? The role of mitochondrial inefficiency in ageing. Exp Geront 35, 811-820.

Brand, M.D, R. Pamplona, M. Portero-Otín, J.R., Requena, S.J., Roebuck, J.A. Buckingham, J.C. Clapham, S. Cadenas, 2002. Oxidative damage and phospholipid fatty acyl composition in skeletal muscle mitochondria from mice underexpressing or overexpressing uncoupling protein 3. Biochem J. 368, 597-603.

Capitanio, D., M. Vasso, C. Fania, M. Moriggi, A. Viganò, P. Procacci, V. Magnaghi and C. Gelfi, 2009. Comparative proteomic profile of rat sciatic nerve and gastrocnemius muscle tissues in ageing by 2-D DIGE. Proteomics 9, 2004-20.

Cioffi, F., R. Senese, P. de Lange, F. Goglia, A. Lanni, A. Lombardi, 2009. Uncoupling proteins: A complex journey to function discovery. Biofactors 35, 417- 428.

Dencher, A., M. Frenzel, N.H. Reifschneider, M. Sugawa and F. Krause, 2007. Proteome alterations in rat mitochondria caused by aging. Ann. NY Acad. Sci. 1100, 291-298.

Ecthay, K.S., 2007. Mitochondrial uncoupling proteins: What is their physiological role? Free Rad. Biol. Med. 43, 1351-1371.

Echtay, K.S., T.C. Esteves, J.L. Pakay, M.B. Jekabsons, A.J. Lambert, M. Portero-Otin, R. Pamplona, A.J. Vidal-Puig, S.J. Wang and M.D. Brand, 2003. A signalling role for 4-hydroxy-2-nonenal in regulation of mitochondrial uncoupling. EMBO J. 22, 4103-4110.

Figueiredo, P.A., M.P. Mota, H.J. Appell and J.A. Duarte, 2008. The role of mitochondria in aging of skeletal muscle. Biogernontology 9, 67-84.

Goglia, F. and V.P. Skulachev, 2003. A function for novel uncoupling proteins: antioxidant defense of mitochondrial matrix by translocating fatty acid peroxides from the inner to the outer membrane leaflet. FASEB J. 17, 1585-1591.

Gong, D.W., S. Monemdjou, O. Gavrilova, L.R. Leon, B. Marcus-Samuels, C.J. Chou, C. Everett, L.P. Kozak, C. Li, C. Deng, M.E Harper and M.L. Reitman, 2000. Lack of obesity and normal response to fasting and thyroid hormone in mice lacking uncoupling protein-3. J Biol Chem. 275, 16251-16257.

Gokey, N., D. McKenzie and J.M. Aiken, 2003. Mitochondrial DNA mutations as a fundamental mechanism in physiological declines associated with aging. Aging Cell 2, 1-7.

Green, D.R. and J.C. Reed, 1998. Mitochondria and apoptosis. Science 281, 1309-13012.

Greenlund, L.J.S. and K.S. Nair, 2003. Sarcopenia – consequences, mechanisms, and potential therapies. Mechanism Ageing Development 124(3), 287-299.

Harman, D., 1956. Aging: a theory based on free radical and radiation chemistry. J Gerontol 11(3), 298-300.

Harper, M.E, L., Bevilacqua, K., Hagopian, R. Weindruch and J.J Ramsey, 2004. Aging, oxidative stress, and mitochondrial uncoupling. Acta Physiol Scand 182, 321-331.

Harper, M.E., K. Green and M.D. Brand, 2008 The efficiency of cellular Energy transduction and its implication for obesity. Ann. Rev. Nutr. 28, 13-33

Jang Y.C., M.S. Lustgarten, Y. Liu, F.L Muller, A., Bhattacharya, H. Liang, A.B. Salmon, S.V. Brooks , L Larkin, C.R. Hayworth, A. Richardson and H. Van Remmen, 2009. Increased superoxide *in vivo* accelerates age-associated muscle atrophy through mitochondrial dysfunction and neuromuscular junction de generation.FASEB J., In press.

Kerner, J., P.J. Turkaly, P.E. Minkler and C.L. and Hoppel, 2001. Aging skeletal muscle mitochondria in the rat: decreased uncoupling protein-3 content. Am J Physiol Endocrinol Metab 281, E1054-E1062.

Korshunov, S.S., O.V. Korkin, E.K. Ruuge, V.P, Skulachev and A.A Starkov, 1998. Fatty acids as natural uncouplers preventing generation of O2.- and H2O2 by mitochondria in the resting state. FEBS Lett 435, 215-218.

Krause, F., 2006. Detection and analysis of protein-protein interactions in organellar and prokaryotic proteomes by native gel electrophoresis: (Membrane) protein complexes and supercomplexes. Electrophoresis 27, 2759-2781.

Lanni, A, L. Beneduce, A. Lombardi, M. Moreno, O. Boss, P. Muzzin, J.P. Giacobino and F. Goglia, 1999. Expression of uncoupling protein-3 and mitochondrial activity in the transition from hypothyroid to hyperthyroid state in rat skeletal muscle. FEBS Lett. 444, 250-254.

Lombardi, A., P. Grasso, M. Moreno, P. de Lange, E. Silvestri, A. Lanni and F. Goglia, 2008. Interrelated influence of superoxides and free fatty acids over mitochondrial uncoupling in skeletal muscle. Biochim Biophys Acta. 1777, 826-833.

Lombardi, A., E. Silvestri, F. Cioffi, R. Senese, A. Lanni, F. Goglia, P. de Lange and M. Moreno, 2009. Defining the transcriptomic and proteomic profiles of rat ageing skeletal muscle by the use of a cDNA array, 2D- and Blue native-PAGE approach. J Proteomics. 72, 708-721.

Lombardi A., R.A. Busiello, L. Napolitano, F. Cioffi, M. Moreno, P. de Lange, E. Silvestri, A. Lanni and F. Goglia, 2010. Uncoupling protein-3 (UCP3) translocates lipid hydroperoxide and mediates lipid hydroperoxide-dependent mitochondrial uncoupling. J Biol Chem, In press.

Luff, A.R., 1998. Age-associated changes in the innervation of muscle fibres and changes in the mechanical properties of motor units. Ann. NY Acad Sci. 854, 92-101.

Mansouri A., F.L. Muller, Y. Liu, R. Ng, J. Faulkner, M. Hamilton, A. Richardson, T. Huang, C.J. Epstein, H.V. Remmen, 2006 Alterations in mitochondrial function, hydrogen peroxide release and oxidative damage in mouse hind-limb skeletal muscle during aging. Mech. Ageing Develop. 68, 1339-1343.

Melton, L.J. III, S. Khosla, C.S. Crowson, M.K. O'Connor, W.M. O'Fallon and B.L. Riggs, 2000. Epidemiology of sarcopenia. Journal American Geriatric Society. 48, 625-630.

Mookerjee S.A., A.S. Divakaruni, M. Jastroch and M.D. Brand, 2010. Mitochondrial uncoupling and lifespan. Mech Ageing Dev. 2010 April.

Nabben, M., J. Hoeks, J.J. Briedé, J.F. Glatz, E. Moonen-Kornips, M.K. Hesselink and P. Schrauwen, 2008. The effect of UCP3 overexpression on mitochondrial ROS production in skeletal muscle of young versus aged mice. FEBS Lett. 582, 4147-4152.

O'Connell, K., J. Gannon, P. Doran and K. Ohlendieck, 2007. Proteomic profiling reveals a severely perturbed protein expression pattern in aged skeletal muscle. Int. J. Mol. Med. 20, 145-153.

O'Connell, K., and K. Ohlendieck, 2009. Proteomic DIGE analysis of the mitochondria-enriched fraction from aged rat skeletal muscle. Proteomic 9, 5509-5552.

Paradies, G., G. Petrosillo, M. Pistolese and F.M. Ruggiero, 2000. The effect of reactive oxygen species generated from the mitochondrial electron transport chain on the cytochrome c oxidase activity and on the cardiolipin content in bovine heart submitochondrial particles. FEBS Lett. 28, 323-326.

Pak, J.W., A. Herbst, E. Bua, J. Kanski, M.A. Alterman and C. Schoneich, 2003. Proteomic identification of age-dependent protein nitration in rat skeletal muscle. Free Radic Biol Med 35, 1229-1239.

Piec, I., A. Listrat, J. Alliot, C. Chambon, R.G. Taylor, and D. Bochet, 2005. Differential proteome analysis of aging in rat skeletal muscle. FASEB J. 19, 1143-1145.

Schagger, H., R. de Coo, M.F., Bauer, S., Hofmann, C. Godinot and U. Brandt, 2004. Significance of respirasomes for the assembly/stability of human respiratory chain complex I. J Biol Chem 279, 36349-36353.

Schagger, H. and K. Pfeiffer, 2000. Supercomplexes in the respiratory chains of yeast and mammalian mitochondria. Embo J 19, 1777-1883.

Shabalina, I.G., T.V. Kramarova, J. Nedergaard and B. Cannon, 2006. Carboxyatractyloside effects on brown-fat mitochondria imply that the adenine nucleotide translocator isoforms ANT1 and ANT2 may be responsible for basal and fatty-acid-induced uncoupling respectively. Biochem. J. 399, 405-414.

Schoneich, C., 2006. Protein modification in aging: an update. Exp Geront 41, 807-812.

Silvestri E, A. Lombardi, P. de Lange, L. Schiavo, A. Lanni, F. Goglia, T.J. Visser and M. Moreno, 2008. Age-related changes in renal and hepatic cellular mechanisms associated with variations in rat serum thyroid hormone levels. Am J Physiol Endocrinol Metab. 294, E1160-1168.

Speakman, J.R., D.A.Talbot, C. Selman, S. Snart, J.S McLaren, P.Redman, E. Krol, D.M. Jackson, M.S. Jonson and M.D. Brand, 2004. Uncoupling and surviving: individual mice with high metabolism have greater mitochondrial uncoupling and live longer. Aging Cell 3, 87-95.

Stadtman, E.R., 2002. Importance of individuality in oxidative phosphorylation stress and aging. Free Radic Biol Med 33, 597-604.

Talbot, D.A, A.J. Lambert and M.D. Brand, 2004. Production of endogenous matrix superoxide from mitochondrial complex I leads to activation of uncoupling protein 3. FEBS Lett. 2; 556, 111-115.

Vidal-Puig, A.J, D. Grujic, C.Y. Zhang, T. Hagen, O. Boss, Y. Ido, A. Szczepanik, J. Wade, V. Mootha, R. Cortright, D.M. Muoio and B.B. Lowell, 2000. Energy metabolism in uncoupling protein 3 gene knockout mice J Biol Chem. 275, 16258-16266.

Transcriptomics of muscle and fat development in growing cattle

J.J. Loor
Mammalian NutriPhysioGenomics, Department of Animal Sciences and Division of Nutritional Sciences, University of Illinois, Urbana, IL 61801, USA

Introduction

As animal agriculture has become highly-competitive with narrow profit margins, research has focused on improving the quality and efficiency of meat and milk production. For the meat-producing industries, emphasis has been on increasing understanding of physiological processes and molecular pathways associated with skeletal muscle tissue growth (Reecy *et al.*, 2006) and how these processes might be affected by management factors such as nutrition and/or weaning age (Myers *et al.*, 1999a,b; Schoonmaker *et al.*, 2003). The growth characteristics of skeletal muscle result in a complex tissue phenotype that comprises muscle fibers, connective tissue, and intramuscular fat, all of which contribute to sensory aspects of meat (Harper, 1999). This very fact makes the *in vivo* study of skeletal muscle tissue growth challenging.

Application of high-throughput sequencing and transcriptomic technologies has dramatically accelerated the rate at which biological and genetic information can be generated. Together with whole-animal level information (e.g. blood metabolites/hormones, tissue chemical composition, performance), large-scale DNA and mRNA information have increased our understanding of mechanisms of metabolic regulation in agricultural species (e.g. Lehnert *et al.*, 2006; Hocquette *et al.*, 2009). Through the use of bioinformatics, clustering, and gene network analysis, transcriptomics is allowing the identification of regulatory mechanisms that are associated with functional development of tissues of agricultural animals. Continued application of genome-enabled (genotyping, microarrays, proteomics, gene silencing) and allied (e.g. phosphoproteomics) technologies will contribute to our understanding of regulatory points at different stages of muscle and fat growth. This information will provide new insights into opportunities for enhancing efficiency of animal production.

Over the last ~10 years there has been considerable effort in using bovine-specific microarrays to gain a better understanding of the genetic control of biochemical and developmental processes that contribute to muscle and fat development not only as they relate to beef quality (e.g. Byrne *et al.*, 2005; Sudre *et al.*, 2005; Bernard *et al.*, 2007; Wang *et al.*, 2009), breed (e.g. Wang *et al.*, 2005a), but also nutrition and stage of growth (e.g. Wang *et al.*, 2005a; Sadkowski *et al.*, 2006; Graugnard *et al.*, 2007; Wang *et al.*, 2009). Large-scale transcriptomics and the candidate gene approach (e.g. Graugnard *et al.*, 2009ab; Waters *et al.*, 2009) have provided insights into the regulation of muscle and fat development in growing and mature cattle. In the following sections, the concept of transcriptional networks in the context of metabolic regulation of skeletal muscle tissue development is presented and used to provide a framework for a brief discussion of recent studies evaluating transcriptomics data (target genes or high-throughput scale) with emphasis on studies dealing with nutritional management during growth. The goal was to provide specific examples from studies where transcriptomics has advanced our understanding of regulation of skeletal muscle tissue development and how information gathered could impact livestock production in the long-term.

The concept of transcriptional networks and metabolic regulation

Extensive work in non-ruminants has provided concrete evidence that metabolic regulation relies partly on transcriptional control as a long-term mechanism affecting the level of expression of several enzymes in a biochemical pathway (e.g. Desvergne *et al.*, 2006). At the transcriptional level, enzymes within classical biochemical pathways can be viewed as a gene network or a collection of DNA segments which can interact either with a regulator such as a transcription factor (TF) or nuclear receptor (NR), but also with each other through their RNA and protein products and with other molecules in the cell (Wittkopp, 2007). These 'global' interactions, thus, can govern the rates at which genes in the network are transcribed into mRNA and eventually into protein which can catalyze a metabolic reaction. Three classical examples of TF and NR controlling lipogenesis and adipogenesis at the transcriptional level are sterol regulatory element binding factor 1 (*SREBF1*), MLX interacting protein-like (MLXIPL; formerly known as carbohydrate-responsive element binding protein or ChREBP), and the ligand-activated nuclear receptor PPARg (*PPARG*), respectively (Fernyhough *et al.*, 2007).

We have proposed recently (Bionaz and Loor, 2008; Graugnard *et al.*, 2009a,b) that temporal examination of most if not all the *known* components of metabolic networks is a suitable means of addressing the issue of transcriptional regulation as it pertains to tissue metabolism in response to growth and/or nutritional management. Because of the complexity and interconnectivity, i.e. the different levels of aggregation from gene-to-protein-to-metabolite-to-function, of metabolic networks in mammals there is a need for implementing systems-level computational approaches that link (at the very least) transcript, metabolite, and phenotype relationships. These approaches will be effective for (1) the discovery of missing information, (2) the formulation of *in silico* models, and (3) as a structured context for analyzing high-throughput aggregated biological data sets (Mo and Palsson, 2008). Integrating transcriptomics data into existing models of metabolism and nutrient use in cattle can go a long way to help understand the critical genetic/environmental interactions in the growing and/or lactating animal (e.g. McNamara, 2010).

Transcriptomics to elucidate novel aspects of intramuscular fat deposition and intracellular energy generation

A cattle muscle/fat cDNA microarray has been used extensively by the CSIRO group in Queensland Australia to evaluate transcript profiles of skeletal muscle tissue across different breeds (Wang *et al.*, 2005a,b), plane of nutrition (Byrne *et al.*, 2005; Lehnert *et al.*, 2006), and during rapid growth (Wang *et al.*, 2005b, 2009). In an initial study to examine breed differences, skeletal muscle profiles of 11 month-old Japanese Black (JB) and Holstein (H) steers were compared (Wang *et al.*, 2005a). The former have a greater inherent capacity for intramuscular fat deposition (i.e. marbling) and have fat with lower melting point (i.e. is greater in monounsaturated fatty acid content). In contrast, H animals have a greater rate of growth and are heavier at slaughter (e.g. Wang *et al.*, 2005b). A more comprehensive evaluation of breed differences as it pertains to growth performance, carcass characteristics, and selected lipogenic enzyme activity has been reported recently (Bonnet *et al.*, 2007).

Genes expressed preferentially (>2-fold expression) in JB vs. H muscle tissue (total of 17) included lipogenic transcription factors (*SREBF1*, thyroid hormone responsive SPOT14 *THRSP*), proteins involved in fatty acid uptake by tissue (lipoprotein lipase, *LPL*), several lipogenic enzymes (stearoyl-CoA desaturase, *SCD*; fatty acid binding protein 4, *FABP4*, cytosolic NADP+-dependent isocitrate dehydrogenase, *IDH1*)), and adipokines (adiponectin, *ADIPOQ*). Skeletal muscle from H was characterized by genes associated with structural proteins (several myosin genes), glycolysis/energy metabolism (pyruvate kinase, *PKM2*; pyruvate dehydrogenase kinase 4, *PDK4*), and muscle contraction (ATPase, Ca++ transporting, cardiac muscle, fast twitch 1, *ATP2A1*). Regarding this difference in structural proteins there also is evidence that genetic selection for muscle growth within a breed can alter differential expression of genes which can in turn determine the metabolic characteristics of muscle types (Sudre *et al.*, 2005). Despite the well-defined effects of age and growth rate on cattle lipid deposition (e.g. Bonnet *et al.*, 2007), which differ across breeds, differences in gene expression profiles observed by Wang *et al.* (2005a,b) provided molecular evidence for the greater propensity of JB to produce beef with greater marbling.

A more recent study (Wang *et al.*, 2009) explored temporal (3, 7, 12, 20, 25, and 30 months of age) gene expression profiles of skeletal muscle from Wagyu × Hereford (WH) and Piedmontese × Hereford cattle (PH), which exhibit contrasting amounts of intramuscular fat content (ca. 10.7% vs. 4.3% for WH and PH). Two animals from each breed with the most extreme intramuscular fat content (16.5% vs. 3.4% for WH and PH) were selected for microarray analysis using the same platform as Wang *et al.* (2005a,b). Microarrays were coupled with quantitative RT-PCR of 17 genes on the entire group of animals (n=6-7). Results revealed contrasting mRNA expression patterns for several genes associated with adipogenesis and lipogenesis between breeds but also across stages of growth. For example, the transcription regulator *PPARG* had greater expression as early as 7 months of age, i.e. close to weaning, in WH vs. PH steers. A battery of genes associated with lipogenesis (e.g. *FABP4*, *SCD*, fatty acid synthase (*FASN*), *ADIPOQ*) and extracellular protein (connective tissue and extracellular matrix) genes [(e.g. collagen, type I, alpha 1 (*COL1A1*), *COL1A2*, fibronectin (*FN1*), secreted protein, acidic, cysteine-rich (osteonectin) (*SPARC*)] also had greater expression in WH steers at 7 months of age.

Unlike previously thought, gene expression patterns suggested (Wang *et al.*, 2009) that the predisposition of WH cattle to accumulate fat is well-developed close to the time of weaning. Such a process also would entail expansion of the extracellular matrix to accommodate intramuscular fat development (Wang *et al.*, 2009). In contrast to WH animals, PH animals had greater coordinated expression of mitochondrial genes associated with oxidative phosphorylation (e.g. mitochondrially encoded cytochrome b (*MTCYB*), cytochrome c oxidase subunit VIIa polypeptide 1 (muscle) (*COX7A*), mitochondrially encoded NADH dehydrogenase 4 (*MTND4*), and *MTND4L*) which authors proposed was suggestive of preferential use of energy in PH animals to support more rapid muscle growth during the period preceding weaning when fractional rates of muscle growth are increased (e.g. Lehnert *et al.*, 2006). Animals with an increased basal energy requirement also may be expected to store less triacylglycerol in adipose tissue (Wang *et al.*, 2009).

Nutritional management and skeletal muscle transcriptomics

Early weaning and nutrition

Weaning calves earlier than the traditional 205 d age, in particular, is a practical management tool that can be used to enhance feed efficiency as well as produce high-quality beef (e.g. Schoonmaker *et al.*, 2002). For example, weaning at ~150 d compared with traditional weaning resulted in 30% more calves grading average choice or above and also improved feed efficiency in the feedlot (Myers *et al.*, 1999b; Graugnard *et al.*, 2009a,b). Very few genes known to be associated with adipogenesis are upregulated in skeletal muscle (i.e. *longissimus lumborum*) of newborn WH compared with PH calves (Lehnert *et al.*, 2007). Those observations coupled with the findings that 11 month JB vs. H cattle (Wang *et al.*, 2005a) had greater expression of genes associated with lipogenesis and adipogenesis suggested that the onset of marbling in animals occurs between birth and the early postweaning phase (Wang *et al.*, 2009). However, the exact timing and development of marbling in cattle are still unknown. It is evident from the above studies that there is a time-frame during which early nutritional intervention could be used as a practical means to initiate precocious intramuscular fat development and, potentially, increase the likelihood of achieving greater marbling at slaughter (Graugnard *et al.*, 2009a,b). Once identified, these putative regulators of adipogenesis could potentially be used as markers of marbling in breeding programs.

Effects of early weaning and level of dietary starch on *longissimus lumborum* transcript profiles

Our research group (Graugnard *et al.*, 2007, 2009b) recently utilized purebred early-weaned (155 d age) Angus steers to examine skeletal muscle transcriptomics in response to feeding a high-starch

(5,980 kJ/kg dry matter) or low-starch (4,970 kJ net energy for gain/kg) diet for 112 d (growing phase) followed by a common high-starch diet (6,030 kJ/kg) for an additional 112 d (finishing phase). *Longissimus lumborum* was biopsied at 0, 56, 112, and 224 d of the start of feeding diets. Tissue RNA was subjected to microarray analysis using the 13,000 gene bovine microarray developed at the University of Illinois. Most of the gene expression changes in our study (FDR-adjusted $P<0.01$) were due to time, i.e. physiological state or growth (Figure 1). However, comparisons of differentially expressed (DEG) between specific time points revealed that most of the changes during growth were observed during the first 56 d of the growing phase (1,471 DEG) and at the mid-point (224 d; 3,200-2,700 DEG) of the finishing phase relative to other points of the growing phase (Figure 1). It was apparent from these data that transcriptional adaptations in skeletal muscle of these early-weaned animals were 'complete' during the first half of the growing phase despite the fact they consumed incremental amounts of feed (~6 kg/d to ~10 kg/d by 112 d) and gained body weight (~1.65 kg/d) through the end of the growing phase (i.e. 112 d on treatments). A second and more robust bout of transcriptional adaptations occurred after all animals consumed the same high-starch/high-energy finishing diet for ~112 d. During that time-frame, there were no differences in dry matter intake (~10 kg/d), estimated energy intake, or feed efficiency (0.15 kg average daily gain/dry matter intake) (Graugnard *et al.*, 2009a).

Our initial bioinformatics evaluation of DEG (using Ingenuity Pathway Analysis®) has revealed that within the set of affected genes (i.e. >3,000) there are several metabolic and signalling pathways that are enriched and whose relative level of activation or inhibition (based on detailed analysis of up- and down-regulated genes within the pathway; see details of the approach in Moyes *et al.*, 2009) changed during the course of the study (Figure 2). For example, a total of 31 and 26 DEG associated with *actin cytoskeleton signalling* and *acute phase response signalling* were found in the comparison of 56 vs. 0 d and, based on their relative up- and down-regulation, it was concluded that there was a *modest* inhibition (gray circle, Figure 2) of these pathways during the first half of the growing phase. Despite changes over time in the significance (Fisher's exact t-test *P*-value) of the association of DEG within these pathways, the analysis also revealed that by 112 d both pathways were *markedly* inhibited.

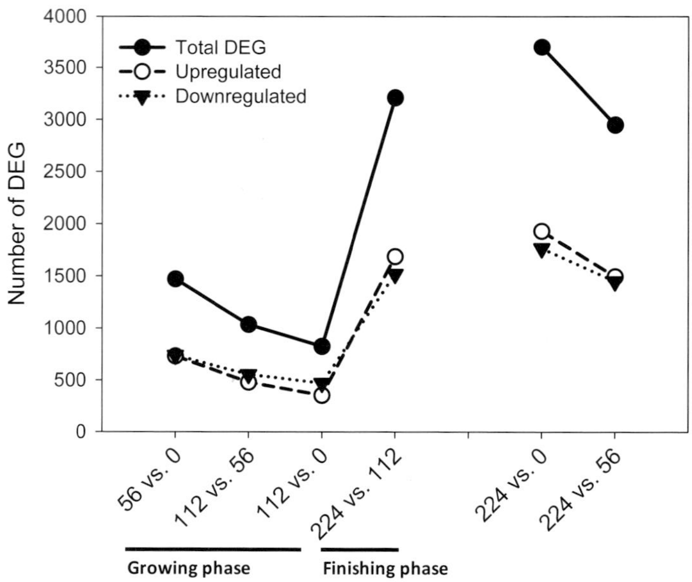

Figure 1. Differentially expressed genes (DEG) in longissimus lumborum *of Angus steers at specific time points during the growing and finishing phase. A total of ~5,700 DEG (FDR* P*<0.01) due to time were found (Graugnard* et al.*, 2007).*

However, at 224 d on study (finishing phase) the *actin cytoskeleton signalling pathway* was *markedly* activated but *acute phase response signalling* was *markedly* inhibited (Figure 2). The behaviour of cytoskeletal signalling genes over time is suggestive of a physiological change in regulation of microtubule and actin dynamics in skeletal muscle as the animals grew, likely related with muscle fibre development (Sudre *et al.*, 2005; Wang *et al.*, 2009). The finding that *acute phase response signalling* was affected during growth also could have important physiological implications in terms of fat and muscle development. Several well-known cytokines and cytokine signalling molecules (e.g. tumor necrosis factor (*TNF*), suppressor of cytokine signalling 1 (*SOCS1*), mitogen-activated protein kinase 1 (*MAPK1*), nuclear factor of kappa light polypeptide gene enhancer in B-cells 1 (*NFKB1*), inhibitor of kappa light polypeptide gene enhancer in B-cells, kinase beta (*IKBKB*), and interleukin 6 receptor (*IL6R*) were among the 37 DEG in this pathway at the end of the study (Figure 2).

Perhaps not surprising, given previous similar work (e.g. Wang *et al.*, 2009), the behaviour of DEG related with *glycolysis/gluconeogenesis*, *oxidative phosphorylation*, and *PPARα/RXRα activation* resulted in an overall activation of these pathways as animals grew, i.e. increased generation of energy via glycolytic flux [(e.g. pyruvate kinase, muscle (*PKM2*) and glyceraldehyde-3-phosphate dehydrogenase (*GAPDH*)] and fatty acid oxidation (e.g. peroxisome proliferator-activated receptor-α (*PPARA*), CD36 molecule (thrombospondin receptor) (*CD36*), acetyl-CoA acyltransferase 1 (*ACAA1*), and carnitine palmitoyltransferase 1B (muscle) (*CPT1B*)] potentially to support skeletal muscle mass deposition. *Oxidative phosphorylation* (with 42 DEG) was the most significantly-enriched pathway and *PPARα/RXRα activation* was among the ones with greatest number of DEG (42) at the end of the study (224 vs. 0 d; Figure 2). Previous studies using macroarrays have observed diet effects on the expression of several genes within the above pathways (i.e. associated with metabolic enzymes and contractile proteins) in steers under different nutritional management (Cassar-Malek *et al.*, 2009). Other studies also have found relationships between energy-related genes and sensory aspects of meat quality including tenderness, juiciness, and flavor [e.g. *CTP1B*, *PDK4*, and NADH dehydrogenase (ubiquinone) 1 beta subcomplex, 4, 15kDa (*NDUFB4*); Bernard *et al.*, 2007]. The above studies underscore the need for more holistic studies of muscle and fat development to gain better understanding of fundamental aspects of tissue physiology as they relate to management and meat quality.

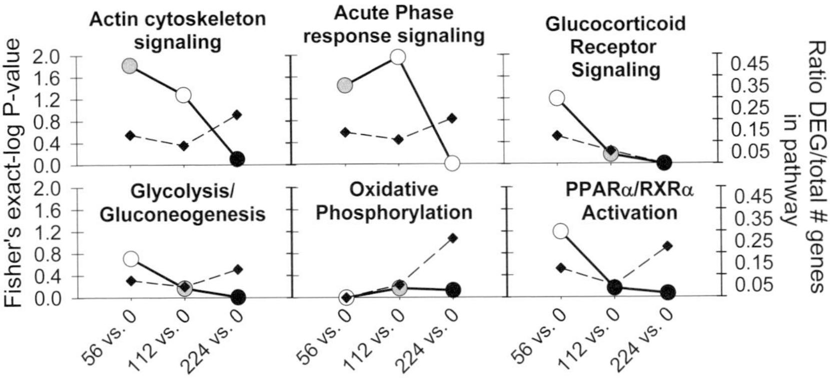

Figure 2. Subset of enriched molecular pathways (Ingenuity Pathway Analysis®) within DEG in longissimus lumborum *of Angus steers at specific time points during the growing and finishing phase (Graugnard* et al., *2007). Symbols denote: marked downregulation (white), modest downregulation (gray), and marked upregulation (black). Y-axis on left, shows Fisher's log-*P *value from Ingenuity analysis and denotes the significance of the genes within each pathway. Y-axis on right, shows the ratio of DEG to the total number of genes in the Ingenuity-curated pathway.*

Adipogenic gene networks during growth: effects of early weaning and level of dietary starch

Previous biochemical studies of adipose tissue depot metabolism in beef cattle have provided basic insights of tissue development as it pertains to substrates used, stage of growth, and/or nutritional management (e.g. Rhoades *et al.*, 2007; Smith *et al.*, 2009). Most current emphasis (e.g. Wang *et al.*, 2005a,b, 2009; Smith *et al.*, 2009), however, has been on examination of selected components of adipogenic and lipogenic gene networks, e.g. *SCD*, *PPARG*, *FASN*, and acetyl-CoA carboxylase-α (*ACACA*) among others. Recent work from our group has used a network approach to study temporal adaptations of adipogenic networks in early-weaned Angus or Angus × Simmental steers (Graugnard *et al.*, 2009a,b). Exposure to high-starch diets during the early growing phase appeared to induce precocious pre-adipocyte differentiation and lipid filling through upregulation of *PPARG* and its target genes (*SCD*, diacylglycerol O-acyltransferase homolog 2 (mouse) (*DGAT2*), and *FABP4*). High-starch also led to greater *SREBF1* and *MLXIPL* during the finishing phase, which coupled with greater glucose-6-phosphate dehydrogenase (*G6PD*) and *ADIPOQ* could help sustain/promote adipogenesis (Graugnard *et al.*, 2009b). The low-starch/high-fibre diet apparently delayed adipocyte differentiation as indicated by increased *THRSP*, *FASN*, *FABP4*, *SCD*, and *DGAT2* during the second half of the growing phase. Quite surprisingly, animals fed low-starch/high-fibre during the growing phase had additional increases in expression of *THRSP*, *PPARG*, *FABP4*, acyl-CoA synthetase medium-chain family member 1 (*ACSM1*), *SCD*, and *DGAT2* during the finishing phase. However, the concomitant increase in nuclear receptor subfamily 2, group F, member 2 (*NR2F2*) expression (a repressor of adipocyte differentiation in rodents; Okamura *et al.*, 2009) might have blunted the expected adipogenic response. Thus, these initial results pointed at the need to not only examine activators of a metabolic pathway but also potential 'inhibitors' or 'controllers'. In addition, from these data we concluded that the sustained upregulation of *MLXIPL* and *SREBF1* due to feeding high-dietary starch was suggestive of 'metabolic imprinting', i.e. long-term upregulation of adipogenic transcription regulators due to exposure of high dietary starch at a young age. Such an effect, if it occurs in cattle intramuscular adipocytes, would lead to further enrichment (Rosen and MacDougald, 2006) of intramuscular adipocytes at a more mature age of the animal. A recent study reported that hypermethylation of PPARγ in rodent preadipocytes prevents transcription of the gene and induction of differentiation (Fujiki *et al.*, 2009). Furthermore, adipocytes from diabetic mice have greater methylation of the *PPARG* promoter coupled with lower mRNA abundance, all of which provided evidence that adipogenic transcription regulators (and potentially enzymes) might be regulated via epigenetic mechanisms. Further detailed studies will be required to determine potential epigenetic regulation of adipogenic and lipogenic genes in cattle fed high-starch diets at an early age. We are currently testing potential metabolic imprinting in *longissimus lumborum* of early-weaned steers (Moisa *et al.*, 2010).

Protein-DNA interactions to study transcriptional networks during muscle and fat development

Current approaches to analyzing temporal transcriptomics data (e.g. *k*-means clustering) can successfully identify groups of co-expressed genes. An important focus in the livestock post-genomic era, however, will be on ways to study the functions of those DEG deemed to be 'key' during muscle and fat development, e.g. TF and NR will be ideal candidates at the outset of these efforts because (as we argued above) their activity is central for regulation of metabolic gene networks. A major first step towards comprehensively understanding the differential control of gene expression in muscle and fat tissue at different stages of growth would be to map the functional regulatory sequences in DNA that are responsible for transcriptional regulation. These regions of DNA are largely composed of transcription binding sites (TBFS) and also NR binding sites (i.e. hormone response elements) (Odom *et al.*, 2004). Identification and categorization of the entire repertoire of TBFS are among the greatest challenges in systems biology (Gerstein *et al.*, 2007). The ENCODE (ENCyclopedia

Of DNA Elements) project was specifically developed to begin the process of defining TBFS in the human genome (Gerstein *et al.*, 2007).

Large-scale, genome-wide TF/NR binding analysis, which identifies physical interactions between regulators and the regulatory DNA regions they bind to, can provide direct evidence of regulatory relationships (Buck and Lieb, 2004). The development of the ChIP (chromatin immunoprecipiation) assay was a defining event in the study of protein-DNA interactions *in vivo* (Buck and Lieb, 2004). The technique allows the identification of DNA elements bound by proteins such as TF or NR under a specific cellular context. The use of the ChIP-real-time quantitative PCR/microarray technique is now considered the gold standard to confirm the presence of a TF or NR at a putative regulatory region (Biddie *et al.*, 2010). It has come to light in recent studies, however, that there is a high degree of discordance between the number of NR binding sites and the number of regulated genes (at least in non-ruminants), which complicates bioinformatic assignment of *binding* to *regulation* and underscores that gene regulation and binding by NR is a highly complex process (Biddie *et al.*, 2010). Despite the possibility for applying this tool to advance the field of cattle physiology, the absence of affordable species-specific reagents (e.g. antibodies) is one important limitation for progress in this area.

In addition to ChIP assays, there are open-source/web-accessible computational tools (e.g. Bailey *et al.*, 2006) that allow researchers to find motifs in DNA or protein sequences that serve as binding sites for TF or NR. This approach could be helpful in analyzing microarray or proteomics data sets, which often uncover large numbers of seemingly co-regulated genes. Software tools can search for statistically-significant motifs within user-provided DNA sequences that may be present in some or all of the input sequences (Bailey *et al.*, 2006). A recent study used this computational approach to search for putative TFBS in microarray data from bovine cardiac muscle (Zadissa *et al.*, 2007).

Ultimately, our complete understanding of TF and NR action will depend on the construction of regulatory networks (e.g. Bionaz and Loor, 2008). While several different approaches are required in order to build complex regulatory networks and connect them with biological functions, the application of ChIP-based approaches, for example, in combination with systematic functional and bioinformatics analysis of the data can help achieve this goal. Future studies in this area will be instrumental to improving our understanding of the molecular mechanisms of gene regulation in muscle and fat, and the direct links with physiological processes in these tissues.

Additional omics technologies in the study of muscle and fat development

Phosphoproteomics: its use in studies of insulin signalling

Mass spectrometry (MS)-based proteomics has become increasingly powerful not only to identify complex protein mixtures but also regulated protein modifications (Aebersold and Mann, 2003). It is well-established in non-ruminants that reversible phosphorylation is a major regulatory mechanism controlling the activity of proteins. Many signalling pathways, including the insulin/IGF-1 signalling pathway transduce signals from the cell surface to downstream targets via tyrosine kinases and phosphatases (e.g. Taniguchi *et al.*, 2006). Insulin or IGF-1 binding initiates a complex cascade of events, starting with phosphorylation of specific tyrosine residues on the insulin and the IGF-1 receptors (Saltiel and Kahn, 2001). Once activated, these receptors phosphorylate a number of docking proteins; the best characterized are the insulin receptor substrate (IRS) proteins 1-4, which upon activation interact with other intracellular signalling molecules primarily through SH2 domains leading to activation of several downstream pathways (e.g. vesicle trafficking, protein synthesis, and glucose uptake) (Saltiel and Kahn, 2001). MS-based proteomics has been recently used to study tyrosine phosphorylation of the insulin pathway in rodent adipocytes (Kruger *et al.*, 2008), allowing the identification of different protein "effector" classes that influence distinct branches of the insulin signalling pathway.

Of 40 identified insulin-induced effectors after a 5 min incubation of differentiated rodent brown adipocytes with insulin (100 nM), 7 were not previously described as components of the insulin signalling pathway including serum deprivation-response protein (SDPR), protein kinase C, delta binding protein (PRKCDBP), low density lipoprotein receptor-related protein 6 (LRP6), and PDZ domain containing 11 (PDZK11), a potential calcium ATPase binding protein. Of the 26 known effectors of the insulin pathway (at least in non-ruminants) insulin receptor (INSR), insulin receptor substrate 1 (IRS1), caveolin 1, caveolae protein, 22kDa (CAV1), and SH2B adaptor protein 2 (SH2B2) were substantially phosphorylated at different tyrosine residues (Kruger *et al.*, 2008). Just as it is often the case with discovery experiments that rely on transcriptomics and proteomics, the phosphoproteomics technique allowed for the discovery of candidate effectors of insulin signalling. Because signalling cascades via hormones and receptor tyrosine kinases occur very rapidly, i.e. represent acute mechanisms of metabolic regulation (contrary to longer-term regulation via changes in transcription), these phosphorylation studies have to be performed using *in vitro* cell/tissue systems often encompassing multiple observations at very early (e.g. 60 s) and 'longer-term' (e.g. 10-30 min) stages. We are currently testing the phosphorylation state of several components of the insulin signalling cascade in subcutaneous adipose tissue from peripartal cows fed different levels of dietary energy *pre partum* (Ji *et al.*, 2010). Those data can easily be complimented with transcriptomics as well as blood hormone and metabolite data to provide several levels of physiological integration. Similar work in growing beef cattle would obviously be valuable.

Concluding remarks

The biological complexity of agricultural animals unavoidably requires a systems biology approach, i.e. a way to systematically study the complex interactions in biological systems using a method of integration instead of reduction. One of the goals of systems biology is to discover new emergent properties that may arise from examining the interactions between all components of a system to arrive at an integrated view of how the organism functions. Work in model organisms during the past 15 years has demonstrated the applicability of high-throughput methods to discern regulatory and metabolic networks (e.g. Mo and Palsson, 2008). We have previously outlined systems biology schemes that would be amenable to high-throughput studies of animal tissue at the level of transcriptomics, proteomics, and metabolomics (see Loor and Cohick, 2009; Loor, 2010). The systems approach might lead to the discovery of regulatory targets that could be tested further (i.e. model-directed discovery) or help address a broader spectrum of basic and practical applications including interpretation of phenotypic data, metabolic engineering, or interpretation of muscle and fat growth phenotypes (Loor and Cohick, 2009). Together with whole-animal level information, both large-scale and network-specific mRNA and protein data are poised to accelerate knowledge of metabolic regulation in agricultural species. Additional technologies to study gene function (gene silencing), protein-DNA interactions, and phosphorylation state of the components of hormonal signalling pathways will compliment the omics' efforts. Ultimately these integrative approaches will lead to new insights into opportunities for enhancing efficiency of animal production.

References

Aebersold, R., and M. Mann, 2003. Mass spectrometry-based proteomics. Nature 422, 198-207.

Bailey, T.L., N. Williams, C. Misleh, and W.W. Li, 2006. Meme: Discovering and analyzing DNA and protein sequence motifs. Nucleic Acids Res 34, W369-373.

Bernard, C., I. Cassar-Malek, M. Le Cunff, H. Dubroeucq, G. Renand and J.F. Hocquette, 2007. New indicators of beef sensory quality revealed by expression of specific genes. 2007. J. Agric. Food Chem. 55, 5229-5237.

Biddie, S.C., S. John, and G.L. Hager, 2010. Genome-wide mechanisms of nuclear receptor action. Trends Endocrinol Metab. 21, 3-9.

Bionaz, M. and J.J. Loor, 2008. Gene networks driving bovine milk fat synthesis during the lactation cycle. BMC Genomics 9, 366.

Bonnet, M., Y. Faulconnier, C. Leroux., C. Jurie, I. Cassar-Malek, D. Bauchart, P. Boulesteix D. Pethick, J.F. Hocquette and Y. Chilliard, 2007. Glucose-6-phosphate dehydrogenase and leptin are related to marbling differences among Limousin and Angus or Japanese Black x Angus steers. J Anim Sci. 85, 2882-2894.

Buck, M.J., and J.D. Lieb, 2004. Chip-chip: Considerations for the design, analysis, and application of genome-wide chromatin immunoprecipitation experiments. Genomics 83, 349-360.

Byrne, K.A., Y.H. Wang, S.A. Lehnert, G.S. Harper, S.M. McWilliam, H.L. Bruce and A. Reverter, 2005. Gene expression profiling of muscle tissue in Brahman steers during nutritional restriction. J. Anim. Sci. 83, 1-12.

Cassar-Malek, I., C. Jurie, C. Bernard, I. Barnola, D. Micol and J.F. Hocquette, 2009. Pasture-feeding of Charolais steers influences skeletal muscle metabolism and gene expression. J Physiol Pharmacol. Suppl 3, 83-90.

Desvergne, B., L. Michalik and W. Wahli, 2006. Transcriptional regulation of metabolism. Physiol. Rev. 86, 465-514.

Fernyhough, M.E., E. Okine, G. Hausman, J.L. Vierck and M.V. Dodson, 2007. PPARg and GLUT-4 expression as developmental regulators/markers for preadipocyte differentiation into an adipocyte. Domest. Anim. Endocrinol. 33, 367-378.

Fujiki, K., F. Kano, K. Shiota, and M. Murata, 2009. Expression of the peroxisome proliferator activated receptor gamma gene is repressed by DNA methylation in visceral adipose tissue of mouse models of diabetes. BMC Biol. 7, 38.

Gerstein, M.B., C. Bruce, J.S. Rozowsky, D. Zheng, J. Du, J. O. Korbel, O. Emanuelsson, Z.D. Zhang, S. Weissman, and M. Snyder, 2007. What is a gene, post-encode? History and updated definition. Genome Res. 17, 669-681.

Graugnard, D., P. Piantoni, M. Bionaz, L.L. Berger, D.B. Faulkner and J.J. Loor, 2009. Adipogenic and energy metabolism gene networks in longissimus lumborum during rapid post-weaning growth in Angus and Angus x Simmental cattle fed high- or low-starch diets. BMC Genomics 10, 142.

Graugnard, D.E., L.L. Berger, D.B. Faulkner and J.J. Loor, 2009. High-starch diets induce precocious adipogenic gene network up-regulation in longissimus lumborum of early-weaned Angus cattle. Br J Nutr. 103, 953-963.

Graugnard, D., S.L. Rodriguez-Zas, D.B. Faulkner, L.L. Berger, R.E. Everts, H.A. Lewin and J.J. Loor, 2007. Temporal longissimus muscle gene expression profiles due to plane of dietary energy in early-weaned Angus steers. J. Dairy Sci. 90, 540.

Harper, G.S, 1999. Trends in skeletal muscle biology and the understanding of toughness in beef. Aust. J. Agric. 50, 1105-1129.

Hocquette, J.F., I. Cassar-Malek, A. Scalbert and F. Guillou, 2009. Contribution of genomics to the understanding of physiological functions. J Physiol Pharmacol. 2009 Suppl 3, 5-16.

Ji, P., J.S. Osorio, J.K. Drackley and J.J. Loor, 2010. Insulin signal transduction in adipose tissue of peripartal dairy cows fed two levels of dietary energy *pre partum*. J. Dairy Sci. 93 (E-Suppl. 1), Abs 870.

Lehnert, S.A., Y.H. Wang, S.H. Tan and A. Reverter, 2006. Gene expression-based approaches to beef quality research. Aust. J. Exp. Ag. 46, 165-172.

Krüger, M., I. Kratchmarova, B. Blagoev, Y.H. Tseng, C.R. Kahn and M. Mann, 2008. Dissection of the insulin signalling pathway via quantitative phosphoproteomics. Proc. Natl. Acad. Sci. USA 105, 2451-2456.

Lehnert, S.A., A. Reverter, K.A. Byrne, Y. Wang, G.S. Nattrass, N.J. Hudson and P.L. Greenwood, 2007. Gene expression studies of developing bovine longissimus muscle from two different beef cattle breeds. BMC Dev Biol. 7, 95.

Loor, J.J., and W.S. Cohick, 2010. ASAS centennial paper: Lactation biology for the twenty-first century. J. Anim. Sci. 87, 813-824.

Loor, J.J., 2010. Genomics of metabolic adaptations in the peripartal cow. Animal (in press).

McNamara, J.P., 2010. Integrating transcriptomic regulation into models of nutrient metabolism in agricultural animals. In: Crovetto, G.M. (eds.) Energy and protein metabolism and nutrition. Wageningen Academic Publishers, Wageningen, the Netherlands, p. 27-28.

Moisa, S.J., D. W.Shike, D. B.Faulkner and J.J. Loor, 2010. Carbohydrate-responsive element binding protein (MLXIPL) and PPARg gene network expression in *longissimus lumborum* of early-weaned and normal-weaned Angus steers fed a high-starch diet during the growing phase. J. Dairy Sci. 93 (E-Suppl. 1), Abs(730).

Mo, M.L., and B.Ø. Palsson, 2009. Understanding human metabolic physiology: a genome-to-systems approach. Trends Biotechnol. 27, 37-44.

Moyes, K.M., J.K. Drackley, D.E. Morin, M. Bionaz, S.L. Rodriguez-Zas, R.E. Everts, H.A. Lewin and J.J. Loor, 2009. Gene network and pathway analysis of bovine mammary tissue challenged with Streptococcus uberis reveals induction of cell proliferation and inhibition of PPARgamma signalling as potential mechanism for the negative relationships between immune response and lipid metabolism. BMC Genomics. 10, 542.

Myers, S.E., D.B. Faulkner, F.A. Ireland and D.F. Parrett, 1999a. Comparison of three weaning ages on cow-calf performance and steer carcass traits. J Anim Sci. 77, 323-9.

Myers, S.E., D.B. Faulkner, T.G. Nash, L.L. Berger, D.F. Parrett and F.K. McKeith, 1999b. Performance and carcass traits of early-weaned steers receiving either a pasture growing period or a finishing diet at weaning. J Anim Sci. 77, 311-22.

Odom, D.T., N. Zizlsperger, D.B. Gordon, G.W. Bell, N.J. Rinaldi, H.L. Murray, T.L. Volkert, J. Schreiber, P.A. Rolfe, D.K. Gifford, E. Fraenkel, G.I. Bell and R.A. Young, 2004. Control of pancreas and liver gene expression by hnf transcription factors. Science 303, 1378-1381.

Okamura, M., H. Kudo, K. Wakabayashi, T. Tanaka, A. Nonaka, A. Uchida, S. Tsutsumi, I. Sakakibara, M. Naito, T.F. Osborne, T. Hamakubo, S. Ito, H. Aburatani, M. Yanagisawa, T. Kodama and J. Sakai, 2009. COUP-TFII acts downstream of Wnt/beta-catenin signal to silence PPARgamma gene expression and repress adipogenesis. Proc. Natl. Acad. Sci USA 106, 5819-5824.

Rhoades, R.D., J.E. Sawyer, K.Y. Chung, M.L. Schell, D.K. Lunt and S.B. Smith, 2007. Effect of dietary energy source on *in vitro* substrate utilization and insulin sensitivity of muscle and adipose tissues of Angus and Wagyu steers. J Anim Sci. 85, 1719-1726.

Reecy, J.M., D.M. Spurlock and C.H. Stahl, 2006. Gene expression profiling: insights into skeletal muscle growth and development. J Anim Sci. 84, 150-154.

Rosen, E.D., and O.A. MacDougald, 2006. Adipocyte differentiation from the inside out. Nat Rev Mol Cell Biol 7, 885-896.

Sadkowski, T., M. Jank, J. Oprzadek and T. Motyl, 2006. Age-dependent changes in bovine skeletal muscle transcriptomic profile. J Physiol Pharmacol. 57 Suppl 7, 95-110.

Saltiel, A.R., and C.R. Kahn, 2001. Insulin signalling and the regulation of glucose and lipid metabolism. Nature 414, 799-806.

Schoonmaker, J.P., V.M. Cecava, D.B. Faulkner, F.L. Fluharty, H.N. Zerby and S.C. Loerch, 2003. Effect of source of energy and rate of growth on performance, carcass characteristics, ruminal fermentation, and serum glucose and insulin of early-weaned steers. J Anim Sci. 81, 843-855.

Schoonmaker, J.P., S.C. Loerch, F.L. Fluharty, H.N. Zerby and T.B. Turner, 2002. Effect of age at feedlot entry on performance and carcass characteristics of bulls and steers. J Anim Sci. 80, 2247-2254.

Smith, S.B., H. Kawachi, C.B. Choi, C.W. Choi, G. Wu and J.E. Sawyer, 2009. Cellular regulation of bovine intramuscular adipose tissue development and composition. J Anim Sci. 87(14 Suppl), E72-82.

Sudre, K., I. Cassar-Malek, A. Listrat, Y. Ueda, C. Leroux, C. Jurie, C. Auffray, G. Renand, P. Martin and J.F. Hocquette, 2005. Biochemical and transcriptomic analyses of two bovine skeletal muscles in Charolais bulls divergently selected for muscle growth. Meat Sci. 70, 267-277.

Taniguchi, C.M., B. Emanuelli, and C.R. Kahn, 2006. Critical nodes in signalling pathways: Insights into insulin action. Nat. Rev. Mol. Cell. Biol. 7, 85-96.

Wang, Y.H., K.A. Byrne, A. Reverter, G.S. Harper, M. Taniguchi, S.M. McWilliam, H. Mannen, K. Oyama and S.A. Lehnert, 2005a. Transcriptional profiling of skeletal muscle tissue from two breeds of cattle. Mamm Genome. 16, 201-210.

Wang, Y.H., A. Reverter, H. Mannen, M. Taniguchi, G.S. Harper, K. Oyama, K.A. Byrne, A. Oka, S. Tsuji and S.A. Lehnert, 2005b. Transcriptional profiling of muscle tissue in growing Japanese Black cattle to identify genes involved with the development of intramuscular fat. Aust. J. Exp. Ag. 45, 809-820.

Wang, Y.H., N.I. Bower, A. Reverter, S.H. Tan, N. De Jager, R. Wang, S.M. McWilliam, L.M. Cafe, P.L. Greenwood and S.A. Lehnert, 2009. Gene expression patterns during intramuscular fat development in cattle. J Anim Sci. 87, 119-130.

Waters, S.M., J.P. Kelly, P. O'Boyle, A.P. Moloney and D.A. Kenny, 2009. Effect of level and duration of dietary n-3 polyunsaturated fatty acid supplementation on the transcriptional regulation of Delta9-desaturase in muscle of beef cattle. J Anim Sci. 87, 244-252.

Wittkopp, P.J., 2007. Variable gene expression in eukaryotes, a network perspective. J Exp Biol 210, 1567-1575.

Zadissa, A., J.C. McEwan, and C.M. Brown, 2007. Inference of transcriptional regulation using gene expression data from the bovine and human genomes. BMC Genomics 8, 265.

Mechanistic analysis of mitochondrial ROS production in chicken skeletal muscle under acute heat stress conditions

M. Kikusato and M. Toyomizu
Laboratory of Animal Nutrition, Division of Life Science, Graduate School of Agricultural Science, Tohoku University, Tsutsumidori-Amamiyamachi 1-1, Aoba-ku, Sendai 981-8555, Japan

Introduction

Heat stress is an environmental factor responsible for oxidative stress. We reported that reactive oxygen species (ROS) production by mitochondria was significantly increased in skeletal muscle of birds after exposure to heat stress (Mujahid *et al.*, 2007), and that muscle mitochondria from heat-stressed birds also had the higher mitochondrial membrane potential ($\Delta\Psi$) (M. Kikusato, unpublished data). These results implied that the overproduction of ROS may depend on $\Delta\Psi$. We also found that heat stress significantly decreased avian uncoupling protein (avUCP) in muscle mitochondria (Mujahid *et al.*, 2009). Therefore, this decrease may contribute to excess ROS production, possibly via a depression of inducible proton leak. To elucidate more precisely the mechanism involved in the overproduction of mitochondrial ROS under heat stress conditions, we have attempted to clarify how heat stress-induced ROS production depends on $\Delta\Psi$, and to evaluate the effect of avUCP downregulation on the inducible proton leak.

Material and methods

Meat-type male chickens obtained from a commercial hatchery were housed in electrically-heated batteries under continuous light. They were provided with access *ad libitum* to water and a commercial starter diet. Chicks (3 weeks old) were heat-stressed by exposure to 34 °C for 12 h (24 °C for controls), before the *pectoralis superficialis* muscles were rapidly excised. Muscle mitochondria were isolated by homogenization, protein digestion and differential centrifugation as previously described (Mujahid *et al.*, 2009). Isolated mitochondria were incubated at 38 °C assay medium, to which 4 mM succinate was used to initiate the respiration. Mitochondrial oxygen consumption and $\Delta\Psi$ were simultaneously measured using electrodes sensitive to oxygen and the potential-dependent probe, triphenylmethyl phosphonium cation ($TPMP^+$), respectively. Modular kinetic analysis was used to measure 'substrate oxidation' by the electron transport chain and 'proton leak', titrating state 4 respiration with sequential additions of FCCP (up to 0.6 μM) and malonate (up to 3 mM), respectively. The mitochondrial ROS (as H_2O_2) production rate was fluorometrically determined in a similar way to measurements of the kinetics. Arachidonic acid (36 μM) and GDP (500 μM) were used to estimate avUCP-mediated proton leak. In all kinetics, rotenone (a complex-I inhibitor) was omitted to induce endogenous superoxide production. Data were presented as means ± SE of 4 replicates of individual mitochondrial measurements.

Results and discussions

$\Delta\Psi$ elevation in state 4 resulted from increase in the substrate oxidation and decrease in basal proton leak (Figure 1A). ROS production at the furthest point (state 4) to the right in the kinetics curve was significantly higher in heat-stressed group than in controls because of the larger $\Delta\Psi$ for heat-stressed group (Figure 1B). Figure 1B also shows that ROS production – regardless of the modular kinetics – was strongly dependent on the $\Delta\Psi$ in both control and heat-stressed groups: the higher the $\Delta\Psi$, the higher the superoxide production in both groups (Figure 1B), even though there were small differences in ROS production at any given $\Delta\Psi$ between the groups (Figure 1B).

Regarding the avUCP-inducible proton leak, basal proton leak was increased by arachidonic acid in both groups, but less in the heat-stressed group than in controls (Figure 2A), indicating that its

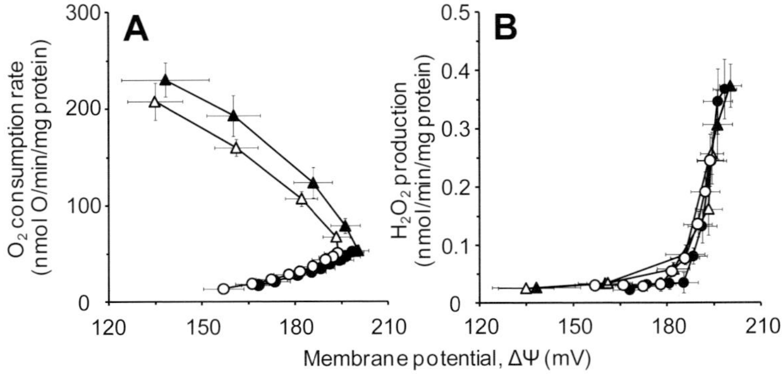

Figure 1. Kinetic response of substrate oxidation (triangles) and basal proton leak (circles) to ΔΨ. (A) ROS production response of both kinetics to ΔΨ. (B) In skeletal muscle mitochondria from control (open symbols) and heat-stressed(closed symbols) birds.

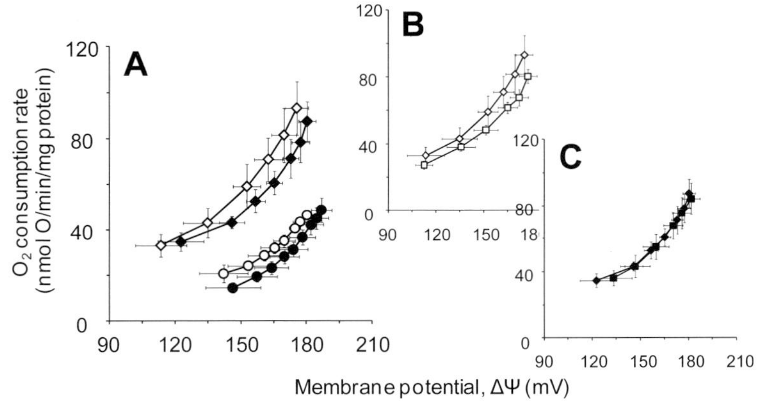

Figure 2. Kinetics of basal (circles) and arachidonic acid-induced (diamonds) proton leak (A), and GDP-inhibited, arachidonic acid-induced proton leak (squares, B and C), in skeletal muscle mitochondria from control (open symbols)and heat-stressed (closed symbols) birds.

uncoupling effect on the muscle mitochondria was diminished in heat-stress. The oxygen consumption rate of the arachidonic acid-induced proton leak was slightly but definitely decreased by GDP in the control group (Figure 2B), but not in the heat-stressed group (Figure 2C), suggesting that avUCP-dependent proton leak may be reduced by heat treatments.

In conclusion, heat stress increased ΔΨ due to enhanced substrate oxidation and decreased proton leak, thereby leading to the overproduction of mitochondrial ROS in muscle mitochondria of birds. The downregulation of avUCP content (data not shown) under heat-stress conditions could also contribute to the overproduction of ROS via a depression in FFA-inducible proton leak.

References

Mujahid A., Y. Akiba, C.H. Warden and M. Toyomizu, 2007. Sequential changes in superoxide production, anion carriers and substrate oxidation in skeletal muscle mitochondria of heat-stressed chickens. FEBS Lett. 581, 3461-3467.

Mujahid A., Y. Akiba and M. Toyomizu, 2009. Olive oil-supplemented diet alleviates acute heat stress-induced mitochondrial ROS production in chicken skeletal muscle. Am. J. Physiol. Regul. Comp. Physiol. 297, R690-698.

Involvement of AMPK, p38 MAPK and PPARα in the regulation of avian uncoupling protein expression: regulation by isoproterenol and fatty acids in chick myoblasts

R. Joubert[1], S. Métayer-Coustard[1], S. Crochet[1], E. Cailleau-Audouin[1], J. Dupont[2,3,4,5], M.J. Duclos[1], S. Tesseraud[1] and A. Collin[1]
[1]INRA, UR83 Recherches Avicoles 37380 Nouzilly, France
[2]INRA, UMR85 Physiologie de la Reproduction et des Comportements, 37380 Nouzilly, France
[3]CNRS, UMR6175 Physiologie de la Reproduction et des Comportements, 37380 Nouzilly, France
[4]Haras Nationaux, 37380 Nouzilly, France
[5]Université François Rabelais de Tours, F-37041 Tours, France

Introduction

The avian uncoupling protein (avUCP) is the unique UCP described in chicken and shares 55% identity of sequence with UCP1 of mammals, involved in the mitochondrial uncoupling that favours thermogenesis in brown adipose tissue. It is orthologous to UCP3 that could transport metabolic anions and limit the mitochondrial production of reactive oxygen species. In chicken, the roles and regulation of avUCP, mostly expressed in muscle, remain to be clarified. Recently, we have shown that avUCP mRNA expression could be under the control of the β-adrenergic system (Joubert *et al.*, 2010), known to be also involved in avian thermoregulation and in UCP1 expression in mammals. This upregulation of avUCP messenger, following intra-muscular isoproterenol injection, was associated with variations in fatty acid metabolism, activation of AMP-activated Protein Kinase (AMPK) and increase of Peroxisome Proliferator-Activated Receptor (PPAR) α, PPARβ/δ and PPARγ coactivator-1α (PGC-1α) mRNA expression.

The aim of this study was to explore *in vitro* the control of avUCP expression by isoproterenol and fatty acids by using primary cell cultures of chick myoblasts, and then to investigate the signalling pathways potentially involved in these regulations.

Material and methods

Primary cultures of myoblasts were derived from *Pectoralis major* muscle satellite cells of 1-day-old chicks. At 80% of confluence, cells were fasted for 3 h in serum-free medium before treatments. Stimulating treatments were of different durations (20 min., 1 h or 2 h) and performed using 100 nM isoproterenol, 0.1% fatty acid supplement (containing linoleic, oleic, myristic, lauric and arachidonic acids; less than 3 mg/l final), 1 mM 5-Aminoimidazole-4-carboxyamide ribonucleoside (AICAR, a pharmacological activator of AMPK) and 100 μM WY-14643 (PPARα agonist) in 0.1% DMSO. Inhibiting treatments were performed using 20 μM Compound C (inhibitor of AMPK) and 10 μM SB202190 (inhibitor of p38 mitogen activated protein kinase, p38 MAPK) in 0.1% DMSO, respectively 2 h and 1 h before stimulations.

Gene expression of avUCP was determined by RT-qPCR (forward: CTACGACCTCATCAAGGACACA; reverse: GAAGGCAGCCACGAAGTGA). Beta-actin was used as the reference gene to standardize mRNA expression of target genes (forward: CTGGCACCTAGCACAATGAA; reverse: CTGCTTGCTGATCCACATCT). Activation of AMPK and p38 MAPK was detected by Western blot using quantification of phosphorylated and total proteins with antibodies previously described by Joubert *et al.* (2010). Vinculin was used as control of loaded proteins.

Results and discussion

Isoproterenol (Iso) increased avUCP expression (+67% after 1 h of stimulation), as previously observed *in vivo* in chicken muscle, but to a greater extent (9 fold-increase; Joubert *et al.*, 2010). This discrepancy could be explained by the concomitant effect of isoproterenol on other factors regulating muscle metabolism *in vivo*, such as plasma thyroid hormones or blood metabolites that could not be investigated here. The use of fatty acid supplement (FA) induced a delayed increase in avUCP expression (+54% after 2 h of stimulation) and was associated with stronger activations of p38 MAPK and AMPK than with isoproterenol.

The AMPK activator AICAR increased avUCP mRNA expression by 63% as compared to controls. The effect of AICAR on avUCP was partially abolished by Compound C, the inhibitor of the AICAR-induced activation of AMPK. Taken together, these results suggest a direct control of AMPK on avUCP expression.

The role of p38 MAPK in avUCP regulation is less clear. On the one hand, Iso and/or fatty acids induced an increase in p38 MAPK activation (+20% 20 min after isoproterenol stimulation). However, SB202190, a p38 MAPK inhibitor, presented a synergic effect with Iso+FA stimulation on avUCP expression (+240% as compared to controls). This huge increase of avUCP mRNA expression was probably due to the AMPK activation (+106%) observed following SB202190 addition. This might represent a peculiarity of avian cells since AMPK is usually not activated by SB202190 in mammalian cells (Jaswal *et al.*, 2007).

To further investigate the regulation of avUCP expression, we studied the effect of WY-14643, a PPARα agonist, on avUCP messenger content. This agonist significantly increased avUCP expression after 2 h of stimulation (+51%), in accordance with the previously observed muscle avUCP upregulation following PPARα overexpression observed after isoproterenol injection *in vivo* (Joubert *et al.*, 2010). These effects of isoproterenol *in vivo* could be due to both b-adrenergic signalling and indirect hormonal or metabolic signals regulating avUCP expression.

In conclusion, AMPK activation and PPARα upregulation are proposed as mechanisms partly mediating b-adrenergic-induced avUCP overexpression. Catecholamines are known as regulators of thermoregulation and lipolysis in avian species. A stimulation of avUCP expression could limit the generation of free radical species when lipid utilization is triggered by the b-adrenergic system in chicken muscle.

References

Jaswal, J.S., M. Gandhi,B.A. Finegan, J.R. Dyck and A.S. Clanachan, 2007. p38 mitogen-activated protein kinase mediates adenosine-induced alterations in myocardial glucose utilization via 5'-AMP-activated protein kinase. Am. J. Physiol. Heart. Circ. Physiol. 292, H1978-H1985.

Joubert, R., S. Métayer Coustard, Q. Swennen, V. Sibut, S. Crochet, E. Cailleau-Audouin, J. Buyse, E. Decuypere, C. Wrutniak-Cabello, G. Cabello, S. Tesseraud, and A. Collin, 2010. The beta-adrenergic system is involved in the regulation of the expression of avian uncoupling protein in the chicken. Domest. Anim. Endocrinol. 38(2),115-125.

Metabolic and signalling pathway alterations in mammary gland of cows fed saturated or unsaturated fat

G. Invernizzi[1,2,3], B. Thering[2,3], M. Bionaz[2,3], G. Savoini[1] and J.J. Loor[2,3]
[1]Department of Sciences and Technologies for Food Safety, University of Milan, 20133 Milan, Italy
[2]Mammalian NutriPhysioGenomics; [3]Department of Animal Sciences and Division of Nutritional Sciences, University of Illinois at Urbana-Champaign, Urbana, 61801 Illinois, USA

Introduction

Milk fat synthesis and milk FA profiles can be greatly affected by nutrition. Several studies have already explored these changes obtained by supplementing in both unsaturated and saturated fat the diet of lactating cows without completely clarifying the systematic adaptations in mammary tissue. A functional genomics approach permits a deeper and broader understanding of the physiological response of the mammary gland to dietary treatments. The goal of this study was to understand the systemic adaptations of biological processes in the mammary gland consequent to dietary supplementation of saturated fat or fish oil.

Material and methods

A 13,257 bovine oligonucleotide microarray in a dye-swap reference design (Loor *et al.*, 2007) was used for transcript profiling of mammary biopsies harvested on d 0, 7 and 21 from the beginning of treatments on 13 Holstein mid-lactation cows (41.7 kg milk/d). The study lasted 4 weeks and the experimental design consisted of three dietary treatments: 4 cows fed a control diet (CTR), 4 cows fed the same diet supplemented with a blend of fish oil and soybean oil (10 g/kg and 25 g/kg DM; FSO) and 5 cows fed the control diet supplemented with 25 g/kg DM of saturated lipid (Energy Booster 100; EB100). Loess normalization and array centering and scaling were used on data obtained from 72 microarrays prior to statistical analysis. Statistics were performed using a mixed effect model on the adjusted ratios of each oligonucleotide with Proc MIXED (SAS®). The model included treatments, time and dye as fixed effects and cow as a random variable. The significant probability values for the treatment and time effects were adjusted for the number of comparisons using Benjamini and Hochberg's false discovery rate (FDR). Data mining was carried out with Ingenuity Pathway Analysis® (IPA) and DAVID (Huang *et al.*, 2009; Dennis *et al.*, 2003). Canonical pathway analysis identified those pathways within the IPA Knowledge Base that were most significantly enriched in the data set. Genes from the data set that met the FDR $\leq$0.05 and post-hoc P-value <0.01 cut-off and were associated with a canonical pathway in the IPA Knowledge Base were considered for the analysis.

Results and discussion

Data on enrichment of most significant canonical signalling and metabolic pathways in the comparison between EB100 and FSO at d 21 are summarized in Table 1. Most enriched metabolic pathways were related to energy metabolism, carbohydrate metabolism and lipid metabolism. The overall effect of EB100 vs. FSO at d 21 on those pathways was a marked inhibition of β-oxidation, glycolysis and TCA cycle. In addition, oxidative phosphorylation as well as sulfur metabolism, all energy-related pathways, was inhibited. The putative inhibition of the pentose phosphate shunt as well that of isocitrate dehydrogenase, the two major sources of NADPH for mammary cells, could be part of the same coordinated mechanism likely directed to a reduction of de novo fatty acids synthesis and, in turn, affect glutathione metabolism which is also dependent on NADPH. Similar to our results, soybean oil supplementation in mice decreased mRNA expression of enzymes of the pentose phosphate shunt, mitochondrial citrate transporter, and enzymes of fatty acid synthesis (Rudolph *et al.*, 2007). Except for aromatase, an essential enzyme for the production of estrogens, the androgen/estrogen metabolism pathway was inhibited. Genes related to cholesterol and HDL uptake (SCARB1) and cholesterol clearance in non-mammary cells (HDLBP) were up-regulated, likely in response to increased availability of this compound in blood. The up-regulation of 11β-hydroxysteroid dehydrogenase 2 (HSD11B2), which converts cortisol to cortisone, could be a response to a local increase of cortisol from exogenous cholesterol. One of the genes most up-regulated (3.3-fold) was FOXO1, a transcription factor (TF) whose function is controlled by insulin signalling. This TF together with PPARGC1A, encoding for the protein PGC-1α (↓), plays a critical role in regulating gluconeogenesis and glycolysis, both of which are classified in the IPA knowledge base as being part of the FXR/RXR activation pathway. Pathways analysis suggested an inhibitory effect of saturated fat on cellular energy production and an enhanced steroid metabolism in mammary tissue.

Table 1. Top metabolic and signalling pathways from Ingenuity Pathways Analysis (IPA) among DEG at d 21 between EB100 and FSO.

Ingenuity canonical pathways	*P*-value	Ratio	DEG	↑/↓	Effect	Class (KEGG orthology)
Metabolic pathways						
Glutathione metabolism	0.0891	0.11	11	3/8	⇓	Metabolism of other amino acids
Oxidative phosphorylation	0.110	0.13	22	0/22	⇓	Energy metabolism
Fructose and mannose metabolism	0.110	0.07	10	3/7	⇓⇓	Carbohydrate metabolism
Citrate cycle	0.119	0.14	8	0/8	⇓⇓	Carbohydrate metabolism
Androgen and estrogen metabolism	0.153	0.06	9	2/7	⇔	Lipid metabolism
Fatty acid metabolism	0.157	0.07	14	1/13	⇓⇓	Lipid metabolism
Glycolysis/gluconeogenesis	0.182	0.09	13	0/13	⇓	Carbohydrate metabolism
Pentose phosphate pathway	0.191	0.08	7	0/7	⇓⇓	Carbohydrate metabolism
Urea cycle and metabolism of amino groups	0.191	0.09	7	0/7	⇓	Amino acid metabolism
Sulfur metabolism	0.191	0.07	4	1/3	↓	Energy metabolism
Signalling pathways						Function (IPA)
FXR/RXR activation	0.0229	0.16	16	3/13		Modulate bile, lipid and glucose homeostasis.
Mitochondrial dysfunction	0.0229	0.14	24	1/23		Affect oxidative stress, apoptosis and mitochondrial DNA damage
Coagulation system	0.0794	0.24	9	1/8		Maintain a fine balance between formation and dissolution of a clot
Xenobiotic metabolism signalling	0.153	0.10	28	9/19		Induction xenobiotic metabolism, elimination and/or detoxification

The *P*-value denotes the significance of the enrichment of a function within the DEG adjusted by Benjamini and Hochberg's FDR ≤0.2. Shown also are the ratio (DEG/number of genes in the pathways), the total number of DEG in the pathway, the number of up-(↑) and down- (↓) regulated DEG in the pathway, the overall effect of the pathways (denoted by ↓ likely inhibited; ⇓ inhibited; ⇓⇓ evidently inhibited; ⇔ equilibrium).

References

Huang D.W., B.T. Sherman and R.A. Lempicki, 2009. Systematic and integrative analysis of large gene lists using DAVID Bioinformatics Resources. Nature Protoc. 4, 44-57.

Dennis G. Jr, B.T. Sherman, D.A. Hosack, J. Yang, W. Gao, H.C. Lane and R.A. Lempicki, 2003. DAVID: Database for Annotation, Visualization, and Integrated Discovery. Genome Biol. 4, P3.

Loor J.J., R.E. Everts, M. Bionaz, H.M. Dann, D.E. Morin, R. Oliveira, S.L. Rodriguez-Zas, J.K. Drackley and H.A. Lewin, 2007. Nutrition-induced ketosis alters metabolic and signalling gene networks in liver of periparturient dairy cows. Physiol. Genomics 32, 105-116.

Rudolph M.C., J.L. McManaman, T. Phang, T. Russell, D.J. Kominsky, N.J. Serkova, T. Stein, S.M. Anderson and M.C. Neville, 2007. Metabolic regulation in the lactating mammary gland: a lipid synthesizing machine. Physiol. Genomics 28, 323-336.

Differential expression of mitochondrial genes in liver from beef calves with divergent phenotypes for feed efficiency

K.A. Johnson[1], H. Neibergs[1], J.J. Michal[1], G.E. Carstens[2], M. Settles[3], A. Hafla[2], T.D. Forbes[2], J.W. Holloway[2] and A. Brosh[4]
[1]Washington State University, USA
[2]Texas A&M University, USA
[3]University of Idaho, USA
[4]Newe-Ya'ar Research Center, Ramat Yishay, Israel

Introduction

Residual feed intake (RFI) is a moderately heritable feed efficiency trait that quantifies the inter-animal variance in observed feed intake that is unexplained by differences in body weight and production and is linked to processes that impact energetic efficiency of beef cattle (Herd and Arthur, 2009). Moore *et al.* (2006) finely mapped the RFI trait to several chromosomes. SNPs associated with RFI but not with feed conversion ratio or feed intake were found indicating these SNPs control other physiological mechanisms (Sherman *et al.*, 2008). Examination of mitochondrial DNA from a small population of steers with divergent RFI phenotypes indicated SNPs were found in the mitochondrial genome (Kolath *et al.*, 2006b). Kolath *et al.* (2006a) and Lancaster *et al.* (2007) reported greater coupling of the electron transport chain in cattle with low RFI phenotypes. Additionally, both studies found mitochondria from efficient animals were better able to carry out oxidative phosphorylation with both NADH- or FADH-linked energy substrates. The aim of this study was to identify differentially expressed genes in liver tissue from beef calves with divergent phenotypes for RFI.

Materials and methods

In study 1, Bonsmara heifers (n=62; initial BW=307±39 kg) were individually fed a roughage-based diet (ME=1.61 Mcal/kg DM) using Calan-gate feeders, whereas, in study 2 crossbred steers (n=170; initial BW=274±26 kg) were individually fed a grain-based diet (ME=2.71 Mcal/kg DM) using a GrowSafe feeding system. For both studies, BW was measured at 14-d intervals and feed intake measured daily for 70 d. Within study, RFI was calculated as the residual from linear regression of DMI on ADG and mid-test $BW^{.75}$, and calves ranked by RFI. For both studies, liver biopsy samples were obtained from calves with the most efficient (n=8) and least efficient (n=8) RFI. Liver biopsies were collected and preserved in RNAlater™ and stored at -80 °C until RNA was extracted. Labelled double-stranded cDNA was hybridized to a custom designed NimbleGen 12x135K array (Roche NimbleGen), washed, scanned and array data extracted. The bovine gene expression array included 34,620 gene transcripts and 38 mitochondrial transcripts. Probe sets with low levels of expression variation across all samples (IQR<0.5) were removed from further analysis. Differential expression was assessed using a linear model with an empirical Bayesian adjustment to the variances and comparisons of interest were extracted using contrasts. The Benjamini and Hochberg method (BH) was used to control for the expected false discovery rate from multiple tests. Probe sets were considered differentially expressed with a BH adjusted *P*-value of <0.05.

Results and discussion

Overall (±SD) ADG and DMI for the heifers were 1.29±0.23 and 8.72±1.14 kg/d, respectively. Heifers with low RFI consumed 24% less (P<0.01) DMI (7.64 vs. 10.03±0.26 kg/d) than heifers with high RFI, even though ADG (1.32 and 1.40±0.07 kg/d) were similar. In study 2, overall (±SD) ADG and DMI for steers were 1.76±0.22 and 9.82±1.03 kg/d, respectively. Steers with low RFI consumed 30% less (P<0.01) DMI (8.34 vs. 11.86±0.26 kg/d) than steers with high RFI, while ADG (1.76 and 1.82±0.07 kg/d) were similar.

Four mitochondrial genes were differentially expressed ($P<0.05$) by the RFI phenotypes in the heifers. NADP-dependent malic enzyme, ATP synthase, and 18 kDa mitochondrial protein were more highly expressed in the inefficient phenotype and L53 39S ribosomal protein more highly expressed in the efficient phenotype. An additional 4 mitochondrial genes tended ($P<0.10$) to be differentially expressed between the heifer RFI phenotypes; TP-53-induced glycolysis and apoptosis regulator, mitochondrion associated apoptosis-inducing factor, methionyl-tRNA formyltransferase, L48 39S ribosomal protein. No other bovine gene transcripts were significantly different between the heifer phenotypes. The mitochondrial genes TP-53-induced glycolysis and apoptosis regulator and cytochrome P450 11A1 had greater expression ($P<0.05$) and monofunctional C1-tetrahydrofolate synthase tended ($P<0.10$) to have greater expression in the efficient steer phenotypes. No other bovine gene transcripts were significantly different between the steers with divergent RFI phenotypes. Lkhagvadorj *et al.* (2010) examined gene expression in liver and adipose from pigs of different RFI phenotypes and found changes in peroxisome proliferator-activated receptor α and γ (PPAR α, γ) and CAMP responsive element binding protein 1 (CREB1) across phenotypes. However, none of these regulatory genes were differentially expressed in our population. The absence of common differentially expressed genes within the RFI phenotypes suggests gender or diet may influence gene expression in heifers and steers with different RFI phenotypes. In heifers and steers exhibiting different RFI phenotypes, genes associated with mitochondrial energy processes are altered.

References

Herd, R.M. and P.F. Arthur, 2009. Physiological basis for RFI. J. Anim. Sci. 87, E64-E71.

Kolath, W.H., M.S. Kerley, J.W. Golden and D.H. Keisler, 2006a. The relationship between mitochondrial function and residual feed intake in Angus steers. J. Anim. Sci. 84:861-865.

Kolath, W.H., M.S. Kerley, J.W. Golden, S.A. Shahid, and G.S. Johnson, 2006b. The relationships among mitochondrial UCP 2 and 3 expression, mitochondrial DNA single nucleotide polymorphisms, and RFI in Angus steers. J. Anim. Sci. 84, 1761-1766.

Lancaster, P.A., G.E. Carstens, J.J. Michal, K.M. Brennan, K.A. Johnson, L.J. Slay, L.O. Tedeschi, and M.E. Davis, 2007. Relationships between hepatic mitochondrial function and RFI in growing beef calves. In: I. Ortigues (ed.) Energy and Protein Metabolism and Nutrition. Wageningen Academic Publishers, Wageningen, the Netherlands, p. 57-58.

Lkhagvadorj, L., Qu, W. Cai, O.P. Couture, C. R. Barb, G.J. Hausman, D. Nettleton, L.L. Anderson, J.C.M. Dekkers, and C.K. Tuggle, 2010. Gene expression profiling of the short-term adaptive response to acute caloric restriction in liver and adipose tissues of pigs differing in feed efficiency. Am. J. Physiol. 298, 494-507.

Moore, S., D.H. Crews, and D. Nkrumah, 2006. Multiple and candidate gene approaches to genetic evaluation of feed efficiency in beef cattle. 8th World Congress on Genetics Appl. to Livestock Prod. 13-18 August, Belo Horizonte, Brazil.

Sherman, E.L., J.D. Nkrumah, B.M. Murdoch and S.S. Moore, 2008. Identification of polymorphisms influencing feed intake and efficiency in beef cattle. Anim. Gen. 39, 225-231.

Hepatic proteome profile of neonatal piglets born to gilts fed diets with differing protein levels

M. Peters, B. Kuhla, I.S. Lang, E.P. Rudolph and C.C. Metges
Leibniz Institute for Farm Animal Biology (FBN), Research Unit Nutritional Physiology, Wilhelm-Stahl-Allee 2, 18196 Dummerstorf, Germany

Introduction

Two dimensional (2D) gel electrophoresis with subsequent mass spectrometry (MS) spot analysis allows the simultaneous determination of changes in steady state levels of numerous proteins in a biological sample (Fuchs *et al.*, 2005) and improves our understanding of regulatory processes and organ function. We used this approach to analyze the effect of maternal protein consumption during gestation on the hepatic proteome profile of newborn pigs. Various studies showed that the effects of an inadequate maternal nutrition during gestation are transfused via maternal metabolic parameters to the growing foetus, and result in developmental adaptations in utero that can permanently change the structure, physiology, and metabolism of the offspring (McMillen and Robinson, 2005). The adaptations often result in an impaired growth and development of the mammalian embryo/foetus during pregnancy (Wu *et al.*, 2006) and are defined as intrauterine growth restriction (IUGR).

Material and methods

Fifty eight gilts (age 241±4 days, body weight [BW] 150.6±10.5 kg) were randomly assigned to three isoenergetic diets with different protein levels fed throughout gestation: adequate protein (AP, 12% CP), high protein (HP, 30% CP) and low protein (LP, 6% CP). Piglet birth BW depended on gilt treatment and were 1.35±0.15 kg (AP), 1.27±0.12 kg (HP) and 1.22±0.11 kg (LP). Liver samples obtained from piglets with high (H, n=42, BW>1.46 kg) and low birth weight (L, n=45, BW<1.1 kg) 24h after term birth were analysed via 2D-SDS-PAGE and MALDI-TOF MS (Kuhla *et al.*, 2007). Relative spot volumes were compared using a three-way ANOVA and SAS procedure MIXED with the main factors gilt diet, gender, and birth weight, and the interaction between diet x birth weight (Tukey-Kramer test). Spots showing significant differences with $P \leq 0.05$ and more than 20% up or down regulation in protein expression were chosen for protein identification by MALDI-TOF MS analysis.

Results and discussion

In neonatal HP as compared to AP offspring, several pathways related to energy homeostasis (GLUD1), protein synthesis (BTF3L2) and degradation (HIBADH) were up regulated, indicating an increased protein turnover in hepatic tissue (Figure 1a), whereas transport proteins (ALB, TF) were less expressed. Low hepatic albumin concentrations affect regulation of oncotic pressure and transport of hydrophobic molecules such as lipid soluble hormones, bile salts, unconjugated bilirubin, and calcium. Decreased levels of serotransferrin limit iron transport, availability and iron-dependent pathways, thereby increasing the risk for anaemia. Interestingly, low maternal protein consumption during gestation and low birth weight (Figure 1A and 1B) predominantly affected pathways related to oxidative stress response (HSP90B1, HSP90AA1, ORM2), energy homeostasis (GLUD1) and cellular proliferation (LMNA). Increased concentrations of several heat shock proteins (HSP90B1, HSP90AA1) and alpha-1 acid glycoprotein (ORM2) suggest systemic oxidative stress which can negatively impact substrate utilization, nutrient homeostasis, immune response and hepatic function. Furthermore, elevated concentrations of heat shock proteins directly modulate apoptotic signalling pathways, supported by the increased abundance of lamin c (LMNA), a nuclear envelop protein which is cleaved by the interleukin-converting enzyme family during apoptosis. In addition to the main effects of diet and birth weight, some proteins showed significant diet × birth weight

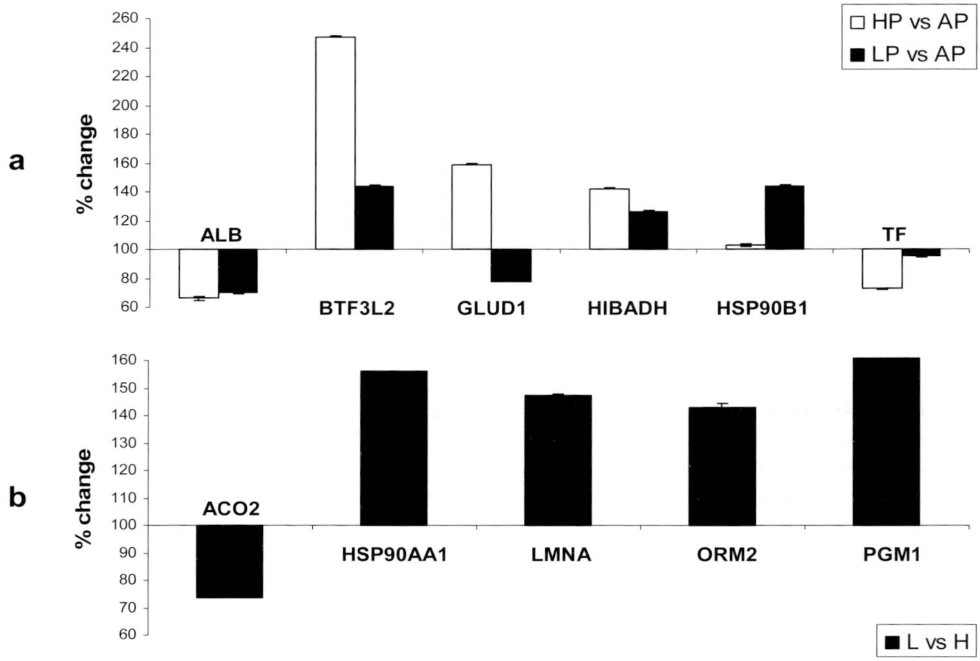

Figure 1. Differentially expressed hepatic proteins of newborn piglets affected by (A) maternal protein consumption during pregnancy or by (B) differing birth weight presented as % change. ALB=albumin, ACO2=aconitase 2, AP=adequate protein, BTF3L2=basic transcription factor 3-like 2, GLUD1=glutamate dehydrogenase 1, H=high birth weight, HIBADH=3-hydroxyisobutyrate dehydrogenase, HP=high protein, HSP90AA1=heat shock protein 90kDa alpha (cytosolic) class A member 1, HSP90B1=heat shock protein 90kDa beta (Grp94) member 1, L=low birth weight, LMNA=lamin A/C, LP=low protein, ORM2=orosomucoid 2, PGM1=phosphoglucomutase 1, TF=transferring.

interaction, e.g. the observed upregulation of GLUD1 or downregulation of ALB in the HP vs. AP group was more pronounced in light piglets. The increase of ORM2 in L vs. H piglets tended to be more apparent in the AP and HP group.

We conclude that some offspring hepatic pathways are heterogeneously affected by maternal protein consumption and/or birth weight during gestation. Further research is underway to validate the results and determine whether the hepatic proteomic profile of foetuses and offspring at later postnatal time points differ from neonatal livers.

References

Fuchs, D., I. Winkelmann, I.T. Johnson, E. Mariman, U. Wenzel, and H. Daniel, 2005. Proteomics in nutrition research: principles, technologies and applications. Br. J. Nutr. 94, 302-314.

Kuhla, B., S. Kuhla, P.E. Rudolph, D. Albrecht, and C.C. Metges, 2007. Proteomics analysis of hypothalamic response to energy restriction in dairy cows. Proteomics 7, 3602-3617.

McMillen, I.C. and J.S. Robinson, 2005. Developmental origins of the metabolic syndrome: prediction, plasticity, and programming. Physiol Rev. 85, 571-633.

Wu, G., F.W. Bazer, J.M. Wallace and T.E. Spencer, 2006. Board-invited review: intrauterine growth retardation: implications for the animal sciences. J. Anim Sci. 84, 2316-2337.

Proteomic investigation of immediate adaptation of adipose tissues to a high protein intake during suckling in the pig

O. Sarr[1,2], *I. Louveau*[1,2], *A. Jamin*[1,2], *I. Le Huërou-Luron*[1,2] *and F. Gondret*[1,2]
[1]*INRA, UMR1079, 35590 Saint-Gilles, France*
[2]*Agrocampus Ouest, UMR1079, 35000 Rennes, France*

Introduction

Proteomics is now recognized as a useful tool in dietary trials to elucidate fine changes in various tissues. However, there is still a paucity of information on how adipose tissue proteome could adapt to changes in dietary protein supply during the neonatal period. In Human, protein contents of infant formula are widely used to promote catch-up growth; however, they are also suspected to increase the risk of later obesity (e.g. Rolland-Cachera *et al.*, 1995). With a natural form of intrauterine growth retardation, the pig is an interesting model to better understand the effects of protein nutrition on adipose tissue development (Morise *et al.*, 2009). This study aims to investigate the immediate effects of milk-replacers providing an adequate or an excess protein supply on early growth and proteomic profiles of adipose tissues in piglets born with a small birth weight.

Material and methods

Crossbred [Pietrain × (Large White × Landrace)] piglets with a small birth weight (0.8 kg to 1.1 kg) were allowed to suckle the dam naturally during the first 48 h after farrowing. Thereafter, they were fed milk-replacers formulated to mimic sow milk (AP, 4.4 g of protein/100 kcal, n=5) or to provide an excess of proteins (HP, 6.2 g of protein/100 kcal, n=5). They were killed at day 28 of age, and perirenal and subcutaneous adipose tissues (PAT and SCAT) were immediately collected. Soluble protein extracts in both tissues were analyzed through 2-dimensional gel electrophoresis (2-DE) and MALDI-TOF/TOF mass spectrometry. The differential abundances of protein spots were analyzed by ANOVA with the fixed effects of dietary group (AP, HP) by using Melanie 7.0 software (GeneBio, Switzerland). Growth data were analyzed by ANOVA using the General Linear Model of SAS (SAS Inst., Cary, NC. USA) with the fixed effects of dietary group.

Results and discussion

Piglets fed the HP milk-replacer had a greater daily weight gain than AP piglets from birth to day 28 (175 vs. 140±4.7 g/day, $P<0.05$). At day 28, the relative weight of PAT was lower in HP piglets than in AP piglets (2.9±0.71 vs. 3.48±0.6 g per kg of body weight) as previously reported (Morise *et al.*, 2009). Moreover, mean adipocyte diameters were lower ($P<0.05$) in the formers compared with the latter piglets in PAT (34±3 vs. 39±4 μm) and in SCAT (36±4 vs. 47±4 μm), however the number of adipocytes in the 2 fat pads did not differ between HP and AP groups. After 2-DE, 34 spots displayed a differential abundance (fold-change > |1.3|, $P<0.10$) in SCAT of HP piglets vs. AP piglets at 28 days of age. Only 7 spots differed in abundance between the 2 dietary groups in PAT. A total of 18 spots corresponding to 16 unique proteins were successfully identified by mass spectrometry. Those proteins are involved in 7 different biological processes, as indicated by Gene ontology classification. Especially, the lactate dehydrogenase (i.e. a glycolytic enzyme), and annexin 2 (i.e. a mediator of GLUT-4 translocation in adipocytes, Huang *et al.*, 2004), were lower in abundance in SCAT of HP piglets than AP piglets. These findings suggest a lower glucose metabolism in SCAT of HP piglets. Similarly, annexin 2 was also down-regulated in PAT of HP piglets compared to AP piglets. In addition, the lower abundance of 2 spots corresponding to glutamate dehydrogenase 1 (GLUD1) in SCAT of HP piglets vs. AP piglets may indicate a reduction in fatty acid synthesis in those piglets. Indeed, besides fatty acid synthesis from glucose, carbon from glutamate (a substrate of GLUD1) may be also incorporated into fatty acids in the adipose tissue (Belfiore and Lannello, 1995).

Alternatively, down-regulation of GLUD1 in HP group may simply sign adaptation of amino-acid metabolism in SCAT to face an excess of amino-acid supply in HP piglets. In support to this latter assumption, up-regulation of 26S subunit proteasome (i.e. the major tool for protein catabolism) and down-regulation of peptidyl-prolyl cis-trans isomerase A (i.e. a catalyzer of protein folding) in SCAT of HP piglets compared to AP animals indicate a modulation of protein turn-over in the formers.

Another important observation was the greater abundance of 2 cytoskeletal proteins (dynactin and cofilin) in SCAT of HP piglets. Non-muscle type cofilin binding to actin has been previously suggested to play a role in lipid accumulation in adipose tissue of pigs (Choi *et al.*, 2003), likely because it controls the mechanical tension of the cells. We suggest that the cytoskeleton of HP piglets vs. AP animals may be involved in regulating adipogenesis in those animals. A greater abundance in glutathione S-transferase (GST)-O, belonging to the multi-gene family of detoxification enzymes, together with a lower abundance in peroxiredoxin 6 were also observed in SCAT of piglets fed the HP formula compared to AP piglets. Both cytoskeleton proteins (including cofilin 2) and other members of GST family have previously identified as having a greater abundance in adipose tissue of fat birds compared to lean birds (Wang *et al.*, 2009). Thus, changes in those proteins may predispose HP piglets to catch-up fat during the fattening period. The lower abundance in annexin 1 in PAT of HP piglets compared with AP piglets is also consistent with the assumption of an early modulation of adipocyte development in response to HP milk. Indeed, annexin 1 gene is generally required for the maintenance of adipocyte and/or preadipocyte numbers (Warne *et al.*, 2006).

In conclusion, our results indicate that neonatal diet enriched in proteins promoted early body growth. It reduced adipose tissue development in the short term, likely due to a lower glucose metabolism in adipocytes and the early modulation of adipogenesis. Our findings represent also novel starting points for elucidating the mechanisms leading to adipose tissue development in the long term.

References

Belfiore, F and S. Lannello, 1995. Fatty acid synthesis from glutamate in the adipose tissue of normal subjects and obese rat. Biochem. Mol. Med. 54, 19-25.

Choi, K.C., S.G. Roh, D. Hishikawa, H. Miyahara, M. Kuno, H. Tsuzuki, A. Tomimatsu, Y.H. Hong, K.K. Cho, K.H. Han and S. Saski, 2003. Differential expression of the non-muscle cofilin gene between subcutaneous and visceral adipose tissue. Biosci. Biotechnol. Biochem. 67, 2262-2265.

Huang, J., S.H. Hsia, T. Imamura, I. Usui and J.M. Olefsky, 2004. Annexin II is a thiazolidinedione-responsive gene involved in insulin-induced glucose transporter isoform 4 translocation in 3T3-L1 adipocytes. Endocrinology 145, 1579-1586.

Morise, A., B. Sève, K. Macé, C. Magliola, I. Luron and I. Louveau, 2009. Impact of intrauterine growth retardation and early protein intake on growth, adipose tissue and the insulin-like growth factor system in piglets. Pediatr. Res. 65, 45-50.

Rolland-Cachera, M.F., M. Deheeger, M. Akrout and F. Bellisle, 1995. Influence of macronutrients on adiposity development: a follow up study of nutrition and growth from 10 months to 8 years of age. Int. J. Obes. Relat. Metab. Disord. 19, 573-578.

Wang, D., N. Wang, and H. Li, 2009. Identification of differentially expressed proteins in adipose tissue of divergently selected broilers. Poult Sci. 88, 2285-2292.

Warne, J.P., C.D. John, H.C. Christian, J.F. Morris, R.J. Flower, D. Sugden, E. Solito, G.E. Gillies and J.C. Burkingham, 2006. Gene deletion reveals roles for annexin A1 in the regulation of lipolysis and IL-6 release in epididymal adipose tissue. Am. J. Physiol. Endocrinol. Metab. 291, E1264-E1273.

The GENOTEND chip: a tool to analyse gene expression in muscles of beef cattle

J.F. Hocquette[1], C. Bernard-Capel[2], M.L. Vuillaume[3], B. Jesson[3], H. Levéziel[4] and I. Cassar-Malek[1]
[1]INRA, UR1213, Unité de Recherches sur les Herbivores, Theix, 63122 Saint-Genès Champanelle, France; jfhocquette@clermont.inra.fr
[2]Institut de l'Elevage, 149 rue de Bercy, 75595 Paris cedex 12, France
[3]IMAXIO, 63360 Saint-Beauzire, France
[4]UMR 1061/Université de Limoges, 87060 Limoges, France

Introduction

There is a need for new knowledge in order to develop animal farming systems that respond to the new and diversified demands by producers and consumers. For producers, it is a question of production of more meat per animal, i.e. of heavy and lean carcasses to ensure the highest potential incomes. Consequently, genetic selection has been directed in favour of muscle development in order to produce lean carcasses with the ultimate objective of increasing the production of muscle quantitatively at the expense of fat. Recent genomic studies have shown that this type of selection may change certain muscle characteristics associated with beef quality (Bernard *et al.*, 2009; Sudre *et al.*, 2005). As for the consumers, they are interested in the production of high-quality meat products. Consumers are indeed looking for bovine meat of high and consistent quality. Intrinsic quality attributes of beef, and especially tenderness, depend on one hand on *post mortem* factors associated with ageing and cooking, and on the other on the muscle characteristics of live animals, which themselves depend on gene expression. Expression levels of these genes and interaction between them can now be assessed thanks to the development of functional genomics (e.g. DNA microarrays and proteomic tools). Functional genomics is also expected to be helpful for the beef industry which is looking for biological or molecular indicators that would identify live animals with desirable quality attributes, in order to direct them towards the most appropriate production system(s) (reviews by Cassar-Malek *et al.*, 2008 and Hocquette *et al.*, 2007). The strategy so far has been to identify genes or proteins that are expressed differentially between extreme animals without any prior knowledge of the processes involved. Previous programs have described new genes associated with growth potential, beef tenderness or flavour (Bernard *et al.*, 2009, 2007). One of these results concerning the DNAJA1 gene was patented (Bernard *et al.*, 2006). DNAJA1 encodes a member of the large 40 kDa heat shock protein family (Hsp40). This protein is a co-chaperone of the 70 kDa heat shock protein (Hsp70) and is believed to play a role in protein folding and mitochondrial protein import. The DNAJA1/Hsp70 complex also directly inhibits apoptosis. The GENOTEND program aims to confirm the relationship between these genes and beef quality including tenderness and flavour. To reach this objective, we developed an Agilent chip with specific probes for the bovine muscular genes known as predictors of muscle growth (including energy and protein metabolism), carcass composition, fat metabolism and beef quality (including marbling).

Material and methods

More than 3000 genes involved in muscle biology or beef quality were selected from genetic or genomic studies. Sometimes, several probes were used for each gene (e.g. 18 probes for DNAJA1). After extraction, the total RNA sample was quantified with a Nanodrop ND.1000 spectrophotometer (ThermoScientific) and evaluated for integrity with the 2100 bioanalyzer (Agilent Technologies) and the RNA 600 LabChip kit. The total RNA was amplified and labelled with Cyanine 3 using Agilent's Low RNA Input Linear Amplification Kit, PLUS, One-Color (Agilent Technologies). Briefly, 500 ng of total RNA was reverse transcribed to double-strand cDNA and cDNA products were then used as templates for *in vitro* transcription to generate fluorescent cRNA. Labelled cRNAs were purified using QIAGEN's RNeasy mini spin columns and eluted in nuclease-free water. Then, cRNA quantity and cyanine incorporation were determined using the Nanodrop spectrophotometer. For each hybridization, 600 ng of Cyanine 3 labelled cRNA were fragmented and hybridised at 65 °C for 17 h to an Agilent 8 × 15K custom Oligo Microarray. After washing, microarrays were scanned using an Agilent DNA G2505B scanner. Feature extraction 9.1 software (Agilent Technologies) was used to assess fluorescent hybridization signals.

Results and discussion

RNA from the *Longissimus thoracis* muscle samples of Charolais young bulls or steers slaughtered in two years (25 young bulls and 22 steers in 2003; 33 young bulls and 19 steers in 2005) was hybridised on the chips. Statistical analyses allowed the genes associated with beef tenderness to be selected. All the DNAJA1 probes gave similar results. For the Charolais animals slaughtered in 2003, DNAJA1 expression was negatively correlated with the initial or global beef tenderness (r=-0.40 to -0.60) in young bulls or steers. However, this was not the case for the animals slaughtered in 2005. This confirms our hypothesis that some markers of beef quality are highly dependant on rearing practices and environmental conditions. However, the expression of other members of the DNAJ family or those linked to other metabolic pathways was associated with beef tenderness. These gene families (including that of heat shock proteins or of muscle metabolism) appeared to be associated with beef quality in all groups of animals. These data are not detailed here because they may be the subject of patent registrations. In conclusion, numerous markers of beef tenderness can be identified but they are often specific to an animal type (steer or young bull), to a breed or to environmental conditions linked to the year. The bovine muscle chip used in this study is available for academic and professional partners[1].

Acknowledgments

This work was funded by APIS-GENE, INTERBEV, the former 'Office de l'Elevage' now 'FranceAgriMer', and the Regional Council of Auvergne in France.

References

Bernard, C., I. Cassar-Malek, M. Le Cunff, H. Dubroeucq, G. Renand and J.F. Hocquette, 2007. New indicators of beef sensory quality revealed by expression of specific genes. J. Agric. Food Chem. 55, 5229-5237.

Bernard, C., I. Cassar-Malek and J.F. Hocquette, 2006. Genomic marker for meat tenderness. Patent 06 300943.5. 12 September 2006.

Bernard, C., I.Cassar-Malek, G. Renand and J.F. Hocquette, 2009. Changes in muscle gene expression related to metabolism according to growth potential in young bulls. Meat Sci. 82, 205-212.

Cassar-Malek, I., B. Picard, C. Bernard and J.F. Hocquette, 2008. Application of gene expression studies in livestock production systems: a European perspective. Austr. J. Exp. Agric. 48, 701-710.

Hocquette, J.F., S. Lehnert, W. Barends, I. Cassar-Malek and B. Picard, 2007. Recent advances in cattle functional genomics and their application to beef quality. Animal 1, 159-173.

Sudre, K., I. Cassar-Malek, A. Listrat, Y. Ueda, C. Leroux, C. Jurie, C. Auffray, G. Renand, P. Martin and J.F. Hocquette, 2005. Biochemical and Transcriptomic analyses of two bovine skeletal muscles in Charolais bulls divergently selected for muscle growth. Meat Sci. 70, 267-277.

[1] The IMAXIO Company is proposing a bovine muscle transcriptomic analysis service.

Phospho-signalling in the liver of periparturient cows

B. Kuhla[1], D. Albrecht[2] and C.C. Metges[1]
[1]Research Unit Nutritional Physiology 'Oskar Kellner', Leibniz Institute for Farm Animal Biology (FBN), Dummerstorf, Germany
[2]Institute of Microbiology, Ernst-Moritz-Arndt-University, Greifswald, Germany

Introduction

Within the periparturient period, feed intake of dairy cows is decreasing from about 6 weeks before parturition, is lowest around calving, and is increasing again in early lactation. During this period, dairy cows mobilize large amounts of muscle protein and body fat which is accompanied by increasing oxidative stress and aberrant release of hormones and cytokines. These endocrine and metabolic changes affect hepatic cellular signalling, metabolic activity and transcription of metabolic enzymes and hepatic cytokines. Which major cellular signalling pathways involved in nutrient sensing, oxidative stress, hormones and cytokines are activated or deactivated by phosphorylation during the periparturient period has not been investigated so far. Therefore, the aim of the study was to explore phospho-related cellular signalling and phospho-regulated metabolic pathways in the liver of periparturient dairy cows.

Material and methods

Four German Holstein cows in 1st lactation were ceased to be milked 10 weeks before expected 2nd calving and were fed a dry-off ration based on grass silage. With the beginning of the close-up period starting from thirty days before expected parturition and during lactation cows received a total mixed ration (TMR). The *ad libitum* feed intake was measured daily until 5 weeks after parturition. Cows were liver biopsied at -60, -40, -21, +1, +14 and +28 days relative to their 2nd parturition. Liver samples were extracted with phosphatase inhibitors and probed using Western blot analysis for total phospho-tyrosins, for the p38 mitogen activated protein kinase (p38 MAPK-Thr180/Tyr182), for the extracellular regulated kinase 1/2 (ERK1/2-Thr202/Tyr204), for the stress-activated protein kinase/Jun-amino-terminal kinase (SAPK/JNK-Thr183/Tyr185), for the AMP-activated kinase-alpha (AMPKα-Thr 172), and for the mammalian target of rapamycin (mTOR-Ser 2448).

In order to identify further activated or deactivated proteins during the transition period, we applied each liver extract (only -21, +1, +14 and +28 days relative to parturition) on a 2-dimensional gel electrophoresis (2DGE) (Kuhla *et al.* 2009). Detection of phospho-proteins was performed by ProQ-Diamond (PQ) staining (Eymann *et al.*, 2007). Subsequently, gels were stained with Colloidal Coomassie (CC) to detect total proteins for normalization (Kuhla *et al.* 2009). The ratio between the total volume (V) in PQ and CC gels ($r = V(PQ)/V(CC)$) was calculated for each gel. In order to differentiate between true and false positive phospho-proteins, a Gaussian-like analysis of log(r) was performed (Eymann *et al.*, 2007). For each gel, proteins were considered to be phosphorylated when log(r) of one spot was >1.2. When at one of the four time points a spot fulfilled the criteria log(r) >1.2 it was considered to be aberrantly phosphorylated and picked for tryptic digestion and subsequent identification by MALDI TOF and MALDI TOF/TOF mass spectrometry.

Results and discussion

With the beginning of TMR feeding average dry matter intake diminished from 16.9 kg (4 weeks before parturition) to 12.7 kg at calving and re-increased to 20.2 kg by the 5th week *post partum*.

Western blot analysis for pTyr revealed several bands aberrantly phosphorylated, for example at ~38, 46, 55 and 290 kDa. Therefore, we first examined phosphorylation of the p38 MAPK signalling pathway known to activate cytokine expression. Activation was highest within the first 14 days after calving, suggesting involvement of oxidative stress and inflammatory cytokines. Surprisingly, phosphorylation of the SAPK/JNK (at ~46 and 55 kDa) pathway which is preferentially activated by a variety of environmental stress factors was diminished after parturition. The energy sensor AMPKα (~64 kDa) activated by an elevated AMP/ATP ratio was not aberrantly phosphorylated before parturition but increasingly phosphorylated after calving and highest 4 weeks *post partum*. Concomitantly, phosphorylation of the downstream target mTOR (~290 kDa), an ATP and branched chain amino acid sensor, was lowest after calving. These results point to a cumulative catabolic state until day 28 *post partum* which is inversely related to the increasing feed intake.

Analysis of phospho-stained proteins by 2DGE revealed 12 phosphorylated protein spots (log ratio >1.2) only occurring at one of the four time points tested. By using MALDI TOF/TOF mass spectrometry we identified 2 enzymes related to oxidative stress: glutathione S-transferase 1 and catalase; 2 enzymes associated with amino acid degradation: arginase 1 and serine hydroxymethyltransferase; 2 enzymes of the glycolysis pathway: 6-phosphofructokinase (known to exert phosphatase activity when phosphorylated) and fructose-bisphosphate aldolase B; 3 enzymes of fat metabolism: acyl-CoA dehydrogenase (showing altered electron transfer and fatty acid beta-oxidation activity when phosphorylated (Kabuyama *et al.*, 2010)), acetyl-CoA acyltransferase, and aldo-keto reductase family 1; one enzyme involved in the oxidative phosphorylation pathway: ATP synthase; and two spots of cytoplasmic skeleton: beta-actin. All these proteins showed minimal phosphorylation either at +1 or +14 days after parturition, reflecting their involvement in metabolic adaptation to parturition.

References

Kuhla, B., D. Albrecht, S. Kuhla and C.C. Metges, 2009. Proteome analysis of fatty liver in feed deprived dairy cows reveals interaction of fuel sensing, calcium, fatty acid and glycogen metabolism. Physiol Genomics 37, 88-98.

Eymann, C., D. Becher, J. Bernhardt, K. Gronau, A. Klutzny and M. Hecker, 2007. Dynamics of protein phosphorylation on Ser/Thr/Tyr in Bacillus subtilis. Proteomics 7, 3509-3526.

Kabuyama, Y., T. Suzuki, N. Nakazawa, J. Yamaki, M.K. Homma and Y. Homma, 2010. Dysregulation of very long chain acyl-CoA dehydrogenase coupled with lipid peroxidation. Am. J. Physiol. Cell Physiol. 298, C107-113.

Part 2. Stable isotopes and other techniques in metabolism and nutrition

Citrulline and urea metabolism in the gut

J.C. Marini
USDA/ARS Children's Nutr. Res. Center, Department of Pediatrics, Baylor College of Medicine, Houston, Texas 77030, USA

Abstract

Citrulline is the endogenous precursor for the synthesis of arginine, and direct functional effects have been recently attributed to this non-protein amino acid. Citrulline is synthesized in the liver, as part of the urea cycle, and in the gut, as part of the small intestine-renal axis for the synthesis of arginine. These two systems are virtually independent and are regulated differently. The enzymes involved in the synthesis and utilization of citrulline and its precursors undergo developmental changes in the gut, which may be geared towards a more efficient utilization of dietary and endogenous arginine during rapid growth. The origin of ornithine utilized for the synthesis of citrulline is controversial, and may reflect the different systems in which it was determined. More research is needed to fully integrate the molecular and cellular processes with the interorgan transaction of citrulline and its precursors involved in the synthesis of citrulline.

Introduction

Citrulline, a non-protein amino acid, was identified independently by Mitsunori Wada in watermelon juice (Wada, 1930) and by Dankwart Ackermann as a product of the degradation of arginine (Ackermann, 1931). Soon afterwards, Hans Krebs and Kurt Henseleit (Krebs and Henseleit, 1932) proposed the first metabolic cycle ever described, in which citrulline was an intermediary in the synthesis of urea. A few years later, the role of aspartate as the donor of the second nitrogen for urea synthesis was established by Sarah Ratner (Ratner and Pappas, 1949).

The five enzymes of the urea cycle (also known as the ornithine cycle) are in close proximity to each other in the hepatocyte, constituting a metabolon spanning two compartments (argininosuccinate synthetase, argininosuccinate lyase, arginase in the cytosol and carbomyl phosphate synthetase and ornithine transcarbamylase in the mitochondria) that channels urea cycle intermediates (Cohen *et al.*, 1987; Cheung *et al.*, 1989). Other enzymes (N-acetylglutamate synthase, glutaminase) and transporters (citrin, ORN1), although not considered part of the cycle per se, are also needed for ammonia detoxification by the liver (see Meijer *et al.*, 1990).

The enzymes of the urea cycle, however, are not only expressed in the liver, but also in other tissues and cell types. In fact, it is believed that the urea cycle evolved from the arginine metabolic pathway present in lower organisms (Takiguchi *et al.*, 1989). This difference in function between urea and arginine synthesis is reflected by the different tissue localization, function and regulation of the enzymes of the urea cycle and other enzymes involved in the metabolism of the urea cycle intermediates, arginine, citrulline and ornithine.

Numerous reviews dealing with different aspect of the metabolism of arginine and citrulline have been published over the last years (Morris, 2006, 2007; Curis *et al.*, 2007; Crenn, 2008; Wu *et al.*, 2009). A renewed interest in arginine metabolism has been sparked due to the functional role of this amino acid in many processes, such as nitric oxide synthesis (Luiking *et al.*, 2005), lipid metabolism (Fu *et al.*, 2005; Hayashi *et al.*, 2005) and diabetes (Kohli *et al.*, 2004).

Citrulline more than the precursor for arginine synthesis

The contribution of the synthesis of citrulline to the economy of arginine in humans is approximately 50% of the average arginine intake (4.4 g/d; (King *et al.*, 2008). Citrulline is not only the precursor for the synthesis arginine, but it is becoming apparent that this amino acid might have additional functional roles. The severe growth retardation seen in rats, consequence of the inhibition of the endogenous synthesis of citrulline, was only partially restored by arginine, but completely prevented by citrulline supplementation (Hoogenraad *et al.*, 1985). It has also been shown that labelled citrulline had a higher rate of protein incorporation than arginine in rats (Featherston *et al.*, 1973); especially, the brain showed a great capability to derive arginine for protein synthesis directly from citrulline. Recent reports show that citrulline supplementation is able to increase protein deposition in a model of ageing (Osowska *et al.*, 2006), and that citrulline seems to be more effective than arginine supplementation (Osowska *et al.*, 2004). Although all these effects may be due to a more efficient utilization of citrulline rather than arginine, recent reports suggest that citrulline may increases protein synthesis through direct activation of the mTOR pathway (Walrand *et al.*, 2008).

Origin of circulating citrulline

Because little or no citrulline is present in the diet (with the notable exception of watermelon; Tedesco *et al.*, 1984), circulating citrulline is almost entirely of endogenous origin. Although there are other sources of endogenous citrulline (e.g. from turnover of citrullinated proteins, by action of nitric oxide synthase on arginine (Hecker *et al.*, 1990) and dimethylarginine dimethylaminohydrolase on dimethylarginine (Ogawa *et al.*, 1989)), it is believed that these sources are quantitatively minor compared to the synthesis of citrulline from ornithine.

A systematic analysis of different tissues showed that activity of the mitochondrial enzymes, carbomyl phosphate synthase I (CPS I) and ornithine transcarbamylase (OTC), was not only present in the liver but also in the small intestine of rats (Jones *et al.*, 1961) and humans (Reichard, 1960). Furthermore, the presence of N-acetylglutamate, the allosteric activator of CPS I, was also found in the small intestine (Raijman, 1974). Thus the enterocyte is equipped to synthesize citrulline from ornithine, ammonia and bicarbonate. Because the enterocytes in adults lack the enzymes that utilize citrulline (for developmental changes in the enzymes, see below), citrulline enters the portal circulation without further metabolism (Windmueller and Spaeth, 1981). This was clearly shown in small intestinal preparations by Windmueller and Spaeth (1974), where citrulline was one of the main products of intestinal metabolism.

It has been shown in rodents that citrulline is neither removed nor released by the liver and thus the plasma entry rate represents citrulline production (Windmueller and Spaeth, 1981). Recently, however, this concept has been challenged in humans undergoing surgery, mostly liver resection (Van de Poll *et al.*, 2007). It is not clear if these human data are applicable to other species or to humans in more physiological conditions. Under some situations the liver releases citrulline, but it seems that this is done only under non-physiological experimental (Drotman and Freedland, 1972) or pathological conditions (argininosuccinic aciduria, Marini and Lee, unpublished data).

For the reasons mentioned above, it is believed that circulating citrulline originates for the most part in the enterocytes, a fact that has been used as a proxy to monitor gut mass and gut function (Crenn *et al.*, 2000; Rhoads *et al.*, 2005).

Fate of circulating citrulline

The ability of the kidney to synthesize arginine from citrulline was shown early on (Borsook and Dubnoff, 1941) which demonstrated that the cytosolic enzymes of the urea cycle, argininosuccinate synthase (AS) and argininosuccinate lyase (ASL), were present in this organ. Arterio-venous differences across the kidney, has shown that >80% of the citrulline produced by the small intestine is taken up by this organ (Windmueller and Spaeth, 1981) and that similar amounts of arginine are released into the circulation (Dhanakoti *et al.*, 1990). This interorgan collaboration for the endogenous synthesis of arginine has been called the intestinal-renal axis (Brosnan and Brosnan, 2004).

It has been recognized that the cytosolic enzymes of the urea cycle are also present in numerous cell types because citrulline can replace arginine when cultured in *in vitro* conditions (Morgan *et al.*, 1958; Tytell and Neuman, 1960; Sun *et al.*, 1979). However, the physiological significance of this local citrulline to arginine synthesis remained unclear until it was discovered that the 'endothelial relaxing factor' was nitric oxide which was derived from arginine, producing citrulline as a co-product (Ignarro *et al.*, 1987). The intracellular recycling of citrulline to arginine, in what has been called the citrulline-nitric oxide cycle, seems to sustain the nitric oxide response (Hecker *et al.*, 1990; Wu and Brosnan, 1992) and it has been proposed that it may be the principal mechanism regulating local arginine availability (Hayashi *et al.*, 2005).

Tracer studies have shown that 50-60% or 75% of circulating citrulline can be accounted for as circulating arginine in humans (Castillo *et al.*, 1993) and mice (Marini *et al.*, 2007), respectively. Because the only known fate of citrulline is conversion to arginine or excretion into the urine (which is negligible (Moinard *et al.*, 2008), it seems that the local utilization of citrulline is responsible for the disposal of a substantial portion of the citrulline flux. This supports the findings that certain cell types are able to utilize citrulline directly to meet their arginine requirements (Featherston *et al.*, 1973).

Regulation of the activity of the hepatic and extrahepatic enzymes of the urea cycle

The different function of the mitochondrial enzymes of the urea cycle in the intestine and liver is reflected by their nutritional (Wraight and Hoogenraad, 1988), hormonal (Ryall *et al.*, 1986) and developmental (Malo *et al.*, 1986; Ryall *et al.*, 1986) regulation. The short term regulation includes an increase in N-acetylglutamate synthase activity and N-acetylglutamate concentration (Kawamoto *et al.*, 1985), which in turn increases CPS I and glutaminase activity (Meijer *et al.*, 1990; Curthoys and Watford, 1995). None of these effects have been described for the intestinal enzymes and short term regulation is very unlikely to occur.

A differential longer term regulation of the mitochondrial enzymes in liver and gut is also evident because high protein diets, glucagon and dexamethasone have shown to increase mRNA expression of the mitochondrial enzymes of the urea cycle in liver, but not in the intestine (Wraight *et al.*, 1985; Hurwitz and Kretchmer, 1986; Ryall *et al.*, 1986; Morris *et al.*, 1987). Likewise, ASS and ASL are regulated differently in kidney and liver. High protein diets, starvation and glucocorticoids greatly increased the expression of these two enzymes in the liver, but had little effect in kidney (Morris *et al.*, 1987, 1989; Goutal *et al.*, 1999).

While urea production can vary widely depending on protein intake, citrulline synthesis is rather constant. No changes in citrulline production have been found in animals (Hartman and Prior, 1992; Prior and Gross, 1995) or humans (Castillo *et al.*, 1993; Tharakan *et al.*, 2008) fed arginine free diets, nor during fed or fasted conditions. The endogenous synthesis of arginine is limited by the supply of citrulline, because the kidney has excess ability to synthesize arginine (Dhanakoti *et al.*, 1990; Moinard *et al.*, 2008).

Ontogeny of the enzymes of the urea cycle

Developmental changes in the activity and expression of the enzymes involved in the metabolism of urea cycle intermediates have been reported in the intestine of rats, mice, piglets and humans (Herzfeld and Raper, 1976; Hurwitz and Kretchmer, 1986; Ryall *et al.*, 1986; Wu, 1995; De Jonge *et al.*, 1998; Kohler *et al.*, 2008). While OTC activity increases progressively from birth (Herzfeld and Raper, 1976), there is a rapid increase in arginase expression and activity concurrent with the ingestion of solid feed (Herzfeld and Raper, 1976; Hurwitz and Kretchmer, 1986; De Jonge *et al.*, 1998), that coincides with a reduction in the expression and activity of AS and ASL (Hurwitz and Kretchmer, 1986; De Jonge *et al.*, 1998). Similar developmental changes for some of these enzymes have been reported recently in human infants (Kohler *et al.*, 2008); however, some noteworthy differences are the presence of arginase II 3 days after birth and that the decline in AS expression occurs at 3-5 years of age (Kohler *et al.*, 2008).

The cytosolic enzymes, AS and ASL, are expressed in kidney at birth albeit at a lower level (~40%) than in adults (Wu and Knabe, 1995; Goutal *et al.*, 1999). In piglets the activity of these enzymes in the small intestine is greater during the neonatal period than later in life (Wu and Knabe, 1995). In mice the specific activity of these enzymes in kidney at birth is approximately a third of the activity found in the intestine, but this relationship changes by day 12 of life, when the renal specific activity is 10 fold greater that the intestinal activity (Hurwitz and Kretchmer, 1986).

Thus, there are important differences in the enzyme expression of citrulline-arginine synthesis between the neonatal period and adulthood, and it is believed that the intestinal-renal axis is not established until after weaning in rodents, piglets and humans. The ability of the neonatal enterocytes to produce arginine was first established *in vitro* in mice (Hurwitz and Kretchmer, 1986) and later in piglets (Blachier *et al.*, 1993). This was corroborated *in vivo* by the net release of arginine by the jejunum in both pre and post weaning piglets (Wu *et al.*, 1994a). However, not all enterocytes present all the enzymes needed for arginine synthesis. Apical enterocytes express AS and ASL and thus are able to make arginine from citrulline, while crypt and lower part of the villus epithelial cells make citrulline from ornithine (De Jonge *et al.*, 1998; Kohler *et al.*, 2008). The implication for the localization of the different enzymes along the intestinal villi is that citrulline has to exit the basal enterocyte where it is synthesized, enter the circulation, and later on be utilized by apical enterocytes to synthesize arginine. This explains the presence of circulating citrulline seen in pre weaning piglets (Flynn *et al.*, 2000; Wilkinson *et al.*, 2004) and human infants (Cavicchi *et al.*, 2009), with concentrations that do not differ from adult values.

Precursors for citrulline synthesis: enzymes involved in the synthesis of citrulline and the supply of precursors

When discussing which precursors are utilized for the synthesis of citrulline, what we really mean is what sources of ornithine are used for citrulline synthesis. Ornithine can be generated 'de novo' from proline and glutamate by action of ornithine amino transferase (OAT) or it can be 'preformed', released by hydrolysis of arginine by arginase at the site of citrulline synthesis (Figure 1). In addition, plasma ornithine can be imported by the enterocyte and utilized for the synthesis of citrulline.

Developmental changes in the enzymes that utilize proline and glutamine to produce glutamate semialdehyde (GSA) have been described in the rat (Herzfeld *et al.*, 1977). Proline oxidase activity in the gut is present in rat pups, but disappears by two weeks of age (Herzfeld and Raper, 1976) and is absent in adult animals (Kawabata *et al.*, 1980). Piglets have shown variable or low levels of proline oxidase activity during their first two months of life (Samuels *et al.*, 1989; Wu, 1997). The enzymatic activity of pyrrolidine-5-carboxylate (P5C) synthase in rodents increases during the preweaning period, but falls to adult levels at around weaning (Riby *et al.*, 1990; Yamada and

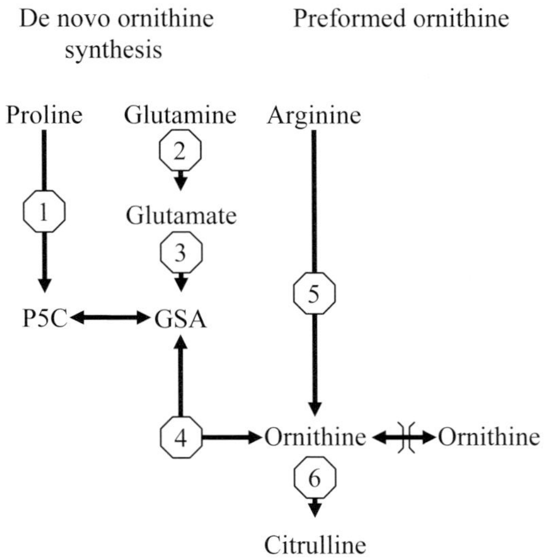

Figure 1. Contribution of different pathways to the ornithine used for citrulline synthesis. De novo *synthesis of ornithine: Proline is oxidised by action of proline oxidase (1) to generate pyrrolidine-5-carboxylate (P5C). P5C interconverts spontaneously with glutamate semialdehyde (GSA). Glutamine is deamidated by action of glutaminase (2) and the resulting glutamate is converted into GSA by pyrrolidine-5-carboxylate synthase (3). GSA is used by ornithine aminotransferase (OAT; 4) to generate ornithine. Note that OAT is used for both synthesis and disposal of ornithine. Preformed ornithine: ornithine is generated from the hydrolysis of arginine by arginase (5), or can be transported from plasma. Ornithine is used for citrulline synthesis by action of ornithine transcarbamylase (OTC; 6).*

Wakabayashi, 1991). In piglets, P5C synthase activity was found soon after birth (Flynn and Wu, 1996), but continuously increased until 40 d of age (Samuels *et al.*, 1989). P5C synthase undergoes a tissue specific differential splicing; the short isoform present in the gut is inhibited by ornithine, but the longer isoform found in multiple tissues is not (Hu *et al.*, 1999).

OAT also undergoes developmental changes and its activity declines towards weaning (Herzfeld and Raper, 1976; Riby *et al.*, 1990). OAT is a bidirectional enzyme that catalyzes the synthesis and disposal of ornithine (Markova *et al.*, 2005; Figure 1). In adults, the catabolic route predominates (McGivan *et al.*, 1977) since the equilibrium strongly favours the formation of GSA (Smith *et al.*, 1967). It has been shown *in vivo* that OAT inhibition in adult mice (Alonso and Rubio, 1989; Seiler *et al.*, 1994) increased plasma ornithine concentration, whereas in suckling piglets OAT inhibition decreased ornithine levels (Flynn and Wu, 1996). The inhibition of OAT in transgenic mice results in a paradoxical neonatal hypo-ornithemia and hyper-ornithemia after weaning, which mimics similar findings reported in humans with gyrate atrophy (Wang *et al.*, 1995). Conversely, the overexpression of OAT results in a decrease in the plasma concentration of ornithine in adult mice (Ventura *et al.*, 2009).

Two different arginase (ARG) isoforms, with different biochemical characteristics and tissue distribution, have been identified (Reddi *et al.*, 1975). These enzymes were shown to be the product of two different, but related genes (see Cederbaum *et al.*, 2004). ARG I (liver type) is cytosolic and the last step of the urea cycle which regenerates ornithine and releases urea. In contrast, ARG II (kidney type) is present in the mitochondria and is involved in providing ornithine for polyamine and proline synthesis (Cederbaum *et al.*, 2004) and in regulating arginine availability for nitric oxide synthesis (Topal *et al.*, 2006). ARG II is present in enterocytes and thus has the potential to generate ornithine for citrulline synthesis at its site of production. However, ARG II is not expressed until a few days before weaning (Herzfeld and Raper, 1976; Hurwitz and Kretchmer, 1986; De Jonge *et al.*, 1998).

All these studies strongly support the notion that in the adult animal, OAT works towards the disposal of ornithine and that it is unlikely that glutamine and proline are major contributors to citrulline synthesis. In the neonate, however, it is likely that 'de novo' ornithine constitutes a major source of the ornithine utilized for citrulline synthesis.

The uptake of plasma ornithine by the enterocyte, and the subsequent utilization for citrulline synthesis, at first view seems redundant since the two different enzyme systems described above are present in the mitochondria to produce ornithine. The mitochondrial ornithine transporter (ORNT1, SLC25A15) has been found in the small intestine of humans and, after liver, is the tissue with the highest expression (Fiermonte *et al.*, 2003; Nishimura and Naito, 2008). In mice, the expression of ORNT1 shows developmental changes increasing early in life and then falling to adult levels at weaning (Begum *et al.*, 2002). In the liver, there is a preferential utilization of cytosolic ornithine for citrulline synthesis (Cohen *et al.*, 1987) and, because the same transporter is present in the enterocyte, it is likely that a similar preference may take place in the gut.

In vitro studies

In vitro studies have shown that enterocytes were able to synthesize citrulline when glutamine, but not ornithine or proline, was the sole substrate (Wu *et al.*, 1994b). However, the addition of NH_4Cl to ornithine (Wu *et al.*, 1994b) or glutamine to proline (Wu, 1997) was able to trigger an increase in the synthesis of citrulline from ornithine and proline, respectively. Likewise, the addition of glutamine increased the production of citrulline from arginine (Guihot *et al.*, 1997). This interaction between possible precursors for citrulline synthesis highlights the difficulty of providing the appropriate mix of precursors and energy substrates to mimic the *in vivo* situation. This is especially true because glutamine may act not only as a precursor for citrulline synthesis, but it is also an important respiratory fuel for the small intestine (Windmueller and Spaeth, 1974). Additionally, glutamine also provides glutamate, which can be used for the synthesis of N-acetylglutamate, and ammonia for the synthesis of carbamylphosphate. Furthermore, the accumulation of glutamate in the medium during *in vitro* culture of enterocytes (Watford, 1994; Wu *et al.*, 1994b) might tilt the OAT balance towards the synthesis of ornithine instead of its degradation (Smith *et al.*, 1967), and thus not reflect *in vivo* metabolism.

Moreover, the isolation of enterocytes for *in vitro* studies results in the loss of the architecture of the intestinal epithelium. The localization of the different enzymes of the urea cycle has been already discussed and in addition it has been shown in rats that OAT is localized in villi, but not in the crypts (Matsuzawa *et al.*, 1994). This crypt-villus localization of enzyme activity adds an extra layer of complexity to the already complex interorgan trafficking of intermediates for citrulline synthesis.

Ex vivo and *in vivo* studies

The central role of the small intestine in the metabolism of glutamine was firmly established by the studies of Windmueller and Spaeth (1974) using isolated perfused small intestinal preparations. Their work revealed that citrulline was one of the many compounds generated by the metabolism of the gut. Since then, glutamine has been considered the main precursor for citrulline synthesis (Fujita and Yanaga, 2007; Deutz, 2008) and tracer studies utilizing 2-^{15}N (amino) glutamine seem to indicate that 60-80% of citrulline originates from glutamine (Boelens *et al.*, 2005; Boelens *et al.*, 2006; Ligthart-Melis *et al.*, 2007; Ligthart-Melis *et al.*, 2008).

The labeling of glutamine with ^{15}N, however, does not follow the carbon skeleton of glutamine and thus no precursor-product relationship can be established using this tracer (Marini *et al.*, 2010b).

Likewise, the incorporation of ^{14}C from glutamine into citrulline (Windmueller and Spaeth, 1974) is not proof that the carbon skeleton of glutamine is used to synthesize citrulline, since CO_2 generated from the oxidation of glutamine, can be incorporated into the ureido group. In fact, it has been shown that U-^{14}C glutamate is incorporated into arginine in neonatal piglets (Murphy *et al.*, 1996), but [3,4] ^{3}H glutamate is not (Wilkinson *et al.*, 2004). It is worthwhile to mention that Windmueller and Spaeth (1974, 1981) explicitly pointed out that citrulline was an end product of glutamine nitrogen metabolism.

Proline seems to be utilized for the synthesis of citrulline and arginine because proline ameliorates arginine deficiency in neonatal piglets (Brunton *et al.*, 1999). The first pass enteral utilization of proline has been shown to be more efficient, since intragastric tracers result in a higher enrichment of arginine than femoral (Murphy *et al.*, 1996) or portal infused tracers (Bertolo *et al.*, 2003). Furthermore, the gut atrophy that results from parenteral nutrition, in combination with the reduced uptake of plasma proline, could explain the low rate of plasma proline utilization for the synthesis of arginine in piglets (Urschel *et al.*, 2007a).

The production of ornithine from arginine is thought not to occur during the neonatal period due to the reported lack of arginase activity in enterocytes (Wu, 1995). This has been confirmed in neonatal piglets because, despite an extensive first pass enteral utilization of arginine (~40%), no labelled urea was generated when an arginine tracer was infused intragastrically (Bertolo *et al.*, 2003). In the adult, however, a similar enteral utilization of arginine is accompanied by the release of urea and ornithine, and the use of arginine for the synthesis of citrulline (Windmueller and Spaeth, 1976).

In both adults and neonates, arginine can generate ornithine in other tissues besides the gut and then serve as precursor for citrulline (and arginine) synthesis. In fact this arginine-arginine cycle has been reported in neonatal piglets (Urschel *et al.*, 2005), humans (Beaumier *et al.*, 1995) and mice (Marini, 2010b,c) and accounts for up to 35% of the arginine flux when arginine is deficient in the diet (Urschel *et al.*, 2005). The utilization of plasma ornithine for citrulline synthesis accounts for 15-30% of the citrulline flux in adult humans (Castillo *et al.*, 1994; Beaumier *et al.*, 1995) and mice (Marini *et al.*, 2010a,c). Likewise, neonatal piglets have shown substantial utilization of plasma ornithine for citrulline and arginine synthesis, as well as intragastrically infused ornithine (Bertolo *et al.*, 2003; Urschel *et al.*, 2007c). In fact, it has been shown that ornithine tracers are more readily used for citrulline synthesis than proline tracers (Urschel *et al.*, 2007b). Therefore, it seems that there is a preferential utilization of 'preformed' rather than '*de novo*' ornithine for citrulline synthesis. The utilization of arginine for the synthesis of the ornithine used in citrulline synthesis seems to be a futile cycle and it has no apparent purpose. It has been shown that citrulline synthesis can be maintained exclusively utilizing endogenous precursors, since a 4 week deprivation of proline, glutamine/glutamate and arginine did not reduce the rate of citrulline synthesis nor increased amino acid oxidation (Tharakan *et al.*, 2008). Which endogenous precursors for citrulline synthesis are used in these conditions remains to be answered.

Since the discovery of citrulline and its functions as intermediate in the urea cycle and as precursor for arginine synthesis, much knowledge has been accumulated on the different aspects of citrulline metabolism. However, the integration of the processes involved at the whole animal and human level has lagged behind. The absence of the intestinal-renal axis for arginine synthesis in neonates, for instance, is based on the presence of the cytosolic enzymes of the urea cycle in the enterocyte. That the kidney presents the same enzymes, albeit at a reduced activity than in adulthood, and that circulating citrulline concentrations are not different than later in life has been ignored. Also, there is no consensus in the literature regarding the precursors for citrulline synthesis. Part of the apparent conflict resides in the fact that the *in vitro* data cannot be directly extrapolated to the whole organism. Moreover, *in vivo* determinations should be considered carefully, since some tracer studies do not take into account the incorporation of the label after tracer oxidation and thus no precursor-product can be established. More research is needed in order to integrate enzymatic expression and activity of different tissues with the interorgan trafficking of precursor and urea cycle intermediates. Urea cycle disorder patients and mouse transgenic models offer the possibility to study the impact of specific enzyme deletions on the metabolism of citrulline *in vivo*. Furthermore, the advent of tissue specific knockout mouse models further expands our ability to probe the role and contribution of different tissues and cell types in the economy of citrulline and arginine.

Acknowledgments

This work was supported by USDA 2533771314 and NIH KO1 RR24173.

References

Ackermann, D., 1931. Über den biologischen abbau des arginins zu citrulline. Z. Physiol. Chem. 203, 66-69.

Alonso, E., and V. Rubio, 1989. Participation of ornithine aminotransferase in the synthesis and catabolism of ornithine in mice – studies using gabaculine and arginine deprivation. Biochem. J. 259, 131-138.

Beaumier, L., L. Castillo, A.M. Ajami, and V.R. Young, 1995. Urea cycle intermediate kinetics and nitrate excretion at normal and 'Therapeutic' Intakes of arginine in humans. Am. J. Physiol. 269, E884-896.

Begum, L., M.A. Jalil, K. Kobayashi, M. Iijima, M.X. Li, T. Yasuda, M. Horiuchi, A. Del Arco, J. Satrústegui, and T. Saheki, 2002. Expression of three mitochondrial solute carriers, citrin, aralar1 and ornithine transporter, in relation to urea cycle in mice. Biochim. Biophys. Acta 1574, 283-292.

Bertolo, R.F., J.A. Brunton, P.B. Pencharz, and R.O. Ball, 2003. Arginine, ornithine, and proline interconversion is dependent on small intestinal metabolism in neonatal pigs. Am. J. Physiol. 284, E915-922.

Blachier, F., H. Mrabettouil, L. Posho, B. Darcyvrillon, and P.H. Duee, 1993. Intestinal arginine metabolism during development- Evidence for de-novo synthesis of l-arginine in newborn pig enterocytes. Eur. J. Biochem. 216, 109-117.

Boelens, P.G., G.C. Melis, P.A. van Leeuwen, G.A. ten Have, and N.E. Deutz, 2006. Route of administration (enteral or parenteral) affects the contribution of l-glutamine to de novo l-arginine synthesis in mice: A stable-isotope study. Am. J. Physiol. 291, E683-E690.

Boelens, P.G., P.A. van Leeuwen, C.H. Dejong, and N.E. Deutz, 2005. Intestinal renal metabolism of l-citrulline and l-arginine following enteral or parenteral infusion of l-alanyl-l-[2,^{15}N]glutamine or l-[2,^{15}N]glutamine in mice. Am. J. Physiol. 289, G679-685.

Borsook, H., and J.W. Dubnoff, 1941. The conversion of citrulline to arginine in the kidney. J. Biol. Chem. 140, 717-738.

Brosnan, M.E., and J.T. Brosnan, 2004. Renal arginine metabolism. J. Nutr. 134, 2791S-2795S.

Brunton, J.A., R.F. Bertolo, P.B. Pencharz, and R.O. Ball, 1999. Proline ameliorates arginine deficiency during enteral but not parenteral feeding in neonatal piglets. Am. J. Physiol. 277, E223-231.

Castillo, L., T.E. Chapman, M. Sanchez, Y. M. Yu, J. F. Burke, A.M. Ajami, J. Vogt, and V.R. Young, 1993. Plasma arginine and citrulline kinetics in adults given adequate and arginine-free diets. Proc. Natl. Acad. Sci. USA 90, 7749-7753.

Castillo, L., M. Sanchez, T.E. Chapman, A. Ajami, J. F. Burke, and V.R. Young, 1994. The plasma flux and oxidation rate of ornithine adaptively decline with restricted arginine intake. Proc. Natl. Acad. Sci. USA 91, 6393-6397.

Cavicchi, C., S. Malvagia, G. La Marca, S. Gasperini, M. A. Donati, E. Zammarchi, R. Guerrini, A. Morrone, and E. Pasquini, 2009. Hypocitrullinemia in expanded newborn screening by LC MS/MS is not a reliable marker for ornithine transcarbamylase deficiency. J. Pharm. Biomed. Anal. 49, 1292-1295.

Cederbaum, S.D., H. Yu, W.W. Grody, R.M. Kern, P. Yoo, and R.K. Iyer, 2004. Arginases I and II: Do their functions overlap? Mol. Genet. Metab. 81 Suppl 1, S38-44.

Cheung, C.W., N.S. Cohen, and L. Raijman, 1989. Channeling of urea cycle intermediates insitu in permeabilized hepatocytes. J. Biol. Chem. 264, 4038-4044.

Cohen, N.S., C.W. Cheung, and L. Raijman, 1987. Channeling of extramitochondrial ornithine to matrix ornithine transcarbamylase. J. Biol. Chem. 262, 203-208.

Crenn, P, 2008. Citrulline and protein metabolism. Nutr. Clin. Metabol. 22, 75-79.

Crenn, P., C. Coudray-Lucas, F. Thuillier, L. Cynober, and B. Messing, 2000. Postabsorptive plasma citrulline concentration is a marker of absorptive enterocyte mass and intestinal failure in humans. Gastroenterol. 119, 1496-1505.

Curis, E., P. Crenn, and L. Cynober, 2007. Citrulline and the gut. Curr. Opin. Clin. Nutr. Metab. Care 10, 620-626.

Curthoys, N.P., and M. Watford, 1995. Regulation of glutaminase activity and glutamine metabolism. Annu. Rev. Nutr. 15, 133-159.

De Jonge, W.J., M.A. Dingemanse, P.A.J. De Boer, W.H. Lamers, and A.F.M. Moorman, 1998. Arginine-metabolizing enzymes in the developing rat small intestine. Pediatr. Res. 43, 442-451.

Deutz, N.E.P, 2008. The 2007 ESPEN Sir David Cuthbertson lecture: Amino acids between and within organs. The glutamate-glutamine-citrulline-arginine pathway. Clin. Nutr. 27, 321-327.

Dhanakoti, S.N., J.T. Brosnan, G.R. Herzberg, and M.E. Brosnan, 1990. Renal arginine synthesis: Studies *in vitro* and *in vivo*. Am. J. Physiol. 259, E437-442.

Drotman, R.B., and R.A. Freedland, 1972. Citrulline metabolism in the perfused rat liver. Am. J. Physiol. 222, 973-975.

Featherston, W.R., Q.R. Rogers, and R.A. Freedland, 1973. Relative importance of kidney and liver in synthesis of arginine by the rat. Am. J. Physiol. 224, 127-129.

Fiermonte, G., V. Dolce, L. David, F.M. Santorelli, C. Dionisi-Vici, F. Palmieri, and J.E. Walker, 2003. The mitochondrial ornithine transporter: Bacterial expression, reconstitution, functional characterization, and tissue distribution of two human isoforms. J. Biol. Chem. 278, 32778-32783.

Flynn, N.E., D.A. Knabe, B.K. Mallick, and G.Wu, 2000. Postnatal changes of plasma amino acids in suckling pigs. J. Anim. Sci. 78, 2369-2375.

Flynn, N.E., and G.Y. Wu, 1996. An important role for endogenous synthesis of arginine in maintaining arginine homeostasis in neonatal pigs. Am. J. Physiol. 40, R1149-R1155.

Fu, W.J.J., T.E. Haynes, R. Kohli, J.B. Hu, W.J. Shi, T.E. Spencer, R.J. Carroll, C.J. Meininger, and G.Y. Wu, 2005. Dietary l-arginine supplementation reduces fat mass in zucker diabetic fatty rats. J. Nutr. 135, 714-721.

Fujita, T., and K. Yanaga, 2007. Association between glutamine extraction and release of citrulline and glycine by the human small intestine. Life Sci. 80, 1846-1850.

Goutal, I., A. Fairand, and A. Husson, 1999. Expression of the genes of arginine-synthesizing enzymes in the rat kidney during development. Biol. Neonate 76, 253-260.

Guihot, G., F. Blachier, V. Colomb, M.T. Morel, P. Raynal, O. Corriol, C. Ricour, and P.H. Duee, 1997. Effect of an elemental vs. a complex diet on l-citrulline production from l-arginine in rat isolated enterocytes. J. Parenter. Enteral Nutr. 21, 316-323.

Hartman, W.J., and R.L. Prior, 1992. Dietary arginine deficiency alters flux of glutamine and urea cycle intermediates across the portal-drained viscera and liver of rats. J. Nutr. 122, 1472-1482.

Hayashi, T., P.A.R. Juliet, H. Matsui-Hirai, A. Miyazaki, A. Fukatsu, J. Funami, A. Iguchi, and L.J. Ignarro, 2005. L-citrulline and l-arginine supplementation retards the progression of high-cholesterol-diet-induced atherosclerosis in rabbits. Proc. Natl. Acad. Sci. U.S.A. 102, 13681-13686.

Hecker, M., W.C. Sessa, H.J. Harris, E.E. Anggard, and J.R. Vane, 1990. The metabolism of l-arginine and its significance for the biosynthesis of endothelium-derived relaxing factor: Cultured endothelial cells recycle l-citrulline to l-arginine. Proc. Natl. Acad. Sci. USA 87, 8612-8616.

Herzfeld, A., V.A. Mezl, and W.E. Knox, 1977. Enzymes metabolizing delta1-pyrroline-5-carboxylate in rat tissues. Biochem. J. 166, 95-103.

Herzfeld, A., and S.M. Raper, 1976. Enzymes of ornithine metabolism in adult and developing rat intestine. Biochim. Biophys. Acta 428, 600-610.

Hoogenraad, N., N. Totino, H. Elmer, C. Wraight, P. Alewood, and R.B. Johns, 1985. Inhibition of intestinal citrulline synthesis causes severe growth-retardation in rats. Am. J. Physiol. 249, G792-G798.

Hu, C.A., W.W. Lin, C. Obie, and D. Valle, 1999. Molecular enzymology of mammalian delta(1)-pyrroline-5-carboxylate synthase – alternative splice donor utilization generates isoforms with different sensitivity to ornithine inhibition. J. Biol. Chem. 274, 6754-6762.

Hurwitz, R., and N. Kretchmer, 1986. Development of arginine-synthesizing enzymes in mouse intestine. Am. J. Physiol. 251, G103-G110.

Ignarro, L.J., G.M. Buga, K.S. Wood, R.E. Byrns, and G. Chaudhuri, 1987. Endothelium-derived relaxing factor produced and released from artery and vein is nitric oxide. Proc. Natl. Acad. Sci. USA 84, 9265-9269.

Jones, M.E., A.D. Anderson, C. Anderson, and S. Hodes, 1961. Citrulline synthesis in rat tissues. Arch. Biochem. Biophys. 95, 499-507.

Kawabata, Y., N. Katunuma, and Y. Sanada, 1980. Characteristics of proline oxidase in rat tissues. J. Biochem. 88, 281-283.

Kawamoto, S., T. Sonoda, A. Ohtake, and M. Tatibana, 1985. Stimulatory effect of arginine on acetylglutamate synthesis in isolated-mitochondria of mouse and rat-liver. Biochem. J. 232, 329-334.

King, D.E., A.G. Mainous Iii, and M.E. Geesey, 2008. Variation in l-arginine intake follow demographics and lifestyle factors that may impact cardiovascular disease risk. Nutr. Res. 28, 21-24.

Kohler, E.S., S. Sankaranarayanan, C.J. Van Ginneken, P. Van Dijk, J.L.M. Vermeulen, J.M. Ruijter, W.H. Lamers, and E. Bruder, 2008. The human neonatal small intestine has the potential for arginine synthesis; developmental changes in the expression of arginine-synthesizing and -catabolizing enzymes. BMC Dev. Biol. 8.

Kohli, R., C.J. Meininger, T.E. Haynes, W.Yan, J.T. Self, and G.Y. Wu, 2004. Dietary l-arginine supplementation enhances endothelial nitric oxide synthesis in streptozotocin-induced diabetic rats. J. Nutr. 134, 600-608.

Krebs, H.A., and K. Henseleit, 1932. Untersuchungen uber die harnstoffbildung im tierkorper. Z. Physiol. Chem, 33-66.

Ligthart-Melis, G.C., M.C.G. van de Poll, P.G. Boelens, C.H.C. Dejong, N.E.P. Deutz, and P.A.M. van Leeuwen, 2008. Glutamine is an important precursor for de novo synthesis of arginine in humans. Am. J. Clin. Nutr. 87, 1282-1289.

Ligthart-Melis, G.C., M.C.G. van de Poll, C.H.C. Dejong, P.G. Boelens, N.E.P. Deutz, and P.A.M. van Leeuwen, 2007. The route of administration (enteral or parenteral) affects the conversion of isotopically labelled l-[2-N-15] glutamine into citrulline and arginine in humans. J. Parenter. Enteral Nutr. 31, 343-348.

Luiking, Y.C., M. Poeze, G. Ramsay, and N.E.P. Deutz, 2005. The role of arginine in infection and sepsis. J. Parenter. Enteral Nutr. 29, S70-S74.

Malo, C., I.A. Qureshi, and J. Letarte, 1986. Postnatal maturation of enterocytes in sparse-fur mutant mice. Am. J. Physiol. 250, G177-G184.

Marini, J.C., I.C. Didelija, L. Castillo, and B. Lee, 2010a. Arginine utilization for citrulline synthesis in arginase II knockout mice. Faseb J. 24, 740.714.

Marini, J.C., I.C. Didelija, L. Castillo, and B. Lee, 2010b. Glutamine: Precursor or nitrogen donor for citrulline synthesis? Am. J. Physiol., In press.

Marini, J.C., I.C. Didelija, L. Castillo, and B. Lee, 2010c. Precursors for the synthesis of citrulline in mice fed arginine free diets. Faseb J. 24, 726-740.

Marini, J.C., A. Erez, L. Castillo, and B. Lee, 2007. Interaction between murine spf-ash mutation and genetic background yields different metabolic phenotypes. Am. J. Physiol. 293, E1764-E1771.

Markova, M., C. Peneff, M.J.E. Hewlins, T. Schirmer, and R. A. John. 2005. Determinants of substrate specificity in omega-aminotransferases. J. Biol. Chem. 280, 36409-36416.

Matsuzawa, T., T. Kobayashi, K. Tashiro, and M. Kasahara, 1994. Changes in ornithine metabolic enzymes induced by dietary-protein in small-intestine and liver-intestine-liver relationship in ornithine supply to liver. J. Biochem. 116, 721-727.

McGivan, J.D., N.M. Bradford, and A.D. Beavis, 1977. Factors influencing activity of ornithine aminotransferase in isolated rat-liver mitochondria. Biochem. J. 162, 147-156.

Meijer, A.J., W.H. Lamers, and R. Chamuleau, 1990. Nitrogen-metabolism and ornithine cycle function. Physiol. Rev. 70, 701-748.

Moinard, C., I. Nicolis, N. Neveux, S. Darquy, S. Benazeth, and L. Cynober, 2008. Dose-ranging effects of citrulline administration on plasma amino acids and hormonal patterns in healthy subjects: The citrudose pharmacokinetic study. Brit. J. Nutr. 99, 855-862.

Morgan, J.F., H.J. Morton, and A.E. Pasieka, 1958. The arginine requirement of tissue cultures. I. Interrelationships between arginine and related compounds. J. Biol. Chem. 233, 664-667.

Morris, S.M, 2007. Arginine metabolism: Boundaries of our knowledge. J. Nutr. 137, 1602S-1609S.

Morris, S.M., Jr, 2006. Arginine: Beyond protein. Am J Clin Nutr 83, 508S-512S.

Morris, S.M., C.L. Moncman, J.S. Holub, and Y. Hod, 1989. Nutritional and hormonal regulation of mrna abundance for arginine biosynthetic enzymes in kidney. Arch. Biochem. Biophys273, 230-237.

Morris, S.M., C.L. Moncman, K.D. Rand, G.J. Dizikes, S.D. Cederbaum, and W.E. O'Brien, 1987. Regulation of messenger-RNA levels for 5 urea cycle enzymes in rat-liver by diet, cyclic-amp, and glucocorticoids. Arch. Biochem. Biophys. 256, 343-353.

Murphy, J.M., S.J. Murch, and R.O. Ball, 1996. Proline is synthesized from glutamate during intragastric infusion but not during intravenous infusion in neonatal piglets. J. Nutr. 126, 878-886.

Nishimura, M., and S. Naito, 2008. Tissue-specific mRNA expression profiles of human solute carrier transporter superfamilies. Drug Metab. Pharmacokinet. 23, 22-44.

Ogawa, T., M. Kimoto, and K. Sasaoka, 1989. Purification and properties of a new enzyme, Ng,Ng-dimethylarginine dimethylaminohydrolase, from rat-kidney. J. Biol. Chem. 264, 10205-10209.

Osowska, S., T. Duchemann, S. Walrand, A. Paillard, Y. Boirie, L. Cynober, and C. Moinard, 2006. Citrulline modulates muscle protein metabolism in old malnourished rats. Am. J. Physiol. 291, E582-E586.

Osowska, S., C. Moinard, N. Neveux, C. Loi, and L. Cynober, 2004. Citrulline increases arginine pools and restores nitrogen balance after massive intestinal resection. Gut 53, 1781-1786.

Prior, R.L., and K.L. Gross, 1995. Dietary arginine deficiency and gut ammonium infusion alter flux of urea cycle intermediates across the portal-drained viscera of pigs. J. Nutr. 125, 251-263.

Raijman, L, 1974. Citrulline synthesis in rat tissues and liver content of carbamoyl phosphate and ornithine. Biochem. J. 138, 225-232.

Ratner, S., and A. Pappas, 1949. Biosynthesis of urea; enzymatic mechanism of arginine synthesis from citrulline. J. Biol. Chem. 179, 1183-1198.

Reddi, P.K., W.E. Knox, and A. Herzfeld, 1975. Types of arginase in rat tissues. Enzyme 20, 305-314.

Reichard, H, 1960. Ornithine carbamyl transferase activity in human tissue homogenate. J. Lab. Clin. Med. 56, 218-221.

Rhoads, J.M., E. Plunkett, J. Galanko, S. Lichtman, L. Taylor, A. Maynor, T. Weiner, K. Freeman, J.L. Guarisco, and G.Y. Wu, 2005. Serum citrulline levels correlate with enteral tolerance and bowel length in infants with short bowel syndrome. J. Pediatr. 146, 542-547.

Riby, J.E., R.E. Hurwitz, and N. Kretchmer, 1990. Development of ornithine metabolism in the mouse intestine. Pediatr. Res. 28, 261-265.

Ryall, J.C., M.A. Quantz, and G.C. Shore, 1986. Rat-liver and intestinal-mucosa differ in the developmental pattern and hormonal-regulation of carbamoyl-phosphate synthetase-i and ornithine carbamoyl transferase gene-expression. Eur. J. Biochem. 156, 453-458.

Samuels, S.E., K.S. Acton, and R.O. Ball, 1989. Pyrroline-5-carboxylate reductase and proline oxidase activity in the neonatal pig. J. Nutr. 119, 1999-2004.

Seiler, N., C. Grauffel, G. Dauneanglard, S. Sarhan, and B. Knodgen, 1994. Decreased hyperammonemia and orotic aciduria due to inactivation of ornithine aminotransferase in mice with a hereditary abnormal ornithine carbamoyltransferase. J. Inherit. Metab. Dis. 17, 691-703.

Smith, A.D., M. Benziman, and H.J. Strecker, 1967. The formation of ornithine from proline in animal tissues. Biochem. J. 104, 557-563.

Sun, N.C., C.R.Y. Sun, R.W. Tennant, and A.W. Hsie, 1979. Selective growth of some rodent epithelial-cells in a medium containing citrulline. Proc. Natl. Acad. Sci. USA 76, 1819-1823.

Takiguchi, M., T. Matsubasa, Y. Amaya, and M. Mori, 1989. Evolutionary aspects of urea cycle enzyme genes. Bioessays 10, 163-166.

Tedesco, T.A., S.A. Benford, R.C. Foster, and L.A. Barness, 1984. Free amino-acids in citrullus-vulgaris (watermelon). Pediatrics 73, 879-879.

Tharakan, J.F., Y.M. Yu, D. Zurakowski, R.M. Roth, V.R. Young, and L. Castillo, 2008. Adaptation to a long term (4 weeks) arginine- and precursor (glutamate, proline and aspartate)-free diet. Clin. Nutr. 27, 513-522.

Topal, J.L.G., A. Brunet, L. Walch, J.L. Boucher, and M. David-Dufilho, 2006. Mitochondrial arginase ii modulates nitric-oxide synthesis through nonfreely exchangeable l-arginine pools in human endothelial cells. J. Pharmacol. Exp. Ther. 318, 1368-1374.

Tytell, A.A., and R.E. Neuman, 1960. Growth response of stable and primary cell cultures to l-ornithine, l-citrulline, and l-arginine. Exp. Cell Res. 20, 84-91.

Urschel, K.L., A.R. Evans, C.W. Wilkinson, P.B. Pencharz, and R.O. Ball, 2007a. Parenterally fed neonatal piglets have a low rate of endogenous arginine synthesis from circulating proline. J. Nutr. 137, 601-606.

Urschel, K. L., M. Rafii, P. B. Pencharz, and R. O. Ball, 2007b. A multitracer stable isotope quantification of the effects of arginine intake on whole body arginine metabolism in neonatal piglets. Am. J. Physiol. 293, E811-E818.

Urschel, K.L., A.K. Shoveller, P.B. Pencharz, and R.O. Ball, 2005. Arginine synthesis does not occur during first-pass hepatic metabolism in the neonatal piglet. Am. J. Physiol. 288, E1244-E1251.

Urschel, K.L., C.W. Wilkinson, P.B. Pencharz, and R.O. Ball, 2007c. Coadministration of ornithine and alpha-ketoglutarate is no more effective than ornithine alone as an arginine precursor in piglets enterally fed an arginine-deficient diet. J. Nutr. 137, 55-62.

van de Poll, M.C.G., M.P.C. Siroen, P.A.M. van Leeuwen, P.B. Soeters, G.C. Melis, P.G. Boelens, N.E.P. Deutz, and C.H.C. Dejong, 2007. Interorgan amino acid exchange in humans: Consequences for arginine and citrulline metabolism. Am. J. Clin. Nutr. 85, 167-172.

Ventura, G., J.P. De Bandt, F. Segaud, C. Perret, D. Robic, O. Levillain, S. Le Plenier, C. Godard, L. Cynober, and C. Moinard, 2009. Overexpression of ornithine aminotransferase: Consequences on amino acid homeostasis. Brit. J. Nutr. 101, 843-851.

Wada, M, 1930. Über citrulline, eine neue aminosäure im preßsaft de wassermelone, citrullus vulgaris schrad. Biochem. Z. 224, 420-429.

Walrand, S., C. Guillet, J. Salles, N. Tardif, A.C. Maurin, P. Fafournoux, N. Cano, and Y. Boirie, 2008. Acides amines et signalisation cellulaire (Amino acids as regulators of cell signalling). Nutr. Clin. Metabol. 22, 161-167.

Wang, T., A.M. Lawler, G. Steel, I. Sipila, A.H. Milam, and D. Valle, 1995. Mice lacking ornithine-delta-amino-transferase have paradoxical neonatal hypoornithinaemia and retinal degeneration. Nat. Genet. 11, 185-190.

Watford, M, 1994. Glutamine-metabolism in rat small-intestine – synthesis of 3-carbon products in isolated enterocytes. Biochim. Biophys. Acta 1200, 73-78.

Wilkinson, D.L., R.F.P. Bertolo, J.A. Brunton, A.K. Shoveller, P.B. Pencharz, and R.O. Ball, 2004. Arginine synthesis is regulated by dietary arginine intake in the enterally fed neonatal piglet. Am. J. Physiol. 287, E454-E462.

Windmueller, H.G., and A.E. Spaeth, 1974. Uptake and metabolism of plasma glutamine by the small intestine. J. Biol. Chem. 249, 5070-5079.

Windmueller, H.G., and A.E. Spaeth, 1976. Metabolism of absorbed aspartate, asparagine, and arginine by rat small intestine *in vivo*. Arch. Biochem. Biophys. 175, 670-676.

Windmueller, H.G., and A.E. Spaeth, 1981. Source and fate of circulating citrulline. Am. J. Physiol. 241, E473-480.

Wraight, C., and N. Hoogenraad, 1988. Dietary-regulation of ornithine transcarbamylase messenger-Rna in liver and small-intestine. Aust. J. Biol. Sci. 41, 435-440.

Wraight, C., K. Lingelbach, and N. Hoogenraad, 1985. Comparison of ornithine transcarbamylase from rat-liver and intestine – evidence for differential regulation of enzyme levels. Eur. J. Biochem. 153, 239-242.

Wu, G., F.W. Bazer, T.A. Davis, S.W. Kim, P. Li, J. Marc Rhoads, M. Carey Satterfield, S.B. Smith, T.E. Spencer, and Y. Yin, 2009. Arginine metabolism and nutrition in growth, health and disease. Amino Acids 37, 153-168.

Wu, G., and D.A. Knabe, 1995. Arginine synthesis in enterocytes of neonatal pigs. Am. J. Physiol. 269; 621-629.

Wu, G.Y, 1995. Urea synthesis in enterocytes of developing pigs. Biochem. J. 312, 717-723.

Wu, G.Y, 1997. Synthesis of citrulline and arginine from proline in enterocytes of postnatal pigs. Am. J. Physiol.-Gastrointestinal and Liver Physiology 35, G1382-G1390.

Wu, G.Y., A.G. Borbolla, and D.A. Knabe, 1994a. The uptake of glutamine and release of arginine, citrulline and proline by the small-intestine of developing pigs. J. Nutr. 124, 2437-2444.

Wu, G.Y. and J.T. Brosnan, 1992. Macrophages can convert citrulline into arginine. Biochem. J. 281, 45-48.

Wu, G.Y., D.A. Knabe, and N.E. Flynn, 1994b. Synthesis of citrulline from glutamine in pig enterocytes. Biochem. J. 299, 115-121.

Yamada, E. and Y. Wakabayashi, 1991. Development of pyrroline-5-carboxylate synthase and N-acetylglutamate synthase and their changes in lactation and aging. Arch. Biochem. Biophys. 291, 15-23.

Intermediary metabolism and neogenesis of nutrients in farm animals

B.J. Bequette[1], *S.W. El-Kadi*[2] *and N.E. Sunny*[3]
[1]*Department of Animal and Avian Sciences, University of Maryland, College Park, MD 20742, USA*
[2]*USDA/ARS, Children's Nutrition Research Center, and Endocrinology and Metabolism Section, Department of Pediatrics, Baylor College of Medicine, Houston, TX 77030, USA*
[3]*Advanced Imaging Research Center, University of Texas Southwestern Medical Center, Dallas, TX 75390, USA*

Abstract

The commercial availability of stable isotope (^{13}C, ^{15}N, ^{2}H)-labelled compounds and highly accurate mass spectrometers has made it possible to probe the details of metabolic pathways involved in macronutrient catabolism and neogenesis. This paper highlights aspects of animal nutrition and metabolism in which uniformly ^{13}C-labelled [U-^{13}C] substrates and ^{13}C-mass isotopomer distribution approaches have been applied to investigations of amino acid and carbohydrate synthesis and catabolism. We will focus on the application of [U-^{13}C] substrates as tracers in chickens, fish, sheep, and cells in culture to quantify rates of macronutrient synthesis, identification of the sources of dietary nutrients that serve as substrates for neogenesis of macronutrients, and investigations of the interconnectivity of the pathways of macronutrient metabolism with those of the Krebs cycle to preserve metabolic flexibility via anaplerotic and cataplerotic sequences.

Introduction

There are several areas of animal nutrition in which our understanding of the routes of macronutrient metabolism and the general contributions these make to neogenesis of nutrients and energy generation is vague at best. In recent years, highly accurate mass spectrometers and the commercial availability of custom stable isotope (^{13}C, ^{15}N, ^{2}H) molecules, in particular multiple-labelled species (e.g. [U-$^{13}C_n$], where n is the number of carbon atoms in the molecule), has made it possible to probe the details and interconnectivity of metabolic pathways involved in macronutrient utilization and synthesis. For years, the biological sciences have exploited the use of uniformly ^{13}C-labelled substrates (e.g. glucose, amino acids, fatty acids, nucleic acids) and mass spectrometric (MS) and nuclear magnetic resonance (NMR) platforms to investigate pathway flux control (Berthold *et al.*, 1994; De Rosiers *et al.*, 1995; Hellerstein, 2003), gluconeogenesis (Lee *et al.*, 1991), ketogenesis (Sunny *et al.*, 2010), lipogenesis (Hellerstein and Neese, 1992), and to assess the dietary essentiality of amino acids (AA) (Berthold *et al.*, 1991) and nucleic acids (Berthold *et al.*, 1995), to highlight a few. While stable isotopes have been employed in studies of farm animal metabolism for decades, most of what is known about intermediary metabolism, and neogenesis and essentiality of dietary macronutrients in farm animals, derives from relatively few studies, largely those from the late Peter Reeds and his colleagues in their work with piglets (e.g. Reeds *et al.*, 1996, 1997; Stoll *et al.*, 1998, 1999; Wykes *et al.*, 1998) and a laying hen (Berthold *et al.*, 1991, 1995).

This paper presents a brief overview of the use of stable isotopes and mass spectrometry, and provides several examples of how these techniques can be applied in farm animals to quantify neogenesis and utilization of macronutrients.

Overview of stable isotopes and mass spectrometry analyses

Gas or liquid chromatography MS and NMR (^{2}H/^{13}C) are the most common experimental approaches employed in investigations of pathway flux analysis. These approaches, when coupled with the application of stable isotope tracers in whole animals, isolated perfused organ systems and cell culture, have allowed researchers to dissect the dynamic processes underlying dietary organic nutrient metabolism, integration of associated metabolic networks and neogenesis of organic nutrients. Here, we will discuss applications of MS analytics to organic nutrient metabolism in farm animals, which is less expensive, requires less time, and offers excellent sensitivity compared to NMR techniques. For applications of NMR techniques to intermediary metabolism the reader is directed to the following papers (Jones *et al.*, 2001; Jin *et al.*, 2004).

A fundamental requirement in the application of stable isotopes and MS techniques is the knowledge of the measurement and interpretation of stable isotope abundances in molecules. For further details, the reader is directed to several excellent reviews on MS techniques (Hellerstein and Neese, 1999), mathematical formalisms for resolving mass spectral data (Lee *et al.*, 1991; Fernandez *et al.*, 1996), and application of stable isotopes and mass isotopomer distribution (**MID**) to metabolic and nutrition research (Hellerstein and Neese, 1992; Berthold *et al.*, 1994; Brunengraber *et al.*, 1997; Reeds *et al.*, 1997; Young and Ajami, 1999; Hellerstein, 2003).

The use of stable isotopes can provide rich metabolic information that differs from that of radioisotopes. For example, quantification of radioisotopes is in terms of energy emitted from the loss of an electron from the nucleus (i.e. isotopic decay), where the energy emitted is directly proportional to the number of radioactive atoms (^{14}C, ^{3}H, ^{35}S) present per sample or molecule weight (i.e. specific activity). Often, specific activity of a whole molecule (e.g. glucose, fatty acids) is determined, and this has three particular disadvantages. First, the concentration of the molecule must be determined and the compound must be isolated for radioactive counting. Both analytical techniques can introduce error in measurement. Second, with ^{14}C or ^{3}H tracers, total molecule specific activity is usually measured. For uniformly ^{14}C-labelled substrates such as glucose and dispensible AA, the disadvantage is that it is not possible to account for ^{14}C recycling back to the parent molecule. Chemical or enzymatic cleavage approaches can be used to yield specific molecular portions of a molecule (e.g. glutamate, Lee *et al.*, 1996; glucose, Xu *et al.*, 2004); however, these methods are labour intensive and for some molecules (e.g. [U-^{14}C] glycerol) it may not be possible. Consequently, use of certain radio labelled compounds may limit information on the flux and activity of metabolic pathways, and the true synthetic rate of the nutrient. Lastly, application of radioisotopes in farm animals presents issues of safety when handling radioactive compounds and in the costs of carcass and sample disposal. By contrast, the stable isotopes (^{13}C, ^{2}H, ^{15}N, ^{18}O) commonly used in organic nutrient metabolism studies are heavier by one or two atomic mass units, and this allows differentiation of molecules (and fragment ions) containing one or more labelled atoms as determined by mass differences on an MS.

Once a sample is introduced into the MS system, ionized fragment ions derived from the analyte are selectively sorted based on their mass-to-charge ratio and recorded. The crude ion abundances are then normalized and corrected for the measured natural abundances (^{13}C, 1.11%; ^{2}H, 0.015%; ^{15}N, 0.365%; ^{18}O, 0.204%; ^{34}S, 4.22%) of stable isotopes present in the original molecule and that contributed by the derivative (if present in the fragment). In most studies of intermediary metabolism and neogenesis of organic nutrients, the administered (intragastric, intravenous, orally or cell media *in vitro*) tracer contains multiple-labelled atoms (e.g. [$^{13}C_6$] glucose, [$^{13}C_5$] glutamate, etc.) which, upon metabolism, leads to the incorporation (synthesis) of ^{13}C-labelled carbon chains into metabolites bearing one or more labelled atoms. Where the metabolite is a mixture of unlabelled and labelled molecules, for example with plasma glucose (Figure 1) where 1, 2, 3 and 6 of the ^{12}C (unlabelled) atoms in glucose molecules have been replaced (enriched) with ^{13}C, mathematical approaches are required to correct (deconvolute) the measured elemental distribution of the stable isotope (Lee *et al.*, 1991; Fernandez *et al.*, 1996). The ^{13}C enrichment data are presented as moles of the isotopomer (M+n) per 100 moles of tracee (M+0). For glucose, M+0 represents the unlabelled glucose and M+n represents the ^{13}C-containing isotopomer species of glucose (n = 1, 2, 3 and 6). Note that the mole ratio representation is analogous to the specific activity formalism used for radioisotopes.

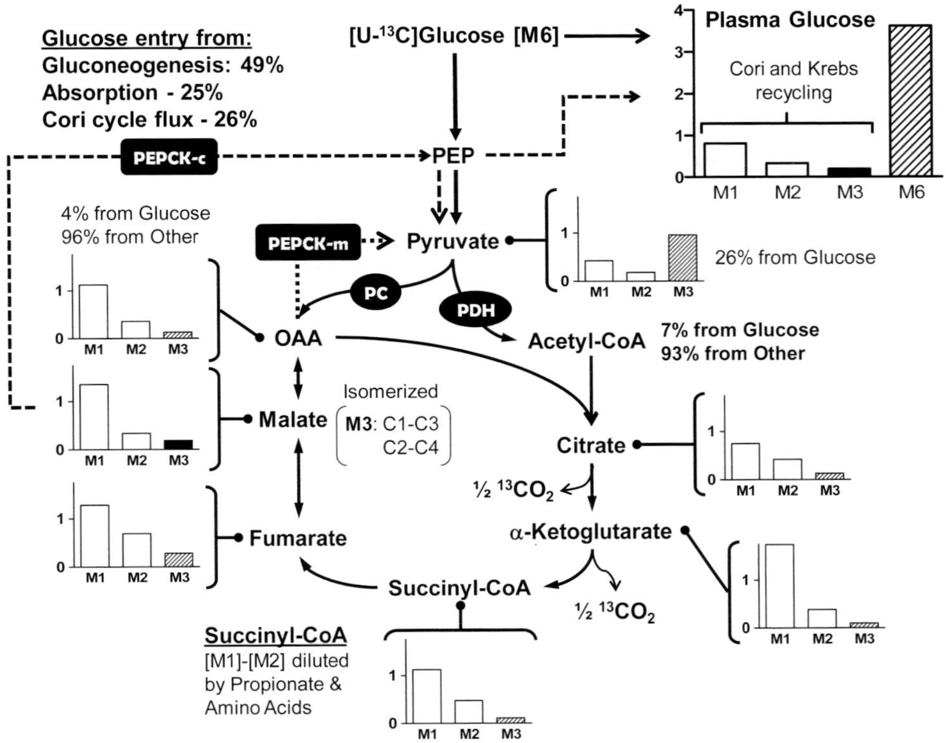

Figure 1. ^{13}C-Mass isotopomer distribution in plasma glucose and of hepatic Krebs cycle intermediates in a sheep given a constant infusion (32 h) of [$^{13}C_6$] glucose into the duodenum (Bequette and El-Kadi, unpubl.). PC, pyruvate carboxylase; PDH, pyruvate dehydrogenase; PEPCK, phosphoenolpyruvate carboxykinase (c, cytosolic; m, mitochondrial).

Apart from the widely used gas chromatography (**GC**)-MS platform, liquid chromatography based MS and MS/MS platforms are now being increasingly used for detecting tracer enrichments (Sunny *et al.*, 2010) and also for mass profiling of metabolites (metabolomics; Millington *et al.*, 1990; Bain *et al.*, 2009) including acylcarnitines, acylCoA and AA. Metabolic flux analysis in combination with metabolomics also allows one to profile alterations in central pathways of glucose, AA and lipid metabolism, which intersect at regular intervals and which often share molecular mediators. The following sections provide examples where the isotopomer patterns resulting from stable isotope based experiments provided metabolic information about organic nutrient utilization and neogenesis of various nutrients.

Gluconeogenesis and glucose carbon utilization

In monogastric and ruminant animals, the dietary requirement for carbohydrates, primarily glucose, is solely based on the metabolism of carbohydrates to generate energy. Thus, there is not a specified dietary requirement for carbohydrates per se, other than as a source of energy. Despite this, there is an obvious metabolic requirement for glucose which necessitates either the inclusion of glucose in the diet or its synthesis *de novo*. To determine the dietary needs for glucose requires answers to several related questions: (1) what is the contribution of dietary glucose to whole body glucose needs; (2) what are the dietary precursors for gluconeogenesis, and (3) other than oxidation (energy), what are the fates of dietary glucose carbon?

Pathways of glucose metabolism have been extensively investigated in a variety of species utilizing ^{13}C and ^{2}H labelled tracers by various MS (Sunny *et al.*, 2010) and NMR (Jones *et al.*, 2001; Jin *et al.*, 2004) isotopomer analytical techniques. The atoms from tracer glucose traverse through multiple pathways before returning as newly synthesized glucose, resulting in a unique metabolic signature (isotopomer pattern) that contains rich information about glucose output, new glucose synthesis and metabolism of glucose carbon through pathways of energy metabolism. The plasma glucose ^{13}C-mass isotopomer data in Table 1 highlights several advantages of using stable isotopes as metabolic probes, and some caveats when interpreting ^{13}C-mass isotopomer data. The data are from a continuous duodenal infusion (32 h) of [$^{13}C_6$] glucose in a sheep, and measurement of plasma glucose ^{13}C-MID by GC-MS. Under the MS conditions that we measured glucose ^{13}C-MID, the MS measures differences in molecular mass only and so it is not possible in this case to determine the position of the ^{13}C label in the molecule. Under different derivatization and MS conditions it is possible to determine the ^{13}C enrichment of the triose molecules that contribute to glucose synthesis (Neese *et al.*, 1995). In the example, there are 3 possible positional ^{13}C-isotopomers for the [M+1] and [M+2] glucose species, whereas for the [M+0], [M+3] and [M+6] glucose species only one isotopomer exists where all carbons contain either ^{12}C or ^{13}C. However, the appearance of plasma [M+6] glucose can only derive from the original [$^{13}C_6$] glucose tracer and the [M+1] to [M+3] glucose species only arise from the recycling of glucose molecules, i.e. Cori and Krebs cycle (Figure 1). During the recycling process, there is loss and exchange of ^{13}C with ^{12}C and dilution of glucose molecules by unlabelled carbon sources (e.g. glycerol, AA) corresponding to 'new' glucose synthesis (i.e. gluconeogenesis), all of which has been estimated on the basis of several

Table 1. Molar tracer:tracee ratios and ^{13}C-enrichments of intestinal mucosa intracellular AA and plasma glucose obtained from 5-day old chicks after 4 days of ad libitum *consumption of a typical starter diet containing [U-^{13}C] algal protein (Sunny and Bequette, unpublished).*[1,2]

	Mass isotopomer tracer:tracee ratio [moles isotopomer per 100 moles tracee]							Source of AA [%]		
	M+1	M+2	M+3	M+4	M+5	M+6	Diet M+n[3]	From diet[4]	Synthesis from AA[5]	Synthesis from other[6]
Leucine	0.00	0.00	0.00	0.00	0.05	4.30	5.70	75.0	-	-
Isoleucine	0.00	0.00	0.00	0.00	0.01	5.29	6.77	78.0	-	-
Glutamine	1.84	1.09	0.41	0.01	0.70	-	3.91	23.9	35.7	46.4
Glutamate	1.47	0.89	0.53	0.00	0.52	-	3.46	20.1	37.2	42.7
Aspartate	0.53	0.49	0.15	1.79	-	-	10.25	23.2	6.3	70.5
Serine	1.50	1.40	1.42	-	-	-	6.83	27.8	27.9	44.3
Plasma glucose	1.96	0.54	0.16	0.00	0.00	0.00	0.00	-	2.6	-

[1] Values are the means of chicks hatched from small (55 g eggs, n=6) and large (70 g eggs, n=6) eggs. Differences were not significant (P>0.05) between small and large eggs. On day 5 post-hatch, small egg chicks weighed 55.7 g (±5.5) and large egg chicks weighed 71.4 g (±3.7).

[2] The starter diet contained (as-fed): 78% ground corn, 13% casein, 2.1% [U-^{13}C] algal protein, 3.5% soy oil, 0.25% L-methionine, 0.1% L-lysine, 0.07% choline (60%) and a complete mineral and vitamin premix.

[3] Isotope enrichment of the AA in the mixed diet, where n is the number of carbon atoms in the AA.

[4] Values for Leucine and Isoleucine were calculated as: (100 × mucosal [M+n])/Diet [M+n]. The dilution factor for Leucine was used to correct other AA for dilution as: (1/0.75)(100 × mucosal [M+n])/Diet [M+n].

[5] AA Synthesis from AA calculated as: % from Diet × (([M+1] + [M+2] × 2 + … + [M+n-1] × (n-1)))/[M+n] × n. Glucose synthesis from AA calculated as: ([M+1] + [M+2] + [M+3])/([M+0] + [M+1] + [M+2] + [M+3]), where [M+0] (unlabelled glucose) is 100.

[6] Synthesis from other calculated as: (100 - % from Diet) minus % from AA.

theoretical models (e.g. Lee *et al.*, 1991; Landau *et al.*, 1998; Katz and Tayek, 1999; Haymond and Sunehag, 2000). Thus, it is possible to calculate Cori cycle flux (glucose carbon recycling) and gluconeogenesis from non-glucose sources separately, with the remainder derived from glucose absorption and glycogen breakdown.

In the study in sheep, we adopted the model of Haymond and Sunehag (2000) to quantify glucose fluxes. Using this model, Cori cycling accounted for 26% (25 g/d), gluconeogenesis 49% (47 g/d) and glucose from absorption and glycogen breakdown (if any) accounted for 25% (24 g/d) of plasma glucose entry. Another advantage of using [$^{13}C_6$] glucose is that it can be used as a metabolic probe to acquire simultaneous information on individual carbon fluxes through glycolysis and the Krebs cycle (e.g. Pascual *et al.*, 1998; Wykes *et al.*, 1998) when combined with ^{13}C-MID analysis of downstream metabolites. The ^{13}C-MID in Krebs cycle intermediates in the sheep liver (Figure 1) yielded ^{13}C-MID patterns that were predictive of known biochemistry, and which confirmed important features of ruminant liver metabolism. For example, glucose metabolism by the ruminant liver made only minor contributions to the carbon fluxes through pyruvate carboxylase (4%) and pyruvate dehydrogenase (7%), as would be expected (Armentano, 1992). Furthermore, it was notable that there was considerable dilution of ^{13}C between α-ketoglutarate and succinyl-CoA, representing the contribution of unlabelled propionate and AA to the flux through succinyl-CoA, and thence their contribution to gluconeogenesis.

We also applied these general principles to determine gluconeogenesis, glucose carbon recycling and the proportion of AA derived from glucose in developing chicken embryos (Sunny *et al.*, 2004; 2010). At the time of lay, the egg contains very little preformed glucose (~250 mg), so it was not surprising that we observed an abrupt, sustained increase in gluconeogenesis as embryonic development advanced (embryonic day 12, 0.12 g/d; day 14, 0.58 g/d; day 16, 0.47 g/d; day 18, 0.43 g/d). This higher glucose synthesis during development was accompanied by a higher flux through the Cori cycle, which conserves 3-carbon precursors for gluconeogenesis and limits total oxygen demands. It was notable that Cori cycling and gluconeogenesis were higher in embryos from small (51 g) compared to typical size (65 g) eggs from day 12 to 16 of embryonic development, and when taken together with the observation that blood concentrations of gluconeogenic AA (threonine, glutamine, arginine, proline, isoleucine, and valine) were lower (25 to 48%) in the small egg embryos, suggests that the small embryos diverted AA away from protein synthesis and towards gluconeogenesis. Perhaps as a consequence, the reduced embryonic growth rate of the embryos from small eggs may have been a consequence of partitioning greater supplies of AA toward gluconeogenesis.

Dispensable amino acid synthesis

In farm and commercial aquatic animals, the indispensable and dispensable AA have been clearly identified. And, for the indispensable AA, their requirements have been determined empirically from dose response and supplementation studies. However, the dietary pattern of indispensable AA for various physiological stages (e.g. growth, lactation and egg laying) is similar, but not identical, to the AA composition of the proteins (e.g. muscle, eggs and milk) being net synthesised (Fuller *et al.*, 1989; Murphy, 1994; Kim *et al.*, 2001). This disparity largely relates to the AA requirement to maintain critical functions and the need to maintain body protein equilibrium, both of which are not well defined. By contrast, although it is known that the dispensable AA are needed in diets for normal growth (Rose *et al.*, 1948, Stucki and Harper, 1961), the dietary pattern of specific dispensable AA for growth and other physiological processes is not known. In the absence of this information, diets are not formulated with targets for dispensable AA, but rather that sufficient dietary (metabolic) nitrogen is available to support their synthesis *de novo*.

In discussing the nutritional adequacy of dispensable AA and the capacity for their synthesis *de novo*, it important to consider that nearly 100% of the dietary supplies of aspartate, glutamate, and glutamine are metabolized by the gastrointestinal tract (Windmueller and Spaeth, 1975, 1980; Stoll *et al.*, 1998, 1999). In consequence, the post-absorptive supplies of these AA for net protein deposition and other functions must derive from synthesis *de novo*. For aspartate, its synthesis occurs solely from the Krebs cycle intermediate oxaloacetate (Figure 1). Here, the source of new carbon skeletons derives from glucose and other substrates (e.g. glycerol, alanine) that are metabolized via pyruvate carboxylase or metabolism of certain indispensable AA (e.g. isoleucine, valine, threonine) prior to oxaloacetate, but not beyond this point (e.g. acetyl-CoA). By contrast, glutamate and glutamine can be synthesized from the Krebs cycle intermediate α-ketoglutarate or from catabolism of proline or arginine. It is clear that, if dispensible AA are involved, this has implications for their dietary requirement and the pattern of AA supply to meet the metabolic demand.

We will discuss several tracer approaches that have been used to determine the sources of carbon for indispensable AA synthesis. Berthold *et al.* (1991) fed a diet containing [U-^{13}C] algal biomass (protein, carbohydrates, and lipids) to a laying hen, collected eggs over the course of 27 d, and harvested tissues from the hen at termination. Their results were in general agreement with the categories of nutritionally indispensable and dispensable, except that all proline and 76% of cysteine deposited in the hen tissues and in the egg had derived solely from the diet (i.e. fully conserved carbon skeleton M+*n*), therefore behaving like an indispensable AA. By contrast, there was virtually no fully ^{13}C-labelled (i.e. [$^{13}C_4$] or [$^{13}C_5$]) aspartate, glutamate, and glutamine in the egg and hen proteins, whereas lower molecular weight ^{13}C-mass isotopomers (i.e. [M+1], [M+2], and [M+3]) were present. Therefore, 89 to 94% of these AA were synthesized *de novo* via operation of the Krebs cycle and entry of ^{13}C-labelled precursors derived from the dietary algal biomass. In this respect, because the [U-^{13}C] biomass contained both [U-^{13}C] AA and [U-^{13}C] carbohydrate (primarily glucose), the source of carbon for dispensable AA synthesis was equivocal.

To further identify the precursors for dispensable AA synthesis, we have taken a similar approach as Berthold *et al.* (1991), except that [U-^{13}C] algal protein (AA), not the complete algal biomass, was fed as a component of the starter diet of chicks from day 1 to 5 post-hatch. Thus, the appearance of less than [M+*n*] ^{13}C-mass isotopomers in a dispensable AA would be clear demonstration of *de novo* synthesis from catabolism of dietary AA. After 4 days of feeding the [U-^{13}C] algal protein, the [M+*n*] ^{13}C enrichment of indispensable AA had not attained the same level of enrichment as the dietary protein-AA, i.e. dilution from unlabelled AA in tissue proteins (Table 1). However, for the indispensable AA leucine, isoleucine and lysine (not shown), the extent of dilution in the intestinal mucosa and plasma free pools, and in tissue protein bound pools (intestines, liver and plasma proteins) was similar. Therefore, the factor for leucine (0.75) was used to adjust each of the dispensable AA for this dilution.

There are several observations we wish to highlight. Firstly, note that for leucine and isoleucine (and other indispensable AA), there was no appearance of lower molecular weight ^{13}C-mass isotopomers, other than low enrichment of the [M+(*n*-1)] isotopomer, which appears because the [U-^{13}C] algal protein is not fully enriched at [M+*n*] (93% [M+*n*], 6% [M+(*n*-1)], and 1% [M+0]). Appearance of only the [M+*n*] ^{13}C-isotopomer confirms the classification of these AA as indispensable, and furthermore that no lower mass ^{13}C-isotopomers appeared indicates that these AA were not synthesized by bacteria in the intestinal lumen of the chicks, at least not from dietary AA. Indeed, for all the biochemically indispensible AA we measured (lysine, arginine, valine), the same labeling patterns were observed.

Second, we estimated that the diet contributed only 20 to 28% of glutamine, glutamate, aspartate and serine in the intestinal mucosa intracellular pool, the remainder synthesized *de novo*. And, because appearance of the lower molecular weight ^{13}C-isotopomers in these AA can only arise from *de novo* synthesis from dietary ^{13}C-labelled AA, we estimated that a considerable proportion of glutamine (36%), glutamate (37%), and serine (28%) was synthesized from dietary AA and the remainder from other sources (e.g. glucose). For aspartate, however, only 6% was synthesized from dietary AA with the majority from other sources. Clearly, it is not possible to determine which AA had been catabolized for synthesis of an indispensable AA. For example, it cannot be ruled out that metabolism of [M+5] proline and [M+6] arginine led to the synthesis of [M+5] glutamate and glutamine, in which case we have underestimated *de novo* synthesis. Similarly, *de novo* synthesis of aspartate would also be underestimated if there was catabolism of [M+4] asparagine and other AA metabolized in the Krebs cycle prior to oxaloacetate. What is clear, however, is that AA that are catabolized for entry into the Krebs cycle between succinyl-CoA and oxaloacetate (Figure 2) did not contribute to glutamate and glutamine synthesis, otherwise there would have been appearance of [M+4] glutamate and glutamine.

Lastly, [M+1] to [M+3] isotopomers were detected in plasma glucose. Again, these isotopomers can only originate from metabolism of dietary ^{13}C-labelled AA. Based on the mole sum of ^{13}C-isotopomers and the mole sum of glucose molecules, we estimated that <3% of glucose entry derived via gluconeogenesis from dietary AA. At most, this value is probably underestimated by a factor of 2, due to loss and exchange of ^{13}C with ^{12}C in the Krebs cycle (Katz and Tayek, 1999). The results suggest that the majority of glucose entry in the chick arises from glucose absorption and glucose carbon recycling (Cori cycle), which is consistent with our estimates in 2-day old chicks (Bequette *et al.*, 2006).

An alternative approach to determine dietary essentiality of AA is based on the use of [$^{13}C_6$] glucose whose subsequent metabolism may lead to appearance of [^{13}C] isotopomers in dispensable AA via cataplerotic sequences of the Krebs cycle; that is, an inside-out approach. This technique has been employed to good effect in human infant studies (Jaksic *et al.*, 1994; Miller *et al.*, 1995), and we have also applied the approach in studies with post-hatch chicks (Sunny *et al.*, 2004) and hybrid striped bass (*Morone saxatilis*) fed [$^{13}C_6$] glucose (Bequette *et al.*, 2005). The comparison between

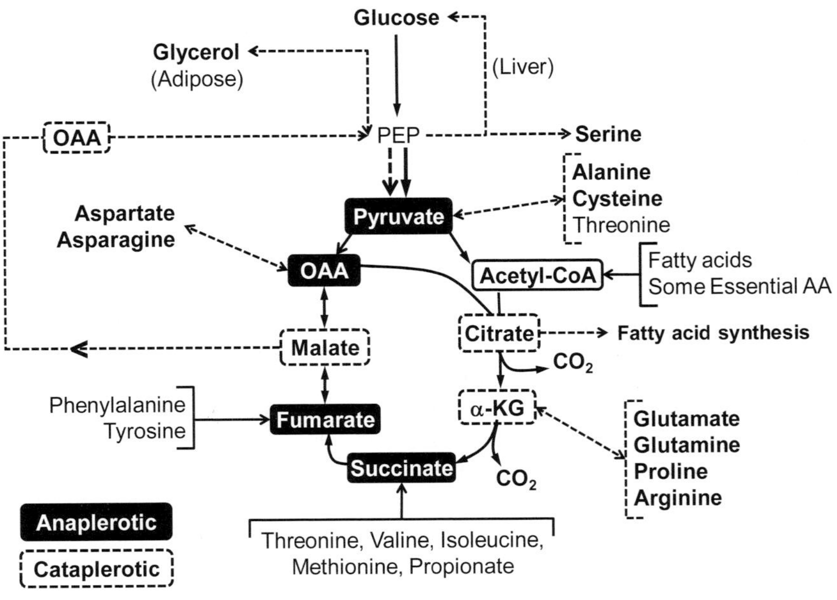

Figure 2. Diagram representing anaplerotic (solid lines and shapes) and cataplerotic (dashed lines and shapes) sequences connecting the Krebs cycle to gluconeogenesis, fatty acid metabolism, and dispensible AA synthesis. Some sequences can serve both anaplerotic and cataplerotic roles, thus linked metabolites (bold) can be catabolized or synthesized. α-KG, α-ketoglutarate; OAA, oxaloacetate; PEP, phosphoenolpyruvate.

the omnivorous chick and the carnivorous fish provides a striking demonstration of the different metabolic strategies these species utilize for AA synthesis, and the sources of substrates and pathways of metabolism they employ for neogenesis of organic nutrients. For all the dispensable AA, there was significant enrichment of the [M+3] isotopomer, which only occurs via operation of the pathway [M+3] or [M+6] glucose ® [M+3] glycerate 3-phosphate ® [M+3] pyruvate ® [M+3] oxaloacetate ® [M+3] a-ketoglutarate ® [M+3] pyrroline-5-carboxylate. In the chicks, which were fed a 65% dextrose diet, very little of serine, aspartate and asparagine had apparently been synthesized in the body from dietary glucose. In contrast, 36% of glutamate and 43% of glutamine had been synthesized from dietary glucose carbon, values similar to the contribution of 'other sources' we estimated using the [U-^{13}C] algal protein approach. It is notable that 23% of proline was synthesized from glucose, a result that is somewhat surprising given that several studies indicate that proline synthesis from glutamate (via α-ketoglutarate) occurs to a limited extent (Graber and Baker, 1973; Wu *et al.*, 1995).

The fish results were in contrast to those of the chick. In the hybrid striped bass (20% dietary dextrin), glucose contributed to only 1 to 6% of aspartate and glutamate synthesis, and there was no synthesis of proline from glucose carbon. Even though as much as 30% of these AA derived from dietary protein, most (64%) of the supply of carbon skeletons for synthesis of these AA derived from other AA, primarily those entering the Krebs cycle prior to oxaloacetate.

Intermediary metabolism: sources of substrates in the Krebs cycle

Two questions central to our understanding of macronutrient synthesis and, relatedly, the basis of dietary energy requirements: (1) What are the pathways that substrates follow? and (2) What are the contributions these substrates make to the flux through central pathways of metabolism? Our group has focused on the gastrointestinal tract, and the metabolic flexibility these tissues possess when substrate supply is varied, particularly in ruminant species where the type and supply of substrates differs from monogastric species. A universal observation has been that a considerable portion of the net removal of AA and glucose by the intestinal tissues involves complete and (or) partial catabolism (Windmueller and Spaeth, 1975, 1980; Stoll *et al.*, 1998; 1999). And, while it has been shown that AA and glucose account for the majority of the CO_2 produced by the gut tissues, it remains unclear how the catabolism of such a varied array of carbon skeletons is coordinated and maintained. For example, complete oxidation in the Krebs cycle necessitates terminal entry of substrates via acetyl-CoA from the pyruvate pool, a process that can only be maintained by balancing substrate flows through anaplerotic and cataplerotic sequences. Both ^{13}C-NMR (Chatham *et al.*, 2003) and GC-MS (Khairallah *et al.*, 2004; Vincent *et al.*, 2004) coupled with stable isotope tracer approaches have been employed to quantify Krebs cycle intermediary metabolite fluxes.

In studies with isolated rumen epithelial and duodenal mucosal cells, we used [U-^{13}C] substrates at physiological concentrations and ^{13}C-MID analysis of Krebs Cycle intermediates to determine the relative contributions of substrates to anaplerotic fluxes. The advantage of [U-^{13}C] tracers over ^{14}C-tracers is that the metabolism of the [U-^{13}C] substrates to end-product flux (or synthesis) can be demonstrated unequivocally by detection of the [U-^{13}C] end-product by GC-MS. For example, when we incubated isolated cells in media containing [$^{13}C_6$] glucose, this lead to the synthesis of [M+3] pyruvate, and thence [M+3] lactate, both labelled forms of which could only arise from direct metabolism [$^{13}C_6$] glucose (El-Kadi *et al.*, 2009). Furthermore, catabolism of this [M+3] pyruvate in the Krebs Cycle can lead to ^{13}C-labelling of Krebs cycle intermediates, in particular [M+2] and [M+3] α-ketoglutarate (Figure 1). Here, [M+3] α-ketoglutarate can only arise from metabolism of glucose via oxaloacetate whereas [M+2] α-ketoglutarate arises mainly from metabolism of glucose via acetyl-CoA. However, in our cell incubations, while glucose metabolism accounted for 5 to 30% of pyruvate and lactate fluxes, there was no appreciable entry of glucose carbon into the Krebs cycle. Rather, glutamate, but not glutamine, was the largest contributor to Krebs cycle intermediate fluxes, with glutamate accounting for as much as 63% of the carbon flux through α-ketoglutarate, a key anaplerotic sequence. Of particular note, when [$^{13}C_5$] glutamate was the tracer, there was significant appearance of [M+1] to [M+4] ^{13}C-isotopomers in all Krebs cycle intermediates, whereas metabolism of other [U-^{13}C] substrates (acetate, propionate, leucine, valine, glutamine and glucose) resulted in only negligible enrichment of the lower molecular weight ^{13}C-isotopomers. This led us to conclude, as did Windmueller and Spaeth (1975, 1980), that glutamate is the major anaplerotic substrate metabolized by the Krebs cycle of rumen epithelia and intestinal mucosal cells. In contrast, metabolism of glucose to lactate occurs to generate some energy and cytosolic NADH, and to allow 3-carbon skeleton recycling to maintain hepatic gluconeogenesis. That the other substrates did not lead to appreciable entry into the Krebs cycle, in particular via acetyl-CoA, was unexpected. However, we should note that the isolated cells derived from the luminal aspect of the gastrointestinal tract, and therefore metabolism by the cells on the serosal aspect of the gut tissues was not included. Indeed, there are several substrates (e.g. glutamine and glucose; Stoll *et al.*, 1999) whose removal and oxidation by the gastrointestinal tract occurs predominantly from the arterial circulation.

Conclusion

Empirical approaches (e.g. feed to tissue conversion, dose titration responses) have been invaluable as the basis for establishing most dietary macronutrient requirements, with particular emphasis on indispensible nutrients. However, empirical approaches do not directly address questions related to the adequacy with which dietary macronutrients support metabolic requirements. Yet, knowledge of rates of organic nutrient catabolism and neogenesis is necessary if the metabolic needs of the animal are to be converted into dietary requirements, hence improving our ability to formulate diets that the animal utilizes most efficiently. Stable isotopes and various MS based protocols are now available to animal scientists to probe multiple central metabolic pathways during different physiological phases. Exploitation of these approaches in animal sciences offers endless opportunities to provide missing details of the biochemical networks of macronutrient metabolism, thus providing dense information about dietary nutrient requirements. We have provided examples of the types of metabolic information that can be derived when multiple-labelled stable isotope tracers and ^{13}C-MID techniques are applied to questions related to the nutritional and metabolic essentiality of macronutrients and neogenesis. In doing so, we also hope to encourage more widespread application of these techniques in the animal sciences, in particular because it is becoming more evident that regulation of cellular metabolism cannot be fully explained based on regulation of molecular mediators and gene function.

References

Armentano, L.E., 1992. Ruminant hepatic metabolism of volatile fatty acids, lactate and pyruvate. J. Nutr. 122, 838-842.

Bain, J.R., R.D. Stevens, B.R. Wenner, O. Ilkayeva, D.M. Muoio and C.B. Newgard, 2009. Metabolomics applied to diabetes research: moving from information to knowledge. Diabetes. 58, 2429-2443.

Bequette, B.J., S.L. Owens, N.E. Sunny, S.W. El-Kadi and L.C. Woods, 2005. Hepatic gluconeogenesis and muscle intermediary metabolism in hybrid striped bass (HSB) determined by ^{13}C-mass isotopomer distribution analysis. J. Anim. Sci. 83 (Suppl. 1), 293.

Bequette, B.J., N.E. Sunny, S.W. El-Kadi, and S.L.Owens, 2006. Application of stable isotopes and mass isotopomer distribution analysis to the study of intermediary metabolism of nutrients. J. Anim. Sci. 84 (E. Suppl.), E50-59.

Berthold, H.K., D.L. Hachey P.J. Reeds O.P. Thomas S. Hoeksema and P.D. Klein, 1991. Uniformly ^{13}C-labelled algal protein used to determine amino acid essentiality *in vivo*. Proc. Natl. Acad. Sci. USA 88, 8091-8095.

Berthold, B.K., L.J. Wykes, F. Jahoor, P.D. Klein and P.J. Reeds, 1994. The use of uniformly labelled substrates and mass isotopomer analysis to study intermediary metabolism. Proc. Nutr. Soc. 53, 345-354.

Berthold, H.K., P. Crain, P.J. Reeds and P.D. Klein, 1995. Dietary pyrimidines but not dietary purines make a significant contribution to nucleic acid synthesis. Proc. Natl. Acad. Sci. USA 92, 10123-10127.

Brunengraber, H., J.K. Kelleher and C. Des Rosiers, 1997. Applications of mass isotopomer analysis to nutrition research. Annu. Rev. Nutr. 17, 559-596.

Chatham, J.C., B. Bouchard and C. Des Rosiers, 2003. A comparison between NMR and GCMS ^{13}C-isotopomer analysis in cardiac metabolism. Mol. Cell. Biochem. 249, 105-112.

Des Rosiers, C., L. Di Donato, B. Comte, A. Laplante, C. Marcoux, F. David, C.A. Fernandez and H. Brunengraber, 1995. Isotopomer analysis of citric acid cycle and gluconeogenesis in rat liver. Reversibility of isocitrate dehydrogenase and involvement of ATP-citrate lyase in gluconeogenesis. J. Biol. Chem. 270, 10027-10036.

El-Kadi, S.W. R.L. Baldwin, VI, K.R. McLeod, N.E. Sunny and B.J. Bequette, 2009. Glutamate is the major anaplerotic substrate in the tricarboxylic acid cycle of isolated rumen epithelial and duodenal mucosal cells from beef cattle. J. Nutr. 139, 869-875.

Fernandez, C.A., C. Des Rosiers, S.F. Previs, F. David and H. Brunengraber, 1996. Correction of ^{13}C mass isotopomer distributions for natural stable isotope abundance. J. Mass Spectrom. 31, 255-262.

Fuller, M.F., R. McWilliams, T.C. Wang and L.R. Giles, 1989. The optimal dietary amino acid pattern for growing pigs. 2. Requirements for maintenance and for tissue protein accretion. Br. J. Nutr. 62, 255-267.

Graber, G. and D.H. Baker, 1973. The essential nature of glycine and proline for growing chickens. Poult. Sci.52, 892-899.

Haymond, M.W. and A.L. Sunehag, 2000. The reciprocal pool model for the measurement of gluconeogenesis by use of [U-^{13}C]glucose. Am. J. Physiol. Endocrinol. Metab. 278, E140-145.

Hellerstein, M.K. and R.A. Neese, 1992. Mass isotopomer distribution analysis: A technique for measuring biosynthesis and turnover of polymers. Am. J. Physiol Endocrinol. Metab. 263, E988-1001.

Hellerstein, M.K. and R.A. Neese, 1999. Mass isotopomer distribution analysis at eight years: Theoretical, analytic, and experimental considerations. Am. J. Physiol Endocrinol. Metab. 276, E1146-1170.

Hellerstein, M.K, 2003. *In vivo* measurement of fluxes through metabolic pathways: The missing link in functional genomics and pharmaceutical research. Annu. Rev. Nutr. 23, 379-402.

Jaksic, T., F. Jahoor, P.J. Reeds and W.C. Heird, 1994. The determination of amino acid synthesis in human neonates using a glucose stable isotope. Surg. Forum 44, 642-686.

Jin, E.S., J.G. Jones, M. Merritt, S.C. Burgess, C.R. Malloy and A.D. Sherry, 2004. Glucose production, gluconeogenesis, and hepatic tricarboxylic acid cycle fluxes measured by nuclear magnetic resonance analysis of a single glucose derivative. Anal. Biochem. 327, 149-155.

Jones, J.G., M.A. Solomon, S.M. Cole, A.D. Sherry and C.R. Malloy, 2001. An integrated ^{2}H and ^{13}C NMR study of gluconeogenesis and TCA cycle Xux in humans, Am. J. Physiol. 281, E848-856.

Katz, J. and J.A. Tayek, 1999. Recycling of glucose and determination of the Cori Cycle and gluconeogenesis. Am. J. Physiol. Endocrinol. Metab. 277, E401-407.

Khairallah, M., F. Labarthe, Bouchard B., Danialou G., B.J. Petrof and C. Des Rosiers, 2004. Profiling substrate fluxes in the isolated working mouse heart using 13C-labelled substrates: focusing on the origin and fate of pyruvate and citrate carbons. Am. J. Physiol. Heart Circ. 286, H1461-1470.

Kim, S.W., D.H. Baker and R.A. Easter, 2001. Dynamic ideal protein and limiting amino acids for lactating sows: The impact of amino acid mobilization. J. Anim. Sci. 79, 2356-2366.

Landau, B. R., J. Wahren, K. Ekberg, S.F. Previs, D. Yang and H. Brunengraber, 1998. Limitations in estimating gluconeogenesis and Cori cycling from mass isotopomer distributions using [U-$^{13}C_6$]glucose. Am. J. Physiol. Endocrinol. Metab. 274, E954-961.

Lee, W.N., S. Sorou and E.A. Bergner. 1991. Glucose isotope, carbon recycling, and gluconeogenesis using [U-^{13}C] glucose and mass isotopomer analysis. Biochem. Med. Metab. Biol. 45, 298-309.

Lee, W.N., J. Edmond, S. Bassilian and J.W. Morrow, 1996. Mass isotopomer study of glutamine oxidation and synthesis in primary culture of astrocites. Dev. Neurosci.18, 469-477.

Miller, R.G., F. Jahoor, P.J. Reeds, W.C. Heird and T. Jaksic, 1995. A new stable isotope tracer technique to assess human neonatal amino acid synthesis. J. Pediatric Surg. 30, 1325-1329.

Millington, D.S., N. Kodo, D.L. Norwood and C.R. Roe, 1990. Tandem mass spectrometry: a new method for acylcarnitine profiling with potential for neonatal screening for inborn errors of metabolism. J. Inherit. Metab. Dis.13, 321-324.

Murphy, M.E., 1994. Amino acid compositions of avian eggs and tissues: Nutritional implications. J. Avian Biol. 25, 27-38.

Neese, R.A., J.M. Schwarz, D. Faix, S. Turner, A. Letscher, D. Vu and M.K. Hellerstein, 1995. Gluconeogenesis and intrahepatic triose phosphate flux in response to fasting or substrate loads. J. Biol. Chem. 270, 14452-14466.

Pascual, M., F. Jahoor and P.J. Reeds, 1998. *In vivo* glucose contribution to glutamate synthesis is maintained while its contribution to acetyl-CoA is lowered in adult mice fed a diet with a high fat:carbohydrate ratio. J. Nutr. 128, 733-739.

Reeds, P.J., D.G. Burrin, F. Jahoor, L. Wykes, J. Henry and E.M. Frazer, 1996. Enteral glutamate is almost completely metabolized in first pass by the gastrointestinal tract of infant pigs. Am. J. Physiol. 270, E413-418.

Reeds, P.J., H.K. Berthold, J.J. Boza, D.G. Burrin, F. Jahoor, T. Jaksic, P.D. Klein, T. Keshen, R. Miller, B. Stoll and L.J. Wykes, 1997. Integration of amino acid and carbon intermediary metabolism: Studies with uniformly labelled tracers and mass isotopomer analysis. Eur. J. Pediatr. 156 (Suppl. 1), S50-58.

Rose, W.C., M.J. Oesterling and M.Womack, 1948. Comparative growth on diets containing ten and nineteen amino acids, with further observations upon the role of glutamic and aspartic acids. J. Biol. Chem. 176, 753-762.

Stoll, B., J. Henry, P.J. Reeds, H. Yu, F. Jahoor and D.G. Burrin, 1998. Catabolism dominates the first-pass intestinal metabolism of dietary essential amino acids in milk-replacer fed piglets J. Nutr.128, 606-614.

Stoll, B, D.G. Burrin, J. Henry, F. Jahoor and P.J. Reeds, 1999. Substrate oxidation by the portal drained viscera of fed piglets. Am. J. Physiol Endocrinol. Metab. 277, E168-175.

Stucki, W.P. and A.E. Harper, 1961. Importance of dispensable amino acids for normal growth of chicks. J. Nutr. 74, 377-383.

Sunny, N.E. and B.J. Bequette, 2010. Gluconeogenesis differs in developing chick-embryos derived from small compared to typical size broiler breeder eggs. J. Anim. Sci. 88, 912-921.

Sunny, N.E., S.W. El-Kadi, M. Oba, S.L. Owens, R. Angel and B.J. Bequette, 2004. Gluconeogenesis and Krebs cycle metabolism in 2-day-old fed and fasted chicks. J. Anim. Sci. 87 (Suppl. 1), 313.

Sunny, N., S. Satapati, X. Fu, T He, R. Medibeigi, C. Spring-Robinson, J. Duarte, M.J. Potthoff, J.D. Browning and S.C. Burgess, 2010. Progressive Adaptation of Hepatic Ketogenesis in Mice Fed a High Fat Diet. Am. J. Physiol. Endocrinol. Metab. [Epub ahead of print] PubMed PMID: 20233938.

Vincent, G., B. Bouchard, M. Khairallah and C. Des Rosiers, 2004. Differential modulation of citrate synthesis and release by fatty acids in perfused working rat hearts. Am. J. Physiol. Heart Circ. Physiol. 286, H257-266.

Windmueller, H.G. and A. E. Spaeth, 1975. Intestinal metabolism of glutamine and glutamate from the lumen as compared to glutamine from blood. Arch. Biochem. Biophys. 171, 662-672.

Windmueller, H.G. and A.E. Spaeth, 1980. Respiratory fuels and nitrogen metabolism *in vivo* in small intestine of fed rats. Quantitative importance of glutamine, glutamate, and aspartate. J. Biol. Chem. 255, 107-112.

Wu, G., N.E. Flynn, W. Yan and D.G. Barstow, Jr., 1995. Glutamine metabolism in chick enterocytes: absence of pyrroline-5-carboxylase synthase and citrulline synthesis. Biochem. J. 306, 717-721.

Wykes, L.J., F. Jahoor and P.J. Reeds, 1998. Gluconeogenesis measured with [U-^{13}C] glucose and mass isotopomer analysis of apoB-100 amino acids in pigs. Am. J. Physiol. 274, E365-376.

Young, V.R. and A. Ajami, 1999. The Rudolf Schoenheimer Centenary Lecture. Isotopes in nutrition research. Proc. Nutr. Soc. 58, 15-32.

Xu, J., V. Chang, S.B. Joseph, C. Trujillo, S. Bassilian, M.F. Saad, W.N. Lee and I.J. Kurland, 2004. Peroxisomal proliferator-activated receptor alpha deficiency diminishes insulin-responsiveness of gluconeogenic/glycolytic/ pentose gene expression and substrate cycle flux. Endocrinology 145, 1087-1095.

Dietary lysine requirements of sows in early- and late-gestation

R.S. Samuel[1], S. Moehn[1], P.B. Pencharz[2] and R.O. Ball[1,2]
[1]Swine Research and Technology Centre, 4-10 Agriculture/Forestry Centre, University of Alberta, T6G 2P5, Edmonton, Alberta, Canada; ron.ball@ualberta.ca
[2]Research Institute, Hospital for Sick Children, M5G 1X8, Toronto, Ontario, Canada

Introduction

Current feeding recommendations specify that pregnant sows be fed a fixed diet through pregnancy. Growth of the placenta declines as pregnancy advances and foetal weight linearly increases after day 70 of gestation (McPherson *et al.*, 2004). Development of the mammary gland occurs very late in gestation, close to parturition (Kim *et al.*, 1999). The relative contribution to dietary amino acid requirements by the products of conception clearly changes during gestation. Thus, a fixed diet during the entire pregnancy seems illogical. Therefore, objective of this study was to determine the lysine requirement of a population of 2nd and 3rd parity sows in early (day 24-45) and late (day 86-110) gestation.

Material and methods

Animals, diets, housing, and feeding

Hypor Hybrid (Hypor Inc) sows, pregnant after their first (n=4) or second (n=3) litters (185.7±9.6 kg BW) were adapted to individual lysine intakes. Three semi-synthetic diets (14.0 MJ ME/kg) based on corn with added amino acids were formulated and mixed to produce a base diet (60% of NRC (1998) requirement) and summit diets for early- and late-gestation (150% and 185% of NRC (1998) requirement, respectively). Each sow received 6 different test diets, in random order, from 60 to 150% of the requirement suggested by NRC (1998) in early- and from 60 to 185% of the requirement suggested by NRC (1998) in late-gestation. Sows were housed individually and fed one-half of their daily feed allowance twice daily, except on study days, when they received 11 meals equivalent to 1/26th of their daily ration every 30 minutes during the 5.5 hour total study period (Moehn *et al.*, 2004). Comparability of this protocol to once daily feeding was previously verified. Individual feed allowances were determined based on body weight and P2 backfat depth at breeding. Phenylalanine, equivalent to the isotope dose, was mixed into individual batches and fed for adaptation periods between study days. Nipple drinkers provided free access to water.

Indirect calorimetry, isotope infusion, and sample collection

After adaptation to each diet, sows were individually housed in respiration chambers for the measurement of indicator amino acid oxidation during a primed, constant infusion of L-[1-^{13}C] phenylalanine and, simultaneously, heat production (HP) over 4 h. The gas exchange was recorded for O_2, CO_2, and CH_4 (Brouwer, 1965) in 1 min intervals. Expired CO_2 and blood plasma were collected in 30 min intervals for determination of ^{13}C enrichment.

Statistics

Statistical analysis was performed using mixed procedure and breakpoint analysis was performed using the non-linear procedure in SAS version 9.1 (SAS Inst. Inc., Cary, NC). The classification variable was lysine intake and individual animals were treated as random variables. Model statements were tested using the Kenward-Roger degrees of freedom method. Least square means were compared using the 'pdiff' option. Data are presented as means ± SEM. Values were considered significant at $P<0.05$.

Results and discussion

The average number of piglets born alive was 13.7±1.9, but ranged from 4 to 20. The average piglet birth weight was 1.5±0.1 kg. Sows gained 600 g/d from breeding and weighed 258.8±8.3 kg at parturition. Actual dietary lysine intakes ranged from 7.5 to 19.3 g/d in early- and 8.1 to 23.7 g/d in late-gestation. Breakpoint analysis of phenylalanine oxidation indicated that the lysine requirement of 2nd parity sows was 13.1 g/d and 18.7 g/d in early and late gestation, respectively. For 3rd parity sows, the dietary lysine requirement was 8.2 g/d and 13.0 g/d for early- and late-gestation, respectively. Similar breakpoints were calculated using HP as the dependent variable. The dietary requirements for lysine in early- and late-gestation are greater than previously reported by NRC (1998) and are similar to values reported by Srichana (2006) and GfE (2008). Phase feeding at least two diets would improve productivity by more correctly providing the necessary nutrients for the growth of the placenta, the piglets, and the mammary gland. Ultimately, phase feeding would provide positive economic returns by reducing feed cost and maximizing lifetime productivity of the sows.

References

Brouwer, E., 1965. Report of sub-committee of the European Association for Animal Production on constants and factors. In: K.L. Blaxter (ed.), Energy metabolism. Academic Press, New York, NY: 441-443.

GfE, 2008. Recommendations for the supply of energy and nutrients to pigs. Frankfurt, Germany, DLG Verlags GmbH.

Kim, S.W., W.L. Hurley, I.K. Han, and R.A. Easter, 1999. Changes in tissue composition associated with mammary gland growth during lactation in sows. J. Anim. Sci. 77, 2510-2516.

McPherson, R.L., F. Ji, G. Wu, J.R. Blanton, and S.W. Kim, 2004. Growth and compositional changes of fetal tissues in pigs. J. Anim. Sci. 82, 2534-2540.

National Research Council, 1998. Nutrient requirements of swine. 10th ed. Washington, D.C., National Academy Press.

Moehn, S., R.F.P. Bertolo, P.B. Pencharz, and R.O. Ball, 2004. Pattern of carbon dioxide production and retention is similar in adult pigs when fen hourly, but not when fed a single meal. BMC Physl. 4, 11-18.

SAS (Statistical Analysis System), 2002. Release 9.1 Edition. SAS Institute, Inc., Cary, NC.

Srichana, P., 2006. SID lysine requirements of gilts during different stages of gestation and the impact of synthetic AA on nitrogen retention. Ajinomoto Animal Nutrition. Ajinomoto Swine Research Report 46.

The metabolic availability of threonine in corn and barley in pregnant sows

C.L. Levesque, S. Moehn, P.B. Pencharz and R.O. Ball
Swine Research and Technology Centre, University of Alberta, Edmonton, AB, Canada

Introduction

Correct estimates of the availability of amino acids (AA) in feedstuffs are important to ensure dietary supply of AA meets the needs of the animal. Recent evidence suggests that digestibility of AA in feedstuffs is greater in gestating sows than growing pigs (Stein *et al.*, 2001); however, published digestibility coefficients for feed ingredients are based on growing pig data. Ileal amino acid digestibility of feedstuffs is a measure of amino acid disappearance from the gut and does not account for the true metabolic utilization of absorbed AA. Metabolic availability (MA) reflects the proportion of dietary AA used for protein synthesis and includes all amino acid losses that occur during digestion, absorption and metabolic utilization (Moehn *et al.*, 2005). It is hypothesized that the availability of AA in feedstuffs will be greater in sows than growing pigs.

Materials and methods

Eight 2nd parity sows ([1-^{13}C] phenylalanine) were used to determine the MA of THR in corn and barley based on the indicator amino acid oxidation method. A paired study was conducted using eight growing pigs ([1-^{14}C] phenylalanine) to allow direct comparison of determined MA values to published ileal digestibility values. A base reference diet was formulated at 50% of the respective daily THR requirement for sows and growers. Crystalline THR was used to create 3 additional reference diets set at 60, 70 and 80% of requirement. All other nutrients were set at 120% of requirement. Test ingredient diets were formulated to supply THR at 80% of requirement.

The trial consisted of 6 consecutive periods (3 d of adaptation + 1 d of expired CO_2 and plasma collection). On collection day, tracer phenylalanine was given orally in 8 ½-hourly meals and expired [1]-labelled CO_2 was quantified. Data was analyzed using the Proc Mixed procedure in SAS (2001) where nesting THR intake within type of THR addition (e.g. free THR or THR in corn or barley) gave the change (slope) in PHE oxidation per g of THR for each type of THR addition. The MA of THR in corn and barley was calculated by dividing the slope for THR from the respective feedstuff by the slope for free THR.

Results and discussion

There was a linear response to increasing THR from free THR and THR in feedstuffs (Figure 1). The MA of THR in corn and barley in sows was 88.0 and 89.3%, respectively (Table 1). The MA of THR in corn in growing pigs was 82.2% (data not shown), similar to the published true ileal digestibility of THR in corn (82.0%). Metabolic availability is therefore, a reliable estimate of the true availability of AA in feedstuffs. Pregnant mature animals have a greater capacity to digest and utilize AA in common feedstuffs than growing animals. Diets formulated for gestating sows using published digestibility coefficients underestimate available AA.

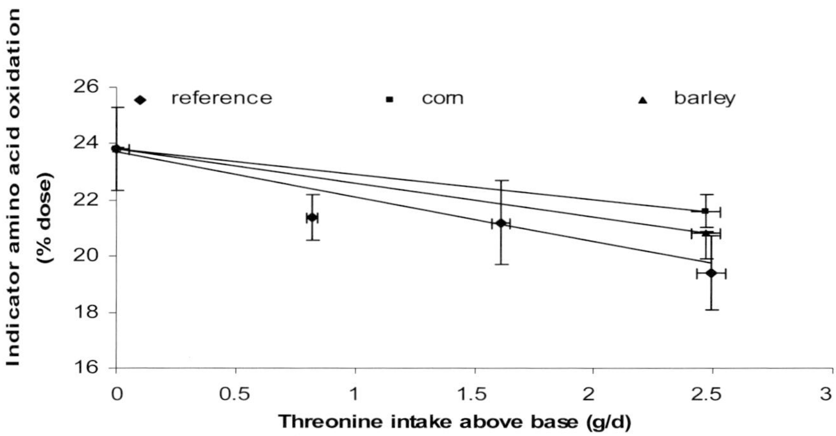

Figure 1. Oxidation of indicator amino acid in response to changes in threonine intake above that provided in the base diet in gestating sows. Regression lines show the response from free threonine compared to threonine provided by the feedstuff.

Table 1. The metabolic availability of threonine in corn and barley in gestating sows based on indicator amino acid oxidation[1].

Item	Threonine intake above base diet[2] g/d	Indicator oxidation % of dose	Oxidation response[3] % of dose per g THR intake	Metabolic availability %
THR-50%	0.00±0.05	23.8±3.6		
THR-60%	0.82±0.02	21.4±2.0		
THR-70%	1.61±0.04	21.2±3.7		
THR-80%	2.49±0.06	19.4±3.2	-1.59±0.53	100.0[4]
Corn	2.47±0.06	21.6±1.4	-1.40±0.62	88.0[5]
Barley	2.47±0.06	20.8±2.2	-1.42±0.81	89.3

[1] Values are means ± SEM, n=6.

[2] Base diet provided THR at 50% of the THR requirement in early gestation as determined by Levesque *et al.*, 2010 (2.5 g/d). Equivalent to ~20% of NRC, 1998 dietary THR requirement for sows in gestation.

[3] Change of indicator oxidation (% of dose) per g additional THR intake. Estimated from mixed model analysis.

[4] Susenbeth *et al.*, 2001.

[5] Oxidation response for corn (or barley) divided by oxidation response for free THR.

References

Moehn, S., R.F.P. Bertolo, P.B. Pencharz and R.O. Ball, 2005. Development of the indicator amino acid technique to determine the availability of amino acids from dietary protein in pigs. J. Nutr. 135, 2866-2870.

Levesque, C.L., S. Moehn, P.B. Pencharz and R.O. Ball. 2010. The requirement for threonine in early and late gestation in sows. Advances in Pork Production, Proc. Banff Pork Seminar 2010, Vol 21. Abstract # 4.

SAS, 2001, Release 8.03 SAS inst. Inc., Gary, NC, USA.

Susenbeth, A., A. Agunbiade and J. Dohms, 2001 Bioavailability of lysine of several protein sources determined by the slope-ratio assay in pigs. In: Fevrier, C., Aumaitre, A., Habe, F., Vares, T., Zjalic, M. (eds.). Protein Feed for Animal Production. EAAP Technical Series 2001: 133-137.

Titration studies to determine amino acid requirements of individual growing pigs

E. van de Hoek[1,2]*, S. Borgijink*[2]*, J.J.G.C. van den Borne*[2]*, W.J.J. Gerrits*[2]*, A.J.M. Jansman*[1] *and C.M.C. van der Peet-Schwering*[1]
[1]*Wageningen UR Livestock Research, Lelystad, the Netherlands*
[2]*Animal Nutrition Group, Wageningen University, Wageningen, the Netherlands*

Introduction

Traditionally, methods for determining amino acid requirements are based on studies in large groups of pigs and estimate the mean requirement of a population. The last decade, new titration approaches for individual pigs have been developed and tested (e.g. the indicator amino acid technique; Pencharz and Ball, 2003). This technique requires the determination of changes in ^{13}C excretion in breath and/or changes in ^{13}C enrichment of plasma amino acids in frequently fed pigs. The sensitivity and applicability of this technique depends a.o. on the length of the titration steps. A shorter step-length is desirable, but absence of the effect of titration step-length on optimal amino acid intakes has not yet been convincingly demonstrated. This can be tested by reversing the direction of the titration in time and by comparing the estimates in individual pigs with a standardized contrast in amino acid requirement. In this paper, a simplified version of the titration technique is tested in two separate experiments, involving measurement of urinary N excretion in meal-fed pigs.

Materials and methods

In Experiment 1, fourteen pigs (initial body weight 27.1±0.2 kg) were assigned to either an increasing or a decreasing lysine supply, with lysine concentrations ranging between 4.7 and 14.0 g/kg in seven equidistant steps at a restricted feed intake of 2.5× the estimated ME requirements for maintenance (ME_m). Each step was applied for a period of 4 d. In Exp. 2, twenty pigs (24.2±0.4 kg) were assigned to a low (LE) or a high level (HE) of energy intake. LE was designed at 2.2× ME_m, HE pigs received an additional amount of energy from starch of 0.5× ME_m. A decreasing lysine titration strategy was applied, with lysine concentrations ranging between 4.5 and 15.9 g/kg in nine equidistant steps of 3 d each. In both experiments total (24 h) urinary N excretion was determined for the last two days of each titration step. For each animal, N excretion in urine was expressed as percentage of N intake. This percentage, reflecting inefficiency of protein utilization, was related to lysine intake using a linear-plateau model. Subsequently, parameter estimates were related to analysis of variance, with titration strategy (Experiment 1) or energy intake level (Experiment 2) as fixed effects.

Results and discussion

In both experiments, a linear-plateau model provided a better (P<0.05) fit of data compared with a linear model (Experiment 1: 11 of the 14 pigs; Experiment 2: 16 of the 20 pigs). In Experiment 1, the increasing and decreasing titration strategy resulted in an estimated lysine requirement of 11.4 g/kg and 10.1 g/kg (P<0.01) respectively, with N excretion in urine at plateau being lower (P<0.01) for the increasing than for the decreasing strategy (23.1% vs. 25.1%; Table 1).

In Experiment 2, an increased energy intake in the form of starch resulted in an increased lysine intake at the point of inflection (14.7 vs. 12.0 g/d; P<0.001) and a lower N excretion in urine at the HE treatment (21.6 vs. 30.1% of N intake, P<0.001). In the lysine limiting phase, the rate of decrease in N excretion in urine was not affected by energy intake (Figure 1).

The different responses between titration strategies in Experiment 1 indicate variation in the rate of adaptation of metabolic processes triggered by changes in amino acid imbalance (e.g. protein

Table 1. Parameter estimates for the linear-plateau model, fitted for the effect of dietary lysine content on N excretion in urine (as % of N intake) at increasing vs. decreasing titration strategies (Experiment 1).

Parameter	Treatment		*P*-value treatment
	Increasing titration	Decreasing titration	
A	60.0±1.3	58.6±1.3	0.464
B	-3.19±0.17	-3.33±0.17	0.563
C	11.4±0.3	10.1±0.2	0.004
Plateau	23.1±0.4	25.1±0.4	0.005

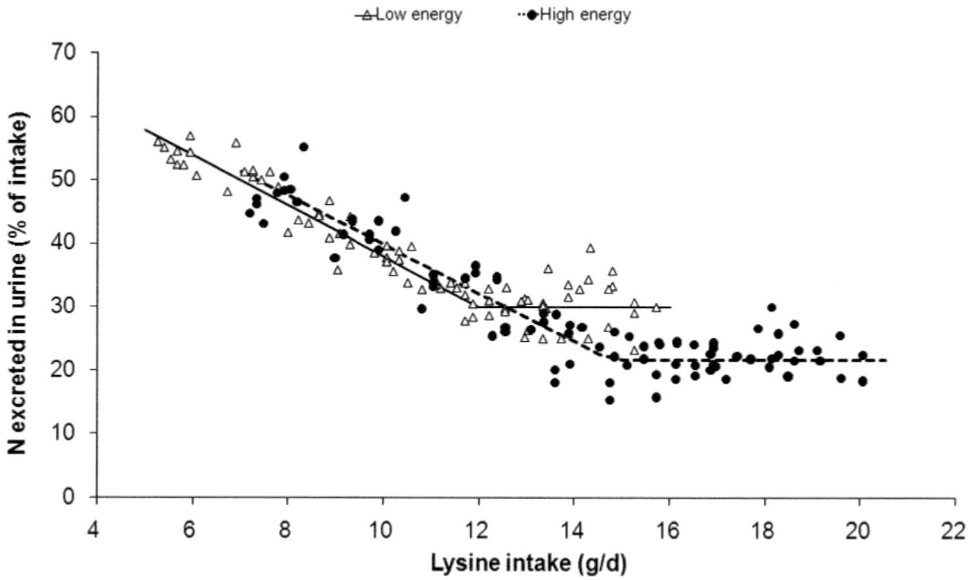

Figure 1. Effect of lysine intake on N excretion in urine (as % of N intake) in growing pigs at low vs. high energy intakes (Experiment 1).

turnover, amino acid oxidation, urea synthesis). The increased inflection point with increasing energy intake from starch (Experiment 2) confirms the generally adopted concept that protein deposition, and thus lysine requirement, increases with energy intake (Bikker *et al.*, 1994). The results of both experiments suggest that this within-animal titration technique is sensitive for determining the requirement of a rate limiting amino acid for growth in individual pigs. Potential applications of this technique include the quantification of effects of various factors, such as health status and genotype, on changes in amino acid requirements of individual pigs.

References

Bikker, P., M.W.A. Verstegen, R.G. Campbell and B. Kemp, 1994. Digestible lysine requirement of gilts with high genetic potential for lean gain, in relation to the level of energy intake. J. Anim. Sci. 72, 1744-1753.

Pencharz, P.B. and R.O. Ball, 2003. Different approaches to define individual amino acid requirements. Ann. Rev. Nutr. 23, 101-116.

Glucose uptake and endogenous glucose production in neonatal calves fed either colostrum or formula

J. Steinhoff[1], S. Görs[1], R.M. Bruckmaier[2], E. Kanitz[1], C.C. Metges[1] and H.M. Hammon[1]
[1]Leibniz Institute for Farm Animal Biology (FBN), Wilhelm-Stahl Allee 2, 18196 Dummerstorf, Germany
[2]Veterinary Physiology, University of Bern, Bremgartenstr. 109a, 3001 Bern, Switzerland

Introduction

Glucose is a major energy source for neonates (Battaglia and Meschia, 1978). Glucose is supplied via glycogenolysis and gluconeogenesis (GNG), in total termed endogenous glucose production (eGP), and by lactose breakdown after milk intake. After birth newborn calves have to adapt from continuous parenteral delivery via placenta to oral glucose absorption and eGP. Previous studies indicate that colostrum with its high concentration of bioactive components may improve glucose status in neonatal calves by promoting oral glucose uptake as well as eGP (Hammon and Blum, 1997; Hammon *et al.*, 2003).

The aim of the study was to determine feeding effects on oral glucose absorption and eGP in neonatal calves fed either colostrum or a milk-based formula with the same nutrient, esp. lactose content as colostrum, but much less bioactive components.

Material and methods

Fourteen neonatal male German Holstein calves were fed either colostrum (C) or formula (F) for 4 days. Composition of colostrum and formula milk were in accordance with previous studies (Hammon *et al.*, 2003; Scheuer *et al.*, 2006). Calves were fed twice daily at 8% of body weight (BW) on day 1 and at 10% of BW from day 2 to 4. The morning meal on day 2 was divided in 10 half-hourly portions. During the interval feeding 5.7 μmol/kg BW [U-$^{13}C_6$]-glucose was added to the milk portions to measure the rate of appearance of intragastric [U-$^{13}C_6$]-glucose (Ra_{ig}). Simultaneously, calves received a primed intravenous infusion of [6.6-$^2H_2$]-glucose (prime: 9.9 μmol/kg BW; infusion: 15.3 μmol/kg BW × h) for 5 h, which started 5 min before first feeding to estimate glucose turnover (Ra_{iv}). The first pass glucose uptake (FPU_{Gluc}) was calculated as follows: $FPU_{Gluc} = (Ra_{ig} - Ra_{iv}) / Ra_{ig}$. To ensure complete uptake of milk diets and oral tracer in some cases tube feeding was necessary. On day 3 GNG was determined by administration of deuterated water. After 16 h without food calves received two oral boli of $^2H_20$ (70%, 10 g/ kg BW per bolus) within 4 h. Fractional GNG was estimated by deuterium enrichments on position C5 related to enrichment on position C2 measured in glucose derivates (Junghans *et al.*, 2010). Glucose turnover during fasting (= eGP) was measured using a primed intravenous infusion of [U-$^{13}C_6$]-glucose (prime: 4.3 μmol/ kg BW; infusion: 6.4 μmol/kg BW × h) according to Junghans *et al.*, (2010).

Blood samples were collected before and during tracer applications on day 2 and 3 to measure plasma appearance of tracers as well as proportion of newly synthesized glucose and plasma concentrations of glucose, NEFA, urea, insulin, glucagon and cortisol. Metabolites were analysed photometrically and hormones by RIA (Hammon *et al.*, 2003; Scheuer *et al.*, 2006).

Data concerning plasma concentrations were analysed by the Mixed Model of SAS with feeding, time of blood sampling and feeding × time interaction as fixed effects and individual calves as random effects. Data related to stable isotope measurements were analysed by General Linear Model with feeding as fixed effect. For analyses of FPU measurement the statistical model included the feeding type and the type of milk application (voluntarily or by tube feeding).

Results and discussion

As expected plasma glucose concentrations in the C group was increased on days 2 and 3 ($P<0.05$) indicating an improved glucose status (Hammon and Blum, 1997). Plasma [U-$^{13}C_6$]-glucose abundance was greater ($P<0.05$) in C than in F and was elevated when calves ingested milk voluntarily. The FPU_{Gluc} tended to be higher ($P<0.1$) in F than in C and was elevated after tube feeding. On the other hand, eGP as well as GNG during fasting on day 3 were not affected by diet, which supports previous findings in calves (Scheuer *et al.*, 2006). Improved glucose status in C was most probably a result of elevated glucose absorption, but not of improved eGP after colostrum feeding. On the other hand, higher FPU in F than C indicated that F calves metabolise more glucose in the splanchnic tissues. Beside diet effects, tube feeding may additionally influence oral glucose absorption due to milk flowing partly into the reticulorumen.

Plasma urea concentrations on day 2 and 3 as well as NEFA concentrations on day 2 were much higher ($P<0.05$) in F than in C, indicating intensified fat and protein breakdown in F calves. Plasma insulin concentrations were not affected by diet but plasma glucagon concentrations were higher ($P<0.05$) on days 3 and 4 in F than in C. Plasma cortisol concentrations decreased ($P<0.05$) on day 2 with feeding in C but not in F calves. Therefore, endocrine changes support intensified fat and protein breakdown, but did not stimulate eGP in F calves.

In conclusion, our data indicate an improved glucose status in C calves that result primarily from enhanced oral glucose absorption rather than colostrum effects on eGP. In addition, metabolic and endocrine changes pointed at an elevated fat and protein breakdown in F calves, obviously to provide alternative substrates to meet energy requirements and to compensate for impaired glucose absorption.

References

Battaglia, F.C. and G. Meschia, 1978. Principal substrates of fetal metabolism. Physiol Rev. 58, 499-527.

Hammon, H.M. and J.W. Blum. 1997. Prolonged colostrum feeding enhances xylose absorption in neonatal calves. J.Anim.Sci. 75, 2915-2919.

Hammon, H.M., S. N. Sauter, M. Reist, Y. Zbinden, C. Philipona, C. Morel and J. W. Blum, 2003. Dexamethasone and colostrum feeding affect hepatic gluconeogenic enzymes differently in neonatal calves. J. Anim. Sci. 81, 3095-3106.

Junghans, P., S. Görs, I.S. Lang, J. Steinhoff, H.M. Hammon and C.C. Metges, 2010. A simplified mass isotopomer approach to estimate gluconeogenesis rate *in vivo* using deuterium oxide. Rapid Commun. Mass Spectrom, in press.

Scheuer, B.H., Y. Zbinden, P. Schneiter, L. Tappy, J.W. Blum and H.M. Hammon, 2006. Effects of colostrum feeding and glucocorticoid administration on insulin-dependent glucose metabolism in neonatal calves. Domest. Anim. Endocrinol. 31, 227-245.

The ^{13}C-bicarbonate tracer (^{13}C-BT) technique for estimation of CO_2 production and energy expenditure in dogs during rest and physical activity

C. Larsson[1], P. Junghans[2] and A.-H. Tauson[1]
[1]*Department of Basic Animal and Veterinary Sciences, Faculty of Life Sciences, University of Copenhagen, Grønnegårdsvej 7, 1870 Frederiksberg C, Denmark*
[2]*Leibniz Institute for Farm Animal Biology (FBN), Research Unit Nutritional Physiology 'Oskar Kellner', Wilhelm-Stahl-Allee 2, 18196 Dummerstorf, Germany*

Introduction

The knowledge of energy requirements in domestic dogs is very limited, and there are several factors that influence their energy demands. Thus, to provide accurate estimates of nutritional requirements and to formulate appropriate guidelines about how to feed dogs, it appears essential to determine the true rate of energy expenditure (EE) in a reliable and feasible way during dogs' daily life, illness, routine jobs or sporting activities, in order to ensure optimal health and performance. To assess EE, previous studies (Junghans *et al.*, 2008) indicate that the completely non-invasive ^{13}C-bicarbonate tracer (^{13}C-BT) technique (i.e. oral administration of $NaH^{13}CO_3$ and samples of breath) may provide correct results, and thus, is an appropriate method for studies with dogs. In the present experiment, the non-invasive ^{13}C-BT technique was used for estimation of CO_2 production (RCO_2) and EE in dogs during rest and physical activity.

Material and methods

RCO_2 and EE were estimated from the ^{13}C kinetics of exhaled breath CO_2 in six dogs of varying breed, age and body weight, after oral administration of (5 mg/kg BW) $NaH^{13}CO_3$. Breath was collected into breath bags by using a mask with a two-way non-rebreathing valve system at -5, 5, 10, 20, 30, 40, 60, 90, 120, 180, 360, 540 and 720 minutes after tracer administration. The measurements were conducted in two periods, the first comprising two days of rest indoors (room temperatures between 18-20 °C), and the second three days with three hours of exercise outdoors per day. The measurements were scattered over five months, where outdoor temperatures at the different times of measurement varied between 8-20 °C.

The ratio $^{13}C/^{12}C$ in collected breath samples was measured by means of an IRIS infrared analyser. The RCO_2 was estimated from the ^{13}C data using the following equation (Elia *et al.*, 1991):

$$RCO_2 = (D / AUC) \times RF \quad (1)$$

where D is the tracer dose administrated, AUC is the area under the $^{13}CO_2$ enrichment-time curve, and RF is the fractional ^{13}C recovery in breath CO_2. Energy expenditure was then calculated as follows (Junghans *et al.*, 2008):

$$EE\ (kJ/\times d) = 4.96 \times RCO_2 + 16.07 \times (RCO_2 / RQ) \quad (2)$$

where RCO_2 is the estimated CO_2 production in litres per day (l/d) and the value of the respiratory quotient (RQ) was assumed to be in accordance with the food quotients (FQ), estimated from the diets of the dogs. The statistical analyses were carried out according to the GLM procedure (SAS, 1989).

Results and discussion

Estimated EE was 483±147 and 1041±336 kJ×kg$^{-0.75}$/d for the six dogs during periods of rest and exercise, respectively. There were significant differences in both RCO_2 and EE between periods (P<0.001), and between dogs within periods (P<0.001) (Table 1).

Variation in EE between the dogs was assumed mainly to be due to individual differences. Related to individual body weight, age and differences in work load during exercise, the results are considered to be reliable and in good agreement with other findings regarding EE in dogs under similar conditions. However, reports from studies on EE in dogs have given fairly varying results (Ballevre *et al.*, 1994; Burger, 1994; Poteau *et al.*, 2002; NRC, 2006), which most likely depends on differences in experimental conditions and methods, as well as the individual differences between the dogs.

In conclusion, it seems possible to use the ^{13}C-BT technique as a minimal restrictive and non-invasive method to obtain reliable estimates of EE in dogs at different activity levels under near natural conditions. However, the accuracy of the estimates depends on the estimates used for the ^{13}C recovery factor (RF) and the respiratory quotient (RQ) in the calculations, and thus, the technique needs to be further standardized and validated.

Table 1. CO_2 production (RCO_2; l/d) and energy expenditure (EE; kJ×kg$^{-0.75}$/d) in the six dogs during two days of rest and during three days with three hours of exercise per day.

	Dog no.					
	1	2	3	4	5	6
Age, (years)	12	5	9	3	1.5	6.5
BW, (kg)	24	24	33	30	11.6	11.8
RCO_2, (l/d)						
Rest	148	184	237	337	186	111
Exercise	284	371	655	758	355	223
EE, (kJ×kg$^{-0.75}$/d)						
Rest	326	405	432	666	664	405
Exercise	613	823	1193	1496	1306	823

References

Ballevre, O., G. Anantharaman-Barr, P. Gicqaello, C. Piquet-Welsh, A.-L. Thielin and E. Fern, 1994. Use of the doubly-labelled water method to assess energy expenditure in free living cats and dogs. J. Nutr. 124, 2594S-2600S.

Burger, I.H., 1994. Energy needs of companion animals: matching food intakes to requirements throughout the life cycle. J. Nutr. 124, 2584S-2593S.

Elia, M., 1991. Estimation of short-term energy expenditure by the labelled bicarbonate method. In: Whitehead RG, Prentice A, (eds.) New techniques in nutritional research. New York, NY: Academic Press: 207-227.

Junghans, P., W. Jentsch and M. Derno, 2008. Non-invasive ^{13}C bicarbonate tracer technique for measuring energy expenditure in men – a pilot study. E-SPEN, the European e-journal of Clinical Nutrition and Metabolism 3, e46-e51.

National Research Council (NRC), 2006. Nutrient requirements of dogs and cats. The National Academies Press. Washington D.C.

Pouteau, E., S.M. Mariot, L.J. Martin, H.J. Dumon, F.J. Mabon, M.A. Krempf, R.J. Robins, D.H. Darmaun, N.A. Naulet and P.G. Nguyen, 2002. Rate of carbon dioxide production and energy expenditure in fed and food-deprived adult dogs determined by indirect calorimetry and isotopic methods. Am. J. Vet. Res. 63, 111-118.

SAS Institute Inc., 1998. SAS/STAT® User's Guide, Version 6, Fourth Edition, Volume 2, Cary, NC: SAS Institute Inc., 846 pp.

A fast, easy, reliable and cheap method to measure the methane production from ruminants

J. Madsen[1], J. Lassen[2], T. Hvelplund[3] and M.R. Weisbjerg[3]
[1]Department of Large Animal Sciences, Faculty of Life Sciences, University of Copenhagen, Groennegaardsvej 2, 1870 Frederiksberg C., Denmark; jom@life.ku.dk
[2]Department of Genetics and Biotechnology, Faculty of Agricultural Sciences, Aarhus University, P.O. Box 50, 8830 Tjele, Denmark
[3]Department of Animal Health and Bioscience, Faculty of Agricultural Sciences, Aarhus University, P.O. Box 50, 8830 Tjele, Denmark

Introduction

Ruminant methane (CH_4) production has for long been in focus by nutritionists. Five to ten percent of the feed gross energy disappears in the air as methane. It is obvious that prevention of this loss of energy by total oxidation in the metabolism of the carbon to CO_2 or by deposition of the carbon in gain or milk would improve the energy efficiency of milk and beef production. Lately, the importance of the ruminant methane production for the global environment has been highlighted and efforts to quantify and reduce the ruminant methane production have been increased. Experiments with respiration chambers are expensive and time consuming and therefore, other methods for individual measurements of methane production have been developed. The most widely used is the SF_6 method. This method has shown to give reasonable results, but with large variation (Pinares-Patino and Clark, 2008). Moreover, the SF_6 is a strong greenhouse gas and its use has been banned in several countries. This paper presents a novel method that can be used on both individual animals and on groups of animals, as example a herd of cattle in a livestock building. In the paper the results of measurements on individual animals are shown. The method is based on simultaneous measurements of CO_2 and CH_4 in air.

Material and method

The method uses CO_2 as internal marker. The CO_2 excretion can be calculated using different data relating the heat production by an animal to the CO_2 excretion. When measuring in a livestock building with several cows, then the CO_2 production can be calculated as 185 l CO_2 per heat producing unit (1 HPU = 1000 W/h) and the number of HPU can be calculated from the body weight (BW) and milk production of the cows. When measuring individual animals the heat production can be estimated as above or from information on individual intake of metabolizable energy (ME) and the milk production and BW gain. The heat produced is the ME intake minus the energy in milk and BW gain and one litre of carbon dioxide is formed per 21.75 kJ heat produced (Madsen *et al.*, 2010). The values for methane production in Figure 1 are calculated according to this method and the methane to carbon dioxide ratio in the breath of the cows was measured when the cows were in an automatic milking system (AMS) to be milked. The numbers are means of the values obtained from 2 to 12 visits in the AMS by 47 Holstein cows. The concentration of CO_2 and CH_4 in the breath was measured using a portable equipment, GASMET 4030 (Gasmet Technologies Oy, Pulttitie 8A, FI-00880 Helsinki, Finland). Almost the same equipment (GASMET 4000) has been tested by Teye *et al.* (2009), and shows precise measurements of different gases.

Results and discussion

The daily milk production by the cows varied between 13 and 43 kg ECM, the feed intake between 103 and 326 MJ ME, and there was a large difference between cows with respect to mobilization and deposition. The cows produced on average 2.5 litres of methane per MJ ME eaten equivalent to 26 litres per kg dry matter or 5.2% of the gross energy excreted as methane. Large individual

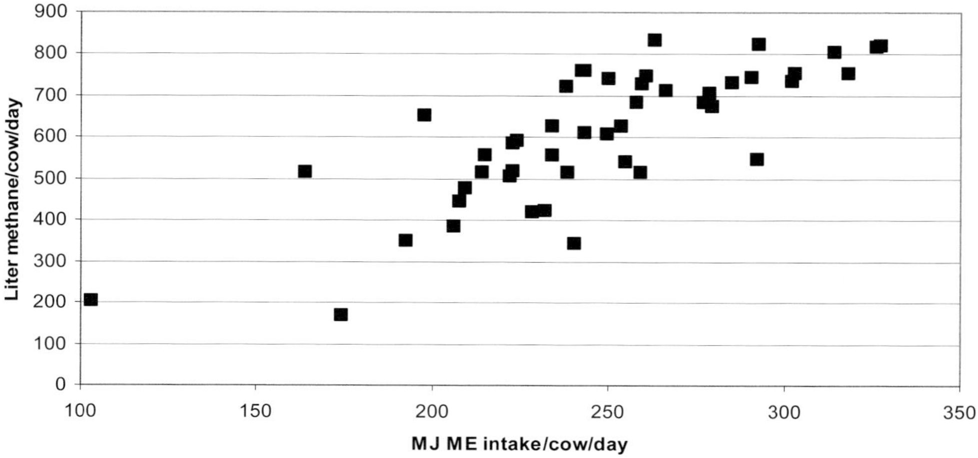

Figure 1. Relation between energy intake and methane production for 47 Holstein cows. Each mark represents one cow.

variation between cows with the same feed intake and production existed with respect to the estimated methane production. This variation is of interest and need to be studied in detail as it may be seen as an opportunity for selection of low methane producing cows. Selection and breeding for low methane production is only an option when a fast, easy and cheap method for measuring the methane production exists as measurements on large number of animals are needed.

References

Madsen, J., B.S. Bjerg, T. Hvelplund, M.R. Weisbjerg and P. Lund, 2010. Methane and carbon dioxide ratio in excreted air for quantification of the methane production from ruminants. Livestock Sci. 129, 223-227.

Pinares-Patino, C.S. and H. Clark, 2008. Reliability of the sulphur hexafluoride tracer technique for methane emission measurement from individual animals: an overview. Aust. J. Exp. Agric. 48, 223-229.

Teye, F.K., E. Alkkiomaki, A. Simojoki, M. Pastell and J. Ahokas, 2009. Instrumentation, measurement and performance of three air quality measurement systems for dairy buildings. Appl. Eng. Agric. 25, 247-256.

Glutamine: precursor or nitrogen donor for the synthesis of citrulline?

J.C. Marini, I.C. Didelija, L. Castillo and B. Lee
USDA ARS CNRC at Baylor College of Medicine, 1100 Bates St., Houston TX 77030, USA

Introduction

Although glutamine (Gln) is considered the main precursor for citrulline (Cit) synthesis, the current literature does not differentiate between the contribution of Gln carbon skeleton versus nonspecific nitrogen (e.g. ammonia) and carbon derived from Gln oxidation. Most of the work to show a precursor-product relationship between Gln and Cit has been done using 2-^{15}N Gln. It is not clear if the ^{15}N follows the carbon skeleton of Gln, and thus indicates a true precursorship for Cit synthesis, or if the ^{15}N is incorporated into Cit during its synthesis by transamination or with the carbamylphosphate group. The aims of the present study were to determine the utilization of (1) Gln nitrogen and (2) dietary precursors for the synthesis of Cit.

Material and methods

Young adult ICR mice (n=10/infusion) were used in the infusion protocols described in Table 1. The Nitrogen Arm of the study was designed to determine the incorporation of both dietary Gln nitrogen groups (and ammonium nitrogen) into Cit. The Carbon Arm was designed to determine the dietary precursors for Cit synthesis. A mixture containing all of the amino acids and glucose was infused intragastrically to achieve a fed steady state. The tracer infusions lasted 4 h and single blood samples were taken at the end from the submandibular bundle.

Amino acid enrichments were determined as their dansyl derivatives by LC MS/MS. By changing the fragmentation collision energy, different Cit fragments were obtained and thus we were able to determine the enrichment of each of the three nitrogen groups of this molecule.

The rate of appearance (Ra) of the amino acids of interest was calculated by the isotopic dilution in plasma of the i.v. infused tracer. The rates of conversion were calculated based on Thompson *et al.* (1989).

Table 1. Tracers infused continuously in mice for 4h to determine the origin of the nitrogen and carbon moieties of citrulline.

	Intragastric		Intravenous
Nitrogen Arm[1]			
Inf. 1 Amino Gln	2-^{15}N glutamine	D_5 glutamine[2]	$^{13}C^{18}O$ urea
Inf. 2 Amido Gln	5-^{15}N glutamine	D_5 glutamine	$^{13}C^{18}O$ urea
Inf. 3 Ammonium	$^{15}NH_4$ acetate	D_5 glutamine	$^{13}C^{18}O$ urea
Carbon Arm			
Inf. 4 Gln	U-$^{13}C_5$ glutamine		^{15}N(ureido) citrulline
Inf. 5 Arg	U-$^{13}C_6$ arginine		^{15}N(ureido) citrulline
Inf. 6 Pro	U-$^{13}C_5$ proline		^{15}N(ureido) citrulline

[1] Nitrogen Arm also included a 5-^{13}C 4,4,5,5 D_4 citrulline intravenous infusion a week prior to the infusion described in the table.

[2] D_5 glutamine is 2,3,3,4,4 D_5 glutamine.

Results and discussion

The RaCit for the Nitrogen Arm was not different among the three infusion protocols (146 μmol/kg/h; P=0.56). The infusion of 2-^{15}N Gln resulted in a similar contribution to the three possible nitrogen groups of Cit (α, δ and ureido; Figure 1A). Both 5-^{15}N Gln and ^{15}N ammonium infusions resulted in a higher enrichment of the ureido group, a smaller enrichment of the α-N and no measurable labeling in the δ position of Cit (not different from zero, P>0.15; Figure 1B). The co-infusion of D_5 Gln resulted in a modest incorporation of the D_5 label into Cit. This indicates that ^{15}N labelled Gln is a poorly suited tracer to determine a precursor-product relationship between Gln and Cit.

The RaCit for the Carbon Arm was not different among the three infusion protocols (165 μmol/kg/h; P=0.86). Dietary arginine (Arg) was the main precursor for circulating Cit (65.8 μmol/kg/h), followed by proline (Pro; 5.5 μmol/kg/h) and Gln (0.7 μmol/kg/h). Dietary Arg was the precursor for almost 40% of the RaCit, while Pro contributed only 3.3% and Gln 0.4%. The contribution of carbon skeleton Gln to circulating Cit in this arm of the study, determined with U-$^{13}C_5$ Gln, was statistically higher (P<0.0001) than in the N arm determined with D_5 Gln (0.4 vs. 0.2% of RaCit).

In conclusion ^{15}N Gln can not be employed to derive precursor-product relationships between Gln and Cit, because it does not trace the carbon skeleton of the Gln. Utilizing U-^{13}C tracers we determined that dietary Arg is the main precursor for Cit synthesis. Our results show that 'preformed' ornithine is the preferred precursor for Cit synthesis, rather than 'de novo' ornithine synthesized from Gln or Pro by action of ornithine aminotransferase.

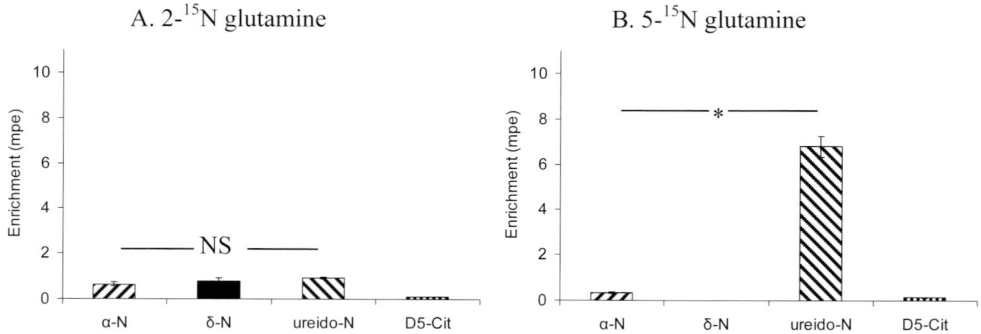

*Figure 1. Contribution of 2- and 5-^{15}N glutamine to the synthesis of citrulline (panels A and B, respectively). The incorporation of the ^{15}N label into each of the 3 positions of citrulline was determined by LC MS/MS. D_5 citrulline was co-infused with the ^{15}N tracers. NS, not significant; * P<0.001; n=10.*

References

Thompson, G.N., P.J. Pacy, H. Merritt, G.C. Ford, M.A. Read, K.N. Cheng and D. Halliday, 1989. Rapid measurement of whole body and forearm protein turnover using a [$^2H_5$]phenylalanine model. Am. J. Physiol. 256, E631-639.

Does endogenous nitrogen contribute to over-estimate bacterial duodenal flow in ruminant estimated by ^{15}N dilution technique?

D.R. Ouellet, D. Valkeners and H. Lapierre
Dairy and Swine R&D Centre, Agriculture and Agri-Food Canada, Sherbrooke, QC, Canada

Introduction

Estimation of duodenal nitrogen (N) flow is essential to adequately estimate protein supply and formulate diets of ruminant. Protein fractions flowing at the duodenum are: bacteria, rumen undegraded feed protein and endogenous N secretions (ENS). Several techniques were developed to estimate duodenal flow of bacterial N. The rumen continuous infusion of ^{15}N-ammonium salt as a marker gives reliable estimates and is often considered the gold-standard method (Broderick and Merchen, 1992). This technique, on the other hand, usually yields higher values than methods based on total purine base, purine derivatives urinary excretion (Reynal *et al.*, 2005) or amino acid profile (Siddons *et al.*, 1982). The basic assumption of the ^{15}N labelling, that all the ^{15}N in the duodenal digesta is of bacterial origin, however, needs to be revalidated. Because samples are harvested after 3-4 d of ^{15}N rumen infusion, the ENS becomes labelled over time: therefore, their release into the forestomach compartment will contribute to ^{15}N flow at the duodenum. As the ^{15}N dilution technique attributes all the ^{15}N in the duodenal flow to bacteria, not accounting for the ^{15}N from ENS may result in an over-estimation of bacterial N. This project was conducted to evaluate the potential contribution of free ENS (not incorporated into bacteria) flowing at the duodenum during a rumen continuous infusion of ^{15}N-ammonium and subsequent over-estimation of the bacterial N flow.

Material and methods

Four cows (557±8 kg) fitted with rumen fistula and a closed-T duodenal cannula were fed 17 kg/d, 12 times a day, a total mixed diet (16% CP and 30% NDF) containing 49% grass silage and 51% of concentrate. The concentrate contained 0.16% of Cr_2O_3 to estimate duodenal DM flow. On day 15, background samples of duodenal digesta, milk and rumen mucosa were taken; then cows were infused intra-ruminally for 96 hours a ^{15}N-ammonium sulfate (98% ape) solution (0.96 g/l) at the rate of 45 ml/h. On day 18 and 19, samples of duodenal digesta (09:00, 11:00, 13:00, and 15:00 h on day 18 and 08:00, 10:00, 12:00, and 14:00 h on day 19) and rumen biopsies (09:00 and 14:00 h on day 18 and 19) were taken daily whereas milk was sampled at milking (08:00 and 20:30 h on day 18., and 08:00 h on day 19). Using the same cows and treatment, the free ENS flow had been determined previously by isotopic dilution of ^{15}N-leucine as described by Ouellet *et al.* (2002). Total duodenal N and ammonia-N in digesta were obtained by Kjeldahl method. The Cr in the digesta was determined by atomic absorption. Isotopic enrichment (IE) was determined using a combustion elemental analyzer coupled with an isotope ratio mass spectrometer. The ammonia in digesta was isolated by diffusion procedure (Brooks *et al.*, 1989) before IE determination. Duodenal bacterial N flows were calculated as follows:

$$\text{Bacterial N flow} = (F_{DUO} \times APE_{DUO} - F_{AMM} \times APE_{AMM}) / APE_{BAC} \quad (1)$$

$$\text{Bacterial } N_{corrected} = (F_{DUO} \times APE_{DUO} - F_{AMM} \times APE_{AMM} - F_{ENS} \times APE_{ENS}) / APE_{BAC} \quad (2)$$

Where F_{DUO}, F_{AMM} and F_{ENS} = duodenal N flow of total N, ammonia, and ENS (estimated previously) and APE_{DUO}, APE_{AMM}, APE_{BAC}, and APE_{ENS} = ^{15}N IE (atom % excess) of the duodenal digesta, ammonia, bacteria, and ENS (rumen mucosa biopsy or alternatively milk protein), respectively. Paired *t*-test comparisons were carried out using the SAS to test for differences in the bacterial flows determined using the standard method and the corrected flows.

Results and discussion

The IE (atom % excess) averaged (SD): 0.0252 (0.0049) for the total duodenal digesta; 0.0373 (0.0074) for the bacteria; 0.0313 (0.0048) for the ammonia; 0.0060 (0.0004) for the rumen mucosa and 0.0186 (0.0026) for the milk protein. The bacteria represented 64.6% (SD=1.6) of total duodenal N flow. Correction of bacteria N flow at the duodenum using the rumen mucosa and milk APE as endogenous pool precursor, respectively, marginally reduced the bacteria flow by 1.1% and 3.7%, respectively. The difference between the standard method and the corrected flow using rumen IE as the precursor pool (mean = 3.9, SD=0.7 gN/d) was greater (P=0.002) than zero. A greater difference was observed for the standard method and the corrected flow using milk as the precursor pool (mean = 12.9, SD=2.2 gN/d, P=0.001).

In conclusion, the contribution of endogenous ^{15}N flow at the duodenum during the determination of duodenal bacterial N flow using the ^{15}N dilution technique does have mathematical significant impact on the estimation of bacterial N flow. However, minimum and maximum differences of estimates were -3 and -16 g of bacterial N per day which may not have a significant biological relevance.

Table 1. Bacterial N flow at the duodenum corrected or not for endogenous N contribution.

Item (g/d)	Number of cows			
	318	327	328	5,313
Total duodenal N flow	541.9	569.8	533.9	584.9
Ammonia-N flow	19.3	26.5	17.2	20.0
ENS flow	25.6	22.0	22.6	22.8
Bacterial N flow	346.5	360.6	357.3	375.4
Bacterial N flow corrected for ENS[rumen mucosa]	343.0	355.8	354.1	371.4
Bacterial N flow corrected for ENS[milk]	331.3	346.6	345.4	365.1

References

Broderick, G.A. and N.R. Merchen, 1992. Markers for quantifying microbial protein synthesis in the rumen. J. Dairy Sci. 75, 2618-2632.

Brooks, P.D., J.M. Stark, B.B. McInteer and T. Preston, 1989. Diffusion method to prepare soil extracts for automated nitrogen-15 analysis. Soil Sci. Soc. Am. J. 53, 1707-1711.

Ouellet, D. R., M. Demers, G. Zuur, G.E. Lobley, J.R. Seoane, J.V. Nolan and H. Lapierre, 2002. Effect of dietary fiber on endogenous nitrogen flows in lactating dairy cows. J. Dairy Sci. 85, 3013-3025.

Reynal, S.M., G.A. Broderick and C. Bearzi, 2005. Comparison of four markers for quantifying microbial protein flow from the rumen of lactating dairy cows. J. Dairy Sci. 88, 4065-4082.

Siddons, R.C., D.E. Beever and J.V. Nolan, 1982. A comparison of methods for the estimation of microbial nitrogen in duodenal digesta of sheep. Br. J. Nutr. 48, 377-389.

Plasma acetate and glucose turnover rates in sheep exposed to cold

M. Al-Mamun[1,2], *M. Sawada*[1] *and H. Sano*[1]
[1]*Dept. of Animal Sciences, Faculty of Agriculture, Iwate University, Japan*
[2]*Dept. of Animal Nutrition, Bangladesh Agricultural University, Mymensingh-2202, Bangladesh*

Introduction

Cold exposure is one of the stresses that results in a variety of negative effects on productivity of livestock through modified digestive, metabolic, and endocrine functions (Sano *et al.*, 2007). Acetate is the most important metabolite being involved in fatty acid and carbohydrate metabolism. Acetate is one of the energy sources in ruminants. Cold exposure enhanced blood glucose metabolism in sheep (Tsuda *et al.*, 1984). There is little information available about the simultaneous study of plasma acetate and glucose metabolism in sheep during cold environment. Therefore, the present study was conducted to determine the plasma acetate and glucose metabolism in sheep simultaneously using stable isotope dilution techniques.

Materials and methods

The experiment was performed using crossbred (Corriedale × Suffolk) sheep (n=5; 3 male and 2 female; approximately 2 yr old, 38±3 kg of initial BW). The animals were offered 92g/kg$^{0.75}$/d of mixed hay (ME 1.79 kcal/g, CP 12.5%) twice daily (08:30 and 20:30) with *ad libitum* water access. The animals were kept in individual pens for a preliminary period of 14 d. Then the sheep were moved to a controlled, thermoneutral (TN, 23±1 °C) environment for 7 d and then temperature was adjusted to 2-4±1 °C cold exposure (CE) for 7 d with relative humidity 70% and lighting present from 08:00 to 22:00.

On the d 21 of TN and on the d 7 of CE two isotope dilution studies using [1-^{13}C]Na-acetate and [U-^{13}C]glucose isotope labelled tracers were performed simultaneously as a primed continuous infusion for 4 h. Blood sampling was performed into sodium heparinised centrifuge tubes immediately before and every 30 min interval during the last 2 h of the isotope infusion. Blood samples were centrifuged at 10,000×*g* for 10 min at 2 °C. Plasma acetate and glucose turnover rates were calculated from the isotopic enrichments of [1-^{13}C]acetate and [U-^{13}C]glucose, respectively using GC/MS according to the methods described before (Al-Mamun *et al.*, 2009; Sano *et al.*, 2007). The statistical analyses were performed using MIXED Procedures of SAS (1996).

Results and discussion

Plasma acetate concentration and turnover rate were numerically higher (P=0.11 & P=0.31, respectively) during CE than during TN environment (Table 1). Plasma glucose concentration was numerically higher (P=0.28), and turnover rate was significantly higher (P=0.02) during CE than during TN environment. Plasma glucose concentration and turnover rate were comparable to the previous findings (Sano *et al.*, 2007). It could be concluded that under present experimental condition CE influenced both plasma acetate and glucose metabolism in sheep.

Table 1. Effect of cold exposure on plasma acetate and glucose concentration, and turnover rates in sheep.

	Treatment[1]		SEM	*P*-value
	TN	CE		
No. of sheep	5	5		
Plasma acetate				
Concentration (mmol/l)	0.9	1.0	0.02	0.11
Turnover rate (mmol/kg$^{0.75}$/d)	151	190	24	0.31
Plasma glucose				
Concentration (mmol/l)	3.3	3.7	0.1	0.28
Turnover rate (mmol/kg$^{0.75}$/d)	38.4	58.4	6.4	0.02

[1] TN, thermoneutral environment (23±1 °C), CE, cold exposure (2-4±1 °C).

References

Al-Mamun, M., K. Goto, S. Chiba and H. Sano, 2009. Responses of plasma acetate metabolism to hop (*Humulus lupulus* L.) in sheep. Int. J. Biol. Sci. 5, 287-292.

Sano, H., H. Sawada, A. Takenami, S. Oda and M. Al-Mamun, 2007. Effects of dietary energy intake and cold exposure on kinetics of plasma glucose metabolism in sheep. J. Anim. Physiol. Anim. Nutr. 91, 1-5.

SAS/STAT® Software, 1996. Changes and Enhancements through Release 6.11. SAS Inst. Inc., Cary, NC, USA.

Tsuda, T., K. Ambo, Y. Shoji, M. Fujita, and K. Sunagawa, 1984. Distribution of energy source expenditure in warm- and cold exposed sheep. Can. J. Anim. Sci. 64 (Suppl.), 265-266.

Plant-species characteristic stable carbon isotope ratios in fatty acids of seed oils: basis for a new tool in ruminant nutrition research?

Eva Katharina Richter[1], Jorge E. Spangenberg[2], Michael Kreuzer[1] and Florian Leiber[1]
[1]ETH Zurich, Institute of Plant, Animal and Agroecosystem Sciences, 8092 Zurich, Switzerland; richtere@ethz.ch
[2]Universite de Lausanne, Institut de Mineralogie et Geochimie, BFSH-2, 1015 Lausanne, Switzerland

Introduction

The digestive and metabolic modification of fatty acids in ruminants has a significant impact on the nutritive quality of lipids in milk and meat and is therefore an important issue of research in ruminant physiology (Lourenço *et al.*, 2008). Studies using labelled nutrients as ruminant metabolic tracers are very expensive. An ongoing study aims to explore the potential use of natural differences in stable isotope composition of individual fatty acids (e.g. Kelly *et al.*, 1997; Spangenberg *et al.*, 1998) as tracers for transformation and fate of fatty acids during digestive, metabolic or technological processes. The major goal of the present study was to investigate the $\delta^{13}C$-variations of individual fatty acids of various seed-producing C_3 plant species suitable for animal feeding. We examined also the discriminatory potential of δ^2H, $\delta^{13}C$ and $\delta^{18}O$ of bulk oil and their fatty acid profiles. The studied plant species included rape *(Brassica napus*; n=25*)*, flax (*Linum usitatissimum;* n=3), false flax (*Camelina sativa;* n=1), poppy (*Papaver somniferum;* n=4), safflower (*Carthamus tinctorius;* n=1), and sunflower (*Helianthus annuus;* n=1).

Material and methods

Seed oils were characterized by fatty acid composition as well as bulk and compound-specific isotopic analyses. After fatty acid saponification and derivatization with methanolic sodium hydroxide and methanolic boron trifluoride, the fatty acid composition was assessed by GC-FID equipped with a 30 m Supelcowax-10 column. For the determination of bulk carbon, oxygen and hydrogen composition ($\delta^{13}C_{bulk}$ values in ‰ VPDB, δ^2H_{bulk} and $\delta^{18}O_{bulk}$ values in ‰ VSMOW), a continuous He-flow combustion and high-temperature conversion elemental analyzer-isotope ratio mass spectrometry (EA-IRMS and TC-EA/IRMS) was used. The $\delta^{13}C$ values of individual fatty acids ($\delta^{13}C_{FA}$) were determined using gas chromatography-combustion-isotope ratio mass spectrometry (GC-C/IRMS). Statistical analysis was performed using SAS software. Average isotopic values of bulk oils and individual fatty acids were subjected to analysis of variance using the general linear model (GLM procedure), considering plant species and individual fatty acids as fixed effects.

Results and discussion

The mean $\delta^{13}C$ values of the main fatty acids in the studied oils ranged between -34.0‰ and -26.1‰. Differences for the four main fatty acids were found within all species; those for rape being most prominent ($P<0.001$). Significant differences between rape, flax and poppy occurred not only for C18:3*n*-3 but also for C16:0 ($P<0.001$). In all species, C18:2*n*-6 was slightly discriminated in $^{13}C/^{12}C$ compared to C18:1*n*-9. For C18:3*n*-3 lower $\delta^{13}C$-values were observed compared to C18:2*n*-6 in oils from rape (-34.0‰) and flax (-32.0‰) but not from poppy oil, which had the highest measured $\delta^{13}C$-value (-26.1‰) for C18:3*n*-3 (Figure 1).

The $\delta^{13}C_{bulk}$ values were typical for C_3 plants and differed significantly ($P<0.001$) between rape, poppy and flax. Rape and flax oil had lower values (-30.2±0.3‰ and -29.9±0.1‰ respectively) than poppy oil (28.8±0.9‰). For the $\delta^{18}O$ values no significant differences between plant species were observed. The safflower oil had the highest $\delta^{18}O$ value (-26.0‰). The δ^2H means of the bulk oils from rape (-175.2±9.5‰) differed from that originating from flax (-167.3±3.6‰) and poppy

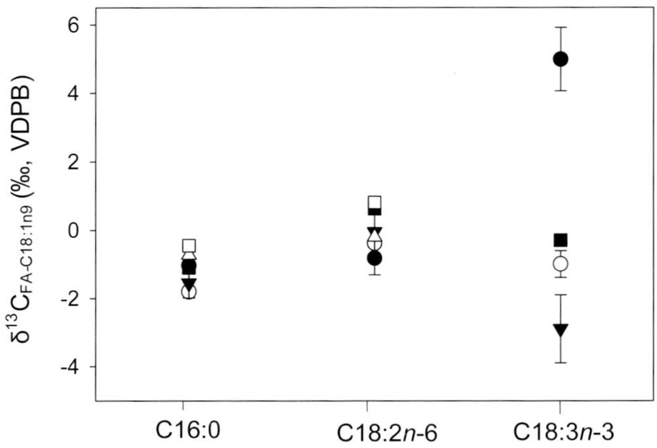

Figure 1. Average differences between $\delta^{13}C$ values (in ‰ VPDB) of fatty acids in flax, poppy, rape, sunflower, false flax and safflower oil relative to $\delta^{13}C$ of C18:1 in the corresponding oils. Vertical bars give the standard deviation of the individual means. ○, flax; ●, poppy; ▼, rape; Δ, sunflower; ■, false flax; □, safflower.

(-160.8±5.6‰) ($P<0.01$), and the overall lowest value was found for the single sample of sunflower bulk oil (-185.0‰).

We have confirmed that carbon isotope composition of individual fatty acids serve to distinguish between different C_3 plant oils. The $\delta^{13}C$ values of C16:0 and C18:3*n*-3 differed significantly between and within species, and from the other fatty acids. In the next step we will explore if the natural $\delta^{13}C_{FA}$ differences in the diet are preserved in some fatty acids generated in the rumen like vaccenic acid and conjugated linoleic acids. If that is possible, this isotopic approach could enable to follow derivatization and fractionation of feed fatty acids in the ruminants' digestive and endogenous metabolism. Further information will provide the fatty acid composition and the bulk $\delta^{13}C$, $\delta^{18}O$ and $\delta^{2}H$ values.

References

Kelly, S., M. Parker, M. Sharman, J. Dennis and I. Goodall, 1997. Assessing the authenticity of single oils using fatty acid stable carbon ($^{13}C/^{12}C$) seed vegetable isotope ratios. Food Chem. 59, 181-186.

Lourenço, M., G. Van Ranst, B. Vlaeminck, S. De Smet and V. Fievez, 2008. Influence of different dietary forages on the fatty acid composition of rumen digesta as well as ruminant meat and milk. Anim. Feed Sci. Technol. 145, 418-437.

Spangenberg, J.E., S.A. Macko and J. Hunziker, 1998. Characterization of olive oil by carbon isotope analysis of individual fatty acids: Implications for authentication. J. Agric. Food Chem. 46, 4179-4148.

Ammonia-N utilization by ruminal bacteria and protozoa from sheep fed diets differing in forage:concentrate ratio and forage type

C. Saro, M.J. Ranilla, M.L. Tejido and M.D. Carro
Instituto de Ganadería de Montaña (CSIC-ULE), Finca Marzanas s/n. 24346 Grulleros, León, Spain

Introduction

Some ruminal microbes are capable of using ammonia as the only nitrogen (N) source, whereas others appear to prefer preformed amino acids or peptides. The degree to which ammonia may constitute a starting point for synthesis of microbial N compounds is expected to vary with the nature of the diet and the microbial fraction considered (Carro and Miller, 1999). The aim of this work was to determine the uptake of ammonia-N by ruminal bacteria and protozoa as affected by diet characteristics.

Materials and methods

Six ruminally cannulated Merino sheep (59.0±4.5 kg body weight) were used in a partially replicated Latin square design with four 16-day periods. The four experimental diets had forage:concentrate ratios (F:C; dry matter basis) of either 70:30 (HF) or 30:70 (HC) being the forage source alfalfa hay (A) or grass hay (G). Diets were offered to the animals twice daily (08:00 and 20:00 h) at a daily rate of 56 g dry matter/kg body weight$^{0.75}$ to minimize feed selection. $^{15}NH_4Cl$ (10% atom excess) was continuously infused into the rumen for 5 days before sampling. On day 16 of each period, about 500 g of rumen content were taken from each sheep at 0, 4, and 8 h after the morning feeding. Rumen contents were squeezed through 4 layers of cheesecloth and the solid digesta was mixed with an equal volume of Coleman solution, homogenized and squeezed again. Both filtrates were pooled and half of the final mixture was used to isolate liquid-associated bacteria (LAB) by differential centrifugation as described by Ranilla and Carro (2003). The rest of the filtrate was transferred to a separation funnel, 5 g of glucose were added, and the mixture was allowed to flocculate for 1 h at 39 °C. The layer containing the protozoa was drawn off, filtered through a 200 μm nylon cloth, and repeatedly washed with Coleman buffer to remove the remaining plant material. Finally, the filtrate was filtered through a 10 μm nylon cloth and washed thoroughly to remove contaminating bacteria. The solid digesta was treated with saline solution (0.9%) containing 0.1% methylcellulose before solid-associated bacteria (SAB) isolation (Ranilla and Carro, 2003). Microbial isolates (MI) were pooled by sheep within each period, lyophilized, grounded to a fine powder, and analysed for N and ^{15}N enrichment. At each sampling time, 50 ml of ruminal fluid were taken, acidified with 5 ml of H_2SO_4 (20%), pooled by sheep and analyzed for ^{15}N enrichment.

Data were analyzed as a mixed model with repeated measures using the PROC MIXED of SAS (SAS Inst. Inc., Cary, NC). The statistical model included MI (SAB, LAB and protozoa), F:C, FOR, period, MI×F:C, MI×FOR, F:C×FOR, MI×F:C×FOR as fixed effects, and sheep as a random effect. Effects were declared significant at $P<0.05$.

Results and discussion

The N content was higher ($P<0.001$) in SAB compared with LAB and protozoa (Table 1). However, MI × F:C and MI × FOR interactions ($P<0.001$) were detected, indicating that N differences among MI were affected by the diet.

The greater ($P<0.001$) ^{15}N enrichment of LAB compared with SAB observed in our study is in accordance with previous *in vivo* (Martin *et al.*, 1994; Reynal *et al.*, 2005) and *in vitro* (Carro and Miller, 1999) studies. Differences in ^{15}N enrichment could be attributed to differential ability of

Table 1. Nitrogen content, ^{15}N enrichment and NH_3-^{15}N capture in ruminal protozoa (PR), solid-associated bacteria (SAB) and liquid-associated bacteria (LAB)[1].

	Nitrogen content (mg/g dry matter)			^{15}N enrichment (atoms % excess)			NH_3-^{15}N capture (%)		
Diet	PR	SAB	LAB	PR	SAB	LAB	PR	SAB	LAB
HFA	63.5	67.9	64.9	0.0566	0.0646	0.0718	30.0	34.4	38.5
HCA	68.5	63.6	60.3	0.0426	0.0570	0.0580	31.3	41.8	42.4
HFG	55.2	69.1	67.6	0.1046	0.1170	0.1323	42.7	47.9	54.3
HCG	62.2	69.2	62.5	0.0636	0.0757	0.0838	35.7	42.7	47.1
Mean values	62.4	67.5	63.8	0.0669	0.0786	0.0865	34.9	41.7	45.6
SEM	1.45			0.00378			1.67		
$P = {}^1$	MI, MI×F:C, MI×FOR			MI, F:C, FOR, F:C×FOR			MI, FOR, F:C×FOR		

[1] MI: microbial isolate ($P<0.001$); FOR = forage ($P<0.001$); F:C = forage:concentrate ratio ($P<0.001$); MI×F:C ($P<0.001$); MI×FOR ($P<0.001$); F:C ×FOR ($P<0.001$).

bacterial groups to use free ammonia as a N source. A higher proportion of species that preferentially incorporate amino acids and peptides in the SAB might explain the different ^{15}N enrichment in the two MI. In fact, the incorporation of N from the ammonia-N fraction was greater ($P<0.001$) in LAB than in SAB for all the diets (Table 1). Whereas LAB are located in free suspension or loosely associated with fibers, SAB are located bound to plant surfaces, where the actual ammonia concentration may be lower than in rumen fluid and could fluctuate widely.

In agreement with previous results (Martin *et al.*, 1994; Reynal *et al.*, 2005), protozoa showed lower ($P<0.001$) ^{15}N enrichment and ammonia-N incorporation compared with LAB and SAB. Protozoa become indirectly enriched with ^{15}N via bacterial predation and their ^{15}N enrichment will be lower than in bacteria as a result of engulfment of unenriched dietary protein (Reynal *et al.*, 2005). The observed differences among MI in ^{15}N enrichment would result in different estimates of microbial N flow to the duodenum depending on the MI used as a reference sample.

Acknowledgements

Financial support from the M.C.Y.T. of Spain (project AGL2004-04755-C02-01) is gratefully acknowledged. C. Saro gratefully acknowledges the receipt of scholarship from the M.E.C. of Spain (AP 2006-03049).

References

Carro, M.D. and E.L. Miller, 1999. Effect of supplementing a fibre basal diet with different nitrogen forms on ruminal fermentation and microbial growth in an *in vitro* semi-continuous culture system (Rusitec). Br. J. Nutr. 82, 149-157.

Martin, C., A.G. Williams and B. Michalet-Doreau, 1994. Isolation and characteristics of the protozoal and bacterial fractions from bovine ruminal contents. J. Anim. Sci. 72, 2962-2968.

Ranilla, M.J. and M.D. Carro, 2003. Diet and procedures used to detach particulate-associated microbes from ruminal digesta influence chemical composition of microbes and estimation of microbial growth in Rusitec fermenters. J. Anim. Sci. 81, 537-544.

Reynal, S.M., G.A. Broderick and C. Bearzi, 2005. Comparison of four markers for quantifying microbial protein flow from the rumen of lactating dairy cows. J. Dairy Sci. 88, 4065-4082.

Precursors for the synthesis of citrulline in mice fed arginine free diets

J.C. Marini, I.C. Didelija, L. Castillo and B. Lee
USDA ARS CNRC at Baylor College of Medicine, 1100 Bates St., Houston TX 77030, USA

Introduction

In our previous study, dietary arginine (Arg) provided the precursor for ~40% of the circulating citrulline (Cit) in adult mice, while dietary proline (Pro) accounted for only 3.4% and the contribution of dietary glutamine (Gln) was negligible (0.4%) (Marini *et al.*, 2010). The aim of the present study was to determine which precursors are utilized for the synthesis of Cit when Arg is absent in the diet.

Material and methods

Young adult ICR mice (n=10/infusion) were used in the infusion protocols described in Table 1. An Arg sufficient (Arg+) or an Arg free (Arg-) diet was infused intragastrically to achieve a fed steady state. The tracer infusions lasted 4 h and single blood samples were taken at the end from the submandibular bundle.

Plasma Cit, Arg, ornithine (Orn) and phenylalanine (Phe) enrichments were determined as their dansyl derivatives by LC MS/MS. Plasma amino acid concentrations were determined by HPLC.

The rate of appearance (Ra) of the amino acids of interest was calculated by the isotopic dilution in plasma of the i.v. infused tracer. First pass extraction of Arg was determined based on the plasma enrichments of the two Arg tracers infused simultaneously i.v. and i.g. The rates of conversion were calculated based on Thompson *et al.* (1989).

Table 1. Infusion protocols followed in conscious mice to determine the contribution of different precursors to the synthesis of citrulline.

	Arg in diet	Intragastric tracer	Intravenous tracers
Infusion 1	+		U-$^{13}C_6$ Arg, D_2 Orn, ^{15}N Cit, D_5 Phe
Infusion 2	-		U-$^{13}C_6$ Arg, D_2 Orn, ^{15}N Cit, D_5 Phe
Infusion 3	-	U-$^{13}C_5$ Pro	D_2 Orn, ^{15}N Cit, D_5 Phe
Infusion 4	-	U-$^{13}C_5$ Gln	D_2 Orn, ^{15}N Cit, D_5 Phe
Infusion 5	+	U-$^{13}C_6$ Arg	1,2$^{13}C_2$ Arg

D_2 Orn is 5,5 D_2 ornithine, ^{15}N Cit is ^{15}N(ureido) citrulline and D_5 Phe is D_5(ring) phenylalanine.

Results and discussion

Feeding an Arg(-) diet did not affect the RaCit in mice (Arg+=164 Arg-= 154 μmol/kg/h; P=0.20) similar to what has been reported previously in other species (Hartman and Prior, 1992; Castillo *et al.*, 1994). No differences in the RaArg (591 vs. 555 μmol/kg/h; P=0.25) or RaPhe (349 vs. 356 μmol/kg/h; P=0.57) were observed for the Arg+ and Arg-, respectively. This indicates that there was no increase in protein degradation as a consequence of feeding a diet devoid of Arg and that dietary Arg was metabolized before reaching the peripheral circulation. This was supported by the observed high first pass extraction of Arg (85.1±0.29%) and the increase in RaOrn (332 vs. 180 μmol/kg/h; P=0.001, Arg+ and Arg-, respectively).

Plasma Orn was the precursor for 30 and 23% of the circulating Cit for the Arg+ and Arg-, respectively (P<0.01). Plasma Arg contributed 24% to the synthesis of Cit independently of the diet (P=0.72). Approximately 70% of this contribution was through plasma Orn and the rest at the site of Cit synthesis. Dietary Pro had a greater (P<0.001) contribution to the synthesis of Cit than Gln (14 and 3% of RaCit, respectively), which agrees with our previous report (Marini *et al.*, 2010). While Pro was utilized as a precursor for plasma Orn, Gln was not, because the labeling of plasma Orn by the Gln tracer was undetectable. Almost 80% of the contribution of dietary Pro to the synthesis of Cit occurred in the small intestine, whereas the remainder was due to the conversion of dietary Pro to Orn elsewhere in the body with the subsequent utilization of this Orn by the small intestine for Cit synthesis.

In conclusion, Cit production was maintained despite the absence of the main dietary precursor (Arg) for its synthesis. Dietary Pro and Gln increased their contribution to Cit synthesis (compared to our previous study), but were not able to completely replace dietary Arg as a precursor. Whereas dietary Pro contributes to the synthesis of circulating Orn, dietary Gln did not, which suggests that Pro is a better precursor for the 'de novo' synthesis of Orn. Plasma Arg was the precursor for 24% of the circulating Cit, regardless of the presence of Arg in the diet. These results support our previous findings (Marini *et al.*, 2010) that 'preformed' Orn is the preferred precursor for Cit synthesis, rather than '*de novo*' Orn generated by action of ornithine amino transferase.

References

Castillo, L., M. Sanchez, T. E. Chapman, A. Ajami, J. F. Burke, and V.R. Young, 1994. The plasma flux and oxidation rate of ornithine adaptively decline with restricted arginine intake. Proc. Natl. Acad. Sci., USA 91, 6393-6397.

Hartman, W.J., and R.L. Prior, 1992. Dietary arginine deficiency alters flux of glutamine and urea cycle intermediates across the portal-drained viscera and liver of rats. J. Nutr. 122, 1472-1482.

Marini JC, I. Cajo, L. Castillo and B. Lee, 2010. Glutamine: precursor or nitrogen donor for the synthesis of citrulline? In: Crovetto, G.M. (ed.) Energy and protein metabolism and nutrition. Wageningen Academic Publishers, Wageningen, the Netherlands, p. 123-124.

Thompson, G.N., P.J. Pacy, H. Merritt, G.C. Ford, M.A. Read, K.N. Cheng and D. Halliday, 1989. Rapid measurement of whole body and forearm protein turnover using a [$^2H_5$]phenylalanine model. Am. J. Physiol. 256, E631-639.

Arginine utilization for citrulline synthesis in arginase II knockout mice

J.C. Marini, I.C. Didelija, L. Castillo and B. Lee
USDA ARS CNRC at Baylor College of Medicine, 1100 Bates St., Houston TX 77030, USA

Introduction

The synthesis of citrulline (Cit) relies mainly on the supply of 'preformed' ornithine (Orn), rather than on the 'de novo' synthesis of Orn from proline (Pro) or glutamine (Gln). Thus utilization of arginine (Arg) for Cit synthesis depends on arginase, which catalyses the hydrolysis of Arg to Orn and urea. The aim of the present study was to determine the precursors utilized for the synthesis of Cit when arginase II, the intestinal isoform, is absent.

Material and methods

Young adult C57BL/6J AII$^{-/-}$ (ARG2; n=17) (Shi *et al.*, 2001) and C57BL/6J (WT; n=19) mice were studied twice. Intragastric infusion of U-$^{13}C_6$ Arg (Inf. 1), U-$^{13}C_5$ Gln (Inf. 2) and U-$^{13}C_5$ Pro (Inf. 3), together with intravenous infusion of 5,5 D_2 Orn and ^{15}N (ureido) Cit (Inf. 1-3) were performed to determine the incorporation of the dietary precursors into Cit. In addition, an intragastric infusion of 1,2$^{13}C_2$ Arg and i.v. infusion of U-$^{13}C_6$ Arg (Inf. 4) was conducted to determine the first pass extraction of Arg.

Plasma Cit, Arg, and Orn enrichments were determined as their dansyl derivatives by LC MS/MS. Plasma amino acid concentrations were determined by HPLC. Arginase II abundances were determined by Western analysis using an anti-arginase II antibody (Santa Cruz Biotechnology Inc., USA).

The rate of appearance (Ra) of the amino acids of interest was calculated by the isotopic dilution in plasma of the i.v. infused tracer. First pass extraction of arginine was determined based on the plasma enrichments of the two arginine tracers infused simultaneously i.v. and i.g. The rates of conversion were calculated based on Thompson *et al.* (1989).

Results and discussion

The lack of arginase II in the small intestine resulted in a reduction in RaCit in ARG2 mice compared to WT animals (121 and 137 μmol/kg/h; $P<0.001$); however, there were no differences ($P=0.326$) in the RaOrn (273 μmol/kg/h). The lower first pass extraction of Arg in ARG2 mice (72 vs. 85%; $P<0.021$), was accompanied by higher RaArg (438 vs. 302 μmol/kg/h; $P<0.002$) and higher plasma Arg concentration (121 vs. 55 μmol/l; $P<0.001$).

Dietary Arg was the main dietary precursor for RaCit in WT mice contributing to the precursor for 33% of the RaCit (44.8 vs. 10 μmol/kg/h; $P<0.001$; Figure 1), which agrees with our previous work on ICR mice (Marini *et al.*, 2010a,b). The lack of ARG2 in the small intestine, and thus the inability to derive Orn from Arg locally by the enterocyte, reduced the total contribution of dietary Arg to ~8% of the circulating Cit. The precursor-intermediate-product model employed to determine the utilization of dietary precursors indicated that no dietary Arg was utilized for Orn production at the site of Cit synthesis in ARG2 mice, consistent with the lack of arginase II shown by western analysis in the intestinal mucosa of these animals.

The contribution to the synthesis of Cit of both dietary Pro and Gln was increased in ARG2 mice, and this occurred at the site of Cit synthesis (Figure 1). Thus the reduction in Orn availability, due

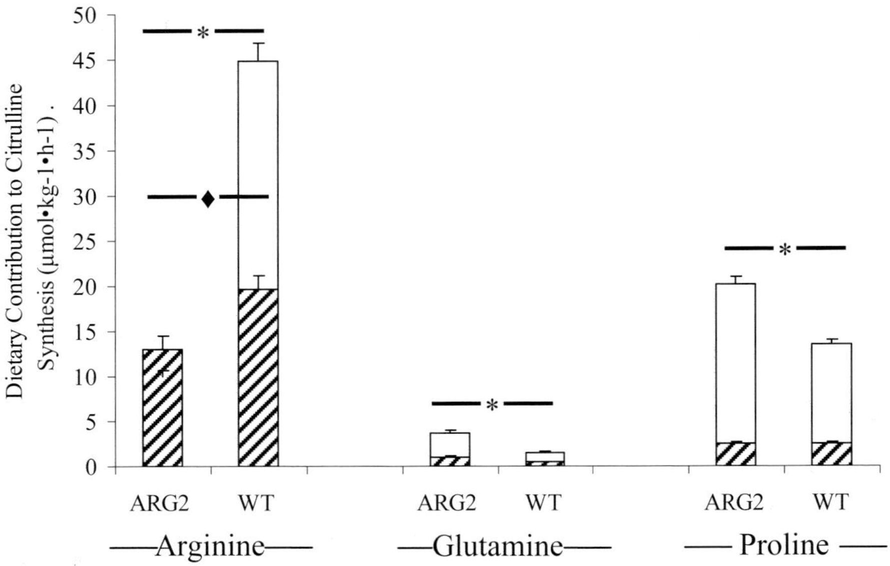

Figure 1. Utilization of dietary arginine, glutamine and proline for the synthesis of citrulline (µmol/kg/h). Bars denote the total contribution of the precursor (P<0.05), shaded portion of the bar the contribution of the precursor through plasma ornithine (♦ P<0.05), and by difference the contribution of the precursor at the site of citrulline synthesis.*

to the lack of ARG2 in enterocytic mitochondria, increased the '*de novo*' synthesis of Orn from Gln and Pro and its utilization for Cit synthesis.

In conclusion, these observations support our previous conclusions (Marini *et al.*, 2010a) that 'preformed' Orn (either plasma Orn or Orn from Arg hydrolyzed locally in the small intestine) is the preferred precursor for the Cit synthesis, rather than '*de novo*' Orn produced from Gln or Pro by action of ornithine aminotransferase.

References

Marini J.C., I. Cajo, L. Castillo and B. Lee, 2010a. Glutamine: precursor or nitrogen donor for the synthesis of citrulline? In: Crovetto, G.M. (ed.) Energy and protein metabolism and nutrition. Wageningen Academic Publishers, Wageningen, the Netherlands, p. 123-124.

Marini J.C., I. Cajo, L. Castillo and B. Lee, 2010b. Precursors for the synthesis of citrulline in mice fed arginine free diets. In: Crovetto, G.M. (ed.) Energy and protein metabolism and nutrition. Wageningen Academic Publishers, Wageningen, the Netherlands, p. 133-134.

Shi, O., S.M. Morris, Jr., H. Zoghbi, C.W. Porter, and W.E. O'Brien. 2001. Generation of a mouse model for arginase II deficiency by targeted disruption of the arginase II gene. Mol Cell Biol 21, 811-813.

Thompson, G.N., P.J. Pacy, H. Merritt, G.C. Ford, M.A. Read, K.N. Cheng, and D. Halliday, 1989. Rapid measurement of whole body and forearm protein turnover using a [$^2H_5$]phenylalanine model. Am J Physiol 256, E631-639.

A large supply of phenylalanine is not oxidised by the mammary gland of dairy cows

S. Lemosquet[1]*, G.E. Lobley*[2]*, R. Koopman*[3]*, L.J.C. van Loon*[3]*, A.K. Kies*[4] *and H. Lapierre*[5]
[1]*INRA UMR1080 Dairy Production, Saint-Gilles, France*
[2]*Rowett Institute of Nutrition and Health, University of Aberdeen, United Kingdom*
[3]*Department of Human Movement Sciences, NUTRIM, Maastricht University, the Netherlands*
[4]*DSM Food Specialities, R&D, Biochemistry and Nutrition Department, Delft, the Netherlands*
[5]*Agriculture and Agri-Food Canada, Sherbrooke, QC, Canada*

Introduction

In the mammary gland (MG) of small ruminants certain amino acids (AA) are reported to be quantitatively transferred into milk protein on a net basis, without detectable oxidation: these AA are classified in group 1 (Mepham, 1982). Phenylalanine (Phe) and its first metabolite, tyrosine (Tyr), are included in this group (as Phe+Tyr). Although Phe can be converted to Tyr within the MG no appearance of $^{14}CO_2$ was detected across the perfused isolated udders of two sheep receiving [^{14}C-U]-L-Phe, suggesting no oxidation of Phe or Tyr (Verbeke *et al.*, 1972). This could be different in cows where the degree of conversion of Phe to Tyr may be higher (see Verbeke *et al.*, 1972). Oxidation of Phe+Tyr across the MG would preclude estimation of mammary plasma flow by the Fick principle as this assumes that net uptake of these AA is quantitatively transferred into milk protein. We hypothesized no oxidation of Phe+Tyr across the MG of dairy cows. A first objective was to ascertain if Phe (plus Tyr) was oxidized across the MG during infusion of large amounts of labelled Phe. A second objective was to quantify the mammary contribution of Phe to Tyr synthesis.

Material and methods

To produce milk with high enrichment of ^{13}C-Phe for a human study (Van Loon *et al.*, 2009), two Holstein dairy cows (23.3 kg/d dry matter (DM) intake; 49.9 kg/d of milk at 27.7 g/kg of true protein content; 34 days in milk) received a diet based on maize silage (Van Loon *et al.*, 2009). They received a large jugular infusion of L[1-^{13}C]Phe (100 g/d) over 2 days while the diet was estimated to supply 125 g/d of intestinal Phe. The cows were equipped with catheters in the tail artery and in the milk vein of the left udder. Blood samples were collected hourly during 3 periods: before the start of the infusion (n=3) and from 6 to 11 h (n=6) and 30 to 35 h (n=6) of the infusion period. Concentrations of CO_2 and isotopic enrichments (IE) were determined in triplicate on blood gas analyser and on a isotope ratio mass spectrometer, respectively. Blood samples (1 ml) were injected into evacuated vacutainers containing 1 ml of lactic acid, immediately mixed and kept at room temperature until analysis. Concentrations and IE of Phe and Tyr were measured on individual plasma samples; IE in milk were determined on milk samples collected at the milkings taken 12, 24 and 36 h after initiation of infusion; analyses were performed by gas chromatography-mass spectrometry.

Results and discussion

Labelled $^{13}CO_2$ (concentration × IE) did not increase between arterial and mammary venous blood (Table 1), i.e. the $^{13}CO_2$ arterio-venous difference (AVD) did not differ from zero (P>0.20) and averaged +0.01 and +0.14±0.11 μM for cow 1 and 2, respectively. These values need to be corrected for sequestration of $^{13}CO_2$ across the MG. This has been reported to range from 1.2 to 4.8% of arterial $^{13}CO_2$ inflow in three dairy cows (Raggio *et al.*, 2006). Based on the extremes of these values, adding this CO_2 sequestration across the MG would result in a $^{13}CO_2$ AVD of between -0.04 to -0.37 μM. Negative values indicate net oxidation but this represents only 0.1 to 0.7% of labelled Phe+Tyr inflow to the MG.

Table 1. Concentrations and enrichments of blood CO_2 during an infusion of L[1-^{13}C]Phe.

Cow #	Time[1]	Concentrations (mM)		Enrichments (ape)[2]		$^{13}CO_2$ concentration (μM)	
		Artery	Vein	Artery	Vein	Artery	Vein
1	6-11 h	22.63	24.45	0.0399	0.0367	9.03	8.94
	30-35 h	21.19	22.94	0.0458	0.0426	9.72	9.78
2	6-11 h	24.80	26.98	0.0390	0.0351	9.70	9.49
	30-35 h	24.35	26.51	0.0466	0.0426	11.34	11.28
	SEM	0.49	0.61	0.0012	0.0011	0.39	0.34

[1] Time after initiation of the infusion.
[2] ape: atom % excess.

Between 1.5% (cow 1) and 3.6% (cow 2) of the Tyr secreted into milk had been synthesized from intracellular Phe within the MG. Total mammary Tyr synthesis from Phe, however, also includes that directed towards synthesis of constitutive protein plus that exported to the mammary vein. Assuming that milk protein secretion represent 75% of MG protein synthesis (Raggio *et al.,* 2006), and that return to the mammary vein can be estimated from the ratios of the IE Tyr/IE Phe in the artery, vein and milk protein, then total conversion of Phe to Tyr within the MG would be the equivalent of 5.2% (cow 1) to 9.4% (cow 2) of Tyr secretion in milk. This is despite the suggestion that hydroxylation of Phe to Tyr is markedly increased at high substrate concentration (Verbeke *et al.*, 1972).

In conclusion, in the current study, even a large and imbalanced supply of Phe did not result in significant oxidation of Phe (+ Tyr) across the MG. This result suggests that under normal feeding conditions, any oxidation of Phe + Tyr is insufficient to preclude the use of these AA to estimate mammary plasma flow using the Fick principle.

Table 2. Enrichments of Phe and Tyr in milk and plasma during an infusion of L[1-^{13}C]Phe.[1]

Cow #	Time[2]	Phe			Tyr		
		Artery	Vein	Milk[3]	Artery	Vein	Milk[3]
1	30-35h	35.13	35.19	33.58	4.04	4.37	4.47
2	30-35h	36.29	35.74	32.77	5.86	7.07	6.84

[1] Atom % excess.
[2] Time after initiation of the infusion.
[3] At the end of the period.

References

Mepham, T.B., 1982. Amino acid utilization by lactating mammary gland. J. Dairy Sci. 65, 287-298.

Raggio, G., S. Lemosquet, G.E. Lobley, H. Rulquin and H. Lapierre, 2006. Effect of casein and propionate supply on mammary protein metabolism in lactating dairy cows. J.Dairy Sci. 89, 4340-4351.

Van Loon, L.J.C., Y. Boirie, A.P. Gijsen, J. Fauquant, A.L. de Ross, A.K. Kies, S. Lemosquet, W.H.M. Sairs and R. Koopman, 2009. The introduction of intrinsically labelled milk protein provides a functional tool for human nutrition research. J. Dairy Sci. 92, 4812-4822.

Verbeke, R., E. Roets, A.M. Massart-Leën and G. Peeters, 1972. Metabolism of [U-^{14}C]-L-threonine and [U-^{14}C]-L-phenylalanine by the isolated perfused udder. J. Dairy Res. 39, 239-250.

Integration of ^{15}N into DNA of liver of chicken embryos

A. Chwalibog[1], E. Sawosz[2], C.C. Metges[3], T. Niemiec[2] and M. Grodzik[2]
[1]University of Copenhagen, Denmark; ac@life.ku.dk
[2]University of Life Sciences, Warsaw, Poland
[3]Leibniz Institute for Farm Animal Biology (FBN), Dummerstorf, Germany

Introduction

It is well known that stable isotopes are not harmful to the organism and recently several new techniques using stable isotope labelled tracers have been successfully applied in measurements of DNA synthesis (Perez and Reeds, 1998; Hellerstein, 2003). Measurements of stable isotopes incorporation into DNA can now be performed with high reproducibility and stability and are powerful tools for quantifying *in vivo* cellular synthesis rate (Fan *et al.*, 2006). Although DNA is synthesised and replicated during the whole life, this process is prevailing and crucial during embryonic life. However, to our knowledge, there is no quantitative data describing the rate of DNA synthesis during prenatal development. Consequently, the objective of the present experiment was to test possibilities of measuring the rate of DNA synthesis in chicken embryos by applying a simple ^{15}N tracer technique.

Materials and methods

Fertilized eggs were divided into 4 groups (4×15): control – not treated, and treated with ^{15}N labelled glycine, ammonium chloride and sodium nitrate. Experimental solutions were given *in ovo* by injection to albumen. The quantity of each compound in the PBS solution was calculated from the content of ^{15}N necessary to provide 5 mg of ^{15}N per kg body weight of embryos at the conclusion of the experiment. After 20 days of incubation eggs were opened and embryos were immediately sacrificed by decapitation. Embryos were weighed and evaluated using Hamburger and Hamilton (1951) standards (HH-standard), including detailed morphological evaluation of dissected organs (heart, liver and spleen). Immediately after decapitation livers were frozen in liquid nitrogen and stored at -80 °C until DNA purification. Determination of ^{15}N enrichment of liver DNA was carried out with an isotope ratio mass spectrometer (Delta S, Finnigan MAT, Bremen, Germany). The ^{15}N enrichments (expressed in atom %) were calculated from the $\delta^{15}N$ values. The $\delta^{15}N$ value is calculated as δ (‰) = $[R_{sample}/R_{standard})-1]\times10^3$, where R is the $[^{15}N]/[^{14}N]$ ratio. The international nitrogen standard is atmospheric N_2 (AIR) with a $[^{15}N]/[^{14}N]$ isotopic ratio $R_{AIR} = 0.0036765$ and has been assigned a $\delta^{15}N$ value of 0‰. The data were analysed by ANOVA, using SAS®.

Results and discussion

After 20 days of incubation, body weight and weights of individual organs from control and treatment groups were not significantly different (P>0.05). Furthermore, all embryos developed normally in accordance with HH-standard. The abundance of ^{15}N was significantly affected by the type of labelled substance (Table 1). The highest level was measured for ammonium chloride, being significantly different from glycine, sodium nitrate and control groups. It is evident that both organic ^{15}N tracers increased ^{15}N atom %, while inorganic $Na^{15}NO_3$ had the same level of ^{15}N as the control group. The level of 0.367 ^{15}N atom % was the same as the atmospheric air standard and the amount naturally occurring in the cells of animal tissue.

The measurements of ^{15}N enrichment have been performed on DNA isolated from chicken embryos' liver. We have chosen chicken embryos because they develop in a 'closed' system without endogenous nutrients and water supply and thus nitrogen metabolism is independent of endogenous substrates, with the exception of exchange with atmospheric air. In this way major drawbacks of calculation and interpretation of results from measurements of DNA synthesis by isotope tracer techniques

Table 1. Content of total nitrogen in dry matter of liver DNA and ^{15}N abundance (^{15}N atom%) in control group (I) and groups treated with ^{15}N labelled glycine (II), ammonium chloride (III) and sodium nitrate (IV).

Group	I	II	III	IV	SE	*P*-value
n	14	11	12	12		
%N in DM	11.20^{b}	11.21^{b}	11.75^{ab}	12.01^{a}	0.017	0.036
^{15}N atom %	0.367^{c}	0.408^{b}	0.432^{a}	0.367^{c}	0.0002	<0.001

^{a,b,c} Values that share no common superscript differ significantly ($P<0.05$).

(Hellerstein, 2003; Fan *et al.*, 2006) could be avoided. Some species of bacteria and fungi can transform inorganic forms of nitrogen into organic precursors (ammonia) of DNA (Acquisti *et al.*, 2009) but inorganic nitrogen assimilation has never been shown for animals. Nevertheless, we have included ^{15}N labelled sodium nitrate to evaluate whether such a phenomenon can occur in chicken embryo. The present results clearly demonstrated that only organic forms of nitrogen could be incorporated into DNA, since content of ^{15}N atom % from $NaNO_3$ was the same as in the control group.

Glycine is one of the main sources of nitrogen atoms for *de novo* synthesis of nucleotides from purine bases and it could be expected that N from glycine was readily incorporated into DNA, to a higher extent than nitrogen from ammonium chloride. However, in this investigation ^{15}N enrichment from NH_4Cl exceeded glycine, probably indicating that released ammonia is quickly fixed by the process of glutamine synthesis and then glutamine amide N is used for synthesis of all nucleotide bases, ie. purines and pyrimidines.

In conclusion, a simple ^{15}N tracer technique injecting ^{15}N in incubating eggs and then measuring ^{15}N enrichment by mass spectrometry can be used to evaluate incorporation of ^{15}N into embryo's DNA. The method is readily applied to the DNA synthesis in the liver of chicken embryos and might be used for other tissues during embryogenesis.

References

Acquisti, C., J.J. Elser and S. Kumar, 2009. Ecological nitrogen limitation shapes the DNA composition of plant genomes. Mol. Biol. Evol. 26, 953-956.

Fan, M.Z., L.I. Chiba, P.D. Matzat, X. Yang, Y.L. Yin, Y. Mine and H.H. Stein, 2006. Measuring synthesis rates of nitrogen-containing polymers by using stable isotope tracers. J. Anim. Sci. 84, E79-83.

Hamburger, V. and H.L. Hamilton, 1951. A series of normal stages in the development of the chick embryo. J. Morphol. 88, 49-92.

Hellerstein, M.K, 2003. *In vivo* measurement of fluxes through metabolic pathways: the missing link in functional genomics and pharmaceutical research. Annu. Rev. Nutr. 23, 379-402.

Perez, J.F. and P.J. Reeds, 1998. A new stable isotope method enables the simultaneous measurement of nucleic acid and protein synthesis *in vivo* in mice. J. Nutr. 128, 1562-1569.

Efficiency of endogenous urea nitrogen incorporation into the ruminal bacteria and milk protein in goats fed diet differentiated in protein level

J. Kowalczyk[1], J. Voigt[2], J.P. Michalski[1] and M. Czauderna[1]
[1]The Kielanowski Institute of Animal Physiology and Nutrition, Polish Academy of Sciences, O5-110 Jabłonna, Poland
[2]Research Institute for the Biology of Farm Animals, 18196 Dummerstorf, Germany

Introduction

Recycling of nitrogen compounds between blood and digesta in ruminants plays an important role in nitrogen utilisation and metabolism (Lapierre and Lobley, 2001; Pfeffer *et al.*, 2009). Accumulation of endogenous blood nitrogen in the protein of ruminal bacteria is accompanied by incorporation into e.g. proteins of blood proteins and other tissues of animals, as well as in milk or secreted in urine and faeces (Beitz, 2004). We hypothesised that the endogenous urea nitrogen of blood can be utilized by ruminants for synthesis of milk proteins. Thus, the purpose of our study was to evaluate the extent of the endogenous urea nitrogen accumulation in ruminal bacteria proteins and proteins of milk of goats fed a low protein (LP), medium protein (MP) or high protein diet (HP).

Material and methods

The investigation was performed on three primiparous Alpine goats (~12 months old) in similar lactation phases. Goats were provided with rumen canulas and permanent catheters into the jugular vein. The experimental design was 3×3 Latin Square. Milked goats of average 35 kg BW were housed in separate metabolic cages, fed every 6 h with three isoenergetic diets containing 11.4 (LP), 13.3 (MP) and 16.9% (HP) of crude protein in DM. The daily amount of the diet was divided into 4 portions (4×550 g); 1 kg of the diet contained 545.5 g hay (CP 11.3%, crude fibre 30.7%; DM 91.5%; 10.6 MJ ME/kg) and 454.5 g concentrate (CP 8.3%; DM 89.4%; 12.5 MJ ME/kg). Water was freely available. The 22-day period consisted of 14 days of adaptation to the diets and 8 days of milk collection. After 14 days of adaptation animals were infused with a continuous infusion of 1.6 g ^{15}N-urea/day (5 mg ^{15}N/day/kg BW) in physiological NaCl solution for 6 days into the jugular vein. Goats were milked twice a day at 06:00 and 18:00, milk was weighed and pooled for 24 h while rumen fluid samples (~200 ml) were collected twice a day at 08:00 and 14:00. All collected samples were stored at -20°C until analyses. The contents of ^{15}N in ruminal bacteria, plasma and milk samples were determined by means of isotope mass spectrometry (Voigt *et al.,* 1980). The concentrations of urea in milk and rumen fluid samples were quantified using high-performance liquid chromatography with pre-column derivation (Czauderna and Kowalczyk, 2009). The data were statistically analysed using the nonparametric Mann-Whitney U test for comparing independent experimental groups. The Statistica v.6 package was used (2002; www.statsoft.pl).

Results and discussion

The average concentration of proteins in milk was similar in LP, MP and HP groups of goats ($P>0.05$) although daily milk productions in the goat groups differed numerically (1,575, 1,492 and 1,770 g/day/goat, respectively). Milk urea concentration was higher in the HP (451 mg/l) than in both the LP group (139 mg/l) and the MP group (342 mg/l) ($P<0.05$), respectively.

Mass spectrometric analyses of milk samples showed that the level of ^{15}N in milk proteins was negatively correlated with the level of protein in the diets (48.3, 31.6 and 15.7 mg/l in the group LP, MP and HP, respectively). The highest enrichment of the ruminal bacteria protein in ^{15}N was found in the animals fed the LP diet. The amount of ^{15}N excreted in urine was greater than in faeces and the least in milk. Incorporation of ^{15}N into the plasma and milk protein was negatively correlated with the level of protein in the diet. The ^{15}N enrichment of plasma protein during the period of isotope infusion was greatest in goats fed the LP diet. Enrichment of ^{15}N in milk protein was correlated ($P<0.01$) with ^{15}N in ruminal bacteria protein (Figure 1). These results showed that urea ^{15}N efficiently accumulated in bacteria protein followed by incorporation of recycled urea ^{15}N into amino acids of milk protein. Our study indicated that endogenous urea nitrogen can be assimilated by ruminants for milk protein synthesis.

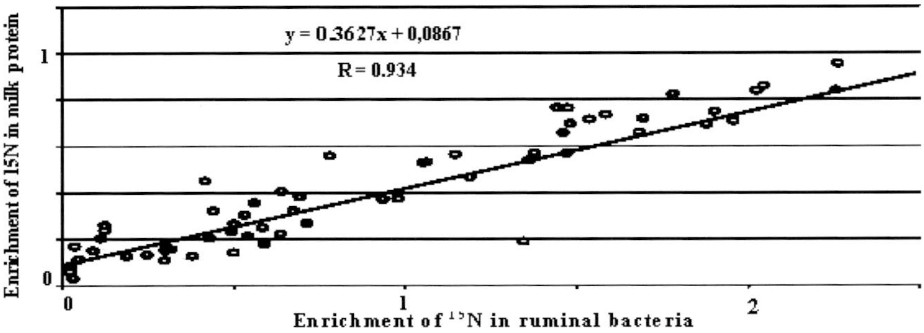

Figure 1. The correlation between enrichment of ^{15}N in milk protein and enrichment of ^{15}N in ruminal bacteria protein

Acknowledgment

Supported in part by the State Committee for Scientific Research, Grant No. 2 P06Z 06329.

References

Beitz, D.C., 2004. Protein and amino acid metabolism. In: Dukes' physiology of domestic animals. W.O. Reece (ed.) Cornell University Press, New York.

Czauderna, M. and K. Kowalczyk, 2009. Easy and accurate determination of urea in milk, blood plasma, urine and selected diets of mammals by high-performance liquid chromatography with photodiode array detection preceded by pre-column derivatization. Chem. Anal. 54, 919-937.

Lapierre, H. and G.E. Lobley, 2001. Nitrogen recycling in the ruminant: A Review. J. Dairy Sci. (E.Suppl.) 84, E223-E236.

Pfeffer, E., H. Speckter, S. Bornemann, A. Holthausen and M. Rodehutscord, 2009. Kinetics of endogenous urea nitrogen in lactating goats and cows fed diets varying in their crude protein concentrations. Arch. Anim. Nutr. 63, 230-242.

Voigt, J., R. Krawielitzki and B. Piatkowski, 1980. Untersuchungen zur Wirkung von Phosphor-saurephenylesrerdiamid als Inhibitor der Pansenurease bei der Milchkuh. 2. Der Umsatz von ^{15}N. Harnstoff. Arch. Anim. Nutr. 30, 825-834.

Studies on amino acid metabolism in rainbow trout: effect of dietary amino acid composition on growth performance and $\delta^{13}C$ of amino acids

J.S.O. McCullagh[1], J. Gaye-Siessegger[2] and U. Focken[3,4]
[1]Chemistry Research Laboratory, Oxford University, Mansfield Rd. Oxford, OX1 3TA, United Kingdom
[2]Fisheries Research Station Baden-Württemberg, Argenweg 50/1, Langenargen, Germany
[3]Department of Aquaculture Systems & Animal Nutrition, University of Hohenheim, Germany
[4]Present address: Johann Heinrich von Thünen-Institut, Institut für Fischereiökologie, Wulfsdorfer Weg 204, Ahrensburg, Germany

Introduction

For fish, proteinogenic amino acids (AAs) are commonly divided into 10 essential (Arg, His, Ile, Leu, Lys, Met, Phe, Thr, Trp and Val) and 10 non-essential AAs (NRC, 1993). The aim of the study was to test whether the dietary non-essential amino acid (NEAA) composition has an effect on growth, protein utilisation and $\delta^{13}C$ of AAs in rainbow trout (*Oncorhynchus mykiss*). In addition, it was investigated whether tissue NEAAs, excluded from the diet, were derived from dietary NEAAs provided in excess or from other carbon sources. ^{13}Cenriched Glu was used to elucidate its metabolic utilisation. Improvements in LC-IRMS (Surveyor HPLC system coupled to a DELTA V isotope ratio mass spectrometer via a Finnigan LC IsoLink interface (Thermo Electron Corporation)) allow the direct determination of $\delta^{13}C$ of individual AAs (McCullagh *et al.*, 2006).

Material and methods

Six diets were made from wheat starch, potato dextrin, wheat germ oil, fish oil, vitamin and mineral premixes, cellulose, betaine as feed attractant and different compositions of synthetic amino acids. Diet 1 was formulated to have the same AA composition as fishmeal. In diet 2, Cys, Gly, Pro and Tyr were isonitrogenously replaced by their precursor AAs Ser, Glu and Phe, and in diet 3, Ala, Asx, Cys, Gly, Pro, Ser and Tyr were isonitrogenously replaced by Glu. Diets 4, 5 and 6 resembled diets 1, 2, and 3, except for Glu, which was enriched in ^{13}C (carboxyl C). In a control diet synthetic amino acids were replaced by defatted fishmeal. All diets were composed to be equal in nitrogen, lipid and energy content. Forty-two rainbow trout (initial body mass 4.7±0.57 g) were divided into 7 groups of 6 fish each. Fish were fed individually at a level of 3.5% body mass equivalent per day in aquaria of a flow-through system. The experiment lasted ten weeks. Data were analysed by two-way ANOVA followed by Tukey's HSD test (STATISTICA 6.0).

Results and discussion

Differences in growth between treatments were not statistically significant due to high individual variability (P=0.059). However, protein gain was significantly influenced by the dietary AA composition (P<0.05) with the highest gain in trout fed the diets with the full AA spectrum (diets 1 and 4). Protein gain did not differ significantly between fish fed diets 2 or 5 and those fed diets 3 or 6. It has been shown that growth and protein utilisation significantly affect bulk $\delta^{13}C$ of fish (Focken, 2001; Gaye-Siessegger *et al.*, 2004). In the present study, dietary AA composition significantly influenced $\delta^{13}C$ of some AAs in rainbow trout (Table 1, Figure 1). The effect of ^{13}C-enriched Glu was small (Table 1). Glu enters the Krebs Cycle via deamination that directly produces α-keto-glutarate. During the step from α-ketoglutarate to succinyl-CoA, the C 1 atom is dissociated. Therefore, only small effects of enriched Glu on $\delta^{13}C$ of individual AAs were expected. The contribution of dietary Glu to tissue Glu was low which is consistent with the fact that Glu has been shown to be used extensively in the gut as an energy source (Reeds *et al.* 1997). Reeds (2000) stated that the classification of AAs into essential and non-essential have become increasingly imprecise. While Arg is classified as an essential AA for birds, carnivores and young mammals, it is a conditionally essential AA for adult mammals (Tapiero *et al.*, 2002). Catfish uses dietary Glu for the synthesis of Arg especially when

Table 1. Effect of dietary NEAA composition and ^{13}C-enriched Glu on $\delta^{13}C$ of AAs in trout (n=36).

Parameter	Ala	Arg	Glu	Gly	His	Lys	Met	Phe	Pro	Ser	Thr	Tyr
diet	***	***	n.i.	***	**	n.s.	n.s.	*	n.i.	n.i.	n.s.	***
enrichment	n.s.	n.s.	n.i.	n.s.	n.s.	n.s.	n.s.	n.s.	n.i.	n.i.	n.s.	n.s.
diet × enrichment	n.s.	n.s.	*	n.s.	n.s.	n.s.	n.s.	n.s.	***	**	n.s.	n.s.

n.s. = not significant; n.i. = non-interpretable due to significant interaction; $^{*}P<0.05$; $^{**}P<0.01$; $^{***}P<0.001$.

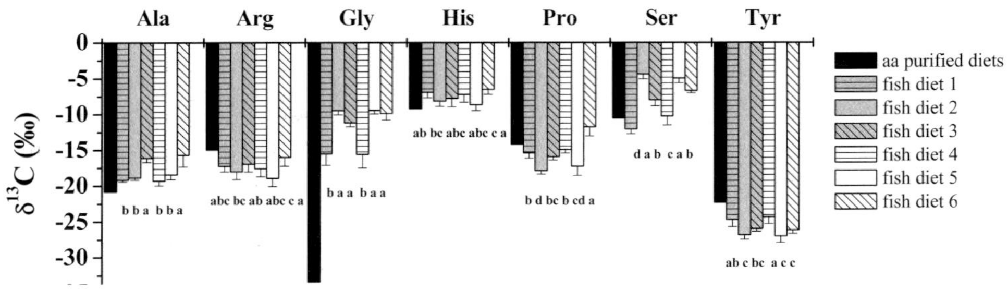

Figure 1. $\delta^{13}C$ of individual AAs in diets and trout (means ± 1 SD).

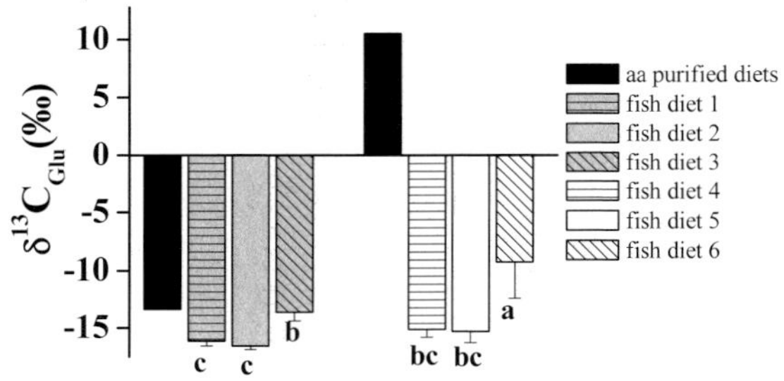

Figure 2. $\delta^{13}C$ of Glu in diets and trout (means ± 1 SD).

Arg is deficient in the diet (Buentello and Gatlin III, 2000). Possibly, trout fed the high Glu diets synthesized Arg from dietary Glu and, thus, Arg in fish tissue was enriched in ^{13}C (-16.9‰ for fish fed diet 3 vs. -16.0‰ for fish fed diet 6). Further research is required to understand the mechanisms that lead to different $\delta^{13}C$ of AAs, however, the results of this study show the importance of dietary NEAA composition for the growth performance of trout.

References

Focken U., 2001. Stable isotopes in animal ecology: the effect of ration size on the trophic shift of C and N isotopes between feed and carcass. Isotop. Environ. Health S. 37, 199-211.

Buentello J.A. and D.M. Gatlin III, 2000. The dietary arginine requirement of channel catfish (*Ictalurus punctatus*) is influ-enced by endogenous synthesis of arginine from glutamic acid. Aquaculture 188, 311-321.

Gaye-Siessegger J., U. Focken, S. Muetzel, Hj. Abel and K. Becker, 2004. Feeding level and individual metabolic rate affect $\delta^{13}C$ and $\delta^{15}N$ values in carp: implications for food web studies. Oecologia 138, 175-183.

McCullagh J., D. Juchelka and R.E.M. Hedges, 2006. Analysis of amino acid ^{13}C abundance from human and faunal bone collagen using liquid chromatography/isotope ratio mass spectrometry. Rapid Commun. Mass Spectrom. 20, 2761-2768.

National Research Council, 1993. Nutrient Requirements of Fish. National Academy Press,Washington, p. 114.

Reeds P.J., 2000. Dispensable and indispensable amino acids for humans. J. Nutr. 130, 1835S-1840S.

Reeds P.J., D.G. Burrin., B. Stoll, F. Jahoor, J. Henry and M.E. Frazer, 1997. Enteral glutamate is the preferential precursor for mucosal glutathione synthesis in the piglet. Am. J. Physiol. 273, E408-E415.

Tapiero H., G. Mathé, P. Couvreur and K.D. Tew, 2002. I. Arginine. Biomed. Pharmacother. 56, 439-445.

Interactive effect of dietary energy concentration and genotype on the nitrogen and energy gain of pigs at a body weight between 20 and 25 kg

L. Permentier[1], D. Maenhout[2], K. Broekman[3], W. Deley[4] and R. Geers[1]
[1]Laboratory for Quality Care in Animal Production, Zootechnical Centre, Catholic University Leuven, Bijzondere weg 12, 3360 Lovenjoel, Belgium
[2]Hendrix Haeck, Zuidkaai 6, 8770 Ingelmunster, Belgium
[3]Hypor B.V., Spoorstraat 69, 5831 CK Boxmeer, the Netherlands
[4]Hypor Belgium NV, Leie Rechteroever 1, 9870 Olsene, Belgium

Introduction

Piétrain crossbreds are commonly known as lean genotypes with a slow growth (Labroue *et al.*, 1999). Nevertheless, it remains unclear whether the lower feed intake capacity of Piétrain crossbreds is the most restrictive factor to realize higher maximal protein deposition (Hermesch *et al.*, 2003). The objective of this study is to increase the energy content in the diet of Piétrain crossbreds to compensate lower intake capacity in order to investigate whether the nitrogen (N) deposition rate can be modified.

Materials and methods

Three Piétrain crossbred types of pigs were selected according to their leanness: a lean (n=78), an intermediate (n=31) and a less lean (n=41) type. As the lean type functions as a pair wise control group to the other two genotypes, the experiment is unbalanced for sample size. At approximately 20 kg, the at random selected animals were housed individually in a pen for a 7-day period. The animals were fed *ad libitum* either a standard energy (9.8 MJ of NE/kg) or a high energy (10.4 MJ of NE/kg) isonitrogenous (18% N) diet. N and net-energy(NE) retention were calculated based on Dual Energy X-ray absorptiometry measurements (DEXA). Daily N and NE retention were analyzed in relation to N intake by using the mixed model procedure within SAS, with the individual pig as the experimental unit and N-intake as covariate. The influence of genotype, dietary energy level and the diet x genotype interaction were investigated.

Results and discussion

Initial body composition and N-intake were similar across genotypes and dietary energy concentrations. The results in Table 1 show two significant effects of diet and genotype. The NE intake is higher if the pigs are fed a higher energy diet. The less lean type reveals a significant higher NE retention, suggesting a higher fat tissue gain, since N retention is equal to the other genotypes. These results are in line with the characteristics of the genotypes. Furthermore, interaction between diet energy level and type of pig was not significant (Quiniou *et al.*, 1996).

Regression analyses indicate that pigs fed a high energy level diet have a higher N retention in comparison to the standard energy level diet at low N intakes, but a lower retention at high intakes. This is demonstrated by the values of the regression equations at 20 and 60 g N intake per day in Table 1, indicating a point between 20 and 60 g N intake per day where the N retention of pigs fed a standard energy level diet becomes higher than the N retentions of the pigs fed a high energy level diet. This point is situated further away from the average N intake for the Piétrain pigs, but around the level of the mean N intake per day for the other two genotypes (not shown), explaining the overall insignificance of the diet in the results, but a trend for the Piétrains' higher average daily gain (ADG) fed the high energy level diet.

Concerning the regression analyses of the NE retention, there are small differences between the genotypes. For the lean type, the high energy level diet has a positive effect on the NE retention.

Table 1. Lsmeans (and standard error) of ADG, N intake, N retention, NE intake, NE retention. Regression parameters of N retention and NE retention of pigs at a body weight of 20 to 25kg.

Genotype (G)	Lean		Intermediate lean		Less lean		*P*-values		
Diet (D)	Standard	High	Standard	High	Standard	High	G	D	G*D
No. of animals	38	40	15	16	21	20			
ADG (g/d)	510.5	584.4	660.2	657.4	710.0	709.5	ns	ns	ns
	(70.6)	(69.4)	(115.6)	(109.4)	(92.4)	(92.6)			
N intake (g/d)	35.23	36.11	40.07	40.98	37.43	36.99	ns	ns	ns
	(1.46)	(1.43)	(2.41)	(2.26)	(1.92)	(1.98)			
N retention (g/d)									
Intercept	-12.00	-4.83	-9.03	-3.19	-8.39	7.76	ns	0.007	ns
Slope	0.760	0.613	0.700	0.518	0.700	0.324	ns	0.010	ns
Mean	16.15	18.07	17.28	15.55	19.57	19.98	ns	ns	ns
	(1.39)	(1.36)	(2.26)	(2.16)	(1.82)	(1.87)			
at 20g N intake	3.2	7.43	4.97	7.17	5.61	14.24			
at 60g N intake	33.6	31.95	32.97	27.89	33.61	27.2			
R-squared	54.31	49.87	49.48	42.85	19.75	39.53			
Root MSE	6.03	5.13	4.74	3.43	7.90	3.46			
NE intake (kJ/d)	12,019	12,983	13,677	14,748	12,776	13,298	ns	0.040	ns
	(511.96)	(499.24)	(844.35)	(790.41)	(673.26)	(694.32)			
NE retention (kJ day^{-1})									
Intercept	-3,626	-3,527	-5,136	-2,228	-2,211	-1,624	ns	ns	ns
Slope	220	224	250	165	249	203	ns	ns	ns
Mean	4,564.83	4,808.25	4,260.71	3,658.84	6,097.95	5,954.30	0.005	ns	ns
	(341.40)	(334.85)	(555.51)	(529.46)	(446.79)	(458.11)			
at 20g N intake	769.6	946.47	-143.09	1,069.4	2,777.19	2,435.2			
at 60g N intake	9,560.8	9,892.87	9,843.31	7,663.4	12,752.79	10,553.6			
R-squared	66.47	68.31	67.49	58.13	39.64	43.25			
Root MSE	1,344.46	1,193.08	1,113.91	747.83	1,689.15	1,705.57			

ns: not significant.

However, for both the intermediate and less lean type there is a point in between 20 and 60 g N intake per day where the NE retention of pigs fed a standard energy level diet becomes higher than the N retentions of the pigs fed a high energy level diet.

In conclusion, increasing the dietary energy level of Piétrain crossbreds at a body weight of approximately 20 kg cannot improve their growth rate, however the less lean type is sensitive to the E-content of diet, resulting in a higher fat tissue gain.

References

Hermesch S., E. Kanis and J.J. Eissen, 2003. Economic weights for feed intake in the growing pig derived from a growth model and an economical model. J. Anim. Sci. 81, 895-903.

Labroue F., R. Guéblez, M.C. Meunier-Salaün and P. Sellier, 1999. Feed intake behaviour of group-housed Piétrain and Large White growing pigs. Ann. Zootech. 4, 247-261.

Quiniou N., J. Noblet and J.-Y. Dourmad, 1996. Effect of energy intake on the performance of different types of pig from 45 to 100kg body weight. 2. Tissue gain. Anim. Sci. 63, 289-296.

Quantitative magnetic resonance (QMR) measurement of changes in body composition of neonatal pigs

A.D. Mitchell[1], T.G. Ramsay[1], G. Taicher[2] and I. Kovner[2]
[1]USDA-ARS, Animal Bioscience & Biotechnology Laboratory, Beltsville, MD 20705, USA
[2]Echo Medical Systems, 14781 Memorial Drive, Suite 222, Houston, TX 77084, USA

Introduction

The survival of low birth weight pigs in particular may depend on energy stores in the body. The growth and composition of the neonatal pig is also of interest because of potential impact on subsequent growth and composition. Studies have found that factors such as birth weight (Powell and Aberle, 1980; Poore and Fowden, 2004) and nutrition (Pond *et al.*, 1965; Campbell and Dunkin, 1983) during the neonatal period can affect both growth and composition. Most studies have relied on chemical analysis of the carcass to measure composition (fat and lean) of the neonatal pig, thus the impact on subsequent composition can only be inferred. QMR is an NMR approach to measuring total body fat, lean and water (Taicher *et al.*, 2005). These measurements are based on quantifying protons associated with lipid and water molecules in the body. Since QMR is very rapid and does not require that the pig be anesthetized, these measurements can be made without interfering with the pig's growth.

Material and methods

Using a QMR instrument (EchoMRIÔ), a total of 60 pigs were scanned an average of 5 times starting at 2.7±1.3 d of age (1.95±0.42 kg) and finally at 13.1±4.3 d (4.14±0.52 kg). Each scan consisted of triplicate measurements. The rates of total body growth and fat and lean deposition were analyzed by linear regression analysis. Subsequently 107 piglets were scanned twice, first at 3±1 d of age and then at 13±4 d of age. Fat, lean and water growth were calculated based on differences between the two measurements. Protein deposition was estimated as 0.25·water deposition (Campbell and Dunkin, 1983). The accuracy of the instrument for measuring total body fat and water had been previously calibrated and validated based on measurements and chemical analysis of 50 piglets weighing 1.2 to 4.3 kg (Kovner *et al.*, 2010).

Results and discussion

Table 1 shows the rates of fat and lean deposition that were observed when pigs were measured by QMR an average of five times during growth from approximately 2 to 4 kg body weight. The mean (±SD) rate of total body growth was 236±76 g/d and was linear throughout the study period (R^2=0.98±0.04). There was a wide range in the rates of both fat and lean deposition and both were linear throughout the study period (R^2=0.97±0.04 and R^2=0.95±0.10, respectively). Consistent with the results by Noblet and Etienne (1987), the rates of both fat and lean deposition were highly correlated (P<0.001) with total body growth rate (r=0.88 and 0.94, respectively). The correlation between the rates of fat and lean deposition was 0.74 (P<0.001).

Table 2 shows the rates of fat and lean deposition that were observed when pigs were measured by QMR only twice during growth from approximately 2 to 4 kg body weight. Similar to the first group, the mean (±SD) rate of total body growth was 230±57 g/d. Likewise, measurements of the rates of fat and lean deposition were similar to the first study, confirming that these rates were linear and that only the two measurements are needed for this growth period. Again, the rates of both fat and lean deposition were highly correlated (P<0.001) with total body growth rate (r=0.88 and 0.97, respectively). In this study, the ratio of total body growth rate to the rate of fat deposition was 5.6, compared to 6.9 reported by Noblet and Etienne (1987). The ratio of protein gain (estimated) to lipid

Table 1. Rates of fat and lean deposition – Linear regression, based on individual measurements (n=60).

Measurement	Mean (g/d) ± SD	Range (g/d)	R^2 (mean ± SD)
Fat deposition rate	32±13	10.6-64.9	0.97±0.04
Lean deposition rate	188±60	39.1-353.6	0.95±0.10

Table 2. Measurement of growth and composition based on two observations per pig (n=107).

Scans	Age (da)	Body Wt (g)	Fat (g)	Fat (%)	Lean (g)	Water (g)
1st Scan	3.0±1.3	2152±428	122±66	5.4±2.3	2014±364	1692±291
2nd Scan	12.9±4.0	4271±601	491±111	11.4±1.7	3767±526	3091±432
Growth	Fat (g)	Fat ($g \cdot d^{-1}$)	Lean (g)	Lean ($g \cdot d^{-1}$)	Water (g)	Water (g/d)
Mean	369±93	41±13	1755±369	191±50	1400±91	152±41
Range	133-631	8-68	810-2532	83-310	600-2033	66-252

gain was 0.93 in this study compared to 1.06 reported by Noblet and Etienne (1987). The correlation between the rates of fat and lean deposition was 0.83 (P<0.001).

The results of these studies demonstrate that QMR is a useful method for measuring changes in body composition in neonatal pigs. Furthermore, the results indicate that during the period of growth from birth to 4 kg, the rates of both fat and lean deposition are linear and highly correlated with total body growth

References

Campbell, R.G. and A.C. Dunkin, 1983. The effects of energy intake and dietary protein on nitrogen retention, growth performance, body composition and some aspects of energy metabolism of baby pigs. Br. J. Nutr. 49, 221-230.

Kovner, I.,Taicher and A. Mitchell, 2009. Calibration and validation of EchoMRI whole body composition analysis based on chemical analysis of piglets. Int. J. Body Comp. Res. (in press).

Noblet, J. and M. Etienne, 1987. Body composition, metabolic rate and utilization of milk nutrients in suckling piglets. Reprod. Nutr. Dev. 27, 829-839.

Pond, W.G., R.H. Barnes, R.B. Bradfield, E. Kwong and L. Krook, 1965. Effect of dietary energy intake on protein deficiency symptoms and body composition of baby pigs fed equalized but suboptimal amounts of protein. J. Nutr. 85, 57-66.

Poore, K.R. and A.L. Fowden, 2004. The effects of birth weight and postnatal growth patterns on fat depth and plasma leptin concentrations in juvenile and adult pigs. J. Physiol. 558, 295-304.

Powell, S.E. and E.D. Aberle, 1980. Effects of birth weight on growth and carcass composition of swine. J. Anim. Sci. 50, 860-868.

Taicher, G.Z., F.C. Tinsley, A. Reiderman and M.L. Heiman, 2003. Quantitative magnetic resonance (QMR) method for bone and whole body composition analysis. Anal. Bioanal. Chem. 377, 990-1002.

The change in threonine requirement from early to late pregnancy in sows

C.L. Levesque, S. Moehn, P.B. Pencharz and R.O. Ball
Swine Research and Technology Centre, University of Alberta, Edmonton, AB, Canada

Introduction

Current NRC (1998) amino acid requirements for sows are based on growing pig data from the 1970's and 1980's; however, genetic selection over the last decade has dramatically increased lean body gain and litter size. The constant values recommended for amino acid requirements during pregnancy (NRC, 1998) assume an equal distribution of nutrient demand throughout pregnancy. However, metabolism in early pregnancy focuses on development of maternal body tissue whereas in late pregnancy foetal and mammary growth makes up the primary demand for nutrients (McPherson *et al.*, 2004). This change in metabolic focus is hypothesized to result in a higher requirement for amino acids during late pregnancy compared to early pregnancy.

Materials and methods

Eight multiparous sows (211±11 kg BW) were used to determine the threonine (THR) requirement during early (25 to 55 d) and late (81 to 111 d) pregnancy using the indicator amino acid oxidation method utilizing L[1-^{13}C]phenylalanine. A corn, cornstarch and sugar base diet was formulated to contain THR at 20 and 60% of NRC (1998) recommendations in early and late pregnancy, respectively. Crystalline THR was used to create 6 diets ranging from 20 to 120% (early pregnancy) and 6 diets from 60 to 180% (late pregnancy) of the current recommended THR intake (10 g/d) based on BW, expected pregnancy gain and litter size.

The trial consisted of 6 consecutive periods (3 d of adaptation + 1 d of expired CO_2 and plasma collection). On collection day, tracer phenylalanine was given orally in 8, half-hourly meals and expired $^{13}CO_2$ was quantified. Plasma samples were taken during tracer infusion and analyzed for free amino acid concentration.

Data were analyzed using a nonlinear Mixed model in SAS (2001).

Results and discussion

Sow reproductive performance was similar to commercial standards (litter size: 14.0±2.6; litter birth weight: 21.3±4.3 kg). Figure 1 shows the response of sows to increasing dietary THR based on L[1-^{13}C]phenylalanine oxidation. The THR requirement in early pregnancy was 5.0±1.0 g/d (R^2=0.71) and in late pregnancy 12.3±2.3 g/d (R^2=0.58) based on indicator amino acid oxidation (Table 1).

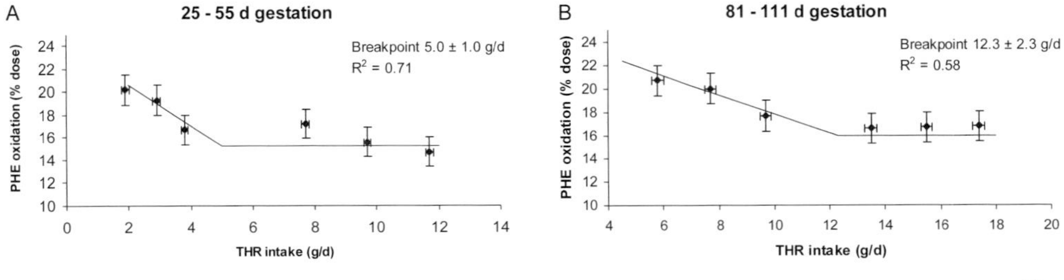

Figure 1. The response of sows to increasing levels of dietary threonine intake based on [1-^{13}C] phenylalanine oxidation (A: early gestation, B: late gestation).

Table 1. Threonine requirement of sows in early and late gestation based on indicator amino acid and plasma threonine concentration.

Stage of gestation	IAAO[1]	se	Plasma THR	se
Early (25-55 d)	5.0	1.0	3.9	0.5
Late (81-111 d)	12.3	2.3	10.5	2.8

[1] IAAO: indicator amino acid oxidation (L[1-^{13}C] phenylalanine.

The THR requirement in early pregnancy was 3.9±0.5 g/d (R^2=0.90) and in late pregnancy was 10.5±2.8 g/d (R^2=0.68) based on plasma THR concentration (Table 1).

The demand for THR increases more than 2-fold in late pregnancy compared to early pregnancy. The requirement for THR was 50 and 125% of the NRC (1998) recommended intake of 10 g/d in early and late pregnancy, respectively.

These results support recently published recommendations for sow feeding that the requirement for amino acids increase in late gestation (Kim *et al.*, 2009; GfE, 2008). Srichana (2006) found that the lysine requirement of sows in early and mid pregnancy was lower than in late pregnancy, 15 and 20 g, respectively.

Feeding a constant level of amino acids during pregnancy does not meet the demand for protein in late pregnancy and results in maternal protein catabolism to support foetal growth. Phase feeding sows during pregnancy will (1) account for changes in amino acid requirements during pregnancy, (2) reduce feed costs by reducing the risk of over feeding AA in early pregnancy and underfeeding amino acids in late pregnancy, and (3) potentially increase sow reproductive performance.

References

GfE, 2008. Recommendations for the supply of energy and nutrients to pigs. DLG Verlags GmbH, Frankfurt, Germany.

Kim, S.W., W.L. Hurley, G. Wu and F. Ji, 2009. Ideal amino acid balance for sows during gestation and lactation. J. Anim. Sci. 87, E123-E132.

McPherson, R., F. Ji, G. Wu, J.R. Blanton and S.W. Kim, 2004. Growth and compositional changes of fetal tissues in pigs. J. Anim. Sci. 82, 2534-2540.

NRC, 1998. Nutrient Requirements for Swine. 10th ed. Natl. Acad. Press. Washington, DC.

SAS, 2001, Release 8.03 SAS inst. Inc., Gary, NC, USA.

Srichana, P, 2006. Amino acid nutrition in gestating and lactating sows. PhD Diss. Univ. of Missouri, Columbia.

Further comparison of direct and indirect estimates of apparent nutrient digestibility with effort to reduce variation by pooling of multi-day fecal samples

J.H. Agudelo[1], C.S. Escobar[1], M.D. Lindemann[1], B.J. Kerr[2], N. Inocencio[1] and G.L. Cromwell[1]
[1]University of Kentucky, Lexington KY, USA
[2]USDA, Ames, IA, USA

Introduction

Total tract apparent digestibility of nutrients is traditionally assessed by either total collection (TC) or by the indicator method (IM). The IM saves labour by measuring the concentration of an indigestible indicator (e.g. Cr_2O_3) in feed and feces, purportedly resulting in similar digestibility values as TC (Schurch *et al.*, 1952). Our previous efforts have demonstrated that the IM methodology may not be as discriminating for the detection of treatment differences as TC, depending on how the IM is conducted. We have reported that a multiday IM sample is better than a single grab sample, and that several days of marker observance in feces are required before the marker content stabilizes in the feces (Agudelo *et al.*, 2010). The objective of this experiment was to further define a more efficacious manner of pooling fecal samples that will provide a digestibility value that most closely approximates that obtained by TC and that minimizes variation and thereby equals the precision of TC.

Materials and methods

Crossbred pigs (n=24; mean BW=46.3 kg) were fed one of 4 diets: (1) low P corn-soybean meal basal diet, (2) basal + virginiamycin (11 mg/kg diet), (3) basal + phytase (300 PU/kg diet), and (4) basal + virginiamycin + phytase. Diets were designed to have differences in P digestibility that would be detectible by TC. Pigs were individually housed in metabolism crates for a 7-day (d) adaptation and 5-d TC. Pigs were fed at 3% of BW divided into 2 equal meals with the beginning and ending meals of the TC marked with indigo carmine. The IM phase immediately followed the TC by including Cr_2O_3 at 0.25% in the feed for 10 days. Collection of fecal samples for IM started the day after Cr_2O_3 was first fed and continued for 9 consecutive d. Feces from collection d 1-9 were kept separated and labelled 'D 1', 'D 2', etc. to 'D 9'. Similar portions of feces from d 5-9 were then composited into 4 additional samples: d 5-6, d 5-7, d 5-8, and d 5-9. The number of samples analyzed was 13/pig for a total of 312 samples for IM digestibility in addition to the 24 samples from TC. All diets and fecal samples were then analyzed for dry matter (DM), gross energy (GE), N, P, acid detergent fiber (ADF), and neutral detergent fiber (NDF). Descriptive statistics of the data were computed in an Excel spreadsheet (Microsoft, Redmond WA, USA). TC data were further analyzed for treatment differences by the GLM procedure (SAS, 2004).

Results and discussion

An extremely large number of samples were analyzed. The information provided by examining 4 diets was similar for many nutrients. In general, the key observations can be summarized by the information for DM and P digestibility from Diets 1 and 3 which provided the largest treatment differences in P digestibility. In general, the highest digestibilities were for DM (>90%), GE (88-89%), and N (89-90%), intermediate for ADF (67-68%) and NDF (62-63%), and lowest for P (33-47%). Table 1 reports the digestibility of DM and P (the extremes of digestibility observed). For all analyzed nutrients, the mean IM digestibility values were less than those by TC (the last row of values in Table 1); the most likely reason for this observation is low marker analysis in the diet because low marker analysis in feces is associated with abnormally high or low digestibility values (as is demonstrated by the values on D 1 of collection). For highly digestible components of the diets, mean values tended to stabilize by D 5 or D 6 and pooling of feces for multiple days (the composited means in the table) improved little the mean or the CV. For the lowest digestible component measured (P), means were never very stable with the IM method and the CV was much greater; in this instance the pooling of samples did improve the CV. Interestingly, the use of phytase (Diet 3 compared to Diet 1) was associated with a lower CV of at least 35% for the composited samples for all measured components except N. While a low digestible component like P had variability for the mean values, a treatment difference of >20% was observed by D 6. In summary, the need for, and potential advantage of, pooling of fecal samples depends on the component of interest.

Table 1. Mean IM digestibility (and associated CV) for dry matter and phosphorus compared to TC based on day of fecal collection.[1]

	Dry matter				Phosphorus			
	Diet 1		Diet 3		Diet 1		Diet 3	
Day	Mean	CV,%	Mean	CV,%	Mean	CV,%	Mean	CV,%
1	-11,275	-208	-32,036	-153	81,067	-214	166,740	-154
2	-37.9	-508	0.1	216,128	-821	-158	-447	-178
3	80.2	11.6	87.3	2.2	-37.1	-205	32.3	39.6
4	87.8	2.3	88.6	2.1	16.2	103.3	42.8	18.9
5	87.5	1.1	89.2	1.4	47.2	87.7	53.8	43.0
6	88.3	1.6	89.2	1.3	23.1	29.0	44.6	16.3
7	87.9	2.0	89.1	1.5	18.2	53.6	40.6	17.0
8	88.4	1.2	89.2	0.7	22.7	43.1	44.0	9.4
9	88.5	2.2	89.0	0.8	22.0	37.6	41.6	12.2
Composited samples								
5-6	88.6	1.5	89.7	0.6	25.4	24.2	45.6	8.2
5-7	88.2	1.7	89.4	0.9	21.1	28.3	42.8	11.9
5-8	88.9	1.2	89.3	0.6	22.8	28.1	43.5	10.1
5-9	88.9	1.6	89.2	0.8	23.1	34.2	43.2	10.0
TC	90.1	0.8	90.0	0.6	33.4[a]	14.3	47.2[b]	9.3

[a,b] $P<0.01$.

[1] Each mean represents values from 6 pigs.

References

Agudelo, J.H., M.D. Lindemann, and G.L. Cromwell, 2010. A comparison of two methods to assess nutrient digestibility in pigs. Livest. Sci., in press.

SAS, 2004. Release 9.1.3. SAS Inst. Inc., Cary, NC, USA.

Schurch, A.F., E.W. Crampton, S.R. Haskell, and L.E. Lloyd, 1952. The use of chromic oxide in digestibility studies with pigs fed *ad libitum* in the barn. J. Anim. Sci. 11, 261-265.

Part 3. Energy/protein tissue metabolism

Non-hepatic glucose production: an update

Alberto Battezzati
International Center for the Assessment of Nutritional Status – DiSTAM, Università degli Studi di Milano, Via G. Colombo, 20133 Milano, Italy

Introduction

In the last century the primary role of the of the liver in producing glucose either by glycogenolysis or gluconeogenesis was acknowledged at a point that the appearance of endogenous glucose in the bloodstream was usually referred to as 'hepatic glucose production'. It had been known for a long time that the kidney has the machinery to synthesize de novo glucose and that it is indeed a source for endogenous glucose in mammalians. Nonetheless, the kidney was considered a gluconeogenic organ only in case of severe acidosis and prolonged fasting.

It was also believed that the gluconeogenic machinery, including four critical enzymes pyruvate carboxylase, PEP carboxykinase, fructose-1,6-diphosphatase, and glucose-6-phosphatase could be fully expressed exclusively in the liver and in the kidney. Specifically, with the exception of hepatocytes and renal cortical cells, it was considered impossible that any glucose synthesized *de novo* or released from local glycogen stores could be released in the systemic circulation, because of the absence of glucose-6-phosphatase.

At the end of the century experimental evidence unequivocally challenged the latter dogma, and a whole family of functionally active glucose-6-phosphatases, variously expressed in several tissues, was discovered. Consequently, many kinetic studies tried to quantify glucose release from various organs besides the liver. However, due to inherent technical problems limiting the accuracy in the estimates of glucose fluxes across organs, controversies still exist about the extent of non-hepatic glucose production in various conditions in humans.

The gluconeogenic potential of extrahepatic organs

Despite conflicting evidences still existing regarding actual gluconeogenic potential of the kidney and other organs (Gerich *et al.*, 2001, Previs *et al.*, 2009), the issue whether humans can produce significant amounts of glucose outside the liver has been conclusively solved. Two critical groups of experiments, although performed after the renal glucose production had already been considered non-negligible in humans with various arterio-venous balance and tracer infusion techniques, definitely and unequivocally demonstrated a significant non-hepatic glucose production. These studies used as an experimental model the anhepatic phase of liver transplantation i.e. the lag-time in which the recipient's liver has been removed but the donor's liver has not yet been replaced. This is a non-physiological, clear-cut model, to demonstrate nonhepatic gluconeogenesis, in the absence of the liver confounding the results with its large output of glucose.

The study published first came from Dr Amiel's group (Joseph *et al.*, 2000). The experimental approach consisted of a primed continuous glucose tracer infusion coupled to an exogenous glucose infusion regulated to maintain moderate hyperglycaemia (7.5-8 mmol/l) through the experiment. The glucose clamp procedure was maintained for the whole duration of the surgical operation, and any variations in the rate of glucose infusion required to maintain the glucose target were recorded. Compared to the other surgical phases, during the anhepatic phase the rate of glucose infusion required only a modest increment in the exogenous glucose infusion, suggesting that either glucose clearance was reduced or endogenous glucose production was maintained. Based on the analysis of tracer derived data, the authors calculated that at the end of the anhepatic phase endogenous glucose was released at a rate approximately 50% of the basal state.

The second group of studies followed a slightly different experimental approach (Battezzati *et al.*, 2002; Battezzati *et al.*, 2004). Subjects undergoing liver transplantation wee studied during the procedure and again several months after its successful completion, and compared to a control group. The experiment begun as soon as the recipient's liver was removed, and ended when the graft circulation was re-established by portal unclamping. In order to obtain all experimental evidence for non hepatic glucose production within this time frame, glucose and amino acid kinetics were investigated by means of stable-isotopes labelled tracers bolus injections followed by frequent arterial glucose sampling. In this way the kinetics of glucose, of the gluconeogenic amino acids alanine and glutamine and of leucine were measured. In addition, the gluconeogenic transfer of 13C of glutamine and alanine tracers to glucose could be determined.

It was found that without the liver, the body can produce and dispose an increased load of gluconeogenic amino acids without net changes in the proteolytic rate, i.e. in their appearance from endogenous proteins catabolism. The plasma and whole-body alanine content and kinetics increased three and two fold respectively compared to controls, with a halved metabolic clearance and a doubled production rate, which was 2% greater than for disposal and caused the increment in concentrations. Free whole-body glutamine decreased 25%, but increased 50% in plasma. Glutamine clearance was halved, and the production decreased by 25%, but remained 2% greater than for disposal. Liver replacement decreased alanine and glutamine concentrations, leaving leucine unchanged.

At the beginning of the anhepatic phase, plasma glucose concentration was increased and then slowly declined, approximately by 15% per hour. Sampling across artery and renal vein showed an arteriovenous gradient of 3.7 mg/dl, proving in line of principle that the right kidney released glucose. In contrast, arterial and portal glucose concentrations were not significantly different, a fact that does not exclude simultaneous glucose uptake and release from the gut. The glucose clearance was 25% reduced, but glucose uptake was similar to that of the control groups. Glucose production without liver was only 30% reduced compared to controls. Glucose became enriched with 13C from glutamine and at a minor extent from alanine, proving conclusively the extrahepatic gluconeogenesis. In addition to alanine and glutamine, the gluconeogenic precursors lactate, pyruvate, and glycerol, and insulin, and the counterregulatory hormones epinephrine, cortisol, growth hormone, and glucagon were all increased severalfold.

The studies during the anhepatic phase of liver transplantation proved that extrahepatic organs synthesize glucose at a rate similar to postabsorptive healthy subjects, when gluconeogenic precursors and counterregulatory hormones are markedly increased. They also suggested that the kidney is the main but possibly not the unique source of the extrahepatic glucose production. The liver proved to be necessary to the disposal of three-carbon compounds as pyruvate, lactate and alanine, but not for glutamine disposal and gluconeogenesis.

Evidence supporting renal glucose production

It is well known that the renal cortex is provided with the full set of enzymes necessary to gluconeogenesis, as evidentiated by the studies of Hans Krebs already in the midst of the last century (Nishiitsutsuji-Uwo *et al.*, 1967; Weidemann and Krebs, 1969). The gluconeogenic potential of the kidney has therefore never been questioned in line of principle, but the quantification of the actual contribution of the kidney to glucose production has been subject of considerable debate. Most of the controversies have arisen from the inaccuracy in measurement of the renal blood flow (which is large), of the glycemic gradient across the renal bed (which is small or even neutral, because the kidney can simultaneously produce and take up glucose), and of the gradients in glucose tracer enrichments or specific activities across the same districts when tracers are simultaneously administered in order to quantify organ glucose uptake and release (Moller *et al.*, 2001).

The estimates of the renal contribution to endogenous glucose production range from almost zero to 30% of the endogenous glucose production (Cersosimo *et al.*, 2000; Ekberg *et al.*, 1999; Moller *et al.*, 2001; Stumvoll *et al.*, 1995). The most consistent contribution of the kidney to the postabsorptive glucose production came from lactate and glutamine, with minor contributions of alanine (Stumvoll *et al.*, 1998a). Reenal glucose production was also suppressed by insulin (Meyer *et al.*, 1998), stimulated by epinephrine and insensitive to glucagon (Stumvoll *et al.*, 1998a, Stumvoll, 1998b).

This issue has been reviewed in 2001 by Gerich *et al.* summarizing the evidence available at that time, providing the following conclusions: (1) the human kidney both consumes and releases glucose in the postabsorptive state. (2) hepatic glucose release cannot account for overall endogenous glucose glucose production. (3) as far as gluconeogenesis is concerned, the kidney may be roughly as important a as the liver. (4) renal glucose release and uptake are under insulin and epinephrine control. (5) during hypoglycemia the kidney can compensate at least partially for impaired hepatic glucose release. (6) in both types 1 and 2 diabetes the kidney releases more glucose contributing to hyperglycemia. Up to now there are unresolved issues concerning the role of the kidney gluconeogenesis in the pathogenesis of renal and hepatic failure, severe infections and aging.

Evidence supporting intestinal glucose production

The existence of intestinal gluconeogenesis in humans is currently matter of debate. In 1978 evidence for gluconeogenic conversion of glutamine carbons in the rat small intestine was provided by Windmueller and Spaeth (Windmueller, 1978), but the idea that fructose and other gluconeogenic substrates could be converted to glucose in the human intestine thanks to a functionally active glucose-6Pase dates back to *in vitro* studies of White and Landau in 1965. It has also been known for a long time that the complete set of gluconeogenic enzymes is expressed in the small intestine of rats, mice, and rabbits at the time they are suckling even though their activities decrease to very low levels thereafter (Hahn and Wei-Ning, 1986; Anderson and Rosendall, 1973; Hahn and Smale, 1982; Watford and Tatro, 1989; Westbury and Hahn, 1984). Recently, when northern blotting and reverse-transcription polymerase chain reaction and highly specific Glc6Pase assay become available, Rajas *et al.* (1999) studied the expression of the Glc6Pase gene in human and in rat tissues and demonstrated that the Glc6Pase gene is expressed in the duodenum, jejunum, and ileum in humans. The same group also showed that both the intestinal rat Glc6Pase and phosphoenolpyruvate carboxykinase (another major regulatory gene of gluconeogenesis) mRNA abundances and protein activities are markedly increased by fasting and insulinopenia (Croset *et al.*, 2001).

The latter findings have substantially contributed to the hypothesis that in mammalians the small intestine might possess the capacity to produce glucose in portal blood and that this production might be augmented in insulinopenic states. This hypothesis requires to be unequivocally proved by kinetic studies quantifying enteral gluconeogenesis.

In suckling rats intestinal mucosa *in vitro*, low-rate gluconeogenesis was quantified by Hahn and Wei-Ning (1986). Critical studies in the adult rat *in vivo* were performed by Croset *et al.* (2001) combining arteriovenous balance and isotopic techniques. They found that in insulinopenia the rat small intestine can contribute up to one fourth of the total endogenous glucose production. Similarly to liver glucose production, small intestine glucose production is acutely suppressed by insulin infusion. The preferred gluconeogenic substrate is glutamine and, to a much lower extent, glycerol. In a subsequent study using this technique, the same group showed that 72 h fasting can increase the contribution of the small intestine up to one third of glucose production (Mithieux *et al.*, 2004). These findings prompt to speculate that the small intestine plays an important regulatory role during prolonged fasting, when increased enteral glucose production is counterparted by decreased hepatic glucose production and increased glycogen stores, hence sparing liver protein catabolism for gluconeogenesis. A second speculation points to the potential role of the small intestine in the production of hyperglycemia in uncompensated diabetes.

Martin *et al.* (2007) challenged this view with *in vitro* and *in vivo* studies published in 2007. They showed that segments from the small intestine of 72 h-fasted rats, incubated with 13C-labelled glutamine, did not synthesize 13C-labelled glucose at all despite a high rate of glutamine utilization and metabolism. In addition they contended that their arteriovenous metabolite concentration difference measurements across the portal vein-drained viscera of 72 h-fasted Wistar and Sprague-Dawley rats couldn't in any way account for the rate of intestinal gluconeogenesis reported by Croset and Mithieux in their 72 h-fasted rats. Finally, they stated that glutamine gluconeogenesis would not be detectable by the currently available methods because the dilution of circulating glucose by newly synthesized glucose would be much smaller than the margins of error of the methods employed. This work and further negative data reported in abstract form from two other groups in piglets, rats and dogs (Burrin *et al.*, 2005; Brunengraber *et al.*, 2007), have led Previs *et al.* (2009) to critically evaluate in a recent review the radioactive and stable isotopic techniques used to measure intestinal gluconeogenesis. In a theoretical work he simulated the impact of unavoidable measurement errors on apparent rates of intestinal gluconeogenesis and concluded that it is currently not possible to provide anything more than rough estimates of intestinal glucose production. In face of the current controversy it is therefore necessary to provide more detailed evidence in order to have the hypothesis generally accepted.

Evidence supporting muscle glucose production

It has been known for years that patients with glycogen storage disease type 1, despite their enzymatic defect, can maintain a residual endogenous glucose production acconting fot up to 50% of healthy subjects, especially when they become older (Kalhan *et al.*, 1982; Tsalikian *et al.*, 1984; Weghuber *et al.*, 2007). The source of this glucose could either derive from residual G6Pase or non specific phosphatases, from amylo-1,6-glucosidase or from acid alpha glucosidase activity (Moses, 2002). Recently, endogenous glucose production was investigated in a patient virtually lacking liver, kidney and intestine glucose production because of an absolute defect of the glucose-6-phosphatase isoform α expressed in these tissues (Huidekoper *et al.*, 2010). In this patient a residual endogenous glucose production was found both from gluconeogenesis and glycogenolysis, accounting for approximately 30% of the rate in healthy subjects.

The authors speculated that this glucose must derive from tissues expressing the glucose-6-phosphatase isoform β, recently proposed by Shieh *et al.* (2003). The expression of this phosphatase, renamed to G6PC3 (glucose-6-phosphatase catalytic subunit) (Hutton *et al.*, 2009) is much lower in kidney and small intestine than in muscle and has been shown to be functionally active in the muscle that, in addition has the largest body reservoir of glycogen. Muscle is therefore the most likely candidate tissue for extrahepatic and extrarenal glucose production. However, direct evidence for the muscular origin of glucose production has to be provided yet.

Evidence supporting brain glucose production

In the brain, neuronal and glial cells are thightly interconnectd from the metabolic standpoint. The human brain contains a substantial amount of glycogen, that can be non invasively detected by *in vivo* with localized 13C-NMR spectroscopy (Oz *et al.*, 2007). At the cellular level, this glycogen is stored in astrocytes and not in neurons even though both cell types use glucose for their energy needs. The glycogen stores are regulated by glucose, hormones, and neurotransmitters, suggesting an important metabolic function: it has been shown that astrocytic glycogen utilization can support neurons during hypoglycemia and during periods of high neuronal activity (Brown and Ransom, 2007). In humans, brain glycogen supports energy metabolism when glucose supply from the blood is inadequate and glycogen levels rebound higher than normal after a single episode of moderate hypoglycemia (Oz *et al.*, 2009). It has been shown that after glycogen breakdown, neurons can export glucose to other cells via gap junctions, or lactate for the use of surrounding neurons (Rouach *et al.*, 2008; Suh *et al.*, 2007).

Astrocytes have recently been found to express the ubiquitous glucose-6Phosphatase also expressed in the muscle, and the glucose-6-P transporter necessary to provide a functional complex (Ghosh *et al.*, 2005). This concept would lead to the consequence that astrocytes could release the stored glycogen as glucose in the brain. Therefore, Shieh *et al.* (2003) hypothesized that in hypoglycemia, astrocytes may provide glucose directly to neurons and possibly to plasma. This issue was addressed with brain microdialysis studies that found evidence of compartmentalisation of brain glucose, suggesting that neurons obtain the majority of glucose reaching via the astrocytic intracellular space and the ECF. Astrocytic glucose-6-phosphatase may therefore permit astrocytes to modulate the trans-astrocytic flux of glucose to adjacent neurons in response to signals reflecting increased neuronal demand. (Forsyth *et al.*, 1996).

Evidence for such release in humans is currently lacking, and it would be difficult to provide by conventional arterio-venous balance techniques coupled to metabolic tracers infusions, because of the complexity of the brain vascularization. However, Eyre *et al.* (1994), documented in 19 children undergoing cardiopulmunary bypass surgery, transitory periods in which the jugular vein glucose concentration exceded the aterial value by up to 5.3 mmol/l. They concluded that the magnitude of the release implies the presence of a glucose reservoir in brain, supporting the idea that that astrocytic glycogen can be exported as glucose, permitting spatial and temporal modulation of glucose delivery to neurons.

Conclusion

In recent years, advances at the molecular level in the knowledge of the gene expression profile of gluconeogenic enzymes in various cells and tissues have opened new perspectives. The idea that endogenous glucose production for the use of other tissues is a function exlusively performed by the liver has been seriously challenged. Specifically a whole family of glucose 6P phosphatase has been discovered, with one subtype ubiquitously distributed. In addition to the liver, the small intestine, the muscle and the brain have been individuated as possible sources of endogenous glucose.

The final proof of these hypothesis depends on the quantification of glucose release from specific organs and tissues in various conditions *in vivo*. Unfortunately, methodologies and studies for substrate kinetics and interorgan metabolic exchanges have not kept the pace with molecular methodologies and studies. The invasiveness of catheterization organ balance studies, the inherent inaccuracy of tracer methodologies, and the complexity of mathematical modelling of biological systems, have not substantially improved in the last two decades. These difficulties continue to be the limiting factors to the full understanding of the metabolic picture at the whole-body level *in vivo*.

The existence and the extent of glucose production from the small intestine, the muscle and the brain, therefore continues to be a matter of debate. Any advancement in this field, however, will have the potential to change the way we consider the concept of liver and kidney reciprocity, the physiological response to exercise and to feeding, and the pathogenesis of many conditions like diabetes, hypoglycecemia, liver and hepatic failure, ischemic and neurodegenerative diseases.

References

Anderson, J.W. and A.F. Rosendall, 1973. Gluconeogenesis in jejunal mucosa of guinea pig. Biochim. Biophys. Acta 304, 384-388.

Battezzati, A., A. Caumo, A. Fattorini, L.P. Sereni, J. Coppa, R. Romito, M. Ammatuna, E. Regalia, V. Mazzaferro and L. Luzi, 2002 Amino acid kinetics during the anhepatic phase of liver transplantation. Diabetes 51, 1690-1698.

Battezzati, A., A. Caumo, F. Martino, L.P. Sereni, J. Coppa, R. Romito, M. Ammatuna, E. Regalia, D.E. Matthews, V. Mazzaferro and L. Luzi, 2004 Nonhepatic glucose production in humans. Am J Physiol Endocrinol Metab. 286, E129-135.

Brown, A.M. and B.R. Ransom, 2007 Astrocyte glycogen and brain energy metabolism. Glia 55, 1263-1271.

Burrin, D., B. Lambert, B. Stoll and X. Guan, 2005. Gastrointestinal gluconeogenesis is absent in the 36-hour fasted piglet. FASEB J. 19, A1675.

Brunengraber, D.Z., C.S. Sayre, F. Allen, A.V. Cendrowski, J.R. Sanabria and S.F. Previs, 2007. Absence of intestinal gluconeogenesis in rats and dogs. FASEB J. 21, A1073.

Cersosimo, E., P. Garlick and J. Ferretti, 2000 Renal substrate metabolism and gluconeogenesis during hypoglycemia in humans. Diabetes 49, 1186-1193.

Croset, M., F. Rajas, C. Zitoun, J.M. Hurot, S. Montano and G. Mithieux, 2001 Rat small intestine is an insulin-sensitive gluconeogenic organ. Diabetes 50, 740-746.

Ekberg, K., B.R. Landau, A. Wajngot, V. Chandramouli, S. Efendic, H. Brunengraber and J. Wahren, 1999 Contributions by kidney and liver to glucose production in the postabsorptive state and after 60 h of fasting. Diabetes 48, 292-298.

Eyre, J.A., A.G. Stuart, R.J. Forsyth, D. Heaviside and K. Bartlett, 1994 Glucose export from the brain in man: evidence for a role for astrocytic glycogen as a reservoir of glucose for neural metabolism. Brain Res. 635, 349-352.

Forsyth, R., A. Fray, M. Boutelle, M. Fillenz, C. Middleditch and A. Burchell, 1996. A role for astrocytes in glucose delivery to neurons? Dev Neurosci. 18, 360-370.

Gerich, J.E., C. Meyer, H.J. Woerle and M. Stumvoll, 2001 Renal gluconeogenesis: its importance in human glucose homeostasis. Diabetes Care 24, 382-91.

Ghosh, A., Y.Y. Cheung, B.C. Mansfield and J.Y. Chou, 2005. Brain contains a functional glucose-6-phosphatase complex capable of endogenous glucose production. J. Biol. Chem. 280, 11114-11119.

Hahn, P. and F.A. Smale, 1982. Phosphoenolpyruvate carboxykinase in the small intestine of developing rodents. J. Nutr. 112, 986-989.

Hahn, P. and H. Wei-Ning, 1986 Gluconeogenesis from lactate in the small intestinal mucosa of suckling rats. Pediatr Res. 20, 1321-1323.

Huidekoper, H.H., G. Visser, M.T. Ackermans, H.P. Sauerwein and F.A. Wijburg, 2010 A potential role for muscle in glucose homeostasis: *in vivo* kinetic studies in glycogen storage disease type 1a and fructose-1,6-bisphosphatase deficiency. J Inherit Metab Dis. 33, 25-31.

Hutton, J.C. and R.M. O'Brien, 2009 Glucose-6-phosphatase catalytic subunit gene family. J Biol Chem. 284, 29241-29245.

Joseph, S.E., N. Heaton, D. Potter, A. Pernet, M.A. Umpleby and S.A. Amiel, 2000 Renal glucose production compensates for the liver during the anhepatic phase of liver transplantation. Diabetes 49, 450-456.

Kalhan, S.C., C. Gilfillan, K.Y. Tserng and S.M. Savin, 1982 Glucose production in type I glycogen storage disease.J Pediatr. 101, 159-160.

Martin, G., B. Ferrier, A. Conjard, M. Martin, R. Nazaret, M. Boghossian, F. Saadé, C. Mancuso, D. Durozard and G. Baverel, 2007. Glutamine gluconeogenesis in the small intestine of 72h-fasted adult rats is undetectable. Biochem. J. 401, 465-473.

Meyer, C., J. Dostou, V. Nadkarni and J. Gerich, 1998 Effects of physiological hyperinsulinemia on systemic, renal, and hepatic substrate metabolism. Am J Physiol. 275(6 Pt 2), F915-921.

Mithieux, G., I. Bady, A. Gautier, M. Croset, F. Rajas and C. Zitoun, 2004 Induction of control genes in intestinal gluconeogenesis is sequential during fasting and maximal in diabetes. Am J Physiol Endocrinol Metab. 286, E370-375.

Moller, N., R.A. Rizza, G.C. Ford and K.S. Nair, 2001 Assessment of postabsorptive renal glucose metabolism in humans with multiple glucose tracers. Diabetes 50, 747-751.

Moses, S.W., 2002 Historical highlights and unsolved problems in glycogen storage disease type 1. Eur J Pediatr. 161 Suppl 1, S2-9.

Nishiitsutsuji-Uwo, J.M., B.D. Ross and H.A. Krebs, 1967. Metabolic activities of the isolated perfused rat kidney. Biochem. J. 103, 852-862.

Oz, G., E.R. Seaquist, A. Kumar, A.B. Criego, L.E. Benedict, J.P. Rao, P.G. Henry, P.F. Van De Moortele and R. Gruetter, 2007 Human brain glycogen content and metabolism: implications on its role in brain energy metabolism. Am J Physiol Endocrinol Metab 292, E946-E951.

Oz, G., A. Kumar, J.P. Rao, C.T. Kodl, L. Chow, L.E. Eberly and E.R. Seaquist, 2009 Human brain glycogen metabolism during and after hypoglycemia. Diabetes 58, 1978-19785.

Previs, S.F., D.Z. Brunengraber and H. Brunengraber, 2009 Is there glucose production outside of the liver and kidney? Annu Rev Nutr. 29, 43-57.

Rajas, F., N. Bruni, S. Montano, C. Zitoun and G. Mithieux, 1999 The glucose-6 phosphatase gene is expressed in human and rat small intestine: regulation of expression in fasted and diabetic rats. Gastroenterol. 117, 132-139.

Rouach, N., A. Koulakoff, V. Abudara, K. Willecke and C. Giaume, 2008 Astroglial metabolic networks sustain hippocampal synaptic transmission. Science 322, 1551-1555.

Shieh, J.J., C.J. Pan, B.C. Mansfield and J.Y. Chou, 2003. A glucose-6-phosphate hydrolase, widely expressed outside the liver, can explain age-dependent resolution of hypoglycemia in glycogen storage disease type Ia. J. Biol. Chem. 278, 47098-470103.

Stumvoll, M., U. Chintalapudi, G. Perriello, S. Welle, O. Gutierrez and J. Gerich, 1995 Uptake and release of glucose by the human kidney. Postabsorptive rates and responses to epinephrine. J Clin Invest. 48, 292-298.

Stumvoll, M., C. Meyer, G. Perriello, M. Kreider, S. Welle and J. Gerich, 1998 Human kidney and liver gluconeogenesis: evidence for organ substrate selectivity. Am J Physiol. 274(5 Pt 1), E817-826.

Stumvoll, M., C. Meyer, M. Kreider, G. Perriello and J. Gerich, 1998 Effects of glucagon on renal and hepatic glutamine gluconeogenesis in normal postabsorptive humans. Metabolism 47, 1227-12232.

Suh, S.W., J.P. Bergher, C.M. Anderson, J.L. Treadway, K. Fosgerau and R.A. Swanson, 2007 Astrocyte glycogen sustains neuronal activity during hypoglycemia: studies with the glycogen phosphorylase inhibitor CP-316, 819 ([R-R*, S*]-5-chloro-N-[2-hydroxy-3-(methoxymethylamino)-3-oxo-1-(phenylmet hyl) propyl]-1H-indole-2-carboxamide). J Pharmacol Exp Ther 321, 45-50.

Tsalikian, E., P. Simmons, J.E. Gerich, C. Howard and M.W. Haymond, 1984 Glucose production and utilization in children with glycogen storage disease type I. Am J Physiol. 247(4 Pt 1), E513-519.

Watford, M. and A.V. Tatro, 1989. Phosphoenolpyruvate carboxykinase of rat small intestine: distribution and regulation of activity and mRNA levels. J. Nutr. 119, 268-272.

Weghuber, D., M. Mandl, M. Krssák, M. Roden, P. Nowotny, A. Brehm, M. Krebs, K. Widhalm and M.G. Bischof, 2007 Characterization of hepatic and brain metabolism in young adults with glycogen storage disease type 1: a magnetic resonance spectroscopy study.Am J Physiol Endocrinol Metab. 293, E1378-384.

Weidemann, M.J. and H.A. Krebs, 1969. The fuel of respiration of rat kidney cortex. Biochem. J. 112, 149-166.

Westbury, K. and P. Hahn, 1984. Fructose-1,6-biphosphatase activity in the intestinal mucosa of developing rats. Am. J. Physiol. 246, G683-686.

White, L.W. and B.R. Landau, 1965 Sugar transport and fructose metabolism in human intestine *in vitro*. J Clin Invest 44, 1200-1213.

Windmueller, H.G. and A.E. Spaeth, 1978. Identification of ketone bodies and glutamine as the major respiratory fuels *in vivo* for postabsorptive rat small intestine. J Biol Chem. 253, 69-76.

From diets to splanchnic released energetic nutrients: empirical predictions in ruminants

I. Ortigues-Marty[1], C. Loncke[1,2], J. Vernet[1], H. Lapierre[3], D. Sauvant[4] and P. Nozière[1]
[1]INRA UR1213 Unité de Recherches sur les Herbivores, Site de Theix, 63122 Saint-Genès Champanelle, France
[2]Present address: Union Invivo ets INZO° SAS, rue de l'Eglise, BP50019, 02407 Chierry Cedex, France
[3]Agriculture and Agri-Food Canada, STN Lennoxville, Sheerbrooke, QC, Canada
[4]INRA-AgroParisTech, UMR 791, Physiologie de la Nutrition et al.imentation, 75231 Paris, France

Abstract

To meet the new constraints and/or objectives of ruminant production systems, feeding systems have to evolve to be more nutrient based. The energy value of feeds being still defined as an aggregated unit, objective of the present work was to evaluate the feasibility of describing energy by the amount of energetic nutrients appearing in the portal vein and released by the splanchnic area towards peripheral tissues. Using the exhaustive Flora database which includes all published data on net splanchnic fluxes of nutrients measured in ruminants, empirical prediction equations were derived by meta-analysis. Equations were validated against other databases established for related parameters and a few unpublished data. Dietary and animal characteristics as defined by INRA proved to be valuable predictors of the net splanchnic release of energetic nutrients. Their use as additional rationing tools to predict animal responses, unaccounted for in present system, is promising. The quantitative coherence between the Metabolisable Energy and Absorbed Energy concepts shows that the necessary condition for energy feeding systems to evolve towards nutrient-based systems will soon be met.

Introduction

For ruminant nutrition research, an improved control of the transformation of feeds into animal tissues and production of economical interest (meat, milk, etc.) has been a long standing goal. Major progresses have been made in the chemical characterization of feeds as well as of their extent and site of digestion. Wells of knowledge have also been accumulated on quantitative *in vivo* tissue and nutrient metabolism. At the same time, ruminant production systems have been facing additional constraints of improved efficiency, product quality, feed security and animal health while limiting wastes to the environment besides productivity.

The present challenge is to select and aggregate the most relevant knowledge into tools that could be used in rationing practices to address the multiple objectives that the ruminant industry is now facing. A number of the newly arising production objectives depend of the adequate prediction of amount and nature of nutrients available for absorption through the gut, following digestion processes, and of their metabolic fate among and within the different body tissues (gut, liver, muscles, adipose tissues, mammary gland, etc.). An evolution towards mechanistic nutrient based systems has been proposed (Reynolds, 2000). Mechanistic digestion and metabolism models exist (as summarised by Nozière *et al.*, 2010b). They have been developed, however, for research purposes and numerous variables and parameters which limit their use in rationing tools. Alternatively, can existing feed evaluation systems integrate nutrient based rationing criteria and thereby accommodate for these multiple objectives?

The French ruminant energy feeding system and objectives of evolution

Existing feeding systems have been developed to adjust rations to the genetic potential of animals, the main issue being productivity and profitability. The INRA feeding system includes a feed evaluation system that defines two nutritional values for each feed, an energy and a protein value, as well as a fill value. It also includes a nutritional requirement system applicable to all types of ruminants. The system is developed for animals as individuals and the metabolic animal unit is considered as a black box. This system is widely used in practice and has proved its robustness (INRA, 2007).

The present paper focuses on energy. Unlike the protein value of feeds, which is now described in terms of Amino Acids Digestible in the Intestine (AADI; INRA, 2007) from feed characteristics, energy is still defined as an aggregated unit. The energy value depends on the chemical composition of feeds and their digestibility (associating *in vivo* and *in situ* data at ruminal and intestinal levels) as well as the efficiency of energy utilisation at reference feeding levels. A specificity of the French energy system was to choose net energy as the unit to define the energy value of feeds and rations.

On these bases, our research objective was to evaluate the feasibility of describing the energy value of feeds and diets in terms of nutrients made available to tissues (in quantity and quality), first at portal and second at splanchnic level. Two constraints were set. The first one was to select predictors of energetic nutrients from the newly revised INRA feed tables (INRA-AFZ, 2004; INRA, 2007). These objectives were set in light of the recent advances towards the prediction of the amount of digestion end-product [starch, fatty acid, volatile fatty acids (VFA)…] from (1) a precise and thorough description of the chemical characteristics of feeds and (2) the introduction of the site and extent of digestion of these individual chemical entities (Offner and Sauvant, 2004). Indeed, only feed characteristics are measurable and available in practice and the objective was to develop prediction models fully compatible with the INRA Feeding System and its associated rationing tool (INRATion). The second one was for the prediction models to apply to the diversity of animal types used in the major ruminant production systems.

Two major questions were formulated:
1. Can net portal absorption of the major energetic nutrients be predicted from dietary characteristics?
2. Can the net splanchnic release of nutrients be predicted from dietary and animal characteristics?

Energetic nutrients considered were the hydrosoluble ones, absorbed into portal blood circulation and contributing for 90 to 95% to metabolisable energy (Journet *et al.*, 1995): major VFA (acetate, propionate and butyrate), glucose and L-lactate, β-hydroxybutyrate (BHBA) and α-amino-nitrogen [α-amino-N, as an indicator of total amino nitrogen (Martineau *et al.*, 2009)]. Long-chain fatty acids absorbed into lymph and not measured with blood trans-organ balances were not considered here.

An empirical meta-analysis modelling approach

An empirical meta-analysis approach was applied in order to develop a quantitative review of the available data and their aggregation towards response laws (Sauvant, 1992). The definite strength of this approach lays in the robustness of the empirically derived response laws, with the limit that the derived laws depend on the physiological and nutritional conditions associated to the data used. Their application is defined by the range of validity of the meta-design. Outside this range of validity, the derived predictions have to be used with caution.

Database and range of validity of the models

The Flora (Flux of nutrients across tissues and Organs in Ruminant Animals; Vernet and Ortigues-Marty, 2006) database was used. It is an exhaustive compilation of the approximately 250 international publications on splanchnic net flux measurements in ruminants (Vernet *et al.*, 2007). It encompasses a wide range of nutritional and physiological conditions.

Intake levels ranged up to 42 g DM / d / kg BW for diets containing up to 100% concentrate (Table 1). The most frequent feeds were the cocks-foot and lucerne hays and corn silage for forages, and corn grain, wheat, corn gluten meal and soya bean meal for concentrates. Lipid supplementation, use of feed additives such as buffers, probiotics, or changes in feed particle size were lacking. Animals included cattle and sheep (no goats), either at or below maintenance, or growing, fattening, gestating or lactating. Milk yields were up to 47.7 kg a day. Body weight gains were up to 1,920 g/d, intensive finishing beef cattle were missing. Low feeding levels and even fasting conditions were represented in sheep.

Considering the work load involved and the public concern on animal experimentation and ethics, it is unlikely that forthcoming research will completely fill the gaps.

Table 1. Range of validity of the prediction models as defined by data available within the Flora database.

Item	Mean	SE	Min	Max
Diet composition (g/kg DM)				
Digestible organic matter	646	94	442	852
Neutral detergent fibre	455	156	133	736
Crude protein	134	35	45	247
Crude fibre	233	94	40	417
Starch	34	36	0	709
Rumen fermentable organic matter	502	43	401	633
Rumen digestible NDF	239	88	85	427
PDI[1]	91	15	28	142
UFL[1]	0.84	0.17	0.44	1.19
% concentrate	37	35	0	100
ME (MJ/ kg DM)	9.95	1.61	6.44	13.12
Intake (g/d/kg body weight)				
DM	21.5	7.64	0	42
Digestible organic matter	13.9	5.02	0	29
Neutral detergent fibre	9.88	5.41	0	27
Starch	3.81	4.06	0	17
Rumen fermentable organic matter	10.9	3.98	0	21
Rumen digestible NDF	5.22	2.72	0	14
Rumen digestible starch	2.47	2.76	0	12

[1] Protein digestible in the intestine and feed unit for lactation as defined by INRA (2007).

Homogeneous characterisation of animals, feeds and diets

Aggregation of nutritional results requires that animals, feeds and diets are described on a common and homogenous basis. Breed, animal type, sex, age and physiological status were the obvious criteria to homogenously describe animals. These qualitative descriptors were completed by quantitative ones: the net energy for production and the energy balance calculated according to INRA (1987, 1988). Tissue energy deposition or mobilisation could thus be estimated. In case of energy mobilisation, the composition of mobilised tissues was estimated and the arterial supply of endogenous nutrients (long chain fatty acids, glycerol and alanine) to splanchnic tissues was estimated (Loncke, 2009).

Major efforts were developed to homogeneously characterise feeds and diets, since publications reporting net splanchnic nutrient fluxes do not systematically nor homogenously report the chemical composition and nutritional value of feeds and diets. We chose to describe all feeds according to INRA feed tables (INRA-AFZ, 2004; INRA, 2007) which give thoroughly detailed, robust and updated information on the chemical and digestion characteristics of an extremely large panel of concentrates (n=160) and forages (n=1,260). After selection, for each publication, of the most representative feed ingredients, the chemical composition and nutritional value of the diets were calculated from Feed Tables by additivity. The calculated values were validated by comparison with the values reported by the authors, when available (Loncke *et al.*, 2009). This strategy allowed testing, as predictors or as secondary variation factors, criteria that were not directly mentioned in the publications. It also allowed comparing the response equations with those derived from other databases using the same principles (Nozière *et al.*, 2010b).

Unit

All quantitative variables of intake, net nutrient fluxes and energy balance were expressed on a body weight basis. Digestive processes are better expressed per kg body weight (Sauvant *et al.*, 2006). This unit was also the only one that allowed normal distributions of the explanatory variables and a full overlap across cattle and sheep data, as already noted by Vernet *et al.* (2005). Loncke *et al.* (2010b) confirmed the relevance of this unit for the energy expenditure of portal drained-viscera.

Within-study meta-analyses, predictors and interfering factors

In order to identify the major driving forces while limiting at best the risk of biases, a within-study approach was applied (Sauvant *et al.*, 2005, 2008). In an iterative process, studies or groups of treatments within publications were selected from Flora if they varied for only one variation factor, chosen as the assumed main predictor of the relationship. The range of nutritional conditions available within Flora could then be fully explored to compare several variation factors and identify the main predictor. The study effect was considered as fixed on the bases that experimental conditions varied across publications first because of clear methodological choices by the research group rather than on a random basis. Consequently, all parameters of the statistical models (slopes, intercepts, LSMeans, within and across study-residues) were tested for influent data and for systematic bias due to secondary variation factors, called interfering factors. Search for interfering factors was systematic and thorough using all quantitative and qualitative variables available in the dataset, namely all variables describing the animals as well as the feeds and diets including the extent and site of digestion or metabolic interactions. Physiological relationships were first established. The predictors and the models that explained the greater proportion of variance and that presented the least interfering factors were considered as being the best, provided they were physiologically relevant. When possible, models were compared with those elaborated for related variables on different databases (Nozière *et al.*, 2010b). When relevant, physiological predictors were then aggregated to identify rationing predictors, easy to use in practice and more generic. In the end, two types of predictors were selected. The first ones represented the 'push' regulations of the mass action law. The second ones illustrated the 'pull' regulations, or more accurately the balance between the genetic potential of the animals and nutrient intake (Ortigues-Marty *et al.*, 2003).

Prediction models of net splanchnic release

In all derived models, the driving forces were similar whatever the species (sheep vs. cattle), nutritional or physiological (non-productive, growing, lactating, gestating) state of the animal. Models could then be applied to all ruminants, only the numerical value of a few parameters sometimes differed.

Table 2. Dietary and animal predictors of the net splanchnic release of energetic nutrients in ruminants (Loncke, 2009).

Nutrient	Predictors	Dietary interfering factors	Mean difference predicted-observed % of observed
Ketogenic nutrients			
Acetate	RFOMI, RdNDF/FOM	none	2.7
	Or RFOMI[1]	dietary NDF	3.4
Butyrate	RFOMI, RdNDF/FOM[1]	none	2.7
BHBA	RFOMI, RdNDF/FOM, EB[1]	none	6.3
Glucogenic nutrients			
Propionate	RFOMI, RdNDF/RFOM[1]	none	8.0
L-lactate	Rd Starch, RFOMI, RdNDF/FOM[1]	none	20
	Or Rd Starch, EB[1]	none	
α-amino-N	PDIE[1]	dietary N content	10
Glucose	Rd and Id Starch, RFOMI, RdNDF/FOM, PDIE[1]	none	14

RFOM(I): rumen fermentable organic matter (intake).
RdNDF: rumen digestible neutral detergent fibre.
Id Starch: intestinal digestible starch.
Rd Starch: rumen digestible starch.
PDIE: protein digestible in the intestine, as limited by the energy supply.
EB: energy balance.
[1] Significant physiological status effect on intercept, LSMeans or slope.

Net splanchnic release of ketogenic nutrients

Splanchnic release of ketogenic nutrients could be predicted (Loncke *et al.*, 2008a, Loncke, 2009; Table 2) from two major dietary predictors: the rumen fermentable organic matter (RFOM) intake and its content in rumen digestible neutral detergent fiber (RdNDF) plus an animal predictor: the energy balance of the animal, for BHBA.

Net splanchnic release of acetate was predicted from its net portal appearance (NPA). As for the two other major VFA, propionate and butyrate,the NPA of acetate was predicted from a combination of two prediction equations: first, the prediction of total VFA-NPA from RFOM intake, and second the prediction of the pattern of portal VFA from the RdNDF concentration in RFOM. Across the liver, the release of acetate responded mainly to mass action law, as no simple relationship could predict endogenous acetate release which accounted for 0 (non productive and gestating adults), 10 (growing) and 24 (lactating) % of NPA. Net splanchnic release of acetate could also be predicted from RFOM intake only but with a lower adjustment and dietary NDF as an interfering factors.

The low net splanchnic release of butyrate was predicted from RFOM intake and the RdNDF concentration in RFOM. The equation combined the prediction of butyrate-NPA and of its net hepatic uptake averaging 75% of incremental butyrate-NPA.

The net splanchnic release of BHBA was based on RFOM intake, the RdNDF concentration in RFOM as well as the energy balance of the animals. Its prediction combined that of BHBA-NPA which depends on acetate and butyrate-NPA, and of its net hepatic release (Loncke *et al.*, 2008a). Basically, the equation combined two distinct responses. The first one applies to animals in positive energy balance where net BHBA release is tightly driven by the dietary supply of precursors. The second one applies to animals in negative energy balance where mobilised body fat becomes the major source of precursors of BHBA. The physiological state of the animal modified the parameters associated to the energy balance term, indicating that fatty acids are not used to the same extent in fat-mobilising animals.

Net splanchnic release of glucogenic nutrients

Splanchnic release of glucogenic nutrients could be predicted (Loncke, 2009; Loncke *et al.*, 2010a; Table 2) from a complementary and wide range of dietary predictors: the RFOM intake and its content in RdNDF, ruminally vs. intestinally digested starch and protein digestible in the small intestine as limited by energy (PDIE), and in some cases energy balance.

The low net splanchnic release of propionate was predicted from RFOM intake and the RdNDF concentration in RFOM. The equation combined the prediction of propionate-NPA and of its net hepatic uptake averaging 91% of incremental propionate-NPA.

The net splanchnic release of L-lactate could be predicted either from RFOM intake and the RdNDF concentration in RFOM, themselves predictors of propionate and L-lactate-NPA, or from rumen digestible starch (to predict L-lactate-NPA) and energy balance (able to predict both net hepatic uptake and release of L-lactate). These predictions will still require improvement to solve the remaining high discrepancy between predicted-observed values.

The, net splanchnic release of α-amino-N was directly derived from PDIE intake, expressing the fact that the mass action law is the first driving force of the net hepatic uptake of α-amino-N. It should be made clear that this equation was strictly established to estimate average gluconeogenic amino acid use in the liver, and will definitely need revision to address fate of individual amino acids.

The net splanchnic release of glucose was then predicted from the glucose-NPA and hepatic supply in gluconeogenic precursors of either dietary or endogenous origin as detailed in the present symposium (Loncke *et al.*, 2010a). This work clearly showed that in growing, finishing or non productive ruminants, net glucose release was directly related to the sum of all potential precursors. Precursor carbons for glucose synthesis were missing only for dairy cows.

Push versus pull type criteria

Two types of prediction criteria of net splanchnic fluxes of nutrients were thus identified. Predictors of nutrient-NPA were under the only control of dietary supply, thus driven by push-type regulations with no interfering factors. However for the liver, both types of predictors were significant. When hepatic metabolism of nutrients responds directly to a mass action law, then the NPA of the nutrient of interest or of its precursors were the best predictors. However, when hepatic metabolism is regulated by an intricate balance between nutrient supply and demand, the energy balance of the animal was an additional appropriate predictor of net hepatic flux. In fact, calculated energy balance was a relevant indicator of the supply of endogenous precursors to the liver, considering the scarcity of quantitative information especially on long chain fatty acid or glycerol. In this respect, NPA and energy balance can both be considered as predictors of nutrient fluxes driven by mass action law. However, it is in these later situations that prediction equations depended most on physiological status of the animals, reflecting additional intricate metabolic regulations and coordination between tissues for nutrient use.

Evaluation of prediction equations

Present models have first been extensively evaluated by comparing the derived predictions with other predictions obtained on related variables (e.g. amount and profile of VFA produced in the rumen) from other databases, characterised according to the same INRA dietary criteria. These evaluation steps have already been published (Nozière *et al.*, 2010a,b).

Absorbed energy versus metabolisable energy

Absorbed energy [AE, Loncke *et al.*, (2010b)] was defined as the difference between digestible energy and energy lost as methane or fermentation heat. It was calculated as:

AE = PDE-EE + PDV-NE + Energy-NPA, and

where PDE-EE: energy expenditure of portal drained-viscera (calculated from oxygen consumption); PDV-NE: net energy deposited in the portal-drained viscera (assumed as negligible); Energy-NPA: NPA of energy nutrient (acetate, propionate, butyrate, glucose, L-lactate, BHBA, total amino acids) + absorbed fatty acids; total amino acid-NPA derived from α-amino-N-NPA according to Martineau *et al.* (2009); absorbed fatty acids being estimated from dietary lipids assuming an apparent digestibility of 80% (Glasser *et al.*, 2008).

Loncke *et al.* (2010b) was able to demonstrate that a global quantitative coherence existed between the classical energetic model of Metabolisable Energy (ME) and the more physiologically based concept of AE. AE was composed of 80.2±4.1% of Energy-NPA and 19.8±4.06% of PDV-EE. It was linearly related to ME_{INRA} expressed either per unit body weight of the animal or per unit dry matter of the diet. Urinary energy and fermentation heat, both calculated according to Sauvant and Giger-Reverdin (2007a), were numerically but not significantly different (urinary energy being higher by 0.75 percentage units). Thus, AE was lower than ME_{INRA} by an average of 9.75% of ME_{INRA}. The difference increased with level of intake probably because of digestive interactions, not accounted for in the INRA calculation of ME. These results suggest that quantification of digestive interactions (Sauvant and Giger-Reverdin, 2007b) will probably improve our models.

INRA criteria selected as predictors

Dietary criteria

The dietary criteria selected from the INRA Feed Tables proved relevant to predict the NPA of all nutrients. Of primary importance were the criteria reflecting the site and extent of digestion of the different chemical fractions of the diet (Loncke *et al.*, 2008b, 2009). The initial and most important predictor was the RFOM intake which predicted the NPA of total volatile fatty acids and of BHBA. Molar proportions of VFA could then be predicted from the composition of RFOM (its content in RdNDF). Diet characteristics which take into account the site and extent of digestion of starch were the best predictors of glucose-NPA (amount of starch digested in the small intestine calculated according to Offner and Sauvant, 2004) and of L-lactate-NPA (amount of starch digested in the rumen). Finally, NPA of α-amino-N was best predicted by PDIE intake, rather than total N intake.

The RFOM is defined within the INRA (2007) PDI system as the sum of digestible organic matter – fat – undegradable crude protein – undegradable starch – fermentation products in the feed. In the INRA feed evaluation system, the digestibility of organic matter content is predicted from NDF, ADF and/or crude fiber contents for concentrates, and from crude protein and crude fiber contents for forages according to plant species and preservation methods. Undegradable crude protein and undegradable starch are estimated by the in sacco method. The RFOM criteria assumes that all digestible fibers and no fat are digested within the rumen. These approximations have probably little impact on predicted VFA-NPA as VFA are also produced in the large intestine and digestion of fat within the rumen is limited. This criteria which had proved its relevance to predict the energy available for microbial protein synthesis, also appeared fully relevant to predict other digestion end-products of ruminal origin. Two limits will have to be overcome; first, it is a calculated entity rather than a measured one. Second, despite the assumed additivity of this criteria, it does not account for digestive interactions.

The INRA Feed Table (2007) defines digestible NDF values for forages only. For concentrates, rumen digestible NDF was calculated as 'NDF - undigestible organic matter - undigestible crude protein - undigestible fat - undigestible starch', assuming a constant fat digestibility of 80% (Doreau and Ferlay, 1994). Digestible NDF from concentrates contributed up to 20% to dietary dNDF and uncertainties associated to this calculation were thus considered to have little impact on the dietary dNDF content. Rumen digestible NDF for all feeds was then defined assuming that 90% of NDF digestion took place in the rumen (Sauvant *et al.*, 2007). Finally, as predictor for VFA profile, RdNDF was expressed as a proportion of RFOM which reduces biases due to digestive interactions as both criteria are positively correlated (Archimède *et al.*, 1997).

Ruminally digested starch was calculated from starch concentration, in sacco degradability (INRA, 2007) and the empirical model of ruminal starch digestion (Offner and Sauvant, 2004). This model accounts for the difference between in sacco and *in vivo* ruminal degradation of starch and the effects of intake level.

The last relevant dietary criteria was the dietary PDIE content which showed a slightly better adjustment than the strict dietary N content or event PDIN. PDIE is the sum of PDIA (dietary protein undegraded in the rumen, but truly digestible in the small intestine) and PDIME (microbial protein that can be synthesized from the energy available in the rumen, when degraded N and other nutrients are not limiting, and which is truly digestible in the small intestine). The PDI system tolerates a deficit in degradable N, evaluated at -20 g maximum (PDIN-PDIE)/Feed Unit (INRA, 2007). Most Flora diets presented a limited degradable N deficit. For 95% of the diets, the (PDIN-PDIE)/Feed Unit value was within the defined boundaries of -20 and +20. It would remain to be tested whether PDIE is also the best predictor in case of higher degradable N deficits.

Animal criteria

The relevant quantitative animal criteria, energy balance, is common to several Feeding Systems. Uncertainties are certainly higher for this criteria compared to dietary criteria because of limited data within Flora on animal performance and measured energy balance. Assumptions had to be made. For growing and non-productive adults, the net energy value of feeds was used rather than the actual net energy of tissues was considered. For lactating cows, the energy and fat content of milk (when not reported) were assumed at 3.1 kJ/l and 40 g/kg milk (INRA, 2007), implying an error in the calculated energy balance of 15% for a difference in fat content of 10 g/kg milk. The amount of endogenous substrates (non esterified fatty acids, glycerol and alanine) made available from body tissue mobilisation was also estimated (Loncke, 2009). Despite these uncertainties, energy balance proved to be a fully relevant predictor and forthcoming efforts to improve the accuracy of theses predictions are warranted.

Use of RFOM in rationing practices

A direct application of our work is the use of the above identified predictors in rationing practices, as indicators of nutrient supply, in addition to the well defined rationing criteria. Indeed, some nutritional situations are not explained by the existing feeding systems. For example, the different growth performances and carcass measured in fattening cattle fed iso-energetic (on a net energy basis) and iso-nitrogenous (PDI basis) hay or maize silage diets (Table 3; Micol *et al.*, 2007). The forage diet was composed of 44% hay, 43% ground maize grain and 13% soybean meal, and the maize silage diet 58% silage, 24% ground maize grain and 18% soybean meal. These iso-energetic and iso-nirogenous diets differed by their RFOM content and their FOM composition. The maize silage diet was associated to a higher RFOM intake and a lower RdNDF content of RFOM. An increased amount of VFA appearing in the portal vein was predicted with little difference in VFA profile and amount of α-amino-N-NPA. The increased ratio of AE/PDI probably explained the higher proportion of adipose tissues in the carcass (Table 3).

Table 3. Measured intakes and performances in Blonde d'Aquitaine fattening steers (as reported by Micol et al., *2007) in relation to predicted dietary criteria and net portal appearance or splanchnic release of nutrients according to INRA (2007) and Loncke (2009).*

	Diets	
	Hay	Maize silage
Intake (g/d/kg body weight)		
UFV	0.018	0.019
PDIE	2.09	2.10
PDIN	2.02	2.00
Predicted FOM	9.78	10.64
Predicted RdNDF/FOM	0.38	0.30
Performances		
Average daily gain (g/d)	1,494	1,709
Adipose tissue (% carcass weight)	8.7	11.3
Predicted net portal appearance (mmol/d/kg body weight)		
Total VFA	59.4	64.5
β-hydroxybutyrate	5.86	6.90
Glucose	0.12	0.12
Lactate	5.58	6.39
α-amino-N	16.1	16.4
Acetate/propionate	2.6	2.3
Predicted net splanchnic release (mmol/d/kg body weight)		
Acetate	44.2	53.3
β-hydroxybutyrate	12.7	13.9
Glucose	16.8	18.5
Lactate	2.64	0.96
α-amino-N	15.5	15.9
Predicted absorbed energy (MJ/d/kg BW)	0.18	0.22
Absorbed energy / PDI	0.086	0.11

Conclusions

Present work resulted in three important outcomes: first, the relevance of the INRA dietary criteria which account for the extent and site of digestion of the different chemical entities as predictors of splanchnic released energetic nutrients and as criteria to be used in rationing practices; second, the relevance of the prediction models for sheep and cattle in different physiological conditions; and third, the quantitative coherence between the ME and AE concepts. It is now likely that after solving the 10% discrepancy between AE and ME_{INRA}, the necessary condition for energy feeding systems to evolve towards nutrient-based systems will be met. The next step will be to widen the range of validity of the models in order to encompass the inclusion of buffers, probiotics ... or fat supplements in the diets or of diet granulometry.

Acknowledgement

The authors wish to thank Limagrain and INZO which partly funded this work.

References

Archimède, H., D. Sauvant and P. Schmidely, 1997. Quantitative review of ruminal and total tract digestion of mixed diet organic matter and carbohydrates. Reprod. Nutr. Dev. 37, 173-189.

Doreau, M., and A. Ferlay, 1994. Digestion and utilization of fatty-acids by ruminants. Anim. Feed Sci. Technol. 45, 379-396.

Glasser F., A. Ferlay, M. Doreau, P. Schmidely, D. Sauvant and Y. Chilliard, 2008. Long-chain fatty acid metabolism in dairy cows: A meta-analysis of milk fatty acid yield in relation to duodenal flows and de novo synthesis. J. Dairy Sci. 91, 2771-2785.

INRA, 2007. Alimentation des bovines, ovins et caprins. Besoins des animaux. Valeurs des aliments. Tables INRA 2007, Versailles, France.

INRA, 1988. Alimentation des bovins, ovins et caprins. R. Jarrige (ed.) INRA pub, Paris, France.

INRA-AFZ, 2004. Tables de composition et de valeur nutritive des matières premières destinées aux animaux d'élevage Porcs, volailles, bovins, ovins, caprins, lapins, chevaux, poissons, 2e édition, 304 p. Sauvant D., J.-M. Perez, and G. Tran (eds.). INRA editions. Versailles, France.

Journet, M., J.B. Huntington and J.-L. Peyraud, 1995. Le bilan des produits terminaux de la digestion. In: Jarrige R., Y. Ruckebusch, C. Demarquilly, M.H. Farce and M. Journet (eds.), Nutrition des Ruminants Domestiques, Ingestion et Digestion: 671-720.

Kraft, G., 2008. Régulation nutritionnelle du métabolisme hépatique des acides aminés chez le ruminant en croissance: Conséquences sur l'apport des nutriments azotés aux muscles. Thèse de Doctorat. AgroParisTech, France.

Loncke, C., 2009. Modélisation des relations entre l'alimentation et les flux splanchniques de nutriments énergétiques chez le ruminant. Thèse de Doctorat. AgroParisTech, Paris France.

Loncke, C., P. Nozière, J. Vernet, H. Lapierre, D. Sauvant and I. Ortigues-Marty, 2008a. Empirical prediction of net splanchnic release of β-hydroxybutyrate in ruminants. In: Chilliard Y., F. Glasser, Y. Faulconnier, F. Bocquier, I. Veissier and M. Doreau (eds.). Ruminant Physiology. Wageningen Academic Publishers, Wageningen: 444-445.

Loncke, C., I. Ortigues-Marty, J. Vernet, H. Lapierre, D. Sauvant and P. Nozière, 2008b. Capacité du système PDI à prédire les quantités d'azote alpha-aminé absorbées en veine porte chez les ruminants. Renc. Rech. Rum. 15, 285.

Loncke, C., I. Ortigues-Marty, J. Vernet, H. Lapierre, D. Sauvant and P. Nozière, 2009. Empirical prediction of net portal appearance of volatile fatty acids, glucose and their secondary metabolites (β-hydroxybutyrate, lactate) from dietary characteristics in ruminants: A meta-analysis approach. J. Anim. Sci. 87, 253-268.

Loncke, C., P. Nozière, J. Vernet, H. Lapierre, D. Sauvant and I. Ortigues-Marty, 2010a. Empirical prediction of net splanchnic release of glucose in Ruminants from dietary and animal characteristics: a meta-analysis approach. ISEP Parma.

Loncke, C., P. Nozière, S. Amblard, S. Léger, J. Vernet, H. Lapierre, D. Sauvant and I. Ortigues-Marty, 2010b. From metabolisable energy to energy of absorbed nutrients: quantitative comparison of models. 7th International Workshop in Modelling Nutrient Digestion and Utilization in Farm Animals, Paris 10-12 Septembre 2009, in press.

Martineau, R., I. Ortigues-Marty, J. Vernet and H. Lapierre, 2009. Technical note: Correction of net portal absorption of nitrogen compounds for differences in methods: Fist step of a meta-analysis. J. Anim. Sci. 87, 3300-3303.

Micol, D., H. Dubroeucq, C. Martin, F. Garcia, M.M. Mialon and J. Agabriel, 2007. Utilisation de rations de valeur contractée pour la finition de jeunes bovines de race Blond d'Aquitaine. Renc. Rech. Rum. 14, 233-236.

Nozière, P., F. Glasser, C. Loncke, I. Ortigues-Marty, J. Vernet and D. Sauvant, 2010a. Modelling rumen volatile fatty acids and its evaluation on net portal fluxes in ruminants. 7th International Workshop in Modelling Nutrient Digestion and Utilization in Farm Animals, Paris 10-12 Septembre 2009, in press.

Nozière, P., I. Ortigues-Marty, C. Loncke and D. Sauvant, 2010b. Carbohydrate quantitative digestion and absorption in ruminants: from feed starch and fiber to nutrients available for tissues. Animal (in press).

Offner, A. and D. Sauvant, 2004. Prediction of *in vivo* starch digestion in cattle from *in situ* data. Animal Feed Science and Technology 111, 41-56.

Ortigues-Marty, I., D. Dardevet and I. Savary-Auzeloux, 2003. Role of the liver in the regulation of energy and protein status. In: Souffrant W.B. and C.C. Metges (eds.), Progress in Research on Energy and Protein Metabolism, EAAP Publ. n° 109. Wageningen Academic Publishers, Wageningen: 83-98.

Reynolds, C.K., 2000. Feed evaluation for animal production. In: Theodorou M.K. and J. France (eds.), Feeding Systems and Feed evaluation models. CABI Publishing, London: 87-108.

Sauvant, D., 1992. La modélisation systémique en nutrition. Reprod. Nutr. Dev. 32, 217-230.

Sauvant D., C. Assoumaya, S. Giger-Reverdin and H. Archimède, 2006. Etude comparative du mode d'expression du niveau d'alimentation chez les ruminants. Renc. Rech. Rum. 13, 103.

Sauvant, D. and S. Giger-Reverdin, 2007a. Empirical modelling by meta analysis of digestive interactions and CH4 production in ruminants. In: Ortigues-Marty I., N. Miraux and W. Brand-Williams (eds.), Energy and Protein Metabolism and Nutrition, EAAP Publication No 124, Wageningen Academic Publishers, the Netherlands: 561-562.

Sauvant, D. and S. Giger-Reverdin, 2007b. Modélisation des interactions digestives et de la production de méthane chez les ruminants. INRA Prod. Anim. 22, 375-384.

Sauvant D., S. Giger-Reverdin, P. Chapoutot and H. Archimède, 2007. Relation entre la digestibilité *in vivo* du NDF de rations et la dégradabilité in sacco du fourrage qu'elle contient. Renc. Rech. Rum. 14, 246.

Sauvant, D. and D. Mertens, 2008. Use of meta-analysis to build a mechanistic model of responses of rumen digestion to dietary fiber in cattle. In: Modeller's Meeting of the ADSA. Can. J. Anim. Sci.

Sauvant, D., P. Schmidely, and J.J. Daudin, 2005. Meta-analysis of experimental data: Application in animal nutrition. INRA Prod. Anim. 18, 63-73.

Vernet, J., J.P. Brun, and I. Ortigues-Marty, 2007. Bibliographical database applied to ruminant nutrition. In: Ortigues-Marty I., N. Miraux and W. Brand-Williams (eds.), Energy and Protein Metabolism and Nutrition EAAP. Publication No. 124, Wageningen Academic Publishers, Wageningen, the Netherlands: 591-592.

Vernet, J., H. Lapierre, P. Nozière, S. Léger, D. Sauvant and I. Ortigues-Marty, 2005. Prediction of blood nutrient supply to issues of economical interest in ruminants: A first step with the prediction of portal blood flow. In: D. R. C. Hill, V. Barra and K. Traoré (eds.) 1st Open International Conference on Modelling and Simulation. OICMS, Clermont Ferrand, France: 163-173.

Vernet, J. and I. Ortigues-Marty, 2006. Conception and development of a bibliographic database of blood nutrient fluxes across organs and tissues in ruminants: data gathering and management prior ot meta-analysis. Reprod. Nutr. Dev. 46, 527-546.

How does increasing protein supply or glucogenic nutrients modify mammary metabolism in lactating dairy cows?

S. Lemosquet[1], J. Guinard-Flament[2], G. Raggio[3], C. Hurtaud[1], J. Van Milgen[4] and H. Lapierre[5]
[1]INRA UMR1080 Dairy Production, 35590 Saint-Gilles, France
[2]Agrocampus Ouest, UMR1080 Dairy Production, 35000 Rennes, France
[3]Organic Dairy Research Centre, Guelph University, Alfred, ON, Canada
[4]INRA UMR1079 SENAH, 35590 Saint-Gilles, France
[5]Agriculture and Agri-Food Canada, Sherbrooke, QC, Canada

Abstract

In lactating dairy cows, yields of milk lactose and protein increase in response to increasing supply of intestinal protein or glucogenic nutrients (GN: ruminal infusion of propionic acid or post-ruminal infusion of glucose), but to a different extent. What are the common and the different changes in mammary metabolism explaining that both type of nutrients (protein or GN) increased both yields of protein and lactose? To answer this question a meta-analysis on mammary uptake data was performed. To analyse intra mammary metabolism and to estimate the requirement and supply of ATP to support the increases in milk synthesis, a modelling approach was performed on one study. Increased protein yield was associated with an increase in mammary net uptake of AA from group I (His, Met, Phe+Tyr), in response to extra protein or extra GN. However, this increased uptake was achieved differently. In response to protein supply, it was linked to an increase in mammary arterio-venous differences whereas in response to GN, it was linked an increase (tendency) in mammary plasma flow. In response to increasing protein supply, the mammary uptake of essential AA from group II (Ile, Lys, Leu and Val) as well as their ratio of mammary uptake to milk output (>1) also increased. This brought extra carbon skeletons and nitrogen for mammary synthesis of other components as non essential AA (NEAA), whereas the mammary uptake of two NEAA, Ala and Glu decreased. By contrast, NEAA uptakes (Ala and Glu) increased in response to GN supply (quadratic effect). Milk fatty acid synthesis was only increased in response to increasing protein supply, supported by an increase in b-hydroxybutyrate (BHBA) uptake. If lactose yield increased in parallel to milk protein yield both in response to protein or GN supply, glucose uptake did not follow lactose yield. Mammary glucose partition between lactose and oxidative pathways seemed a key mechanism to explain the increased lactose yield, in particular in response to GN supply. The increased protein synthesis was the only synthesis that required a substantial increase in ATP production, based on the modelling approach. The extra ATP required could be supplied by the catabolism of BHBA and group II AA in response to protein or the catabolism of extra glucose and lactate in response to GN supply. This meta-analysis confirms that the increase in uptake of AA from group I is a common mechanism to support an increase in milk protein yield. It also confirms that the regulation of the partition of their precursors (acetate, glucose, BHBA, group II AA and NEAA) towards milk component synthesis or oxidation is probably one of the key factors in the regulation of milk volume and composition.

Introduction

Yields of milk, protein and lactose increase curvilinearly in response to increasing either energy or protein supply in dairy cows (Vérité and Delaby, 2000; Brun-Lafleur *et al.*, 2010). By contrast, increasing protein supply by post-ruminal infusions of casein (CN) increased milk fat yield (Lemosquet *et al.*, 2007) whereas increasing the supply of glucogenic nutrients (GN: ruminal propionic acid (C3) or intestinal glucose (GLC)) decreases milk fat yield (Rigout *et al.*, 2003). The main nutrients metabolised by the mammary gland are amino acids (AA), glucose, acetate and b-hydroxybutyrate (BHBA). They mainly contribute to the synthesis of protein, lactose and fatty acids (FA) with short and medium carbon chains (C4 to half of C16), respectively, in the mammary gland (Smith *et al.*, 1983). Increasing protein or energy supply differently modifies the types of nutrients absorbed and available for mammary metabolism. Accordingly, the total splanchnic flux of AA (Raggio *et al.*, 2004) increases with increasing protein supply while whole body glucose rate of appearance increases in response to GN and protein supply (Lemosquet *et al.*, 2007, 2009a,b). Given such different patterns in nutrient availability, how does mammary metabolism adjust to increase yields of milk protein and lactose and to modify yields of synthesised FA with increased supplies of protein and/or GN?

The literature offers no clear answer to this question, probably because the mammary metabolism can adapt differently between studies, depending of the nutritional status of the animal and its great metabolic flexibility to produce milk (Lobley, 2007). Most nutrients taken up by the mammary gland do not only contribute to milk synthesis but can also be oxidized to supply ATP (Smith *et al.*, 1983), which probably contribute to metabolic flexibility of this organ. Interestingly, when energy or GN supply result in increased milk protein yield, assumed ATP availability for the mammary gland could be a limiting factor for milk protein synthesis because synthesis of a peptide bond is energetically costly (4 to 5 ATP per link: Lobley, 1990; Van Milgen, 2002).

In this work, two complementary approaches were applied. First, a meta-analysis approach was used to determine how the uptakes of major nutrients by the mammary gland were modified to support increased secretion of milk components in response to increased supplies of protein and/or GN. This approach was chosen to discern the common and the differing metabolic pathways used by the mammary gland to increase milk protein and lactose yields. Second, the contributions of nutrients to the synthesis of milk components *per se* and the ATP needed for these syntheses by the mammary were both estimated. For this purpose, the biochemical model proposed by Van Milgen (2002) adapted to the mammary gland of ruminants (Lemosquet *et al.*, 2010a) was used in one example.

Materials and methods

A database was built to analyse mammary gland metabolism through the net balance technique. It comprised 4 small sub data-basis. The effect of increasing the protein supply (7 through CN infusion and 3 through diets) was analysed using 10 studies (13 redesigned experiment nt=31; indicated [1] in the reference list and also analysed by Lemosquet *et al.*, 2007). The effect of GN infusion was analysed using 5 studies (9 redesigned experiments, nt=21 indicated [2] in the reference list) including 3 studies analysing the combined effect of increasing protein and GN. Only 3 studies in two publications reported ruminal C3 infusions. The data in the database did not allow analysing the partitioning of acetate and BHBA for FA synthesis or oxidation because the milk FA profile was not reported in most experiments. Therefore, the database was completed with 7 studies reporting the FA profile of milk fat in response to protein 2 studies, nt=11, indicated [3] in the reference list) or GN supply (6 studies; nt=29 indicated [4] in the reference list), although not reporting mammary net flux.

Diet composition and dry matter intake were required to calculate metabolisable protein supply (i.e. the amount of AA absorbed in the small intestinal) using the INRA (1989) feeding system and expressed in PDIE (protein truly digestible in the small intestine as allowed by available energy in the rumen). *By pass* starch and total C3 produced in rumen were estimated using the INRA (1988) feed tables and the calculations of Rigout *et al.* (2003) to compare the present response curves to this previous work. The molar proportion of C3 in the rumen reported in each publication on GN supply was required (Rigout *et al.*, 2003).

In the database, dry matter intake (DMI) ranged between 10.4 kg/d and 28 kg/d (mean ± SE = 17.8±3 kg/d) in link with milk yield that ranged between 11.1 kg/d to 38.2 kg/d (25.4±5.7 kg/d). Milk protein, fat and lactose contents ranged between 23.8 g/d and 38.5 g/d (31.0±3.0 kg/d), 27.1 and 54.3 g/d (40.5±5.9 g/d) and between 44 g/d to 51.2 g/d (48.0±2.0 g/d), respectively. Net energy supply of lactation (NE_L by diet plus infusion) ranged between 16 Mcal/d to 46 Mcal/d (28.8±5.5 g/d) and PDIE supply (by diet plus infusion) ranged between 837 g/d and 3335 g/d (1764±507 g/d).

Intestinal glucose (diet plus infusion) ranged between 0.2 kg/d to 3.0 kg/d (1.0±0.8 kg/d). The molar proportion of ruminal C3 in and the estimated quantity of C3 produced in the rumen ranged between 15.6% and 37.2%, (21.0±5.6%), and between 8.5 mol/d and 35.4 mol/d (18.6±5.2 mol/d), respectively.

The total amounts of GN (i.e. total intestinal GLC or ruminal C3 produced from diet plus infusion) and PDIE (from diet plus infusion) were expressed as the flow of carbon units (mmol C/h) using the Rulquin *et al.* (2001) tables to calculate the AA (AADI) composition of each diet (in of %PDIE). Lactose, protein and FA synthesised by the mammary gland were also expressed as the flow of carbon units (mmol C/h) using the calculations of Raggio *et al.* (2006a) for protein and of Lemosquet *et al.* (2009d) for other nutrients. For simplicity, the calculations for FA, all C4 to C14 FA were considered to be synthesised by the mammary gland while 55% of the C16 FA was considered to be synthesised by the mammary gland (Smith *et al.*, 1983). The longer chain FA (45% of C16, all C18) were not considered since they are not synthesised within the mammary gland (Smith *et al.*, 1983). Mammary uptake was calculated from measurement on the half-udder by multiplying half-udder plasma flow (MPF) by arterio venous (AV) concentration differences. Mammary plasma flow (MPF) was measured in one half-udder using an ultrasound probe (2 studies, nt=7), *p*-amino hippurate dilution (2 studies nt=5), or estimated by the Fick principle using Phe+Tyr (nt=30) as AA markers (Mepham, 1982). The ratio of mammary uptake and milk protein output of AA (U:O) was calculated to analyse mammary metabolism using the classification of Mepham (1982). For this, His, Met and Phe+Tyr were included in group I because these AA are taken up by the mammary gland, on a net basis, at a level close to their secretion in milk protein (i.e. their U:O ratio is close to 1). On the other hand, Ile, Leu, Lys and Val were included in group II because their U:O ratio can be significantly greater than 1. Data underwent analysis of covariance using the GLM procedure of SAS (2004) according the following model:

$$y_{ijk} = \text{Experiment}_i + a + (b \times x_j) + (c \times x_j^2) + \varepsilon_{ijk}$$

where y is the variable to study and x is the supply of either PDIE or GN (diet + infusions). Experiment was included as a fixed effect (Loncke *et al.*, 2009). Significance was declared at $P \leq 0.05$ and tendency at $0.05 < P \leq 0.1$. Equations were included in Table 1 only if the slope (the quadratic term) and the experiment effects were significant.

The ATP cost of milk synthesis was calculated using the stoichiometry model of Van Milgen (2002) adapted to the ruminant mammary gland (Lemosquet *et al.*, 2010a). This biochemical model is based on the concept that there is no accumulation of any intermediary metabolites (i.e. pyruvate, a-ketoglutarate) in the mammary epithelial cell (i.e. pyruvate, a-ketoglutarate). It was used in one experiment (Raggio *et al.*, 2006a,b; Lemosquet *et al.*, 2009c) increasing protein supply through duodenal CN infusions and increasing energy supply through ruminal C3 infusions. This experiment was chosen because enough nutrients were taken up to account for the nitrogen and carbon output in milk and as blood carbon dioxide (without any excess).

Results and discussion

Supply of both protein and GN increases milk protein yield, but to a different extent

In the database, milk protein yield increased in response to increasing protein and GN supply in agreement with results reported in the literature (Vérité and Delaby, 2000; Rigout *et al.*, 2003; Brun-Lafleur *et al.*, 2010). Milk protein yield increased in a quadratic curve (ranged: 501 g/d to 1,014 g/d) in response to increasing PDIE (867 g/d to 3,335 g/d and 38 mmol C/g):

Protein yield (g/d) = 191 (±86) + 0.41 (±0.09) × PDIE (g/d) - $4.9 \times 10^{-5} (\pm 2.3 \times 10^{-5})$ × PDIE^2 (g/d); n=42 from sub data-bases [1] and [3], Sxy=35.2, adjusted $R^2 = 0.96$; $^{*}P<0.05$.

Protein yield (mmol C/h) = 342 (±146)* + 0.46 (±0.1)* × PDIE (mmol C/h) - 3.4×10^{-5} ($\pm1.5\times10^{-5}$)*× $PDIE^2$ (mmol C/h); n=42, Sxy=60.7, adjusted R^2 = 0.97, *$P<0.05$.

This efficiency of PDIE utilization (first slope in g/d: 0.41) for milk protein yield was probably due to the fact that the PDI balance (intake – requirement) was higher than -100 g/d in most (33 on 42) experiments (Vérité and Delaby, 2000; Brun-Lafleur *et al.*, 2010).

The responses of milk yield and protein content to GN were close to that reported by Rigout *et al.* (2003), with a linear increase in milk protein yield to GN nutrient supply when expressed as in Rigout *et al.* (2003) in NE_L (ranging between 2,080 mmol C/d and 4,000 mmol C/d or 4.5 and 8.5 Mcal/d using 0.495 and 0.266 Mcal of NE_L per mol of GLC and 0.266 Mcal of NE_L per mol of C3):

Protein yield (g/d) = 700 (±21)* + 13.7 (±3.1)* × GN (Mcal/d of NE_L); n=50 from sub data-bases [2] and [4], Sxy=24.0; adjusted R^2 = 0.97; *$P<0.05$.

Protein yield (mmol C/h) = 1,228 (±36)* + 0.048 (±0.11)* × GN (mmol C/h); n=50, Sxy=42.0, adjusted R^2 = 0.97; *$P<0.05$.

However, the increase in milk protein yield in response to increasing GN supply was limited to one tenth of the increase (slope 0.048 vs. 0.048 in mmol C/h) observed with increasing protein supply because GN supply ranged from 2,122 mmol C/h to 5,500 mmol C/h and protein supply 1,500 mmol C/h to 5,500 mmol C/h. In cows producing 800 g/d of milk protein, increasing milk protein yield by 30 g/d (52 mmol C/h) required an large increase in ruminal C3 supply of 1,130 mmol C/h (i.e. an increase in C3 molar proportion in the rumen from 17% to 27% in Lemosquet *et al.*, 2009c). By contrast, increasing milk protein yield about 30 g/d (52.4 mmol C/h) required only about 120 g/d (200 mmol C/h) of PDIE.

Milk lactose increases in parallel to milk protein but not milk fatty acid

The variations in yields of protein, lactose and FA synthesised by the mammary gland induced by the supply of protein and GN were compared on a carbon unit basis. Milk lactose yield increased linearly in parallel with milk protein yield (Figure 1) in response to increasing both protein supply and GN supply, with slopes significantly lower than 1 (0.64±0.05 for protein and 0.76±0.15 for GN; Figure 1). Conversely, yield of FA synthesised only increased in response to protein supply (Figure 1) with a slope of 0.49. Long chain FA (C18), not synthesised in the mammary gland, also increased in response to protein supply (not shown). At the opposite, the amount of carbon utilized in FA synthesis did not change in response to GN supply due to a decrease of short chain FA and a greater elongation of FA (Hurtaud *et al.*, 1998). The decreased amount of carbon from milk fat was thus explained by a decrease in long chain FA (not shown).

Increasing milk protein yield requires increased uptake of group I but not group II AA

In response to protein and GN supplies, the increase in milk protein was linked in both cases to an increased uptake of AA from group I (His, Met, Phe+Tyr) whereas the uptake of other AA varied differently depending on the type of nutrients supplied (Table 1). Interestingly, the increase in uptake of AA from group I occurred through different mechanisms according to whether protein or GN was supplied. In response to an increasing protein supply, group I AA uptake can be explained by an increase in arterial concentrations and AV differences probably linked to the overall increase in intestinal AA supply while MPF did not change (not shown). By contrast, in response to increasing GN supply, the arterial concentration of AA from group I did not increase and the AV difference decreased (not shown). The increased uptake of AA from group I was therefore linked to a tendency of MPF to increase linearly (Table 1) in response to increasing GN supply. In individual experiments,

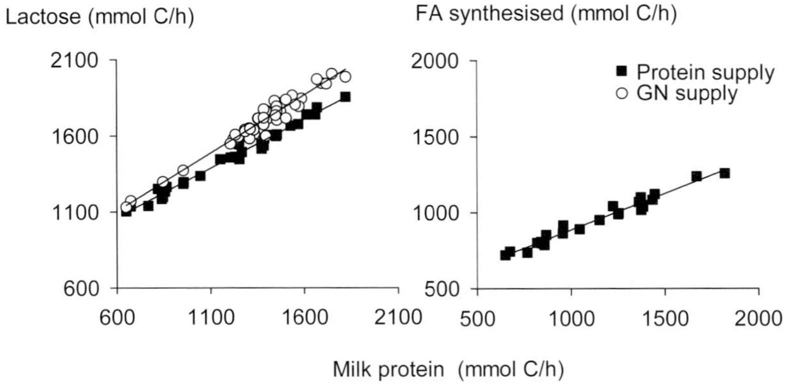

Figure 1. Within experiment relationships between increased milk lactose yield and milk fatty acid (FA) synthesised in the mammary gland and increased milk protein yield. (For increasing protein supply from sub data-bases [1] and [3]: lactose yield = 676 (±69) + 0.64 (±0.05)* × x, n=32, Sxy=32.2, adjusted R^2 (adj.R^2)=0.99, *P<0.05; FA synthesised = 405 (±81)* + 0.49 (±0.07)* × x + Experiment*, n=23, Sxy=33.4, adj.R^2 = 0.98, *P<0.05; for increasing glucogenic nutrients (GN: ruminal propionic acid or post-ruminal glucose) from sub data-bases [1] and [4]: lactose yield = 646 (±81)* + 0.76 (±0.15)* × x, n=46, Sxy=47.3, adj.R^2=0.98, *P<0.05; x = milk protein yield, all values in mmol c/h).*

the increase was only observed (Clark *et al.,* 1977; Rigout *et al.*, 2002; Raggio *et al.*, 2006) with an increase in GN higher than 2500 mmol C/h (>17% of the lower GN level). An increased splanchnic blood flow and cardiac frequency was observed in response to increasing energy (Huntington, 1990, personal results in Lemosquet *et al.*, 2009a) suggesting an increased blood flow at whole body level.

In this database, the MPF was estimated using the Fick principle in 8 out of the 12 studies. This principle is based an U:O of Phe+Tyr equal to 1 whatever the situation which can be questionable when protein supply varied. Several arguments validate the utilisation of the Fick principle. First, MPF measured using an ultrasound flow did not change in response to increased protein supply (Guinard *et al.*, 1994b) and increased in response to increased GN or energy supplies (Rulquin *et al.*, 2004) as measured in a recent experiment (Table 2; Lemosquet *et al.*, 2009a). Second, this hypothesis of an U:O of Phe+Tyr equal to 1 was recently supported in an 'extreme' situation, as almost no oxidation of Phe+Tyr was observed in the mammary gland of two cows given a large amount of Phe (Lemosquet *et al.*, 2010b).

Among group I AA, it was also possible to study the mammary uptake of Met+His and the U:O ratio of Met and His (Table 1). The Met+His uptake significantly increased in response to both treatments and the U:O ratio did not differ from 1 for Met and His suggesting that milk protein secretion was mostly linked to mammary net uptake of almost all Met and His. Overall, the increased mammary uptake of AA from group I is a necessary mechanism to support an increase in protein synthesised by the mammary gland, which basically fits with the (almost) complete conversion of AA taken up for milk protein output.

Among the other AA (Table 1), the uptake of AA from group II (Ile, Leu, Lys, Val) and the uptake of Ala and Glu (NEAA) did not vary in the same way in response to increasing protein and GN supplies. The mammary uptake of AA from group II increased in response to increasing protein supply as did their AV differences (one component of the uptake). The AV difference increased in relation with the increased arterial concentration, probably due to their increased intestinal supply. In addition, the mammary U:O ratio of group II AA increased. This excess uptake relative to milk protein output could supply extra carbon for oxidative pathways or as carbon skeletons for the synthesis of other

Table 1. Half-udder uptake of nutrients in response to increasing protein supply (PDIE) or glucogenic nutrients (intestinal glucose and ruminal propionic acid): only significant equations were reported.

	n	Equation[1] $a + (b \times X + (c \times X^2) + \text{Experiment}$	Sxy	adj.R[2]
PDIE[3] (mmol C/h) from sub data-base [1] (Range: 837 to 3,335 mmol C/h)				
AA				
Group I[4] A[5] (µmol C/l)	31	$252(\pm138)^{\dagger} + 0.32(\pm0.05)^{*} \times X$	92.8	0.77
Group I[4] AV[6] (µmol C/l)	31	$78(\pm121) + 0.11(\pm0.04)^{*} \times X$	81.4	0.31
Group I[4] uptake[7] (mmol C/h)	31	$53(\pm12)^{*} + 0.039(\pm0.006)^{*} \times X$	8.23	0.93
Group II[8] A[5] (µmol C/l)	31	$-1505(\pm541)^{*} + 1.45(\pm0.17)^{*} \times X$	364	0.86
Group II[8] AV[6] (µmol C/l)	31	$-43(\pm160) + 0.36(\pm0.05)^{*} \times X$	108	0.84
Group II[8] uptake[7] (mmol C/h)	31	$-3.5(\pm39) + 0.10(\pm0.01)^{*} \times X$	26.2	0.88
Group II[8] U:O[9] ratio	31	$0.50(\pm0.14)^{*} + 2.7\times10^{-4}\pm(5\times10^{-4})^{*} \times X$	0.10	0.84
Ala+Glu[10] AV[6] (µmol C/L)	27	$378(\pm60)^{*} -0.042(\pm0.02)^{*} \times X$	39.6	0.81
Ala+Glu[10] uptake[7] (mmol C/h)	27	$114(\pm18)^{*} - 0.013(\pm0.006)^{*} \times X$	11.7	0.86
Ala+Glu[10] U:O[9] ratio	27	$4.95(\pm0.58)^{*} - 8.4\times10^{-4}\pm(1.9\times10^{-4})^{*} \times X$	0.28	0.74
Glucose AV[6] (µmol C/l)	24	$2.33(\pm0.38)^{*} + 7.4\times10^{-4}\pm(1.3\times10^{-4})^{*} \times X$	0.24	0.88
BHBA[11] A[5] (µmol C/l)	20	$2.39(\pm0.42)^{*} + 3.0\times10^{-4}\pm(1.5\times10^{-4})^{\dagger} \times X$	0.26	0.93
BHBA[11] AV[6] (µmol C/l)	20	$0.16(\pm0.41) + 3.8\times10^{-4}\pm(1.5\times10^{-4})^{*} \times X$	0.26	0.81
BHBA[11] uptake[7] (mmol C/h)	20	$80(\pm27)^{*} + 0.02\pm(0.01)^{*} \times X$	17.1	0.69
Glucogenic nutrients (mmol C/h)[12] from sub data-base [2] (Range: 2,122 to 3,947 mmol C/h)				
MPF (l/h)	21	$140(\pm80)^{\dagger} + 0.055(\pm0.028)^{\dagger} \times X$	45.9	0.31
AA				
Group I[4] uptake (mmol C/h)	21	$103(\pm15)^{*} + 9.6\times10^{-3}\pm(5.3\times10^{-4})^{\dagger} \times X$	8.43	0.87
Group II[8] A[5] (µmol C/l)	21	$4321(\pm353)^{*} - 0.58(\pm0.13)^{*} \times X$	203	0.91
Group II[8] AV[6] (µmol C/l)	21	$1493(\pm153)^{*} - 0.17(\pm0.06)^{*} \times X$	88.0	0.87
Ala+Glu[10] uptake[7] (µmol C/h)	18	$314(\pm104)^{*} - 0.17(\pm0.07)^{*} \times X + 3.1\times10^{-5}(\pm1.1\times10^{-5})^{*} \times X^2$	10.2	0.77
Ala+Glu[10] U:O[9] ratio	18	$11.2(\pm3.3)^{*} - 6\times10^{-3}(\pm2\times10^{-3})^{*} \times X + 1.1\times10^{-6}(\pm0.3\times10^{-6})^{*} \times X^2$	0.32	0.75
Glucose A[5] (µmol C/l)	17	$17.3(\pm0.94)^{*} + 1.1\times10^{-3}(\pm0.3\times10^{-3})^{*} \times X$	0.515	0.77
Glucose AV[6] (µmol C/l)	17	$6.0(\pm0.4)^{*} - 6.4\times10^{-4}(\pm1.3\times10^{-4})^{*} \times X$	0.207	0.84
Lactose: glucose uptake ratio	17	$-0.8(\pm0.7) + 1.1\times10^{-3}(\pm0.5\times10^{-3})^{*} \times X - 1.8\times10^{-7}(\pm0.8\times10^{-7})^{*} \times X^2$	0.074	0.56

$^{*}P<0.05$, $^{\dagger}P<0.01$; [1]Equations of analysis of covariance (experiment effect: $P<0.05$); [2]adjusted R^2; [3]X = Protein Digestible in the Intestine (PDIE; INRA, 1989); [4]His+Met+Phe+Tyr; [5]Arterial concentration; [6]Arterio venous differences; [7]Half-udder net uptake; [8]Ile+Lys+Leu+Val; [9]Uptake to output in milk protein ratio; [10]Alanine plus glutamate; [11]b-Hydroxybutyrate; [12]X = intestinal glucose and ruminal propionic acid (infusions plus diet).

nutrients. Similarly, extra nitrogen, could be used for synthesis (Lapierre *et al.*, 2009) of non essential AA (NEAA). In parallel, the uptake of Ala plus Glu (Table 1), two NEAA that can be synthesised by the mammary gland, decreased when milk protein yield increased in response to protein supply, leading to a significant decrease in their U:O ratio.

Conversely, the uptake of group II AA was not affected by GN supply because the two components of uptake varied in opposite direction: the AV difference decreased as arterial concentration while MPF increased. In the absence of any significant increase in net uptake of group II AA by the mammary gland, the increased uptake of nitrogen required for the increased milk protein yield could be supplied

Table 2. Comparison of the mammary (half-udder) plasma flow (MPF; L/h) accuracy observed in lactating cows receiving two levels (Low: L and High: H) of energy (E) in combination with two levels of protein (P).

Type of measurement of MPF	Treatments				Root MSE[1]	P<		
	LELP n=4	LEHP n=4	HELP n=4	HEHP n=3		E	P	E × P
Ultrasound probe[2]	260	280	353	305	34.6	0.02	0.49	0.13
Fick principle[3] GC-MS[4]	221	254	309	279	34.3	0.03	0.95	0.15
Cationic[5]	262	296	370	283	74.0	0.29	0.55	0.19

[1] Standard error of the GLM model (SAS, 2004).

[2] MPF measured using an ultra sound probe (Probe A; f = 20 mm; transonic).

[3] MPF estimated using the Fick principle with Phe+Tyr as AA markers.

[4] Phe+Tyr concentrations measured by isotopic dilution (GC-MS; Calder *et al.*, 1999).

[5] Phe+Tyr concentrations measured with a cationic exchange column as in Guinard *et al.* (1994b).

by an increased NEAA uptake. The uptake of Ala plus Glu tended to increase (quadratic curve) in response to increased GN supply, and their U:O increased providing extra nitrogen and carbon for synthesis of other NEAA.

Taken together, this meta-analysis indicates that the increased uptake of AA from group I is the common mechanism needed to support an increase in milk protein secretion in response to increasing protein or GN supplies. However, the mechanisms by which the uptake of AA from group I increased were different, probably linked to the nature of nutrients absorbed and available and by characteristics that affect the nutrient supply, like MPF. In addition, the uptake of group II AA and of two NEAA by the mammary gland as well as the metabolism by the mammary gland varied differently in response to protein and GN supply. The increase in milk protein yield, was more limited in response to GN supply than to milk protein supply probably because less nitrogen was provided to the mammary gland with GN supply through the increased uptake of AA from group I AA and NEAA than through the increased uptake of AA from groups I and II with protein supply (Raggio *et al.*, 2006a). These different variations reflect the metabolic flexibility of the mammary gland concerning AA metabolism.

Increasing protein supply increases β-hydroxybutyrate uptake and fatty acid synthesis

The increase in milk FA synthesis in response to increasing protein supply was not linked to changes in uptake of acetate carbon but to an increase in BHBA uptake due to an increase in AV difference (Table 1). There also was a tendency to an increase in arterial BHBA concentration. As milk C18 FA (directly taken up by the mammary gland) also increased, it suggested an increased body fat mobilisation. Conversely, increasing GN decreased the uptake of BHBA by the mammary gland in two experiments (Rigout *et al.*, 2002; Raggio *et al.*, 2006) although the synthesis of FA was not affected.

Increasing milk lactose yield does not require an increased uptake of glucose

No clear relationship was observed between increased lactose yield and uptake of glucose in response to increased protein supply. With increasing protein supply, AV differences of glucose increased but not the net uptake of glucose did not as MPF. Arterial concentration of glucose significantly increased in response to GN supply but AV difference decreased, whereas MPF increased and therefore glucose net uptake did not change with GN supply. However, the lactose-to-glucose uptake ratio increased curvilinearly in response to GN supply (Table 1) suggesting a change in glucose partitioning in the mammary gland. The utilization of glucose for lactose synthesis increased at low GN supply (<2,500 mmol C/h when MPF did not increase). Above 2,500 mmol C/h of GN supply, the ratio decreased suggesting an increased utilization of glucose in other pathways than lactose synthesis.

Can metabolic flexibility of the mammary gland explain the increased lactose synthesis?

It can not be completely excluded that the absence of an increase in glucose uptake might be for a part explained by the limitations of this meta-analysis approach. The uptakes of energy-yielding nutrients were less documented than the uptake of EAA and the precision of the AA analysis could set a limit to accurate estimate of MPF as shown by the standard errors of the model in Table 2. However, several points strongly suggested a high metabolic flexibility of the mammary gland, utilising different nutrients in different pathways to achieve milk synthesis (Lobley, 2007) as observed for AA since uptakes of group II AA, Ala and Glu were altered differently in response to supply of protein or GN. The extent of glucose utilisation for lactose synthesis and in oxidative pathways varied widely between experiments based on the lactose-to-glucose uptake ratio representing between 0.49 and 1.03 of mammary gland glucose uptake (on a carbon basis: mean ± SE: 0.69±0.15). When GN supply was limited, a greater proportion of glucose uptake was utilized for lactose synthesis whereas with a greater GN supply a greater proportion of glucose was utilised in other pathways including oxidation. Indeed, acetate, BHBA and also AA (from group II and NEAA) as well as glucose can be oxidised to provide ATP (Smith *et al.,* 1983; Bequette *et al.*, 1996; Raggio *et al.*, 2006a). In addition, these nutrients can also provide 3-carbon intermediates that could substitute for glucose in a variety of metabolic pathways (Bequette *et al.*, 2006). Tracer studies conducted *in vitro* with mammary gland explants from goats (Roets *et al.*, 1983; Bequette *et al.*, 2006) have shown incorporation of AA-carbon into lactose (most EAA) and milk fat (Leu). All these findings point to a metabolic flexibility of the mammary gland (Lobley, 2007) and may explain part of the variation between experiments in glucose utilization for lactose synthesis.

The next question concerns the origin of ATP required by the mammary gland to support milk production. In particular, in response to the increased protein supply, can the uptake of group II AA and BHBA supply enough ATP to support the increase in yields of protein, lactose and FA? When increasing GN supply does not significantly increase GLC uptake how can milk protein and lactose yield increase?

Much ATP is required for milk protein synthesis

In the experiment (Raggio *et al.*, 2006a,b) utilised in the model (Van Milgen, 2002), the increase in milk protein yield was lower than predicted by the meta-analysis (166 g/d vs. 250 g/d) in response to CN infusion into the duodenum and was similar to the prediction (36 g/d vs. 30 g/d) in response to increasing C3 infusions (Control: Ctrl vs. C3). In addition, uptakes of nutriments (Lemosquet *et al.*, 2009c) varied similarly than in the present meta-analysis.

In this experiment, the increased lactose yield induced by the CN treatment required only an additional 22 mmol ATP/h/half-udder (Table 3). Milk FA synthesis did not necessarily require ATP and could even produce ATP depending on which biochemical pathways were taken account to estimate the ATP cost. In fact, synthesis of milk triglycerides (FA and glycerol-3-Phosphate) required a lot of ATP (Table 3) and also a lot of NADPH and glycerol-3-phospate. If the acetate and BHBA required for NADPH synthesis and the amount of glucose required for glycerol-3P for triglycerides are both used in the biochemical pathways, this will generate intermediary metabolites (in particular a-ketoglutarate, Van Milgen, 2002). Their subsequent oxidation in the tricarboxylic acid cycle will generate NADH and ATP. The ATP balance for FA synthesis will then correspond to a net gain (Table 3).

Table 3. Comparison of ATP synthesis and utilisation in response to increasing protein supply or increasing ruminal propionic acid supply in one experiment.[1]

ATP (mmol/h/half-udder):	Treatments		
	Ctrl[2]	CN[3]	C3[4]
Cost for lactose synthesis	-149	-171	-149
Cost of FA[5] + glycerol-3-P[6] synthesis[7]	-1,066	-1,296	-1,083
Gain by acetate+ BHBA[8] + glucose required for NADPH and glycerol-3-P[6,9]	+1,362	+1,588	+1,337
Balance for FA + glycerol-3-P synthesis	+296	+292	+254
Cost of total protein synthesis and turnover	-857	-1,027	-998
Gain by AA[10] catabolism	+280	+391	+389
Gain[7] by BHBA[8] catabolism	+1,138	+2,647	+934
Gain by glucose and lactate oxidation	+1,429	+125	+3,650

[1] Raggio *et al.* (2006a,b); Lemosquet *et al.* (2009d).
[2] Control (diet + water infusion).
[3] CN infusion in the duodenum.
[4] Ruminal propionic acid infusion.
[5] Fatty acids synthesised in the mammary gland.
[6] Glycerol-3- phosphate required for milk triglycerides.
[7] Using the hypothesis that half of the primer carbon chain for FA came from acetate and a half from b-hydroxybutyrate (the hypothesis was different in Lemosquet *et al.*, 2010a).
[8] b-hydroxybutyrate maximal oxidation through isocitrate dehydrogenase (ICDH) pathway.
[9] using the hypothesis of 66% of NADPH produced through pentose phosphate pathway and 33% through ICDH pathway.
[10] Amino acids.

Overall, only the increased protein yield required an increase in ATP production in the mammary gland. The extra ATP production required for the increased milk protein yield was 170 mmol/h/half-udder in response to increased CN supply and 141 mmol/h/half-udder in response to increased C3 supply (Table 3). In fact, the amount of protein synthesised in the mammary gland is always higher to the amount secreted in milk because of the protein turnover (Lobley, 1990; Hanigan *et al.*, 2009). Protein turnover is extremely costly in ATP since synthesis of each peptide bond required 5 ATP in the present model (Van Milgen, 2002) and the subsequent protein degradation required 1 ATP (Lobley *et al.*, 1990). In the present experiment, total mammary protein synthesis (including constitutive proteins) was calculated to amount to 130, 130 and 140% of milk protein secretion in Ctrl, CN, and C3 treatment, using L[1-^{13}C]Leu leading to a mammary protein degradation estimated at between 30 and 40% of milk protein synthesis. However, the calculation of the rate of protein turnover depends of the labelled AA used and the precursor pool chosen to analyse the enrichment (venous plasma, or intra cellular). In lactating goats, mammary tissue protein synthesis could represent up to 300% of milk protein synthesis (Hanigan *et al.*, 2009). The real ATP cost lies probably between the two calculated values because with Hanigan *et al.* (2009) estimation, there was not enough ATP to account for NEAA synthesised in the mammary gland that returned in the mammary vein (Lapierre *et al.*, 2009).

In this experiment, the catabolism of AA taken up in excess could explain a part of the increase in ATP requirement for milk protein synthesis in CN and C3 treatments. The increased catabolism of AA (mainly from Group II) in CN and of NEAA in C3 compared with Ctrl increased the ATP production by +111 mmol/h/half-udder and +109 mmol/h/half-udder, respectively (Table 3). With the Hanigan estimates, additional ATP was required to support the increased protein yield in the CN treatment. However, in this experiment, there were other additional sources of ATP to explain the increase in milk synthesis: in addition to AA catabolism, mammary uptake of BHBA highly increased in response to CN (compared to other experiment of the databases) and BHBA oxidation could provide until +1,519 mmol/h/half-udder of ATP. A higher glucose and lactate oxidation could provide until +2,221 mmol/h/half-udder of ATP in response to C3 that largely cover the increased ATP cost needed to cover the increased protein synthesis.

Overall, protein synthesis in the mammary gland is costly in term of ATP because of the cost of peptide bonds and because protein synthesis is greater than protein secretion in the milk because of the protein turnover. However several other nutrients (glucose, lactate, BHBA and AA from group II) could provide these additional ATP required, confirming the metabolic flexibility of the mammary gland to increase milk protein yield.

Conclusion

Whether induced by increasing the supply of protein or of GN, the increase in milk protein yield is linked to an increase in group I AA because their net uptakes are close to the secretion in milk protein. The mammary gland seems to use different mechanisms to support the increased protein yield and to provide the additional ATP. Several nutrients can be used to synthesise this ATP, thereby, allowing mammary gland metabolic flexibility. When extra protein is supplied, the catabolism of AA, especially group II AA and of BHBA seemed to be able to meet a large proportion of the increased ATP cost of protein synthesis. Overall, the regulation of milk lactose and fat yield seems to depend on the partitioning of their precursors (acetate, glucose, BHBA and also AA from group II and NEAA) towards milk component synthesis or oxidation. Further work is required to determine the rate of protein turnover in the mammary gland and to identify the driving forces regulating the partitioning of nutrients in the mammary gland between synthesis of milk component and oxidative pathways.

References

Note: numbers between [] at the end of a reference designate the literature used for the sub-database of that number.

Bequette, B.J., F.R.C. Backwell, J.C. MacRae, G.E. Lobley, L.A. Crompton, J.A. Metcalf, and J.D. Sutton, 1996. Effect of intravenous amino acid infusion on leucine oxidation across the mammary gland of the lactating goat. J. Dairy Sci. 79, 2217-2224.

Bequette, B.J., N.E. Sunny, S.W. El-kadi and S.L. Owens, 2006. Application of stable isotope and mass isotopomer distribution analysis to the study of intermediary metabolism of nutrients. J. Anim. Sci. 84 (E.Suppl.), E50-E59.

Brun-Lafleur, L., L. Delaby, F. Husson and and P. Faverdin, 2010. Predicting energy × protein interaction on milk yield and milk composition in dairy cows. J Dairy Sci. (accepted for publication).

Calder, A.G., K.E. Garden, S.E. Anderson and G.E. Lobley, 1999. Quantification of blood and plasma amino acids using isotope dilution electron impact gas chromatography_mass spectrometry with U_^{13}C amino acids as internal standards. Rapid Commun. Mass Spectrom. 13, 2080-2083.

Choung, J.J. and D.G. Chamberlain, 1993. The effects of abomasal infusions of casein or soya-bean-protein Isolate on the milk-production of dairy-cows in midlactation. Br. J. Nutr. 69, 103-115. [3]

Clark, J. H., H.R. Spires, R.G. Derrig and M.R. Bennink, 1977. Milk production, nitrogen utilization and glucose synthesis in lactating cows infused postruminally with sodium caseinate and glucose. J. Nutr. 107, 631-644. [1,2]

Guinard, J. and H. Rulquin, 1994a. Effect of graded levels of duodenal infusions of casein on mammary uptake in lactating cows. 2. Individual amino acids. J. Dairy Sci. 77, 3304-3311. [1]

Guinard, J., H. Rulquin and R. Vérité, 1994b. Effect of graded levels of duodenal infusions of casein on mammary uptake in lactating cows. 1. Major nutrients. J. Dairy Sci. 77, 2221-2231. [1]

Hanigan, M.D., J. France, S.J. Mabjeesh, W.C. McNabb and B.J. Bequette, 2009. High rates of mammary tissue protein turnover in lacxtating goats are exnergetically costly. J. Nutr 139, 1118-1127.

Huhtanen, P., A. Vanhatalo and T. Varvikko, 2002. Effects of abomasal infusions of histidine, glucose, and leucine on milk production and plasma metabolites of dairy cows fed grass silage diets. J. Dairy Sci. 85, 204-216. [2]

Huntington, G.B., 1990. Energy metabolism in the digestive tract and liver of cattle: influence of physiological state and nutrition. Reprod. Nutr. Dev. 30, 35-47.

Hurtaud, C., S. Lemosquet and H. Rulquin, 2000. Effect of graded duodenal infusions of glucose on yield and composition of milk from dairy cows.2. Diets based on grass silage. J. Dairy Sci. 83, 2952-2962. [4]

Hurtaud, C., H. Rulquin and R. Vérité, 1998. Effect of level and type of energy source (volatile fatty acids or glucose) on milk yield, composition and coagulating properties in dairy cows. Reprod. Nutr. Dev. 38, 315-330. [4]

INRA, 1989. Ruminant Nutrition: Recommended Allowances and Feed Tables. In: Jarrige, R. (ed.). John Libbey Eurotext, London, England, 303 pp.

Korhonen, M., A. Vanhatalo and P. Huhtanen, 2002. Effect of protein source on amino acid supply, milk production and metabolism of plasma nutrients in dairy cows fed grass silage. J. Dairy Sci. 85, 3336-3335. [1]

Lapierre, H., L. Doepel, E. Milne and G.E. Lobley, 2009. Responses in mammary and splanchnic metabolism to altered lysine supply in dairy cows. Animal 3, 360-371.

Lemosquet, S., O. Abdou Arbi, A. Siegel, J. Guinard-Flament, J. Van Milgen and J. Bourbon, 2010a. A generic stoichiometric model to analyse flexibility of the mammary gland in lactating dairy cows. In: Sauvant, D., J. Agabriel, P. Faverdin, P. Lescoat and J. Van Milgen (eds.). Modelling Nutrient Digestion and Utilization in Farm Animals. 10-12 september, Paris. Wageningen academic Publisher, Wageningen, the Netherlands. 10 pp. (accepted for publication).

Lemosquet, S., F. Bardey, H. Rulquin, H. Lapierre and J. Guinard-Flament, 2009a. Mammary glucose metabolism in response to energy ad/or protein supply in lactating dairy cows. J. Dairy Sci. 92, E-Suppl.1, 149 (abstr.).

Lemosquet, S., E. Delamaire, H. Lapierre, J. W. Blum and J. L. Peyraud, 2009b. Effects of glucose, propionic acid and non-essential amino acids on glucose metabolism and milk yield in Holstein dairy cows. J. Dairy Sci. 92, 1-14. [4]

Lemosquet, S., G.E. Lobley, R. Koopman, L.J.C. Van Loon, A.K. Kies and H. Lapierre, 2010b. 3rd EAAP International Symposium on Energy and Protein Metabolism and Nutrition, Parma (Italy), 6-10 September 2010. short communication n°7763.

Lemosquet, S., G. Raggio, H. Lapierre, J. Guinard-Flament and H. Rulquin, 2007. Effects of protein supply on whole body glucose rate of appearance and mammary gland metabolism of energy nutrients in ruminants. In: Ortigues-Marty I. (ed.), Energy and Protein Metabolism and Nutrition. Wageningen Academic Publishers, Wageningen, the Netherlands, EAAP publication N° 124: 581-582.

Lemosquet, S., G. Raggio, G.E. Lobley, H. Rulquin, J. Guinard-Flament and H. Lapierre, 2009c. Whole-body glucose metabolism and mammary energetic nutrient metabolism in lactating dairy cows receiving digestive infusions of casein and propionic acid. J. Dairy Sci. 92, 6068-6082. [1,2]

Lobley, G.E., 1990. Energy-metabolism reactions in ruminant muscle - responses to age, nutrition and hormonal status. Reprod. Nutr. Dev. 30, 13-34.

Lobley, G.E., 2007. Protein-energy interaction: horizontal aspects. In: Ortigues-Marty I. (ed.), Energy and Protein Metabolism and Nutrition. Wageningen Academic Publishers, Wageningen, the Netherlands, EAAP publication N° 124: 445-461.

Loncke, C., I. Ortigues-Marty, J. Vernet, H. Lapierre, D. Sauvant and P. Noziere, 2009. Empirical prediction of net portal appearance of volatile fatty acids, glucose, and their secondary metabolites (beta-hydroxybutyrate, lactate) from dietary characteristics in ruminants: A meta-analysis approach. J. Anim. Sci. 87, 253-268.

Mackle, T.R., K. Dwyer, K.L. Ingvartsen, P.Y. Chouinard, D.A. Ross and D. E. Bauman, 2000. Effects of insulin and postruminal supply of protein on use of amino acids by the mammary gland for milk protein synthesis. J. Dairy Sci. 83, 93-105. [1]

Mepham, T.B., 1982. Amino acid utilization by lactating mammary gland. J. Dairy Sci. 65, 287-298.

Metcalf, J.A., D.E. Beever, J.D. Sutton, C. Wray, R.T. Evans and D.J. Humphries, 1994. The effect of supplementary protein on *in vivo* metabolism of the mammary gland in lactating dairy cows. J. Dairy Sci. 77, 1816-1827. [1]

Metcalf, J. A., C. Wray, E.E. Chettle, J.D. Sutton, D.E. Beever, L.A. Crompton, J.C. McRae, B.J. Bequette and F. R. C. Backwell, 1996. The effect of dietary crude protein as protected soybean meal on mammary metabolism in the lactating dairy cow. J. Dairy Sci. 79, 603-611. [1]

Miettinen, H. and P. Huhtanen, 1996. Effects of the ratio of ruminal propionate to butyrate on milk yield and blood metabolites in dairy cows. J. Dairy Sci. 79, 851-861. [2]

Miettinen, H. and P. Huhtanen, 1997. Effects of silage fermentation and post_ruminal casein supplementaion in lactating dairy cows, 2-energy metabolites and plasma amino acids. J. Sci. Food Agric 74, 459-468. [4]

Raggio, G., S. Lemosquet, G.E. Lobley, H. Rulquin and H. Lapierre, 2006a. Effect of casein and propionate supply on mammary protein metabolism in lactating dairy cows. J. Dairy Sci. 89, 4340-4351. [1,3]

Raggio, G., G.E. Lobley, S. Lemosquet, H. Rulquin and H. Lapierre, 2006b. Effect of casein and propionate supply on whole body protein metabolism in lactating dairy cows. Can. J. Anim. Sci. 86, 81-89. [1,3]

Raggio, G., D. Pacheco, R. Berthiaume, G.E. Lobley, D. Pellerin, G. Allard, P. Dubreuil and H. Lapierre, 2004. Effect of level of metabolizable protein on splanchnic flux of amino acids in lactating dairy cows. J. Dairy Sci. 87, 3461-3472. [1]

Rigout, S., C. Hurtaud, S. Lemosquet, A. Bach and H. Rulquin, 2003. Lactational effect of propionic acid and duodenal glucose in cows. J. Dairy Sci. 86, 243-253. [4]

Rigout, S., S. Lemosquet, A. Bach, J.W. Blum and H. Rulquin, 2002a. Duodenal infusion of glucose decreases milk fat production glucose in grass silage-fed dairy cows. J. Dairy Sci. 85, 2541-2550. [2]

Rigout, S., S. Lemosquet, J.E. Van Eys, J.W. Blum and H. Rulquin, 2002b. Duodenal glucose increases glucose fluxes and lactose synthesis in grass silage-fed dairy cows. J. Dairy Sci. 85, 595-606. [2]

Roets, E., A.M. Massart leen, G. Peeters and T. Veneman, 1983. Metabolism of leucine by isolated perfused goat udder. J. Dairy Res. 50, 413-424.

Rulquin, H., 1982. Effects of ruminal infusion of volatile fatty-acids and duodenal infusion of caseinate on digestion and metabolism in the dairy-cow. 1. Production and Digestion. Reprod. Nutr. Dev. 22, 905-921. [3, 4]

Rulquin, H., 1983. Effects of ruminal infusion of volatile fatty-acids and duodenal infusion of caseinate on digestion and metabolism in the dairy-cow. 2. General and mammary metabolisms. Reprod. Nutr. Dev. 23, 1029-1042. [3,4]

Rulquin, H., S. Rigout, S. Lemosquet, and A. Bach, 2004. Infusion of glucose directs circulating amino acids to the mammary gland in well-fed dairy cows. J. Dairy Sci. 87, 340-349. [2]

Rulquin, H., R. Vérité and J. Guinard-Flament, 2001. Tables AADI des Aliments. INRA Prod. Anim. 14 (Suppl.), 1-16.

SAS, 2004. SAS/STATÒ 9.1. User's Guide. SAS Institute, Cary, NC.

Smith, G.H., B. Crabtree, and R. Smith, 1983. Energy metabolism in the mammary gland. In: Mepham, T.B. (ed.): Biochemistry of Lactation. Elsevier, Amsterdam, the Netherlands: 121-141.

Vanhatalo, A., T. Varvikko and P. Huhtanen, 2003. Effects of various glucogenic sources on production and metabolic responses of dairy cows fed grass silage-based diets. J. Dairy Sci. 86, 3249-3259. [1,2]

Van Milgen, J., 2002. Modeling biochemical aspects of energy metabolism in mammals. J. Nutr. 132, 3195-3202.

Vérité, R. and L. Delaby, 2000. Relation between nutrition, performances and nitrogen excretion in dairy cows. Ann. Zootech. 49, 217-230.

Nutrition restriction in late-gestation and postnatal overfeeding change metabolic preferences in skeletal muscle *longissimus dorsi* of sheep

L. Hou[1,3], E. Nilsson[3], K. Dixen[2], J.B. Hansen[2], A. Vaag[3] and M.O. Nielsen[1]
[1]Department of Basic Animal and Veterinary Sciences, Faculty of Life Sciences, University of Copenhagen, Denmark
[2]Department of Biomedical Sciences, Faculty of Health Sciences, University of Copenhagen, Denmark
[3]Steno Diabetes Center, Gentofte, Denmark

Introduction

Maternal nutrition restriction during late gestation influences foetal development, and impacts glucose-insulin homeostasis in adult offspring. Additionally, intrauterine under-nutrition followed by postnatal over-nutrition leads to rapid catch-up growth, altered adiposity, and impaired glucose tolerance in young age (Ozanne and Hales, 2005; Hernandez-Valencia and Patti, 2006). Skeletal muscle is an important site for glucose and fatty acid catabolism, and reduced glucose uptake and intra-muscular lipid accumulation are related to impaired insulin sensitivity in skeletal muscle (DeFronzo and Tripathy, 2009; Kraegen and Cooney, 2008). The aim of this experiment was to demonstrate if foetal programming leads to changes in catabolic preferences in skeletal muscle.

Materials and methods

Twenty twin-pregnant Shropshire ewes were fed either an Norm (~requirements) or Low (50% of requirements) diet the last 6 weeks of gestation (term=147d). From 3-days until 6-months of age, twin lambs were assigned to each their feeding: CONV (moderate hay feeding) or HCHF (high-fat-high-carbohydrate) supplemented with milk replacer between 0-8 weeks of age. From 6 to 24 months of age, sheep were raised on the same moderate grass based diet. Biopsies of *m. longissimus dorsi* were obtained *post mortem* at either 6 or 24 months of age. Target genes mRNA levels were measured by qPCR, and were grouped into four categories: (1) glucose metabolism: *INSRβ*, *IRS1*, and *GLUT4*, (2) regulators of oxidation: *PPARα* and *PPARδ*, (3) coordinators: *PGC1α* and *PGC1β*, and (4) mitochondrial FA oxidation regulators: *CPT1b* and *UCP3*. *ACTB* served as an endogenous control. All samples were run in triplicate, and expression data were calculated by using the ΔΔCt method and expressed as a ratio to the *ACTB* reference. The data were analyzed by the PROC MIXED procedure in SAS (v9.2, SAS institute, USA). Variables in the statistical models included age at slaughtering (6 or 24 months old), maternal diet (Norm or Low), and postnatal diet (CONV or HCHF), and their interactions as fixed effects; and interaction of the factors ewe_no and lamb_no was included as random effect. Normality of data was achieved by logarithm-transformation and outliers identified based on residual plots. Presented results are expressed as least squares means (LSM) with standard error of mean (SEM) of logarithm-transformed data. The PDIFF option in SAS was used to generate comparisons between treatment means.

Results

Postnatal HCHF feeding significantly up-regulated the mRNA level of genes involved in both glucose uptake (*INSRβ, IRS1* and *GLUT4*), mitochondrial FA oxidation regulators (*CPT1b* and *UCP3*), as well as the metabolic coordinator *PGC1β*. Expression of these genes were not affected by age (except for a downregulation of *INSRβ* with age), or by prenatal diet despite the fact that the animals fed the HCHF diet in early postnatal life had been switched back to a low-fat moderate diet for 18 months before the sampling in young adult life (24 months of age). Increased mRNA levels were found also for skeletal muscle *PGC1 β*, *PPARα* and *PPARδ* in response to the HCHF diet and the postnatal dietary effect appeared to be more pronounced in young as compared to older animals. However, this pattern of development was influenced also by the prenatal diet, although not in a very clear pattern (see Table 1).

Table 1. Sheep m. longissimus dorsi *mRNA expression of target genes as affected by late gestation maternal diet (Low or Norm), postnatal diet (conventional moderate CONV or high-carbohydrate-high-fat HCHF) and age (6 or 24 months).*

Postnatal[1]	CONV				HCHF			
IRS1	0.12±0.04^{a}				0.26±0.04^{b}			
GLUT4	-0.12±0.04^{a}				0.06±0.04^{b}			
CPT1b	-0.10±0.04^{a}				0.08±0.04^{b}			
UCP3	-0.13±0.06^{a}				0.05±0.06^{b}			
Postnatal-age[2]	CONV-6		CONV-24		HCHF-6		HCHF-24	
INSRβ	-0.03±0.04^{c}		-0.17±0.04^{d}		0.09±0.04^{abd}		-0.02±0.04^{b}	
Postnatal[3]	CONV				HCHF			
Maternal-age:	Low 6	Low 24	Norm 6	Norm 24	Low 6	Low 24	Norm 6	Norm 24
PGC1β	-0.11± 0.07^{abc}	-0.16± 0.06^{abc}	-0.24± 0.06^{ac}	-0.02± 0.07^{bc}	-0.12± 0.07^{abc}	0.04± 0.07^{b}	0.03± 0.07^{bc}	0.00± 0.07^{bc}
PPARα	-0.18± 0.07^{a}	0.02± 0.06^{b}	-0.03± 0.06^{ab}	-0.13± 0.07^{ab}	-0.01± 0.07^{ab}	-0.12± 0.07^{ab}	0.03± 0.07^{b}	0.06± 0.07^{b}
PPARδ	0.02± 0.09^{bc}	0.04± 0.08^{bc}	0.22± 0.08^{ab}	-0.12± 0.09^{bc}	0.32± 0.09^{a}	-0.07± 0.09^{bc}	0.19± 0.08^{a}	0.06± 0.09^{abc}

Data are presented as LSM ± SEM of logarithm-transformed data. Values within a row with different superscripts differ significantly (P<0.05). HCHF up-regulated *IRS1, GLUT4, CPT1b, UCP3 (1) and INSRβ* (P<0.05); *INSRβ* was also up-regulated with age (P<0.05); PGC1β: Lowest value in CONV-Norm-6, lower than HCHF-Norm-6, HCHF-Norm-24, HCHF-Low-24, and CONV-Norm-24 (P<0.05), highest value in HCHF-Low-24, higher than HCHF-Norm-6 and CONV-Low-6 (P<0.05); *PPARα*: Lowest value in CONV-Low-6, lower than HCHF-Norm-6, CONV-Low-24 and HCHF-Norm-24 (P<0.05), highest value in HCHF-Norm-24, higher thanCONV-Low-6 (P<0.05). *PPARδ*: Lowest value in CONV-Norm-24, lower than HCHF-Low-6, HCHF-Norm-6, and CONV-Norm-6 (P<0.05), highest value in HCHF-Low-6, higher than CONV-Low-6, CONV-Low-24, HCHF-Low-24, and CONV-Norm-24 (P<0.05).

Conclusion

Postnatal nutrition is a determinant of the metabolic pattern of skeletal muscle during early life and has long-term implications for the metabolic pattern later in life even after dietary correction. The increased expression of glucose uptake related genes would be a compensatory mechanism to counteract lowered insulin sensitivity (results not shown). The effects of intrauterine nutrition restriction were observed on the metabolic regulators (*PPAR*s) and *PGC1β* but not other genes analyzed. Prenatal nutrition did not appear to have major influence on gene expression in early postnatal life, but the metabolic long-term adaptive response to an adverse diet early in postnatal life was apparently influenced. Foetal nutrition may have implications for the metabolic flexibility to adapt to a postnatal over-nutrition environment.

References

DeFronzo, R.A. and D. Tripathy, 2009. Skeletal muscle insulin resistance is the primary defect in type 2 diabetes. Diabetes. Care. 32, Suppl 2, S157-163.

Hernandez-Valencia. M. and M.E. Patti, 2006. A thin phenotype is protective for impaired glucose tolerance and related to low birth weight in mice. Arch. Med. Res. 37, 813-817.

Kraegen, E.W. and G.J. Cooney, 2008. Free fatty acids and skeletal muscle insulin resistance. Curr. Opin. Lipidol. 19, 235-241.

Ozanne, S.E. and C.N. Hales, 2005. Poor fetal growth followed by rapid postnatal catch-up growth leads to premature death. Mech. Ageing. Dev. 126, 852-854.

Effect of weaning and milk replacer on aminotransferase activity in plasma and liver of Saanen goat kids

D. Magistrelli, A.A. Aufy and F. Rosi
Università degli Studi, Dipartimento di Scienze Animali, via Celoria 2, 20133 Milano, Italy

Introduction

Weaning is the process of switching young mammals from milk to solid feed. In ruminants, weaning is not just an event, but a period in which milk is progressively substituted by forage and concentrate or grain-based diets, which are processed in the developing rumen. During the weaning period, the diet switches from a mixture of casein, lactose and triglycerides to a more complex nutrient mixture and the pattern of nutrients absorbed shifts from glucose, long-chain fatty acids, and milk-derived amino acids to volatile fatty acids, ketones and amino acids from feed and microbial sources. This change requires the adaptation of liver metabolism and activity, that shifts from glycolytic to gluconeogenic pathways. In particular, amino acids from the breakdown of dietary proteins can be used either for synthesis of new proteins and in the production of glucose and/or ketone bodies (Baldwin *et al.*, 2004). For this latter purpose, amino acids must be transaminated and deaminated so that the carbon skeleton can be utilized in gluconeogenic and/or ketogenic pathways (Hagopian *et al.*, 2003). In light of this, the aim of the present study was to investigate the effect of weaning on plasmatic and hepatic activity of aminotransferases in goats. Moreover, considering that in modern intensive farms weaning is often conducted with the use of a milk replacer, instead of whole milk (since it is cheaper and gives the opportunity to save milk for human consumption), the effect of milk replacer feeding was also investigated in goat kids.

Materials and methods

Thirty-six three-day old Saanen goat kids were divided into four groups (GM, WGM, MR and WMR). GM kids received goat milk to age 48 days. WGM group was initially fed goat milk, but started weaning at 25 days and was completely weaned by 40 days. MR kids were fed a milk replacer to age 48 days. WMR kids were initially fed the milk replacer and then subjected to the same weaning program used for WGM kids. The weaning program was based on the progressive replacement of milk by a weaning mixture constituted by grass hay (30%), dehydrated alfalfa hay (10%), steam-flaked corn (19%), corn gluten meal (3%), dried sugar beet pulp (8%), soybean meal (15%), sunflower seeds (4%), sugar cane molasses (4%) and mineral/vitamin supplement (7%). The experimental feeds were administered twice a day (9:00 am and 7:00 pm). Weekly, body weight (BW) was recorded and plasma samples were taken before the first meal of the day and analyzed for alanine aminotransferase (ALT) and aspartate aminotransferase (AST) activity (Boehringer Mannheim GmbH). On day 48, all the kids were slaughtered, liver weight was recorded and liver samples were taken and analyzed for DNA and RNA content (Munro and Fleck, 1969), ALT and AST activity. Data obtained were analyzed within each observation day by a 2-way ANOVA, using the GLM procedure of SAS (SAS Institute Inc.), considering the effect of weaning and type of milk as independent variables, with a 2×2 factorial arrangement of treatments.

Results and discussion

Body weight did not differ among groups (Table 1). No difference was observed in plasma analysis until the completion of weaning for WGM and WMR kids (data not shown), but 8 days after (on day 48 of age) plasma ALT activity was higher in weaned kids (WGM and WMR) compared to milk-fed ones (GM and MR) (Table 1).

Table 1. Body weight (BW), liver weight, plasma and liver analysis on day 48 of age.

	GM	WGM	MR	WMR	SEM	*P*-value weaning	*P*-value type of milk
BW (kg)	15.0	13.0	14.6	13.3	0.52	ns	ns
Plasma ALT (U/l)	8.54	15.1	9.69	13.4	0.68	<0.001	ns
Plasma AST (U/l)	62.4	64.5	60.1	60.3	2.69	ns	ns
Liver weight (g)	404	221	424	220	28.2	0.001	ns
Liver DNA (mg/g)	2.88	7.57	3.35	8.25	0.79	<0.001	ns
Total liver DNA (g)	1.22	1.67	1.18	1.70	0.16	ns	ns
Liver RNA (mg/g)	9.66	10.2	9.27	9.61	0.15	ns	ns
Liver ALT (U/g)	6.29	8.83	10.0	11.7	1.99	ns	ns
Liver AST (U/g)	169	224	191	210	6.11	<0.001	ns

ns = not significant.

At slaughtering, liver weight was lower in weaned kids (Table 1). Liver DNA, expressed as mg/g of fresh tissue was higher in weaned than milk-fed kids (Table 1), but total liver DNA did not differ among the experimental groups (Table 1). Liver ALT activity was 30% higher in WGM than GM group and 17% higher in WMR than MR group, but these differences were not significant, because of the great variability within groups (Table 1). Liver AST activity was higher in weaned than milk-fed kids (Table 1).

Obtained data suggest that nor milk replacer feeding, neither weaning practiced under the conditions of the present study, result in any negative effect on the performance of the kids. Milk replacer feeding has no consequence on aminotransferase activity in both plasma and liver of goat kids. By contrast, weaning alters liver weight and activity, without affecting the number of cells within the tissue. The transition from milk to solid feed triggers ruminal development (Baldwin *et al.*, 2004), reducing the availability of carbohydrate for post-ruminal digestion. At weaning therefore the liver is stimulated to fulfil a larger portion of the animal's glucose requirement, by increasing hepatic activity of gluconeogenic enzymes (Baldwin *et al.*, 2004) and by increasing amino acid transamination, which is the first step in the catabolism of most amino acids for glucose production (Hagopian *et al.*, 2003). According to this, in the goat kids of the present study, hepatic activity of AST increased after the transition to solid feed, but ALT did not seem to be affected by weaning. Whereas the most of liver AST is localized into the mitochondria, hepatic ALT is limited to the cytosol (Rosenthal and Haight, 1990) of the hepatocyte, which is the principal site of ALT synthesis in the body. In light of this, it is possible to hypothesize that hepatic activity of ALT does not change at weaning, because of its secretion into the bloodstream, thereby raising its plasmatic activity. Further studies will verify this hypothesis, even investigating the linkage between hepatic expression of aminotransferases and their level and activity in both liver and plasma of weaning kids.

References

Baldwin, R.L. VI, K.R. McLeod, J.L. Klotz and R.N. Heitmann, 2004. Rumen development, intestinal growth and hepatic metabolism in the pre- and postweaning ruminant. J. Dairy Sci. 87, E55-E65.

Hagopian, K., J.J. Ramsey and R. Weindruch, 2003. Caloric restriction increases gluco-neogenic and transaminase enzyme activities in mouse liver. Exp. Gerontol. 38, 267-278.

Munro H.N. and A. Fleck, 1969. Analysis of tissues and body fluids for nitrogenous constituents. In: Munro H.N. (ed.), Mammalian Protein Metabolism, Vol. 3., Academic Press, New York, NY, USA: 481-483.

Rosenthal P. and M. Haight, 1990. Aminotransferase as a prognostic index in infants with liver disease. Clin. Chem. 36, 346-348.

High protein milk formula and amoxicillin induced changes in protein metabolism of porcine intrauterine growth restricted neonate

A. Jamin, J.-N. Thibault, N. Mézière, P. Ganier, B. Sève and N. Le Floc'h
INRA, UMR1079 Systèmes d'Elevage, Nutrition Animale et Humaine (SENAH), 35590 Saint Gilles, France

Introduction

Intra-uterine growth retarded (IUGR) individuals represent approximately 10% of human infants worldwide. Growth restriction is an adaptation to either oxygen or nutrients intra-uterine deprivation. Despite neonatal intensive care and use of antibiotics to prevent infection, IUGR is a major cause of mortality and morbidity in the perinatal period (Aucott *et al.,* 2004). When not breast fed, IUGR neonates are usually fed a high protein formula (HP) to ensure catch-up growth (Thureen and Heird, 2005). Nevertheless, the effects of such a practice on whole-body and tissue protein metabolism and development in IUGR neonates are unknown.The IUGR pig is a natural model which is similar to human asymmetric IUGR (Bauer *et al.*, 1998). This animal model can be used to understand the impact of nutritional strategies on the development of IUGR neonates. Our objective was to determine if feeding IUGR piglets with a HP formula supplemented (HPA) or not with an antibiotic in comparison to normal protein (NP) formula may modify tissue protein metabolism. We measured protein synthesis and the activities of enzymes involved in proteolysis in the jejunum, ileum, liver and *semi-tendinosus* muscle of artificially reared IUGR piglets.

Material and methods

The experiments were performed on 36 full term IUGR Piétrain × (Large White × Landrace) piglets (average birth weight 0.96±0.11 kg). Piglets were separated from the sows 2 days after birth. From this moment, 3 experimental groups were constituted. The first group (NP: normal protein content) received an artificial milk formulated to supply the same proportions of protein, fat and carbohydrates as in sow milk. A second group (HP: high protein content) was fed an artificial milk formulated to provide a 50% higher amount of protein of similar composition. A third group received the HP formula supplemented with amoxicillin between d 2 and d 7 (HPA). All piglets were placed in incubators and bottle-fed 10 times a day between d 2 and d 7. At d 7, piglets were transferred into individual cages equipped with automated devices delivering milk 10 times a day. Piglets were weighted daily from d 2 to d 6 then weekly until d 28. Feed intake was recorded daily.

Within each group, 18 piglets were slaughtered at d 7 and at d28. The fractional protein synthesis rate (Ks; % per day) was measured in the mucosa of the jejunum and ileum, the liver and the *semitendinosus* muscle after piglets were injected with a flooding dose of ^{15}N-valine (Hamard *et al.* 2009). Proteasome, cathepsine L and calpain activities were determined in those tissues. The data were analysed with the GLM procedure of SAS. The model included the age, the diet and the age * diet interaction.

Results and discussion

NP, HP and HPA IUGR piglets had similar birth weight (0.95 kg). Piglet average body weights were 1.38 kg at day 7 and 5.51 kg at day 28. Energy intake of NP, HP and HPA pigs did not differ during the experimental period. As expected, protein intake of HP and HPA piglets was higher than in NP piglets (P<0.0001). Tissue protein content increased (P<0.05) between d7 and d28 in the jejunum (10.5 vs. 12.0 g/100 g) and the liver (11.9 vs. 13.3 g/100 g) but was modified neither by the dietary protein content, nor by antibiotic. Tissue Ks decreased with age, except in jejunum (Table 1). Ileum and liver Ks were higher in HP than in NP piglets at d7. Ks of HPA piglets were lower than those of NP piglets in muscle at d28. Calpain and cathepsin L activities increased with age in jejunum. In ileum, calpain activity increased with age and cathepsin L activity was higher in HPA than in NP and HP piglets. Liver calpain activity decreased with age and was lower in HP and HPA than in NP piglets at d28. In muscle, calpain, cathepsin L and proteasome activities decreased with age.

The effects of age reflecting the tissue maturation process largely dominated the impact of diets. Nevertheless our results indicate that excess protein with or without antibiotic treatment may differentially impact liver and muscle protein metabolism according to their maturation level. This questions the effects of nutrition on short term tissue maturation and long term consequences.

Table 1. Fractional rates (K_s, %/d) of protein synthesis, proteasome, calpaïn and cathepsin L activities (RFU/min/mg protein) measured in tissues of piglets fed NP, HP and HPA diets at d 7 and d 28 (n=6/group).

Age	d 7			d 28				*P*		
Diets	NP	HP	HPA	NP	HP	HPA	SEM	diet	age	diet × age
K_s										
jejunum	55.5[a]	58.9[a]	57.6[a]	60.9[a]	50.6[b]	56.1[ab]	2.39	0.35	0.45	0.02
ileum	58.5	64.2	54.5	53.3	53.5	48.2	2.8	0.03	0.003	0.57
liver	66.5[a]	80.7[b]	73.4[ab]	60.2[ac]	56.3[c]	54.8[c]	3.05	0.20	<0.0001	0.01
muscle	18.6[a]	19.7[a]	20.7[a]	12.2[b]	10.2[bc]	9.2[c]	0.98	0.88	<0.0001	0.04
Proteasome										
jejunum	25908	25569	26586	31343	25827	26137	1986	0.30	0.28	0.26
ileum	21828	23123	21803	26196	23846	22263	2793	0.72	0.40	0.72
liver	12499	12848	12767	13352	11808	11117	682	0.33	0.27	0.16
muscle	634	783	691	497	376	578	66	0.53	0.0002	0.60
Calpain										
jejunum	6469	6842	6794	9307	8640	9460	819	0.89	0.0007	0.79
ileum	8944	9962	7819	12427	14015	14552	2452	0.86	0.02	0.77
liver	11505[a]	11649[a]	12054[a]	11045[a]	7451[b]	6884[b]	902	0.11	0.0002	0.04
muscle	529	505	561	382	308	524	83	0.25	0.06	0.61
Cathepsin L										
jejunum	1286	989	1136	2511	2943	4125	819	0.60	0.003	0.54
ileum	3041	2971	5642	2965	3019	6085	1160	0.02	0.88	0.97
liver	2333	3358	2415	3889	2301	2530	801	0.71	0.75	0.26
muscle	965	2237	1068	732	572	632	111	0.10	<0.0001	0.34

[a,b,c] LS means having a different superscript letter differ at $P \leq 0.05$.

References

Aucott, S.W., P.K. Donohue, and F.J. Northington, 2004. Increased morbidity in severe early intrauterine growth restriction. J. Perinatol. 24, 435-440.

Bauer, R., B. Walter, A. Hoppe, E. Gaser, V. Lampe, E. Kauf, and U. Zwiener, 1998. Body weight distribution and organ size in newborn swine - a study describing an animal model for asymmetrical intrauterine growth retardation. Exp. Toxicol. Pathol. 50, 59-65.

Hamard, A., B. Seve, and N. Le Floc'h, 2009. A moderate threonine deficiency differently affects protein metabolism in tissues of early-weaned piglets. Comp Biochem. Physiol A Mol. Integr. Physiol 152, 491-497.

Thureen, P. and W.C. Heird, 2005. Protein and energy requirements of the preterm/low birthweight (LBW) infant. Pediatr. Res 57, 95R-98R.

Energy and protein-related metabolisms in liver, skeletal muscle and adipose tissue of pigs in a divergent selection experiment for residual feed intake

T. Le Naou[1,2], N. Le Floc'h[1,2], L. Lefaucheur[1,2], I. Louveau[1,2], P. Sellier[3], Y. Billon[4], H. Gilbert[3] and F. Gondret[1,2]
[1]INRA, UMR1079 Systèmes d'Elevage, Nutrition Animale et Humaine (SENAH), 35590 Saint Gilles, France
[2]Agrocampus Ouest, UMR1079, 35000 Rennes, France
[3]INRA, UMR1313 Génétique Animale et Biologie Intégrative (GABI), Domaine de Vilvert, 78352 Jouy-en-Josas Cedex, France
[4]INRA, UE967 Génétique Expérimentale en Productions Animales (GEPA), St-Pierre d'Amilly, BP52, 17700 Surgères, France

Introduction

Feed intake and feed efficiency are economically important traits because feed is the greatest variable cost in pig production. Feed efficiency can be measured as residual feed intake (RFI), which is the difference between the actual and expected feed intake based on the rate and tissue composition of body weight (BW) gain over a given period. The physiological basis of RFI generally includes intake and digestion of feed, physical activity, thermoregulation, and anabolic/catabolic metabolisms underlying variations in whole-body and tissue compositions (Herd and Arthur, 2009). A previous study has indicated minimal genetic correlation between RFI and growth rate but a positive genetic correlation between RFI and the amount of fat depots (Gilbert *et al.*, 2007). Lefaucheur *et al.* (2008) have also shown a higher muscle glycolytic potential and an impaired meat quality in the most efficient pigs. The aim of the present experiment was to determine the main changes in tissue energy and protein metabolisms associated with a genetic difference for RFI in growing pigs.

Material and methods

Weaned female pigs from 2 lines that underwent six generations of a divergent selection for RFI (high [RFI^+] and low [RFI^-]) were compared. A third group (RFI^+_R) composed of high-RFI pigs restricted to the same level of feed per $BW^{0.60}$ than RFI^- pigs during the growing-finishing period was also included. This should allow dissociating effects of genetic variation in RFI from those of feed intake. All pigs were slaughtered at the same BW (114.7±0.6 kg), 2 hours after the last meal. Liver and *Longissimus* muscle (LM) as organs involved in active energy and protein metabolisms, and dorsal subcutaneous adipose tissue (SCAT) responsible for the long-term storage of energy into lipids, were sampled. Performance during growth and carcass composition traits at slaughter were recorded on a total of 38 animals from 17 litters in 2 successive repeats. Biochemical data were obtained from a subset of 24 animals from the first repeat out of the 38 pigs. Data were analyzed by ANOVA, with the fixed effects of RFI group, repeat, and RFI group × repeat, and dam nested within group as a random effect.

Results and discussion

Only RFI^+ pigs displayed an accelerated growth during the growing-finishing period (Table 1). However, both RFI^+ and RFI^+_R pigs have a deteriorated food conversion ratio as compared to RFI^- pigs. At slaughter, there were no difference in lipid contents of SCAT, liver and LM between the 3 groups (Table 1). Specific activity of fatty acid synthase (FAS), i.e. the key lipogenic enzyme, was similar in SCAT of the 3 groups, whereas it was the greatest in the liver of RFI^+ pigs. The physiological significance of the latter observation remains to be assessed, because liver is not considered as the main lipogenic organ in pigs at this slaughter weight. A greater activity of calpain enzyme was observed in the liver of both RFI$^+$ and RFI$^+$_R pigs compared to RFI$^-$ pigs (P=0.01, data not shown), suggesting an impact of selection on hepatic protein catabolism. The activity level of the same enzyme did not change in the LM. Greater specific activities of enzymes representative of glycolysis (lactacte dehydrogenase, LDH), fatty acids oxidation (beta-hydroxylacyl Coenzyme-A dehydrogenase, HAD) or terminal oxidative catabolism (citrate synthase, CS) were shown in the liver and LM of RFI$^+$ pigs compared to RFI$^-$ pigs (Table 1). When compared to RFI$^+$ group, restricted feeding in RFI$^+$_R pigs had no effects in the liver, but decreased the activity levels of the same enzymes in the LM. This suggests an accelerated glycolysis and a higher fatty acid oxidation in the liver as a direct response to selection for high RFI in pigs. The modifications observed in the LM were likely more associated to variations in feed intake between pigs.

Table 1. Performance, lipid content and enzyme activities in tissues.

Trait	RFI$^-$	RFI$^+$	RFI$^+$_R	P_{group}
Performance in growing-finishing	n=16	n=11	n=11	
Average daily gain, g/d	911^a	1,010^b	869^a	0.004
Food conversion ratio	2.57^a	2.85^b	2.87^b	0.003
Biochemical data	n=10	n=7	n=7	
Dorsal subcutaneous adipose tissue				
Lipid content, % fresh tissue	81.6	83.5	82.1	0.449
FAS activity, nm/min/mg proteins	23.8	17.4	18.9	0.178
Liver				
Lipid content, % fresh tissue	4.33	4.76	4.34	0.212
FAS activity, nm/min/mg proteins	0.90^b	1.46^a	0.71^b	0.039
LDH activity, µm/min/mg proteins	5.69^a	7.46^b	6.96^{ab}	0.083
CS activity, µm/min/mg proteins	1.71^a	2.00^b	1.92^b	0.028
HAD activity, µm/min/mg proteins	0.23^a	0.33^b	0.30^b	0.002
Longissimus muscle				
Lipid content, % fresh tissue	1.59	1.97	1.83	0.254
LDH activity, µm/min/mg proteins	557^a	620^b	542^a	0.026
CS activity, µm/min/mg proteins	1.25^a	1.54^b	1.36^a	0.006
HAD activity, µm/min/mg proteins	0.76^a	1.01^b	0.89^c	0.003

^{a,b,c} LS means having a different superscript letter differ at $P \leq 0.05$.

Conclusion

Selection for high RFI was suggested to accelerate glucose and fatty acid catabolisms in the liver, possibly to provide energy to sustain a greater maintenance and an accelerated growth.

References

Gilbert H., J-P. Bidanel, J. Gruand, J-C. Caritez, Y. Billon, P. Guillouet, H. Lagant, J. Noblet, and P. Sellier, 2007. Genetic parameters for residual feed intake in growing pigs, with emphasis on genetic relationships with carcass and meat quality traits. J. Anim. Sci. 85, 3182-3188.

Herd R.M., and P.F. Arthur, 2009. Physiological basis for residual feed intake. J. Anim. Sci. 87 (E. suppl), E64-E71.

Lefaucheur L., B. Lebret, P. Ecolan, M. Galian, M. Damon, I. Louveau, A. Prunier, P. Sellier, and H. Gilbert, 2008. Divergent selection on 'residual feed intake' in pigs: impact on growth performance, muscle compositional traits and meat quality. Journées Rech. Porcine 40, 83-84.

Selection for muscling increases adipose tissue response to adrenaline

G.E. Gardner[1,2,3], K.M. Martin[2], D.W. Pethick[1,2,3], M.B. Ferguson[2,3], J.M. Thompson[1,2] and P. McGilchrist[1,3]
[1]Australian Cooperative Research Centre for Beef Genetic Technologies, Australia
[2]Cooperative Research Centre for Sheep Industry Innovation, Australia
[3]School of Veterinary & Biomedical Science, Murdoch University, 6150, WA, Australia

Introduction

Carcass lean meat yield is a key profit driver within the sheep and beef industries, and is improved by selection for muscling. This trait can be improved through visual selection or with the use of estimated breeding values for yearling eye muscle depth (YEMD) in sheep, and retail beef yield in cattle. The resulting muscle hypertrophy impacts upon carcass composition, proportionately increasing muscle and decreasing subcutaneous fat (Perry *et al.*, 1993; Hegarty *et al.*, 2006). At a physiological level the mechanisms under-pinning these compositional differences are unclear, with a number of key hormonal axis likely to be impacted. The key regulatory stress hormone adrenaline, which causes mobilisation of adipose triacylglycerol releasing non-esterified fatty acids (NEFA) for energy production could have a large effect. However, Gilson *et al.*, (1986) indicated that NEFA response to adrenaline would simply reflect whole body fatness, and therefore its responsiveness is likely to be less in sheep and cattle selected for muscling. Thus, there is no known physiological shift which alters the catabolic mechanism in adipose of high muscled animals causing their reduced phenotypic leanness. In this study we test the hypothesis that adipose response to adrenaline will reflect whole body fatness in sheep and cattle selected for muscling.

Material and methods

This paper details two experiments where adrenaline challenges were administered to a group of 20 sheep at 4 and 16 months of age, as well as 21 steers at 18 months of age. The sheep were the progeny of Merino and Poll Dorset sires selected for a diverse range in Australian Sheep Breeding Values for yearling eye muscle depth (YEMD). The steers were the progeny of an Angus herd selected for divergence in muscling over 15 years. Prior to administering adrenaline challenges, animals were habituated in individual pens for 2 weeks on an *ad-libitum* grain-based diet. Adrenaline challenges were then administered via indwelling jugular catheters at 7 levels (2/day) ranging between 0.1-3.0 μg/kg liveweight. Sixteen blood samples were taken between -30 & 130 minutes relative to adrenaline administration. Plasma was analysed for NEFA concentration which reflects the adipose response to adrenaline. In both experiments plasma NEFA area under curve (AUC) for the first 10 minutes following adrenaline challenge was analysed using a linear mixed effects model. Fixed effects included muscling genotype (cattle experiment only) and age (sheep experiment only), covariates were adrenaline challenge and YEMD (sheep experiment only), and animal within sire was used as the random term.

Results and discussion

In the sheep, at 4 months of age the plasma NEFA AUC following all levels of adrenaline challenge was about 0.3 mM/10 min greater ($P<0.01$) in the progeny of high YEMD sires (Figure 1A). At 16 months of age this difference was not evident, and the adrenaline response had reduced by about 5-fold. Similarly, in beef cattle, the NEFA AUC was about 0.2 mM/10 min greater ($P<0.01$) in the high muscled genotype across all levels of adrenaline challenge (Figure 1B).

In both sheep and cattle the progeny of more heavily muscled sires had adipose tissue that was more responsive to adrenaline, contrary to our initial hypothesis which assumed that NEFA response to

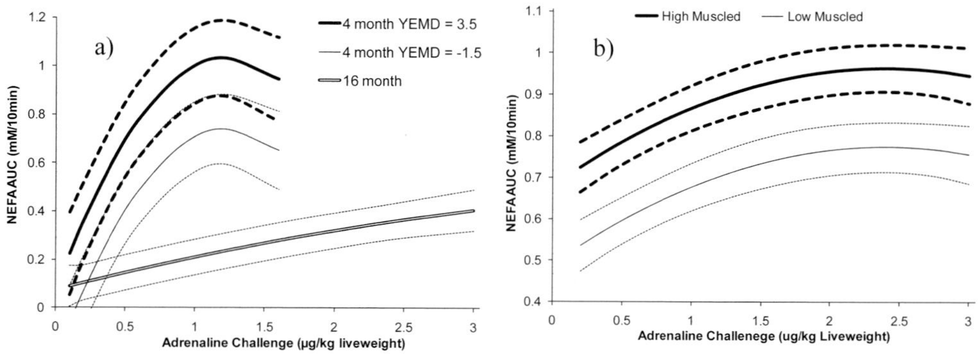

Figure 1. Effect of adrenaline challenge on plasma NEFA area under the curve (mM/10 min) for (A) sheep at 4 and 16 months of age with high and low YEMD and (B) cattle selected for high or low muscling. Values are least square means; dashed lines are s.e.m.

adrenaline would simply reflect total adiposity. To further emphasise the significance of this result, the high muscled cattle used in this study had a hind-limb that was composed of 11% less fat than the low muscled cattle (data not shown), thus the greater adipose response to adrenaline has been demonstrated in animals with proportionately less fat. While the physiological mechanisms contributing to these results are unclear, the adipose tissue of lean sheep has been shown to have greater levels of vascularisation which may enhance perfusion and thus response to adrenaline (Gregory *et al.*, 1986). In sheep the differences between the high and low muscled animals diminished as they approached maturity, along with total adipose responsiveness to adrenaline, thus lipolysis may be a less important determinant of overall adiposity and carcase composition in older animals. None-the-less, greater adrenaline induced lipolysis in high muscled sheep and cattle may partly explain why they are leaner, particularly at a younger age.

References

Gilson, T.L., A.D. Kennedy, and T., Rampersand, 1996. Effects of breed and adipose depot location on responsiveness and sensitivity to adrenergic stimulation in ovine adipose tissue. Comp. Biochem. Physiol. I 115C, 19-26.

Gregory, N.G., R.J. Christopherson, and D. Lister, 1986. Adipose tissue capillary blood flow in relation to fatness in sheep. Res. Vet. Sci. 40, 352-356.

Hegarty R.S., D.L. Hopkins, T.C. Farrell, R. Banks and S. Harden, 2006. Effects of available nutrition and sire breeding values for growth and muscling on the development of crossbred lambs. 2:Composition and commercial yield. Aus. J. Agri. Res. 57, 617-626.

Perry, D., A.P. Yeates and W.A. McKiernan, 1993. Muscle score: Its usefulness in describing the potential yield of saleable meat from live steers and their carcasses. Aust. J. Exp. Ag. 33, 275-281.

Effects of dietary nitrogen content and intravenous urea infusion on ruminal and portal-drained visceral extraction of arterial urea in lactating Holstein cows

N.B. Kristensen, A.C. Storm and M. Larsen
Department of Animal Health and Bioscience, Aarhus University, Tjele, Denmark

Introduction

There exists a mismatch between the general agreement that urea transport across gut epithelia is directed by mass action (Harmeyer and Martens, 1980; Sunny *et al.*, 2007) and the curvilinear relationship between arterial urea concentration and the extraction ratio of arterial urea across the PDV in cattle showing declining urea extraction with increasing blood urea concentration (Calsamiglia *et al.*, 2010; Røjen *et al.*, 2008). Two overall hypotheses can explain the observed relationship between epithelial urea supply and extraction ratio in cattle: (a) mass action driven transport combined with down-regulation of epithelial urea permeability with increasing nitrogen (N) status and (b) fixed transport capacity for urea across the epithelia independent of blood urea supply to the epithelium. The objective of the present experiment was to investigate the ruminal and portal-drained visceral (PDV) extraction of arterial urea at low and high N intake and to investigate the interaction between dietary N status and the blood urea level on PDV and ruminal extraction of arterial urea.

Materials and methods

Nine second parity Danish Holstein cows implanted with a ruminal cannula and permanent indwelling catheters in an artery, mesenteric vein, hepatic portal vein (n=8), hepatic vein, and ruminal vein (n=8) were used. Cows were randomly allocated to two experimental diets in a cross over design (balanced for 8 cows) with 21 d periods. The dietary treatments were LOW-N (12.9% crude protein) and HIGH-N (17.1% crude protein). The N content of the diets was changed by replacing soybean meal with soyhulls and molasses. The rations were based on corn silage (30%), rolled barley (30%) and grass clover silage (25%). Cows were sampled on d 14 of each period. The interaction between adaptation to dietary treatments and blood concentrations of urea was studied by dividing samplings into a 2.5 h period without urea infusion followed by a 2.5 h period with primed continuous jugular vein infusion of urea (0.493±0.012 mmol/kg BW per h; approx. 1.7 and 0.7 times net hepatic flux with LOW-N and HIGH-N, respectively). Five sets of blood samples were collected at 30 min intervals before initiation of urea infusion and five sets of blood samples were collected during urea infusion. Data on blood concentrations and extractions were analyzed by the MIXED procedure (SAS) as mean values (cow × dietary treatment × infusion state) by contrasts defining before infusion LOW-N versus before infusion HIGH-N, with-infusion LOW-N versus with-infusion HIGH-N, and the interaction. The model had sequence, period, dietary treatment, urea infusion (before or with), and the interaction dietary treatment × urea infusion as fixed effects. Cow within sequence and cow within treatment were included as random effects. Intake and milk yield data were analyzed using a reduced model.

Results and discussion

Dry matter intake (18.3 vs. 20.5±0.5 kg/d) and yield of energy corrected milk (36 vs. 42±1 kg/d) were less (P<0.01) in LOW-N compared with HIGH-N. The blood urea concentration was lower (P<0.01) in LOW-N compared with HIGH-N, but no interaction (P=0.47; Table 1) between treatment and infusion was observed indicating a similar increase in blood urea with intravenous infusion independent of dietary treatment. Both the PDV and ruminal extraction ratios of arterial urea were greater (P<0.01) in LOW-N compared with HIGH-N and the differences persisted also during urea infusion i.e. no interaction between treatment and infusion (P=0.33 to P=0.35). The net portal uptake

Table 1. Portal-drained visceral and ruminal extraction of arterial urea in lactating Holstein cows fed a low or high N diet with and without intravenous infusion of urea.

	LOW-N		HIGH-N		SEM	Trt × Infusion
	Before infusion	With infusion	Before infusion	With infusion		
Arterial blood urea[1], mmol/l	1.37	3.38	4.09	5.91	0.18	0.47
Portal extraction of arterial urea[1], %	5.4	5.2	2.7	3.0	0.5	0.35
Ruminal extraction of arterial urea[1], %	23.8	22.6	7.1	9.1	2.1	0.33

[1] Test for effect of LOW-N vs. HIGH-N (P<0.05).

of arterial urea was greater with HIGH-N compared with LOW-N before infusion (P<0.01) and the portal uptake increase with infusion with both treatments although the increase was less with HIGH-N (interaction P=0.06).

The present study demonstrated that urea transport across gut epithelia is directly proportional to the blood concentration of urea in lactating dairy cows i.e. in the short term the extraction ratio upon passage of the epithelial bed is unaffected by increased blood urea concentration. The epithelial permeability for urea was markedly affected by the ration offered and the extraction ratio of urea was up-regulated with LOW-N at the level of the rumen and the whole PDV. However, the relative increase in extraction was greatest for the rumen. The data obtained in the present study implies that the curvilinear relationship observed between blood urea concentration and PDV urea extraction in lactating dairy cows is caused by adaptation of the epithelial urea transport to the nutritional state of the cow and not an artifact of a transport system for urea running at a constant rate (i.e. not zero-order regulation).

The mechanims underlying the changes in gut epithelial urea permeability as affected by dietary composition remain to be indenfied and candidates may be found among proteins expressed by the cow (even though UT-B seems unlikely to be involved) and proteins expressed by bacteria colonizing the rumen epithelium (Stewart and Smith, 2005). Resolving the mechanisms of urea transport across gut epithelia is of major importance for evaluating the potential role of urea recycling (urea salvaging mechanisms) in improving N efficiency of lactating dairy cows and other ruminants.

References

Calsamiglia, S., A. Ferret, C.K. Reynolds, N.B. Kristensen and A.M. Van Vuuren, 2010. Strategies for optimizing nitrogen use by ruminants. Animal, In press.

Harmeyer, J. and H. Martens, 1980. Aspects of urea metabolism in ruminants with reference to the goat. J. Dairy Sci. 63, 1707-1728.

Røjen, B.A., P. Lund and N.B. Kristensen, 2008. Urea and short-chain fatty acids metabolism in Holstein cows fed a low-nitrogen grass-based diet. Animal 2, 500-513.

Stewart, G.S., C. Graham, S. Cattell, T.P.L. Smith, N.L. Simmons and C.P. Smith, 2005. UT-B is expressed in bovine rumen: portential role in ruminal urea transport. Am. J. Physiol. Regul. Integr. Comp. Physiol. 289, R605-R612.

Sunny, N.E., S.L. Owens, R.L. Baldwin, VI, S.W. El-Kadi, R.A. Kohn and B.J. Bequette, 2007. Salvage of blood urea nitrogen in sheep is highly dependent upon plasma urea concentration and the efficiency of capture within the digestive tract. J. Anim. Sci. 85, 1006-1013.

Glucagon-like peptide-2 alters amino acid fluxes across the portal-drained viscera of ruminant calves

C.C. Taylor-Edwards[1], D.G. Burrin[2], N.B. Kristensen[3], K.R. McLeod[1] and D.L. Harmon[1]
[1]Department of Animal and Food Sciences, University of Kentucky, Lexington, KY, USA
[2]USDA/ARS Children's Nutr. Res. Center, Baylor College of Medicine, Houston, TX, USA
[3]Department of Animal Health and Bioscience, Aarhus University, Tjele, Denmark

Introduction

In ruminant calves, glucagon-like peptide-2 (GLP-2) increased small intestinal mass and epithelial mass, villus height, crypt depth, and bromodeoxyuridine (BrdU) labeling in the small intestine without affecting total body mass (Taylor-Edwards *et al.*, 2009). The gastrointestinal tract uses amino acids (AA) as an energy source and for protein synthesis (Bertolo and Burrin, 2008; Wu, 2009). Thus, intestinal growth caused by GLP-2 would likely also alter portal drained viscera (PDV) utilization of AA. Furthermore, previous research (Guan *et al.*, 2003) has demonstrated that GLP-2 increases PDV glucose utilization in parenterally-fed piglets, but no information is known in ruminants, which often differ in their regulation and utilization of glucose compared to non-ruminants. Thus, the aim of this experiment was to examine the effects of exogenously-administered GLP-2 on blood AA concentrations and fluxes across the PDV and utilization of glucose by the PDV.

Materials and methods

Eight weaned Holstein calves with catheters in the carotid artery, portal vein, and mesenteric vein were paired by age and randomly assigned to treatment: Control (0.5% bovine serum albumin [BSA] in saline; n=4) or GLP-2 (50 μg/kg BW bovine GLP-2 in BSA; n=4). Intake was 2.75% of BW (DM-basis). Treatments were injected subcutaneously every 12 h for 10 d. On day 10, after a 12 h fast to avoid prandial effects on blood flow, calves were continuously infused with para-aminohippuric acid and [U-^{13}C] glucose to measure portal blood flow and PDV utilization of glucose, respectively. Blood samples were taken every 15 min for 150 min. Plasma was analyzed for para-aminohippuric acid, for total free AA by HPLC (phenylisothiocyanate derivitization), and for [U-^{13}C] glucose enrichment (Kristensen *et al.*, 2002). Net PDV flux was calculated as portal plasma flow × (portal – arterial nutrient concentration), with positive values indicating net release and negative values indicating net uptake. Data were analyzed using the MIXED procedure of SAS with treatment as a fixed effect and block as a random effect.

Results and discussion

Treatment with GLP-2 for 10 d reduced (P<0.05) arterial concentrations of the essential AA Leu, Lys, Phe, and Val (Table 1), and the non-essential AA Gln (Table 1) and Ala, Asn, and Pro (not shown). Likewise, 10-d treatment with GLP-2 reduced net PDV release of the essential AA Arg, Ile, Leu, Lys, and Phe (Table 1), and the non-essential AA Ala, Asn, Asp, Gly, Pro, and Ser (not shown). Moreover, negative correlations were found between small intestinal growth measures and arterial AA concentrations and PDV flux. For example, jejunal crypt cell BrdU labeling was negatively correlated with PDV release of Leu, Lys, Phe, Asp, Gly, Pro, Ser, and ornithine (P<0.01). These results suggest GLP-2 increased growth and resulted in greater sequestration of AA in the PDV, perhaps for protein synthesis.Glutamate and Gln are used for gut energy metabolism and growth, including synthesis of other AA such as Arg, Pro, ornithine, and citrulline (Bertolo and Burrin, 2008). Arterial Gln concentrations were 23% lower with GLP-2 treatment and net PDV uptake of Gln was unchanged (Table 1); thus, the net PDV extraction of Gln was greater after GLP-2 (0.20 vs. 0.10, P<0.0001). Greater arterial Gln extraction and reduced PDV ornithine flux may have contributed to the increased citrulline export from the gut and likely resulted in greater renal conversion of citrulline to Arg, and thus greater Arg concentrations in GLP-2-treated calves. Despite greater Arg concentrations, PDV Arg release was reduced, suggesting PDV retention, perhaps to support small intestinal mucosal growth. Despite changes in AA metabolism, whole body glucose irreversible loss (GIL) did not differ between treatments (117.4 mmol/h). Glucose utilization by the PDV and non-PDV tissues was 0.41 and 0.63, respectively, of GIL and was not affected by treatment. This contrasts with TPN-fed piglets, where GLP-2 increased PDV glucose uptake and extraction (Guan *et al.*, 2003). The effects of GLP-2 on PDV glucose utilization may be a result of species adaptations in non-ruminants versus ruminants. Although PDV use of glucose is significant, ruminants also rely on energy substrates besides glucose to spare glucose for vital functions. Availability of substrates such as Glu, Gln, propionate, and butyrate reduced glucose oxidation in ovine duodenal mucosal cells without reducing glucose uptake (Oba *et al.*, 2008). Thus, GLP-2 may have increased Gln extraction rather than glucose to provide the energy to support growth. Flux of AA across the PDV is affected by GLP-2, potentially by increased small intestinal epithelial growth (Taylor-Edwards *et al.*, 2009) and thus energy and AA requirements of this tissue. Increased PDV extraction of Gln and alterations in PDV metabolism of Arg, ornithine, and citrulline support the concept that intestine-specific AA metabolism is affected by GLP-2. However, unchanged glucose metabolism suggests GLP-2 effects on PDV glucose metabolism in ruminants are more transient or less significant than for non-ruminants.

Table 1. Small intestinal epithelia percentage, crypt cell BrdU labeling (% of total cells), arterial amino acid concentrations and PDV fluxes in calves treated with GLP-2.

	Control	GLP-2	SEM	$P\leq$	Control	GLP-2	SEM	$P\leq$
	Jejunum				Ileum			
Epithelia, %	57.8	70.3	4.00	0.10	52.0	65.2	3.51	0.04
BrdU, %	16.8	23.1	0.86	0.01	16.9	20.0	1.60	0.05
	Arterial concentration, µM				PDV Flux, mmol/h			
Arg	117.3	166.1	5.11	<0.0001	3.72	1.90	0.540	0.03
Ile	117.7	102.1	8.67	0.10	4.16	1.56	0.846	0.05
Leu	134.5	102.5	9.25	0.006	4.86	2.13	0.662	0.004
Lys	64.6	52.3	4.20	0.01	3.71	1.64	0.649	0.01
Phe	47.2	38.2	2.84	0.005	3.08	1.48	0.332	0.003
Val	247.4	199.6	17.73	0.02	3.12	2.13	1.500	0.65
Gln	208.7	160.6	10.92	0.0003	-5.23	-6.87	1.043	0.12
Cit	46.9	77.4	3.30	<0.0001	2.02	1.79	0.529	0.76
Orn	21.1	23.1	1.48	0.15	1.04	0.062	0.296	0.002

References

Bertolo, R.F. and D.G. Burrin, 2008. Comparative aspects of tissue glutamine and proline metabolism. J. Nutr. 138, 2032S-2039S.

Guan, X., B. Stoll, X. Lu, K.A. Tappenden, J.J. Holst, B. Hartmann, and D.G. Burrin, 2003. GLP-2-mediated up-regulation of intestinal blood flow and glucose uptake is nitric-oxide dependent in TPN-fed piglets. Gastroenterology 125, 136-147.

Kristensen, N.B., A. Danfaer, B.A. Rojen, B.-M.L. Raun, M.R. Weisbjerg, and T. Hvelplund, 2002. Metabolism of propionate and 1,2-propanediol absorbed from the washed reticulorumen of lactating cows. J. Anim. Sci. 80, 2168-2175.

Oba, M., R.L. Baldwin, VI, and B.J. Bequette, 2004. Oxidation of glucose, glutamate, and glutamine by isolated ovine enterocytes *in vitro* is decreased by the presence of other metabolic fuels. J. Anim. Sci. 82, 479-486.

Taylor-Edwards, C.C., D.G. Burrin, K.R. McLeod, and D.L. Harmon, 2009. Glucagon-like peptide-2 increases small intestinal mass of calves. J. Anim. Sci. 87 (E-Suppl. 2), 234 (Abstr.).

Wu, G., 2009. Amino acids: metabolism, functions, and nutrition. Amino Acids 37, 1-17.

Methionine hydroxy analogue conversion in intestinal Caco-2 cells

R. Martín-Venegas[1], Y. Mercier[2], P.A. Geraert[2] and R. Ferrer[1]
[1]Departament de Fisiologia, Facultat de Farmàcia, Universitat de Barcelona, av. Joan XXIII s/n, 08028 Barcelona, Spain
[2]Adisseo France S.A.S., 42, avenue Aristide Briand, 92160 Antony, France

Introduction

The DL-2-hydroxy-(4-methylthio)butanoic acid (HMTBA) is a synthetic source of dietary Met widely used in poultry nutrition. Upon absorption, the biological utilization of HMTBA relies on its conversion to L-Met a process which is similar in avian and mammal species (Dibner and Knight, 1984; Fang *et al.*, 2010). In a first step, HMTBA is oxidized through a stereospecific enzymatic process. L-HMTBA is substrate of L-2-hydroxy acid oxidase (L-HAOX), present mainly in chicken hepatic and renal peroxisomes (Dibner and Knight, 1984; Gordon and Sizer, 1965). D-HMTBA is oxidized by D-2-hydroxy acid dehydrogenase (D-HADH), found in the mitochondrial fraction of different tissues (Knight and Dibner, 1984). Subsequent to the formation of the common intermediate 2-keto-(4-methylthio)butanoic acid (KMB), the second step is its conversion to L-Met by transamination, which is ubiquitous (Rangel-Lugo and Austic, 1998). Taking into account the importance of HMTBA in animal production, the determination of the role of the intestinal epithelium in HMTBA conversion is of high interest. Therefore, the aim of the present study was to investigate in intestinal Caco-2 cultures HMTBA oxidation and transamination. Cultures of Caco-2 cells form a highly polarized epithelium with many of the properties of the intestinal villous absorptive cells and constitute an *in vitro* experimental model, currently used to evaluate intestinal transport and metabolism (Quaroni and Hochmann, 1996).

Materials and methods

Cell culture. Caco-2 cells were kindly provided by Dr. David Thwaites at the School of Cell and Molecular Biosciences, University of Newcastle upon Tyne (UK) and cultured during 21 days as previously described (Martín-Venegas *et al.*, 2007).

D-HADH and L-HAOX assay. The enzymatic activities were estimated following Schreiner and Jones (1988) using D-lactate as substrate for D-HADH, and glycolic acid and L-2-hydroxyisocaproic acid for L-HAOX-A and B isozymes, respectively (Duley and Holmes, 1976).

Transamination study. The capacity of the cells to convert KMB to L-Met was assayed following Rangel-Lugo and Austic (1998) using different amino acids as amino group donors. Met was quantified by ion exchange chromatography as previously described (Martin-Venegas *et al.*, 2006).

Statistical analysis. Results were given as means ± SEM. Significant differences were detected by one-way ANOVA followed by Student's t test using the SPSSâ software (SPSS Inc. Chicago, IL, USA). $P<0.05$ was considered to denote significance.

Results and discussion

The results indicate that Caco-2 cells showed higher D-HADH (3.38±0.63 mU/μg protein, n=6, $P<0.05$) than L-HAOX activities. In the case of L-HAOX, the data reveal a higher capacity to convert glycolic acid than L-2-hydroxyisocaproic acid to KMB (0.28±0.04 vs. 0.12±0.01 mU/μg protein, n=4 cultures, $P<0.05$), thus suggesting that, although both L-HAOX-A and B isozymes are present, the isozyme A is more active. As for KMB transamination, the results indicate that all the amino acids tested (L-Leu, L-Val, L-Ala, L-Phe, L-Pro, L-Asn, L-Gln, L-Ser, L-Tyr, L-Asp, L-Glu,

Table 1. KMB transamination to L-Met.

nmol/µg protein		nmol/µg protein		nmol/µg protein	
L-Leu	12.08±0.65[a]	L-Asn	1.13±0.02[e]	L-Asp	0.08±0.02[g]
L-Val	4.62±0.35[b,c]	L-Gln	5.89±0.39[b]	L-Glu	1.99±0.07[d]
L-Ala	3.10±0.97[c,d,e]	L-Ser	1.30±0.10[e]	L-Arg	1.01±0.03[e]
L-Phe	2.71±0.26[d]	L-Tyr	0.68±0.01[f]	L-Lys	1.13±0.07[e]
L-Pro	1.30±0.06[e]				

Mean values labelled with different letters are significantly different ($P<0.05$) (mean ± SEM of n=4 monolayers).

L-Arg and L-Lys) participate in the formation of L-Met from KMB, the branched-chain amino acid L-Leu being the preferred amino group donor (Table 1).

In conclusion, these results confirm D-HADH activity in the intestinal epithelium and describe for the first time the presence of both isozymes of L-HAOX in this tissue, thus indicating the capacity of the intestinal epithelium to convert both isomers of HMTBA to L-Met. The results also reveal that transamination is not specifically linked to an amino group donor, thus suggesting that this is not the limiting step in HMTBA conversion to L-Met.

References

Dibner J.J. and C.D. Knight, 1984. Conversion of 2-hydroxy-4-(methylthio)butanoic acid to L-methionine in the chick: a stereospecific pathway. J. Nutr. 114, 1716-1723.

Duley J.A. and R.S. Holmes, 1976. L-alpha-Hydroxyacid oxidase isozymes. Purification and molecular properties. Eur. J. Biochem. 63, 163-173.

Fang Z., H. Lou, H. Wei, F. Huang, Z. Qi, S. Jiang and J. Peng, 2010. Methionine metabolism in piglets fed DL-methionine or its hydroxy analogue was affected by distribution of enzymes oxidizing these sources to keto-methionine. J. Agric. Food Chem. 58, 2008-2014.

Gordon R.S. and I.W. Sizer, 1965. Conversion of methionine hydroxy analogue to methionine in the chick. Poult. Sci. 44, 673-678.

Knight C.D. and J.J. Dibner, 1984. Comparative absorption of 2-hydroxy-4-(methylthio)-butanoic acid and L-methionine in the broiler chick. J. Nutr. 114, 2179-2186.

Martín-Venegas R., P.A. Geraert and R. Ferrer, 2006. Conversion of the methionine analogue, DL-2-hydroxy-(4-methylthio)butanoic acid, to sulfur-containing amino acids in the chicken intestine. Poult. Sci. 85, 1932-1938.

Martín-Venegas R., M.J. Rodríguez-Lagunas, P.A. Geraert and R. Ferrer, 2007. Monocarboxylate transporter 1 mediates DL-2-hydroxy-(4-methylthio)butanoic acid transport across the apical membrane of Caco-2 cell monolayers. J. Nutr. 137, 49-54.

Quaroni A. and J. Hochmann, 1996. Development of intestinal cell culture models for drug transport and metabolism studies. Adv. Drug Deliv. Rev. 22, 3-52.

Rangel-Lugo M. and R.E. Austic, 1998. Transamination of 2-oxo-4-(methylthio)butanoic acid in chicken tissues. Poult. Sci. 77, 98-104.

Schreiner C.L. and E.E. Jones, 1988. Basis of the stereospecific preference of porcine kidney fibroblasts for D-2-hydroxy-4-methylthiobutanoic acid as a source of methionine. J. Nutr. 118, 818-828.

Effects of changing dietary lysine level from a deficient to a sufficient level on expression of proteolytic-related genes in growing rats

A. Ishida, K. Nakashima and M. Katsumata
Molecular Nutrition Research Team, National Institute of Livestock and Grassland Science, Tsukuba, 305-0901, Japan

Introduction

We reported that changing dietary lysine level from a deficient to a sufficient level induced increment of growth rate in growing rats (Ishida *et al.*, 2007). This increment was considerable during the next three days after the changing dietary lysine levels. The rates of muscle protein degradation were lower in the rats exhibiting increment of growth rate during this period. We inferred this lower degradation of muscle protein at least partly contributed to the considerable increment of the growth rate. In order to investigate underlying mechanisms of suppression of muscle proteolysis induced by change of dietary lysine levels, we examined the time-course of changes of expression of proteolytic-related genes.

Material and methods

Thirty six male 4-wk-old rats were randomly assigned to one of two diets; a control diet and a low lysine (LL) diet. The control diet contained all essential amino acids in the recommended amounts, including 13 g lysine/kg on fed basis while the LL diet was similar but contained only 4.5 g lysine/kg on fed basis. The animals were subjected to these diets during the first 14 days. From day 14, the rats fed the low-lysine diet were given the control diet until the end of experiment (day 17) (LC group) while the rats subjected to the control diet were continuously fed the control diet throughout the experiment period (control group). On days 14, 15 and 17, serum and gastrocnemius muscle samples were taken from six rats from each group. The levels of expression of proteolytic-related genes (u-calpain, m-calpain, cathepsin L, caspase-3, proteasome alpha subunit, Atrogin-1, MuRF1) were determined by Real-time RT-PCR method. We measured concentrations of corticosterone in the serum samples. The effects of the treatments were assessed with a two-way ANOVA. When the effects were significant, the differences between the means were compared by Student's t-test or Tukey's test. All the calculations were carried out using the GLM procedure of SAS (version 9.1, 2003).

Results and discussion

Although the mRNA expression of capase-3 in gastrocnemius muscle on day 14 was not affected by dietary lysine levels, the expression of its mRNA of the LC group one day after the changing dietary lysine levels (on day 15) was lower than the control group ($P<0.05$; Table 1). Although the mRNA expression of atrogin-1 in gastrocnemius muscle on day 14 was not affected by dietary lysine levels, its expression of the LC group on days 15 and 17 was lower than the control group ($P<0.05$; Table 1). Further, atrogin-1 mRNA expression of the LC group on day 15 was lower than that of the LC group on day 14 ($P<0.05$; Table 1). The treatments did not affect mRNA expression of the other proteolytic-related genes (data not shown). The concentrations of serum corticosterone of the LC group tented to be higher than those of the control group on day 14 ($P=0.059$; data not shown), whereas the concentrations rapidly declined after the changing dietary lysine levels. The concentrations of serum corticosterone were not different between the two groups on day 17 (data not shown).

The results suggest that suppression of muscle proteolysis after the changing dietary lysine levels may be attributed to down-regulation of caspase 3 and atrogin-1 mRNAs expression. Since corticosterone induces expression of atrogin-1 mRNA (Sacheck *et al.,* 2004), change of the concentrations of the

Table 1. Relative expression of caspase3 and atrogin-1 mRNA in the muscle of rat on compensatory growth after dietary lysine deficiency.

	Caspase3 (relative expression)			Atrogin-1(relative expression		
Day[1]	Control	LC	RMSE	Control	LC	RMSE
Day 14	1.00	0.96[b]	0.09	1.00	1.22[a]	0.35
Day 15	1.10	0.90[*b]	0.12	1.21	0.62[*b]	0.27
Day 17	1.00	1.14[a]	0.13	1.36	0.85[*ab]	0.28

[a,b] Statistically different ($P<0.05$), between different letters. *$P<0.05$ vs. control rats.
[1] Day 14: lysine deficient status, day 15: one day after lysine sufficiency, day 17:three day after lysine sufficiency.

hormone after the changing dietary lysine levels can be involved in the mechanisms regulating expression of atrogin-1 mRNA. However, further investigations are necessary to establish the mechanisms in this respect.

References

Ishida, A., K. Nakashima and M. Katsumata, 2007. Changing dietary lysine level from a deficient to a sufficient level greatly enhances growth rate of growing rats. In: Ortigues-Marty I. (ed.), Energy and Protein Metabolism and Nutrition, Wageningen Academic Publishers, Wageningen, the Netherlands, EAAP Series 124, 531-532.

Sacheck, J.M., A. Ohtsuka, S.C. McLary and A.L. Goldberg, 2004. IGF-I stimulates muscle growth by suppressing protein breakdown and expression of atrophy-related ubiquitin ligases, atrogin-1 and MuRF1. Am J Physiol Endocrinol Metab. 287, 591-601.

Dietary tryptophan downregulates the gene expression of lipopolysaccharide-induced genes in healthy pigs susceptible to enterotoxigenic *E. coli* K88

P. Trevisi[1], E. Corrent[2], S. Messori[1], D. Priori[1] and P. Bosi[1]
[1]University of Bologna, DIPROVAL, Via Rosselli 107, Reggio Emilia, Italy
[2]Ajinomoto Eurolysine S.A.S., 75017 Paris, France

Introduction

An important mechanism in regulating immunological and neurological responses is the catabolism of tryptophan (Trp). The pool of enzymes that break down Trp through the kynurenine pathway (so called from one of the final products) are found in numerous cell types, including cells of the immune system. Some of these enzymes are induced by some inflammatory signals. Thus, tissue inflammatory condition can cause depletion of tryptophan. This raises the hypothesis that intestinal infections could increase the requirement of Trp. Supplementary L-Trp limited the decrease of feed intake and growth in Enterotoxigenic *E. coli* K88 (ETEC) susceptible pigs, upon oral challenge with this pathogen (Trevisi *et al.*, 2009). Our aim was to assess if genes involved in the intestinal barrier are changed by dietary Trp in healthy pigs susceptible or not to ETEC.

Materials and methods

A polymorphism on Mucin 4 (MUC4) gene has been identified as a marker for susceptibility to ETEC. Sows and boars were screened for this polymorphism, following the protocol of Jensen *et al.* (2006), then litters from parents heterozygous were tested, to obtain piglets homozygous for the resistant and sensitive alleles ($2^{-/-}$ and $2^{+/+}$ pigs within each litter). So 36 piglets were selected, weaned at 24 d, and assigned to two diets within each genotype per litter. The pigs were *ad libitum* fed the same basal diet (18.7% CP, 2.52 Mcal/kg of NE, no antimicrobial), but the Trp to Lysine (Lys) ratio (17% or 22%, standardized ileal digestible basis) was varied with or without the 0.5 g/kg L-Trp addition. Thus two factors, Trp to Lys ratio (17%, 22%) and genotype ($MUC4^{-/-}$, $MUC4^{+/+}$) were arranged in a factorial design. After 21±1 days the pigs were anaesthetized and euthanized, the mid jejunum of each pig was collected and the *in vitro* ETEC adhesion test (Bosi *et al.*, 2004) was performed. Pigs were considered susceptible with values ≥6 bacteria adhering per 250 μm length of villous, non-susceptible with no adhesion, and mildly susceptible with values <6 and >0.

A oligonucleotide-Microarray preliminary test was done to highlight the most differentially expressed genes in jejunum tissue from 3 resistant and 3 susceptible pigs selected from two litters. All the samples passed the evaluation by Agilent Bioanalyzer 2100 and mRNA was labelled and hybridised to GeneChip® Porcine Genome Array. CEL files were background-subtracted, normalized with the GC-RMA algorithm in the Affymetrix package of Bioconductor. The presence of artifacts and the consistency of normalisation across arrays were checked by *affy* and *affyPLM* packages available on Bioconductor, and a cluster representation of mRNA sequences was produced obtained by *affy*. Model fitting and hypothesis testing were done by the *limma* package from Bioconductor. Three contrasts were fitted: $MUC4^{-/-}$ vs. $MUC4^{+/+}$; Low Trp vs. High Trp; ($MUC4^{-/-}$ and ($MUC4^{+/+}$, High Trp)) vs. $MUC4^{+/+}$, Low Trp. The last contrast was generated to assess one interaction. The empirical Bayes moderated *t* test was used to assess differential expression, the corresponding *P*-values were adjusted to control the false discovery rate, and the genes with adjusted *P*-values <0.1 were declared differentially expressed (Jensen *et al.*, 2007).

Then the gene expression of the most interesting genes was assessed on the whole sample, with the procedure reported in our previous papers (Bosi *et al.*, 2006). Data of gene expression were analyzed by analysis of variance, considering Trp to Lys ratio, susceptibility to ETEC, the interaction, and the litter. When the interaction was statistically significant, linear contrasts were used.

Results and discussion

The cluster representation of mRNA sequences preliminary tested by Affymetrix Microchips on the six pigs was examined (graph not shown). It was evident from the figure that there was a cluster association related to the polymorphism for MUC4. Furthermore while in the group MUC4$^{-/-}$ there was no association with the diet, in the MUC4$^{+/+}$ the sub-cluster did differentiate the subjects for the Trp addition.

On total subjects in trial, within susceptible pigs, three genes involved in the intestinal barrier response and induced by lipopolysaccharide were less expressed in jejunum samples from Trp-supplemented pigs ($P<0.05$) (Reg3g, Regenerating islet-derived 3 gamma; SFTPD, Surfactant; pulmonary-associated protein D; CFB, Complement factor B). Within unsupplemented pigs, the gene expression of Reg3g and SFTPD increased with the susceptibility to ETEC ($P<0.05$), while the Interleukine-8 tended to be less expressed with the Trp supplementation whatever was the ETEC susceptibility ($P=0.09$). Reg3g and SFTPD are C-type lectins often increased in bacteria induced infections. Particularly it was already observed that Reg3 (also called Pancreatitis-associated protein) mRNA increased in porcine small intestinal loops infected with *E. coli* K88, as compared with non infected loops. Factor B is a protein that plays a central role in the 'alternative pathway' of complement activation. It is also stimulated by various bacteria products, including LPS. Glutathione peroxidase 2 gene expression was not significantly affected. No DNA from ETEC was detected in the jejunum content of pigs.

Conclusions

Tryptophan favourably interacts to reduce the induction of lipopolysaccharide on genes involved in the intestinal barrier in ETEC susceptible pigs, but the causative mechanism needs to be established.

Acknowledgements

Ajinomoto Eurolysine financially supported this work.

References

Bosi, P., C. Gremokolini, P. Trevisi, M. Mazzoni, P. Bonilauri, G. Sarli and L. Casini, 2004. Oral challenge with *E. coli* K88 as a tool to assess growth and health performance in feeding trials of weaned pigs. J. Rech. Porcine 36, 125-132.

Bosi, P., M. Mazzoni, S. De Filippi, L. Casini, P. Trevisi, G. Petrosino and G. Lalatta-Costerbosa, 2006. Continuous dietary supply of free calcium formate negatively affects parietal cell population and gastric RNA expression for H+/K+-ATPase in weaning pigs. J. Nutr. 136, 1229-1235.

Golkar, L., X.Z. Ding, M.B. Ujiki, M.R. Salabat, D.L. Kelly, D. Scholtens, A.J. Fought, D.J. Bentrem, M.S. Talamonti, R.H. Belland and T.E. Adrian, 2007. Resveratrol inhibits pancreatic cancer cell proliferation through transcriptional induction of macrophage inhibitory cytokine-1. J Surg Res.138, 163-169.

Jensen, G.M., K. Frydendahl, O. Svendsen, C.B. Jorgensen S. Cirera, M. Fredholm, J.P. Nielsen and K. Møller, 2006. Experimental infection with *Escherichia coli* O149:F4ac in weaned piglets, Vet. Microbiol. 115, 243-249.

Trevisi, P., D. Melchior, M. Mazzoni, L. Casini, S. De Filippi, L. Minieri, G. Lalatta-Costerbosa and P. Bosi, 2009. A tryptophan-enriched diet improves feed intake and growth performance of susceptible weanling pigs orally challenged with *E. coli* K88. J. Anim Sci. 87, 148-156.

Goblet cells in the small intestine of young pigs fed low-threonine diet supplemented with threonine or with different sources of nonessential amino acids

E. Swiech, M. Boryczka, M. Taciak, A. Tusnio and L. Buraczewska
The Kielanowski Institute of Animal Physiology and Nutrition, Instytucka street 3, 05-110 Jablonna, Poland

Introduction

The intestinal epithelium is protected by mucous layer composed mainly of mucins secreted by goblet cells (GC). Mucins are classified into acidic (sialomucins and sulfomucins) and neutral types, and their proportions vary along the gut. An optimal protection of epithelium depends on both quantitative (thickness) and qualitative properties of mucins. Acidic mucins increase the resistance of mucus against bacterial enzymes (Deplancke and Gaskins, 2001). The increase in GC number in the small intestinal villi (V) may potentially increase the mucin secretion capacity of the mucosa and, consequently, improve gut health. All mucins are rich in serine, proline and particularly in threonine (Thr). The aim of the study was to estimate a Thr-sparing effect of nonessential amino acids (AA), added in the form of wheat gluten (WG) or monosodium glutamate (MSG) to a diet deficient in Thr and balanced for other essential AA, on GC producing different types of mucins along the small intestine of young pigs.

Material and methods

The experiment was carried out on five groups, each of six male pigs (from 15 to 40 kg of BW), fed during 28 days on Thr-deficient (80% of requirement) diet supplemented with 0 (negative control), 20 and 40 g/kg WG protein or 17.6 g/kg MSG (equivalent to the sum of glutamic and aspartic acid contents in 40 g WG protein), and on Thr-adequate diet (positive control). All diets were supplemented with crystalline AA to cover the AA requirement of pigs. The tissue samples of the duodenum (D), mid-jejunum (J) and ileum (I) were taken. The number of GC containing neutral, acidic, and sulfomucins per the left side of V and per crypt (C) was estimated in intestinal sections stained with periodic acid-Schiff, Alcian blue pH 2.5, and Alcian blue pH 1.0, respectively. Data were statistically evaluated using one-way Kruskal-Wallis ANOVA followed by the Mann-Whitney test.

Results and discussion

The number of GC containing neutral, acidic and sulfomucins in three segments of the small intestine are presented in Table 1. In D and J, Thr supplementation significantly increased number of GC containing all types of mucins, with exception of neutral mucin-containing GC in C. In I, only number of GC containing sulfomucins and neutral mucins was greater in Thr-adequate than Thr-deficient group in C. This is in agreement with results obtained by Law *et al.* (2007), who found an increased number of GC, containing all types of mucins, in D and I of neonatal piglets fed intragastrically Thr-adequate diet, in comparison with Thr-deficient diet. On the contrary, Hamard *et al.* (2007) did not find positive effect of Thr intake on a number of GC in the ileum and colon in piglets.

The effect of supplementing the Thr-deficient diet with WG or MSG differed among the intestinal segments and parameters. In D, supplementation with 40 g/kg WG had no effect on the GC number in V and C while in J it increased or tended to increase the number of all types GC in C and only acidic ones in V. On the contrary, in I, diet with 40 g/kg WG decreased or tended to decrease the number of all GC in V and had no effect in C. In D, supplementing with MSG increased or tended to increase the number of GC in V and decreased sulfomucin GC in C. In J, MSG affected GC number

Table 1. Number of goblet cells containing neutral or acidic mucins in villus and crypt in the small intestine of pigs.

	Diet[1]					
Type of mucin	B	B+WG20	B+WG40	B+MSG	B+Thr	SEM
Duodenum						
neutral in villi	5.8^{a}	5.4^{a}	6.0^{ab}	6.8^{b}	6.8^{b}	0.35
acidic in villi	5.4^{ab}	4.9^{a}	6.2^{ab}	6.5^{b}	8.1^{c}	0.35
sulfomucins in villi	5.9^{b}	4.8^{a}	6.2^{b}	7.6^{c}	7.3^{c}	0.36
neutral in crypts	6.5	6.5	6.3	7.2	7.2	0.28
acidic in crypts	6.7^{a}	6.9^{a}	6.7^{a}	6.5^{a}	11.2^{b}	0.33
sulfomucins in crypts	6.4^{b}	7.0^{b}	7.2^{b}	5.6^{a}	12.6^{c}	0.41
Mid-jejunum						
neutral in villi	3.3^{a}	4.0^{ab}	3.7^{a}	3.6^{a}	4.3^{b}	0.27
acidic in villi	3.7^{a}	3.5^{a}	4.8^{b}	3.3^{a}	4.9^{b}	0.27
sulfomucins in villi	4.4^{a}	4.7^{a}	4.5^{a}	4.1^{a}	5.5^{b}	0.36
neutral in crypts	4.6^{a}	6.0^{c}	5.4^{ab}	5.2^{bc}	4.7^{ab}	0.26
acidic in crypts	4.3^{a}	5.0^{b}	5.0^{b}	4.7^{ab}	8.0^{c}	0.27
sulfomucins in crypts	4.5^{a}	5.3^{ab}	5.4^{b}	4.7^{a}	8.1^{c}	0.27
Ileum						
neutral in villi	7.9^{d}	6.2^{bc}	6.0^{b}	4.4^{a}	7.3^{cd}	0.37
acidic in villi	6.6^{cd}	4.7^{ab}	5.5^{bc}	4.6^{a}	7.6^{d}	0.38
sulfomucins in villi	7.9^{bc}	6.4^{b}	7.1^{bc}	5.0^{a}	8.4^{c}	0.42
neutral in crypts	13.2^{a}	12.2^{a}	13.7^{a}	13.8^{a}	16.3^{b}	0.40
acidic in crypts	15.3^{bc}	14.6^{ab}	15.3^{abc}	13.6^{a}	16.9^{c}	0.60
sulfomucins in crypts	14.4^{bc}	14.8^{bc}	14.9^{cd}	12.3^{a}	17.4^{d}	0.65

[1] B: Thr-deficient (negative control); WG: wheat gluten; MSG: monosodium glutamate; B+Thr: Thr-adequate (positive control).

[a,b,c,d] Means in the same row with different superscripts differ at $P \leq 0.05$.

neither in V nor in C (except increase of neutral GC in C) while in I it considerably decreased all GC in V and C except neutral GC in C.

It may be concluded that feeding Thr-deficient diet decreases mostly the number of acidic GC, particularly sulfomucin and that supplementing this diet with MSG and WG affects GC positively but in a very inconsistent way. This suggests that increase in nonessential AA supply of Thr-deficient diets may have beneficial effects in practical feeding of young pigs.

References

Deplancke, B., and H.R. Gaskins, 2001. Microbial modulation of innate defense: goblet cells and the intestinal mucus layer. Am. J. Clin. Nutr. 73, 1131S-1141S.

Hamard, A., B. Séve, and N. le Floc'h, 2007. Intestinal development and growth performance of early-weaned piglets fed a low-threonine diet. Animal 1, 1134-1142.

Law, G.K., R.F. Bertolo, A. Adjiri-Awere, P.B. Pencharz, and R.O. Ball, 2007. Adequate oral threonine is critical for mucin production and gut function in neonatal piglets. Am. J. Physiol. Gastrointest. Liver Physiol. 292, G1293-G1301.

Dietary nucleotides in broilers: Effects on productive performances and intestinal morphometry

L. Liotta[1], F. Marino[2], G. Savoini[3], V. Chiofalo[1], G. Lanteri[2], E. D'Alessandro[1] and B. Chiofalo[1]
[1]Dipartimento di Morfologia, Biochimica, Fisiologia e Produzioni Animali, Università degli Studi di Messina, Polo Universitario Annunziata, Messina, Italy
[2]Dipartimento di Sanità Pubblica Veterinaria, Università degli Studi di Messina, Polo Universitario Annunziata, Messina, Italy
[3]Dipartimento di Scienze e Tecnologie Veterinarie per la Sicurezza Alimentare, Università degli Studi di Milano, Via Celoria, 10, Milano, Italy

Introduction

Nucleotides are normal components of the diet and the body provides mechanism for their absorption and incorporation into tissue (Sanchez-Pozo and Gil, 2002). However, during periods of rapid growth, certain disease states, limited nutrient intake or disturbed endogenous synthesis of nucleotides, their availability could limit the maturation of fast dividing tissues with low biosynthetic capacity, such as the intestine (Van Buren and Rudolph, 1997). Regarding chickens, dramatic changes occur in the development of the small intestinal mucosa after hatching, including enterocyte maturation, intensive cryptogenesis and villous growth (Geyra *et al.*, 2001). This intestinal development influences the growth rate, since intestinal maturation plays a rate determining role in providing the substrates for growth. Aim of this study was to evaluate the effect of nucleotide supplementation on growth performance and intestinal morphometry of male broiler chickens.

Material and methods

Male broiler chickens (n=60,000), ROSS 508 genotype, were divided into two homogeneous groups: Control (C) and Nucleotides (N); the latter received, from the birth (1 day of age) to the slaughtering age (52 days), the basal diet, which were the same for both groups, supplemented with 0.1% of a Nucleotide pool containing adenosine, guanosine, cytidine and uridine 5'-monophosphates in similar quantities, extracted by yeasts (*Saccharomyces cerevisiae*) through an enzymatic hydrolysis.

From each of the two groups, 130 animals were randomly selected. Body weights (BW) were determined and used to calculate Average Daily Gain (ADG). Feed intake was calculated at the end of the trial (52 d) on 2×30,000 animals in order to determine the Feed Conversion Ratio (FCR) of the entire experimental period. Mortality rate and Litter quality were determined on 2×30,000 animals.

At slaughtering, 15 chickens were again randomly selected among the 130 per group. Intestinal tract were removed; small pieces of descendent duodenum of each animal were taken at the level of the medium portion, using as reference point the pancreas between the ascendant and descendent segments of duodenum, sampled and processed. Five μm thick sections were obtained and stained. The following morphometric analysis using a Zeiss Axiophot was performed: total number of villi (TNV) per the whole circumference transverse sections; full length profiles of villi (FLV), continuous with the submucosa, and villi width (VW), at the medium length of each one, were measured on three villi per each broiler.

Data was subjected to ANOVA (Proc. GLM, SAS, 2001) considering the variable: diet.

Results and discussion

Performance during the experiment is shown in Table 1. During the first 10 days of age no significant differences were observed for the BW of the two groups, whereas, at 20, 40 and 52 days the N

Table 1. Effects of dietary nucleotides on technical performance of broilers.

Item	C	N	SEM	*P*-values
Body weight at 1 day (g)	53.56	44.77	3.16	0.6703
Body weight at 10 days (g)	342.40	355.50	4.06	0.5202
Body weight at 20 days (g)	746.06	895.01	10.8	0.0001
Body weight at 40 days (g)	2,173.99	2,376.33	58.2	0.0001
Body weight at 52 days (g)	3,149.91	3,300.10	62.6	0.0001
Average daily gain 1 to 52 days (g/d)	59.54	62.60	4.17	0.0012
Feed intake 1 to 52 days (kg)	7.24	7.29	0.63	0.8921
FCR 1 to 52 days (kg/kg)	2.34	2.24	0.35	0.5525
Litter score: 1-10 / very wet - very dry litter	5.5	5.6	0.56	0.7332
Mortality 1 to 52 days (%)	3.20	3.25	0.62	0.5221

group showed a higher BW; moreover, broiler chickens of the N group, have shown higher ADG. These results are in agreement with Tipa (2002) in chickens receiving nucleotides. No significant differences were observed for FCR, Litter score and Mortality.

Dietary nucleotides have determined a significant increase of the TNV (N 49.33 vs. C 39.67; P=0.0014) as well as the VW (N 0.19 vs. C 0.12 mm; P=0.0001), whereas a significant increase for the C group were observed for the FLV (N 0.98 vs. C 1.58 mm; P=0.049).

Differences observed for the BW and ADG as well as for the intestinal morphology in the N group could be related to an increased growth and maturation of intestinal epithelial cells and, consequently, an increased total mucosal area as observed by Domeneghini *et al.* (2004) in swine, with positive functional effects, e.g. on feed utilisation (Zhang *et al.* 2008).

References

Domeneghini, C., A. Di Giancamillo, G. Savoini, R. Paratte, V. Bontempo and V. Dell'Orto, 2004. Structural patterns of swine ileal mucosa following L-glutamine and nucleotide administration during the weaning period. An histochemical and histometrical study. Histol. Histopathol. 19, 49-58.

Geyra, A., Z. Uni, and D. Sklan., 2001. Enterocyte dynamics and mucosal development in the posthatch chick. Poultry Sci. 80, 776-782.

Sanchez-Pozo, A. and A. Gil. 2002. Nucleotides are semiessential nutritional components. Brit. J. Nutr. 87, S135-S137.

SAS, 2001. User's Guide Version 8.2 Statistical Analysis System Inst. Cary, NC, USA.

Tipa, C. 2002. Production performance of broilers fed yeast extract as a partial or total replacement for fishmeal. DVM Thesis, Univ. of Philippines at Los Banos, Laguna.

Van Buren, C.T. and F. Rudolph. 1997. Dietary nucleotides: a conditional requirement. Nutrition. 13, 470-472.

Zhang, G.Q., Q.G. Ma and C. Ji. 2008. Effects of dietary Inosinic Acid on carcass characteristics, meat quality, and deposition of Inosinic Acid in broilers. Poultry Sci. 87, 1364-1369.

Effect of early feeding on intermediary metabolism of broiler chicks differing in hatching time

M. Debonne, A. Witters, Y. Li, H. Willemsen, E. Decuypere, J. Buyse and N. Everaert
Department of Biosystems, K.U. Leuven, Leuven, Belgium

Introduction

Chicks hatch over a hatching window of 48 hours. In practice, after 21 days, incubators are opened and chicks are removed from the machine (hour 0 of chronological age). The batch of chicks is never homogenous and it is wondered whether chicks that hatch early (around ED20) are physiologically different in comparison with chicks that hatch late (ED21). It is not exceptional that chicks that hatched first are deprived from feed and water for 72 hours after they hatched. Several studies however have shown that early feeding after hatching has beneficial effects on subsequent performance of chicks (Noy and Sklan, 1999; Gonzales *et al.*, 2003). Moreover, Careghi *et al.* (2005) found that late hatchers benefit more from early feeding than early and mid hatchers in terms of growth rate when the chronological age was considered. The aim of this study was to further investigate the effect of immediate or delayed (48 hours after hatching time) feeding on the metabolism of the early, mid and late hatchers.

Material and methods

1,200 hatching eggs were collected from Ross breeders of 47 weeks old and set for incubation. The eggs were incubated under standard conditions. From the 468th hour of incubation until the 512th hour of incubation, hatch was followed. All eggs were checked every hour and hatchlings were weighed and numbered immediately after hatch. Three groups were distinguished: early, mid and late hatchers. Half of the early, mid and late group received feed immediately after hatch, whereas the other half was fasted for 48 hours. At their biological age (age calculated from the precise moment of hatch) of 48 hours, the fasted chicks of the early, mid and late group were allowed access to feed and water. Samples were taken when chicks had a biological age of 15, 39, 53, 101 and 149 hours. Fifteen chicks per group were weighed, and blood was taken from the neck vein. After decapitation, livers were collected and glycogen levels were determined. Plasma T_3, triglycerides, and non-esterified fatty acids (NEFA) concentrations were measured. A general linear model was used to study the effect of age, hatching time (early, mid or late), time of feeding (immediate or delay of 48 hours) and their interactions on measured parameters. When there was a significant effect of a variable, then these means were further compared by Tukey's test, per day. Significance was set at a level of 5% level. Values are expressed as mean ± SEM.

Results

In general, there was no interaction between hatching time and time of feeding. However, on certain time points, the time of hatching influenced the delay in feed differently. Results are shown in the table. Chicks fasted for 48 hours after hatch were significantly lighter than chicks that were fed immediately after hatch, irrespective of hatching time. In the late group, the difference in chick weight remained until 149 hours of biological age even when the previously fasted chicks received feed (data not shown). Fasting decreased plasma T_3 levels in all groups. When fasted chicks were fed, plasma T_3 values were restored and even a slight overcompensation could be observed in the late hatchers at 53 hours. Hepatic glycogen reserves were completely depleted in chicks fasted for 48 hours after hatching, irrespective of hatching time. At 53 hours, hence after feeding, glycogen concentrations exceeded those of the chicks that had access to feed immediately after hatch. At 39 hours, plasma levels of NEFA were only significantly decreased in the late hatchers that were feed deprived compared to their fed counterparts. Feeding restored plasma NEFA levels in this group. Plasma concentrations of triglycerides were significantly lower in all groups as a result of fasting but when the chicks were fed, overcompensation could be observed in all groups which was more pronounced in the late hatching group.

Discussion

Our results show that feed deprivation after hatch impairs postnatal growth at least until 5 days of age. Gonzales *et al.* (2003) even reported decreased performance at slaughter weight when chicks were fasted for more than 24 hours immediately after hatching. Interestingly, late hatchers needed more time as their body weight remained lower at 7 days of age. When chicks were fasted after hatch, hepatic glycogen reserves were depleted and circulating plasma triglyceride levels were lower, suggesting that carbohydrate catabolism increased and lipid synthesis were decreased compared to their fed counterparts. The rebound in plasma triglycerides after feeding was more pronounced for the late hatchers and this might be due to a higher feed consumption (Careghi *et al.*, unpublished data), which might also be the cause for the postprandial increase in plasma T_3 levels in this group. In conclusion, immediate feeding after hatch is a prerequisite for optimal postnatal performance. Late hatchers seem to be affected more by feed deprivation, notwithstanding their more pronounced rebound in postprandial metabolism.

Table 1. Chick weight (g), hepatic glycogen content (mg/g), plasma metabolite (mmol/l) and thyroid (ng/ml) concentrations at 39 and 53 hours of biological age in early, mid and late hatchers with immediate or delayed (48 hours) access to feed.

		Early	Mid	Late
39 hours of biological age				
Chick weight (g)	Feed	55.8±1.5[a]	54.0±1.1[a]	55.1±1.8[a]
	No feed	42.5±1.3[b]	44.8±1.1[b]	42.1±0.9[b]
Glycogen (mg/g liver)	Feed	11.23±0.64[a]	11.37±0.72[a]	7.78±0.84[a]
	No feed	0.06±0.05[b]	1.40±0.05[b]	0.12±0.08[b]
NEFA (mmol/l)	Feed	2.35±0.53	0.99±0.32	1.41±0.35[a]
	No feed	1.76±0.46	1.14±0.22	0.55±0.12[b]
Triglycerides (mmol/l)	Feed	1.12±0.16[a]	1.26±0.24[a]	1.71±0.13[a]
	No feed	0.42±0.03[b]	0.48±0.13[b]	0.50±0.04[b]
T_3 (ng/ml)	Feed	1.83±0.08[a]	1.99±0.14[a]	2.17±0.08[a]
	No feed	0.94±0.03[b]	1.12±0.11[b]	1.10±0.08[b]
53 hours of biological age				
Chick weight (g)	Feed	61.5±1.1[a]	61.1±1.7[a]	64.5±1.6[a]
	No feed	54.3±1.5[b]	57.6±0.8[b]	52.9±1.3[b]
Glycogen (mg/g liver)	Feed	3.20±0.63[b]	7.21±2.11[b]	5.13±1.20[b]
	No feed	9.17±0.83[a]	12.15±1.25[a]	12.94±1.15[a]
NEFA (mmol/l)	Feed	1.88±0.48	1.61±0.49	2.38±0.59
	No feed	1.80±0.45	1.03±0.21	2.15±0.68
Triglycerides (mmol/l)	Feed	0.94±0.10	1.12±0.14	0.82±0.10[b]
	No feed	1.20±0.08	1.54±0.14	1.30±0.16[a]
T_3 (ng/ml)	Feed	1.73±0.14	1.65±0.13	1.37±0.07[b]
	No feed	1.81±0.11	1.98±0.14	1.79±0.09[a]

[a,b] Means between time of feeding differ (P<0.05) per hatching time.

References

Careghi, C., K. Tona, O. Onagbesan, J. Buyse, E. Decuypere, and V. Bruggeman, 2005. The effects of the spread of hatch and interaction with delayed feed access after hatch on broiler performance until seven days of age. Poult. Sci. 84, 1314-1320.

Gonzales, E., N. Kondo, E.S.P.B. Saldanha, M.M. Loddy, C. Careghi, and E. Decuypere, 2003. Performance and physiological parameters of broiler chickens subjected to fasting in the neonatal period. Poult. Sci. 82, 1250-1256.

Noy, Y. and D. Sklan, 1999. The importance of early nutrition. WPSA 12th European Symposium on Poultry Nutrition. Veldhoven, Nederland: 143-151.

Effect of nanoparticles of silver and gold on gas exchange and heat production of chicken embryos

L. Pineda[1], A. Chwalibog[1], E. Sawosz[2], A. Hotowy[1], J. Elnif[1], F. Sawosz[1], T. Niemiec[2] and A. Ali[1]
[1]University of Copenhagen, Denmark
[2]University of Life Science, Warsaw, Poland

Introduction

Despite the potential application of noble nanometals in animal production, limited information is available concerning the impact of nanoproducts in animal nutrition (Fondevilla *et al.*, 2009; Sawosz *et al.*, 2007). Beside antimicrobial and immuno-stimulating properties, it was hypothesized that nanometals increases anabolic activity. Metal like silver is a carrier of easy available oxygen atoms that can be used in cellular respiration or oxidative metabolism (Sawosz *et al.*, 2007). Using chicken embryo as a model, an experiment was conducted to determine the effects of nanoparticles of silver and gold on gas exchange and heat production.

Materials and methods

A total of 72 fertile Lohman eggs were weighed and randomly distributed into 3 experimental groups consisting of injection of a colloidal solution of nanoparticles of silver (Ag) or gold (Au) together with a control uninjected group (C). Eggs were set in 3 batches, 24 per batch, and incubated for 21 days under standard conditions (37.8 °C, humidity 55%, turn once per hour during first 18 d, and later at 37 °C and humidity 60%). The eggs were injected at day one of incubation with 0.3 ml of colloidal Ag and Au nanoparticles at the concentration of 50 ppm.

Gas exchange was measured in an open-air-circuit respiration unit (Micro-Oxymax calorimeter from Columbus Instruments, Columbus, Ohio, USA) equipped with four respiration chambers at days 10, 13, 16 and 19 of incubation. Measurements were made on four eggs at a time. They were candled prior to measurement to check the presence of embryo and then set into each respiration chamber for 3 hours from 9:00 to 12:00 followed by another group measured from 13:00 to 16:00.

Data were analyzed using GLM procedure of SAS (SAS Institute Inc., 2009) considering the main effects of treatment and age; and their interaction.

Results and discussion

The parameters measured were normalized to 50 g egg weight to eliminate differences in weight. The weight of eggs and chicks was not affected by the treatments. The egg weight at the start of incubation was similar in all groups, 60.0 (SEM 0.29) g; while the hatching weight was 41.3 (SEM 0.42) g.

Pattern of gas exchange and heat production was consistent in all the groups, following the model for typical avian embryos (Janke *et al.*, 2004) and in accordance with the findings of Chwalibog *et al.* (2007) who observed a similar slow increase until day 10 and then rapid increase at days 13 to 16 of incubation and then slowing down until 1 to 2 days before hatching (Figure 1).

Interestingly, the nanoparticles did not elicit the same significant effect on all parameters measured (Table 1) except for RQ (data not shown), which was calculated to be around 0.7 in all the groups. Nano-silver significantly increased oxygen consumption and carbon dioxide production; and consequently the rate of heat production. No significant difference was observed on the interaction between treatment and age.

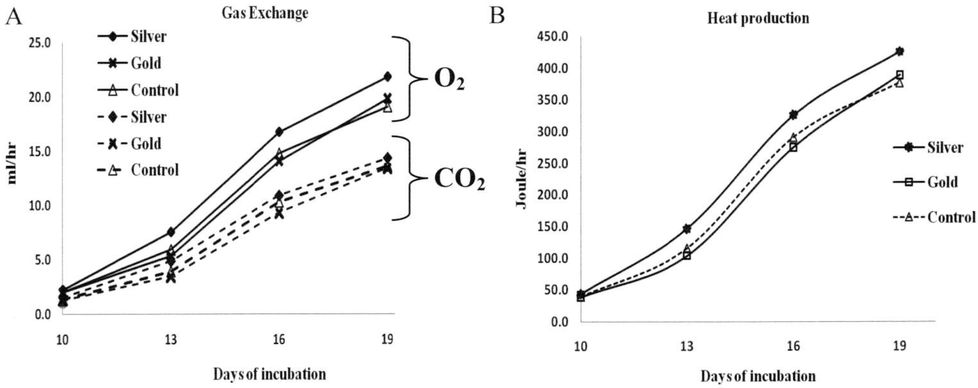

Figure 1. (A) Oxygen consumption and CO_2 production. (B) Heat production during incubation period.

Table 1. Gas exchange and heat production.

Indices	Treatment			Age, d				Root MSE
	C	Ag	Au	10	13	16	19	
Egg, g				57.7	56.2	55.3	54.2	1.527
Per 50 g egg								
O_2, ml/h	10.3^b	11.7^a	9.82^b	2.02	5.85	15.8	19.8	1.705
CO_2, ml/h	7.19^{ab}	7.71^a	6.47^b	1.36	3.81	10.6	13.3	1.621
HE, J/h	202.3^b	228.4^a	191.4^b	39.5	13.9	309.4	386.9	33.54

[a,b] Values that share no common superscript differ significantly ($P<0.05$).

The results suggest that nanoparticles of silver increased rate of metabolism of chicken during embryonic development, however, more research should be done to elucidate the mode of action of nano Ag.

References

Chwalibog, A., A-H. Tauson, A. Ali, C. Matthiesen, K. Thorhauge and G. Thorbek. 2007. Gas exchange, heat production and oxidation of fat in chicken embryos from a fast or slow growing line. Comparative Bio. and Physio.Part A 146, 305-309.

Fondevila, M., R. Herrer, M.C. Casallas, L. Abecia and J.J. Ducha. 2009. Silver nanoparticles as potential antimicrobial additive for weaned pigs. Anim. Feed Sci. and Tech. 150, 259-269.

Janke, O., B. Tzschentke and M.L. Boerjan. 2004. Comparative investigations of heat production and body temperature in embryos of modern chicken breeds. Avian Poult. Biol. Rev. 15, 191-196.

SAS Institute Inc., 2009. SAS[Ò] Procedure Guide. Version 9.2. SAS Institute Inc., Cary, NC, USA.

Sawosz, E., M. Binek, M. Grodzik, M. Zielinska, P. Sysa, M. Szmidt, T. Niemiec and A. Chwalibog. 2007. Influence of hydrocolloidal silver nanoparticles on gastrointestinal microflora and morphology of enterocytes of quails. Arch.Anim.Nutr. 6, 444-451.

Effects of birth weight and dietary caloric density on growth, voluntary intake and body composition of newborn lambs

J.M. Ramos-Nieves[1], G. Bernal-Santos[2], A. Faciola[1], M.E. Van Amburgh[1] and Y.R. Boisclair[1]
[1]Department of Animal Science, Cornell University, Ithaca, NY, 14853, USA
[2]Universidad Autónoma de Querétaro, Facultad de Ciencias Naturales, Medicina Veterinaria y Zootecnia, Querétaro, Qro. CP 76000, México

Introduction

For more than a decade, it has been recognized that foetuses mount physiological adaptations to survive adverse events that occur during uterine life. These physiological adaptations become maladaptive and increase disease risk when sustained after birth (Nijland *et al.*, 2008).

Ewes gestating multiple offspring in a single pregnancy give birth to lambs that suffer from Intra Uterine Growth Retardation (IUGR) during foetal life. IUGR lambs exhibit hyperphagia, lower energy expenditure and increased fat deposition after birth. Using the IUGR sheep model described by Greenwood *et al.* (2002), we tested the hypothesis that IUGR lambs are unable to regulate their voluntary intake when offered a high fat, hypercaloric diet, as seen in IUGR rodents (Vickers *et al.*, 2000).

Methods

Thirty two neonatal male lambs were selected on the basis of their birth weight (below 3 kg for IUGR lambs and above 3.9 kg for Normal size lambs) and studied between birth and postnatal day 14. Lambs used in this experiment were Finn × Dorset genotype. Lambs were collected from the Cornell T&R Sheep Center at birth before suckling and offered a dose of artificial colostrum and kept in an enclosed and controlled environment for the rest of the experiment.

IUGR and Normal size lambs were randomly allocated to receive *ad libitum* amounts of either a Low or High fat milk replacer (20% or 40% Fat/kg Dry Matter [DM]) starting on postnatal day 1. Both milk replacers contained 28% Crude Protein [CP]/kg of DM).

DM intake and body weight were measured daily. On day 14, animals were euthanized and tissues were collected. The whole body was fractionated into carcass and viscera (visceral fat depots and blood were included in the viscera fraction). Proximate analyses were conducted on both fractions to determine fat, crude protein, ash and DM content.

Data were analyzed with a model accounting for Birth Weight Category (BWC, IUGR vs. Normal), Dietary Fat (DF, LF vs. HF) and their interaction (BWC×DF) as fixed effects and animal as random effect. Statistical significance was set at $P<0.05$ for main effects.

Results and discussion

At birth, IUGR lambs were lighter (2.5±0.06 vs. 4.1±0.06 kg, $P<0.001$), consumed less DM (2.8±0.1 vs. 3.7±0.1 kg, $P<0.001$), and grew at a slower rate (254±10 vs. 321±10 g/d, $P<0.001$) than Normal lambs. When expressed relative to body weight (BW), however, both feed intake and growth rate were higher for IUGR than Normal lambs (feed intake, 50.7±1.3 vs. 45.2±1.3 g DM/kg BW, $P<0.01$; relative growth rate, 66±1.3 vs. 56±1.3 g/kg BW, $P<0.001$). Dietary Fat did not affect absolute or relative DM intake or growth rate.

Energetic efficiency was higher in IUGR than Normal lambs (222±5 vs. 205±5 g empty BW gain/Mcal intake energy, P=0.01). Proximate composition of carcass and viscera are shown in Table 1. IUGR lambs had more fat and less protein in the carcass than Normal lambs but only on the Low fat diet. IUGR lambs had significantly more visceral fat than Normal lambs irrespective of the fat content of the diet.

In conclusion, we did not observe any evidence that voluntary feed intake of newborn lambs is regulated by caloric density irrespective of size at birth. Evidence of a predisposition for visceral adiposity was observed in the IUGR lambs.

Table 1. Proximate composition of carcass and viscera fractions of Intra Uterine Growth Retarded (IUGR) and Normal size lambs at 14 d of age fed a milk replacer of either Low (LF, 20% DM) or High fat (HF, 40% DM) content.

	IUGR		Normal			Significance		
	LF	HF	LF	HF	SEM	BWC	DF	BWC×DF
Carcass								
CP, %DM	60.9	56.6	64.3	56.3	1.1	NS	<0.001	0.11
Fat, %DM	28.0	32.5	24.2	33.4	1.4	NS	<0.001	0.11
Ash, %DM	10.5	9.6	11.4	10.0	0.5	NS	0.01	0.58
Viscera								
CP, %DM	55.4	51.4	61.9	54.3	1.9	0.02	<0.01	0.35
Fat, %DM	34.0	38.6	26.6	36.9	2.1	0.04	0.001	0.18
Ash, %DM	3.9	3.7	4.4	3.7	0.2	NS	NS	0.21

References

Greenwood, P.L., A.S. Hunt, J.W. Hermanson and A.W. Bell, 1998. Effects of birth weight and postnatal nutrition on neonatal sheep: I. Body growth and composition, and some aspects of energetic efficiency. J Anim. Sci. 76, 2354-2367.

Nijland, M.J., S.P. Ford and P.W. Nathanielsz, 2008. Prenatal origins of adult disease. Curr. Opin. Obstet. Gynecol. 20, 132-138.

Vickers, M.H., B.H. Breier, W.S. Cutfield, P.L. Hofman and P.D. Gluckman, 2000. Fetal origins of hyperphagia, obesity and hypertension and postnatal amplification by hypercaloric nutrition. Am J Physiol. Endocrinol. Metab. 279, E83-87.

Effect of abomasal infusion of oligofructose on portal-drained visceral ammonia and urea fluxes in lactating Holstein cows

N.B. Kristensen and M. Larsen
Department of Animal Health and Bioscience, Aarhus University, Tjele, Denmark

Introduction

Increasing supply of fermentable carbohydrate to the hindgut is know to increase fecal nitrogen (N) excretion, decrease urinary N excretion, and decreases the blood urea concentration in ruminants (Gressley and Armentano, 2007; Thornton *et al.*, 1970). These effects have generally been attributed to increasing transfer of blood urea to the hindgut with increasing carbohydrate fermentation. However, the large proportion of total portal-drained visceral (PDV) urea uptake accounted for by forestomach uptake (Theurer *et al.*, 2002), the incapability of dairy cattle to increase urea transfer to gut tissues with decreasing N intake (Calsamiglia *et al.*, 2010), and the very low PDV uptake of arterial urea in growing pigs (Kristensen *et al.*, 2009) lead us to question that changes in PDV urea fluxes is the causative link between increased fermentation of carbohydrate in the hindgut and the observed effects on urea status and N excretion. The aim of the present study was to investigate if changes in ammonia absorption or changes in urea transport to the gut were primarily affected by abomasal oligofructose infusion in lactating Holstein cows. Resolving this question will be of importance to modeling urea metabolism.

Materials and methods

Nine lactating Holstein cows fitted with ruminal cannulas, permanent indwelling catheters in an artery and mesenteric, hepatic portal (n=8), hepatic (n=8), and ruminal veins (n=7) as well as a transruminal abomasal infusion device were allocated to two experimental treatments in a cross over design with 14 d periods. Cows were fed the same ration with both treatments (total mixed ration based on 30% corn silage, 25% grass silage, and 30% rolled barley containing 13.9% crude protein). Treatments differed in the abomasal infusate: control (CON) 10 l/d of water and oligofructose (OLIGO) 1,500 g/d of Orafti P95 (Alisano, Birkerød, Denmark). Feed intake and milk yield were registered daily. Blood (arterial, portal, hepatic, and ruminal), ruminal fluid, and milk samples were obtained on d 14 of each period. Blood flow in portal and hepatic veins were determined by downstream dilution of *p*-aminohippuric acid. Data were analyzed using the MIXED procedure (SAS, 2002) using a model containing the fixed effects of treatment, sequence, period, sampling time, and the interaction between sampling time and treatment. Cow within sequence was considered a random effect and sampling time considered a repeated measure. A reduced model was used to analyse feed intake and milk data.

Results and discussion

Oligofructose infusion decreased ($P<0.01$) dry matter intake with 0.74±0.20 kg/d equivalent to half the infused amount of oligofructose (Table 1). The yield of energy corrected milk was not affected by treatment ($P=0.42$) and neither were any milk composition variables (data not shown) except for urea content that decreased with OLIGO ($P=0.04$; Table 1). OLIGO decreased the arterial blood concentration of urea ($P=0.03$), but did not affect ($P=0.71$) the arterial ammonia concentration. The net PDV uptake of arterial urea (negative net portal flux) decreased ($P<0.01$) with OLIGO compared to CON indicating that the reduced blood concentration of urea with OLIGO cannot be explained by an increased PDV uptake of urea from the arterial blood. On the contrary OLIGO decreased ($P<0.01$) the net absorption of ammonia by 88±19 mmol/h. The decrease in ammonia absorption equals 123 g crude protein/kg infused oligofructose. Neither ruminal ammonia concentration nor ruminal vein – arterial concentration differences for urea or ammonia were affected by treatment

Table 1. Dry matter intake, milk production, and ammonia and urea flux variables in lactating Holstein cows abomasally infused with water (CON) or 1,500 g/d of oligofructose (OLIGO).

	Treatment (Trt)		SEM	*P*-values	
	CON	OLIGO		Trt	Trt × time
Dry matter intake, kg/d	19.7	19.0	0.6	<0.01	-
Yield of energy corrected milk, kg/d	31.5	31.9	1.1	0.42	-
Milk urea, mM	2.17	1.69	0.14	0.04	-
Arterial urea, mM	2.39	1.88	0.15	0.03	0.46
Arterial ammonia, mM	0.116	0.115	0.004	0.71	0.63
Net portal flux of urea, mmol/h	-232	-192	20	<0.01	0.09
Net portal flux of ammonia, mmol/h	476	388	28	<0.01	0.63
Ruminal ammonia, mM	3.17	2.68	0.26	0.19	0.89
Ruminal vein – arterial urea, mM	-0.38	-0.35	0.04	0.28	0.38
Ruminal vein – arterial ammonia, mM	0.62	0.61	0.07	0.93	0.49

(P=0.19 to P=0.93) indicating that up-regulation of urea transfer to the hindgut in expense of ruminal urea uptake was unlikely to have happened.

It is concluded that increasing supply of fermentable carbohydrate to the hindgut did not increase urea transfer from the blood to the hindgut. Reduced arterial blood and milk urea concentrations with abomasal oligofructose infusion was likely caused by a reduced ammonia input to hepatic urea synthesis due to increased sequestration of ammonia in the hindgut.

References

Calsamiglia, S., A. Ferret, C.K. Reynolds, N.B. Kristensen and A.M. Van Vuuren, 2010. Strategies for optimizing nitrogen use by ruminants. Animal, In Press.

Gressley, T.F. and L.E. Armentano, 2007. Effects of low rumen-degradable protein or abomasal fructan infusion on diet digestibility and urinary nitrogen excretion in lactating dairy cows. J. Dairy Sci. 90, 1340-1353.

Kristensen, N.B., J.V. Nørgaard, S. Wamberg, M. Engbæk, J.A. Fernández, H.D. Zacho and H.D. Poulsen, 2009. Absorption and metabolism of benzoic acid in growing pigs. J. Anim. Sci. 87, 2815-2822.

SAS, 2002. Release 9.1. SAS Inst. Inc., Cary, NC, USA.

Theurer, C.B., G.B. Huntington, J.T. Huber, R.S. Swingle and J.A. Moore, 2002. Net absorption and utilization of nitrogenous compounds across ruminal, intestinal, and hepatic tissues of growing beef steers fed dry-rolled or steam-flaked sorghum grain. J. Anim. Sci. 80, 525-532.

Thornton, R.F., P.R. Bird, M. Somers and R.J. Moir, 1970. Urea excretion in ruminants III. The role of the hind-gut (caecum and colon). Aust. J. Agric. Res. 21, 345-354.

Effect of glucogenic and ketogenic feeding strategies on splanchnic metabolism in *post partum* transition dairy cows

M. Larsen and N.B. Kristensen
Faculty of Agricultural Sciences, Aarhus University, 8830 Tjele, Denmark

Introduction

Recent production trials have shown beneficial effects of feeding glucogenic diets to *post partum* transition cows. However, in a recent study we found a dramatic negative impact on the *post partum* adaptation of metabolism and feed intake with infusion of 1,500 g/d of glucose to the small intestine (Larsen and Kristensen, 2009). To which extent the negative effect was related to the amount of glucose supplied or the use of glucose and not starch is unknown. The aim of the study was to investigate the effect of glucogenic and ketogenic feeding strategies on metabolic adaptation and splanchnic metabolism in *post partum* transition cows.

Material and methods

Six Holstein cows implanted with ruminal cannulas and permanent indwelling catheters in the major splanchnic blood vessels during the dry period preceding second calving were used. Cows were blocked according to expected calving date and randomly assigned to 1 of 2 feeding strategies initiated at the day of calving: glucogenic strategy (GLUCO; 56.5% NaOH treated wheat grain, 25.8% grass-clover silage and 17.7% concentrate mix; 7.56 NE_L MJ/kg DM, 18.6% CP, 4.0% crude fat, 39.6% starch, 5.1% sugar, 17.8% NDF), or ketogenic strategy (KETO; 40.5% fodder beets, 25.8% grass-clover silage, 15% NaOH treated wheat grain and 18.7% concentrate mix; 6.98 NE_L MJ/kg DM, 18.2% CP, 3.2% crude fat, 11.3% starch, 31.3% sugar, 18.4% NDF). In both diets, NaOH treated wheat provided the vast majority of starch, of which the rumen degradability has been found to 59% (Larsen *et al.*, 2009). The concentrate mixes were composed of rumen protected soybean meal, maize gluten, vegetable fat and mineral/vitamins. All cows received the same non-lactation diet in the dry period. Diets were fed 3 times daily at 08:00, 16:00 and 24:00 h in equally sized meals. Cows were milked 3 times daily at 06:00, 14:00, and 22:00 h. Lactation diets were offered *ad libitum*. Feed intake and milk yield were recorded daily. Milk and blood were sampled on 4, 15, 29 days in milk (DIM). Eight hourly sets of arterial, portal vein, and hepatic vein samples were collected simultaneously beginning 30 min before feeding at 08:00 h. Blood plasma flow was measured by down stream dilution of para-aminohippuric acid infused into a mesenteric vein. The portal recovery of glucose equivalents from feed starch was estimated by correcting the net portal flux of glucose for the portal drained visceral extraction rate of arterial glucose (2.4%; Larsen and Kristensen, 2009). Means within cow and DIM were subjected to statistical analysis with a model including treatment (Trt), DIM and Trt×DIM, where DIM within cow was considered a repeated measure (autoregressive order 1) using the MIXED procedure of SAS.

Results and discussion

Dry matter intake was unaffected by treatment (P=0.95; Table 1). The arterial glucose concentration was affected by treatment and DIM (interaction; P=0.03), reflecting *post partum* increasing arterial glucose with GLUCO as compared with a continual *post partum* decrease in KETO. The arterial concentration of β-OH-butyrate (BHBA) increased with KETO to very high levels at 29 DIM (4.2 mM) as compared with consistently moderate levels with GLUCO (interaction; P=0.01). Overall, low arterial glucose concentrations did not seem to drive feed intake, as could be hypothesised from the dramatic lower rate in *post partum* increases in feed intake induced by high levels of glucose in arterial blood (Larsen and Kristensen, 2009). Milk lactose was lower (P=0.01) with KETO compared with GLUCO.

Table 1. Effect of glucogenic and ketogenic feeding strategies to post partum *transition cows.*

	Treatment (Trt) 1 to 29 DIM		SEM	*P*-value		
	GLUCO	KETO		Trt	DIM	Trt×DIM
Dry matter intake, kg/d	16.6	16.5	0.7	0.95	<0.01	0.24
Starch intake, kg/d	6.6	1.1	0.2	<0.01	0.02	0.08
Milk, kg/d	38.0	32.5	1.8	0.07	<0.01	0.52
Milk composition, g/kg						
Fat	40	47	1.5	0.02	<0.01	0.13
Protein	33	36	1.7	0.25	<0.01	0.99
Lactose	48	46	0.3	0.01	0.43	0.16
Arterial concentration, mM						
Glucose	3.4	2.7	0.2	0.05	<0.01	0.03
BHBA	0.822	2.514	0.261	0.02	0.01	0.01
Glycerol	0.022	0.023	0.003	0.74	0.03	0.75
Glucose fluxes, mmol/h						
Net portal	128	23	12	<0.01	0.07	<0.01
Corrected net portal	226	115	20	0.02	0.12	<0.01
Net hepatic	572	599	49	0.71	0.50	0.46
Portal recovery of starch, %	16	43	5	0.02	0.02	0.02
BHBA fluxes, mmol/h						
Net portal	159	160	51	0.98	0.87	0.94
Net hepatic	320	554	81	0.13	0.12	0.20

The net portal release (positive fluxes) of glucose was greater with GLUCO ($P<0.01$; Table 1) compared with KETO. The net hepatic release of glucose was unaffected ($P=0.71$) by the varying supply of glucose from the portal vein. The net portal recovery of glucose equivalents from dietary starch was estimated to 16% with GLUCO containing 56.5% NaOH treated wheat compared with 43% with KETO containing only 15.0% NaOH treated wheat. The decreased efficacy in transfer of glucose from dietary starch to portal blood in response to increased small intestinal entry is likely related to limitations in small intestinal starch digestion, as it previously was observed that the portal drained visceral extraction ratio of arterial U-^{13}C labelled glucose was unaffected by glucose supply in the *post partum* period (Larsen and Kristensen, 2009).

The net hepatic release of BHBA was numerically higher ($P=0.13$; Table 1) with KETO as compared to GLUCO, indicating that liver metabolism of butyric acid and of acetoacetate from rumen epithelial metabolism of butyric acid may have been responsible for the greater levels of BHBA in arterial blood with KETO. Hepatic oxidation of NEFA could have contributed to greater arterial BHBA concentrations with KETO, but arterial glycerol concentrations did not differ between treatments ($P=0.74$) indicating no differences in fat mobilisation.

References

Larsen, M. and N.B. Kristensen, 2009. Effect of abomasal glucose infusion on splanchnic and whole body glucose metabolism in periparturient dairy cows. J. Dairy Sci. 92, 1071-1083.

Larsen, M., P. Lund, M.R. Weisbjerg and T. Hvelplund, 2009. Digestion site of starch from cereals and legumes in lactating dairy cows. Anim. Feed Sci. Technol. 153, 236-248.

Foetal umbilical venous and arterial plasma amino acid concentrations are depending on the protein level of gestation diets fed to gilts

S. Görs, I.S. Lang, B. Stabenow, U. Hennig, K. Brüssow, W. Otten, C. Rehfeldt and C.C. Metges
Leibniz Institute for Farm Animal Biology (FBN), Wilhelm-Stahl-Allee 2, 18196 Dummerstorf, Germany

Introduction

Nutrient intake during gestation affects the intrauterine environment, and thus developmental conditions for the foetus. Amino acids and glucose are foetal nutrients of paramount importance. We have shown that diets with protein levels below and above recommendations fed to gilts throughout gestation lead to intrauterine growth retardation (Metges *et al.,* 2009). Therefore, we investigated whether these diets alter plasma amino acid concentrations in late gestation (day 93) were foetal body mass accretion is highest. We show results for umbilical venous and arterial, and maternal venous plasma.

Material and methods

German Landrace gilts (n=26) were fed isoenergetic diets (~13.6 MJ ME/ kg) with control (CP, ~12%), low (LP, ~6%) or high (HP, ~30%) protein levels throughout gestation. At day 93 of gestation umbilical venous and arterial blood samples and simultaneously from maternal *vena jugularis* were collected during Caesarean section. The concentrations of 20 free proteinogenic amino acids (AA) were analyzed by HPLC (Series 1200; Agilent Technologies, Waldbronn, Germany) with automated pre-column derivatization and fluorescence detection (Krömer *et al.*, 2005). Separation was carried out at a flow rate of 0.8 ml/min within 45 min on a 250×4 mm Hyperclone ODS (C18) 120 Å column protected by a 4×3 mm C18 pre-column (Phenomenex, Aschaffenburg, Germany) using a gradient with 40 mM phosphate buffer (pH 7.8) and acetonitrile/methanol/water (v:v:v, 45:45:10) ranging from 6-100%.

Preliminary statistical comparison of diet group means was done by t-tests. A statistical model with consideration of mothers and body birth mass is currently developed.

Results and discussion

Evaluation of plasma AA (gilts: n=9 CP, 8 LP, 9 HP) indicates that total amino acid concentrations in maternal vein were numerically lower in gilts fed low and high protein diets. In LP compared to CP gilts Asp, Cys, Arg, His, Thr, Trp, Phe, Tyr, Val, Ile, and Leu concentrations were decreased ($P<0.043$) whereas Gly and Ala were increased ($P<0.010$). Plasma concentrations of Asp, Glu, Gln, Ala, Cys, Ser, His, Arg, Trp, and Phe were reduced in HP vs. CP gilts ($P<0.035$), while Val and Lys were on a higher level ($P<0.036$) in comparison to the CP group. This is in agreement with the known alterations of plasma AA concentrations in response to low or high protein intake (e.g. Moundras *et al.*, 1993.).

Blood is transferred from the mother to the foetus via the umbilical vein. We investigated umbilical vein AA concentrations in a subset of 22-24 foetuses per dietary group (n=1-4 per gilt). In agreement with Wu *et al.* (1995) some AA like Lys were released by the placenta probably due to degradation of proteins, Others like Leu were extracted (Table 1). Gln was the dominant AA in umbilical vein accounting for 20-22% of total proteinogenic AA and the umbilical/maternal vein ratios 1.9-2.3 were greater than those in other species like humans or sheeps (Wu *et al.*, 1995). The total AA concentration did not differ between maternal and umbilical vein for CP animals but there was a decrease of total AA between maternal and umbilical vein in the LP (P=0.065) and HP group (P=0.023) suggesting

Table 1. Plasma amino acids in maternal vein and umbilical cord in dependence of protein supply in German Landrace gilts at day 93 of gestation (means ± SEM).

Amino acid	Maternal vein			Umbilical vein			Umbilical artery		
(AA, μM)	CP[2]	LP[2]	HP[2]	CP	LP	HP	CP	LP	HP
n	9	8	9	22	22	24	22	22	24
8 indispensable[1]+Arg	1311^{A}	887^{B}	1381^{Aa}	1231^{A}	973^{B}	1061^{Bb}	1152^{A}	921^{B}	1014^{ABb}
	±90	±32	±47	±70	±36	±39	±62	±36	±39
Lys	75^{Aa}	80^{Aa}	108^{Ba}	221^{Ab}	152^{Bb}	172^{Cb}	205^{Ab}	143^{Bb}	163^{Bb}
	±9	±9	±10	±18	±11	±11	±16	±11	±10
Leu	226^{Aa}	161^{Ba}	215^{Aa}	120^{Ab}	101^{Bb}	104^{ABb}	106^{b}	92^{b}	96^{b}
	±12	±5	±6	±8	±4	±5	±7	±4	±5
Arg	159^{Aa}	120^{B}	134^{Ba}	144^{Aab}	119^{B}	125^{Bab}	132^{Ab}	113^{B}	118^{ABb}
	±7	±11	±5	±7	±5	±6	±6	±4	±6
Glu	240^{Aa}	239^{Aa}	146^{Ba}	252^{a}	199^{a}	243^{b}	482^{b}	425^{b}	455^{c}
	±17	±11	±11	±38	±20	±27	±29	±39	±39
Gln	449^{Aa}	437^{Aa}	358^{Ba}	835^{b}	823^{b}	837^{b}	677^{c}	688^{c}	714^{c}
	±16	±11	±10	±43	±40	±33	±29	±31	±31

[1] Thr, Val, Met, Trp, Phe, Ile, Leu, Lys.

[2] CP: control protein; LP: low protein; HP: high protein.

[a,b,c] Means carrying different lower case letters differ between vessels within diet.

[A,B,C] Means carrying different upper case letters differ between diets within vessel; (P<0.05).

AA utilization in the placenta. In the umbilical vein all indispensable AA and arginine were lower in the LP than in the CP group (P<0.035). In addition, HP umbilical vein concentrations of Lys, Leu, Trp, Thr, Phe, and Arg were lower than in controls (P<0.085). This suggests that protein synthesis and therefore growth should be restricted in LP as well as in HP foetuses as compared to the control counterparts. The reduced Arg in LP and HP might point toward a reduced blood flow. Irrespective of the maternal diet only Gln and Glu levels differed between umbilical vein and artery (P<0.011) indicating an involvement of Gln in foetal nitrogen supply (extraction: CP 158, LP 135, HP 123 μM) whereby Glu was formed (release: CP 230, LP 226, HP 212 μM). With the exception of Asp and Glu all other AA were numerically taken up by the foetuses. Since foetal body mass at day 93 of gestation was independent of the maternal diet our results suggest a restricted amino acid availability in foetuses of LP and HP fed gilts in the last month of gestation indicating diet dependent intrauterine growth retardation in late gestation.

This study was supported by the Deutsche Forschungsgemeinschaft (PAK 24, ME 1420/8-1).

References

Krömer, J.O., M. Fritz, E. Heinzle E and C. Wittmann, 2005. *In vivo* quantification of intracellular amino acids and intermediates of the methionine pathway in Corynebacterium glutamicum. Anal. Biochem. 340, 171-173.

Metges, C.C., I.S. Lang, S. Goers, P. Junghans, U. Hennig, B. Stabenow, F. Schneider, W. Otten and C. Rehfeldt, 2009. Inadequate protein levels during gestation in gilts affect gestation body mass and fatness as well as offspring birth weight and insulin sensitivity at 10 wk of age. J. Anim. Sci. 87, E-Suppl 2 / J. Dairy Sci. 92, E-Suppl 1, 503-504.

Moundras, C., C. Remesy and C. Demigne, 1993. Dietary protein paradox: deacrease of amino acid availability induced by high-protein diets. Am. J. Physiol. 264 (Gastrointest. Liver Physiol. 27), G1057-G1065.

Wu, G., F.W. Bazer and W. Tou, 1995. Developmental changes of free amino acid concentrations in fetal fluids of pigs. J. Nutr. 125, 2859-2868.

Change of plasma arginine flux around weaning period in young calves

T. Obitsu, T. Mita, R. Fukumori, T. Sugino and K. Taniguchi
Graduate School of Biosphere Science, Hiroshima University, Kagamiyama 1-4-4, Higashihiroshima-shi, 739-8528, Japan

Introduction

Arginine (Arg) is removed by splanchnic tissues and converted to urea and other compounds such as nitric oxide. In young calves, the gastrointestinal mass and hepatic urea production increase after weaning (Klein *et al.*, 1987; Obitsu and Taniguchi, 2009). Such developmental change may result in an increase of Arg removal by the splanchnic tissues and a reduction of peripheral Arg supply. In this study, we measured plasma Arg flux in young calves at pre- and post weaning stages with intravenous or oral administration of the tracer to elucidate the change of the flux and first-pass splanchnic extraction of Arg around weaning period.

Materials and methods

Four newborn Holstein male calves were fed, at 09:00 and 17:00, whole milk at 10% of body weight with a calf starter (crude protein: 200g/kg dry matter) and hay (crude protein: 124 g/kg dry matter) *ad libitum* till 6 weeks of age. After weaning (47 days of age), calves were fed the calf starter and hay in a ratio of 70:30 to fulfill their energy need for 0.7 kg/d of body weight gain. Warm water (0.5 l) was also given twice per day by bottle feeding to maintain the esophageal groove reflex. At 2, 6 (pre-weaning) and 10 weeks (post-weaning) of age, faeces and urine were collected for 8 days to determine digestible nitrogen (N) and urinary urea. On day 3 of each collection period, calves received a 6-h primed (7 μmol/kg) constant (7 μmol/kg/h) infusion of [guanido-$^{15}N_2$]arginine into the jugular vein from 12:00 to 18:00. On day 7 of each collection period, calves were orally given [guanido-$^{15}N_2$]arginine (7 μmol/kg) by bottle feeding with milk (pre-weaning, 200 ml) or water (post-wenanig, 100 ml) every 1 h from 12:00 to 18:00. During tracer administration, calves received milk or solid feeds every hour in equal portions and blood was collected from the jugular vein every 15 min from 17:00 to 18:00. Plasma arginine fluxes were calculated for both tracer administrations (IV and oral) based on plasma [$^{15}N_2$]arginine enrichments. First-pass splanchnic extraction of Arg was calculated by the difference of the fluxes determined by IV and oral administration of the tracer. Additionally, blood was collected from the jugular vein before feeding (08:50) every two weeks from 1 to 13 weeks of age to measure plasma concentrations of free amino acids.

Results and discussion

Calves ingested small amounts of starter and hay before weaning, whereas the solid feed intake dramatically increased after weaning (Table 1). Although N intake increased after weaning (495 to 693 mgN/kg/d), digestible N intake per body weight was not different between pre- and post-weaning due to a decrease in N digestibility after weaning. Oral arginine flux per body weight was not changed with age, whereas IV arginine flux decreased with advancing age (Table 1). Consequently, estimated first-pass extraction by the splanchnic tissues increased with advancing age. This increased arginine catabolism after weaning may be caused by elevated hepatic urea production (Obitsu and Taniguchi, 2009) indicated by a trend of increased urinary urea-N excretion (P=0.14).

Plasma Arg concentration was constant from 3 to 13 weeks of age (Figure 1). In contrast, plasma concentration of citrulline (Cit, cubic, P<0.003) and ornithine (Orn, linear, P=0.101) increased with advancing age. The elevated plasma Cit may enhance renal Arg synthesis from Cit (Windmueller and Spaeth, 1981). These results indicate that circulating plasma Arg level in calves is maintained around weaning period even though splanchnic extraction of Arg increases after weaning.

Table 1. Feed intake and plasma arginine flux determined by intravenous infusion or oral administration of [$^{15}N_2$] arginine in pre- (2 and 6 weeks) and post-weaning (10 weeks) calves.

	Age			SEM
	2 weeks	6 weeks	10 weeks	
Feed intake				
Whole milk, kg/d	5.1	5.1	-	-
Calf starter, gDM/d	54^{c}	247^{b}	1,623^{a}	49.3
Hay, g DM/d	25^{c}	87^{b}	242^{a}	17.7
Digestible N intake, mgN/kg/d	445	441	501	25.5
Urinary urea-N excretion, mgN/kg/d	63	67	85	7.1
Plasma arginine flux, µmol/kg/h				
IV infusion	205^{a}	166^{b}	152^{b}	7.4
Oral administration	270	243	267	9.5
First-pass splanchnich extraction				
µmol/kg/h	65^{b}	77^{ab}	114^{a}	9.9
% of oral flux	24.1^{b}	31.7^{ab}	42.9^{a}	3.31

[a,b,c] Means with different superscript differ (P<0.05). SEM: Standard error of means.

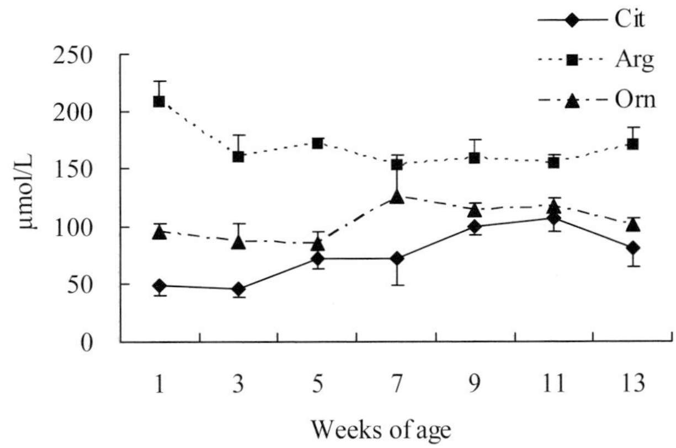

Figure 1. Changes of plasma concentrations of Arginine (Arg), Ornitine (Orn) and Citrulline (Cit) in calves from 1 to 13 weeks of age. Data are means ± SE, n=4.

References

Klein, R.D., R.L. Kincaid, A.S. Hodgson, J.H. Harrison, J.K. Hillers and J.D. Cronrath, 1987. Dietary fiber and early weaning on growth and rumen development of calves. J. Dairy Sci. 70, 2095-2104.

Obitsu, T. and K. Taniguchi, 2009. Quantitative comparison of diversity and conformity in nitrogen recycling of ruminants. Asian-Aust. J. Anim. Sci. 22, 440-447.

Windmueller, H.G. and A.E. Spaeth, 1981. Source and fate of circulating citrulline. Am. J. Physiol. 241 (Endocrinol. Metab.5), E473-E480.

Empirical prediction of net splanchnic release of glucose in Ruminants from dietary and animal characteristics: a meta-analysis approach

C. Loncke[1,4], P. Nozière[1], J. Vernet[1], H. Lapierre[2], D. Sauvant[3] and I. Ortigues-Marty[1]
[1]INRA UR 1213 Herbivores, Site de Theix, 63122 St Genès Champanelle, France
[2]Dairy and Swine R&D Centre, Agriculture and Agri-Food Canada, Sherbrooke, QC J1M1Z3 Canada
[3]INRA UMR 791, Physiologie de la Nutrition et al.imentation, INAPG, 75231 Paris, France
[4]Present address: Union Invivo ets. INZO° SAS, 02407 Chierry, France

Introduction

Feed evaluation systems have to evolve towards nutrient-based systems. In ruminants, glucose is an essential nutrient both for milk synthesis, muscle growth and meat quality. Glucose availability to peripheral tissues depends on its net splanchnic release (NSR) which combines net portal absorption (NPA) of glucose and hepatic gluconeogenesis. An empirical prediction model of NPA-glucose has been established from the dietary starch intake and its intestinal and ruminal digestibility (Loncke *et al.*, 2009). Mechanistic prediction models of hepatic glucose release exist. However, they are limited to dairy cows, they are based on detailed metabolic pathways and are difficult to use in rationing practices. Thus, the objective of the present study was to develop an empirical prediction model of the NSR-glucose, which (1) could be applicable to ruminants in different physiological statuses and (2) is based on simple and generic predictors chosen from diet (composition and intake) and animal characteristics.

Material and methods

Meta-analytical methods (Sauvant *et al.*, 2008) were applied to all international publications reporting net splanchnic glucose fluxes in sheep or cattle, growing, lactating, gestating or non-productive ruminants extracted from the exhaustive FLORA database (Vernet and Ortigues Marty, 2006). Diets were characterised according to INRA Feed Tables (2007). Predictors of animal characteristics were physiological status (phy) and energy balance (EB, MJ/d/kg BW). Predictors (mmol C/h/kg BW) of glucose-NSR reflected (a) glucose-NPA [P1], (b) precursor's availability for hepatic gluconeogenesis of dietary origin [P2 = NPA of propionate (C3) + glucogenic amino acids (glAA calculated as 45% of NPA-total AA)], or (c) from tissue metabolism including mobilisation [P3 = L-lactate-NPA + hepatic glycerol + alanine supply, the later being estimated using EB and assuming that mobilised energy was composed of 7% protein (4.7% alanine in protein) and 93% fat (10.4% glycerol in fat)] (Loncke, 2009). When nutrient-NPA were missing (in 41, 29 and 4% of the publications for C3, L-lactate and glAA-NPA respectively), they were predicted from diet composition and intake and validated according to Loncke *et al.* (2009). Response equation of NSR glucose ($Y = \alpha + \beta X + \text{phy} + \alpha i(\text{phy}) + \text{phy} \times X + e$, with α = overall intercept; αi = effect of the experimental group I on the intercept α, nested within phy; β = slope; e = error) to its precursors was investigated using a GLM model where the within experiment effect was considered as fixed. The interfering factors were systematically checked. A first evaluation step was then applied whereby glucose-NSR was predicted for all publications and compared to reported values, by regression analysis. Predictions used combinations of dietary predictors of nutrient-NPA (Loncke *et al.*, 2009) and the response equation of NSR glucose presented here.

Results and discussion

43 publications (95 treatments) were used. 57, 22, 4, 17% of the data were on non productive adults, growing/finishing, gestating animals, and lactation (most early lactating) cattle, respectively). EB ranged from -159 to 109, 0.7 to 91 -159 to 15 and -80 to 59 kJ/d kg/BW in the same respective order. Intakes averaged 35 g DMI/d.kg/BW of diets containing 70 g concentrate/100 g DM, extreme levels

being 0 to 41 g DMI/d.kg/BW and 0 to 100 g concentrate /100 g DM. Contribution of P1, P2 and P3 to the sum of all glucogenic precursors (P1+P2+P3) averaged respectively 2.62, 74.47 and 22.91%. Results were within the range of 0.9 to 8.2 mmolC.h^{-1}.kg BW^{-1}, NSR-glucose was linearly related to (P1+P2+P3), as follows: NSR-glucose = 0.750* (0.34) + 0.828*** (0.10) × (P1+P2+P3); RMSE = 0.54 and R^2_{adj} = 0.89. Linearity of the model was probably due to the range of NPA-precursors. On average 83 (± 10)% of precursors were used for gluconeogenesis. The intercept was significantly affected by phy. At equal precursor supply, the NSR-glucose was at mean 6.6% higher for dairy cattle and 1.7% lower for growing animals, as compared to non-productive or late gestating adults (Figure 1).

Evaluation of the predicted glucose-NSR against reported values showed a slope not significantly different from 1 (0.96±0.10; P=0.73) and an intercept not different from 0 (0.12±0.36; P=0.74). In conclusion, NSR-glucose can be predicted from dietary characteristics using the nutrient-NPA prediction equations and the EB of ruminants. It will contribute to the improvement of rationing tools for ruminants of different species and physiological status.

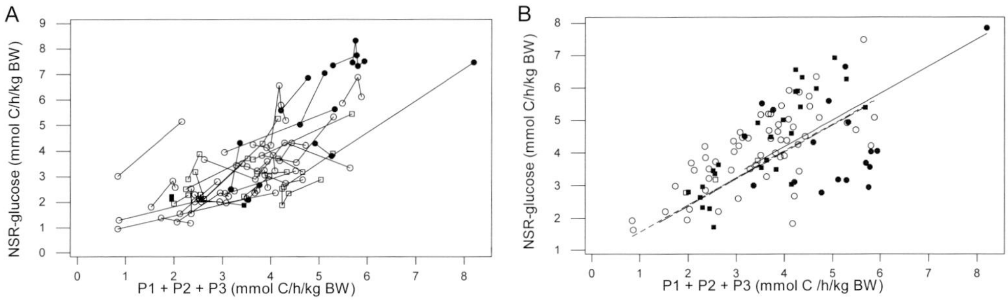

Figure 1. Within-study (A) and adjusted (B) relationship between NSR-glucose and the glucogenic precursors (P1+P2+P3); the line represents the predicted values and symbols represent the residuals.

Acknowledgement

Financial support from INZO° and Limagrain is acknowledged.

References

INRA, 2007. Alimentation des bovins, ovins et caprins - Tables INRA 2007, Versailles, France.

Loncke, C., 2009. Modélisation des relations entre l'alimentation et les flux splanchniques de nutriments énergétiques chez le ruminant, Thèse, AgroParisTech - ABIES, Paris, France.

Loncke, C., I. Ortigues-Marty, J. Vernet, H. Lapierre, D. Sauvant and P. Nozière. 2009. Empirical prediction of net portal appearance of volatile fatty acids, glucose, and their secondary metabolites (β-hydroxybutyrate, lactate) from dietary characteristics in ruminants: A meta-analysis approach. J. Anim. Sci. 87, 253-268.

Sauvant, D., P. Schmidely, J.J. Daudin and N.R. St-Pierre, 2008. Meta-analyses of experimental data in animal nutrition Animal 2, 1203-1214.

Vernet, J. and I. Ortigues-Marty, 2006. Conception and development of a bibliographic database of blood nutrient fluxes across organs and tissues in ruminants: Data gathering and management prior to meta-analysis. Reprod. Nutr. Dev. 46, 527-546.

Energy expenditure of portal drained viscera relative to the total in Iberian vs. Landrace growing gilts

I. Fernández-Fígares, J.M. Rodríguez-López, L. González-Valero and M. Lachica
Animal Nutrition Department, Estación Experimental del Zaidín (CSIC), Camino del Jueves s/n, 18100 Armilla, Granada, Spain

Introduction

The gastrointestinal tract plays an essential role in nutrient distribution to peripheral tissues and considering its total mass, represents a great part of the energy requirements of the whole animal. Rivera-Ferre *et al.* (2005) have shown that the splanchnic tissues contribution to total BW is greater in Iberian (native) than in Landrace (modern) pigs (10.2 and 8.3%, respectively) with no differences in protein content of viscera, therefore, the metabolic activity of these tissues in Iberian could imply a greater proportion of total energy requirements compared to Landrace pigs.

There is only one study which compares energy expenditure of PDV in modern breed vs. Iberian pigs (Lachica *et al.*, 2010). The objective of this work was to determine the PDV heat production (PDVHP) relative to the total (THP) in Iberian vs. Landrace growing gilts fed with a diet with optimal protein content for Iberian pigs.

Material and methods

Twelve (6/breed) gilts (27.8±0.90 kg BW) were used. Three catheters were placed: in carotid artery and portal vein for blood sampling, and in mesenteric vein for para-aminohippuric acid (PAH) infusion to measure blood flow. Pigs were on metabolic cages and restrictively fed along the whole experiment twice a day 25 and 75% of the daily ration (1 kg) at 9 am and 3 pm, respectively, an equilibrated diet (88.5% DM, 140 g/kg CP and 14-14.5 MJ ME/kg DM). Two trials were carried out. First trial began when animals recovered their pre-surgery appetite for at least 7 days, two weeks on average. Forty five min prior blood sampling a 15 ml pulse dose of PAH (2% w/v) was infused into mesenteric vein, followed by continuous infusion of 0.8 ml/min. A 4.5 ml blood sample was anaerobically taken simultaneously from carotid artery and portal vein 5 min before and 0.5, 1, 1.5, 2, 2.5, 3, 3.5, 4, 5 and 6 h after feeding 25% of daily ration, into an heparinized tube for instantaneous measurement of O_2 concentration and saturation of haemoglobin (by an hemoximeter). The rest of blood was centrifuged for packed cell volume and plasma harvested and stored at -20 °C until PAH analysis. The remaining daily ration was given immediately after sampling. Whole-blood flow and O_2 consumption rates were based on Fick principle. The assumed energy equivalent for O_2 was 20.4 kJ/l. For the second trial, after blood sampling pigs were individually moved on its own metabolic cage to a respirometry chamber. After 18 h of adaptation, THP was measured over 6 h following the same feeding protocol of first trial.

Data were analyzed using a two way ANOVA where breed and time of sampling were fixed factors. Because the interaction was not statistically significant, data were pooled and subjected to one way ANOVA and multiple range comparisons between breeds by means of the Tukey's t-test.

Results and discussion

Results are displayed in Table 1. Blood flow was relatively low in both breeds in comparison with data in the literature (e.g. Yen *et al.*, 2004) where feeding level during the sampling period was greater than in the present study. Blood flow and O_2 consumption were lower ($P<0.05$) in Iberian pigs compared to Landrace. Similar results were found by Lachica *et al.* (2010) when a 16% protein content diet was used instead. The lower PDV flow rate observed in Iberian compared to Landrace contrasts with the inferior growth energy efficiency for protein deposition reported for Iberian pigs (Nieto *et al.*, 2002). On the contrary, Yen *et al.* (2004) reported no differences in portal flow while PDV fraction of whole-body O_2 consumption was greater in Meishan than in a modern crossbred pig; in our study, however, accounted for 12 vs. 8.5% for Landrace and Iberian, respectively. Iberian pigs fed with similar intake of a 16% CP diet, accounted for 9.5 of the THP (Lachica *et al.*, 2009) vs. 8.5% in the present study, and PDVHP were similar to that obtained by Lachica *et al.* (2010) with pigs of comparable weights fed the same amount of an equivalent diet (same ingredients and ME, 16% CP). The higher ($P<0.05$) THP observed in Iberian than in Landrace agrees with the low growth energy efficiency of Iberian pigs.

In conclusion, energy expenditure associated to PDV and its proportion relative to THP was greater in Landrace than in Iberian pigs of the same weight.

Table 1. Blood flow, PDV oxygen consumption, PDV heat production (PDVHP) and total HP (THP) in Iberian (n=6) and Landrace (n=6) pigs (values are mean ± SE for ten postprandial measurements).

	Iberian	Landrace	SE
Portal blood flow (ml/min)			
Preprandial[1]	629[a]	878[b]	67
Postprandial	747[a]	1,110[b]	56
Blood oxygen consumption (mmol/min)	1.30[a]	1.77[b]	0.07
PVDHP (kJ/kg$^{0.75}$ per day)[2]	75[a]	92[b]	3.3
THP (kJ/kg$^{0.75}$ per day)	887[a]	730[b]	37.6

[a,b] For the comparisons made, values within a row with unlike superscript letter were significantly different ($P<0.05$).
[1] Values are mean ± SE for one preprandial measurement.
[2] Calculated from: PDVHP=20.4 (kJ/l) × O_2 (l/kg$^{0.75}$ per day).

References

Lachica, M., J.M. Rodríguez-López, L. González-Valero and I. Fernández-Fígares, 2009. Energy expenditure of splanchnic tissues relative to the total in Iberian growing gilts. In: 60th Annual Meeting of the European Association for Animal Production. Wageningen Academic Publishers, the Netherlands, EAAP Book of abstracts No. 15: 585.

Lachica, M., J.M. Rodríguez-López, L. González-Valero and I. Fernández-Fígares, 2010. Energy expenditure of splanchnic tissues in Iberian and Landrace growing gilts. Livest. Sci. (in press).

Nieto, R., Miranda, A., García, M.A., Aguilera, J.F., 2002. The effect of dietary protein content and feeding level on the rate of protein deposition and energy utilization in growing Iberian pigs, from 15 to 50 kg body weight. Br. J. Nutr. 88, 39-49.

Rivera-Ferre, M.G., J.F. Aguilera and R. Nieto, 2005. Muscle fractional protein synthesis is higher in Iberian than in Landrace growing pigs fed adequate or lysine-deficient diets. J. Nutr. 135, 469-478.

Yen, J.T., V.H. Varel and J.A. Nienaber, 2004. Metabolic and microbial responses in western crossbred and Meishan growing pigs fed a high-fiber diet. J. Anim. Sci. 82, 1740-1755.

Energy expenditure of portal-drained viscera in Iberian gilts fed acorn

M. Lachica, J.M. Rodríguez-López, L. González-Valero and I. Fernández-Fígares
Animal Nutrition Department, Estación Experimental del Zaidín (CSIC), Camino del Jueves s/n, 18100 Armilla, Granada, Spain

Introduction

The production system of the Iberian pig is orientated towards a final grazing period where the acorn is the main component of the diet. This fattening period -named 'montanera'- follows the intensive rearing and produces a radical shift in type of diet, providing products of peculiar organoleptical properties and extraordinary high price. The acorn has a very low protein content with unbalanced amino acids profile, where lysine is the first limiting amino acid, and contains significant amount of condensed tannins (particularly in the hulls) bound to dietary and endogenous proteins. Conversely, it is rich in carbohydrates (starch) and lipids (Nieto *et al.*, 2002).

There are only two reports on the energy expenditure associated to portal-drained viscera (PDV) in Iberian pigs. The aim of this work was to determine PDV heat production (PDVHP) in Iberian growing gilts fed acorns from evergreen oak (*Quercus rotundifolia*), and to find out if there is an adaptation to acorn feeding over time associated with PDVHP.

Material and methods

Two periods of sampling, both under identical conditions, were carried out with six gilts. Three catheters were placed in each pig (approximately 26 kg average BW): in carotid artery and portal vein for blood sampling, and in mesenteric vein for para-aminohippuric acid (PAH) infusion to measure blood flow. Pigs were on metabolic cages and fed twice a day at 9 am and 6 hours later (25 and 75% of the daily ration, respectively) an equilibrated diet (140 g/kg CP and 14-14.5 MJ ME/kg DM) based on barley and soya bean meal, at 85% *ad libitum* calculated as a function of BW. First period of sampling began at least 7 days after the animals had recovered its pre-surgery appetite (34 kg average BW). The day previous to sampling, pigs were given 2.4 kg of acorn (85% *ad libitum* calculated as a function of BW, corresponding to 1.405 kg, 23.27 MJ and 71.0 g of DMI, MEI and CP intake, respectively) as the only feed source and following the same feeding schedule described above. Forty five min prior to blood sampling a 15 ml pulse dose of PAH (2% w/v) was infused into mesenteric vein, followed by continuous infusion of 0.8 ml/min. A 4.5 ml blood sample was anaerobically taken simultaneously from carotid artery and portal vein 5 min before and 0.5, 1, 1.5, 2, 2.5, 3, 3.5, 4, 5 and 6 h after feeding 25% of their total daily acorn ration, into an heparinized tube for instantaneous measurement of O_2 concentration and saturation of haemoglobin (by an hemoximeter). The rest of blood was centrifuged for packed cell volume and plasma harvested and stored at -20 °C until PAH analysis. Whole-blood flow and O_2 consumption rates were based on Fick principle. The assumed energy equivalent for O_2 was 20.4 kJ/l. For the second period of sampling, after one week in which the pigs were exclusively fed acorn, the same fed and sampling schedule was repeated.

Time of sampling was not statistically significant and data were pooled and subjected to one way ANOVA and multiple range comparisons between trials by means of a Tukey´s t-test.

Results and discussion

Table 1 shows the results obtained. Blood flow and PDVHP were relatively low in the 1st period of sampling but similar to another study in Iberian pigs with analogous feeding level (85% *ad libitum*; Lachica *et al.*, 2010). Similarly, Yen and Nienaber (1992) observed in pigs switched to an antimicrobial-supplemented diet a reduction in portal blood flow and oxidative demand by the PDV.

Table 1. Blood flow, net flux of oxygen across the PDV and its heat production (PDVHP) in Iberian pigs (n=6) fed acorns at identical intake level in the 1st and 2nd period of sampling (values are mean ± SE for ten postprandial measurements).

	1st period	2nd period	SE
Portal blood flow (ml/min)			
Preprandial[1]	660[a]	1,021[b]	75
Postprandial	887[a]	1,304[b]	64
Blood oxygen consumption (mmol/min)	1.7[a]	2.3[b]	0.12
PDVHP (kJ/kg$^{0.75}$ per day)[2]	79[a]	102[b]	5.3

[a,b] For the comparisons made, values within a row with unlike superscript letter were significantly different (P<0.05).

[1] Values are mean ± SE for one preprandial measurement.

[2] Calculated from: PDVHP=20.4 (kJ/l) × O_2 (l/kg$^{0.75}$ per day).

However, in the present study pre and postprandial PDV blood flow were greater (P<0.05) in 2nd period of sampling, indicating a sharp change induced by the adaption to acorn, even when the intake level was identical in both sampling periods. This effect greatly affected (P<0.05) the PDVHP. We speculate that the increase in PDVHP could be associated with an adaptation of intestinal microflora to acorn nutrients. In this sense Rerat (1993) pointed out a 'time factor' in the relationships between digestive and metabolic use of nutrients. It seems that after one week, there could be an increment in acorn diet metabolicity.

In conclusion, after one week of feeding exclusively with acorn, there was an increase in PDVHP, which indicates that under practical feeding conditions, a gradual change to acorn may be advisable. Further research on net portal absorption of nutrients is warranted.

References

Lachica, M., J.M. Rodríguez-López, L. González-Valero and I. Fernández-Fígares, 2010. Energy expenditure of splanchnic tissues in Iberian and Landrace growing gilts. Livest. Sci. (in press).

Nieto, R., M. Rivera, M.A. García and J.F. Aguilera, 2002. Amino acid availability and energy value of acorns in the Iberian pig. Livest. Prod. Sci. 77, 227-239.

Rerat, A., 1993. Nutritional supply of proteins and absorption of their hydrolysis products: consequences on metabolism. Proc. Nutr. Soc. 52, 335-344.

Yen, J.T. and J.A. Nienaber, 1992. Influence of carbadox on fasting oxygen consumption by portal vein-drained organs and by the whole animal in growing pigs. J. Anim. Sci. 70, 478-483.

Mammary energy consumption and its relation to whole body energy metabolism in lactating cows fed high-concentrate diets

K. Higuchi[1], *F. Ohtani*[1], *Y. Kobayashi*[1], *I. Nonaka*[1], *K. Yayou*[2], *M. Sutoh*[1] *and O. Enishi*[1]
[1]*National Institute of Livestock and Grassland Sciences, 2 Ikenodai, Tsukuba, Ibaraki 305-0901, Japan*
[2]*National Institute of Agrobiological Sciences, 2 Ikenodai, Tsukuba, Ibaraki 305-8602, Japan*

Introduction

To evaluate the efficiency of feed for milk production, measuring whole body energy metabolism is of interest in dairy nutrition. In addition to whole body level, measuring energy metabolism at a local organ, such as mammary gland, gives us valuable information because mammary gland requires a large amount of nutrient and metabolism during lactation. We have started a series of work to estimate the effect of roughage-concentrate ratio of feed on energy metabolism of lactating cow. Thus, in this experiment, the energy efficiency for milk production of cows fed high-concentrate diets was measured by estimating energy efficiency at mammary gland and whole-body levels.

Material and methods

Three Holstein lactating cows (average BW 623 kg, average milk yield 27 kg/day, average DIM 162 days) fitted with an ultrasound flow probe (14A, Transonic Systems Inc., New York, USA) around the left external pudic artery were housed in a temperature and humidity controlled-room and fed 3 high-concentrate diets: a control diet (Control, with 30.0% Italian ryegrass silage, 12.0% soybean meal, 56.6% concentrate mix, and 1.4% vitamin and mineral mix), a bypass-protein diet (Bypass, replacing soybean meal of the control diet with bypass-protein), and a starch diet (Starch, adding 7% starch to the control diet) over 21-days periods according to a 3×3 Latin square design. Composition of the Control and Bypass diets was iso-nitrogenous and iso-energetic, and the amount to meet 110% of TDN requirement (NARO, 2007) was fed. Milking was held twice daily (8:30 and 18:00). From the day 10 to 14, whole body energy metabolism was determined by using open circuit respiratory chamber with digestion trial apparatus (Iwasaki *et al.*, 1982). The efficiency of metabolizable energy (ME) utilization for lactation (k_l) was calculated assuming ME requirement for maintenance at 486.6 kJ/kg $BW^{0.75}$ (NARO, 2007). At the day 19, blood samples were simultaneously collected into heparinized syringes from left carotid artery and mammary vein via catheters at almost every 3 hours for 8 times. The blood dioxygen (O_2) concentration was immediately measured by a fully automated blood gas analyzer (CHIRON 840, Siemens Healthcare Diagnostics Co. Ltd., Tokyo, Japan). The data for blood flow rate were collected every 50 msec via blood flow monitor (T201, Transonic Systems Inc., New York, USA), recorded by computers, and averaged for each 3-hours sample. The O_2 consumption of left udder was calculated from the difference in O_2 concentration between carotid artery and mammary vein multiplied by mammary blood flow. The O_2 consumption of whole udder was calculated from the O_2 consumption of left udder multiplied by the ratio of whole milk yield to left milk yield. The energy consumption was calculated by using O_2-energy conversion coefficient from the respiratory trial. The data were analyzed by the GLM procedures of SAS 9.2 with the diet as a factor and the experimental period and the cow as blocks.

Results and discussion

There was no significant differences were detected between experimental diets in all items (Table 1). Gross energy intake was bit higher in Starch diet, however, the fecal energy also tended to be higher in Starch diet, thus the ME intake was similar. The higher heat production (HP) and similar milk energy in Starch diet resulted in lower k_l value. The average metabolizability and k_l was calculated as 0.61 and 0.62, respectively, and the value of k_l was very similar to 0.62 (NARO, 2007) and 0.63 (ARC, 1980) when metabolizability was 0.60. Mammary energy expenditure was calculated as 94 kJ/kgBW$^{0.75}$ and this accounts for about 10% of HP and 5% of ME intake. We roughly estimated the energy expenditure of whole udder from reports in recent years as 87 (Guinard-Flament *et al.*, 2007) and 129 kJ/MBS/day (Lemosquet *et al.*, 2009), the present result was almost similar level. When an energy uptake of mammary gland was calculated as milk energy output plus mammary energy consumption (Guinard *et al.*, 1994), the ratio of milk energy output to energy uptake of mammary gland was calculated as 0.88 and this value was very similar to the data described by Guinard *et al.*, (1994). In conclusion, there was no significant difference between experimental diets, but the collected data from mammary gland were comparable to the data for whole body energy metabolism and other experiments. Further studies are also needed to evaluate the effect of feed on mammary gland metabolism and relationships between energy and other nutrients metabolism in mammary gland.

Table 1. Performance, whole body and mammary gland energy metabolism.

	Control	Bypass	Starch	SEM
Body weight, kg	582	588	593	2
Dry matter intake, kg/day	18.8	18.9	19.8	0.2
Milk yield, kg/day	26.4	26.4	26.5	0.4
Gross energy intake, kJ/kg BW$^{0.75}$/day	2,956	2,942	3,058	33
Fecal energy, kJ/kg BW$^{0.75}$/day	879	878	964	16
Urinary energy, kJ/kg BW$^{0.75}$/day	74	71	69	3
Methane energy, kJ/kg BW$^{0.75}$/day	199	197	199	9
Heat production, kJ/kg BW$^{0.75}$/day	962	950	990	23
Milk energy, kJ/kg BW$^{0.75}$/day	686	678	688	13
Retained energy, kJ/kg BW$^{0.75}$/day	156	170	147	21
ME intake, kJ/kg BW$^{0.75}$/day	1,805	1,797	1,825	14
k_l	0.62	0.63	0.60	0.02
Mammary blood flow, l/min	5.9	6.1	6.4	0.3
Mammary energy expenditure, kJ/kg BW$^{0.75}$/day	101	103	77	10
Mammary energy uptake, kJ/kg BW$^{0.75}$/day	787	781	765	8
Mammary energy efficiency	0.87	0.87	0.90	0.01

References

Agricultural Research Council, 1980. The Nutrient Requirements of Ruminant Livestock. CAB International, UK.

Guinard, J., H. Rulquin and R. Verite, 1994. Effect of graded levels of duodenal infusions of casein on mammary uptake in lactating cows. 1. major nutrients. J. Dairy Sci., 77, 2221-2231.

Guinard-Flament, J., E. Delamaire, P. Lamberton, and J.L. Peyraud, 2007. Adaptations of mammary uptake and nutrient use to once-daily milking and feed restriction in dairy cows. J. Dairy Sci. 90, 5062-5072.

Iwasaki, K., T. Haryu, R. Tano, F. Terada, M. Itoh and K. Kameoka, 1982. New animal metabolism facility especially the description of respirational apparatus. Bull. Natl. Inst. Anim. Indust. (Jpn.) 39, 41-78.

Lemosquet, S., G. Raggio, G.E. Lobley, H. Rulquin, J. Guinard-Flament and H. Lapierre, 2009. Whole-body glucose metabolism and mammary energetic nutrient metabolism in lactating dairy cows receiving digestive infusions of casein and propionic acid. J. Dairy Sci. 92, 6068-6082.

National Agriculture and Food Research Organization (NARO), 2007. Japanese Feeding Standard for Dairy Cattle, Japan Livestock Industry Association, Japan.

Part 4. Systemic and local regulation mechanisms

Could 'minor' nutrients such as B vitamins alter major metabolic pathways in lactating dairy cow?

C.L. Girard[1], A. Preynat[1], B. Graulet[2] and H. Lapierre[1]
[1]Dairy and Swine Research and Development Centre, Agriculture and Agri-Food Canada, Sherbrooke, Québec, J1M 1Z3, Canada
[2]INRA, Unité de Recherches sur les Herbivores, Equipe Digestion Microbienne et Absorption, 63122 Saint-Genès Champanelle, France

Abstract

Knowledge on dairy cow requirements for major nutrients increased substantially over the last decades and was integrated into programs used to balance dairy rations to substantially increase milk production and efficiency. However, over that period where milk production increased, little consideration has been given on the importance and the roles of B vitamins. The supply in those vitamins from synthesis by the ruminal microflora is generally sufficient to avoid deficiency symptoms and as such, had often been assumed to be sufficient to fulfill the cow's requirement under most circumstances. However, the sole synthesis from rumen microflora could be suboptimal when it comes to optimize metabolic efficiency, production, composition and nutritional quality of milk of high-producing dairy cows. Folic acid, one of the B vitamins, has the single, important biochemical function of accepting and releasing one-carbon units, which gives it a major role in DNA synthesis and *de novo* formation of methyl groups for the methylation cycle. On the other hand, vitamin B_{12} is involved in only two metabolic pathways: regeneration of methionine and tetrahydrofolate in the remethylation cycle and in the entry of propionate in the Krebs cycle for gluconeogenesis, through the vitamin B_{12}-dependent enzyme, methylmalonyl-coenzyme A (CoA) mutase. As these reactions are linked to carbohydrate, protein and lipid metabolism, it is then likely that the demand for these cofactors increases with milk yield. Indeed, results from recent studies highlight the fact that folic acid and vitamin B_{12} need to be supplemented to fully achieve their metabolic influence through their role as facilitators (coenzyme, cofactor) of enzymatic reactions and significantly affect performance of dairy cows. For example, results from one study shows that, when folic acid was given in combination with vitamin B_{12}, metabolic efficiency was improved as suggested by similar lactational performance and dry matter intake than cows fed folic acid supplements alone but increased plasma glucose and decreased hepatic concentrations of lipids. In a recent study, folic acid needed to be combined to vitamin B_{12} to increase milk and milk component yields over a whole lactation. More specifically, in these cows at their 12th week of lactation, a combined supplement of folic acid and vitamin B_{12} increased milk production as well as milk lactose, protein and total solid yields as compared to unsupplemented cows. Whole body rate of appearance of glucose also tended to increase with the vitamin supplementation with a similar quantitative magnitude as the milk lactose yield increment. This suggests that the effects of the combined supplements of folic acid and vitamin B_{12} on lactational performance were not mainly explained by methionine economy through a more efficient methylneogenesis as first hypothesized but were rather related to increased glucose availability. This also suggests that although folic acid is not known to participate in the methylmalonylCoA mutase pathway, it could possibly interfere with this pathway. These results confirm the important role of these B vitamins in the high productive dairy cows and the needs to define and integrate their appropriate requirements in modern predictive schemes.

Introduction

In a dairy industry context in most Western countries, production and associated research are no longer oriented towards increasing only production per se but instead focus on meeting consumers' demand for safe and high quality dairy products that are produced in a manner that is respectful of the environment and animal well-being, but still cost-effective. Improving metabolic efficiency is an approach that helps reducing the environmental impact of dairy production while taking into account dairy cow well-being. In that respect, nutrients considered 'minor' until recently can exert mechanistic regulations on metabolic pathways. Therefore, a better knowledge of their mode of action and related requirements would certainly have a positive influence on the efficiency of the cow, taking more into account the subtlety of her high demanding metabolism. The vitamins would certainly fit in that category of not too well known but nevertheless essential nutrients.

For centuries and until the end of the 19th century, food was considered important as a source of protein, energy and 'ash'. Raising the concept that low intakes of specific nutrients can cause certain diseases was considered revolutionary. The term 'vitamin', which appears at the very end of the 19th century, was born from this new 'revolutionary' concept that diet and health were closely interrelated, concept which allows emergence of nutrition as a science (Combs, 1998). Unlike other classes of nutrients except minerals, vitamins have no structural functions. Indeed, as B-vitamins play highly specific roles, such as coenzymes/cofactors, they are requested only in small amounts.

Bechdel *et al.* (1928) demonstrated that bacteria present in the rumen of a cow produced high levels of B vitamins, even if the animal's diet provided very small amounts of those vitamins. Furthermore, over the years since the discovery of B vitamins, it appears that true deficiency of these vitamins is rare in animals with a functional rumen.

> 'Because the gross functional and morphological changes cause by deprivation of the vitamins were the source of their discovery as important nutrients, these signs have become the focus of attention for many with interests in humans and/or veterinary health' (Combs, 1998).

As a result, traditionally, the impact of subclinical deficiency on maintenance of normal and efficient metabolism has been disregarded. Therefore, it became the dogma generally accepted that, in dairy cows, B-vitamin requirements can be met solely through synthesis by ruminal bacteria and dietary sources that escape from the rumen (NRC, 2001). Consequently, very little research effort has been directed at defining dairy cow requirements for B vitamins. As the majority of those vitamins act as essential cofactors in reactions linked to carbohydrate, protein and lipid metabolism, it is then likely that the demand for these cofactors increases with milk yield. Thenceforth, even if the supply in those vitamins from synthesis by the ruminal microflora is generally sufficient to avoid apparition of deficiency symptoms, it may not be sufficient to optimize metabolic efficiency, production, composition and nutritional quality of milk of high-producing dairy cows. This hypothesis is supported by studies reporting beneficial effects of supplementation of thiamin (Shaver and Bal, 2000), niacin (Fronk and Schultz, 1979; Riddell *et al.*, 1981), biotin (Zimmerly and Weiss, 2001), folic acid (Girard *et al.*, 1995; Girard and Matte, 1998) and vitamin B_{12} (Girard and Matte, 2005). The present paper aims to illustrate how supply in some B vitamins, specifically folic acid and vitamin B_{12}, could affect major metabolic pathways and improve the metabolic efficiency of dairy cows.

Metabolic roles of folic acid

In mammals folic acid has the single, important biochemical function of accepting and releasing one-carbon units (Choi and Mason, 2000). Cellular tetrahydrofolate (THF), originating from dietary folic acid, accepts one-carbon units from donors such as serine, glycine, histidine or formate and transfers them for purine and pyrimidine synthesis for DNA synthesis (Figure 1) (Bailey and Gregory, 1999). Reactions in this pathway are reversible. Tetrahydrofolate, after being reversibly converted to 5,10-methylene-THF could be directed towards the methylation cycle after its irreversible conversion to 5-methyl-THF which provides de novo methyl groups for remethylation of homocysteine to methionine. This reaction is mediated by an enzyme, the methionine synthase for which vitamin B_{12} is a coenzyme. The major function of the methylation cycle is to provide S-adenosylmethionine (SAM) which is the major donor of methyl groups in mammals. After having given its methyl group,

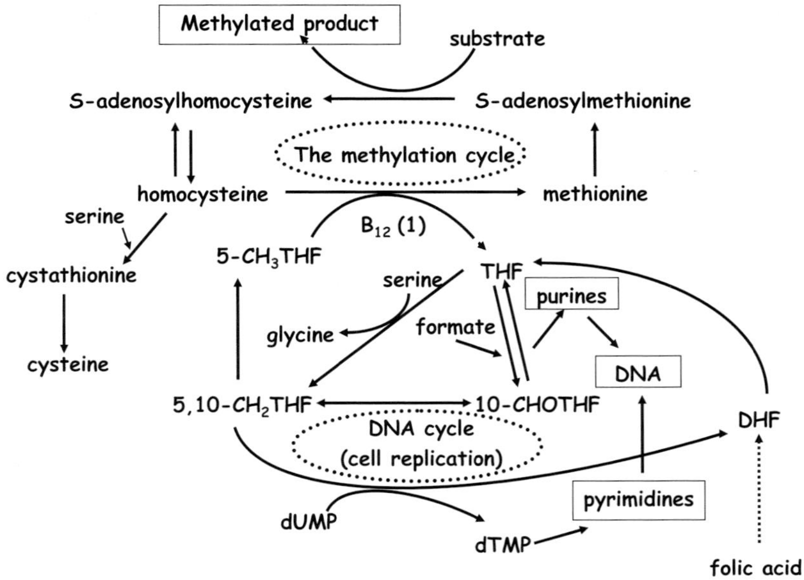

Figure 1. Roles of folic acid in DNA synthesis and methylation cycle.
$B_{12}(1)$: vitamin B_{12}-dependent enzyme, methionine synthase; DHF: dihydrofolate; THF: tetrahydrofolate, 5-CH_3-THF: methyl-THF; 5, 10-CH_2-THF: methylene-THF; 10-CHO-THF: formyl-THF; dUMP: desoxyuridylic acid; dTMP: thymidylic acid.

SAM becomes S-adenosylhomocysteine and then, homocysteine which could be catabolized through the transsulfuration pathway or remethylated again to methionine (Lucock, 2000).

Therefore, folic acid is crucial for *de novo* DNA synthesis and replication, and thus, cell division. It is also essential to DNA repair; folic acid deficiency causes an imbalance in DNA precursors, uracil misincorporation and chromosome breakage (Duthie, 1999; Kim, 1999). In the methylation cycle, the role of folate coenzymes is to provide one-carbon units for *de novo* formation of methyl groups (methylneogenesis) to ensure a constant supply in SAM which is the primary methylating agent. Among the numerous enzymatic reactions mediated by SAM, there are DNA methylation, which controls gene transcription and genetic stability, and synthesis of phosphatidylcholine, choline, creatine and neurotransmitters, such as serotonin, dopamine, norepinephrin and epinephrine.

The methylation cycle is especially important in ruminant animals. In non ruminant animals, methionine, choline and creatine are the major dietary sources of methyl groups (Snoswell and Xue, 1987). However, in ruminants, dietary choline is destroyed by ruminal microflora, plant material in the ruminant diet does not contain significant amount of creatine or carnitine and methionine supply is frequently lower than requirements, especially in high-producing lactating dairy cows fed a maize-based diet (NRC, 2001). To survive under those conditions of continuous low intake of methyl nutrients, ruminants developed adaptive mechanisms: a marked reduction of methyl group catabolism and an increased capacity for methylneogenesis (Snoswell and Xue, 1987). Moreover, lactation increases the demands for both methylated compounds, such as milk choline and creatine and for methionine to support milk protein synthesis (Xue and Snoswell, 1985). As an example, in order to synthesize choline to compensate for the loss in milk without depleting the body choline pool, the activity of phospholipid methyltransferases is increased by 34% in liver of lactating ewes as compared to non-lactating sheep (Xue and Snoswell, 1985). Inhibition of methionine synthase activity by nitrous oxide reduces by 70% methionine concentration in plasma and by 55% SAM concentration in liver of sheep demonstrating that sheep depend heavily on this metabolic pathway for provision of labile methyl groups (Snoswell and Xue, 1987). Therefore, *de novo* synthesis of methyl groups becomes of predominant importance for ruminant and this happens through folate metabolism (Snoswell and Xue, 1987).

Folic acid supplements for dairy cows

Girard *et al.* (1995) reported that weekly intramuscular injections of folic acid (160 mg of pteroylmonoglutamic acid) given from day 45 of gestation to 6 weeks after calving tended, as lactation progressed, to increase milk production and milk protein content from day 45 of gestation until drying off. After calving, however, injections of folic acid for 6 weeks had no effect on milk production. Supplementary folic acid, during that period, had no effect on milk protein content in primiparous cows, 33.2 and 33.7 g/kg for cows injected or not with folic acid but for cows in their second or greater lactation, milk protein content was 35.1 and 32.3 g/kg for cows injected or not with folic acid, respectively.

By comparison with unsupplemented cows, dietary supplements of folic acid (2 or 4 mg per kg of body weight) given daily from 4 weeks before expected time of calving until day 305 of lactation decreased milk production in primiparous cows during the first 100 d of lactation although this effect was no longer significant over a 305-day lactation period. By contrast, in multiparous cows, milk production increased linearly with the dose of folic acid supplemented, with the largest effect being observed between 100 and 200 d of lactation. Milk production over the whole lactation (3 to 305 days) averaged 8,284, 8,549 and 8,953 kg for the control, 2 mg and 4 mg groups, respectively. As dietary supplements of folic acid did not change milk composition in multiparous cows, milk component yields exhibited the same upward trend as milk yield (Girard and Matte, 1998). Non protein nitrogen content in milk of primiparous cows was not affected by the supplements but it was decreased during the first 100 d of lactation in multiparous cows fed supplementary folic acid giving an indication that nitrogen was used more efficiently. Dry matter intake did not differ among treatments (Girard and Matte, 1998). Incidentally, in that experiment, in spite of a sufficient dietary supply in cobalt, it was observed that serum vitamin B_{12} was low in early lactation, more so in primiparous than multiparous cows (Girard and Matte, 1999). Results from these first two experiments show that supplementary folic acid might improve the efficiency of nutrient utilization but only in cows in second or greater lactation.

Attempts to elucidate the modes of action of supplementary folic acid

Given the metabolic roles of folic acid, the effects of supplementary folic acid on lactational performance could be related to its action on DNA synthesis and/or the methylation cycle. Cellular concentration of SAM is function of the availability of methyl groups which are provided as preformed labile methyl groups by methionine, betaine and choline or by *de novo* synthesis from folate metabolism. If the effects of supplementary folic acid on lactational performance of dairy cows are mediated through the methylation cycle, then supplementary methionine, a major source of preformed methyl groups for SAM synthesis, should reduce the demand for folic acid. Moreover, one should keep in mind that folic acid and vitamin B_{12} share a critical interface (Figure 1). As the production of 5-methyl-THF is irreversible and as this form of folates can not be retained in the cells, a vitamin B_{12} deficiency can trap folic acid under its methylated form and reduce its cellular utilization (Scott and Weir, 1981; Bässler, 1997). In summary, a vitamin B_{12} deficiency leads to a secondary folate deficiency by interfering with folate utilization in cells (Scott, 1999).

Therefore, a third experiment was undertaken in an attempt to determine if the effects of supplementary folic acid on lactational performance previously observed could be explained by an improved methylneogenesis. Multiparous dairy cows were fed a diet estimated to supply methionine at 1.75%

of metabolisable protein supply, equivalent to 70% of methionine requirement; half of them received also a supplement of rumen-protected methionine in order to cover methionine requirement (NRC, 2001). Within each level of methionine supplements, the cows received daily a dietary supplement of 0, 3 or 6 mg folic acid per kg body weight from 3 weeks before the expected calving date until 305 d of lactation. Total milk production for the 305-day lactation was unaffected by folic acid or rumen-protected methionine supplements and averaged 10,584 kg. Milk component yields and dry matter intake were also unaltered by treatments (Girard *et al.*, 2005). Overall, 8 weeks after calving, serum concentrations of folates and cysteine were increased whereas concentrations of vitamin B_{12}, homocysteine and methionine were decreased as compared to later in lactation. As previously described in Figure 1, homocysteine has two metabolic fates: remethylation into methionine under the action of the vitamin B_{12}-dependent enzyme, methionine synthase, or catabolism to form cysteine (Selhub, 1999). Serum concentrations of vitamin B_{12} increased through lactation according to a pattern similar to the one observed by Girard and Matte (1999), although the concentrations were lower than previously reported for multiparous cows and, instead, were similar to those reported for primiparous cows (Girard and Matte, 1999). These data, coupled with the slower serum clearance of folates measured following an intravenous bolus of folic acid at that time, suggest that low supply in vitamin B_{12} during early lactation could interfere with folate utilization. It could explain the limited lactational response to supplementary folic acid observed in the present experiment.

In a meantime, primiparous cows were fed the same basal diet than in the experiment described previously and supplemented daily with folic acid and rumen-protected methionine. The effects of weekly intramuscular injections of saline (0.95 NaCl) or 10 mg of vitamin B_{12} on milk production were monitored from 4 to 18 weeks of lactation. Supplementary vitamin B_{12} tended to increase milk yield from 28.5 to 31.1 kg/d and increased energy-corrected milk as well as milk yields of solids, fat and lactose but had no effect on dry matter intake or milk composition. Furthermore, packed cell volume and blood haemoglobin increased and serum methylmalonic acid decreased in cows that received vitamin B_{12} injections. The first observation suggests that low vitamin B_{12} supply interfered with folate metabolism because folic acid deficiency, through its role in DNA synthesis, affects hematopoiesis (Bills *et al.*, 1992). The second one indicates that low vitamin B_{12} supply interfered with the other vitamin B_{12}-dependent enzyme, methylmalonyl-CoA mutase. These findings supported the hypothesis that vitamin B_{12} supply is suboptimal in early lactation (Girard and Matte, 2005) and may have limited the potential role of folic acid in these first studies.

Vitamin B_{12} supplements: restoration of folate metabolism efficiency or improvement of propionate utilization?

Besides its role in the methylation cycle (Figure 1), vitamin B_{12} plays a major role for the entry of propionate in the Krebs cycle and gluconeogenesis, through the vitamin B_{12}-dependent enzyme, methylmalonyl-coenzyme A (CoA) mutase (McDowell, 2000). Propionate, originating from rumen fermentation of carbohydrates, provides propionyl-CoA which is carboxylated to methylmalonyl-CoA by propionyl-CoA carboxylase. This pathway can also participate to catabolism of some amino acids (isoleucine, valine, threonine, methionine). Methylmalonyl-CoA is then isomerized in succinyl-CoA under the action of the vitamin B_{12}-dependent enzyme, methylmalonyl-CoA mutase (Le Grusse and Watier, 1993). Succinyl-CoA finally enters into the Krebs cycle where it can be used for gluconeogenesis. A deficiency of vitamin B_{12} increases concentrations of methylmalonyl-CoA and its catabolites, methylmalonic acid (Combs, 1998).

It was not before the late seventies, that the concept that vitamin B_{12} supply for lactating dairy cows could not be optimal was raised. However, supplementary vitamin B_{12} generally fails to affect milk and component yields in cows (Frobish et Davis, 1977; Elliot *et al.*, 1979; Croom *et al.*, 1981). Elliot *et al.* (1979) reported that bi-weekly intramuscular injections of vitamin B_{12} given from weeks 2 to 8 *post partum* had no effect on milk fat content but increased milk fat yield due to a non-statistically significant increase in milk yield. Using a cross-over design, Croom *et al.* (1981) observed that weekly intramuscular injections of vitamin B_{12} given during weeks 9 to 11 or weeks 13 to 15 *post partum* had no effect on milk fat content or yield as compared to their respective control period. None of these studies measured milk protein content and yield.

Indeed, the lack of effect of vitamin B_{12} supplements given alone on lactational performance of dairy cows was observed again in a factorial experiment studying the effects of dietary supplements of folic acid and/or vitamin B_{12} given from 3 weeks before the expected time of calving until 8 weeks of lactation. During the first 8 weeks of lactation, dietary supplements of folic acid, given alone or in combination with vitamin B_{12}, increased milk production by 3.4 kg/d, from 38.0 to 41.4±1.0 kg/d and milk protein by 75 g/d without any effect on dry matter intake. In cows fed supplementary folic alone, plasma concentration of glucose was 10% lower and hepatic concentrations of lipids during the first 4 weeks of lactation were twice those of cows fed the combined supplement of folic acid and vitamin B_{12} (Graulet *et al.*, 2007). Lipid accumulation in liver was also observed in vitamin B_{12} deficient sheep (Smith *et al.*, 1974; Kennedy *et al.*, 1994). In ruminants, methylmalonyl-CoA plays a unique regulating role in gluconeogenesis and β-oxidation of fatty acids (Kennedy *et al.*, 1994). Indeed, in sheep liver, methylmalonyl-CoA inhibits β-oxidation of fatty acids (Brindle *et al.*, 1985). If the entry of propionate in the Krebs cycle is slowed down by a lack of vitamin B_{12} (reduction of the isomerisation of methylmalonyl-CoA to succinyl-CoA), concentrations of methylmalonyl-CoA, and then, methylmalonic acid, its catabolite, are increased, reducing β-oxidation, even if gluconeogenesis is also reduced (Zammit, 1990). Therefore, in ruminants, vitamin B_{12} deficiency increases energy deficit by interfering with propionate utilization which increases mobilization of body reserves but then blocks fatty acid oxidation in liver, leading to their accumulation. Moreover, vitamin B_{12} deficiency also affects synthesis of VLDL (Kennedy *et al.*, 1994; Stangl *et al.*, 1999), then reducing liver ability to export lipids. All together, results from this study indicate that when folic acid is given in combination with vitamin B_{12}, metabolic efficiency is improved as compared to folic acid alone as suggested by similar lactational performance and dry matter intake but increased plasma glucose and decreased hepatic concentrations of lipids (Graulet *et al.*, 2007).

Further to this experiment, another study was conducted aiming to precise the importance of an increased provision of newly formed methyl groups (methylneogenesis) through the folate metabolism on lactational performance of dairy cows and if the supply in vitamin B_{12} could affect this metabolic pathway. In the eventuality that the effects of supplementary folic acid are mostly due to an improved methylneogenesis, then supplementary methionine, a major source of preformed methyl groups, should reduce the requirements for these vitamins (Figure 1). Multiparous Holstein cows were fed or not a supplement of rumen-protected methionine and within each level of methionine (1.83 and 2.23% of metabolisable protein supply), cows received either no vitamin supplement, weekly intramuscular injections of folic acid alone or combined with vitamin B_{12}, from 3 weeks before to 16 weeks after calving. There was no treatment effect on dry matter intake. Folic acid supplements alone or supplementary dietary methionine had no effect on milk yield but the combined supplement of folic acid and vitamin B_{12} increased milk production by 1.4 kg/d during the complete experimental period. This effect was more pronounced during the first 4 weeks of lactation; milk lactose, crude protein and total solid yields followed the same trend than milk production. Intramuscular injections of folic acid alone or combined with vitamin B_{12} tended to decrease plasma concentrations of homocysteine. Rumen-protected methionine and vitamin supplements both affected expression of genes related to the methylation cycle. However, because supplying preformed labile methyl groups as rumen-protected methionine had no effect either on milk and milk component yields, it is unlikely that the effect of the combined supplement of vitamins on lactational performance were due to an improvement in methyl group supply through methylneogenesis. Nevertheless, folic acid and vitamin B_{12} given together increased milk and milk component yields without effect on dry matter intake or plasma non-esterified fatty acids. These results suggest that the effects of the combined supplement of vitamins on lactational performance were due to an effect on energy metabolism rather than on methylneogenesis (Preynat *et al.*, 2009a, 2010).

Further measurements were made, within this last study, to confirm these hypotheses. At 12 weeks of lactation, glucose and methionine kinetics were measured by isotope dilution using continuous infusions of D[U-^{13}C]glucose, [^{13}C]NaHCO_3 and L[1-^{13}C,$^{2}H_3$]methionine. Due to the cost of isotopes, this part of the study was limited to the cows receiving no vitamin injections or the combined injections of folic acid and vitamin B_{12}, with or without the rumen-protected supplement of methionine. On the week on these measurements, supplementary B-vitamins increased milk production from 34.7 to 38.9±1.0 kg/d as well as milk lactose, protein and total solid yields. Whole body glucose rate of appearance, representing the sum of glucose absorption, gluconeogenesis and glycogen turnover, tended to increase with the vitamin supplementation with a similar quantitative magnitude as the milk lactose yield increment, whereas methionine supplementation had no effect on glucose kinetics. Supplementary folic acid plus vitamin B_{12} tended to increase the irreversible loss rates (ILR) of methionine by approximately 20% in cows fed no methionine supplements but had no effect on cows fed rumen-protected methionine. Folic acid plus vitamin B_{12} supplements decreased methionine entry in the transmethylation pathway (Figure 1) in cows fed rumen-protected methionine, whereas it had no effect when the cows did not receive methionine supplementation. Vitamin supplementation increased methionine utilization for protein synthesis through increased protein turnover when methionine was deficient and through decreased methionine oxidation when rumen-protected methionine was fed. In the current study, the increases of both milk lactose and protein yields with folic acid plus vitamin B_{12} supplements were related to the positive effect of vitamins on glucose rate of appearance and methionine ILR, the later being mostly due to an increase of methionine directed towards protein synthesis. Milk protein yield was not modified by supplementary methionine, despite an increment in methionine ILR and the use of methionine for protein synthesis, suggesting that supplementary methionine increased protein turnover. In conclusion, the effects of the supplements of folic acid and vitamin B_{12} on lactational performance were not mainly explained by methionine economy through a more efficient methylneogenesis as first hypothesized but were rather related to increased glucose availability and changes in methionine metabolism (Preynat *et al.*, 2009b).

As described previously, it is well-known that a lack of vitamin B_{12} interferes with folate metabolism even leading to a secondary deficiency in folic acid in spite of an adequate supply in the vitamin. Although folic acid is not known to participate in the methylmalonylCoA mutase pathway, recent results, however, seem to indicate that folic acid supply could also interfere with this pathway. In summary, Graulet *et al.* (2007) observed that the requirement of the methylmalonylCoA mutase for its vitamin B_{12} cofactor decreased when the two vitamins were given together as compared to folic acid or vitamin B_{12} given separately, improving the efficiency of this metabolic reaction. Preynat *et al.* (2010) observed that gene expression of methylmalonylCoA mutase was increased by folic acid with or without vitamin B_{12} supplements as compared to unsupplemented cows but this increase was greater in cows injected with the two vitamins together. A similar observation was reported in humans recently (Selhub *et al.*, 2007). Little is known about the transport mechanisms of vitamin B_{12} from cytosol to mitochondria. Observations in cell cultures suggest that the effect of folic acid supply on methylmalonylCoA mutase is possibly linked to coordination of the distribution of vitamin B_{12} between cytosol (methionine synthase) and mitochondria (methylmalonylCoA mutase; Riedel *et al.*, 1999).

Conclusion

Results reported in the present paper described the long but still incomplete journey towards the understanding of the roles of two B vitamins, folic acid and vitamin B_{12}, on metabolism and related lactational performance of dairy cows. This description is only one example of how 'minor' nutrients, in this case two B vitamins, can alter major metabolic pathways and affect metabolic utilization of nutrients provided by the diet, increasing yields of milk components at a similar intake. The objective of the present paper was to demonstrate how, in the quest towards a better understanding of nutrient requirement and metabolism, inadequate supply of generally neglected minor nutrients, such as B vitamins, could distort the whole picture and possibly explain some of the discrepancies in results from experiments otherwise similar. Minor nutrient requirements cannot only be assessed by the absence of deficiency symptoms. An adequate supply should target optimization of the metabolic efficiency, and in the case of these 2 B vitamins, we cannot assume anymore that the supply from rumen microflora is sufficient to maximize the efficiency of the high producing dairy cows. As knowledge is built on these requirements, recommendations should be included in models used to balance dairy rations, for the benefits of the cow, the producer and the environment.

References

Bailey, L.B. and J.F. Gregory III, 1999. Folate metabolism and requirements. J. Nutr. 129, 779-782.

Bässler, K.H., 1997. Enzymatic effects of folic acid and vitamin B_{12}. Internat. J. Vitam. Nutr. Res. 67, 385-388.

Bechdel, S.I., H.E. Honeywell, R.A. Dutcher and M.H. Knutsen, 1928. Synthesis of vitamin B in the rumen of the cow. J. Biol. Chem. 80, 231-238.

Bills, N.D., M.J. Koury, A.J. Clifford and E.N. Dessypris, 1992. Ineffective hematopoiesis in folate-deficient mice. Blood. 79, 2273-2280.

Brindle, N.P.J., V.A. Zammit and C.I. Pogson. 1985. Inhibition of sheep liver carnitine palmitoyltransferase by methylmalonyl-CoA. Biochem Soc. Trans. 13, 880-881.

Choi, S.-W. and J.B. Mason, 2000. Folate and carcinogenesis: an integrated scheme. J. Nutr. 130, 129-132.

Combs, G.F.Jr. 1998, The vitamins. Fondamental aspects in nutrition and health. 2nd edition. Academic Press. San Diego. CA, USA.

Croom, W.J.Jr., A.H. Rakes, A.C. Linnerud, G.A. Ducharme and J.M. Elliot, 1981. Vitamin B_{12} administration for milk fat synthesis in lactating dairy cows fed a low fiber diet. J. Dairy Sci. 64, 1555-1560.

Duthie, S. J., 1999. Folic acid deficiency and cancer: mechanisms of DNA instability. Br. Med. Bull. 55, 578-592.

Elliot, J.M., E.P. Barton and J.A. Williams, 1979. Milk fat as related to vitamin B_{12} status. J. Dairy Sci. 62, 642-645.

Frobish, R.A. and C.L. Davis, 1977. Theory involving propionate and vitamin B_{12} in the low-milk fat syndrome. J. Dairy Sci. 60, 268-273.

Fronk, T.J. and L.H. Schultz, 1979. Oral nicotinic acid as a treatment for ketosis. J. Dairy Sci. 62, 1804-1807.

Girard, C.L. and J. J. Matte, 1998. Dietary supplements of folic acid during lactation; effects on the performance of dairy cows. J. Dairy Sci. 81, 1412-1419.

Girard, C.L. and J.J. Matte, 1999. Changes in serum concentrations of folates, pyridoxal, pyridoxal-5-phosphate and vitamin B_{12} during lactation of dairy cows fed dietary supplements of folic acid. Can. J. Anim. Sci. 79, 107-113.

Girard, C.L. and J.J. Matte, 2005. Effects of intramuscular injections of vitamin B_{12} on lactation performance of dairy cows fed dietary supplements of folic acid and rumen-protected methionine. J. Dairy Sci. 88, 671-676.

Girard, C.L., Matte, J.J. and G.F. Tremblay, 1995. Gestation and lactation of dairy cows: a role for folic acid? J. Dairy Sci. 78, 404-411.

Girard, C.L., H. Lapierre, J.J. Matte and G.E. Lobley, 2005. Effects of dietary supplements of folic acid and rumen-protected methionine on lactational performance and folate metabolism of dairy cows. J. Dairy Sci. 88, 660-670.

Graulet, B., J.J. Matte, A. Desrochers, L. Doepel, M.F. Palin and C.L. Girard, 2007. Effects of dietary supplements of folic acid and vitamin B_{12} on metabolism of dairy cows in early lactation. J. Dairy Sci. 90, 3442-3455.

Kennedy, D.G., P.B. Young, W.J. Blanchflower, J. M. Scott, D.G. Weir, A.M. Molloy and S. Kennedy, 1994. Cobalt-vitamin B_{12} deficiency causes lipid accumulation, lipid peroxidation and decreased α-tocopherol concentrations in the liver of sheep. Internat. J. Vit. Nutr. Res. 64, 270-276.

Kim, Y. I., 1999. Folate and carcinogenesis: evidence, mechanisms, and implications. J. Nutr. Biochem. 10, 66-88.

Le Grusse, J. and B. Watier. 1993. Les vitamines. Données biochimiques, nutritionnelles et cliniques. Centre d'étude et d'information sur les vitamines, Produits Roche. Neuilly-sur-Seine, France.

Lucock, M., 2000. Folic acid: nutritional biochemistry, molecular biology, and role in disease processes. Mol. Genet. Metab. 71, 121-138.

McDowell LR., 2000. Vitamins in animal and human nutrition. 2nd ed. Iowa State University Press, Ames, Iowa, USA.

National Research Council, 2001. Nutrient Requirements of Dairy Cattle. 7th rev. ed. Natl. Acad. Sci., Washington, DC, USA.

Preynat, A., H. Lapierre, M.C. Thivierge, M.F. Palin, J.J. Matte, A. Desrochers and C.L. Girard, 2009a. Influence of methionine supply on the responses of lactational performance of dairy cows to supplementary folic acid and vitamin B_{12}. J. Dairy Sci. 92, 1685-1695.

Preynat, A., H. Lapierre, M.C. Thivierge, M.F. Palin, J.J. Matte, A. Desrochers and C.L. Girard, 2009b. Effects of supplements of folic acid, vitamin B_{12} and rumen-protected methionine on whole body metabolism of methionine and glucose in lactating dairy cows. J. Dairy Sci. 92, 677-689.

Preynat, A., H. Lapierre, M.C. Thivierge, M.F. Palin, J.J. Matte, A. Desrochers and C.L. Girard, 2010. Effects of supplementary folic acid and vitamin B_{12} on hepatic metabolism of dairy cows according to methionine supply. J. Dairy Sci. 93, in press.

Riddell, D.O., Bartley, E.E. and A.D. Dayton, 1981. Effect of nicotinic acid on microbial protein synthesis *in vitro* and on dairy cattle growth and milk production. J. Dairy Sci. 64, 782-791.

Riedel, B., T. Fiskerstrand, H. Refsum and P.M. Ueland, 1999. Co-ordinate variations in methylmalonyl-CoA mutase, and the cobalamin cofactors in human glioma cells during nitrous oxide exposure and the subsequent recovery phase. Biochem. J. 341, 133-138.

Scott, J.M., 1999. Folate and vitamin B_{12}. Proc. Nutr. Soc. 58, 441-448.

Scott, J.M., and D.G. Weir, 1981. The methyl trap. A physiological response in man to prevent methyl group deficiency in kwashiorkor (methionine deficiency) and an explanation for folic-acid induced exacerbation of subacute combined degeneration in pernicious anaemia. Lancet. 2, 337-340.

Selhub, J., 1999. Homocysteine metabolism. Annu. Rev. Nutr. 19, 217-246.

Selhub, J., M.S. Morris and P.F. Jacques, 2007. In vitamin B_{12} deficiency, higher serum folate is associated with increased total homocysteine and methylmalonic acid concentrations. Proc. Natl. Acad. Sci. USA. 104, 19995-20000.

Shaver, R.D. and M.A. Bal, 2000. Effect of dietary thiamin supplementation on milk production by dairy cows. J. Dairy Sci. 83, 2335-2340.

Smith, R.M., W.S. Osborne-White and J.M. Gawthorne, 1974. Folic acid metabolism in vitamin B_{12}-deficient sheep. Effects of injected methionine on liver constituents associated with folate metabolism. Biochem. J. 142, 105-117.

Snoswell, A.M. and G.-P. Xue, 1987. Methyl group metabolism in sheep. Comp. Biochem. Physiol. 88B, 383-394.

Stangl, G. I., F. J. Schwarz and M. Kirchgessner, 1999. Moderate long-term cobalt-deficiency affects liver, brain and erythrocyte lipids and lipoproteins of cattle. Nutr. Res. 19, 415-427.

Xue, G.-P. and A.M. Snoswell, 1985. Regulation of methyl group metabolism in lactating ewes. Biochem. Internat. 11, 381-385.

Zammit, V.A., 1990. Ketogenesis in the liver of ruminants – adaptations to a challenge. J. Agric. Sci. (Camb.). 115, 155-162.

Zimmerly, C.A. and W.P. Weiss, 2001. Effects of supplemental dietary biotin on performance of Holstein cows during early lactation. J. Dairy Sci. 84, 498-506.

Renal blood flow increase induced by amino acids in sheep: role of nitric oxide and prostaglandins

A. Cirio[1], I. Tebot[1], C. Paquet[2], J.-Y. Ayoub[2] and J.-M. Bonnet[2]
[1]Area de Fisiología, Facultad de Veterinaria, Montevideo, Uruguay
[2]Université de Lyon, VetAgro Sup Campus vétérinaire de Lyon, EA 4173 Inserm ERI 22, Marcy l'Etoile, France

Introduction

As largely stated in the literature, renal blood flow (RBF) increases after a protein-rich meal or during systemic infusion of amino acids (AA) in different species. However, the mechanisms involved are still not fully elucidated. Nitric oxide (NO) has been proposed as a step in the renal hemodynamic response to AA (Chen *et al.*, 1992). Nevertheless, we recently reported that blocking NO synthesis with N^G-nitro-L-arginine methyl ester (L-NAME) reduced renal perfusion in sheep, but only partially prevented RBF increase induced by insulin, suggesting the existence of complementary vasodilatory agents (Cirio *et al.*, 2009b). Local vasodilatory prostaglandins (PG) have been invoked to explain the renal hemodynamic response to AA, but some findings are contradictory (reviewed by Woods, 1993). More recently, Tuttle *et al.* (2002) reported, in normal subjects, that the RBF increase induced by AA infusion, is not PG-dependent. Moreover, we recently reported in sheep that blocking PG synthesis with ketoprofen reduced renal perfusion but failed to prevent the RBF increase induced by insulin, suggesting the involvement of vasodilatory agents other than PG (Cirio *et al.*, 2009a). The aim of the present work was to evaluate the participation of NO and PG in the renal vasodilation induced by AA infusion, using a sheep model chronically equipped with a flow-metering system.

Material and methods

Under general anesthesia, 6 adult Ile de France ewes (54-70 kg BW) were bilaterally implanted with transit-time ultrasonic flow-metering probes (4 mm, R-series, Transonic Systems, Ithaca, NY, USA) around renal arteries for RBF measurement, as previously described (Tebot *et al.*, 2009). After recovery, sheep were housed in individual pens for chronic studies and fed on alfalfa pellets, hay and *ad libitum* water. Data were transmitted to a processing system (Acqknowledge III for MP150WSW, Biopac Systems Inc., Sta. Barbara, CA, USA), and mean values of RBF (right + left flows) calculated every 10 min. On experimental days, 16 h-fasted sheep received constant rate i.v. infusions (0.5 ml/min, 09:00-11:00) of (a) mixed AA (Vintene, Baxter, Maurepas, France) 6 mg/kg/min, (b) L-NAME (Sigma-Aldrich, Lyon, France), 0.22 mg/kg/min in 0.9% saline, + AA solution during the second hour, and (c) ketoprofen (Ketofen, Merial, Lyon, France), 0.2 mg/kg/min primed with 6 mg/kg in saline + AA solution during the second hour. Each sheep received one perfusion per diem, on alternate days and random order. Significance of RBF changes during and after infusions, related to 1 h pre-infusion mean values, was determined by a *t*-test for dependent samples (Statistica 6.0, Statsoft, USA).

Results and discussion

RBF increased up to 12% ($P<0.05$) above basal values during AA infusion (Figure 1). This result is in opposition to that of Boyce *et al.* (2004) who, using the para-aminohippurate clearance and perfusing 4 AA, did not observe any change in RBF. As expected from the literature, during the first hour of L-NAME and ketoprofen infusions, RBF decreased (maximum of 11% and 4%, $P<0.05$, respectively). The AA administration during the second hour of L-NAME raised the RBF, but only after 20 min of perfusion. The AA administration during the second hour of ketoprofen raised the RBF to values similar to those observed with the perfusion of AA alone. The high values of RBF after the end of perfusions could be attributed to the known renal vasodilatatory effect of the AA-induced

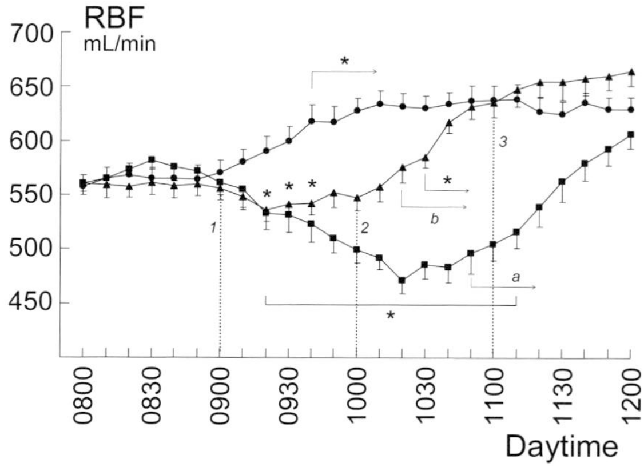

*Figure 1. Renal blood flow (RBF) during and after 2 h perfusions of mixed AA (●, dotted lines 1-3) and of L-NAME (■) and ketoprofen (▲) (dotted lines 1-3) with mixed AA during the second hour (dotted lines 2-3). Data are 10 min means ± SE. * = P<0.05 vs. 1 h pre-infusion mean value; a,b = P<0.05 vs. the minimum value in L-NAME (1020) and ketopro-fen (0920) perfusions; n=6 sheep.*

insulin or glucagon release. It is concluded that NO seems to play a minor role in AA-induced renal vasodilatation (as also proposed by Chen *et al.* (1992) in anesthetized rats), and that PG are not involved in this effect. Both facts suggest the participation of complementary vasodilatory agents accounting for the effect of AA on renal hemodynamics, and strengthen our previous findings (Cirio *et al.*, 1990a,b) on the RBF response to insulin after blockade of NO and PG synthesis in sheep.

Acknowledgement

Financial support: Comité ECOS-*Sud*, action U05S02 and the Conseil Régional Rhône-Alpes, Coopéra 2009.

References

Boyce, A.C., S. Chinchen and K.J. Gibson, 2004. Is the renal haemodynamic response to amino acids present in sheep? Proc. of the Sydney 2004 Meeting of the Aust. Physiol. Soc. 35, 34P.

Chen, C., K.D. Mitchell and L.G. Navar, 1992. Role of endothelium-derived nitric oxide in the renal hemodynamic response to amino acid infusion. Am. J. Physiol. 263, R510-516.

Cirio, A., I. Tebot, J-Y. Ayoub, C. Paquet, S. Junot and J-M. Bonnet, 2009a. Blocking vasodilatory prostaglandins synthesis by ketoprofen fails to prevent the renal blood flow increase induced by insulin in conscious sheep. In: Chilliard Y., F. Glasser, Y. Faulconnier, F. Bocquier, I. Veissier and M. Doreau (eds.), Ruminant Physiology. Wageningen Acad. Publishers, Wageningen, the Netherlands, Proc. XIth ISRP: 488-489.

Cirio, A., I. Tebot, C. Paquet, J-Y. Ayoub and J-M. Bonnet, 2009b. Blocking NO synthesis by L-NAME perfusion partially prevents the renal blood flow increase induced by insulin perfusion in conscious sheep. In: Chilliard Y., F. Glasser, Y. Faulconnier, F. Bocquier, I. Veissier and M. Doreau (eds.), Ruminant Physiology. Wageningen Acad. Publishers, Wageningen, the Netherlands, Proc. XIth ISRP: 490-491.

Tebot, I., J-M. Bonnet, S. Junot, J-Y. Ayoub, C. Paquet and A. Cirio, 2009. Roles of eating, rumination, and arterial pressure in determination of the circadian rhythm of renal blood flow in sheep. J. Anim. Sci. 87, 554-561.

Tuttle, K.R., M.E. Puhlman, S.K. Cooney and R.A. Short, 2002. Effects of amino acids and glucagons on renal hemodynamics in type 1 diabetes. Am. J. Physiol. 282, F103-112.

Woods, L, 1993. Mechanisms of renal hemodynamic regulation in response to protein feeding. Kidney Int. 44, 659-675.

m-TOR independent model of protein synthesis regulation by essential amino acids in mammary epithelial cells

S. Arriola, W.A.D. Nayananjalie, J.A.D.A. Appuhamy and M.D. Hanigan
Dept. of Dairy Science, Virginia Polytechnic Institute and State University, Blacksburg, VA 24061, USA

Introduction

Essential amino acids (EAA) stimulate protein synthesis through the m-TOR pathway (Kimball and Jefferson, 2002). A TOR-independent pathway through which AA may play a key role in protein synthesis regulation has been recently described in yeast. The model proposes that an increase of uncharged tRNA, due to EAA starvation, would promote free t-RNA binding to the histidyl-tRNA synthetase homologous domain in the GCN2 kinase of eIF2a subunit at Ser 51 (Dever *et al.*, 2007). The phosphorylation of eIF2a down-regulates the nucleotide exchange activity of eIF2B on eIF2. Exchange of GDP for GTP is necessary for the initiation factor-guanine nucleotide complex to bind met-tRNA and initiate a new translation.

The objective of this study was to explore the effect of individual EAA in the phosphorylation of eIF2a at Ser 51 in epithelial mammary cells after EAA depletion.

Materials and methods

The experiment was conducted using an immortalized mammary epithelial cell line (MAC-T; Huynh *et al.*, 1991). Cells were grown in DMEM/F12 medium (Invitrogen, Carlsberg, CA, USA) supplemented with 7% foetal bovine serum, 100 U/ml Penicilin, 100 ug/ml Streptomycin, and 0.25 ug/ml Amphotericin B (HyClone Laboratories Inc., Logan, UT, USA). Before the application of the treatments, cells were starved for 2 h in DMEM/F12, with EAA concentrations at 5% of normal DMEM/F12 EAA concentrations. The experimental design was a randomized block design with three replications per treatment. Treatments were the starvation media, complete DMEM/F12 media, and the starvation media supplemented with individual EAA at 100% of normal DMEM/F12 concentrations. Cells were incubated in treatment media for 1 h. After incubation, plates were placed on ice, and ice-cold stop buffer was added at 2 ml per well. Stop buffer was removed, 1 ml of ice-cold lysis buffer was added to the wells, and cells were removed from the plate by manual scraping. Cell lysates were centrifuged at 12,000×g for 10 min at 4 °C, and supernates were recovered and analyzed by Western immunoblotting for total and phosphorylated forms of eIF2a (Ser 51). The ratio of phosphorylated to total protein was analyzed statistically using the MIXED procedure of SAS (2009) with treatment specified as a fixed effect. Means were separated by Dunnett test, with starvation media used as the control. Significance was declared at $P<0.05$.

Results and discussion

Table 1 shows the phosphorylation ratios and significance level of complete and individual EAA media compared with starvation media. Supplementation of Phe, His, and Val significantly affected ($P<0.05$) the eIF2a phosphorylation ratio. Supplementation of Trp and Arg tended to reduce the phosphorylation ratio as compared to that observed in cells grown in starvation media. The effect of His on phosphorylation of eIF2a has been reported in yeast with His-synthesis knock out genes (Zaborske *et al.*, 2009). Zaborske *et al.* (2009) also observed reduced charging level in His-tRNA, indicating that this and other tRNA may be regulating the phosphorylation of eIF2a at Ser 51. In the same experiment, it was reported a positive effect of starvation with Trp on eIF2a phosphorylation, similar to the trend found in our study. Arginine has not been observed to affect eIF2a phosphorylation. Conversely, Leu has been reported to stimulate protein synthesis and inhibit phosphorylation of

Table 1. Phosphorylation ratio of eIF2α in Mac-T cells grown in complete DMEM/F12, essential amino acid starved media1, or starved media plus individual amino acids. The pooled SE was 0.039.

Media	Starv.	Complete	+Ile	+Leu	+Val	+Met	+Phe	+Trp	+Thr	+His	+Arg
Ratio	0.311	0.214	0.227	0.297	0.188	0.250	0.166	0.212	0.232	0.176	0.211
P-value	-	0.132	0.24	0.99	0.034	0.57	0.010	0.12	0.29	0.017	0.11

[1] Starvation media (starv.) contained 5% of normal essential amino acid concentrations; amino acid supplemented media was starvation media plus the specified individual amino acid at 100% of DMEM/F12 concentrations.

eIF2a in mouse liver cells (Anthony *et al.*, 2004), but no effect of Leu on the phosphorylation ratio of eIF2a was observed in these mammary epithelial cells.

These results demonstrate that short-term EAA starvation of mammary epithelial cells alters the phosphorylation state of eIF2a, and the effects can be reversed by the addition of at least Phe, His, or Val. This was an exploratory experiment. More research is needed to determine the long-term effects of individual EAA on the regulation of translation initiation and protein synthesis in mammary epithelial cells.

References

Anthony, T.G., B.J. McDaniel, R.L. Byerley, B.C. McGrath, D.R. Cavener, M.A. McNurlan and R.C. Wek, 2004. Preservation of Liver Protein Synthesis during Dietary Leucine Deprivation Occurs at the Expense of Skeletal Muscle Mass in Mice Deleted for eIF2 Kinase GCN2. J. Biol. Chem. 279, 36553-36561.

Dever, T.E., A.C. Dar and F. Sicheri, 2007. The eIF2α Kinases. In: Mathews, M.B., Sonenberg, N., Hershey, J.W.B. (eds.), Translational Control in Biology and Medicine. Cold Spring Harbor, New York.

Huynh, H.T., G. Robitaille and J.D. Turner, 1991. Establishment of bovine mammary epithelial cells (MAC-T): An *in vitro* model for bovine lactation. Experimental Cell Research 197, 191-199.

Kimball, S.R. and L.S. Jefferson, 2002. Control of protein synthesis by amino acid availability. Current Opinion in Clinical Nutrition & Metabolic Care 5, 63-67.

Zaborske, J.M., J. Narasimhan, L. Jiang, S.A. Wek, K.A. Dittmar, F. Freimoser, T. Pan and R.C. Wek, 2009. Genome-wide analysis of tRNA charging and activation of the eIF2 kinase Gcn2p. J. Biol. Chem. 284, 25254-25267.

Developmental changes in insulin- and amino acid-induced mTOR signalling regulate muscle protein synthesis in neonatal pigs

T.A. Davis, A. Suryawan, M.L. Fiorotto, R.A. Orellana and D.G. Burrin
USDA/ARS Children's Nutrition Research Center, Critical Care Medicine Division, Department of Pediatrics, Baylor College of Medicine, 1100 Bates Street, Houston, Texas 77030, USA

Introduction

The enhanced efficiency with which dietary protein is used for growth in the neonate is due to the ability of neonatal muscle to markedly increase protein synthesis in response to feeding (Davis *et al.*, 1996). The stimulation of protein synthesis by feeding in neonatal muscle is independently modulated by the post-prandial rise in insulin and amino acids (Davis *et al.*, 2002; O'Connor *et al.*, 2003). These responses decrease with development.

We have previously demonstrated that feeding increases the activation of the insulin and amino acid signalling pathways leading to translation initiation in skeletal muscle of neonatal pigs and these responses decrease with development (Davis *et al.*, 2000; Suryawan *et al.*, 2001; Suryawan *et al.*, 2006). Both insulin and amino acids increase protein synthesis in skeletal muscle of the neonatal pig by increasing the activation of mammalian target of rapamycin (mTOR) and its downstream signalling proteins, ribosomal protein S6K1 and eukaryotic initiation factor binding protein 1, and promoting the binding of eIF4E to eIF4G. Insulin, but not amino acids, increases the activation of protein kinase B (Suryawan *et al.*, 2007). All of these changes decrease with development (Davis and Fiorotto, 2009).

To better understand the intracellular mechanisms involved in the postprandial stimulation of protein synthesis in skeletal muscle of neonates, we examined the independent effects of amino acids and insulin on the activation of potential mTOR regulators that had been recently identified in cell culture systems.

Material and methods

Overnight-fasted 6 and 26 day old pigs were studied during: (1) euinsulinemic-euglycemic-euaminoacidemic conditions (controls), (2) euinsulinemic-euglycemic-hyperaminoacidemic clamps (AA), and (3) hyperinsulinemic-euglycemic-euaminoacidemic clamps (INS) (n=4-6/treatment/age). Fractional rates of protein synthesis and the abundance and activation of purported regulatory proteins of mTOR were determined in longissimus dorsi muscle. Data were analyzed by two-way ANOVA.

Results and discussion

Raising either insulin or amino acids from the fasting to the fed level increased the fractional rates of protein synthesis in skeletal muscle ($P<0.05$) and the responses were higher in 6 day than 26 day old pigs ($P<0.05$). Both insulin and amino acids independently increased ($P<0.05$) the phosphorylation, and thus, activation of mTOR, and these responses decreased with development ($P<0.05$). Insulin, but not AA, enhanced TSC2 ($P<0.05$) and PRAS40 phosphorylation $P<0.05$), and thus reduced the activation of these inhibitors. These effects decreased with development ($P<0.05$). Neither INS nor AA altered the composition of the mTOR complex 1 (raptor, mTOR, and GβL) or mTOR complex 2 (rictor, mTOR, and GβL) and there were no changes with development. PLD1 abundance, phosphorylation, and association with Rheb, which have been postulated to regulate mTOR activation, were higher in 6 than in 26 day old pigs ($P<0.05$) but were unaffected by insulin or amino acids. RagB abundance and the association of RagB with Raptor, which has been postulated to mediate amino acid-induced mTOR activation, were higher in 6 day than in 26 day old pigs ($P<0.05$), but were unaffected by insulin or amino acids. The abundance of another purported activator, Vps34, and inhibitor, FKBP38, of mTOR activation was unaffected by insulin, amino acids, or age. The activation of the mTOR inhibitor, AMP-activated protein kinase, and the regulator of elongation, eukaryotic elongation factor 2, were also unaffected by insulin, amino acids or age.

The results suggest that TSC2, PRAS40, Rheb, PLD1 and Rag proteins modulate mTORC1 activation in skeletal muscle of neonatal pigs. The activation of these signalling components leading to translation initiation is developmentally regulated and likely contributes to the high rate of protein synthesis and rapid gain in skeletal muscle mass in neonates.

Acknowledgements

Supported by NIH grants AR-44474 and USDA/ARS 6250510000-33.

References

Davis, T.A., D.G. Burrin, M.L. Fiorotto and H.V. Nguyen, 1996. Protein synthesis in skeletal muscle and jejunum is more responsive to feeding in 7- than 26-day-old pigs. Am. J. Physiol. 270, E802-E809.

Davis, T.A. and M.F. Fiorotto, 2009. Regulation of muscle growth in neonates. Curr. Opin. Clin. Nutr. 12,78-85.

Davis, T.A., M.L. Fiorotto, D.G. Burrin, P.J. Reeds, H.V. Nguyen, P.R. Beckett, R.C. Vann and P.M. O'Connor, 2002. Stimulation of protein synthesis by both insulin and amino acids is unique to skeletal muscle in neonatal pigs. Am. J. Physiol. 282, E880-E890.

Davis, T.A., H.V. Nguyen, A. Suryawan, J.A. Bush, L.S. Jefferson and S.R Kimball, 2000. Developmental changes in the feeding-induced stimulation of translation initiation in muscle of neonatal pigs. Am. J. Physiol. 279, E1226-E1234.

O'Connor, P.M., S.R. Kimball, A. Suryawan, J.A. Bush, H.V. Nguyen, LS. Jefferson and T.A. Davis, 2003. Regulation of translation initiation by insulin and amino acids in skeletal muscle of neonatal pigs. Am. J. Physiol. 285, E40-E53.

Suryawan, A., J. Escobar, J..W Frank, H.V. Nguyen and T.A. Davis, 2006. Developmental regulation of the activation of signalling components leading to translation initiation in skeletal muscle of neonatal pigs. Am. J. Physiol. 291, E849-59.

Suryawan, A., H.V. Nguyen, J.A. Bush and T.A. Davis, 2001. Developmental changes in the feeding-induced activation of the insulin-signalling pathway in neonatal pigs. Am. J. Physiol. 281, E908-E915.

Suryawan, A., R.A. Orellana, H.V. Nguyen, A.S. Jeyapalan, J.R. Fleming and T.A. Davis, 2007. Activation by insulin and amino acids of signalling components leading to translation initiation in skeletal muscle of neonatal pigs. Am. J. Physiol. Endocrinol. Metab 293, E1597-E1605.

Nano-nutrition in experiments with chicken embryos

M.K. Zielińska[1], E. Sawosz[1], A. Chwalibog[2], M. Grodzik[1], T. Ostaszewska[3] and M. Kamaszewski[3]
[1]Department of Animal Nutrition and Feed Sciences, Warsaw University of Life Sciences, Poland; Zielinska.marlena@gmail.com
[2]Department of Basic Animal and Veterinary Sciences, University of Copenhagen, Denmark
[3]Division of Ichthyobiology and Fisheries, Warsaw University of Life Sciences, Poland

Introduction

Intensity of poultry selection increased percentage of lean meat, especially mass of breast muscles, but at the same time decreased quality of meat, via enhance hypertrophy of myocytes. To prevent this mechanism, scientists try to find methods stimulating increase number of muscles cells, to maintain proper balance between size and number of myocytes. Nutrition '*in ovo*' can provide embryos with nutrients being not adequate for optimal growth and development of embryos and for development after hatching (Foye *et al.*, 2006). Taurine can be a donor of sulphur-compounds, being necessary to forming sulphate glycoaminoglycanes, which as a membrane receptors participate in communication between cells (Page, 2000) and support cellular nutrition via low density lipoprotein binding (Lewis and Xu, 2008). Nanoparticles of gold show affinity to –SH groups and are used as a carrier of sulphur-containing pharmaceutics (Han, 2007). The objective of presented experiments was to reveal influence of taurine, nanoparticles of gold or nanoparticles of gold conjugated with taurine on molecular and morphological status of the proliferation of muscle's cells activity at the end of embryogenesis.

Material and methods

Fertilized eggs (n=120, 66.9±3.6 g) from Ross Line 308 hens, obtained from hatchery Dembowka, Poland, were divided into 4 groups: I (control), group II (hydrocolloid of Au nanoparticles), group III (taurine), group IV (hydrocolloids of Au nanoparticles with taurine). Experimental solutions were given *in ovo*, at day 3 of incubation, by injection 0.5 ml of experimental solution to air sack using 1ml tuberculin syringe in sterile conditions. After 20 days of incubation eggs were taken from the incubator, opened and embryos were sacrificed by decapitation and immediately samples of muscles were pull in RNAlater Tissue Collection: RNA Stabilization Solution (Ambion Inc.) and stored at 8 °C for PCR Real Time analyses or pull in Bouin's solution for immunohistochemical analyses. The data regarding expression of Proliferating Cell Nuclear Antigen (PCNA) were analysed according to the factorial design, using analysis of variance ANOVA by Statgraphics Plus 4.1. Differences with $P<0.05$ were considered significant.

Results and discussion

The results of the experiment are presented in Table 1. Taurine decreased ($P<0.01$) PCNA at mRNA level in pectoral muscle, increased ($P<0.01$) % PCNA positive cells in total cells, and the average number of nuclei/mm^3 within the muscle fibre fasciculus in cross sectional image, whereas gold had no influence on the measured parameters.

Assuming that an increase of PCNA positive cells in the muscles reflects state when more cells are involved in non-dividing structure of myotubes, this may suggest improved balance between hypertrophy to hyperplasia in chicken after hatching.

In the present experiment reduced expression of PCNA mRNA in pectoral muscle of chicken embryo at 20 day of incubation was observed with taurine. We can suspect that down regulation mechanism could be a reason of this phenomenon. Higher concentration of protein can decline gene expression,

Table 1. Effect of taurine and gold on PCNA at mRNA level gene expression (PCR Real Time) in pectoral muscle, % PCNA positive cells in total cells (immunochemistry) and the average number of nuclei/mm³ within the muscle fibre fasciculus in cross sectional image (immunochemistry).

Without taurine		Taurine			ANOVA Effect of		
Au-	Au+	Au-	Au+	SEM	Au *P*	Taurine *P*	Au × Taurine *P*
mRNA expression, normalized ratio to reference gene β-actin							
3.27	3.60	0.99	1.01	0.72	0.81	0.003	0.84
protein expression, %PCNA positive cells/total cells							
0.402	0.478	0.576	0.555	0.374	0.47	0.003	0.22
average number of nuclei within the muscle fibre fasciculus in cross sectional image							
21,781	20,589	27,096	26,888	1,526	0.66	0.001	0.75

being a mechanism which protects cells against over-activity of proliferation process (Kelman, 1997). However, further studies are required to explain present results.

Conclusion

Taurine, supplemented *in ovo* to chicken embryo, enhanced organization of breast muscle by increasing number of muscles cells, by activating molecular mechanisms evaluated by expression of PCNA. Nanoparticles of gold did not influence examined parameters.

Acknowledgements

A part of presented results is included in doctoral thesis by Marlena Zielinska. This study was partially supported by Grand MNiSW N N311 303836 from Polish Ministry of Science.

References

Foye, O.T., Z. Uni, P.R. Ferket, 2006. Effect of in ovo feeding egg white protein, beta-hydroxy-beta-methylbutyrate, and carbohydrates on glycogen status and neonatal growth of turkeys. Poult. Sci. 85, 1185-1192.

Han, G., P. Ghosh, M, De and V.M, Rotello, 2007. Drug and gene delivery using gold nanoparticles. NanoBioTechnology 3, 40-45.

Kelman, Z, 1997. PCNA: structure, functions and interactions. Oncogene 14, 629-640.

Lewis, E.J. and X. Xu, 2008. Abnormal glomerular permeability characteristics in diabetic nephropathy. Diabetes Care 31, 202-207.

Page, C, 2000. The role of proteoglycans in the regulation of airways inflammation and airways remodelling. J. Allergy Clin. Immunol. 105, S518-521.

Prolonged stimulation of protein synthesis by leucine is dependent on amino acid availability

T.A. Davis, A. Suryawan, F.A. Wilson, M.L. Fiorotto, M.C. Gazzaneo, R.A. Orellana and D.G. Burrin
USDA/ARS Children's Nutrition Research Center, Critical Care Medicine, Department of Pediatrics, Baylor College of Medicine, 1100 Bates Street, Houston, Texas 77030, USA

Introduction

Leucine is unique among the amino acids in its ability to enhance protein synthesis by activating translation initiation (Kimball and Jefferson, 2005). Our laboratory has shown that raising leucine to postprandial levels whilst keeping all other amino acids at the postabsorptive level acutely stimulates protein synthesis in skeletal muscle of neonatal pigs (Escobar *et al.*, 2005). However, this response cannot be maintained unless the resulting leucine-induced fall in amino acids is prevented (Escobar *et al.*, 2007). The leucine-induced stimulation of protein synthesis is mediated by activation of the mammalian target of rapamycin complex (mTORC1) pathway but the mechanism by which amino acids activate mTORC1 is not understood. To assess the potential value of leucine supplementation to optimize the nutritional management of neonates, we examined the effect of prolonged leucine infusion on protein synthesis in skeletal muscles of different fiber types and in visceral tissues of the neonatal pig and the role of other amino acids in the response. We further examined the role of leucine in the activation of signalling components of translation.

Material and methods

Overnight fasted neonatal pigs (5 d old; n=18) were infused for 24 h with saline, leucine (400 μmol/kg/h), or leucine with replacement amino acids (Leu+AA) to prevent leucine-induced hypoaminoacidemia. Fractional rates of protein synthesis in longissimus dorsi, gastrocnemius, and masseter muscles, heart (right and left ventricular walls), liver, pancreas, kidney, and jejunum were determined. Activation of signalling components that have been implicated in the regulation of protein synthesis by amino acids were examined in detail in the longissimus dorsi muscle and the downstream effectors of mTORC1 were examined in other tissues. Data were analyzed by GLM and Tukey multiple comparisons procedures.

Results and discussion

Prolonged leucine administration stimulated protein synthesis in the longissimus dorsi, a fast-twitch glycolytic muscle, only when hypoaminoacidemia was prevented by replacement amino acid infusion (Table 1). Leucine had no effect on the phosphorylation of PKB, AMPK, TSC2, or eEF2, the association of mTOR with raptor, GßL, or rictor; or the phosphorylation of raptor or PRAS40 in muscle. Leucine supplementation increased phosphorylation of mTOR, 4EBP1 and S6K1, and eIF4E•eIF4G association whereas eIF2α phosphorylation was reduced and these responses were not further altered by correcting for the leucine-induced hypoaminoacidemia. In muscles of mixed and oxidative fiber types, the liver and pancreas, leucine also increased protein synthesis but only in the presence of replacement amino acids (Table 2). Leucine increased mTOR effectors in these tissues, in the presence of both hypoaminoacidemia and euaminoacidemia. Leucine had no effect on protein synthesis in the heart, kidney or small intestine.

Thus, although the prolonged infusion of leucine activates mTOR and its downstream targets in skeletal muscle and some visceral tissues of neonatal pigs, higher protein synthesis rates are dependent upon amino acid availability.

Table 1. Protein synthesis and nutrient signalling in longissimus dorsi *muscle of neonatal pigs infused for 24 h with saline or leucine, with and without replacement amino acids.*

	Saline	Leucine	Leu + AA
Protein synthesis (%/d)	12.43±0.80[b]	13.40±0.75[b]	16.51±0.49[a]
PKB phospho Ser[473] (AU)	1.58±0.42	1.57±0.42	1.68±0.42
AMPK phospho Thr[172] (AU)	3.38±1.16	3.56±0.70	3.83±0.82
TSC2 phospho Thr[1462] (AU)	3.73±0.86	4.09±0.38	3.99±0.68
mTOR phospho Ser[2448] (AU)	0.20±0.05[b]	0.37±0.07[ab]	0.40±0.04[a]
PRAS40 phospho Thr[246] (AU)	1.50±0.14	1.52±0.33	1.63±0.25
Raptor phospho Ser[792] (AU)	1.30±0.22	1.10±0.18	1.03±0.14
S6K1 phospho Thr[398] (AU)	0.06±0.02[b]	0.27±0.08[a]	0.29±0.09[a]
4E-BP1 phospho Thr[70] (AU)	0.63±0.15[b]	1.27±0.25[a]	1.47±0.32[a]
eIF4G phospho Ser[1180] (AU)	0.43±0.15[b]	0.95±0.20[a]	1.26±0.36[a]
eIF2α phospho Ser[51] (AU)	4.40±1.26[a]	1.63±0.52[b]	1.71±0.48[b]
eEF2 phospho Thr[56] (AU)	4.47±0.31	4.34±0.32	3.98±0.48
eIF4E • eIF4G (AU)	0.54±0.18[b]	3.83±1.22[a]	4.49±1.01[a]
eIF4E • 4E-BP1 (AU)	1.80±0.20[a]	0.77±0.17[b]	0.77±0.16[b]

[a,b] Means in a row with superscripts without a common letter differ, $P<0.05$. AU = arbitrary units.

Table 2. Protein synthesis rates in muscle and visceral tissues of neonatal pigs infused for 24 h with saline or leucine with and without replacement amino acids

	Saline	Leucine	Leu + AA
Gastrocnemius (%/d)	12.62±1.85b	11.69±1.10b	17.19±1.27a
Masseter (%/d)	12.53±0.74 b	13.48±0.78 b	16.49±0.82 a
Right heart (%/d)	9.55±1.81	9.32±0.87	9.99±0.74
Left heart (%/d)	14.78±1.35	14.07±0.84	17.29±1.83
Liver (%/d)	78.01±3.20b	79.03±1.59 b	92.34±1.22 a
Pancreas (%/d)	81.68±5.29b	70.24±1.38b	114.9±7.8a
Kidney (%/d)	34.88±2.51	34.36±3.20	40.40±2.14

[a,b] Means in a row with superscripts without a common letter differ, $P<0.05$.

Acknowledgements

Supported by Ajinomoto Amino Acid Research Program, NIH R01 AR-44474, and USDA/ARS Cooperative Agreement no. 58-6250-6-001.

References

Escobar, J., J.W. Frank, A. Suryawan, H.V. Nguyen, S.R. Kimball, L.S. Jefferson and T.A. Davis, 2005. Physiological rise in plasma leucine stimulates muscle protein synthesis in neonatal pigs by enhancing translation initiation factor activation. Am. J. Physiol. Endocrinol. Metab. 288, E914-E921.

Escobar, J., J.S. Frank, A. Suryawan, H.V. Nguyen and T.A. Davis, 2007. Amino acid availability and age affect the leucine stimulation of protein synthesis and eIF4F formation in muscle. Am. J. Physiol. Endocrinol. Metab. 293, E1615-E621.

Kimball, S.R. and L.S. Jefferson, 2005. Role of amino acids in the translational control of protein synthesis in mammals. Sem. Cell. Dev. Biol. 16, 21-27.

Relationship between ruminal environment and chewing behavior in dairy cattle

A. Arieli[1], *P.J. Weimer*[2] *and M.B. Hall*[2]
[1]*The Hebrew University of Jerusalem, Rehovot 76100, Israel*
[2]*U. S. Dairy Forage Research Center, 1925 Linden Drive, Madison, WI 53706, USA*

Introduction

Dairy cows require an optimal amount of dietary fiber for effective gut functioning. The term effective fiber integrates physical and chemical fiber traits, that stimulate chewing, saliva- buffering in the rumen, and ruminal retention and passage. In dairy cattle effective fiber correlates with ruminal pH, and chewing activity (Mertens, 1997). Recently, dairy facilities have begun using rumination sensors to track rumen functioning and cow's health. These sensors display a negative correlation between pH and temperature, due to a close relationship between microbial acid production and fermentation heat in the rumen (AlZahal *et al.*, 2008). Though rumen temperature may vary transiently with water intake, its long-term monitoring may further aid in exploring the cow's energy budget and heat increment (Beatty *et al.*, 2008). Information on the relationship between 24-h variation in ruminal environment traits and the chewing behavior of dairy cows is incomplete. The aim of this trial was to explore the relationship between cows' feeding behavior and dynamics of ruminal pH and temperature.

Materials and methods

Four rumen-cannulated late-lactation Holstein cows were assigned to a switchback design, for three periods of 7 days each. Cows were fed with a total mixed ration (TMR) at 8 h. Diet contained 32% NDF, 16.5% CP and 6.3 MJ NE per kg DM (CON). Feed allotment was adjusted to 8% orts. In the last 48 h of each period, two cows received an extra 2 kg of fine meadow fescue grass hay (TRT), top-dressed on the TMR at 11 h. Rumen sensors (Kahne Limited, Auckland, New Zealand) were inserted into the rumen, and dynamics of pH and temperature were measured at 5-min intervals. During the same 48 h period, except during milkings, cow behavior was observed every 5 min and was noted as standing, lying, drinking, eating, or ruminating. Cows were milked at 6 and 17 h. Rectal temperature and respiration rate were measured at 7 and 18 h; air temperature and humidity were monitored every hour. Results were analyzed by ANOVA using the GLM procedure of SAS (2001) to examine the effect of treatment, cow, and period; day in period, and hours from feeding were analyzed as repeated measures.

Results and discussion

Mean ambient temperature and relative humidity during the trial were 23 °C and 55%, respectively. Mean rectal temperature and respiration rate were 38.7 °C (SD 0.5) and 45 (SD 14) breaths per min. Daily feed intake was 19.5 kg of DM in both treatments. Intake of NDF was 0.92 and 0.99% of BW in CON and TRT diets, respectively ($P<0.01$). Mean pH was higher, 6.35 and 6.25, ($P=0.05$) and rumen temperature was lower, 39.55 and 39.89 °C, ($P<0.01$) in CON and TRT cows, respectively. Feeding-behavior parameters did not differ between diets. All cows spent 11.4 (SD 13.5), 25.2 (SD 21.3), and 3.8 (SD 5.3) min/h eating, ruminating and drinking, respectively, and 34 (SD 20.4) min/h in the lying position. CON diet cows had two eating peaks (30 to 40 min/h), one immediately after feeding and the other immediately after evening milking. With TRT, cows had an extra eating peak following the hay addition. About two-thirds of the rumination time was spent in the lying position.

We hypothesized that hay supplementation would increase chewing activity, salivation and pH. Relative to CON, an increase of 0.2 unit in pH ($P=0.020$) was indeed observed during first 3 h

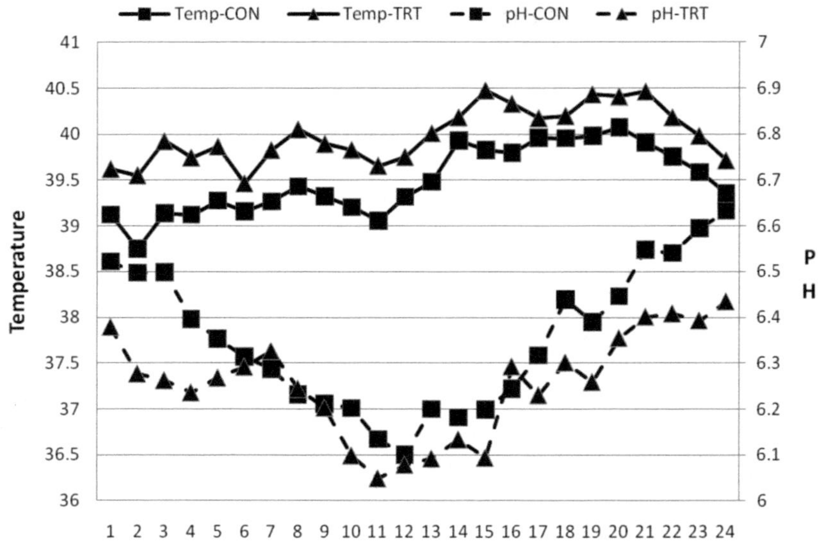

Figure 1. Ruminal temperature and pH.
CON = control diet; TRT = added 2 kg hay per day; ↑ = time of hay supplementation.

post-hay feeding (Figure 1). The lower pH for TRT may reflect a greater retention of feed particles in the rumen, and perhaps a greater extent of ruminal fermentation of the diet. In both diets, minimal pH was recorded 12 to 13 h after feeding, whereas peak ruminal temperature was observed 6 to 7 h later. A negative correlation between ruminal pH and temperature was depicted during the first 12 h after feeding. Decreased fermentation, increased acid absorption and saliva buffer flow (Allen, 1997) apparently caused the later recovery. The slow upturn of rumen temperature during the subsequent 12 h (Figure 1) might reflect reduced nutrient availability, longer rumination time, postural modification, and/or decreased drinking bouts throughout the evening and night hours.

Continuous and combined monitoring of ruminal pH and temperature appears to depict the fluctuation in fermentation. Under certain conditions, thermal balance may become a metabolic constraint for high-yielding cows (Illius and Jessop, 1996). Under those conditions designing low thermogenic diets might aid in energy intake and productivity. Further trials are needed to determine if this tool can also be used to assess of fermentative heat production.

References

Allen, M.S., 1997. Relationship between fermentation acid production in the rumen and the requirement for physically effective fiber. J. Dairy Sci. 80, 1447-1462.

AlZahal, O., E. Kebreab, J. France, M. Froetschel and B.W. McBride, 2008. Ruminal temperature may aid in the detection of subacute ruminal acidosis. J. Dairy Sci. 91, 202-207.

Beatty, D.T., A. Barnes, E. Taylor and S.K. Maloney. 2008. Do changes in feed intake or ambient temperature cause changes in cattle rumen temperature relative to core temperature? J. Therm. Biol. 33, 12-19.

Illius, A.W. and Jessop N.S. 1996. Metabolic constraints on voluntary intake in ruminants. J. Anim. Sci. 74, 3052-3062.

Mertens, D. R. 1997. Creating a system for meeting the fiber requirements of dairy cows. J. Dairy Sci. 80, 1463-1481.

SAS, Release 8.01, SAS Inst. Inc., Cary, NC, USA.

Effects of energy level and change of feed intake during late pregnancy on responses to glucose in dairy cows

T. Kokkonen, S. Salin, P. Pursiainen, J. Taponen, K. Elo and A. Vanhatalo
University of Helsinki, FIN-00014 University of Helsinki, Finland

Introduction

Voluntary dry matter intake (DMI) of dairy cows decreases during the final 2 to 3 weeks before parturition (Ingvartsen and Andersen, 2000). In order to reduce fatty acid mobilisation from adipose tissue and the consequent hepatic lipid accumulation, there has been considerable interest to maximize *pre partum* feed intake. However, there is evidence that maximizing intake with low NDF diets has not efficiently prevented metabolic disorders (Grummer *et al.,* 2004). Furthermore, the decrease of feed intake during the last week of pregnancy is more pronounced with high-energy diets than with low-energy diets. On the other hand, several studies have shown that feed-intake dip near calving can be avoided by feeding the cows at the level of requirements (Agenäs *et al.,* 2003; Kokkonen *et al.,* 2005). High-energy diet fed to *pre partum* cows may also increase insulin resistance of peripheral tissues during transition period (Holtenius *et al.,* 2003), thus promoting tissue mobilisation. The current study was carried out to test the hypothesis that high energy allowance during dry period, in combination with a decrease of feed intake near calving, induces whole-body insulin resistance in dairy cows, and this effect remains during early lactation.

Material and methods

Sixteen multiparous cows were used in a randomized complete block design. During the dry period, the cows were divided into two treatments: restricted feeding according to the energy requirements of pregnant cow (RESTR) or high energy feeding (HIGH) combined with decreasing feed allowance during the last three weeks of pregnancy. The diet consisted solely of grass silage during early dry period (weeks 6 to 4 before calving), and average energy allowances were 95 MJ/d in RESTR and 144 MJ/d in HIGH group. During close-up period the diet comprised of grass silage and concentrate (30% of dietary energy). Average energy allowances during close-up period were 107 MJ/d in RESTR and 135 MJ/d in HIGH group. After parturition, cows in both groups were offered grass silage *ad libitum*, and a similar concentrate ration. Average CP and ME contents of the diet during the second lactation week were 174 g/kg DM and 11.6 MJ/kg DM.

Intravenous glucose tolerance test (IVGTT) was performed at 7±1 d before the expected calving day and 10±1 d after parturition, by administering 0.25 g/kg body weight of glucose. Blood samples were collected at -15, -5, 5, 10, 15, 20, 30, 40, 50, 60, 90, 120, 150 and 180 min relative to glucose infusion. Data were analysed by mixed procedure of SAS (version 9.1) with a model that included the random effect of cow and fixed effect of treatment. The fixed effect of time between IVGTT and actual calving date was included in the model that was used to analyse *pre partum* data.

Results and discussion

During the *pre partum* IVGTT, the total DMI was 25% higher in the HIGH group than in the RESTR group (11.8 vs. 9.5 kg/d, $P<0.001$). The interval between IVGTT at 7 d before the expected calving date and the actual calving averaged 10 d (range 2 to 18 d) in the HIGH group, and 9 d (range 1 to 15 d) in the RESTR group. Glucose peak concentration, clearance rate or area under the curve were not different between treatments (Table 1). Holtenius *et al.* (2003) observed that high energy allowance (178% of energy requirement) increased glucose clearance rate 3 weeks *pre partum*, which they attributed to higher hepatic gluconeogenesis, as a consequence of increased gastrointestinal uptake of glucose precursors.

Table 1. Effect of dry period feeding level on plasma glucose responses to glucose challenge at 7 d pre partum.

	Treatment		SEM	*P*-value
	RESTR	HIGH		
Basal (mmol/l)	4.5	4.4	0.16	0.39
Peak (mmol/l)	15.8	16.9	0.60	0.21
CR_{120} (%/min)	1.3	1.4	0.08	0.25
$AUC_{180'}$ (mmol/l × 180 min)	429	432	27.2	0.90

CR: Clearance rate during the first 120 min; AUC: Area under the curve during 180 min.

At the time of *post partum* IVGTT, the total DMI were 18.9 and 18.4 kg/d (*P*=*0*.50) in HIGH and RESTR, respectively. Average milk yields were 37.4 and 37.3 kg/d (*P*=*0.94*). Glucose peak concentration, clearance rate or area under the curve did not differ between treatments (Table 2). In contrast, Holtenius *et al.* (2003) reported lower glucose clearance rate 3 wk *post partum* in the cows fed high energy diets during dry period.

In conclusion, high energy allowance during dry period, combined with decreased feed intake during close-up period, did not affect glucose clearance in dairy cows.

Table 2. Effect of dry period feeding level on plasma glucose responses to glucose challenge at 10 d post partum.

	Treatment		SEM	*P*-value
	RESTR	HIGH		
Basal (mmol/l)	3.2	3.1	0.23	0.85
Peak (mmol/l)	14.7	15.9	0.61	0.13
CR_{120} (%/min)	2.3	2.3	0.13	0.49
$AUC_{180'}$ (mmol/l × 180 min)	299	310	18.0	0.69

CR: Clearance rate during the first 120 min; AUC: Area under the curve during 180 min.

References

Agenäs, S., E. Burstedt and K. Holtenius, 2003. Effects of feeding intensity during the dry period. 1. Feed intake, body weight, and milk production. J. Dairy Sci. 86, 870-882.

Grummer, R.R., D.G. Mashek and A. Hayirli, 2004. Dry matter intake and energy balance in the transition period. Vet. Clin. North Am. Food Anim. Pract. 20, 447-470.

Holtenius K., S. Agenäs, C. Delavaud and Y. Chilliard, 2003. Effects of feeding intensity during the dry period. 2. Metabolic and hormonal responses. J. Dairy Sci. 86, 883-891.

Ingvartsen, K.L. and J.B. Andersen, 2000. Integration of metabolism and intake regulation: A review focusing on periparturient animals. J. Dairy Sci. 83, 1573-1597.

Kokkonen, T., J. Taponen, T. Anttila, L. Syrjälä-Qvist, C. Delavaud, Y. Chilliard, M. Tuori, A.T. Tesfa, 2005. Effect of body fatness and glucogenic supplement on lipid and protein mobilization and plasma leptin in dairy cows. J. Dairy Sci. 88, 1127-1141.

Mechanisms regulating the peripheral utilisation of glucose: involvement of AMPK

H. Mameri[1], J. Dupont[2,3,4,5], R. Joubert[1], A. Collin[1], S. Crochet[1], E. Cailleau-Audouin[1], S. Tesseraud[1] and S. Métayer-Coustard[1]
[1]*INRA, UR83 Recherches Avicoles, 37380 Nouzilly, France*
[2]*INRA, UMR85 Physiologie de la Reproduction et des Comportements, 37380 Nouzilly, France*
[3]*CNRS, UMR6175, 37380 Nouzilly, France*
[4]*Université François Rabelais de Tours, 37000 Tours, France*
[5]*Haras Nationaux, 37380 Nouzilly, France*

Introduction

Understanding the mechanisms regulating growth and meat quality requires studying glucose utilization by muscle. Several studies highlighted the major role of adenosine monophosphate (AMP)-activated protein kinase (AMPK) in regulating the level of muscle glycogen (Carling and Hardie, 1989; Jorgensen *et al.*, 2004). Indeed, AMPK plays a key role in controlling whole body energy metabolism, including the regulation of plasma glucose levels, fatty acid oxidation and glycogen metabolism (Carling, 2004). However, its regulation by nutritional factors is still poorly known. To better understand the underlying mechanisms, we used a model of sequential feeding, which consists of giving several diets of different nutritional values, for several cycles. Recently, the use of a 48 h cycle sequential feeding varying in protein, energy contents, or both led to similar growth performance and carcass composition as the complete diet (Bouvarel *et al.*, 2008). In this model, the recorded ultimate pH (pHu) of breast muscle was lower with diets varying in protein contents than other treatments. This decrease in breast pHu suggests higher muscle glycogen content at death (Berri *et al.* 2005) and, thus, higher energy storage in the muscles. The present study aimed (1) to evaluate the main effects of energy and protein contents on AMPK activation and thus analyse potential consequences of sequential feeding on glucose metabolism, (2) to appreciate the potential adaptation of animals to sequential feeding and (3) to precise *in vitro* mechanisms involved in short-term regulation of the peripheral utilization of glucose.

Material and methods

One-d-old male broiler chicks (Ross PM3) were raised from 1 day old until 3 weeks of age in an environmentally-controlled poultry house. All the chickens received the same starter diet (2,900 kcal/kg of ME, 21% CP). At 10 d of age, chickens were given control or sequential treatments. Seven diets were used during the sequential feeding period. The control treatment was complete feed (3,000 kcal/kg of ME, 19.5% CP). Two diets were isoenergetic (3,000 kcal/kg of ME) with differences in protein and essential amino acid contents: the high-protein feed (P^+ = 23.4% CP) and the low-protein feed (P^- = 15.6% CP). Two diets were isoproteic (19.5% CP) with differences in ME content: the low-energy feed (E^- = 2,800 kcal/kg of ME) and the high-energy feed (E^+ = 3,200 kcal/kg of ME). Two diets had differences in both protein and energy contents: the high-protein and low-energy feed (P^+E^-) and the low protein and high energy feed (P^-E^+). Sequential feeding was carried out during 48-h cycles between 10 to 14-15 d of age. Three different sequential feeding alternations were studied and compared with the control group: (1) alternation of diets varying in CP (P^+ followed by P^-), (2) in energy (E^- followed by E^+) or (3) in protein and energy contents (P^+E^- followed by P^-E^+). At 14-15 d of age, blood was withdrawn and animals (n=6 per diet; 7 diets) were sacrificed to collect the *Pectoralis major* (PM) muscle 5 hours after the change of food to study effect of different diets within each sequence (P^- versus P^+; E^- versus E^+, E^-P^+ versus E^+P^-) on glucose metabolism. Plasma samples were stored at -20 °C until the determination of glucose, insulin, amino acids levels and muscle samples at -80 °C for further western blot analysis and glycogen content measurement. Statistical analysis was performed by analysis of variance (ANOVA) using the Statview Software program, version 5.0 (SAS Institute, 1992-1998, Cary, NC) and followed by a Fisher test to detect significant differences between treatments. Values are expressed as means ± SEM.

QT6, stable and permanent cell line from methylcholanthrene-induced pectoralis fibrosarcomas of Japanese quail *Coturnix coturnix japonica* developed by Antin and Ordahl, was used in this study. QT6 cells were grown in McCoy medium supplemented with 10% foetal calf serum and 1% chicken serum to 80-90% confluence, fasted 16 h in serum-free medium, washed once with PBS and incubated for 2 h in different media varying in glucose (1 to 3 g/l of glucose) and/or amino acid concentrations (0.5× to 2×). Cell samples were stored at -80 °C until further analysis.

Results and discussion

The model of sequential feeding used in this study with varying daily protein and energy supply showed a short-term regulation of intracellular signalling pathways involved in carbohydrate metabolism, in particular that of AMPK (P<0.005). For example, activation of this kinase was higher with the E^-P^+ compared to E^+P^- treatment (1.50±0.24 and 0.52±0.11 in arbitrary unit pAMPK/Tubuline, respectively, P<0.001). This activation state of AMPK was associated with modifications in glycogen store (P<0.005). Indeed, E^-P^+ treatment showed a reduction (P<0.0001) in glycogen store compared to glycogen content measured in E^+P^- treatment (79.1±3.4 and 99.9±4.3 µmol/g eq lactate, respectively). However, this model did not allow us to determine if the modifications observed (cell signalling, glycogen metabolism) are mediated directly by amino acids themselves or by hormones.

We completed our study with an *in vitro* approach. Incubating QT6 fibroblasts in media varying in amino acids and/or glucose concentrations, we showed for example a higher activation of AMPK with medium mimicking 'E^-P^+ treatment' (glucose 1 g/l and amino acid 2×) compared to 'E^+P^-' (glucose 3 g/l and amino acid 0.5×) or 'control' (glucose 2 g/l and amino acid 1×) treatments. More generally, we demonstrated that a variable supply of amino acids and/or D-glucose can directly modify the activation of AMPK, independently of any hormonal regulation.

In conclusion, our preliminary results suggest that nutrients, in particular amino acids, seem to be involved in the regulation of the AMPK pathway. Optimizing their supply could be a tool for controlling muscle glycogen store.

References

Berri, C., M. Debut, V. Santé-Lhoutellier, C. Arnould, B. Boutten, N. Sellier, E. Baéza, N. Jehl, Y. Jégo, M.J. Duclos and E. Le Bihan-Duval, 2005. Variations in chicken breast meat quality: implications of struggle and muscle glycogen content at death. Br Poult Sci. 46, 572-579.

Bouvarel, I., A.M. Chagneau, P. Lescoat, S. Tesseraud and C. Leterrier, 2008. Forty-eight-hour cycle sequential feeding with diets varying in protein and energy contents: adaptation in broilers at different ages. Poult Sci. 87, 196-203.

Carling, D., 2004. The AMP-activated protein kinase cascade-a unifying system for energy control. Trends Biochem Sci. 29, 18-24.

Carling, D. and D.G. Hardie, 1989. The substrate and sequence specificity of the AMP-activated protein kinase. Phosphorylation of glycogen synthase and phosphorylase kinase. Biochim Biophys Acta. 1012, 81-86.

Jørgensen, S.B., J.N. Nielsen, J.B. Birk, G.S. Olsen, B. Viollet, F. Andreelli, P. Schjerling, S. Vaulont, D.G. Hardie, B.F. Hansen, E.A. Richter and J.F.Wojtaszewski, 2004. The alpha2-5'AMP-activated protein kinase is a site 2 glycogen synthase kinase in skeletal muscle and is responsive to glucose loading. Diabetes 53, 74-81.

The role of feeding regimens in regulating metabolism of sexually mature broiler breeders

R.D. Ekmay[1], M. de Beer[2], R.W. Rosebrough[3], M.P. Richards[3], J.P. McMurtry[3] and C.N. Coon[1]
[1]University of Arkansas, Center of Excellence for Poultry Science, Fayetteville, Arkansas 72701, USA
[2]Aviagen, 5015 Bradford Dr., Huntsville, Alabama 35805, USA
[3]Animal Bioscience and Biotechnology Laboratory, USDA-ARS, 10300 Baltimore Ave., Beltsville, Maryland 20705, USA

Introduction

The metabolic consequences of the different restriction feeding regimens have only recently been elucidated (De Beer *et al.*, 2008). A large amount of evidence has arisen for the programming of body composition and metabolism by early nutrition. This study attempts to determine whether the metabolic patterns established during the growing phase of broiler breeder hens remain at the onset of sexual maturity.

Material and methods

A flock of Cobb 500 pullets were divided into two groups at 4 wks and randomly placed on either an everyday (ED) or skip-a-day (SKP) feeding regimen. At 24 wk, all hens were switched over to an ED regimen. At 26.4 wk, hens were randomly selected and euthanized at intervals after feeding. Livers were sampled from 4 hens at 4-h intervals for 24 h for a total of 28 samples per treatment. Blood was sampled from 4 hens per sampling time; sampling times were 0, 30, and 60 min and 2 and 4 h after feeding and then every 4 h up to 24 h for a total of 36 samples per treatment. Liver samples were analyzed for mRNA expression of acetyl CoA carboxylase (ACC), aspartate aminotransferase (AAT), isocitrate dehydrogenase (ICDH), malic enzyme (ME) and β-actin. Plasma concentrations of insulin, glucagon, IGF-I and II, T_3, T_4, corticosterone, and leptin were determined. At 28, 40, and 65 wk, 15 hens per regimen were euthanized, eggs and abdominal fat pad collected and the fatty acid profile determined.

Results and discussion

Time and feeding interaction effects were found (P=0.013 and 0.0005, respectively) for ACC and ME mRNA expression. The pattern of ACC expression in SKP reared hens was more consistent and was similar to results seen at 16 wk (De Beer *et al.*, 2007). ED reared hens showed large surges in ACC expression different from those in pullets. Malic enzyme was found to be elevated in 16 wk old pullets on a SKP feeding regimen (De Beer *et al.*, 2008). Significant interaction effect and a higher peak in SKP-reared hens suggest that the previous findings may still hold true in mature breeder hens. Overall plasma levels of IGF-II were higher in SKP reared birds (P=0.0005). Sixteen wk-old pullets showed no difference in IGF-II levels, however, the ED pullets showed a significantly higher IGF-I level. Significant interaction effects were seen (P=0.006) in corticosterone levels. Corticosterone was shown to be significantly elevated in pullets on a SKP regimen (De Beer *et al.*, 2008). Despite a normalized feeding schedule, SKP reared breeder hens were still showing increased corticosterone with a peak over threefold greater than ED reared breeders. Total monounsaturated fatty acids in abdominal fat were reduced in hens reared on an ED regimen. These same hens had higher levels of 15, 17, 18:3n6, and 20:3n6 in abdominal fat. Higher levels of polyunsaturated fatty acids were also found in eggs produced by hens reared on an ED regimen: especially 18:3n3, 20:4n6, and 20:2. In conclusion, SKP-reared hens show characteristics of increased lipogenesis: a pattern that was established during the pullet stage. These same SKP-reared hens also had higher levels of mono-unsaturated fatty acids in the abdominal fat pad and eggs. The carry-over of certain traits may be due to early development in SKP birds.

Table 1. Mean levels of plasma hormones at 16 wk and 26 wk in broiler breeders reared on either an everyday (ED) or skip a day (SKP) regimen.

	26 wk			16 wk		
Hormones	ED	SKP	*P* value	ED	SKP	*P*-value
Leptin (ng/ml)	4.49	4.75	0.05	10.07	8.23	<0.01
Glucagon (pg/ml)	278.18	319.78	0.08	253.84	279.57	0.59
IGF-2 (ng/ml)	63.71	70.56	<0.01	117.58	111.00	0.27
Corticosterone (ng/ml)	2.73	4.38	<0.01	1.20	2.41	<0.01

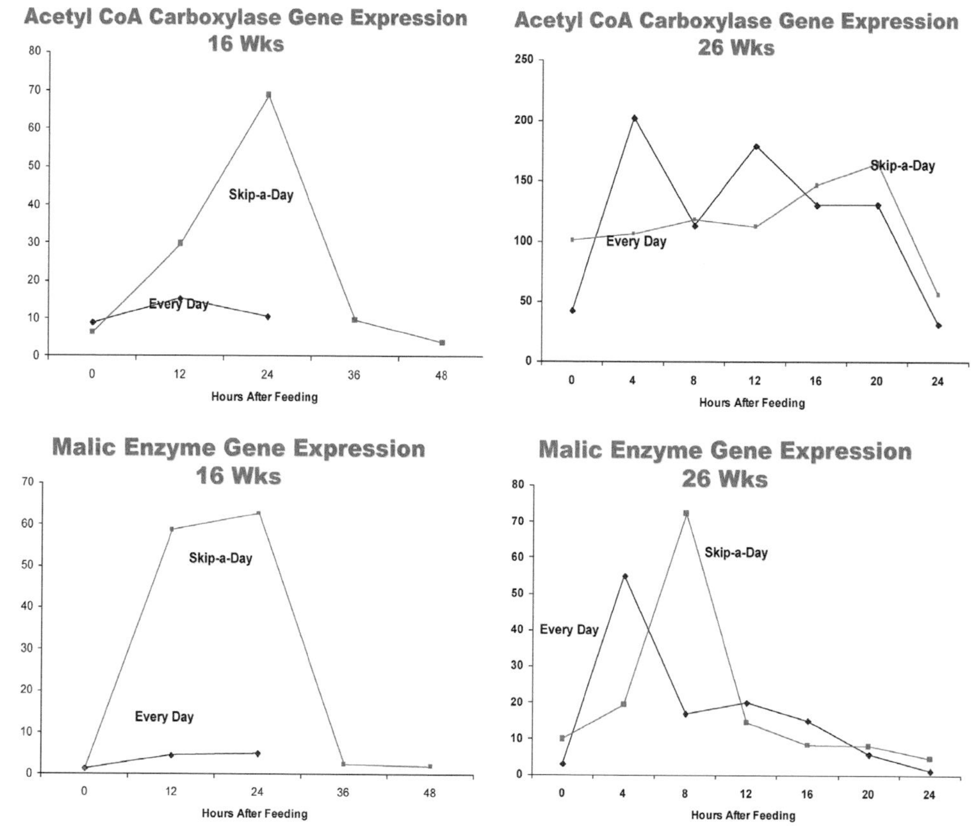

Figure 1. Gene expression (relative to β-actin) of ACC and ME in 16 wk and 26 wk broiler breeders reared on either ED or SKP feeding regimen.

References

De Beer, M., R.W. Rosebrough, B.A. Russell, S.M. Poch, M.P. Richards, and C.N. Coon, 2007. An examination of the role of feeding regimens in reglating metabolism during the broiler breeder grower period. 1. Hepatic lipid metabolism. Poult. Sci. 86, 1726-1738.

De Beer, M., J.P.McMurtry, D.M.Brocht, and C.N. Coon, 2008. An examination of the role of feeding regimens in regulating metabolism during the broiler breeder grower period. 2. Plasma hormones and metabolites. Poult. Sci. 87, 264-275.

The effects of leptin on adipocyte and liver metabolism in leptin-responsive and leptin-resistant rats

W.P. Flatt, D.P. Brewer and R.B.S. Harris
Department of Foods and Nutrition, University of Georgia, Athens, GA 30602-3632, USA

Introduction

Both central and peripheral leptin reduce body fat while sparing lean tissue (Harris *et al.*, 1998). Central injections of leptin (Gullicksen *et al.*, 2002) are associated with increased fatty acid oxidation, lipolysis, and energy expenditure. In contrast, continuous peripheral infusions of physiological concentrations leptin produce less dramatic changes in body fat that are not necessarily associated with an inhibition of energy intake (Bowen *et al.*, 2003) or an increase in energy expenditure (Harris *et al.*, 2007), therefore it is unclear whether peripheral and central leptin activate the same mechanisms. The following experiments measured energy expenditure, and glucose and fatty acid utilization in white adipose tissue from rats receiving chronic peripheral infusions of a physiological dose of leptin or physiological buffered saline (PBS).

Materials and methods

In Experiment l, 30 male 250 g Sprague Dawley rats were fed chow. Daily body weight and food intake were measured and the rats were adapted to calorimetry cages (Oxymax, Columbus Instrument Co., OH) described previously (Wang *et al.*, 1999). Energy expenditure and respiratory exchange ratio (RER) were measured for two baseline days and then the rats were fitted with intraperitoneal (i.p.) Alzet pumps delivering either PBS or 40 μg leptin/day (recombinant rat leptin, R&D Systems, MN). A third group of rats was infused with PBS, but fed the intake of leptin-infused rats. Energy expenditure and RER were used to test for treatment effects. On the fourth day of infusion, trunk blood was collected for measurement of serum hormones and metabolites. Glucose and fatty acid utilization were measured in isolated epididymal (EPI) adipocytes (Harris and Martin, 1986). Glucose oxidation and incorporation into fatty acids (FAS) were measured with or without 100 μU/ml insulin. Palmitate oxidation (FAO) and incorporation into fatty acids (FAE) were measured, and adipocyte glycerol release was measured as an index of lipolysis (Harris *et al.*, 1995). Data were compared by analysis of variance and post-hoc Duncan's Multiple Range test.

In Experiment 2, 18 male Wistar 230 g rats were fed either a LF (12% kcal fat: D02041902 Research Diets) or a HF diet (60% kcal fat; D12492 Research Diets). Leptin responsiveness was tested by measuring 12 hour food intake following an i.p. injection of 2.5 mg/kg leptin. Leptin inhibited 12 hour food intake of both LF and HF-fed rats at 5 and 9 weeks, but after 11 weeks leptin resistance was confirmed in HF-fed rats. Daily food intakes and body weights were then measured for 4 days before the rats were fitted with i.p. Alzet pumps that delivered either 70 μg leptin/day or an equal volume of PBS. On day 2 the rats were moved to the indirect calorimeters and data were collected from the last three days. On day 11 the rats were killed and measures of serum metabolites and EPI adipocytes and liver metabolism were made.

Results and discussion

In Experiment 1 leptin inhibited food intake of chow-fed Sprague Dawley rats (65^a, 56^b, 53^c g/4 days for PBS, leptin and pair-fed respectively. Values with different superscripts are different at $P<0.05$). At the end of the study inguinal, EPI (2.9^a, 2.6^b, 2.3^b g), retroperitoneal and mesenteric white fat depots and intrascapular brown adipose tissue (0.19^a, 0.12^b, 0.15^b g) were smaller in leptin-infused and pair-fed rats than controls. Leptin inhibited EPI cell basal FAS by 55% and prevented an insulin response. There was a non-significant 36% inhibition of FAS in pair-fed rats. FAE was 44% lower in leptin-treated than control rats, but adipocyte FAO and glycerol release were not different. Adipocyte glucose oxidation was stimulated by insulin, but there was no effect of leptin. Pair-fed rats had lower concentrations of serum insulin (1.0^{a}, 0.7^{b}, 0.4^{b} ng/ml) and higher free fatty acids (0.5^{a}, 0.4^{a}, 0.6^{b} mEq/l) than PBS or leptin-treated rats.

In Experiment 2, HF-fed rats weighed 50 g more and had higher energy expenditures, but lower RERs than LF-fed rats. There was no effect of leptin on energy expenditure or RER in either dietary group. White fat depots were twice as big in HF-fed as LF-fed rats. Leptin did not decrease fat pad size, but there was a trend for an increase in LF-fed rats. Neither diet nor leptin modified adipocyte FAS, FAE, FAO, glucose oxidation or lipolysis. Leptin increased liver FAS (166 vs.73 nmol/liver/hr; $P<0.05$) and lipid (3.6 vs. 2.6 g/liver; $P<0.05$) content in LF-fed rats, but had no effect in HF-fed rats. Serum glucose was higher in HF-fed than LF-fed rats (145 vs. 116 mg/dl; $P<0.05$), but there were no differences in serum insulin, adiponectin, triglycerides, free fatty acids or glycerol.

In leptin-responsive rats low physiological doses of leptin reduced body fat by inhibiting lipogenesis rather than stimulating lipolysis or FAO. Leptin prevented a drop in energy expenditure even when food intake was inhibited. None of these leptin responses were apparent in Experiment 2 and body fat was not reduced in leptin-treated animals. Because it took 11 weeks to induce resistance to leptin injections in HF-fed rats both groups were relatively obese and older than those in Experiment 1. HF-fed rats were fatter, had higher energy expenditures, but lower RERs than LF-fed rats. Leptin stimulated liver lipogenesis in LF-fed rats, increased liver lipid and there was a trend for leptin to increase fat pad weight. These results suggest that as Wistar rats become more obese there is an intermediate stage of leptin resistance during which leptin promotes lipid deposition.

References

Bowen, H., T.D. Mitchell and R.B. Harris, 2003 Method of leptin dosing, strain, and group housing influence leptin sensitivity in high-fat-fed weamling mice. Am J Physiol Regul Integr Comp Physiol 284, R87-100.

Gullicksen, P.S., W.P.Flatt, R.G. Dean, D.L. Hartzell and C.A. Baile, 2002. Energy metabolism and expression of uncoupling proteins 1,2, and 3 after 21 days of recovery from intracerebroventricular mouse leptin in rats. Physiol Behav 75, 473-482.

Harris, R.B., T.J. Bartness and H.J.Grill, 2007. Leptin responsiveness in chronically decerebrate rats. Endocrinology 148, 4623-4633.

Harris, R.B. and R.J.Martin, 1986. Metabolic response to a specific lipid-depleting factor in parabiotic rats. Am J Physiol 250, R276-286.

Harris, R.B., R.J. Martin and R.C. Bruch, 1995. Dissociation between food intake, diet composition and metabolism in parabiotic partners of obese rats. Am J Physiol 268, R874-883.

Harris, R.B., J. Zhou, S.M. Redmann, Jr., G.N. Smagin, S.R. Smith, E. Rodgers and J.J. Zachwieja, 1998. A leptin dose-response study in obese (ob/ob) and lean (+/?) mice. Endocrinology 139, 8-19.

Wang, T., D.L. Hartzell, B.S. Rose,W.P. Flatt, M.G. Hulsey, N.K. Menon, R.A. Makula and C.A. Baile, 1999. Metabolic responses to intracerebroventricular leptin and restricted feeding. Physiol Behav 65, 839-848.

Effects of nutrient restriction on mammary cell turnover in lactating dairy cows

F. Dessauge[1,2], V. Lollivier[1,2,3], L. Finot[1,2], S. Wiart[1,2], E. Cutullic[1,2], C. Disenhaus[1,2,3], S. Barbey[4], S. Lemosquet[1,2] and M. Boutinaud[1,2]
[1]INRA UMR 1080 Dairy Production, 35000 Rennes, France
[2]AGROCAMPUS UMR 1080 Dairy Production, 35000 Rennes, France
[3]Université européenne de Bretagne, France
[4]INRA UE 326 Domaine expérimental du Pin au Haras, 61310 Le Pin au Haras, France

Introduction

Feeding level acts on lactation performances in dairy cows: more precisely a nutrient restriction results in decreased milk yield (MY; Friggens *et al.*, 1995; Andersen *et al.*, 2003; Norgaard *et al.*, 2005). However, little is known about the effects of nutrient restriction on mammary epithelial cells and tissue remodelling. The aims of the study were to investigate the effects of nutrient restriction on mammary tissue morphology and on mammary epithelial cell turnover in lactating dairy cows. We hypothesized that nutrient restriction reduces MY by affecting mammary cell turnover (determined as the balance between cell proliferation and apoptosis).

Material and methods

We used 15 Holstein × Normande crossbred dairy cows, divided into 2 groups submitted to different feeding levels. From calving to week 11 *post partum*, the cows were fed a total mixed ration composed either of 55% maize silage, 15% dehydrated alfalfa and 30% concentrate (Basal diet-group as a control group, n=7) or of 60% grass silage and 40% hay (Restricted diet-group, n=8). Cows were milked twice daily. MY was recorded at each milking (twice a day) and milk composition was monthly measured. After 11 weeks of lactation, cows were slaughtered and mammary glands were removed and weighed. The mammary DNA concentration was measured by a fluorometric method, in order to evaluate the total amount of DNA in the mammary gland and to estimate the total number of mammary cells (Boutinaud *et al.*, 2003). Expression of proteins involved in proliferation and cell death (DNA fragmentation) were evaluated on mammary tissue by real-time qPCR, Western Blotting and immunohistochemical staining as previously reported (Benchedly *et al.*, 2009; Dessauge *et al.*, 2009). Analysis of metalloprotease activity by the gelatin-containing zymogram was performed from mammary gland extracts. The data were analysed using the SAS MIXED procedure.

Results and discussion

The cows of the Restricted diet-group had a lower 11-week average daily MY from calving to slaughter than Basal diet-group cows (20.5 vs. 33.5 kg/d, $P<0.001$, Table 1). Feed restriction decreased milk protein content ($P<0.001$) and milk lactose content ($P<0.002$), without modifying milk fat composition (Table 1). Mammary glands from Restricted diet-group cows were less heavy than those of Basal diet-group (16.7 vs. 26.9 kg, $P<0.001$). At the mammary tissue level, the total amount of DNA in the mammary gland and the size of the mammary acini were lower ($P<0.01$, -20%, and -41%, respectively) in the Restricted diet-group compared to the Basal diet-group. Mammary cell proliferation did not vary significantly with feed restriction at the time of slaughtering whereas TUNEL assay showed a higher level of apoptosis in the mammary gland of Restricted diet-group ($P<0.001$). Gelatin zymography showed a higher level of matrix metalloproteases MMP9 and MMP2 proteolytic activity in the Restricted diet-group compared to the Basal diet-group, suggesting a remodelling of the extracellular matrix in the restricted diet-group.

Table 1. Main zootechnical and physiological results.

	Basal diet	Restricted diet	SEM	Significance *P*
Milk yield (kg/d)	33.5	20.5	0.41	<0.0001
Fat content (%)	4.2	4.2	0.007	0.95
Protein content (%)	3.0	2.7	0.02	<0.0001
Lactose content (%)	5.0	4.8	0.02	<0.001
Total mammary DNA content (mg)	52	42	2.6	0.003
Proliferation (%)	22.8	23.8	2.31	0.80
Apoptosis (%)	0.14	0.3	0.06	0.02
MMPs activity (Integrated Density)	1.97	2.32	0.11	0.004
Acini size (μm^2)	22,505	10,587	3,643	0.02

In conclusion, the lower MY induced by nutrient restriction in lactating dairy cows was partly due to a lower number of mammary cells. This was a consequence of a higher level of apoptosis but also to a mammary remodelling by the extracellular matrix. The future objective will be the study of the influence of nutrient restriction on mammary cell activity.

References

Andersen, J.B., N.C. Friggens, K. Sejrsen, M.T. Sørensen, L. Munksgaard and K.L. Ingvartsen, 2003. The effects of low vs. high concentrate level in the diet on performance in cows milked two or three times daily in early lactation. Livest. Prod. Sci. 81, 119-128.

Ben Chedly, H., P. Lacasse, P-G. Marnet, S. Wiart-letort, L. Finot and M. Boutinaud, 2009. Cell junction disruption after 36h milk accumulation was associated with changes in mammary secretory activity and dynamics in lactating dairy goats. J Physiol Phamacol 60, Suppl 2, 105-111.

Boutinaud, M., Rousseau C., Keisler D. H., Djiane J. and Jammes H., 2003. Growth hormone and milking frequency act differently on goat mammary gland in late lactation. J. Dairy Sci 86, 509-520.

Dessauge, F., Finot L., Wiart S., Aubry JM., Ellis SE. (2009) Effects of ovariectomy in prepubertal goats. Journal of Physiology and Pharmacology Aug 60, Suppl 2, 127-133.

Friggens, N., G.C. Emmans, S. Robertson, D.G. Chamberlain, C.T. Whittemore, and J.D. Oldham, 1995. The lactational responses of dairy cows to amount of feed and to the source of carbohydrate energy. J Dairy Sci 78, 1734-1744.

Nørgaard, J., A. Sørensen, M.T. Sørensen, J.B. Andersen, and K. Sejrsen, 2005. Mammary cell turnover and enzyme activity in dairy cows: effects of milking frequency and diet energy density. J Dairy Sci 88, 975-982.

Expression of ghrelin, leptin and leptin receptor genes in the GIT of newborn calves fed with different types of liquid feed

P. Górka[1], M. Jaworski[1], J. Flaga[1], U. Kaczor[2], Z.M. Kowalski[1] and R. Zabielski[3]
[1]Department of Animal Nutrition and Feed Management, University of Agriculture in Krakow, Cracow, Poland
[2]Department of Sheep and Goat Breeding, University of Agriculture in Krakow, Cracow, Poland,
[3]Department of Physiological Sciences, Faculty of Veterinary Medicine, Warsaw University of Life Sciences, Warsaw, Poland

Introduction

Butyric acid was shown to be a potential stimulator of gastrointestinal tract (GIT) development in various mammal species, including calves (Kotunia *et al.*, 2004; Guilloteau *et al.*, 2009). Its dietary supplementation is especially beneficial in newborn calves fed with milk replacer (MR). A lack of bioactive components and/or a presence of antinutritional factors in MR is known to slow down the calf small intestine development (Blättler *et al.*, 2001). Supplementation of sodium butyrate (SB) in MR positively affected mitotic/apoptotic ratio in the small intestine epithelium, as well as villus high and brush border enzymes activity in calves (Górka *et al.*, 2009a; Guilloteau *et al.*, 2009). Dietary supplementation of SB showed also a stimulatory effect on rumen development in calves (Górka *et al.*, 2009b). However, controlling mechanisms of SB on calf GIT development are still unknown. Such effects may be mediated by the action of ghrelin and leptin. Thus the aim of the study was to determine the effect of liquid feed type on ghrelin and leptin mRNA expression in calf GIT.

Materials and methods

Twenty one bull calves (5 day old) were randomly allocated to three groups (7 per treatment): fed with whole milk (WM), MR, or MR supplemented with SB (0.3% as fed; MR+SB). Liquid feed DMI was equal in all treatments and amounted 1% of initial BW. Up to 25% of MR (22% CP, 17% fat) was based on soy protein concentrate. Animals were slaughtered at 26 days of age. One cm^2 of whole thickness samples from the rumen (dorsal and ventral sac), abomasum, duodenum and middle jejunum were taken for analyses. Expression of mRNA of ghrelin, leptin and long form of leptin receptor (ObRb) were evaluated using the semi-quantitative RT-PCR method. The β-actin mRNA expression was taken as a housekeeping gene. The PCR products were run in a 2.0% agarose gel and stained with ethidium bromide for visualization and the density of the gel band was determined using the Scion Image for Windows (Scion Corporation, Maryland, USA). Data were subjected to one-way analysis of variance using PROC GLM (SAS 8.01). Pre-planed orthogonal contrast (WM vs. MR and MR+SB, MR vs. MR+SB) were used. The significance was declared at $P \leq 0.05$ and tendencies at $P \leq 0.10$.

Results and discussion

Calves fed with WM had higher mRNA expressions of ghrelin in the abomasum, leptin receptor in the dorsal (P=0.02) and ventral sac of the rumen (P=0.10), and in the abomasum (P=0.03) as compared to calves fed with MR and MR+SB (Table 1). On the other hand, mRNA expression of leptin receptor in duodenum was lower in calves fed with WM as compared to calves fed with MRs (P=0.03) and calves fed with MR+SB had higher mRNA expression of leptin receptor in the duodenum as compared to calves fed with MR (P=0.10). Liquid feed type had no effect on leptin mRNA expression.

Results of the present study showed a potential effect of liquid feed type on ghrelin and ObRb mRNA expression. Local down-regulation of ghrelin and ObRb mRNA expression in calves fed with MR

Table 1. Effect of liquid feed type on ghrelin and long form of leptin receptor (ObRb) mRNA expression.

Item	Group[1]			SE[3]	Contrasts[2]	
	WM	MR	MR+SB		1	2
Ghrelin/ β-actin						
Rumen (dorsal sac)	0.59	0.49	0.87	0.11	NS	NS
Rumen (ventral sac)	0.62	0.50	0.44	0.07	NS	NS
Abomasum	6.94	1.83	2.15	0.82	0.04	NS
Duodenum	2.52	1.24	2.91	0.48	NS	NS
Jejunum	0.28	0.29	0.27	0.02	NS	NS
ObRb/ β-actin						
Rumen (dorsal sac)	2.46	1.06	1.01	0.28	0.02	NS
Rumen (ventral sac)	2.30	1.19	1.61	0.30	0.10	NS
Abomasum	2.34	0.94	0.70	0.29	0.03	NS
Duodenum	1.27	0.63	1.46	0.18	NS	0.10
Jejunum	0.48	0.67	0.61	0.09	0.03	NS

[1] WM = whole milk, MR = milk replacer, MR+SB = milk replacer with sodium butyrate (0.3% as fed); mean values given.

[2] 1 = WM vs. MR and MR+SB; 2 = MR vs. MR+SB; NS = not significant.

[3] Standard error.

suggest negative effect of MR on those genes expression and potential implication of ghrelin and leptin in calf GIT development. Effect of SB on calf GIT development seems not to be exerted by means of ghrelin and leptin gene expressions.

Acknowledgement

Grant no. 1164/P01/2006/31, Ministry of Science and Higher Education, Poland

References

Blättler, U., H.M. Hammon, C. Morel, C. Philipona, A. Rauprich, V. Romé, I. Le Huërou-Luron, P. Guilloteau and J.W. Blum, 2001. Feeding colostrum, its composition and feeding duration variably modify proliferation and morphology of the intestine and digestive enzyme activities of neonatal calves. J. Nutr. 131, 1256-1263.

Górka, P., Z.M. Kowalski, P. Pietrzak, A. Kotunia, R. Zabielski, R. Kiljańczyk and J. Flaga, 2009a. The effect of sodium butyrate supplementation in milk replacer and starter diet on rumen development in calves. J. Physiol. Pharmacol. 60 (Suppl. 3), 47-53.

Górka, P., P. Pietrzak, A. Kotunia, J. Flaga, Z.M. Kowalski and R. Zabielski, 2009b. The effect of type of liquid feed on small intestine development in newborn calves. In: Ruminant physiology Digestion, metabolism, and effects of nutrition on reproduction and welfare. Wageningen Academic Publisher, Wageningen, the Netherlands. 192-193.

Guilloteau, P., R. Zabielski, J.C. David, J.W. Blum, J.A. Morisset, M. Biernat, J. Woliński, D. Laubitz and Y. Hamon, 2009. Sodium-butyrate as a growth promoter in milk replacer formula for young calves. J. Dairy Sci. 92, 1038-1049.

Kotunia, A., J. Woliński, D. Laubitz, M. Jurkowska, V. Rome, P. Guilloteau and R. Zabielski, 2004. Effect of sodium butyrate on the small intestine development in neonatal piglets feed by artificial sow. J. Physiol. Pharmacol. 55 (Suppl. 2), 59-68.

Fasting induces an increase in AMPKα2 mRNA gene expression in the hypothalamus of broiler chicks

Z. Song[1], L. Liu[2], H. Lin[2], E. Decuypere[1] and J. Buyse[1]
[1]Laboratory of Physiology, Immunology and Genetics of Domestic Animals, Department of Biosystems, Catholic University Leuven, Kasteelpark Arenberg 30, 3001 Leuven, Belgium
[2]College of Animal Science and Technology, Shandong Agricultural University, Taian, Shandong 271018, China

Introduction

The control and regulation of voluntary feed intake in humans and animals is a very complex system involving central and peripheral regulatory mechanisms. Recently, 5'-adenosine monophosphate-activated protein kinase (AMPK) was reported to be implicated in the central control of food intake in mammals (Andersson *et al.*, 2004; Xue and Kahn, 2006). However, such information in avian species is lacking. The aim of the present experiment was therefore to investigate if AMPKα1 and -α2 subunit gene expressions are modulated by fasting and refeeding in meat-type chicks and to determine the interrelationship with hypothalamic (an)orexigenic neuropeptide and fatty acid synthetase mRNA levels.

Material and methods

Forty Arbor Acres broiler chicks with similar body weight were raised in wired cages. At the age of 7 days, feed was deprived for 24 h. The chicks were refed for 24 h. Eight chickens were euthanized at day 7 (before fasting), at day 8 (after 24 h fasting) and at day 9 (after 24 h of refeeding). Hypothalami were collected for quantification of mRNA levels by RT-PCR (calculations according to Livak and Schmittgen, 2001) of AMPKα1, AMPKα2, upstream Serine/threonine AMPK kinase (LKB1), agouti-related peptide (AgRP), neuropeptide Y (NPY), corticotrophin-releasing hormone (CRH), proopiomelanocortin (POMC), and fatty acid synthase (FAS). The mRNA levels of these genes were normalized to 18s rRNA levels (delta Ct). Data were analyzed by ANOVA and differences between treatment means were discerned with post-hoc tukey test (SAS, 2001).

Results and discussion

The effects of fasting/refeeding on hypothalamic AMPK α-subunit, LKB1, (an)orexigenic neuropeptides and FAS mRNA levels are illustrated in Figure 1. Fasting did not affect hypothalamic AMPKα1 and LKB1 mRNA levels but significantly increased hypothalamic AMPKα2 mRNA levels. In contrast, hypothalamic CRH gene expression was significantly reduced after 24 h of feed deprivation. On the other hand, fasting induced a significant increase in hypothalamic AgRP, POMC and FAS mRNA levels. Refeeding for 24 h caused a normalization of the expression of most genes to prefasting levels except that the mRNA levels of AMPKα1 and LKB1 of refed chicks were significantly lower compared to pre-fasting levels. Hypothalamic NPY gene expression was not influenced by the fasting/refeeding cycle. It is clear that the nutritional status does affect the hypothalamic gene expression of the energy sensor AMPK catalytic subunits α1 and α2 in a different way as previously observed in skeletal muscle (Fujii *et al.*, 2000). The causal interrelationship between AMPK, FAS and some (an)orexigenic neuropeptides in the hypothalamus of chickens in relation to energy homeostasis needs to be established.

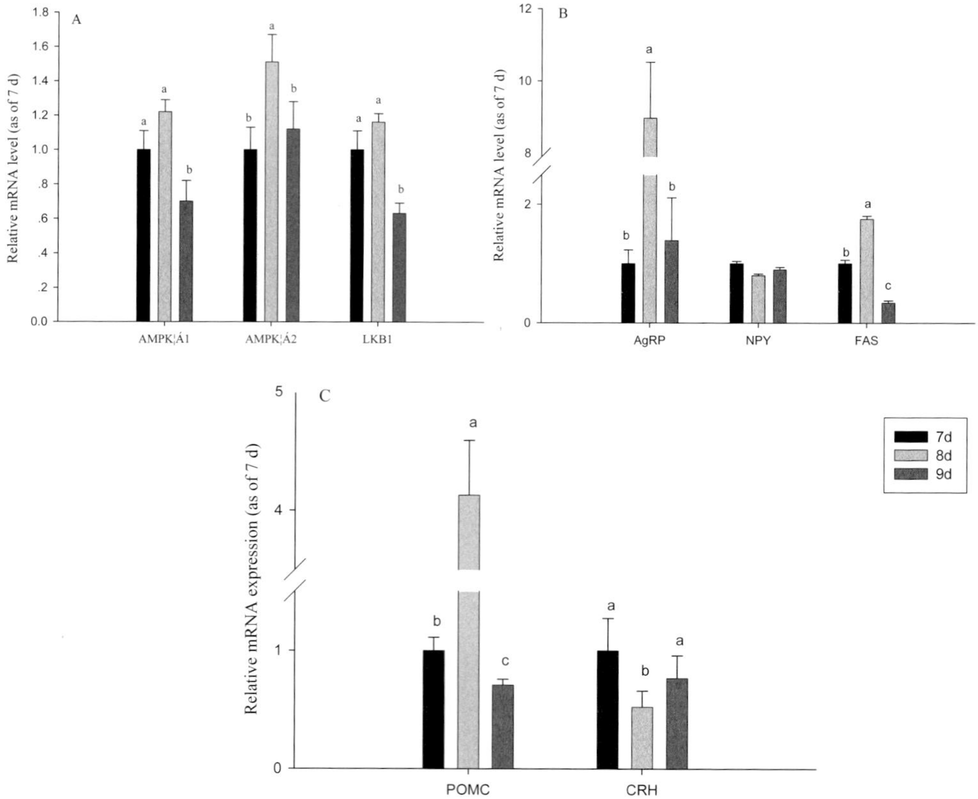

Figure 1. Effects of feeding ad libitum *(7 d), fasting (8 d) and refeeding (9 d) on hypothalamic AMPK and (an)orexigenic peptide mRNA levels. A. Effect on AMPK mRNA expression; B. Effect on orexigenic neuropeptides expression; C. Effect on anorexigenic neuropeptides expression. AMPK: 5'-adenosine monophosphate-activated protein kinase, LKB1: upstream serine/threonine AMPK kinase, AgRP: agouti-related peptide, NPY: neuropeptide Y, CRH: corticotrophin-releasing hormone, POMC: proopiomelanocortin, FAS: fatty acid synthase; mRNA levels normalized to 18s rRNA level. Bars, within a gene, with a different letter differ significantly (*P*<0.05).*

References

Andersson, U., K. Filipsson, C.R. Abbott, A. Woods, K. Smith, S.R. Bloom, D. Carling and C.J. Small, 2004. AMP-activated protein kinase plays a role in the control of food intake. J. Biol. Chem. 279, 12005-12008.

Fujii, N., T. Hayashi, M.F. Hirshman, J.T. Smith, S.A. Habinowski, L. Kaijser, J. Mu, O. Ljungqvist, M.J. Birnbaun, L.A. Witters, A. Thorell and L.J. Goodyear, 2000. Exercise induces isoform-specific increase in 5'AMP-activated protein kinase activity in human skeletal muscle. Biochem. Biophys. Res. Commun. 273, 1150-1155.

Livak, K.J. and T.D. Schmittgen, 2001. Analysis of relative gene expression data using real-time quantitative PCR and the $2^{-\Delta\Delta CT}$ method. Methods. 25, 402-408.

SAS, 2001. Release 8.01. SAS Inst. Inc., Cary, NC, USA.

Xue, B. and B.B. Kahn, 2006. AMPK integrates nutrient and hormonal signals to regulate food intake and energy balance through effects in the hypothalamus and peripheral tissues. J. Physiol. 574, 73-83.

Comparison of expression of proteolytic-related genes in skeletal muscles of layer and broiler chickens

K. Nakashima, A. Ishida and M. Katsumata
National Institute of Livestock and Grassland Science, Tsukuba 305-0901, Japan

Introduction

Broiler chickens have been selected for rapid growth and large muscle mass, while layer chickens have been selected for egg production. The growth rates of broilers are considerably faster than those of layers. Further, the muscle protein degradation rates of broilers are lower in those of layers (Hayashi *et al.*, 1985). This lower muscle protein degradation might play a pivotal role in causing the large differences in growth rates of broilers and layers. It is likely that the differences between layers and broilers in muscle protein degradation rates are related to the expression of proteolytic-related genes in the muscles. Here, we compared the expression of proteolytic-related genes in the skeletal muscles of broiler and layer chicks.

Materials and methods

Layer and broiler chicks (1 day old) from commercial hatchery were individually housed in a continuously lighted room in wire-floored cages. The chicks were allowed free access to food and water. All chicks were reared under these conditions until 7 or 14 days of age. When the chicks were 7 or 14 days of age (n=6), pectoralis muscle was excised, frozen in liquid nitrogen and stored at -80 °C. Total RNA was extracted from the muscle tissue, and complementary DNA was synthesized from the total RNA. Real-time PCR primers were designed for atrogin-1/MAFbx, proteasome C2 subunit (C2 subunit) and glyceraldehyde-3-phosphate dehydrogenase (GAPDH). Sequences of the forward primers and reverse primers are as follows: atrogin-1/MAFbx, 5'-CCA ACA ACC CAG AGA CCT GT-3' - 5'-GGA GCT TCA CAC GAA CAT GA-3'(NM_001030956), C2 subunit, 5'-AAC ACA CGC TGT TCT GGT TG-3' - 5'-CTG CGT TGG TAT CTG GGT TT-3'(AF027978) and GAPDH, 5'-CCT CTC TGG CAA AGT CCA AG-3' - 5'-CAT CTG CCC ATT TGA TGT TG-3'(K01458). The levels of mRNA were measured by real-time PCR analysis. GAPDH expression was measured as an internal control.

Results and discussion

In the present study, the expression of proteolytic-related genes was compared in the muscle of layer and broiler chickens. The data in Table 1 show that pectoralis muscle weight at 7 and 14 days of age were significantly greater for broilers than for layers. Protein accumulation in growing muscle represents the balance between synthesis and degradation. Fractional rates of protein synthesis has been reported to be slightly greater in broiler than in layer chickens (Maeda *et al.*, 1984), but also not differ between the two strains (Hayashi *et al.*, 1985). At ages of 7 to 14 day, rates of protein degradation in the fast-growing chickens were significantly lower than those in the slow-growing chickens (Tesseraud *et al.*, 2000). The expression of atrogin-1/MAFbx and C2 subunit of skeletal muscle in layer and broiler chickens are also given in Table 1. The mRNA expression of atrogin-1/MAFbx and C2 subunit in skeletal muscle was lower in broilers than in layers both at 7 and 14 days of ages. The present study is the first to demonstrate that an important component of difference of muscle protein degradation between layers and broilers is its ability to affect the expression of ubiquitin ligase and proteasome subunit mRNA.

In conclusion, this study shows that the expression of atrogin-1/MAFbx and C2 subunit mRNA in skeletal muscle of broilers was lower than in layers, suggesting lower muscle protein degradation in broiler than in layer chicken, responsible for the increased muscle mass. These results indicate that

Table 1. Pectoralis muscle weight and relative expression of atrogin-1/MAFbx in the muscle of layer and broiler chicken at 7 and 14 days of age.

	7 d		14 d	
	Layer	Broiler	Layer	Broiler
Pectoralis muscle weight (g)	3.7±0.6	8.0±0.8**	7.1±1.0	37.0±3.3**
Atrogin-1/MAFbx (fold change)	1.0±0.3	0.6±0.2**	1.2±0.2	0.3±0.1**
C2 subunit (fold change)	1.0±0.2	0.8±0.2*	1.2±0.2	1.0±0.2*

Values represent the means±SD of 6 animals. Values differ from those for layer chickens. Data were analyzed by the Student's *t*-test. A *P*-value of <0.05 was considered statistically significant. * $P<0.05$, ** $P<0.01$.

ubiquitin-proteasome pathway is different between the strains, and may be one of factor responsible for the difference in muscle protein accretion through alteration of protein degradation between broiler and layer chickens.

References

Hayashi K., Y. Tomita, Y. Maeda, Y. Shinagawa, K. Inoue and T. Hashizume, 1985. The rate of degradation of myofibrillar proteins of skeletal muscle in broiler and layer chickens estimated by N^{τ}-methylhistidine in excreta, Br. J. Nutr. 54, 157-163.

Maeda Y., K. Hayashi, S. Toyohara and T. Hashiguchi, 1984. Variation among chicken stocks in the fractional rates of muscle protein synthesis and degradation. Biochem. Genet. 22, 687-700.

Tesseraud S., AM. Chagneau and J. Grizard, 2000. Muscle protein turnover during early development in chickens divergently selected for growth rate. Poult. Sci. 79, 1465-1471.

Part 5. Energy/protein and their interaction on productive functions: ruminants

Protein supply, glucose kinetics and milk yield in dairy cows

H. Lapierre[1], C.E. Galindo[2], S. Lemosquet[3], I. Ortigues-Marty[4], L. Doepel[5] and D.R. Ouellet[1]
[1]Agriculture and Agri-Food Canada, Sherbrooke, QC, Canada
[2]Université Laval, Québec, QC, Canada
[3]UMR1080 Dairy Production, INRA, Saint-Gilles, France
[4]UR1213 URH, INRA, Theix, 63122 St Genès Champanelle, France
[5]University of Calgary, Alberta, Canada

Abstract

Despite the traditional approach of thinking of energy and protein as two separate entities when balancing dairy rations, the response of the cow to increased supply of nutrients often crosses these artificial compartments. For example, supplementation of protein as casein infusions increases not only protein yield but also milk and lactose yields. As part of the quest to enhance the biological and economic efficiency of the cow, we need to understand how and where such interactions occur. The response in lactose yield to protein supplementation is almost identical (g/g) to the response in protein yield. An obvious candidate for the link between lactose output and protein supply is glucose, as increasing protein supply through casein infusion increases the whole body (WB) rate of appearance (Ra) of glucose (sum of real portal absorption, gluconeogenesis and glycogenolysis). In ruminants, utilization of arterial glucose by the portal-drained viscera averages 22% of WB glucose utilization and real portal absorption contributes, on average for 14% of WB Ra of glucose, with variations from 0 to 37%, in relation with starch digested in the intestine. Most of the remainder of WB Ra originates from gluconeogenesis, with a liver contribution of at least 85% and the kidneys contribution at most 15% to non-detectable levels in animals fed a concentrate diet. Infusion of casein did not alter real portal absorption of glucose and neither did it decrease portal-drained viscera utilization of arterial glucose. Indeed, in one study where measured simultaneously, the increased WB Ra of glucose induced by post-ruminal casein infusion originated from an increment in liver net flux. In dairy cows, the efficiency of WB transfer of the carbons of casein into glucose is on average higher than the maximal theoretical synthesis of glucose from protein supplementation, once the increment in milk protein yield is taken into account. Stimulation of utilization of other glucose precursors or recycling of glucose could explain this high efficiency. Despite a strong relationship between casein supply, WB Ra of glucose and lactose yield, increased WB Ra of glucose does not seem to be the driving force stimulating lactose yield. Increased lactose yield is only observed when the increment in WB Ra of glucose is due to increased protein supply; the same relationship does not exist when glucose Ra is increased through energy supply. It may be that some of the essential AA are playing a key role in some metabolic pathways or are simply stimulating protein synthesis. The latter hypothesis might result in increased milk protein synthesis, 'pulling' lactose synthesis due to its osmotic role, or in stimulating 'enzyme machinery' involved in gluconeogenesis, milk protein and/or lactose synthesis. Overall, this 'simple' example demonstrates the intricate integration between protein and energy metabolism.

Introduction

Most of the models used to balance dairy rations consider protein and energy as two separate entities. Despite this dichotomy that we have imposed on the predictive models, 'cross-reactions' between energy and protein supply and the milk component responses occur. One of the 'simplest' examples is the response of milk and lactose yields to protein supplementation, in addition to the protein response. Using from the database of Doepel *et al.* (2004) only the studies where casein, casein hydrolysates or amino acids (AA) with a casein profile had been infused, the response (g/g) of lactose yield to the increased protein supply is almost identical to the response of protein yield (Figure 1).

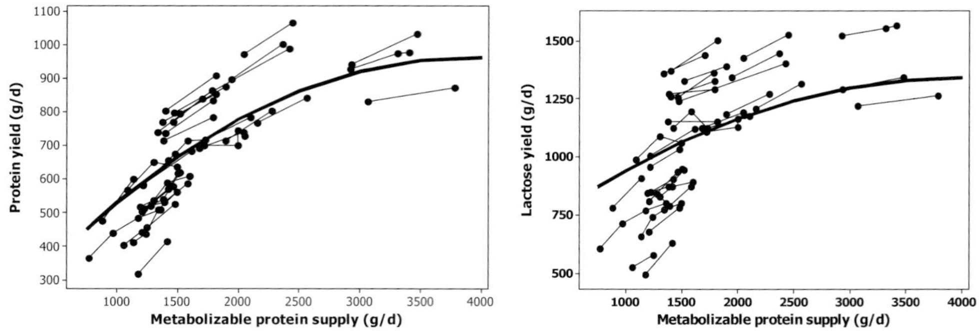

Figure 1. Milk protein and lactose yield response to increased supply of protein from post-ruminal infusions of casein, casein hydrolysates or amino acids with whole protein profile.
$MPY = 196^{**}(\pm 32) + 0.38^{**}(\pm 0.03) \times MP - 0.000043^{**}(\pm 0.000008) \times MP^2 + Exp.$, R^2adj: 98.8%,
$MLY = 630^{**}(\pm 53) + 0.36^{**}(\pm 0.05) \times MP - 0.000045^{**}(\pm 0.000013) \times MP^2 + Exp.$, R^2adj: 98.6%,
*with 32 experiments, n=80; where all units are in g/d and MPY=milk protein yield, MLY= milk lactose yield and MP=metabolizable protein supply (NRC, 2001);** indicates* $P<0.01$.

How does protein supplementation, or more correctly the absorbed AA, affect lactose yield? Already in 1906, the role of glucose for lactose secretion was proposed (Kaufmann and Magne, 1906) and later, it was observed that, in ruminants, more than 85% of the lactose originates from glucose (Bickertaffe *et al.*, 1974). Therefore, increased glucose availability from protein supply would be a first obvious route between protein supply and lactose output. But, where and how, in the body, and with which efficiency does an increased protein supply affect the glucose availability to the dairy cow? And with which efficiency does the dairy cow use this increased glucose availability to export milk lactose? This presentation does not pretend to be an exhaustive literature review of the whole complexity of energy and protein interaction, but rather illustrates from a simple situation that protein and energy metabolism are so closely linked that we should really integrate both their supply and requirements in our predictive models used to balance rations.

What is whole body rate of appearance of glucose?

Increasing protein supply is known to increase whole body (WB) rate of appearance (Ra) of glucose in ruminants (see review Lemosquet *et al.*, 2007). The measurement of WB Ra requires the utilization of isotopes, earlier studies being conducted with radioactive isotopes and more recent studies conducted with stable isotopes. Only studies infusing proteins post-ruminally will be considered in this review, allowing precise determination of the extra supply of AA without relying on any predictive models. Table 1 summarizes studies conducted on the effect of post-ruminal supply of casein on WB Ra of glucose in dairy cows. All these studies involved continuous infusions of labelled glucose and WB Ra of glucose was calculated using the rate of infusion and the specific activity or the isotopic enrichment of glucose in peripheral plasma. For reasons described below, the study of Hanigan *et al.* (2004) reporting glucose hepatic net flux was added to the database. The position of the isotope on the labelled molecule of glucose alters the absolute value of Ra due to variable measurements of recycling (Wolfe, 1992), but as we will be looking at increments, reported values will be directly used without correction.

First, all studies reported an increase, at least numerically, of the WB Ra of glucose when casein was provided to dairy cows. But, what does an increased WB Ra of glucose mean? The WB Ra of glucose is the sum of all the inflows of glucose in the plasma, i.e. in a fed ruminant, gluconeogenesis plus glucose intestinal absorption reaching blood circulation plus glucose from the turnover of glycogen returning to the plasma pool. Intestinal absorption of glucose plus gluconeogenesis from absorbed nutrients represent a net supply of glucose to the animal. On the other hand, glucose originating from

Table 1. Effect of post-ruminal infusion of casein (CN) and/or energy on whole body glucose rate of appearance (Ra), liver net flux and yields of milk, protein and lactose.

Treatment	Glucose Ra (g/d)	Liver net flux (g/d)	Milk yield (kg/d)	Lactose yield (g/d)	Protein yield (g/d)	Reference
Control	2,760		28.8	1,404[1]	963	Clark *et al.*, 1977
+ CN (450 g/d Na-CN)	2,890		31.0	1,510[1]	1,072	
+ Glucose (450 g/d)	2,850		28.9	1,407[1]	957	
+ CN & Glucose	3,150		31.6	1,539[1]	1,104	
Control	1,931		22.1	992	607	König *et al.*, 1984
+ CN (240 g/d Na-CN)	2,169		23.8	1,085	697	
+ CN (460 g/d Na-CN)	2,200		23.5	1,065	683	
Control	2,540		25.5	1,276	752	Lemosquet *et al.*, 2009a
+ CN (743 g/d Ca-CN)	2,830		29.8	1,448	904	
+ Propionate (rumen 1,043 g/d)	2,925		25.7	1,288	784	
+ CN & propionate	3,270		30.5	1,464	996	
Control	2,756	1,970	30.9	1,420	951	Galindo, 2010
+ CN (647 g/d Na-CN)	3,110	2,402	32.9	1,484	1,058	
Control		2,022	14.8	671	611	Hanigan *et al.*, 2004
+ CN (200 g/d)		2,411	16.9	768	679	
+ CN (400 g/d)		2,125	16.1	724	654	
+ CN (600 g/d)		2,773	15.9	727	656	

[1] Estimated from milk yield with a lactose concentration of 47.0 g lactose/kg milk (average of other references).

glycogenolysis, either reaching directly the plasma pool or being shuttled to the liver as lactate and then being resynthesized in glucose does not represent a net supply to the animal if the glycogen pool is not depleting but just being turned over. The relative contribution of each function to WB Ra obviously depends on the status of the animal.

Real portal absorption

In ruminants, net portal appearance of glucose is relatively limited as shown by an average 4±13 g/d, not significantly different from 0, obtained from 306 measurements (Reynolds, 2006), but with individual values bouncing below and above 432 g/d. Net appearance across the portal-drained viscera (PDV) is the result of two opposite actions: absorption of glucose from starch small intestinal digestion (+) and utilization of glucose from arterial supply by the PDV (-). First, intestinal digestion of starch and subsequent absorption of glucose yields a positive balance of glucose across the PDV. A recent meta-analysis estimated that for 1 g of starch digested in the intestine, 0.43 g of glucose is absorbed (Loncke *et al.*, 2009), in the range of values reported when starch was infused (Reynolds, 2006). Therefore, the magnitude of the real portal absorption of glucose depends, partly, on the amount of starch intake by-passing the rumen. On the other hand, the PDV uses glucose both from arterial and lumen supply. These utilizations can be estimated from uptake of lumen or systemic labelled glucose by the PDV. Very limited data exist in ruminants on utilization from lumen supply and El-Kadi *et al.* (2003) reported that most of the glucose used by the PDV would be of arterial supply. Glucose utilization from plasma arterial supply by the PDV is more documented and averaged 22% of WB utilization or 0.7 g/d/kg BW, excluding treatments involving post-ruminal glucose or starch infusions (Table 2) but showed large variations from 0.3 to 2.2 g/d/kg BW. In recent studies in lactating dairy cows (Larsen and Kristensen, 2009; Galindo, 2010), glucose PDV utilization from plasma supply averaged 423 and 397 g/d, i.e. 0.7 and 0.8 g/d/kg BW, respectively. From the meta-analysis quoted

Table 2. Glucose whole body rate of appearance (WB Ra) and tissue kinetics in ruminants (g/d).

Species (BW, kg)	Treatment	WB	Portal-drained viscera			Liver			Reference
		Ra	Net flux	Utilization	Real	Net flux	Utilization	Real	
Sheep (57)	Non-pregnant; hay	118	-17	14	-5	94	9	104	Bergman *et al.*, 1970
	Non-pregnant; grain + hay	134	-12	18	0	82	31	114	
	Non-pregnant; hay	84	-34	22	-13	79	-4	76	
	Pregnant; hay	199	-31	45	11	170	0	169	
	Pregnant; fasted	134	-17	28	11	115	-11	102	
Steers (235)	Starch (rumen; 800 g/d)	782	-130	121	-9	540	91	631	Harmon *et al.*, 2001
	Starch (abomasum (Ab); 800 g/d)	1,140	276	281	557	592	-60	531	
Dairy cows (626)	Control	2,756	337	402	739	1,970	95	2,065	Galindo, 2010
	+ Casein (CN; Ab, 647 g/d Na-CN)	3,110	281	393	674	2,402	104	2,506	
Sheep (59)	Pregnant	234	26	38	57	142	5	155	van der Walt *et al.*, 1983
	Lactating	345	20	68	111	242	-3	241	
Sheep (46)	Control	132	5	44	49				Balcells *et al.*, 1995
	+ Glucose (jugular, 65 g/d)	194	-5	47	41				
	+ Glucose (jugular, 130 g/d)	283	-13	86	73				
Sheep (33)	Control	98	-16	36	21				El-Kadi *et al.*, 2006
	+ CN (duodenum; 35 g/d)	104	-16	35	19				
	+ CN (duodenum; 70 g/d)	112	-18	43	25				
	+ CN (duodenum; 105 g/d)	116	-31	46	15				
Sheep (29)	Hay	123	-12	8	-4				Huntington *et al.*, 1980
	85% concentrate	230	35	22	57				
Dairy cows (536)	*Pre partum* (595 kg)	2,201	-134	292	158				Larsen and Kristensen, 2009
	Lactating (mean of 4, 15, 29 DIM)	3,826	-204	408	203				
	Lactating + Glucose (Ab; 1.5 kg/d)	3,624	795	439	1,234				
Sheep (44)	Grass pellets	148	-26	65	39				Piccioli Cappelli *et al.*, 1997
	+ Glucose (jugular; 130 g/d)	207	-70	54	-16				
	+ Glucose (duodenum; 130 g/d)	246	-3	96	93				
Steers (127)	Control	511	-106	139	32				Seal and Parker, 1994
	+ Propionate (rumen; 0.5 mol/d)	509	-74	58	-16				
	+ Propionate (rumen; 1.0 mol/d)	580	-122	65	-58				

above, glucose utilization by the PDV was estimated to be in a similar range, and averaged 0.5 g/d/kg BW in ruminants (346 g/d for a 650 kg cow) but also presented a fairly wide variation from 0.4 to 1.3 g/d/kg BW (Loncke *et al.*, 2009). It is not clear at this point if variations observed between studies are technical or real. In sheep, arterial utilization of glucose by the PDV increased with WB Ra of glucose (altered with intrajugular infusions of glucose), but remained a fixed proportion, 28%, of WB Ra (Balcells *et al.,* 1995). In steers, switching the site of starch digestion from the rumen to the small intestine increased both WB Ra of glucose and the proportion of WB Ra used by the PDV from 15 to 28%.(Harmon *et al.* 2001). However, in dairy cows, increasing starch intake increased WB Ra of glucose but did not affect glucose utilization by the PDV (Kristensen *et al.*, 2006). Also, estimated PDV utilization did not differ between diets providing duodenal starch and starch-free diets (Loncke *et al.*, 2009). An adequate quantification of the PDV utilization of glucose, including factors that might affect it, is much needed to determine the real amount of glucose absorbed from the diet. Nevertheless, the limited data available clearly indicate that the real absorption of glucose and its true contribution to the WB Ra is more important than usually acknowledged using only net portal appearance. Actually, from studies reported in Table 2 with no post-ruminal infusion of starch or glucose, the real contribution of absorption to WB Ra averaged 14% of glucose WB Ra in ruminants, varying from 0 to 37% in relation with intestinal starch digestion.

Gluconeogenesis

If absorption contributes, on average, to 14% of WB Ra of glucose, another important inflow has to be made. It has long been recognized that in ruminants glucose has to be provided through gluconeogenesis, due to limited absorption of glucose compared with monogastrics (Lindsay, 2006). It was only in 1970, however, that a study was conducted measuring simultaneously the WB Ra of glucose and fluxes of glucose across the PDV and the liver (Bergman *et al.*, 1970; Table 2). This confirmed the important role of the liver which contributed, on a net basis, to 82% of WB Ra in these studies, in animals mainly fed hay. As for the PDV, hepatic utilization of glucose derived from plasma should be added to this net flux. In sheep, glucose utilization by the liver from plasma (portal and arterial) supply was very small, increasing by 4% the real glucose hepatic production to 86% of WB Ra. Such a small utilization of glucose by the liver had been confirmed later in steers (Harmon *et al.*, 2001) and in cows (Kristensen *et al.*, 2006; Galindo, 2010), where it averaged between 3 and 5% of WB Ra. These very low measurements of hepatic utilization of glucose from plasma supply are in agreement with the fact that the liver has been reported to mainly use fatty acids as an energy source (Krebs, 1972). This estimation of glucose utilization by the liver might be, however, slightly underestimated because of the recycling of secondary labelled metabolite (e.g. lactate) into glucose in the liver or if cells of the liver are using newly synthesized, unlabelled glucose.

In the study of Bergman *et al.* (1970), the splanchnic contribution to WB Ra varied from 81 to 91%, still leaving slightly more than 10% of the Ra to be provided by another source. From measurements of renal net fluxes, Bergman *et al.* (1974) estimated that in sheep renal gluconeogenesis contributed on average 12% (from 8 to 16%) of WB Ra of glucose. These sheep were fed good quality leguminous hay. However, in growing beef heifers, the renal net flux of glucose averaged only between 4% (high intake) to 6% (low intake) of liver net glucose flux when fed a 75% alfalfa diet, whereas it was even were negative when the heifers were fed a 75% concentrate diet (Reynolds *et al.*, 1991). Similarly, in steers, estimation of peripheral glucose production decreased from 242 to -17 g/d, when the site of starch digestion was switched from the rumen to the small intestine (Harmon *et al.*, 2001). In lactating dairy cows, contribution from peripheral tissues was estimated to be almost null (Galindo, 2010). All together these data suggest that the contribution of the kidneys to WB Ra of glucose is related to the type of feeding and would be fairly limited in animals well fed a concentrate diet providing starch to the small intestine.

Glycogenolysis

Information on the magnitude of the contribution of the turnover of the pool of glycogen to WB Ra of glucose is scarce. In ruminants, Annisson *et al.* (1963) reported that 'There was very little incorporation of ^{14}C into glycogen during the infusion of [^{14}C]lactate or [^{14}C]glucose'. Therefore, the variation of the turnover of the pool of glycogen in response to protein supply will be considered minimal and not contributing to variations of WB Ra to increased protein supply.

Where can amino acids alter glucose kinetics?

As mentioned previously, the WB Ra of glucose is the sum gluconeogenesis plus glucose intestinal absorption plus the turnover of glucose from glycogenolysis returning to the plasma pool. Therefore, simplistically, there are three sites where an increased protein supply may affect WB Ra of glucose: across the small intestine through increased real absorption and/or across the liver or across the kidney through increased gluconeogenesis (as previously mentioned, potential alterations of the pool size and the turnover of glycogen will be considered minimal and not included in this discussion).

Absorption

Increased real portal absorption through increased protein supply could occur through two different mechanisms: 1) increased intestinal digestion of the starch and/or 2) epithelial cells of the small intestine using additional AA from the extra supply (from the lumen or from arterial supply) as an energy source thereby decreasing their utilization of glucose absorbed from the lumen. The first route is supported by the fact that increased presence of casein into the duodenum has been shown to increase post-ruminal digestion of starch in sheep, steers and dairy cows, maybe related to increased secretion of α-amylase (see review from Larsen, 2009). To support the second hypothesis, it is known that the PDV are using glucose from both arterial and lumen source (El-Kadi *et al.*, 2003) and that some AA are used as an energy source by the PDV. In young pigs the contribution of glutamate plus glutamine to CO_2 production across the PDV (54%) exceeded that derived from the sum of enteral and systemic glucose utilisation (44%; Stoll *et al.*, 1999). Such data, however, do not exist for ruminants, but there is substantial evidence that the digestive tract of the ruminants has the ability to catabolize AA to generate energy (see review by Lobley and Lapierre, 2003). In dairy cows, a substantial utilization of glutamate and glutamine by the PDV was also observed, with net portal appearance of glutamine being negative and the absorption of glutamate representing less than 10% of the small intestinal disappearance of glutamate plus glutamine (Berthiaume *et al.*, 2001). Such demand certainly is linked, in part, to use of the AA carbon as an energy source. The PDV also oxidizes the EAA. For example, leucine oxidation across the PDV has been observed in sheep (Lobley *et al.*, 2003) and increased with protein supply in dairy cows (Lapierre *et al.*, 2002). The situation is not as clear for the other EAA (see review from Lobley and Lapierre, 2003). From comparison between predicted digestive flows of EAA and measured net portal appearance in dairy cows, it has been proposed that the PDV would mainly oxidize, in addition to leucine, isoleucine and valine (Pacheco *et al.*, 2006). However, despite these two possible mechanisms to increase real portal absorption, in two studies measuring it specifically (net portal appearance plus PDV arterial utilization), casein infusion did not alter real portal absorption (El-Kadi *et al.*, 2006; Galindo, 2010).

A decreased utilization of glucose by the PDV from arterial supply alters the net flux but not the real absorption and WB RA of glucose. A decreased utilization of arterial supply would, however, spare glucose for other tissues, including the mammary gland. Recent studies have tried to determine if an extra supply of protein or candidate AA, such as glutamine, might effectively spare glucose utilization by the PDV. Indirectly, a sparing effect of glucose arterial utilization by the PDV could be observed through an increased net portal appearance of glucose. In dairy cows abomasal infusions of casein did not increase net portal appearance of glucose (Hanigan *et al.*, 2004) neither did glutamine abomasal infusion (Doepel *et al.*, 2007). Direct measurements of glucose utilization using labelled glucose lead to the same conclusions. In sheep, the PDV appeared to metabolize increasing amounts of the branched-chain AA, glutamate and glutamine with infusion of casein, but glucose utilization by the PDV was not affected (El-Kadi *et al.*, 2006). Similarly, in a study in dairy cows, abomasal infusions of casein did not alter glucose utilization by the PDV (Galindo, 2010: Table 2). This does not preclude a contribution of AA extracted by the PDV to energy expenditure but, if it does happen, this does not increase the glucose availability to post-PDV tissues.

Altogether these data suggest that the increment in WB Ra of glucose observed with increased protein supply did not originate from increased real portal absorption. In addition, there is no evidence that, albeit some AA are oxidized by the PDV, increased supply of protein would spare glucose for peripheral tissues.

Gluconeogenesis

If the increment of WB Ra of glucose with increased casein supply does not originate from the PDV, options are for increments in gluconeogenesis from the liver or from the kidneys. Although the major glucose precursor is propionate (see reviews Huntingon *et al.*, 2006; Reynolds, 2006), most AA, except leucine and lysine, can make a net contribution to gluconeogenesis, and most of them through entry in the Krebs cycle. The contribution of AA to gluconeogenesis varies with the feeding status. In well fed sheep, AA contribution to gluconeogenesis was estimated to range between 13 to 15% (Ford and Reilly, 1969) or from 11 to 29% (Wolf and Bergman, 1972). In cattle, AA were also estimated to contribute from 11 to 30% of glucose hepatic release (Huntington *et al.*, 2006). These would be maximal estimates as it has been shown in dairy cows that part of the AA extracted by the liver is used for the synthesis of export protein: up to 20% of phenylalanine removal was used for this purpose (Raggio *et al.*, 2007). The AA can also contribute to renal gluconeogenesis in sheep. Kaufman and Bergman (1974) estimated that between 20 and 30% of renal gluconeogenesis originated from AA. Indeed in fed sheep, alanine, aspartate and glutamine were removed by the kidneys and their maximal contribution to renal gluconeogenesis averaged 12%. Also, as evidence for contribution of AA to renal gluconeogenesis, Wolf and Bergman (1972) reported that: 'conversion of ^{14}C-labelled aspartate or glutamate to glucose was 1.5-4 times greater in the whole body than calculated from hepatic uptake of the ^{14}C-labelled amino acid'. However, overall, this indicates a relatively small contribution (less than 2%) of renal neoglucogenesis from AA to WB Ra of glucose in a fed animal.

Only one study reported concurrent measurement of the response of glucose WB Ra and liver net production to increased protein supply in lactating dairy cows: the response of WB Ra of glucose to casein infusion was totally covered by the increased in hepatic net flux (Galindo, 2010; Table 1). As the increment in WB Ra of glucose was associated with the increment in liver net flux and as the data on the response of WB Ra of glucose to increased protein supply are limited in dairy cows, intention was to include studies reporting the net liver flux of glucose in response to casein infusion. However, only one additional study was added to the database (Hanigan *et al.*, 2004; Table 1). One study in non-lactating dairy cows could not be used because the profile of AA was very different from casein (Wray-Cahen *et al.*, 1997).

The advantage of studies reporting net flux across the liver is that in addition to glucose, information on the relative contribution of AA can be estimated from the liver uptake of AA. In the study of Hanigan *et al.* (2004), increasing supply of casein from 0 to 600 g/d increased liver release of glucose (Table 1), but liver removal of AA did not significantly increase. For the highest supplementation treatment, however, numerical increased removal of AA could account for about half of the increased release of glucose. In late gestation, dry dairy cows, infusion of a commercial mixture of AA increased liver net flux of glucose and if the liver uptake of alanine, glycine and glutamine was assumed to be directed totally to gluconeogenesis, these AA would have provided nearly all the 3-C units needed to account for the extra glucose produced (Wray-Cahen *et al.*, 1997). Unfortunately, data are not yet available to estimate if the increased removal of AA by the liver matched the increased glucose hepatic release in the study of Galindo (2010).

This close relationship observed between the increment in WB Ra of glucose and increased net liver release of glucose (Galindo, 2010) suggests very limited contribution of increased renal gluconeogenesis to support the increment of glucose WB Ra that has been observed with increased protein supply.

Efficiency of transfer

Glucose formed from AA can then be used for lactose synthesis and, already in 1964, it has been reported in a dairy cow that after administration of a mixture of [^{14}C]AA, 'the milk lactose also became radioactive to an extent indicating that about 12 per cent of the lactose was formed by gluconeogenesis from protein' (Hunter and Milsson, 1964). A recent *in vitro* study, however, suggests that part of the contribution of AA to lactose synthesis might also occur within the mammary gland (Bequette *et al.*, 2006). Figure 2 illustrates the relationship between the metabolizable protein supply (MP, NRC, 2001) and the glucose 'availability' (WB Ra or liver net flux). The limited dataset did not yield a quadratic relationship between yields of protein or lactose and MP supply, as observed from the database of Doepel *et al.* (2004). However, a first striking observation is the similarity of the response between the yields of protein (P<0.01, 0.22±0.04 g/g; 7 experiments, n=17; R^2adj=95.7%) and lactose (P<0.01, 0.20±0.04 g/g; R^2adj=98.4%) to casein infusion. Therefore, in terms of milk output, casein infusion has an effect almost as strong on lactose yield as on protein yield.

Another interesting observation relies on the very high transfer of the infused casein into glucose WB Ra. According to metabolic pathways, the maximal theoretical conversion of casein into glucose would be approximately 0.60 g of glucose per g of casein. The relationship between MP supply and WB glucose Ra was linear (P<0.01) with a slope of 0.60±0.14 (7 experiments, n=17, R^2adj=86.7%, Figure 2), with, therefore, an observed efficiency of transfer similar to the maximal theoretical efficiency. In addition, if the increased milk protein yield is removed from the increased protein supply, the slope of the relationship increased to 0.76±0.18 (P<0.01). The reason for this high efficiency is not clear. It is possible that the supply of protein has positively stimulated the cow' metabolism and that in such conditions, nutrients other than the additional supply of AA (including dietary AA) have been diverted towards glucose synthesis, if that is recognized as a metabolic priority. Indeed, in cows maintained on a constant feed intake, initiation of lactation increased WB Ra of glucose (Bennink *et al.*, 1992). In view of the very small direct contribution of the glycogen pool turnover to WB Ra of glucose, it is unlikely that it would be stimulation of glycogenolysis that

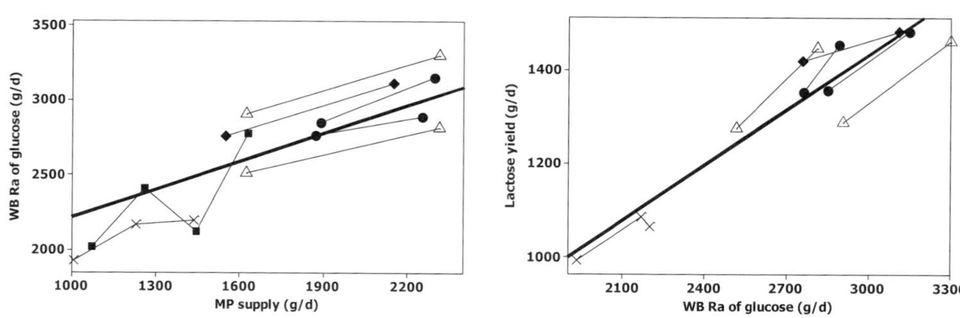

Figure 2. Relationship between whole body rate of appearance (WB Ra) of glucose, metabolizable protein (MP) supply and lactose yield in studies involving casein infusion. Data from: ● Clark et al., *1977, ♦ Galindo, 2010, ■ Hanigan* et al., *2004, × König* et al., *1984 and Δ Lemosquet* et al., *2009a.*

would be responsible for the high 'recovery' of glucose from protein supply. Another option could also be that part of the increased WB Ra and net hepatic flux of glucose would be the result of an increased recycling from glucose to peripheral tissue and lactate coming back to the liver being used for gluconeogenesis.

If we then look at the transfer of this increment of WB Ra of glucose into the increment in lactose yield, it averaged 0.40 (±0.06) g/g ($P<0.01$, 6 experiments, n=13, R^2adj=96.2%; Figure 2), excluding the milk lactose response from Hanigan *et al.* (2004). This value is slightly smaller than the average 0.49 value of lactose yield/WB Ra calculated for Control treatments of Table 1. This might indicate a lower marginal efficiency at higher glucose availability or also the fact that part of this increment in WB Ra of glucose is not a net increment but originates from recycling of metabolites from intermediate metabolism.

The key question then becomes: 'Is this increment in WB Ra of glucose the driving force for the increased lactose yield?' Overall, a good relationship has been reported between WB Ra of glucose and lactose output (Danfær *et al.*, 1995) and this also holds for the present dataset. However, in two studies where casein infusion increased both WB Ra of glucose and lactose yields, glucose or propionate infusions to the same cows also increased WB Ra of glucose but failed to increase yields of milk and lactose (Clark *et al.*, 1977; Lemosquet *et al.*, 2009a). This suggests that, at least in these studies, with cows still having the potential to improve their lactose yield, availability of glucose was not the critical driver of lactose and milk yields. The flexibility of the mammary gland in using different nutrients to support increased yields of lactose and protein is discussed in another contribution in the present symposium (Lemosquet *et al.*, 2010). Along the same line of reasoning, Lemosquet *et al.* (2009b) reported recently that infusion of glucose, propionate or non-EAA increased glucose WB Ra, but did not increase milk and lactose yields. This non-response of milk yield to non-EAA is in agreement with previous observations where the supply of non-EAA had little influence on milk yield (Metcalf *et al.*, 1996; Schwab *et al.*, 1996; Doepel *et al.*, 2010). Overall, this suggests that although increased glucose availability with casein supply might support the increased lactose yield, glucose availability per se is not the driving force.

Tentatively, some explanations can be proposed to explain this apparent discrepancy, but at this point, they are speculation as no direct observations are available to support or contradict them. One possibility is that stimulation of protein synthesis through increased protein supply has an overall positive effect on all proteins, including enzymes related to gluconeogenesis and lactose synthesis as well as milk protein per se. Increased milk protein synthesis would 'pull' the demand for lactose, due to its major osmotic role (Mepham, 1993), and therefore stimulate gluconeogenesis. That would imply that glucose precursors other than infused AA are used more efficiently towards glucose synthesis (through a reduced oxidation, for example) when extra protein is supplied. Another explanation relies on the potential effect of AA in stimulating protein synthesis through cell signalling pathways (Davis *et al.*, 2003). However, recently, Rius *et al.* (2009) reported that not only casein but also starch was involved in the regulation of the mTOR pathway in the mammary gland of dairy cows.

Conclusion

In summary, increasing supply of protein (casein infusion) directly increases WB Ra of glucose, mainly though an increased hepatic gluconeogenesis. However, the extra supply of AA from infused casein can not entirely explain the increase of WB Ra of glucose. This would suggest sparing of AA or other glucose precursors when metabolism is stimulated by extra protein, but demonstration of that hypothesis is still to come. The increment in WB Ra of glucose induced by casein infusion does not seem, however, to be the driving force behind the observed increased lactose and milk yields. Some essential AA are probably required to stimulate protein synthesis, either as constituent of the protein per se or playing a key role in some metabolic pathways. Overall, this 'simple' example demonstrates the intricate integration between protein and energy metabolism in a lactating dairy cow. This clearly indicates the requirement of considering them together, rather than as two distinct entities, either when conducting trials or when balancing dairy rations.

References

Annison, E.F., D.B. Lindsay and R.R. White, 1963 Metabolic interrelations of glucose and lactate in sheep. Biochem. J. 88, 243-248.

Balcells, J., C.J. Seal and D.S. Parker, 1995. Effect of intravenous glucose infusion on metabolism of portal-drained viscera in sheep fed a cereal/straw-based diet. J. Anim. Sci. 73, 2146-2155.

Bennink, M.R., R.W. Mellenberger, R.A. Frobish and D.E. Bauman, 1972. Glucose oxidation and entry rate as affected by the initiation of lactation. J. Dairy Sci. 55 (Suppl. 1), 712.

Bequette, B.J., N.E. Sunny, S.W. El-Kadi and S.L. Owens, 2006. Application of stable isotopes and mass isotopomer distribution analysis to the study of intermediary metabolism on nutrients. J. Anim. Sci. 84 (E. Suppl.), E50-E59.

Bergman, E.N., M.L. Katz and C.F. Kaufman, 1970. Quantitative aspects of hepatic and portal glucose metabolism and turnover in sheep. Am. J. Physiol. 219, 785-793.

Bergman, E.N., C.F. Kaufman, J.E. Wolff and H.H. Williams, 1974. Renal metabolism of amino-acids and ammonia in fed and fasted pregnant sheep. Am. J. Physiol. 226, 833-837.

Berthiaume, R., P. Dubreuil, M. Stevenson, B.W. McBride and H. Lapierre, 2001. Intestinal disappearance, mesenteric and portal appearance of amino acids in dairy cows fed ruminally protected methionine. J. Dairy Sci. 84, 194-203.

Bickerstaffe, R., E.F. Annison and J.L. Linzell, 1974. The metabolism of glucose, acetate, lipids and amino acids in lactating dairy cows. J. Agric. Sci. 82, 71-85.

Clark, J.H., H.R. Spires, R.G. Derrig and M.R. Bennink, 1977. Milk production, nitrogen utilization and glucose synthesis in lactating cows infused postruminally with sodium caseinate and glucose. J. Nutr. 107, 631-644.

Danfær, A., V. Tetens and N. Agergaard, 1995. Review and an experimental study on the physiological and quantitative aspects of gluconeogenesis in lactating ruminants. Comp. Biochem. Physiol. 111B, 201-210.

Davis, T.A., A. Suryawan, J.A. Bush, P.M. O'Connor, and M.C. Thivierge, 2003. Interaction of amino acids and insulin in the regulation of protein metabolism in growing animals. Can. J. Anim. Sci. 83, 357-364.

Doepel, L., M.D. Hanigan, J.J. Kennelly, D. Pacheco, I.F. López-Campbell, and H. Lapierre, 2004. Milk protein synthesis as a function of amino acid supply. J. Dairy Sci. 87, 1279-1297.

Doepel, L., G.E. Lobley, J.F. Bernier, P. Dubreuil and H. Lapierre, 2007. Effect of glutamine supplementation on splanchnic metabolism in lactating dairy cows. J. Dairy Sci. 90, 4325-4333.

Doepel, L. and H. Lapierre, 2010. Effect of post-ruminal supplementation of amino acids on production performance of lactating dairy cows. J. Dairy Sci., In press.

El-Kadi, S.W., N.E. Sunny, M. Oba, S.L. Owens and B.J. Bequette, 2003. Glucose metabolism by the gastrointestinal tract of sheep as affected by protein supply. EAAP plublication No.109. Ed. W.B. Souffrant and C.C. Metges, Wageningen Press, the Netherlands, 401-404.

El-Kadi, S. W., R.L. Baldwin VI, N.E. Sunny, S.L. Owens and B.J. Bequette, 2006. Intestinal protein supply alters amino acid, but not glucose, metabolism by the sheep gastrointestinal tract. J. Nutr. 136, 1261-1269.

Ford, E.J.H. and P.E.B. Reilly, 1969. The utilization of plasma free amino acid and glucose carbon in sheep. Res. Vet. Sci. 10, 409-418.

Galindo, C.E. 2010. Effet des acides amines sur le métabolisme du glucose chez la vache laitière. Mémoire de maîtrise. Université Laval, Québec, QC Canada, 98 p.

Hanigan, M., C.K. Reynolds, D.J. Humphries, B. Lupoli and J.D. Sutton, 2004. A model for net amino acid absorption and utilization by the portal-drained viscera of the lactating dairy cow. J. Dairy Sci. 87, 4247-4268.

Harmon, D.L., C.J Richards, K.C Swanson, J.A. Howell, J.C. Matthews, A.D. True, G.B. Huntington, S.A. Gahr and R.W. Russell, 2001. Influence of ruminal or postruminal starch on visceral glucose metabolism in steers. In Energy metabolism in animals, A. Chwalibog and K. Jakobsen, (eds.) EAAP Publication No. 103. Wageningen Academic Publishers, the Netherlands: 273-276.

Hunter, G.D. and G.C. Millson, 1964. Gluconeogenesis in the lactating dairy cow. Res. Vet. Sci. 5, 1-6.

Huntington, G.B., R.L. Prior and R.A. Britton, 1980. Glucose and lactate absorption and metabolic inter-relationships in lambs switched from low to high concentrate diets. J. Nutr. 110, 1904-1913.

Huntington, G.B, D.L. Harmon and C.J. Richards, 2006. Sites, rates, and limits of starch digestion and glucose metabolism in growing cattle. J. Anim. Sci.. 84 (E. Suppl.), E14-E24.

Kaufmann, M. and H. Magne, 1906. Sur la consommation du glucose du sang par le tissu de la glande mammaire. Comptes rendus de l'Académie des sciences. 143, 779-782.

Kaufman, C.F. and E.N. Bergman, 1974.Renal ketone body metabolism and gluconeogenesis in normal and hypoglycemic sheep. Am. J. Physiol. 226, 827-832.

König, B.A., J.D. Oldham and D.S. Parker, 1984. The effect of abomasal infusion of casein on acetate, palmitate and glucose kinetics in cows during early lactation. Br. J. Nutr. 52, 319-328.

Krebs, H.A. 1972. Some aspects of the regulation of fuel supply in omnivorous animals. Adv. Enzyme Regul. 10, 397-420.

Kristensen, N.B., B.M.L. Raun and B.A. Røjen, 2006. Splanchnic metabolism of [U-^{13}C]glucose in lactating dairy cows. J. Dairy Sci. 89 (Supp.1), 73.

Lapierre, H., J.P. Blouin, J.F. Bernier, C.K. Reynolds, P. Dubreuil, and G.E. Lobley, 2002. Effect of supply of metabolizable protein on whole body and splanchnic leucine metabolism in lactating dairy cows. J Dairy Sci 85, 2631-2641.

Larsen, M. 2009. Starch to transition dairy cows. Ph.D. Thesis. University of Copenhagen, Denmark. 113 p.

Larsen, M. and N.B. Kristensen, 2009. Effect of abomasal glucose infusion on splanchnic and whole-body glucose metabolism in periparturient cows. J. Dairy Sci. 92, 1071-1083.

Lemosquet, S., G. Raggio, H. Lapierre, J. Guinard-Flament, and H. Rulquin, 2007. Effect of protein supply on whole body glucose rate of appearance and mammary gland metabolism of energy nutrients in ruminants. EAAP publication No.124, I. Ortigues-Marty (ed.), Wageningen Academic Publishers, the Netherlands: 581-582.

Lemosquet, S., G. Raggio, G. E. Lobley, H. Rulquin, J. Guinard-Flament, and H. Lapierre, 2009a. Whole-body glucose metabolism and mammary energetic nutrient metabolism in lactating dairy cows receiving digestive infusions of casein and propionic acid. J. Dairy Sci. 92, 6068-6082.

Lemosquet, S., E. Delamaire, H. Lapierre, J. W. Blum, and J. L. Peyraud, 2009b. Effects of glucose, propionic acid, and nonessential amino acids on glucose metabolism and milk yield in Holstein dairy cows. J. Dairy Sci. 92, 3244-3257.

Lemosquet, S., G. Raggio, P. Faverdin, J. Guinard-Flament, C. Hurtaud, J Van Milgen and H. Lapierre, 2010. How increasing protein supply or glucogenic nutrients modify mammary metabolism in lactating dairy cows. 3rd EAAP International symposium on energy and protein metabolism. Parma, Italy.

Lindsay, D.B., 2006. Ruminant metabolism in the last 100 years. J. Agric. Sci. 144, 205-219.

Lobley, G.E. and H. Lapierre, 2003. Post-absorptive metabolism of amino acids. EAAP plublication No.109. W.B. Souffrant and C.C. Metges (eds.), Wageningen Academic Publishers, the Netherlands: 737-756.

Lobley, G.E., X. Shen, G. Le, D.M. Bremner, E. Milne, C.A. Graham, S.E. Anderson and N. Dennison, 2003. Oxidation of essential amino acids by the ovine gastrointestinal tract. Br. J Nutr. 89, 617-630.

Loncke, C., I. Ortigues-Marty, J. Vernet, H. Lapierre, D. Sauvant and P. Nozière, 2009. Empirical prediction of net portal appearance of volatile fatty acids, glucose and their secondary metabolites (β-hydroxybutyrate, lactate) from dietary characteristics in Ruminants: a meta-analysis approach. J. Anim. Sci. 87, 253-268.

Mepham, T.B. 1993. The development of ideas on the role of glucose in regulating milk secretion. Aust. J. Agric. Res. 44, 509-522.

Metcalf, J.A., L.A. Crompton, D. Wray-Cahen, M.A. Lomax, J.D. Sutton, D.E. Beever, J.C. Macrae, B.J. Bequette, F.R.C. Backwell and G.E. Lobley, 1996. Responses in milk constituents to intravascular administration of two mixtures of amino acids to dairy cows. J Dairy Sci. 79, 1425-1429.

National Research Council (NRC), 2001. Nutrient Requirements of Dairy Cattle. 7th rev. ed. Natl. Acad. Sci., Washington, DC.

Pacheco, D., C.G. Schwab, R. Berthiaume G. Raggio and H. Lapierre, 2006. Comparison of net portal absorption with predicted flow of digestible amino acids, a scope for improving current models? J. Dairy Sci. 89, 4747-4757.

Piccioli Cappelli, F.P., C.J. Seal and D.S. Parker, 1997. Glucose and [^{13}C]leucine metabolism by the portal-drained viscera of sheep fed on dried grass with acute intravenous and intraduodenal infusions of glucose. Br. J. Nutr. 78, 931-946.

Raggio, G., G.E. Lobley, R. Berthiaume, D. Pellerin, G. Allard, P. Dubreuil and H. Lapierre, 2007. Effect of protein supply on hepatic synthesis of plasma and constitutive proteins in lactating dairy cows. J. Dairy Sci. 90, 352-359.

Reynolds, C.K. 2006. Production and metabolic effects of site of starch digestion in dairy cattle. Anim. Feed Sci. Tech. 130, 78-94.

Reynolds, C.K., H.F. Tyrrell and P.J. Reynolds, 1991. Effect of diet forage-to-concentrate ratio and intake on energy metabolism in growing beef heifers: net nutrient metabolism by visceral tissues. J. Nutr. 121, 1004-1015.

Rius, A.G., J. Escobar, O. Becvar, D. Kirovski and M.D. Hanigan, 2009. Mammary cell signalling responses to abomasal starch and casein infusions in lactating dairy cows. J. Dairy Sci. 92 (E-Suppl.1), 469.

Schwab, C.G., L.D. Satter and A.B. Clay, 1976. Response of lactating dairy cows to abomasal infusion of amino acids. J. Dairy Sci. 59, 1254-1270.

Seal, C.J. and D. S. Parker, 1994. Effect of intraruminal propionic acid infusion on metabolism of mesenteric- and portal-drained viscera in growing steers fed a forage diet: I. Volatile fatty acids, glucose, and lactate. J. Anim. Sci. 72, 1325-1334.

Stoll, B., D. G. Burrin, J.F. Henry, F. Jahoor and P.J. Reeds, 1999. Dietary and systemic phenylalanine utilization for mucosal and hepatic constitutive protein synthesis in pigs. Am. J. Physiol. Gastrointest. Liver Physiol. 276, G49-G57.

Van Der Walt, J.G., G.D. Baird, and E.N. Bergman, 1983. Tissue glucose and lactate metabolism and interconversions in pregnant and lactating sheep. Br. J. Nutr. 50, 267.

Wolfe, R.R. 1992. Radioactive and Stable Isotope tracers in Biomedicine, Principles and Practice of Kinetic Analysis. John Wiley & Sons Inc, New York.

Wolf, J.E. and E.N. Bergman, 1972, Gluconeogenesis from plasma amino acids in fed sheep. Amer. J. Physiol. 223, 455-460.

Wray-Cahen, D., J.A. Metcalf, F.R.C. Backwell, B.J. Bequette, D.S. Brown, J.D. Sutton and G.E. Lobley, 1997. Hepatic response to increased exogenous supply of plasma amino acids by infusion into the mesenteric vein of Holstein-Friesian cows in late gestation. Br. J. Nutr. 78, 913-930.

Effects of energy and protein supply on milk protein yield responses in dairy cows

P. Huhtanen[1] *and A.N. Hristov*[2]
[1]*Department of Agriculture for Northern Sweden, Swedish University of Agricultural Sciences, Umeå, Sweden*
[2]*Department of Dairy and Animal Sciences, Pennsylvania State University, University Park, Pennsylvania, USA*

Abstract

Consumers demand production of animal protein with minimal environmental impact, which requires improving the efficiency of conversion of feed nitrogen into milk and meat protein. Accurate determination of animal protein requirements and evaluation of dietary protein supply are critically important for optimizing production with minimum nitrogen input in dairy production systems. As milk nitrogen efficiency is inversely related to dietary nitrogen input, diets with greater protein concentration predictably result in lower milk protein conversion efficiency. Most current feeding systems, however, assume a constant efficiency of conversion of metabolizable protein, above maintenance requirements, into milk protein. Thus, milk protein yield is overestimated at high and underestimated at low protein intakes. This analysis discusses milk protein responses to dietary energy and protein supply in dairy cows and suggests improvements to current milk protein yield-predicting models. A meta-analysis of studies with post-ruminally infused casein showed transfer efficiency of casein into milk protein of 0.29, which is clearly below default values of conversion of metabolizable protein into milk protein in protein evaluation systems. This decreased efficiency can be partially explained by enhanced liver metabolism and use of absorbed amino acids as an internal energy source by the mammary gland. Previous meta-analysis showed that milk protein yield can be predicted better from energy than from metabolizable protein supply, which along with substantial and consistent production responses to post-ruminal casein infusions suggests that inaccurate estimation of metabolizable protein supply, particularly from feed, contributes to variable efficiency of metabolizable protein utilization. Data from milk production and nutrient infusion studies demonstrate an important role of energy supply in transferring metabolizable protein into milk protein. An empirical model based on energy intake and feed metabolizable protein predicted milk protein yield responses better than models based on the supplies of metabolizable protein or metabolizable energy, demonstrating the significant role of energy in protein utilization. Increased energy supply provides other nutrients, glucose and acetate, that are needed to support increased milk production in addition to amino acids. Future models should integrate the concepts of nitrogen metabolism in the digestive tract and tissue metabolism of amino acids and translate this into mechanistic models that are capable of predicting milk protein yield and efficiency of nitrogen utilization for wide ranges of dietary inputs and genetic potential of the cows. To develop such models that are better than current empirical models, not only understanding of whole body amino acid metabolism must be improved, but also more accurate estimates of the supply of absorbed, especially feed, amino acids are needed.

Introduction

With the increased consumer concerns about the impact of animal production on the environment, more attention has been directed in recent years to the nitrogen (N) efficiency of dairy production. Proper determination of animal protein requirements and evaluation of protein supply, usually expressed as metabolisable protein (MP), are critically important for optimizing production with minimum N input in dairy production systems. Although, depending on the basal diet, increasing N input may produce an increase in milk protein yield (MPY), the efficiency of conversion of dietary N into milk protein will predictably decrease (Colmenero and Broderick, 2006). Dietary CP concentration was the best single predictor of milk N efficiency, while CP degradability apparently had only a minor effect (Huhtanen and Hristov, 2009).

To optimize milk protein production, requirements must be accurately determined and matched with dietary supply. Through microbial digestion in the reticulo-rumen, ruminants have a capacity to utilize forages that are indigestible in non-ruminants. Dietary proteins are degraded in the rumen and used in microbial protein synthesis, which modifies dietary supply of protein both quantitatively and qualitatively making predictions of the amount and profile of amino acids (AA) absorbed from the small intestine difficult.

Feed protein evaluation systems (e.g. Vérité and Peyraud, 1989; NRC, 2001) assume a constant efficiency of conversion of MP above maintenance requirement to milk protein. However, most prediction schemes overestimate milk protein yield from dairy cows at high protein intakes and underestimate it at low protein intakes (Doepel *el al.*, 2004; Lapierre *et al.*, 2007). The decreased efficiency of the transfer of absorbed AA into milk protein at high protein diets can be biologically explained by intermediary metabolism of AA between the duodenum and the mammary gland (Lapierre *et al.*, 2006b). Because energy and bacterial MP are better predictors of MPY than total MP (Huhtanen and Hristov, 2009), it can be suggested that inaccurate estimates of the supply of absorbed AA, especially feed AA, by the current protein evaluation systems can also have an impact on variable efficiencies of transfer of MP into milk protein. The objective of this paper is to discuss MPY responses to energy and protein supply in milk production trials. We will focus on specific aspects of the evaluation of the supply of absorbed AA (MP) from the small intestine and suggest improvements to current models predicting milk protein yield.

Casein infusion

Post-ruminal casein infusions have often been used to study milk and protein yield responses to dietary protein. Casein infusion studies have some methodological benefits in examining potential MPY responses, since the change in input of rumen undegraded protein (RUP) can accurately be determined and the effects on rumen metabolism and feed intake are marginal.

For estimating the marginal protein utilisation of post-ruminally infused casein (mean 0.34 kg/d; SD = 0.16) a meta-analysis of published data (75 diets, 40 comparisons) was conducted. The mean DM intake and milk yield were 15.9 and 22.8 kg/d, respectively. Dietary CP concentration in the control treatments was 152 (SD=16) g/kg DM and milk N efficiency, calculated as MPY/CP intake, was 0.282. The regression equations included a random intercept and slope effect for each study in the database (St-Pierre, 2001). Casein infused was converted to MP by assuming values of 90 and 100% for the CP concentration and intestinal digestibility, respectively. The latter value was estimated by the Lucas test from the data of Whitelaw *et al.* (1986), who infused casein at four graded levels.

The linear transfer efficiency of casein into milk protein was 0.285 (Figure 1) that is clearly below the default values of the conversion efficiency of MP to milk protein (e.g. 0.67 in NRC, 2001). The quadratic model improved the fit of the data indicating diminishing efficiency of MP utilization with increased supply. The maximum MPY response of 160 g/d was predicted at 0.94 kg casein per day. The CP concentration of the basal diet had no influence on marginal MPY response to infused casein, but the response was greater in studies with low yielding cows compared with studies with high yielding cows (0.35 vs. 0.25). Because of its ideal AA composition, the marginal MPY responses to post-ruminal casein infusions may be considered as the maximum that can be attained in practical feeding conditions by increasing the supply of RUP. A smaller MPY response to abomasal infusion of soya protein isolate compared with casein (Choung and Chamberlain, 1993) suggests that with practical diets the responses to increased supply of RUP are likely to be smaller than those observed with casein.

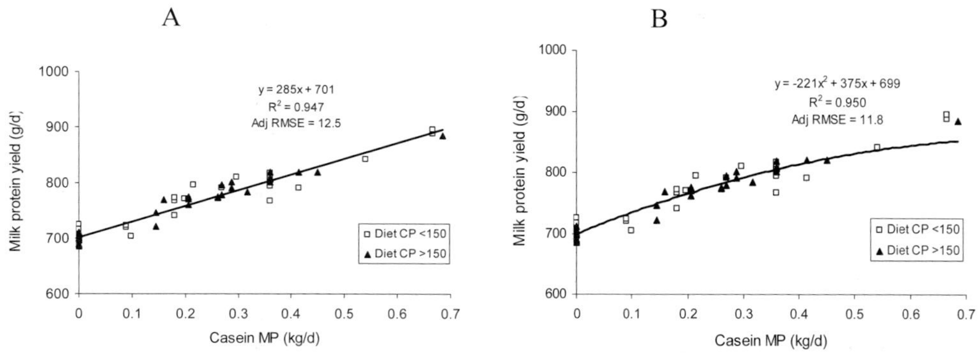

Figure 1. The linear (A) and quadratic (B) effects of post-ruminal casein infusions on milk protein yield. The symbols (□ and ▲) represent studies with dietary CP concentration below or above 150 g/kg DM, respectively.

Relatively small MPY response to increased protein supply can most likely be related to the fact that other nutrients are limiting milk production. In the present dataset, milk energy output increased 14.4 MJ per kg incremental MP from casein. Assuming that the carbon chain was entirely used for ATP synthesis, ME concentration of casein would be 18 MJ/kg DM (Guinard *et al.*, 1994) and the efficiency of incremental ME utilisation was approximately 0.80. This value is higher than utilization of ME above maintenance for milk production (0.60-0.64), and suggests that incremental ME is either used with a higher efficiency or body stores provided additional nutrients required to support production. Increased MPY with infused casein was associated with at least similar increase in lactose yield that requires additional glucose as a substrate. Liver extraction of AA relative to portal absorption is increased at higher AA supply, such that the increment in post-liver supply is smaller than the increment in portal absorption (Guerino *et al.*, 1991; Bruckental *et al.*, 1997) that in turn is smaller than the increment in MP supply (Lapierre *et al.*, 2006a). It appears that the close link between milk protein and lactose output sets a limit for marginal protein utilization when the supply of AA increased without concomitant increases in energy and other nutrients.

Milk protein responses protein supply

The transfer efficiency of MP to milk protein was 0.21 and 0.33 in a meta-analysis of datasets derived from North American (n=736) and North European (n=998) milk production trials (Huhtanen and Hristov, 2009). These values are also markedly below the default value of the NRC (2001) system (0.68) of the utilization of MP above maintenance for milk protein synthesis. There was a strong quadratic relationship between MP supply and MPY, indicating diminishing returns to increased protein supply, supporting the view that the efficiency of MP transfer to milk protein is not a constant (Doepel *et al.*, 2004; Lapierre *et al.*, 2006b; 2007). The supply of MP was calculated according to NRC (2001), but most likely the same conclusion could be drawn with most other feed protein evaluation system. Decreased transfer coefficient can be explained by incomplete recovery of absorbed AA in the liver and increased AA uptake by the liver with increased MP intake. With casein infusion, mammary uptake of Group 2 AA (Ile, Leu, Lys, Val) increased to a greater extent than required for the synthesis of non-essential AA and oxidation of leucine was increased (Raggio *et al.*, 2006a), suggesting that these AA can also be used as an internal energy source in the mammary gland.

In addition to intermediary metabolism of AA between the duodenum and the mammary gland, biologically explaining the decreased efficiency of the transfer of absorbed AA into milk protein (Lapierre *et al.*, 2006b), errors in estimating the supply can also influence the estimates of the utilization MP for milk protein synthesis. Overestimating the increase in the supply of MP would result in underestimating the transfer efficiency of MP to milk protein and *vice versa*. Our meta-analysis (Huhtanen and Hristov, 2009) suggested that the protein evaluation systems are not very accurate in estimating differences between the diets in MP supply. Energy (expressed as total digestible nutrients) and bacterial MP predicted MPY responses at least as precisely as the total MP. On the other hand, consistent and predictable MPY responses to casein infusion (Figure 1) suggest that when the change in feed MP supply is accurately estimated, the total MP should improve MPY predictions compared with models based on energy or bacterial MP. Indeed, when the total MP was split into bacterial and feed MP components, the precision of MPY predictions were clearly improved for both datasets (Huhtanen and Hristov, 2009). The regression coefficients (g milk protein per g MP) were approximately 5-fold greater for bacterial MP than feed MP: 0.49 and 0.10 for the North American and 0.75 and 0.14 for the North European dataset, respectively.

The lower value for feed MP can be expected, since bacterial MP is associated with increased supply of other nutrients (energy), whereas feed MP provides only absorbed amino acids that may also have poorer profile compared with bacterial MP. The transfer efficiency of infused casein was clearly higher than that of feed MP in production trials (0.29 vs. 0.10 and 0.14). Higher efficiency can be expected for post-ruminally infused casein than for feed MP due to its better AA profile. In the North American dataset, the low coefficient can partly be due to overfeeding of MP, but in the North European dataset the diets supplied on average 93% of MP requirement calculated according to NRC (2001) and the cows should be responsive to increased MP supply.

In addition to the greater residual variance of MPY prediction model in the North European dataset (Huhtanen and Hristov, 2009), variances of the regression intercept and slope were also greater for the total MP model compared with bacterial MP and ME models. This can be due to: (1) a higher transfer efficiency of microbial protein compared with feed protein, (2) variable AA profile in feed MP, or (3) inaccurate estimates of feed MP. In the analysis of data from production trials, the effect of CP degradability estimated using tabulated degradation parameters (NRC, 2001) on MPY was highly significant when it was used with TDN and CP intake in MPY prediction models (Huhtanen and Hristov, 2009). However, this model predicted a transfer efficiency of only 0.06 for additional MP derived from reduced ruminal protein degradability. Because energy and bacterial MP predicted MPY better than total MP, it seems that the tabulated degradation parameters reflect qualitative rather than quantitative differences between feeds. The small effect of degradability is consistent with the results of Ipharraguerre and Clark (2005), who did not find any difference in milk yield when solvent extracted soybean meal was compared with treated soybean meal or other low degradability protein supplements. The small, or even non-existing responses to reduced ruminal feed CP degradability are surprising, as a number of studies have demonstrated that there are differences in duodenal feed protein flow and milk protein yield that are associated with differences in ruminal protein degradability (Reynal and Broderick, 2003; Reynal *et al.*, 2007).

Determination of MP supply

For optimization of protein nutrition within economic and environmental constraints, it is essential that the supply of AA absorbed from the small intestine can be determined accurately and precisely. The first step in determining the protein supply to the ruminant is estimating protein flow into the small intestine. Traditionally, measurement of compartmental nutrient flows within the digestive tract and microbial protein synthesis in the rumen have relied on sampling through simple T-cannulas fitted either in the abomasum or proximal duodenum (Harmon and Richards, 1997). Obtaining unrepresentative samples that may be contaminated with endogenous secretions into the abomasum are among the problems associated with this approach. Duodenal CP flow includes three major fractions: microbial CP, rumen undegraded protein and endogenous protein. The contribution of these fractions is variable, but usually microbial protein provides the majority of AA flowing to the small intestine of dairy cows (Clark *et al.* 1992). The flow of feed protein is usually calculated as a difference between total and microbial protein flow, but to determine the true net protein flow endogenous protein must be subtracted. Determination of endogenous protein flow is technically challenging and the estimates are derived from studies conducted in rather artificial conditions such as intragastric nutrient infusions (Ørskov *et al.*, 1986), or feeding N-free diets (Hart and Leibholtz, 1991). More recently, Quellet *et al.* (2002) have used ^{15}N-labelled leucine that allows estimation of endogenous protein that is free or incorporated into microbial protein.

The omasal sampling technique (Huhtanen *et al.,* 1997) allows digesta sampling from the omasal canal via a ruminal cannula without repeated entry into the omasum. This technique requires less surgical intervention than using abomasal or duodenal cannulas and it avoids potentially confounding contamination from abomasal secretions. The slope between NRC (2001) predicted and observed microbial CP flow was 1.26 (Broderick *et al.*, 2010) suggesting a greater contribution of microbial protein to total MP supply than the current models predict. If true, this could partly explain the much higher transfer coefficient of bacterial MP compared with feed MP in the Huhtanen and Hristov (2009) meta-analysis. A triple-marker system was used to measure DM flow and in most cases microbial protein synthesis was estimated with ^{15}N as a microbial marker. The efficiency of microbial N synthesis increased significantly with increased DM intake (Broderick *et al.*, 2010). Improved efficiency of microbial N synthesis is not taken into account in the NRC (2001) system, whereas the amount of fermentable substrate is discounted for feeding level effects. Taking these factors into account, it is possible that the range in microbial MP supply is greater than the predicted range. If true, this can also contribute to a high transfer coefficient of microbial MP to milk protein.

Calculations of feed MP are usually based on tabulated values for ruminal protein degradability estimated by incubating feed samples *in situ* for different length of periods and fitting the data to kinetic models for estimating degradation parameters. Our meta-analysis (Huhtanen and Hristov, 2009) indicated that using tabulated values of ruminal protein degradability and digestibility of RUP did not improve predictions of MPY. Actually, total MP calculated using the mean degradability and RUP digestibility values resulted in a smaller prediction error of MPY than MP calculated using tabulated values in the North European dataset. It could be argued that tabulated values are less accurate compared to the values determined in each study. However, considering the large between-laboratory variability in protein degradability determined by the *in situ* method (Madsen and Hvelpund, 1994), this is unlikely. In the evaluation of protein systems, a model with a constant degradability for all feeds resulted in a smaller variation in the efficiency of MP utilization within study than using the values determined *in situ* (Tuori *et al.* 1998), even though the values were determined in two laboratories using the same procedures.

Technical details of the *in situ* procedure are discussed in several reviews (e.g. Nocek, 1988; Nozière and Michalet-Doreau, 2000; Hvelplund and Weisbjerg, 2000) that provide insight into the sources of variation and methodological aspects of the procedure. Even though the technical problems could be solved, two false assumptions of the kinetic model used in computing effective ruminal protein degradability (Ørskov and McDonald, 1979; NRC, 2001) can influence the degradability values, and consequently change the relative differences in degradability and MP values of the feedstuffs. One model assumption is that the soluble A-fraction disappearing at zero time from the bags is degraded at infinite rate without any escape from the rumen. However, there is considerable experimental evidence based on different methodologies demonstrating that this assumption is not correct. Hristov and Broderick (1996) estimated the flow of N fractions from rumen pool sizes and fractional passage rates of rumen solid and liquid phase and found that a considerable fraction of feed N escaped rumen as soluble non-ammonia N (SNAN). Similarly, studies using the omasal sampling technique (Choi *et al.*, 2002; Reynal *et al.*, 2007) demonstrated an escape of feed SNAN from the rumen with peptides being quantitatively the most important component. Volden *et al.* (2002) used a single dose of silage soluble N fraction and estimated that 8-10% of this fraction escaped as amino-N. Furthermore, studies based on intrinsically ^{15}N-labelled forages reported that an important fraction of feed N escaped the rumen with the liquid phase as amino-N (Hristov *et al.*, 2001; Ahvenjärvi *et al.,*

2007). Finally, the effect of the proportion of silage soluble non-ammonia N (soluble N - ammonia N) was insignificant, when MPY responses were predicted from MP supply that was calculated using a constant degradability for forage CP. This approach assumes that the solubility does not influence degradability and the true MP value (Huhtanen *et al.*, 2008a).

The *in situ* kinetic models (Ørskov and McDonald, 1979; NRC, 2001) assume the rumen as a single compartment system with random passage of feed particles irrespective of their size or specific gravity. However, the marker kinetics studies based on duodenal sampling have strongly indicated that the passage of feed particles cannot be described by these models (Ellis *et al.,* 1994; Huhtanen *et al.*, 2006). Both intrinsically (ADF-^{15}N) (Huhtanen and Hristov, 2001) and extrinsically (Lund *et al.*, 2006) labelled forages showed an ascending phase of marker excretion curves, and using the passage rate estimated from the descending phase of marker curves clearly underestimates the residence time in the rumen during which the feed is subjected to degradation. In the meta-analysis of Huhtanen and Hristov (2009), the mean passage rate of forages in the NE data averaged 0.05 per h when calculated using the NRC (2001) equations. The corresponding rumen residence time (20 h) is markedly shorter than estimated from duodenal marker excretion curves (Huhtanen and Hristov, 2001; Lund *et al.*, 2006), or from indigestible NDF passage rate estimated by rumen evacuation technique (Huhtanen *et al.*, 2006).

Using a biologically more correct kinetic model to calculate protein degradability from the kinetic data will influence the degradability values with the extent of change depending on the type of feed. Potential escape of the soluble A-fraction will decrease more degradability of feeds with a large A-fraction, whereas including retention in the first rumen compartment (large particle pool) will increase estimated degradability more for feeds with low soluble A-fraction. Overall, using a biologically more correct kinetic model will reduce the differences in estimated ruminal protein degradability, and consequently will have an impact on the transfer efficiency of feed and total MP into milk protein. Indeed, in the meta-analysis of omasal flow data, the slope between NRC (2001) predicted and observed feed protein flow varied between 0.70 to 0.78, depending on how endogenous protein was taken into account (Broderick *et al.*, 2010).

Effects of energy supply on milk yield responses

The models based on energy supply predicted milk protein responses at least as well as estimated MP supply (Huhtanen and Hristov, 2009). In ruminants, energy intake influences both absorbed AA and energy supplies, mainly because microbial protein synthesis is closely related to the supply of fermentable substrates. Microbial protein synthesis was directly related to OM truly digested in the rumen (Broderick *et al.,* 2010), whereas composition of the diet (CP, RDP, NDF, NFC) had only marginal, if any, effects on the efficiency of microbial protein synthesis. The supply of glucose and acetate available for the mammary gland are also directly related to energy intake, and consequently, to microbial protein flow.

Interactions between energy and protein have been investigated using post-ruminal casein or AA infusion and either ruminal propionate or abomasal glucose infusion. No interactions between glucose and casein infusion on MPY were reported (Clark *et al.*, 1977; Vanhatalo *et al.* 2003). Similarly, the combinations of casein and propionate (Raggio *et al.,* 2006b) or histidine and glucose (Huhtanen *et al.*, 2002) showed additive responses (Figure 2). These results demonstrate that the efficiency of AA utilization can be improved by increasing the supply of energy substrates while maintaining protein supply. These effects can be due to a reduced amount of AA used for gluconeogenesis and, in consequence, more AA available for milk protein production.

Grouping the North European dataset (Huhtanen and Hristov, 2009) according to study type (protein supplementation, concentrate supplementation, silage digestibility and silage fermentation quality)

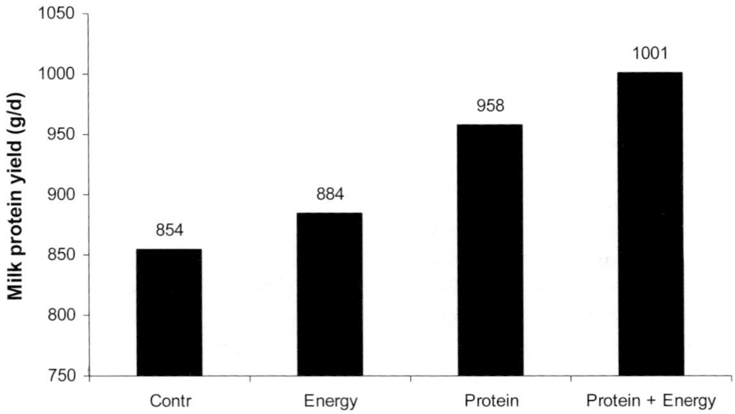

Figure 2. The effects of energy [glucose (3 studies) or propionate (1 study)] and protein [casein (3) or histidine (1)] infusions on milk protein yield (data from Clark et al.*, 1977; Huhtanen* et al.*, 2002; Vanhatalo* et al.*, 2003; Raggio* et al.*, 2006b).*

showed large variation in the transfer efficiency of calculated MP (NRC, 2001) into milk protein (Table 1). The transfer coefficient of MP was the smallest in protein supplementation studies, where energy components (usually grain) in the concentrate were replaced with protein supplements such as soybean, rapeseed and fishmeal. Considering that the AA profile of protein supplements is generally poorer than that of casein and the possible overestimation of the differences in protein degradability, the MP response in protein supplementation studies was relatively high compared with casein infusion data. However, protein supplementation has increased feed intake and diet digestibility (Oldham, 1980; Huhtanen *et al.*, 2008b; Nousiainen *et al.*, 2009), and consequently energy supply, thereby facilitating a greater transfer efficiency of MP. The highest transfer efficiency of MP was observed when silage digestibility was changed by the stage of maturity at harvest. Intermediate transfer efficiency was found when all data were analyzed together. Milk protein yield responses both to ME (3.27 vs. 4.47 g/MJ ME) and MP (189 vs. 308 g/kg MP) were smaller in the North American

Table 1. Milk protein yield responses (g/d) to ME (MJ/d) and MP supply (kg/d) when nutrient supply is manipulated by different strategies (data from Huhtanen and Hristov, 2009).

	N[1]	Intercept	Slope	SE	Adj. RMSE[2]	AIC[3]
ME models						
All data	982	177	4.47	0.12	19.1	9,788
Concentrate level	217	141	4.80	0.14	14.5	2,210
Protein level	335	-8	5.86	0.17	18.9	3,375
Silage digestibility	79	162	4.61	0.38	12.2	823
Silage fermentation	237	170	4.32	0.19	13.1	2,315
MP models						
All data	982	270	308	11.4	20.7	10,057
Concentrate level	217	135	402	16.0	12.3	2,264
Protein level	335	388	254	13.2	11.7	3,488
Silage digestibility	79	21	454	38.5	15.1	833
Silage fermentation	237	99	408	17.9	12.9	2,320

[1] N = Number of diets.

[2] Adj. RMSE = Residual mean square error adjusted for random study effect.

[3] AIC = Akaike's information criteria.

studies compared with the North European studies indicating a higher plane of nutrition and lower responsiveness to increased nutrient supply.

The MPY responses to increased ME supply varied depending on how the nutrient supply was modified, but variation was much smaller than that observed for MP models suggesting that ME is a more uniform 'currency' in predicting MPY responses than MP with different diets. A greater coefficient in protein supplementation studies is probably associated with positive effects of increased protein supply on milk yield as the higher transfer efficiency of ME above maintenance for energy corrected milk production (0.180 vs. 0.125 kg/MJ ME) suggests.

The data in Figure 2 and Table 1 indicate that the transfer efficiency of MP into milk protein is strongly influenced by associated changes in the supply of other nutrients, or more generally energy. To test that hypothesis, the transfer efficiency of MP was plotted against the ME/MP ratio in incremental nutrient supply. There was a strong positive relationship between the transfer efficiency of MP and the nutrient ratio (Figure 3). The intercept of regression suggests that transfer efficiency of MP to milk protein is only 0.08 when the increase in MP supply is not associated with any increase in energy supply.

Previous meta-analysis (Doepel *et al.*, 2004; Huhtanen and Hristov, 2009) have demonstrated decreasing transfer efficiency of MP into milk protein with increased MP supply. Therefore, variable, rather than fixed, factors for transfer efficiencies must be incorporated into future predictive models (Lapierre *et al.*, 2006b). Developing such empirical models using only MP supply as an independent variable may not be very successful due to the strong influence of associated changes in energy supply on the transfer efficiency of MP. In the future, mechanistic models taking into account both intermediary metabolism of AA between duodenum and mammary gland and the supply of energy yielding nutrients are required for accurate predictions of MPY responses for different diets. Meanwhile, empirical models based on energy and ME supplies as variables predicting MPY responses can improve the accuracy of MPY predictions compared with the current linear models. The following MPY (g/d) models were developed from the North European dataset (Huhtanen and Hristov, 2009) using the supplies of ME available for production (MJ/d) and feed MP (kg/d) as independent variables:

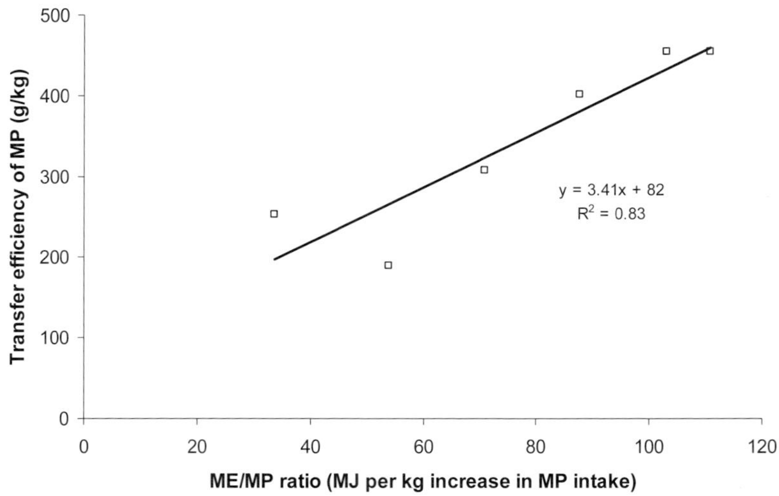

Figure 3. Relationship between ME/MP ratio in nutrient supply and transfer efficiency of MP into milk protein. The ME/MP ratio was estimated by a mixed model regression analysis (random study effect) with a model: ME (MJ/d) = MP (kg/d)

$$MPY = 201(\pm 16) + 3.73(\pm 0.13) \times ME + 102(\pm 10) \times Feed\ MP\ (Adj.\ RMSE = 17.8) \quad (1)$$

$$MPY = 157(\pm 18) + 3.84(\pm 0.13) \times ME + 278(\pm 34) \times Feed\ MP - 124(\pm 23) \times Feed\ MP^2\ (Adj.\ RMSE = 17.5) \quad (2)$$

Prediction errors of these models were smaller than of models based only on ME or MP supplies (Table 1). The quadratic effect of feed MP was significant and the prediction error was slightly smaller than for the linear model. The small regression coefficient for feed MP suggests that increases in MPY are marginal, when the maximum ME intake response to supplementary protein is achieved. The quadratic equation predicted that the maximum MPY response above the mean feed MP intake was 34 g/d and to reach it, about 0.5 kg/d of additional feed MP was needed. In omasal flow data (Broderick *et al.*, 2010) the transfer efficiency of feed MP was 0.144 when analyzed with the ME + MP model. Feed N flow was corrected for endogenous protein according to Ørskov *et al.* (1986) and a value of 0.75 was used for digestibility of RUP. The higher transfer efficiency in the omasal flow data compared with the data from production trials (0.144 vs. 0.102) is consistent with the flow data that showed smaller differences in feed N flow (Broderick *et al.*, 2010) than predicted by NRC (2001) from tabulated values based on the *in situ* data.

When the data was grouped by mean milk yield in the study into three groups of about equal size (<24, 24-29 and >29 kg energy corrected milk/d), MPY responses to increased feed MP supply were similar among the groups levels (Figure 4). In these data, the increased milk yield is both due to increased genetic potential and improved diets. The relationship between feed MP intake and MPY was similar among the three groups and does not support the view that transfer efficiency of MP would be related to the genetic potential of the cows. An increased efficiency of microbial protein synthesis may explain why high producing cows do not benefit more from increased feed MP supply than low producing cows.

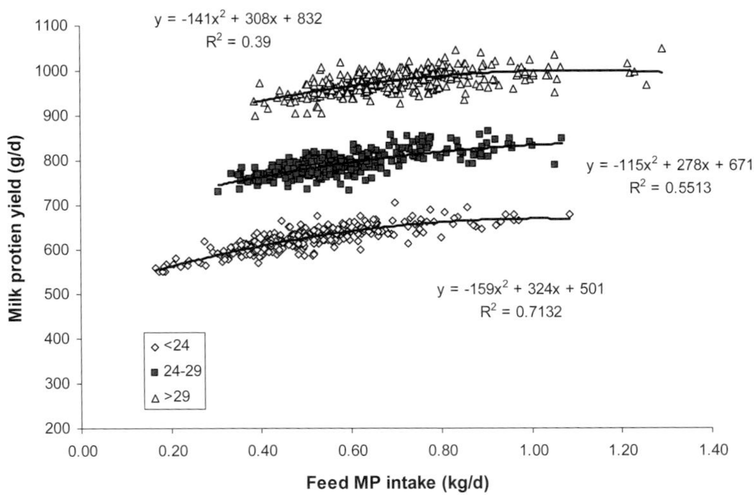

Figure 4. Relationship between feed MP supply and milk protein yield in cows at different production levels. The values are adjusted for the random study effect and differences in ME intake within each group (data from Huhtanen and Hristov, 2009).

Conclusions

Transfer efficiency of MP into milk protein is not constant, and variable efficiencies must be used to predict more accurately milk protein yield responses to increased MP supply. Increased AA metabolism in the body tissues biologically explains the reduced transfer efficiency with increased MP supply. Milk protein yield was predicted better from energy than MP supply, which together with substantial and consistent production responses to post-ruminal casein infusions suggests that inaccurate estimation of MP supply, especially of the feed MP component, contributes to the variable efficiency of MP utilization. Data from both milk production and nutrient infusion studies demonstrated an important role of energy supply in transferring MP into milk protein. An empirical model based on energy intake and feed MP, predicted MPY responses better than model based on the supplies of MP or ME, demonstrating the significant role of energy in protein utilization. An increased energy supply provides other nutrients, glucose and acetate, that are needed to support increased milk production in addition to AA. Future models should integrate the concepts of N metabolism in the digestive tract and metabolism of AA in the body tissues to mechanistic models that are capable in predicting MPY and efficiency of N utilization for wide ranges of dietary inputs and genetic potential of the cows. To develop such models that are better than current empirical models, not only understanding of whole body AA metabolism must be improved, but also more accurate estimates of the supply of absorbed AA, especially feed AA, are needed.

References

Ahvenjärvi, S., A. Vanhatalo, P. Huhtanen and A. Hristov, 2007. Ruminal metabolism of 15N labelled ammonium-N and grass silage soluble non-ammonia N. J. Dairy Sci. 90 (Suppl. 1), 232. (Abstr.).

Broderick, G.A., P. Huhtanen, S. Ahvenjärvi, S.M. Reynal and K.J. Shingfield, 2010. Quantifying ruminal nitrogen metabolism using the omasal sampling technique in cattle – A meta-analysis. J. Dairy Sci., in press.

Bruckental, I., G.B. Huntington, C. Kirk Baer, and R.A. Erdman. 1997. The effect of abomasal infusion of casein and recombinant somatotropin hormone injection on nitrogen balance and amino acid fluxes in portal-drained viscera and net hepatic and total splanchnic blood in Holstein steers. J. Anim. Sci. 75, 1119-1129.

Choi, C.W., S. Ahvenjärvi, A. Vanhatalo, V. Toivonen and P. Huhtanen, 2002. Quantitation of the flow of soluble non-ammonia nitrogen entering the omasal canal of dairy cows fed grass silage based diets. Anim. Feed Sci. Technol. 96, 203-220.

Choung, J.J.and D.G. Chamberlain, 1993. The effects of abomasal infusions of casein or soya-bean-protein isolate on the milk production of dairy cows in mid lactation. Br. J. Nutr. 69, 103-115.

Clark, J.H., H.R, Spires, R.G. Derrig and M.R. Bennink, 1977. Milk production, nitrogen utilization and glucose synthesis in lactating cows infused postruminally with sodium caseinate and glucose. J. Nutr. 107, 631-644.

Clark J.H., T.H Klushmeyer and M.R. Cameron, 1992. Microbial protein synthesis and flow of nitrogen fractions to the duodenum of dairy cows. J. Dairy Sci. 75, 2304-2323.

Colmenero, J.J. and G.A. Broderick, 2006. Effect of dietary crude protein concentration on milk production and nitrogen utilization in lactating dairy cows. J. Dairy Sci. 89, 1704-1712.

Doepel, L., D. Pacheco, J.J. Kennelly, M.D. Hanigan, I.F. López, and H. Lapierre, 2004. Milk protein synthesis as a function of amino acid supply. J. Dairy Sci. 87:1279-1297.

Guerino, F., G.B. Huntington and R.A. Erdman. 1991. The net portal and hepatic flux of metabolites and oxygen consumption in growing beef steers given postruminal casein. J. Anim. Sci. 69, 387-395.

Guinard, J., H. Rulquin and R. Verite. 1994. Effect of graded levels of duodenal infusions of casein on mammary uptake in lactating cows. 1. Major nutrients. J. Dairy Sci. 77, 2221-2231.

Harmon, D.L. and C.J. Richards, 197. Considerations for gastrointestinal cannulation in ruminants. J. Anim. Sci. 75, 2248-2255.

Hart, F.J. and J. Leibholz, 1990. A note on the flow of endogenous protein to the omasum and abomasums of steers. Anim. Prod. 51, 217-219.

Hristov, A.N. and G.A. Broderick, 1996. Synthesis of microbial protein in ruminally cannulated cows fed alfalfa silage, alfalfa hay, or corn silage. J. Dairy Sci. 79, 1627-1637.

Hristov, A.N., P. Huhtanen, L.M. Rode, T.A. McAllister and S.N. Acharya, 2001. Comparison of the ruminal metabolism of nitrogen from ^{15}N-labelled alfalfa preserved as hay or as silage. J. Dairy Sci. 84, 2738-2750.

Huhtanen, P., P.G. Brotz and L.D. Satter, 1997. Omasal sampling technique for assessing fermentative digestion in the forestomach of dairy cows. J. Anim. Sci. 75, 1380-1392.

Huhtanen, P. and A.N. Hristov, 2001. Estimating passage kinetics using fibre-bound ^{15}N as an internal marker. Anim. Feed Sci. Technol. 94: 29-41.

Huhtanen, P., A. Vanhatalo and T. Varvikko, 2002. Effects of abomasal infusions of histidine, glucose, and leucine on milk production and plasma metabolites of dairy cows fed grass silage diets. J. Dairy Sci. 85, 204-216.

Huhtanen, P., S. Ahvenjärvi, M.R. Weisbjerg and P. Nørgaard, 2006. Digestion and passage of carbohydrates. In: Sejrsen, K., T. Hvelplund and M.O. Nielsen (editors). Ruminant physiology: digestion, metabolism and impact of nutrition in gene impression, immunology and stress. Wageningen Academic Publishers, Wageningen, the Netherlands. Pp. 87-135.

Huhtanen, P., M. Rinne and J. Nousiainen, 2008a. Effects of silage soluble N components on metabolizable protein concentration: a meta-analysis of dairy cow production experiments. J. Dairy Sci. 91, 1150-1158.

Huhtanen, P., M. Rinne and J. Nousiainen, 2008b. Evaluation of the concentrate factors affecting silage intake of dairy cows; a development of the relative total diet intake index. Animal 2, 942-953

Huhtanen, P. and A.N. Hristov, 2009. A meta-analysis of the effects of protein concentration and degradability on milk protein yield and milk N efficiency in dairy cows. J. Dairy Sci. 92, 3222-3232.

Hvelplund, T. and M.R. Weisbjerg. 2000. *In situ* techniques for the estimation of protein degradability and postrumen availability. In: Givens, D.I. E. Owen, R.F.E. Axford and H.M. Omed (editors) Forage evaluation in ruminant nutrition.. CABI Publishing, Wallingford, UK: 233-258.

Ipharraguerre, I.R. and J.H. Clark, 2005. Impacts of the source and amount of crude protein on the intestinal supply of nitrogen fractions and performance of dairy cows. J. Dairy Sci. 88, E Suppl., E22-E37.

Guinard, J., H. Rulquin and R. Verite. 1994. Effect of graded levels of duodenal infusions of casein on mammary uptake in lactating cows. 1. Major nutrients. J. Dairy Sci. 77, 2221-2231.

Lapierre, H., D. Pacheco, R. Berthiaume, D.R. Ouellet, C.G. Schwab, P. Dubreuil, G. Holtrop and G.E. Lobley, 2006b. What is the true supply of amino acids for a dairy cow? J. Dairy Sci. 89, E. Suppl., E1-E14.

Lapierre, H., G.E. Lobley, D.R.Ouellet, L. Doepel and D. Pacheco, 2007. Amino acid requirements for lactating dairy cows: Reconciling predictive models and biology. In: Proc. Cornell Nutr. Conf. 2007.

Lapierre, H., R. Berthiaume, G. Raggio, M.C. Thiverge, L. Doepel, D. Pacheco, P. Dubreuil, G.E. Lobley, 2006a. The route of absorbed nitrogen to milk protein. Anim. Sci. 80, 11-22.

Lund, P., M.R. Weisbjerg and T. Hvelplund, 2006. Passage kinetics of fibre in dairy cows obtained from duodenal and faecal ytterbium excretion. Effect of forage type. Anim. Feed Sci. Technol. 128, 229-252.

Madsen, J. and T. Hvelplund, 1994. Prediction of *in situ* protein degradability in the rumen. Results of a European ringtest. Livest. Prod. Sci, 39:201-212.

Nocek, J.E., 1988. *In situ* and other methods to estimate ruminal protein and energy digestibility: a review. J. Dairy Sci. 71, 2051-2069.

Nousiainen, J., M. Rinne and P. Huhtanen, 2009. A meta-analysis of feed digestion in dairy cows. 1. The effects of forage and concentrate factors on total diet digestibility. J. Dairy Sci. 92, 5019-5030.

Nozière, P. And B. Michalet-Doreau, 2000. *In sacco* methods. In: J.P.F. D´Mello (ed.) Farm Animal Metabolism and Nutrition. CAB International: 233-253.

NRC. 2001. Nutrient requirements of dairy cattle. 7th rev. ed. Natl. Acad. Press, Washington, DC.

Oldham, J.D., 1984. Protein energy relationships in dairy cows. J. Dairy Sci. 67, 1090-1114.

Ørskov E.R. and I. McDonald, 1979. The estimation of protein degradability in the rumen from incubation measurements weighted according to rate of passage. J. Agric. Sci., Camb. 92, 499-503.

Ørskov, E.R., N.A. Macleod,and D.J. Kyle, 1986. Flow of nitrogen from the rumen and abomasum in cattle and sheep given proteinfree nutrients by intragastric infusion. Br. J. Nutr. 56, 241-248.

Ouellet, D.R., M. Demers, G. Zuur, G.E. Lobley, J.R. Seoane, J.V. Nolan and H. Lapierre, 2002. Effect of dietary fiber on endogenous nitrogen flows in lactating dairy cows. J. Dairy Sci. 85, 3013-3025.

Raggio, G., S. Lemosquet, G.E. Lobley, H. Rulquin and H. Lapierre. 2006a. Effect of casein and propionate supply on mammary gland protein metabolism. J. Dairy Sci. 89, 4340-4351.

Raggio, G., G.E. Lobley, S. Lemosquet, H. Rulquin and H. Lapierre, 2006b. Effect of casein and propionate supply on whole body protein metabolism in lactating dairy cows. Can. J. Anim. Sci. 86:81-89.

Reynal, S.M. and G.A. Broderick, 2003. Effects of feeding dairy cows protein supplements of varying ruminal degradability. J. Dairy Sci. 86:835-843.

Reynal, S.M, I.R. Ipharraguerre, M. Liñeiro, A.F. Brito, G.A. Broderick and J.H. Clark, 2007. Omasal flow of soluble proteins, peptides, and free amino acids in dairy cows fed diets supplemented with proteins of varying ruminal degradabilities. J. Dairy Sci. 90, 1887-1903.

St-Pierre, N.R., 2001. Invited review: Integrating quantitative findings from multiple studies using mixed model methodology. J. Dairy Sci. 84, 741-755.

Tuori, M., K. Kaustell and P. Huhtanen, 1998. Comparison of the protein evaluation systems of feed for dairy cows. Livest. Prod. Sci. 55, 33-46.

Vanhatalo, A., T. Varvikko and P. Huhtanen, 2003a. Effects of casein and glucose on responses of cows fed diets based on restrictively fermented grass silage. J. Dairy Sci. 86, 3260-3270.

Vérité R. and J.-L, Peyraud, 1989. Protein: the PDI system. In: R. Jarrige (ed.). Ruminant Nutrition. Recommended Allowances and Feed Tables. INRA: 33-48.

Volden H., L.T. Mydland and V. Olaisen, 2002. Apparent ruminal degradation and rumen escape of soluble nitrogen fractions in grass and grass silage administered intraruminally to lactating dairy cows. J. Anim. Sci. 80, 2704-2716.

Whitelaw, F.G., J.S. Milne and E.R. Ørskov, 1986. The nitrogen and energy metabolism of lactating cows given abomasal infusions of casein. Br. J. Nutr. 55, 537-556.

Providing an 'ideal' profile of essential amino acids in the intestine increased milk protein yield in lactating dairy cows fed both below and above protein requirements

M.N. Haque, H. Rulquin, A. Andrade, P. Faverdin, J.L Peyraud and S. Lemosquet
INRA UMR 1080 Dairy Production, 35590 Saint-Gilles, France

Introduction

The efficiency of amino acid (AA) conversion into milk protein in lactating dairy cows can be improved by manipulating the AA profile of metabolisable protein, as demonstrated by post-ruminal supplementation of methionine (Met) and lysine (Lys). An ideal profile for intestinally absorbed essential AA (EAA) was proposed by Rulquin *et al.* (2007) including 10 EAA. However, this profile was never tested and was also not compared against a more simple mixture of EAA. Rulquin *et al.* (1994) also suggested that milk protein response could not be increased by adjusting the AA profile of Lys and Met in cows fed below their protein requirements. The first objective of this study was to analyse the effect of the 'ideal' EAA profile proposed by Rulquin *et al.* (2007) on milk yield and composition in dairy cows fed either a low protein (LP) or a high protein (HP) diet and to compare its effects with an infusion of 4 EAA.

Material and methods

Six multiparous Holstein dairy cows at 26±7 weeks of lactation received six treatments combining two levels of protein supply by the diet and 3 different mixtures of AA supplied by duodenal infusions. The experiment was divided in two principal 3 week periods where cows received either LP or HP treatment (diet plus AA infusions) according to a switchback design. Within each 3 week period, cows received one of the three AA infusions for a week according 3×3 Latin square designs. The LP and HP diets provided 90 and 130 g/kg of DM of crude protein (CP) respectively corresponding to 77.6 vs. 95.8 g/kg DM of PDIE (Protein digested in the small intestine expected from ruminal energy supply; INRA, 2007) or 12.2 vs. 14.9 g/MJ of PDIE/NE_L (Net Energy of lactation). They were composed (% on DM basis) of 70.5 vs. 64.4 maize silage, 0 vs. 9.5 soybean meal, 25.4 vs. 24.0 energy concentrate, 1.4 *vs*. 0.5 urea and 2.7 vs. 2.6 minerals for LP and HP, respectively. The diets were estimated to provide (in % of PDIE; INRA, 2007): 6.8% of Lys, 1.8% of Met, 8.4% of leucine (Leu), 2.1% of histidine (His), 4.8% of arginine (Arg), 5.3% of isoleucine (Ile), 4.9% of phenylalanine (Phe), 5.1% of threonine (Thr), 4.3% of tyrosine (Tyr), 5.8% of valine (Val) and 14.5% of glutamate (Glu). Three iso-PDI AA mixtures were duodenally infused to supplement LP and HP, providing 10 g/kg of DM of CP and PDIE, each. The three AA mixtures were 1) Glu (181 g/d) as control; 2) '4EAA': Lys (20 g/d), Met (14 g/d), Leu (19 g/d), His (21 g/d) and Glu (107 g/d); 3) 'Ideal' a mixture that contained similar amounts of Lys, Met, Leu, His as 4EAA plus Ile (7 g/d), Val (11 g/d), Phe (1 g/d), Trp (4 g/d), Tyr (22.9 g/d) and Glu (55 g/d). Results were analysed by using MIXED procedure of SAS (2004) with a random effect of cow according the model:

$$y_{ijklmo} = \mu + Cow_i + Period_j + Protein_k + Week(Period)_l + AA_m + Protein_k \times AA_m + Protein_k \times Week(Period)_l + \varepsilon_{ijklmo}$$

Two orthogonal contrasts were performed to compare the three AA treatments: 1) Contrast EAA (Glu *vs*. Ideal and 4EAA) and 2) and contrast Ideal (4EAA vs. Ideal).

Results and discussion

The interaction between the level of protein supply and AA infusions was never significant. The intake of DM was lower when LP diets were fed in agreement with Faverdin *et al.* (2003). Total

PDIE intakes (diet plus infusion) were 1650 g/d and 2089 g/d in LP and HP, respectively. Cows were in negative PDI balance in LP treatments (-90 g/d) and in positive PDI balance (96 g/d) in HP treatments, as expected. However in all treatments, NE_L balance was not significantly different from zero. Protein supply increased milk yield by 14% and protein yield by 19%, and tended to decrease lactose content (Table 1). Supply of EAA (contrast EAA) increased milk yield, protein yield and protein content by 3%, 8% and 5% respectively, however this increase was similar whether cows were infused with the '4EAA' or with the 'ideal' EAA mixture (contrast Ideal not shown). Milk fat content was decreased with EAA supplies (4EAA and Ideal vs. Glu). Gross efficiency of protein utilisation for milk protein and the net milk PDIE efficiency were both significantly improved during 4EAA and ideal treatments compared to Glu treatment.

In conclusion, our results clearly showed that a correction of the EAA profile as proposed by Rulquin *et al.* (2007) improved the protein efficiency independently of the level of protein supply. In addition, the supply of the four most limiting EAA in the diet (Lys, Met, His, Leu) was sufficient for increasing the protein efficiency similarly as adjusting the complete profile.

Table 1. Effect of AA profiles on dry matter intake (DMI) and milk production.

	LP[1]			HP[2]			P<		
	Glu	4EAA	Ideal	Glu	4EAA	Ideal	SEM	Prot[3]	EAA[4]
DMI[1], kg/d	18.7	18.8	19.2	19.9	20.0	19.9	0.20	0.01	NS
Milk yield, kg/d	28.8	29.5	29.6	32.7	33.7	33.7	0.81	0.02	0.02
Protein yield, g/d	802	854	869	955	1,037	1,023	20.8	0.03	0.01
Protein content, %	2.78	2.89	2.95	2.93	3.08	3.05	0.39	0.02	0.01
Fat content, %,	3.52	3.28	3.22	3.33	3.18	3.08	1.01	NS	0.01
Lactose content, %	4.96	4.81	4.80	4.91	4.75	4.75	0.27	0.07	NS
Gross PDIE Eff.[5]	0.49	0.52	0.52	0.46	0.50	0.49	0.01	NS	0.01
Milk PDIE Eff.[6]	0.63	0.68	0.68	0.56	0.61	0.60	0.01	NS	0.01

[1] Low protein supply; [2] High protein supply; [3] Protein supply contrast: LP vs. HP; [4] Contrast (4 EAA and Ideal) vs. Glu; [5] Gross efficiency of milk protein synthesis (milk protein yield / PDIE intake); [6] Milk PDIE efficiency of utilization (milk protein yield/[PDIE intake-PDIE requirement for maintenance]).

References

Faverdin, P., D. M'hamed and R. Verite. 2003. Effects of metabolizable protein on intake and milk production of dairy cows independent of effects on ruminal digestion. Anim. Sci. 76, 137-146.

INRA, 2007. Alimentation des bovines, ovins et caprins. Besoins des animaux- Valeurs des aliments. Tables INRA 2007. Editions Quae, Versailles, France. 305 pp.

Rulquin, H., C. Hurtaud and L. Delaby, 1994. Effects of dietary protein level on lactational responses of dairy cows to rumen-protected methionine and lysine. Ann Zootech. 43, 245 (Abstr.).

Rulquin, H., G. Raggio, H. Lapierre and S. Lemosquet, 2007. Relationship between intestinal supply of essential amino acids and their mammary metabolism in the lactating dairy cow. In: Ortigues-Marty I. (ed.), Energy and Protein Metabolism and Nutrition. Wageningen Academic Publishers, Wageningen, the Netherlands, EAAP publication N° 124: 587-588.

Balancing carbohydrates in dry rations for dairy cows

M. Fustini, A. Palmonari, G. Canestrari and A. Formigoni
DIMORFIPA, University of Bologna, Bologna, Italy

Introduction

Diets for cows producing milk for Parmigiano Reggiano cheese are not allow to contain silages. Thus the primary forage which can be fed is dry hay, alfalfa, grass and straw. Moreover the hay must be finely chopped to promote a well blended diet. The challenge which then arises is to provide enough peNDF (physically effective neutral detergent fiber), to allow chewing and rumination, and obtaining energy from concentrate without creating ruminal disorders. Several researchers described effects of different amount of energy and fiber on milk production and quality (Sutton *et al.* 2003; Zebeli *et al.* 2009), some others evaluated the importance of forage chopping length on ruminal pH and chewing activity (Yang and Beauchemin, 2009). The objective of this study was to evaluate two diets having the same content of rumen fermentable carbohydrates but originating from different sources.

Material and methods

The trail was conducted at the Faculty of Veterinary Medicine, Bologna, upon approval by the University Animal Care and Use Committee. Forages were chopped to a particle length of 3cm using a Zago Mixer (Padova, Italy) and included in a TMR offered daily. Fifty lactating Holstein cows were allocated to matched pairs of cows on the basis of calving date, milk yield, milk composition, live weight and parity. One cow from each pair was allocated randomly one to the two groups fed the experimental diets in a cross-over design during 18 d with 14 d of adaptation. Diets were formulated using CPM Dairy v.3.0.8 to meet the energy and protein requirement, allowing the same milk quality and production. The two diets were different for forage:concentrate ratio, which was 55:45 for the HAY diet (grass hay as only forage source), and 45:55 for STRAW diet (34% grass hay and 11% wheat straw). Straw was included in STRAW diet to cover the requirement for peNDF. The HAY diet was characterized by a high amount of fermentable structural carbohydrates, while the STRAW diet was rich in concentrate (Non Structural Carbohydrates) plus the addition of straw and more by-pass protein than the HAY diet. Milk composition was analyzed daily in the collection period for lactose, protein and fat, and the amount of milk produced was recorded.

Using the RO Tap system peNDF was measured daily during the collection period as percentage of diet NDF greater than 1.18 mm (Mertens, 1997). Eating and ruminating activities were recorded on d 14 through d 18 of each period (two 24-h periods/cow) using the Institute of Grassland and Environmental Research Behavior Recorders and Graze Jaw Movement Analysis Software (Ultra Sound Advice, London, UK; Kononoff *et al.*, 2002). Rumen fluid and fecal samples were collected on d 15 and 18 of each period right before feeding and 6 h later. Diet digestibility was calculated using acid insoluble ash. Feed samples were collected daily and composited weekly for analysis.

Statistical analysis was conducted using PROC MIXED of SAS (2006).

Results and discussion

Dry matter intake was similar between diets (24.6 kg with HAY versus 24.4 kg with STRAW). Total peNDF content (Mertens, 1997), was 21.5% for the HAY diet and 15.5% for the STRAW diet. Milk production did not differ across treatments (28.0 kg with HAY vs. 28.9 kg with STRAW), but the STRAW diet significantly increased protein (3.37% with HAY vs. 3.48% with STRAW) and lactose percentage. Chewing data (Table 1) show a longer time spent eating of the STRAW diet per kg of DM, NDF and peNDF compare to the HAY diet. No differences were found in ruminating

Table 1. Effects of different carbohydrates balance on chewing activity of lactating dairy cow.

			Hay	Straw	P-value
DM	Eating	min/kg SS	12.81±2.16	15.34±2.02	0.1
	Ruminating	min/kg SS	17.72±2.51	17.88±3.57	NS
NDF	Eating	min/kg NDF	29.2±4.92	37±5.59	<0.05
	Ruminating	min/kg NDF	40.4±5.73	43.1±8.60	NS
peNDF	Eating	min/kg peNDF	59.6±10.04	98.92±14.96	<0.01
	Ruminating	min/kg peNDF	82.45±11.69	115.25±23.00	<0.05

time. The STRAW diet was potentially under estimated by the peNDF system for total chewing. *In vivo* digestibility of the diets revealed a higher total NDF and ADF digestibility for the HAY diet (65.5% and 59.7% vs. 55.2% and 49.2% for the STRAW diet). VFA profile was similar for both diets, suggesting no differences in carbohydrate balance, while the pH decreased after feeding, as a consequence of fermentation. However, pH was not affected by treatment (6.38 with HAY vs. 6.50 with STRAW). In conclusion, it seems to be possible to balance diets by good quality forages thus using a lower amount of concentrate. Moreover, it appears to be possible to obtain a similar result, in terms of milk production and quality, when using straw to cover the requirement of peNDF and concentrate to fulfill the need for fermentable carbohydrates, and thus energy.

References

Kononoff, P.J., H.A. Lehman and A.J. Heinrichs, 2002. A comparison of methods used to measure eating and ruminating activity in cattle. J. Dairy Sci. 85, 1801-1803.

Mertens, D.R., 1997. Creating a System for Meeting the Fiber Requirements of Dairy Cows, J. Dairy Sci. 80, 1463-1481.

Sutton, J.D., M.S. Dhanoa, S.V. Morant, J. France, D.J. Napper and E. Schuller, 2003. Rates of Production of Acetate, Propionate, and Butyrate in the Rumen of Lactating Dairy Cows Given Normal and Low-Roughage Diets. J. Dairy Sci. 86, 3620-3633.

Yang, W.Z. and K.A. Beauchemin, 2009. Increasing physically effective fiber content of dairy cow diets through forage proportion versus forage chop length: Chewing and ruminal pH. J. Dairy Sci. 92, 1603-1615.

Zebeli, Q.J., D. Mansmann, H. Steingass and B.N. Ametaj, 2009. Balancing diets for physically effective fiber and ruminally degradable starch: A key to lower the risk of sub-acute rumen acidosis and improve productivity of dairy cattle. Livest. Sci. 127, 1-10.

Intake, milk production and nitrogen balance of goats fed diets with differing energy content in early lactation

S. Colombini, G. Galassi, L. Malagutti, M. Zucali, G.M. Crovetto and L. Rapetti
Università degli Studi, Dipartimento di Scienze Animali, via Celoria 2, 20133 Milano, Italy

Introduction

In ruminants, dry matter intake (DMI) increases gradually during the early *post partum* period and reaches the maximum value several weeks after the peak milk yield with a consequently negative energy balance. In early lactation, the limited DMI requires to formulate diets with a high crude protein (CP) concentration to meet the CP requirements for milk protein yield (Schmidely *et al.*, 2002). However, the efficiency of utilization of nitrogen (N) by ruminants is generally low and it may be partially related to an impaired utilization of N in the rumen. In fact, a major determinant of microbial protein synthesis is the availability of energy-yielding substrates (Dijkstra *et al.*, 1998). Consequently, in early lactation, the use of diets with high energy and medium CP concentration may improve N efficiency and energy balance.

The aim of the study was to determine the effect on DMI, milk production and N balance of goats fed diets with different energy content in early lactation.

Material and methods

Twelve multiparous Saanen goats were divided into two groups and fed diets with differing energy content: high (H) or low (L). In order to differentiate the diets for the energy content a different forage:concentrate ratio was used. Composition (on DM basis) of H and L diets were respectively: 9.0 and 11.9% wheat straw, 36.4 and 47.9% dehydrated lucerne hay, 21.4 and 14.6% maize meal, 9.0 and 6.7% barley meal, 9.4 and 6.4% soybean meal, 7.2 and 4.9% dried sugarbeet pulp, 2.5% sugarcane molasses, 3.2% Megalac® and 2.1% mineral and vitamin supplement. The calculated dietary net energy of lactation (NEL) contents (INRA, 2007) were 6.98 and 6.42 (MJ NEL/kg DM) for diets H and L. Starch, NDF and CP concentrations (% DM) were: 20.7, 32.2 and 16.0 (H) and 14.6, 37.0 and 15.8 (L). Goats were confined in individual metabolic cages to allow total collection of excreta. Milk yield and composition, and N balance were measured in two different periods (7 days each) at the 4th week of lactation (WOL) (W4) (average days in milk: 24) and at the 8th WOL (W8) (average days in milk: 57). Live weights were as follows: W4: 60.5 and 59.4 kg, W8: 64.3 and 62.0 kg, for H and L, respectively. Samples of rumen fluid were taken during W8 using an esophageal probe at 0, 3 and 6 h after the morning feeding. Data were analyzed by procedure GLM of SAS.

Results and discussion

Two goats of treatment H were removed from the experiment due to udder health problems. The DMI was affected by the period (2,395 vs. 2,776 g/d for W4 and W8, respectively, $P<0.001$) whilst the effect of the diet was significant only for W4 (2,509 vs. 2,281 g/d for H and L, respectively, $P<0.05$) (Table 1). Milk yield (g/d) was higher ($P<0.001$) for diet H (4,317) than for diet L (3,731). Milk urea was slightly lower for diet H ($P=0.07$) due to the higher NFC/NDF ratio. Milk urea content is related to N utilization, and the higher milk protein content of treatment H (3.11 vs. 3.00%, $P<0.05$) confirms a better N utilization at rumen level for this diet. The higher concentration of NFC in diet H compared with diet L (39.4 vs. 34.5% on DM) determined a higher acidity of the rumen fluid: the nadir values observed 3 h post-feeding were 6.18 and 6.57 ($P<0.001$), for H and L treatments, respectively.

N intake was higher for W8 than for W4 (Table 2). Total N excretion was affected only by the period (35.1 vs. 40.7 g/d, for W4 and W8, respectively, $P<0.001$) due to the greater N intake in W8 which

Table 1. Dry matter intake, milk yield and composition of the goats fed H and L diets during the 4th and the 8th week of lactation.

		W4		W8		SE	P-value	
		H	L	H	L		Diet	W
DMI	g/d	2,509[b]	2,281[c]	2,722[a]	2,831[a]	54.3	0.265	<0.001
Milk yield (MY)	g/d	4,382[a]	3,570[c]	4,252[a]	3,893[b]	109	<0.001	0.358
FCM[1]	g/d	4,547[a]	3,719[bc]	3,996[b]	3,645[c]	101	<0.001	0.010
Milk fat	%	3.71	3.80	3.11	3.13	0.18	0.744	0.005
Milk protein	%	3.23	3.09	2.99	2.90	0.04	0.024	<0.001
Milk urea	mg/dl	28.3	33.0	39.3	41.7	1.85	0.070	<0.001
Feed efficiency	MY/DMI	1.78	1.56	1.58	1.38	0.04	<0.001	<0.001

[1] FCM = Milk yield × (0.1375 × Fat % + 0.0825 × Protein % + 0.263) (INRA, 2007).

[a,b,c] Least squares means within row with different superscripts differ (*P*<0.05) when the *P*-value for the interaction of main effects was <0.05.

increased the excretion of urinary N (18.4 vs. 23.0 g/d, *P*<0.001) for both diets. The excretion of faecal N was not affected by period and diet. N balance was already positive in W4, and, as expected, it increased significantly in W8; retained N was also higher for treatment L compared to treatment H. Schmidely *et al.* (2002) reported in goats at the 6th week of lactation an average value of retained N of 10.2% of N intake. In conclusion, a high energy diet increases dairy efficiency and N utilization for milk production.

Table 2. N balance of the goats fed H and L diets during the 4th and the 8th week of lactation.

		W4		W8		SE	P-value	
		H	L	H	L		Diet	W
N intake (NI)	g/d	63.1	60.5	70.8	73.9	1.49	0.887	<0.001
Faecal N	g/d	18.4	17.3	18.7	19.8	1.00	0.988	0.166
	% NI	29.2	28.7	26.6	26.8	1.16	0.902	0.065
Urine N	g/d	17.8	19.1	23.3	22.6	0.95	0.720	<0.001
	% NI	28.1	31.5	32.9	30.4	1.33	0.763	0.163
Milk N	g/d	22.2[a]	17.3[c]	20.0[b]	17.7[c]	0.53	<0.001	0.102
	% NI	35.8	28.5	28.6	24.0	0.86	<0.001	<0.001
N Retention	g/d	4.8	6.8	8.9	13.8	1.01	0.006	<0.001
	% NI	6.9	11.3	11.9	18.9	1.49	0.003	0.002

[a,b,c] Least squares means within row with different superscripts differ (*P*<0.05) when the *P*- value for the interaction of main effects was <0.05.

References

Dijkstra, J., J. France and D.R. Davies, 1998. Different mathematical approaches to estimating microbial protein supply in ruminants. J. Dairy Sci. 81:3370-3384.

INRA, 2007. Alimentation des bovins, ovins et caprins- Besoins des animaux – Valeurs des aliments. Ed. Quae.

Schmidely, P., F. Meschy, J. Tessier and D. Sauvant, 2002. Lactation response and nitrogen, calcium, and phosphorus utilization of dairy goats differing by the genotype for {alpha}S1-casein in milk, and fed diets varying in crude protein concentration. J. Dairy Sci. 85, 2299-2307.

Performance and metabolism during physiological energy deficiency in early lactation and during energy restriction at 100 DIM in high-yielding dairy cows

J. Groß[1], H.A. van Dorland[2], R.M. Bruckmaier[2] and F.J. Schwarz[1]
[1]Technische Universität München, Department Animal Science, Chair of Animal Nutrition, Liesel-Beckmann-Strasse 6, 85350 Freising-Weihenstephan, Germany
[2]University of Bern, Veterinary Physiology, Bremgartenstrasse 109a, 3001 Bern, Switzerland

Introduction

A negative energy balance (NEB) usually occurs in dairy cows after calving (Fenwick *et al.*, 2008; Van Dorland *et al.*, 2009), but can also appear later in lactation during periods of insufficient feed supply. The aim of this study was to investigate and compare effects of the physiological energy deficiency in early lactation and a following nutritionally induced negative energy balance at 100 days in milk (DIM) on changes in performance and metabolism. Furtheron, responses of cows were investigated in a subsequent refeeding period.

Material and methods

Fifty multiparous Holstein dairy cows (3.2±1.4 lactations) were allocated according to their energy balance (EB) in wks 1-12 p.p. (=experimental period 1) at around 100 DIM to either a control (n=25) or a restriction group (n=25) for 3 weeks (=period 2). Thereafter cows were further monitored during a subsequent 8-wk realimentation period (=period 3). Throughout the experiment all cows obtained a partial mixed ration (PMR; based on maize silage, grass silage, hay and concentrate) *ad libitum*, except for restricted cows during the restricted feeding period. For restricted cows additional hay was mixed into the PMR to decrease the energy content of the ration. In order to induce an energy deficiency of at least 30% of the calculated energy requirements, the amount of PMR was individually limited. Additional concentrate was fed individually when milk yield was above 21 kg/d, except during the restriction period, where it was set by 0.5 kg FM/d. Feed intake and milk yield were recorded daily, live weight (LW) weekly. Blood samples were taken once a week, milk samples twice a week. EB of each cow was calculated from daily feed intake, maintenance requirement and milk yield. Data were analyzed with the Proc Mixed procedure from SAS, including wk and group as fixed effects, and cow as repeated subject. A covariate based on period 1 was additionally used for analysis of data from the restriction and realimentation period.

Results

Feed intake increased steadily from 14.9±0.2 kg DM/d (wk 1 p.p.) to over 22 kg DM/d in wks 7-12 of period 1. NEB had a nadir in wk 1 p.p. with -46.1 MJ NEL/d and cows covered only 70% of their energetic requirements by feed intake. EB turned positive in wk 9 p.p. over all cows and reached a level of 103% of the calculated demand before experimental period 2. During period 2 restricted cows had a mean DMI of 10.3 kg/d and thereof covered 51% of their energy requirements, whereas control cows had a DMI of 21.1 kg/d and an energy balance of 104%. In wk 1 of realimentation control cows still had a higher feed intake than restricted cows (20.4 vs. 18.7 kg DM/d; $P<0.01$). EB for restricted cows turned positive again in wk 2 of realimentation and averaged 109% of the calculated demand until the end of the study (control cows 108%).

Milk yield started at 27.5±0.7 kg/d in wk 1 p.p., peaked in wk 6 p.p. (39.5±0.8 kg/d) and decreased to 32.8±0.8 kg/d in wk 12 p.p. During period 2 restricted cows (27.4 kg/d) had a lower milk yield than control cows (30.5 kg/d; $P<0.05$). Milk yield of restricted cows increased already in wk 1 of the realimentation period from 27.6 to 28.6 kg/d and was thereafter even between 0.5 and 1.0 kg/d above the control group (wks 2-8 in phase 3; $P>0.05$).

Milk fat percentage was highest in wk 1 p.p. (5.48%) and dropped to 4.00% in wk 7 p.p. Within 1 wk of feed restriction, milk fat content raised from 4.30 to 4.63% ($P<0.10$). However, there were no differences between the groups during the rest of period 2 and 3. Milk protein content was highest in wk 1 p.p. (4.09%) and decreased to 3.03% in wk 4 p.p. Protein content decreased significantly in restricted cows from initially 3.37% to a mean value of 3.19% in period 2. In wk 1 of realimentation restricted cows had a lower protein percentage (3.33 vs. 3.39%, $P<0.05$), but recovered completely thereafter.

The LW decreased after parturition from 668 kg (wk 1 p.p.) to 647 kg (wk 4 p.p.). In period 2 feed restricted cows showed a lower LW than control animals (627 vs. 655 kg; $P<0.05$). During the realimentation period restricted cows gained LW and were equal to control cows in wk 2.

Plasma glucose had a nadir in wk 2 p.p. (3.30 mmol/l) and increased to 4.13 mmol/l in wk 12 p.p. For restricted cows glucose was lower (3.85 vs. 4.06 mmol/l; $P<0.05$) in period 2 and reached control cows again in wk 4 of period 3. NEFA concentration was highest in period 1 in wk 2 p.p. (0.90±0.06 mmol/l) and decreased to 0.13 mmol/l at the end of phase 1. Restricted cows had higher values of NEFA in period 2 (0.23 mmol/l) than control cows (0.14 mmol/l) ($P<0.05$). In period 3 there were no more detectable differences. BHBA increased from 0.70 mmol/l (wk 1 p.p.) to a maximum in wk 3 p.p. (0.98 mmol/l). Thereafter values declined to 0.50 from wk 7-12 of period 1. In period 2 BHBA was slightly higher for restricted cows (0.62 vs. 0.52 mmol/l) and decreased in period 3 to the levels of control animals.

Conclusions

Despite a nutritional NEB that was even higher than in early lactation, covering about 50% of energetic demands, the relative decline in milk yield was low in comparison to the high mobilization rate (equivalent in energy for about 20 kg milk). The extent of changes in plasma metabolites was also lower than expected during the induced tremendous mobilization compared to the amplitudes in early lactation. The responses of the restricted cows during the realimentation period included a compensation in performance and metabolic parameters. Milk yield was even higher in realimentation period for restricted than control cows, although EB was at the same level.

References

Fenwick, M.A., R. Fitzpatrick, D.A. Kenny, M.G. Diskin, J. Patton, J.J. Murphy and D.C. Wathes, 2008. Interrelationships between negative energy balance (NEB) and IGF regulation in liver of lactating dairy cows. Domestic Animal Endocrinology 34, 31-44.

Van Dorland, H.A., S. Richter, I. Morel, M.G. Doherr, N. Castro, and R.M. Bruckmaier, 2009. Variation in hepatic regulation of metabolism during the dry period and in early lactation in dairy cows. J. Dairy Sci. 92, 1924-1940.

Predicting the effect of energy × protein interaction on milk yield and composition in dairy cows

L. Brun-Lafleur[1,2,3], L. Delaby[1,2], F. Husson[4] and P. Faverdin[1,2]
[1]*INRA, UMR1080 Production du Lait, 35590 Saint-Gilles, France*
[2]*Agrocampus-Ouest, UMR1080 Production du Lait, F-35000, Rennes, France*
[3]*Institut de l'Elevage, 35652, Le Rheu, France*
[4]*Agrocampus-Ouest, UMR6625 Mathématiques Appliquées, 35000, Rennes, France*

Introduction

Dietary supplies of energy and protein to dairy cows are 2 important levers by which milk yield and composition can be modulated in the short term. When one of those supplies increases, milk yield increases curvilinearly and reaches plateaus (Coulon and Rémond, 1991; Vérité and Delaby, 2000). However, energy and protein supplies often vary simultaneously, but only few studies have addressed their possible interaction (Gordon and Forbes, 1970; Krohn and Andersen, 1980; Broderick, 2003). The objective of this trial was thus to build response equations of milk yield and composition to joint variations of energy and protein supplies, and to determine whether those responses depend on parity and production level.

Material and methods

Forty-eight Holstein cows were blocked into 4 homogeneous groups according to parity and production level: high- or low-producing and multiparous (41.7 and 34.0 kg of milk/d at the beginning of the trial, respectively) or primiparous (32.5 and 28.5 kg of milk/d, respectively). According to a balanced incomplete block design including 3 4-wk periods, 9 dietary treatments were compared for each group. A reference treatment was calculated to cover the predicted requirements of each group and was applied to each cow. The other 8 treatments corresponded to fixed supplies of energy (in UFL = French forage unit, 1 UFL = 1.7 Mcal of net energy for lactation) and protein (in g PDIE = grams of protein digested in the small intestine), higher or lower than those of the reference treatment. The treatments were named by their level of supply of energy (E--, E-, E0, E+, E++, from the lowest to the highest) and of protein (P--, P-, P0, P+, P++) (Figure 1). Thirty-six diets were thus formulated and distributed as total mixed rations in limited quantity to control the supplies.

Results and discussion

There were energy × protein interactions ($P<0.05$) not only in milk yield but also in protein content and yield with a similar pattern (Table 1). Conversely, energy and protein had additive effects on fat yield and no effect on fat content. Milk yield response to energy supply was nil with low protein supply and increased with higher protein supply (Figure 2). This interaction was similar for protein yield and content. Using partial derivatives of those equations, it is possible to test the hypothesis that the results were consistent with a response corresponding to the more limiting of the 2 factors: energy and protein. Milk yield response supports this concept because it was well predicted by the minimum of the response permitted either by energy supply if protein supply was not limiting or by protein supply if energy supply was not limiting. This experiment showed that milk yield response to variations in dietary supply depended on the cows' production level. High producing cows were more sensitive in their response in milk, especially with low energy and protein supplies. Genetic potential may impact nutrient partitioning between the different functions, namely lactation and control of body reserves (Friggens and Newbold, 2007). Protein content response to dietary supply changes was greater for primiparous than for multiparous cows, especially with low protein supply (E+P- and E0P--). This confirms that primiparous cows have a greater protein requirement for growth than multiparous cows, and then give priority to this function if protein supply is low.

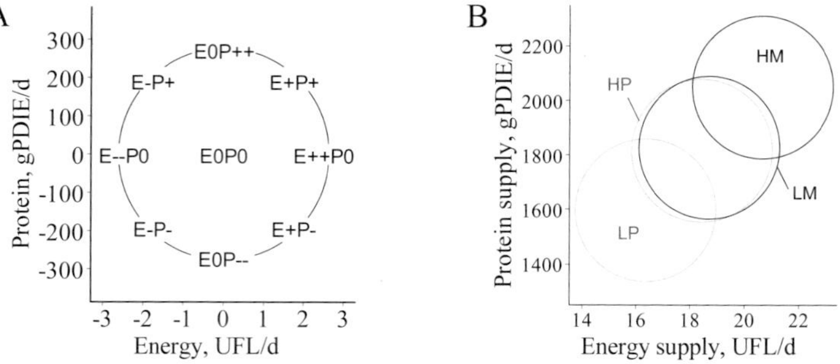

Figure 1. Definition of dietary treatments in terms of (A) difference between supplies and requirements of energy and protein and (B) the corresponding supplies for each group: high- (H) or low- (L) producing and multiparous (M, in black) or primiparous (P, in grey).

Table 1. Response of milk yield and composition to variations of energy (1 UFL/d) and protein (100 g of PDIE/d) supplies compared to requirements.

	Effect: coefficient (P(>F))					RMSE
	Energy		Protein		Energy × protein	
	Linear	Quadratic	Linear	Quadratic		
Milk yield, kg/d	0.25 (0.005)	-0.097 (0.045)	0.49 (<0.001)	-0.052 (0.048)	0.12 (0.023)	0.99
Fat yield, g/d	8.4 (0.015)		18 (<0.001)			42
Protein yield, g/d	12 (<0.001)	-4.2 (0.010)	22 (<0.001)	-2.5 (0.006)	5.1 (0.004)	33
Protein content, g/kg	0.16 (0.004)	-0.060 (0.051)	0.25 (<0.001)	-0.054 (0.002)	0.078 (0.017)	0.63

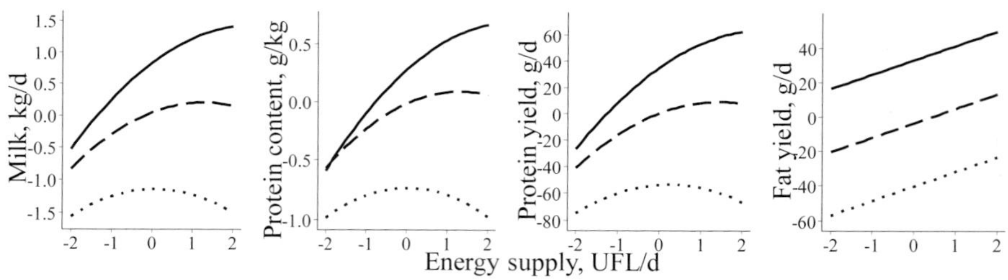

Figure 2. Response of milk yield, protein yield, protein content, and fat yield to variations of energy and protein supplies compared to requirements (dotted line = -200 g of PDIE/d, long-dashed line = 0 g of PDIE/d, and solid line = +200 g of PDIE/d).

References

Broderick, G.A., 2003. Effects of varying dietary protein and energy levels on the production of lactating dairy cows. J. Dairy Sci. 86, 1370-1381.

Coulon, J.B. and B. Rémond, 1991. Variations in milk output and milk protein content in response to the level of energy supply to the dairy cow: a review. Livest. Prod. Sci. 29, 31-47.

Friggens, N.C. and J.R. Newbold, 2007. Towards a biological basis for predicting nutrient partitioning: the dairy cow as an example. Animal 1, 87-97.

Gordon, F.J. and T.J. Forbes, 1970. The associative effect of level of energy and protein intake in the dairy cow. J. Dairy Res. 37, 481-491.

Krohn, C.C. and P.E. Andersen, 1980. Different energy and protein levels for dairy cows in the early weeks of lactation. Livest. Prod. Sci. 7, 555-568.

Vérité, R. and L. Delaby, 2000. Relation between nutrition, performances and nitrogen excretion in dairy cows. Ann. Zootech. 49, 217-230.

Effects of various levels of crude protein and metabolizable energy intake on heat production of Brahman cattle fed under humid tropical conditions

A. Chaokaur[1], T. Nishida[2] and K. Sommart[3]
[1]Faculty of Animal Science and Agriculture, Silpakorn University, 76120, Thailand
[2]Japan International Research Center for Agricultural Sciences, Japan
[3]Department of Animal Science, Khon Kaen University, 40002, Thailand

Introduction

The energy expenditure for heat production widely varies with feed intake (Freetly *et al.*, 2006). Recent reviews are available about the interaction between protein and energy supply on metabolism and growth performance in growing beef cattle (NRC, 2000; Schroeder *et al.*, 2007). However, the effect of interaction between protein and energy on energy expenditure in to heat has not been elucidated in growing beef cattle. Therefore, the objective of this study was to determine heat production of Brahman cattle offered varying levels of crude protein and metabolizable energy intake under humid tropical conditions. The measurement of heat production requires direct or indirect calorimetry. A respiration apparatus for large animals on the Indochina peninsula is only available in Thailand up to now (Chaokaur *et al.*, 2007).

Materials and methods

Sixteen growing male Brahman cattle (BW 226.9±43.5 kg; age 1 year old) were housed in individual pens. The animals were assigned to a 2×2 factorial arrangement in a randomized complete block design. Dietary treatments contained CP intake either at 5 or at 15 g/kg $BW^{0.75}$ (L, low and H, high) and ME intake either at 694 or at 1,007 kJ/kg $BW^{0.75}$/d (L, low and H, high). The animals were assigned randomly to one of four feeding treatments (LL, LH, HL and HH). Temperature and humidity during experiment were 28.1±1.8 °C and 69.2±4.0%, respectively. The animals were standing or lying, and the equipment for heat production measurements (3-day sampling) consisted of respiration head boxes. Nutrient digestibilities were determined in a 6-day sampling of the complete feces. The oxygen consumed, carbon dioxide and methane produced were determined by an open respiration head boxes system, and heat production (HP) was calculated using Brouwer's equation (Brouwer, 1965). The animals were fasting 5-day for heat production measurements over the last 2-day. The heat increment (HI) was calculated as the difference between heat production obtained in fed cattle minus heat production obtained in fasting cattle. The data were analyzed as a 2×2 factorial arrangement in a randomized block using GLM procedure (SAS, 1999). Multiple comparisons among means were carried out with Duncan news's multiple range tests at the level of $P<0.05$.

Results and discussion

Average heat production in fasting cattle was 289 kJ/kg $BW^{0.75}$/d. Table 1 shows the results of HP and HI of cattle subjected to the dietary treatments. The interaction of CP×ME was not significant ($P>0.01$) for nutrient digestibilities. An increased intake of ME increased digestibilities of dry matter, organic matter and crude protein. An increased intake of CP increased the digestibility of crude protein. The results indicated that the interaction of CP×ME was not significant ($P>0.01$) for energy expenditure in terms of HP and HI. Cattle fed high levels of ME showed a higher ($P<0.01$) percentage of dietary GE lost as heat increment. This result agreed with the report by Freetly *et al.* (2006) who showed that an increased recovered energy is associated with a rapid increase in heat production.

It is concluded that an increase in intake of crude protein and metabolizable energy intake independently of each other affect heat energy loss of beef cattle fed under humid tropical conditions.

Table 1. Effects of dietary crude protein and metabolizable energy on nutrient digestibility and heat production of growing Brahman cattle[1].

Item	Level of CP intake		Level of ME intake		SE	P-value[2]		
	Low	High	Low	High		CP	ME	CP×ME
DM intake, g/kg $BW^{0.75}$	83.4	80.9	82.8	81.5	0.85	0.06	0.29	0.29
CP intake, g/kg $BW^{0.75}$	4.94^b	14.58^a	8.37^b	11.15^a	0.43	<0.01	<0.01	<0.05
Digestibility, %								
DM	61.8^b	63.2^a	55.5^b	69.5^a	1.21	0.41	<0.01	0.45
OM	65.0^b	65.8^a	58.3^b	72.4^a	1.03	0.60	<0.01	0.55
CP	40.8^b	71.4^a	50.7^b	61.5^a	3.07	<0.01	<0.05	0.13
NDF	58.2^b	57.7^a	56.7^b	59.1^a	1.28	0.76	0.22	0.33
Energy intake, kJ/kg $BW^{0.75}$								
GE	1,274	1,262	1,261	1,276	22.8	0.71	0.64	0.23
DE	915	930	797	1048	28.4	0.72	<0.01	0.35
ME	795	787	666^b	917^a	29.1	0.85	<0.01	0.36
Heat production								
kJ/kg $BW^{0.75}$/d	628	620	599	649	19.5	0.76	0.10	0.51
% of GEI	49.6	49.2	47.5	51.3	1.57	0.84	0.12	0.88
Heat increment								
kJ/kg $BW^{0.75}$/d	399	331	310	360	19.5	0.76	1.00	0.51
% of GEI	23.4^b	29.9^a	22.2^b	31.1^a	1.56	<0.05	<0.01	0.66

[1] SE, standard errors; DM, dry matter; OM, organic matter; NDF, neutral detergent fiber; CP, crude protein; GE, Gross energy; DE, digestible energy; ME, metabolizable energy; GEI, Gross energy intake.
[2] Probability of a significant effect of crude protein intake (CP) or metabolizable energy intake (ME) or crude protein intake and metabolizable energy intake interaction (CP×ME).
[a,b] Least square means with different superscripts among treatments significantly differ ($P<0.05$).

References

Brouwer, E., 1965. Report of subcommittee on constants and factors in energy metabolism. In: Blaxter, K.L. (ed.), Energy Metabolism. EAAP Publication No. 11, New York Academic Press, New York: 441-443.

Chaokaur, A., T. Nishida, I. Phaowphaisal, P. Pholsen, R. Chaithiang and K. Sommart. 2007. Energy metabolism and energy requirement for maintenance of Brahman steers in tropical conditions. In: I., Ortigues-Marty, N. Miraux and W. Brand-Williams (eds.), Energy and Protein Metabolism and Nutrition. Wageningen Acad. Publ., Wageningen, the Netherlands. EAAP No. 124: 505-506.

Freetly, H.C., J.A. Nienaber, and T. Brown-Brandl, 2006. Changes in heat production by mature cows after changes in feeding level. J. Anim. Sci. 84, 1429-1438.

NRC. 2000. Nutrient Requirements of Beef Cattle. 7th ed. Natl. Res. Counc., Natl. Acad. Press, Washington, D.C.

SAS. 1999. SAS User's Guide: Statistics, Version 6.12th ed. SAS Institute Inc.Cary, NC.

Schroeder, G.F. and E.C. Titgemeyer, 2008. Interaction between protein and energy supply on protein utilization in growing cattle: A review. Livest. Sci. 114, 1-10.

Analysis of liveweight and growth performance in Australian lambs

C.G. Jose, C.F. Hansen, K.L. Pearce, S.I. Mortimer, A.J. Ball, K.G. Geenty and G.E. Gardner
Cooperative Research Centre for Sheep Industry Innovation, School of Veterinary and Biomedical Sciences, Murdoch University, Murdoch, Western Australia 6150, Australia

Introduction

The age-weight relationship of growing lambs is of major economic importance to the Australian sheep industry. Birth, weaning and slaughter weights can be used to derive important growth traits such as relative and absolute growth rates by the use of nonlinear growth functions such as Brody (1945). These functions have been used to model liveweight gain, providing a relatively precise description of the growth pattern of animals with a clear biological interpretation. Analysis of animal growth performance is essential to establish precise feeding strategies as well as providing accurate age estimates for animals to reach target slaughter weights. Early estimation of these traits is vital to optimise production and genetic selection. The objectives of this study were to determine environmental and genetic factors affecting post-weaning weight gains in Australian sheep flocks with known genetic potential for growth.

Material and methods

Details of the design of the Sheep CRC's Information Nucleus Flock were presented by Fogarty *et al.* (2007). Briefly, about 2000 lambs were produced in 2007 from Merino and crossbred ewes located at seven research sites across Australia (Katanning WA, Cowra NSW, Kirby NSW, Struan SA, Turretfield SA, Hamilton VIC, and Rutherglen VIC) which represented a broad cross section of Australian production systems. These ewes were artificially inseminated with semen from 93 key industry sires and were measured and sampled for carcass, meat and growth traits. A Brody curve was fitted to the weight data (collected fortnightly) of each lamb slaughtered via the proc NLIN function in SAS (SAS 9.0 Inst. Inc., Cay, NC) using the Gauss-Newton method. Individual curves and their first derivatives were used to compute the predicted weights (Wt) in kg at day 100 (weaning), growth rates (Gr) in g/d at day 100. The Brody function is described as:

$$Wt[t] = A\cdot(1-b\cdot e^{-k\cdot t})$$

where *y* is liveweight, *a* is asymptotic (mature) weight, *k* is exponential constant, *t* is time (age in days). The Gr function (GR[t]) was equal to the first derivative of the Brody function:

$$GR[t] = dWt[t] / dt = k\cdot A\cdot b\cdot e^{-k\cdot t}$$

Gr at a particular time point describes gain in weight that depends not only on previous weight points but also on curve points in later stages of growth. Wt and Gr data at day 100 was analysed using a linear mixed effects model including fixed effects (and their relevant interactions) for site, sex, birth type-rear type, dam breed within sire type, and sire type. Sire and dam identification were included as random terms. The sire estimates from this model were subsequently analysed for their association with Australian Sheep Breeding Values (ASBVs) for post weaning weight (PWWT), C-site fat depth (PFAT), and eye muscle depth (PEMD) using a general linear model including sire type as fixed effect, sire ASBV for PWWT, PFAT and PEMD as covariates, and the first order interaction between sire type and ASBVs.

Results and discussion

Aligning well with previous studies (Rodríguez *et al.* 2008; Borg *et al.* 2009), the Wt and Gr at day 100 were greater (P<0.01) for male, terminal sired lambs (versus Merino or Border-Leicester sires), and maternal dams (i.e. Border-Leicester vs. Merino) that were born and raised as singles. Lambs at Katanning, Cowra and Turretfield were the heaviest (P<0.01; 31.1±0.34, 30.9±0.34 and 30.0±0.38 kg) and lambs at Katanning grew faster than at any other site (P<0.01; 191±2.6 g/d). In contrast lambs at Hamilton were the lightest and grew the slowest (17.0±0.40 kg, 113±3.0 g/d). Sire estimates demonstrated a positive association with PWWT ASBV for both Wt and Gr at day 100 and 150 (P<0.01). For the progeny of Terminal sires, an increase of one ASBV unit at day 100 increased Gr by about 1.4g/d (Figure 1A) and liveweights by about 200g (data not shown). Sire estimates for PEMD ASBV demonstrated a positive association (P<0.01) with liveweight at day 100 (Figure 1B) and 150, but only for terminal and maternal sires. However there was no effect on growth rate at these time points, indicating that the growth path for the High and Low PEMD lambs must have diverged at some point prior to 100 days to generate the liveweight differences. Future analysis of the effect of PEMD on birth weights may help to elucidate this. This study highlighted the effectiveness of selecting sires with high PWWT and PEMD ASBVs to increase growth rates and animal weights in production lambs.

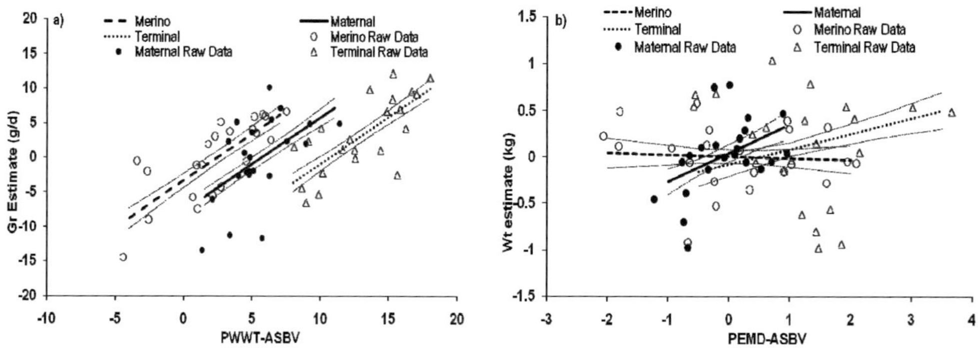

Figure 1. Effect of merino, maternal and terminal sires estimates for (a) post weaning weight (PWWT) Australian Sheep Breeding Values (ASBVs) on Gr (g/d) at 100 days and (b) eye muscle depth (PEMD) ASBV on predicted Wt (kg) at 100 days.

References

Borg, R.C., D.R. Notter and R.W. Kott, 2009. Phenotypic and genetic associations between lamb growth traits and adult ewe body weights in western range sheep. J. Anim Sci. 87(11): 3506-3514.

Brody, S., 1945. Bioenergetics and Growth. New York, Reinhold.

Fogarty, N.M., R.G. Banks, J.H.J. van der Werf, A.J. Ball and J.P. Gibson, 2007. The information nucleus – a new concept to enhance sheep industry genetic improvement. Proc. Ass. Adv. Anim. Breed. & Genet. 17, 29-32.

Rodríguez, A.B., R. Bodas, N. Prieto, R. Landa, A.R. Mantecón and F.J. Giráldez, 2008. Effect of sex and feeding system on feed intake, growth, and meat and carcass characteristics of fattening Assaf lambs. Livestock Science 116, 118-125.

Hepatic transcriptome of beef steers is differentially modulated by composition of energy-substrate supply in growing beef steers

R.L. Baldwin VI[1], J. Sumner-Thomson[2], E.E. Connor[1], R.W. Li[1], K.R. McLeod[3] and B.J. Bequette[2]
[1]Bovine Functional Genomics Laboratory, Animal and Nutrition Resources Institute USDA-ARS, Beltsville, MD 20705, USA
[2]Department of Animal and Avian Sciences, University of Maryland, College Park 20742, USA
[3]Department of Animal Science, University of Kentucky, Lexington KY 40506, USA

Introduction

Development of improved carcass quality characteristics (higher protein deposition concomitantly with lower fat) while maintaining or improving the efficiency of production is fundamental to the economic stability of the beef industry. Recently, Connor *et al.* (2009) using a compensatory gain model identified specific mitochondrial gene expression patterns implicating them as putative hepatic arbiters and/or biomarkers for increased nutrient use efficiency in compensating steers. Selection for and manipulation of gene expression will likely be a management tool implemented in future production paradigms requiring improved understanding of putative metabolic regulators of tissue accretion. Nutrient delivery, both form (e.g. carbohydrate, fat, and protein) and site of absorption (rumen vs. small intestines) elicits effects on carcass composition (lean vs. fat; McLeod *et al.*, 2006) concomitant with alterations in expression of specific genes that results in altered rates of tissue accretion (Baldwin *et al.*, 2007). Elucidation of nutrient-driven mechanisms in ruminants and clarification of physiological and metabolic mechanisms underlying growth responses has been difficult to attain when macronutrients are manipulated by diet ingredients. Thus, our objective was to characterize genomic responses in economically important and metabolically important tissues when macronutrient supply is altered by direct ruminal (diet) or abomasal (infusion) administration of specific nutrients. This paper reports specific hepatic gene expression responses to abomasally administered starch (partially-hydrolyzed) and casein, and dietary inclusion of propionate in growing beef steers over 42 days.

Materials and methods

Nineteen Wye Black Angus beef steers (272.5±17.6 kg initial body weight (BW)) were fed a forage-based diet and infused per abomasum with either water (WC; n=4), casein (AC; n=5) or starch (ASH; n=5), or fed sodium-propionate (PRO; n=5) for 42 d. Treatments were administered on an equal energy basis (167.4 kJ metabolizable energy/kg metabolic BW). On day 42, steers were euthanized and hepatic tissue was collected for transcriptome profiling. Total RNA was isolated for hybridization to a custom bovine high-density, whole-genome 60-mer array (Roche NimbleGen). Quality and concentration of RNA were determined using an Agilent 2100 Bioanalyzer and ND-1000 spectrophotometer, respectively. Probes were labelled per standard procedures outlined by Roche NimbleGen Systems for hybridization to the high-density oligonucleotide microarray at the Microarray Core Facility in Reykjavik, Iceland in a single batch of 19 arrays. Two samples did not meet the control requirements for cDNA synthesis yield and quality. Using first principal component analysis verified by cluster analysis, two additional samples were excluded from ANOVA and Permutation Analysis for Differential Expression (PADE) that adds elements to minimize chances for false discovery rate. Remaining samples were used for final ANOVA and PADE (3 for WC, 3 for PRO, 4 for ASH and 5 for AC). Contrasts used were individual treatment vs. all other treatments (e.g. PRO vs. WC plus AC plus ASH). Identification, gene ontology, and pathway analysis was determined using the online resources provided by BioRag (2010).

Result and discussion

Infused steers gained more weight (3.6 to 5.2%, $P<0.05$) compared to WC. There were 4713 probes identified as differentially expressed ($P<0.05$) by ANOVA. Subsequent analysis using a conservative (false discovery rate set to 0) PADE approach yielded 110 transcripts for ASH, 6 for AC, and 1 for PRO and 3 for WC. The number of transcripts statistically identified may reflect the loss in power associated with sample size (n=3 for WC and PRO treatment) and the higher variability of AC microarray results observed.

The one transcript for PRO has not yet been annotated to either human or bovine gene annotations and AC results yielded only a single gene *Bos taurus adenylate kinase 3-like 1 (AK3L1)*, a nuclear gene encoding mitochondrial protein, and the WC transcript identified was *Bos taurus major histocompatibility complex, class I, A (BoLA)*, mRNA. In contrast, ASH elicited changes in the expression of 42 annotated genes which encode for proteins in a number of signalling pathways including G protein signalling (*AKAP7* and *ADCY3*), transforming growth factor Beta signalling (*SMAD2* and *BAMBI*), and the cell cycling (*GNG2* and *ADCY3*). Using a combined filter of ANOVA $P<0.05$ and absolute fold change >2 yielded greater numbers of differentially-expressed genes which will enhance the power of the data set to identify pathways regulated by these alterations in nutrient delivery. Identification of uniquely differentially-expressed genes due to macronutrient infusion and pathway analysis is fundamental to identification of causative factors regulating efficient nutrient use in ruminants.

References

Baldwin, R.L., VI, K.R. McLeod, J.P., McNamara, T.H. Elsasser and R.G. Baumann, 2007. Influence of abomasal carbohydrates on subcutaneous, omental, and mesenteric adipose lipogenic and lipolytic rates in growing beef steers. J. Anim. Sci. 85, 2271-2282.

BioRag (Bioresource for array genes) Available at: www.biorag.org.

Connor E.E., S. Kahl, T.H. Elsasser, J.S. Parker, R.W. Li, C.P. Van Tassell, R.L. Baldwin 6th, and S.M. Barao, 2009. Enhanced mitochondrial complex gene function and reduced liver size may mediate improved feed efficiency of beef cattle during compensatory growth. Funct. Integr. Genomics 10, 39-51.

McLeod, K.R., R.L. Baldwin, VI, M.B. Solomon, and R.G. Baumann, 2006. Influence of ruminal and postruminal carbohydrate infusion on visceral organ mass and adipose tissue accretion in growing beef steers. J. Anim. Sci. 85, 2256-2270.

Energy requirements of double-muscled Piemontese bulls

S. Schiavon, F. Tagliapietra, L. Bailoni and G. Bittante
Dipartimento di Scienze Animali, viale dell'Università 16, 25020 Legnaro (PD), Italy

Introduction

Robelin and Daenicke (1980) suggested that energy and protein requirements of growing cattle may differs among breeds because of divergent amounts of protein and adipose tissue deposited. The NRC (1996) reported daily net energy requirements for maintenance (NE_m) of 0.322 $MJ \cdot EBW^{0.75}$, where EBW is the average empty body weight (kg). Energy requirements of double muscled (dbm) bulls are uncertain because of the extremely lean composition of weight gain and of body mass. De Campeneere *et al.* (2001) found NE_m values to be surprisingly higher (+ 57%) in Belgian Blue dbm bulls compared to those given by NRC (1996), in apparent contradiction with other reports (Fiems *et al.*, 1999). Piemontese is a dbm breed with few information being available on production traits. In this paper data from a trial conducted on Piemontese dbm bulls were analyzed to estimate net energy retained in the body tissues (NE_R) and these were used to estimate maintenance requirements (NE_m).

Material and methods

The trial (Dal Maso *et al.*, 2009) involved 48 bulls being 229±18 days old, divided in four groups and housed in 12 pens, and fed four corn silage and cereal based rations (T) differing in crude protein content (145 or 108 g/kg dry matter), with or without the addition of rumen protected conjugated linoleic acid. Each group was further divided in three sub groups (W) with initial BW of 208±11, 240±9 and 263±19 kg, respectively. Dry matter (DM) intake was recorded daily on pen basis and the bulls were weighed monthly. Faecal grab samples, collected during the first (0 to 120 d), the central (120 to 233 d) and the final (233 to 332 d) periods (P), and acid detergent lignin were used to estimate energy digestibility coefficients. Empty BW (EBW) was computed as 0.89·BW (De Campeneere *et al.*, 2001). The average metabolic EBW over the course of each period was computed as (Foster, 1983): $(EBW_f^{0.75} - EBW_i^{0.75})/[1.75 \cdot (EBW_f - EBW_i)]$, where $_f$ and $_i$ indicate the end and the start of each period, respectively. All the bulls were slaughtered at the same day (average final BW: 678±56 kg). Empty body composition: at the start of the trial was estimated using the De Campeneere *et al.* (2001) equations; at the end of trial it was estimated from the chemical composition of muscles, fat and bones tissues dissected from the V rib cut collected at slaughter, at intermediate periods linear changes were assumed. Energy retained (NE_R) was computed assuming 23.02 and 38.90 MJ/kg of retained protein (Pr) and lipid (Lr), respectively. ME_{intake} and ME use efficiencies (k_m and k_g for maintenance and growth, respectively) were computed following Sauvant *et al.* (2004). NE for maintenance, computed as $NE_m = (ME_{intake} - NE_R/k_g) \cdot k_m$, was adjusted for NE_m used for thermoregulation as the environmental temperatures were known (Johnson, 1986). Data, averaged by pen, were analyzed with a model considering the effects of P, T, W, and interactions W×P and W×T.

Results and discussion

From the first to the last period, Pr significantly decreased and Lr was kept at about 0.1 kg/d (Table 1); ME intake significantly increased but the values of metabolizability (q=ME/GE), and those of k_m and k_g did not change; consequently NE_R significantly decreased. NE_m spent for thermoregulation was high in the first and in the third periods (cold seasons) and zero in the central one, as the ambient temperatures averaged 11, 8 and 21 °C, respectively. No significant influence of period was found on NE_m, which averaged 0.520 MJ/d·per kg $EBW^{0.75}$. Treatment did not influence NE_m. The value of NE_m found in this trial was close to that of 0.507 MJ/d·per kg $EBW^{0.75}$ found by De Campeneere *et al.* (2001) on Belgian Blue dbm bulls and much higher (+61%) than the value commonly used for cattle of conventional breeds (NRC, 1996). The high NE_m value of dbm bulls is likely due to the high

Table 1. Energy requirements of Piemontese double-muscled bulls.

		Age mo.			Root MSE	*P*-value	
		7-11	11-15	15-18		P	T
$EBW^{0.75}$ [1]	$kg^{0.75}$	76.5^C	99.5^B	115.4^A	2.66	<0.01	0.02
Body protein[2]	kg/kg EBW	0.193^C	0.200^B	0.202^A	0.006	<0.01	0.43
Body lipid[2]	kg/kg EBW	0.052^C	0.065^B	0.075^A	0.002	<0.01	0.68
Pr[3]	$kg·d^{-1}$	0.273^A	0.219^B	0.167^C	0.026	<0.01	0.47
Lr[3]	$kg·d^{-1}$	0.102	0.113	0.102	0.028	0.51	0.57
ME_{intake}	$MJ/kg\ EBW^{0.75}$	1.111^A	0.907^B	0.956^B	0.063	<0.01	0.11
q (ME/GE)		0.57	0.58	0.58	0.03	0.60	0.49
k_m		0.72	0.72	0.72	0.01	0.60	0.49
k_g		0.45	0.46	0.46	0.03	0.60	0.49
NE_R	$MJ/kg\ EBW^{0.75}$	0.134^A	0.095^B	0.068^C	0.011	<0.01	0.51
NE_m for cold[4]	$MJ/kg\ EBW^{0.75}$	0.047^B	0.00^C	0.064^A	<0.01	<0.01	0.77
NE_m [5]	$MJ/kg\ EBW^{0.75}$	0.536	0.505	0.519	0.060	0.49	0.18

P = period; T = treatment; A,B,C $P<0.01$. [1] EBW = 0.89·BW (De Campeneere *et al.*, 2001); [2] From equations of De Campeneere *et al.* (2001) and the chemical composition of the V rib; [3] Pr = Protein retention, Lr = lipid retention; [4] According to Johnson (1986); [5] $NE_m = (ME_{intake} - NE_R/k_g)·k_m$.

proportion of body protein mass, but it can also include amounts of energy not fully accounted by k_g. The k_g estimates are based on conventional cattle data where the Pr:Lr ratio was lower compared to that of dbm bulls. As the efficiency of ME utilization is lower for Pr compared to Lr (Geay, 1984), k_g assessed following Sauvant *et al.* (2004) could be overestimated for dbm cattle.

References

Dal Maso, M., S. Schiavon, F. Tagliapietra, A. Simonetto and G. Bittante, 2009. Growth Performance and N excretion of double muscled Piemontese bulls fed low protein rations with or without the addition of rumen protected conjugated linoleic acid. Ital. J. Anim. Sci. 8, 175-177.

De Campeneere, S., L.O. Fiems and C.V. Bouqué, 2001. Energy and protein requirements of Belgian Blue double-muscled bulls. Anim. Feed Sci. Technol. 90, 153-167.

Fiems, L.O., R. J. Moermans, Ch.V. Bouquè, B.G. Cottyn and S. De Campeneere, 1999. Voluntary feed intake in dual purpose and double-muscled Belgian White-blue bulls. In: Van der Heide, D., Huisman E.A., Kanis E., Osse J.W.M., Verstegen M.W.A. (eds.), Proceedings of the 5th zodiac Symposium on Regulation of Feed intake, Wageningen, NL.

Foster, W.H., D.J. Kilpatrick and I.H. Heaney, 1983. Genetic variation in the efficiency of energy utilization by the fattening pig. Anim. Prod. 37, 387-393.

Geay, Y., 1984. Energy and protein utilization in growing cattle. J. Anim. Sci. 58, 766-778.

Johnson, D.E., 1986. Climatic stress and production efficiency. Limiting the effects of stress on cattle. Western Regional Res. Publ. n. 009 and Utah Agric. Exp. Sta. Res. Bull. 512, 17.

NRC, 1996. Nutrient requirements of beef cattle. Seventh revised edition. Natl. Acad. Press, Washington, D.C.

Robelin, J. and R. Daenicke, 1980. Variations of net requirements for cattle growth with liveweight, liveweight gain, breed and sex. Ann. Zootech. 29, 99-118.

Sauvant, D., J.M. Perez, and G. Tran, 2004. Tables of composition and nutritional value of feed materials: pigs, poultry, cattle, sheep, goats, rabbits, horses and fish. INRA Editions, Versailles, France.

Energy expenditure and physical activity of grass-fed dairy cows on pasture and in the barn measured at different time points in lactation

L.D. Kaufmann[1], A. Münger[1], P. Junghans[2], S. Görs[2], C.C. Metges[2] and F. Dohme[1]
[1]Agroscope Liebefeld-Posieux, Research Station ALP, Posieux, Switzerland
[2]Leibniz Institute for Farm Animal Biology (FBN), Dummerstorf, Germany

Introduction

There is evidence that the energy requirements for maintenance of grass-fed dairy cows are higher than presently assumed (Bruinenberg *et al.*, 2002). According to Agnew and Yan (2000) the higher dry matter (DM) intake associated with high fibre diets results in greater gut fill, larger strain of rumination and digestion and consequently contributes to a higher maintenance metabolic rate. Cattle fed forage und especially fresh grass produced more heat during ingestion of one kg DM than cattle fed rolled barley (Susenbeth *et al.,* 2004). In addition, feeding of high fibre diets, such as grass-based diets, may cause a deficiency of fermentable organic matter (energy) in the rumen leading to an inefficient utilization of crude protein and consequently to a higher urea concentration in blood and urine (Reynolds *et al.*, 1991; Bruinenberg *et al.*, 2002). However, although differences between net energy output and predicted net energy intake in grazing systems may partly be attributed to the energy costs for metabolism and excretion of excess nitrogen, other important factors such as walking and grazing activity must be involved. The higher energy expenditure (EE) during grazing may also result from a greater physical effort for eating feed from the sward. As data substantiating this assumption are scarce the objective of the present study was to compare EE as well as physical activity and feeding behaviour of dairy cows fed grass either on pasture or in the barn at different time points in lactation.

Material and methods

Fourteen Holstein cows (body weight (BW): 660±65.0 kg; milk yield: 44±2.7 kg/d) were subjected to two treatments in a repeated cross over design with three experimental series. At the beginning of each series, cows were on average 38, 94 and 171±11.1 d in milk (DIM), respectively. Each series consisted of a 7 d adaptation and a 7 d collection period. Cows either grazed on pasture or had *ad libitum* access to grass cut from the same paddock in a free-stall barn. All cows were supplemented with a cereal-based concentrate to meet their predicted nutrient requirements. The concentrate was offered in two equal meals at 06:30 and 16:30 h after milking in the free-stall barn using weighing troughs (Insentec B.V., Marknesse, the Netherlands). Grass intake was estimated using the double alkane technique. Throughout the collection periods milk yield and milk components were recorded daily. On each day of the collection period, CO_2 production (RCO_2) of one cow in the barn and of one cow on pasture was determined by the ^{13}C bicarbonate dilution technique (Junghans *et al.*, 2007). The EE was estimated using a modified version of the equation of Brouwer (1965):

$$EE = 4.96 \times RCO_2 + 16.07 \times RCO_2 / 1.$$

After administration of the tracer (0.7 mg $NaH^{13}CO_3$/kg BW) into the jugular vein, blood was sampled from 0700 to 1300 h either manually in the barn or with an automatic blood sampling system (IceSampler®, IceRobotics, Roslin, UK) on pasture. During the same time period cows' physical activity and feeding behaviour were recorded over three days with a pedometer (IceTag3D®, IceRobotics) and the IGER behaviour recorder (Rutter *et al.*, 1997), respectively. Data were evaluated by analysis of variance with production system, experimental series and their interaction as fixed effects in the model.

Results and discussion

Milk (36.7 kg/d), fat (1.37 kg/d) and protein yield (1.15 kg/d) did not differ ($P>0.05$) between treatments but decreased with increasing days in milk ($P<0.05$). Grass intake was lower ($P<0.01$) for grazing cows (16.8 kg DM/d) compared to those fed grass in the barn (18.9 kg DM/d). The lowest intake was observed in the first experimental series and the highest in the second series while in the third series intake was intermediate ($P<0.001$). In contrast, concentrate intake which was the same ($P>0.05$) in both treatment groups (5.14 kg DM/d) decreased with increasing stage of lactation ($P<0.05$). Within the 6-h measurement period, grazing cows produced more ($P<0.001$) CO_2 and consequently expended more ($P<0.001$) energy than cows fed grass in the barn (273 vs. 231 kJ/kg $BW^{0.75}$). Differences in EE between the two treatments did not ($P>0.05$) change with increasing DIM. Cows on pasture were more active which means that they spent proportionally more ($P<0.001$) time walking (30 vs. 9%) and less ($P<0.001$) time standing (41 vs. 54%) and lying (30 vs. 36%) than cows in the barn. Feeding behavior of the treatment groups changed in such a manner that the proportion of time spent eating (50 vs. 44%) was higher ($P<0.001$) and that of time spent ruminating (18 vs. 23%) was lower ($P<0.001$) for cows on pasture compared to cows in the barn. This is in line with Graf *et al.* (2005) who found that cows grazed full time on a pasture providing good quality herbage required more time to consume one kg of DM than cows that grazed part time und were supplemented with hay or corn silage.

In conclusion, higher physical and feeding activity may at least partly cause higher EE of grazing cows. The unchanged milk production along with a lower feed intake indicates that these cows mobilised body reserves to cover additional energy requirements.

References

Agnew, R.E. and T. Yan, 2000. Impact of recent research on energy feeding systems for dairy cattle. Livest. Prod. Sci. 66, 197-215.

Brouwer, E., 1965. Report of sub-committee on constants and factors. In: Blaxter, K.L. (ed.) Energy Metabolism. Academic Press, London: 441.

Bruinenberg, M.H., Y. Van der Honing, R.E. Agnew, T. Yan, A.M. van Vuuren and H. Valk, 2002. Energy metabolism of dairy cows fed on grass. Livest. Prod. Sci. 75, 117-128.

Graf, C.M., M. Kreuzer and F. Dohme, 2005. Effects of supplemental hay and corn silage versus full-time grazing on ruminal pH and chewing activity of dairy cows. J. Dairy Sci. 88, 711-725.

Junghans, P., J. Voigt, W. Jentsch, C.C. Metges and M. Derno, 2007. The ^{13}C bicarbonate dilution technique to determine energy expenditure in young bulls validated by indirect calorimetry. Livest. Sci. 110, 280-287.

Reynolds, C.K., H.F. Tyrrell and P.J. Reynolds, 1991. Effects of diet forage-to-concentrate ratio and intake on energy metabolism in growing beef heifers: whole body energy and nitrogen balance and visceral heat production. J. Nutr. 121, 994-1003.

Rutter, S.M., R.A. Champion and P.D. Penning, 1997. An automatic system to record foraging behaviour in free-ranging ruminants. Appl. Anim. Behav. Sci. 54, 185-195.

Susenbeth, A., T. Dickel, K.H. Südekum, W. Drochner and H. Steingass, 2004. Energy requirements of cattle for standing and for ingestion, estimated by a ruminal emptying technique. J. Anim. Sci. 82, 129-136.

Phenotypic variation of energy intake and heat production in lactating cows fed *ad libitum* and restricted

M. Derno, A. Schwarm, B. Kuhla and M. Schweigel
Leibniz-Institute for Farm Animal Biology (FBN), Research Unit Nutritional Physiology 'Oskar Kellner', 18196 Dummerstorf, Germany

Introduction

In dairy cows, pregnancy is a critical time for metabolic adaptation to the following lactation and for distinct changes in feed intake. Based on the state of the art and own previous work it seems that *pre partum* dry matter intake (DMI) and body condition have a major effect on early *post partum* feed intake and (lipid) metabolism and thus on the seriousness of negative energy balance experienced by high-yielding dairy cows during early lactation. The ability to cope with energy deficiency is differently expressed in individual animals. For further investigations reaching the genetic level it is necessary to detect such phenotypical variations. Therefore, we performed a pilot respiration study with lactating Holstein Friesian cows and induced a situation of negative energy balance by restricted feeding. We predicted that cows differ inter-individually both in energy expenditure (EE) and in EE relative to DMI. Implications for their respective energy status *post partum* are proposed.

Materials and methods

The experiment was performed with four randomly selected Holstein Friesian dairy cows being in the state of mid-gestation with an initial body mass of 568±34 kg and an initial milk yield of 19.2±1.2 l/d (mean ± SD) (Table 1). Animals were halter-trained and adapted to handling and the respiration chambers. They were fed a total mixed ration (TMR) which consisted of, on a DM basis, 70%

Table 1. Data of selected parameters of the respiration experiment with lactating Holstein cows in mid gestation fed ad libitum *(AL) or restricted (RE).*

Item	Unit		Animal 1	Animal 2	Animal 3	Animal 4
Live weight	kg		607	584	536	592
Milk	l/d	AL	18.2	19.6	18.2	20.6
		RE	17.9	19.6	20.7	19.8
Water intake	l/d	AL	58	67	57	50
		RE	28	44	42	24
ME intake	kJ $kg^{-0.75}$/d	AL	1,600	1,557	1,651	1,577
		RE	930	788	897	811
Heat production	kJ $kg^{-0.75}$/d	AL	925	966	1,072	1,103
		RE	826	837	909	858
Min Heat production[1]	kJ $kg^{-0.75}$/d	AL	814	762	909	875
		RE	665	617	641	648
Standing time	h/d	AL	7.7	7.2	13.5	9.6
		RE	6.7	4.9	8.7	7.3
NEFAFFA[2]	µmol/l	AL	84.0	79.45	84.3	63.8
		RE	180.6	138.6	185.2	222.3
BHBA[2]	mmol/l	AL	0.59	0.71	0.67	0.88
		RE	0.49	0.48	0.41	0.59

[1] Data are means of the 10 lowest heat production values during the time course of the day.
[2] Data are means of 24 values of samples taken hourly over the day.

corn silage, 4% grass hay, 26% concentrate, and a mineral mixture (DM=28%; CP=13.6%; crude fibre=16.5%; ME=10.6 MJ/kg). TMR was adjusted with additional concentrates of 1 kg per 2 litres of milk produced above 17 l. The cows were milked at 06:30 and 14:30 h, fed meals of equal size at 07:00 and 15:00 h and had free access to water. On the day before the experiments started, cows were transferred to open-circuit respiration chambers having an ambient temperature of 15 °C and 60% relative humidity (Derno *et al.* 2009). Feed intake was monitored daily and on the first day of the respiration measurements the cows were fed *ad libitum* (AL), while on the second day they only received 50% of this amount (RE).

Gas exchange and physical activity were measured continuously throughout the trial period. Standing and lying times were registered by a photoelectric cell. This allowed for monitoring of the effects of activity and posture on heat production (HP). Daily HP (24 h) was estimated based on measurements of O_2 consumption, CO_2 and CH_4 production..

Blood samples were drawn by a jugular vein catheter in 1 h-intervals and plasma concentrations of non-esterified fatty acids ([NEFA]) and of β-hydroxybutyrate ([BHBA]) were determined. The data were analyzed by paired t-test using SPSS (version 15.0).

Results and discussion

The animals differed in the level of feed intake and physical activity, and showed a corresponding variability in HP, with animals 1 and 2 showing lower HP than animals 3 and 4 (Table 1). The decline of mean daily HP induced by the 50% reduction of energy intake also differed between animals and ranged from 11 to 22%. Standing was the predominant component of physical activity. In all animals standing time significantly decreased during food deprivation (P=0.047) and thus, it seems to be a major strategy to reduce EE under such conditions.

Significantly higher [NEFA] (P=0.015) during food deprivation and significantly lower [BHBA] (P=0.013) can be explained by mobilization of body fat reserves and by a diminished ruminal production of SCFA leading to a reduced delivery of ketone bodies from rumen mucosa to blood. The intensity of fat mobilization varies between animals and thus, an 1.7 to 3.5-fold increase of the blood [NEFA] was observed.

Animal 4 showed a negative energy balance (-11 kJ $kg^{-0.75}$/d) at the high intake level already. At both intake levels energy balance was highest in animal 1 (high intake: 269 kJ $kg^{-0.75}$/d; low intake: -311 kJ $kg^{-0.75}$/d). With the exception of animal 1, energy balance was very similar (in a range of -514 and -545 kJ $kg^{-0.75}$/d) for animals 2 to 4. To conclude, animals differed in their ability to cope with metabolic imbalances. It is expected that the energy status of animal 1 *post partum* will be most stable.

References

Derno M., H.G. Elsner, E.A. Paetow, H. Scholze and M. Schweigel, 2009. Technical note: A new facility for continuous respiration measurements in lactating cows. J. Dairy Sci. 92, 2804-2808.

A meta-analysis of the effects of rumen protected choline supplementation on milk production in dairy cows

L. Pinotti[1], C. Polidori[2], A. Campagnoli[2], V. Dell'Orto[1] and A. Baldi[1]
[1]Department of Veterinary Sciences and Technology for Food Safety, Università degli Studi di Milano, Via Celoria 10, 20133, Milano, Italy
[2]UNITEL-Università Telematica Internazionale, Via Fantoli, 16/15, 20138, Milano, Italy

Introduction

Choline occupies a key position between energy and protein metabolism. Two types of choline functions are known: as choline per se, for which the choline moiety is required, and in the function of a methyl donor. Choline per se plays a major role in lipid metabolism, particularly in lipid transport, as lipotropic agent, but it is also an important source of labile methyl groups for the biosynthesis of other methylated compounds. Based on this second function, choline and methionine pathways partially overlap in providing methyl groups in a variety of reactions. Based on these assumptions we investigated the effects of rumen protected choline administration on milk production in dairy cows. To achieve this pourpose a meta-analysis was carried out to summarize available scientific evidence for the effect of oral rumen protected choline (RPC) supplementation in dairy cows.

Materials and methods

The effects of RPC administration on milk production was investigated by a regression model based on the results of 11 different studies published between 1991 and 2008 (Erdman *et al.*, 1991; Deuchler *et al.*, 1998; Hartwell *et al.*, 2000; Pinotti *et al.*, 2003, 2004; Pipenbrink and Overton, 2003; Janovick *et al.*, 2006; Xu *et al.*, 2006; Zahra *et al.*, 2006; Davidson *et al.*, 2008; Elek *et al.*, 2008). Mean and standard error (SEM) of milk production of the 42 experimental groups (on the basis of RPC amount administered) were considered, and data were analyzed by the PROC MIXED procedure of SAS (SAS Institute, 2001). According to St-Pierre (2001), the analysis included RPC supplementation (control/RPC) as fixed effect, the variability among experiments as random effect and their interaction. Treatments schedule, dry matter intake and dietary composition, accounted for most of the variability among experiments and were highly correlated in a preliminary data analysis; accordingly, these variables were considered as experimental effects and treated as random components in the mixed model, assuming that the results of different experiments are affected by different experimental conditions. For each study, data were also weighted by the reciprocal of the squared standard error of the mean value, in order to consider the unequal variance among studies.

Results and discussion

Milk yield responses to RPC administration are reported in Table 1. In general an improved milk yield is evident as a consequence of RPC administration, even though, as expected, the dose of choline administered is also relevant. Considering three supplementation classes, i.e. control (RPC: 0 g/d), low level (RPC: <10 g/d) and high level (RPC: >10 g/d), the meta-analysis indicated that a low level (less then 10 g per day) of RPC supplementation tended (P=0.15) to increase milk yield and a high level of supplementation of RPC significantly (P<0.001) increased milk yield (Table 1).

Overall these results suggest that greater choline supply seems to be helpful in optimising milk production in dairy cows, even though other factors such as dietary composition, dietary crude protein content and post-ruminal methionine supply merit further investigations.

Table 1. Results from mixed model when the different amounts of RPC administrated were grouped into three concentrations levels

Choline amount administered	Milk yield estimate (kg/d)	SEM	RPC effect
Control (0 g/d)	32.35	1.53	-
RPC low level (<10 g/d)	33.40	1.59	0.15
RPC high level (> 10 g/d)	35.18	1.58	<0.001

References

Davidson, S., B.A. Hopkins, J. Odle, C. Brownie, V. Fellner, and L.W. Whitlow, 2008. Supplementing limited methionine diets with rumen-protected methionine, betaine, and choline in early lactation holstein cows. J. Dairy Sci. 91, 1552-1559.

Deuchler, K.N., L.S. Piperova and R.A. Erdman, 1998. Milk choline secretion as an indirect indicator of postruminal choline supply. J. Dairy Sci. 81, 238-242.

Elek, P., J.R. Newbold, T. Gaal, L. Wagner, and F. Husveth, 2008. Effects of rumen-protected choline supplementation on milk production and choline supply of periparturient dairy cows. Animal 2, 1595-1601.

Erdman, R.A. and B.K. Sharma, 1991. Effect of dietary rumen-protected choline in lactating dairy cows. J. Dairy Sci. 74, 1641-1647.

Hartwell, J.R., M.J. Cecavqa and S.S. Donkin, 2000. Impact of dietary rumen undegradable protein and rumen-protected choline on intake, peripartum liver triacylglyceride, plasma metabolites and milk production in transition dairy cows. J. Dairy Sci. 83, 2907-2917.

Janovick Guretzky, N.A., D.B. Carlson, J.E. Garrett and J.K. Drackley, 2006. Lipid metabolite profiles and milk production for holstein and jersey cows fed rumen-protected choline during the periparturient period. J. Dairy Sci. 89, 188-200.

Piepenbrink, M.S. and T.R. Overton, 2003. Liver metabolism and production of cows fed increasing amounts of rumen-protected choline during the periparturient period. J. Dairy Sci. 86, 1722-1733.

Pinotti, L., A. Baldi, I. Politis, R. Rebucci, L. Sangalli and V. Dell'Orto, 2003. Rumen protected choline administration to transition cows: effects on milk production and vitamin E status. J. Vet. Med., Series A 50, 1821.

Pinotti, L., A. Campagnoli, L. Sangalli, R. Rebucci, V. Dell'Orto, and A. Baldi, 2004. Metabolism in periparturient dairy cows fed rumen-protected choline. J. Anim. Feed. Sci., 13, Suppl. 1, 551-554.

SAS, 2001. Release 9.01. SAS Inst. Inc., Cary, NC, USA.

St-Pierre, N.R., 2001. *Invited review*: Integrating quantitative findings from multiple studies using mixed model methodology. J. Dairy Sci. 84, 741-755.

Xu, G., J. Ye, J. Liu, and Y. Yu, 2006. Effect of rumen-protected choline addition on milk performance and blood metabolic parameters in transition dairy cows. Asian-Austr. J. Anim. Sci. 19, 390-395.

Zahra, L.C., T.F. Duffield, K.E. Leslie, T.R. Overton, D. Putnam and S.J. LeBlanc, 2006. Effects of rumen-protected choline and monensin on milk production and metabolism of periparturient dairy cows. J. Dairy Sci. 89, 4808-4818.

Interaction diet energy level x genotype at alpha s1 casein locus in lactating goats fed *ad libitum*: effects on metabolic and endocrinal response

R.I. Pagano[1], A. Di Trana[2], B. Valenti[1], A. De Angelis[1], M. Avondo[1], P. Di Gregorio[2] and P. Pennisi[1]
[1]DACPA, University of Catania, Via Valdisavoia 5, 95123 Catania, Italy
[2]Dip. SPA, University of Basilicata, Viale dell'Ateneo Lucano 10, 85100 Potenza, Italy

Introduction

In *Girgentana* goats homozygous for strong alleles (AA) at αs1 casein, a high energy intake improves the efficiency of transformation of the diet into milk and casein yield, compared to goats homozygous for weak alleles (FF) (Pagano *et al.*, 2010). This result needs to be studied from the metabolic and hormonal perspective in order to elucidate differences in the partition of the nutrients between goats of different genetic variants at αs1 casein. The objective of this study was to investigate this in conjunction with three levels of hay inclusion in the diet following the metabolic and hormonal profile in lactating goats.

Material and methods

Eighteen goats homogeneous for milk production (1.3±0.3 kg/d), days of lactation (94±12 d) and body weight (36.9±4.7 kg) were selected on the basis of their genotype at the αs1 casein locus (G), as follows: nine goats homozygous for strong (AA) alleles and nine goats homozygous for weak alleles (FF). The goats were used in a 2×3 factorial arrangement of treatments, with the two genotypes (AA, FF) and three diets at different energy level. The diets were: a pelleted alfalfa hay (H100, 1,099 kcal/kg DM) and two pelleted complete feeds including alfalfa hay at 65% (H65, 1,386 kcal NEl/kg DM) and 30% (H30, 1,590 kcal NEl/kg DM). The three different diets had the same crude protein content (15.4% in dry matter). All animals were housed in individual pens. During the pre-experimental period (12 d), the animals received *ad libitum* a mix of the three experimental diets. Each trial consisted of 15 d for adaptation and 8 d for data and sample collection, during which the goats received *ad libitum* the scheduled diet.

Blood samples were taken from all goats at the end of pre-experimental period and of each experimental period. An automated analyzer was used to determine glucose, cholesterol and urea contents in plasma samples. Non-esterified fatty acids (NEFA) and beta-hydroxy butyric acid (BHBA) were analyzed by using commercial kits. Insulin, free triiodothyronine (fT3) and free thyroxine (fT4) were measured in duplicate by ELISA kits

Changes in plasma concentration of metabolites and hormones were analysed by means of ANOVA procedure and analysis included the main effect of the αs1 casein genotype (AA, FF), the diet (100H, 65H, 30H) and the interaction genotype x diet. Insulin, fT3 and fT4 data were transformed into log, ln (y+1) and ln (y+10) values, respectively, before statistical analysis. Data of the pre-experimental period were used as a covariates.

Results and discussion

Plasma concentrations of glucose and NEFA did not significantly change during the trial. The BHBA levels were significantly (P<0.05) lower in the 100H group compared with the 65H group (Table 1). Plasma cholesterol concentrations were significantly higher (P<0.01) in the 30H and the 65H groups than in the 100H group. An opposite trend was noted for plasma urea levels, with its levels significantly (P<0.05) increasing in the 100H and the 65H group and decreasing in the 30H group relative to the initial value. The insulin concentration was affected (P<0.05) by the αs1 casein genotype. As indicated in Table 1 an increase in plasma insulin was observed when animals were homozygous for strong (AA) alleles, whereas no differences (P>0.05) were detected between diets. In previous studies on goats at different genotype at the αs1 casein locus, the authors did not report on differences in insulin levels between strong and weak alleles which would be significant. Schmidely *et al.*, (1997) showed insignificantly higher insulin levels for the AA genotype compared to the FF genotype at the αs1 casein locus. A significant effect of the genotype (P<0.01) on plasma concentrations of fT3 was detected. The concentration of fT3 was higher by 25% in the AA genotype than in the FF genotype, whereas fT4 was not significantly affected by genotype. In adult small ruminants, fT3 and fT4 are a small fraction of the thyroid hormones responsible for the biological activity. These fractions are available for utilization by peripheral tissues and they control the rates of secretion of TRH and TSH. The well established role of thyroid hormones for the maintenance of lactation might explain why high levels of fT3 were observed in the AA genotype characterized by higher milk yield, compared to FF goats (1,419 g vs. 1,014 g; P<0.05) as reported in our experiment (Pagano *et al.,* 2010). A tendency was found for the genotype x diet interaction in fT3 being higher in AA goats than FF goats but only when fed the diet 65H. The results suggest a major role of the hormonal component than the metabolic component in the genotype. The high concentration of insulin and fT3 in AA genotype may be related to a more efficient milk synthesis in subjects with strong alleles and fed *ad libitum*.

Table 1. The effect of diet and genotype on metabolic and endocrinal response

Parameters	Hay inclusion (HI)			Genotypes (G)		SEM	Significance of P		
	100H	65H	30H	FF	AA		HI	G	HI x G
Glucose, mmol/l	2.77	2.70	2.82	2.88	2.71	0.09	ns	ns	ns
NEFA, mmol/l	0.14	0.12	0.13	0.12	0.15	0.04	ns	ns	ns
BHBA, mmol/l	0.24^{b}	0.44^{a}	0.35^{ab}	0.41	0.29	0.07	*	ns	ns
Cholesterol, mmol/l	1.48^{b}	1.79^{a}	1.84^{a}	1.73	1.66	0.09	**	ns	ns
Urea, mmol/l	8.68^{a}	8.55^{a}	7.66^{b}	8.58	8.03	0.41	*	ns	ns
Insulin (log ng/l)	2.36	2.34	2.41	2.29^{b}	2.45^{a}	0.05	ns	*	ns
fT3 (ln+1 pg/ml)	0.93	0.95	0.89	0.82^{b}	1.03^{a}	0.05	ns	**	&
fT4 (ln+10 ng/dl)	2.39	2.39	2.39	2.39	2.41	0.01	ns	ns	ns

* P<0.05; ** P<0.01; & P=0.10; ns = not significant; ^{a,b} P<0.05.

Acknowledgement

The research was funded by MIUR, project of High National Interest PRIN 2007.

References

Pagano R.I., P. Pennisi, B. Valenti, M. Lanza, A. Di Trana, P. Di Gregorio and M. Avondo, 2010. Effect of αs1-casein genotype and its interaction with diet energy level on milk production and quality in Girgentana goats fed *ad libitum*. J. Dairy Res (in press).

Schmidely P., P. Morand-Fehr and D. Sauvant, 1997. Effects of genetic variant for alphaS1 casein and diet nitrogen content on the response of some plasma characteristics after alanine infusion in early and mid lactation goats. Options Méditerranéennes, Series A, 155-158.

Size-dependent effect of metabolic hormones on milk fat globules

H. Malka and N. Argov
Department of Animal Science, the Faculty of Agriculture, Food and Environmental Quality Sciences, the Hebrew University of Jerusalem, Israel

Introduction

Milk fat is secreted in a unique structure termed milk fat globules (MFG) which consist of a triglyceride (Tg) core covered with three layers of phospholipids (Pl) (Heid and Keenan, 2005). In milk the MFG diameter spans over three orders of magnitude; from 200 nm to more than 15 microns (Couvreur and Hurtaud, 2007). Surface area-to-volume ratio determine the dominant lipid component of the MFG: large MFG (lMFG) will be dominated by their Tg core whereas small MFG (sMFG) will be dominated by their Pl envelope. Thus MFG of different sizes will have different lipid composition (Argov and Wochsmann, 2008). Insulin is a metabolic hormone. Its contribution to milk fat composition is well established especially under negative energy balance or under milk fat depression induced by glucogenic dietary regime (Corl and Butler, 2006). Insulin effect on milk fatty acid composition during established lactation is not well described. Furthermore, the questions whether the various mechanisms involved in lipid synthesis and fatty acid derivations in the mammary gland that are subjected to insulin regulation, differ between small and large MFGs were never addressed. The aim of this experiment was to study endogenous insulin concentration and metabolic status effect on milk fat composition. Furthermore, this study aimed at elucidating the differential effect of metabolic hormones, such as insulin, on MFG of various sizes.

Materials and methods

Multiparous cows at 60-70 days *post partum* (n=37) were screened for endogenous insulin level by blood sampling after morning and evening milking. A high positive correlation ($r>0.96$, $P<0.05$) was found between morning and evening insulin concentrations. Milk was collected from cows with the highest and lowest endogenous insulin concentration (n=7). From the initial 37 cows, cows were randomly assigned to 2 groups (n=14) that were drenched daily, for 10 days, 2 different solutions: 500 ml of saline solution (control; c) and 500 ml of propylene glycol solution (treatment; PG). Radioimmunoassay confirmed an approximate 3 fold increase in plasma insulin that peaked approximately 12 h after PG administration in the PG group, while no change in the insulin plasma concentration was observed in the control group. During the experimental period, milk (15 ml) was collected every morning and half was immediately frozen and half was first separated to sMFG and lMFG according to their density. Lipid species, fatty acid concentration and composition of total milk, lMFG and sMFG were determined by liquid and gas chromatography, respectively.

Results and discussion

In the whole milk, a strong association was found prior to the beginning of the study between endogenous insulin plasma concentration and milk fatty acid profile. Apparently high or low endogenous insulin concentration is associated differently with various metabolic processes in the mammary gland that determine milk fatty acid composition such as desaturation and elongation processes (Table 1). Testing the insulin plasma levels in treatment and control cows, and its effect on milk fatty acids composition revealed size dependent influence (Figure 1). Increased plasma insulin concentration was associated with increased omega 3 fatty acid concentration only in sMFG and had no affect on the omega 3 fatty acid concentration in lMFG. Furthermore increased insulin plasma concentration was associated with decreased saturated fatty acid concentration in sMGF but not in lMFG (Figure 1). The results imply that endogenous and induced insulin concentrations have different effect on milk lipid composition. Furthermore, induced insulin plasma concentration has a distinct effect on sMFG composition.

Table 1. Milk fatty acid composition in cows with high and low endogenous plasma insulin concentration. Cows (n=37) were screened for endogenous insulin plasma concentration and the cows with the highest and lowest insulin concentration (P<0.05; n=7) were divided into two groups (high and low). Differences in various fatty acid concentrations were found between cows with different endogenous plasma insulin concentrations. Additionally, markers for flux of fatty acids through the elongase (20:1n9/18:1n9, 22:1n9/20:1n9) and desaturase pathways (20:1/20;0, 16:1n7/16:0) were significantly different between the two groups.

Variable	High (mean)	Low (mean)	*P*-value
16:1n7	1.98	2.14	0.0819
c18:0	9.08	8.34	0.015
c18:1n7	0.73	1.09	0.0274
c18:3n6	0.045	0.028	0.0009
c20:0	0.026	0.052	0.1727
c20:1n9	0.11	0.08	0.0013
c20:4n6	0.19	0.166	0.0038
c22:4n6	0.0098	0.029	0.1577
c20:1n9/c18:1n9	0.005	0.0034	0.0135
c22:1n9/c20:1n9	0.002	0.006	0.0156
c20:1/c20:0	6.4	1.99	0.0014
c16:1n7/16:0	0.047	0.055	0.0436

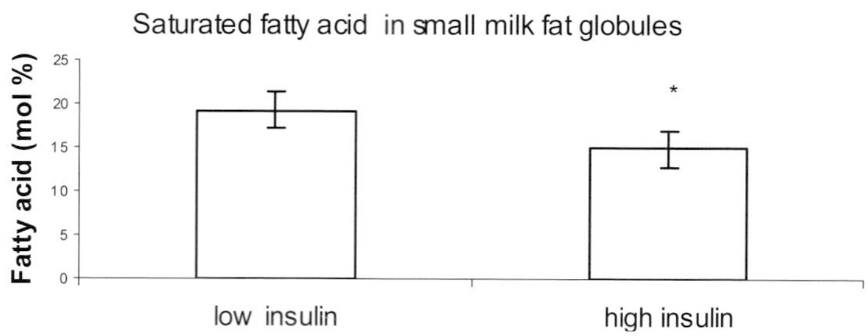

Figure 1. Size specific insulin effect on milk fat globule composition. Milk was collected from cows (n=12) after 4 days of propylene glycol drenching (high insulin) or control cows (low insulin). Milk was separated to small and large milk fat globules. Insulin plasma concentration was positively associated with saturated fatty acid concentration only in the small but not in lMFG (P<0.05).

References

Argov, N., D. Lemay and J.B. German, 2008. Milk fat globule structure and function: nanoscience comes to milk production.. Trends Food Sci Tech 19, 617-623.

Argov, N., S. Wachsmann-Hogiu, S.L. Freeman, T. Huser, C.B. Lebrilla and J.B. German, 2008. Size-dependent lipid content in human milk fat globules. J Agri Food Chem. 56, 7446-7450.

Corl, B.A., S.T. Butler, W.R. Butler and D.L. Bauman, 2006. Short communication: Regulation of milk fat yield and fatty acid composition by insulin. J Dairy Sci. 89, 4171-4175.

Couvreur, S., C. Hurtaud, P.G. Marnet, P. Faverdin and J.L. Peyraud, 2007. Composition of ilk fat from cows selected for milk fat globule size and offered either fresh pasture or a corn silage- based diet. J Dairy Sci. 90, 392-403.

Heid, H.W and T.W. Keenan, 2005. Intracellilar origin and secretion of milk fat globules. Eur J Cell Biol. 84, 245-258.

Pfeuffer, M. and J. Schrezenmeir, 2006. Milk and the metabolic syndrom. Obesity Rev 8, 109-118.

Development of a mechanistic metabolic model of regulation of reproductive processes in dairy cattle

P. Celi, I. Lean, J. McNamara, H. Raadsma and A. Rabiee
Camden, NSW, Australia and Pullman, WA, Australia

Introduction

Reproductive efficiency continues to be suboptimal in many animals, even in very well managed dairy herds. There is no single point issue to fertility, as it is a multifaceted process including regulation of luteinizing hormone and follicle stimulating hormone, response of ovarian cells, development of follicles, ovulation, signs of visible estrus, timely insemination, fertilization, embryonic development, implantation and foetal growth. Each one of these processes can be sub-optimal and can be affected by a number of factors. In general, genotype of the animal, and phenotype, including processes affected by amount and type of nutrient intake all can affect one or more of these processes.

Therefore, for good reason, approaches to describing the system of reproduction and solving the suboptimal performance has been somewhat fragmentary and focused on general animal or environmental characteristics such as body condition score and energy balance. While there is not necessarily anything wrong with this approach, by its nature this approach is limited for determining underlying physiological mechanisms. If there are genetic markers for some aspects of reproduction, there must be specific genes and proteins involved. If nutrient type or amount affects fertility, there must be some specific cellular mechanism. Although a massive amount of data has been collected on many aspects of reproduction, there have been few attempts to take a systems biology approach, a quantitative approach to describing the overall system, especially for research. Good meta-analytical and farm level research has been done, and in fact, reproductive fertility continues to improve. Yet there is no systematic, physiological, mechanistic approach to the problem.

Therefore, the objective was to construct and begin evaluation of a deterministic, mechanistic, dynamic model of nutritional and genetic control of reproductive processes in the dairy cow. The objective of this conceptual research model is to describe control of reproductive processes in dairy cattle at the metabolic level; and to be suitable for evaluation of data, concepts and hypotheses regarding underlying genetic, nutritional and physiological control of reproduction.

Methodology and description

The project was started with an existing model of metabolism in the cow, published and validated (Baldwin *et al.*, 1987; referred to as Molly), which describes utilization of glucose, amino acids and fatty acids by muscle, adipose, visceral and mammary tissues at an aggregated metabolic pathway level. Elements of genetic background, response to nutritional environment and metabolic hormones are explicitly embodied in equation forms and parameter values, such as maximal velocity, substrate sensitivity and control by anabolic and catabolic hormones.

Next, a model of reproductive processes was developed that included flux of follicle stimulating hormone, luteinizing hormone, estrogen and progesterone in cycling and pregnant animals, insemination, fertilization and development of the calf to term. The model links glucose with LH release (with a low sensitivity, that is glucose availability must be severely restricted before affecting LH release); glucose and IGF1 with follicular growth; and also describes effects of feed intake, metabolic rate and milk production on liver metabolism of estrogen and progesterone. The models were then integrated into one system to link genetic elements (for example, genetic merit for milk); nutrient intake and use with reproductive processes, with an integration interval of one day. The model is aggregated at the pathway level and tracks pools in moles (e.g. nutrients, hormones) or kg (milk, body fat) and fluxes in moles/d or kg/d. For example, in cycling animals: Delta progesterone (moles/d) = production of progesterone - degradation of progesterone; Where production of progesterone = $0.0066 \times t^4 - 0.3074 \times t^3 + 3.6504 \times t^2 - 1.5237 \times t$; util_Progesterone = Progesterone × K_util_prog × Liver_Clear_Prog; where Liver_Clear_Prog is a function of metabolic rate.

All parameters are values that can vary depending on the animal being modeled or hypothesis tested (one could alter the parameter for metabolic rate to compare new data to the model, for example). Once fertilization and embryonic development begins the production and degradation of progesterone are changed to describe the usual pattern of concentration during pregnancy.

Results and discussion

The nutritional elements of the model have been extensively evaluated and validated for the objective previously (Hanigan *et al.*, 2006). Included in the nutritional model are genetically controlled elements as noted above. For the reproductive elements, description of hormonal concentrations during cycling and pregnancy simulated literature values within one standard deviation of the mean based on several publications. The relation of nutrient use to reproductive functions was consistent with current data and concepts; for example, increasing metabolic rate increased degradation of estrogen and progesterone, decreased peak concentrations and days of elevated concentration, and decreased progesterone in early pregnancy. Progesterone concentrations are related to embryonic mortality by using a minimally required concentrations that can be varied to test alternative hypotheses.

The behavior of the model was in the proper direction and magnitude for hormone concentrations, timing of ovulation, maintenance of pregnancy, growth of the calf, using a standard that simulated rates and concentrations were within one standard deviation of the observed means, and adequacy was decided if the output was within one least significant difference (LSD) using a paired comparison based on $P<0.05$. The model can be useful to frame specific hypotheses on control of reproductive processes by genetic and nutritional mechanisms, and to form a framework of more specific models.

References

Baldwin, R.L., J. France, J.H.M. Thornley and D.E. Beever, 1987. Metabolism of the lactating cow. Animal elements of a mechanistic model. J. Dairy Res. 54, 99-106.

Hanigan, M.D., H.G. Bateman, J.G. Fadel, and J.P. McNamara, 2006.Metabolic Models of Ruminant Metabolism: Recent Improvements and Current Status J. Dairy Sci. 89, E52-64E.

Rabiee, A.R., K.L. Macmillan and F. Schwarzenberger, 2001.The effect of level of feed intake on progesterone clearance rate by measuring faecal progesterone metabolites in grazing dairy cows. Anim. Repr. Sci. 67, 2005-2014.

Chemical composition of the foetuses in dairy goats in the final third of pregnancy

L.D. Lima, K T. Resende, I.A.M.A. Texeira, H.G. Silva, O. Boaventura Neto and B. Biagioli
UNESP, São Paulo State University, Jaboticabal, SP 14870-000, Brazil

Introduction

The knowledge of body composition is important to estimate nutritional requirements. From the methods used for determining the concentration of nutrients in the body of the animal, the direct method is the most suitable and reliable (ARC, 1980). Thus, the determination of the foetus composition is useful to calculate the demand for nutrients during pregnancy, allowing the formulation of appropriate diets for individual stages of pregnancy. The metabolism during pregnancy passes by profound changes, especially in the last six weeks, when 70% of foetal growth occurs (Russell, 1982). According to NRC (1985), goat pregnancy is divided into two distinct phases; the first before the 120 days is the anabolic phase and the last 30 days which is a phase of nutrients mobilization to supply the pregnant uterus demand. The aim of this study was to evaluate development and chemical composition of foetuses in pregnant of dairy goats with 1 and 2 foetuses at 80, 110 and 140 days of pregnancy.

Material and methods

The data for this study were obtained from 44 pregnant dairy goats (average body weight = 50.6±7.71 kg, age = 4 years, body condition score = 2.58±0.59) fed a diet based on hay and concentrate (roughage: concentrate ratio of 60:40) formulated according to their requirements. Once pregnancy was confirmed, the goats were allocated to treatments as follows: two types of pregnancy (single and twin) and three pregnancy ages (80, 110 and 140 days). At the pre-established pregnancy age goats were slaughtered and the reproductive tract was removed and separated from the cervix, dissected into uterus, foetus, placenta and placental fluid. After the withdrawal of the foetuses, they were weighed and frozen. Contents of dry matter (DM), ether extract (EE) and crude protein (CP) of the foetuses were determined according to AOAC (1995). Gross energy (GE) was measured with a bomb calorimeter. The experiment was designed as a completely randomized 2×3 factorial design. The procedure MIXED of SAS (2002) was applied. An exponential growth model was applied, using the slice option in MIXED procedure, as $\text{Ln}(Y) = (b_0 + b_1T + b_2T^2) + \varepsilon$; NID $(0; s^2)$, where: Y = the value as a function of pregnancy days; b_0 = intercept, b_1 and b_2 define the variation of Y in terms of days of pregnancy, and T = days after mating.

Results and discussion

The amount of nutrients retained in the foetuses did not differ significantly ($P>0.05$) between number of foetus, and there was no interaction between number of foetus and stages of pregnancy. Thus, for determining best fitting equation, this factor number of foetus was removed from the model. It was observed very low nutrient deposition in the foetus body from mating to 80 days of pregnancy, mainly because water represents 85% of their total mass, from 110 days of gestation, the amount of EE, CP and GE in foetus tissues increased exponentially, which followed the fast foetus development. It is also important to highlight that the amount of protein in the foetus at 140 days of pregnancy was 93.9% higher compared to foetus at 80 days of pregnancy and foetal growth in this interval was 89.4%. Our findings confirm the importance of protein in the foetus development, showing that protein demand is higher at the end of pregnancy when the gain of foetus body weight is exponentially increased. The deposition of energy in the body of the foetus followed the same pattern observed for the protein. Both, protein and energy, are important in this stage and their needs increase exponentially with foetal development.

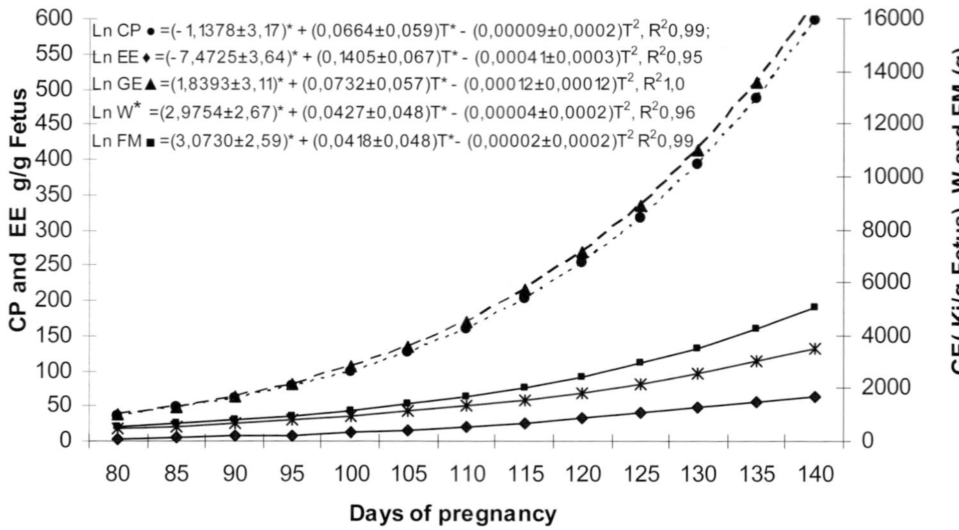

*Figure 1.Composition of gross energy (GE, k;, ▲), crude protein (CP, g; ●), ether extract (EE, g; ♦), water (W, g; *) and foetuses mass (FM, g; ■) as a function of days of pregnancy (T) observed in non-lactating dairy goats,* * = P<*0.001.*

References

Agricultural Research Council (ARC), 1980. The nutrient requirements of farm livestock London, 351 p.

Association of Official Analytical Chemists (AOAC), 1995. Official Methods of Analysis. 16.ed. AOAC, Washington, DC.

National Research Council, 1985. Nutrient requirement of sheeps: 6 ed. Washington: National Academy Press, 99 p.

Russel, A.J.F., 1982. Nutricion de las ovejas gestantes. In: Maluenda, P.D. (ed.) Manejo e enfermedades de las ovejas. Zaragoza: Acribia: 225-242.

SAS – Statistical Analysis Systems Institute, 2002. SAS user's guide: Statistics Version 2002. SAS, Cary, NC.

Early life nutritional management and effects on long term productivity of dairy calves

F. Soberon, E. Raffrenato, R.W. Everett and M.E. Van Amburgh
Department of Animal Science, Cornell University, Ithaca, NY, USA

Introduction

Recent studies in dairy calves have demonstrated the effects of early life nutrient status on long term performance. Studies comparing the effects of suckling versus controlled intakes or *ad-libitum* feeding of calves from birth to 42 or 56 days of life have found that increasing the nutrient intake prior to 56 days of life resulted in increased milk yield during their first lactation ranging from 450 to 1,300 kg of milk compared to the restricted fed calves during the same period (Foldager *et al.*, 1994; 1997; Bar-Peled *et al.*, 1997; Drackley *et al.*, 2007). It is difficult to conduct many of these studies using direct comparisons due to the time and resources needed. We believe there are other methods by which this effect can be evaluated and those methods involve the use of genetic evaluation algorithms already used for genetic evaluation of populations of cattle. The objectives of this study were to better understand the relationship between growth rate prior to weaning and from birth to breeding and milk production, and to determine what factors from the early rearing period have an impact on future performance of dairy calves. We hypothesized that increased energy intake above maintenance is involved in some metabolic programming that results in greater mammary cell activity, an increased efficiency of use of absorbed nutrients or an associate increase in voluntary dry matter intake that remains with the animal throughout its life. A test day model was used to develop residuals for individuals within herds and the residuals were then subjected to further statistical evaluation. It is assumed, since the residuals are used to make genetic evaluations, the residuals provide a less biased analysis of yield effects and help account for the temporal and environmental effects of such an evaluation of lactating cattle.

Materials and methods

Calf records were collected from two New York dairy farms from the computerized farm records. For Herd A, birth weight, birth height, weaning weight, weaning height, age at first calving (AFC), monthly average temperatures and milk replacer intake were collected from available records; average daily gain (ADG) pre-weaning was calculated. For Herd B, birth weights, weaning weights, breeding weights, and AFC were collected and ADG pre-weaning and ADG to breeding were calculated. Milk production and milk composition records were collected for both farms through Dairy Herd Improvement (DHI) and analyzed with a Test Day Model (TDM) (Bauman *et al.*, 1999). The TDM residuals were analyzed using both Mixed model and GLM procedures (SAS, 2002) and significance was declared at $P<0.05$. Models evaluating milk yield included linear and quadratic evaluations of ADG within farm and among farm and non-significant relationships were dropped from the model. Milk yield data was analyzed within farm and not among farms since it was determined that the environmental effects from birth through calving were farm specific and independent such that ADG should not be analyzed among farms even if farm is characterized as a separate effect.

Results and discussion

Significant positive correlations with first lactation milk yield were found for season of birth, year of birth, weaning weight, and ADG from birth to weaning on both farms. When accounting for effect of season and year of birth, for every additional kg of ADG pre-weaning, heifers produced 1,188 kg more milk ($P<0.01$) during first lactation in Herd A. Calves in Herd A were fed the same amount of nutrients every day despite variation in ambient temperature which led to the observation of altered pre-weaning growth rates due to changes in maintenance requirements and subsequent energy available for growth. Thus, the amount of MegaJoules (MJ) consumed above maintenance decreased with lower ambient temperatures. A strong and highly significant relationship was observed in Herd A between intake over maintenance and TDM residuals. For every additional MJ over maintenance consumed from milk replacer during the pre-weaning period, heifers produced 62.81 kg of milk during the first lactation and a total of 237 kg more milk per MJ above maintenance during three lactations (P<0.01). Similarly, in Herd B, for every additional kg of ADG, pre-weaning heifers produced 1,507 kg more milk (P<0.01). For cattle in Herd B, positive correlations with first lactation milk yield were also found for breeding weight, ADG from birth to breeding and ADG from weaning to breeding. When the entire pre-breeding period was considered, for every additional kg of ADG from birth to breeding, heifers produced 1,736 kg more milk (P<0.01) during first lactation. Within this dataset, pre-weaning growth rate accounted for 25% of the variation in first lactation milk production, which is greater than any genetic component. Conclusions from this analysis suggest that the productive life of an animal can be influenced by nutrient availability during an early phase of life and that some programming event is being set through early life nutrient intake. A better understanding of the mechanisms involved in this effect could have pronounced implications in production agriculture as well as in the understanding of long-term metabolic regulations in the animal.

References

Bar-Peled U., B. Robinzon, E. Maltz, H. Tagari, Y. Folman, I. Bruckental, H. Voet, H. Gacitua, and A.R. Lehrer, 1997. Increased weight gain and effects on production parameters of Holstein heifer calves that were allowed to suckle from birth to six weeks of age. J. Dairy Sci. 80, 2523-2528.

Bartlett, K.S., F.K. McKeith, M.J. VandeHaar, G.E. Dahl, and J.K. Drackley, 2006. Growth and body composition of dairy calves fed milk replacers containing different amounts of protein at two feeding rates. J. Anim. Sci. 84, 1454-1467.

Bauman, D.E., R.W. Everett, W.H. Weiland, and R.J. Collier, 1999. Production responses to bovine somatotropin in Northeast dairy herds. J. Dairy Sci. 82, 2564-2573.

Drackley, J.K., B.C. Pollard, H.M. Dann and J.A. Stamey, 2007. First lactation milk production for cows fed control or intensified milk replacer programs as calves. J. Dairy Sci. 90, 614 (Abstr.).

Foldager, J. and C.C. Krohn, 1994. Heifer calves reared on very high or normal levels of whole milk from birth to 6-8 weeks of age and their subsequent milk production. Proc. Soc. Nutr. Physiol. Page 3 (Abstr.)

Foldager, J., C.C. Krohn and L. Morgensen, 1997. Level of milk for female calves affects their milk production in first lactation. Proc. European Assoc. Animal Prod. 48th Annual Meeting. Secion C-3.77 (Abstr.)

Effect of metabolizable energy intake on partition of body fat in dry Pelibuey ewes fed high fibre rations in the tropics

A.J. Chay-Canul, A.J. Ayala-Burgos, J.G. Magaña-Monforte and J.C. Kú-Vera
Facultad de Medicina Veterinaria y Zootecnia, Universidad Autónoma de Yucatán, México; aayala@uady.mx

Introduction

Adipose tissue represents an important energy reserve (39 kJ/g) when forage availability is scarce (Ermias *et al.*, 2002). Breed, physiological state and environmental conditions play a significant role in the amount and distribution of body fat depots (Sprinkle *et al.*, 1998). Pelibuey is the most extensively employed maternal breed under practical conditions in tropical Mexico and little is known concerning its ability for accumulation and mobilization of body fat. This work evaluated the effect of metabolizable energy intake (MEI) on changes in fat depots of mature Pelibuey ewes fed high fibre rations in the tropics.

Materials and methods

Eighteen adult Pelibuey ewes (non pregnant, non lactating) with a mean body weight (BW) of 37±4.4 kg, of the same body condition (BC, 2.5) and being 3 years of age, were divided into 3 groups of 6 animals and assigned to a completely randomized experimental design. Ewes were individually penned in metabolic crates, and fed at three levels of MEI: Low (L), Medium (M) and High (H) during 65 days. The feed offered was adjusted every two weeks based on body weigth. Feeding was based on chopped fresh *Pennisetum purpureum* grass, supplemented with a mixture based on the fruit of the legume *Mucuna pruriens*, corn and cane molasses. Chemical composition of the forage was: Dry matter (DM): 283 g/kg, crude protein (CP): 31 g/kg DM, neutral detergent fibre (NDF): 693 g/kg DM. it was provided at a rate of 44 g DM/kg $BW^{0.75}$/d. Chemical composition of the supplement was: 140 g CP/kg DM and 11.5 MJ ME/kg DM, and was given at a rate of 0, 16 and 32 g DM/kg $BW^{0.75}$/d for treatments L, M and H respectively. Rations were kept isonitrogenous by supplying a mixture of urea and cane molasses diluted in water. Urea was given at a rate of 1.8, 0.6 and 0 g/kg $BW^{0.75}$/d for L, M and H respectively. Dry matter intake (DMI) was measured every second week during five consecutive days. MEI was estimated according to DMI and the Digestible Organic Matter in the DM (DOMD), using the McDonald *et al.* (2002) equation (ME MJ/kg DM: % DOMD × 0.16). Ewes were weighed every two weeks. At the end of the experiment ewes were fasted for 24 h, weighed and slaughtered. Data recorded at slaughter were weights of viscera and carcass. Adipose tissue was dissected, weighed and grouped in two main depots: carcass and internal. Internal fat depot consisted of pelvic (surrounding kidneys and pelvic region), mesenteric and that from inter-intestinal regions. Carcass was split at the dorsal midline in two equal halves, weighed, and chilled to 6 °C for 24 h. After refrigeration, the left half of the carcass was completely dissected into subcutaneous and intermuscular fat (carcass fat), muscle, bone and each component was weighed separately. Two ewes were removed at the end of the experiment because of illness. Data were analysed as a completely randomized design using analysis of variance and Tukey test to verify the significance of differences among treatments, and testing the linear (*L*) or quadratic (*Q*) effects of MEI on response variables. ME requirement for maintenance was assumed to be 426 kJ ME/kg $BW^{0.75}$/d (AFRC, 1993). Huxley's allometric growth equations were used on slaughter and carcass dissection data to assess the differential growth of body fat depots relative to total body fat (TBF).

Results and discussion

Daily MEI was equivalent to approximately 58, 110 and 125% of the maintenance energy requirement (AFRC, 1993) for L, M and H respectively (Table 1). Ewes in the L and M groups lost 7.3 and 1.4 kg respectively, while those in group H gained 2.1 kg which represented 19, 4 and 5% of the initial BW for L, M and H, respectively. TBF, carcass (CF) and internal fat (IF) were different ($P<0.05$) among treatments, and had a linear relationship with MEI ($P<0.0001$). At low levels of MEI, proportion of IF and CF was approximately 50%, however, as the MEI was increased, the proportion of IF was increased up to 57 and 60% for M and H respectively. On the other hand, the mesenteric and pelvic fat deposits were those which increased in a larger proportion with respect to the interintestinal fat deposit. The slopes of relative growth of fat depots in relation to TBF were: for CF at all feeding levels: b=0.78; and for IF at all feeding levels: b=1.23. Regarding IF, the pelvic fat depot had a higher slope (b=1.40), followed by mesenteric (b=1.31) and interintestinal fat (b=1.03).

Table 1. Intake and body fat partition in dry mature Pelibuey ewes fed three levels of ME.

Item	Low (n=5)	Medium (n=6)	High (n=5)	P	
				L	Q
Intake					
Total DM, g/kg$^{0.75}$/d	36.1^{a}	56.2^{b}	65.7^{c}	<0.0001	0.08
ME, MJ/kg$^{0.75}$/d	0.247^{a}	0.472^{b}	0.532^{b}	<0.0001	0.06
Fat partition (kg/animal)					
Total fat (kg)	1.60^{a}	3.41^{b}	5.55^{c}	<0.0001	0.70
Carcass fat (kg)	0.80^{a}	1.46^{b}	2.19^{c}	<0.0001	0.85
Internal fat (kg)	0.80^{a}	1.95^{b}	3.35^{c}	<0.0001	0.64
Mesenteric fat (kg)	0.36^{a}	0.73^{a}	1.63^{b}	<0.0001	0.09
Interintestinal fat (kg)	0.29^{a}	0.70^{b}	0.85^{b}	<0.0001	0.09
Pelvic fat (kg)	0.16^{a}	0.52^{b}	0.87^{c}	<0.0001	0.92

$^{a-c}$ Means with different superscript letters in a row differ ($P<0.05$).

Conclusion

Mature Pelibuey ewes stored a considerable proportion of MEI in the IF depots rather than in the CF. The pattern of accumulation-mobilization of fat in this breed probably represents an adaptive strategy for energy storage during times of plenty in prevision for periods of feed scarcity. The dynamic interplay among fat storage and mobilization, depending on availability or scarcity of food, may have an effect on energy utilization in mature Pelibuey ewes.

References

AFRC, 1993. Energy and Protein Requirements of Ruminants. CAB International, Wallingford, UK.

Ermias, E., A. Yami and J.E.O. Rege, 2002: Fat deposition in tropical sheep as adaptive attribute to periodic feed fluctuation. J. Anim. Breed. Genet. 119, 235-246.

MacDonald, P., R.A. Edwards, J.F.D. Greenhalgh and C.A. Morgan, 2002. Animal Nutrition. 6th Ed. Prentice Hall. pp. 693.

Sprinkle J.E., C.L. Ferrell, J.W. Holloway, B.G. Warrington, L.W. Greene, G. Wu and J.W. Stuth, 1998. Adipose tissue partitioning of limit-fed beef cattle and beef cattle with *ad libitum* access to feed differing in adaptation to heat. J. Anim. Sci.. 76, 665-673.

Effect on animal performance of the source of energy (glucogenic vs. lipogenic) of early lactating dairy cow diets with similar content of metabolizable energy

M. Danesh Mesgaran[1], G. Koolabadi[2], A. Banikamalil[2] and S. Danesh Mesgaran[1]
[1]Dept. Animal Science, Excellence Centre for Animal Science, Faculty of Agriculture, Ferdowsi University of Mashhad, P O Box 91775-1163, Mashhad, Iran
[2]Research Department, Dashat Co., Nayshabour, Iran

Introduction

High-producing dairy cows are challenged *post partum* with large metabolic demands caused by the sudden increase in energy requirements. Two approaches to increase the energy content of diets fed to cows in early lactation are to increase the grain content of the diet or to supplement the diet with fat (Gong, 2002). The characteristics of metabolic problems of this lactation phase suggest a role for the balance in availability of lipogenic and glucogenic nutrients (Van Knegsel *et al.*, 2005, 2007). In ruminants, lipogenic nutrients originate either from fibre that stimulates the ruminal production of acetate and butyrate or from dietary fat, or are derived from body reserves. Glucogenic nutrients originate from starch escaped from rumen degradation or gluconeogenesis (Gaynor *et al.*, 1995). However, the contribution of intestinally digested starch to metabolic glucose is highly variable (Mills *et al.*, 1999). The aim of the present study was to determine the effect of early lactating dairy cow diets containing different energy sources (glucogenic vs. lipogenic) with equal metabolizable energy (ME) content on early lactation performance of Holstein cows.

Materials and methods

Three early lactating dairy cow diets were provided containing different energy sources including glucogenic source [barley grain (BG)] or lipogenic sources [sugar beet pulp (SBP) or protected palm fat (PF)]. The sources of energy provided 18% of the total daily requirements for metabolizable energy (312 MJ/head/day). Diets were formulated to support 43 kg milk/d and consisted of forge (alfalfa hay and corn silage) and concentrates (maize grain, soybean meal, cottonseed meal and BG or SBP or PF) in a ratio of 1:1 (CP= 180 g/kg DM). Diets were fed as TMR to 120 multiparous early lactating Holstein dairy cows (DIM= 16±3, 40 cows per each diet) for 10 weeks as *ad libitum*. Feed intake was monitored daily, milk yield and composition were recorded weekly, and body condition score (BCS, five-point scale where 1=thin to 5=obese) was determined monthly. Data were analyzed using the MIXED procedure of SAS (2001) for a completely randomized design with repeated measures. The model included the energy sources, time, and 2-way interactions of energy sources with time. Duncan's test was used to determine the significance of means at $P<0.05$.

Results and discussion

Data of feed intake, milk production and composition, and BCS are presented in Table 1. Results indicated that the animals fed SBP had a lower ($P<0.05$) daily feed intake compared with the other groups. Milk yield of the cows receiving the diets of BG and PF was considerably higher ($P<0.05$) than those fed the SBP diet. However, milk fat concentration (g/kg) of the cows fed SBP was higher ($P<0.05$) than that of the other animals. There was no significant effect of the diet on BCS.

The results of the present study did not confirm the findings of the previous studies that observed no increase in milk fat percentage after feeding a lipogenic diet (Van Knegsel *et al.*, 2007). Extra lipogenic nutrients, as digestible fiber, often increase the milk fat percentage, as also observed in this study, which is not the effect of the lipogenic character of the diet but of its effects on ruminal fermentation. The present study was designed to feed diets of similar ME content but contrast in

Table 1 Dry matter intake (DMI, kg/d), milk production (kg/d), milk composition (g/kg) and body condition score (BCS) of dairy cows fed diets containing different energy sources including glucogenic source [barley grain (BG)] or lipogenic sources [sugar beet pulp (SBP) or protected palm fat (PF)] during weeks 2 to 12 of lactation.

Variable	Diet			SEM	*P*-value		
	BG	SBP	PF		Diet (D)	Week (W)	DxW
DMI	24.2[a]	22.1[a]	23.7[a]	0.6	<0.050	<0.050	NS
Milk yield	41.9[a]	37.8[a]	40.4[a]	1.1	<0.05	<0.050	NS
Milk fat	35[a]	37[a]	34[a]	1.2	<0.05	<0.050	NS
Milk protein	30.2	30.4	30.3	1.3	NS	<0.050	NS
Milk lactose	47.4	48.3	47.5	1.7	NS	NS	NS
BCS	2.9	3.0	3.0	0.3	NS	-	-

[a,b] The difference between means with different letter is significant at $P<0.05$. NS = not significant.

lipogenic (in two different types as fat or digestible fiber) and glucogenic nutrient supply. Despite a similar quantity of available ME, cows fed the FP as lipogenic diet and BG as glucogenic diet partitioned the same amount of energy to milk. Therefore, the present data did not confirm the hypothesis that energy partitioning between milk and body tissue can be altered by feeding isoenergetic diets that differ in lipogenic and glucogenic nutrient supply (Van Knegsel *et al.*, 2007).

References

Gaynor, P.J., D.R. Waldo, A.V. Capucco, R.A. Erdman, L.W. Douglass and B.B. Teter, 1995. Milk fat depression, the gluconeogenic theory, and trans-C18:1 fatty acids. J. Dairy Sci. 78, 2008-2015.

Gong, J.G., 2002. Influence of metabolic hormones and nutrition on ovarian follicle development in cattle: Practical implications. Domest. Anim. Endocrinol. 23, 229-241.

Mills, J.A.N., J. France, and J. Dijkstra, 1999. A review of starch digestion in the lactating dairy cow and proposals for a mechanistic model: 2. Postruminal starch digestion and small intestinal glucose absorption. J. Anim. Feed Sci. 8, 451-481.

Van Knegsel, A.T.M., H. van den Brand, J. Dijkstra, S. Tamminga, and B. Kemp, 2005. Effect of dietary energy source on energy balance, production, metabolic disorders and reproduction in lactating dairy cattle. Reprod. Nutr. Dev. 45, 665-688.

Van Knegsel, A.T.M., H. van den Brand, J. Dijkstra, W.M. van Straalen, R. Jorritsma, S. Tamminga, S. and B. Kemp, 2007. Effect of glucogenic vs. lipogenic diets on energy balance, blood metabolites, and reproduction in primiparous and multiparous dairy cows in early lactation. J. Dairy Sci. 90, 3397-3409.

Heart rate in dairy cows grazing in an alpine area

S. Colombini, A. Tamburini, M. Guerci, L. Rapetti and G.M. Crovetto
Università degli Studi di Milano, Dipartimento di Scienze Animali, via Celoria 2, Milan, Italy

Introduction

In dairy cows grazing activity increases the energy requirements. Heart Rate (HR) measurements are a practicable method for estimating accurately the short-term and long-term variations of energy expenditure in free-range cattle (Brosh *et al.*, 1998, Brosh, 2007). There are few studies on dairy cows grazing alpine pastures.

The purpose of this study was to evaluate the relation between daily grazing behavior and milk yield on HR of grazing cows and to estimate the energy expenditure of cows supplemented with different levels of concentrate.

Materials and methods

The trial was conducted over two summers on a high alpine pasture (Alpe Culino, SO, Valtellina, 1660-2000 m.a.s.l.). In each year, 45 dairy Brown Swiss cows were controlled for daily milk yield for a period of 21 d. Three cows were monitored for dry matter intake (*n*-alkane method) and for HR (Polar Equine) for 10 d in each year (mean every 5 sec). Heart rate data (averaged every 5 min) and energy expenditure (computed with a O_2 pulse of 320 µl/beat/kg Body Weight$^{0.75}$) were analyzed (proc GLM procedure of SAS, 2000) considering four different time periods per each day obtained by direct observation (milking, grazing or travelling, daily and night rest) and three different concentrate level (0, 1.5 and 3 kg/d as fed). Concentrate was offered twice a day at milking.

Results and discussion

The HR was different among cows fed with a different concentrate level (Table 1), showing a numerically higher value for the cows of the Group 0 probably due to higher grazing activity to cover energy requirements. The difference in HR between years may be explained by the varying quality of mountain pastures and by the different milk yield (11.4, 14.1, 14.7 kg/d of FCM in 2007 and 10.2, 13.5, 12.2 kg/d of FCM in 2008 for 0, 1.5 and 3 kg/d of concentrate, respectively). The highest HR values were registered during grazing (Table 2), whilst no significant differences were detected between milking or resting. These results, and the diurnal patterns of HR (Figure 1), were similar to Brosh *et al.* (2003, 2006) that have shown that these patterns for grazing cows also depend on other activities (travelling, moving, grazing), while HR for confined ruminants mainly depends on hourly patterns of feed intake.

Table 1. Heart Rate (HR, beats/min) for different levels of concentrate in diet (LS means).

	Concentrate level			SEM	Contrasts		
	0 kg/d	1.5 kg/d	3 kg/d		1 vs. 2	1 vs. 3	2 vs. 3
HR (beats/min)	72.6	68.4	70.6	1.45	0.03	0.31	0.27
HR and year							
2007	68.9	70.3	73.5	2.09	0.62	0.09	0.18
2008	76.3	66.5	67.7	2.37	0.0002	0.0027	0.70

Table2. Heart Rate (HR, beats/min) for different daily activities (LS means).

	Daily activity				SEM	Contrasts					
	grazing	milking	daily rest	night rest		1 vs. 2	1 vs. 3	1 vs. 4	2 vs. 3	2 vs. 4	3 vs. 4
HR (beats/min)	75.7	67.2	70.7	68.6	1.82	0.001	0.036	0.001	0.14	0.47	0.36
HR and concentrate level											
0 kg/d	77.1	68.3	72.7	72.4	2.95	0.01	0.26	0.16	0.25	0.19	0.94
1.5 kg/d	73.6	64.9	68.9	66.0	3.20	0.03	0.28	0.04	0.34	0.74	0.45
3 kg/d	76.4	68.3	70.4	67.4	3.31	0.03	0.15	0.02	0.62	0.81	0.48

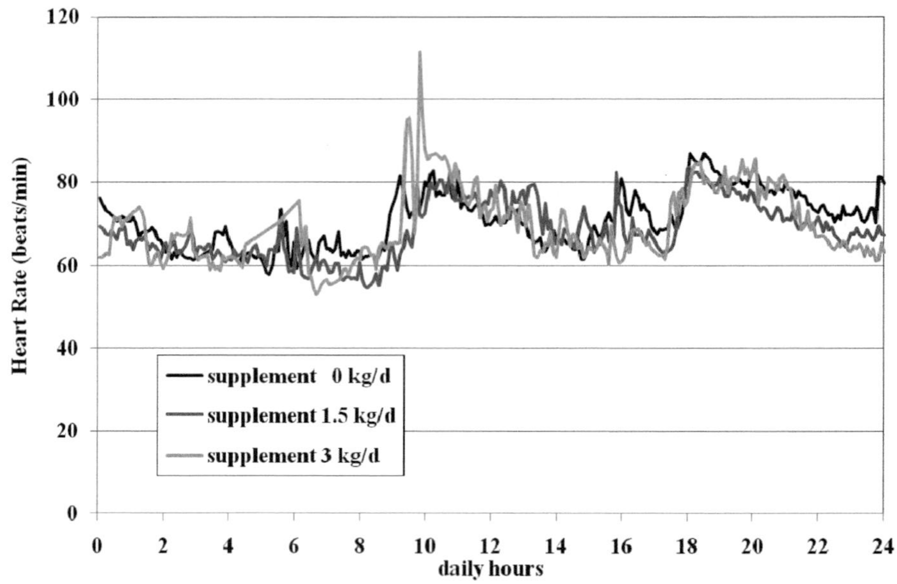

Figure 1. Heart rate (beats/min) for different concentrate levels in daily hours.

References

Aharoni, Y., A. Brosh, and Y. Harari, 2005. Night feeding for high yielding dairy cows in hot weather: Effects on intake, milk yield and energy expenditures. Livest. Prod. Sci. 92, 207-219.

Brosh, A., 2007. Heart rate measurements as an index of energy expenditure and energy balance in ruminants: A review. J. Anim. Sci. 85, 1213-1227.

Brosh, A., Y. Aharoni, A.A. Degen, D. Wright and B. Young, 1998. Estimation of energy expenditure from heart rate measurements in cattle maintained under different conditions. J. Anim. Sci. 76, 3054-3064.

Brosh, A., Z. Henkin, E.D. Ungar, A. Dolev, A. Orlov, Y. Yehuda, and Y. Aharoni. 2006. Energy cost of cows' grazing activity: The use of heart rate GPS methods for direct field estimation. J. Anim. Sci. 84, 1951-1967.

SAS Institute, 2000. Release 8.1, USA. Cary, NC: SAS Institute Inc.

The effects of hydrolyzable tannins on rumen fluid traits and production performances in dairy sheep fed on pasture

G. Pulina[1], G. Battacone[1], A. Mazzette[1], M. Acciaro[2], M. Decandia[2], M. Sitzia[2] and A. Nudda[1]
[1]Dipartimento di Scienze Zootecniche, University of Sassari, Italy
[2]Dipartimento per la Ricerca nelle Produzioni Animali, AGRIS Sardegna, Sassari, Italy

Introduction

In the Mediterranean area, feeding strategies of dairy sheep are based on grazing pastures that offer different herbage quality in relation to both botanical composition and season. Particularly in spring, a high level of soluble proteins characterizes pastures dominated by legume species. However, in this season, due to the high herbage availability, shepherds tend to reduce the concentrate/forage ratio. As a consequence the animal diet results to be unbalanced between nitrogen and carbohydrate intake (Cannas, 2004).

The inclusion of tannins in the diet can affect the efficiency of rumen fermentation, possibly resulting in relevant positive effects on the nutritional status of grazing ruminants (Min *et al.*, 2003). In fact, the use of tannins in grazing ruminants reduces OM digestibility in the rumen and increases protein escape. The latter is particularly desirable when animals graze on grass with a high content of soluble protein, because tannins can reduce the production of ammonia in the rumen and increase the protein potentially available for intestinal digestion. However, tannins could become ineffective when herbage nitrogen content falls rapidly in late spring.

The objective of this work was to evaluate the effects of different levels of tannins on rumen fluid traits and production performances in dairy sheep grazing a pasture initially rich in nitrogen and receiving restricted supplementation (250 g/d of concentrate per head).

Material and methods

Thirty-six 2-4-year-old Sarda ewes (12 per experimental group) in mid lactation (90-120 days in milking, DIM) were used in this 6 weeks trial. The ewes grazed on pasture constituted by 76% of *Medicago polymorpha* L., 20% of *Lolium multiflorum* L. and 4% of other species (composition: DM 25%; CP 17.3%, NDF 47.5% in DM). The animals were supplemented with 300 g/d of alfalfa hay and 250 g/d of concentrate per head (87.1% DM, 17.6% CP, 29.4% NDF in DM) containing three levels (0, 6% and 12%, on a DM basis; T0, T6 and T12, respectively) of a commercial chestnut hydrolizable tannin (Nuova Rivart Spa, Radicofani, Italy). Individual intake of pasture grass (once, Dove and Mayes, 1991), concentrate (daily) and alfalfa hay (daily, on a group basis) were measured. Daily milk production was recorded and milk samples were collected weekly for analysis of fat, protein, lactose, SCC and urea. Rumen fluid samples were collected every 10 days with a stomach tube from five animals/group to measure traits of rumen biochemistry: pH, VFA (acetic, propionic and butyric acids), ammonia N (NH3-N) and long chain fatty acid profile. Data were analyzed by ANOVA including tannin level, sampling and their interaction as fixed factors.

Results and discussion

Alfalfa hay DM intake was 258 g/d on average. Concentrate intakes with T6 and T12 were 12.6% and 27.2%, respectively, lower than that of T0 (246 g/d), probably because of the taste of tannins. Herbage intake on pasture was not different between groups (997, 1159 and 924 (SEM ±110) g of DM for T0, T6 and T12, respectively). Tannins did not influence milk yield and milk fat and protein content (Table 1), which fell linearly throughout the period without differences between groups. Milk urea increased with time in all groups independently of tannin addition (data not shown). The lack of tannin effect was unexpected and is hard to explain. The SCC was lowered by the inclusion of tannin in the diet (Table 1).

Rumen biochemistry traits are reported in Table 2. Proportions of C18:2 n6 and C18:3 n3 were increased by the inclusion of tannins. Ammonia tended to be reduced by the inclusion of tannin in the diet and markedly decreased from the first to the second sampling following the CP content in the pasture. In conclusion, tannins could be useful to modulate nitrogen utilization in sheep grazing forages in the growing phase. However, further studies are necessary to verify why the pattern of rumen fluid ammonia and milk urea did not correspond.

Table 1. Milk yield and its composition in ewes fed concentrate with different doses of tannins[1]

	T0	T6	T12
Milk yield, g/d	1,311	1,316	1,303
Fat, %	5.6	5.4	5.5
Protein, %	5.14	5.07	5.08
Lactose, %	4.62[c]	4.67[cd]	4.71[d]
Urea, mg/dl	51.1	51.6	52.6
SCC, log10 (*1000/ml)	3.03[a]	2.67[b]	2.73[b]

[1]T0 = 0%, T6 = 6%, T12 = 12% of commercial chestnut hydrolyzable tannin, on a DM basis.
[a,b] Different superscripts within a row differ for $P<0.01$ and [c,d] $P<0.1$.

Table 2. Rumen fluid traits of ewes fed concentrate with different doses of tannins[1].

	T0	T1	T2
pH	6.22[c]	6.34[d]	6.27[cd]
Acetic, % of VFA	61.7[a]	59.4[b]	62.6[a]
Propionic, % of VFA	22.0	21.9	21.1
Butiric, % of VFA	16.2[a]	18.8[b]	16.3[a]
Ammonia, mg/L	2.41[a]	2.12[ab]	1.91[b]
C18:2 n6, % of FA	4.5[a]	4.8[ab]	5.5[b]
C18:3 n3, % of FA	2.3[c]	2.9[d]	2.9[d]

[1] T0 = 0%, T6 = 6%, T12 = 12% of commercial chesnut hydrolyzable tannin, on a DM basis.
[a,b] Different superscripts within a row differ for $P<0.01$ and [c,d] $P<0.1$.

Acknowledgements

Research was supported by the Ministero dell'Istruzione dell'Università e della Ricerca (Project PRIN 2008).

References

Cannas, A., 2004. Feeding of lactating ewes. In: Dairy Sheep Nutrition. G. Pulina (ed.) CAB International, Wallingford, Oxon, UK: 79-108.

Dove, H. and Mayes, R.W., 1991. The use of plant wax alkanes as marker substances in studies of the nutrition of herbivores: A review. Aust. J. Agric. Res. 42, 913-952.

Min, B.R., T.N. Barry, G.T. Attwood and W.C. McNabb, 2003. The effect of condensed tannins on the nutrition and health of ruminants fed fresh temperate forages: a review. Anim. Feed Sci. Tech. 106, 3-19.

Effect of a phytogenic feed additive on rumen fermentation and bacterial protein reaching the small intestine

M. Veit[1], A. Mader[1], K. Wendler[1] and J. Schmidt[2]
[1]Delacon Phytogenic Feed Additives, Steyregg, Austria
[2]University of West Hungary, Mosonmagyaróvár, Hungary

Introduction

In dairy production, rumen function, feed intake, and health status have to be optimized to achieve profitable milk production. Furthermore, the demand for emission reductions in livestock production is becoming more important. This is because methane is one of the primary greenhouse gases and livestock production is the major source of anthropogenic methane (Wood and Knipmeyer, 1998). Methane is produced as a by-product of digestive processes and represents a serious loss of feed energy (2-12%) (Pen *et al.*, 2006). Specific plant substances, like saponins may inhibit the growth of protozoa, resulting in a reduction of gaseous emissions such as methane. Thereby, metabolic energy and nitrogen losses may be reduced and feed efficiency can be increased. The objective of the present study was to evaluate the effects of the phytogenic feed additive Rumex (Delacon Phytognic Feed Additives, Steyregg, Austria) on rumen fermentation, microbial activity and bacterial protein in the duodenum. Rumex consists mainly of a mixture of essential oils (≥1.5%) with standardized thymol and of saponins extracted from *Quillaja saponaria* and herbs.

Material and methods

A trial in two phases (control, treatment) of 26 days with four rumen-fistulated steers (n=4) was conducted. Steers were between 600-650 kg in bodyweight and were fed on a roughage based diet according to maintenance requirements. The diet (10.2 MJ net energy (NE)/kg dry matter (DM); 13.1% crude protein (CP) and 21.8% crude fiber CF)) consisted of 8 kg corn silage, 4 kg hay and 3 kg concentrate (75% maize, 14.5% wheat, 8% sunflower meal). During the control phase, the animals received no additive, while in the testing phase a total of 5 g/head/day of Rumex were applied directly via the rumen fistula (2.5 g during morning and 2.5 g during afternoon feeding). In both, control and testing phase, the adaptation period was 21 days, followed by a 5-day trial period in which the following parameters were measured every day in samples from rumen fluid, at 0, 2, 4, 6 and 8 hours after morning feeding: pH, volatile fatty acids (VFA) contents, and microbial activity (MA). The MA was measured by the Nitrite Reduction Method (Horváth, 1979) in rumen fluid. In this test, the time needed by rumen bacteria for the degradation of nitrite is measured, allowing for an estimation of the activity of nitrite-reducing microbes. The concentration of VFA in the rumen fluid was analyzed by HPLC.

In order to evaluate the bacterial protein synthesized in the rumen and reaching the small intestine, a trial with the same design as above but only two steers (n=2) (550-600 kg) per phase with rumen and duodenal fistulas was conducted. The diet (10.2 MJ/kg NE; 13.6% CP and 21.6% CF) consisted of 6 kg corn silage, 4 kg hay and 3 kg concentrate (as described above). Samples were taken on days 2 and 4 of the trial period, between 6 am and 4 pm, at an interval of every 2 h. The parameters measured were: pH, DM, CP, CF, diaminopimelic acid (DAPA) and TiO_2 content. DAPA is a marker for bacterial protein occurring from rumen fermentation and was measured by the method of Krawielitczki and Piatkowski (1977). TiO_2 was used as a marker for DM, to calculate the quantity of chymus passing the duodenum. The evaluation of CF degradation was made by measuring the CF content in the feed and in the chymus, relative to the quantity of chymus passing the duodenum. The data were analyzed by the GLM procedure with SAS 9.1, considering the effect of treatment.

Results and discussion

The additive had a positive influence on the MA in the rumen. In Table 1, the results of the nitrite reduction test indicate a significant increase in MA during the testing phase.

As a result of the higher MA, the content of VFA increased in the testing phase. Significantly higher concentrations of the following VFA were observed in the Rumex group: propionic acid, acetic acid and iso-butyric acid. The acetic:propionic acid ratio was decreased by Rumex as the concentration of propionic acid was increased more than the concentration of acetic acid. The quantity of bacterial protein passing the duodenum was 935±257 g/d in the control and 1298±86 g/d in the test phase. This numerical increase might be the result of a higher MA. A further reason for this numerical increase in bacterial protein might be the anti-protozoal activity of the saponins present in Rumex. The investigation on CF intake and CF reaching the duodenum showed that numerically more CF was degraded by the rumen's microbes when Rumex was added to the feed (44.8±17.4% and 46.28±16.7% in the control and testing phase, respectively).

Table 1. Microbial activity (nitrite concentration: KNO_2 per ml) in rumen fluid after feeding (means ± standard deviation).

	Time in minutes		
	0.2	0.5	0.7
Control	4.90±2.31[a]	9.35±5.33[a]	12.90±6.27[a]
Rumex	3.30±0.73[b]	7.80±2.57[a]	10.70±2.39[a]

[a,b] Means with different superscripts within a row differ significantly ($P<0.05$).

Conclusion

The application of the phytogenic feed additive tested shows some potential for optimizing rumen function. Positive effects were especially observed for the microbial activity and the concentration of volatile fatty acids in the rumen. Furthermore, bacterial protein reaching the small intestine was slightly increased and crude fiber digestibility was numerically improved.

References

Horváth, Z., 1979. Veterinary Clinical Laboratory Tests (in Hungarian). Mezőgazdasági Kiadó Budapest.

Krawielitzki, R. and B. Piatkowski, 1977. Studies on the determination of the part of bacterial nitrogen in relation to total nitrogen content in the duodenum of cows. Archiv für Tierernährung 27, 309-314.

Pen, B., C. Sar, B. Mwenya, K. Kuwaki, R. Morikawa and J. Takahashi, 2006. Effects of *Yucca schidigera* and *Quillaja saponaria* extracts on *in vitro* ruminal fermentation and methane emission. Anim. Feed Sci. Tech. 129, 175-186.

Wood, C. and C.K. Knipmeyer, 1998. Global climate change and environmental stewardship by ruminant livestock producers. Agricultural Education, University of Missouri and the National Council for Agricultural Education.

Microbial protein flow in lactating and dry cows fed grass silages of different maturities

S.J. Krizsan[1,3], *G. Broderick*[2], *S.K. Nes*[1] *and H. Volden*[1]
[1]*Dept. of Anim. and Aquacultural Sci., Norwegian University of Life Sciences, Ås, Norway*
[2]*Agricultural Research Service, USDA, US Dairy Forage Research Center, Madison, WI 53706, USA*
[3]*Department of Agricultural Research for Northern Sweden, Swedish University of Agricultural Sciences, Umeå, Sweden*

Introduction

Dietary protein escaping ruminal degradation and microbial protein synthesised in the rumen are the main sources that provide the ruminant with absorbed and metabolisable protein to support maintenance and production requirements. Two experiments were conducted to study microbial protein yield in dairy cows fed grass silage of different maturities at high and low level of feed intake.

Materials and methods

Six lactating Norwegian Red cows producing 34±4.5 kg milk/d at the start of the experiment were used in an incomplete Latin square design with a 3×2 factorial arrangement of treatments and four 24-d periods to test effects at a high intake level (Trial 1). Treatments were three grass silages fed *ad libitum and* harvested at different maturity, each combined with one of two levels of concentrate. Silage-specific chemical composition was (mean ± SD): G1 (310±14.3 g dry matter (DM)/kg, 412±9.7 g neutral detergent fibre (NDF)/kg of DM and 215±6.1 g crude protein (CP)/kg of DM), G2 (349±10.0 g DM/kg, 530±11.8 g NDF/kg of DM and 127±4.7 g CP/kg of DM), and G3 (242±9.0 g DM/kg, 639±21.0 g NDF/kg of DM and 125±5.2 g CP/kg of DM). Four dry Norwegian Red cows were fed two grass silages, supplemented with three kg (as fed) of concentrate or without any supplementation to test effects at a low level of intake (Trial 2). Intake of DM was restricted to 7.5 kg/d. The experiment was conducted as a 4×4 Latin square with 22-d periods. Silage-specific chemical composition was (mean ± SD): G4 (437±3.5 g DM/kg, 440±3.3 g NDF/kg of DM and 149±5.1 g CP/kg of DM), and G5 (341±22.5 g DM/kg, 540±5.6 g NDF/kg of DM and 123±1.8 g CP/kg of DM). Silages in the experiments were produced or purchased locally; the leys consisted mainly of timothy and meadow fescue. Concentrate was composed mainly of wheat starch, contained no NDF and CP concentration was 168±14.3 g/kg of DM. Nutrient flow was quantified by the omasal sampling technique with the use three markers (Co + Yb + iNDF). Continuous infusion of +10 atom% $(^{15}NH_4)_2SO_4$ was used to label microbial protein. All data were analysed using the MIXED procedure of SAS (2003), and diet effects were separated using orthogonal contrasts.

Results and discussion

During Trial 1, intake of DM and N were higher for cows fed G1 diets (17.0 kg/d and 546 g/d, respectively) than those fed G2 (15.4 kg/d and 338 g/d, respectively) and G3 diets (15.4 kg/d and 337 g/d, respectively) (P=0.01). Intake of N was highest for cows fed G4 silage and supplemented with concentrate (191 g/d) and lowest for cows fed only G5 silage (148 g/d) (P=0.03). Microbial flow parameters for Trial 1 and 2 are given in Tables 1 and 2, respectively. In Trial 1 concentrate level did not affect any of the traits presented (P≥0.25). Consistent with the results of Vanhatalo *et al.* (2009) nonammonia N (NAN) and microbial NAN flow decreased when the grass silage was harvested at a more mature growth stage (P≤0.09; Table 1 and 2). Microbial NAN flow was higher with concentrate in Trial 2 (P=0.01; Table 2); this effect had not been observed in the lactating cows. A moderate increase in the amount of concentrate has been shown to enhance microbial outflow (Jaakkola and Huhtanen, 1993). It seems that the proportion of microbial N as a fraction of total

NAN flow was comparable in the present trials. A lower microbial efficiency for dry cows than for lactating cows might have been expected.

Table 1. Effect of silage source on microbial flow parameters and OM truly digested in the rumen of lactating dairy cows (trial 1).

	Silage[1]			SE[4]	*P*-value[2]
	G1	G2	G3		L_{NDF}
Nonammonia N (NAN) flow, g/d	329	301	271	25	0.09
Microbial NAN flow, g/d	217	190	195	17	0.04
Microbial efficiency, g of NAN/kg of OMTDR	17.0	17.7	18.8	0.9	0.12
OMTDR[3], %	82.0	75.0	71.6	1.9	<0.01

[1] Chemical characterisation of the silages given in text.
[2] Probability of a significant effect of L_{NDF} = linear effect of silage NDF concentration.
[3] OMTDR = OM truly digested in the rumen.
[4] Unbalanced design; highest SE among treatments is given.

Table 2. Effect of concentrate level and silage source on microbial flow parameters and OM truly digested in the rumen of dry cows (trail 2).

	Silage[1]				SE	*P*-value[2]	
	G4		G5			L_C	L_{NDF}
Concentrate level[4]	0	3	0	3			
Nonammonia N (NAN) flow, g/d	143	159	123	129	7	0.14	0.01
Microbial NAN flow, g/d	89.7	117	76.4	88.6	5.6	0.01	0.01
Microbial efficiency, g of NAN/kg of OMTDR	17.9	21.1	15.5	16.1	1.3	0.16	0.03
OMTDR[3], %	73.6	80.1	71.4	79.0	9.0	<0.01	0.16

[1] Chemical characterisation of the silages given in text.
[2] Probability of a significant effect of L_C = linear effect of concentrate level and L_{NDF} = linear effect of silage NDF concentration.
[3] OMTDR = OM truly digested in the rumen.
[4] Concentrate level in kg/d as fed.

References

Jakkola, S. and P. Huhtanen, 1993. The effects of the forage preservation method and the proportion of concentrate on nitrogen digestion and rumen fermentation in cattle. Grass Forage Sci. 48, 146-154.

SAS Institute, 2003. User's Guide: Statistics. Version 9.1 ed. SAS Inst. Inc., Cary, NC.

Vanhatalo, A., K. Kuoppala, S. Ahvenjärvi and M. Rinne, 2009. Effects of feeding grass or red clover silage cut at two maturity stages in dairy cows. 1. Nitrogen metabolism and supply of amino acids. J. Dairy Sci. 92, 5620-5633.

The effects of rumen ammonia concentration on feed intake, fibre digestibility and milk yield in dairy cows

S. Ahvenjärvi[1] *and P. Huhtanen*[2]
[1]*MTT Agrifood Research Finland, Jokioinen, Finland*
[2]*Swedish University of Agricultural Sciences, Umeå, Sweden*

Introduction

Supplementation of dairy cow diets based on forage and carbohydrate enriched concentrates with protein feeds has often resulted in good production responses in terms of milk yield (Rinne *et al.* 1999). In spite of good economical returns a large proportion of feed protein entering the rumen is degraded to ammonia or incorporated directly into the rumen microbial protein and only a small proportion enters the omasal canal undegradable (Broderick *et al.* 2010). Extensive ruminal degradability of feed protein results in poor marginal utilisation of supplementary N and consequently the efficiency of N utilisation inevitably decreases as the crude protein (CP) concentration of the diet increases. Since the efforts to improve the efficiency of protein utilisation through decreasing the ruminal degradability of feed protein have failed or been unsatisfactory the most optimal N utilisation is achieved when the diet supplies sufficient rumen degradable protein to meet but not exceed the rumen microbial requirements. The objective of the current study was to identify the critical ammonia levels in the rumen below which the ruminal fibre digestibility and microbial N synthesis will be decreased relative to optimal conditions (Hoover, 1986).

Materials and methods

To study the effects of low ruminal ammonia levels on feed intake, fibre digestibility, N utilisation and milk yield in dairy cows the basal diet was formulated to supply inadequate amounts of rumen degradable CP. The diet was comprised of highly digestible grass silage (707 g digestible OM per kg DM) with moderate CP concentration (141 g/kg DM). Forage and concentrates were fed as a total mixed ration that contained 511 g of grass silage, 156 g of wheat, 152 g of oats, 147 g of molassed sugar beet pulp, 20 g of solvent extracted rapeseed meal, and 15 g of mineral and vitamin premix per kg DM. The CP concentration of the basal diet was 127 g/kg DM. The experimental animals were multiparous dairy cows equipped with a rumen cannula, 77 DIM and 620 kg BW at the beginning of the study. The experimental treatments were five levels of urea infused into the rumen at a rate of 0, 17, 33, 49 and 66 g/d of N. Treatments were allocated to each cow over five 14 d periods according to 5×5 Latin square change over design. Nutrient digestibility and N excretion in urine was determined based on total collection of faeces and urine conducted on d 11-13 of each period. Rumen ammonia concentrations were determined on d 14 based on 8 samples collected at 1.5 h intervals between morning and evening meals. Milk yield was measured daily throughout the experiment and milk composition and urea concentrations were analysed from samples collected from four consecutive milkings at the end of each period.

Results and discussion

The level of urea infusion had no effect on DM, digestible OM or NDF intake or NDF digestibility (Table 1). Nitrogen intake from the basal diet was not affected by the treatment but urea infusions increased the total N intake ($P<0.01$), milk yield ($P=0.08$) and milk protein yield ($P=0.08$) in a linear manner. As a response to urea infusions ruminal ammonia concentrations increased in a curvilinear manner because there were no effects at the lower levels (U1-U4) but at the highest level of urea infusion the ammonia concentration sharply increased. The responses observed in milk urea concentrations were not entirely consistent with those in ruminal ammonia because milk urea concentrations increased in a linear manner ($P<0.01$). In agreement with milk urea concentration

Table 1. Effects of urea infusion on feed intake, milk yield and composition and nutrient degestibility

Item	Urea infusion level					SEM[1]	Orthogonal contrasts	
	U1	U2	U3	U4	U5		Linear	Quadratic
DM intake, kg/d	22.5	22.7	22.5	22.6	23.4	0.54	0.35	0.46
Digestible OM intake, kg/d	14.2	14.5	14.3	14.4	14.9	0.41	0.31	0.62
NDF intake, kg/d	8.20	8.26	8.18	8.24	8.52	0.197	0.35	0.44
NDF digestibility	0.549	0.559	0.562	0.559	0.564	0.0088	0.31	0.61
N intake, g/d								
Basal diet	458	461	457	460	475	11.0	0.35	0.46
Total	458	478	490	508	541	11.1	<0.01	0.46
Milk, kg/d	32.2	32.9	33.0	34.8	33.7	0.82	0.08	0.61
Milk protein, kg/d	1.06	1.10	1.09	1.16	1.15	0.038	0.08	0.98
Milk urea, mg/100 ml	9.5	12.7	12.6	13.7	16.9	0.77	<0.01	0.67
Ruminal ammonia, mmol/l	2.7	2.6	2.2	2.4	3.6	0.29	0.09	0.01
N in urine, g/d	109	119	131	145	151	5.1	<0.01	0.81
N in faeces, g/d	177	173	177	182	184	3.3	0.05	0.36
N efficiency[2]	0.364	0.359	0.346	0.358	0.333	0.0106	0.08	0.69

[1] Standard error of the mean.

[2] Milk protein N/N intake.

N excretion in urine exhibited linear increases, whereas N efficiency (milk protein N/N intake) decreased in a linear manner with increasing levels of urea infusion.

Based on a meta-analysis of 32 studies Broderick *et al.* (2010) concluded that the omasal canal non ammonia-N flows were equal to N intake at ruminal ammonia concentration of 5.1 mmol/l and milk urea concentration of 17.8 mg/100 ml. Under such circumstances there is no net absorption of ammonia from the rumen but the amount of protein degraded in the rumen equals the microbial N synthesis. Based on the above mentioned criteria ruminal ammonia concentrations were critically low on all urea infusion levels in the current study. In spite of low ruminal ammonia-N concentrations the major proportion (0.67) of urea-N infused into the rumen was excreted in urine suggesting that the rumen microbes couldn't completely utilize extracellular ammonia-N under limiting conditions. The current findings indicate that the microbial N synthesis was more sensitive to moderately low ruminal ammonia N concentrations than fibre digestion because intraruminal urea infusions seemed to increase microbial N synthesis but no effect was observed in fibre digestion or feed intake.

References

Broderick, G.A., P. Huhtanen, S. Ahvenjärvi, S.M. Reynal and K.J. Shingfield, 2010. Quantifying ruminal nitrogen metabolism using the omasal sampling technique in cattle – a meta-analysis. J. Dairy Sci. Accepted.

Hoover, W.H., 1986. Chemical factors involved in ruminal fiber digestion. J. Dairy Sci. 69, 2755-2766.

Rinne, M., S. Jaakkola, T. Varvikko and P. Huhtanen, 1999. Effects of type and amount of rapeseed feed on milk production. Acta Agric. Scand. Sect A, Animal Sci. 49, 137-148.

Nitrogen efficiency of early lactating dairy cows as affected by dietary carbohydrates and protein

P. Bani, F. Piccioli Cappelli, L. Calamari and G. Bertoni
Istituto di Zootecnica, Facoltà di Agraria, Via Emilia Parmense, 84, 29122 Piacenza, Italy

Introduction

Dairy farming is considered to be an important contributor to nitrogen pollution of atmosphere and soil and recently there has been a general effort toward an optimization of nitrogen nutrition of dairy cows. The complex effects of diet, besides those on rumen microflora output, include also the modification of absorbed nutrient and of the endocrine system that, in turn, influences the nutrients partitioning (Hart, 1983). The literature is not consistent about the possibility to avoid a high protein concentration in dairy rations without affecting milk yield and its protein content (Broderick, 2003; Olmos Colmenero and Broderick, 2006). Milk N efficiency, expressed as the percentage of dietary N secreted in milk, is generally low but in commercial dairy herds a wide range, from 20 up to 35%, has been observed (Recktenwald and Van Amburgh, 2009). A correct balance and synchronization of carbohydrates and protein in the rumen can maximize protein yield by the rumen microflora and reduce the body utilization of amino acids for energy purpose (Seal and Reynolds, 1993). Nevertheless, excessive feeding of fermentable energy can lower milk fat and induce some metabolic problems. The aim of this study was to verify the effects of two levels of protein combined with two levels of fermentable carbohydrates on lactating cow performance, with particular regard to the efficiency of dietary nitrogen utilization.

Material and methods

The trial was carried out on four Italian Friesian dairy cows in the 2nd month of lactation and fed a diet based on corn silage, alfalfa and ryegrass hay. According to a 4x4 Latin square design with periods of 24 days (17 d of adaptation and 7 days of controls) cows were fed four experimental diets obtained combining two levels of NSC with two levels of protein. The four diets were: high NSC and high protein (HSHP= 30.3% NSC and 16.0% CP in DM), high NCS and low protein (HSLP = 30.6% NSC and 15.1% CP in DM); low NSC and high protein (LSHP = 26.6% NSC and 16.4% CP in DM), low NSC and low protein (LSLP = 26.7% NSC and 15.3% CP on DM). Dietary NDF was kept at 32.8±1.1% of DM and fat content at 4.4±0.2% DM. Feed intake was monitored daily and milk yield at any milking. Twice a week a representative sample from morning milking was analyzed for fat, protein, lactose, casein and urea content. Live weight and BCS were evaluated fortnightly. At the end of each period urine was collected at 07:00 and feces were collected at 07:00 and 14:00. Dry matter and pH were measured in feces; creatinine, total nitrogen, urea nitrogen and pH in urine. Urine volume was calculated assuming a daily creatinine excretion as high as 22.8 mg/ d^{-1} kg LW^{-1}. Data processing was performed using the MIXED and GLM procedures of SAS, assuming as statistical threshold $P<0.05$.

Results and discussion

DMI was slightly higher in the LS diets (25.7 vs. 25.4 kg h^{-1} d^{-1}, $P<0.05$). Fecal traits did not change according to NSC or CP levels and did not provide evidence for any unfavourable fermentation in the large intestine. Milk yield was as high as 45.0 kg h^{-1} d^{-1} on average. Cows on HP diets produced more milk (45.5 vs. 44.5 kg h^{-1} d^{-1}), and a NSCxCP interaction was evident ($P<0.05$) with the highest yield being recorded for the HSHP diet. Milk composition was not affected by NSC or CP levels, and the same was true for total nitrogen secreted in milk (on average 226 g h^{-1} d^{-1} on average).

Levels of NSC did not influence urea level either in blood (4.24 vs. 4.37 mmol/L for HS and LS diets, respectively) or in milk where it was some 15% lower (3.82 vs. 3.73 mmol/L), but CP levels raised this metabolite in blood (4.67 vs. 3.95 mmol/L for HP and LP diets, respectively) as well as in milk (4.03 vs. 3.49 mmol/L). Urea in urine was higher in HP diets (189 vs. 148 mmol/L, $P<0.05$) and it represented also the main portion of total urinary nitrogen (75.2 vs. 70.9%, $P<0.05$). The urinary excretion of nitrogen was on average 31.3% of nitrogen intake and was not significantly influenced by NSC or CP levels, though numerically lower in diets with higher NSC (29.5 vs. 32.7%; $P<0.09$) or lower CP levels (29.6 vs. 32.9%; $P<0.06$). Based on BCS, variations in body N appeared to have been small and therefore were not taken into account for further calculations. The average efficiency of transferring dietary N to milk nitrogen was 34.7%. Dietary NSC levels did not modify this efficiency, but the high CP levels lowered it (33.7 vs. 35.6% for HP and LP diets, respectively; $P<0.01$). For any kg of nitrogen excreted (N intake minus N in milk), cows on LP diets yielded 112 kg d^{-1} of milk but only 103 kg d^{-1} on HP diets. An inverse and linear relationship was obvious between milk yield and nitrogen efficiency: for each increase in daily milk yield level, a reduction of 24 g of nitrogen excreted/kg of milk ws estimated, confirming that well managed high yielding cows are more efficient than cows with lower production. The possibility to optimize dairy cattle protein nutrition without major effects on milk yield and quality was confirmed, by that improving MNE and reducing the environmental impact of dairy farming.

References

Broderick G.A., 2003. Effects of varying dietary protein and energy levels on the production of lactating dairy cows. J. Dairy Sci. 86, 1370-1381.

Hart, I.C., 1983. Endocrine control of nutrient partition in lactating ruminants. Proc. Nutr. Soc. 42:181-212.

Olmos Colmenero, J.J. and G.A. Broderick, 2006. Effect of dietary crude protein concentration on milk production and nitrogen utilization in lactating dairy cows. J. Dairy Sci. 89, 1704-1712.

Recktenwald, E.B. and M.E. Van Amburgh, 2009. Refining nitrogen feeding using current knowledge of recycled urea and microbial nitrogen uptake. Proc. Cornell Nutr. Conf. Feed Manufacturers, 66-76.

Seal, C.J. and C.K. Reynold, 1993. Nutritional implications of gastrointestinal and liver metabolism in ruminants. Nutr. Res. Rev. 6, 185-208.

Blood methionine and lysine concentration in lactating dairy cows supplemented with commercial rumen-proteced methionine and lysine products

A. Gallo[1], G. Fusconi[1], L. Fiorentini[1], E. Grilli[2], P. Fantinati[3] and F. Masoero[1]
[1]Istituto di Scienze degli Alimenti e della Nutrizione, Facoltà di Agraria, Università Cattolica del Sacro Cuore, Piacenza, Italy
[2]DIMORFIPA, Facoltà di Veterinaria, Bologna, Italy
[3]Vetagro S.p.A., Reggio Emilia, Italy

Introduction

It is well known that methionine (Met) and lysine (Lys) are two of the most limiting amino acids (AA) for milk and protein production in lactating dairy cows fed corn-based diets (Schwab *et al.*, 2003; Rulquin, 2004). The NRC (2001) suggested concentrations of Met and Lys of 2.4% and 7.2% on a metabolizable protein (MP) basis, respectively, in order to maximize the use of MP for milk and milk protein yield by lactating dairy cows. However, it should be taken in consideration that these concentrations are hardly achieved. Nutritionists have two methods to fed lactating dairy cows with AA balanced diets: one is to incorporate feeds with a high level of rumen undegradable protein, and the other is to supplement the diet with ruminally protected (RP) synthetic AA (Schwab *et al.*, 2003). The aim of this work was to assess the bioavailability of a RP D,L-Met and a RP L-Lys HCl products by a standardized blood test (Südekum *et al.*, 2004).

Materials and methods

Two commercial products, a RP D,L-Met and a RP L-Lys HCl (Timet® and Relys®, respectively; Vetagro S.p.A., Reggio Emilia, RE, Italy) coated with hydrogenated triglycerides were tested. Four multiparous lactating dairy cows (body weight 605±23 kg and days in milk 125±25) were fed a basal diet typical for many areas of Po valley (Table 1).

The first week animals received the basal diet plus a supplementation of 500 g/d of soybean meal twice a day at the beginning of morning feeding (8:00) and after the evening milking (16:00). During the second week, from day 1 to day 5, each cow received daily the equivalent of 25 g of D,L-Met and 58 g L-Lys HCl, respectively. Based on manufacturer's specification, this meant 48 g/cow of Timet® and 132 g/cow of Relys®. For this purpose, the two products were simultaneously mixed with 500 g of soybean meal and supplemented to the animals twice a day as described previously. Blood samples were obtained by jugular venipunctures the day before (day 0) and 5 days after (day 5) the treatment,

Table 1. Ingredients of the basal diet.

Ingredients (g/kg dry matter)	Basal diet
Corn silage	312
Corn meal	249
Alfalfa hay, dehydrated	208
Protein supplement[1]	103
Cotton seed, whole with lint	85
Soybean meal	34
Calcium soap[2]	9

[1] Contents per kg of premix: Soybean meal 600 g; sunflower meal 300 g; mineral and vitamin supplement 100 g; 120,000 IU of vitamin A; 9,000 IU of vitamin D_3; 90 mg of vitamin E; 3.6 mg of Co; 19.2 mg of I; 1.44 mg of Se; 600 mg of Mn; 62.4 mg of Cu; 2,240 mg of Zn; 1.92 mg of Mo; 360 mg of Fe.
[2] Megalac.

at 8:00 h and 14:00 h, and directly filled into 10 ml tubes containing Li-heparin. Plasma was separated and directly frozen at -20 °C until being analyzed. Met and Lys contents were quantified by HPLC as described by Rossi *et al.* (2003). The plasma urea level was determined according to Sampson *et al.* (1980). Data were subjected to analysis of variance by the GLM procedure of SAS (2003).

Results and discussion

Plasma Met level was increased ($P<0.01$) by Timet® supplementation by 56%. Similarly, plasma Lys was increased ($P<0.05$) by Relys® supplementation by 41%. These results are assumed to indicate an effective rumen protection of the two tested products. Plasma urea content was reduced ($P<0.01$) by 13% after administration of the RP AA, showing a more efficient utilization of dietary nitrogen when these two products were administered to animals. Similar results were reported by Gao *et al.* (2008): the authors showed a reduction of the blood urea nitrogen after administration of RP-AA to lactating dairy cows. This probably because a balanced AA profile cause a better absorption and utilization of dietary nitrogen and a reduction of nitrogen loss in gastro-intestinal tract of lactating dairy cows. Concluding, the supplementation of the two RP AA products reduced the blood urea nitrogen content and improved the protein utilization efficiency by lactating dairy cows.

Table 2. Blood plasma methionine (Met, mg/100 ml), lysine (Lys, mg/100 ml) and urea (nmol/l) the day before (day 0) and 5 days after start of feeding ruminal protected D,L-Met and L-Lys HCl products (Timet® and Relys®, respectively).

	Days of treatment		SEM	*P*-value
	Day 0	Day 5		
Met	0.325±0.050	0.508±0.095	0.025	<0.01
Lys	0.786±0.286	1.107±0.176	0.079	<0.05
Urea	5.88±0.569	5.11±0.945	0.092	<0.01

Acknowledgements

This research was supported by the Research program VETAGRO, PNR 2005-2007 'LATTIERO/CASEARIO' n. 27778.

References

Gao, Y., J. Li, W. Jia, Q. Li, and Y. Cao, 2008. Response of lactating cows to supplemental rumen protected methionine and niacin. Front. Agric. China 2, 121-124.

NRC (National Research Council), 2001. Nutrient Requirements of Dairy Cattle, 7th revised ed. National Academic Science, Washington, DC, USA.

Rossi, F., M. Moschini, F. Masoero, G. Capanna and G. Piva, 2003. Rumen degradation and intestinal digestibility of rumen protected amino acids: comparison between *in situ* and *in vitro* data. Anim. Feed Sci. Technol. 108, 223-229.

Rulquin, H., S. Rigout, S. Lemosquet and A. Bach, 2004. Infusion of glucose directs circulating amino acids to the mammary gland in well-fed dairy cows. J. Dairy Sci. 87, 340-349.

Sampson, E.J., M.A. Baird, C.A. Burtis, E.M. Smith, D.L. Witte and D.D. Bayse, 1980. A couple-enzyme equilibrium method for measuring urea in serum. Clin. Chem. 26, 816-826.

SAS, 2003. Version 9.13 Edition. SAS Institute Inc., North Carolina, USA.

Schwab, C.G., R.S. Ordway and N.L. Whitehouse, 2003. The Latest on Amino Acid Feeding. In: Proc. Southwest Nutrition and Management Conference, Phoenix, AZ: 27-41.

Südekum, K.H., S. Wolffram, P. Ader and J.C. Robert, 2004. Bioavailability of three ruminally protected sources in cattle. Anim. Feed Sci. Technol. 113, 17-25.

Part 6. Energy/protein and their interaction on productive functions: monogastrics

Update on endocrine mechanisms regulating nutrient partitioning in growing animals

Y.R. Boisclair, S.L. Giesy and J.M. Ramos-Nieves
Department of Animal Science, Cornell University, Ithaca, NY 14853, USA

Introduction

In landmark papers in the 1980's, Bauman and colleagues reviewed determinants of productive efficiency in lactating dairy cows (Bauman and Currie, 1980; Bauman *et al.*, 1984). They concluded that nutrient partitioning accounted for much of the variation between animals and proposed that endocrine systems played a major role in regulating this process. Since, a wealth of evidence has been collected supporting a role for endocrine systems in prioritizing nutrient use among tissues, not only during lactation but also during growth (Dunshea *et al.*, 2005; Etherton and Bauman, 1998).

Most of the research in the area of growth focused on three broad systems acting predominantly on peripheral tissues, namely estrogenic and androgenic steroids, β-agonists and lastly growth hormone (GH). The bulk of this work was performed in cattle and swine and has been reviewed in detail elsewhere (Dunshea *et al.*, 2005; Sillence, 2004). Over the last decade, however, there has been a growing recognition that centrally-regulated mechanisms play a crucial role in regulating appetite and more importantly the disposition of nutrients between tissues (Nogueiras *et al.*, 2009). Our goal is to illustrate actions by both levels of regulation. We will do so first by reviewing recent work answering some long-standing questions on the peripheral actions of GH and its essential mediator insulin-like growth factor-I (IGF-I), with both considered as members of a single system (GH-IGF-I system). We will then discuss the leptin signalling system to illustrate the role of the central nervous system in regulating nutrient allocation.

The GH-IGF-I system

Core elements of the GH-IGF-I system

Signalling by the GH-IGF-I system is initiated when pituitary-derived GH binds to its cognate receptor. The GH receptor (GHR) belongs to the cytokine receptor superfamily (Vijayakumar *et al.*, 2010). The GHR is abundant in liver but can be found at significant level in tissues accounting for a major portion of growth such as skeletal muscle and adipose tissue. Upon binding, GH induces a series of biochemical events leading to the formation of a fully active GHR signalling complex, i.e. homodimerization of the GHR, recruitment of the cytoplasmic tyrosine kinase JAK2 and phosphorylation of key tyrosine residues (Lanning and Carter-Su, 2006; Vijayakumar *et al.*, 2010). This is followed by the activation of intracellular signalling cascades such as the signal transducers and activators of transcription (STAT)-5, the phosphatidylinositol 3'-kinase (PI3K) pathway and the Ras-mitogen-activated protein kinase (MAPK) pathway (Lanning and Carter-Su, 2006; Vijayakumar *et al.*, 2010).

One of the signature events of GH action is induction of IGF-I gene transcription (Lanning and Carter-Su, 2006). The mechanism underlying this transcriptional response is now understood to involve STAT5 (Vijayakumar *et al.*, 2010). Upon phosphorylation by the activated GHR complex, STAT5 dimerizes and translocates to the nucleus where it binds to its DNA binding sites. STAT5 binding sites have been found in the IGF-I gene of many species (Lanning and Carter-Su, 2006; Vijayakumar *et al.*, 2010).

IGF-I is an extracellular signal with significant homologies to insulin (Vijayakumar *et al.*, 2010). Not surprisingly, IGF-I acts via a receptor (IGFIR) that has also significant homology with the insulin receptor and its signalling (Vijayakumar *et al.*, 2010). The IGF-I produced by most tissues behaves as a growth factor, which by definition acts either on the cell of origin (autocrine action) or on a neighboring cell (paracrine action). Locally produced IGF-I is usually bound to one member of the IGFBP family to yield binary complexes of 40-50 kDa (Boisclair *et al.*, 2001; Le Roith *et al.*, 2001).

The liver is a major site of IGF-I synthesis in postnatal life. This IGF-I is unable to act on liver because hepatocytes do not express the IGFIR at meaningful level. Instead, liver-derived IGF-I is secreted in plasma where it acts in an endocrine manner. During postnatal life, this IGF-I circulates in ternary complexes of 150 kDa composed of one molecule each of IGF-I, IGF binding protein (IGFBP)-3 or IGFBP-5 and the acid labile subunit (ALS) (Boisclair *et al.*, 2001).

Role of the GH-IGF-I system in growth

Normal postnatal growth requires the combined actions of GH and IGF-I. This has been unambiguously demonstrated by the growth deficit seen in mice lacking both GH and IGF-I actions (Lupu *et al.*, 2001). The contribution of each signal was derived by comparing the postnatal growth of mice harboring null GH receptor or IGF-I genes (Lupu *et al.*, 2001). They found that the GH-IGF-I system accounted for 83% of postnatal growth, leaving a mere 17% for any other system. Specifically, GH and IGF-I account respectively for 14% and 35% of postnatal growth whereas the contribution of GH mediated by IGF-I is 34%. At maturity, mice lacking GH-IGF-I signalling weighed a mere 5 grams, which is only twice the weight of the second smallest known mammal. Recent data in cattle confirm the importance of the GH-IGF-I system for postnatal growth. For example, a line of miniature Bos indicus cattle was identified suffering from an ~30% deficit in growth rate, height and weight (Liu *et al.*, 1999). This impairment is associated with a single mutation in the GH gene, mutating an amino acid residue that is essential for high affinity binding of GH to the GHR (McCormack *et al.*, 2009). A line of Holstein cattle was recently reported with normal levels of plasma GH, IGF-I and other metabolic hormones but with a 75% reduction in IGFIR expression in skeletal muscle (Blum *et al.*, 2007). At 265 days of age, these animals were 59% lighter that their normal half-siblings (101 vs. 245 kg).

Chronic GH treatment of growing animals including pigs and cattle increases plasma IGF-I and liver IGF-I expression (Etherton and Bauman, 1998). GH treatment also increases IGF-I expression in extra-hepatic tissues such as skeletal muscle and adipose tissue (Ueki *et al.*, 2009). This leads to the following question: What is the relative contribution of plasma vs. locally derived IGF-I for postnatal growth? The first attempt to answer this question involved mouse models devoid of either source of IGF-I. On one hand, liver IGF-I deficient mice grow normally after birth suggesting little functional significance for plasma IGF-I (Yakar *et al.*, 1999). On the other hand, IGF-I knockout mice suffer from a 70% growth deficit that can be completely corrected by exclusive re-expression of IGF-I in the liver (Wu *et al.*, 2009). These experiments show that each source can compensate for absence of the other but do not resolve their relative role in postnatal growth.

To address the physiological significance of plasma IGF-I for GH-dependent growth, we developed a mouse model lacking ALS. In these mice, liver-derived IGF-I is no longer recruited in long-lived ternary complexes of 150 kDa and as a consequence, plasma IGF-I is reduced by ~70% throughout postnatal life (Ueki *et al.*, 2000). Null ALS mice and their wild type counterparts were treated with exogenous GH between day 35 and 63 of age (Ueki *et al.*, 2009). GH treatment doubled plasma IGF-I concentration in wild type mice but had no effect whatsoever in null ALS mice. This difference in the effect of GH on plasma IGF-I occurred despite identical stimulation of hepatic IGF-I mRNA in both wild type and null ALS mice. During the treatment period, GH stimulated body weight gain in both genotypes, but less so in null than wild type mice (Figure 1A). The gain dependent exclusively on exogenous GH was reduced by 50% in null ALS mice relative to their wild type counterparts (Figure 1B).

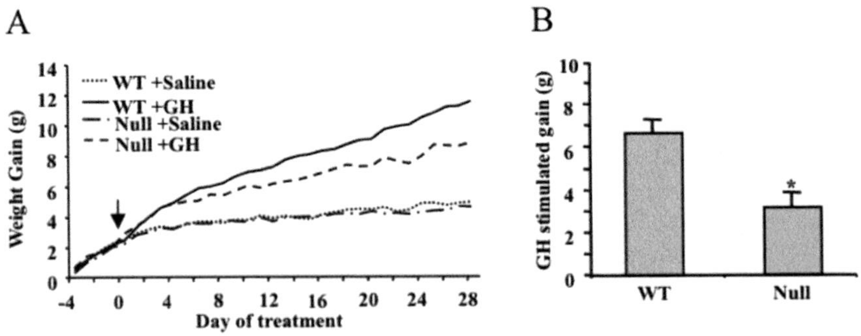

Figure 1. Effect of chronic GH treatment in wild type (WT) and null ALS (null) mice. (A) Cumulative gain was calculated by difference from body weight start of saline or GH injections on day 35 of age (shown with vertical arrow). (B) The growth depending uniquely on GH was calculated for both WT and null mice. Adapted from Ueki et al.*, 2009.*

The remaining growth effects of exogenous GH must be accounted for by extra-hepatic actions, which can be independent or dependent on local IGF-I production. The balance between these 2 classes of GH actions varies across tissues. So the first category is likely dominant in liver because it lacks IGFIR. Accordingly, the ability of GH to stimulate liver growth was undiminished in null ALS mice even though plasma IGF-I failed to rise (Ueki *et al.*, 2009). Finally, GH actions that are dependent on local IGF-I synthesis also appear intact in null ALS mice because we found equally effective GH stimulation of IGF-I expression in muscle and adipose tissue.

Overall, these data are consistent with a physiological significant role for liver-derived IGF-I. The liver is a major peripheral sensor of anabolic inputs represented by the prevailing combination of nutrients and hormones. Increased hepatic IGF-I synthesis that occurs with high anabolic inputs supplements locally produced IGF-I to bring about the full growth response.

Role of the GH-IGF-I system in metabolism

GH is known to have major effects on glucose metabolism (Etherton and Bauman, 1998, Moller and Jorgensen, 2009). In brief, GH attenuates the ability of insulin to reduce hepatic glucose production in liver and to stimulate glucose uptake in muscle and adipose tissue. This GH action accounts at least in part for the reduction in the expression of the lipogenic program in growing pigs treated chronically with GH and their lower rates of lipid accretion (Etherton and Bauman, 1998). A recently identified mechanism involving the PI3K pathway could explain a portion of the GH-dependent attenuation of insulin action (Barbour *et al.*, 2005). PI3K is a heterodimer of a p110 catalytic subunit and a regulatory p85 subunit providing the binding site for recruitment to the activated insulin receptor. GH specifically increases the abundance of p85 relative to p110, such that it competes with the heterodimeric PI3K for binding to the insulin receptor complex. This mechanism has been described in both muscle and adipose tissue in the mouse (Barbour *et al.*, 2005; Del Rincon *et al.*, 2007), and we recently showed that it also occurs in adipose tissue of GH-treated cattle (Rhoads *et al.*, 2007). This raises the possibility that it accounts for the antagonistic effects of GH on insulin action in domestic animals as well.

GH also stimulates protein accretion, and this action in skeletal muscle appears to be mediated predominantly by IGF-I (Clemmons, 2009; Etherton and Bauman, 1998). Three specific IGF-I effects culminate to increase protein accretion in skeletal muscle, i.e. stimulation of proliferation of satellite cells, promotion of their assembly into myotubes and finally stimulation of hypertrophy of the myofiber by increasing protein synthesis and reducing protein degradation (Clemmons, 2009). It is important to mention that the hyperinsulinemia that is seen in GH-treated animals does not play a role in the positive effects of GH on protein synthesis (Wilson *et al.,* 2008). Finally, a recent report also suggested that GH stimulates fiber formation by itself, independently of IGF-I (Sotiropoulos *et al.,* 2006).

Leptin as an example of centrally mediated mechanisms

The importance of the central nervous system (CNS) for the regulation of energy intake was established long ago (Gao and Horvath, 2007). This regulation involves the hypothalamus, a region located at the basis of the brain. The hypothalamus consists of neurons organized in anatomically defined clusters such as the arcuate nucleus (ARC), the paraventricular nucleus (PVN), the dorsomedial area (DMA), the lateral hypothalamic area (LHA) and the ventromedial area (VMA) (Gao and Horvath, 2007; Schwartz *et al.,* 2000). Targeted destruction of the VMA in rodents resulted in unchecked appetite and obesity whereas destruction of the LHA caused the opposite (Gao and Horvath, 2007).

Despite such data, understanding the mechanistic basis for these phenotypes progressed slowly until the discovery of the hormone leptin in 1994. Humans and mice devoid of leptin are morbidly obese due to their inability to regulate appetite (Farooqi and O'Rahilly, 2007; Schwartz *et al.,* 2000). The discovery of leptin provided the entry point into the central mechanisms used by peripheral hormones to regulate energy intake. Recent work shows that central leptin signalling also alters the pattern of nutrients used by innervated tissues (Nogueiras *et al.,* 2009).

Core elements of the leptin system

Leptin is synthesized almost exclusively by adipose tissue in most species, including cattle, sheep and pig (Boisclair *et al.,* 2006). It is also expressed at very low level in other tissues at specific time of development, such as the brain and liver in the foetal sheep and mammary epithelial cells in late pregnant and lactating ruminants. The two most important factors regulating plasma leptin level in postnatal animals are adiposity and nutrition. In mature sheep and cattle, adiposity and the plane of nutrition have been estimated to account for ~35 and 17% of the variation in plasma leptin concentration (Boisclair *et al.,* 2006). Finally, a number of hormones have been shown to regulate the synthesis of leptin, with insulin being a positive regulator and β-adrenergic signalling having the opposite effect. Results on the effect of GH are equivocal, with different studies reporting no effects as well as positive or negative effects (Boisclair *et al.,* 2006).

Leptin binds to a family of transmembrane receptors produced by transcription of a single leptin receptor gene (Ob-R) (Villanueva *et al.,* 2009). The multiple isoforms are the result of alternate splicing of the primary Ob-R transcript. One of the isoforms, known as the long receptor or Ob-Rb, contains an extracellular binding domain, a transmembrane domain and a complete cytoplasmic signalling domain. As the GHR, Ob-Rb belongs to the cytokine receptor superfamily, and not surprisingly, activates series of signalling events reminiscent of GHR signalling (i.e. JAK2 activation followed by signalling via STAT3, STAT5, MAPK and PI3K) (Villanueva *et al.,* 2009). Ob-Rb gene transcripts have been identified in most domestic species, including chicken despite a continued failure to identify the leptin gene in that species (Sharp *et al.,* 2008).

Five truncated receptors have been identified in rodents (Ahima and Flier, 2000). These isoforms have the same exact extracellular binding domain as Ob-Rb, but differ in lacking all of the other 2 domains (Ob-Re) or most of the cytoplasmic signalling domain (Ob-Ra, Ob-Rc, Ob-Rd and Ob-Rf). Accordingly, they all bind leptin perfectly well but fail to initiate significant signalling. Information on these truncated isoforms is limited in domestic species, except for the Ob-Ra isoform which has been found in most of these animals.

In rodents, Ob-Rb is expressed almost exclusively in the brain with particularly strong expression in the hypothalamus (Ahima and Flier, 2000; Villanueva *et al.,* 2009). The implication of this observation is that the brain is the main leptin target tissue. A wealth of evidence supports this model of operation in rodents. For example, the obese and diabetic phenotypes of mice suffering from leptin signalling deficiencies are completely abolished by correcting the genetic defect only in the hypothalamus (De Luca *et al.,* 2005). Moreover, knockout of the leptin receptor in neural tissues phenocopies the *db/db* mouse which lacks Ob-Rb (Cohen *et al.,* 2001). Finally, *db/db* mice retain normal expression of Ob-Ra, but are nevertheless as obese as leptin-deficient *ob/ob* mice (Ahima and Flier, 2000).

Compelling evidence also exists for central action in ruminants and other domestic animals (Boisclair *et al.,* 2006), but whether leptin also has broad actions in the periphery has been the subject of controversy. We have addressed this question indirectly by surveying Ob-R expression in growing dairy heifers with independent real time PCR assays measuring all leptin receptor isoforms (Ob-R_{TOTAL}), the long isoform (Ob-Rb) and the most abundant short isoform (Ob-Ra) (Thorn *et al.,* 2007). Ob-R_{TOTAL} is detected in all peripheral tissues, with most having more abundant expression than the hypothalamus (Figure 2A). Liver, which has the highest abundance, exceeds hypothalamic expression by over a 1000 fold. However, the signalling competent isoform Ob-Rb accounts for the majority of Ob-R_{TOTAL} transcripts only in the hypothalamus and is nearly absent in peripheral tissues, including liver (Figure 2B). Ob-Ra, on the other hand, accounts for nearly all of Ob-R_{TOTAL} in peripheral tissues. Overall, these findings are in full agreement with the rodent data and suggest that the hypothalamus is the major site of leptin action in cattle. We are not aware of similar surveys of Ob-R expression in other agricultural animals but a different scenario appears very unlikely.

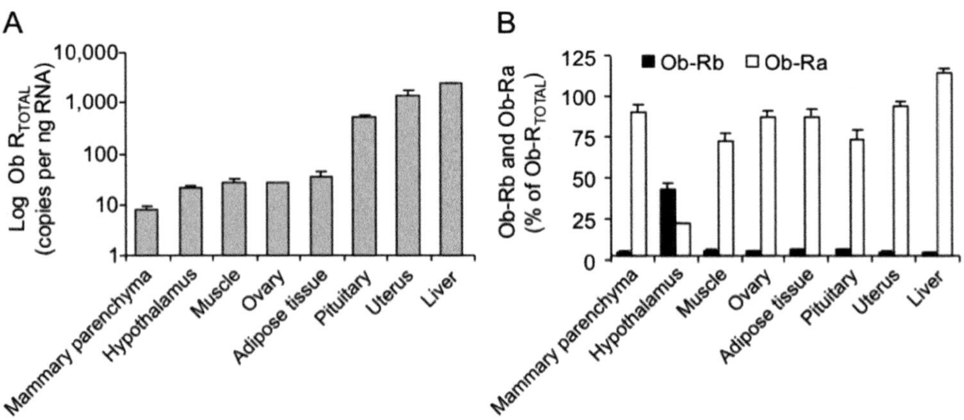

Figure 2. Expression of leptin receptor isoforms in dairy heifers. (A) Transcript copy number for Ob-R_{total} were measured. (B) Transcript copy number were measured for OB-Ra and OB-Rb and expressed as a percentage of Ob-R_{total}. Adapted from Thorn et al., *2007.*

Leptin actions in the hypothalamus

Neurons located in the ARC have strong Ob-Rb expression and connect to neurons located in other clusters via an extensive network of axons (Figure 3). This feature gave rise to a model where ARC neurons sense variation in plasma leptin and then communicate this information to neurons located in other hypothalamic centers (Gao and Horvath, 2007; Schwartz *et al.,* 2000). According to this model, ARC neurons are first order (or sensory) neurons whereas the other hypothalamic centers contain second order (or effector) neurons. It is now obvious that monitoring of plasma leptin is not restricted to the ARC. Indeed, the majority of Ob-Rb containing neurons are distributed at lower abundance across other hypothalamic centers and other central nervous tissue (e.g. brain stem) and their activation is needed for the full complement of leptin actions (Myers *et al.,* 2009). The broad

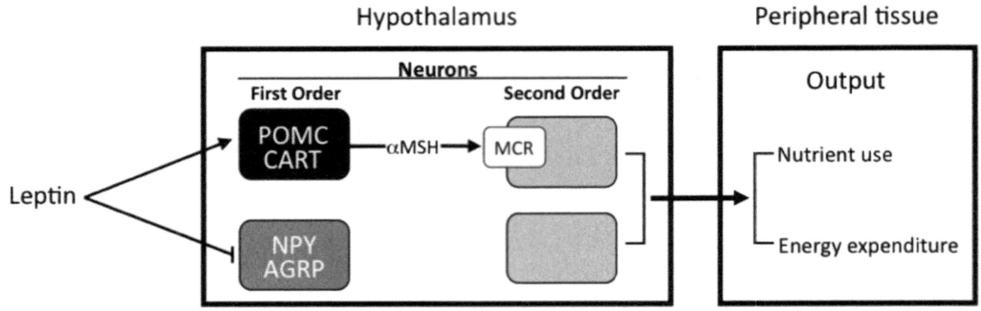

Figure 3. Model of leptin action. Leptin stimulates (line with arrow) or inhibit (blunt line) first order of neurons. This results in the release of αMSH near second order neurons, activation of MCR and finally peripheral actions via the sympathic nervous system.

significance of this observation was confirmed by retention of leptin-mediated activation of PVN, LHA and DMA neurons after surgical destruction of the ARC in sheep (Qi *et al.,* 2010).

Even if an over-simplification, this 'first order-second order' model is useful in understanding how leptin signalling in the brain is translated into whole animal responses. The ARC contains two major neuronal populations (Gao and Horvath, 2007, Schwartz *et al.,* 2000). The first population produces the peptides pro-opiomelanocortin (POMC) and cocaine and amphetamine related transcript (CART). Leptin increases POMC synthesis and processing into anorexic neuropeptides such as the melanocortin α-melanocyte stimulating hormone (αMSH). The second population of neurons is the diametrical opposite to POMC neurons: They produce the orexigenic agouti related protein (AGRP) and neuropeptide Y (NPY) and leptin inhibits their production (Schwartz *et al.,* 2000).

Upon leptin activation, POMC neurons release αMSH at their axonal terminals, in close proximity to the second order neurons located in the other hypothalamic clusters (Figure 3). There, αMSH activates its own receptors, which consist of melanocortin-3 and -4 receptors (MCR) (Gao and Horvath, 2007). Interestingly, AGRP also binds MCR, but unlike αMSH, blocks rather than activates signalling. As discussed in the next section, activation (or inhibition) of MCR located on second order neurons initiates a number of responses impacting nutrient utilization by peripheral tissues.

Alteration of peripheral nutrient utilization by central leptin signalling

Positive associations between plasma leptin concentration, obesity and Type II diabetes initially suggested that leptin impacted negatively glucose utilization. However, experimental evidence indicates the exact opposite. Leptin therapy stimulates whole body glucose turnover and glucose uptake by skeletal muscles in the absence of any change in insulin levels (Kamohara *et al.,* 1997). Leptin injection into the VMA recapitulates these positive effects in the heart, skeletal muscles and brown adipose tissues, but not in white adipose tissue (Minokoshi *et al.,* 1999). Streptozotocin diabetic rats provide additional evidence implicating leptin in peripheral glucose utilization. Chronic leptin therapy normalized plasma glucose and peripheral insulin responsiveness in the face of severe hypoinsulinemia (Chinookoswong *et al.,* 1999), and again, these effects were recapitulated by central leptin administration (Hidaka *et al.,* 2002).

Leptin also exerts major effects on the disposal and storage of fat independently of its effects on intake, again by acting predominantly in the brain. Elevating plasma leptin in normal rodents reduces fat depots more rapidly and extensively than seen in pair-fed controls (Sarmiento *et al.,* 1997). Remarkablely, leptin-treated rodents lose only fat while pair-fed controls lose both fat and protein (Levin *et al.,* 1996).

How can leptin acting in the brain have such profound effects on the fate of glucose and lipids in the periphery? A first clue is that a majority of these effects are dependent on innervation and more specifically on the sympathetic autonomic outflow to liver, skeletal muscle and adipose tissue (Buettner *et al.,* 2008; Nogueiras *et al.,* 2007). Second, a subset of PVN, DMA and LHA neurons expresses MCR and control sympathetic tone to peripheral tissues (Bartness and Song, 2007). This suggests a model whereby leptin induces first order ARC neurons, which then release αMSH to activate second order neurons controlling sympathetic outflow to peripheral tissues (Figure 3).

In support of this model, central infusion of αMSH amplified insulin responses (i.e. greater inhibition of hepatic glucose production in liver and stimulation of glucose disposal in peripheral tissues) during euglycemic hyperinsulinemic clamps (Obici *et al.,* 2001). Central MCR signalling also limits lipid deposition by promoting lipolysis in adipose tissue and simultaneously inducing the β-oxidative program in skeletal muscle (Nogueiras *et al.,* 2007). Importantly, these effects are independent of variation in feed intake, thus representing true partitioning effects between muscle and adipose tissue (Nogueiras *et al.,* 2007, 2009). The physiological relevance of this system is supported by reciprocal effects when MCR receptors are blocked (i.e. reduced leptin-induced glucose uptake in muscle, reduced β-oxidation in skeletal muscle, increased lipogenesis in adipose tissue) (Nogueiras *et al.,* 2007; Toda *et al.,* 2009).

Conclusion

The conceptual framework proposed by Bauman and colleagues whereby hormonal systems play a pre-eminent role in regulating nutrient partitioning remains valid in year 2010. This model can be updated by adding leptin and perhaps other hormones working predominantly in the brain (e.g. ghrelin, glucagon-like peptide-1, etc.) to the list of regulatory players (Nogueiras *et al.,* 2009). As obvious from this review, all the evidence involving leptin and the central system in peripheral nutrient allocation has been obtained in rodents. Accordingly, the challenge over the next few years will be to demonstrate that the central nervous system plays a role in nutrient allocation of domestic animals as it does in rodents. Given the conservation of mechanisms regulating growth and nutrient partitioning, as illustrated by the GH-IGF-I system, we have little doubt that this will prove true. This will give animal scientists a complete new set of avenues to optimize nutrient utilization during growth and improve sustainability of animal agriculture.

References

Ahima, R.S. and J.S. Flier, 2000. Leptin. Annu Rev Physiol. 62, 413-437.

Barbour, L.A., S. Mizanoor Rahman, I. Gurevich, J.W. Leitner, S.J. Fischer, M.D. Roper, T.A. Knotts, Y. Vo, C.E. McCurdy, S. Yakar, D. Leroith, C.R. Kahn, L.C. Cantley, J.E. Friedman, and B. Draznin, 2005. Increased P85alpha is a potent negative regulator of skeletal muscle insulin signalling and induces *in vivo* insulin resistance associated with growth hormone excess. J Biol Chem. 280, 37489-37494.

Bartness, T.J. and C.K. Song, 2007. Brain-adipose tissue neural crosstalk. Physiol Behav. 91, 343-351.

Bauman, D.E. and W.B. Currie, 1980. Partitioning of nutrients during pregnancy and lactation: A review of mechanisms involving homeostasis and homeorhesis. J Dairy Sci. 63, 1514-1529.

Bauman, D.E., S.N. McCutcheon, W.D. Steinhour, P.J. Eppard, and S.J. Sechen, 1984. Sources of variation and prospects for improvements of productive efficiency in the dairy cow: a review. J Anim Sci. 60, 583-592.

Blum, J.W., T.H. Elsasser, D.L. Greger, S. Wittenberg, F. de Vries, and O. Distl, 2007. Insulin-like growth factor type-1 receptor down-regulation associated with dwarfism in Holstein calves. Domest Anim Endocrinol. 33, 245-268.

Boisclair, Y.R., R.P. Rhoads, I. Ueki, J. Wang, and G.T. Ooi, 2001. The acid-labile subunit (ALS) of the 150 kDa IGF-binding protein complex: an important but forgotten component of the circulating IGF system. J Endocrinol. 170, 63-70.

Boisclair, Y.R., S.R. Wesolowski, J.W. Kim, and R.A. Ehrhardt, 2006. Roles of growth hormone and leptin in the periparturient dairy cow. Pages 327-346 *in* Proceedings of the Xth International Symposium on Ruminant Physiology. K. Sejrsen, T. Hvelplund, and M. O. Nielsen, eds. Wageningen Academic Publishers, Wageningen.

Buettner, C., E.D. Muse, A. Cheng, L. Chen, T. Scherer, A. Pocai, K. Su, B. Cheng, X. Li, J. Harvey-White, G.J. Schwartz, G. Kunos, L. Rossetti, and C. Buettner, 2008. Leptin controls adipose tissue lipogenesis via central, STAT3-independent mechanisms. Nat Med. 14, 667-675.

Chinookoswong, N., J.-L. Wang, and Z.-Q. Shi, 1999. Leptin restores euglycemia and normalizes glucose turnover in insulin-deficient diabetes in the rat. Diabetes. 48, 1487-1492.

Clemmons, D.R., 2009. Role of IGF-I in skeletal muscle mass maintenance. Trends Endocrinol Metab. 20, 349-356.

Cohen, P., C. Zhao, X. Cai, J.M. Montez, S.C. Rohani, P. Feinstein, P. Mombaerts, and J.M. Friedman, 2001. Selective deletion of leptin receptor in neurons leads to obesity. J Clin Invest. 108, 1113-1121.

De Luca, C., T.J. Kowalski, Y. Zhang, J.K. Elmquist, C. Lee, M.W. Kilimann, T. Ludwig, S.M. Liu, and S.C. Chua, Jr, 2005. Complete rescue of obesity, diabetes, and infertility in db/db mice by neuron-specific LEPR-B transgenes. J Clin Invest. 115, 3484-3493.

Del Rincon, J.P., K. Iida, B.D. Gaylinn, C.E. McCurdy, J.W. Leitner, L.A. Barbour, J.J. Kopchick, J.E. Friedman, B. Draznin, and M.O. Thorner, 2007. Growth hormone regulation of p85alpha expression and phosphoinositide 3-kinase activity in adipose tissue: mechanism for growth hormone-mediated insulin resistance. Diabetes. 56, 1638-1646.

Dunshea, F.R., D.N. D'Souza, D.W. Pethick, G.S. Harper, and R.D. Warner, 2005. Effects of dietary factors and other metabolic modifiers on quality and nutritional value of meat. Meat Sci. 71, 8-38.

Etherton, T.D. and D.E. Bauman, 1998. Biology of somatotropin in growth and lactation of domestic animals. Physiol Rev. 78, 745-761.

Farooqi, I.S. and S. O'Rahilly, 2007. Genetic factors in human obesity. Obes Rev. 8 Suppl 1, 37-40.

Gao, Q. and T.L. Horvath, 2007. Neurobiology of feeding and energy expenditure. Annu Rev Neurosci. 30, 367-398.

Hidaka, S., H. Yoshimatsu, S. Kondou, Y. Tsuruta, K. Oka, H. Noguchi, K. Okamoto, H. Sakino, Y. Teshima, T. Okeda, and T. Sakata, 2002. Chronic central leptin infusion restores hyperglycemia independent of food intake and insulin level in streptozotocin-induced diabetic rats. Faseb J. 16, 509-518.

Kamohara, S., R. Burcelin, J.L. Halaas, J.M. Friedman, and M.J. Charron, 1997. Acute stimulation of glucose metabolism in mice by leptin treatment. Nature. 389, 374-377.

Lanning, N. J. and C. Carter-Su, 2006. Recent advances in growth hormone signalling. Rev Endocr Metab Disord. 7, 225-235.

Le Roith, D., C. Bondy, S. Yakar, J.L. Liu, and A. Butler, 2001. The somatomedin hypothesis, 2001. Endocr Rev. 22, 53-74.

Levin, N., C. Nelson, A. Gurney, R. Vandlen, and F. de Sauvage, 1996. Decreased food intake does not completely account for adiposity reduction after ob protein infusion. Proceedings of the National Academy of Sciences USA 93, 1726-1730.

Liu, J., C.K. Boyd, Y. Kobayashi, C.C. Chase, Jr., A.C. Hammond, T.A. Olson, T.H. Elsasser, and M.C. Lucy, 1999. A novel phenotype for Lardon dwarfism in miniature Bos indicus cattle suggests that the expression of growth hormone receptor 1A in liver is required for normal growth. Domest Anim Endocrinol. 17, 421-437.

Lupu, F., J.D. Terwilliger, K. Lee, G.V. Segre, and A. Efstratiadis, 2001. Roles of growth hormone and insulin-like growth factor 1 in mouse postnatal growth. Dev Biol. 229, 141-162.

McCormack, B.L., C.C. Chase, Jr., T.A. Olson, T.H. Elsasser, A.C. Hammond, T.H. Welsh, Jr., H. Jiang, R.D. Randel, C.A. Okamura, and M.C. Lucy, 2009. A miniature condition in Brahman cattle is associated with a single nucleotide mutation within the growth hormone gene. Domest Anim Endocrinol. 37, 104-111.

Minokoshi, Y., M.S. Haque, and T. Shimazu, 1999. Microinjection of leptin into the ventromedial hypothalamus increases glucose uptake in peripheral tissues in rats. Diabetes. 48, 287-291.

Moller, N. and J.O. Jorgensen, 2009. Effects of growth hormone on glucose, lipid, and protein metabolism in human subjects. Endocr Rev. 30, 152-177.

Myers, M.G., Jr., H. Munzberg, G.M. Leinninger, and R.L. Leshan, 2009. The geometry of leptin action in the brain: more complicated than a simple ARC. Cell Metab. 9, 117-123.

Nogueiras, R., M. Lopez, and C. Dieguez, 2009. Regulation of lipid metabolism by energy availability: a role for the central nervous system. Obes Rev. 11(3), 185-201.

Nogueiras, R., P. Wiedmer, D. Perez-Tilve, C. Veyrat-Durebex, J.M. Keogh, G.M. Sutton, P. T. Pfluger, T.R. Castaneda, S. Neschen, S.M. Hofmann, P.N. Howles, D.A. Morgan, S.C. Benoit, I. Szanto, B. Schrott, A. Schurmann, H.G. Joost, C. Hammond, D.Y. Hui, S.C. Woods, K. Rahmouni, A.A. Butler, I.S. Farooqi, S. O'Rahilly, F. Rohner-Jeanrenaud, and M.H. Tschop, 2007. The central melanocortin system directly controls peripheral lipid metabolism. J Clin Invest. 117, 3475-3488.

Obici, S., Z. Feng, J. Tan, L. Liu, G. Karkanias, and L. Rossetti, 2001. Central melanocortin receptors regulate insulin action. J Clin Invest. 108, 1079-1085.

Qi, Y., B.A. Henry, B.J. Oldfield, and I.J. Clarke, 2010. The action of leptin on appetite-regulating cells in the ovine hypothalamus: demonstration of direct action in the absence of the arcuate nucleus. Endocrinology. 151, 2106-2116.

Rhoads, R.P., J.W. Kim, M.E. Van Amburgh, R.A. Ehrhardt, S.J. Frank, and Y.R. Boisclair, 2007. Effect of nutrition on the GH responsiveness of liver and adipose tissue in dairy cows. J Endocrinol, 195, 49-58.

Sarmiento, U., B. Benson, S. Kaufman, L. Ross, M. Qi, S. Scully, and C. DiPalma, 1997. Morphologic and molecular changes induced by recombinant human leptin in the white and brown adipose tissues of C57BL/6 mice. Lab. Invest. 77, 243-256.

Schwartz, M.W., S.C. Woods, D. Porte, Jr., R.J. Seeley, and D.G. Baskin, 2000. Central nervous system control of food intake. Nature. 404, 661-671.

Sharp, P.J., I.C. Dunn, D. Waddington, and T. Boswell, 2008. Chicken leptin. Gen Comp Endocrinol. 158, 2-4.

Sillence, M.N, 2004. Technologies for the control of fat and lean deposition in livestock. Vet J. 167, 242-257.

Sotiropoulos, A., M. Ohanna, C. Kedzia, R.K. Menon, J.J. Kopchick, P.A. Kelly, and M. Pende, 2006. Growth hormone promotes skeletal muscle cell fusion independent of insulin-like growth factor 1 up-regulation. Proc Natl Acad Sci USA. 103, 7315-7320.

Thorn, S.R., M.J. Meyer, M.E. Van Amburgh, and Y.R. Boisclair, 2007. Effect of estrogen on leptin and expression of leptin receptor transcripts in prepubertal dairy heifers. J Dairy Sci. 90, 3742-3750.

Toda, C., T. Shiuchi, S. Lee, M. Yamato-Esaki, Y. Fujino, A. Suzuki, S. Okamoto, and Y. Minokoshi, 2009. Distinct effects of leptin and a melanocortin receptor agonist injected into medial hypothalamic nuclei on glucose uptake in peripheral tissues. Diabetes. 58(12), 2757-2765.

Ueki, I., S.L. Giesy, K.J. Harvatine, J.W. Kim, and Y.R. Boisclair, 2009. The acid-labile subunit is required for full effects of exogenous growth hormone on growth and carbohydrate metabolism. Endocrinology. 150, 3145-3152.

Ueki, I., G.T. Ooi, M.L. Tremblay, K.R. Hurst, L.A. Bach, and Y.R. Boisclair, 2000. Inactivation of the acid labile subunit gene in mice results in mild retardation of postnatal growth despite profound disruptions in the circulating insulin-like growth factor system. Proc Natl Acad Sci USA. 97, 6868-6873.

Vijayakumar, A., R. Novosyadlyy, Y. Wu, S. Yakar, and D. LeRoith, 2010. Biological effects of growth hormone on carbohydrate and lipid metabolism. Growth Horm IGF Res. 20, 1-7.

Villanueva, E.C., H. Munzberg, D. Cota, R.L. Leshan, K. Kopp, R. Ishida-Takahashi, J.C. Jones, D.C. Fingar, R.J. Seeley, and M.G. Myers, Jr, 2009. Complex regulation of mammalian target of rapamycin complex 1 in the basomedial hypothalamus by leptin and nutritional status. Endocrinology. 150, 4541-4551.

Wilson, F.A., R.A. Orellana, A. Suryawan, H.V. Nguyen, A.S. Jeyapalan, J. Frank, and T.A. Davis, 2008. Stimulation of muscle protein synthesis by somatotropin in pigs is independent of the somatotropin-induced increase in circulating insulin. Am J Physiol Endocrinol Metab. 295, E187-194.

Wu, Y., H. Sun, S. Yakar, and D. LeRoith, 2009. Elevated levels of insulin-like growth factor (IGF)-I in serum rescue the severe growth retardation of IGF-I null mice. Endocrinology. 150(9), 4395-4403.

Yakar, S., J.L. Liu, B. Stannard, A. Butler, D. Accili, B. Sauer, and D. LeRoith, 1999. Normal growth and development in the absence of hepatic insulin-like growth factor I. Proc Natl Acad Sci USA. 96, 7324-7329.

Biochemical and physiological limitations to efficiency of amino acid utilization for animal growth

Guoyao Wu[1,2]
[1]*Department of Animal Science, Texas A&M University, College Station, TX 77843, USA; g-wu@tamu.edu*
[2]*State Key Laboratory of Animal Nutrition, China Agricultural University, 100191 Beijing, China*

Abstract

Multiple pathways exist for the utilization of amino acids by the animal. They are building blocks of proteins and essential precursors for the production of biologically important nitrogenous substances (including creatine, dopamine, glutathione, heme, nitric oxide, polyamines, serotonin, and taurine). These physiological processes are obligatory for the reproduction, health and survival of organisms. Additionally, syntheses of large amounts of arginine, proline, aspartate, glutamine, and glycine (whose provision from conventional diets is inadequate for tissue protein synthesis) via inter-organ cooperation necessitate the degradation of nutritionally essential amino acids (EAA). Thus, when diets do not contain adequate NEAA, EAA catabolism fulfils an important purpose in growing animals. Moreover, intestinal microbes extensively utilize both EAA and nutritionally nonessential amino acids (NEAA) for protein synthesis to maintain their normal population and activity. Due to these biochemical events, efficiency of utilization of dietary proteins for animal growth is far below 100%. For example, under current feeding programs, only 70% and 55% of dietary amino acids are deposited as proteins in 14-day-old pigs reared by sows and 30-day-old pigs weaned at 21 days of age, respectively. The remaining amino acids are degraded to CO_2, NO, CO, H_2S, methane, H_2O, ammonia, urea, nitrate, and other nitrogenous metabolites. Because current diets contain substantially more EAA, but less NEAA, than needed for protein accretion in growing pigs, supplementing NEAA and prebiotics to a low-protein diet may be effective in improving efficiency of amino acid utilization and supporting optimal growth performance of pigs.

Introduction

Protein is quantitatively the most expensive nutrient in swine production. The conversion of food proteins into animal proteins requires complex biochemical and physiological processes, which include digestion, absorption, metabolism (involving intestinal-lumen microorganisms, the splanchnic bed, and other organs), and the regulation of these events via multiple signalling pathways (Wu, 2009). This constitutes dynamic utilization of both nutritionally essential amino acids (EAA) and nonessential amino acids (NEAA) (Figure 1). Except for glutamate, glutamine, and aspartate, dietary amino acids are primarily used for protein accretion in young pigs (Wu *et al.*, 2010). Although little is known about amounts of amino acids utilized for the synthesis of non-protein substances in the body, it has been estimated that 10-40% of dietary EAA and some of the NEAA (asparagine, cysteine, serine, and tyrosine) that enter into the portal circulation are degraded in extra-intestinal tissues (Wu *et al.*, 2010).

Under current feeding programs, efficiency of utilization of dietary proteins for animal growth remains suboptimal. For example, in 14-day-old pigs reared by sows and 30-day-old pigs weaned at 21 days of age, 70% and 55% of dietary amino acids are deposited in body proteins, respectively (Wu *et al.*, 2010). The remaining amino acids must be degraded to CO_2, NO, CO, H_2S, methane, H_2O, ammonia, urea, nitrate, and other nitrogenous metabolites (Wu and Meininger, 2002; Li *et al.*, 2009). Excretion of these products in urine and feces is a source of environmental pollution and can contribute to global climate changes. Therefore, there are urgent needs to better understand biochemical and physiological limitations to efficiency of amino acid utilization in swine. The major objective of this article is to address the question of whether catabolism of EAA fulfils an important purpose in growing animals.

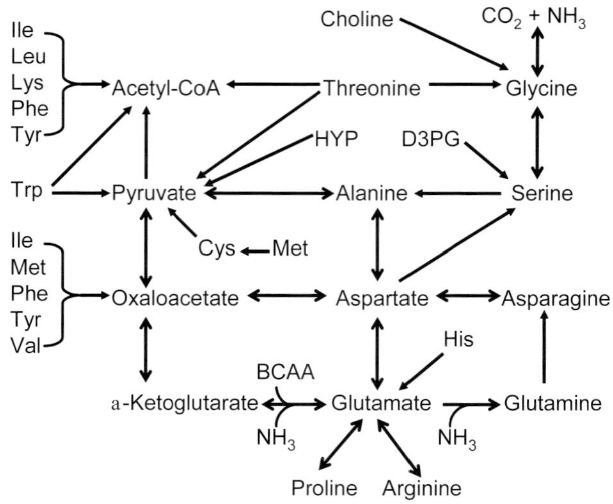

Figure 1. Dynamic utilization of amino acids in pigs. Degradation of essential amino acids via inter-organ cooperation results in synthesis of nonessential amino acids. BCAA, branched-chain amino acids; D3PG, D-3-phosphoglycerate (an intermediate of glucose metabolism); HYP, hydroxyproline. Synthesis of serine from its carbon skeleton (D3PG) requires amino acids (e.g. aspartate and glutamate) as donors of the amino group.

Obligatory use of EAA and NEAA for the production of nonprotein nitrogenous substances

Amino acids are essential precursors for the synthesis of a wide array of nitrogenous substances with enormous biological importance. Some of these bioactive molecules are summarized in Table 1, which include neurotransmitters (e.g. γ-aminobutyrate, dopamine, and serotonin), hormones (e.g. epinephrine and nor epinephrine), vasodilators, signalling gases (NO, CO and H_2S), antioxidants (glutathione, creatine, melatonin, melanin, and taurine), methyl donors, as well as key regulators of metabolism, growth, development, immune response, and health (Wu, 2009). Of particular note, large amounts of glutamine and arginine are used by cells of the immune system to defend the host from infectious pathogens, particularly in response to immunological challenges (e.g. consumption

Table 1. Major nitrogenous metabolites produced from amino acids in animals.[1]

Amino acid	Metabolite	Amino acid	Metabolite	Amino acid	Metabolite
Arginine	NO and ornithine	Lysine	Hydroxylysine	Arg & Met	Polyamines
Cysteine	Taurine and H_2S	Methionine	BET, CHO and SAM	Arg, Met & Gly	Creatine
Glutamate	γ-Aminobutyrate	Phenylalanine	Tyrosine	Cys, Glu & Gly	Glutathione
Glutamine	NH_3 and GlcN-6P	Proline	Hydroxyproline	AGGS	Nucleic acids
Glycine	Heme and CO	Serine	Glycine	Lys, Met & Ser	Carnitine
Histidine	Histamine and UCA	Tryptophan	STN, MLT and ANS	β-Ala & His	Carnosine
BCAA	Glutamine	Tyrosine	DOP, MEL and EPN	β-Ala & MH	Balenine

[1] Adapted from Wu (2009). AGGS, aspartate, glutamine, glycine and serine; BCAA, branched-chain amino acids; BET, betaine; CHO, choline; CO, carbon monoxide; DOP, dopamine; EPN, epinephrine and nor epinephrine; GlcN-6P, glucosamine-6-phosphate; MEL, melanin; MLT, melatonin; MH, 3-methylhistidine; NO, nitric oxide; SAM, S-adenosylmethionine; STN, serotonin, UCA, urocanic acid.

of 50% free glutamine and 50% free arginine in plasma within 1 h and 2.5 h, respectively; Li *et al.*, 2007). Additionally, 17% of dietary arginine is used for creatine synthesis alone in piglets (Wu *et al.*, 2004). Furthermore, 25% of dietary methionine is utilized by methylation reactions via the formation of S-adenosylmethionine in young pigs (Brosnan *et al.*, 2009). These metabolic transformations are essential for the reproduction, health and survival of organisms.

Degradation of dietary EAA for NEAA synthesis in the small intestine

Recent studies reveal that both EAA and NEAA in the diet are extensively degraded by the pig small intestine (Wu, 1998; Stoll and Burrin, 2006). Results of our studies indicate that nearly all of glutamate and aspartate, 67-70% of glutamine, and 30-40% of proline in the enteral diet are catabolized by the small intestine of neonatal, weaned, and gestating swine (Wu *et al.*, 2010). Similar results were obtained by Reeds and coworkers for glutamate and aspartate (Reeds *et al.*, 1996, 1997). Thus, only 5% of glutamate and aspartate, 30-33% of glutamine, as well as 60% of proline and arginine in the enteral diet enter the portal circulation. The rate of degradation of glutamate in the enteral diet by the small intestine is the greatest among amino acids, followed by glutamine, aspartate, and proline. It is now known that bacteria in the lumen of the small intestine can degrade these amino acids (Bergen and Wu, 2009; Dai *et al.* 2010). Furthermore, absorptive epithelial cells of the neonatal gut (enterocytes) extensively catabolize glutamate, aspartate, glutamine and proline (Wu an Morris, 1998).

Enterocytes of both pre- and post-weaning pigs do not express key enzymes for degrading histidine, lysine, methionine, phenylalanine, threonine or tryptophan and, therefore, have little ability to oxidize them to CO_2 (Chen *et al.*, 2007, 2009). However, some evidence shows that 40% leucine, 30% isoleucine, 40% valine, 50% lysine, 50% methionine, 45% phenylalanine, and 60% threonine) may be extracted by the small intestine of weaned piglets in first pass (Stoll and Burrin, 2006). Using bolus oral or intravenous administration of the same dose of a test amino acid, we obtained lower values of first-pass intestinal use for histidine, lysine, methionine, phenylalanine, threonine, and tryptophan in 14-day-old suckling pigs and 35-day-old weaned pigs. Data on the net balance of amino acids across the portal-drained viscera in milk protein-fed young pigs are summarized in Table 2. Less than 20% of the extracted EAA are utilized for intestinal mucosal protein synthesis (Stoll and Burrin, 2006), and greater than 80% of the extracted EAA are presumably degraded by

Table 2. Net balance of amino acids across the portal-drained viscera in milk protein-fed young pigs (6-10 kg).[1]

Amino acid	Percentage of dietary intake	Amino acid	Percentage of dietary intake
EAA		NEAA	
Arginine	147	Alanine	154
Histidine	71	Asparagine	74
Isoleucine	66	Aspartate	5
Leucine	64	Cysteine	69
Lysine	55	Glutamate	3
Methionine	69	Glutamine	-16
Phenylalanine	63	Glycine	69
Proline	59	Serine	66
Threonine	50	Tyrosine	71
Tryptophan	75		
Valine	65		

[1] Vaues (percentage of dietary intake) are adapted from Stoll and Burrin (2006) and Wu *et al.* (2010).

microorganisms in the intestinal lumen (Dai *et al.*, 2010). Dietary amino acids likely play an important role in regulating microbial composition and activity in the gut.

Metabolism of amino acids in the intestine is of nutritionally and physiological significance. For example, these nutrients play an important role in maintaining gut integrity and function, regulating endogenous synthesis of amino acids (citrulline, arginine, proline and alanine), and modulating the availability of dietary amino acids to extra-intestinal tissues (Wu, 2009). Additionally, because elevated levels of glutamate and aspartate in the circulation exert a neurotoxic effect (Meldrum and Garthwaite, 1990; Stout *et al.*, 1998), their extensive catabolism by the small intestine is essential to the survival of organisms. However, irreversible loss of dietary amino acids (particularly EAA) from the gut results in reduced efficiency of their utilization for growth. Therefore, inhibiting microbial activity through dietary supplementation with antibiotics or prebiotics can enhance the entry of dietary amino acids into the portal circulation (Yin *et al.*, 2010) and growth performance of young pigs (Kong *et al.*, 2009).

Degradation of dietary EAA for NEAA synthesis in extraintestinal tissues

Glycine

Milk is the sole source of nutrients for suckling piglets. The analysis of glycine content in sow milk and the piglet body reveals that glycine is the most deficient amino acid in this neonatal food (Table 3). Strikingly, milk-derived glycine meets only 23% of the need for protein synthesis in 14-day-old pigs (Wu *et al.*, 2010). Similarly, glycine is the second most deficient amino acid in a typical corn- and soybean meal-based diet. Although biochemistry textbooks state that glycine is synthesized from serine, 81% of milk-born serine is utilized for protein deposition in sow-reared piglets and the diet provides at most 0.32 g serine for glycine formation (Table 3). Therefore, in both milk-fed and post-weaning pigs, approximately 90% of glycine must be synthesized from precursors other than serine, which likely include threonine, choline, and ammonia plus CO_2. At present, little is known about pathways (including substrates and reactions) for glycine synthesis in pigs.

Arginine

It is now known that sow milk is markedly deficient in arginine and provides at most 40% of arginine for protein deposition in the 14-day-old pig (Wu and Knabe, 1995; Wu *et al.*, 2004). The supply of arginine from a corn- and soybean meal-based diet is also inadequate for maximal growth of post-weaning pigs (Wu *et al.*, 1997). Thus, endogenous synthesis of arginine from glutamine, glutamate and proline must take place to meet the basic physiological needs for protein accretion and other pathways (e.g. generation of NO, polyamines, and creatine; Wu and Morris, 1998). The arterial source of glutamine for the intestinal synthesis of citrulline (the immediate precursor of arginine in kidneys and extra-renal cells) is the catabolism of branched-chain amino acids (BCAA; leucine, isoleucine, and valine) primarily in skeletal muscle (Figure 2). Therefore, degradation of BCAA derived from the small-intestinal lumen ensures endogenous provision of arginine to compensate for its deficiency in diets for young pigs.

Proline

In addition to arginine, sow milk does not supply adequate amounts of proline for protein synthesis (Wu *et al.*, 2010). Based on a degradation rate of 0.93 g/kg body weight per day for proline in young pigs (Murphy *et al.*, 1996), the *de novo* synthesis of proline must occur at a rate of at least 1.11 g/kg body weight per day (or at least 60% of the proline needed for protein accretion) in sow-reared piglets. The relative deficiency of proline is even more severe for post-weaning pigs fed a corn- and soybean meal-based diet (Table 3). Indeed, the amount of dietary proline that enters into the portal

Table 3. Degradation or minimal synthesis of amino acids for protein accretion as absolute or percentage of amounts entering the portal vein from the small-intestinal lumen in fed young pigs.[1]

Amino acid	Sow-reared pigs[2]		Weaned pigs[3]	
	Absolute amount (g/day)	Percentage (%)	Absolute amount (g/day)	Percentage (%)
EAA				
Arginine	+1.09	+103	+0.26	+10
Histidine	-0.10	-13	-0.58	-37
Isoleucine	-0.29	-21	-0.39	-20
Leucine	-0.62	-22	-0.84	-23
Lysine	-1.18	-38	-0.68	-22
Methionine	-0.25	-29	-0.13	-15
Phenylalanine	-0.40	-27	-0.97	-41
Proline	+0.71	+23	+2.15	+74
Threonine	-0.21	-16	-0.36	-20
Tryptophan	-0.17	-33	-0.15	-25
Valine	-0.17	-11	-0.22	-11
NEAA				
Alanine	+0.70	+51	-0.82	-23
Asparagine	-0.88	-44	-0.97	-40
Aspartate	+1.24	+1127	+1.52	+661
Cysteine	-0.08	-16	-0.37	-41
Glutamate	+2.26	+1076	+3.24	+1542
Glutamine	+0.17	+12	+0.30	+17
Glycine	+2.75	+335	+2.58	+117
Serine	-0.32	-19	-0.23	-11
Tyrosine	-0.59	-41	-0.78	-41

[1] Adapted from Wu *et al.* (2010).
[2] Fourteen-day-old piglets reared by sows.
[3] Thirty-day-old pigs weaned at 21 days of age to a corn- and soybean meal-based diet.
The signs '-' and '+' denote degradation and minimal synthesis. Total synthesis of amino acids should include amino acids deposited in tissue proteins and those oxidized to urea, CO_2 and H_2O.

vein meets only 58% of the requirement for protein synthesis in weaned pigs. The ultimate sources of the nitrogen and carbon for proline synthesis are primarily BCAA and other amino acids (e.g. methionine, phenylalanine, tryptophan, tyrosine, and ornithine).

Alanine, aspartate, glutamate and glutamine

Almost none of the dietary aspartate and glutamate enters the portal circulation and only 30% of dietary glutamine is absorbed into the portal vein, because of their extensive degradation by the portal-drained viscera (Stoll and Burrin, 2006). Sow milk provides at most 66%, 8%, and 9% of alanine, aspartate, and glutamate for protein deposition in suckling pigs, respectively (Kim and Wu, 2004). Most of essential amino acids can be metabolized to form these three amino acids (Figure 1). Additionally, there is a high rate of whole-body glutamine utilization in the piglet (at least 1.65 g/kg body weight per day) for oxidation (0.47 g glutamine/kg body weight per day; Stoll *et al.*, 1999) and protein synthesis (0.41 g glutamine/kg body weight per day; Wu *et al.*, 2010). Because only a small amount of dietary glutamine enters the systemic circulation (0.36 g/kg body weight per day), the rate of *de novo* synthesis of glutamine is likely the greatest in the suckling piglet (at least 0.88

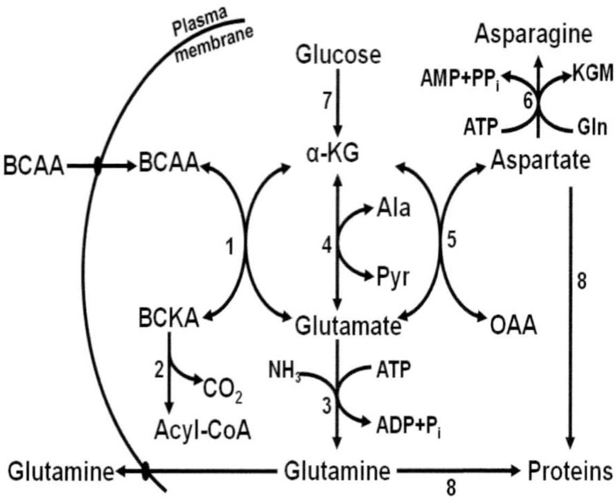

Figure 2. Pathways of BCAA catabolism for the synthesis of glutamate, alanine, glutamine, aspartate and asparagine in pigs. Enzymes that catalyze the indicated reactions are: 1) BCAT; 2) BCKAD; 3) GS; 4) GOT; 5) GPT; 6) AS; 7) glucose metabolism via glycolysis and the Krebs cycle; 8) protein synthesis. The corresponding α-ketoacids of leucine, isoleucine and valine are α-ketoisocaproate, α-keto-β-methylvalerate, and α-ketoisovalerate, respectively. Insulin-sensitive tissues (e.g. skeletal muscle, heart, and adipose tissue) take up BCAA from the arterial circulation and release glutamine through specific transporters on the plasma membrane. Reprinted from Journal of Nutrition (Li et al.*, 2009) with permission from American Society of Nutrition.*

g glutamine/kg body weight per day). In post-weaning pigs, the enteral diet provides sufficient alanine, but not aspartate, glutamate or glutamine. As noted above, BCAA provide the amino group for glutamine synthesis primarily in insulin-sensitive organs (e.g. skeletal muscle, heart, and adipose tissue) of animals (Figure 2).

It is clear that large amounts of arginine, proline, aspartate, glutamine, and glycine must be synthesized in both milk-fed and post-weaning pigs. This is also true for alanine in suckling piglets. Syntheses of some amino acids necessitate the degradation of EAA (whose carbon skeletons cannot be synthesized) in pigs. Thus, when diets do not contain adequate NEAA, EAA catabolism fulfils an important purpose in growing animals and should not be considered as futile events inherited from ancient ancestors. Recent advances on NEAA nutrition have important implications for formulating new swine diets with the goal of reducing EAA content. For example, current diets contain substantially more EAA, but less NEAA, than needed for protein accretion in growing pigs (Table 2). Supplementing NEAA to a low-protein diet may be effective in improving the efficiency of metabolic transformations and supporting optimal growth performance of pigs.

Changes in gene expression and signalling pathways with postnatal growth

Efficiency of utilization of dietary amino acids by animals critically depends on the regulation of gene expression in a cell-specific manner (Wu, 2009). Amino acids are not only substrates for protein synthesis but also affect one or more of the following steps: modification of chromatin, transcription, post-transcriptional modification, RNA transport, mRNA degradation, translation, and post-translational modifications (Bruhat *et al.*, 2009). The mammalian target of rapamycin (mTOR), a highly conserved serine/threonine protein kinase, is the master regulator of translation (Figure 3) and consists of mTOR complexes 1and 2 (Shaw, 2008). Amino acids (e.g. glutamine, arginine, and leucine) stimulate the phosphorylation of mTOR in a cell-specific manner, leading to phosphorylation of p70S6 kinase 1 and eIF4E-binding protein-1 (4EBP1) proteins and, therefore,

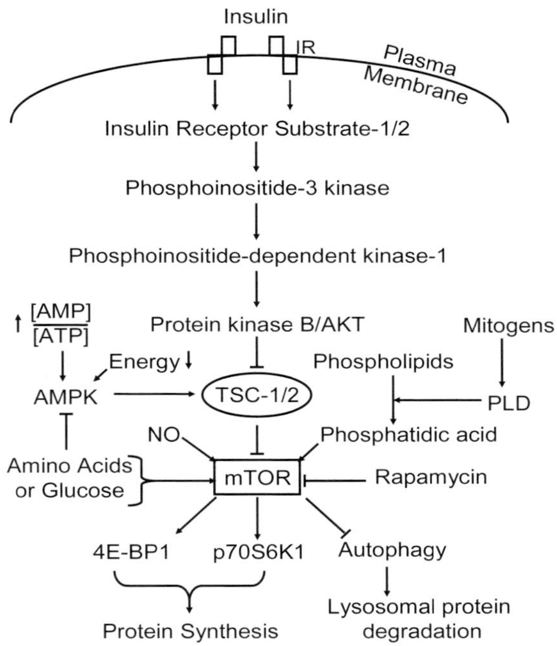

Figure 3. Activation of protein synthesis by amino acids and growth factors through the mTOR signalling pathway. mTOR (a protein kinase) phosphorylates eIF4E-binding protein-1 (4E-BP1) and ribosomal protein S6 kinase-1 (S6K1), thereby stimulating protein synthesis and inhibiting autophagy (a key step in lysosomal proteolysis). mTOR is inhibited by TSC-1/2 (tuberous sclerosis complex-1/2) whose activity is enhanced by AMPK (AMP-activated protein kinase) but suppressed by protein kinase B (also known as AKT). Phosphorylation of AKT in response to insulin and other growth factors relieves an inhibitory effect of TSC-1/2 on mTOR. Additionally, certain nutrients (e.g. glutamine, arginine, leucine, and glucose) and phosphatidic acid produced by phospholipase D (PLD) stimulate mTOR phosphorylation and thus increase its activity. Oxidation of AA, glucose and fatty acids increases cellular ratios of ATP:AMP, therefore reducing AMPK activity. Reprinted from Livestock Science (Wu et al.*, 2007) with permission from Elsevier.*

the formation of translation initiation complex (Yao *et al.*, 2008; Davis and Fiorotto, 2009; Rhoads and Wu, 2009). Besides mTOR signalling, arginine can regulate gene expression and activity of AMPK, another important protein in nutrient sensing that modulates oxidation of energy substrates and insulin sensitivity (Jobgen *et al.*, 2006). Additionally, glutamine is known to activate several signalling pathways, including extracellular signal-related kinase, Jun kinase, mitogen-activated protein kinase, protein kinase A, nuclear receptors, zinc fingers proteins, and helix-turn-helix proteins (Brasse-Lagnel *et al.*, 2009; Rhoads and Wu, 2009). Furthermore, nitric oxide (NO), carbon monoxide (CO), and hydrogen sulfide (H_2S), which are products of arginine, glycine, and cysteine catabolism, respectively, beneficially regulate diverse physiological and immunological processes in animals via cGMP- and cAMP-dependent pathways (Li *et al.*, 2007; 2009).

Increasing evidence shows that expression of genes as well as anti-oxidative and regulatory proteins in pig tissues changes with postnatal growth (Wang *et al.*, 2008, 2010). Notably, protein levels for positive regulators of the mTOR signalling pathway are higher in skeletal muscle of young than older pigs (Suryawan *et al.*, 2009), resulting in a decrease in the rate of protein synthesis with age (Davis and Fiorotto, 2009). Thus, efficiency of dietary EAA and NEAA for protein synthesis is 35% and 26% greater, respectively, in suckling piglets than in post-weaning pigs (Wu *et al.*, 2010).

Enhancing efficiency of utilization of dietary amino acids for growth

New knowledge about regulation of gene expression and signalling pathways has made it possible to develop novel nutritional management for ameliorating the biochemical and physiological limitations to efficiency of amino acid utilization in swine. For example, oral administration of L-glutamine (1 g/kg body weight per day) to low-birth-weight piglets can improve their survival and growth, particularly in response to endotoxin treatment (Haynes *et al.*, 2009). Also, dietary supplementation with 1% L-glutamine to early-weaned pigs prevents jejunal atrophy during the first week postweaning and increases the gain:feed ratio by 25% during the second week postweaning (Wu *et al.*, 1996; Wang *et al.*, 2008). Furthermore, dietary supplementation with 0.2% and 0.4% L-arginine to 7- to 21-day-old milk-fed pigs (artificially reared on a liquid-milk feeding system) dose-dependently enhances plasma arginine concentrations (30% and 61%), reduces plasma ammonia levels (20% and 35%), and increases weight gain (28% and 66%) (Kim and Wu, 2004). Finally, dietary L-arginine supplementation increases tissue anti-oxidative capacity (Ma *et al.* 2010) and lean tissue growth (Tan *et al.* 2009) in growing-finishing pigs.

Conclusion

There have been rapid advances in amino acid nutrition over the past decade. Despite much research, efficiency of the utilization of dietary protein for swine growth and production remains suboptimal due to both biochemical and physiological limitations. Among the constraining factors are: (1) the extensive degradation of both EAA and NEAA by the small intestine and extra-intestinal tissues, (2) the obligatory use of amino acids for the production of nonprotein nitrogenous substances, and (3) age-dependent decline in muscle mTOR activity. Based on new knowledge about intestinal microbial metabolism as well as the regulation of gene expression and signalling pathways in mammalian cells, practical methods involving dietary supplementation with prebiotics, glutamine and arginine have be developed to enhance the net absorption of dietary amino acids into the systemic circulation, stimulate cellular protein synthesis, and improve the efficiency of amino acid utilization in pigs.

Acknowledgments

The author thanks all of his collaborators, postdoctoral fellows, graduate students, and technicians for important contributions to the studies described here, as well as Dr. Niels Kristensen for helpful discussion. This work was supported by the National Research Initiative Competitive Grant Animal Reproduction Program (No. 2006-35203-17199 and 2008-35203-19120) and Animal Growth and Nutrient Utilization Program (2008-35206-18762, 2008-35206-18764 and 2009-35206-05211) from the USDA Cooperative State Research, Education, and Extension Service, and Texas AgriLife Research (H-82000).

References

Bergen, W.G. and Wu, G., 2009. Intestinal nitrogen recycling and utilization in health and disease. J. Nutr. 139, 821-825.

Brasse-Lagnel, C., Lavoinne, A. and Husson, A., 2009. Control of mammalian gene expression by amino acids, especially glutamine. FEBS Journal 276, 1826-1844.

Brosnan, J.T., Wijekoon, E.P., Warford-Woolgar, L., Trottier, N.L., Brosnan, M.E., Brunton, J.A. and Bertolo, R.F.P., 2009. Creatine synthesis is a major metabolic process in neonatal piglets and has important implications for amino acid metabolism and methyl balance. J Nutr. 139, 1292-1297.

Bruhat, A., Cherasse, Y., Chaveroux, C., Maurin, A.C., Jousse, C. and Fafournous, P., 2009. Amino acids as regulators of gene expression in mammals: molecular mechanisms. BioFactors 35, 249-257.

Chen, L., Yin, Y., Jobgen, W.J., Jobgen, S.C., Knabe, D.A., Hu, W. and Wu, G., 2007. *In vitro* oxidation of essential amino acids by intestinal mucosal cells of growing pigs. Livest. Sci. 109, 19-23.

Chen, L., Li, P., Wang, J., Li, X., Gao, H., Yin, Y., Hou, Y. and Wu, G., 2009. Catabolism of essential amino acids in developing porcine enterocytes. Amino Acids 37, 143-152.

Dai, Z.L., Zhang, J., Wu, G. and W.Y. Zhu., 2010. Utilization of amino acids by bacteria from the pig small intestine. Amino Acids. doi: 10.1007/s00726-010-0556-9.

Davis, T.A. and Fiorotto, M.L., 2009. Regulation of muscle growth in neonates. Curr. Op. Nutr. Metab. Care 12, 78-85.

Haynes, T.E., Li, P., Li, X.L., Shimotori, K., Sato, H., Flynn, N.E., Wang, J.J., Knabe, D.A. and Wu, G., 2009. L-Glutamine or L-alanyl-L-glutamine prevents oxidant- or endotoxin-induced death of neonatal enterocytes. Amino Acids 37, 131-142.

Jobgen, W.S., Fried, S.K., Fu, W.J., Meininger, C.J. and Wu, G., 2006. Regulatory role for the arginine-nitric oxide pathway in metabolism of energy substrates. J. Nutr.Biochem. 17, 571-588.

Kim, S.W. and Wu, G., 2004. Dietary arginine supplementation enhances the growth of milk-fed young pigs. J Nutr. 134, 625-630.

Kong, X.F., Yin, Y.L., He, Q.H., Yin, F.G., Liu, H.J., Li, T.J., Huang, R.L., Geng, M.M., Ruan, Z., Deng, Z.Y., Xie, M.Y. and Wu, G., 2009. Dietary supplementation with Chinese herbal powder enhances ileal digestibilities and serum concentrations of amino acids in young pigs. Amino Acids 37, 573-582.

Li, P., Yin, Y.L., Li, D.F., Kim, S.W. and Wu, G., 2007. Amino acids and immune function. Br. J Nutr. 98, 237-252.

Li, X., Bazer, F.W., Gao, H., Jobgen, W., Johnson, G.A., Li, P., McKnight, J.R., Satterfield, M.C., Spencer, T.E. and Wu, G., 2009. Amino acids and gaseous signalling. Amino Acids 37, 65-78.

Ma, X., Lin, Y., Jiang, Z., Zheng, C., Zhou, G., Yu, D., Cao, T., Wang, J. and Chen, F., 2010. Dietary arginine supplementation enhances antioxidative capacity and improves meat quality of finishing pigs. Amino Acids 38, 95-102.

Meldrum, B. and Garthwaite, J., 1990. Excitatory amino acid neurotoxicity and neurodegenerative disease. Trends Pharm Sci 11, 379-387.

Murphy, J.M., Murphy, S.J. and Ball, R.O., 1996. Proline is synthesized from glutamate during intragastric infusion but not during intravenous infusion in neonatal piglets. J Nutr 126, 878-886.

Reeds, P.J., Burrin, D.G., Jahoor, F., Wykes, L., Henry, J. and Frazer, M.E., 1996. Enteral glutamate is almost completely metabolized in first pass by the gastrointestinal tract of infant pigs. Am J Phys 270, E413-E418.

Reeds, P.J., Burrin, D.G., Stoll, B., Jahoor, F., Wykes, L., Henry, J. and Frazer, M.E., 1997. Enteral glutamate is the preferential source for mucosal glutathione synthesis in pigs. Am J Phys 273, E408-E415.

Rhoads, J.M. and Wu, G., 2009. Glutamine, arginine, and leucine signalling in the intestine. Amino Acids 37, 111-122.

Shaw, R.J., 2008. mTOR signalling: RAG GTPases transmit the amino acid signal. Trends Biochem Sci 33, 565-568.

Stoll, B., Burrin, D.G., Henry, J., Yu, H., Jahoor, F. and Reeds, P.J., 1999. Substrate oxidation by the portal drained viscera of fed piglets. Am J Phys 277, E168-E175.

Stoll, B. and Burrin, D.G., 2006. Measuring splanchnic amino acid metabolism *in vivo* using stable isotopic tracers. J Anim Sci 84, E60-E72.

Stout, A.K., Raphael, H.M., Kanterewicz, B.I., Klann, E. and Reynolds, I.J., 1998. Glutamate-induced neuron death requires mitochondrial calcium uptake. Nature Neurosci 1, 366-373.

Suryawan, A., O'Connor, P.M.J., Bush, J.A., Nguyen, H.V. and Davis, T.A., 2009. Differential regulation of protein synthesis by amino acids and insulin in peripheral and visceral tissues of neonatal pigs. Amino Acids 37, 105-110.

Tan, B., Yin, Y.L., Liu, Z.Q., Li, X.G., Xu, H.J., Kong, X.F., Huang, R.L., Tang, W.J., Shinzato, I., Smith, S.B. and Wu, G., 2009. Dietary L-arginine supplementation increases muscle gain and reduces body fat mass in growing-finishing pigs. Amino Acids 37, 169-175.

Wang, J.J., Chen, L.X., Li, P., Li, X.L., Zhou, H.J., Wang, F.L., Li, D.F., Yin, Y.L. and Wu, G., 2008. Gene expression is altered in piglet small intestine by weaning and dietary glutamine supplementation. J Nutr 138, 1025-1032.

Wang, X.Q., Wu, W.Z., Lin, G., Li, D.F., Wu, G. and J.J. Wang., 2010. Temporal proteomic analysis reveals continuous impairment of intestinal development in neonatal piglets with intrauterine growth restriction. J Prot Res 9, 924-935.

Wu, G., 1998. Intestinal mucosal amino acid catabolism. J Nutr 128, 1249-1252.

Wu, G., 2009. Amino acids: metabolism, functions, and nutrition. Amino Acids 37, 1-17.

Wu, G. and Knabe, D.A., 1995. Arginine synthesis in enterocytes of neonatal pigs. Am J Phys 269, R621-R629.

Wu, G., Meier, S.A. and Knabe, D.A., 1996. Dietary glutamine supplementation prevents jejunal atrophy in weaned pigs. J Nutr 126, 2578-2584.

Wu, G. and Morris, S.M., Jr., 1998. Arginine metabolism: nitric oxide and beyond. Biochem J 336, 1-17.

Wu, G., Davis, P.K., Flynn, N.E., Knabe, D.A. and Davidson J.T., 1997. Endogenous synthesis of arginine plays an important role in maintaining arginine homeostasis in postweaning growing pigs. J Nutr 127,2342-2349.

Wu, G. and Meininger, C.J., 2002. Regulation of nitric oxide synthesis by dietary factors. Ann Rev Nutr 22, 61-86.

Wu, G., Knabe, D.A. and Kim, S.W., 2004. Arginine nutrition in neonatal pigs. J Nutr 134, 2783S-2390S.

Wu, G., Bazer, F.W., Burghardt, R.C., Johnson, G.A., Kim, S.W., Knabe, D.A., Li, X.L., Satterfield, M.C., Smith, S.B. and Spencer, T.E., 2010. Functional amino acids in swine nutrition and production. In: Dynamics in Animal Nutrition (J. Doppenberg, ed.), Wageningen Academic Publishers, the Netherlands (In Press).

Yao, K., Yin, Y.L., Chu, W.Y., Liu, Z.Q., Deng, D., Li, T.J., Huang, R.L., Zhang, J.S., Tan, B.E., Wang, W. and Wu, G., 2008. Dietary arginine supplementation increases mTOR signalling activity in skeletal muscle of neonatal pigs. J Nutr 138, 867-872.

Yin, Y.L., Huang, R.L., Li, T.J., Ruan, Z., Xie, M.Y., Deng, Z.Y., Hou, Y.Q. and Wu, G., 2010. Amino acid metabolism in the portal-drained viscera of young pigs: effects of dietary supplementation with chitosan and pea hull. Amino Acids. DOI: 10.1007/s00726-010-0577-4.

The effect on growing pig performance of changes in energy intake achieved through restriction of feed intake versus changes in dietary energy concentration

A.D. Beaulieu[1], J.P. Marriott[1], J.E. Pettigrew[2], C.M. Nyachoti[3] and J.F. Patience[4]
[1]Prairie Swine Centre, Inc., Saskatoon, SK S7H 5N9, Canada
[2]University of Illinois, Urbana, IL 61801, USA
[3]University of Manitoba, Winnipeg, MB R3T 2N2, Canada
[4]Iowa State University, Ames, IA 50011, USA

Introduction

Energy intake in the pig can be manipulated through restriction of his feed intake or by altering the energy density of his diet. The former approach is commonly taken in a research setting while the latter is generally the more accessible approach in commercial pork production.

Restriction of feed intake in the growing pig results in decreases in energy intake and in average daily gain (Weis *et al.*, 2004). Some authors report analogous findings when dietary energy concentration is manipulated (Wu *et al.*, 2007), while others report that energy concentration does not affect energy intake or growth performance (Beaulieu *et al.*, 2009). The objective of the present experiment was to compare the response of the pig to changes in energy intake brought about by changes in feed intake with the response of the pig to changes in energy intake brought about by changes in dietary energy concentration.

Materials and methods

Dietary treatments were arranged in a 3×3 factorial design with 3 feeding levels (80, 90 and 100% *ad libitum*) and 3 concentrations of net energy (2.18, 2.29 and 2.40 Mcal/kg). Net energy concentrations were adjusted through proportional changes in the inclusion levels of wheat (15.00, 39.55 and 64.51% as-fed), barley (55.45, 31.33 and 6.80% as-fed) and canola oil (1.00, 2.25 and 3.50% as-fed) in the experimental diets.

Seventy-two individually-housed barrows (initial body-weight 30±2 kg) each received one of nine dietary treatments. On a weekly basis the pigs were weighed, the feed allowances of the restricted-fed pigs were adjusted and the feed intake (disappearance) of the *ad libitum*-fed pigs was determined. Pigs were removed from the experiment at a body-weight of 60±2 kg.

Data were analyzed by analysis of covariance, with initial weight as the covariate. The MIXED procedure of SAS (SAS Institute, 1996) was employed; the model account for the fixed effects of dietary energy concentration, energy intake and their interaction and the random effects of block and pen.

Results and discussion

No interactive effects between feeding level and dietary energy concentration were found ($P>0.10$). The main effects of feeding level and of energy concentration on growth performance are presented in Table 1. Pigs offered feed at 80, 90 and 100% of *ad libitum* had NE intakes of 3.7, 4.3 and 4.7 Mcal/d ($P<0.0001$), while dietary energy concentration had no effect on NE intake ($P>0.10$).

Both the quantity of energy used for maintenance, expressed on a per day basis, and the efficiency of NE utilization for growth were unaffected by dietary treatment ($P>0.10$); however, when expressed on a per pig basis, the quantity of energy required for maintenance was observed to increase as feeding level decreased ($P<0.0001$). This observation can be attributed to the increased number of days spent on test as feeding level decreased (Table 1; $P<0.0001$).

Analysis showed that pigs fed at 80, 90 and 100% of *ad libitum* had carcass CP contents of 54.48, 52.86 and 49.16% (as-is; $P<0.0001$) and accrued protein at rates of 135.3, 152.9 and 172.1 g/d ($P<0.0001$), respectively. These parameters were unaffected by dietary energy concentration ($P>0.10$).

This study demonstrates that the responses of growing pigs to changes in dietary energy concentration differ from their responses to changes in feed allowance. This finding challenges the applicability of those principles of energy metabolism developed through restriction of feed intake to situations, most critically commercial pork production, in which the energy concentration of the diet is to be manipulated.

Table 1. Effects of feeding level and energy concentration on the performance of growing pigs.

	Feeding level, % *ad libitum*				NE concentration, Mcal/kg			
Item	80	90	100	SEM	2.18	2.29	2.40	SEM
N	24	23	23	-	24	23	23	-
Initial BW, kg	30.4	30.4	29.6	0.43	30.2	30.2	30.0	0.13
Final BW, kg	59.6	60.3	60.2	0.37	60.3	59.7	60.0	0.31
Days on test[1]	41	35	29	5.96	35	35	35	0.12
ADG, kg[1]	0.72	0.85	1.06	0.17	0.89	0.85	0.88	0.02
ADFI, kg[1]	1.61	1.87	2.05	0.22	1.93	1.81	1.78	0.08
ADFI, % NRC[1,3]	76	87	96	10.08	86	85	87	1.04
G:F, kg/kg[2]	0.45	0.46	0.52	0.04	0.46	0.47	0.50	0.02

[1] Significant main effect of feeding level ($P<0.0001$).
[2] Significant main effect of feeding level ($P=0.0006$).
[3] Significant main effect of NE concentration ($P=0.0016$).

References

Beaulieu, A.D., N.H. Williams, and J.F. Patience, 2009. Response to dietary digestible energy concentration in growing pigs fed cereal grain-based diets. J. Anim. Sci. 87, 965-976.

SAS Institute Inc. 1996. SAS/STAT User's Guide. Release 6.12. SAS Institute Inc., Cary, NC USA.

Weis, R.N., S.H. Birkett, P.C.H. Morel, and C.F.M. de Lange, 2004. Effects of energy intake and body weight on physical and chemical body composition in growing entire male pigs. J. Anim. Sci. 82, 109-121.

Wu, Z., D. Li, Y. Ma, Y. Yu and J. Noblet, 2007. Evaluation of energy systems in determining the energy cost of gain of growing-finishing pigs fed diets containing different levels of dietary fat. J. Anim. Sci. 61, 1-9.

Effects of xylanase and β-glucanase supplementation to cereal diets on nitrogen balance of growing pigs

U. Dickhoefer[1], M. Naatjes[1], B. Blank[1], R. Kühl[1], P. Ader[2] and A. Susenbeth[1]
[1]Institute of Animal Nutrition and Physiology, Christian-Albrechts-University, 24118 Kiel, Germany
[2]BASF SE, 67056 Ludwigshafen, Germany

Introduction

Common cereal grains used in pig nutrition contain considerable amounts of non-starch polysaccharides (NSP), which reduce diet digestibility due to a higher digesta viscosity (Campbell and Bedford, 1992) and increase endogenous losses of protein and amino acids (AA; Jansman *et al.*, 2002). Hence, in a recent experiment by Blank (2009) the requirements of pigs for precaecal digestible threonine (Thr) increased by 0.37 g per 100 g of ingested neutral detergent fibre from wheat bran. The current study therefore tested the hypothesis whether xylanase and β-glucanase supplementation to a cereal diet reduces these anti-nutritional effects and consequently improves nitrogen (N) retention in growing pigs.

Material and methods

In animals with a high protein accretion potential, absorption and endogenous losses of the first limiting AA largely determine protein retention by the animal. Since Thr concentrations in endogenous proteins are high (Jansman *et al.*, 2002), a decrease in endogenous protein losses will lead to a pronounced effect on available Thr. Thus, N retention from a Thr limited diet (Thr:lysine ratio of 0.56) was used as an indicator for possible effects of NSP-degrading enzymes on absorption and endogenous losses of protein and AA (Blank, 2009). In 3 consecutive balance periods (7 d adaptation + 7 d faeces and urine collection), 3 experimental diets were fed to 12 castrated male pigs (Pietrain x dbNaima; initial body weight (BW) 31 kg; final BW 66 kg) in a cross-over design: a cereal-based diet (37% wheat, 16% rye, 15% barley, 17% soybean meal) without enzyme supplementation (T1), supplemented with 40 mg/kg (T2) and 80 mg/kg (T3) of an enzyme preparation. The enzyme preparation contained 7,600 XU/g of endo-1,4-β-xylanase and 3,250 GU/g endo-1,4-β-glucanase. Each animal was offered 1.5 kg of feed per day. In the second and third balance period, pigs additionally received 150 g and 300 g of corn starch, respectively. Faeces and urine were collected quantitatively and stored at -20 °C. Urine was collected in bottles containing 20 ml H_2SO_4 (20%, v/v) to keep pH<2. Least squares means, standard deviations from the means and standard errors were calculated using the mixed model procedure (Proc Mixed) of SAS (1996), where diet, period and animal were treated as fixed effects, since no significant interactions between these variables were found. Significant differences ($P<0.05$) between treatment groups were determined by the Tukey-Kramer-Test.

Results and discussion

Daily urinary and faecal N excretion averaged 11.3 g (SD 1.5) and 6.5 (SD 1.2) for animals of all treatment groups (Table 1; $P=0.960$; $P=0.580$). N retention was 23.3 g (SD 2.0), 22.9 (SD 1.7) and 23.7 (SD 2.1) in T1, T2 and T3 pigs, respectively, equivalent to 0.568 (SD 0.047), 0.557 (SD 0.042) and 0.576 (SD 0.051) of their daily N intake ($P=0.624$). N excretion and retention were identical between treatments groups, stressing that enzyme supplementation did not improve precaecal N absorption and/or reduce endogenous protein losses neither at medium nor at high supplementation level.

Similarly, ileal digestibilities of DM, CP, lipid and starch components of un-supplemented barley-based diets fed to growing pigs (20 and 27 kg initial BW) were similar to those with β-glucanase

Table 1. Intake, excretion and retention of nitrogen (N) of growing pigs fed a cereal diet supplemented with different levels of a 1,4-β-xylanase and 1,4-β-glucanase mixture (least squares means ± standard deviation; SEM standard error of the mean).

		Diet			SEM	*P*-value
		T1	T2	T3		
N intake	g/d	41.1	41.1	41.1		
Faecal N		6.6±1.6	6.6±1.0	6.5±1.0	0.196	0.960
Urinary N		11.2±1.4	11.6±1.6	11.0±1.5	0.259	0.580
N retention		23.3±1.9	22.9±1.7	23.7±2.1	0.318	0.624
Faecal N	g/g N intake	0.160±0.039	0.161±0.023	0.157±0.024	0.005	0.960
Urinary N		0.272±0.033	0.282±0.039	0.267±0.036	0.006	0.580
N retention		0.568±0.047	0.557±0.042	0.576±0.051	0.008	0.624

supplementation in two experiments by McCann *et al.* (2006). In pigs of a similar BW than in our study, about 20% of dietary neutral detergent fibre are already digested before the terminal ileum (Schulze *et al.*, 1994). Although pigs do not produce endogenous enzymes to hydrolyze dietary fibre, enzymes of microbes present in the proximal part of the stomach and the small intestine allow for this precaecal fibre fermentation (Galassi *et al.*, 2004). Moreover, higher activities of pancreatic proteases, amylases and lipases secreted into the ileum in 56 d- than in 28 d-old pigs (Jensen *et al.*, 1997) reduce the potential for responses to exogenous enzyme supplementation with advancing age of the pig.

Conclusion

Supplementation of NSP-degrading enzymes to cereal diets fed to pigs with a BW of above 30 kg does not reduce the negative effects of dietary fibre on precaecal protein digestibility and on endogenous protein and AA losses.

References

Blank B., 2009. Studies on the threonine requirement in growing pigs. PhD thesis, Christian-Albrechts-University of Kiel, Kiel, Germany, 55 pp.

Campbell G.L. and M.R. Bedford, 1992. Enzyme Applications for Monogastric Feeds – a Review. Can. J. Anim. Sci. 72, 449-466.

Galassi G., G.M. Crovetto, L. Rapetti and A. Tamburini, 2004. Energy and nitrogen balance in heavy pigs fed different fibre sources. Livest. Prod. Sci. 85, 253-262.

Jansman A.J.M., W. Smink, P. van Leeuwen and M. Rademacher, 2002. Evaluation through literature data of the amount and amino acid composition of basal endogenous crude protein at the terminal ileum of pigs. Anim. Feed Sci. Technol. 98, 49-60.

Jensen M.S., S.K. Jensen and K. Jakobsen, 1997. Development of digestive enzymes in pigs with emphasis on lipolytic activity in the stomach and pancreas. J. Anim. Sci. 75, 437-445.

McCann M.E.E., J.D.G. McEvoy and K.J. McCracken, 2006. Factors affecting digestibility of barley-based diets for growing pigs. Livest. Sci. 102, 51-59.

SAS. 1996. SAS/STAT Change and Enhancements through Release 6.11. SAS Inst. Inc. Cary, NC..

Schulze H., P. Vanleeuwen, M.W.A. Verstegen, J. Huisman, W.B. Souffrant and F. Ahrens, 1994. Effect of level of dietary neutral detergent fiber on ileal apparent digestibility and ileal nitrogen losses in pigs. J. Anim. Sci. 72, 2362-2368.

Effect of body composition on diet selection in finishing pigs

M. Azain, S. A. Meers, T. Dean Pringle and Rick D. Jones
University of Georgia, Athens, GA, USA

Introduction

Animals will regulate intake of energy and other nutrients, including protein and essential AA, if they are given a choice between diets (Henry, 1987; Webster, 1993; Roberts and Azain, 1997). The present study was designed to test the hypothesis that pigs of similar age and weight, but differing in body composition, would select dietary protein based on their individual needs. It was assumed that pigs with greater 10th rib fat thickness would have higher rates of lipid accretion and lower rates of protein accretion than leaner pigs of the same BW. Furthermore, pigs with greater amounts of body fat, and thus a lower percent lean mass, would have a lower protein requirement and self-select a diet with less protein than leaner pigs.

Materials and methods

The first experiment was conducted with gilts and was designed in a 2×2 factorial arrangement with main effects of body fat (phenotype, Lean vs. Fat) and diet (Single vs. Choice). Crossbred gilts (n=32) with an initial wt. of 79 kg were selected from a larger pool of animals based on BW and 10th rib fat thickness. Selected gilts were classified into one of 2 phenotypes (Fat, 2.24 cm; or Lean, 1.51 cm). In experiment 2, castrated male terminal crossbred pigs (n=32 in two replicate trials) with an initial BW of 69.2±1.2 kg were selected from a larger pool of animals and sorted into Fat (2.33 cm) and Lean (1.75 cm) phenotypes.

Pigs were individually penned (0.86×1.80 m), were randomly assigned to one of two feeding programs: Single or Choice. Pigs in the Choice feeding program were allowed to self-select between diets with high and low protein, while those in the Single program were fed a 50:50 blend of the 2 diets. Pelleted diets were: (1) corn based diet supplemented with crystalline AA (8.5% CP, 0.58% lysine) and (2) a high protein, corn-soybean meal based diet (22.7% CP, 1.275% lysine). During wk 1 (d0-7), all pigs were fed a 50/50 mix of diets 1 and 2. Beginning on d7, half of the pigs in each phenotype were assigned to the Choice feeding program which continued until d 28.

Differences in BW and total feed intake in each experiment were analyzed using the PROC GLM procedure in SAS (2004) with main effects of phenotype, feeding program, trial, and their interactions included in the model. Intake of the HP and LP diets in pigs in the Choice treatment, were also analyzed as repeated measures using the PROC Mixed procedure in SAS (2004). The model included fixed effects of phenotype, day, and day x phenotype and the random effects of pig within phenotype and trial.

Results and discussion

Experiment 1

In the gilts, initial (79.4±1.0 kg), average daily gain (ADG, 1.06 kg/d) or total feed intake (ADFI 2.64 kg/d) were not different across treatment groups for the period from d 7-28. Initial differences in fat thickness were maintained (d 28: Fat, 3.02 cm vs. Lean, 2.42 cm, $P<0.01$). There was no effect of diet on fat thickness or on the calculated change in fat depth. Although there was no difference in total ADFI between phenotypes, the pattern of selection differed between Lean and Fat pigs. From d 7-28, Fat/Choice pigs consumed 69.0% of their feed as the LP diet and 31.0% from the HP. In contrast, Lean/Choice pigs consumed 35.6% LP feed and 64.4% HP feed ($P<0.01$). Over the period of 7-28 d and as result of differences in selection patterns, Lean pigs selected a 16.88% CP diet whereas Fat pigs selected a 12.82% CP diet ($P<0.01$).

Experiment 2

In the barrows, initial BW (69.2±1.2 kg), ADG (1.27 kg/d) ADFI (3.09 kg/d) were not different across treatment groups. The initial differences in fat thickness were maintained throughout the study (d 28: Fat, 2.82 cm vs. lean, 2.42 cm, $P<0.01$). However, it should be noted that the magnitude of the difference in fat thickness in the barrows (8-14%) was less than that in the gilts (20-28%). Although the differences in selection were not statistically significant, Lean barrows selected a 17.4% CP diet, whereas Fat barrows selected a 15.9% CP diet ($P>0.20$).

The objective of the present work was to determine if pigs of similar genetic background and age selected differently and if the pattern of selection was related to composition of gain. For this work, it was assumed that pigs of similar weight, but differing in 10th rib fat thickness would represent differences in composition of gain that may result in diet selection differences. The underlying assumption was that fatter pigs, that is animals with greater amounts of body fat and thus a lower percent lean mass, would self-select a diet with less protein than lean pigs (those with higher amounts of lean mass and a lower percent of body fat). This was clearly the case in experiment 1 with the female pigs where there were significant differences in diet selection between animals classified as fat or lean, but was not seen in the barrows in experiment 2. The differences between pigs classified as fat and lean in the barrow study were less obvious than those in the gilts. This may have contributed to the lack of a significant difference. The implication of this work is that diet selection can be used to allow pigs to more closely meet their individual nutrient requirements.

References

Henry, Y., 1987. Self-selection by growing pigs of diets differing in lysine content. J. Anim. Sci. 65, 1257-1265.

Roberts, T.J. and M.J. Azain, 1997. Somatotropin treatment reduces energy intake without altering protein intake in pigs selecting between high and low protein diets. J. Nutr. 127, 2047-2053.

SAS Institute, 2004. SAS / STAT User's Guide. Release 9.1. Cary, NC: SAS Institute.

Webster, A.J., 1993. Energy partitioning, tissue growth and appetite control. Proc. Nutr. Soc. 52, 69-76.

Milk intake and protein and energy deposition in suckling Iberian piglets

M.A. Aguinaga, F. Gómez-Carballar, A.I. Navarro, R. Nieto and J.F. Aguilera
Instituto de Nutrición Animal, Estación Experimental del Zaidín (CSIC), Cno. del Jueves s/n, 18100 Armilla, Granada, Spain

Introduction

Under similar commercial conditions growth rate of the Iberian suckling piglet is lower than those observed for conventional or lean genotypes. The Iberian sow's milk yield and composition could explain such a low performance. However, a low biological utilization of protein and energy in the milk by the sow-reared piglet cannot be discarded. Within a programme aiming at improving litter growth before weaning, an experiment was made to relate the supply of nutrients from milk with the retention of protein and energy in the body of the Iberian suckling piglet.

Materials and methods

The experimental protocol was approved by the Bioethical Committee of CSIC. Two trial replicates (TR) were made, each with eight pure-bred Iberian sows of similar genetic background, age and live weight. Four sows were used for milk production, litter growth and piglet nutrient balance measurements, and the other four sows, for milk sample collection. The environmental temperature of the farrowing room was 27±2 °C in the first replicate and 22±2 °C, in the second. In both replicates deliveries took place within a 2-day period. Shortly after birth, litter size was equalized to seven piglets by cross-fostering when necessary. Milk intake was determined weekly, from day 5 of lactation, by the weigh-suckle-weigh technique. Eight determinations at 75 minutes intervals were made on each day of measurement and the total amount of milk was extrapolated to a 24-hour period. Corrections were made for piglets' weight losses during suckling due to activity metabolism, as described by Klaver *et al.* (1981), and fluids and faecal losses. Piglets were weighed individually at birth and every seven days, prior to initiating a milk intake determination. The comparative slaughter procedure was used to determine nutrient and energy retention. Within each litter, one piglet at birth and four piglets at 35 d of age were anaesthetized and subsequently bled. Body components were kept at -20 °C until analysis. Samples were analysed separately for crude protein (CP; N × 6.25), gross energy (GE) and ash. Body fat was calculated assuming that the energy content of body protein and fat was 23.8 and 39.8 kJ/g, respectively. Within replicate, average data from the initial slaughter group were used to estimate the body composition of the other 16 piglets at birth. Milk samples were collected from four sows on days 5, 12, 19, 26 and 34 *post partum*. After suckling, piglets were separated from the dam. Two hours later, the sows were injected with 10 IU oxytocin and the milk was collected from all functional glands by hand-milking. Samples were stored at -20 °C until analysed (total solids, CP (total N × 6.38), total fat, GE, ash) by standard procedures. The effect of TR was assessed by one-way ANOVA by a computer software package (StatGraphics Centurion XV, version 15.2.06, StatPoint Inc.). Linear and multiple regressions relating ME intake (MEI) and energy retained (ER) or MEI and ER as protein (ER_p) and ER as fat (ER_f) were used to estimate maintenance needs (ME_m), the net efficiency of use of ME for growth (k_g) and the partial efficiencies of utilization of ME for protein (k_p) and fat (k_f) deposition.

Results and discussion

No significant differences between the two trial replicates were observed for most of the parameters studied, so values presented are means of both replicates. Total milk yield did not differ between trials, being on average 5,163±215 g/day. This value is well below that of 6.57 kg/d for a 5-week lactation adopted by ARC (1981) for conventional sows. However, this yield results in a daily intake of 821 g milk for an average litter of 8 piglets, a figure comparable to our observations. Milk intake per piglet showed a tendency to be higher in the second trial (832 vs. 893 g/d, respectively; P=0.066). An average protein to energy ratio of 11.3 g/MJ milk GE across lactation was observed. The average BW at birth was 1.410 kg and was not influenced by TR (P=0.740). The IB piglets grew at an average rate of 168 g/d, irrespective of TR (P=0.378); therefore, at the lowest end of the range of growth rate values reported by the literature. Milk conversion ratio and growth rate per MJ milk GE intake did not differed statistically between TR, being on average 5.15 and 42.1, respectively. Mean energy cost of body gain was estimated as 28.6 kJ milk GE/g gain, a 55.4% higher than the figure of 18.4 MJ milk energy calculated as an average from several surveys (Noblet *et al.*, 1998). The overall efficiency of protein accretion was rather poor and differed between TR (0.619 and 0.571; P=0.016). Between birth and weaning after 34-d lactation, 27.4 g protein, 22.7 g fat and 1,615 kJ energy were deposited daily. ME_m was calculated as 404 kJ/kg $BW^{0.75}$; k_g as 0.584; k_p and k_f, as 0.373 and 0.696, respectively. In Iberian growing pigs given an adequate supply of ideal protein we obtained for ME_m an estimate of 422 kJ/kg $BW^{0.75}$ per d and an identical value for k_g (0.582; Nieto *et al.*, 2002). While our k_f value approaches those reported in the literature for early weaned piglets, our estimate for k_p is far below corresponding values found in these studies (0.74-0.83). In Iberian growing pigs we have also obtained very low k_p values (0.303; Nieto *et al.*, 2002), corroborating that in this breed considerable more energy is invested per unit of protein deposited, presumably as a result of a comparatively higher muscle protein turnover (Rivera-Ferre *et al.*, 2005). It is concluded that the poor efficiency of utilization of sow's milk nutrients may explain the low rate of growth of the Iberian suckling piglet in comparison with conventional or lean genotypes.

References

ARC, 1981. The Nutrient Requirements of Pigs. Agricultural Research Council. Commonwealth Agricultural Bureaux, Slough, UK.

Klaver, J., G.J.M. van Kempen, P.G.B. de Lange, M.W.A. Verstegen, and H. Boer, 1981. Milk composition and daily yield of different milk components as affected by sow condition and lactation/feeding regimen. J. Anim. Sci. 52, 1091-1097.

Nieto, R., A. Miranda, M.A. García, and J.F. Aguilera. 2002. The effect of dietary protein content and feeding level on the rate of protein deposition and energy utilization in growing IB pigs from 15 to 50 kg body weight. Br. J. Nutr. 88, 39-49.

Noblet, J., M. Etienne, and J.-Y. Dourmand. 1998. Energetic efficiency of milk production. In The lactating sow. M.W.A. Verstegen, P. J. Moughan, and J. W. Scharama, eds. Wageningen Pers, Wageningen, the Netherlands: 113.

Rivera-Ferre, M.G., J.F. Aguilera, and R. Nieto. 2005. Muscle fractional protein synthesis is higher in IB than in Landrace growing pigs fed adequate or lysine-deficient diets. J. Nutr. 135, 469-478.

High dietary leucine levels are performance depressive in diets for pigs limiting in isoleucine or valine

M.K. Wiltafsky[1], M.W. Pfaffl[2] and F.X. Roth[2]
[1]Evonik Degussa GmbH, Rodenbacher Chaussee 4, 63457 Hanau-Wolfgang, Germany
[2]Technische Universität München, Hochfeldweg 6, 85350 Freising, Germany

Introduction

The branched-chain amino acids (BCAA: Ile, Leu, Val) share the first two steps of their catabolism. The first step is a reversible transamination. The resulting branched-chain α-keto acids (BCKA) are then oxidatively decarboxylized by the branched-chain α-keto-acid dehydrogenase (BCKDH). This step is irreversible and regulates the BCAA catabolism. The BCKDH activity is highly regulated and Leu seems to have the greatest impact on its activity (Harper *et al.*, 1984). The aim of the present study was to investigate the impact of excess levels of Leu on animal performance and BCAA catabolism.

Material and methods

In each of the two experiments, 48 pigs (8-25 kg of body weight) were allotted to 6 dietary groups. Diets contained 13.5 MJ metabolizable energy and 11.0 g/kg standardized ileal digestible (SID) lysine. Treat 1 was a positive control (SID Ile 6.3 g/kg; SID Leu 11.7 g/kg; SID Val 7.3 g/kg). In the 5 experimental groups the supply of either Ile (Exp. 1; SID Ile 5.0 g/kg) or Val (Exp. 2; SID Val 6.2 g/kg) was reduced to avoid surplus. The dietary Leu content was increased by supplementation of L-Leu to reach 100, 150, 175, and 200% of the control levels (Treat 2-5). In Treat 6 the Leu and Val (Exp. 1) or Leu and Ile (Exp. 2) supply were set at 200% of the control levels. Samples of blood (Treat 1-6) and liver (Treat 2, 3, 5) were taken at the termination of the trial (after 2.5 hours of fasting). Levels of plasma amino acids, serum BCKA and liver BCKDH activity were measured (Wiltafsky *et al.*, 2009; Pailla *et al.*, 2000; Nakai *et al.*, 2000). Data were analyzed by the MIXED procedure (SAS, 2004) considering treatment and sex as fixed and litter as random effect.

Results and discussion

In Exp. 1, the decreased Ile supply showed little effect on average daily gain (ADG; Table 1) whereas in Exp. 2 ADG was significantly reduced as the Val supply decreased (Table 2). Increasing the supply of Leu from Treat 2 to 5 resulted in a linear decrease of performance in both experiments. In Treat 6, the additional oversupply of Val (Exp. 1) reduced ADG by another 10% (n.s.) whereas the additional oversupply of Ile in Exp. 2 was without effect. Average daily feed intake (ADFI) paralleled ADG. Increasing the dietary Leu level resulted in both trials in increasing levels of plasma Leu and its BCKA. The basal activity of the BCKDH increased linearly with increasing Leu supply, indicating increased BCAA catabolism. This could be confirmed by decreased levels of plasma Val and Ile and their BCKA. In contrast to the basal BCKDH activity, the total activity of the completely activated (dephosphorylated) BCKDH complex was not influenced by the fed diets.

This study demonstrated that a dietary Leu content above the requirement increases the activity of the BCAA degrading enzyme BCKDH. In consequence the catabolism of all three BCAA was increased and the animal's performance was reduced in the present study. Moreover, oversupply of Leu might increase the requirement for Ile and Val. To assure optimal supply of BCAA and for correct determination of BCAA requirement this antagonism should be taken into account.

Table 1. Results of experiment 1.

Treat	1	2	3	4	5	6	SEM	t-Test		Contrasts 2 to 5	
								1 vs. 2	5 vs. 6	Linear	Quadratic
Leu, %	100	100	150	175	200	200					
Ile, %	100	80	80	80	80	80					
Val, %	100	100	100	100	100	200					
ADFI, g/d	660	647	611	572	558	510	17	0.752	0.383	0.037	0.918
ADG, g/d	478	468	439	408	397	358	12	0.742	0.331	0.028	0.916
G:F, g/kg	724	726	722	716	715	706	5	0.881	0.602	0.352	0.990
Serum BCKA, μmol/l											
Keto-Ile	26.9	5.4	5.9	5.8	5.0	4.5	1.3	<0.001	0.686	0.914	0.611
Keto-Leu	28.6	33.4	43.7	44.4	48.5	45.0	1.6	0.171	0.335	<0.001	0.473
Keto-Val	8.6	11.6	5.1	4.1	3.3	19.3	0.9	0.054	<0.001	<0.001	0.006
BCKDH activity, nmol NADH/g/min											
Basal	-	26	46	-	70	-	7	-	-	0.002	0.186
Total	-	204	205	-	249	-	10	-	-	0.170	0.170

Table 2. Results of experiment 2.

Treat	1	2	3	4	5	6	SEM	t-Test		Contrasts 2 to 5	
								1 vs. 2	5 vs. 6	Linear	Quadratic
Leu, %	100	100	150	175	200	200					
Ile, %	100	100	100	100	100	200					
Val, %	100	85	85	85	85	85					
ADFI, g/d	783	563	441	414	394	383	26	0.005	0.708	<0.001	0.172
ADG, g/d	563	396	296	275	247	252	20	0.001	0.812	<0.001	0.219
G:F, g/kg	722	706	670	662	630	651	8	0.339	0.306	0.007	0.821
Serum BCKA, μmol/l											
Keto-Ile	40.1	33.2	16.8	12.7	11.2	26.8	1.9	0.238	<0.001	<0.001	<0.001
Keto-Leu	33.0	29.0	41.0	49.4	43.3	41.6	1.8	0.593	0.791	0.002	0.065
Keto-Val	12.2	0.8	1.9	1.4	1.9	1.5	0.7	0.002	0.376	0.042	0.339
BCKDH activity, nmol NADH/g/min											
Basal	-	14	31	-	39	-	4	-	-	0.003	0.774
Total	-	167	167	-	163	-	10	-	-	0.857	0.815

References

Harper, A.E., R.H. Miller and K.P. Block, 1984. Branched-chain amino acid metabolism. Ann. Rev. Nutr. 4, 409-454.

Nakai, N., R. Kobayashi, K.M. Popov, R.A. Harris and Y. Shimomura, 2000. Determination of branched-chain α-keto acid dehydrogenase activity state and branched-chain α-keto acid dehydrogenase Kinase activity and protein in mammalian tissues. Methods Enzymol 324, 48-62.

Pailla, K., F. Blonde-Cynober, C. Aussel, J.P. DeBandt and L. Cynober, 2000. Branched-chain keto-acids and pyruvate in blood: measurement by HPLC with flourimetric detection and changes in older subjects. Clin. Chem. 46, 848-853.

SAS, 2004. Release 9.1. SAS Inst. Inc., Cary, NC, USA.

Wiltafsky, M.K., B. Schmidtlein and F.X. Roth, 2009. Estimates of the optimum ratio of standardized ileal digestible valine to lysine for eight to twenty-five kilograms of body weight pigs. J. Anim. Sci. 87, 2544-2553.

Inadequate protein intake of pregnant gilts affects body composition and skeletal muscle properties of newborn piglets

C. Rehfeldt[1], J. Block[1], L. Lefaucheur[2], B. Stabenow[1], W. Otten[1], C.C. Metges[1] and C. Kalbe[1]
[1]Leibniz Institute for Farm Animal Biology (FBN), 18196 Dummerstorf, Germany
[2]Institut National de la Recherche Agronomique (INRA), Unité Mixte de Recherche sur les Systèmes d'Elevage, la Nutrition Animale et Humaine, Saint Gilles, France

Introduction

Intrauterine growth retardation (IUGR) in the litter-bearing pig is a known phenomenon with economic impact. IUGR manifests as low body weight at birth that in turn is incident with decreased survival, restricted postnatal growth, poorer carcass and meat quality (Gondret *et al.*, 2006; Rehfeldt *et al.*, 2008). Imbalanced maternal nutrition may be one reason of IUGR and little is known about the impact of the quality of maternal diets on the offspring. Here we present results on the effects of low and high protein intake of pregnant gilts on offspring birth weight, body composition and cellular properties of skeletal muscle and adipose tissue.

Materials and methods

An experiment with German Landrace gilts was conducted over eight temporally successive replicates. From the day prior to insemination to parturition gilts were fed 2.3 to 2.9 kg/d of corn-barley/soybean meal diets (~13.7 MJ ME/kg) with target crude protein levels of 60 (low protein, LP; n=19), 120 (adequate protein, CP; n=22), and 300 (high protein, HP; n=21) g/kg throughout gestation. After farrowing 166 piglets (regularly the heaviest, 2 of medium weight, the lightest, but >800 g, per litter) from 15 CP, 15 LP and 17 HP gilts were sacrificed at d 1 of age. Body composition was determined by dissection and chemical analysis (Rehfeldt *et al.*, 2001a). Histological and biochemical characteristics (myofibre numbers, DNA, RNA, protein contents, CK activity, myosin heavy chain (MyHC) isoform proportions) of *semitendinosus* (ST), *longissimus* (LD), and *biceps femoris* (BF) muscles were analyzed (see Lefaucheur *et al.*, 2001; Rehfeldt *et al.*, 2001b). Fat cell size and DNA content were measured in subcutaneous adipose tissue (SC). Data were analyzed by the MIXED procedure (SAS, 2006) with diet (D), sex, litter size (L1<13; L2≥13), birth weight class - BW (LW: ≤1.08 kg; MW: 1.09 to 1.45 kg; HW: ≥1.46 kg), replicate (R) and interactions as fixed factors and the gilt (D×R×L) as random factor. The Tukey post hoc test was used to separate LSMeans.

Results and discussion

Birth weight was lower both in LP and HP compared with CP piglets (Table 1) indicating that not only insufficient (e.g. Davis *et al.*, 1997), but also excessive (only shown in rodents, e.g. Daenzer *et al.*, 2002) maternal protein supply impairs intrauterine growth. Adjusting to equal birth weight by including BW as fixed factor, HP piglets exhibited the highest live weight at d 1 of age because of highest gain. Internal fat depots were not significantly affected by dietary treatment. However, HP piglets had heavier brains and throats than CP piglets, while tongue, throat, diaphragm, colon, and kidneys were heavier compared with LP piglets (data not shown). The relative kidney weights were reduced in LP piglets of low birth weight originating from large litters (0.93 vs. 1.14%; $P<0.07$ for D×L×BW) compared with CP piglets suggesting disturbed kidney development in response to low maternal nutrient plus protein supply. The proportion of skeletal muscle tissue was higher in HP than in CP and LP piglets (Table 1), which was mainly pronounced in the light BW class ($P<0.01$ for D×BW). Weight and circumference (3.95 vs. 3.64 cm; $P<0.001$) of ST muscle were greater in HP than in LP, but both were not different from CP piglets. This was associated with greater total DNA, protein, primary (P) and secondary (S) myofibre numbers at unchanged S/P ratio in ST, and CK activity (ST, LD, BF) but lower proportion of the embryonic MyHC isoform (LD) denoting advanced muscle development and higher maturity of the piglets in response to the maternal HP diet at comparable BW. While SC fat mass remained unchanged, total DNA (P=0.08) and concentration were reduced in SC fat of LP compared with CP piglets. The results suggest that both an increase (250%) and a reduction (50%) in protein supply to gilts during pregnancy retard intrauterine growth, whereas body and tissue composition of the offspring is divergently affected.

Table 1. Body composition, skeletal muscle and subcutaneous fat (SC) properties of piglets born to gilts fed adequate (CP), high (HP) or low (LP) protein levels during gestation.

Item	CP	HP	LP	SEM	P			
					D	Sex	L	BW
N total (d 0)	259	248	238					
Birth weight, g	1,405 ^{a}	1,211 ^{b}	1,186 ^{b}	45	0.001	0.001	0.01	-
N sample (d 1)	55	56	55					
Live weight, g	1,337^{ab}	1,395 ^{a}	1,287 ^{b}	30	0.04	ns	ns	<0.0001
Muscle tissue, %	42.8 ^{a}	44.2 ^{b}	43.1 ^{a}	0.34	0.01	<0.01	ns	<0.0001
ST weight, g	2.94^{ab}	3.11^{a}	2.73^{b}	0.08	<0.01	ns	ns	<0.0001
ST protein, mg	273.1^{ab}	288.8^{a}	250.5^{b}	8.6	<0.01	0.02	ns	<0.0001
ST DNA, mg	6.47^{ab}	6.79^{a}	5.96^{b}	0.17	<0.01	<0.01	ns	<0.0001
CK, IU/mg^{c}	6.22^{ab}	6.42^{a}	5.90^{b}	0.13	0.02	ns	ns	<0.0001
MyHCemb, %^{d}	7.29^{ab}	6.92^{a}	8.18^{b}	0.32	0.01	ns	ns	Ns
ST fibre number	440,725^{a}	511,292^{b}	401,467^{a}	15,513	<0.0001	0.03	ns	<0.0001
ST primary fibres	19,389^{ab}	21,725^{a}	17,298^{b}	774	0.001	ns	ns	0.0001
ST second. fibres	421,445^{a}	489,554^{b}	384,180^{a}	15,069	<0.0001	0.03	ns	<0.0001
S/P ratio	22.2	22.9	22.5	0.8	ns	0.07	ns	0.05
Loin SC fat, g	5.36	5.29	5.08	0.25	ns	0.01	ns	<0.0001
SC adipocyte area, μm^2	275	316	339	35	ns	ns	ns	<0.0001
SC DNA, µg/g	1,321^{a}	1,133^{ab}	1,077^{b}	62	0.02	ns	ns	0.01

D: diet; L: litter size group; BW: birth weight class; ^{a,b} P<0.05; ns: not significant; ^{c} ST, LD, BF; ^{d} LD.

Acknowledgements

The authors are grateful to the DFG for financial support (KA 1844/1-1, PAK 24).

References

Daenzer, M., S. Ortmann, S. Klaus and C.C. Metges, 2002. Prenatal high protein exposure decreases energy expenditure and increases adiposity in young rats. J. Nutr. 132, 142-144.

Davis, T.A, M.L Fiorotto, D.G. Burrin, W.G. Pond and H.V. Nguyen, 1997. Intrauterine growth restriction does not alter response of protein synthesis to feeding in newborn pigs. Am. J. Physiol. 272, E877-884.

Gondret, F., L. Lefaucheur, H. Juin, I. Louveau and B. Lebret, 2006. Low birth weight is associated with enlarged muscle fiber area and impaired meat tenderness of the longissimus muscle in pigs. J. Anim. Sci. 84, 93-103.

Lefaucheur, L., P. Ecolan, G. Lossec, J.C. Gabillard, G. Butler-Browne and P. Herpin, 2001. Influence of early cold exposure on myofiber maturation in pig skeletal muscle. J. Muscle Res. Cell Motil. 22, 439-452.

Rehfeldt, C., G. Kuhn, G. Nürnberg, E. Kanitz, F. Schneider, M. Beyer, K. Nürnberg and K. Ender, 2001a. Effects of exogenous somatotropin during early gestation on maternal performance, fetal growth, and compositional traits in pigs. J. Anim Sci. 79, 1789-1799.

Rehfeldt, C., G. Kuhn, J. Vanselow, R. Furbass, I. Fiedler, G. Nürnberg, A.K. Clelland, N.C. Stickland and K. Ender, 2001b. Maternal treatment with somatotropin during early gestation affects basic events of myogenesis in pigs. Cell Tissue Res. 306, 429-440.

SAS, 2006. Release 9.2 SAS Inst. Inc., Cary, NC, USA.

Conjugated linoleic acids in gestating and lactating sow diets affect offspring body composition

R. Gerritsen[1], P. Bikker[1] and A.-M. Pfeiffer[2]
[1]Schothorst Feed Research, P.O. Box 533, 8200 AM Lelystad, the Netherlands
[2]BASF, Berliner Buro, Charlottestr. 59, 1011 Berlin, Germany

Introduction

Conjugated linoleic acids (CLA) are isomers of linoleic acid with many unique biological properties compared to the common unsaturated fatty acids, e.g. linoleic acid. These properties have been extensively studied in dairy cattle, growing pigs and rodents. Only a small number of studies have been conducted in sows. Inclusion of CLA in sow diets increased the CLA content in colostrum and milk (Bontempo *et al.*, 2004). Results on the influence of CLA in sow diets on offspring growth performance and tissue deposition are scarce and variable. Harrell *et al.* (2002) did not find an effect on piglet growth performance, whilst a tendency towards growth reduction was found in the study of Poulos *et al.* (2004). As results on offspring performance are variable and effects on body composition of suckling piglets have not been reported, the aim of this study was to determine the effects of incremental levels of CLA in diets of gestating and lactating sows on offspring performance and tissue deposition.

Materials and methods

The experiment was comprised of four treatments: a control diet and three incremental levels of Luta-CLA (1%, 2% or 3%). Luta-CLA consists of conjugated vegetable fatty acids of methyl ester containing minimally 56% C18:2. Diets were fed during the last 6 weeks of gestation and throughout lactation. Up to 110 days of gestation, sows were fed gestation diets and from day 110 onwards lactation diets were fed. In total, 80 sows (Great Yorkshire x Finnish Landrace) were blocked based on parity. Sows from each block were randomly allocated to the treatments.

The experimental diets were formulated according to Dutch CVB and SFR standards for diets of gestating (20% wheat, 15% barley, 15% wheat middlings, 15% soybean hulls) and lactating sows (35% wheat, 20% barley, 10% rapeseed meal, 9% wheat middlings). In both the gestation and lactation diets, Luta-CLA was exchanged with soybean oil.

Observations included milk composition and milk production by the weigh-suckle-weigh method on days 5, 13 and 25 post-farrowing, calculated energy balance of the sow, offspring performance, body composition of piglets at birth and weaning, lipid and protein deposition in the suckling piglets. In order to determine body composition of the piglet, one newborn piglet per litter (representative in median body weight of the litter) was sacrificed within 16 h after birth and another piglet per litter was sacrificed at weaning.

Data were analyzed by ANOVA followed by a least significant difference-test using GenStat software in the following model $Y_{ijk} = \mu + parity_i + month_j + treatment_k + e_{ijk}$, where month is the month the sow entered the trial. In addition, CLA effects were tested for linearity using the following model $Yi = \beta_0 + \beta_1 \times Parity + \beta_2 \times Month + \beta_3 \times CLA_i + \beta_4 \times CLA_i^2$.

Results and discussion

Dietary CLA linearly increased birth weight of piglets (Table 1; *P*=0.02). Also piglet body gain until 14 days of age tended to increase linearly with dietary CLA content (*P*=0.06). Growth rate during the entire 4-wk suckling period, however, was not significantly affected by dietary treatments. Dietary

CLA content did not influence tissue composition at birth. Dietary CLA reduced fat deposition during the suckling period and the fat content at weaning (Table 1; $P<0.01$). The decrease in carcass fat was more than 30% compared to the control treatment. This could be caused by the lower daily milk lipid (decrease by 20-30%) and energy supply to suckling piglets or a direct effect of CLA supply via the milk. CLA may decrease lipid synthesis processes in adipose tissue, although these effects have not been consistently reported (Mersmann, 2002). Furthermore, piglet body protein and moisture at weaning linearly increased with increasing dietary CLA ($P<0.01$). Piglet growth rate tended to increase ($P=0.07$) from weaning to 10 weeks of age. This is in agreement with results of Bee (2000). We hypothesize that the lower energy intake, lipid deposition and body lipid content have resulted in compensatory growth and feed intake in the post-weaning period.

In conclusion, dietary CLA supplementation in gestating and lactating sow diets resulted in an improved energy balance of the sow, an increased birth weight, a reduced carcass fat content in piglets at weaning and a tendency to an increased growth rate during the post-weaning period until 10 weeks of age.

Table 1. Effect of CLA content in the diet on offspring performance and tissue composition at weaning.

	CLA- content				SEM
	0%	1%	2%	3%	
Piglet performance during suckling period					
Birth weight (g)	1,395[ab]	1,362[a]	1,485[b]	1,502[b]	40
Gain week 1-2 (g)	226	226	229	245	7.4
Gain week 3-4 (g)	236	237	229	229	7.9
Standardized litter size[1]	11.2	11.7	11.2	11.7	0.28
Carcass composition at weaning					
Moisture (g/kg)	646.2[a]	672.4[b]	687.4[c]	693.0[c]	4.86
Protein (g/kg)	160.2[a]	164.9[b]	170.8[c]	170.7[c]	1.30
Lipid (g/kg)	155.0[c]	123.9[b]	109.0[a]	104.6[a]	4.95

[a,b,c] Different superscripts within one row indicate significant differences between treatments ($P<0.05$).

[1] Litters were standardized before day 3 post-farrowing to 10-11 suckling piglets for primiparous sows and 11-12 piglets for multiparous sows.

References

Bee, G., 2000. Dietary conjugated linoleic acid consumption during pregnancy and lactation influences growth and tissue composition in weaned pigs. J. Nutr. 130, 2981-2989.

Bontempo, V., D. Sciannimanocp, G. Pastorelli, R. Rossi, F. Rosi and C. Corino, 2004. Dietary conjugated linoleic acid positively affects immunologic variables in lactating sows and piglets. J. Nutr. 134, 817-824.

Harrell, R.J., O. Philips, D.L. Jerome, R.D. Boyd, D.A. Dwyer and D.E. Bauman, 2002. Effects of conjugated linoleic acid on milk composition and baby pig growth in lactating sows. NC State University, Annual Swine report.

Mersmann, H.J., 2002. Mechanisms for conjugated linoleic acid-mediated reduction in fat deposition. J. Anim. Sci. 80 (E. suppl. 2), E126-134.

Poulos, S.P., M.J. Azain and G.J. Hausman, 2004. Conjugated linoleic acid (CLA) during gestation and lactation does not alter sow performance or body weight gain and adiposity in progeny. Anim. Res. 53, 275-288.

Effects of dietary protein content on milk composition of large white sows in tropical climate

B.A.N. Silva[1,2], Y. Primot[3], J. Mourot[4], J.L. Gourdine[1] and D. Renaudeau[1]
[1]INRA UR 143 Unité de Recherches Zootechniques, 97170 Petit Bourg, Guadeloupe, France
[2]IPG, Institute for Pig Genetics, P.O. Box 43, 6640 AA Beuningen, the Netherlands
[3]Ajinomoto Eurolysine S.A.S. 153, rue de Courcelles, 75817 Paris Cedex 17, France
[4]INRA, UMR 1079 Systèmes d'Elevage, Nutrition Animale et Humaine 35590 St Gilles, France

Introduction

The growth rate of nursing piglets is mainly determined by milk nutrient output. In consequence, the quantity and the composition of milk produced by sows are key factors in successful piglet production. Milk production appears to be highly variable and depends on many factors. Under tropical conditions, milk yield was reduced by at least 30 to 50% in comparison with data obtained in temperate countries (Gourdine *et al.*, 2004). This result is mainly connected to the combined negative effect of high ambient temperature and relative humidity resulting in a concomitant reduction of voluntary feed consumption and milk production. Data on milk composition obtained in tropical countries are scarce and limited to the general composition (total solids, protein, fat and ash contents) (Renaudeau *et al.*, 2003; Gourdine *et al.*, 2005). The aim of this study was to evaluate the amino acids and fatty acids compositions of sow's milk composition in relation to the stage of lactation and dietary protein content under tropical climatic conditions.

Materials and methods

Milk composition was studied in a total of eighteen Large White multiparous sows in 3 successive batches of 6 sows. The study was conducted at the INRA facilities in Guadeloupe, French West Indies (latitude 16°N, longitude 61°W) characterized as having a tropical humid climate. At farrowing, sows were allocated to two dietary protein treatments. The experimental diets were formulated using corn, wheat middlings and soybean meal to contain 17.3 or 14.3% crude protein (CP). Both diets supplied the same concentration of standardized digestible lysine (0.80 g/MJ NE). Milk samples were collected manually from all the active glands on each sow on d 7, 14, 21 and 27 of lactation. Sow performance data were submitted to an analysis of variance (Proc GLM, SAS) with diet, batch, and interaction as main effects. Milk composition data were submitted to a linear mixed model (SAS) including the fixed effect of diet, batch, and lactation stage. The random effect of the sow was added in the model to take into account of the correlations among measurements made on the same subject. These correlations were modeled using an UN covariance structure.

Results and discussion

General results are shown in Table 1. Milk production and composition were not affected by dietary treatment ($P>0.05$). The dry matter and ash contents in milk linearly increased with the advancement of lactation (17.6 to 19.9%, and 0.72 vs. 0.97%, respectively from d 7 to d 27; $P<0.01$). Lactose content increased from d 7 to d 14 (3.95 vs. 4.91; $P<0.01$) and thereafter remained constant. Fat content did not change during lactation and averaged 7.5%. Expressed in g/16 g N, the amino acid concentrations in milk protein were affected by the stage of lactation: methionine, threonine, tryptophan, valine, and alanine concentrations decreased ($P<0.05$) but glycine and glutamic acid contents increased ($P<0.05$) from d 7 to d 27. Similar results were reported by Csapó *et al.* (1996). As the AA are derived from milk proteins, changes in AA pattern during lactation reflects a change in the relative distribution of milk proteins with different AA pattern. Fatty acids composition in milk fat was not influenced ($P>0.05$) by the stage of lactation. In agreement with data previously published in the literature more than 80% of the fatty acids in sow's milk fat were palmitic (16:0),

Table 1. Effect of lactation stage on chemical, amino acid, and fatty acid composition of sow's milk protein (Least Square Means).

	Lactation day, d				STD[1]	Statistical analysis[2]
	7	14	21	27		
No. observations	18	18	18	18		
Chemical composition, %						
Dry matter	17.6^{a}	19.3^{b}	18.7^{b}	19.9^{b}	1.2	S**, B**
Ash	0.72^{a}	0.80^{b}	0.88^{c}	0.97^{d}	0.06	S**
Nitrogen	0.82^{ab}	0.80^{a}	0.83^{ab}	0.87^{b}	0.07	S*, B**
Lipids	7.12	8.00	7.07	7.65	1.06	
Lactose	3.95^{a}	4.91^{b}	4.88^{b}	4.90^{b}	0.36	S**
Total essential amino acids, g/16 g N	44.4	44.0	43.9	43.5	1.4	
Total non essential amino acids, g/16 g N	53.8	54.2	54.6	54.8	1.3	
Total saturated fatty acids, %	37.1	38.0	38.9	38.0	2.6	
Total monounsaturated fatty acids, %	38.3	36.8	37.1	35.6	4.4	
Total polyunsaturated fatty acids, %	24.8	25.4	24.4	26.7	3.2	

[1] Residual Standard Variation.

[2] From an analysis of variance with a general linear model including the effect of diet (D) and batch (B), lactation stage (S) and their interactions as fixed effects. Repeated measurements of milk chemical composition were analysed using an unstructured covariance structure with sows within batch as a subject. Statistical significance: * $P<0.05$, ** $P<0.01$.

oleic (18:1) and linoleic acids (18:2) (Csapó *et al.*, 1996; Gerfault *et al.*, 1999). Significant residual correlations were reported between performance during lactating and milk composition. Maternal BW loss during lactation was negatively correlated with the average daily feed intake (r=-0.76) and positively correlated with backfat thickness loss (r=0.55). In addition, there was a positive correlation between milk production and body reserves mobilisation (r=0.82). According to our findings, changes in milk production can affect the AA composition of milk proteins.

References

Csapó, J., T.G. Martin, Z.S. Csapó-kiss and Z. H¦zas, 1996. Protein, fats, vitamin and mineral concentrations in porcine colostrum and milk from parturition to 60 days. Int. Dairy J. 6, 881-902.

Gerfault, V., J. Mourot, M. Etienne and A. Mounier, 1999. Influence of the source of lipids added to the sow pregnancy diet on reproductive performance and on body composition of piglets at birth. [French]. Journées de la Recherche Porcine en France 31, 191-197.

Gourdine, J.L., J.P. Bidanel, J. Noblet and D., Renaudeau, 2005. Effects of breed and season on performance of lactating sows in a tropical humid climate. J. Anim. Sci. 84, 360-369.

Gourdine, J.L., D. Renaudeau, J. Noblet and J.P. Bidanel, 2004. Effects of season and parity on performance of lactating sows in a tropical climate. Anim. Sci. 79, 273-282.

Renaudeau, D., C. Anaïs and J. Noblet, 2003. Effects of dietary fiber on performance of multiparous lactating sows in a tropical climate. J. Anim. Sci. 81, 717-725.

Energy and nutrient retention in the weaned Iberian piglet

J.A. Conde-Aguilera, M.A. Aguinaga, J.F. Aguilera and R. Nieto
Instituto de Nutrición Animal (IFNA), Estación Experimental del Zaidín (CSIC), Camino del jueves s/n, 18100 Armilla, Granada, Spain

Introduction

Energy and protein requirements of the growing and fattening Iberian pig, a lipogenic, slow-growing porcine breed, differ markedly (lower ME_m requirements and dietary protein levels) from those of conventional modern pig genotypes, as we have previously shown (Nieto *et al.*, 2002; Barea *et al.*, 2007). Nevertheless, the optimum dietary protein/energy ratio for maximum protein deposition in the young Iberian piglet remains unknown. Nutrient requirements during this period of growth might differ from those determined for older pigs as the capacity for lean tissue deposition rapidly decreases as the animal ages. Therefore, the objective of the present work was to define how the young Iberian piglet utilizes the energy and protein of the diet for the maintenance of body functions and tissue growth. This information is essential to design experimental diets with optimum protein and energy concentration for maximizing lean growth and minimizing N environmental losses during this growing stage.

Materials and methods

A comparative slaughter experiment was performed with forty-eight castrated male Iberian piglets from shortly after weaning (9.9±0.09 kg) up to 25.1±0.09 kg BW. Six additional piglets were slaughtered to estimate initial body composition. The treatments followed a factorial design with 4 dietary protein concentrations (CP, 201, 176, 149 and 123 g CP/kg DM) and 2 feeding levels (FL, 0.95 and 0.70 × *ad libitum*) with six piglets per combination of treatments. Diets were isoenergetic (14.6 kJ ME/g DM) and the amino acid (ideal) profile was maintained constant by diluting the more concentrated diet with a protein-free mixture. Pigs were placed in individual pens (27±1.5 °C) and randomly allocated to experimental treatments. They were fed in two-daily equal meals and water was freely available. BW was recorded weekly. Digestibility trials were performed at approximately 13 kg BW. At 25 kg BW piglets were slaughtered. Four components were obtained for each pig (carcass; head, feet and tail; viscera and blood) which were kept at -20 °C until analysis. The right half of the carcass and the rest of components were separately ground, homogenized and sub-samples taken for freeze-drying and subsequent analysis (crude protein (N × 6.25), GE and ash). Body fat was calculated assuming the energy content of body protein and fat as 23.9 and 39.8 kJ/g. The experimental protocol was approved by the Bioethical Committee of the CSIC.

Treatment effects were assessed by two-way ANOVA according to a randomized design with CP, FL as fixed effects, including their interactions, by a computer software package (StatGraphics Centurion XV, version 15.2.06, StatPoint Inc.). Piglet was considered as the experimental unit. Linear and multiple regressions approaches were used to estimate ME_m requirements, net ME efficiencies for energy deposition and its partition as protein and fat.

Results and discussion

Daily feed intake did not differ between CP treatments. Average values were 801 and 602 g DM/d for animals fed 0.95 and 0.70 × *ad libitum*. Animal performance was enhanced (P<0.001) in piglets fed diets containing 201 and 176 g CP/kg DM (P<0.001) with no significant differences between these two CP levels. The highest values for daily gain (g/d), gain:feed (g/g) and gain:ME intake (g/MJ) were obtained in piglets fed 201 and 176 g CP/kg DM at 0.95 × *ad libitum* (on average, 413 g/d; 0.518 and 35.3 g/MJ, respectively). These values are considerably lower than those observed in modern porcine genotypes (Le Bellego *et al.,* 2002). Body retention of nutrients was affected significantly by CP and FL ($P<0.001$, Table 1). Protein deposition was higher in diets containing 201 and 176 g CP/kg DM ($P<0.001$). Fat accretion increased as CP decreased, and mineral and water paralleled the pattern of protein retention ($P<0.001$). Energy and nutrient retention were higher in piglets fed at 0.95 × *ad libitum* ($P<0.001$). CP × FL interactions were not significant.

N retained/N intake (g/g) decreased from 0.413 to 0.370 as CP increased ($P<0.05$) and was not affected by FL. Protein retention / ME intake (g/MJ) increased in diets containing 201 and 176 g CP/kg DM ($P<0.001$) and averaged 4.92 g, a value considerably higher than that found for Iberian pigs growing from 15 to 50 kg BW (Nieto *et al.,* 2002). Highly significant linear and multiple regressions were established relating ME intake and total energy retained (ER), or ME intake and ER as protein and as fat, respectively. ME_m requirements were estimated from this approach as 391 $kJ/kg^{0.75}d$, which is below to those established for conventional pigs. Our estimations of the efficiency of utilization of ME for growth (k_g) and of the partial efficiencies of ME utilization for protein (k_p) and fat deposition (k_f) (0.53; 0.27 and 0.61, respectively) are clearly lower than the values reported for conventional pigs.

This work underlines differential metabolic characteristics in the Iberian piglet and suggests the convenience of formulating specific practical diets to maximize lean growth. According to the present results, a diet containing 176 g ideal CP/kg DM (12.3 g CP/MJ ME or 9.22 g digestible CP/MJ ME) provides a balanced protein/energy ratio in the post-weaning stage.

Table 1. Effects of protein and energy intake on nutrient and energy deposition in Iberian piglets growing from 10 to 25 kg BW.

	CP (g/kg DM)				SEM	FL (x *ad libitum*)			*P*-value	
	201	176	149	123		0.95	0.70	SEM	CP	FL
Protein (N× 6.25, g/d)	51.6^a	48.1^a	41.2^b	35.8^c	1.30	50.7	37.7	0.9	***	***
Fat (g/d)	58.4^c	62.2^{bc}	71.8^{ab}	79.0^a	2.99	82.3	53.4	2.11	***	***
Ash (g/d)	10.2^a	10.1^{ab}	8.6^{bc}	7.9^c	0.41	10.1	8.3	0.29	***	***
Water (g/d)	207^a	203^a	175^b	156^b	5.2	221	149	3.7	***	***
Energy (MJ/d)	3.6	3.6	3.8	4.0	0.13	4.5	3.0	0.09	0.068	***

*** $P<0.001$.

References

Barea, R., R. Nieto, and J.F. Aguilera, 2007. Effects of the dietary protein content and the feeding level on protein and energy metabolism in Iberian pigs growing from 50 to 100 kg body weight. Animal 1, 357-365.

Le Bellego, L. and J. Noblet, 2002. Performance and utilization of dietary energy and amino acids in piglets fed low protein diets. Livest. Prod. Sci. 76, 45-48.

Nieto, R., A. Miranda, M.A. García and J.F. Aguilera, 2002. The effect of dietary protein content and feeding level on the rate of protein deposition and energy utilization in growing Iberian pigs from 15 to 50 kg body weight. Br. J. Nutr. 88, 39-49.

Iberian vs. Landrace × Large White pigs: nutrient utilization, animal performance and relative organ weights at two stages of growth

R. Barea, R. Nieto and J.F. Aguilera
Instituto de Nutrición Animal (IFNA), Estación Experimental del Zaidín (CSIC), Camino del Jueves s/n, 18100 Armilla, Granada, Spain

Introduction

The effects of pig genotype on total-tract digestibility (TTD) of nutrients and energy have been widely reported in literature; nevertheless, there is controversial information on the digestive capacity of indigenous pig breeds as compared with lean-type pigs (Freire *et al.*, 2003). The proportional weights of viscera seem to be affected by genotype, decreasing in lean-type pigs as compared with indigenous genotypes (Noblet *et al.*, 1999). This fact might have consequences in nutrient absorption and overall efficiency of energy use. The strategy of the current study was to test the effects of pig genotype on growth rate, efficiency of nutrient utilization, nitrogen retention and relative organ weights at growing and finishing stages of growth.

Materials and methods

A total of 19 Iberian (IB) and 19 Landrace × Large White (LD) pigs were used. Three pigs per genotype were slaughtered at the start of the experiment (15 kg bodyweight, BW). The remaining pigs were fed one of two diets differing in crude protein (CP) content (12 or 17% of dry matter) using a pair-fed procedure (8 pigs per treatment and breed). The feeding level was restricted at 0.8× *ad libitum* of the IB pigs, which have higher intake capacity. Four pigs per treatment and genotype were slaughtered at 50 kg BW, and the remaining pigs were kept on the trial until slaughter at 120 kg BW. Nutrient digestibility and N balance trials were performed at 30 and 80 kg BW. After slaughter, the gastrointestinal tract was carefully emptied and weighed as well as the rest of viscera organs. Data were subjected to analysis of variance using the GLM procedure of SAS in a factorial arrangement that included 2 genotypes, 2 levels of CP and their interaction. The effects of BW at slaughter on proportional weights of viscera were also tested for each genotype. Experimental procedures were approved by the Bioethical Committee of the CSIC.

Results and discussion

Daily gain was significantly higher in LD than in IB pigs (741 vs. 611 g/d, respectively) and, thus, the IB pigs presented on average a 30% reduction in gain-to-feed ratio when compared with LD. This lower performance was in agreement with the lower capacity for growth of obese-type compared with leaner pigs. Nitrogen TTD was significantly higher for LD pigs at 30 kg BW, whereas at 80 kg BW we observed higher values for the digestibility of DM, organic matter and energy in IB than in LD pigs (Figure 1). Both N retention and efficiency of N retention was higher in LD pigs (+30% as mean value), irrespective of the experimental period studied. The higher digestive capacity of IB pigs at 80 kg BW differs from the findings of Morales *et al.* (2002) who compared finishing IB vs. Landrace pigs. At 30 kg BW, pigs fed the higher dietary CP content presented higher DM, organic matter, N and energy digestibilities and higher N retained (g/d) ($P<0.01$).

Only the small intestine weight (as related to empty BW) was significantly higher in LD than in IB pigs; however, this parameter was not necessary related to a better overall nutrient absorption because of the higher TTD coefficients observed in IB pigs at 80 kg BW. Increasing BW resulted in a dramatic decrease of the proportional weight of the GIT in both LD and IB pigs from 15 to 120 kg BW.

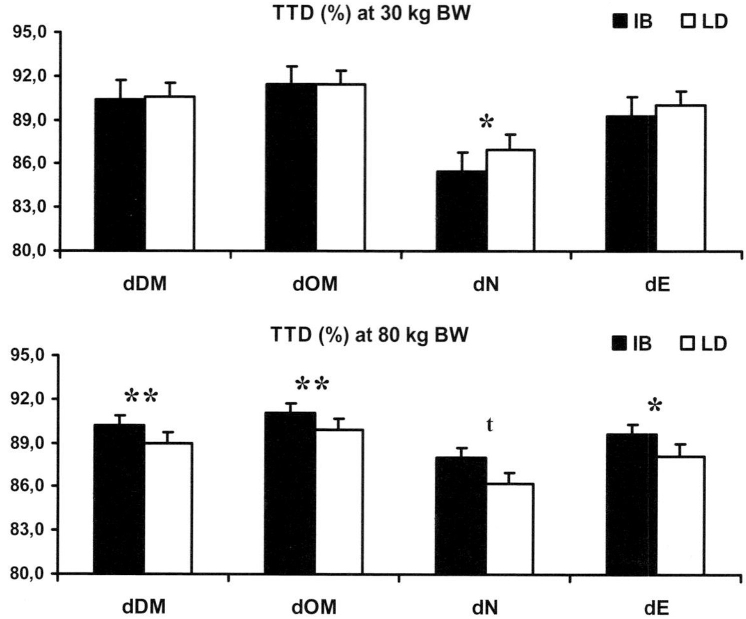

Figure 1. Total-tract digestibility coefficients (TTD, %) for dry matter (dDM), organic matter (dOM), N (dN) and energy (dE) of Iberian (IB) and Landrace × Large White (LD) pigs at 30 and 80 kg BW. Differences between breeds: ** P<*0.01;* * P<*0.05; t* P<*0.10.*

Our results show important differences that can affect protein and energy metabolism in IB pigs as compared with modern genotypes. However, complementary research is required to better understand physiological and biochemical mechanisms responsible for these variations and to unravel differences in growth potential between these pig types.

References

Freire, J.P.B., R.I.M., Dias, L.F. Cunha, and A. Aumaitre, 2003. The effect of genotype and dietary fibre level on the caecal bacterial enzyme activity of young piglets: digestive consequences. Anim. Feed Sci. Technol. 106, 119-130.

Morales, J., J.F. Pérez, M.D. Baucells, J. Mourot and J. Gasa, 2002. Comparative digestibility and lipogenic activity in Landrace and Iberian finishing pigs fed *ad libitum* corn- and corn-sorghum-acorn-based diets. Liv. Prod. Sci. 77, 195-205.

Noblet, J., C. Karage, S. Dubois and J. van Milgen, 1999. Metabolic utilization of energy and maintenance requirements in growing pigs: effects of sex and genotype. J. Anim. Sci. 77, 1208-1216.

Foetal protein restriction of male mink (*Neovison vison*) kits leads to lower protein oxidation and lower abundance of hepatic fructose-1,6-bisphosphatase mRNA during post-weaning growth

C.F. Matthiesen[1], D. Blache[2], P.D. Thomsen[1] and A.-H. Tauson[1]
[1]Department of Basic Animal and Veterinary Sciences, Faculty of Life Sciences, University of Copenhagen, Denmark
[2]Faculty of Natural and Agricultural Sciences, School of Animal Biology, University of Western Australia, Australia

Introduction

The importance of adequate nutrition during foetal life for lifetime health is increasingly recognized, and related research has developed rapidly in recent years. Several animal models (e.g rats and sheep) and approaches, one being maternal protein restriction, have been used (McMillen and Robinson, 2005). Due to the high protein requirement, this might be of particular importance among strict carnivore species e.g. mink and cat. A low protein allowance to mink during late-gestation have led to lower abundance of some key hepatic enzymes in the foetuses and lower kit birth weight (Matthiesen *et al.*, 2010). Our objectives were to investigate if low protein during foetal life led to adaptive changes in the protein oxidation and to better nutrient utilization during post-weaning growth. Further, to study if changes in the mRNA abundance of key hepatic gluconeogenic and glycolytic enzymes detected in protein restricted foetal hepatic tissue were detectable during post-weaning growth.

Materials and methods

Thirty two male mink kits were used, out of which 16 (FLP1) were exposed to foetal life low protein allowance by feeding their mothers a low protein diet the last 16.3±1.8 days of pregnancy, whereas the dams of the remaining 16 kits (FAP1) had been adequately fed, according to NRC (1982), during gestation. The kits were divided into two feeding groups during post-weaning growth, each comprising 8 FLP1 and 8 FAP1. One group was given an adequate level of protein (AP; 32% of metabolizeble energy (ME) from protein) and the other was given an insufficient level of protein (LP; 18% of ME from protein) during a period of three weeks, starting at weaning when the kits were 7 weeks old. All animals were fed *ad libitum* and feed intake was recorded. Respiration and balance experiments were performed by means of indirect calorimetry in an open-air circulation system. The male kits were euthanized at the end of the experiment for tissue collection for determination of gene expression of key hepatic gluconeogenic and glycolytic enzymes. All data were statistically analysed using the MIXED procedure in SAS (Littell *et al.*, 2006) according to the following model: $Y_{ijk} = \mu + \alpha_i + \beta_j + (\alpha\beta)_{ij} + \varepsilon_{ijk}$, Y_{ijk} is the Y_{ijk}*th* observation, μ is the general mean, α_i is the fixed effect of foetal protein allowance (FAP1, FLP1), and β_j is the fixed effect of postnatal dietary treatment (AP, LP); $(\alpha\beta)_{ij}$ is the interaction between the fixed effects and ε_{ijk} is the residual error $\sim N(0,\sigma_i^2)$. The normality of the residuals was tested. The results are presented as least square means (LSmeans), whereas the variance of each model is presented as the square root of the residuals (RR). Pairwise comparisons of LSmeans were made using the PDIFF option and differences were denoted as significant if $P<0.05$.

Results and discussion

Metabolic traits and post-weaning growth were mainly affected by post-weaning dietary treatment (Table 1). A tendency towards higher heat production among the foetal protein restricted than adequate treated male kits indicated a reduced metabolic efficiency which is similar to findings among foetal restricted lambs (Husted *et al.*, 2007). However, there was also a tendency for interaction between maternal treatment and post-weaning diet. Body composition data revealed no differences between treatments. The lower protein allowance during foetal life led to a significantly (*P*=0.006) lower protein oxidation suggesting an expression of metabolic flexibility and indicating an ability to conserve nitrogen due to foetal life experience. Of the hepatic traits measured the lower abundance of hepatic fructose-1,6-bisphosphatase mRNA in 9.5 weeks old kits confirm that adaptation in gene expression occurring during foetal life still is detectable during post-weaning growth. The consequences of this difference are unknown, though it might affect glucose homeostasis. In conclusion, effects of foetal life nutrient supply were still detectable post-weaning.

Table 1. The energy metabolism measured as metabolizable energy (ME), heat production (HE), retained energy (RE), retained fat (RF) and body weight as well as the oxidation of protein (OXP), fat (OXF), and carbohydrates (OXCHO) in male mink kits exposed to adequate (FAP1) or low (FLP1) protein levels during foetal life combined with adequate (AP) or low (LP) protein allowance from 7 to 9.5 weeks of age.

n=32	Treatment				RR	*P*-value		
	FAP1-AP	FAP1-LP	FLP1-AP	FLP1-LP		M	P	M × P
ME [kJ $kg^{-0.75}$/day]	1,676[a]	1,064[b]	1,554[a]	1,094[b]	110	NS	<0.001	NS
HE [kJ $kg^{-0.75}$/day]	770[a]	675[b]	776[a]	755[a]	38	0.07	0.02	0.10
RE [kJ $kg^{-0.75}$/day]	905[a]	340[b]	749[c]	338[b]	87	0.12	<0.001	0.12
RF [g $kg^{-0.75}$/day]	13.7[a]	5.9[b]	9.7[c]	5.4[b]	2.0	0.07	0.002	0.14
Substrate oxidation [% of HE]								
OXP	32[a]	15[b]	25[c]	12[b]	3	0.006	<0.001	NS
OXF	39	30	42	31	11	NS	NS	NS
OXCHO	29[a]	55[b]	33[a]	57[b]	10	NS	<0.001	NS
Body weight [g]	865[a]	690[b]	873[a]	645[b]	56	NS	<0.001	NS

RR: Root of residuals. Values with different superscripts [a,b,c] differ significantly within treatment, maternal or postnatal diet. NS: not significant. M: Maternal diet applied during foetal life. P: postnatal diet fed from 7 to 9.5 weeks of age.

References

Husted S.M., M.O. Nielsen, M.P. Tygesen, A. Kiani and D. Blache, 2007. Programming of intermediate metabolism in young lambs affected by late gestational maternal undernourishment. Am J Physiol Endocrinol Metab. 293, E548-E557.

Littell R.C., M.A. Milliken, W.W. Stroup, R.D. Wolfinger and O. Schabenberger, 2006. SAS for mixed models, 2nd ed. Cary, NC: SAS Institute Inc.

Matthiesen C.F., D. Blache, P.D. Thomsen and A. Tauson, 2010. Effect of late gestation low protein supply to mink (*Mustela vison*) dams on reproductive performance and metabolism of dam and offspring. Arch Anim Nutr. 64, 56-76.

McMillen I.C. and J.S. Robinson, 2005. Developmental origins of the metabolic syndrome: prediction, plasticity, and programming. Physiol Rev. 85:571-633.

NRC, 1982. Nutrient requirements of mink and foxes, National Research Council. Washington (DC): National Research Council, National Academy Press.

Effect of feed intake level on the contents of digestible and metabolisable energy in diets for piglets

M. Goerke[1], M. Eklund[1], D. Jezierny[1], N. Sauer[1], H.-P. Piepho[2], U. Messerschmidt[1] and R. Mosenthin[1]
[1]Institute of Animal Nutrition, University of Hohenheim, 70593 Stuttgart, Germany
[2]Bioinformatics Unit, University of Hohenheim, 70593 Stuttgart, Germany

Introduction

In Europe, energy evaluation systems, based on digestible (DE), metabolisable (ME) or net (NE) energy values, are used to describe energy contents in feed ingredients for pig nutrition (GfE, 2006). Most energy values in feed ingredients for growing pigs were obtained in studies in which the feed intake was restricted, although *ad libitum* feeding regime represents on-farm feeding conditions. Moreover, in post weaning piglets, the voluntary feed intake is highly variable (Brooks and Tsourgiannis, 2003). There is evidence, that high levels of feed intake can reduce ileal protein and amino acid digestibility in growing pigs and piglets (Moter and Stein, 2004; Goerke *et al.*, 2009), due to a limited capacity of the gastrointestinal tract for protein and amino acid digestion and absorption (Li *et al.*, 1993). The effect of a wide range in feed intake levels on the contents of digestible and metabolisable energy in diets for piglets has not yet been determined.

Materials and methods

In total, 36 barrows (17 days old) with an average initial body weight of 5.6±0.68 kg were used in 3 consecutive experiments with 3 periods and 12 piglets each. Every period included 5 days for adaptation to the assay diets followed by 5 days for collection of faeces and urine. The piglets were kept in metabolic crates which allowed for quantitative collection of urine. Adhesive collection bags attached to the pigs' anus were used for quantitative faeces collection. The 18 different assay diets consisted mainly of corn starch, 1 of 18 different batches of soybean meal and casein. The soybean meal batches originated either from Argentina (n=6), Brazil (n=6) or the US (n=6). The gross energy content in the 18 different assay diets ranged from 17.20 to 17.41 MJ/kg (as is). The assay diets were fed at 3 graded levels of feed intake corresponding to 30, 45 and 60 g assay diet (as is)/kg average body weight of the pigs at the beginning of each experimental period.

The data were analysed using the MIXED procedure of SAS (2003), considering level of feed intake (n=3), experiment (n=3), origin of soybean meal (n=3) and the carry over effects of feed intake level (n=4) and of soybean meal origin (n=4) as fixed effects. Period (n=3) and animal (n=12) within an experiment and soybean meal batch (n=6) within a soybean meal origin were considered to be random. The difference between planned and actual feed intake level within each animal and period, and the initial bodyweight of each piglet were included as covariates in the model, and have been set equal to 0 and 5.6 kg, respectively.

Results and discussion

The total tract digestibility of dry matter (dDM) ranged from 92.0% for the 60 g/kg body weight level to 92.3% for the 30 and 45 g/kg body weight level and did not differ between the levels of feed intake (P>0.05) (Table 1). The energy digestibility in the assay diets averaged 92.0%. The contents of DE and ME in the assay diets were similar for all three levels of feed intake (P>0.05). The DE contents in the assay diets amounted to 15.88, 15.93, and 15.94 MJ/kg assay diet (as is) and the ME contents were 15.32, 15.40 and 15.42 MJ/kg assay diet (as is) for the 30, 45, and 60 g/kg body weight feed intake level, respectively. In agreement with the results of the present study, Kelly *et al.* (1991) failed to show any effect of feed intake level on energy digestibility in piglets when feeding a highly digestible assay diet with an average energy digestibility of 90%. In contrast, Ball and

Table 1. Effect of level of feed intake on total tract digestibility of dry matter (dDM) and contents of digestible energy (DE) and metabolisable energy (ME) in the assay diets[1].

	Feed intake level (g/kg body weight)			*P*-value[2]
	30	45	60	
Observations	32	31	31	
dDM (%)	92.3±0.19	92.3±0.23	92.0±0.23	0.293
DE (MJ/kg as is)	15.88±0.076	15.93±0.075	15.94±0.070	0.578
ME (MJ/kg as is)	15.32±0.084	15.40±0.082	15.42±0.074	0.299

[1] LSmeans ± SEM.

[2] *P*-value for Wald-type F-tests for treatment differences.

Aherne (1987) reported for piglets considerable lower energy digestibility values amounting to 80% for cereal-based assay diets. Digestibility values decreased by 1.6 percentage units as the level of intake increased from 80% of *ad libitum* intake to *ad libitum* intake. In conclusion, the results of the present study show, that the feed intake level does not affect the contents of DE and ME in highly digestible assay diets for piglets. Further studies with piglets are warranted to investigate the effect of a large range in feed intake levels on energy digestibility of diets containing feed ingredients with lower digestibilities such as cereals and their by-products.

References

Ball, R.O. and F.X. Aherne, 1987. Influence of dietary nutrient density, level of feed intake and weaning age on young pigs. II. Apparent nutrient digestibility and incidence and severity of diarrhea. Can. J. Anim. Sci. 67, 1105-1115.

Brooks, P.H. and C.A. Tsourgiannis, 2003. Factors affecting the voluntary feed intake of the weaned pig. In: Pluske, J.R., J. Le Dividich and M.W.A. Verstegen (eds.), Weaning the pig: Concepts and Consequences. Wageningen Academic Publishers, Wageningen, the Netherlands: 81-116.

GfE, 2006. Empfehlungen zur Energie- und Nährstoffversorgung von Schweinen. DLG Verlag, Frankfurt am Main.

Goerke, M., M. Eklund, F. Rink, N. Sauer, M. Rademacher and R. Mosenthin, 2009. Effect of feed intake level on standardized ileal digestibility of crude protein and amino acids in piglets. Proceedings of the 18th international scientific symposium on nutrition of domestic animals 'Zadravec-Erjavec days', Radenci, Slovenia: 195-203.

Kelly, D., J.A. Smyth and K.J. McCracken, 1991. Digestive development of the early-weaned pig. 2. Effect of level of food intake on digestive enzyme activity during the immediate post-weaning period. Brit. J. Nutr. 65, 181-188.

Li, S., W.C. Sauer and M.Z. Fan, 1993. The effect of dietary crude protein level on amino acid digestibility in early-weaned pigs. J. Anim. Physiol. Anim. Nutr. 70, 26-37.

Moter, V. and H.H. Stein, 2004. Effect of feed intake on endogenous losses and amino acid and energy digestibility by growing pigs. J. Anim. Sci. 82, 3518-3525.

SAS, 2003. SAS User's Guide: Statistics, Version 9. SAS Inst. Inc., Cary, NC, USA.

Impact of dietary markers on fecal microbial ecology, fecal VFA, and nutrient digestibility coefficients in finishing pigs

B.J. Kerr, T.E. Weber and C.J. Ziemer
USDA-Agricultural Research Service, Ames, IA, USA

Introduction

Utilization of markers in nutrition studies has a long history (Lloyd *et al.*, 1955; Kotb and Luckey, 1972; Van Soest *et al.*, 1983; Kavanagh *et al.*, 2001) with the most common markers including chromic oxide, ferric oxide, and titanium dioxide. Recently, Jagger *et al.* (1992) and Thompson and Wiseman (1998) suggested that titanium dioxide was one of the better markers for determination of fecal digestibility values in the pig. Although markers are commonly utilized in nutrition experiments, no data could be found on their impact on microbial ecology. The objective of the current experiment was to determine the impact of three commonly utilized digestibility markers on fecal microbial ecology and VFA, and nutrient digestibility coefficients.

Materials and methods

Gilts (48) were randomly allotted to 4 dietary treatments: control or the control diet with 0.5% Cr_2O_3, Fe_2O_3, or TiO_2. Pigs were fed diets for 33 d, prior to being weighed (BW = 115 kg) and moved to metabolism crates where they were fed their respective diets at 3% of their BW for 3 d followed by a 4 d total fecal collection period. For nutrient digestibility, feces were collected twice daily and stored at 0 °C until analysis. For microbial ecology, 50 g of feces was collected from each pig on d-38 and stored at 0 °C until analysis. Bacterial counts were determined using fluorescence *in situ* hybridization and DGGE used to profile of dominant bacteria. Fecal nucleic acids were extracted with PCR primers targeting the variable part of 16S ribosomal RNA gene. Gel image processing, band numbers per sample, and cluster analysis of DGGE banding patterns were performed using BioNumerics^{TM} software. Fecal ammonia was analyzed colorimetrically, fecal VFA using GC-SPME, DM by oven drying, C, N, and S by thermocombustion, Ca, Cr, Cu, Fe, P, S Ti, and Zn by ICP, ADF and NDF by filter-bag technology, EE by accelerated solvent extraction (petroleum ether), and GE by bomb calorimeter. With the individual pig as the experimental unit, data were subjected to ANOVA with an overall model SE and P value, means reported as LSMEANS, and preplanned contrasts provided between pigs fed each dietary marker and pigs fed the control diet.

Results and discussion

Total microbial and bacterial counts in pig feces did not differ due to any markers; however, Fe_2O_3 resulted in decreased numbers of Archaea in the feces (P<0.10, Table 1). Inclusion of Cr_2O_3 and Fe_2O_3 in diets increased bacterial diversity (P<0.05) and altered evenness of bacterial species in the feces (P<0.10, Table 1). Fecal ammonia or VFA were not affected by dietary marker inclusion, except for isocaproic acid which was higher in pigs fed the diet with no marker relative to pigs fed Cr_2O_3, Fe_2O_3, and TiO_2 (P<0.01, data not shown). Digestibilities of most nutrients were not affected by dietary treatment (Table 1). Pigs fed the diet containing Cr_2O_3 had a higher Ca, Cu, P (P<0.05), and C and P (P<0.10) digestibility compared to pigs fed no marker. Pigs fed the diet containing Fe_2O_3 had higher Ca and S digestibility compared to pigs fed no marker (P<0.10). Pigs fed TiO_2 had higher C and EE (P<0.10) and Zn (P<0.05) digestibility compared to pigs fed no marker. Overall, the data suggest that TiO_2 may be the best overall marker when considering both microbial ecology and nutrient digestibility.

Table 1. Microbial ecology and apparent total tract nutrient digestibility of pigs fed different dietary markers for 38 d.

	Dietary marker[1]				Statistics[2]				
	None	Cr	Fe	Ti	SE	M	Cr	Fe	Ti
Counts, cfu × 10^9									
Total	6.27	6.16	6.25	5.98	0.45	0.97	0.86	0.98	0.65
Bacteria	4.89	5.51	5.72	6.13	0.45	0.27	0.33	0.19	0.06
Archaea	0.69	0.74	0.58	0.68	0.06	0.08	0.47	0.07	0.83
Diversity indices									
Richness	11.25	15.50	15.75	14.42	1.34	0.08	0.03	0.02	0.10
Evenness	1.04	1.21	1.15	1.13	0.06	0.29	0.06	0.23	0.30
S-W index[3]	2.51	3.32	3.15	3.03	0.26	0.16	0.03	0.09	0.17
Apparent nutrient digestibility									
ADF	86.6	86.0	84.9	88.3	0.9	0.07	0.66	0.20	0.18
Calcium	41.0	52.4	48.0	40.8	2.7	0.01	0.01	0.08	0.96
Carbon	86.9	88.2	87.8	88.1	0.5	0.25	0.07	0.18	0.09
Copper	10.5	19.8	11.8	17.0	2.7	0.06	0.02	0.76	0.12
Dry matter	87.2	88.1	87.8	87.8	0.5	0.70	0.24	0.46	0.44
Ether extract	61.4	65.0	63.1	65.0	1.5	0.25	0.08	0.40	0.09
Gross energy	86.2	87.3	87.1	87.2	0.5	0.37	0.12	0.19	0.17
NDF	80.8	80.1	81.5	82.3	0.9	0.36	0.60	0.56	0.24
Nitrogen	84.4	85.1	85.5	85.6	0.6	0.54	0.42	0.21	0.21
Phosphorus	47.2	54.7	49.9	48.0	2.2	0.09	0.02	0.39	0.79
Sulfur	77.1	78.7	79.2	78.0	0.8	0.32	0.19	0.08	0.41
Zinc	21.8	26.3	26.3	36.7	2.8	0.01	0.28	0.27	0.01

[1] Cr, Fe, and Ti refer to Cr_2O_3, Fe_2O_3, and TiO_2, respectively. There were 12 pigs per dietary treatment with a d-33 BW of 115.1 kg (SD = 10.0).

[2] Model standard error (SE) and probability level (M), with Cr, Fe, and Ti representing the preplanned comparison of pigs fed each marker to pigs fed the control diet.

[3] Shannon-Wiener index.

References

Jagger, S., J. Wiseman, D.J.A. Cole and J. Craigon, 1992. Evaluation of inert markers for the determination of ileal and faecal apparent digestibility values in pigs. Br. J. Nutr. 68, 729-739.

Kavanagh, S., P.B. Lynch, F. O'Mara and P.J. Caffrey, 2001. A comparison of total collection and marker technique for the measurement of apparent digestibility of diets for growing pigs. Anim. Feed Sci. Tech. 89, 49-58.

Kotb, A.R. and T.D. Luckey, 1972. Markers in nutrition. Nutr. Abstr. Rev. 42, 813-845.

Lloyd, L.E., B.E. Rutherford and E.W. Crampton, 1955. A comparison of titanic oxide and chromic oxide as index materials for determining apparent digestibility. J. Nutr. 56, 265-271.

Thompson, J.E., and J. Wiseman. 1998. Comparison between titanium dioxide as an inert marker and total collection in the determination of digestible energy of diets fed to pigs. In: Proc. Br. Soc. Anim. Sci., Scarborough, UK: 157 (Abstr.).

Van Soest, P.J., P. Uden, and K.F. Wrick, 1983. Critique and evaluation of markers for use in nutrition of humans and farm and laboratory animals. Nutr. Rep. Int. 27, 17-28.

Modification of caecal fermentation by tannic acid and protein in rats

M. Barszcz, M. Taciak and J. Skomiał
The Kielanowski Institute of Animal Physiology and Nutrition, Polish Academy of Sciences, Instytucka 3, 05-110 Jabłonna, Poland

Introduction

Tannic acid (TA) is a polyphenolic compound, belonging to the class of hydrolysable tannins. Depending on the dose it can exert both antinutritional and health-promoting effects. TA can decrease the digestibility of nutrients, reduce liveweight gains and have toxic effects (Khan and Hadi, 1998). On the other hands it has antioxidant and anticarcinogenic properties (Lopes *et al.*, 1999). TA forms complexes with nutrients, especially with dietary protein, which becomes indigestible (Mueller-Harvey, 2006). Undigested protein reaches the large intestine and undergoes bacterial fermentation leading to the production of volatile fatty acids (VFA) and also many potentially toxic compounds such as ammonia. Microflora of the large intestine plays an important role in the metabolism of endo- and exogenic substances. The level of some bacterial enzymes is thought to be a factor affecting colon carcinogenesis. One of these enzymes is β-glucuronidase (βG). The aim of this study was to examine the effect of TA and protein level in the diet on fermentation processes in the caecum of rats and activity of bacterial βG.

Material and methods

Twelve groups of six ten-week-old male Wistar rats were given diets containing 10% or 18% of crude protein (CP; casein) free of TA (control diets) or supplemented with 0.25%, 0.5%, 1%, 1.5% and 2% of TA. Animals were kept in individual metabolic cages. After 21 days of the experiment rats were sacrificed and VFA concentrations in caecal digesta were determined using a HP 5890 Gas Chromatography System fitted with Supelco NUKOL column. Caecal digesta samples for βG assay were homogenized in potassium phosphate buffer and then prepared according to the method of Jenab and Thompson (1996), and analysed using spectrophotometer UNICAM UV 300. The effect of TA and protein, and their interactions were determined by two-way ANOVA according to the model: $y_{ijk}=\mu+\alpha_i+\beta_j+(\alpha\beta)_{ij}+e_{ijk}$. Differences between treatments were analysed *post hoc* by Least Significant Difference test using the SPSS 14.0 PL statistical package. The effects were considered to be significant at $P\leq0.05$.

Results and discussion

Higher CP level and TA supplementation significantly increased caecal pool of VFA and acetic acid concentration (Table 1). Differences were between control group and groups with 1%, 1.5% and 2% of TA. Acetic acid was the major metabolite and constituted from 62.1% to 69.5% of the sum of VFA. The increase of caecal VFA pool may be attributed both to positive effect on fermentation intensity and to slower absorption of these compounds. Higher CP level in the diet increased branched-chain fatty acids (BCFA) concentration in the caecum of rats. These acids derived from breakdown of branched-chain amino acids: valine, leucine and isoleucine. They are markers of proteolytic fermentation leading to production of many toxic compounds (Hughes *et al.*, 2000). Probably, higher amount of protein, that reaches the caecum and stimulates the development of proteolytic microflora, is the reason of increased concentration of valeric, isovaleric and isobutyric acids. The results indicate that 1% and 1.5% of TA additive increases concentration of BCFA, in case of diets with higher level of CP.

The activity of bacterial βG decreased significantly, when diets supplemented with TA were fed, which may point the smaller risk for colon cancer development. The lowest βG activity, 68% lower

Table 1. Indicators of caecal fermentation in rats.

Group	Experimental factor [%]		Volatile fatty acids [μM/caecum]							β-glucuronidase acitivity [nM of phenolphthalein/ g /min.]
	TA	CP	C_2	C_3	i-C_4	C_4	i-C_5	C_5	Σ	
I	0.0	10	94.3	34.6	3.1	15.8	2.1	1.9	152.0	69.2
II	0.25	10	123.0	29.9	3.6	17.8	2.9	2.7	180.1	37.0
III	0.5	10	124.6	35.3	4.1	20.2	3.6	3.4	191.2	49.7
IV	1.0	10	164.2	34.8	3.9	24.3	3.0	3.2	233.4	20.8
V	1.5	10	141.8	31.0	3.5	16.6	2.7	3.6	199.1	31.5
VI	2.0	10	136.8	35.5	3.9	15.8	3.7	2.3	198.0	32.8
VII	0.0	18	115.6	36.1	3.7	13.6	3.2	2.9	175.1	78.7
VIII	0.25	18	148.3	43.1	4.2	18.0	3.2	3.6	220.5	64.0
IX	0.5	18	126.9	39.9	4.2	15.5	4.2	3.5	194.2	28.0
X	1.0	18	158.2	42.4	4.1	15.9	3.4	3.5	227.5	26.8
XI	1.5	18	188.2	43.8	6.4	23.4	6.5	5.8	274.2	22.1
XII	2.0	18	178.2	41.3	6.8	22.5	7.0	5.5	261.3	39.7
SEM			6.0	1.5	0.2	0.9	0.3	0.2	8.0	4.0
TA effect			*	ns	**	ns	**	ns	*	**
CP effect			ns	*	**	ns	**	**	*	ns
Interaction			ns	ns	**	ns	ns	ns	ns	ns

* $P \leq 0.05$; ** $P \leq 0.01$; ns: not significant; C2: acetic acid; C3: propionic acid; i-C4: isobutyric acid; C4: butyric acid; i-C5: isovaleric acid; C5: valeric acid.

compared with the control group, was observed in groups receiving diets with 1% of TA. Observed inhibition is probably caused by the formation of tannin-enzyme complex (Aerts *et al.*, 1999).

It can be supposed that 1% or 1.5% of TA supplementation may be advantageous because of the positive effect on VFA profile and the most potent inhibition of bacterial βG activity.

References

Aerts, R.J., T.N. Barry and W.C. McNabb, 1999. Polyphenols and agriculture: beneficial effects of proanthocyanidins in forages. Agric. Ecosyst. Environ. 75, 1-12.

Hughes, R., E.A.M. Magee and S. Bingham, 2000. Protein degradation in the large intestine: relevance to colorectal cancer. Curr. Issues Intest. Microbiol. 2, 51-58.

Jenab, M. and L.U. Thompson, 1996. The influence of flaxseed and lignans on colon carcinogenesis and β-glucuronidase activity. Carcinogenesis 17. 1343-1348.

Khan, N.S. and S.M., Hadi, 1998. Structural features of tannic acid important for DNA degradation in the presence of Cu(II). Mutagenesis 13, 271-274.

Lopes, G.K.B., H.M. Schulman and M. Hermes-Lima, 1999. Polyphenol tannic acid inhibits hydroxyl radical formation from Fenton reaction by complexing ferrous ions. Biochim. Biophys. Acta 1472, 142-152.

Mueller-Harvey, I., 2006. Unravelling the conundrum of tannins in animal nutrition and health. J. Sci. Food Agic. 86, 2010-2037.

Effect of energy intake level, body weight and pig genotype on the dynamics and composition of growth and maintenance energy requirement in growing pigs between 25 and 125 kg BW

H.R. Martínez-Ramírez and C.F.M. De Lange
Centre for Nutrition Modelling, Department of Animal and Poultry Science, University of Guelph, Guelph, ON, N1G 2W1, Canada

Introduction

For a factorial estimation of energy requirements of growing pigs, information is required about the partitioning of retained energy between body protein deposition (PD) and body lipid deposition (LD), as well energetic efficiencies of using dietary nutrients for PD and LD (Schinckel and de Lange, 1996; Birket and de Lange, 2001). Maintenance energy requirements may be estimated from physical or chemical body composition (Whittemore, 1983; van Milgen *et al.*, 1998). The objective of this study was to determine the effect of energy intake level (EI) on chemical and physical body composition, as well as the dynamics of PD, LD and maintenance metabolizable energy requirements (MEm), in growing gilts of two pig genotypes (PG: purebred Yorkshire - Y, and commercial cross bred - C) between 25 and 125 kg body weight (BW), using the serial slaughter technique.

Materials and methods

At approximately 25 kg BW, 32 pigs for each PG (Y and C) were allotted to one of two EI (70 or 90% of calculated voluntary daily digestible energy intake according to NRC, 1998). For each treatment, chemical body composition (represented by the ratio between whole body lipid to whole body protein mass; BL/BP) was determined at 25 (n=2), 50 (n=3), 75 (n=4), 100 (n=3) and 125 (n=4) kg BW according to established procedures in our laboratory (Martínez-Ramírez *et al.*, 2008). Corn, wheat and soybean meal based diets were formulated to ensure that essential nutrient intake exceeded requirements for maximum PD. Retained energy (RE) was calculated as PD × 23.7 kJ/g + LD × 39.6 kJ/g. The MEm was calculated as the difference between ME intake and ME requirements for growth (PD × 43.9 kJ/g + LD × 52.8 kJ/g; Birket and de Lange, 2001). Both RE and MEm were calculated for each 25 BW range.

Results were subjected to analysis of variance using the Proc Mixed of SAS v9.2 (SAS, Inst., Cary, NC). Pigs and blocks (group of pigs upon arrival, n=2) were used as random effects. The EI, PG and their interaction were considered as fixed effects. The effect of litter was deemed not significant (P>0.10). Initial BW was used as a covariate when growth performance was evaluated. Differences among treatments means were assessed using the Tukey-Kramer Honestly Significant Difference Test.

Results and discussion

There were no interactive effects of EI and PG on growth performance, BL/BP, Ld, Pd, RE and MEm (P>0.10). Across various BW ranges, C pigs grew faster than Y pigs (P<0.05) and daily BW gain increased with EI (P<0.01). Treatments effects on feed efficiency were similar to those on daily BW gain. Throughout the experiment, BP content was higher and BL was lower for C pigs that for Y pigs (P<0.01). Body composition (BL/BP) was lower for C pigs than for Y pigs and increased with EI and BW for both PG (P<0.05). The body water to BP ratio, expresses per kg BP^0.855 (De Lange *et al.* 2003) and the body ash to BP ratio were not affected by BW, EI, and PG, indicating that LB/PB is the main determinant of variation in chemical body composition in pigs. Observed treatment effects on growth performance and BL/BP were consistent with effects on PD and LD, even though the effect of EI and PG were not significant for some of the 25 kg BW ranges. Both

PD and LD increased with EI ($P<0.05$), whereas PD was higher and LD lower for C pigs than for Y pigs ($P<0.10$).

For all BW ranges, except 25 to 50 kg BW, RE was not influenced by genotype (25 to 125 kg BW: 10.85 vs. 11.35 MJ/d for C and Y, respectively; $P>0.10$), even though the LD/PD ratio was lower in C pigs compared to Y pigs (25 to 125 kg BW: 1.29 vs. 1.64 g/d; $P<0.002$). Higher EI yielded higher RE ($P<0.05$). Daily MEm between 25 and 100 kg BW, but not between 25 and 125 kg BW, was influenced by both PG (13.2 vs. 11.7 MJ/d for C and Y, respectively; $P=0.065$) and EI (11.7 vs. 13.2 MJ/d for 70 and 90% of voluntary DEi, respectively; $P=0.058$). Across treatments, higher MEm coincided with lower BL/BP, indicating that BL/BP should be considered when estimating MEm. The latter implies that a simple ME partitioning model, relating MEm to BW, is insufficient to represent EI and PG effects on energy partitioning.

References

Birkett, S. and K. de Lange, 2001. Calibration of a nutrient flow model of energy utilization by growing pigs. Brit. J. Nutr. 86, 675-689.

De Lange, C.F.M., P.C.H. Morel, and S.H. Birkett, 2003. Modeling chemical and physical body composition of the growing pig. J. Anim. Sci. 81(E Suppl. 2), E159-E165.

Martínez-Ramírez, H.R., E. A. Jeaurond, and C. F. M. de Lange, 2008. Dynamics of body protein deposition and changes in body composition after sudden changes in amino acid intake: I. Barrows. J. Anim. Sci. 86, 2156-2167.

NRC, 1998. Nutrient Requirements of Swine. 10th rev. ed. Natl. Acad. Press, Washington, DC.

SAS Institute Inc., 2009. SAS OnlineDoc®. Version 9.2. SAS Institute Inc., Cary, N.C. USA.

Schinckel, A.P. and C.F.M. de Lange, 1996. Characterization of growth parameters needed as inputs for pig growth models. J. Anim. Sci. 74, 2021-2036.

Van Milgen, J., J.F. Bernier, Y. Lecozler, S. Dubois, and J. Noblet, 1998. Major determinants of fasting heat production and energetic cost of activity in growing pigs of different body weight and breed/castration combination. Br. J. Nutr. 79, 1-9.

Whittemore, C.T., 1983. Development of recommended energy and protein allowances for growing pigs. Agric. Syst. 11, 159-186.

Xylanase and benzoic acid in the fattening heavy pig: effects on growth performance and on nitrogen and energy balance

L. Malagutti, S. Colombini, L. Rapetti, M. Pirondini, D. Magistrelli and G. Galassi
Università degli Studi, Dipartimento di Scienze Animali, via Celoria 2, 20133 Milano, Italy

Introduction

Organic acids are largely used in pig nutrition as inhibitors of microbial bacteria. Organic acids are effective in reducing the incidence of porcine post-weaning diarrhoea and improving animal growing performances (Partanen and Mroz, 1999; Tsiloyiannis *et al.* 2001; Partanen, 2001). Benzoic acid is an organic carboxylic acid, approved in the European Union for use in fattening pig diets with an inclusion concentration of 0.5 to 1.0% as fed considering 88% DM content. This level, according to Torrallardona *et al.* (2007) can improve performance of weanling piglets. Also exogenous enzymes, such as xylanase and β-glucanase, capable to influence the utilization of dietary fermentable carbohydrates, are largely used in pig feeding in order to increase nutrient digestibility and decrease ammonia emission by optimizing microbial activity (Leek *et al.* 2007; Garry *et al.* 2007).

Aim of the study was to test benzoic acid and xylanase as supplements in the diet for the typical Italian heavy pig and study their effects on the growing performance, nitrogen and energy balance in heavy pigs, from 40 to 170 kg bodyweight (BW).

Material and methods

Eighty Landrace × Large White barrows of about 40 kg BW were randomly divided into 4 groups of 20 animals; the pigs of each group was divided into 4 pens of 5 animals each. Pigs were individually weighed to calculate the average daily gain (ADG) and feed intake was registered per each pen, in order to compute the feed conversion ratio (FCR, kg DM intake/kg weight gain). In comparison with a traditional dietary treatment (C), three treatments were tested: in diet A, the C diet was added 0.5% benzoic acid (Vevo Vitall®, DSM Nutritional Products); in diet X, C diet was added 0.02% xylanase (Ronozyme WX®, DSM Nutritional Products), whereas in diet AX both 0.5% benzoic acid and 0.02% xylanase were added. During the performance trial feeding was restricted accordingly to bodyweight (DM fed from to 9.7 to 6.1% $BW^{0.75}$) for diet C, whereas diets A, X and AX, taking into account the positive effect of xylanase and benzoic acid on digestibility, were fed at 95% in comparison with C to have similar digestible energy content. Dietary treatment C consisted of 3 diets (Phase 1: 40-90 kg BW, Phase 2: 90-130 kg BW, Phase 3: 130-160 kg BW), to meet the requirements of the animals. At the weight of approximately 135 kg, sixteen pigs, 4 for each dietary treatment, were used for the digestibility trial. The animals were housed individually in metabolism cages throughout the study period. During each test period the animals were placed individually in an open-circuit respiration chamber to measure respiratory exchange over three 24-h cycles. During the digestibility trial feeding was restricted accordingly to bodyweight (DM fed=6.7% $BW^{0.75}$) for all the dietary treatments; the diet C included 50% maize meal, 30% barley meal, 10% soybean meal, 6.5% wheat bran, 1.8% calcium carbonate. 0.5% sodium chloride, 0.5% dicalcium phosphate, 0.5% vit/minerals, 0.15% L-Lys HCl, 0.05% DL-Met. Crude Protein (CP) content was 13.9, 13.4, 13.4 and 13.5% on DM for diets C, A, X and AX, respectively; Metabolizable Energy (ME) content was 15.32, 15.45, 15.44 and 15.39 MJ/kg DM for diets C, A, X and AX, respectively. The data were analysed by the GLM procedure (SAS, 2001). For some dependent variables (ADG and FCR) data were covariated for the initial body weight. The model applied was the following: $y_{ij} = \mu + t_i + \beta (x_{ij} - x) + e_{ij}$ where: y_{ij} = dependent variable; μ = general mean; t_i = effect of the i-diet; $\beta (x_{ij} - x)$ = effect of the covariate Initial BW; e_{ij} = residual error. Data on nutrient digestibility, N balance and energy utilization were analysed using the following model: $y_{ij} = \mu + t_i + e_{ij}$.

Results and discussion

The performance trial did not show differences among treatments: for C, A, X and AX diets the ADG were 756, 740, 738 and 751 g and the FCR 3.11, 3.03, 3.05 and 3.04, respectively. Digestibilities were similar among dietary treatments: for diets C, A, X and AX nitrogen digestibility was 88.2, 89.2, 87.7 and 88.8% (P=0.82) and organic matter digestibility was 86.7, 88.3, 87.0 and 87.4% (P=0.53), respectively. No statistical differences were observed for nitrogen balance and daily energy utilization and partition (Table 1). The higher nitrogen intake observed for diet C was due to the larger content of CP of C diet. The inclusion of benzoic acid at 0.5% and xylanase at 0.02% in the feed for fattening heavy pigs did not determine statistically significant differences for digestibility of nutrients, nitrogen balance and energy utilization.

Table 1. Nitrogen balance and daily energy utilization and partition.

		C	A	X	AX	SE
Intake N (IN)	g/d	65.2[a]	63.1[b]	62.6[b]	62.9[b]	0.29
	g/kg $BW^{0.75}$	1.55[a]	1.48[b]	1.45[b]	1.46[b]	0.015
N in feaces	g/kg $BW^{0.75}$	0.18	0.16	0.18	0.16	0.015
	% IN	11.8	10.8	12.3	11.2	1.16
N in urine	g/kg $BW^{0.75}$	0.79	0.78	0.78	0.79	0.029
	% IN	51.0	52.6	53.8	54.0	1.91
N retained	g/kg $BW^{0.75}$	0.58	0.54	0.49	0.51	0.036
	% IN	37.2	36.5	33.9	34.8	2.47
Intake energy (IE)	MJ/d	54. 7	54.8	54.3	54.3	0.25
	kJ/$BW^{0.75}$	1,297	1,270	1,276	1,258	13.1
Digestible energy	kJ/$BW^{0.75}$	1,106	1,085	1,105	1,079	11.5
	% IE	85.3	85.5	86.6	85.8	0.77
Metabolizable energy	kJ/$BW^{0.75}$	1067	1048	1068	1040	11.5
	% IE	82.3	82.6	83.7	82.7	0.82
Heat production	kJ/$BW^{0.75}$	518	520	533	500	9.38
	% IE	39.9	40.9	41.8	39.7	0.61
Retained energy	kJ/$BW^{0.75}$	549	528	535	540	12.1
	% IE	42.4	41.6	41.9	42.9	1.01

[a,b] With different letter on the same row P<0.05.

References

Garry, B.P., M. Fogarty, T.P. Curran and J.V. O'Doherty, 2007. Effect of cereal type and exogenous enzyme supplementation in pig diets on odour and ammonia emissions. Livest. Sci. 109, 212-215.

Leek, A.B.G., J.J. Callan, P. Reilly, V.E Beattie and J.V. O'Doherty, 2007. Apparent component digestibility and manure ammonia emission in finishing pigs fed diets based on barley, maize or wheat prepared without or with exogenous non-starch polysaccharide enzymes. Anim. Feed Sci. Technol. 135, 86-99.

Partanen, K. and Z. Mroz, 1999. Organic acids for performance enhancement in pig diets. Nutr. Res. Rev. 12, 117-145.

Partanen, K., 2001. Organic acids – their efficacy and modes of action in pigs. In: Piva, A., Bach Knudsen, K.E., Lindberg, J.E. (eds.), Gut Environment of Pigs. Nottingham University Press, Nottingham, UK: 201-217.

SAS, 2001. Release 8.01. SAS Inst. Inc., Cary, NC, USA.

Torrallardona D., I. Badiola and Z. Mroz, 2007. Effects of benzoic acid on performance and ecology of gastrointestinal microbiota in weanling piglets. Livest. Sci. 108, 210-213.

Tsiloyiannis, V.K., S.C Kyriakis, J. Vlemmas and K. Sarris, 2001. The effect of organic acids on the control of porcine post-weaning diarrhoea. Res. Vet. Sci. 70, 287-293.

Effects of benzoic acid on nitrogen, phosphorus and energy balance in heavy pigs

G. Galassi, L. Malagutti, L. Rapetti, S. Colombini, D. Magistrelli and G.M. Crovetto
Università degli Studi, Dipartimento di Scienze Animali, via Celoria 2, 20133 Milano, Italy

Introduction

Benzoic acid is an organic carboxylic acid used in pig nutrition as antimicrobial growth promoter particularly for weanling piglets. *In vitro* studies have shown that benzoic acid has a strong antimicrobial effect (Knarreborg *et al.*, 2002), but little is known about its efficacy *in vivo*. Torrallardona *et al.* (2007) observed that the addition of benzoic acid at 0.5% to the diet of weanling pigs improves performance by influencing ileal and caecal microbiota of the piglets. The aim of the study was to test the effects of two dietary levels of benzoic acid (Vevo Vitall®, DSM Nutritional Products) on nitrogen, phosphorus and energy balance in the typical Italian heavy pig during the last phase of growth.

Material and methods

Three dietary treatments were tested in six Landrace × Large White barrows of 124±16 kg body weight (BW) on average in a repeated Latin Square design (3×3×2). The pigs were housed in individual metabolic cages to collect feaces and urine separately, in 3 collection periods of 7 days, after 7 days of adaptation. The animals were individually housed in open circuit respiration chambers to study the energy metabolism. The dietary treatments were as follows (% on DM): (1) diet C (maize meal 55%, barley meal 30%, wheat bran 10%, soybean meal 6.5%, calcium carbonate 1.8%, sodium chloride 0.5%, dicalcium phosphate 0.5%, vit/minerals 0.5%, L-Lys HCl 0.15%, DL-Met 0.05%); (2) diet B1: diet C plus 0.5% benzoic acid; (3) diet B2: diet C plus 1.0% benzoic acid. Benzoic acid was added in substitution of maize meal. DM fed was fixed at 6.8% $BW^{0.75}$. The average metabolic body weight was 37.0, 37.1 and 36.9 for pigs fed diets C, B1 and B2, respectively. The data were analysed by the GLM procedure (SAS, 2001). The model was: $Y_{ijk(t)} = \mu + S_i + A_{ij} + P_k + T_{(t)} + e_{ijk}$ where: $Y_{ijk(t)}$ = dependent variable; μ = general mean; S_i = square effect (i=1,2); A_{ij} = effect of animal within each square (j=1,3); P_k = effect of period (k=1,3); $T_{(t)}$ = effect of treatment (t=1,3); e_{ijk} = residual error.

Results and discussion

On DM, diets C, B1 and B2 resulted to have 14.1, 14.1 and 14.5% crude protein (CP), 15.38, 15.44 and 15.38 MJ ME/kg and 0.41, 0.41 and 0.38% total phosphorus, respectively.

No significant differences among treatments were registered, in terms of digestibility, for any parameter studied: 88.3, 87.9 and 88.6% for OM; 86.0, 86.6 and 86.7% for CP and 54.9, 50.5 and 56.6% for NDF for C, B1 and B2, respectively. Urinary nitrogen excretion (Table 1) was numerically lower for diets added with benzoic acid (50.3, 48.9, 45.5% of intake N, for C, B1 and B2, respectively) although differences were not statistically significant because of unexpected large variability; as a result N retention was 35.8, 37.4, 40.8% of intake N for C, B1 and B2, respectively. Concerning P balance, the effects of benzoic acid in pigs have been studied for long (Mroz *et al.*, 1998; Partanen and Mroz, 2000), however uncertainties on its action remain. For example the retention of P decreased in one study of Mroz *et al.* (1997) but it increased in a study of Sauer *et al.* (2009). In our study no differences were observed among diets and the amounts retained were 40.7, 43.1 and 42.1% of P intake for C, B1 and B2, respectively. Benzoic acid did not show any effect on energy metabolism (Table 2).

Table 1. Nitrogen and phosphorus balances.

		C	B1	B2	SE
Intake N (IN)	g/d	46.4	48.5	46.2	2.83
	g/kg $BW^{0.75}$	1.28	1.33	1.26	0.077
N in feaces	g/kg $BW^{0.75}$	0.18	0.18	0.17	0.008
	% IN	13.8	13.7	13.1	0.59
N in urine	g/kg $BW^{0.75}$	0.63	0.62	0.56	0.035
	% IN	50.4	48. 9	45.3	4.92
N retained	g/kg $BW^{0.75}$	0.47	0.53	0.53	0.079
	% IN	35.8	37.4	41.6	5.12
Intake P (IP)	g/d	8.4	8.7	7.7	0.51
	g/kg $BW^{0.75}$	0.23	0.24	0.21	0.013
P in feaces	g/kg $BW^{0.75}$	0.12	0.12	0.11	0.006
	% IP	53.5	51.7	52.4	1.54
P in urine	g/kg $BW^{0.75}$	0.012	0.011	0.011	0.001
	% IP	5.75	5.17	5.44	0.68
P retained	g/kg $BW^{0.75}$	0.09	0.09	0.11	0.009
	% IP	40.7	43.1	42.1	1.67

Table 2. Daily energy utilization and partition.

		C	B1	B2	SE
Intake energy (IE)	MJ/d	37.79	39.79	37.96	2.31
	kJ/$BW^{0.75}$	1042	1087	1032	63.5
Digestible energy	kJ/$BW^{0.75}$	892	932	890	58.8
	% IE	85.9	85.4	86.3	0.57
Metabolizable energy	kJ/$BW^{0.75}$	966	903	861	58.6
	% IE	83.3	82.6	83.4	0.76
Heat production	kJ/$BW^{0.75}$	430	441	416	13.3
	% IE	42.9	42.8	40.7	2.31
Retained energy	kJ/$BW^{0.75}$	435	462	445	52.6
	% IE	40.4	39.8	42.7	2.89

References

Knarreborg, A., N. Miquel, T. Granli, B.B Jensen, 2002. Establishment and application of an *in vitro* methodology to study the effects of organic acids on coliform and lactic acid bacteria in the proximal part of the gastrointestinal tract of piglets. Anim. Feed Sci. Technol. 99, 131-140.

Mroz, Z., W. Krasucki and E. Grela, 1997. Prevention of bacteriuria and ammonia emission by adding sodium benzoate to diets for pregnant sows. Proceed.of the 6th Biennial Conf. of the Australasian Pig Science Association. APSA. Canberra, Australia:. 294.

Mroz, Z., A.W. Jongbloed, K. Partanen, J.Th.M. Diepen,.K. van Vreman and J. Kogut, 1998. Ileal digestibility of amino acids in pig diets of different buffering capacity and with supplementary organic acids. J. Anim. Feed Sci. 7, 191-197.

Partanen, K.H. and Z. Mroz, 1999. Organic acids for performance enhancement in pig diets. Nutr. Res. Rev. 12, 117-145.

SAS, 2001. Release 8.01. SAS Inst. Inc., Cary, NC, USA.

Sauer W., M. Cervantes, J. Yanez, B. Araiza, G. Murdoch, A. Morales and R.T. Zijlstra, 2009. Effect of dietary inclusion of benzoic acid on mineral balance in growing pigs. Livest. Sci. 122, 162-168.

Torrallardona D., I. Badiola and Z. Mroz, 2007. Effects of benzoic acid on performance and ecology of gastrointestinal microbiota in weanling piglets. Livest. Sci. 108, 210-213.

Nanoparticles of silver in broiler production: effects on energy metabolism and growth performance

L. Pineda[1], A. Chwalibog[1], E. Sawosz[2], A. Hotowy[1], F. Sawosz[1], J. Elnif[1] and A. Ali[1]
[1]University of Copenhagen, Denmark
[2]Warsaw University of Life Sciences, Poland

Introduction

The application of alternative feed additives, which do not generate antibiotic-resistant microbial strains, and which may increase resistance to stress and promote growth is a scientific challenge of the present time. Over the years, research has been going on to find alternative growth promoting feed additives as a means to improve health and growth of animals. Numerous nutritional additives are already available in the market; however, more are still under development (Mellor, 2000). Recently, research focus has been directed toward the application of nanotechnology as alternative method of protecting animals (Salata, 2004).

Metallic silver in the form of nanoparticles is making a remarkable comeback as a potential antimicrobial agent (Rai *et al.*, 2009). The application of nanotechnology allows silver nanoparticles to be produced with unique biological properties and corresponding reduction in toxicity. The use of nanoparticles of silver in animal production has potential, however, there are only few investigations regarding applicability in animal nutrition (Sawosz *et al.*, 2009; Fondevilla *et al.*, 2009).

The objective of the experiment was to evaluate the effects of different levels of colloidal nanoparticles of silver on energy metabolism and growth performance of broiler chicken.

Materials and methods

A total of 54, one-day-old male broiler chicks (Ross × Ross 308) were randomly allocated into three treatments with 18 birds per group replicated 3 times. The birds were fed *ad libitum* with a standard diet for broiler chicken and with free access to drinking water containing 0, 10 and 25 ppm of hydrocolloidal nanoparticles of silver (nano-Ag) offered for four weeks. Four balance experiments were carried out at 7-14, 15-21, 22-28 and 29-35 days of age to determine nitrogen and energy balances. Each balance period included 4-day collection period and 22-hour gas exchange measurements by indirect calorimetry in open-air-circuit respiration unit.

Data from the last 2 balance periods were pooled and analyzed using one way ANOVA procedure of SAS (SAS Institute Inc., 2009), considering effect of treatment. The data from the first 2 weeks were excluded due to technical problems.

Results and discussion

There were no significant differences with regard to all the parameters measured, except for energy content in droppings and water intake. The levels of nano-Ag in the drinking water did not contribute to a significant increase nor decrease on the rate of metabolism and growth performance of broiler chicken (Table 1). Consistent with the findings of Sawosz *et al.* (2009), colloidal nano-Ag did not influence development of chicken embryos; but in contrary to the work of Fondavilla *et al.* (2009); who recorded significant improvement on the growth of weaned piglet given Ag nanoparticles as dietary additive.

Table 1. Growth performance, nitrogen and energy metabolism[1] of chickens supplied with 0, 10 and 25 ppm of nano-Ag.

Treatment	0 ppm	10 ppm	25 ppm	SEM[2]	*P*-value
Growth performance					
Body weight gain [g]	304	326	325	19.8	0.89
Dry matter intake [g]	514	564	587	45.6	0.81
Water intake [ml/d]	251	314	410	23.1	0.008
Feed conversion ratio	1.65	1.73	1.80	0.083	0.37
Energy metabolism					
Intake energy [kJ/d]	2,162	2,381	2,478	141.9	0.67
Energy in droppings [kJ/d]	218	217	318	15.6	0.003
Metabolizable energy [kJ kg/d]	1,943	2,164	2,160	149.0	0.80
Heat production [kJ/d]	1,340	1,683	1,646	132.0	0.55
Retained energy [kJ/d]	573	656	733	272.1	0.98

[1] Values are means of 6 cages, each containing 2 chickens.

[2] Pooled standard error of the mean.

Intake of ME and HE were numerically higher in nano-Ag treated groups, but not sufficient enough to cause significant variation. Similarly, retention of energy was not affected by the treatments and the difference was not significant between groups.

The result implies that nanoparticles of silver did not affect energy metabolism and did not stimulate nor depress growth of chicken. However, it has to be kept in mind that the present results included only two weeks of measurements and so far may have only preliminary character.

References

Fondevila, M., R. Herrer, M.C. Casallas, L. Abecia and J.J. Ducha, 2009. Silver nanoparticles as potential antimicrobial additive for weaned pigs. Anim. Feed Sci. and Tech. 150, 259-269.

Mellor, S., 2000. Alternatives to antibiotics. All About Feed.Net. [online]. Available at http://www.allaboutfeed.net/article-database/five-largest-companies-id17.html (accessed 05 April 2010).

Rai, M., A. Yadav and A. Gade, 2009. Silver nanoparticles as a new generation of antimicrobials. Biotechnology Advances 27:76-83

Salata, O.V. 2004. Application of nanoparticles in biology and medicine. J. Nanotechnology. 2(1), 6pp.

Sawosz, E., M. Grodzik, M. Zielinska, T. Niemiec, B. Olszanska and A. Chwalibog, 2009. Nanoparticles of silver do not affect growth, development and DNA oxidative damage in chicken embryos. Arch. Geflugelk. 73: 208-213.

SAS Institute Inc., 2009. SAS® Procedure Guide. Version 9.2. SAS Institute Inc., Cary, NC, USA.

Energy metabolism in young mink kits (*Neovison vison*) affected by protein and carbohydrate level in the diet

A.L. Frydendahl Hellwing, N.E. Hansen and A.-H. Tauson
Department of Basic Animal and Veterinary Sciences, Faculty of Life Sciences, University of Copenhagen, Groennegaardsvej 3, 1870 Frederiksberg C, Denmark

Introduction

The mink is a strict carnivore and mink diets usually have a high content of protein. The energy metabolism in young minks in the transition period from milk to solid food is not investigated in detail, and the protein requirement is poorly defined. The substrate oxidation can give useful information about the relative contribution of different nutrients to the total heat production (HE; Tauson *et al.,* 1997). The aim of the study was to examine the effect of different provision of protein and carbohydrate on the energy metabolism and substrate oxidation of mink kits between 6 and 12 weeks of age.

Material and methods

Thirty-two pairs of male mink kits were allocated to four different diets four weeks *post partum*. The diets contained either 30% or 45% of the metabolisable energy (ME) from protein and 15% or 25% of ME from carbohydrate. The diet codes were HPHC (high protein, high carbohydrate; 45:25), LPHC (low protein, high carbohydrate; 30:25), HPLC (45:15) and LPLC (30:15). The fat content in the diets accounted for 30%, 45%, 40% and 55% of ME for diets HPHC, LPHC, HPLC and LPLC, respectively. The diets were mainly based on fish offal, whole fish, poultry offal, heat treated barley and wheat, soy oil and lard. Heat production was calculated from nitrogen and carbon balances which were measured when the kits were 6, 9 and 12 weeks old. Each 4 days balance period included a 22 h measurement of carbondioxid production (VCO_2) by indirect calorimetry in an open-air circulation system. The oxygen consumption was calculated from HE, CVO_2, and urinary N using the equation by Brouwer (1965). Substrate oxidation was calculated according to Chwalibog *et al.* (1998). The statistical analyses were made in proc GLM in SAS (SAS Institute Inc., 1990) with diet and period as main effects. Results are presented as least square means (LSmeans). Pair wise comparisons of LSmeans, and differences were considered significant if $P<0.05$.

Results and discussion

Mink kits on LPLC gained significantly less weight than kits on the other diets. Although the intake of dry matter was significantly lower on LPLC, the intake of gross energy was the same (Table 1) because the gross energy content of LPLC was higher than that of the other diets. Intake of ME was, because of low fat digestibility, significantly reduced on LPLC (Hellwing *et al.*, 2009). The low ME intake also resulted in a significantly lower HE on this diet, but retained energy (RE) was not significantly different from the other diets although there was a clear difference in weight gain between diets. This might be explained by measurements of RE and energy retained in fat having a considerably larger variance than live weight gain. The retention of protein was significantly lower on LPLC than on other diets, also reflected in a lower weight gain. The percentage of heat from oxidation of protein (OXP/HE) was highest on the two high protein diets (Figure 1). OXP/HE was below 30% on LPHC which is rather low for the mink (Tauson *et al.*, 1997). This reflected the high protein retention in these animals that were growing rapidly, and with the low protein provision from this diet little protein was left for oxidation. As expected the percentage of heat from oxidation of fat (OXF/HE) was lowest on HPHC, which had the lowest fat content. In conclusion, animals on diet LPLC had a lower ME intake, RN and HE, but not RE, than animals on the other diets, whereas results from the LPHC diet were not different from those on the other diets, suggesting that the protein requirement was sustained. Substrate oxidation generally reflected dietary composition.

Table 1. Energy metabolism in mink kits fed different levels of protein and carbohydrate.

	HPLC	LPHC	HPHC	LPLC	SEM	*P*-value
N	24	24	24	23		
Live weight [g]	1,016[a]	995[a]	939[b]	752[c]	16.4	<0.001
Live weight gain [g/day]	24[ab]	29[a]	22[b]	14[c]	2.1	<0.001
Intake of dry matter [kJ/kg$^{0.75}$/day]	63.7[ab]	65.7[a]	68.9[a]	59.2[b]	2.0	0.01
Intake of gross energy [kJ/kg$^{0.75}$/day]	1,503	1,545	1,519	1,499	48	0.91
Metabolisable energy [kJ/kg$^{0.75}$/day]	1,167[a]	1,218[a]	1,131[a]	1,011[b]	33.1	<0.001
Heat production [g/kg$^{0.75}$/day]	778[a]	775[a]	761[a]	705[b]	14.8	0.004
Retained energy [kJ/kg$^{0.75}$/day]	376	439	370	305	35.6	0.10
Retained energy in protein [kJ/kg$^{0.75}$/day]	190[a]	215[a]	202[a]	130[b]	9.7	<0.001
Retained energy in fat [kJ/kg$^{0.75}$/day]	173	230	173	130	29.4	0.15

HP: High protein, LP: Low protein, HC: High carbohydrate, and LC: Low carbohydrate.
[a,b,c] Values in a row with different superscript differ significantly ($P<0.05$).

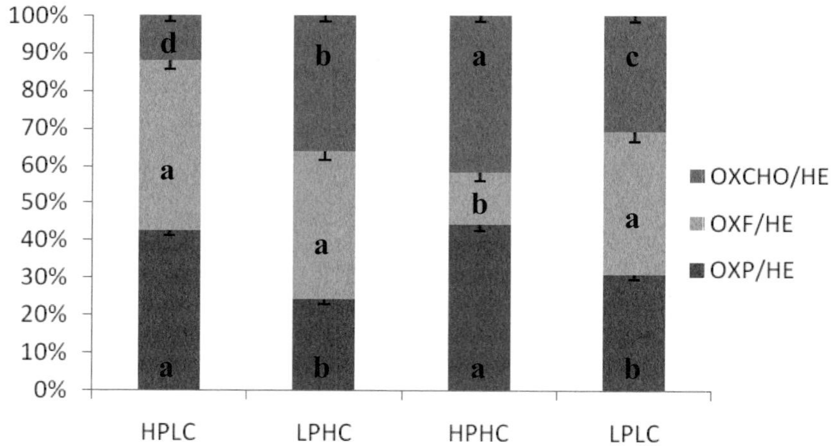

Figure 1. Heat from oxidation of protein (OXP/HE) P<0.001, *fat (OXF/HE)* P<0.001 *and carbohydrate (OXCHO/HE)* P<0.001 *in relation to total heat production.*

References

Brouwer, E., 1965. Report of sub-committee on constants and factors. In: Blaxter K.L. (ed.), Energy metabolism. Academic Press, London. EAAP No. 11: 441-443.

Chwalibog, A., A.-H. Tauson, R. Fink, and G. Thorbek, 1998. Oxidation of substrates and lipogenesis in pigs *(Sus scrofa)*, mink (*Mustela vison*) and rats (*Ratus norvegicus*). Thermochimica Acta 309, 49-56.

Hellwing, A.L.F., A.-H. Tauson and N.E. Hansen, 2009. Apparent and true digestibility of protein and amino acids during the early growth period in mink fed different levels of carbohydrates and protein. NJF Semeinar No. 427, Vaasa, Finland.

SAS Institute Inc., 1990. SAS/STAT® User's guide, Version 6, 4th edition, Cary NC: SAS institute Inc. 943 pp.

Tauson, A.-H., R. Fink and A. Chwalibog, 1997. Can gas exchange measurements be used for calculation of nutrient oxidation in mink (*Musteal vison*) exposed to short-term changes in energy supply? Z. Ernährungswiss 36, 317-320.

Comparison of the slaughtering and the T-cannulation methods on ileal digestibility of nutrients in pigs

C. Pedersen, E. Vinyeta and R. Gerritsen
Schothorst Feed Research, P.O. Box 533, 8200 AM Lelystad, the Netherlands

Introduction

Traditionally, the slaughter method was used for collection of ileal samples, in order to determine crude protein (CP) and amino acids (AA) digestibility in pigs. Later T-cannulation became the dominating method, however, in the literature, few comparison studies between the slaughter method and T-cannulation method are reported (Donkoh *et al.*, 1994; Moughan and Smith, 1997). In both studies (same laboratory) no significant difference between the two methods was found, however, greater variation for slaughter method was found. Moughan and Smith (1997) fed barley as sole ingredients, while Donkoh *et al.* (1994) fed a synthetic diet, with meat and bone meal (200 g/kg; starch 621 g/kg) as only protein source.

The aim of the current experiment was to compare the variation in ileal digestibility between the slaughter and T-cannulation methods, when a balance diet and a high fibre diet are used. This will be used for calculating the number of replicates for each method and investigate if type of diets should be take into account the type of diet tested.

Materials and methods

In total two trials were performed. In trial 1 the slaughter method was used, while in trial 2, the T-cannulation method was used after which the animals were slaughtered.

In trial 1, twenty growing pigs (31.2±1.95) were used. After an 18-d adaptation period pigs were slaughtered and ileal samples were collected. Pigs were fed 50% of their morning feeding at 6AM, followed by feeding the remaining 50% 2.5 h prior to slaughter. Pigs were slaughtered at an interval of 10 minutes. Ileal content from the last 100 cm, excluding the last 10 cm, of the ileum was collected. Pigs were euthanized by an overdose of T61® in the heart, after having been put to sleep with a combination of stressnil (azaperone) and zoletil (tiletamine). Samples from eight pigs per treatment were analysed.

In trial 2, sixteen T-cannulated pigs were used. A T-cannula was surgically placed at the end of the small intestine when the animals weighed about 25kg. After a five day adaptation period, ileal samples were collected on days 6 and 7 between 7.30 and 3.00 PM, covering the time between morning and afternoon feeding. Pigs were fed at 7.30 and 3.15 PM.

Two different diets were used in both experiments. Diet A was based on barley (29%), wheat (29%), peas (10%) and maize (8%), while diet B was based on wheat DDGS (30%), barley (20%), wheat (20%), peas (7%) and maize (5%). Cr_2O_3 (0.3%) was added as a marker to the diets. Apparent ileal digestibility was calculated using the marker concentration in feed and ileal content, in combination with the concentration of nutrients.

A two way Anova comparison between the two diets for all nutrients was performed. For calculating the number of replicates needed using the slaughter and T-cannulation methods, values obtained for diet A and B were used. Calculations were done with accuracy of 80 and 90%, difference in responses were detected at 1.0, 1.5 and 2.0 units. Statistical analyses were performed in GenStat.

Results and discussion

Diet A contained 149 g CP/kg and 186 g NSP/kg, while diet B contained 204 g CP/kg and 274 g NSP/kg, respectively. The body weight of the pig on the day of slaughter was 40.3±2.3 kg in trial 1 and 68.2±3.3 kg in trial 2.

No significant (P>0.39) differences between the slaughter (trial 1) and T-cannulation (trial 2) methods on ileal digestibility of DM, OM, ash and CP for diet A or B were found (data not shown).

In Table 1, the calculated number of replicates needed is listed. The data shows that fewer replicates are needed when the T-cannulation method is used in comparison to the slaughter method. The data are in agreement with data reported by Moughan and Smith (1997) and Donkoh *et al.* (1994). It can therefore be concluded that using the T-cannulation method is a better option, as on the one hand fewer replicates are needed, and on the other hand the same animal can be used several times and in that way becomes its own control. The data also shows that number of replicates for diet B is higher than diet A. A possible reason for the greater number of replicates found for diet B, can be the higher NSP content.

Table 1. Calculation of the number of replicates needed using the slaughter and T-cannulation methods, using accuracy of OM ileal digestibility as response parameter.

Diet	Probability	Response to be detected	Slaughter method	T-cannulation method
A	80	1.0	18	13
		1.5	10	7
		2.0	7	5
	90	1.0	23	16
		1.5	12	9
		2.0	8	6
B	80	1.0	22	17
		1.5	11	9
		2.0	7	6
	90	1.0	28	22
		1.5	14	11
		2.0	9	8

References

Donkoh A., P.J. Moughan and W.C. Smith, 1994. Comparison of the slaughter method and simple T-piece cannulation of the terminal ileum for determining ileal amino acid digestibility in meat and bone meal for the growing pig. Anim. Feed Sci. Technol. 49, 43-56.

Moughan P.J. and W.C. Smith, 1987. A note on the effect of cannulation of the terminal ileum of the growing pigs on the apparent ileal digestibility of amino acids in ground barley. Anim. Prod. 44, 319-321.

Effects of the addition of succinic acid to the diet on the zootechnical performance and the ileal digestibility of the pig

P. Guggenbuhl, A. Piñón Quintana and C. Simões Nunes
DSM Nutritional Products France SA, Centre de Recherche en Nutrition Animale (CRNA), BP 170, 68305 Saint-Louis cedex, France

Introduction

The use of organic acids in pig nutrition is increasing. Their use can improve animal performance especially in young animals (Guggenbuhl *et al.* 2007; Partanena *et al.*, 2007) and for some, antimicrobial activities have been shown (Gong *et al.*, 2008). The interest of succinic acid (SA) lies to its anti-bacterial effects (Séon and Simões Nunes, 2009) and to its role in the tri-carboxylic acid cycle resulting in the release of energy. Thus, the aim of the present studies was to evaluate the effects of the addition of SA to the diet on the zootechnical performance and the ileal apparent digestibility (IAD) of some elements in the pig.

Materials and methods

The trials were conducted at DSM Nutritional Products France, according to the French legislation on the protection of vertebrate animals used for experimental purposes. In a first experiment, 96 28-day old Large-White × Landrace weaner piglets having an initial body weight of 7.79±0.84 kg were used. Piglets were randomly allocated into four equally sized groups and housed in 4 pens per group in an environmentally controlled room. They were fed for 32 days either a basal diet (Ctrl-a), or the basal diet with addition of SA at the levels of 0.5% (SA-0.5a), 1% (SA-1a) and 2% (SA-2a) respectively. The basal diet was formulated to meet the animals' requirements according to the NRC (1998). Performance was evaluated for the 32 days of the trial duration. In a second experiment, 8 Large-White × Landrace pigs having an initial body weight around 45 kg were subjected to an ileo-rectal anastomosis according to the techniques described by Laplace *et al.* (1994). Pigs were housed individually in stainless-steel metabolic cages in an environmentally controlled room. After a 10-day post-surgical recovery period, pigs were fed, based on a double 3×3 Latin square design, the previous mentioned basal diet (Ctrl-b) or the basal diet supplemented with SA at 0.5% (SA-0.5b) or 1.0% (SA-1b) respectively. Chromium oxide was added at a concentration of 0.4% to all the diets in order to measure IAD of total nitrogen (N), total energy (E), total fat (F) and amino acids (AA).

Statistical treatment of the results involved the calculation of the mean and of the standard deviation of the mean as well as a two-factor hierarchical analysis of variance. The mathematical model was: $Y(ij) = m + treatment_i + e_{ij}$ for experiment 1 and $Y_{ijkl} = m + treatment_i + period_j + square_k + pig(square)_{kl} + e_{ijkl}$ for experiment 2. The analysis of variance was followed by a Duncan multiple range test when a significant i effect without j effect was observed (Snedecor and Cochran, 1989).

Results and discussion

The animals did not present any symptoms of illness during the experiments. In the first experiment the general performance was relatively good. The daily weight gain, feed intake and feed conversion ratio were not modified by any of the dietary treatments (Table 1). In the second experiment the ingestion of SA dose dependently improved the IAD of all AA. The IAD responses were significant ($P<0.05$) with 1% SA for Lys (+5.3%), Thr (+10.6%), Ala (+9.9%), Asp+Asn (+5.6%), Ile (+14.6%), Leu (+6.0%), Phe (+5.6%), Tyr (+4.6%), Val (+14.8%), branched chain AA (+9.4%), essential AA (+6.7%) and total AA (+5.7%; Table 2). It is well known that reducing the microbial gut population results in a higher nutrient availability (Partanena *et al.*, 2007), but the present results, especially on the IAD of AA, suggested that the effects observed for SA were only in part due to its antibacterial

Table 1. Effects of succinic acid on the piglet performance.

Treatment groups	Ctrl-a	SA-0.5a	SA-1a	SA-2a
Daily weight gain (g)[1]	298±67	321±48	340±59	318±62
Feed intake (g/day)[2]	543±74	532±22	550±35	521±28
Feed conversion ratio (g/g)[2]	1.842±0.168	1.673±0.041	1.624±0.027	1.642±0.062

[1] Mean ± σ of 4 observations of 24 individual weights each.

[2] Mean ± σ on of 4 determinations.

Table 2. Effects of succinic acid on the pig ileo-rectal apparent digestibility (mean ± σ).

Group	Ctrl-b	SA-0.5b	SA-1b	Group	Ctrl-b	SA-0.5b	SA-1b
Total N	63.1±6.5	63.4±6.6	65.2±6.0	Asp+Asn	67.2^{b}±5.5	68.8^{ab}±5.8	70.9^{b}±4.1
Total F	56.3±9.3	57.9±9.7	58.2±12.6	Ile	66.9^{b}±9.7	72.3^{ab}±7.9	76.7^{b}±5.1
Total E	57.2±10.2	58.7±11.6	60.6±10.9	Leu	73.2^{b}±5.4	75.6^{ab}±5.5	77.6^{b}±3.7
Total AA	70.0^{b}±5.5	72.3^{ab}±6.2	74.0^{a}±4.4	Lys	74.7^{b}±5.5	76.5^{ab}±4.6	78.7^{b}±3.2
Essential AA	72.1^{b}±5.7	74.8^{ab}±5.8	77.0^{a}±3.8	Phe	74.5^{b}±5.2	76.7^{ab}±5.2	78.7^{b}±3.4
BCAA	69.8^{b}±7.1	73.5^{ab}±6.6	76.3^{a}±4.3	Thr	64.3^{b}±7.8	67.8^{ab}±9.7	71.1^{b}±6.4
Sulfured AA	77.1±4.7	79.6±5.5	79.5±4.0	Tyr	72.5^{b}±4.9	74.5^{ab}±5.3	75.8^{b}±4.0
Ala	55.5^{b}±8.0	58.3^{ab}±8.5	61.0^{a}±6.1	Val	63.8^{b}±9.6	69.3^{ab}±8.6	73.2^{b}±5.4

[a,b] Mean values within a row with unlike superscript letters were significantly different *P<0.05;* BCAA = branched chain aa.

effect. As intermediary of the tricarboxylic acid cycle, the intrinsic properties of SA in providing energy could explain part of the observations. This is particularly true for the AA absorption which is an active transport but also for the fat digestibility. It appeared that SA can be a basis for the development of new organic acids.

References

Gong, J., H. Yu, T. Liu, M. Li, W. Si, C.F.M. de Lange and C. Dewey, 2008. Characterization of ileal bacterial microbiota in newly-weaned pigs in response to feeding lincomycin, organic acids or herbal extract. Livest. Sci 116, 318-322.

Guggenbuhl, P., A. Séon, A. Piñón Quintana and C. Simões Nunes, 2007. Effects of dietary supplementation with benzoic acid (VevoVitall®) on the zootechnical performance, the gastrointestinal microflora and the ileal digestibility of the young pig. Livest. Sci. 108, 218-221.

Laplace, J.P., W.B. Souffrant, U. Hennig, E. Chabeauti and C. Février, 1994. Measurement of precaecal dietary protein and plant cell wall digestion in pigs comparison of four surgical procedures for ileo-rectal anastomosis. Livest. Prod. Sci. 40, 313-328.

NRC, 1998. Nutrient requirements of swine, 10th revised edition, National Academic Press, Washington.

Partanena, K., T. Jalavaa and J. Valajaa, 2007. Effects of a dietary organic acid mixture and of dietary fibre levels on ileal and faecal nutrient apparent digestibility, bacterial nitrogen flow, microbial metabolite concentrations and rate of passage in the digestive tract of pigs. Animal 1, 389-403.

Séon, A. and C. Simões Nunes, 2009. Évaluation *in vitro* des effets de différents acides organiques sur trois serovars de *Salmonella enterica* d'origine porcine. Journées Rech. Porcine en France 41, 53-54.

Snedecor, G.W. and W.G. Cochran, 1989. Statistical methods, 8th edition, Iowa University Press, Ames.

Part 7. Energy/protein nutrition and environmental sustainability

Energy and protein interactions and their effect on nitrogen excretion in dairy cows

E. Kebreab[1], A.B. Strathe[1], J. Dijkstra[2], J.A.N. Mills[3], C.K. Reynolds[3], L.A. Crompton[3], T. Yan[4] and J. France[4]
[1]Department of Animal Science, University of California, Davis, Davis CA, USA
[2]Animal Nutrition Group, Wageningen University, Wageningen, the Netherlands
[3]Animal Science Research Group, School of Agriculture, Policy and Development, University of Reading, Whiteknights, P.O. Box 237, Reading RG6 6AR, United Kingdom
[4]Agri-Food and Biosciences Institute, Hillsborough Co. Down, Northern Ireland BT26 6DR, United Kingdom
[5]Department of Animal and Poultry Science, University of Guelph, Guelph ON, Canada

Abstract

The principal driver of nitrogen (N) losses from the body including excretion and secretion in milk is N intake. However, other covariates may also play a role in modifying the partitioning of N. This study tests the hypothesis that N partitioning in dairy cows is affected by energy and protein interactions. A database containing 470 dairy cow observations was collated from calorimetry experiments. The data include N and energy parameters of the diet and N utilization by the animal. Univariate and multivariate meta-analyses that considered both within and between study effects were conducted to generate prediction equations based on N intake alone or with an energy component. The univariate models showed that there was a strong positive linear relationship between N intake and N excretion in faeces, urine and milk. The slopes were 0.28 faeces N, 0.38 urine N and 0.20 milk N. Multivariate model analysis did not improve the fit. Metabolizable energy intake had a significant positive effect on the amount of milk N in proportion to faeces and urine N, which is also supported by other studies. Another measure of energy considered as a covariate to N intake was diet quality or metabolizability (the concentration of metabolizable energy relative to gross energy of the diet). Diet quality also had a positive linear relationship with the proportion of milk N relative to N excreted in faeces and urine. Metabolizability had the largest effect on faeces N due to lower protein digestibility of low quality diets. Urine N was also affected by diet quality and the magnitude of the effect was higher than for milk N. This research shows that including a measure of diet quality as a covariate with N intake in a model of N excretion can enhance our understanding of the effects of diet composition on N losses from dairy cows. The new prediction equations developed in this study could be used to monitor N losses from dairy systems.

Introduction

Increasing feed efficiency and reducing environmental pollution, particularly nitrogen (N) has been the subject of several previous studies. In ruminants, dietary N is partitioned into proteinaceous products such as milk and meat or excreted in faeces and urine. The amount of N lost in urine and faeces by the high-producing dairy cow is greater than in any other ruminant or non-ruminant production system and as a consequence, most studies have been carried out in dairy cattle (e.g. Castillo *et al.*, 2001a, b). Nitrogen intake has been identified as the principal driver of N excretion and studies have shown that reducing N intake reduces N in faeces and urine. Castillo *et al.* (2000) and Kebreab *et al.* (2001) reported a linear relationship between N intake and N excretion in faeces and N secretion in milk; urine N, however, was exponentially related to N intake. A limitation of the study was that the authors did not take study effect into account, and some of the urine N data was calculated by difference and was not directly measured. A more rigorous statistical analysis is therefore required to establish the relationships between N intake and excretion.

Another shortcoming of the previous empirical analyses of N utilization was the failure to investigate the effect of covariates. Reynolds and Firkins (2005) identified that N excretion is related not only to N intake but to other parameters such as the energy content of the diet. Milk production, milk N and apparent N utilization have been reported to increase and urinary N excretion to decrease by supplementing high digestibility, high protein dairy cow diets with cereal grains such as barley (Cohen *et al.*, 2006). The authors attributed the higher milk production in supplemented cows to the higher metabolizable energy intakes. The key energy parameters that may influence N excretion are metabolizability, i.e. feed quality defined as the proportion of metabolizable energy (ME) to gross energy of feed (AFRC, 1993), and ME intake (MJ/d).

A meta-analytical approach is best suited to test the hypothesis that energy variables have an effect on N utilization and excretion rates. The aim of the present study was to investigate the effects of diet quality and ME intake on the efficiency of utilizing N intake for milk production and subsequent excretion of N in faeces and urine.

Materials and methods

Database

A database containing N and energy balance data of 470 dairy cow observations was assembled from calorimetry studies conducted at the Centre for Dairy Research at the University of Reading, the Agricultural Research Institute for Northern Ireland, Queens University of Belfast and Grassland Research Institute, Hurley. For details of diet composition, experimental design and references see Kebreab *et al.* (2003). Nitrogen intake, faecal and urinary N excretion, milk N secretion and metabolizable energy intakes were all measured in the experiments.

Statistical analysis

Nitrogen balance data was analyzed in three fundamental ways using N intake as the primary covariate. First, the response variables faeces, urine and milk N were modelled using univariate normal mixed models for identifying potential covariates. Secondly, an index of N efficiency was produced by expressing the ratio of milk N to excreta N and analyzing this as a function of the potential covariates. Finally, a multivariate analysis was undertaken to generate new prediction equations for N partitioning in dairy cows. The multivariate approach was chosen because faeces, urine and milk N are correlated biologically and statistically.

Let y_{ij} and x_{ij} denote N output for the *j*th observation from the *i*th study. Then the following normal linear mixed model may be fitted:

$$y_{ij} = \beta_0 + b_{0_i} + (\beta_1 + b_{1_i}) \times x_{1ij} + (\beta_2 + b_{2_i}) \times x_{2ij} + e_{ij}$$

where $(\beta_0, \beta_1, \beta_2)^T$ and $(b_{0_i}, b_{1_i}, b_{2_i})$ are fixed and random effects regression coefficients related to N intake (x_{1ij}) and metabolizability or ME intake (x_{2ij}). The standard assumptions were $(b_{0_i}, b_{1_i}, b_{2_i})^T \sim N(0,\Omega)$ and $e_{ij} \sim N(0,\sigma^2)$. The matrix Ω was an unstructured positive-definite variance-covariance matrix.

If y_{ij} denote the ratio of milk to excreta N for the *j*th observation from the *i*th study then the following mixed effect model may be fitted:

$$y_{ij} = \beta_0 + b_{0_i} + (\beta_1 + b_{1_i}) \times x_{ij} + e_{ij}$$

where $(\beta_0, \beta_1)^T$ and $(b_{0_i}, b_{1_i})^T$ are fixed and random effects regression coefficients related the covariates metabolizability, and ME intake (x_{ij}). The standard assumptions were $(b_{0_i}, b_{1_i})^T \sim N(0,\Omega)$ and $e_{ij} \sim N(0,\sigma^2)$. All univariate models were implemented in the linear mixed effects function in R (lmer(); Bates and Maechler, 2009). The best performing model was identified based on the Bayesian information criterion (BIC) as the goodness-of-fit indicator.

The multivariate model can be written in the following way and centring the mean, i.e.

$$y_{1ij} = \beta_1 + \beta_2 \times (x_{1ij} - \bar{x}_1) + \beta_3 \times (x_{2ij} - \bar{x}_2) + b_{1_i} + e_{1ij}$$

$$y_{2ij} = \beta_4 + \beta_5 \times (x_{1ij} - \bar{x}_1) + \beta_6 \times (x_{2ij} - \bar{x}_2) + b_{2_i} + e_{2ij}$$

$$y_{3ij} = \beta_7 + \beta_8 \times (x_{1ij} - \bar{x}_1) + \beta_9 \times (x_{2ij} - \bar{x}_2) + b_{3_i} + e_{3ij}$$

In the multivariate case, correlations were introduced in both the random effects and residual variances, i.e. $(b_{1_i}, b_{2_i}, b_{3_i})^T \sim N(0,\Omega)$ and $(e_{1ij}, e_{2ij}, e_{3ij})^T \sim N(0,\Sigma)$ where T denotes the transpose. The matrices Σ and Ω were unstructured positive-definite variance-covariance matrices. The populations' effects were denoted by parameters $(\beta_1,\ldots, \beta_{12})$. The multivariate model was implemented in OpenBUGS and the parameters where estimated using a Gibbs sampling scheme (Lunn *et al.*, 2000; Thomas *et al.*, 2006). The non-informative priors were specified because the likelihood should dominate the posterior. Convergence was established using the Gelman-Rubin statistic as the main determinant and running three chains. Convergence was established after 10,000 samples, i.e. the burn-in period. Inference was based on an additional 100,000 samples from the posterior distribution where every 10th sample was used. The deviance information criterion (DIC) was used to compare models with varying complexity. This can be considered an adaptation of the Akaike Information Criteria to Bayesian models incorporating prior information, whether through fully specified proper prior distributions or hierarchical models (Spiegelhalter *et al.*, 2002). The notion that 'smaller is better' is preserved in the DIC.

Results and discussion

The univariate analysis confirmed the previous reports of a linear relationship between N intake (*NI*; g/d) and N excretion (g/d).

$$N\ excretion = 30\ (SE{=}20) + 0.67\ (SE{=}0.044) NI \quad (1)$$

This is in broad agreement with Kebreab *et al.* (2001) who reported that N in urine and faeces accounted for approximately 62% of N intake. There was also a linear relationship between *NI* and N in faeces, urine and milk (all in g/d):

$$N\ faeces = 10\ (SE{=}9.0) + 0.28\ (SE{=}0.023) NI \quad (2)$$
$$N\ urine = 20\ (SE{=}20) + 0.38\ (SE{=}0.039) NI \quad (3)$$
$$N\ milk = 30\ (SE{=}10) + 0.20\ (SE{=}0.03) NI \quad (4)$$

Kebreab *et al.* (2001) also reported a similar positive linear relationship between N intake and faeces N. However, the authors reported that the slope for N excreted in faeces was 0.16, but the intercept was higher. For a cow consuming 400 g N/d, Equation 2 estimates that 31% would be excreted in faeces, compared to 35% in Kebreab *et al.* (2001) equation estimate. The relationship between N intake and milk N was almost identical with the slope of 0.19. In the analysis of Castillo *et al.* (2000), based on data for 91 diets and 580 cows taken from the literature, the slopes for faeces and milk N were 0.21 and 0.17 of the N intake, respectively. There was also a positive linear relationship between N intake and urinary N using the univariate analysis method. The results agree with Weiss *et al.* (2009) who found that increasing protein concentration in the diet linearly increased N excretion in faeces and urine. Mills *et al.* (2009) also reported that the increase in urine N was positively and linearly related to N intake. However, Castillo *et al.* (2000) and Kebreab *et al.* (2001) reported an exponential relationship between N intake and urinary N excretion. One explanation of this finding could have been, in cases of positive N balance, inclusion of data where urine N was not measured directly but calculated by difference might have led to an overestimation of urinary N excretion.

A multivariate analysis was conducted because the excretions were considered to be correlated with each other. The only difference from the univariate analysis was a slight change in the slope of faeces N (from 0.28 to 0.29). Slopes for urine and milk N were unchanged and the standard deviations were slightly lower for urine N and slightly higher for milk N (Figure 1). Therefore, in the absence of other covariates (as will be discussed later), either univariate or multivariate models can be used to predict N excretions.

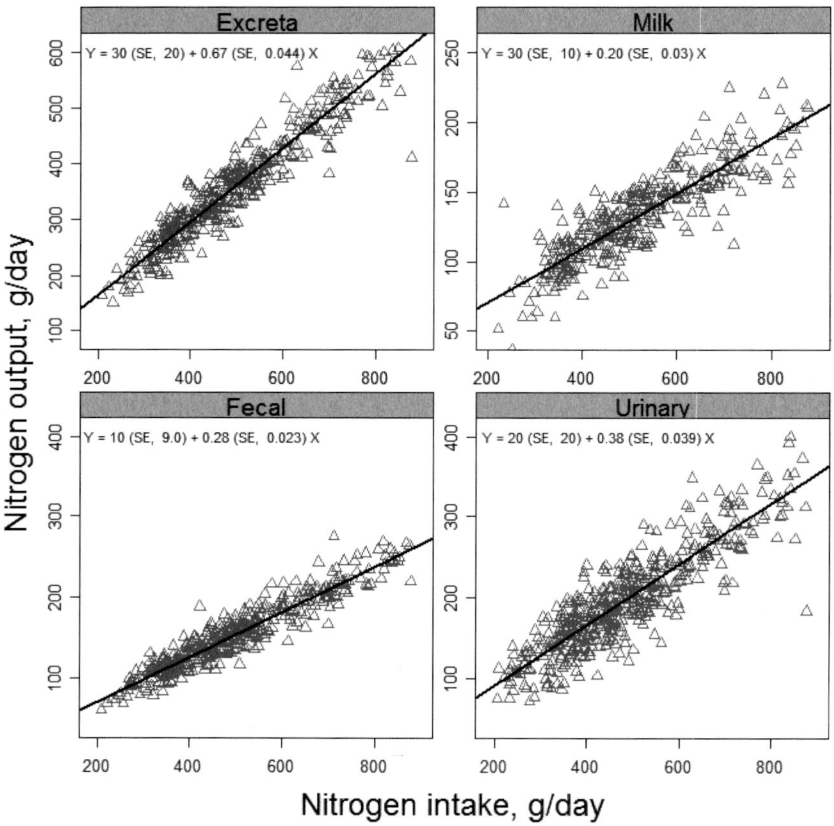

Figure 1. Univariate relationships between nitrogen intake and nitrogen excretion in faeces, urine and milk. Excreta are the combination of faecal and urinary nitrogen. The responses have been adjusted for study effect.

Effect of ME on nitrogen excretion

The effect of ME on N utilization was analyzed by regressing ME intake against the ratio of milk N to that excreted in faeces and urine (Figure 2). There was a positive relationship indicating that there was more N secreted in milk as ME intake increased. Cohen *et al.* (2006) also reported that milk N increased at the expense of urinary N as ME increased. In the same report, the authors also found that

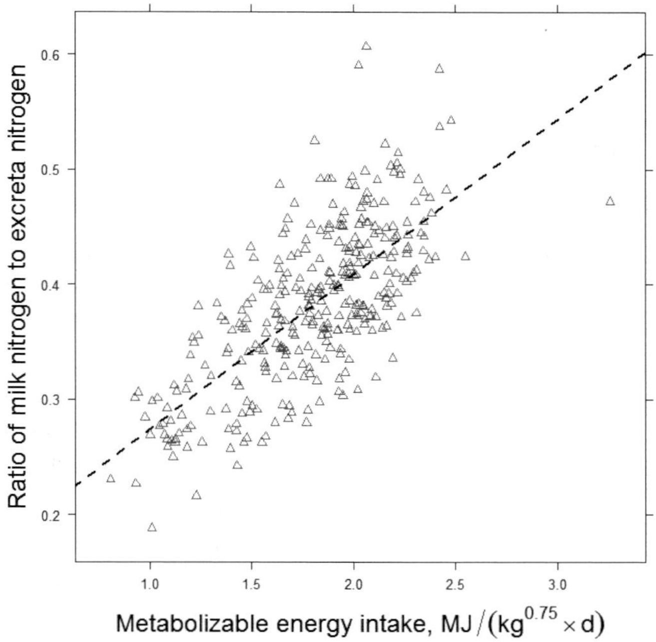

Figure 2. The effect of metabolizable energy intake on the proportion of nitrogen in milk and that excreted in faeces and urine. The responses have been adjusted for study effect.

increasing ME intake did not affect N intake, faeces N excretion, and overall N balance and microbial protein synthesis. Nitrogen excretion patterns may be affected by the site of ME supplementation. Urinary N excretion was decreased and faecal N increased with abomasal infusions of up to 1 kg/d of pectin (Gressley and Armentano, 2005); however, there were no significant differences in milk yield in dairy cows fed molasses or abomasally infused with pectin, but milk fat was decreased.

Cohen *et al.* (2006) cautioned that DMI and milk yield could be suppressed, and any benefits derived from supplemental carbohydrates overshadowed, if the level of cereal grain supplementation caused the neutral detergent fibre (NDF) concentration of the diet to reach a point where digestion was impaired or acidosis prevailed. The authors also reported that the cost of the supplement and the practices needed to feed the supplement should not outweigh the marginal responses to supplementation. However, under European Union (EU) circumstances the level of production per unit area may be limited by environmental constrains. This limitation of production or income may become more important if the EU ends the milk quota system.

The type of energy fed to dairy cows has been reported to have the potential to affect N excretion patterns. For example, a study by Castillo *et al.* (2001a) found that feeding a concentrate with high starch degradability resulted in 30% more N excretion in urine compared to the other study diets, which the authors attributed to an increased amount of rumen-fermentable energy supporting higher rates of microbial protein synthesis. Although urinary N was increased, faecal and milk N remained relatively constant in relation to N intake across all treatments. Milk yield was not affected by concentrate type, but milk protein concentration was highest for diets containing low or high degradability starch. Castillo *et al.* (2001a) concluded it was possible to achieve lower N losses and improve N use efficiency without adverse effects on milk production by using mixtures of energy sources that include low degradability starches and balancing diets for animal requirements and level of milk production. In cows fed diets differing in their content of non-fibrous carbohydrates (NFC), cows fed higher levels of NFC had a lower milk fat content and higher milk protein than the cows receiving low levels of NFC (Cherney *et al.*, 2003). Faecal N excretion was constant across all treatments. Urinary N excretion, however, was higher for the low NFC diets, possibly due to less efficient rumen microbial N conversion. Cows fed the high NFC diets had greater N intake and milk N, as well as a higher N use efficiency. In the same study, Cherney *et al.* (2003) also compared diets containing sucrose or maize. When 10% of maize was replaced with sucrose, there was little impact on milk production and composition but N use efficiency was lower for the diets containing higher sucrose. Cherney *et al.* (2003) concluded that replacing maize with sucrose was not to be recommended as the decrease in N use efficiency and N retained would represent an increased environmental impact of dairy farming. Diet also affects the composition of urinary N. For example, van Vuuren and Smits (1997) showed that changing levels of fermentable organic matter and hence the yield of microbial protein, may result in changes in the proportion of urea, uric acid and allantoin-N. Therefore, this may be part of the reason for higher urinary N excretion observed by Castillo *et al.* (2001a) when cows were fed diets supplemented with high degradable starch sources.

A multivariate analysis was conducted that included N intake and ME intake as covariates in predicting N output in faeces, urine and milk. Metabolizable intake had a significant effect on predicting urinary N excretion and milk N secretion but not faecal N excretion. It should be noted that ME intake already includes faecal and urinary energy as part of the calculation, therefore, correlations are expected. The residual standard deviations from the multivariate analysis including ME intake as a covariate were 13.8, 27.9 and 15.0 g/d. The prediction equations for urinary N excretion and milk N secretion (which had significant effect of ME intake) were:

$$N\ urine = 47.8\ (SE{=}20.1) + 0.56\ (SE{=}0.03) NI - 71.4\ (SE{=}12.2)\ ME \quad (5)$$

$$N\ milk = 2.04\ (SE{=}11.7) + 0.10\ (SE{=}0.023) NI + 45.9\ (SE{=}5.43) ME \quad (6)$$

Effect of metabolizability on nitrogen excretion

Several studies have highlighted the wide variation in N excretion, particularly urine N. Weiss *et al.* (2009) reported that the variation in urinary N was 3.5 times greater than for faecal N excretion and Mills *et al.* (2009) also found that there was greater variation in N excretion relative to milk yield. The residual standard deviations for N losses in faeces, urine and milk in the present study were estimated to be 14.6, 32.5 and 17.5 g/d, respectively from the univariate analysis. The higher residual standard deviation for urine N compared to faeces or milk output presents an opportunity to manipulate diets to reduce urine N excretion and may be, in part, explained by metabolizability of the diet.

There was a positive linear increase in milk N relative to urine and faeces as the metabolizability of the diet increased from 0.45 to 0.70 (Figure 3). There was a significant effect of metabolizability when explicitly added as a covariate to the relationship between N intake and faecal N (Table 1). Lower metabolizability values indicated lower digestibility of energy in the diet, and therefore, greater hindgut fermentation of fibre or starch, which would increase microbial protein in faeces. In addition, low quality diets might have lower digestible protein, increasing the amount of N excreted in faeces. The effect was not as large but still significant when metabolizability was considered in the relationship between N intake and urinary N. The effect of metabolizability appears to be lowest for the relationship between N intake and milk N. There was an apparent negative linear relationship between diet quality and N excretion, but the magnitude varied between faeces, urine and milk. Therefore, for prediction of N utilization and partitioning in excreta, it is recommended that metabolizability be included as a covariate.

A multivariate analysis was conducted that included N intake and metabolizability as covariates in predicting N output in faeces, urine and milk. Metabolizability had a significant effect on predicting

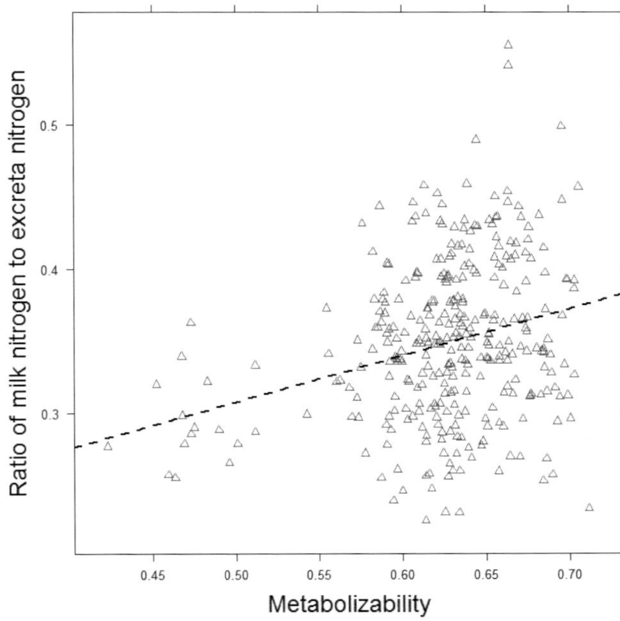

Figure 3. The effect of metabolizability on the proportion of nitrogen in milk and that excreted in faeces and urine. The responses have been adjusted for study effect.

Table 1. Effect of covariates (ME intake and metabolizability (q)) on the estimate of nitrogen (N) excretion in faeces, urine and milk.

Response	Covariate	Estimate (SE)	P-value	BIC[1]
FaecalN	Nitrogen intake	0.28 (0.022)	<0.001	3,496
	Nitrogen + ME[2]	0.27 (0.023)	<0.001	
	ME	1.14 (6.1)	0.85	3,546
	Nitrogen + q[3]	0.29 (0.023)	<0.001	
	q	-305 (31.5)	<0.001	3,410
UrinaryN	Nitrogen intake	0.38 (0.038)	<0.001	4,651
	Nitrogen + ME	0.56 (0.031)	<0.001	
	ME	-71.4 (12.2)	<0.001	4,606
	Nitrogen + q	0.38 (0.037)	<0.001	
	q	-174 (80.6)	0.03	4,660
MilkN	Nitrogen intake	0.20 (0.031)	<0.001	3,051
	Nitrogen + ME	0.10 (0.024)	<0.001	
	ME	45.9 (5.4)	<0.001	2,986
	Nitrogen + q	0.20 (0.033)	<0.001	
	q	-96.1 (41.2)	0.02	3,069

[1] BIC = Bayesian Information Criterion.

[2] Nitrogen + ME = Shows the coefficient for N intake when ME was added as covariate in a multivariate setting.

[3] Nitrogen + q = Shows the coefficient for N intake when q was added as covariate in a multivariate setting.

faecal and urinary N excretion but not milk N secretion when all outputs were considered to be correlated. The residual standard deviations from the multivariate analysis including energy as a covariate were 14.3, 32.5 and 17.5 g/d which was an improvement on the univariate analysis. The DIC for the univariate method of analysis was calculated to be 8876, and for the multivariate method it was 8800. This indicates a substantial improvement in the fit to data; therefore, generating prediction equations for whole animal N flows are best accomplished using multivariate statistical models because these account for correlated random study effects and residual variances. The new prediction equations for faecal and urinary N excretion (which had significant effect of metabolizability) were:

$$N\,faeces = 244\ (SE=23.3) + 0.25\ (SE=0.01)NI - 346\ (SE=36.5)q \quad (7)$$

$$N\,urine = 150\ (SE=53.6) + 0.46\ (SE=0.02)NI - 257\ (SE=85.4)q \quad (8)$$

Figure 4 illustrates the effect of modifying diet quality on faecal N excretion. Although, N intake is the primary driver for N excretion, the predictive model was improved for faecal and urinary N by including representation of diet quality.

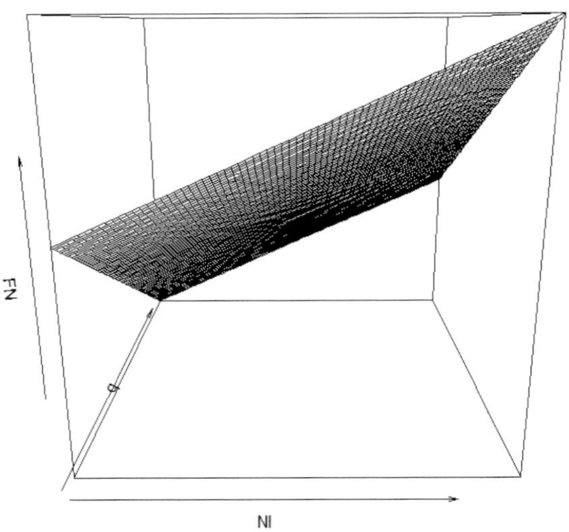

Figure 4. A 3-dimensional representation of a multivariate relationship between nitrogen intake (NI), faeces nitrogen excretion (FN) and diet quality or metabolizability (q).

In conclusion, there is a clear interaction between N and energy and affects the relative partitioning of N in faeces, urine and milk. Therefore when considering the effect of reducing N concentration in feed through for example, reduced use of fertilizers, the changes in energy (either ME content or diet quality) should also be taken into account to avoid overestimating N availability to the animal.

References

Bates, D. and M. Maechler, 2009. lme4: Linear mixed-effects models using S4 classes. R package version 0.999375-32. http://CRAN.R-project.org/package=lme4. Accessed March 25, 2010.

Castillo, A.R., E. Kebreab, D.E. Beever and J. France, 2000. A review of efficiency of nitrogen utilisation in dairy cows and its relationship with the environmental pollution. J. Anim. Feed Sci. 9, 1-32.

Castillo, A.R., E. Kebreab, D.E. Beever, J.H. Barbi, J.D. Sutton, H.C. Kirby and J. France, 2001a. The effect of energy supplementation on nitrogen utilization in lactating dairy cows fed grass silage diets. J. Anim. Sci. 79, 240-246.

Castillo, A.R., E. Kebreab, D.E. Beever, J.H. Barbi, J.D. Sutton, H.C. Kirby and J. France, 2001b. The effect of protein supplementation on nitrogen utilisation in grass silage diets by lactating dairy cows. J. Anim. Sci. 79, 247-253.

Cherney, D.J.R., J.H. Cherney, and L.E. Chase, 2003. Influence of dietary nonfiber carbohydrate concentration and supplementation of sucrose on lactation performance of cows fed fescue silage. J. Dairy Sci. 86, 3983-3991.

Cohen, D.C., C.R. Stockdale and P.T. Doyle, 2006. Feeding an energy supplement with white clover silage improves rumen fermentation, metabolisable protein utilisation, and milk production in dairy cows. Aust. J. Agric. Sci. 57, 367-375.

Firkins, J. L. and C.K. Reynolds, 2005. Whole-animal nitrogen balance in cattle. In: Pfeffer, E. and A.N. Hristov (eds.), Nitrogen and phosphorus nutrition of cattle. CAB International, Cambridge, MA, USA: 167-186.

Gressley, T.F. and L.E. Armentano, 2005. Effect of abomasal pectin infusion on digestion and nitrogen balance in lactating dairy cows. J. Dairy Sci. 88, 4028-4044.

Kebreab, E., J. France, D.E. Beever and A.R. Castillo, 2001. Nitrogen pollution by dairy cows and its mitigation. Nutr. Cycl. Agroecosyst. 60, 275-285.

Kebreab, E., J. France, R.E. Agnew, T. Yan, M.S. Dhanoa, J. Dijkstra, D.E. Beever and C.K. Reynolds, 2003. Alternatives to linear analysis of energy balance data from lactating dairy cows. J. Dairy Sci. 86, 2904-2913.

Lunn, D.J., A. Thomas, N. Best and D. Spiegelhalter, 2000. WinBUGS – a Bayesian modelling framework: concepts, structure, and extensibility. Statist. Comp. 10, 325-337.

Mills, J.A.N., L.A. Crompton, A. Bannink, S. Tamminga, J.M. Moorby and C.K. Reynolds, 2009. Predicting methane emissions and nitrogen excretion from cattle. In: Crompton, L.A. and T.R. Wheeler (eds.), Proceedings of the forty-first meeting of the agricultural research modellers' group, J. Agric. Sci. 147, 741.

Spiegelhalter D.J., N. Best, B.P. Carlin and A. van der Linde, 2002. Bayesian measures of model complexity and fit (with discussion). J. Roy. Statist. Soc. B. 64, 583-640.

Thomas, A., B. O'Hara, U. Ligges and S. Sturtz, 2006. Making BUGS Open. R News 6:12-17.

Van Vuuren, A.M. and M.C.J. Smits, 1997. Effect of nitrogen and sodium chloride intake on production and composition of urine in dairy cows. In: Jarvis, S.C. and B.F. Pain (eds.), Gaseous nitrogen emissions from grasslands. CAB International, Wallingford, UK: 95-99.

Weiss, W.P., L.B. Willet, N.R. St-Pierre, D.C. Borger, T.R. Mckelvey and D.J. Wyatt. 2009. Varying forage type, metabolizable protein concentration, and carbohydrate source affects manure excretion, manure ammonia, and nitrogen metabolism of dairy cows. J. Dairy Sci. 92, 5607-5619.

Ruminant nutrition regimes to reduce greenhouse gas emissions in dairy cows

C.K. Reynolds[1], J.A.N. Mills[1], L.A. Crompton[1], D.I. Givens[1] and A. Bannink[2]
[1]Animal Science Research Group, School of Agriculture, Policy and Development, University of Reading, Whiteknights, P.O. Box 237, Reading RG6 6AR, United Kingdom
[2]Wageningen UR Livestock Research, P.O. Box 65, 8200 AB Lelystad, the Netherlands

Abstract

At present there is considerable concern about the emissions of methane and nitrous oxide attributable to milk and meat production by ruminants. Numerous meta-analyses of measurements of methane emissions from cattle obtained from studies of energy metabolism have demonstrated that methane emission is determined primarily by feed dry matter intake (DMI), but variation attributable to animals and diet composition are also very important. In addition, the relative amount of methane emitted per unit DMI is greater at lower intakes of the same diet. As regards diet composition, the type of carbohydrates in the basal diet and their digestibility have repeatedly been shown to be important, with the relative amounts of starch, soluble carbohydrates, hemicellulose, and cellulose exerting a significant influence on the amount of methane arising for a given level of DMI. Supplemental fat has long been known to reduce methane emission per unit DMI. The effects of supplemental fat are multifactorial, and thus the effect varies with fat source. In general, sources of longer chain polyunsaturated fatty acids are most efficacious, but any supplemental fat may also reduce fibre degradation, DMI and milk yield, or alter milk composition. As regards nitrogen excretion into manure, which is a determinant of nitrous oxide emissions, the efficiency of utilization of dietary nitrogen consumed can be increased by reducing total nitrogen intake, but the extent to which dietary nitrogen can be reduced without sacrificing DMI and milk production remains to be determined, and will depend on the basal diet and level of production.. As for methane emissions, benefits of supplemental starch for the efficiency of dietary nitrogen utilization are evident. Whilst many of these regimes may have more relevance to more intensive production systems, forages remain the principal component of dairy diets and there remains an important role for milk and meat production from grazing ruminants. Thus there is also a need for dietary regimes and sward management practices that minimize their greenhouse gas emissions.

Introduction

In recent years there has been increasing concern regarding the contribution of livestock production to greenhouse gas emissions (FAO, 2007), and particularly the anthropogenic emissions of methane and nitrous oxide attributable to domestic ruminants. Whilst estimates of the total 'carbon footprint' of ruminant meat and milk production relative to other anthropogenic sources (FAO, 2007 and 2010) are the subject of considerable debate and discussion (Gill *et al.*, 2010), and vary geographically (Pitesky *et al.*, 2009), there is now a commitment on the part of many governments to reduce greenhouse gas emissions, including those from animal agriculture. In this regard production of milk and meat by ruminants are estimated to account for substantial portions of national methane and nitrous oxide emissions of many countries. Similarly, the emission of other nitrogenous compounds (nitrates and ammonia) from animal feeding operations is now regulated in many countries, and there is increasing consumer concern about these and other purported environmental impacts of milk and meat production. As methane emitted and nitrogen excreted in urine and faeces represent an economic cost to the producer in terms of the efficiency of feed component utilization, any management practices that reduce their emission and increase the capture of energy and nitrogen in milk and meat conceivably represent a 'win-win' scenario in terms of reduced environmental impact and economic return. However, many mitigation options under consideration, such as bioactive plant components or other novel feed additives, may have a cost that outweighs any benefit in terms of productive response. In this regard 'best practice' in terms of diet formulation and management approaches that improve feed conversion efficiency and thereby reduce methane and nitrogen excretion may be the most practical mitigation options in the immediate future. Our paper will briefly consider effects of diet composition in terms of carbohydrate, protein, and lipid components on methane and nitrogen emissions, with a focus on lactating dairy cows.

Methane production

As regards methane emission from ruminants, there have been numerous summarizations of data from individual measurements of energy balance of lactating dairy cows from laboratories in the USA, UK and other parts of Europe (e.g. Moe and Tyrrell, 1979; Holter and Young, 1992; Kirchgeßner *et al.*, 1995; Wilkerson *et al.*, 1995; Yan *et al.*, 2000; Mills *et al.*, 2001, 2003; Kebreab *et al.*, 2008; Ellis *et al.*, 2009). In addition, there have been a number of recent reviews and meta-analyses of published data (typically treatment means), from beef and dairy cattle and sheep (e.g. Boadi *et al.*, 2004; Beauchemin *et al.*, 2008; Ellis *et al.*, 2007; Martin *et al.*, 2010). These summarizations have been used to explore the major dietary factors that determine methane emission from dairy cows and develop regression equations and other models that predict methane emission. For all of the models developed the major determinant of total methane excretion is the amount of dry matter intake (DMI), or more precisely the fermentable organic matter consumed. In a recent meta-analysis of 1,335 measurements of methane emissions from energy metabolism studies with cattle, Mills *et al.* (2009) reported that DMI was the major determinant of methane emission (Figure 1). This relationship between DMI and methane emission reflects the relationship between DMI and fermentable organic matter, but in the analysis of Mills *et al.* (2009) the prediction of methane emission was only slightly improved when digestible energy intake was used in place of DMI. Although much of the variation in methane emission by cattle can be explained by differences in their feed DMI, there is still considerable variation in methane emission at a given level of DMI (Figure 1), which reflects the influence of other factors including animal variation and diet composition. At a practical level, variation in DMI for a given level of production is of less relevance to the mitigation of methane emission than diet composition and other management options subject to manipulation 'on farm'.

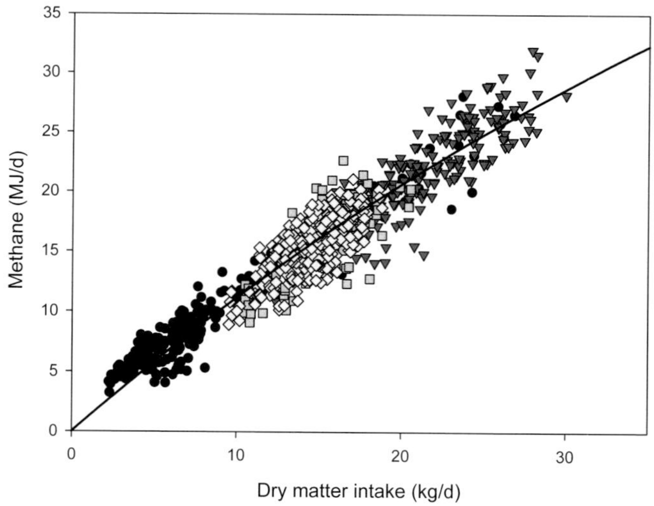

Figure 1. Relationship between dry matter intake and methane energy emission in beef and dairy cattle (Mills et al.*, 2009).*

Effects of diet composition

In addition to total DMI, numerous other dietary factors have significant effects on methane excretion, including the amount and type of fibre, starch and sugars included in the diet (Kebreab *et al.*, 2006). As reviewed by Kebreab *et al.* (2006), predictions of methane production by ruminants have been based on DMI (Kriss, 1930; Bratzler and Forbes, 1940), intake of digestible carbohydrates (Bratzler and Forbes, 1940; Moe and Tyrrell, 1979), diet digestibility and level of feed intake (Blaxter and Clapperton, 1965; Murray *et al.*, 1978), as well as diet intake and relative composition (Holter and Young, 1992; Yan *et al.*, 2000). In addition, an effect of rate of rumen fermentation and rumen pH has been described (Bannink *et al.*, 2010). In one summarization of calorimetry studies that included measurements of intake of diet components and methane production by lactating dairy cattle, Kirchgeßner (1995) reported that when considered across a range of basal diet types and levels of production the primary determinant of methane emission was amount of feed DMI, but the consumption of individual diet components, based on proximate analysis, gave a better prediction of methane emission. In this regard the amounts of crude fibre and to a lesser extent nitrogen free extract had a positive effect on methane emission, whilst consumption of crude protein had little effect. These findings agree well with a previous analysis of data from lactating dairy cows by Moe and Tyrrell (1979), who found that in addition to total DMI, the major dietary components determining methane production were the intake of cellulose, hemicellulose, and 'soluble residue' (nonfibre carbohydrate or NFC), and that the prediction of methane emission was improved when the amounts of these components digested was used in their regression.

Effects of intake level

Moe and Tyrrell (1979) also reported that the effect of carbohydrate type was greater in lactating dairy cows than in dry cows fed at lower levels of DMI, and presumably lower rates of rumen turnover, greater extent of digestion of diet components, and differences in fermentation profile and rumen pH. In this regard, Tyrrell *et al.* (1990) found that methane emission (L/kg DMI) was lower in lactating than nonpregnant dry dairy cows fed the same diet (27.2 vs. 35.1, respectively), and that this difference was associated with differences in dry matter (DM) digestibility of the diet (645 vs. 720 g/kg, respectively). This difference between lactating and dry cows was observed in Jersey and Holstein cows fed diets based on maize silage and alfalfa haylage as forage sources, but it is not certain if similar effects of level of intake and/or physiological state on relative methane emission apply to diets based on other forages or components. However, a recent meta-analysis found that for beef and dairy cattle methane emission as a percentage of gross energy intake generally declined with increasing DMI (Figure 2; Mills *et al.*, 2009), which likely reflects similar effects of level of intake on rumen dynamics and diet component digestion. For the measurements in Figure 2, lower levels of methane emission relative to gross energy intake at lower levels of DMI were generally observed for growing beef cattle fed pelleted, high concentrate diets (e.g. Reynolds *et al.*, 1991).

Effects of carbohydrate type

Subsequent analyses of dietary factors determining methane emission have to a large extent agreed with the conclusions of Moe and Tyrrell (1979) and Kirchgeßner *et al.* (1995) regarding the influence of relative amounts of fibrous and non-fibrous carbohydrates consumed by cattle on their methane emission (e.g. Ellis *et al.*, 2007, 2009). Based on the results of a dynamic model of methane excretion, Mills *et al.* (2001) suggested replacing soluble sugars with starch, and replacing grass silage with maize silage, as approaches that would reduce methane excretion from lactating dairy cows. The conclusions of Mills *et al.* (2001) were founded on the relationship between the stoichiometry of volatile fatty acid (VFA) production during fermentation and the reduction of carbon dioxide by surplus hydrogen. Production of propionate and branched chain VFA acts as a hydrogen sink whereas production of acetate and butyrate is a hydrogen source. The stoichiometry was derived

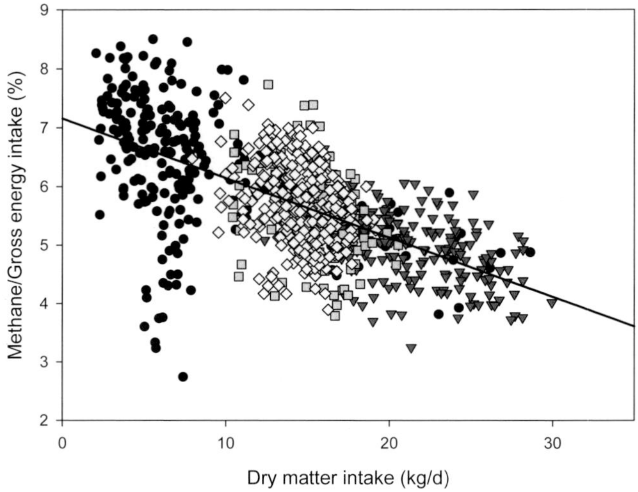

Figure 2. Relationship between daily dry matter intake (kg/d) and methane excretion as percentage of gross energy intake in beef and dairy cattle (Mills et al.*, 2009).*

from rumen observations in lactating cows by Bannink *et al.* (2006) and describes the effects of diet type (proportion of dietary forage) and fermented substrate type on VFA production. More recently, Bannink *et al.* (2010) derived pH-dependent relationships for the easily-fermented carbohydrates (sugars and starch). At high pH (6.5), this stoichiometry generates the highest methane yield from fermented soluble carbohydrates. At low pH (5.5), methane yield for fermented soluble carbohydrates was strongly reduced and yield from fermented fibre was highest. They concluded that the least methane is formed from starch and protein fermentation. In a subsequent non-linear meta-analysis of methane emission measurements from calorimetry studies, Mills *et al.* (2003) reported that methane emission was best predicted on the basis of DMI and an adjustment for the ratio of starch to acid detergent fibre (ADF; cellulose and lignin) concentrations in the diet. A subsequent analysis of a more comprehensive data set that included a broader range of dietary starch concentrations (Mills *et al.*, 2009) found that the effect of starch concentration on methane concentrations reached a plateau at higher concentrations typically found in finishing beef cattle diets in the USA.

The predictions and conclusions of Mills *et al.* (2001) have been supported by more recent experimental observations. For example, studies in dairy cows (Hindrichsen *et al.*, 2005) and *in vitro* (Hindrichsen and Kreuzer, 2009) have shown that sugars have a higher methanogenic potential than starch, particularly at high pH, which confirms the conclusions from *in vivo* data by Bannink *et al.* (2010) More recent experimental work has also shown that methane emission per kg DMI is lower in lactating dairy cows fed diets based on maize silage compared to grass silage (McCourt *et al.*, 2007; Reynolds *et al.*, 2010), and even lower when whole-crop wheat silage is fed (McCourt *et al.*, 2007). In the study of Reynolds *et al.* (2010), diets were formulated to be similar in starch and neutral detergent fibre (NDF) concentrations by manipulating the concentrate portion of the diets, thus differences in methane emission per kg DMI between the maize and grass-based diets may have been attributable to differences in the rate and extent of degradation of carbohydrate components. In this regard the extent of carbohydrate degradation in the rumen and hindgut may vary with forage maturity, such that the methane potential may be reduced as the extent of lignification increases (Hindrichsen *et al.*, 2005). In lactating dairy cows, methane emission per unit DMI was not affected by maize silage maturity (Cammell *et al.*, 2000), but in growing beef cattle there was a linear reduction in methane/kg DMI with increasing maturity of maize silage fed *ad libitum* that was associated with a linear decrease in NDF digestibility (McGeough *et al.*, 2010). As stated previously,

Moe and Tyrrell (1979) concluded that the relative amounts of digestible hemicellulose, cellulose, and NFC consumed determined methane production as opposed to the absolute amounts consumed.

Effects of dietary protein

As regards the effect of dietary protein on methane excretion, the summarizations of Moe and Tyrrell (1979) and Kirchgeßner *et al.* (1995) found that the amount of total or digestible protein consumed did not have a large effect on methane emission compared to the effects of digested carbohydrate. A significant effect of diet crude protein concentration was noted in some of the regression equations reported by Holter and Young (1995). As noted by Kirchgeßner *et al.* (1995), the majority of the measurements summarized to date have been obtained at protein intakes at or above requirements, and at lower protein concentrations an imbalance of rumen degradable protein and fermentable energy could influence microbial growth and methane emissions. However, subsequent analyses have found no significant effects of dietary crude protein intake or concentration on methane emissions from cattle (Ellis *et al.*, 2009; Mills *et al.*, 2009). In a recent experiment, lactating dairy cows were fed high maize- or high grass-silage diets differing in concentrations of crude protein (145, 165 and 185 g/kg of DM) and rumen degradable protein, and there was no effect of dietary protein on methane emissions (Reynolds *et al.*, 2010). The preceding discussion applies to the protein concentration of the total diet fed. As regards the nitrogen fertilization rate and resulting protein concentration of forage fed to lactating dairy cows, Bannink *et al.* (2010) reported a negative correlation between grass protein concentration and methane emissions. This may relate to a lower methane production from fermented protein compared to most carbohydrates, or associated changes in the carbohydrate composition and degradability of the herbage fed.

Effects of dietary fat

In their summarization of measurements of methane emissions from studies of energy metabolism in cattle, Moe and Tyrrell (1979) found that diet ether extract intake did not have a significant effect on methane emission. In a more recent meta-analysis of measurements from calorimetry trials, Mills *et al.* (2009) also found that diet ether extract concentration or ether extract intake was not significantly related to methane emission. In contrast, Kirchgeßner *et al.* (1995) found that intake of ether extract had a large negative effect on methane emissions from dairy cattle. In the studies summarized by Moe and Tyrrell (1979) and Mills *et al.* (2009) there were relatively few experiments where supplemental fat was fed, and as discussed by Ellis *et al.* (2009), the influence of dietary fat concentration on methane emission is of greater relevance when diets are supplemented with fat, as the variation in basal diet fat concentration is normally small and doest not exert a strong effect on methane emission. In this regard it has long been known that feeding supplemental fat reduces methane emissions from ruminants (e.g. Czerkawski *et al.*, 1966; Andrew *et al.*, 1991; Tyrrell *et al.*, 1990). However, the current global concern regarding the need to reduce anthropogenic sources of greenhouse gas emissions has recently led to a plethora of more recent studies of the effects of a myriad of supplemental dietary fats on methane emissions (for recent reviews see e.g. Beauchemin *et al.*, 2008 and Martin *et al.*, 2010).

Several *in vitro* and *in vivo* studies have demonstrated marked reductions in methanogenesis as levels of supplemental fat are increased. The effects vary according to basal diet, type of fat and level of inclusion in the diet. Effects of supplemental fat on methane emissions are multifactorial. Firstly, fat supplies digestible energy that is not fermented, thus reduces the amount of methane emitted relative to DMI or energy consumption. Secondly, the use of hydrogen to saturate any dietary unsaturated fatty acids acts as an alternative sink for hydrogen that reduces the need for methane production. Finally, the inhibitory effects of specific fatty acids on rumen microbes and fibre degradation, and associated changes in fermentation dynamics that together can have pronounced inhibitory effects on methane emission. Recent research has focussed specifically on the methane suppressing effects of the medium chain fatty acids due to their apparent ability to achieve a suppression in methanogenesis whilst exerting lesser influences on DMI and animal performance than longer chain alternatives containing polyunsaturated fatty acids. Many studies have demonstrated large reductions in methane emission for C12:0 and C14:0 rich oils added to ruminant diets (Beauchemin *et al.*, 2008), but many of the conclusions drawn thus far are dependent on purely *in vitro* work or studies conducted with sheep. In a recent study with lactating dairy cows (Crompton *et al.*, 2010), decreases in methane emission with supplemental coconut oil were associated with decreases in DMI.

In practice supplemental fats and oils in various forms are often fed to dairy cows in order to increase the dietary energy density and also to manipulate milk composition. In particular there has been much interest, for well over 50 years, in the effects of supplemental fats on milk fatty acid profile (e.g. Kirchgeßner *et al.*, 1967). It can generally be assumed that when supplemental fat is fed, there will be a reduction in methane emission per kg DMI or per kg milk yield and fatty acid composition may be altered (Beauchemin *et al.*, 2008; Martin *et al.*, 2010). However, recent studies suggest that dietary fats containing C18:3 and longer chain polyunsaturated fatty acids, such as linseed, marine algae, and fish oils, are particularly potent in suppressing methane emissions (Clapperton, 1974; Fievez *et al.*, 2003; Czerkawski *et al.*, 1966; Martin *et al.*, 2010). This may relate to their degree of unsaturation, their reactivity in the rumen, and their effects on specific rumen microbes (e.g. fibrolytic bacteria and protozoa).

Other supplements and additives

There is a long list of other dietary supplements and feed additives that have been shown to reduce the total amount of methane produced relative to DMI or milk yield, at least during shorter term studies (Kebreab *et al.*, 2006; Beauchemin *et al.*, 2008; Martin *et al.*, 2010). These include antibiotic compounds and plant bioactives, yeasts, organic acids, and other hydrogen acceptors such as nitrates (Bozic *et al.*, 2009). In many cases compounds found to be effective *in vitro* have had limited practical efficacy *in vivo*, and similarly compounds effective at reducing methane emissions in sheep have had little or no effect in beef cattle or lactating dairy cows (McAllister and Newbold, 2008; Foley *et al.*, 2008; McCourt *et al.*, 2008). These differences in response may reflect differences between *in vitro* and *in vivo* ecosystems, and differences in rate of passage, microbial populations, and rumen ecology between sheep and cattle. In addition, in some cases compounds such as ionophores may be more effective in the short term than in the longer term, presumably due to microbial and rumen adaptations (Beauchemin *et al.*, 2008). In many cases, the efficacy of many of these additives, including supplemental fats, in reducing methane emissions *in vivo* have not been tested in the longer term. Considering the global scale of the research effort in this area, and the potential economic and environmental benefits to be gained, it is highly likely that effective and practical (and economically viable) additives will be discovered, developed and marketed in the foreseeable future.

Practical implications

Without a viable additive, vaccine, or similar 'magic bullet' that can be practically applied to abolish enteric methane emissions by ruminants, what practical management alternatives are available to mitigate methane emissions? Current assessments of the extensive data bases describing methane emissions during measurements of energy balance suggest that DMI explains some 90% of the variation in methane emission, but also that the relative amount of feed energy lost as methane reduces as level of DMI increases. This suggests that management strategies that improve feed conversion to milk and meat will reduce associated methane emissions, and that management systems that achieve higher DMI will also reduce relative methane losses. In this regard, there is currently considerable interest in the potential heritability of feed conversion efficiency and relative amount of methane emission per unit of milk or meat produced. Numerous studies have also suggested that increasing dietary starch at the expense of cellulose or soluble carbohydrates will reduce methane emission per unit DMI, and this is particularly true for starch sources with a slower rate of degradation in the rumen (Mills *et al.*, 2003). This again suggests a benefit of more intensive production systems reliant on cereal grains, and maize grain in particular. However, the potential negative effects of excess dietary starch, and the implications of cereal grain use in ruminant production systems in terms of carbon footprint and competition for resources must also be considered. The potential trade offs between mitigation options for reducing methane versus nitrogen emissions must also be considered. For example if increases fertilization rate reduces the methane production from grass (Bannink *et al.*, 2010), there may be negative consequences for associated nitrous oxide emissions. Finally, feeding supplemental fat is a widespread practice in ruminant production systems, and especially in the dairy industry, and there is no question that supplement fat will reduce methane emission relative to DMI. More recent research has also shown that specific fats, and thus fatty acids, are more potent at reducing methane emission than other fats. As with any dietary supplement or feed additive, the decision to use dietary fats or not is often based on cost and potential detrimental effects on diet utilization relative to benefit effects, but in future the benefit of reduced methane emission may also weigh into the decision as a benefit of feeding fat.

Nitrogen excretion

There is no question that the most important factor determining total nitrogen excretion as manure (faeces plus urine) in lactating dairy cows, and beef cattle, is total dietary nitrogen intake (James *et al.*, 1999; Castillo *et al.*, 2000; Kebreab *et al.*, 2001; Yan *et al.*, 2006). This reflects the fact that across a range of dietary types and levels of production, milk nitrogen excretion as a percentage of nitrogen intake decreases with increasing nitrogen intake (Figure 3; Mills *et al.*, 2009). However, as for the relationship between methane emission and DMI, there is variation in the excretion measured at a given level of nitrogen intake, which may arise due to effects of experimental or animal variation, or dietary and other environmental variables affecting the supply of nitrogenous compounds relative to requirements (Kebreab *et al.*, 2010). In this regard, the type of forage and concentrates included in the ration also has a small but important effect on total nitrogen excretion within individual experiments (Kebreab *et al.*, 2001, 2010). This implies that in spite of the overriding effect of total nitrogen intake, there is scope for reducing nitrogen excretion by reducing the amount fed, but also through other

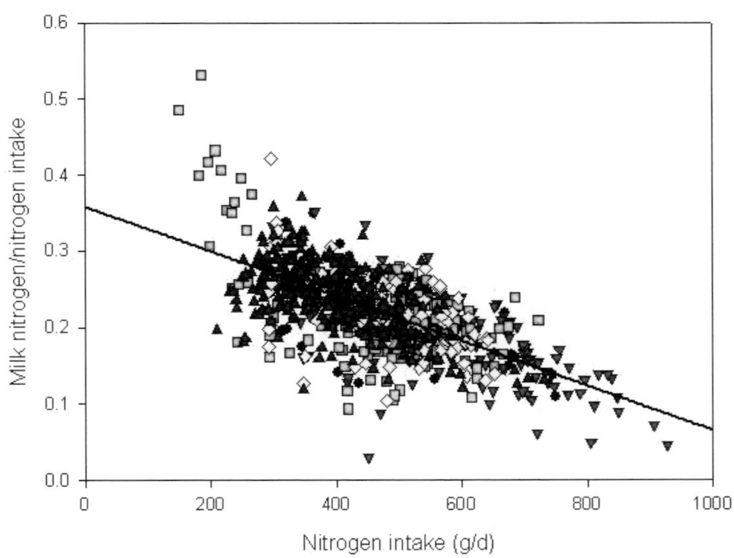

Figure 3. Milk nitrogen excretion (relative to nitrogen intake) and nitrogen intake in beef and dairy cattle (Mills et al.*, 2009).*

dietary management strategies. In addition, whilst the total amount of nitrogen excreted as manure increases linearly with increasing nitrogen intake, the proportion excreted in urine increases at an accelerating rate as nitrogen intake increases (Castillo *et al.*, 2000; Kebreab *et al.*, 2010). This may to a large extent reflect increasing absorption of nitrogenous compounds in excess of requirement, and an increasing proportion of N excretion as urea. This is a particular concern for ammonia emissions, as urinary urea is the primary source of nitrogen for ammonia generation in manure slurry, and to a large extent reflects the relatively low efficiency of dietary nitrogen conversion to proteinaceous product in ruminants (Figure 3), which seldom exceeds 30% in practice (Frank and Swensson, 2002).

More precise nutritional management of growing and lactating ruminants will reduce the amount of manure nitrogen excreted, as well as the portion excreted as more volatile urinary nitrogen. In addition to a reduction in the amount of nitrogen fed (Castillo *et al.*, 2000), strategies suggested for reducing nitrogen excretion and ammonia losses from cattle include increasing the metabolizable energy concentration of the ration fed (Kebreab *et al.*, 2001, 2010), particularly through the use of maize based concentrates containing slowly degraded starch (Castillo *et al.*, 2000). In a recent study, reducing dietary crude protein concentration from 145 to 185 g/kg DM to provide metabolizable protein at levels below, near, and above requirements for milk protein yield increased the gross efficiency of dietary nitrogen utilization for milk nitrogen production by 30% (Reynolds *et al.*, 2010). All of the approaches for reducing nitrogen excretion described above will impact the amount and profile of absorbable amino acids provided for absorption from the small intestine, which are an important determinant of the overall efficiency of dietary nitrogen utilization (Noftsger and St-Pierre, 2003). The increased dietary inclusion of starch, and particularly maize starch, may also reduce methane emission as discussed above. In addition, there is currently considerable interest in identifying plant bioactive compounds, such as essential oils or tannins, which slow the degradation of protein in the rumen and thereby reduce ammonia absorption and urinary nitrogen excretion (e.g. Calsamiglia *et al.*, 2007).

Nitrous oxide emission

Nitrous oxide is a greenhouse gas with an estimated potential some 12 to 14 times greater than the greenhouse gas potential of methane. The majority of nitrous oxide emissions attributable to ruminant production systems arise from soil following applications of inorganic fertilizer or manure. Losses from manure can be mitigated in part through manure management systems (Monteny *et al.*, 2006), but any dietary approach that reduces nitrogen excretion into manure will ultimately contribute to a reduction in nitrous oxide emissions attributable to milk and meat production by ruminants. Current assessments of the effect of dietary composition and intake on nitrogen excretion are based on measurements of nitrogen balance using total collection of faeces and urine. There has long been concern about the potential loss of gaseous nitrogen, including ammonia and nitrous oxide, during measurements of nitrogen balance, even when urine is acidified to reduce ammonia loss (Martin, 1966; Spanghero and Kowalski, 1997; Reynolds and Kristensen, 2008). In a recent study for which energy and nitrogen balance was measured in lactating dairy cows (Reynolds *et al.*, 2010), respiration chambers were equipped for measurements of ammonia and nitrous oxide emission in exhaust air, and ammonia capture in air conditioner condensate. There was no loss of nitrous oxide measured during measurements of nitrogen and energy balance and losses of ammonia and exhaust air plus ammonia captured in air conditioner condensate were minimal, and only accounted for at most less than 4 g of daily nitrogen balance. The amount of ammonia lost in exhaust air increased with increasing dietary nitrogen concentration, presumably due to increased concentration of nitrogen in faeces and urine. This confirms that direct emissions of ammonia by ruminants are minimal, and emissions of nitrous oxide are not attributable to ruminants per se, but to their manure (and ensiled forages; Monteny *et al.*, 2006). In this regard there are numerous management and environmental factors that influence the extent of nitrogen loss from manure as nitrous oxide, including manure composition, storage and handling, as well as method, timing and site of application to soils (Monteny *et al.*, 2006).

Conclusions

At the present time, 'best practice' in terms of diet formulation and management represents one of the most practical approaches to mitigating methane and nitrous oxide emissions attributable to milk and meat production. Although DMI is the overriding determinant of methane emission by ruminants, considerable variation in methane emission at a given level of DMI is apparent. This variation can be attributed to animals and their diets, and represents opportunities for reducing methane emissions through diet formulation and perhaps genetic selection. As regards diet composition, meta-analysis of existing data suggests that increasing concentrations of starch, and in particular maize starch, at the expense of cellulose (ADF) and/or soluble carbohydrates will reduce methane emission per unit DMI. Supplemental fats also reduce methane emissions, whilst increased milk yield will similarly reduce methane emission per unit DMI or milk yield. A number of summarizations of extant data bases have found no relationship between protein concentration of the total diet and methane emission, but there is evidence of a reduction in methane emission from lactating cows fed less mature forages with higher protein concentrations and associated changes in carbohydrate composition and quality. As regards nitrogen and nitrous oxide emissions, improvements in the efficiency of dietary nitrogen utilization for milk protein production have been associated with reduced dietary nitrogen concentration, and increased dietary starch concentrations, but the extent to which dietary protein can be reduced and fed to more precisely meet amino acid requirements without sacrificing milk yield remains to be demonstrated. Finally, many of these approaches are more suited to so-called 'intensive' production systems, and the wider implications of cereal use for milk and meat production, and the value of ruminants in extensive systems, must also be considered. In this regard there remains an important role of grazing ruminants for food production and thus a need for dietary and grazing approaches and sward management practices that minimize their greenhouse gas emissions.

References

Andrew, S.M., H.F. Tyrrell, C.K. Reynolds and R.A. Erdman, 1991. Net energy value for lactation of a dietary fat, calcium salts of long chain fatty acids, for cows fed silage based diets. J. Dairy Sci. 74, 2588-2600.

Bannink, A., J. Kogut, J. Dijkstra, J. France, E. Kebreab, A.M. Van Vuuren and S. Tamminga, 2006. Estimation of the stoichiometry of volatile fatty acid production in the rumen of lactating cows. J. Theoretical Biol. 238, 36-51.

Bannink, A., M.C.J. Smits, E. Kebreab, J.A.N. Mills, J.L. Ellis, A. Klop, J. France and J. Dijkstra, 2010. Simulating the effects of grassland management and grass ensiling on methane emission from lactating cows. J. Agric. Sci. (Cambr.) 148, 55-72.

Beauchimin, K.A., M. Kreuzer, F. O'Mara and T.A. McAllister, 2008. Nutritional management for enteric methane abatement: a review. Austr. J. Exper. Agric. 48, 21-27.

Blaxter, K.L. and J.L. Clapperton, 1965. Prediction of the amount of methane produced by ruminants. Brit. J. Nutr. 19, 511-522.

Boadi, D., C. Benchaar, J. Chiquette and D. Massé, 2004. Mitigation strategies to reduce enteric methane emissions from dairy cows: Update review. Can. J. Anim. Sci. 84, 319-335.

Bozic, A.K., R.C. Anderson, G.E. Carstens, S.C. Ricke, T.R. Callaway, M.T. Yokoyama, J.K. Wang and D.J. Nisbet, 2009. Effects of methane-inhibitors nitrate, nitroethane, lauric acid, Lauricidin, and the Hawaiian marine algae *Chaetoceros* on ruminal fermentation *in vitro*. Bioresource Technology 100, 4017-4025.

Bratzler, J.W., and E.B. Forbes, 1940. The estimation of methane production by cattle. J. Nutr. 19, 611-615.

Calsamiglia, S., M. Busquet, P.W. Cardozo, L. Castillejos and A. Ferret, 2007. Invited review: Essential oils as modifiers of rumen microbial fermentation. J Dairy Sci. 90, 2580-2595.

Cammell, S.B., J.D. Sutton, D.E. Beever, D.J. Humphries, and R.H. Phipps, 2000. The effect of crop maturity on the nutritional value of maize silage for lactating dairy cows. 1. Energy and nitrogen utilization. Anim. Sc. 71, 381-390.

Castillo, A.R., E. Kebreab, D.E. Beever and J. France, 2000. A review of efficiency of nitrogen utilisation in dairy cows and its relationship with the environmental pollution. J. Anim. Feed Sci. 9, 1-32.

Clapperton, J.L., 1974. The effect of trichloroacetamide, chloroform and linseed oil given into the rumen of sheep on some of the end-products of rumen digestion. Br. J. Nutr. 32:155-161.

Crompton, L.A., Mills, J.A.N. and Reynolds, C.K, 2010. Effect of feeding frequency and replacing calcium salts of palm oil with crushed rapeseed or coconut oil on methane emissions in lactating dairy cows. Proc. Nut. Soc., in press.

Czerkawski, J.W., K.L. Blaxter and F.W. Wainman, 1966. The effect of linseed oil and of linseed oil fatty acids incorporated in the diet on the metabolism of sheep. Br. J. Nutr. 20, 485-494.

Ellis, J.L., E. Kebreab, N.E. Odongo, B.W. McBride, E.K. Okine and J. France, 2007. Prediction of methane production from dairy and beef cattle. J Dairy Sci. 90, 3456-3466.

Ellis, J.L., E. Kebreab, N.E. Odongo, K. Beauchemin, S. McGinn, J.D. Nkrumah, S.S. Moore, R. Christopherson, G. K. Murdoch, B.W. McBride, E.K. Okine and J. France, 2009. Modeling methane production from beef cattle using linear and nonlinear approaches. J Anim Sci. 87,1334-1345.

FAO, 2007. Livestocks Long Shadow. Environmental Issues and Options. Food and Agriculture Organization of the United Nations.

FAO, 2010. Greenhouse Gas Emissions from the Dairy Sector. A Life Cycle Assessment. Food and Agriculture Organization of the United Nations, Animal Production and Health Division.

Fievez, V., F. Dohme, M. Danneels, K. Raes and D. Demeyer, 2003. Fish oils as potent rumen methane inhibitors and associated effects on rumen fermentation *in vitro* and *in vivo*. Anim. Feed Sci. Tech. 104, 41-58.

Foley, P.A., D.A. Kenney, D.K. Lovett, J.J. Callan, T.M. Boland and F.P. O'Mara, 2008. Effect of DL-malic acid supplementation on feed intake, methane emissions, and performance of lactating dairy cows at pasture. J. Dairy Sci. 2008, 3258-3264.

Frank, B. and C. Swensson, 2002. Relationship between content of crude protein in rations for dairy cows and milk yield, concentration of urea in milk and ammonia emissions. J. Dairy Sci. 85, 1829-1838.

Gill, M., P. Smith and J.M. Wilkinson, 2010. Mitigating climate change: the role of domestic livestock. Animal 4, 323-333.

Hindrichsen, I. K., H. -R. Wettstein, A. Machmuller, B. Jorg, and M. Kreuzer, 2005. Effect of the carbohydrate composition of feed concentrates on methane emission from dairy cows and their slurry. Environ. Monitoring Assessment 107, 329-350.

Hindrichsen, I.K. and M. Kreuzer, 2009. High methanogenic potential of sucrose compared with starch at high ruminal pH. J. Anim. Physiol. 93, 61-65.

Holter, J.B. and A.J. Young, 1992. Methane prediction in dry and lactating Holstein cows. J. Dairy Sci. 75, 2165.

Kebreab, E., J. France, D.E. Beever and A.R. Castillo, 2001. Nitrogen pollution by dairy cows and its mitigation. Nutr. Cycl. Agroecosyst. 60, 275-285.

Kebreab, E., K. Clark, C.Wagner-Riddle and J. France, 2006. Methane and nitrous oxide emissions from Canadian animal agriculture: A review. Can. J. Anim. Sci. 86, 135-158.

Kebreab E, K. A. Johnson, S. L. Archibeque, D. Pape and T. Wirth, 2008. Model for estimating enteric methane emissions from United States dairy and feedlot cattle. J Anim Sci. 86, 2738-2748.

Kebreab, E., A.B. Strathe, J. Dijkstra, J.A.N. Mills, C.K. Reynolds, L.A. Crompton, T. Yan and J. France, 2010. Energy and protein interactions and their effect on nitrogen excretion in dairy cows. Proceedings of the 3rd EAAP Energy and Protein Symposium, in press.

Kirchgeßner, M., H. Friesecke and G. Koch, 1967. Nutrition and the Composition of Milk. Crosby Lockwood & Son, London.

Kirchgeßner, M., W. Windisch and H.L. Müller, 1995. Nutritional factors for the quantification of methane production. In: Englehardt, W. v., S. Leonhard-Marek, G. Breves, and D. Giesecke (eds.), Ruminant Physiology: Digestion, Metabolism, Growth and Reproduction. Proceedings of the Eighth International Symposium on Ruminant Physiology. Ferdinand Enke Verlag, Stuttgart, Germany, pp 333-348.

Kriss, M., 1930. Quantitative relations of the dry matter of the food consumed, the heat production, the gaseous outgo, and the insensible loss in body weight of cattle. J. Ag. Res. 40, 283-295.

Martin, A.K., 1966. Some errors in the determination of nitrogen retention of sheep by nitrogen balance studies. Br. J. Nutr. 20, 325-337.

Martin, C., D.P. Morgavi and M. Doreau, 2010. Methane mitigation in ruminants: from microbe to farm scale. Animal 4, 351-365.

McAllister, T. A. and C.J. Newbold, 2008. Redirecting rumen fermentation to reduce methanogenesis. Austr. J. Exper. Agric. 48, 7-13.

Mc Court, A., T. Yan, and C.S. Mayne, 2007. Effect of forage type on methane production from dairy cows. Proc. Brit. Soc. Anim. Sci., 48.

McCourt, A.R., T. Yan, S. Mayne, and J. Wallace, 2008. Effect of inclusion of encapusulated fumaric acid on methane production from grazing dairy cows. Proc. Brit. Soc. Anim. Sci., 64.

McGeough, E.J., P. O'Kiely, P.A. Foley, K.J. Hart, T.M. Boland and D.A. Kenny, 2010. Methane emissions, feed intake, and performance of finishing beef cattle offered maize silages harvested at 4 different stages of maturity. J Anim Sci. 88, 1479-1491.

Mills, J.A.N., J. Dijkstra, A. Bannink, S.B. Cammell, E. Kebreab and J. France, 2001. A mechanistic model of whole-tract digestion and mathanogenesis in the lactating dairy cow: Model development, evaluation and application. J. Anim. Sci. 79, 1584-1597.

Mills J.A.N., E. Kebreab, C.M. Yates, L.A. Crompton, S.B. Cammell, M.S. Dhanoa, R.E. Agnew and J. France, 2003. Alternative approaches to predicting methane emissions from dairy cows J. Anim. Sci. 81, 3141-3150.

Mills, J.A.N., L.A. Crompton, A. Bannink, S. Tamminga, J.M. Moorby and C.K. Reynolds, 2009. Predicting methane emissions and nitrogen excretion from cattle. In: Crompton, L.A. and T.R. Wheeler (editors), Proceedings of the forty-first meeting of the agricultural research modellers' group, J. Agric. Sci. 147, 741.

Moe, P.W. and H.F. Tyrrell, 1979. Methane production in dairy cows. J. Dairy Sci. 62, 1583-1586.

Monteny, G.-J., A. Bannink and D. Chadwick, 2006. Greenhouse gas abatement strategies for animal husbandry. Agriculture, Ecosystems and Environment 112, 163-170.

Murray, R.M., Bryant, A.M. and Leng, R.A, 1978. Methane production in the rumen and lower gut of sheep given lucerne chaff: effect of level of intake. Br. J. Nutr. 39, 337-345.

Noftsger, S. and N.R. St-Pierre, 2003. Supplementation of methionine and selection of highly digestible rumen undegradable protein to improve nitrogen efficiency for milk production. J. Dairy Sci. 86, 958-969.

Pitesky, M.E., K.R. Stackhouse and F.M. Mitloehner, 2009. Clearing the air: Livestock's contribution to climate change. Adv. Agronomy 103, 1-222.

Reynolds, C.K., H.F. Tyrrell and P.J. Reynolds, 1991. Effects of diet foragetoconcentrate ratio and intake on energy metabolism in growing beef heifers: Whole body energy and nitrogen balance and visceral heat production. J. Nutr. 121, 9941003.

Reynolds, C.K., and N.B. Kristensen, 2008. Nitrogen recycling through the gut and the nitrogen economy of ruminants – an asynchronous symbiosis. J. Anim. Sci. 86 (E. Suppl.), E293-E305.

Reynolds, C.K., L.A. Crompton, J.A.N. Mills, D.J. Humphries, P. Kirton, A.E. Relling, T.H. Misselbrook, D.R. Chadwick and D.I. Givens, 2010. Effects of diet protein level and forage source on energy and nitrogen balance and methane and nitrogen excretion in lactating dairy cows. Proceedings of the EAAP Energy and Protein Symposium, submitted.

Spanghero, M. and Z.M. Kowalski, 1997. Critical analysis of N balance experiments with lactating cows. Livestock Prod. Sci. 52, 113-122.

Tyrrell, H.F., C.K. Reynolds and H.D. Baxter, 1990. Energy metabolism of Jersey and Holstein cows fed total mixed diets with or without whole cottonseed. J. Dairy Sci. 73 (supplement 1), 192.

Wilkerson, V.A., D.P. Casper and D.R. Mertens, 1995 The prediction of methane production of Holstein cows by several equations. J. Dairy Sci. 78, 2402-2414.

Yan, T., R.E. Agnew, F.J. Gordon and M.G. Porter, 2000. Prediction of methane energy output in dairy and beef cattle offered grass silage-based diets. Livestock Prod. Sci. 64, 253-263.

Yan, T., J.P. Frost, R.E. Agnew, R.C. Binnie and C.S. Mayne, 2006. Relationships among manure nitrogen output and dietary and animal factors in lactating dairy cows. J. Dairy Sci. 89, 3981-3991.

Milk production and nitrogen excretion of cows fed diets with differing starch/protein ratios

S. Colombini[1], G. Galassi[1], G. Colombari[2], D. Magistrelli[1], M. Guerci[1], L. Rapetti[1] and G.M. Crovetto[1]
[1]Università degli Studi, Dipartimento di Scienze Animali, via Celoria 2, 20133 Milano, Italy
[2]ERSAF, Struttura 'Sperimentazione Agronomica e Lattiero Casearia e Gestione Aziende Agrozootecniche', Italy

Introduction

One of the major objectives of the European Common Agricultural Policy is the development of a sustainable farming system with environment-friendly production management aimed at reducing nitrogen (N) excretion. The reduction of urinary N excretion is a promising way to decrease N losses from dairy farms and ammonia emission. Balancing diets for CP content may represent a useful tool for reducing total urinary N excretion. In a previous study (Crovetto *et al.*, 2009) a dietary CP content of 17.2% on DM increased milk yield in comparison with a content of 15.4%, but it also resulted in a higher content of milk urea nitrogen (MUN), probably due to an inadequate ratio between ruminal fermentable energy and protein. The aim of this work was to evaluate the effects of feeding dairy cows diets with differing energy/protein ratio on milk production and N excretion.

Materials and methods

Twenty lactating Holstein cows were divided into 2 groups on the basis of milk yield (kg/d): low (L) or high (H). Within each group, cows were randomly divided in two subgroups and fed, in a change-over design of 2 periods (14 d of adaptation and 7 d of sample collection), 2 diets (Table 1) with a different starch to protein ratio: low (L; 1.4) or high (H; >1.6). Corn silage, Italian ryegrass and alfalfa hay, corn, barley and soybean meal and soybean flakes were included in each diet to give the formulated CP and starch contents. Individual milk yield was measured daily and individual milk samples were collected at each milking and analysed for fat, protein and milk urea N (MUN) (MilkoScan™ FT6000, FOSS, Hillerod, Denmark). Diets and orts were collected daily to determine dry matter (DM) content and diet chemical composition. Individual urine and faeces samples were collected after morning and before evening milking on 2 different days. Acidified urine samples were analysed for total N and creatinine. Faecal output was estimated using indigestible ADF as an internal marker and samples were analysed for total N. Data were analysed by proc mixed of SAS.

Results and discussion

Within each productive level, milk and FCM yields were not affected by dietary treatment (Table 2). Within the high yielding cows a milk yield higher than 35 kg/d was achieved with a dietary CP

Table 1. Chemical composition (% DM) of the experimental diets.

Milk yield	L		H	
Starch/protein ratio	L	H	L	H
Starch	21.0	23.6	23.4	26.1
Crude protein	15.5	14.9	16.6	15.3
aNDFom	36.0	33.7	32.7	31.6
ADFom	26.7	24.8	24.1	23.0
NFC	40.1	43.0	42.2	44.8

content of 15.3% on DM, lower than the values (16.5-17%) reported in other studies (Ipharraguerre and Clark, 2005; Colmenero and Broderick, 2006). Milk fat and protein contents were lower for the highest production level (Table 2) and, overall, were not affected by diet. Milk urea N was higher for the diets with lower starch/protein ratio and it was not affected by the productive level. The diets with the higher starch/protein ratio decreased urinary N excretion, expressed as either daily excretion or as a percentage of N-intake (Table 3). Faecal N excretion (% N intake) was increased by the diets with the higher starch/protein ratio, although the amount of faecal N was affected only by the production level, due to the difference in N-intake. Nitrogen efficiency (N milk/N intake) was higher for the high producing cows and, within this group, was significantly higher for the diet with the higher starch/protein ratio (Table 3). Overall N efficiency was 25.5% and similar to the value of 25.8% reported in a study conducted in Spain over 76 farms (Arriaga *et al.*, 2009). The diet with the higher starch/protein ratio reduced urinary N excretion in both groups, but improved N efficiency only at the higher production level.

Table 2. Dry matter intake, milk yield and quality of cows fed the experimental diets.

Milk yield		L		H		SE	Probability		
Starch/protein ratio		L	H	L	H		Production	Diet	Prod. × Diet
DMI	kg/d	23.0	23.8	26.6	26.9	0.57	<0.001	0.007	0.175
Milk	kg/d	25.0	24.5	37.5	37.0	1.48	<0.001	0.250	0.963
FCM	kg/d	26.6	25.9	33.5	33.9	1.07	<0.001	0.193	0.151
Fat	%	4.18	4.10	3.36	3.45	0.13	<0.001	0.935	0.052
Protein	%	3.41	3.48	3.14	3.18	0.09	0.003	0.162	0.834
MUN	mg/dl	12.5	11.2	13.0	9.6	0.42	0.294	<0.001	0.059

Table 3. N excretion with urine and faeces of the experimental diets fed to the dairy cows.

Milk yield		L		H			Probability		
Starch/protein ratio		L	H	L	H	SE	Production	Diet	Prod. × Diet
N-intake (NI)	g/d	569	569	708	657	14.2	<0.001	<0.001	<0.001
N-faeces	g/d	223	233	300	304	10.9	<0.001	0.390	0.660
N-urine	g/d	244	217	285	235	12.3	0.049	0.002	0.275
N-faeces	% NI	39	41	42	46	1.32	0.019	0.002	0.428
N-urine	% NI	43	38	40	36	1.68	0.158	0.015	0.993
N efficiency	%	24	24	26	28	0.60	0.002	0.025	0.045

References

Arriaga, H., M. Pinto, S. Calsamiglia and P. Merino, 2009. Nutritional and management strategies on nitrogen and phosphorus use efficiency of lactating dairy cattle on commercial farms: An environmental perspective. J. Dairy Sci. 92, 204-215.

Colmenero, J.J., and G.A. Broderick, 2006. Effect of dietary crude protein concentration on milk production and nitrogen utilization in lactating dairy cows. J. Dairy Sci. 89, 1704-1712.

Crovetto, G.M., S. Colombini, G. Colombari and L. Rapetti, 2009. Effects of constant vs variable dietary protein content on milk production and N utilization in dairy cows. Ital. J. Anim. Sci. 8, 292-294.

Ipharraguerre, I.R. and J.H. Clark, 2005. Varying protein and starch in the diet of dairy cows. II. Effects on performance and nitrogen utilization for milk production. J. Dairy Sci. 88, 2556-2570.

Nitrogen metabolism of diets differing in forage type and forage to concentrate ratio in sheep and goats

G. Cantalapiedra-Hijar[1], E. Molina-Alcaide[1], M.J. Ranilla[2], D.R. Yáñez-Ruiz[1], M.L. Tejido[2] and M.D. Carro[2]
[1]Estación Experimental del Zaidín (CSIC), Profesor Albareda, 1, 18008 Granada, Spain
[2]Departamento de Producción Animal, Universidad de León, 24071 León, Spain

Introduction

Nitrogen represents an important source of the environmental contamination from animal production. Excretion of N in faeces and urine depends mainly on the N content of the diet and its digestibility, but it also may vary with animal species (Devendra, 1990). The aim of this study was to compare N utilization in sheep and goats fed four medium-good quality diets at similar levels of intake. The diet ingredients were selected to represent those most frequently used under practical feeding conditions of both animal species in Spain.

Materials and methods

Four rumen-fistulated Granadina goats (46.5±1.43 kg) and 4 Merino sheep (52.5±3.10 kg) were used in 4 experimental periods. Four total mixed diets were formulated according to a 2×2 factorial arrangement of treatments. The diets had forage:concentrate (FC; dry matter (DM) basis) ratios of 70:30 (H) or 30:70 (L) with either lucerne hay (A) or grass hay (G) as forage, and were designated as HA, LA, HG and LG. Nitrogen content was 26.9, 28.3, 19.4 and 25.6 g/kg DM and neutral-detergent fibre content was 426, 374, 499 and 401 g/kg DM for HA, LA, HG and LG diets, respectively. Diets were offered to the animals twice daily (08:00 and 14:00 h) at a daily rate of 56 g DM/kg body weight$^{0.75}$ to minimize feed selection. Each period consisted of 10 d of adaptation, 7 d for determination of N excretion (animals in metabolism cages equipped for quantitative collection of faeces and urine) and 1 d for determining ruminal NH_3-N concentrations (0 and 4 h after the morning feeding). Data were analyzed using the MIXED procedure (SAS Inst. Inc., Cary, NC). The effects of animal species (AS), FC ratio, type of forage (FOR), period, and the interactions AS×FC, AS×FOR were considered fixed, and animal within species effect was considered random.

Results and discussion

There was no difference between AS in N intake (Table 1). The percentage of N intake excreted in urine was higher in goats than in sheep (averaged across diets, 59.7 and 40.1%, respectively), but the percentage of N intake in faeces was lower in goats compared with sheep (averaged across diets, 23.2 and 27.7%, respectively). Protein digestibility was greater in goats than in sheep, which was consistent with the greater NH_3-N concentrations observed in the rumen of goats. These results are in accordance with others (Molina *et al.*, 2000) and would indicate either a greater proteolytic activity, more intense chewing activity or both in goats compared with sheep. A significant AS×FC interaction was detected for N digestibility and ruminal NH_3-N concentrations but there were no AS×FOR interactions for any parameter.

Goats showed lower N retention than sheep for all diets either expressed in absolute terms (g/d) or as g per kg of digestible organic matter intake (DOMI), which is in agreement with previous studies (Al Jassim *et al.*, 1991; Molina *et al.*, 2000).

These results indicate different N utilization in goats and sheep with similar N intakes. The interspecies differences in N digestibility and ruminal ammonia concentrations seem to be modulated by the composition of the diet. The average percentage of total N intake excreted in both the urine and faeces

Table 1. Nitrogen intake (NI), excretion and digestibility (ND), ruminal NH_3-N concentrations and N retained in sheep and goats fed four different diets.

Item	Animal species	Diet				SEM	Effects[1]				
		HA	LA	HG	LG		AS	FC	FOR	AS×FC	AS×FOR
N intake	Sheep	26.8	31.5	18.6	25.5	1.21	0.16	<0.001	<0.001	0.06	0.16
(g/d)	Goat	27.8	28.5	16.6	20.3						
Urinary N	Sheep	41.2	44.5	33.7	40.9	1.78	<0.001	<0.001	0.02	0.95	0.10
(% NI)	Goat	58.2	62.2	55.7	62.5						
Faecal N	Sheep	24.6	22.5	37.4	26.2	0.74	0.007	<0.001	<0.001	0.17	0.13
(% NI)	Goat	18.9	20.8	35.6	17.3						
ND	Sheep	75.4	77.5	62.6	73.8	3.04	0.006	<0.001	<0.001	0.02	0.36
(% NI)	Goat	81.1	79.2	64.4	82.7						
NH3-N	Sheep	252	239	93	170	16.8	0.01	<0.001	<0.001	0.01	0.95
(mg/l)	Goat	227	306	139	304						
Retained	Sheep	34.2	33.1	28.9	32.9	1.56	<0.001	0.07	0.002	0.53	0.24
N (% NI)	Goat	22.9	17.0	8.7	20.2						
Retained	Sheep	9.57	10.5	5.86	8.94	0.541	<0.001	0.003	<0.001	0.10	0.49
N (g/kg DOMI)	Goat	8.11	5.77	1.90	5.60						

[1] AS = animal species; FC = forage:concentrate ratio; FOR = forage.

was 67.8 and 82.8% for sheep and goats, respectively, which could have important environmental implications.

Acknowledgements

Financial support from the M.C.Y.T. of Spain (projects AGL2004-04755-C02-01 and C02-02) is gratefully acknowledged. G. Cantalapiedra-Hijar gratefully acknowledges the receipt of scholarship from the M.E.C. of Spain (AP2004-1908).

References

Al Jassim, R.A.M., S.A. Hassan, A.N. Al-Ani and T.K. Dana, 1991. Effects of undegradable protein supplementation on digestion and nitrogen balance in sheep and goats. Small Rum. Res. 5, 57-63.

Devendra, C., 1990. Comparative aspects of digestive physiology and nutrition in goats and sheep. In: Devendra, C. and E. Mazumi (eds.), Ruminant Nutrition and Physiology in Asia. IDRC, Singapore, SG: 45-60.

Molina, M.E., A.I. Martín. and J.F. Aguilera, 2000. A comparative study of nutrient digestibility, kinetics of degradation and passage and rumen fermentation pattern in goats and sheep offered good quality diets. Livest. Prod. Sci. 64, 215-223.

Nitrogen losses per unit of nitrogen deposition as derived from modelling of protein utilization depending on dietary protein quality parameters and age of growing barrows

F. Liebert and C. Wecke
Department of Animal Sciences, Division Animal Nutrition Physiology, Göttingen University, Kellnerweg 6, 37077 Göttingen, Germany

Introduction

Optimization of protein metabolism in growing animals is achieved, when the aimed level of growth is yielded with both minimized amino acid (AA) losses during absorption and during post-absorptive utilization process. However, N losses via faeces play a minor role in growing pigs when compared to N-losses via urine. Consequently, assessing requirements in terms of quantity, efficacy and optimal dietary AA ratio to meet the metabolic needs is required to decrease urine-N losses within the physiological limitations. Modelling procedures may provide additional information to quantify and to predict the effects of modulating the dietary protein quality on efficiency of N-deposition and resulting N-losses. Our investigation aimed to apply a verified modelling procedure for prediction of N-losses per unit N-deposition of growing barrows depending on age.

Materials and methods

Totally 64 barrows of modern genotype (Pietrain × (Large white × German landrace)) were utilized for N-balance studies (n=4) in five age periods, corresponding to 30, 50, 70, 90, and 110 kg BW. Within each of the age periods (24 balance periods), the influence of graded dietary protein supply on N-excretion pattern was observed. Six graded levels of dietary protein supply with equal protein composition was achieved by dilution of a high protein mixture (33.9% CP; Lys:Met+Cys:Thr:Try =1:0.77:0.84:0.26) with barley (25%), wheat (25%), potato protein (16.5%), wheat gluten (14%) and soybean protein concentrate (6%) as main ingredients, fortified with DL-Met and L-Try. A protein free mixture based on wheat starch was utilized to yield lysine limiting diets with graded protein supply. Statistical analyses of N-balance data made use of the N-utilization model as described for growing animals (Liebert, 2008; Wecke and Liebert, 2009) and its model parameter b, expressing the slope of the exponential function dependent on dietary protein quality. N-maintenance requirement (NMR) and threshold values for daily N-retention ($NR_{max}T$) dependent on age were taken from Wecke and Liebert (2009) for further modelling of N-losses, at least per unit of N-deposition.

Results and discussion

According to earlier observations (Thong and Liebert, 2004), daily faecal N-excretion remained on a rather low level below 500 $mgN/BW_{kg}^{0.67}$ when daily protein supply markedly increased the N-intake up to a level near 6,000 $mgN/BW_{kg}^{0.67}$. The potential of increasing dietary protein quality to lower faecal N-excretion is very limited. At highest protein supply, daily N-excretion via urine exceeded 2,500 $mgN/BW_{kg}^{0.67}$. As expected, the observed response to graded levels of dietary protein supply was robust and of quantitative importance. However, improving dietary protein quality (AA-balance) significantly reduced urine N-losses. As an example, results for an average BW of 70 kg are presented in more detail.

N-excretion (NEX) and N-deposition (ND) dependent on N-intake (NI) are approximated by the following equations:

$$NEX = NI - 2{,}856 \times (1 - e^{-0.000319 \times NI}) + 375$$
$$ND = 2{,}856 \times (1 - e^{-0.000319 \times NI}) - 375$$

Relating NEX : ND from both of the equations, the relationship between NI and NEX : ND ratio dependent on dietary protein quality is given in Figure 1.

It is demonstrated that NEX : ND ratio yields a minimum which is far from acceptable NI to yield ND according to the potential of fast growing pigs. However, the applied procedure allows identification of a dietary protein quality which provides a compromise from ecological point of view. Consequently, further investigations will have to find a context for modelling between protein quality, minimal dietary protein supply, individual amino acid efficiency and optimal dietary amino acid ratio to couple adequate growth performance and minimal ecological load for growing pigs.

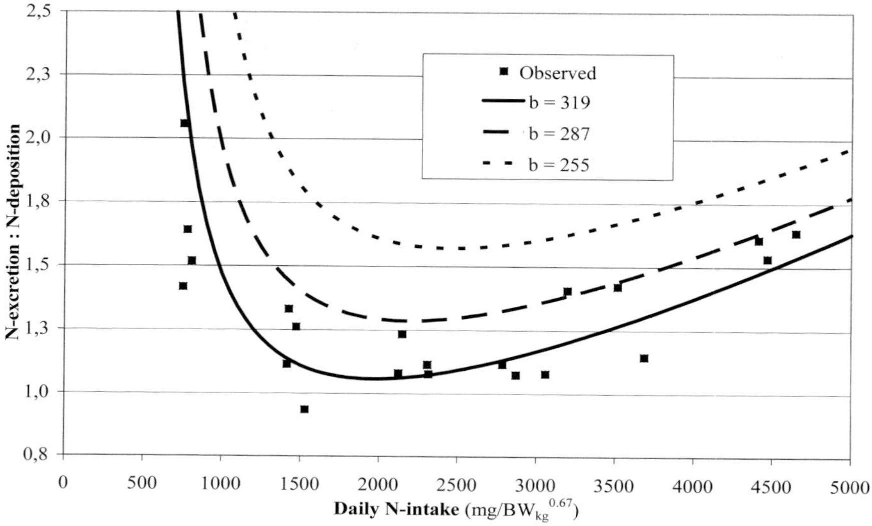

Figure 1. Relationship between NI and NEX:ND ratio in growing pigs (70 kg BW) at 3 different levels of dietary protein quality as described by model parameter (b).

References

Liebert, F., 2008. Modelling of protein metabolism yields amino acid requirements dependent on dietary amino acid efficiency, growth response, genotype and age of growing chicken. Avian Biol. Res. 1, 101-110.

Thong, H.T. and F. Liebert, 2004. Potential for protein deposition and threonine requirement of modern genotype barrows fed graded levels of protein with threonine as the limiting amino acid. J. Anim. Physiol. a. Anim. Nutr. 88, 196-203.

Wecke, C. and F. Liebert, 2009. Lysine requirement studies in modern genotype barrows dependent on age, protein deposition and dietary lysine efficiency. J. Anim. Physiol A. Anim. Nutr. 93, 295-304.

Redefining nutrient protein energy requirements of fish in sub-optimum environments

C.G. Carter, R.S. Katersky, J. Barnes, R.C. Hauler and A.R. Bridle
Tasmanian Aquaculture and Fisheries Institute, University of Tasmania, Private Bag 49, Hobart, Tasmania 7001, Australia

Introduction

Nutrient requirements of fish have mainly been defined for small juvenile animals held under optimum environmental conditions. In contrast, the majority of aquaculture is conducted outdoors and where fish are subject to periods of sub-optimum environmental conditions (Carter *et al.*, 2005), which are likely to increase due to climate change and threaten the sustainability of aquaculture. To further understand environment-nutrient interactions in relation to protein and energy requirements two model fish species were selected (Carter *et al.*, 2005; Katersky and Carter, 2007). Temperate Atlantic salmon (*Salmo salar*) and tropical barramundi or Asian sea bass (*Lates calcarifer*) perform well in aquaculture over a wide temperature range (Katersky and Carter, 2007).

Materials and methods

Four extruded salmon diets of varying protein and energy combinations were produced on a commercial extruder and vacuum coated with fish oil. Each diet was fed to satiation to triplicate groups of 9 salmon held at 19 °C and two dissolved oxygen (DO) levels of >90 and 70% saturation. In addition and at both DO levels single groups of 9 salmon were fed at 80, 60 and 30% of satiation. Relationships between intake for digestible energy (DE) and protein (DP) and whole carcass protein and energy retention were used to predict optimum dietary DP to DE for growing salmon. Barramundi experiments on protein synthesis were as described by Katersky and Carter (2007). Anabolic stimulation efficiency (100% × protein synthesis/protein intake) and synthesis retention (100% × protein retention/protein synthesis) (Carter and Houlihan, 2001) provided further analysis of protein utilisation. Statistical analyses were carried out with SPSS.

Results and discussion

At 19 °C relationships between nutrient retention and digestible nutrient intake were described by: digestible energy intake ($kJ.kg^{-0.8}/d$) = 185.8+1.5 Energy retention ($kJ.kg^{-0.8}/d$) (R^2=0.98; F=1019; P<0.001); digestible protein intake ($g.kg^{-0.7}/d$) = 2.19+1.87 protein retention ($g.kg^{-0.7}/d$) (R^2=0.96; F=514; P<0.001); and the optimum DP:DE was 19.77 g DP/MJ DE. Compared to salmon at 90% DO maximum feed intake and maximum growth rate were lower at 70% DO (Figure 1). At 70% DO energy retention was similar and protein retention lower than at 90% DO. Consequently, the optimum DP:DE increased to 20.3 g DP/MJ DE. Under optimum conditions of <15 °C and 100% DO the optimum DP:DE is 18.5 g DP/MJ DE, illustrating the effect of sub-optimal environmental conditions in increasing protein metabolism and requirements.

For salmonids increasing dietary protein:energy stimulates protein synthesis (Figure 2A) to regulate excess amino acids and consequently synthesis retention decreases (Figure 2B). Preliminary analysis of barramundi suggested more efficient protein metabolism at high dietary protein:energy under optimum temperatures, where as at a sub-optimum temperature anabolic stimulation and synthesis retention were lower and similar to salmonids (Figure 2). Several factors, including a higher protein requirement, faster growth and higher protein retention efficiency, may explain the differences. Whilst sub-optimum environmental conditions test the adequacy of nutrient supply and impact negatively on growth, these two fish species had robust physiological systems for maintaining efficient growth across a wide range of environmental conditions and that make them ideal for sustainable aquaculture.

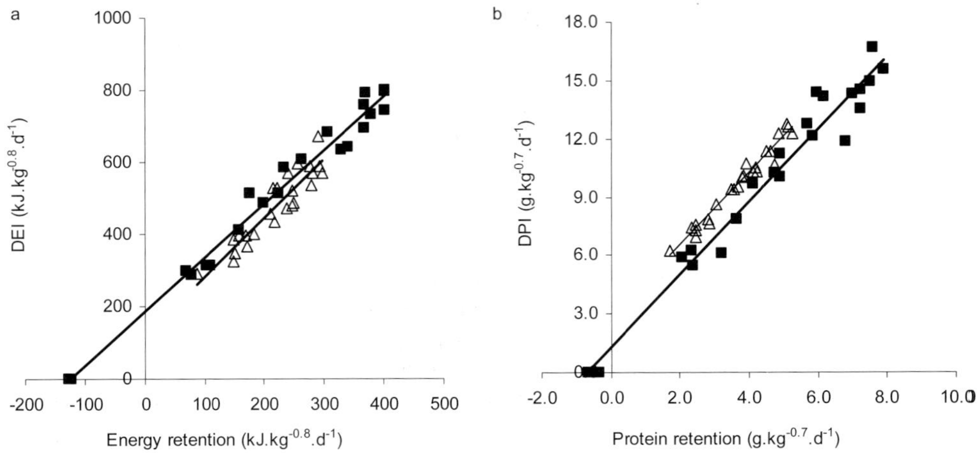

Figure 1. Relationships between (A) energy retention and digestible energy intake (DEI) and (B) protein retention and digestible protein (DPI) intake for Atlantic salmon at 19 °C and >90% (■) or 70% (Δ) dissolved oxygen.

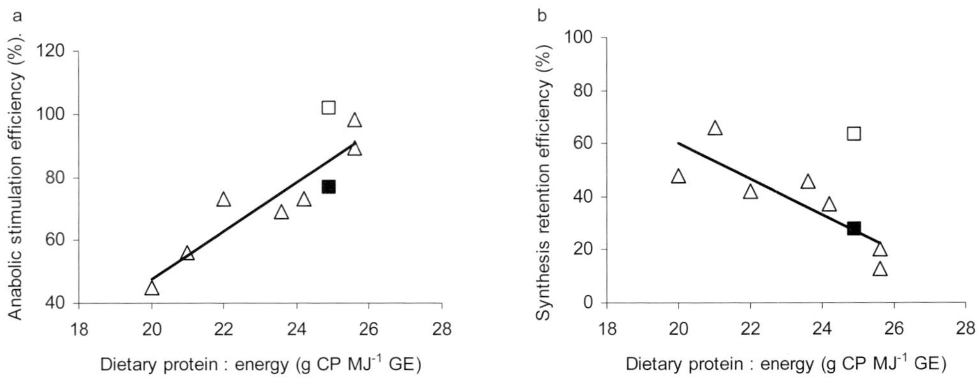

Figure 2. Relationships between dietary protein:energy and (A) anabolic stimulation efficiency and (B) synthesis retention efficiency for salmonid species (Δ) (adapted from Carter and Houlihan, 2001) and for barramundi at optimum (□) and sub-optimum (■) temperatures (Katersky and Carter, 2007).

References

Carter, C.G. and D.F. Houlihan, 2001. Protein synthesis. In: P.A. Wright and P.M. Anderson (eds.). Nitrogen excretion. Academic Press, New York: 31-75.

Carter, C.G., R.S. Katersky, J. Barnes, K. Irwin and R.C. Hauler, 2005. Interactions between temperature and nutrition in Atlantic salmon *Salmo salar* and barramundi *Lates calcarifer*. In: Howell, B. and R. Flos (eds.), Lessons from the past to optimise the future. European Aquaculture Society Special Publication 35: 156-157.

Katersky, R.S., and C.G. Carter, 2007. High growth efficiency occurs over a wide temperature range for juvenile barramundi *Lates calcarifer* fed a balanced diet. Aquaculture 272, 444-450.

Emissions of hydrogen gas from sheep fed fresh perennial ryegrass and pelleted diets

C.S. Pinares-Patiño, S. Muetzel, G. Molano, C. Hunt and H. Clark
AgResearch Grasslands, Private Bag 11008, Palmerston North, New Zealand

Introduction

Ruminant methane (CH_4) emissions are the result of microbial fermentation of feed components, mainly in the rumen. Primary digestive microorganisms (bacteria, protozoa and fungi) hydrolyse plant cell walls, starch and protein into sugars and amino acids, which are then fermented, producing volatile fatty acids (VFA: chiefly acetate, propionate and butyrate), hydrogen (H_2), CO_2, ammonia and heat by both primary and secondary digestive microorganisms (McAllister *et al.*, 1996). As a last step in the process, methanogens reduce CO_2 to CH_4 using H_2 as a source of energy. The VFA are absorbed and utilised by the animal, whereas CH_4 and the heat of fermentation represent a loss of dietary energy. It is believed that under the normal ruminal environment, the H_2 balance resulting from the net H_2 production from fermentation of carbohydrates and amino acids to VFA and that used for biosynthesis of microbial cells and biohydrogenation of unsaturated fats, is used for CH_4 production (Czerkawski, 1972; Benchaar *et al.*, 1998). It also seems that emissions of enteric H_2 have been mostly neglected or attributed to abnormal rumen conditions. On a molecular basis, the heat of combustion of H_2 is 3.1 times that of CH_4, and hence emissions of H_2 may add to the gaseous waste of energy. Preliminary observations at our institute indicated that sheep fed on diets containing grains had non-negligible concentrations of H_2 in their breath. Hydrogen emissions were measured in a study planned to corroborate consistency of sheep CH_4 emission rankings (rank) under contrasting diets. This report discusses H_2 emissions only.

Materials and methods

Twenty sheep (average liveweight = 45 kg), ranked low (n=10) and high (n=10) CH_4 emitters (emissions per unit of feed intake), were fed at 1.3 times maintenance energy requirements first on a fresh perennial ryegrass diet (grass) and then on a pelleted diet (pellet) containing 60% lucerne hay and 40% wheat grain (dry matter basis). Feed was offered in equal portions twice a day, with water available *ad libitum*. The sheep were staggered in three groups of 8, 8 and 4 animals in order to match the availability of the 8-chamber sheep respiration facility. Each measurement group consisted of equal numbers of low and high CH_4 ranking animals. On each diet, the sheep were acclimatised for 21 days before a two-day measurement of gas exchange was conducted using respiration chambers.

On a dry matter basis, the crude protein (g/kg), neutral detergent fibre (g/kg) and estimated metabolisable energy (MJ ME/kg) contents of ryegrass were 216, 409 and 12.8, respectively, whereas the corresponding values for pellet were 156, 269 and 9.8, respectively. The 8-chamber animal respiration system was that described by Pinares-Patiño *et al.* (2008), with CH_4 concentrations measured using a Servomex 4900 gas analyser (Servomex Group Ltd., East Sussex, UK), which was fitted with a electrochemical H_2 detector. The detection ranges for CH_4 and H_2 were 0-200 and 0-50 ppm, with accuracies of 0.5 and 5 ppm, respectively.

Data were analysed using the mixed effects model of the REML procedure of GenStat, with 'measurement group' and 'chamber' as random effects and 'rank', 'diet' and 'rank × diet' (R×D) as fixed effects.

Results and discussion

Overall, high CH_4 emitting sheep had larger CH_4 yields (emissions per unit of intake) than their low emitting counterparts, and this difference was much greater on the pellet diet than on grass (Table 1). Sheep CH_4 emission ranks did not differ in H_2 yield, but whereas little H_2 emission was observed on the grass diet, emissions of H_2 on the pellet diet were important, with individual animals emitting up to 0.52 g/kg DMI. When emissions of H_2 were added to those of CH_4, the share of H_2 on the total gaseous energy emissions were almost insignificant on the grass diet, but represented up to 1.9% on the pellet diet, with low CH_4 emission-ranking sheep shoving tendency for a higher share of H_2 than their high CH_4 emission counterparts. Individuals on the pellet diet had up to 10% of their gaseous emissions represented by H_2.

Contrary to what was observed on the grass diet, sheep on the pellet diet ate their feed allowance within 30 min of feeding and that may have resulted in a sudden increase in digestion rate which was not equalled by the rate of methanogenic activity. Findings of this study suggest that under particular feeding conditions, emissions of H_2 may be non-negligible and hence may challenge the accuracy of current feed evaluation systems. Hydrogen is an indirect greenhouse gas and in the current study, such an effect was negligible.

Table 1. Methane and hydrogen emissions.

	Low rank		High rank		SEM	*P*-value		
	Grass	Pellet	Grass	Pellet		Rank	Diet	R×D
CH_4 (g/d)	22.7a	18.6b	26.4a	26.1a	1.69	<0.01	0.02	0.06
CH_4 (g/kg DMI)	22.1a	17.8b	24.9a	24.2a	1.21	<0.01	<0.001	0.01
CH_4 energy (KJ/d)	1263a	1034b	1470a	1453a	93.9	<0.01	0.02	0.06
H_2 (g/d)	0.01a	0.15b	0.02a	0.19b	1.10	0.43	<0.001	0.59
H_2 (g/kg DMI)	0.01a	0.14b	0.02a	0.18b	1.08	0.49	<0.001	0.59
H_2 energy (KJ/d)	1.7a	18.9b	2.2a	23.5b	1.27	0.72	<0.001	0.90
H_2 energy (% of gaseous emissions)	0.13a	1.90b	0.16a	1.70b	1.17	0.81	<0.001	0.21

a,b $P<0.05$.

Acknowledgements

This study was funded by the New Zealand Pastoral Greenhouse Gas Research Consortium.

References

Benchaar, C., J. Rivest, C. Pomar and J. Chiquette, 1998. Prediction of methane production from dairy cows using existing mechanistic models and regression equations. J. Anim. Sci. 76, 617-627.

Czerkawski, J.W., 1972. Fate of metabolic hydrogen in the rumen. Proc. Nutr. Soc. 31, 141-146.

McAllister, T.A., E.K. Okine, G.W. Mathison and K.J. Cheng, 1996. Dietary, environmental and microbiological aspects of methane production in ruminants. Can. J. Anim. Sci. 76, 231-243.

Pinares-Patiño, C.S., H. Clark, G.C. Waghorn, C. Hunt, R. Martin, P. Lovejoy and J. West, 2008. Ruminant methane – Extension of the animal calorimetry facility at AgResearch Grasslands (MAF POL_2008-28 139-4). Available at http://www.maf.govt.nz/climatechange/slm/grants/research/2007-08/pdf/2008-28-animal-calorimetry-facility.pdf.

Plant additives and their effectiveness in mitigating methane emitted by fattening bulls

S.M. Staerfl, M. Kreuzer and C.R. Soliva
ETH Zurich, Institute of Plant, Animal and Agroecosystem Sciences, 8092 Zurich, Switzerland

Introduction

Enteric fermentation of feed in ruminants results in methane (CH_4) production. Methane is not only a greenhouse gas with a high global warming potential but also a loss of feed energy to the animal (Johnson and Johnson, 1995). Studies showed that feeding strategies might be useful to decrease ruminants' CH_4 emissions (i.e. Carulla *et al.*, 2005; Kamra *et al.*, 2006) whereby specific plant additives may be especially worthwhile to be taken into consideration (i.e. Carulla *et al.*, 2005; Patra *et al.*, 2006) Therefore the aim of the present experiment was to investigate the CH_4 mitigating effect of three plant additives when being fed to fattening bulls.

Material and methods

Twenty-four growing fattening bulls (Brown Swiss × Limousin; initial body weight of 350±30 kg) were divided into four groups (n=6), housed together and fed individually with diets consisting of maize silage *ad libitum* and concentrate (2.0 kg dry matter (DM) per day and bull). A basal diet served as control while for the remaining three groups the concentrate was supplemented either with an extract rich in condensed tannins (*Acacia mearnsii*, Weibull black, TANAC S.A., Montenegro, Brazil; targeted: 3% of daily DM intake (DMI)), or dried and milled garlic bulbs (*Allium sativum*; targeted: 1.5% of DMI/day), or dried and milled hypocotyls of maca (*Lepidium meyenii*; targeted: 1.5% of DMI/day). After feeding the bulls their individual diet for 4 months, they were kept in tie stalls with individual troughs for 8 days. During this collection period, DM intake as well as fecal and urinary output (non-separated and non-acidified; the resulting N losses were assumed to have been proportionate to excretion in all treatments) were determined on a daily basis. Within this 8-day period, the animals also spent 2 days (2 × 22.5 h) in open circuit respiratory chambers to measure gas exchange including methane emission. Nutrients and energy contents were analyzed in feeds and excreta samples. Data was analyzed using the MIXED procedure of SAS (version 9.1, SAS Institute inc., Cary, NC, USA).

Results and discussion

Daily nutrient intakes with the four diets did not differ ($P>0.05$) and were on average (kg/d): 6.7 for organic matter, 1.0 for crude protein, and 1.9 for neutral detergent fiber. Selected dietary effects on the energy and nitrogen balance are presented in Table 1. There were no significant differences regarding energy intake and excretion through feces, urine and heat among the four diets. Energy loss with CH_4 was significantly lower with tannin supplementation, compared to control (-23.8%). This is similar to the findings of previous experiments (Carulla *et al.*, 2005; Grainger *et al.*, 2009) where *Acacia mearnsii* extract supplemented diets decreased CH_4 formation. In the present study, the addition of maca and garlic numerically increased the energy loss with CH_4 (+5.2 and +6.4%, respectively, compared to control).

Tannins significantly influenced N utilization in the fattening bulls. The estimated proportion of N lost with manure relative to N intake seemed to have been increased with the tannin diet by 22%. This can be mainly explained by the ability of tannins to form complexes with proteins, thus protecting feed protein from ruminal degradation (Barry and McNabb, 1999). Although tannin-protein complexes are partially dissolved at low pH, total tract protein digestibility may still be decreased resulting in increased endogenous N losses (Norton, 2000).

Table 1. Energy and nitrogen balance of fattening bulls fed four different diets (n=6).

	Control	Tannin	Maca	Garlic	SEM	*P*-value
Energy intake (MJ/day)						
Gross energy	108.7	108.1	110.7	111.9	2.32	0.66
Metabolizable energy	72.3	70.3	74.0	74.5	1.55	0.28
Energy excretion (MJ/day)						
Feces & urine	30.6	33.3	30.4	31.1	1.61	0.61
Methane	5.87[a]	4.47[b]	6.19[a]	6.27[a]	0.29	0.002
Heat	34.3	34.1	34.9	35.3	0.76	0.69
Total	70.7	71.9	71.6	72.7	2.24	0.94
Body energy retention (MJ/day)	38.0	36.2	39.1	39.2	1.17	0.30
k_m[c]	0.74	0.73	0.74	0.74	0.004	0.64
k_f[c]	0.53	0.51	0.53	0.53	0.008	0.64
Nitrogen intake and excretion						
N-intake (g/day)	142.3	142.1	147.9	146.6	1.84	0.09
Feces & urine N (% of N-intake)[d]	31.5[b]	38.6[a]	31.4[b]	33.9[ab]	1.77	0.046

[a,b] Means differ at $P<0.05$.

[c] $k_m=0.35$ (ME/GE)+0.503; $k_f=0.78$ (ME/GE)+0.006 (AFRC, 1993).

[d] Estimated proportions assuming that N losses during storage had been proportionate to excretion in all treatments.

It can be concluded that the tannin extract tested has the potential to lower CH_4 formation when fed to fattening bulls. Although there was no reduction in average daily gains (ADG; data not shown) in the period around the balance measurements, across the entire fattening period there was a reduction in ADG by 8.5% compared to control which is to be considered in feasibility calculations of this strategy.

References

AFRC, 1993. Energy and protein requirements of ruminants. Agricultural and Food Research Council Technical Committee on Responses to Nutrients. CAB International, Wallingford. U.K.

Barry, T.N. and W.C. McNabb, 1999. The implications of condensed tannins on the nutritive value of temperate forages fed to Ruminants. Br. J. Nutr. 81, 263-272.

Carulla, J.E., M. Kreuzer, A. Machmüller and H.D. Hess, 2005. Supplementation of *Acacia mearnsii* tannins decreased methanogenesis and urinary nitrogen in forage-fed sheep. Austr. J. Agr. Res. 56, 961-970.

Grainger, C., T. Clarke, M.J. Auldist, K.A. Beauchemin, S.M. McGinn, G.C. Waghorn and R.J. Eckard, R.J., 2009. Potential use of *Acacia mearnsii* condensed tannins to reduce methane emissions and nitrogen excretion from grazing dairy cows. Can. J. Anim. Sci. 89, 241-251.

Johnson, K.A. and D.E. Johnson, 1995. Methane emissions from cattle. J. Anim. Sci. 73, 2483-2492.

Kamra, D.N., N. Agarwal and L.C. Chaudhary, 2006. Inhibition of ruminal methanogenesis by tropical plants containing secondary compounds. In 'Greenhouse Gases and Animal Agriculture: An Update', C.R. Soliva, J. Takahashi, M. Kreuzer, (eds.). Elsevier, Amsterdam, the Netherlands: 156-163.

Norton, B.W., 2000. The significance of tannins in tropical animal production. In: (ed.) J.D. Brooker, Tannins in livestock and human nutrition. Proceedings of an International Workshop', Adelaide, Australia, 31 May-2 June 1999: 14-23.

Patra, A.K., D.N. Kamra and N. Agarwal, 2006 Effect of spices on rumen fermentation, methanogenesis and protozoa counts in *in vitro* gas production test. In 'Greenhouse Gases and Animal Agriculture: An Update', C.R. Soliva, J. Takahashi, M. Kreuzer (eds.). Elsevier, Amsterdam, the Netherlands: 176-179.

A screening test of the use of essential oils compounds on ryegrass silage for preventing nitrogen losses in sustainable dairy production systems.

A. Foskolos, S. Cavini, M. Rodriquez-Prado, A. Ferret and S. Calsamiglia
Departament de Ciéncia Animal i dels Aliments, Universitat Autónoma de Barcelona, 08193 Bellaterra, Spain

Introduction

Nitrogen (N) has been proposed as the main element to be balanced in sustainable dairy production due to its key role in determining yield and consequently its over application that leads to environmental pollution (Goulding, 2007). Silages make an important contribution to the total N intake. However, most of the N in fresh forage is transformed into non protein nitrogen (NPN) during the ensiling processes (Slottner and Bertilsson, 2006). NPN is rapidly transformed into ammonia in the rumen, which is the main source of N losses in ruminants (Tamminga, 1996). Essential oils (EO) have antimicrobial activities and its application in the rumen environment has resulted in a reduction of deamination and inhibition of peptidolysis (Calsamiglia *et al.*, 2007) Therefore, our aim was to evaluate the effect of EO compounds on protein degradation of ryegrass silage.

Materials and methods

Five different studies were conducted. Microsilos (n=144) were prepared in polyester bags with 2.0 kg of fresh chopped ryegrass forage, sprayed according to treatments and sealed with an automated vacuum machine. The EO compounds tested were: thymol (THY), eugenol (EUG), cinnamaldehyde (CIN), capsaicin (CAP) and carvacrol (CAR), at 4 doses: 0, 50, 500 and 2,000 mg/kg of fresh forage. Silages were opened after 35 days and sampled. Samples were analyzed for pH, dry matter, lactic acid, lactic acid bacteria (LAB) and clostridia. Silage juice (SJ) was extracted after blending 25 g of silage with 225 ml of distilled water for 30 sec in a high speed blender and filtering the slurry through 2 layers of cheesecloth. Samples of SJ were precipitated with trichloroacetic acid or tungstic acid for the determination of NPN, large peptides and small peptides as described by Winter *et al.* (1964). N was analyzed by the Kjeldahl method. Ammonia was determined by spectrophotometry. Statistical analyses were conducted using SAS (v. 9.1) and the addition of each EO was analysed separately.

Results and discussion

Ryegrass silages produced in this experiment had extensive proteolysis, with NPN being as high as 60% of the total N (see CTR group of CAR). The main component of the NPN fraction was small peptides, demonstrating an extensive peptidolysis. CAP did not affect any of the variables and thus results are not shown. The amount of small peptides was affected only by the addition of CIN at 2,000 mg/kg. That was in accordance with the overall decrease of the NPN (53.5 vs. 43.8% TN, at 0 vs. 2,000 mg/kg) and the lower concentrations of ammonia-N, indicating the inhibition of proteolysis and deamination caused by CIN in the high dose. All EO at the highest dose decreased the concentration of ammonia indicating inhibition of deamination. Among the EO, CAR also reduced ammonia concentration at 500 mg/kg of forage.

Generally, the addition of EO compounds tended to increase the pH of silages, even though the effect was not significant due to the high standard error, with the exception of CAR and THY where a dose of 2,000 mg/kg of forage significantly increase the pH. The addition of 2,000 mg/kg CAR, and THY decreased the concentration of lactic acid of silages. However, LAB only decreased in THY.

Table 1. The effect of essential oils compounds on ryegrass silage characteristics[1,2].

		pH	DM	LAB	Lactic acid	TN	NPN	NH_3-N	Small Pep.	Large Pep.
EUG	CTR	5.81	13.38	8.57[a]	51.97	38.86	21.82	1.74[a]	20.17	-0.09
	2000	6.48	13.95	7.89[b]	40.05	37.87	21.07	0.96[b]	17.6	2.48
	SE	0.317	0.323	0.124	7.240	1.201	1.507	0.108	1.990	1.005
	P	NS	NS	**	NS	NS	NS	***	NS	NS
CIN	CTR	5.63	13.03	8.58[a]	50.70	42.01	22.38[a]	1.85[a]	19.19[a]	1.35
	2000	6.02	12.35	8.21[b]	41.47	40.29	18.40[b]	1.23[b]	15.87[b]	1.29
	SE	0.279	0.282	0.097	6.47	0.970	0.908	0.134	1.074	1.122
	P	NS	0.1	0.1	NS	NS	**	**	*	NS
CAR	CTR	5.77[a]	13.26	8.04	58.71[a]	38.61	23.38	1.80[a]	21.17	0.40
	500	6.46[b]	14.21	8.25	36.38[b]	37.02	22.13	1.20[c]	19.88	1.05
	2000	6.57[b]	14.09	7.11	12.22c	37.83	21.96	1.08c	21.22	-0.34
	SE	0.171	0.345	0.395	3.902	1.300	0.864	0.154	1.26	1.131
	P	*	NS	NS	***	NS	NS	**	NS	NS
THY	CTR	5.60[a]	12.58	7.88[a]	45.87[a]	48.88[a]	23.17	1.60[a]	19.78	1.79
	2000	6.32[b]	12.93	5.66[b]	3.54[b]	42.29[b]	26.19	0.85[b]	24.14	1.19
	SE	0.283	0.351	0.412	7.366	1.990	1.474	0.136	1.483	1.452
	P	*	NS	**	***	†	NS	**	NS	NS

[1] Parameters: pH, dry matter (DM; %), lactic acid bacteria (LAB; log cfu), lactic acid (mg 7), total nitrogen (TN; g/kg of DM), non protein nitrogen (NPN; g/kg of DM), ammonia nitrogen (NH_3-N; g/kg of DM), small peptides (small pep; g/kg of DM) and large peptides (large pep; g/kg of DM).
[2] *P*-values: † = $P<0.1$; * = $P<0.05$; ** = $P<0.01$; *** = $P<0.001$.

In conclusion, the addition of EO compounds on ryegrass silage did affect protein degradation by inhibiting deamination and decreasing peptidolysis. However the dose required was probably too high, making its practical application difficult.

Acknowledgements

This study has been carried out with financial support from EU FP7 REDNEX.

References

Calsamiglia, S., M. Busquet, P.W. Cardozo, L. Castillejos and A. Ferret, 2007. Essential oils as modifiers of rumen microbial fermentation: a review. J. Dairy Sci. 90, 2580-2595.

Goulding, K.W.T., 2007. Perspective. nutrient management on farms or 'you get out what you put in'. J. Sci. Food Agri. 87, 177-180.

Slottner, D. and J. Bertilsson, 2006. Effect of ensiling technology on protein degradation during ensilage. Anim. Feed Sci. Technol. 127, 101-111.

Tamminga, S., 1996. A review on environmental impacts of nutritional strategies in ruminants. J. Anim. Sci. 74, 3112-3124.

Winter, K.A., R.R. Johnson and B.A. Dehority, 1964. Metabolism of urea nitrogen by mixed cultures of rumen bacteria grown on cellulose. J. Dairy Sci. 47, 793-797.

Increased protein utilization for milk in cows fed buckwheat silage compared to silage made from ryegrass or chicory

T. Kälber, M. Kreuzer and F. Leiber
ETH Zurich, Institute of Plant, Animal and Agroecosystem Sciences, Universitätsstrasse 2, 8092 Zurich, Switzerland; tkaelber@ethz.ch

Introduction

Buckwheat (*Fagopyrum esculentum*) plays a role in crop rotation systems for weed reduction and regeneration of the soil (Bjorkman, 2008). Besides its high agronomic and ecological value it might also be useful as an alternative feed for livestock. In moderate climatic regions, buckwheat can be sown in late summer as a catch crop making this plant a livestock feed which is not in competition for acreage with human food cereals. Feeding whole ensiled buckwheat to dairy cows (about 100 g/kg of diet; replacing maize silage) was found to remain without consequences on performance of cows yielding >35 milk/day (Amelchanka *et al.*, 2010), and fresh buckwheat was found to promote the transfer rate of beneficial a-linolenic acid from feed to cow's milk to twofold values compared to ryegrass (Kälber *et al.*, 2009). Further, during an *in vitro* fermentation experiment, fresh buckwheat reduced ruminal ammonia production and increased apparent bacterial protein synthesis per unit of dietary N (Amelchanka *et al.*, 2010). This led to the hypothesis that feeding buckwheat could enhance N utilization by ruminants and reduce N losses, thus making respective buckwheat diets a contribution to the establishments of environmentally friendly animal husbandry systems.

Materials and methods

In the present experiment, buckwheat was compared to a forage herb (chicory, *Cichorium intybus*) and a common forage grass (ryegrass, *Lolium multiflorum* var. *westerwoldicum*). Buckwheat and chicory were sown in mixture with ryegrass (eventually containing 68% buckwheat and 50% chicory of fresh matter in the respective cultures). The cultures were cut 8 weeks after sowing when buckwheat was already in the flowering state. The material was ensiled in small round bales in order to have a constant quality during the feeding experiment. Eighteen dairy cows of the breeds Holstein-Frisian and Brown Swiss (average: 233 days in milk and a milk yield of 19.3 kg/d) were allocated to three feeding treatments: buckwheat silage (FE), chicory silage (CI) and pure ryegrass silage (LM). Groups were balanced for breed, milk yield and milk composition. Buckwheat and chicory diets were blended with ryegrass silage (25% of dietary dry matter [DM]). Silages were offered *ad libitum* and cows were supplemented daily with 1 kg each of a protein concentrate and of an energy concentrate to avoid major imbalances. After an adaptation period of 7 days, cows were moved to barns equipped for metabolism experiments for 6 days. During this period, urine and faeces were collected separately and completely every day. Milk samples were taken twice a day and feed samples three times during the experiment. Milk samples were analyzed by near-infrared technique (Milkoscan 4000, Foss Electric, Hillerød, Denmark) for fat, protein, lactose and urea. Feed, faeces and urine samples were analyzed for N contents with a C/N analyzer (Analysator CN-2000, Leco, St. Joseph, MI, USA; AOAC index no. 977.02). Nutrient and organic acid composition of the silages were analyzed by standard procedures as described in Amelchanka *et al.* (2010). Data was evaluated by analysis of variance using a monofactorial model. Significances of differences among groups were tested with Tukey's method using the general linear model of SPSS.

Results and discussion

Table 1 reports parameters of silage quality and performance of the cows. Ensiling of all plants resulted in good quality silages, with ammonium being at the borderline for chicory. The crude protein (CP) content was low in the buckwheat silage, which led to a lower N intake compared to the other diets. Dry matter intake was numerically highest with chicory. Contrary to this, NDF intake was lowest with chicory. Although total milk yield as well as fat and protein yield were numerically lower in the group fed with buckwheat silage, this was not statistically significant. The ratio of milk N to ingested N was higher with buckwheat silage compared to the CI treatment; there was also a similar tendency relative to the LM treatment. This observation might be explained by the lower CP concentration of the FE diet. However, Leiber *et al.* (2004) showed that in cows with a shortage of dietary CP, milk protein content may decrease, which was not the case in the present study. Further, with FE milk N in relation to excreta N (sum of N faeces and N urine) was highest and urine N related to excreta N was lowest. Therefore it appears that buckwheat silage, compared to CI and LM, leads to a comparably high N-conversion into milk protein and to a reduction of N losses to the environment in cows at the given level of performance.

Table 1. Parameters of silage quality and performance of cows.

Parameter	Treatment			SEM	*P*-value
	FE	CI	LM		
Crude protein (g/kg DM)	122	157	152		
NH_4^+ (g/kg DM)	0.5	6.6	4.7		
Lactate (g/kg DM)	34.2	48.9	76.3		
Acetate (g/kg DM)	8.7	56.6	26.9		
Dry matter intake (kg/d)	11.4	13.3	12.7	0.43	0.17
NDF intake (kg/d)	6.44	5.65	6.49	0.179	0.093
Milk yield (kg/d)	15.8	18.3	17.6	0.69	0.32
Milk fat yield (g/d)	765	833	787	34.0	0.7
Milk protein yield (g/d)	567	625	626	20.2	0.4
N intake (g/d)	320^{b}	436^{a}	405^{a}	0.14	<0.001
Milk N/N intake	0.28^{a}	0.23^{b}	0.24^{ab}	0.008	0.009
Milk N/N excreta	0.29^{a}	0.24^{b}	0.27^{ab}	0.008	0.036
Urine N/N excreta	0.58^{b}	0.63^{ab}	0.65^{a}	0.013	0.047

^{a,b} Means within row without common superscript are significantly different (P<0.05).

References

Amelchanka, S.L., M. Kreuzer and F. Leiber, 2010. Utility of buckwheat (*Fagopyrum esculentum* Moench) as feed: Effect of forage and grain on *in vitro* ruminal fermentation and performance of dairy cows. Anim. Feed Sci. Technol. 155, 111-121.

Bjorkman, T., 2008. Buckwheat cover crops for vegetable rotations to reduce weeds and improve soil condition. HortScience 43, 1139.

Kälber, T., M. Kreuzer, H.R. Wettstein and F. Leiber, 2009. Recovery of a-linolenic acid in milk fat of dairy cows fed flowering forage plants. In: Ruminant physiology. (eds.) Y. Chilliard *et al.*, Wageningen Academic Publishers, Wageningen, the Netherlands: 436-437.

Leiber, F., M. Kreuzer, B. Jörg, H. Leuenberger and H.R. Wettstein, 2004. Contribution of altitude and alpine origin of forage to the influence of alpine sojourn of cows on intake, nitrogen conversion, metabolic stress and milk synthesis. Anim. Sci. 78, 451-466.

A calorimetry system for metabolism trials: methane production in goats

C. Fernández, M.C. López, L. Ródenas, E. Martínez, J.J. Pascual, E. Blas and C. Cervera
Instituto de Ciencia y Tecnología Animal, Universidad Politécnica de Valencia, 46022 Valencia, Spain

Introduction

Global warming, caused by increasing atmospheric concentrations of greenhouse gases, is a major worldwide environmental, economic and social threat. Mitigating methane (CH_4) losses from goats has economic as well as environmental benefits. Modifying nutrition is one of the strategies to reduce CH_4 emissions. Feeding diets with higher grain content is one way to reduce CH_4 losses (Johnson and Johnson, 1995; Beauchemin *et al.* 2009). Spanish productive system is based on a high use of concentrate and less forage. Respiration calorimetry has been a usual method to determine heat production and CH_4 in energy metabolism in farm animals. The main objective of the present study was to design a mobile respiration mask device to determine *in vivo* CH_4 production and, their CH_4 conversion factor to greenhouse gas inventories. The device has been tested by goats fed mixed diets based on either barley or corn.

Materials and methods

Three Murciano-Granadina goats (33.08±2.1 kg body weight) were selected from the Experimental Farm at the University. Goats were fed at about 1.2× maintenance level. Diets were balanced according to Lachica and Aguilera (2003). Diets were a mixed diet with 60% alfalfa hay and 40% concentrate. Diet B had barley grain as cereal (30%) and diet C corn grain (30%). Sunflower meal (10%) was added to each diet as protein source. Both diets contained (per kg of dry matter) 19.30 MJ of gross energy (GE), 170 g of crude protein and 497 g of neutral-detergent fiber. The trial included 2 consecutive sequential dietary treatments: The first treatment was diet C; the second treatment was diet B. The design consisted of 7 days for adaptation +5 experimental days for digestibility determination +1 day for air collection and energy balance (goats in metabolic cages individually confined and open circuit respiration mask). Respiration calorimetry device was installed on a trolley in order to shift from one cage to another. The data were analyzed by general linear model from SPSSv16 (2008) considering the effect of diet as fixed effect.

Results and discussion

The portable respiration device built at the Polytechnic University of Valencia was used as an indirect open calorimeter, since it is equipped with an air entry and a mass flow meter that automatically totalizes the air volume and send an aliquot sample to the gas impermeable collection bag. The CO_2 and CH_4 analyzers ranged between 0.0 and 1.5% and 0.0 and 0.15%, respectively, operating on infrared principle; whereas O_2 analyzer is working on a paramagnetic principle with a range of 19 to 21%.

Table 1 reports the daily energy balance. Dry matter intake (DMI) and energy digestibility were not significantly different. Average heat production was 555 kJ/ kg $BW^{0.75}$, being similar to values observed by Aguilera *et al.* (1990) with lactating Granadina goats using open-circuit respiration chambers (540 kJ/kg $BW^{0.75}$). We observed a significant ($P<0.05$) reduction on CH_4 production from 60 (diet C) to 39 (diet B) kJ/kg $BW^{0.75}$. Prieto *et al.* (1990), with Granadina goats at maintenance, found average values of 45 kJ/kg $BW^{0.75}$(alfalfa hay based diet) and 61 kJ/kg $BW^{0.75}$ (pelleted alfalfa and barley). Beauchemin and McGinn (2005) in feedlot cattle showed that substitution of barley by corn grain reduces CH_4 production. Casper *et al.* (1999), with cows fed barley and soybean meal found highest fractional passage rate and shorter retention times than cows fed corn and soybean meal diet.

Table 1. Daily energy utilization by goat.

Diet	B	C	SEM	P-Value
Dry matter intake, kg/day	0.72	0.80	0.05	0.414
Gross energy intake (GEI), kJ/kg $BW^{0.75}$	996	1124	42.6	0.144
Digestible energy intake, kJ/kg $BW^{0.75}$	676	781	88.0	0.156
CH_4 production, kJ/kg $BW^{0.75}$	39	60	5.3	0.018
Ym (CH_4 production, % of GEI)	4.0	5.4	0.40	0.072
Metabolisable energy intake, kJ/kg $BW^{0.75}$	615	699	33.3	0.241
Heat production, kJ/kg $BW^{0.75}$	579	531	28.5	0.460
Retained energy, kJ/kg $BW^{0.75}$	36	169	53.0	0.249

Enteric CH_4 emission from ruminants is proportional to DMI (Blaxter and Clapperton, 1965), and thus is usually normalized by expressing it as a percentage of GE intake (Ym). In studies with Japanese goats, CH_4 emission ranged from 5.0 to 8.2% with ten different mixed rations (Bhatta *et al.*, 2008). When we expressed CH_4 as a percentage of GEI, no significant difference in CH_4 production was observed. The average value obtained in our study was of 4.7%, which is in the range proposed by Johnson and Johnson (1995) and IPCC (2006) [values from 2 (90% grain diets) to 11% (forage diets)]. Beauchemin and McGinn (2005) observed a loss of 4% of GE as CH_4 for feedlot cattle fed barley based diets.

Our study showed that the mobile open circuit mask design is a useful system to measure the CH_4, CO_2, and O_2 in the air breathing of animals and offers the possibility to measure CH_4 production under practical conditions. Such data (Ym) could contribute to more accurate estimates of emission factors used in greenhouse gas inventories.

References

Aguilera, J.F., C. Prieto and J. Fonollá, 1990. Protein and energy metabolism of lactating Granadina goats. Br. J. Nutr. 63, 165-175.

Beauchemin, K., T. McAllister and S. McGinn, 2009. Dietary mitigation of enteric methane from cattle. CAB Reviews: Perspectives in Agriculture, Veterinary Science, Nutrition and Natural Resources 4: 035.

Beauchemin, K.A. and S.M. McGinn, 2005. Methane emissions from feedlot cattle fed barley or corn diets. J. Anim. Sci. 83, 653-661.

Bhatta, R., O. Enishi, N. Takusari, K. Higuchi, I. Nonaka and M. Kurihara, 2008. Diet effects on methane production by goats and a comparison between measurement methodologies. J. Agri. Sci. 146, 705-715.

Blaxter, K.L. and J.L. Clapperton, 1965. Prediction of the amount of methane produced by ruminants. Br. J. Nutr. 19, 511-522.

Casper, D.P., Maiga, H.A., Brouk, M.J. and D.J. Schingoethe, 1999. Synchronization of carbohydrate and protein sources on fermentation and passage rates in dairy cows. J. Dairy Sci. 28, 1779-1790.

International Panel on Climate Change (IPCC), 2006. Working Group 1 to the Fourth IPCC Assessment Report. http://www.ipcc.ch/pdf/assessment-report/ar4.

Johnson, K.A. and D.E. Johnson, 1995. Methane emissions in cattle. J. Anim. Sci. 73, 2483-2492.

Lachica, M. and J.F. Aguilera, 2003. Estimation of energy needs in the free-ranging goat with particular reference to the assessment of its energy expenditure by the ^{13}C-bicarbonate method. Small Rum. Res. 49, 303-318.

Prieto, C., J.F. Aguilera, L. Lara and J. Fonollá, 1990. Protein and energy requirements for maintenance of indigenous Granadina goats. Br. J. Nutr. 63, 155-163.

Statistical Package for the Social Sciences, 2008. SPSS v16 for Windows.

In vitro fermentation and methane production of fava and soy beans

A. Guglielmelli, S. Calabrò, M. Cutrignelli, O. Gonzalez, F. Infascelli, R. Tudisco and V. Piccolo
Dipartimento di Scienze Zootecniche e Ispezione degli alimenti, Università di Napoli Federico II, Via F. Delpino 1, 80137 Napoli, Italy

Introduction

Cultivation of legumes (fava bean, pea, lupine) can improve soil fertility, reduce supply of nitrogen (N) fertiliser and, due to the tannins presence (Woodward *et al.*, 2001), restrict CH_4 production. In addition, legume grain, in some agricultural condition, such as Mediterranean area, can replace partially or totally soybean in ruminant diets (Cutrignelli *et al.*, 2008). In this study the *in vitro* fermentation characteristics of six diets for ruminants with fava bean (*Vicia faba minor*) or solvent extracted (s.e.) soybean as protein sources were studied.

Materials and methods

For each protein source (fava bean: FD and soybean s.e. SD), three different diets [CP: 10, 13 and 15% DM; NE_l (MJ)/CP: 0.50, 0.46 and 0.41] were formulated with wheat straw and corn meal. The estimated protein digestible at intestinal level (PDI) of all the diets ranged between 8.3 and 16.2% DM. Fermentation characteristics were studied using the *in vitro* gas production technique, by incubating two series of bottles for each diet (1.015±0.012 g) in triplicate under anaerobic condition with buffered rumen fluid collected at the slaughterhouse from six cows fed a standard diet (Calabrò *et al.*, 2006). The gas produced during the incubation was recorded using a manual system and the cumulative gas production was related to the incubated organic matter (OMCV). For the 1st series, the fermentation was stopped at 72 h and the fermentation liquor was analyzed for NH_3 by spectrophotometer and for volatile fatty acids (VFA) by gas chromatography; the organic matter disappearance (dOM) was measured by filtering and successively burning at 550 °C the bottles residues. CH_4 was determined by a chemical method (Pell and Schofield, 1993) using NaOH (10%) at 39 °C to separate CH_4 from CO_2. CH_4 production was also estimated from VFA using stoichiometric equations (Van Soest *et al.*, 1994). In order to study the fermentation kinetics of the diets, the 2nd series was stopped at 120 h and the gas production profiles were fitted to a sigmoid model. Data were statistically processed with SAS (2000).

Results and discussion

The trend of fermentation rate was similar for all the diets and the process was almost completed at 72 h (Figure 1). However, FD showed always higher gas production compared to SD; FD15 had the highest gas production and fermentation rate. Protein source and level significantly ($P<0.01$) affected most of fermentation characteristics recorded at 72 h (Table 1). In particular, FD showed higher dOM and OMCV, CH_4, NH_3, and total VFA production compared to SD. The higher acetic and butyric acid production and higher AB/P ratio in FD explains the higher CH_4 production. As expected, dOM and VFA increased with the decrease of NE_l/CP in the diets. CH_4 showed an opposite trend ($P<0.01$); probably the shift in protein to carbohydrates changed the rumen H balance. Significant correlations ($P<0.05$) were found between CH_4 (ml/g of degraded OM) and NE_l/CP or PDI (0.838 and -0.877, respectively) confirming that the optimisation of rumen conditions for microbial growth is related to a decrease of CH_4 production (Moss, 1994). The mean value of the chemically determined CH_4 registered for all the diets (41.8 ml/g of incubated DM) was consistent with the data reported by Getachew *et al.* (2005) and similar to that estimated stoichiometrically (42.3 ml/g). In addition, the percentage of CH_4 on the total gas (Figure 2) was similar to the average value (18%) reported by Blümmel and Ørskov (1993). According to Fievez *et al.* (2005), the results concerning CH_4 confirm the suitability of the chemical method which has the advantage to be very simple and inexpensive

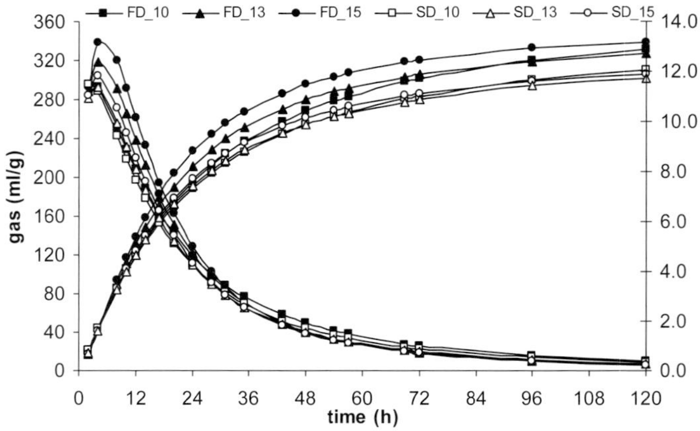

Figure 1. Gas production and rate over time.

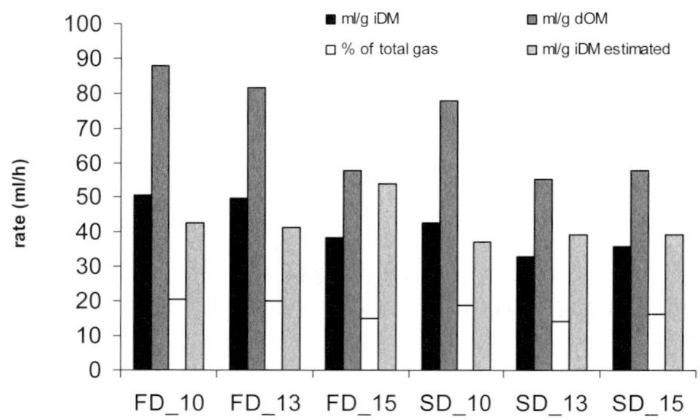

Figure 2. Methane production at 72 h.

Table 1. In vitro *fermentation parameters at 72 h.*

Diets	dOM %	OMCV ml/g OM	NH_3 mg/g DM	CH_4 ml/g DM	CH_4 ml/g dOM	Ace	Prop	But	VFA	AB/P
						Mm /g litre of incubated DM				
FD	73.2^{A}	310^{A}	44.5	46.3^{A}	75.7^{A}	45.9	27.3	10.4^{A}	96.2	2.11
SD	69.8^{B}	284^{B}	39.3	37.2^{B}	63.6^{B}	40.7	27.5	9.59^{B}	84.2	1.79
CP 10	66.8^{C}	294	42.8^{B}	46.6^{A}	82.9^{A}	41.2	26.8	8.93^{B}	84.1	1.87
CP 13	71.4^{B}	295	36.9^{C}	41.4^{AB}	68.3^{Ba}	41.9	25.7	9.13^{AB}	91.3	2.05
CP 15	76.3^{A}	302	46.1^{A}	37.3^{B}	57.8^{Bb}	46.8	29.7	10.4^{A}	95.2	1.94
MSE	1.86	41.3	2.17	21.7	61.6	31.5	9.14	0.654	79.5	0.089
D*CP	NS	NS	***	***	NS	NS	NS	NS	NS	NS

VFA: ace + prop + but + isobut + valeric + isovaleric; AB/P: (acetate + butyrate)/propionate.
NS: not significant. ^{a,b} P<0.05; ^{A,B,C} and *** P<0.01; MSE: mean square error.

and compatible in those laboratories where GC equipments are not available. However, some error can affect the data: not all CO_2 may be dissolved in NaOH, and CH_4 is the main, but not the only, gas of fermentation which remain insoluble in alkaline solution. FD had higher CH_4 production than SD, which is negative from the environmental point of view. The more intensive fermentation process with FD led to a higher dOM and VFA production than SD, which may be beneficial to the energy supply of ruminants.

References

Blümmel, M. and E.R. Ørskov, 1993. Comparison of *in vitro* gas production and nylon bag degradability of roughages in predicting of food intake in cattle. Anim. Feed. Sci. Technol. 40, 109-119.

Calabrò, S., F. Carone, M.I. Cutrignelli, S. D'Urso, G. Piccolo, R. Tudisco, G. Angelino and F. Infascelli, 2006. The effect of haymaking on the neutral detergent soluble fraction of two intercropped forages cut at different growth stages. Ital. J. Anim. Sci., 327-339.

Cutrignelli, M.I., G. Piccolo, F. Bovera, S. Calabrò, S. D'Urso, R. Tudisco and F. Infascelli, 2008. Effects of protein source and energy value of diet on the performance of young Marchigiana bulls: 1 *Infra vitam* performance and carcass quality. Ital. J. Anim. Sci. 3, 271-286.

Fievez, V., O.J. Babayemi and D. Demeyer, 2005. Estimation of direct and indirect gas production in syringes: a tool to estimate short chain fatty acid. Anim. Feed Sci. Technol. 123-124, 197-210.

Getachew, G., P.H. Robinson, E.J. DePeters, and S.J. Taylor, 2005. Methane production from commercial dairy rations estimated using an *in vitro* gas technique. Animal Feed Science and Technology. 123-124, 391-402.

Moss, A.R., 1994. Methane production by ruminants: Literature review of I. Dietary manipulation to reduce methane production and II. Laboratory procedures for estimating methane potential of diets, Nutr. Abstr. Rev. Ser. B 64, 786-806.

Pell, A.N. and P. Schofield, 1992. Computerized monitoring of gas production to measure forage digestion *in vitro*. J Dairy Sci. 76, 1063-1073.

SAS\STAT, 2000. User's Guide (Release 6.03). SAS Inst. Inc., Cary, NC.

Van Soest, P.J., 1994. Nutritional Ecology of the Ruminant (2nd Ed.). Cornell University Press, Ithaca, NY.

Woodward, S.L., G.C. Waghorn, M.J. Ulyatt and K.R. Lassey, 2001. Early indications that feeding Lotus will reduce methane emission from ruminants. Proc. N.Z. Anim. Prod. 61, 2326.

Influence of dietary tannin levels on methane production from ruminant livestock: a meta-analysis

A. Jayanegara, C.R. Soliva, S. Marquardt, M. Kreuzer and F. Leiber
ETH Zurich, Institute of Plant, Animal and Agroecosystem Sciences, Universitaetstrasse 2, 8092 Zurich, Switzerland

Introduction

Tannins represent a class of plant secondary metabolites and are produced by plants in their intermediary metabolism. Tannins are considered to be a promising group of substances to decrease methane (CH_4) emission from ruminants by dietary means. However, there is no common agreement whether tannins generally decrease CH_4 formation *in vivo* or not and to which extent (Beauchemin *et al.*, 2008). Therefore, research on this particular topic is needed to summarize and to quantify the tannin effects on CH_4 production from ruminants and its associated variables. The objective of this study was to estimate the relationship between dietary tannin levels and CH_4 emission from ruminant animals by compiling available literature data of respective experiments using a statistical meta-analysis approach.

Materials and methods

A database was created from *in vivo* experiments testing tannins as dietary supplements (both condensed and hydrolysable tannins) and giving CH_4 production as a measured variable. The database contained experimental factors (animal type, body weight, tannin source, tannin level) and various observed variables (digestibility, CH_4, volatile fatty acids (VFA) and protozoal counts). Results from nine publications reporting results from *in vivo* experiments with both small (goat, sheep) and large ruminants (cattle) were used (Sliwinski *et al.*, 2002; Carulla *et al.*, 2005; Puchala *et al.*, 2005; Oliveira *et al.*, 2007; Beauchemin *et al.*, 2007; Tiemann *et al.*, 2008; Animut *et al.*, 2008a,b; Grainger *et al.*, 2009). In total 25 dietary tannin treatments were pooled in the database. The variables being dependent on body size (CH_4 production) were standardized by relating them to metabolic body weight ($BW^{0.75}$) to counterbalance the variation associated with body weight among and within ruminant species.

The analysis of the data assembled in the database was made by a statistical meta-analysis approach (St-Pierre, 2001; Sauvant *et al.*, 2008). Experiments were treated as random effects whereas tannin levels were treated as fixed effects using the following model:

$$Y_{ijk} = \mu + Tannin_i + Exp_j + E_{ijk},$$

where Y_{ijk} = observations, μ = overall mean, $Tannin_i$ = fixed effect of tannin levels (in g/kg feed dry matter), Exp_j = random effect of experiments, and E_{ijk} = random residual error. The model was applied without weighting the observations for standard errors and the procedure MIXED of SAS version 9.1 was used.

Results and conclusion

The evaluation showed that increasing dietary tannin levels linearly decreased daily CH_4 emission per kg $BW^{0.75}$ (Table 1). This was accompanied by linear decreases of log protozoal counts and of organic matter digestibility. Tannin level did not significantly affect VFA profiles. Still a linear decrease was also observed for CH_4 production per unit of organic matter digested with increasing tannin levels. These findings suggest that increasing dietary tannin levels appears to decrease CH_4 production from ruminants. However, the effect is accompanied by a reduction in organic matter

Table 1. Meta-analysis of tannin levels (X; in g/kg feed dry matter) and dependent variables (Y).

Variable	n exp	n treat	Intercept	Slope	*P*-value		R^2
					Intercept	Slope	
Organic matter digestibility (%)	6	17	66.7	-0.122	<0.001	<0.001	0.94
CH_4 (l/kg $BW^{0.75}$ per day)	9	25	1.97	-0.005	<0.001	0.008	0.73
CH_4/OM digested (l/kg)	6	17	2.9	-0.148	<0.001	<0.001	0.93
Total VFA (mmol/l)	7	18	93.8	-0.089	<0.001	ns	0.24
C_2 (% of total)	7	18	69.8	-0.004	<0.001	ns	0.05
C_3 (% of total)	7	18	18.2	0.0004	<0.001	ns	0.00
C_2:C_3	7	18	4.24	0.0005	<0.001	ns	0.00
log protozoal counts (10^5/ml)	4	11	6.07	-0.003	<0.001	0.003	0.93

n exp = number of experiments; n treat = number of treatments; BW = body weight; C_2 = acetate; C_3 = propionate.

digestibility though at a lower magnitude than CH_4 abatement. Methane mitigation seems to be associated with a decrease in protozoal growth where part of the methanogens is attached.

References

Animut, G., R. Puchala, A.L. Goetsch, A.K. Patra, T. Sahlu, V.H. Varel and J. Wells, 2008a. Methane emission by goats consuming different sources of condensed tannins. Anim. Feed Sci. Technol. 144, 228-241.

Animut, G., R. Puchala, A.L. Goetsch, A.K. Patra, T. Sahlu, V.H. Varel and J. Wells, 2008b. Methane emission by goats consuming diets with different levels of condensed tannins from lespedeza. Anim. Feed Sci. Technol. 144, 212-227.

Beauchemin, K.A., S.M. McGinn, T.F. Martinez and T.A. McAllister, 2007. Use of condensed tannin extract from quebracho trees to reduce methane emissions from cattle. J. Anim. Sci. 85, 1990-1996.

Beauchemin, K.A., M. Kreuzer, F. O'Mara T.A. McAlister, 2008. Nutritional management for enteric methane abatement: a review. Aust. J. Exp. Agric. 48, 21-27.

Carulla, J.E., M. Kreuzer, A. Machmüller and H.D. Hess, 2005. Supplementation of *Acacia mearnsii* tannins decreases methanogenesis and urinary nitrogen in forage-fed sheep. Aust. J. Agric. Res. 56, 961-970.

Grainger, C., T. Clarke, M.J. Auldist, K.A. Beauchemin, S.M. McGinn, G.C. Waghorn and R.J. Eckard, 2009. Potential use of *Acacia mearnsii* condensed tannins to reduce methane emissions and nitrogen excretion from grazing dairy cows. Can. J. Anim. Sci. 89, 241-251.

Oliveira, S.G., T.T. Berchielli, M.S. Pedreira, O. Primavesi, R. Frighetto and M.A. Lima, 2007. Effect of tannin levels in sorghum silage and concentrate supplementation on apparent digestibility and methane emission in beef cattle. Anim. Feed Sci. Technol. 135, 236-248.

Puchala, R., B.R. Min, A.L. Goetsch and T. Sahlu, 2005. The effect of a condensed tannin-containing forage on methane emission by goats. J. Anim. Sci. 83, 182-186.

Sauvant, D., P. Schmidely, J.J. Daudin and N.R. St-Pierre, 2008. Meta-analyses of experimental data in animal nutrition. Animal 2, 1203-1214.

Sliwinski, B.J., M. Kreuzer, H.R. Wettstein and A. Machmüller, 2002. Rumen fermentation and nitrogen balance of lambs fed diets containing plant extracts rich in tannins and saponins, and associated emissions of nitrogen and methane. Arch. Anim. Nutr. 56, 379-392.

St-Pierre, N.R., 2001. Integrating quantitative findings from multiple studies using mixed model methodology. J. Dairy Sci. 84, 741-755.

Tiemann, T.T., C.E. Lascano, H.R. Wettstein, A.C. Mayer, M. Kreuzer and H.D. Hess, 2008. Effect of the tropical tannin-rich shrub legumes *Calliandra calothyrsus* and *Flemingia macrophylla* on methane emission and nitrogen and energy balance in growing lambs. Animal 2, 790-799.

Effects of diet protein level and forage source on energy and nitrogen balance and methane and nitrogen excretion in lactating dairy cows

C.K. Reynolds[1], L.A. Crompton[1], J.A.N. Mills[1], D.J. Humphries[1], P. Kirton[1], A.E. Relling[2], T.H. Misselbrook[3], D.R. Chadwick[3] and D.I. Givens[1]
[1]Department of Agriculture, University of Reading, P.O. Box 237, Earley Gate, RG6 6AR, United Kingdom
[2]Department of Animal Sciences, The Ohio State University, Wooster, OH 44691, USA
[3]North Wyke Research, Okehampton, Devon, EX20 2SB, United Kingdom

Introduction

Increasing concerns regarding nitrogen and methane emissions from dairy cattle are bringing pressures on producers to develop feeding strategies for reduced nitrogen and methane excretion relative to milk yield. The objective of the present study was to determine the effect of forage source (maize or grass silage) and dietary protein supply on energy and nitrogen balance and methane and nitrogen excretion in lactating dairy cows.

Materials and methods

Six Holstein-Friesian dairy cows in mid-lactation were fed *ad libitum* total mixed rations consisting of a 50:50 mixture (dry matter [DM] basis) of forage:concentrate, with the forage comprised on a DM basis of either 25:75 or 75:25 grass:maize silage. Concentrates were formulated to provide diet CP levels of approximately 140, 160 and 180 g per kg ration DM, in a 2×3 factorial experiment, giving 6 treatments. Diets were formulated to give increments of estimated metabolizable protein (MP) and rumen degradable protein (RDP) relative to predicted requirements using the Feed into Milk (FiM) and NRC rationing systems, in part using rumen-protected soya bean protein. For the 3 increments of dietary protein concentration supply of MP averaged 79, 95, and 109 and 101, 112, and 122% of estimated requirements based on NRC and FiM, respectively. Dietary supply of RDP for the 3 increments of dietary protein averaged 100, 102, and 105 and 88, 97 and 107% of NRC and FiM estimates, respectively. Cows were randomly assigned to diets in a 6×5 incomplete Latin Square design with 4 week periods. Measurements of respiratory exchange and energy and nitrogen digestion, excretion, and balance were obtained in the last week of each period. Measurements of N_2O and NH_3 emissions from respiration chambers were obtained to account for possible losses of N during balance measurements, although urine was acidified to prevent N loss. Data were analyzed statistically using the Mixed procedure of SAS® (SAS Institute Inc.) and a model testing fixed effects of forage source, protein level, and their interaction and random effects of period and cow. The main effect of diet protein level was partitioned into linear and quadratic effects using orthogonal contrasts.

Results and discussion

Feed DMI was increased by feeding the high-maize silage (MS) diets and by increasing dietary protein, but the effect of protein differed with forage type (Table 1). Digestibility of feed DM was increased linearly by increasing dietary protein and was greater for the high-grass silage (GS) diets, thus faecal DM excretion was reduced. In contrast, urine volume was increased linearly by dietary protein and was greater for the GS diets. Milk energy yield was greater for the MS diets. Milk protein yield was increased by feeding the MS diets, whilst milk fat yield increased linearly with increasing dietary protein (data not shown). Dietary protein concentrations were slightly higher than formulations, and there was no effect of dietary protein on milk protein yield, but urine N excretion increased linearly with increasing dietary protein and was greater for the GS diets. Faecal N excretion was increased by dietary protein for MS diets, but not GS diets, perhaps due to differences in total DMI, the ingredients used for diet formulation, and the extent of hindgut fermentation between the 2 forages.

Table 1. Components of energy and nitrogen balance in mid-lactation dairy cows fed diets containing two ratios of maize silage:grass silage (MS:GS) and 3 levels of dietary protein.

	75:25 MS:GS			75:25 GS:MS			SEM
Protein, g/kg DM…	140	160	180	140	160	180	
DM intake, kg/d[1,b,***]	21.92	22.30	23.10	19.79	21.35	19.76	0.84
DM digested, g/kg[2,a]	0.697	0.713	0.723	0.706	0.729	0.739	0.008
Urine, kg/d[1,a]	18.63	20.17	23.62	22.91	25.68	27.12	1.35
Milk yield, kg/d	32.2	32.8	33.5	30.9	31.3	32.4	2.44
Milk energy, MJ/d[c]	100.5	101.5	103.0	94.4	95.8	99.8	6.25
CH_4, L/d	636.5	671.0	652.9	643.0	650.5	659.8	37.2
CH_4, L/kg DMI[1]	28.38	28.94	28.09	32.94	31.78	33.36	1.54
CH_4, L/kg milk[3]	19.78	19.73	19.94	20.89	22.30	19.89	1.23
N balance, g/d							
Intake[1,a,**]	510	597	689	462	548	588	21.6
Faecal[3,**]	204	232	256	218	215	212	13.9
Urine[3,a]	95	119	160	107	138	165	8.2
Milk[1,**]	168	169	173	160	171	159	8.4
NH_3[a]	0.81	1.65	2.41	1.32	1.19	1.93	0.26
Condensate NH_3[1]	1.05	1.14	1.24	1.45	1.35	1.51	0.15
Milk N/Intake N[1,a]	0.327	0.284	0.250	0.344	0.309	0.270	0.094

[1,2,3] Forage effect at $P<0.01$, $P<0.05$, and $P<0.10$, respectively.
[a,b,c] Protein effect at $P<0.01$, $P<0.05$, and $P<0.10$, respectively.
[***,**,*] Forage by protein interaction at $P<0.01$, $P<0.05$, and $P<0.10$, respectively.

Methane excretion was not affected by diet, but methane excretion per kg DMI and per kg milk yield were greater when the GS diet was fed. There was no emission of N_2O from the chambers, and on average there was a net uptake of 88 mg N_2O N/d from incoming air. Emission of NH_3 in exhaust air increased linearly with increasing dietary protein concentration and NH_3 in air conditioner condensate was greater for GS diets.

Conclusions

Although diet had no effect on methane excretion by lactating dairy cows in the present study, feeding the MS diets increased DMI and tended to increase milk yield, thus methane excretion per kg DMI and per kg milk yield was reduced. The incremental response of DMI and milk protein yield to dietary protein supply differed with forage source, but milk yield was not affected by dietary protein. The efficiency of dietary N utilization for milk N production was lower for the MS diets and increased linearly with decreasing dietary protein concentration. There was a small emission of NH_3N from the cows or their feed and excreta during measurements of N balance, but no measureable emission of N_2O.

Acknowledgements

Funding by DEFRA (project AC0209) is gratefully acknowledged.

Methane emission in livestock and diet characteristics

A. Tamburini, S. Colombini, C. Penati, M. Zucali, P. Roveda, L. Rapetti and G.M. Crovetto
Università degli Studi di Milano, Dipartimento di Scienze Animali, via Celoria 2, Milan, Italy

Introduction

Methane is the second greenhouse gas after carbon dioxide. Although its concentration in the atmosphere is much lower than CO_2, its potential in climate change is higher. Enteric fermentation and manure produce 80% of agricultural methane emissions and 35 to 40% of the total anthropogenic methane emissions (Steinfeld, 2006). The production of methane represents an energy loss for ruminants (2-12% of Gross Energy intake; Johnson and Johnson, 1995). The two primary dietary factors involved in the pathway of methane production are the amount and type of dietary carbohydrates fermented in the rumen. Particularly, the amount determines the quantity, while both factors determine the proportions of the different volatile fatty acids produced. Therefore, diet characteristics can be used to predict methane production from ruminants through equations (Mills *et al.*, 2003; Ellis *et al.*, 2007). The purpose of this study was to evaluate the effect on methane production of diet characteristics in different species of ruminants.

Materials and methods

A database was created, consisting of 241 observations from 10 digestibility trials, performed between 1992 and 2008 at the Department of Animal Science at Milan (Italy). Each observation represented weekly average data for each animal, for a total of 35 diets. Methane production was recorded using open-circuit respiration chambers. The studies involved dairy goats, dairy cows, beefs and wethers (at maintenance status with forage diets). The database was divided in two groups: Small ruminants (n=214) and Large ruminants (n=27) (Table 1). Multiple regression analyses (PROC REG; SAS Institute, 2000) were carried out in order to determine relationships among methane production and dietary variables. Equations developed in this study were evaluated using R^2, root mean square prediction error expressed as a proportion of the observed mean (RMSPE) and Sawa's Bayesian information criterion (BIC) (Ellis *et al.*, 2007).

Table 1. Small and large ruminants database.

Variable		Small ruminants		Large ruminants			Small ruminants		Large ruminants	
		Mean	SEM	Mean	SEM		Mean	SEM	Mean	SEM
$BW^{\wedge 0.75}$	kg	22.9	0.35	119	1.20					
Methane	$kJ/BW^{0.75}$	103	3.48	143	5.26	% GEI	6.37	0.08	5.83	0.19
GEI	$kJ/BW^{0.75}$	1,665	54	2,578	155					
DMI	$g/BW^{0.75}$	89.1	2.87	139	8.12					
OMI	"	81.0	2.63	129	7.25	% DMI	90.6	0.21	93.0	0.45
NDF	"	36.1	1.11	49.4	3.24	"	42.1	0.55	35.1	0.40
ADF	"	20.8	0.57	26.2	1.79	"	25.4	0.51	18.5	0.27
ADL	"	4.17	0.34	5.99	0.23	"	4.10	0.23	3.63	0.03
NSC	"	25.6	1.08	50.7	2.29	"	27.9	0.81	37.6	0.92
Hemicellulose	"	15.3	0.62	23.3	1.46	"	16.7	0.34	16.6	0.14
Cellulose	"	17.4	0.65	26.0	0.93	"	20.8	0.51	15.8	0.07
Ether extract	"	3.90	0.18	6.18	0.31	"	4.27	0.11	4.57	0.18

BW = body weight, GEI = gross energy intake; OMI = organic matter intake.

Table 2. Evaluation of linear diet-based equations developed.

	Equation	R^2	BIC	RMSPE
Small ruminants (n=214)				
SA	CH_4 (kJ/BW$^{0.75}$) = 9.573(±3.57) + 1.058(±0.04) × DMI (g/BW$^{0.75}$)	0.80	1280	20.7
SB	CH_4 (kJ/BW$^{0.75}$) = 13.94(±3.30) + 0.077(±0.003) × DE (kJ/BW$^{0.75}$)	0.82	1267	20.1
SC	CH_4 (kJ/BW$^{0.75}$) = 16.84(±3.35) + 0.086(±0.003) × ME (kJ/BW$^{0.75}$)	0.80	1280	20.7
SD	CH_4 (kJ/BW$^{0.75}$) = 8.86(±4.29) + 2.63(±0.11) × NDF (g/BW$^{0.75}$)	0.74	1339	23.9
SE	CH_4 (kJ/BW$^{0.75}$) = 13.41(±6.34) + 4.32(±0.28) × ADF (g/BW$^{0.75}$)	0.53	1460	32.0
SF	CH_4 (kJ/BW$^{0.75}$) = 46.74(±4.61) + 2.19(±0.15) × NFC (g/BW$^{0.75}$)	0.50	1472	33.0
SG	CH_4 (kJ/BW$^{0.75}$) = 43.02(±9.27) + 0.957(±0.04) × DMI (g/BW$^{0.75}$) - 0.958(±0.25) × ADF (g/BW$^{0.75}$)	0.82	1267	20.0
Large ruminants (n=27)				
LA	CH_4 (kJ/BW$^{0.75}$) = 60.43(±7.25) + 0.597(±0.05) × DMI (g/BW$^{0.75}$)	0.85	133	10.4
LB	CH_4 (kJ/BW$^{0.75}$) = 56.67(±7.52) + 0.049(±0.004) × DE (kJ/BW$^{0.75}$)	0.85	132	10.4
LC	CH_4 (kJ/BW$^{0.75}$) = 62.64(±7.61) + 0.051(±0.005) × ME (kJ/BW$^{0.75}$)	0.83	136	11.1
LD	CH_4 (kJ/BW$^{0.75}$) = 69.04(±6.33) + 1.50(±0.12) × NDF (g/BW$^{0.75}$)	0.86	131	10.1
LE	CH_4 (kJ/BW$^{0.75}$) = 72.51(±6.33) + 2.71(±0.23) × ADF (g/BW$^{0.75}$)	0.85	133	10.5
LF	CH_4 (kJ/BW$^{0.75}$) = 45.78(±12.94) + 1.93(±0.25) × NFC (g/BW$^{0.75}$)	0.71	151	14.6
LG	CH_4 (kJ/BW$^{0.75}$) = 33.77(±4.36) + 0.545(±0.09) × DMI (g/BW$^{0.75}$) + 1.84(±0.81) × ADF (g/BW$^{0.75}$)	0.85	134	10.8

Results and discussion

In this study the best predictors of methane emission for Small ruminants (Table 2) were DMI (g/BW$^{0.75}$), DE and ME (kJ/BW$^{0.75}$), which is in agreement with Mills *et al.* (2003) and Ellis *et al.* (2007) on large ruminants studies. Overall, for Small ruminants, using DE as predictor of CH_4 production had a higher R^2 and lower BIC and RMSPE than using DMI or ME. For Large ruminants (Table 2), NDF intake (g/BW$^{0.75}$) was the best predictor of CH_4 production; however, all the other equations, excluded the equation based on NFC intake, gave satisfactory results. Moe and Tyrrell (1979) found fermentation of soluble carbohydrate to be less methanogenic than cell wall carbohydrates and this explains the lower precision for this equation. Concentrations of NDF, ADF and NFC (expressed as % DMI, Table 1) were also tested, but they did not affect methane emission significantly (r^2=0.14, 0.40, 0.02). This can be explained by the lower variability (SEM/mean in Table 1) of concentrations in comparison with the intakes data. Therefore in this database, diet characteristics had less influence on methane emission than ingestion level. In all databases, increasing the complexity of the equation by multiple regressions analysis did not improve predictions significantly.

References

Ellis J.L., K. Kebreab, N.E. Odongo, B.W. McBride, E.K. Okine and J. France, 2007. Prediction of methane production from dairy and beef cattle. J. Dairy Sci. 90, 3456-3466.

Johnson, K.A. and D.E. Johnson, 1995. Methane emissions from cattle. J. Anim. Sci. 73, 2483-2492.

Mills J.A.N., E. Kebreab, C.M. Yates, L.A. Crompton, S.B. Cammell, M.S. Dhanoa, R.E. Agnew and J. France, 2003. Alternative approaches to predicting methane emissions from dairy cows. J. Anim. Sci. 81, 3141-3150.

Moe, P.W. and H.F. Tyrrell. 1979. Methane production in dairy cows. J. Dairy Sci. 62, 1583-1586.

SAS Institute, 2000. Release 8.1, USA. Cary, NC, SAS Institute Inc.

Steinfeld H., P. Gerber, T. Wassenaar, V. Castel, M. Rosales and C. de Haan., 2006. Livestock's long shadow. Environmental issues and options. FAO. ftp//ftp.fao.org/docrep/fao/010/a0701e/A0701E00.pdf.

Part 8. Energy/protein metabolism and nutrition, and health in farm and companion animals

Advantage of complementary liver transcripts to understand metabolic biodiversity in dairy cows?

R.M. Bruckmaier and H.A. van Dorland
Veterinary Physiology, Vetsuisse Faculty University of Bern, Switzerland

Nutrient partitioning and metabolic challenges in dairy cows

Homeostatic regulation keeps a metabolic equilibrium in the cow's body for maintenance, deposition and mobilization of energy and nutrient reserves.

Due to their pivotal role for the maintenance of the species, pregnancy and milk secretion have developed a high priority of energy and nutrient supply within the mammalian organism during evolution. This directed way of energy and nutrient partitioning was coined as homeorhesis (Bauman and Currie, 1980). Homeorhesis has been defined as the orchestrated or coordinated control of metabolism of body tissues necessary to support a dominant physiological process or state (Bauman and Currie, 1980). Any nutrient deficiency is aimed to be fully compensated by the mobilization of body reserves at least for a limited period of time. Homeorhesis does not guarantee overall metabolic equilibrium if the energy and nutrient resources are limited. Therefore, if there is a gap between maintenance and production requirements, homeorhetic regulation can lead to metabolic disorders even to the point of severe metabolic diseases.

The metabolic priority of either conceptus or mammary gland changes throughout all functional stages of these tissues. Immediately after conception, metabolic stress of the dam results in impaired embryonic development as evidenced by reduced embryo quality in high yielding dairy cows as compared with non-lactating animals (Leroy *et al.*, 2005). Until the last trimester, when foetal growth reaches exponential rates, the additional nutrient requirements of the conceptus are relatively small. However, deficient supply or overfeeding of the dam during this time can affect body composition (rather than growth rates) of the foetus, as shown in sheep (Symonds *et al.*, 2004; Wallace *et al.*, 2005). The metabolic dominance of the conceptus thus increases towards the last weeks of pregnancy. However, the importance of competition for energy and nutrients between the conceptus and the mammary gland even at the end of pregnancy is small because energy and nutrient requirements for maintenance and pregnancy increase only by 20-25% (Moe *et al.*, 1972). This does not interfere with the needs for milk secretion because cows do not yield very much milk during this period or are dried off. In addition, animal breeding for high milk production did not considerably change the needs of the developing foetus in dairy cows.

In contrast, the mammary gland has obviously the highest priority at the start of lactation. Despite energy deficiency the milk production shortly after parturition is maintained and even further increases at the risk of metabolic disorders, whereas feeding induced deficiency in later lactational stages cause an immediate reduction of milk production (Sejrsen *et al.*, 1982; Gross *et al.*, 2010). Milk in ruminants is essential for the survival of the newborn during the colostral period and the following weeks. With increasing age of the offspring, the uptake of solid feed becomes more and more important for the development of the ruminant's digestion and metabolism while the importance of milk as the only nutrient source diminishes (Guilloteau *et al.*, 2009). The metabolic priority of the early lactating mammary gland, that developed during evolution, combined with dairy cow breeding for high milk production has, however, increased the demands of the mammary gland steadily throughout decades (Oltenacu and Algers, 2005). Feed intake did not adequately follow this increase in yield. The peripartal period, i.e. the time of metabolic changes from pregnancy to lactation and the first weeks of lactation require therefore an enormous adaptive performance of the cow's metabolism.

Specific metabolic adaptations during early lactation

The high demands of energy, glucose, lipids, proteins, and minerals for milk synthesis cannot be covered by feed intake during early lactation (Bauman and Currie, 1980; Kessel *et al.* 2008). Feed intake is rather low during the final *pre partum* period and adapts only slowly to the higher lactational demands while milk production is already high immediately after parturition and increases further during the first weeks of lactation (Ingvartsen and Andersen, 2000; Kessel *et al.*, 2008). In particular there is mostly a depression of feed intake during the periparturient period (Grummer, 1993). Consequently, a transiently negative balance of energy (NEB) and nutrients during the *post partum* period is mostly observed. The mobilization of body reserves - mainly lipids and to some extent proteins, and some minerals - allows transient adaptation and occurs in cows as well as in many other mammalian species, but the extent of NEB is particularly increased with increased milk yield. The required metabolic adaptations can be successful but can also lead to metabolic disorders (Hachenberg *et al.*, 2007; Kessel *et al.*, 2008). Dairy cows therefore need sooner or later to fill the gap between nutrient demand and nutrient uptake.

A key nutrient for milk synthesis is glucose. Most of the glucose turnover, more than 80%, is used by the mammary gland for the synthesis of lactose during peak lactation (Bauman and Currie, 1980). Therefore, hepatic gluconeogenesis and glycogenolysis are increased to a maximum at the start of lactation. Only a relatively small amount of glucose, is not needed for lactose synthesis, and is available for oxidation as an energy source for the mammary gland and other tissues (Lemosquet *et al.*, 2009). Lipids therefore become the main energy source. To cover the increased needs of fatty acids as an energy source and for milk fat synthesis, body lipid stores are mobilized and circulate in blood as non-esterified fatty acids (NEFA), while lipogenesis is simultaneously reduced. Further adaptations are the mobilization of amino acids and peptides from protein reserves in skeletal muscle tissues and other body tissues (such as the uterus) for use in protein synthesis and energy metabolism, mobilization of calcium from bones, and the increased gastrointestinal absorption (Bauman and Currie, 1980).

Hormones are essential for the regulation and coordination of the metabolic adaptations. A most important mechanism of adaptation is the decrease of plasma insulin concentrations, combined with insulin resistance, and simultaneously rather high plasma concentrations of glucagon. Along with enhanced sensitivity to effects of catecholamines, this is one of the main triggers to activate hormone-sensitive lipases in adipose tissue and hence fat mobilization. Low circulating insulin is involved in the uncoupling of the somatotropic axis in the liver via down-regulation of the hepatic growth hormone receptor (Kobayashi *et al.*, 1999; Rhoads *et al.*, 2004). Consequently plasma growth hormone (GH) levels increase, while plasma insulin-like growth factor-I (IGF-1) concentrations remain low. A direct effect of GH is the stimulation of lipolysis from adipose tissue and this effect is supported by enhanced sentivity to beta-adrenergic effects of catecholamines. While plasma insulin concentrations are low, the high levels of GH and NEFA induce an additional insulin resistance in the peripheral tissues. NEFA can be used as a priority source of oxidation for energy delivery in the liver and other tissues. In addition the liver releases the NEFA after transformation to VLDL to other tissues via the blood. A surplus of NEFA is esterified in the liver, accumulates there, and may cause fatty liver.

Due to the ruminal fermentation, glucose is scarcely absorbed directly from feed, except from some feeds, such as maize. However, most glucose even then does not reach the peripheral circulation because it is metabolized in the gastrointestinal tract. The major part of glucose is derived from hepatic gluconeogenesis. Hepatic gluconeogenesis depends on overall Krebs cycle activity, especially the oxalaoactetate availability and particularly on the availability of substrates, i.e. propionate, lactate, amino acids and glycerol. A high level gluconeogenesis pulls on oxaloacetate reserves, therefore reduces the amounts and thus the availability of oxaloacetate in hepatocyte mitochondria.

Consequently the oxidation of fatty acids via the Krebs cycle is limited, and an alternative metabolic pathway is activated through enhanced synthesis of ketone bodies from acetyl-CoA in the liver.

While most tissues are under control of insulin, glucagon, and other hormones, the nutrient uptake of the mammary gland is independent of insulin via hormone-independent glucose transporters (GLUT 1, GLUT 3). Thus the uptake of nutrients by the mammary gland from the blood circulation is not inhibited while the other body tissues release metabolic reserves, thus causing a catabolic status of the cow.

Failure of adaptation

During the transition period, the success of metabolic adaptation to the changed demands for energy and nutrients in early lactation depends of the intensity and dimension of the challenge and of the physiological adaptation of the cow to the metabolic challenge. Therefore, there seems to be an increased risk of impaired metabolic adaptation with increasing milk production levels (Simianer *et al.,* 1991; Fleischer *et al.,* 2001). However, some studies show that high milk production does not necessarily lead to metabolic diseases (Gröhn *et al.,* 1995; Ingvartsen *et al.,* 2003), i.e. a successful metabolic adaptation is possible despite a negative energy balance which cannot be avoided in early lactation. Interestingly, the highest susceptibility for production-related diseases like ketosis, lameness, mastitis or displaced abomasum occurs already before the peak of milk yield (Goff and Horst., 2006) and concurrently with the greatest rate of increase of milk production (Ingvartsen *et al.*, 2003), which is also the period of the nadir of energy balance and the greatest rate of body tissue mobilization and highest plasma levels.

If the lipase activity causes a mobilization of fat from adipose tissue beyond the oxidation capacity, too much NEFA infiltrate the liver, causing the development of fatty liver, and impaired function also of other important metabolic activities of the liver including gluconeogenesis (Rukkwamsuk *et al.*, 2000).

A consequence of unsuccessful adaptation is the occurrence of production-related diseases such as hypocalcaemia, retained placenta, fatty liver and ketosis, and impaired immune function causing increased risk of mastitis, and metritis. Reduced reproductive performance based on different pathophysiological dysfunctions during the *post partum* period is another consequence of high metabolic load.

The role of the liver in metabolic regulation

Because the liver plays a key role in metabolic adaptations during the transition period from late pregnancy to early lactation, and is important for metabolic homeostasis in the dairy cow, understanding hepatic regulation of metabolism during the periparturient period and early lactation has during recent years increasingly gained importance.

Hepatic regulation of metabolism during the transition period may vary between cows, and may be an underlying cause why some cows are more susceptible to metabolic and related disorders than others.

Individual differerences of metabolic adaptations

It is know from practical farming that individual cows adapt differently to metabolic challenges despite similar performance levels and similar feeding and housing conditions. While most of the cows are able to adapt successfully to milk production even at high production levels, some react with metabolic disorders which cause an increased occurrence of production-related diseases. This is confirmed by a number of scientific investigations which showed that the endocrine and metabolic adaptations to support milk production in early lactation vary between individual cows (Jorritsma *et al.*, 2000; Hachenberg *et al.*, 2007; Kessel *et al.*, 2008; Van Dorland *et al.*, 2009). It is not possible to avoid the metabolic challenge as such because milk production at the start of lactation is too high to have the requirements of energy and nutrients covered via feed intake during this period.

Cows with an optimal adaptive performance, with no occurrence of health disorders, would be most suitable to select for breeding programs aimed at breeding for metabolic robustness in dairy cows.

In this respect, a number of studies were performed that focused on identification of indicators for increased risk of metabolic and related disorders. Cameron *et al.* (1998) identified a negative energy balance *pre partum* and a high body condition score as important risk factors for displaced abomasum. In another study high concentrations of NEFA and beta-hydroxybutyrate (BHBA), as a response to the peripartum negative energy balance, were associated with increased risk for abomasal displacement (LeBlanc *et al.*, 2005). In addition, high plasma concentrations of NEFA and urea, and concomitantly low blood glucose concentrations between 6 and 17 days *post partum* were discovered as major risk factors for hepatic lipidosis (Jorritsma *et al.*, 2001). Hammon *et al.* (2009) found a negative relationship between liver fat content and hepatic glucose metabolism. The mRNA expression of the glucogenogenic enzyme pyruvatcarboxylase (PC) was higher in cows with a high than with a low liver fat content. Low plasma glucose concentrations during the first week after parturition were indicative for increased risk of ketosis in the weeks thereafter (Reist *et al.*, 2003).

The variation between cows to adapt successfully to lactation may have a genetic basis (Ingvartsen *et al.*, 2003; Drackley *et al.*, 2005). It is, however, not known which genes are or may be responsible. Relatively enhanced feed intake capacity and relatively increased efficiency of digestion may explain, at least under practical conditions, why some cows do not suffer from production-related diseases.

The use of functional transcripts in the liver

Molecular biology approaches have increasingly provided new insights into the regulation of metabolism in relation to dairy cattle nutrition (Drackley *et al.*, 2006). Because of the central importance of the liver in metabolic regulation it is useful to determine hepatic factors with key functions in metabolism. In particular key enzymes of the various metabolic pathways are of interest.

Repeated sampling at different metabolic stages is very promising. However, in this case the sample size can be limiting for methods at a protein level such as Western blotting. Alternatively the quantitative determination of functional factors at a transcript (mRNA) level has advantages. Although mRNA does not necessarily reflect the activity of an enzyme or receptor, correlations between mRNA expression and activity or concentration of the protein are close in many cases, as shown below. In addition, the transcript is expectedly less under short-term environmental influence is therefore more stable and thus provides a more constant information about a certain factor. Most importantly, RT-qPCR allows the quantitative determination of a huge number of factors in a very small sample (20 mg).

A close correlation was shown between hepatic pyruvate carboxylase (PC) enzyme activity and PC mRNA abundance (Greenfield *et al.*, 2000). Both, enzyme activity and mRNA abundance increased during early transition (Greenfield *et al.*, 2000) or feed restriction (Velez and Donkin, 2005). A number of further hepatic key enzymes of different metabolic pathways, transcription factors, and nuclear receptors respond to negative energy balance with changes of their mRNA abundance (Loor *et al.*, 2007). Also the periparturient down-regulation of the hepatic GH receptor, and of hepatic IGF-1 could be demonstrated at the mRNA level (Radcliff *et al.*, 2003; Rhoads *et al.*, 2008). Due to the lacking feedback via IGF-1, plasma GH concentrations increase, and exert mainly a direct lipolytic effect in adipose tissue. In addition, metabolic enzymes such as phosphoenolpyruvate carboxykinase

(PEPCK), shown at the level of mRNA abundance, are stimulated by exogenous GH (Velez and Donkin, 2004). Thus, also the repartitioning effect of GH for nutrients toward milk synthesis (hepatic gluconeogenesis) could be demonstrated at the level of mRNA abundance. Although the changes of the mRNA abundance of the various factors are rather small (up to 3-fold), in most of the performed studies (Loor *et al.*, 2005, 2007; Hammon *et al.*, 2009) these changes are obviously of significant importance for the activity or concentration of the respective hepatic protein. It has been shown in liver biopsies from cows after slaughter that the mRNA abundance of a number of factors is very similar in the entire organ, however vary between different livers (Van Dorland and Bruckmaier, 2009). This is an important precondition for repeated biopsy sampling of the same organ to compare the mRNA abundance at different time points. The approach to measure a number of selected functional transcripts in repeated biopsy samples at different metabolic states of dairy cows is promising to show timely changes of these factors, and in addition, different strategies of metabolic adaption to lactation in individual animals.

Liver transcripts for describing metabolic biodiversity in dairy cows

A number of studies was performed recently to evaluate metabolic and endocrine changes in dairy cows during the transition period. This was performed through measurements of metabolites and hormones in blood plasma and, in addition, through measurements of mRNA abundance from a number of hepatic key enzymes of various metabolic pathways, transcription factors and nuclear receptors (Van Dorland *et al.*, 2009; Graber *et al.*, unpublished results). The focus of these studies was to investigate the dynamics of liver transcripts underlying the observed variation in adaptation among cows as described by metabolic and endocrine plasma parameters. In addition, the relationship between plasma parameters and mRNA expression of parameters in the liver was evaluated. In these studies no experimental treatments were applied, i.e. merely the indivual adaptation to metabolic transition under given feeding and housing conditions was recorded. Therefore a large number of cows could be followed over time with calving as the metabolic challenge to which the cows had to adapt.

By Graber *et al.* (unpublished results) a large-scale field study was carried out that included 232 dairy cows (parity >3) from 64 farms in Switzerland. This study appears to be the first study to include a large number of cows for evaluation of liver metabolism by means of mRNA abundance of genes encoding key parameters involved in metabolic processes around parturition. Blood and liver samples were collected 3 weeks before, 3 weeks after, and 12 weeks after parturition. Blood plasma was assayed for concentrations of glucose, NEFA, BHBA, cholesterol, triglycerides, urea, albumin, protein, insulin, IGF-1, leptin, 3,5,3`-triiodthyronine (T_3), and thyroxine (T_4). Liver samples were obtained at the same time-points, and were measured for mRNA abundance of 26 candidate genes encoding enzymes and nuclear receptors that are involved in gluconeogenesis, fatty acid β-oxidation, fatty acid and triglyceride synthesis, ketogenesis, citric acid cycle, cholesterol synthesis, and the urea cycle (Figure 1).

From the significant changes in concentrations of the measured plasma parameters across the transition period from 3 weeks before to 3 weeks after parturition, it was concluded that the cows in the study experienced a marked and individually very different metabolic load in early lactation (Figure 2), demonstrated by changes of plasma NEFA concentrations. Similarly, significant changes in mRNA abundance were observed across the transition period for most of hepatic candidate genes (PEPCKm and c, PC, G6PC, ACSL1, CPT1A, CPT2, ACADM, ACADVL, ACLY, GPAM, GPD2, CS, ASS1, OCT, PPARα, SREBF1, and LXRα). Even though these changes were significant, the difference in mRNA abundance between time-points and cows for all hepatic parameters was small, illustrated by mRNA abundance of ACADVL (Figure 3 showing the individual mRNA abundance of ACADVL in cows across sampling time-points).

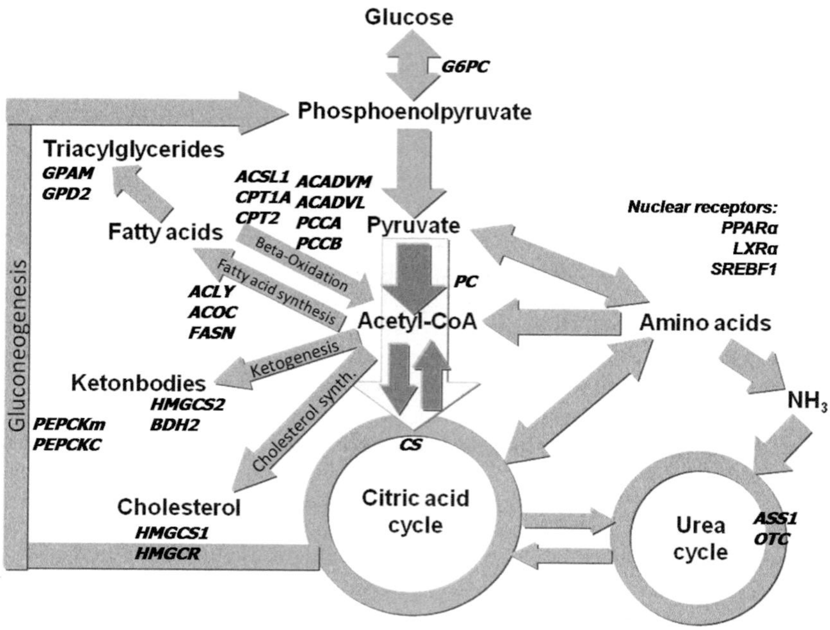

Figure 1. Schematic representation of key metabolic processes in the liver including the measured target genes (ACADM, acyl-CoA dehydrogenase, medium chain; ACADVL, acyl-coenzyme A dehydrogenase, very long chain; ACoC, Acetyl-CoA-Carboxylase; ACSL1, acyl-CoA synthetase long-chain 1; ASS1, argininosuccinate synthetase 1; BDH2, 3-hydroxybutyrate dehydrogenase 2; CPT1A, carnitine palmitoyltransferase 1A; CPT2, carnitine palmitoyltransferase 2; CS, citrate synthase; FASN, fatty acid synthase; GPAM, Glycerol-3-phophate acyltransferase; GPD2, glycerol-3-phosphate dehydrogenase 2; G6PC, Glucose-6-phosphatase; HMGCR, 3-hydroxy-3-methylglutaryl-coenzyme A reductase; HMGCS1, 3-hydroxy-3-methylglutaryl-coenzyme A synthase 1; HMGCS2, 3-hydroxy-3-methylglutaryl-coenzyme A synthase 2; LXRα, liver X receptor α; OCT, ornithine transcarbamylase; PCCA, propionyl-CoA carboxylase alpha; PCCB, propionyl-CoA carboxylase beta; PPARα, peroxisome proliferators-activated receptor α; PC, pyruvate carboxylase; PEPCKc, cytosolic phosphoenolpyruvate carboxykinase; PEPCKm, mitochondrial phosphoenolpyruvate carboxykinase; SREBF1, sterol regulatory element binding factor 1).

The above observations confirm a previous study (Van Dorland *et al.*, 2009), that included 28 dairy cows kept on an experimental farm, followed and sampled from drying off to week 14 of lactation. As described in the previous paragraph, timely changes in mRNA abundance were small (up to 1.5 cycles, i.e. up to 2.8-fold change of expression), but significant for most of the hepatic parameters measured (PEPCKm and c, CPT1A, CPT2, CS, ACLY, HMGCS1 and 2, and PPARα). This implies that the physiological state relative to parturition can be well characterized by the selected and measured target liver transcripts. However, liver transcripts failed to explain differences in the adaptive performance in dairy cows during moderate metabolic challenge, because van Dorland *et al.* (2009) did not observe differences between animals in early lactation that were retrospectively classified into 2 groups according to their plasma concentration of BHBA in week 4 p.p. with 0.75 mmol/l being the cut-off point that differentiated the two groups. This may partly be explained by the moderate challenge the cows experienced in early lactation as the energy balance in both cow groups turned positive in wk 7 p.p. already. Nevertheless, the number and type of significant correlations between plasma and hepatic parameters at each time-point and between groups were not similar, which suggests that even during a moderate metabolic challenge, differences in adaptation reactions in plasma are associated with variation in metabolic regulation in the liver (Figure 3 shows significant correlations among measured plasma and hepatic parameters in week 4 p.p.).

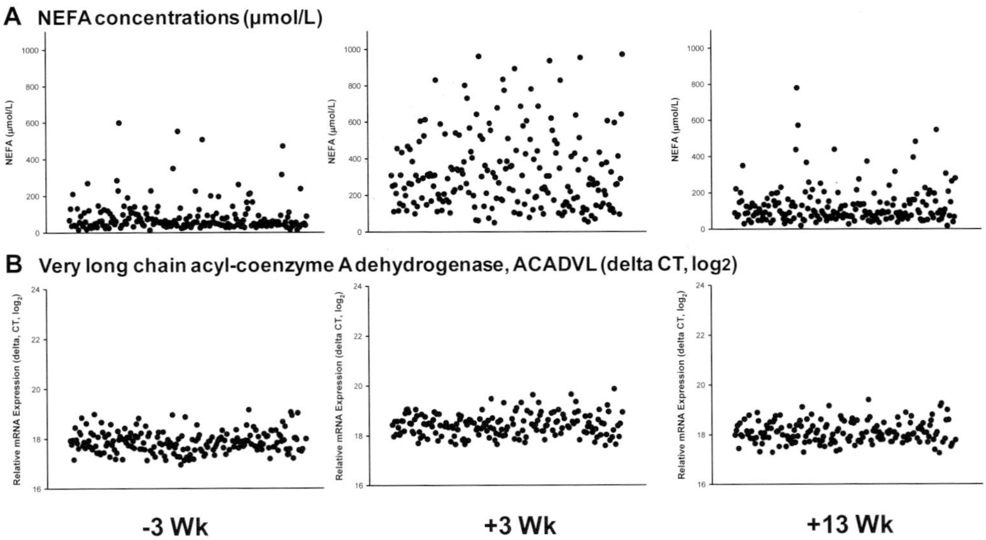

Figure 2. Individual adaptation reactions in cows for NEFA concentration and mRNA abundance of ACADVL across time-points. (A) NEFA concentrations (µmol/l). (B) Very long chain acyl-coenzyme A dehydrogenase, ACADVL (delta CT, log2).

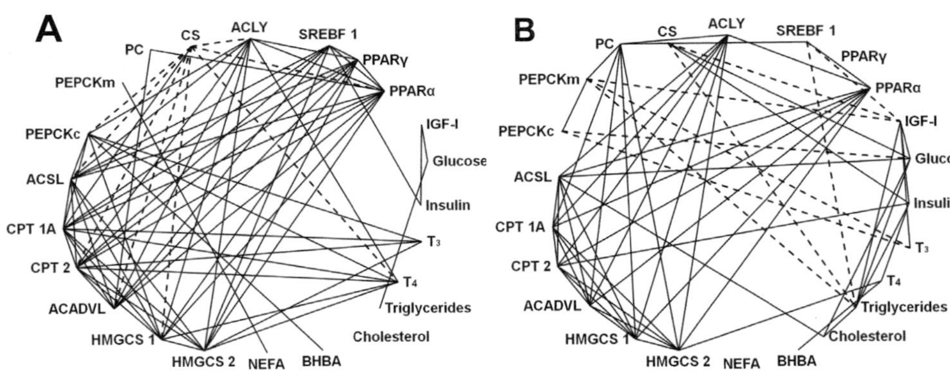

Figure 3. Correlations (Spearman rank correlation coefficients) among mRNA abundance of hepatic genes and concentrations of blood plasma in cows with BHBA concentration of >0.75 mmol/l (A) and in cows with BHBA concentration of <0.75 mmol/l. (B) In week 4 p.p. Only correlations are shown with P*<0.05. Full lines represent positive, and dotted lines represent negative correlations (ACADVL, acyl-coenzyme A dehydrogenase very long chain; ACLY, ATP citrate lyase; ACSL, acyl-CoA synthetase long-chain; CPT1A, carnitine palmitoyltransferase 1A; CPT2, carnitine palmitoyltransferase 2; CS, citrate synthase; HMGCS1, 3-hydroxy-3-methylglutaryl-coenzyme A synthase 1; HMGCS2, 3-hydroxy-3-methylglutaryl-coenzyme A synthase 2; PC, pyruvate carboxylase; PEPCKc, cytosolic phosphoenolpyruvate carboxykinase; PEPCKm, mitochondrial phosphoenolpyruvate carboxykinase; PPARα, peroxisome proliferators-activated receptor α; PPARγ, peroxisome proliferators-activated receptor γ; SREBF 1, sterol regulatory element binding factor 1; T_3, Triiodothyronine; T_4, Thyroxine).*

To test if mRNA abundance of the target genes is suitable information for explaining the variation in adaptive performance during a marked metabolic challenge, data from the field study (Graber *et al.*, unpublished results) were used to investigate cows experiencing a high versus a low metabolic load in early lactation.

For this study, two groups of cows were formed based on the plasma concentrations of glucose, NEFA, and BHBA in week 4 p.p. Cows in the first group had a glucose concentrations of <3.0 mmol/l, NEFA concentrations of >300 µmol/l, and BHBA concentrations of >1.0 mmol/l, and were referred as metabolically challenged (GRP-; n=33). The second group included cows with glucose concentrations of >3.0 mmol/l, NEFA concentrations of <300 µmol/l, and BHBA concentrations of <1.0 mmol/l, which were referred as metabolically not challenged (GRP+; n=40). Cows that did not fall into either group were excluded from this evaluation. Results showed that mRNA abundance of parameters regulating gluconeogenesis (PEPCKm and PC) in GRP- cows were higher than in GRP+ cows ($P<0.05$). The mRNA abundance of parameters regulating fatty acid oxidation (ACSL, CPT2, ACADVL) in GRP- cows were also higher than in GRP+ cows ($P<0.05$). Furthermore, mRNA abundance of PPARα was lower for GRP- compared to GRP+ ($P<0.01$). No group differences in mRNA levels of the other hepatic parameters were detected. From this evaluation it was evidenced that metabolically challenged dairy cows (on the basis of concentrations of glucose, BHBA, and NEFA) show differences in liver transcripts, compared to metabolically unchallenged dairy cows with a similar milk production in early lactation. These findings indicate that liver transcripts do aid in characterizing differences in adaptive performance in dairy cows during early lactation. Whether or not the differences in liver transcripts between cows are a result (environmentally affected) from or are causal (underlying genetic component such as polymorphisms) to the observed differences in plasma concentrations of parameters has to be investigated.

Conclusion

The hepatic mRNA abundance of metabolic enzymes, nuclear receptors and transcription factors is highly reproducible. This confirms the assumption that on a mRNA level these factors are not susceptible to short-term environmental changes, and may therefore be a useful tool to characterize metabolic biodiversity. Small differences of mRNA expression my indicate polymorphisms at a DNA level which needs to be further investigated. On the other hand, the regulation of key factors of hepatic metabolism cannot only be based on the small differences of mRNA abundance. Most likely, short-term regulation is mainly based on post-transcriptional mechanisms.

References

Bauman, D.E., and W.B. Currie, 1980. Partitioning of nutrients during pregnancy and lactation: a review of mechanisms involving homeostasis and homeorhesis. J. Dairy Sci. 63, 1514-1529.

Cameron, R.E.B., P.B. Dyk, T.H. Herdt, J.B. Kaneene, R. Miller, H.F. Bucholtz, J.S. Liesman, M.J. Vandehaar, and R.S. Emery, 1998. Dry cow diet, management, and energy balance as risk factors for displaced abomasum in high producing dairy herds. J. Dairy Sci. 81, 132-139.

Drackley, J.K., T.R. Overton, and G.N. Douglas. 2001. Adaptations of glucose and long-chain fatty acid metabolism in liver of dairy cows during the periparturient period. J. Dairy Sci 84 (E. Suppl.), E100-E112.

Drackley, J.K., M. Heather, G. Dann, N. Douglas, N.A. Janovick Guretzky, N.B. Litherland, J.P. Underwood, and J.J. Loor, 2005. Physiological and pathological adaptations in dairy cows that may increase susceptibility to periparturient diseases and disorders. Ital. J. Anim. Sci. 4, 323-344.

Drackley, J.K., S.S. Donkin, and C.K. Reynolds, 2006. Major advances in fundamental dairy cattle nutrition. J. Dairy Sci. 89, 1324-1336.

Fleischer, P., M. Metzner, M. Beyerbach, M. Hoedemaker, and W. Klee, 2001. The relationship between milk yield and the incidence of some diseases in dairy cows. J. Dairy Sci. 84, 2025-2035.

Goff, J.P., and R.L. Horst, 1997. Physiological changes at parturition and their relationship to metabolic disorders. J. Dairy Sci. 80, 1260-1268.

Greenfield, R.B., M.J. Cecava, and S.S. Donkin, 2000. Changes in mRNA expression for gluconeogenic enzymes in liver of dairy cattle during the transition to lactation. J. Dairy Sci. 83, 1228-1236.

Gröhn, Y.T., S.W. Eicker, and J.A. Hertl, 1995. The Association between previous 305-day milk yield and disease in New York State dairy cows. J. Dairy Sci. 78, 1693-1702.

Groß, J., H.A. van Dorland, R.M. Bruckmaier, and F.J. Schwarz, 2010. Produktionsparameter von hochleistenden Milchkühen zu Laktationsbeginn, während einer Energierestriktion nach 100 Laktationstagen und anschließender Realimentation. Proc. Soc. Nutr. Physiol. 19, 55.

Grummer, R.R., 1993. Etiology of lipid-related metabolic disorders in periparturient dairy cows. J. Dairy Sci. 76, 3882-3896.

Guilloteau, P., R. Zabielski, and J.W. Blum, 2009. Gastrointestinal tract and digestion in the young ruminant: ontogenesis, adaptations, consequences and manipulations. J. Physiol. Pharmacol. 60, 37-46.

Hachenberg, S., C. Weinkauf, S. Hiss, and H. Sauerwein, 2007. Evaluation of classification modes potentially suitable to identify metabolic stress in healthy dairy cows during the peripartal period. J. Anim. Sci. 85, 1923-1932.

Hammon, H.M., G. Stürmer, F. Schneider, A. Tuchscherer, H. Blum, T. Engelhard, A. Genzel, R. Staufenbiel, and W. Kanitz, 2009. Performance and metabolic and endocrine changes with emphasis on glucose metabolism in high-yielding dairy cows with high and low fat content in liver after calving. J. Dairy Sci. 92, 1554-1566.

Ingvartsen, K.L., and J.B. Andersen, 2000. Integration of metabolism and intake regulation: A review focusing on periparturient animals. J. Dairy Sci 83, 1573-1597.

Ingvartsen, K.L., R.J. Dewhurst, and N.C. Friggens, 2003. On the relationship between lactational performance and health: is it yield or metabolic imbalance that cause production diseases in dairy cattle? A position paper. Livest. Prod. Sci. 83, 277-308.

Jorritsma, R., H. Jorritsma, Y.H. Schukken, and G.H. Wentink. 2000. Relationship between fatty liver and fertility and some periparturient diseases in commercial dutch dairy herds. Theriogenology 54, 1065-1074.

Jorritsma, R., H. Jorritsma, Y.H. Schukken, P.C. Bartlett, Th. Wensing, and G.H. Wentink. 2001. Prevalence and indicators of *post partum* fatty infiltration of the liver in nine commercial dairy herds in the Netherlands. Livest. Prod. Sci. 68, 53-60.

Kessel, S., M. Stroehl, H.H.D. Meyer, S. Hiss, H. Sauerwein, F.J. Schwarz, and R.M. Bruckmaier, 2008. Individual variability in physiological adaptation to metabolic stress during early lactation in dairy cows kept under equal conditions. J. Anim. Sci. 86, 2903-2912.

Kobayashi, Y., C.K. Boyd, C.J. Bracken, W.R. Lamberson, D.H. Keisler, and M.C. Lucy, 1999. Reduced growth hormone receptor (GHR) messenger ribonucleic Acid in liver of periparturient cattle is caused by a specific down-regulation of GHR 1A that is associated with decreased insulin-like growth factor I. Endocrinology 140, 3947-3954.

LeBlanc, S.J., K.E. Leslie, and T.F. Duffield, 2005. Metabolic predictors of displaced abomasum in dairy cattle. J. Dairy Sci. 88, 159-170.

Lemosquet, S., E. Delamaire, H. Lapierre, J.W. Blum, and J.L. Peyraud, 2009. Effects of glucose, propionic acid, and nonessential amino acids on glucose metabolism and milk yield in Holstein dairy cows. J. Dairy Sci. 92, 3244-57.

Leroy, J.L., G. Opsomer, S. De Vliegher, T. Vanholder, L. Goossens, A. Geldhof, P.E. Bols, A. de Kruif and A. Van Soom, 2005. Comparison of embryo quality in high-yielding dairy cows, in dairy heifers and in beef cows. Theriogenology 64, 2022-2036.

Loor, J.J., H.M. Dann, R.E. Everts, R. Oliveira, C.A. Green, N.A. Janovick Guretzky, S.L. Rodriguez-Zas, H.A. Lewin, and J.K. Drackley, 2005. Temporal gene expression profiling of liver from periparturient dairy cows reveals complex adaptive mechanisms in hepatic function. Physiol. Genomics 23, 217-226.

Loor, J.J., H.M. Dann, R.E. Everts, R. Oliveira, C.A. Green, N.A. Janovick-Guretzky, N. Litherland, S.L. Rodriguez-Zas, H.A. Lewin and J.K. Drackley, 2006. Plane of nutrition *pre partum* alters hepatic gene expression and function in dairy cows as assessed by longitudinal transcript and metabolic profiling. Physiol. Genomics 27, 29-41.

Loor, J.J., R.E. Everts, M. Bionaz, H.M. Dann, D.E. Morin, R. Oliveira, S.L. Rodriguez-Zas, J.K. Drackley, and H.A. Lewin, 2007. Nutrition-induced ketosis alters metabolic and signalling gene networks in liver of periparturient dairy cows. Physiol. Genomics 32, 105-116.

Moe, P.W., and H.F. Tyrrell, 1972. Metabolizable energy requirements of pregnant dairy cows. J. Dairy Sci. 55, 480-483.

Oltenacu, P.A. and B. Algers, 2005. Selection for increased production and the welfare of dairy cows: are new breeding goals needed? Ambio Vol. 34, 4-5.

Radcliff, R.P., B.L. McCormack, B.A. Crooker and M.C. Lucy, 2003. Plasma hormones and expression of growth hormone receptor and insulin-like growth factor-I mRNA in hepatic tissue of periparturient dairy cows. J. Dairy Sci. 86, 3920-3926.

Reist, M., D. Erdin, D. von Euw, K. Tschuemperlin, H. Leuenberger, C. Delavaud, Y. Chilliard, H.M. Hammon, N. Kuenzi, and J.W. Blum, 2003. Concentrate feeding strategy in lactating dairy cows: metabolic and endocrine changes with emphasis on leptin. J. Dairy Sci. 86, 1690-1706.

Rhoads, R.P., J.W. Kim, B.J. Leury, L.H. Baumgard, N. Segoale, S.J. Frank, D.E. Bauman, and Y.R. Boisclair, 2004. Insulin increases the abundance of the growth hormone receptor in liver and adipose tissue of periparturient dairy cows. J. Nutr. 134, 1020-1027.

Rhoads, M.L., J.P. Meyer, W.R. Lamberson, D.H. Keisler, and M.C. Lucy, 2008. Uterine and hepatic gene expression in relation to days *post partum*, estrus, and pregnancy in *post partum* dairy cows. J. Dairy Sci. 91, 140-150.

Rukkwamsuk, T., M.J.H. Geelen, T.A.M. Kruip, and T. Wensing, 2000. Interrelation of fatty acid composition in adipose tissue, serum, and liver of dairy cows during the development of fatty liver *post partum*. J. Dairy Sci. 83, 52-59.

Sejrsen K., 1994. Relationships between nutrition, puberty and mammary development in cattle. Proc. Nutr. Soc. 53, 103-111.

Sejrsen, K., J.T. Huber, H.A. Tucker, and R.M. Akers, 1982. Influence of nutrition of mammary development in pre- and postpubertal heifers. J. Dairy Sci. 65, 793-800.

Simianer, H., H. Solbu, and L.R. Schaeffer, 1991. Estimated genetic correlations between disease and yield traits in dairy cattle. J. Dairy Sci. 74, 4358-4365.

Symonds, M.E., S. Pearce, J. Bispham, D.S. Gardner, and T. Stephenson, 2004. Timing of nutrient restriction and programming of fetal adipose tissue development. Proc. Nutr. Soc. 63, 397-403.

Van Dorland, H.A., and R.M. Bruckmaier. Short communication: Regional mRNA expression of key gluconeogenic enzymes in the liver of dairy cows, 2009. J Anim Physiol Animal Nutr. (in press, DOI: 10.1111/j.1439-0396.2009.00935.x).

Van Dorland, H.A., M. Graber, S. Kohler, T. Kaufmann, and R.M. Bruckmaier, 2009. Expression of genes involved in different metabolic pathways in the liver of metabolically challenged dairy cows during early lactation: a field study. In: Y. Chilliard, F. Glasser, Y. Faulconnier, F. Bocquier, I. Veissier and M. Doreau (eds.) Ruminant physiology 2009: Digestion, metabolism, and effects of nutrition on reproduction and welfare. Wageningen Academic Publishers, the Netherlands: 682-683.

Van Dorland, H.A., S. Richter, I. Morel, M.G. Doherr, N. Castro, and R.M. Bruckmaier. 2009. Variation in hepatic regulation of metabolism during the dry period and in early lactation in dairy cows. J. Dairy Sci. 92, 1924-1940.

Velez, J.C., and S.S. Donkin, 2005. Feed restriction induces pyruvate carboxylase but not phosphoenolpyruvate carboxykinase in dairy cows. J. Dairy Sci. 88, 2938-2948.

Wallace, J.M., T.R. Regnault, S.W. Limesand, W.W. Hay Jr., and R.V. Anthony, 2005. Investigating the causes of low birth weight in contrasting ovine paradigms. J. Physiol. 565, 19-26.

Interactions between energy and protein status, immunity and infections in farm animals

N. Lacetera, U. Bernabucci and B. Ronchi
Dipartimento di Produzioni Animali, Università degli Studi della Tuscia, via San Camillo De Lellis, Viterbo, Italy

Abstract

Interactions between metabolism and immunity involve a group of biomolecules (insulin, leptin, heat shock proteins, etc.), which can regulate both functions. Throughout its action on nutrients uptake, insulin promotes lymphocyte activation and responsiveness, and cytokines action; leptin, a well known regulator of lipid metabolism, acts as an acute-phase reactant during inflammation, stimulates chemotaxis, phagocytosis and release of oxygen radicals, stimulates proliferation of immune cells and promotes secretion of cytokines; heat shock proteins may protect cells from excessive lipid accumulation, regulate inflammatory events and activate immunity. Several studies pointed out cause-and-effect relationships between negative energy balance, impairment of the immune response and infection outbreaks in periparturient dairy ruminants, and suggested that changes of circulating metabolites may be potential markers for the risk of infections in early lactating animals. Prolonged activation of the immune system (disease challenge or poor hygienic environment) is responsible for metabolic effects (synthesis of new proteins, anorexia, nutrients diversion, etc.), which may cause reduction of productive and reproductive performances in farm animals. Numerous studies pointed out that dietary energy and protein may affect immune system and that their manipulation in terms of quantity and/or quality may help to modulate immune functions, increase resistance to infectious agents and avoid the negative consequences of chronic immune activation. Further research is needed to elucidate the intimate mechanisms through which metabolism and immunity interact and to establish the best immunonutritional interventions that may help in keeping farm animals healthy and capable to perform at their full potential.

Mechanisms of host defence and resistance to infections

The immune system has evolved as a complex of mechanisms to protect the host from invasion by pathogenic organisms. Before the development of an acquired immune response to a pathogen, there are innate immune mechanisms, which provide some degree of protection from disease. However, infectious organisms are often able to evade these defences. The innate immune mechanisms are non-specific, in that they are effective against a variety of pathogens, do not require prior exposure for their induction, and include mechanical barriers (skin and mucosae), secreted products (lysozyme, lactoferrin, tumor necrosis factor, etc.), inflammatory cells (neutrophils and macrophages) and physiologic functions (cough reflex, mucociliary blanket activity, etc.) (Gershwin *et al.*, 1995).

The ability of the immune system to respond specifically to a foreign substance, commonly named antigen, forms the basis for acquired immunity. Initial exposure to antigen (virus, bacteria, parasite, etc.) elicits a primary immune response, which serves to prime the immune system and to create a system of immunologic memory, such that subsequent exposure to the same antigen will elicit a faster, greater and longer-lasting immune response than the initial exposure. The adaptive immune response may be considered as divided in three major phases: the antigen recognition and presentation phase, the activation phase of lymphocytes, and the effector phase (Tizard, 1992). The latter phase may be distinguished in humoral immunity, which deals with infectious agents in the blood and body tissues, and cell-mediated immunity, which deals with body cells that have been infected. In general, the humoral system is managed by B-lymphocytes (with help from specific subsets of T-lymphocytes), and the cell-mediated system is managed by T-lymphocytes. Both systems exhibit fascinating complexity and interrelationships that allow them to fine-tune immune reactions to almost any antigen, or molecule that stimulates an immune response.

The ability of the host to counteract microbial invasion and resist to infections may be influenced by virulence features, environmental factors, including quantity and quality of nutrients, and/or defects in host defence mechanisms (Tizard, 1992).

Biomolecules at the interface between metabolic and immune functions

Several studies recognized a series of biomolecules, which can regulate both metabolism and immunity (Matarese and La Cava, 2004). Concentration of these molecules in the body compartments varies due to physiological state, nutritional, immune or environmental challenge, and their changes may affect both metabolic and immune functions. Some of these molecules are secreted only from non immune cells (insulin, a-melanocyte stimulating hormone and Zinc-α2-glycoprotein), whereas others derive either from non immune or immune cells (leptin, corticotropin-releasing hormone, tumor necrosis factor (TNF)-α, interferon (IFN)-γ, interleukin (IL)-1 and IL6, etc.).

For long time, insulin has been recognized only as a metabolic hormone. In the last few years, it has been reported that throughout its action on nutrients uptake, insulin raises the levels of intermediary cellular metabolism while promoting lymphocyte activation and responsiveness, and cytokines action. Antigens or mitogens activated T cells express insulin receptors, and after binding to its receptor on activated T cells, insulin stimulates glucose uptake and oxidation, amino acid transport, lipid metabolism and protein synthesis (Stentz and Kitabchi, 2003). Furthermore, defects in insulin action (e.g. in conditions such as insulin resistance) result in immunesuppression, and during hyperglycaemia and ketoacidosis, insulin promotes the secretion of inflammatory cytokines and favours T-cell activation (Dandona *et al.*, 2004).

Leptin is another hormone capable of significant influence on energy balance and both innate and adaptive immunity. The major metabolic role of leptin is to gauge the amount of body fat, to stop food intake and increase basal metabolism (Friedman and Halaas, 1998). Together with IL-1, IL-6 and TNF-α, leptin acts as an acute-phase reactant during inflammation (Fantuzzi and Faggioni, 2000). In natural immunity, leptin stimulates chemotaxis, phagocytosis and release of oxygen radicals in polymorphonuclear (PMN) cells (Caldefie-Chezet *et al.*, 2003; Moore *et al.*, 2003). In adaptive immunity, leptin stimulates proliferation of naive T cells and promotes the secretion of T helper 1 (Th1) cytokines (Lord *et al.*, 1998; Farooqi *et al.*, 2002).

A series of recent studies indicated that also heat shock proteins (Hsp) may play a role in the regulation of metabolism and immunity. Splenic lymphocytes from mice kept on a lipidic diet causing hyperilipidemia have, compared to those isolated from non hyperlipidemic mice, a higher predisposition to apoptosis, which is preceded by the appearance of intracellular Hsp (Carratelli *et al.*, 1999). Furthermore, a recent study demonstrated that heat shock-induced upregulation of intracellular Hsp72 alleviated insulin resistance and reduced liver lipidosis (Morino *et al.*, 2008). On the other hand, several studies documented that Hsp are components of the network involved in the regulation of inflammatory events (Pockley *et al.*, 2008), and that they activate immune defences by optimising antigen processing and presentation (Campisi *et al.*, 2003; Asea, 2005).

Two recent studies testified that the dual role of Hsp as regulators of metabolism and immunity is likely to be applicable also to bovine. Eitam *et al.* (2009) reported that low-energy diet promoted cell-specific Hsp response with an increase of Hsp90 but unchanged levels of Hsp70 mRNAs in white blood cells, and with a lower expression of Hsp70 in milk cells of beef cows. In a second study, Catalani *et al.* (2010) described an increase of plasma and intracellular Hsp72 in post-calving dairy cows, which was associated with transient increase of TNF-α and plasma nonesterified fatty acids

(NEFA), and decrease of the ability of peripheral blood mononuclear cells (PBMC) to proliferate following lypopolysaccharide (LPS) stimulation (Figure 1). These authors hypothesized that the post-calving increase of Hsp72 may be linked to metabolic, inflammatory and immunological conditions of early lactation, and that Hsp72 upregulation may have an adaptive role in limiting the consequences of an excessive lipid mobilization and inflammation. Furthermore, results from that study suggested the hypothesis that the post-calving increase of Hsp72 at PBMC level may contribute to establish an endotoxin tolerance-like condition in early lactating dairy cows.

Many questions remain to be answered on the exact role of the molecules at the interface between metabolism and immune tuning. Although significant knowledge has been gained in this field, further research is needed and guaranteed. A series of *in vitro* studies are being carried out at our site to establish possible cause-and-effect relationships between changes of intra and extracellular Hsp72, metabolic, and immune parameters in dairy cows.

Relationships between energy balance, immunity and resistance to infections

A state of energy balance is that in which the total energy intake of a living organism equals total energy needs (http://nutritionanddiet.medical-dictionaries.org/Energy_balance). In early lactating dairy cows, energy intake is usually lower than energy required for maintenance and milk production, which results in negative energy balance (NEB) and mobilization of body reserves (Van Knegsel *et al.*, 2005).

Several studies also demonstrated that dairy cows, sheep and goats are likely to suffer from immunodepression around parturition (Kehrli *et al.*, 1998; Lacetera *et al.*, 2004a, 2006; Goff, 2006), and hypothesized that this may at least partially explain the higher incidence of infections in this critical period.

Several studies have thus been performed in the attempt to establish possible cause-and-effect relationships between NEB and impairment of the immune response in periparturient dairy ruminants. Recently, Wathes *et al.* (2009) reported that several inflammatory response genes were upregulated in cows undergoing conditions of severe NEB, that this may explain why these cows undergo a long-lasting active uterine inflammatory response *post partum*, and also that this may prevent cows from mounting an effective immune response to the microbial challenge experienced after calving. Additionally, in a study carried out to better understand the mechanisms associated with NEB and risk of mastitis during the transition period, Moyes *et al.* (2010) compared mammary tissue gene expression profiles during a *Streptococcus uberis* mastitis challenge between cows subjected to dietary-induced NEB and cows fed to maintain positive energy balance. The main finding from this study was that the majority of genes involved in immune response were down-regulated in NEB cows. Previous studies were based on the observation that also ketotic ruminants are immunesuppressed (Erb and Grohn, 1988; Andersson, 1993; Suriyasathaporn *et al.*, 1999), and were aimed at verifying the hypothesis that immunesuppression under conditions of NEB might be due to increased concentration of ketone bodies (Targowski and Klucinski, 1983; Franklin *et al.*, 1991; Suriyasathaporn *et al.*, 1999). However, results of those studies were conflicting and did not permit to support this hypothesis. More recently, other authors (Hoeben *et al.*, 2000; Lacetera *et al.*, 2001) reported negative relationships between immune functions and plasma concentrations of NEFA, which also increase sharply in ketotic ruminants, and indicated plasma concentrations of these metabolites as possible diagnostic markers of impaired immunity around parturition. Following these observations, several studies have thus been performed to verify more accurately the hypothesis that increased plasma NEFA due to the energy deficit-related mobilization of body reserves might play a role in the immunesuppression taking place around parturition. A first series of studies were carried out *in vitro* and demonstrated that a mixture of NEFA reflecting composition of ruminants'

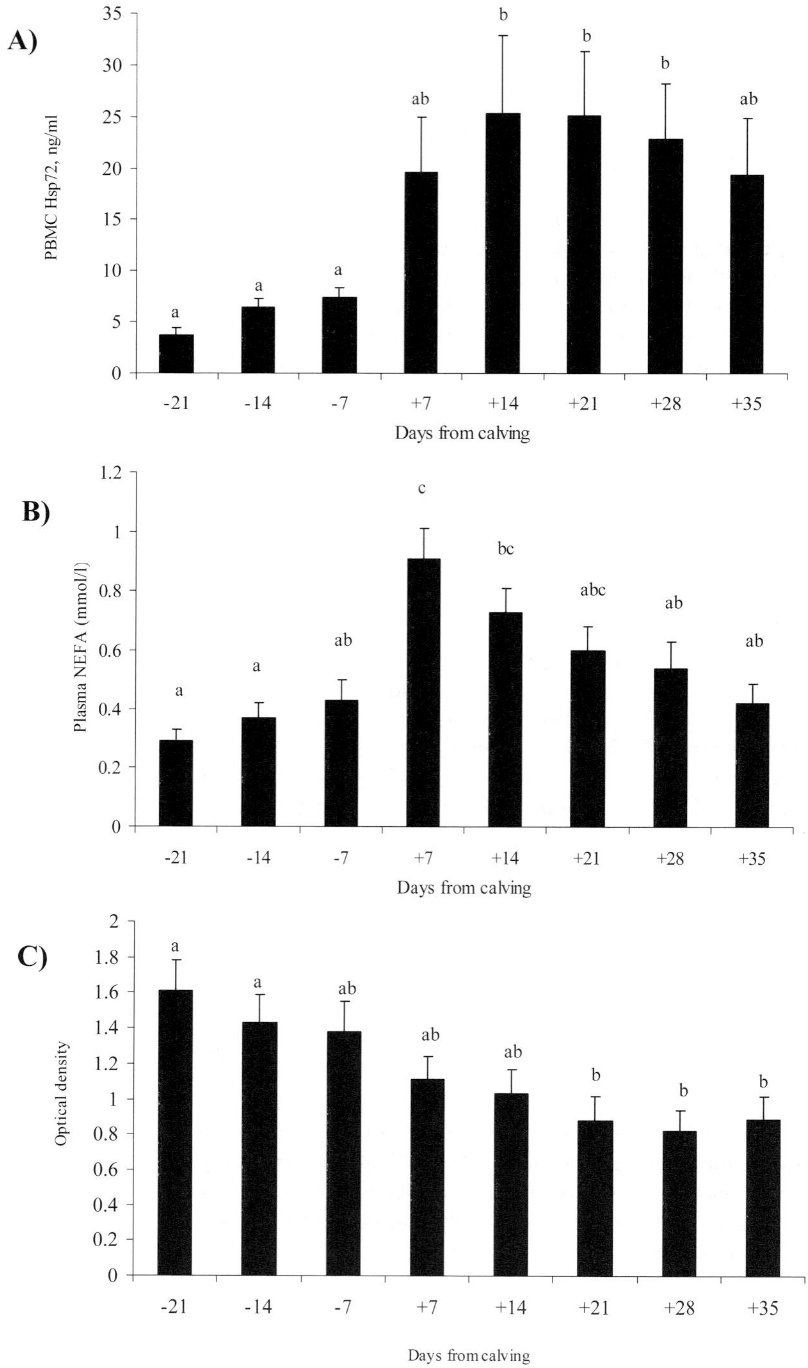

*Figure 1. Changes of inducible heat shock protein 72 kDa molecular weight (Hsp72) in peripheral blood mononuclear cells (PBMC) (panel A), of plasma nonesterified fatty acids (NEFA) (panel B) and of the proliferative response of lipopolysaccharide-stimulated PBMC (panel C) in peri-parturient dairy cows. Data are reported as least square means ± standard error. Different letters indicate significant differences between time points (*P<0.05*). Modified from Catalani* et al. *(2010).*

plasma NEFA at concentrations mimicking intense lipomobilization, altered the ability of PBMC to proliferate or secrete immunoglobulin (Ig)M and IFN-γ in response to polyclonal stimuli (Lacetera *et al.*, 2002, 2004) or viability and oxidative burst of PMN cells (Scalia *et al.*, 2006). Finally, an *ex vivo* study demonstrated that the larger extent of lipomobilization taking place in overconditioned cows compared to normal or thin cows was associated with a more pronounced immunedepression around calving (Lacetera *et al.*, 2005).

Therefore, findings from these trials encouraged realization of several epidemiological studies aimed at establishing whether the amount of body fat and/or the intensity of lipomobilization around calving were associated with the incidence of infections in the same physiological period. On this regard, Hammon *et al.* (2006) reported that an intense lipomobilisation around calving was associated with high incidence of metritis, and Rezamand *et al.* (2007) observed that cows identified with a new subclinical intrammammary infection had greater *pre partum* body condition score, body weight, and greater body weight loss compared with cows that did not develop a new intrammammary infection. Finally, Moyes *et al.* (2009) reported that substances in blood, especially NEFA, may be potential markers for the risk of mastitis in early lactating dairy cows, and Ospina *et al.* (2010) indicated that *post partum* serum NEFA concentration was associated with the risk of developing metritis, or retained placenta during the first 30 days in milk.

Nutritional and metabolic consequences of immune activation and infections

While nutritional deficiency or imbalance lead to immune dysfunction, it is becoming increasingly clear that an excessive immune activation may alter metabolic status and nutrients requirement (Husband and Bryden, 1996). Important consequences of immune stimulation include production of the pro-inflammatory cytokines, activation of the acute phase response, fever, inappetence, amino acid resorption from muscle, and redirection of nutrients towards liver anabolism of acute phase proteins (Colditz, 2002). Therefore, does not surprise whether under chronic activation of the immune system, as may occur during disease challenge or under conditions of a high microbial burden (poor hygienic environment), a reduction of productive and reproductive performances may occur.

Different models have been used to test the effects of immune activation on nutrition and metabolism. In a chicken study, Klasing and Calvert (1999) reported that immune activation was responsible for a significant increase of the cost for maintenance in terms of lysine utilization: percentage of cost for immunity changes from 1.17% to approximately 7% of the total cost following LPS challenge. Additionally, Lobley *et al.* (2001) indicated that immune activation is responsible for utilization of glutamine (GLN) and treonine as testified by reduction of plasma concentrations of these two aminoacids following LPS administration in sheep. Finally, Waldron *et al.* (2003) reported that immune activation by intravenous administration of LPS resulted in significant dose-dependent changes in metabolic measurements in dairy cows. In particular, these authors noticed that aspects of energy metabolism important to metabolic health are altered following immune activation and that these alterations may suggest possible mechanisms to explain the association between metabolic disorders and infectious diseases in dairy cattle.

As already reported above, cytokines are important biomolecules at the interface between metabolic and immune function, and play a central role in those situations where the immune activation may have nutritional/metabolic consequences (Elsasser *et al.*, 2008). In this context, the experimental models based on cytokines administration provided interesting results. Kushibiki *et al.* (2002) reported that intravenous administration of TNF-α was responsible for a significant increase of plasma NEFA in dairy cows, which would testify the need for massive utilization of body reserves. In pigs, very low physiological concentrations of TNFα interfere with protein synthesis and muscle cell development by inducing a state of insulin-like growth factor 1 receptor resistance (Broussard *et al.*, 2003). Finally, Trevisi *et al.* (2009) reported that compared with a non-treated group, cows orally administered with low doses of IFN-α showed larger decrease of BCS along with decreased milk yield, increased haptoglobin and ceruloplasmin, slower increase of negative acute phase proteins (albumin, cholesterol, paraoxonase, vitamin A), larger decrease of plasma glucose and greater values of plasma NEFA and β-hydroxybutyrate after calving.

Altogether, research findings reviewed within these paragraphs suggest the existence of intricate and multidirectional relationships between energy balance, immune activation, inflammatory and Hsp72 responses, immunosuppression and infections in early lactating dairy cows, which have been summarized in Figure 2.

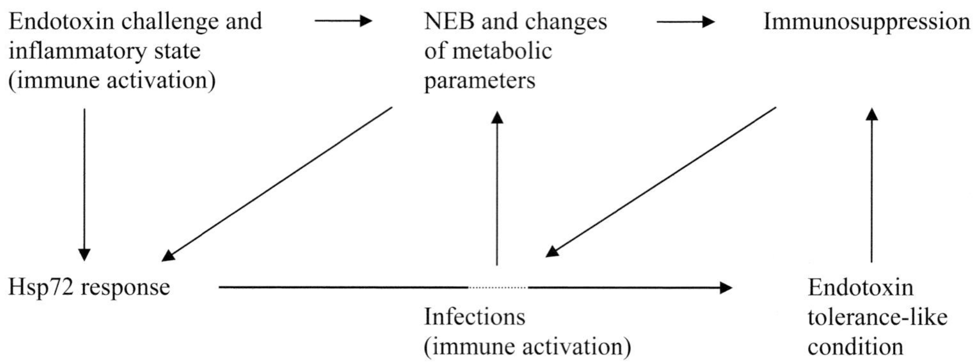

Figure 2. Diagram showing possible relationships between negative energy balance (NEB), immune activation, heat shock proteins (Hsp)72 kDa and inflammatory responses, immunosuppression and infections in early lactating dairy cows. These hypothesis have been already presented and discussed extensively elsewhere (Catalani et al.*, 2010). Briefly, translocation of bacterial endotoxin from the reproductive and digestive tracts to general circulation, the inflammatory state and the high incidence of infections in the early lactation period represent conditions, which may account for an endotoxin tolerance-like condition in early post-calving dairy cows. Hyporesponsiveness to lypopolysaccharide (LPS) following repeated exposure to endotoxin has been termed endotoxin tolerance and is a mechanism for preventing excessive inflammatory conditions, even if a persisting hyporesponsiveness bears the risk of causing immunesuppression and, hence, increased risk of secondary infections. In this scenario, it has been suggested that the gradual post partum decline of the ability of PBMC to proliferate in response to LPS might be interpreted as an ancillary sign of hyporesponsiveness compatible with a state of endotoxin tolerance. Finally, it is also suggested that the negative energy balance itself, due to the high demands of milk synthesis and to a certain extent also to the immune activation, and related changes of metabolic parameters may contribute to explain immunesuppression and upregulation of Hsp72.*

Nutritional modulation of the immune response and resistance to infections with special reference to energy and protein

In the context of this review, the role of energy and protein as immunomodulators will be briefly considered by two different points of view. First of all, energy and protein will be regarded as tools to enhance the ability of farm animals to mount an immune response, which may help to counteract invasion by pathogenic organisms. Secondly, energy nutrients will be also considered as components of strategies that may reduce the risk of a chronic activation of the immune system.

Several studies have been carried out in the last decades to identify nutrients capable to stimulate immune response in humans, laboratory animals, and farm animal species. With regard to domestic food animals, Blecha and Charley (1990) indicated that utilization of immunostimulant nutrients would be particularly useful when animals may be immunologically compromised, like during the neonatal period, and prior to or during exposure to stressors or pathogenic organisms. Oral treatment of ruminants with a mixture of glycerol and propylene glycol improved energy metabolism during the stress of feed restriction and pre-slaughter transport (Pethick *et al.*, 2000) and may provide a means for improving glucose availability to production tissues during immune activation. Experimental studies have shown that feeding GLN to young animals enhances immune functions and reduces infections or infectious morbidity (Johnson *et al.*, 2006). These effects have been primarily attributed to the effect of GLN on the health of the intestinal epithelial cell, but GLN is also an energy substrate for lymphocytes and macrophages, important for the optimal function of T and B cells and required for neutrophil bactericidal function. These observations support the concept that this amino acid may become essential in the diet during periods of immune stress. Finally, in a recent study it was reported that arginine requirements of starting broiler chicks for optimal immune functions are higher than those for maximum growth performance or feed efficiency and are dependent on dietary protein concentration (Jahanian, 2009).

As already reported above, the inflammatory response to microbial challenge and infection, although an essential part of immune function, carries the risk of severe tissue depletion and immunosuppression (Grimble, 2001). Fatty acids can impact host inflammatory responses in several ways. An adequate supply of fatty acids is essential as a key energy source where they can be oxidized to produce Acyl-CoA and yield ATP. However, when present in excess, high levels of circulating NEFA are associated with increased systemic inflammatory, oxidative stress and immunesuppression (Bernabucci *et al.*, 2005; Sordillo *et al.*, 2009). However, numerous studies carried out in humans and laboratory animals testified that increased intake of n-3 polyunsaturated fatty acids results in lower production of proinflammatory cytokines (Grimble, 2001), and few recent studies demonstrated that nutritional-induced changes of plasma NEFA composition may differentially affect immunoresponsiveness of periparturient dairy cows undergoing moderate/intense mobilization of lipid depots (Lessard *et al.*, 2003; Lacetera *et al.*, 2007).

The precise mechanisms through which fatty acids may differentially affect inflammation and immune cell functions remain to be ascertained in ruminants, although it is likely that they may largely reflect those already described in other species (changes in the plasma membrane fatty acid composition, biosynthesis of lipid mediators, susceptibility of cells to oxidative stress, etc.).

Concluding remarks

Metabolic balance is essential to ensure the correct functioning of the immune system and guarantee maximum resistance to invading pathogens. On the other hand, excessive activation of defensive mechanisms may have metabolic consequences, which are likely to cause impairment of reproductive and productive efficiency in farm animals.

Further research is needed and guaranteed to elucidate the intimate mechanisms through which metabolism and immunity interact and to establish appropriate and sustainable immunonutritional interventions that may help in keeping farm animals healthy and capable to perform to their maximum potential.

Acknowledgements

The authors would like to thank MIPAF (CLIMANIMAL and SELMOL Projects), MINSAL (IZSLER PRF 2006201 Project), MIUR (PRIN), and Università della Tuscia (RSA) for financial support of the dairy cow studies on the relationships between NEFA or Hsp72 and immune response, reported in this paper.

References

Andersson, R., 1993. Ketone bodies as risk factors for metabolism disturbances and mastitis. Milch Praxis 31, 200-201.

Asea, A., 2005. Stress proteins and initiation of immune response: chaperokine activity of HSP72. Exerc. Immunol. Rev. 11, 34-45.

Bernabucci, U., B. Ronchi B., N. Lacetera and A. Nardone, 2005. Influence of body condition score on relationships between metabolic status and oxidative stress in periparturient dairy cows. J. Dairy Sci. 88, 2017-2026.

Blecha, F. and B. Charley, 1990. Rationale for using immunopotentiators in domestic food animals. Adv. Vet. Sci. Comp. Med. 35, 3-19.

Broussard, S.R., R.H. McCusker, J.E. Novakosfski, K. Strle, W.H. Shen, R.W. Johnson, G.G. Freund, R. Dantzer and K.W. Kelly, 2003. Cytokine-hormone interactions: Tumor necrosis factor α impairs biologic activity and downstream activation of signals of the insulin-like growth factor I receptor in myoblasts. Endocrinology 144, 2988-2996.

Caldefie-Chezet, F., A. Poulin and M.P. Vasson, 2003. Leptin regulates functional capacities of polymorphonuclear neutrophils. Free Radic. Res. 37, 809-814.

Campisi, J., T. Leem and M. Fleshner, 2003. Stress-induced extracellular HSP72 is a functionally significant danger signal to the immune system. Cell Stress Chaperones 8:272-286.

Carratelli, C.R., I. Nuzzo, T. Vitiello, E. Galdiero and F. Galdiero, 1999. The effect of dietary lipid manipulation on murine splenic lymphocytes apoptosis and heat shock protein over expression. FEMS Immunol. Med. Microbiol. 24, 19-25.

Catalani, E., M. Amadori, A. Vitali, U. Bernabucci, A. Nardone and N. Lacetera, 2010. The Hsp72 response in periparturient dairy cows: relationships with metabolic and immunological parameters. Cell Stress Chaperones, doi:10.1007/s12192-010-0186-x.

Colditz, I.G., 2002. Effects of the immune system on metabolism: implications for production and disease resistance in livestock. Liv. Prod. Sci. 75, 257-268.

Dandona, P., A. Aljada and A. Bandyopadhyay, 2004. Inflammation as the link between insulin resistance, obesity and diabetes. Trends Immunol. 25, 4-7.

Eitam, H., A. Brosh, A. Orlov, I. Izhaki and A. Shabtay, 2009. Caloric stress alters fat characteristics and HSP70 expression in somatic cells of lactating beef cows. Cell Stress Chaperones 14, 173-182.

Elsasser, T.H., T.J. Caperna, C-J. Li, S. Kahl and J.L. Sartin, 2008. Critical control points in the impact of the proinflammatory immune response on growth and metabolism. J. Anim. Sci. 86 (E. Suppl.), E105-E125.

Erb, H.N. and Y.T. Grohn, 1988. Epidemiology of metabolic disorders in the periparturient dairy cow. J. Dairy Sci. 71, 2557-2571.

Farooqi, S., G. Matarese, G.M. Lord, J.M. Keogh, E. Lawrence, C. Agwu, V. Sanna, S.A. Jebb, F. Perna, S. Fontana, R.I. Lechler, A.M. DePaoli and S. O'Rahilly, 2002. Beneficial effects of leptin on obesity, T cell hyporesponsiveness, and neuroendocrine/metabolic dysfunction of human congenital leptin deficiency. J. Clin. Invest. 110, 1093-1103.

Franklin, S.T., J.W. Young and B.J. Nonnecke, 1991. Effects of ketones, acetate, butyrate and glucose on bovine lymphocyte proliferation. J. Dairy Sci. 74, 2507-2514.

Friedman, J.M. and J.L. Halaas, 1998. Leptin and the regulation of body weight in mammals. Nature 395, 763-770.

Gershwin, L.J., S. Krakowka and R.G. Olsen, 1995. Principles of immunology. Innate immunity. In: L.L. Duncan (ed.), Immunology and immunopathology of domestic animals. Mosby-Year Book, St. Louis, Missouri, USA: 3-5.

Goff, J.P., 2006. Major advances in our understanding of nutritional influences on bovine health. J. Dairy Sci. 89, 1292-1301.

Grimble, S.F., 2001. Nutritional modulation of immune function. Proc. Nutr. Soc. 60, 389-397.

Hammon, D.S., I.M. Evjen, T.R. Dhiman, J.P. Goff and J.L. Walters, 2006. Neutrophil function and energy status in Holstein cows with uterine health disorders. Vet. Immunol. Immunopathol. 113, 21-29.

Hoeben, D., C. Burvenich, A.M. Massart-Leen, M. Lenjou, G. Nijs, D. Van Bockstaele and J.F. Beckers, 1999. *In vitro* effect of ketone bodies, glucocorticosteroids and bovine pregnancy-associated glycoprotein on cultures of bone marrow progenitor cells of cows and calves. Vet. Immunol. Immunopathol. 68, 229-240.

Husband, A.J. and W.L. Bryden, 1996. Nutrition, stress and immune activation. Proc. Nutr. Soc. Austral. 20, 60-70.

Jahanian, R. 2009. Immunological responses as affected by dietary protein and arginine concentrations in starting broiler chicks. Poult. Sci. 88, 1818-1824.

Johnson, I.R., R.O. Ball, V.E. Baracos and C.J. Field, 2006. Glutamine supplementation influences immune development in the newly weaned piglet. Dev. Comp. Immunol. 30, 1191-1202.

Kehrli, M.E. Jr., K. Kimura, J.P. Goff, J.R. Stabel and B.J. Nonnecke, 1999. Immunological dysfunction in periparturient cows – What role does it play in *post partum* infectious diseases? In: Proc. Annu. Conf. Am. Assoc. Bovine Pract: 24-28.

Klasing, K.C. and C.C. Calvert, 1999. The care and feeding of the immune system: an analysis of lysine needs. In: Proc. VIIIth Int. Symp. Protein Metabol. Nutr., Wageningen Press, Wageningen: 253-264.

Kushibiki S., K. Hodate, H. Shingu, T. Hayashi, E. Touno, M. Shinoda and Y. Yokomizo, 2002. Alterations in lipid metabolism induced by recombinant bovine tumor necrosis factor-alpha administration to dairy heifers. J. Anim. Sci. 80, 2151-2157.

Lacetera N., U. Bernabucci, B. Ronchi and A. Nardone, 2001. Effects of subclinical pregnancy toxemia on immune responses in sheep. Am. J. Vet. Res. 62, 1020-1024.

Lacetera N., O. Franci, D. Scalia, U. Bernabucci, B. Ronchi and A. Nardone, 2002. Effects of nonesterified fatty acids and β-hydroxybutyrate on functions of mononuclear cells obtained from ewes. Am. J. Vet. Res. 63, 414-418.

Lacetera N., U. Bernabucci, A. Nardone and B. Ronchi, 2004a. Immunological parameters in periparturient sheep. J. Anim. Feed Sci. 13, 601-604.

Lacetera N., D. Scalia, O. Franci, U. Bernabucci, B. Ronchi and A. Nardone, 2004b. Effects of nonesterified fatty acids on lymphocyte functions in dairy heifers. J. Dairy Sci. 87, 1012-1014.

Lacetera N., D. Scalia, U. Bernabucci, B. Ronchi, D. Pirazzi and A. Nardone, 2005. Lymphocyte functions in overconditioned cows around parturition. J. Dairy Sci. 88, 2010-2016.

Lacetera N., U. Bernabucci and B. Ronchi, 2006. Proliferative response of peripheral blood lymphocytes in periparturient dairy goats: a preliminary study. In: Proc. 2nd Eur. Vet. Immunol. Workshop: 85-86.

Lacetera N., D. Scalia, D.G. Mashek, U. Bernabucci and R.R. Grummer, 2007. Effects of intravenous triacylglycerol emulsions on lymphocyte responses to mitogens in fasted dairy cows undergoing intense lipomobilization. J. Dairy Res. 74, 323-328.

Lessard M., N. Gagnon and H.V. Petit, 2003 Immune response of *post partum* dairy cows fed flaxseed. J. Dairy Sci. 86, 2647-2657.

Lobley G.E., S.O. Hoskin and C.J. McNeil, 2001. Glutamine in animal science and production. J. Nutr. 131, 2525S-2531S.

Lord G.M., G. Matarese, J.K. Howard, R.J. Baker, S.R. Bloom and R.I. Lechler, 1998. Leptin modulates the T-cell immune response and reverses starvation-induced immunosuppression. Nature 394, 897-901.

Matarese G. and A. La Cava, 2004. The intricate interface between immune system and metabolism. Trends Immunol. 25, 193-200.

Moore S.I., G.B. Huffnagle, G.H. Chen, E.S. White and P. Mancuso, 2003. Leptin modulates neutrophil phagocytosis of Klebsiella pneumoniae. Infect. Immun. 71, 4182-4185.

Morino S., T. Kondo, K. Sasaki, H. Adachi, M.A. Suico, E. Sekimoto, T. Matsuda, T. Shuto, E. Araki and H. Kai, 2008. Mild electrical stimulation with heat shock ameliorates insulin resistance via enhanced insulin signalling. PLoS one. doi:10.1371/journal.pone.0004068.

Moyes K.M., T. Larsen, N.C. Friggens, J.K. Drackley and K.L. Ingvartsen, 2009. Identification of potential markers in blood for the development of subclinical and clinical mastitis in dairy cattle at parturition and during early lactation. J. Dairy Sci. 92, 5419-5428.

Moyes K.M., J.K. Drackley, D.E. Morin, S.L. Rodriguez-Zas, R.E. Everts.3, H.A. Lewin and J.J. Loor, 2010. Mammary gene expression profiles during an intramammary challenge reveal potential mechanisms linking negative energy balance with impaired immune response. Physiol. Genomics 41, 161-170.

Ospina P.A., D.V. Nydam, T. Stokol and T.R. Overton, 2010. Evaluation of nonesterified fatty acids and β-hydroxybutyrate in transition dairy cattle in the northeastern United States: Critical thresholds for prediction of clinical diseases. J. Dairy Sci. 93, 546-554.

Pethick D.W., L. Cummins, G.E. Gardner, R.H. Jacobs, B.W. Knee, M. McDowell, B.L. McIntyre, G. Tudor, P.J. Walker and R.D. Warner, 2000. The regulation of glycogen level in the muscle of ruminants by nutrition. Proc. N. Z. Soc. Anim. Prod. 60, 94-98.

Pockley A.G., M. Muthana and S.K. Calderwood, 2008. The dual immunoregulatory roles of stress proteins. Trends Biochem. Sci. 33:71-79.

Rezamand P., T.A. Hoagland, K.M. Moyes, L.K. Silbart and S.M. Andrew, 2007. Energy status, lipid-soluble vitamins, and acute phase proteins in periparturient Holstein and Jersey dairy cows with or without subclinical mastitis. J. Dairy Sci. 90, 5097-5107.

Scalia D., N. Lacetera, U. Bernabucci, K. Demeyere, L. Duchateau, C. Burvenich, 2006. *In vitro* effects of non-esterified fatty acids on bovine neutrophils oxidative burst and viability. J. Dairy Sci. 89, 147-154.

Sordillo L.M., G. A. Contreras and S.L. Aitken, 2009. Metabolic factors affecting the inflammatory response of periparturient dairy cows. Animal Health Res. Rev. 10, 53-63.

Stentz F.B. and A.E. Kitabchi, 2003. Activated T lymphocytes in type 2 diabetes: implications from *in vitro* studies. Curr. Drug Targets 4, 493-503.

Suriyasathaporn W., A.J.J.M. Daemen, E.N. Noordhuizen-Stassen, S.J. Dieleman, M. Nielen and Y.H. Schukken, 1999. β-hydroxybutyrate levels in peripheral blood and ketone bodies supplemented in culture media affect the *in vitro* chemotaxis of bovine leukocytes. Vet. Immunol. Immunopathol. 68, 177-186.

Targowski S.P. and W. Klucinski, 1983. Reduction in mitogenic response of bovine lymphocytes by ketone bodies. Am. J. Vet. Res. 44, 828-830.

Tizard I., 1992. Veterinary immunology. An introduction. I. Tizard (ed.), Saunders Company, Philadelphia, Pennsylvania, USA: 1-7.

Trevisi E., M. Amadori, A.M. Bakudila and G. Bertoni, 2009. Metabolic changes in dairy cows induced by oral, low-dose interferon-alpha treatment. J. Anim Sci. 87, 3020-3029.

Van Knegsel A.T.M., H. Van den Brand, J. Dijkstra, S. Tamminga and B. Kemp, 2005. Effect of dietary energy source on energy balance, production, metabolic disorders and reproduction in lactating dairy cattle. Reprod. Nutr. Dev. 45, 665-688.

Waldron M.R., T. Nishida, B.J. Nonnecke and T.R. Overton, 2003. Effect of lipopolysaccharide on indices of peripheral and hepatic metabolism in lactating cows. J. Dairy Sci. 86, 3447-3459.

Wathes D.C., Z. Cheng, W. Chowdhury, M.A. Fenwick, R. Fitzpatrick, D.G. Morris, J. Patton and J.J. Murphy, 2009. Negative energy balance alters global gene expression and immune responses in the uterus of *post partum* dairy cows. Physiol. Genomics 39, 1-13.

An additional study on the relationship between the inflammatory condition at calving time and net energy efficiency in dairy cows

E. Trevisi, A. Ferrari, F. Piccioli-Cappelli, P. Grossi and G. Bertoni
Istituto di Zootecnica, Università Cattolica del Sacro Cuore, Piacenza, Italy

Introduction

Inflammation is a complex response of the organism to injuries or infections (Calder, 2002). Unfortunately, this response can be overburden, and produce several negative effects. For example, in dairy cows during the peripartum period inflammatory phenomena are very common (Bertoni *et al.*, 2008). They can negatively affect the net energy efficiency, likely due to the increased energy demand by the immune system (Elsasser *et al.*, 1995). A reduction of energy efficiency in the *post partum* cows affected by serious inflammations has been already observed by Trevisi *et al.*, 2007. Aim of this work is to provide further evidences, about the relationships between the severity of the peripartum inflammation and the net energy efficiency in dairy cows.

Materials and methods

Twenty multiparous dairy cows were investigated one month before and after calving. Cows were individually and daily monitored for health status, rectal temperature, milk yield, water consumption and dry matter intake (DMI). Body weight and body condition score (BCS) were measured once a week; blood samples were collected two times a week or daily from 10 days before calving to 10 days in milk (DIM). Hematochemical profile, included positive and negative acute phase reactants, was determined according to Bertoni *et al.* (2008). Milk samples were collected weekly from 2 consecutive milkings to determine fat, protein, lactose content (Milkoscan^{TM} FT120, Foss Electric, Hillerød, Denmark) and somatic cell count (Fossomatic 180, Foss Electric). Cows were retrospectively ranked according to liver functionality index (LFI; Bertoni *et al.*, 2006), that includes three liver indices: albumin, lipoproteins (measured as total cholesterol) and total bilirubin (its excretory enzyme is synthesized by the liver). The 6 top (HI-LFI) and the 6 worst (LO-LFI) cows were further investigated assessing retinol (as index of retinol binding protein = RBP), non esterified fatty acids (NEFA), β-hydroxybutyrate (BHB), thiol groups and insulin in blood. RBP allowed the evaluation of an additional liver activity index (LAI, Bertoni *et al.*, 2008). Moreover, weekly energy balance [(feed - (maintenance + milk production)] as well as the net energy efficiency [milk energy/ (feed energy - maintenance energy - reserves energy)] were evaluated in the first 30 DIM. Statistical analysis was performed by repeated measures variance test, using a mixed model (MIXED procedure, SAS Inst. Inc., Cary, NC), that included LFI and DIM as fixed effects and their interaction.

Results and discussion

All the studied cows underwent to an inflammatory state around calving as indicated by the raise of positive acute phase proteins (APP), however followed by faster (HI-LFI) or slower (LO-LFI) recovery of negative APP. In particular HI-LFI displayed a lower increase of haptoglobin (a positive APP; NS vs. LO-LFI) and a quicker recovery of the pre-calving levels (Table 1). Thus, HI-LFI (1.3±0.6 vs. -3.3±1.5 of LO-LFI) represents cows with minor negative consequences on the usual liver synthesis activity. HI-LFI group displayed half number of cows with at least one disease (2 vs. 4) as well as cows with lower severity of health disorders in the peripartum, confirming previous results (Bertoni *et al.*, 2006). Furthermore, HI-LFI cows displayed a lower rectal temperature after calving. Considering these clinical data, our major interest focused on the different consequences of inflammation, as suggested by variations of negative APP. In the first 30 DIM (average ± standard deviation), HI-LFI cows showed the highest levels of albumin (36.2±2.3 vs. 33.5±2.1 g/l, $P<0.01$), cholesterol (2.9±1.1 vs. 2.6±1.1 mmol/l, NS) and the lowest level of bilirubin (3.38±2.09 vs.

Table 1. Average (± Standard Error of Mean) DMI and blood parameters recorded in the month before and after calving in dairy cows grouped according to high (HI) or low (LO) LFI.

	DMI (kg/d)		Haptoglobin (g/l)		Retinol (µg/100 ml)		NEFA (mmol/l)	
LFI index	HI	LO	HI	LO	HI	LO	HI	LO
-28/-1 DIM	11.9±0.4	11.1±0.6	0.12±0.15	0.13±0.10	53.8±1.8	41.0±1.8	0.20±0.09	0.33±0.11
+1/+28 DIM	17.9±0.8	15.5±0.7	0.45±0.16	0.54±0.13	45.8±2.7	32.9±1.7	0.54±0.07	0.73±0.07
P (factor LFI)	<0.10		0.29		<0.01		<0.05	

5.60±3.18 µmol/l, $P<0.05$). Interestingly, LAI index which includes also retinol (index of RBP, another negative APP) showed an highly significant correlation (r=0.87; $P<0.001$) with LFI index. The similar trend of the two indices supports the reliability of the easier-assayed LFI. In the first 30 DIM, the negative energy balance (NEB) appears less pronounced in HI-LFI respect to LO-LFI cows (-10.3±3.4 vs. -13.2±4.5 Mcal/d, $P<0.10$). This is confirmed by the lower BCS loss observed in HI-LFI cows (-0.31±0.1 vs. -0.46±0.1 scores, $P<0.01$). Moreover, these results are supported by differences in blood parameters implying a lower lipomobilization: i.e. HI-LFI cows showed lower levels of NEFA (Table 1) and BHB (1.2±0.7 vs. 1.7±1.8 mmol/l), while slightly higher values of insulin (2.3±1.0 vs. 2.1±0.9 µU/ml). Conversely, NEB data apparently conflict with the higher milk yield recorded for HI-LFI cows (38.3±4.9 vs. 34.4±6.5 kg/d of LO-LFI, $P<0.14$). However, both higher DMI (Table 1) and higher net energy efficiency (120.3±19.5 vs. 102.3±13.2, $P<0.06$ from 7 to 21 DIM) could explain the data observed.

In conclusion, our results showed that cows with minor inflammatory consequences during the peripartum period were also characterized by a less severe negative energy balance, likely due to the higher DMI and to the lower energy requested by the immune system (higher efficiency), as suggested also by the better health status. Such observations confirm that inflammation in early lactation is a physiological expensive and dangerous process to prevent by suitable and effective health care and management strategy.

References

Bertoni G., E. Trevisi, A. Ferrari and A. Gubbiotti, 2006. The dairy cow performances can be affected by inflammations occurring around calving. 57th EAAP Meeting, 17-20 September 2006, Antalya, Turkey, 325.

Bertoni G., E Trevisi, X Han and M Bionaz, 2008. Effects of inflammatory conditions on liver activity in puerperium period and consequences for performance in dairy cows. J. Dairy Sci. 91, 3300-3310.

Calder, P.C., 2002. Dietary modification of inflammation with lipids. Proc Nutr Soc. 61, 345-358.

Elsasser T.H., N.C. Steele and R. Fayer, 1995. Cytokines, stress, and growth modulation. In: Myers, M. J. and M. P. Martaugh (eds.), Cytokines in animal health and disease. Marcel Dekker, Inc., New York.

Trevisi E., A. Gubbiotti and G. Bertoni. 2007. Effects of inflammation in peripartum dairy cows on milk yield, energy balance and efficiency. Wageningen Academic Publisher, Wageningen, the Netherlands. EAAP publication n. 124. Energy and Protein Metabolism and Nutrition: 395-396.

Metabolites in cerebrospinal fluid and plasma of feed-restricted dairy cows: implications on appetite regulation

T. Laeger, S. Görs and B. Kuhla
Leibniz Institute for Farm Animal Biology (FBN), Research Unit Nutritional Physiology 'Oskar Kellner', Wilhelm-Stahl-Allee 2, 18196 Dummerstorf, Germany; b.kuhla@fbn-dummerstorf.de

Introduction

Nutritional metabolites in blood such as fatty acids, amino acids as well as hormones and cytokines act as hunger or satiety signals in the brain and play an important role in the regulation of food intake. These substances are also components of the cerebrospinal fluid (CSF) which surrounds the hypothalamus and brain stem, two major centers of feed intake regulation. The contribution of orexigenic and anorexigenic signals in CSF of cows has not been studied so far. Thus, the aim of this study was to elucidate associations existing between both plasma and CSF metabolites each in response to different feeding levels.

Materials and methods

Six German Holstein dairy cows between 87 and 96 days of the 2nd lactation were fed *ad libitum* a total mixed ration (TMR) 2-times daily (7 am and 4 pm) for four days. After local anesthesia, CSF from spinal cord was obtained by lumbar puncture and blood EDTA-plasma was withdrawn from the jugular vein before morning feeding. Afterwards, animals were feed restricted to 50% of the previous *ad libitum* intake for four days and subsequently sampled again for CSF and plasma. Body weight, feed intake, water intake, and milk production were determined daily. Concentrations of non-esterified fatty acids (NEFA), glucose, cholesterol, urea, and β-hydroxybutyric acid (BHBA) were determined by routine analyses (Cobas Mira). The concentrations of plasma volatile fatty acids (VFA) were determined by GC-FID on a 0.25 μM FFAP column according to Ryan (1980). Amino acids (AA) were analyzed by HPLC (Series 1200, Agilent Technologies; 250x4 mm Hyperclone ODS (C18) 120 Å column protected by a 4×3 mm C18 pre-column) with automated pre-column derivatization and fluorescence detection (Krömer *et al.*, 2005). The data were statistically analyzed using the paired t-test.

Results and discussion

As compared to the *ad libitum* feeding (mean ± SD, 123.8±10.2 MJ/day) period, restricted feeding for four days (69.1±5.4 MJ/day) resulted in significant loss of body weight, decreased water intake, and reduced milk yield. Urea concentrations were not affected by reduced feed intake, neither in plasma nor in CSF (Table 1). Likewise glucose concentrations did not change in response to feed restriction but tended to be reduced in CSF as compared to plasma concentrations. At the lower intake level, plasma but not CSF BHBA concentrations tended to be elevated when compared to *ad libitum* intake. After feed restriction plasma cholesterol was significantly elevated whereas CSF cholesterol tended to be lower. After feed restriction, plasma NEFA concentrations were significantly increased but decreased (not detectable) in the CSF of feed restricted cows. This might be due to the fact that long-chain NEFA may hardly enter the blood-brain and the blood-CSF barrier. The analyses of VFA plasma concentrations revealed a tendency for decreased acetic acid after feed restriction (Table 2). With the exception of glutamine, all AA were up to 30-fold higher concentrated in plasma than in CSF. Glutamine was almost doubled in CSF compared with plasma in *ad libitum* fed animals (451.2±49.4 μM vs. 250.5±42.9 μM). After restrictive feeding CSF glutamine was diminished (403.6±58.4 μM) while plasma glutamine was not affected (284.6±41.7 μM). The branched-chain AA (BCAAs) leucine, isoleucine, and valine which are supposed to be involved in satiety signalling did not change in response to feed restriction in CSF. Plasma leucine was significantly increased

Table 1. Concentrations of plasma (P) and CSF metabolites after ad libitum *and 50% feed restriction.*

Item	Fluid	Diet		*P*-value
		Ad libitum	50% restrictive	
BHBA (mmol/l)	P	0.35±0.09	0.62±0.22	0.088
	CSF	0.13±0.06	0.19±0.09	0.307
Cholesterol (mmol/l)	P	4.87±0.43	5.44±0.37	<0.001
	CSF	0.11±0.11	0.06±0.05	0.577
NEFA (mmol/l)	P	0.889±0.227	1.989±0.713	0.008
	CSF	0.010±0.005	-	-
Glucose (mmol/l)	P	4.04±0.26	3.85±0.59	0.323
	CSF	2.38±0.10	2.19±0.19	0.083
Urea (mmol/l)	P	4.36±0.62	4.53±0.78	0.221
	CSF	4.43±0.81	4.13±1.32	0.822

Table 2. Concentrations of jugular plasma (P) VFA concentrations (µmol/l) after ad libitum *and 50% feed restriction.*

Item	Fluid	Diet		*P*-value
		Ad libitum	50% restrictive	
Acetic acid	P	662.65±135.84	458.13±133.48	0.057
n-Butyric acid	P	13.19±16.54	0.00	0.149
n-Caproic acid	P	4.70±8.34	0.00	0.277
i-Valeric acid	P	9.84±15.97	0.00	0.240
Total	P	690.38±156.96	458.13±133.48	0.056

(110.3±11.2 µM to 147.8±26.7 µM, $P<0.05$), whereas isoleucine tended to rise (97.7±13.6 µM to 127.9±36.9 µM) while valine was not affected (213.8±28.6 µM to 230.1±54.8 µM) by reduced feed intake. Moreover, plasma arginine, a precursor for NO production and thus involved in appetite regulation, was significantly elevated after restriction by 32% while plasma tryptophan, a precursor of serotonin, was decreased by 32%. CSF serine and threonine shown to be related to reduced food intake (Veldhorst *et al.*, 2009), were significantly diminished after feed restriction (Ser: 53.3±7.8 µM to 44.5±7.8 µM; Thr: 23.1±4.5 µM to 18.3±3.8 µM; $P<0.05$). Earlier studies deduced the orexigenic and anorexigenic effects of metabolites primarily from plasma data. Here we demonstrate that numerous metabolites reveal different concentration responses in plasma as compared to CSF after feed restriction. Therefore, studies assigning plasma metabolites a negligible or anorexigenic role *per se* should be interpreted with more caution.

References

Krömer, J.O., M. Fritz, E. Heinzle and C. Wittmann, 2005. *In vivo* quantification of intracellular amino acids and intermediates of the methionine pathway in *Corynebacterium glutamicum*. Anal Biochem 340, 171-173.

Ryan, J.P., 1980. Determination of volatile fatty acids and some related compounds in bovine rumen fluid, urine, and blood plasma, by gas-liquid chromatography. Anal. Biochem. 108, 374-384.

Veldhorst, M.A.B., A.G. Nieuwenhuizen, A. Hochstenbach-Waelen, K.R. Westerterp, M.P.K.J. Engelen, R.J.M. Brummer, N.E.P. Deutz and M.S. Westerterp-Plantenga, 2009. Effects of complete whey-protein breakfasts versus whey without GMP-breakfasts on energy intake and satiety. Appetite 52, 388-395.

Effect of rumen-protected choline on performance and hepatic triacylglycerol concentrations in early-lactating dairy cattle

A.M. van Vuuren[1], R.M.A. Goselink[1], R. Zom[1], M.J. de Veth[2] and J. Van Baal[1]
[1]Wageningen UR Livestock Research, P.O. Box 65, 8200 AB Lelystad, the Netherlands
[2]Balchem Corporation, 52 Sunrise Park Rd., New Hampton, NY 10958, USA

Introduction

In the first weeks of lactation, dairy cows mobilize body fat resulting in an increased release of non-esterified fatty acids (NEFA) into circulation. The NEFA are taken up by the liver and here they are oxidized or re-esterified into triacylglycerols (TAG), which are released through packaging in very low density lipoproteins (VLDL). When TAG formation is not balanced by VLDL synthesis, TAG may accumulate in the liver. TAG accumulation is regarded as a predisposing factor for other metabolic disorders and infection diseases and impaired reproductive performance (Drackley and Andersen, 2006). Choline is an essential component for phosphatidylcholine synthesis, an important component of VLDL. However, 79 to 99% of dietary choline is degraded in the rumen (Sharma and Erdman, 1989). Encapsulation of choline to reduce ruminal degradation will increase choline supply, supporting VLDL synthesis. Indeed, Cooke *et al.* (2007) observed a reduction in hepatic TAG when supplementing 15 g/d of rumen-protected choline (RPC) to non-lactating cows. In the present trial we studied the effect of RPC supplementation on feed intake, milk production and liver TAG concentrations of dairy cows during the periparturient period.

Materials and methods

In a complete randomized blocked designed experiment, 38 dairy cows were paired in 19 blocks. Cows within each block were randomly allocated to treatment group 'RPC' or 'Control'. Cows on 'RPC' were daily supplemented with a mixture of 60 g of RPC (ReaShure™, Balchem Corporation, New Hampton, NY, USA) and 540 g of soybean meal (SBM). Cows on 'Control' received a mixture of SBM and palm oil instead (582 and 18 g/d, respectively), to obtain equal protein and energy supplementation. Mixtures were fed with a concentrate dispenser, starting 3 weeks *ante partum* (*ap*) until 6 weeks *post partum* (*pp*). Cows *ap* received a mixed ration of corn silage (30% of mix DM), wilted grass silage (30%), wheat straw (32%), SBM (7%), and vitamins and minerals (1%). Cows *pp* received a mixed ration of corn silage (35% of mix DM), wilted grass silage (52%), grass seed straw (6%), SBM (6%), and vitamins and minerals (1%). Additionally, cows received a concentrate mixture, increasing gradually up to 1 kg/d on day -1 *ap*, further increasing *pp* by 0.5 kg/day up to the maximum level of 9 kg/d from d16 *pp* until the end of the trial. Individual feed intake and milk production were recorded daily and averaged by week. Milk was sampled weekly for analyses of fat, protein and lactose. Additionally, 8 blocks of cows were selected for liver biopsies and blood sampling. Liver biopsies were taken in wk3 *ap* and in wk1, 3 and 6 *pp*, for TAG analyses. Blood samples were taken in wk3 *ap* and on d1, d4, d7, d10, wk2, wk3 and wk6 *pp* for glucose, NEFA, and β-hydroxybutyrate (BHB) analyses. Results of hepatic TAG concentrations were tested statistically by ANOVA; all other treatment effects were tested by mixed model analysis using the REML procedure.

Results and discussion

No significant treatment effects were observed for DM intake (DMI) *pp*, which was 20.3 and 21.0 kg/d for Control and RPC, respectively. Directly after calving, DMI was significantly higher for RPC than for Control; 16.0 and 14.4 kg/d, respectively. This indicates that cows supplemented with RPC had a significantly improved DMI directly after calving (Figure 1). Because similar amounts of concentrates were fed to both treatment groups, the higher DMI at the beginning of lactation for RPC was due to higher voluntary DMI of the mixed ration.

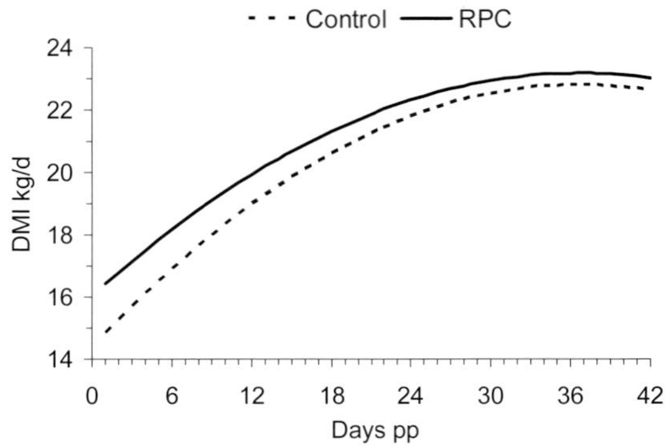

Figure 1. Modelled effect of RPC on DMI.

No treatment effects were observed for yields of milk, milk fat and milk protein and were 40.0, 1.81 and 1.14 kg for Control and 41.2, 1.85 and 1.37 for RPC, respectively. In the first week of lactation, milk protein yield was significantly higher for RPC than for Control; 1.26 and 1.12 kg/d, respectively.

Plasma concentrations of glucose, NEFA and BHB were also not affected by treatment. Values for NEFA and BHB were always below cut-off values for (sub)clinical ketosis (NEFA <0.5 to 0.6 mmol/l and BHB <1 mmol/l).

Supplementing RPC resulted in lower liver TAG concentrations in wk1 (36% reduction; P=0.04) and wk3 (32% reduction; P=0.12) *pp* (Figure 2). The reduction in liver TAG is consistent with the mode of action of choline and the effect seen in other livestock animals when choline is supplemented (Cooke *et al.*, 2007).

In conclusion, RPC supplementation reduced hepatic TAG concentration. Supplementing RPC coincided with improved DMI and increased production of milk protein at the onset of lactation.

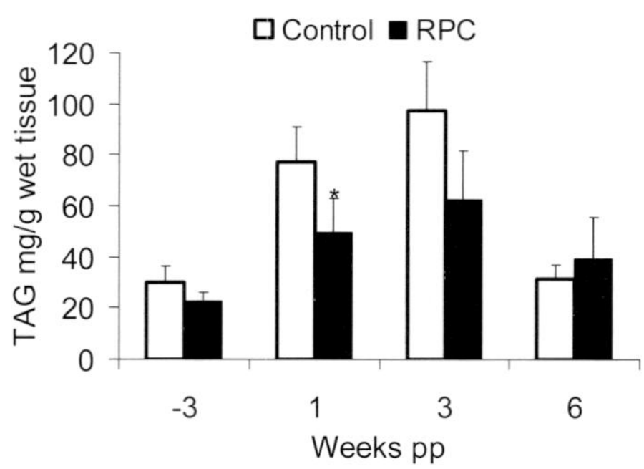

Figure 2. Effect of RPC on hepatic TAG concentrations.

References

Cooke, R.F., N.S. del Rio, D.Z. Caraviello, S.J. Bertics, M.H. Ramos and R.R. Grummer, 2007. Supplemental choline for prevention and alleviation of fatty liver in dairy cattle. J. Dairy Sci. 90, 2413-2418.

Drackley, J.K. and J.B. Andersen, 2006. Splanchnic metabolism of long-chain fatty acids in ruminants. In: Sejrsen, K., T. Hvelplund and M.O. Nielsen (eds.), Ruminant Physiology. Wageningen Academic Publishers, Wageningen, the Netherlands: 199-224.

Sharma B.K. and R.A. Erdman, 1989. Degradation of choline from selected foodstuffs and choline supplements. J. Dairy Sci. 72, 2772-2776.

Foetal protein-malnutrition of the F1-generation in mink (*Neovison vison*) causes changes in F2 progeny

C.F. Matthiesen[1], D. Blache[2], P.D. Thomsen[1] and A.-H. Tauson[1]
[1]Department of Basic Animal and Veterinary Sciences, Faculty of Life Sciences, University of Copenhagen, Denmark
[2]Faculty of Natural and Agricultural Sciences, School of Animal Biology, University of Western Australia, Australia

Introduction

Protein malnutrition during foetal life can induce metabolic changes in offspring, due to metabolic and structural adaptation to the available nutrient supply in order to maximize the outcome (Lucas, 1991). It is known that for offspring exposed to a postnatal environment different from that predicted in foetal life, the resulting mismatch can put them at increased risk of developing metabolic and cardiovascular diseases later in life (Gluckman and Hanson, 2004). More specifically, protein restriction occurring during pregnancy has been found to reduce birth weight, affect glucose homeostasis (Ozanne and Hales, 2002), elevate blood pressure (Langley and Jackson, 1994), cause structural changes in the liver of adult rat offspring (Burns *et al.*, 1997), and to reduce the islet and β-cell mass in the endocrine pancreas leading to reduced insulin secretion (Snock *et al.*, 1990). Of considerable interest, however, is the finding that modification and consequences of a period of malnutrition during pregnancy are not necessarily limited to first-generation offspring (i.e. the F_1 generation) but can be transgenerational (Pinheiro *et al.*, 2008). An adequate protein provision during pregnancy might therefore be of particular importance for a strict carnivore species e.g. mink and cat, due to their high requirement for dietary protein provision. Our objectives were to investigate if changes detected in foetal hepatic tissue of the F_1-generation caused by low protein provision during foetal life can become transgenerational and be detected in the F_2-generation offspring. Further, to study postnatal growth of offspring (F_2-generation) of foetal life protein restricted mothers.

Materials and methods

Fifty-six yearling female mink of the standard black genotype were used, 23 of which had been exposed to low protein provision (14% of metabolizable energy – ME – from protein; FLP1) and 33 to adequate protein provision (29% of ME from protein; FAP1) the last 17.9±3.6 days of foetal life (F_1-generation). The yearlings had been given adequate dietary protein from birth onwards. Forty-eight dams were used to study reproductive performance and to record kit body weight at birth and until 28 days of age of the F_2-generation offspring (FLP2 and FAP2). Sixteen animals, eight dams from each foetal life treatment, were used in balance and respiration experiments in the first and last thirds of true gestation, and after parturition the dams and their offspring (six kits in each litter) were measured in the second and fourth week of lactation. Blood was drawn once a week from the animals for hormone analyses. Eight dams were euthanized in late gestation for tissue collection for determination of gene expression of key hepatic gluconeogenic and glycolytic enzymes.

Results and discussion

The quantitative metabolic traits were largely unaffected by protein supply during foetal life, but ME intake was significantly higher, and the retained energy tended to be higher among FLP1 than FAP1 during the fourth week of lactation. The F_2 offspring of FLP1 dams had significantly (P=0.003) higher birth weight than those of FAP1 dams. Plasma concentrations of leptin, insulin and IGF-1 were not affected by foetal life protein provision but insulin and IGF-1 tended to be lower among FLP1 than FAP1 dams. The tendency towards lower plasma insulin could indicate gestational hyperglycemia in the FLP1 dams supported by higher foetal growth among their offspring. However, this remains to be confirmed by glucose measurements. No significant differences in the relative abundances of glucose-6-phosphatase (G-6-Pase), fructose-1,6-bisphosphatase (Fru-1,6-P_2ase), phosphoenol-pyruvate carboxy-kinase (PEPCK), and pyruvate kinase (PKM_2) mRNA in hepatic tissue between FLP1 and FAP1 dams were found. Despite this, the relative abundance of PKM_2 mRNA was significantly (P=0.007) lower, and that of Fru-1,6-P_2ase mRNA tended (P=0.08) to be lower in hepatic tissue of F_2 foetuses of FLP1 than FAP1 dams (Figure 1).

Our study confirms that changes in hepatic enzymes affecting glucose homeostasis, induced by malnutrition during foetal life of the F_1-generation, might be transgenerational and be transmitted from the F_1 to the F_2 generation in mink. These findings indicate that even though most metabolism traits display no effect of protein provision during early life, nutritionally induced effects still appear at the molecular level in the subsequent generation.

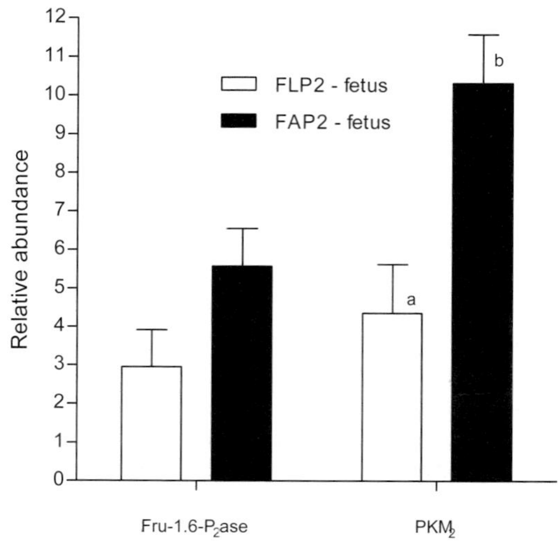

Figure 1. The relative abundance of fructose-1,6-bisphosphatase (Fru-1,6-P_2ase) and pyruvate kinase (PKM_2) mRNA normalized to 18s rRNA in the hepatic tissue of foetuses from mothers exposed to adequate (FAP2) or low protein provision (FLP2) during early life. [a,b] *Mean values within gene with unlike lower-case superscript letters were significantly different.*

References

Burns S.P., M. Desai, R.D. Cohen, D. Robert, C.N. Hales, R.A. Iles, J.P. Germain, T.C.H. Going and R.A. Bailey, 1997. Gluconeogenesis, glucose handling, and structural changes in livers of the adult offspring of rats partially deprived of protein during pregnancy and lactation. J. Clin. Invest. 100, 1768-1774.

Gluckman, P.D. and M.A. Hanson, 2004. Living with the past: evalution, development, and patterns of disease. Science 305, 1733-1736.

Lucas, A., 1991. Programming by Early Nutrition in Man. In: Bock, G.R. and Whelan, J. (eds.) The Childhood Environment and Adult Disease – Ciba Foundation Symposium 156. Chichester, England: John Wiley & Sons Ltd: 38-55.

Ozanne, S.E. and C.N. Hales, 2002. Early programming of glucose-insulin metabolism. Trends Endocrinol Metab 13, 368-373.

Pinheiro, A.R., I.D.M. Salvucci, M.B. Aguila and C.A. Mandarim-de-Lacerda, 2008. Protein restriction during gestation and/or lactation causes adverse transgenerational effects on biometry and glucose metabolism in F1 and F2 progenies of rats. Clin Sci 114, 381-392.

Snoeck, A., C. Remacle, B. Reusens and J.J. Hoet, 1990. Effect of a low protein diet during pregnancy on the fetal rat endocrine pancreas. Biol. Neonate. 57, 107-118.

Digestive efficiency, metabolism of nitrogen and methane emission in sheep, goats and llamas fed grass based diets differing in protein content

M.O. Nielsen, E. Tejada, A. Chwalibog, A.H. Tauson and L. Nielsen
Department of Basic Animal and Veterinary Sciences, Faculty of Life Sciences, University of Copenhagen, DK-1870 Frederiksberg C, Denmark

Introduction

Llamas are so-called pseudo-ruminant animals which have their evolutionary origin in the arid high Andean mountains in South America. They ruminate and are forestomach fermenters. But their forestomachs are not as fully compartmentalized and do not retain feed particles and liquid in the same way as in the true ruminants (Clemens and Stevens, 1980).

It has been claimed that llamas should be superior to the true ruminants in their ability to digest and utilize low protein feeds (Van Saun, 2006; Hinderer and von Engelhardt, 1975). The aim of this study was to investigate (1) if this superiority of the llama can be ascribed to an altered strategy for nitrogen metabolism and excretion from the body, and (2) whether the anatomical distinctive features of the llama foregut is associated with less synthesis and emission of methane from fermentation than in true ruminants.

Materials and methods

Six non-pregnant adult female animals of each of the species sheep, goats and llamas were used in this experiment in a change over design with two different diets: HP: high protein artificially dried hay (crude protein content CP=17.7%) and LP: low protein grass seed hay (CP=6.5%). The experiment was divided into two periods of 28 days. Half the animals from each species received one diet in one period and the other diet in the next period. Both diets were equally represented in each period. The diets were fed ad lib during days 1-20 of each period, and at 80% of ad lib during days 21-28.

Animals were placed in metabolic cages for 5 days (d 23-28) for collection of faeces and urine, and quantitative energy metabolism was determined over 22 h by indirect calorimetry in open-air respiration chambers. Blood samplings were obtained from indwelling catheters in a jugular vein on day 28 of each period at -0.5, +1, +2.5 and +4 hours from the morning feeding. Catheterisations were performed the day before blood samplings.

Data were analyzed statistically in a linear model in R version 2.7.0 (Copyright (C) 2008 The R Foundation for Statistical Computing). The model included systematic effects of species, diet and period, with a random effect of animal number and animal number within the period.

Results and discussion

Llamas had lower dry matter (DM) intake per kg metabolic body weight (MBW) than small ruminants (~35.9 in sheep and goat vs. 26.2 kg/MBW in llama) and a lower energy expenditure/MBW (0.38-0.41 in sheep and goats vs. 0.29 MJ/MBW in llama) on the LP diet.

Small ruminants had superior digestibilities of DM (62-64% in sheep and goat vs. 55% in llama) and N (56-58% in sheep and goat vs. 48% in llama) on the HP diet. But digestibilities were more negatively affected in small ruminants by the LP diet (decreased to 27-38% for energy in sheep and goats and to 48% in llama). Consequently llamas had the most efficient digestion on the LP diet. Digestibilities for N became negative for sheep (-21%) and goats (-58%) but were not significantly different from 0 (-5%) in llamas on the LP diet.

Llamas efficiently reduced N excretion in urine on the LP diet (to 0.12 g/MBW from 0.3 on the HP diet), although only a minor reduction was observed in plasma urea (not shown). Sheep were most efficient in reducing urinary N excretion (0.75 on HP to 0.12 g/MBW on LP diet) and also reduced levels of N in urine to low levels on the LP diet (0.44% N compared to 1.12% N on the HP diet). N content in urine was low (0.52-0.56% N) and not affected by diet in the goat, whilst plasma urea decreased markedly in both sheep and goat on the LP diet.

Creatinine levels in plasma (Figure 1) were very high in llamas, and increased markedly on the LP diet, despite lower protein supply. Only small increases were seen in sheep and goats.

The present study lend support to theories about more efficient recycling of N to the foregut in the llama compared to the small ruminants explaining the superior feed utilization on poor quality diets - except that sheep were even more efficient than the llama in reducing N excretion in urine on the low protein diet.

The finding that llamas had much higher creatinine levels in plasma *per se*, and responded with much higher increases in creatinine when shifted to a low protein diet indicate that llamas through evolution may have evolved a capacity to adapt also intermediary nitrogen metabolism to cope with protein deficiency.

Methane emission/kg DMI was of the same magnitude (29 l/kg DMI) in sheep and goats irrespective of the diet, and the methane emission from llamas on the LP diet was in the same range. But the llama emitted significantly less methane (19.6 l/kg DMI) on the HP diet. This indicates that anatomical features of the foregut may impact methanogenesis.

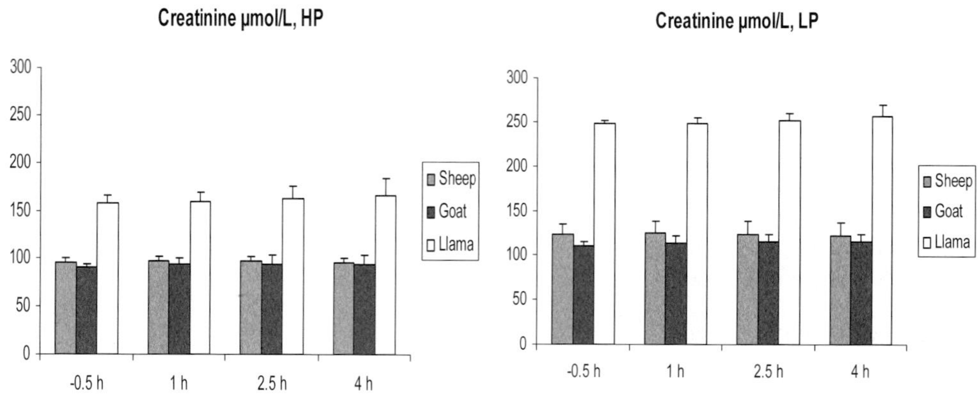

Figure 1. Plasma creatinin levels at different time points relative to morning feeding in sheep, goat and llama fed a high protein (HP; left panel) or low protein (LP) diet.

References

Clemens, E.T. and C.E. Stevens, 1980. A comparison of gastrointestinal transit time in ten species of mammal. J. Agr. Sci. 94, 735-737.

Hinderer, S. and W. von Engelhardt, 1975. Urea metabolism in the llama. Comp. Biochem. Physiol. 52A, 619-622.

Van Saun, R.J., 2006. Nutrient requirements of South American Camelids: A factorial approach. Small Ruminant Res. 61, 165-186.

Periparturient energy status and its implications for lymphocyte responsiveness in high-yielding dairy cows

A. Schwarm, M. Derno and M. Schweigel
Leibniz Institute for Farm Animal Biology (FBN), Department of Nutritional Physiology, 'Oskar Kellner', Wilhelm-Stahl-Allee 2, 18196 Dummerstorf, Germany

Introduction

Immunosuppression associated with gestation and parturition has been reported in many species. In cows various parameters of lymphocyte function, such as antibody and cytokine production or proliferative response, have been shown to be impaired around calving (Kashiwazaki *et al.*, 1985; Goff and Horst, 1997; Lacetera *et al.*, 2005). However, the strongly energy-dependent process of lymphocyte activation (G_1 cells, ~ first 24 hrs of cell cycle) has not been studied so far. The aim of this study was to investigate a possible relationship between reduced blood lymphocyte responsiveness in cows and physiological changes, especially hyperketonemia, related to energy deficiency attributed to insufficient energy intake relative to milk production.

Materials and methods

To do this, three experiments were performed with peripheral blood mononuclear cells (PBMC) during which the respiration rate of stimulated versus control cultures was studied. PBMC of dry cows were incubated (1) with the ketone body β-hydroxybutyrate (BHBA) at concentrations of 1.2 mM or 4 mM, typically found in subclinical and clinical ketosis and (2) in serum obtained from cows fed *ad libitum* (*ante partum*, dry-off period) or restricted (50%, 2 weeks *post partum*), respectively, with the latter reflecting a physiological state of negative energy balance. Serum contained 0.48 and 1.04 mmol BHBA/l, respectively. (3) In the final experiment cells isolated from high-yielding dairy cows ante- and *post partum* were incubated. Cell number, vitality and purity of the PBMC preparation were detected using a cell counter, the trypan blue exclusion test and flow cytometry, respectively. Isolated mononuclear cells consisted of 70-85% lymphocytes and 5-10% monocytes, with a viability always higher than 95%. Phytohemagglutinine (PHA, 4 μg/ml) was used for stimulation of lymphocytes. Oxygen consumption rate (OCR) of non-stimulated and stimulated lymphocytes was assessed fluorometrically (BD Oxygen Biosensor System). Interleukin-2 production (BD IL-2 ELISA Kit II) after 24 hrs of culture was used as a measure of functional lymphocyte activity. The data were analyzed by paired t-test using SPSS (version 15.0).

Results and discussion

Until now, data on respiration rates of bovine lymphocytes have not been published. Compared to the PHA-induced OCR of dry cow PBMC incubated for 24 hrs in media containing 1.2 mM BHBA, incubation in media containing 4 mM BHBA led to a significant 19% reduction (n=4, P=0.046, Figure 1). Correspondingly, we found a tendency towards a reduced PHA-induced OCR (33 pmol/min/10^5 viable cells) in dry cow PBMC cultivated in serum obtained from early lactating, restrictive-fed cows compared to that observed in PBMC cultivated in serum from dry cows fed *ad libitum* (71 pmol/min/10^5 viable cells; n=4). The reduction of the lymphocyte responsiveness to antigenic stimulation after exposure to elevated concentrations of BHBA is in accord with e.g. Targowski and Klucinski (1983) who measured the incorporation of [^{3}H]thymidine into the DNA of lymphocytes. Relative high NEFA and ketone body concentrations observed in serum of early lactating cows presumably interfere with lymphoid cell functions due to perturbation of properties of cellular membranes, suppression of cytokine synthesis, formation of lipid peroxides, gene regulation, and induction of necrosis or apoptosis (De Pablo and De Cienfuegos, 2000). Presumably a prolonged exposition of lymphocytes to high [BHBA] or serum from energy-deficient cows prior to PHA-stimulation, e.g. 24 hrs instead of 1.5 hrs, would have led to a more distinct difference in oxygen consumption within the two treatment

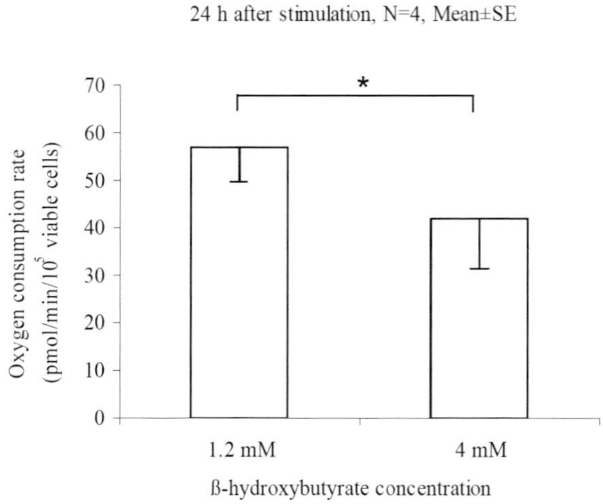

Figure 1. Oxygen consumption rate (pmol/min/10^5 viable cells) of cow peripheral blood mononuclear cells at 24 hours after phytohaemagglutinine-stimulation (PHA, 4 μg/ml). Cells were incubated in media containing either 1.2 mM or 4 mM β-hydroxybutyrate (BHBA) corresponding to plasma levels observed under conditions of subclinical and clinical ketosis. Bars represent means ± SE of 4 animals. *P=*0.046.*

groups. Basal OCR (non-stimulated) in PBMC from cows *post partum* was on average higher than *ante partum* (17 vs. 5 pmol/min/10^5 viable cells; n=3), reflecting higher energy expenditure *post partum*. PHA-induced OCR reached maximum levels of 180 pmol/min/10^5 viable cells after 24 hrs. These results are in agreement with basal and post stimulation OCRs (8 and 108 pmol/min/10^5 total cells) measured amperometrically (Clarke electrode) in rat thymocytes and human PBMC (Krauss *et al.,* 1999; Schmid *et al.,* 2000). Our higher post stimulation OCR most possibly results from a longer duration (24 hrs) of the experiment in comparison to the short term stimulation (1 to 20 min) used by the other authors. A lower stimulation index (SI = OCR of stimulated / OCR of control cultures after 24 hrs) was observed with PBMC from *ante partum* cows (-5 weeks, mean±SD, 3.0±1.5; n=6) compared to that obtained two weeks *post partum* (5.6±4.3; n=5). The higher variability observed in the latter may reflect inter-individual differences in the capability to cope with energy deficiency after parturition. In addition, the reduced interleukin-2 production of stimulated *post partum* compared to *ante partum* PBMC (3.2 vs. 4.1 pg/ml; n=2) is indicative of an impaired functional immune response after parturition, e.g. the abortion of the cell cycle during the early G_1 phase.

References

De Pablo, M.A. and G.A. De Cienfuegos, 2000. Modulatory effects of dietary lipids on immune system functions. Immunol Cell Biol 78, 31-39.

Goff, J.P. and R.L. Horst, 1997. Physiological changes at parturition and their relationship to metabolic disorders. J Dairy Sci 80, 1260-1268.

Kashiwazaki, Y., Y. Maede and S. Namioka, 1985. Transformation of Bovine Peripheral-Blood Lymphocytes in the Perinatal-Period. Jap J Vet Sci 47, 337-339.

Krauss, S., F. Buttgereit and M.D. Brand, 1999. Effects of the mitogen concanavalin A on pathways of thymocyte energy metabolism. Biochim Biophys Acta Bioenerg 1412, 129-138.

Lacetera, N., D. Scalia, U. Bernabucci, B. Ronchi, D. Pirazzi and A. Nardone, 2005. Lymphocyte functions in overconditioned cows around parturition. J Dairy Sci 88, 2010-2016.

Schmid, D., G.R. Burmester, R. Tripmacher, A. Kuhnke and F. Buttgereit, 2000. Bioenergetics of human peripheral blood mononuclear cell metabolism in quiescent, activated, and glucocorticoid-treated states. Bioscience Reports 20, 289-302.

Targowski, S.P. and W. Klucinski, 1983. Reduction in mitogenic response of bovine lymphocytes by ketone-bodies. Am J Vet Res 44, 828-830.

Immune system stimulation increases reduced glutathione synthesis rate in growing pigs

A. Rakhshandeh and C.F.M. De Lange
Department of Animal and Poultry Science, University of Guelph, N1G 2W1, Guelph, Canada

Introduction

Reduced glutathione (GSH) is a tripeptide that is ubiquitous in mammalian cells and has many important protective and metabolic functions, including defense against oxidative stress, detoxification of xenobiotics, and acting as the cysteine (CYS) storage and transporter. Glutathione homeostasis is crucial for normal cellular processes in defined physiological conditions such as disease (Breuillé and Obled, 2000). The availability of CYS is considered to be the rate limiting factor for synthesis of GSH (Grimble *et al.*, 1992). As skeletal muscle become the main source of CYS during disease (Jahoor *et al.*, 1999) increases in GSH pool size or synthesis rate may contribute to muscle protein mobilization. To better understand *in vivo* aspects of GSH metabolism during immune system stimulation (ISS) concentration and synthesis rate of reduced GSH were measured in various tissues in growing pigs.

Material and methods

Ten gilts (BW 11.0±0.24 kg) were fed-restricted and assigned to a sulfur amino acid (SAA) limiting diet (SAA intake 2.0 g/d). Animals were surgically catheterized in the jugular vein and injected intramuscularly with either saline (n=6; ISS-) or increasing amounts of *Escherichia coli* lipopolysaccharide (LPS): dose I, 60 μg/kg BW; dose II, given 48 h later, 67 μg/kg BW (n=8; ISS+). The day after the 2nd injection pigs were constantly infused with ^{35}S-cysteine for 5 hours at a rate of 23.1 μCi/kg BW/h. Serial blood samples were taken for determining plasma specific radioactivity (SRA). Daily blood samples were taken for plasma acute-phase protein measurement. Immediately following isotope infusion, pigs were sacrificed by infusion of sodium pentobarbital for tissue sampling. Plasma and tissue free CYS and reduced GSH were isolated and quantified as the monobromobimane derivatives by HPLC. Radioactivity of the compounds was determined by Beckman Coulter LS6000 beta counter. Rate of GSH synthesis was determined by measurement of the specific radioactivity of GSH-bound CYS and tissue free CYS at the end of the infusion period (Table 1). Eye temperature was monitored once daily for ISS- and ISS+, using infrared imaging technique, to ensure the effectiveness of ISS. Repeated measurement analysis (PROC MIXED in SAS) was used to evaluate change in plasma SRA, acute-phase protein level and eye temperature over time.

Results and discussion

Relative to ISS-, repeated injection with LPS increased eye temperature (36.5 vs. 37.4 °C, SE 0.19; *P<0.01*), plasma levels of haptoglobin (1.1 vs. 1.4 g/l, SE 0.06; *P<0.05*) and fibrinogen (2.1 vs. 3.0 g/l, SE 0.16; *P<0.01*) but reduced plasma albumin (26.4 vs. 23.4 g/l, SE 0.91; *P<0.01*), confirming successful ISS. Concentration of GSH (mmol/kg ± SE, ISS- vs. ISS+) was lower (*P<0.05*) in spleen (1.4±0.08 vs.1.16±0.03), kidneys (2.6±0.08 vs. 2.17±0.09) and erythrocytes (mmol/l; 5.9±1.07 vs. 3.6±1.10), and was higher in muscle (mmol/kg; 0.2±0.01 vs. 0.3±0.01) and plasma (μmol/l; 10.2±1.07 vs. 13.7±1.30) for ISS+. There was no ISS effect on GSH concentrations in liver, small and large intestine, heart and lung (2.1±0.17, 1.1±0.09, 0.6±0.04, 0.6±0.09 and 0.6±0.07 mmol/kg, respectively). Immune system stimulation increased (*P<0.05*) fractional synthesis rate (FSR) of GSH in liver, small and large intestine, heart, lung and kidneys, but had no effect (*P>0.05*) on FSR in spleen and erythrocyte (Table 1). Absolute synthesis rate of GSH was higher in spleen and liver, but not in erythrocyte of ISS animals (Table 1; *P<0.03*). These results are in general agreement with those observed in septic rats (Malmezat *et al.*, 2000). Change in GSH pool size is function of balance

Table 1. Impact of immune system stimulation (ISS) on fractional synthesis rate (FSR) and absolute synthesis rate (ASR) of glutathione (GSH) in selected tissues.[3]

Tissue	ISS-	ISS+	*P*
FSR %/d[1]			
Liver	263±16.2	355±14.1	0.001
Spleen	257±17.9	377±53.8	0.067
Small intestine	299±44.5	426±36.8	0.042
Large intestine	110±41.8	219±36.3	0.035
Heart	35.2±13.8	80.6±11.3	0.035
Lung	197±23.2	287±16.3	0.016
Kidney	150±18.2	278±54.9	0.050
Muscle	85.9±6.7	118±11.8	0.049
Erythrocyte	59.3±3.0	73.8±7.2	0.123
ASR[2]			
Liver, mmol/tissue/d	1.30±0.2	2.00±0.2	0.030
Spleen, μmol/tissue/d	61.0±30.0	131±28.5	0.020
Erythrocyte, mmol/l/d	0.80±0.4	1.10±0.3	0.597

[1] Fractional synthesis rate = [(product SRA / precursor SRA) × (24 h / infusion period)] × 100.

[2] Absolute synthesis rate = FSR × Pool size.

[3] Data are least square mean ± SE.

between its synthesis rate and uptake from circulation and rate of its disappearance (Breuillé and Obled, 2000). Although responses varied among GSH pools, the current findings suggest that GSH synthesis and utilization increase during ISS, contributing to increased CYS requirement during ISS.

References

Breuillé, D., and C. Obled, 2000. Cysteine and glutathione in catabolic state 2000. In: Fürst, P., Young, V., (eds.) Protein, peptides and amino-acids in enteral nutrition. Vevey: Nestlé Nutrition Services: 173-197.

Grimble, R. F., A.A. Jackson., C. Persaud, M.J. Wride, F. Delers and R. Engler, 1992. Cysteine and glycine supplementation modulate the metabolic response to tumor necrosis factor alpha in rats fed a low protein diet. J Nutr. 122, 2066-2073.

Jahoor, F., L. Wykes, M. Del Rosario, M. Frazer, and P.J. Reeds, 1999. Chronic protein undernutrition and an acute inflammatory stimulus elicit different protein kinetic responses in plasma but not in muscle of piglets. J. Nutr. 129, 693-699.

Malmezat, T., D. Breuille, P. Captain, P. Minard, and C. Obled. 2000. Glutathione turnover is increased during acute phase of sepsis in rats. J. Nutr. 130, 1239-1246.

Modulation of goat monocyte immune functions by ω-3 polyunsaturated fatty acids EPA and DHA

G. Savoini[2,4], C. Lecchi[1], A. Agazzi[2], S. Modina[3], M. Ferroni[2], L.F. Pisani[1], G. Invernizzi[2], P. Sartorelli[1,4] and F. Ceciliani[1,4]
[1]Dipartimento di Patologia Animale, Igiene e Sanità Pubblica Veterinaria. Università di Milano, via Celoria 10, 20133 Milano, Italy
[2]Dipartimento di Scienze e Tecnologie Veterinarie per la Sicurezza Alimentare. Università di Milano, via Celoria 10, 20133 Milano, Italy
[3]Dipartimento di Scienze Animali, Università di Milano, via Celoria 10, 20133 Milano, Italy
[4]CISMA Centro Interdipartimentale di Studi sulla Ghiandola Mammaria, Università di Milano, via Celoria 10, 20133 Milano, Italy

Introduction

Fatty acids are key nutrients for animal cells. They may be oxidized as energy source or stored as triglycerides or glycogen. Alternatively, they may be taken up by cells and used for synthesis of phospholipids, cholesterol esters, prostaglandins and other powerful signalling agents (Calder, 2006). Fatty acids metabolism plays a significant role in immune cells functions, both in humans and in animals (Yaqoob and Calder, 2007). The most powerful immunomodulatory effect is achieved by two w-3 polyunsaturated fatty acids (PUFAs), namely eicosapentaenoic acid (EPA, C20:5[ω3]) and docosahexaenoic acid (DHA C22:6[ω3]) (Calder, 2006). Polyunsaturated fatty acids modulate the immune response in humans and livestock, though the mechanisms involved are only partially characterized.

In the present study, the *in vitro* effects of two ω-3 PUFAs, namely EPA and DHA, on some inflammatory functions of goat monocytes including phagocytosis, extracellular respiratory burst, as well as the capability to modulate the expression of some pro-inflammatory cytokines, were investigated. Their apoptosis rate and the cytoplasmic accumulation of lipid droplets were also studied.

Materials and methods

The purification of monocytes was carried out by magnetic sorting of $CD14^+$ cells, as previously described (Ceciliani *et al.*, 2007). Six clinically healthy multiparous lactating goats (Saanen) between 5 and 6 years of age and between 30 and 33 weeks after calving were used for blood collection.

Phagocytosis assay was carried out by measuring the fluorescence of fluorescein-labelled *E. coli* bioparticles according to the method of Pisani *et al.* (2009). The modulation of oxidative burst activity of isolated goat monocytes was studied by cytochrome *c* reduction assay as described by Sartorelli *et al.* (2000). Cytokine mRNA expression was determined by quantitative PRC (Real Time PCR). Apoptosis was determined by measuring the modification of the activities of two enzymes that play effector roles in apoptosis: caspase-3 and caspase-7 as previously described in bovine (Ceciliani *et al.*, 2007). Intracellular lipid droplet content was determined after Nile red staining.

Statistical analysis was carried out by ANOVA using the General Linear Model of SAS (SAS/STAT, Version V8, 1999, SAS Inst., Inc., NC, USA). Significance was declared for $P<0.01$ and $P<0.05$.

Results and discussion

Both EPA and DHA up-regulate phagocytosis of monocytes, as reported. Due to the relatively brief incubation timeframe, it seems unlikely that these fatty acids have been already integrated in the plasma membrane of monocytes. Therefore, the up-regulation of phagocytosis activity was likely to be induced by a direct interaction with cell-surface proteins involved in phagocytosis.

Results on cytokine regulatory activity are inconsistent. Treating of monocytes with EPA and DHA does not apparently affect apoptosis rate, except when used at very high concentration. Both EPA and DHA treatment induced a different accumulation of lipid droplets in goat monocytes. Remarkably, profound effects of EPA and DHA on cellular lipid accumulation were discovered, as clearly shown in Figure 1. EPA treatment generated preferentially large droplets, whereas DHA mainly produced smaller ones. This is, at least to the best knowledge of the authors, the first report that shows that EPA and DHA can differentially affect the size of lipid droplets within the cytoplasm. The biological significance of this finding remains elusive, since the pathways that drive to the formation of lipid bodies are not fully understood. In conclusion, this study demonstrated that both EPA and DHA have important and specific roles in modulating monocyte immune functions and therefore support their use in animal diets not only to increase the nutritional value of milk for human consumption, but also because they can improve animal health. Several aspects of the potential immunomodulatory effects of ω-3 fatty acids are still elusive, the most evident being the molecular basis of the sometime divergent mechanism of action between EPA and DHA. The involvement of lipid droplets formation due to ω-3 fatty acids treatment in immune modulation of monocyte function is a final aspect that needs to be further characterized.

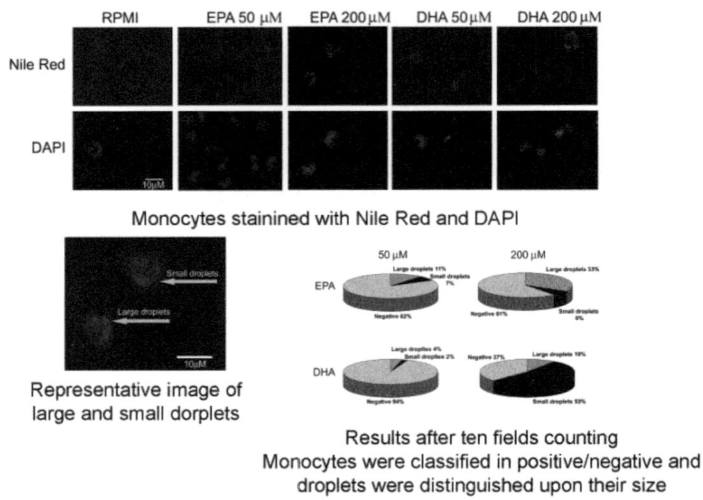

Figure 1. EPA and DHA regulate lipid droplet formation.

References

Calder, P.C., 2006. N-3 polyunsaturated fatty acids, inflammation, and inflammatory diseases. Am J Clin Nutr. 83 (6 Suppl), 1505S-1519S.

Ceciliani, F., V. Pocacqua, A. Miranda-Ribera, V. Bronzo, C. Lecchi and P. Sartorelli, 2007. Alpha(1)-acid glycoprotein modulates apoptosis in bovine monocytes. Vet. Immunol. Immunopathol. 116 (3-4), 145-152.

Pisani L.F., C. Lecchi, G. Invernizzi, P. Sartorelli, G. Savoini and F. Ceciliani, 2009. *In vitro* modulatory effect of omega-3 polyunsaturated fatty acid (EPA and DHA) on phagocytosis and ROS production of goat neutrophils. Vet. Immunol. Immunopathol. 131(1-2), 79-85.

Sartorelli P., S. Paltrinieri and S. Comazzi, 2000. Non-specific immunity and ketone bodies. II: *In vitro* studies on adherence and superoxide anion production in ovine neutrophils. J. Vet. Med. A. Physiol. Pathol. Clin. Med. 47, 1-8.

Yaqoob P. and P.C. Calder, 2007. Fatty acids and immune function: new insights into mechanisms. Br. J. Nutr. 98 Suppl 1, S41-45.

Dietary protein and methionine intakes affect glutamate dehydrogenase and alanine aminotransferase activities in the juvenile marine shrimp *Penaeus monodon*

L. Richard[1,2], C. Vachot[1], J. Brèque[1], P.-P. Blanc[2], V. Rigolet[2], S.J. Kaushik[1] and I. Geurden[1]
[1]INRA, UMR1067 Nutrition, Aquaculture and Genomics, 64310 St-Pée-sur-Nivelle, France
[2]AQUALMA, BP 93 Immeuble SCIM, 4 rue Galliéni, Mahajanga 401, Madagascar

Introduction

Penaeus monodon (Decapod, crustacean) was found to have a high (approximately 35% of diet dry weight) requirement for dietary protein (Richard *et al.*, 2010). Aquatic crustaceans are ammoniotelic (Regnault, 1987) and have amino acid (AA) catabolic pathways (transamination, deamination) similar to those of vertebrates (Claybrook, 1983). The high protein requirements of teleosts is considered to be related to their incapacity to down-regulate AA catabolism when fed suboptimal protein levels (Rumsey, 1981; Walton and Cowey, 1982). The objectives of this study were to characterize the activity of two key catabolic enzymes (glutamate dehydrogenase and alanine aminotransferase) in different tissues of *P. monodon* and to examine whether the activity of both enzymes is regulated by dietary protein and methionine intake.

Material and methods

Juvenile *P. monodon* (initial weight 2.4 g) were fed during six weeks one of six semi-purified diets (based on casein and coated crystalline AA as nitrogen source) formulated to contain a crude protein level (CP, % DM) of 14.4% (CP10), 34% (CP30), or 54% (CP50) with two levels of methionine (normal or 30% reduced). Hepatopancreas, gills and tail abdominal muscle were sampled four hours after the last feeding for enzyme analyses and stored at -80 °C until analyses of alanine aminotransferase (ALAT, EC 2.6.1.2.) and glutamate dehydrogenase (GDH, EC 1.4.1.3.) activities. All parameters were analysed using a two-way ANOVA (dietary protein level x dietary methionine level) (Statistica 5.0, StatSoft Inc., Tulsa OK, USA).

Results and discussion

The high protein requirement of *P. monodon* does not seem to be related to an enzymatic disorder in the regulation of AA deamination since *P. monodon* appears able not only to upregulate (+35%), but also to downregulate (-26%) muscle GDH activity according to changes in dietary protein supply (Table 1, $P<0.05$). GDH appears to be much more sensitive to dietary protein level than ALAT (Figure 1), which tends to indicate that this enzyme plays a central role in AA catabolism of shrimp. Regarding the relative methionine deficiency, GDH activity increased 4-fold in the gills but only at the protein intake level close to requirement (CP30m vs. CP30), as indicated by the significant interaction between dietary protein and methionine levels (Table 1, $P<0.05$). This suggests that gills are involved in ammonia formation and not only in ammonia excretion.

References

Claybrook, L., 1983. Nitrogen metabolism. In: The Biology of Crustacea, vol. 5, L.H. Mantel (ed.), Academic Press, New York, USA: 163-213.

Marte, C.L., 1980. The food and feeding habit of *Peaneus monodon* Fabricius collected from Makato River, Aklan, Philippines (*Decapoda natantia*). Crustaceana 38, 225-236.

Regnault, M., 1987. Nitrogen excretion in marine and freshwater crustacean. Biol. Rev. 62, 1-24.

Table 1. Total activity of GDH and ALAT in muscle and gills of Penaeus monodon *fed semi-purified diets with one of three dietary protein levels deficient or not in methionine.*

	Muscle				Gills			
	GDH (U/g tissue)		ALAT (U/g tissue)		GDH (U/g tissue)		ALAT (U/g tissue)	
	Mean	SD	Mean	SD	Mean	SD	Mean	SD
CP10	0.34	0.17	6.8	1.8	0.3 [b]	0.2	8.8	2.2
CP10m	0.33	0.15	7.9	1.3	0.5 [b]	0.3	7.4	1.5
CP30	0.48	0.14	8.9	2.3	0.3 [b]	0.3	8.6	2.7
CP30m	0.55	0.14	8.7	2.2	1.2 [a]	0.4	10.3	0.7
CP50	0.65	0.17	9.8	1.1	0.5 [b]	0.5	8.5	3.9
CP50m	0.65	0.11	9.5	2.4	0.3 [b]	0.4	8.6	4.0
Two-way ANOVA analysis[1]								
Protein	**<0.0001**		**0.0230**		0.0568		0.4659	
Methionine	0.7291		0.7861		**0.0280**		0.8948	
Protein × methionine	0.7302		0.6025		**0.0031**		0.3896	

[1] Differences were significant for *P*<0.05 (bold).

[a,b] In each column, different letters depict significantly different groups (*P*<0.05, 1-way ANOVA).

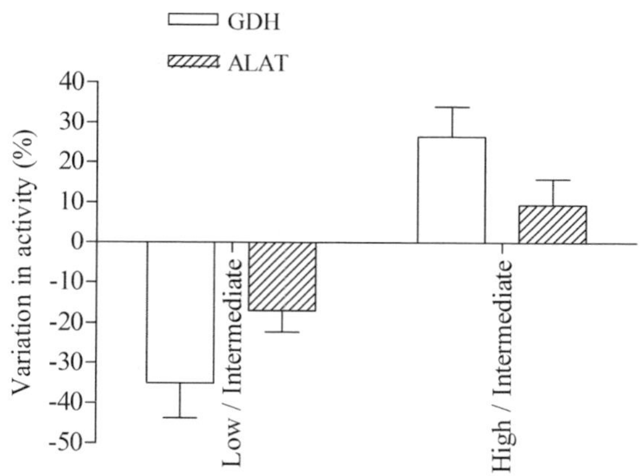

Figure 1. Variation in GDH and ALAT activities when lowering or increasing dietary protein content.

Richard, L., P.-P. Blanc, V. Rigolet, S. Kaushik and I. Geurden, 2010. Maintenance and growth requirements for nitrogen, lysine and methionine and their utilisation efficiencies in juvenile black tiger shrimp, *Penaeus monodon*, using a factorial approach. Brit J Nutr., doi:10.1017/S0007114509992844.

Rumsey, G.L., 1981. Significance of nitrogen metabolism: Why does the salmonid require a high protein diet? Salmonid Nov-Dec, 20-24.

Walton, M.J. and C.B. Cowey, 1982. Apects of intermediary metabolism in salmonid fish. Comp. Biochem. Physiol. 73B, 59-79.

Adaptive metabolic responses in sheep, goats and llama when fed grass based diets differing in protein content

M.O. Nielsen, E. Tejada, V.G. Christensen and L. Nielsen
Department of Basic Animal and Veterinary Sciences, Faculty of Life Sciences, University of Copenhagen, 1870 Frederiksberg C, Denmark

Introduction

Llamas are so-called pseudo-ruminant animals. They have the ability to digest and utilize forage fibre in their forestomachs and they ruminate as do the true ruminants, but their forestomachs are not as fully compartmentalized and do not retain feed particles and liquid in the same way as in the true ruminants. Nevertheless it has been claimed that llamas should be superior to ruminants like sheep, goats and cattle in their ability to digest and utilize poor quality feeds with low nitrogen content (Van Saun, 2006, Hinderer and von Engelhardt, 1975) as a result of their evolutionary origin in the arid high Andean mountains in South America.

The aim of this study was to investigate if the evolutionary adaptation of the llamas to low protein diets also involve metabolic adaptation of e.g. the glucose-insulin homeostasis.

Materials and methods

Six non-pregnant adult female animals of each of the species sheep, goat and llama were used in this experiment in a change over design with two different diets. The diets consisted of a high protein artificially dried hay (HP; crude protein content CP=17.7%) and a low protein grass seed hay (LP; CP=6.5%). The experiment was divided into two periods of 28 days, and half the animals within each species received one diet in one period and the other diet in the next period so that both diets were equally represented in each period. The diets were fed ad lib during days 1-28 of each period, and at 80% of ad lib from day 21-28.

Blood samplings were obtained from indwelling catheters in a jugular vein on day 28 of each period at -½, +1, +2.5 and +4 hours after the morning feeding. Catheterisations were performed the day before samplings. Blood parameters were analyzed statistically in a linear model in R version 2.7.0 (2008-04-22) (Copyright (C) 2008 The R Foundation for Statistical Computing). The model included systematic effects of species, diet, time and period, with a random effect of animal number and animal number within the period.

Results and discussion

Llamas had significantly higher glucose and insulin levels on the HP diet compared to the true ruminants (Figure 1). Plasma glucose in llamas on the HP diet was high even for monogastric animals, although only little glucose should be absorbed in the llama due to forestomach fermentation, and despite substantially higher insulin levels compared to sheep and goats. This shows that the llama can be considered to have an intrinsic insulin insensitivity. Insulin but not glucose levels decreased in response to a LP diet, but the differences between the true ruminants and the pseudo-ruminant llama persisted. The metabolic implications of this insulin insensitivity in the llama is not known, but it must have metabolic implications and may possibly contribute to explain why energy expenditure per kg metabolic bodyweight is lower in llamas as compared to small ruminants, particularly in LP diets (Nielsen *et al.*, 2010).

BOHB levels in llamas were remarkably low for a foregut fermenter on the HP diet, but increased on the LP relative to the HP diet presumably reflecting increased ketogenesis, whereas BOHB levels in

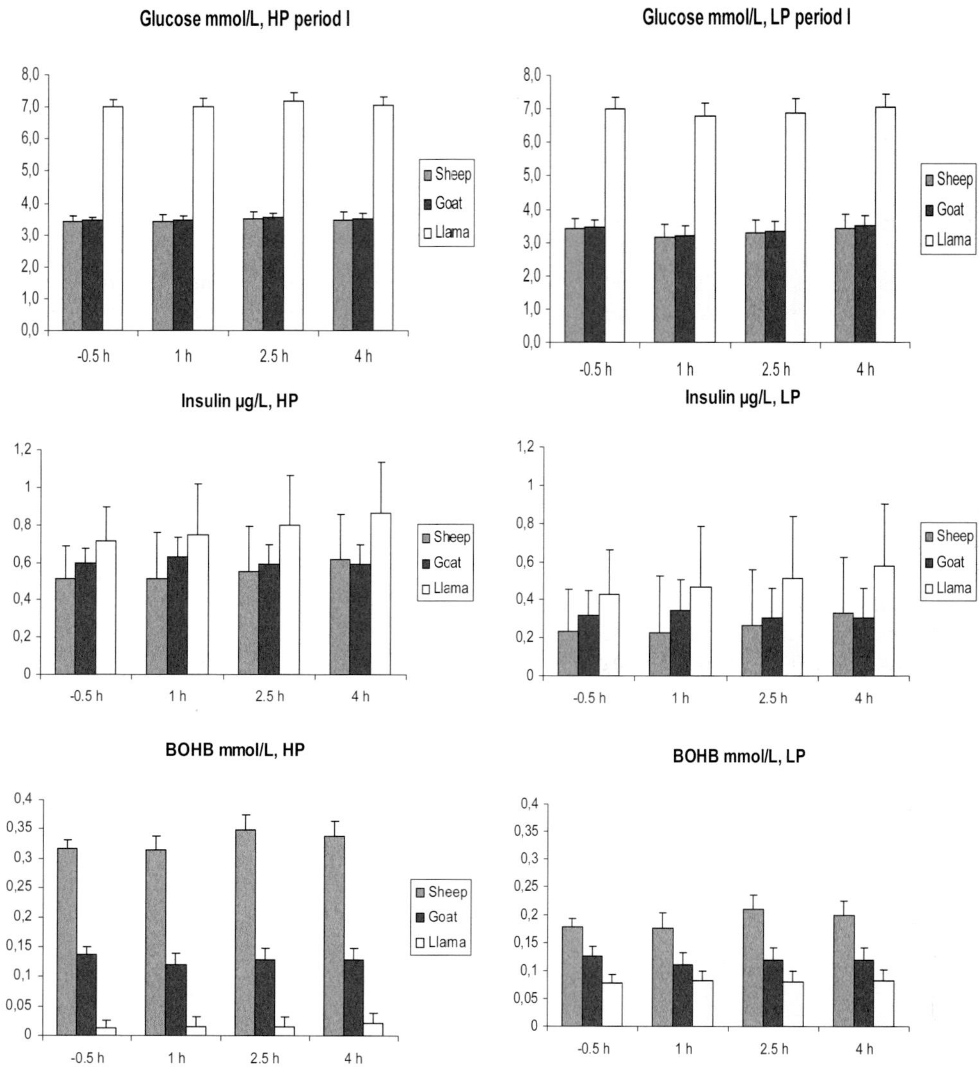

Figure 1. Postprandial glucose, insulin and BOHB responses in sheep, goats and llamas fed diets with high (HP) or low (LP) protein content.

the sheep decreased on the LP diet due to lowered DM intake. The underlying reason for this species difference with respect to adaptation of SCFA production in the forestomachs remains to be elucidated.

References

Hinderer, S. and W. von Engelhardt, 1975. Urea metabolism in the llama. Comp. Biochem. Physiol. 52A, 619-622.

Nielsen, M.O., E. Tejada, A. Chwalibog, A.H. Tauson and L. Nielsen, 2010. Digestive efficiency, metabolism of nitrogen and methane emission in sheep, goats and llamas fed grass based diets differing in protein content. In: Crovetto, G.M. (ed.) Energy and protein metabolism and nutrition. Wageningen Academic Publishers, Wageningen, the Netherlands, p. 497-498.

Van Saun, R.J., 2006. Nutrient requirements of South American Camelids: A factorial approach. Small Ruminant Res. 61, 165-186.

Immune system stimulation alters plasma cysteine kinetics in growing pigs

A. Rakhshandeh[1], *K. de Ridder*[1], *J.K. Htoo*[2] *and C.F.M. De Lange*[1]
[1]*Department of Animal and Poultry Science, University of Guelph, N1G 2W1, Guelph, Canada*
[2]*Evonik Degussa GmbH, 63457 Hanau, Germany*

Introduction

Changes in plasma free amino acid (AA) kinetics reflect modification of AA metabolism in different metabolic states (Melchior *et al.*, 2004). Cysteine (CYS) is considered to be conditionally essential during immune system stimulation (ISS) due to its increased utilization for synthesis of protein and metabolites such as albumin, glutathione (GSH), defensins. (Jahoor *et al.*, 1999; Malmezat *et al.*, 2000; Tang and Selsted, 1993). Plasma CYS flux, GSH level and rates of appearance of sulfate (SO_4; end product of CYS catabolism) and taurine (TAU) were determined to evaluate the impact of ISS on CYS kinetics.

Material and methods

Ten gilts (BW 11.0±0.24 kg) were fed restricted and assigned to a sulfur amino acid (SAA) limiting diet (SAA intake 2.0 g/d). Animals were surgically catheterized in the jugular vein and injected intramuscularly with either saline (n=4; ISS-) or increasing amounts of *Escherichia coli* lipopolysaccharide (LPS): dose I, 60 μg/kg BW; dose II, given 48 h later, 67 μg/kg BW (n=8; ISS+). The day after the 2nd injection pigs were constantly infused with ^{35}S-cysteine for 5 hours at a rate of 23.1 μCi/ kg BW/h. Serial blood samples were taken for determining plasma specific radioactivity (SRA). Daily blood samples were taken for plasma acute-phase protein measurement. Immediately following isotope infusion, pigs were sacrificed by infusion of sodium pentobarbital. Plasma CYS, as the monobromobimane derivative, TAU as the phenylisothiocyanate derivative, and SO_4 were isolated and quantified by HPLC. Rates of plasma SO_4 and TAU appearance were determined by measurement of SRA of ^{35}SO4 and ^{35}S- TAU, relative to that of CYS at the end of the infusion period (Table 1). Radioactivity of the compounds was determined by Beckman Coulter LS6000 beta counter. Eye temperature was monitored once daily for ISS- and ISS+ using infrared imaging technique, to ensure the effectiveness of ISS. Repeated measurement analysis (PROC MIXED in SAS) was used to evaluate changes in plasma SRA, acute-phase protein level and eye temperature over time.

Results and discussion

Relative to ISS-, repeated injection with LPS increased eye temperature (36.5 vs. 37.4 °C, SE 0.19; *P<0.01*), plasma levels of haptoglobin (1.1 vs. 1.4 g/l, SE 0.06; *P<0.05*) and fibrinogen (2.1 vs. 3.0 g/l, SE 0.16; *P<0.01*) but reduced plasma albumin (26.4 vs. 23.4 g/l, SE 0.91; *P<0.01*), confirming successful ISS. Reduced plasma CYS level and increased plasma CYS flux (Table 1) indicate modification in CYS metabolism during ISS. These results are in general agreement with those observed in septic patients and rats (Lyons *et al.*, 2001; Malmezat *et al.*, 2000). Immune system stimulation increased plasma GSH level, but it had no effect on plasma levels or fractional and absolute rate of appearance (FRA and ARA) of TAU and SO_4 (Table 1), suggesting preferential utilization of CYS, mainly for synthesis of GSH, immune proteins and other metabolites during ISS. We have shown an increase in synthesis rate of GSH in several tissues of ISS pigs (Rakhshandeh *et al.* 2010; this meeting). This study suggests that reduced plasma CYS level and increased CYS flux during ISS are due to enhanced utilization of CYS for synthesis of GSH, immune proteins and other metabolites, and not as result of enhanced irreversible loss of CYS. Enhanced CYS utilization during ISS may increase dietary CYS requirements.

Table 1. Impact of immune system stimulation (ISS) on plasma cysteine (CYS) kinetics in gilts.[6]

Parameter	ISS-	ISS+	*P*
CYS concentration, µM	67.0±4.5	56.1±4.40	0.012
CYS SRA[1], dpm/nmol	342±54.0	360±43.2	0.635
CYS flux [2], µmol/kg/h	133±14.7	167±10.6	0.010
SO_4 concentration, mM	0.90±0.1	1.00±0.1	0.585
SO_4 SRA, dpm/nmol	7.80±1.4	9.80±1.0	0.276
SO_4 FRA[3], %/d	11.8±1.8	12.7±1.6	0.692
SO_4 ARA[4], µ/kg/h	19.0±2.8	24.8±2.9	0.189
% of CYS flux catabolized	13.4±1.8	16.0±1.5	0.300
TAU concentration[5], µM	36.7±4.5	54.4±7.9	0.109
TAU SRA, dpm/nmol	5.70±1.9	7.20±1.1	0.521
TAU FRA, %/d	7.80±2.7	10.0±2.5	0.580
TAU ARA, µ/kg/h	0.90±0.2	0.60±0.1	0.318
GSH concentration, µM	10.2±1.3	13.7±1.3	0.005

[1] Specific radioactivity.

[2] Cysteine flux = infusion rate of ^{35}S-CYS (dpm/kg/h) / CYS SRA (dpm/µmol) at plateau.

[3] Fractional rate of appearance = [(product SRA / precursor SRA) × (24 h / infusion period)] × 100.

[4] Absolute rate of appearance = [(FRA × product concentration)/24] × plasma volume. It was assumed that blood volume is 6.5% of the body weight.

[5] Taurine concentration.

[6] Data are least square mean ± SE.

References

Jahoor, F., L. Wykes, M. Del Rosario, M. Frazer, and P.J. Reeds, 1999. Chronic protein undernutrition and an acute inflammatory stimulus elicit different protein kinetic responses in plasma but not in muscle of piglets. J. Nutr. 129, 693-699.

Lyons, J., A.R. Pfeiffer, Y. Ming-Yu, X. Lu, D. Zurkaowski, M. Curley, C. Duggan, J. Thomson, A. Ajami, S. Borgonha, V.R. Young and L. Castillo, 2001. Cysteine metabolism and whole blood glutathione synthesis in septic pediatric patients. Crit. Care. Med. 29, 870-877.

Malmezat, T., D. Breuille, P. Captain, P. Minard, and C. Obled, 2000. Glutathione turnover is increased during acute phase of sepsis in rats. J. Nutr. 130, 1239-1246.

Malmezat, T., D. Breuillé, C. Pouyet, C. Buffière, P. Denis, P.P. Mirand and C. Obled, 2000. Methionine transsulfuration is increased during sepsis in rats. Am J Physiol Endocrinol Metab. 279, 1391-1397.

Tang, Y.Q. and M.E. Selsted, 1993. Characterization of the disulfide motif in in BNBD-12, an antimaicrobial β-defensin peptide from bovine neutrophil. J. Bio. Chem.268, 6649-6653.

Piglets adaptation during weaning: physiological changes and energy metabolism consequences

A. Minuti, L. Bomba, M. Lizier, R. Mazza, F. Lucchini, F. Rossi, B. Paganini, F. Chegdani, P. Ajmone Marsan and E. Trevisi
Nutrigenomic Research Center (NUTRIGEN), Università Cattolica del Sacro Cuore, via Emilia Parmense 84, 29122 Piacenza, Italy; erminio.trevisi@unicatt.it

Introduction

Weaning is a very critical period for piglets (Lallès *et al.*, 2007), it involves separation from the sow and adaptation to several novelties: new environment, social condition, housing, gut microbiota and nutrition system. All these factors contribute to increase piglet susceptibility to gut disorders (e.g. diarrhea), intestinal infections and, consequently, mucosa inflammation with possible systemic consequences (Pié *et al.*, 2004). To overcome these problems, in past years, antibiotics have been used in the weaning prophylaxis. Since 2006 they are banned in EU to reduce the risk of spreading of antibiotic resistance. However, alternative strategies are still not completely satisfactory. Aim of this study was to investigate the cellular and physiological mechanisms of piglet stress response and adaptation to a risky diet (high starch) during the weaning period.

Material and methods

Eighteen female piglets of similar age and weight were investigated during the experiment. Six 28-day-old control piglets (CTR0) were slaughtered just before weaning, the remaining 12 test piglets were moved in a new farm and slaughtered 6 days after transfer. In the new farm piglets were fed a high starch diet (crude protein 21.9%; starch 43.6%; net energy 2,595 kcal/kg on dry matter basis), without antibiotic addition (to increase the digestive disorder risk). Parameters collected just before or after slaughtering included body weight; wide hema-tochemical profile (ILAB 600, Instrumentation Laboratory, Lexington, MA, USA) as described by Bertoni *et al.*, (2008); serum amiloid A (SAA, by ELISA, Tridelta, Ireland) and intestinal fatty acid binding protein concentration (I-FABP, by ELISA, HyCult Biotechnology B.V., the Netherlands); pH and volatile fatty acids (VFA) in digesta; ileum, caecum and colon tissue histological evaluation by optical microscopy. The 12 test piglets were divided in two groups according to presence (DR; 7 animals) or absence (NODR; 5 animals) of diarrhea. For the statistical evaluation, data were subjected to ANOVA using the GLM procedure (SAS Inst. Inc., Cary, NC, USA, release 9.1) including the group variable (CTR0, DR, NODR) in the model.

Results and discussion

Piglets submitted to the same stress agents during weaning (transport, new environment, new diet, etc.) showed wide differences in their response. Six days after transfer, the DR group showed a reduction while NODR a small rise of body weight (-149.3 vs. 38.3 g/d; $P<0.01$). In accordance with body weight change, differences emerged in the use of body reserves. Subject with diarrhea showed higher level of triglycerides, NEFA, urea and creatinine, all markers of adipose and muscle catabolism (Table 1). In the ileum, pH of digesta (Table 1) was higher in DR compared to NODR ($P<0.05$), while the total VFA concentration was lower (6.43 vs. 23.98 mM/l; $P<0.05$), contributing to explain the difference in pH level. Villus/crypt ratio (Table 1) was lower in DR group, versus both CTR0 ($P<0.05$) and NODR piglets (NS), suggesting intestinal atrophy and damage in the DR group. In these animals, digestive disorders, indicated by higher pH of digesta, seem to trigger ileum inflammatory events and cause the decrease of the villus/crypt ratio (Spreeuwenberg *et al.*, 2001). This hypothesis is confirmed also by the higher level of I-FABP (NS) in DR piglets, a marker of intestinal epithelial cell damage. DR animals also showed higher level of total proteins and albumin (Table 1),

Table 1. Average values of some digestive tract features and blood parameters in weaned piglets grouped according to clinical signs of diarrhea (CTR0=control; DR=diarrhea; NODR=no diarrhea).

		CTR0	DR	NODR	PSE
Ileum pH		$7.13^{a,b}$	7.42^{b}	6.48^{a}	0.175
Ileum villus/crypt ratio		2.41^{b}	1.90^{a}	$2.16^{a,b}$	0.291
Plasma I-FABP	pg/ml	135.9	202.0	97.2	57.916
Plasma NEFA	mmol/l	$0.40^{a,b}$	0.67^{b}	0.22^{a}	0.591
Plasma creatinine	mmol/l	85.0^{a}	137.5^{b}	$97.1^{a,b}$	28.929
Plasma albumin	g/l	30.14^{a}	43.51^{c}	38.18^{b}	0.776
Plasma ceruloplasmin	μmol/l	13.87^{a}	17.42^{b}	14.13^{a}	0.685
Plasma SAA	μg/ml	35.1	102.0	53.7	49.132

^{a,b,c} $P<0.05$; PSE=Pooled Standard Error.

likely due to hemoconcentration, and of inflammatory markers: ceruloplasmin ($P<0.05$), haptoglobin and SAA (NS). Moreover, DR had a higher level of total bilirubin compared to NODR (2.85 vs. 1,41 μmol/l; $P<0.06$), indicating a reduction of its clearance. Therefore, the local inflammation of the intestine seems to induce systemic consequences (Rowlands and Gardiner, 1998), as suggested by changes in hepatic synthesis of inflammatory markers (raise of ceruloplasmin and total bilirubin).

In conclusion, during weaning several animals showed symptoms of digestive disorders together with ileal villous atrophy, indicating local inflammation and signals of systemic effects mediated by the liver. Likely, the feed intake reduction and/or a severe impairment of the intestinal barrier function, justify the reduction of nutrient absorption, as indicated by differences in energy and protein metabolism and body weight losses. A gene expression analysis of ileal tissue is presently in progress, to understand the molecular basis of the observed metabolic responses.

References

Bertoni, G., E. Trevisi, X. Han, and M. Bionaz, 2008. Effects of inflammatory conditions on liver activity in puerperium period and consequences for performance in dairy cows. J. Dairy Sci. 91, 3300-3310.

Lallès, J.P., P. Bosi, H. Smidt, and C.R. Stokes, 2007. Nutritional management of gut health in pigs around weaning. Proc Nutr Soc. May 66, 260-268.

Pie, S., J.P. Lalles, F. Blazy, J. Laffitte, B. Seve, and I.P. Oswald, 2004. Weaning is associated with an upregulation of expression of inflammatory cytokines in the intestine of piglets. J. Nutr. 134, 641-647.

Rowlands, B.J., and K.R. Gardiner, 1998. Nutritional modulation of gut inflammation. Proc Nutr Soc. 57, 395-401.

Spreeuwenberg, M.A.M., J.M.A.J. Verdonk, H.R. Gaskins, and M.W.A. Verstegen, 2001. Small intestine epithelial barrier function is compromised in pigs with low feed intake at weaning. J. Nutr. 131, 1520-1527.

Seasonal effects on energy requirements in senior cats in a temperate environment

E.N. Bermingham[1], K. Weidgraaf[2], M. Hekman[2], M.H. Tavendale[1], N.C. Roy[1] and D.G. Thomas[2]
[1]Food Nutrition Genomics Team, Food Metabolism & Microbiology Section, AgResearch Grasslands, Private Bag 11008, Palmerston North, New Zealand
[2]Institute of Food, Nutrition and Human Health, Massey University, Private Bag 11222, Palmerston North, New Zealand

Introduction

Recently, Bermingham *et al.* (2010) identified the lack of data concerning the energy requirements in older cats. As obesity and related illnesses (e.g. diabetes, joint problems, etc.) are increasing in companion animals, it is important that the energy requirements of older cats are established. The National Research Council (2006) highlights the energy requirements for cats at high (36 °C) or room temperatures (25 °C), however, no information currently exists on energy requirements in temperate climates, such as those experienced in New Zealand where cats typically have outdoor access all year round. Therefore, this study aimed to determine the maintenance energy requirements of healthy older cats (c. 10 years of age), in an outdoor environment during summer and winter.

Materials and methods

Eight mixed-sex, neutered cats (c. 9.9 years of age) were housed in individual pens (80×80×110 cm) for 5 weeks (1 week adaptation, 4 week trial) during summer and winter. Cats were in visual contact with each other, and were familiar with the single pens. The cats were fed *ad libitum* AAFCO tested, commercially available fresh wet food daily, had water available *ad libitum* and were exposed to natural light regimes. The cats were weighed and body-condition scored weekly.

During the third week of the feeding study, energy expenditure was estimated in the cats using 0.7 g/kg BW dose of deuterium oxide (D_2O) and 0.13 g/kg BW of oxygen-18 (O^{18}) water (Backus *et al.*, 2000). Food and water was withheld 24 h beforehand. A baseline blood sample (day 0) was taken by jugular venepuncture prior to the intravenously injection with the double-labelled water. Blood samples were collected on day 3, 6 and 12 after injection to determine energy expenditure during this period (Kanchuk *et al.*, 2003). Plasma was separated from each blood sample and frozen at -20 °C until analysis. Enrichment of D_2O and O^{18} were determined by gas chromatography-isotope ratio mass spectrometry.

Energy intakes were determined on week 4 of the experimental period. Energy expenditure was determined using the equations outlined by Ballevre *et al.* (1994), assuming a corrected food quotient of 0.89 (Ballevre *et al.*, 1994).

Energy intake was analysed using a Repeated Measures procedure whilst other variables (e.g. energy expenditure, bodyweight) were analysed using ANOVA (Genstat, version 11). Results are reported as mean and standard error of difference (SED).

Results and discussion

The average temperature during the 4 week periods in summer and winter were 18.5±0.5 °C and 8.5±0.4 °C, respectively. Season had no effect on the bodyweight of the cats (P>0.05; Table 1). Energy intakes and energy expenditure during the two seasons were unchanged (P>0.05; Table 1).

Table 1. Bodyweight and daily energy intake and energy expenditure of senior cats during summer and winter.

	Summer	Winter	SED	*P*-value
Body weight (kg)	4.6	4.5	0.34	0.72
Lean mass (kg)	3.3	3.2	0.20	0.68
Energy intake (kcal/kg BW/d)[1]	63.9	69.0	3.40	0.30
Energy intake (kcal/kg lean mass/d)[1]	84.4	93.5	7.00	0.25
Energy expenditure (kcal/kg BW/d)[2]	55.9	50.7	4.40	0.26
Energy expenditure (kcal/kg lean mass/d)[2]	76.3	73.7	5.64	0.65

[1] Energy intake calculated from average daily intakes during week 4 of the experimental period.
[2] Determined from the injection of deuterated water and oxygen-18 on week 3 of the experimental period.

The cats in this study had higher energy intakes (averaging 67 kcal/kg BW/d) than those reported in the literature for senior cats (51.1 kcal/kg BW/d; Bermingham *et al.*, 2010), which may reflect differences between cats kept outside in temperate climates compared to cats studied in indoor environments.

References

Backus R.C., P.J. Havel, R.L. Gingerich and Q.R. Rogers, 2000. Relationship between serum leptin immunoreactivity and body fat mass as estimated by use of a novel gas-phase Fourier transform infrared spectroscopy deuterium dilution method in cats. Am. J. Vet. Res. 61, 796-801.

Ballevre O., G. Anantharaman-Bar, and P. Gicquello, 1994. Use of doubly-labelled water method to assess energy expenditure in free living cats and dogs. J. Nutr. 124, 2594S-2600S.

Bermingham E.N., D.G. Thomas, P.J. Morris and A.J. Hawthorne, 2010. Energy requirements in cats: a meta-analysis. Bri. J. Nutr. 103, 1083-1093.

Kanchuk M.L., R.C. Backus, C.C. Calvert, J.G. Morris, and Q.R. Rogers, 2003. Weight gain in gonadectomized normal and lipoprotein lipase-deficient male domestic cats results from increased food intake and not decreased energy expenditure. J. Nutr. 133, 1866-1874.

National Research Council (NRC), 2006. Energy. In: Nutrient requirements of dogs and cats, Washington DC: National Academy Press: 28-48.

Energy metabolism of growing blue foxes, *Alopex lagopus*

N. Koskinen[1] *and A-H. Tauson*[2]
[1]*MTT Agrifood Research Finland, Animal Production Research, Tervaniementie 179, Hyvinkää, Finland*
[2]*Faculty of Life Sciences, University of Copenhagen, Frederiksberg C, Denmark*

Introduction

Among blue fox (*Alopex lagopus*) selective breeding has focused on producing heavy animals and meanwhile reproductive performance has declined among yearling vixens. At present blue foxes are fed *ad libitum* during the growing – furring period both those aimed for pelting and the future breeding stock. The blue fox exhibits seasonal fluctuations in feed intake and accretion of body fat, feed intake and body fat retention being very high during autumn and early winter if fed unrestrictedly. It has been indicated that energy supply, body condition and reproductive processes are closely connected (Blache *et al.*, 2004). The main objective of this project was to establish baseline data on the energy requirement of growing blue foxes by measuring feed intake, energy expenditure, and protein and fat retention. The data will be used to develop practical farming feeding recommendations for young future breeding blue fox vixens, in order to avoid excessive fat retention during the autumn, and then severe weight reduction during the winter.

Materials and methods

The experiment was accomplished from August to December 2007 in five different periods, carried out at monthly intervals. Sixteen juvenile blue fox vixens were used and they were fed the same conventional diet with the chemical composition of 37.8% dry matter, 1.9% ash, 11.7% crude protein, 8.7% fat and 15.5% carbohydrate as fed. The treatments with different energy supply were: (1) *ad libitum*, target body condition 'very high'; (2) 20-30% below the *ad libitum* group, target body condition 'high'; (3) 35-45% below *ad libitum*, target condition 'ideal mating condition'; (4) 50-60% below the *ad libitum* group, target condition 'lean'. The treatments were used in order to establish a population of animals with different body condition. The experiment was performed in five 7 d balance periods (3 d adaptation and 4 d quantitative collection of faeces and urine), starting when the animals were 10 weeks old and ending when the animals were 30 weeks old. Each period included a 22 h respiration experiment by means of indirect calorimetry in an open-air circulation system. Heat production (HE) was calculated according to the formula by Brouwer (1965).

Statistical analyses were carried out using the MIXED procedure for repeated measures in SAS (SAS release 9.1) with treatment group (1-4), and period (1-5) as the main effects, and interactions between the main effects. The covariance structure was UN (unstructured) and was chosen according Akaike's information criterion.

Results and discussion

The feed consumption was close to target in the four treatment groups and the increase in animal live weight reflected the energy supply. Mean live weights were similar (4,880-5,087 g) among groups at the beginning of the trial, but at the end of the experiment the group mean weights were significantly different, as intended. Because the experiment was performed during the growth period, animal live weights, metabolizable energy, heat production, retained energy and retained nitrogen were strongly affected by measurement period. Also the effect of the treatment on the metabolizable and retained energy and effect of the period on the heat production were found (Table 1). Feeding recommendations for the breeding and production blue fox females have been assigned according to the results. Recent data suggest that blue fox vixens being selected for breeding purposes during the summer, and then

Table 1. Animal live weights (g), metabolizable energy (ME), heat production (HE), retained energy (RE) kJ/kg$^{0.75}$ and retained nitrogen (RN, g/kg$^{0.75}$ day) for blue fox vixens raised on different energy supply and measured in five periods from 10 to 30 weeks of age.

	Group	Period					RR	*P*-value; effect of		
		I	II	III	IV	V		Group	Period	G×P
Live weight										
	1	5,087	7,752	10,154	11,307	12,839				
	2	4,880	7,175	9,335	11,359	12,525				
	3	4,990	7,197	8,569	10,024	10,930				
	4	4,891	6,907	8,021	8,968	9,238	467	0.0007	<0.0001	<0.0001
ME										
	1	1,616	1,392	1,074	835	819				
	2	1,634	1,392	1,017	820	711				
	3	1,502	1,250	855	753	598				
	4	1,458	1,133	752	593	462	137	0.0017	<0.0001	0.0044
HE										
	1	570	388	291	479	489				
	2	567	407	345	487	421				
	3	568	378	289	454	461				
	4	466	368	232	397	363	68	0.0661	<0.0001	0.9033
RE										
	1	1,046	1,004	783	356	330				
	2	1,067	985	673	332	290				
	3	934	872	554	298	137				
	4	992	764	529	196	99	139	0.0086	<0.0001	0.1698
RN										
	1	1.09	0.94	0.46	0.21	0.15				
	2	1.14	0.77	0.33	0.21	0.15				
	3	1.06	0.61	0.33	0.22	0.07				
	4	1.11	0.58	0.34	0.29	0.07	0.13	0.1703	<0.0001	0.1462

rearing them on a restricted feeding regimen, has resulted in improved reproductive results when compared to vixens reared on *ad libitum* feeding (Koskinen *et al.*, 2008). Future investigations will focus on the reproductive performance in vixens raised on different levels of energy supply.

References

Blache, D., S. Zhang and G.B. Martin, 2003. Fertility in male sheep: modulators of the acute effects of nutrition on the reproductive axis of male sheep. Reproduction Supplement 61, 387-402.

Brouwer, E., 1965. Report of sub-committee on constants and factors. In: Blaxter, K.L. (ed.) Energy metabolism of farm animals. EAAP Publication No. 11. London: Academic Press: 441-443.

Koskinen, N., A-H. Tauson, J. Sepponen and T.Rekilä, 2008. Feeding history affects cub survival of young breeding blue foxes (*Alopex lagopus*), a field study. In: International Fur Animal Scientific Association 2008 Congress: August 19-23: Halifax, Nova Scotia, Canada. Scientifur 32 (4), 16-17.

Part 9. Evaluation and modelling of feed value and requirements: ruminants

The partial efficiency of use of metabolisable energy for growth in ruminants

L.O. Tedeschi[1], D.G. Fox[2], G.E. Carstens[1] and C.L. Ferrell[3,4]
[1]Texas A&M University, College Station, TX 77843-2471, USA; luis.tedeschi@tamu.edu
[2]Cornell University, Ithaca, NY 17853, USA
[3]USDA, ARS, US Meat Animal Research Centre, Clay Centre, NE 68933-0166, USA
[4]Current address: F & J Livestock Consulting, Saronville, NE 68975, USA

Abstract

Published information on animal bioenergetics, and energy partitioning and utilization is vast and contains broad scientific knowledge but lacks proper integration. The main objective of this paper is to discuss the principal factors affecting the partial efficiency of use of metabolisable energy (ME) for growth (k_g). The development of the partial efficiencies of use of ME to net energy (NE) for maintenance and growth has dramatically increased our understanding of energy partition for growing cattle in the last century. Maintenance and growth and their partial efficiencies of use of ME are intrinsically connected and interdependent, suggesting they should be analyzed together and cannot be studied apart. There are several factors that can affect k_g; the key ones are dietary (energy concentration and end product volatile fatty acid – VFA – profile) and composition of the gain. We hypothesized that dietary factors provide the initial base of the energetic efficiency given the proportion of VFA and maybe glucose availability, but animal genetics and stage of maturity ultimately control the proportion of fat and protein deposited. Therefore, the maximization of the efficiency of use of ME can be achieved if diets are accurately formulated to meet the desired composition of the gain under specific production scenarios. Asynchrony between these two factors (and maybe others) is likely to decrease the energy efficiency and be limited by the least efficient one. A more integrated system of maintenance and growth nutritional energetics is needed. The inability to completely separate maintenance from growth requirements for different levels of energy intake likely leads to inaccurate assignments of energy requirements, causing errors in the prediction of k_g. Efforts should be focused on developing methods to integrate dietary and animal factors to more accurately predict k_g and specific factors should be determined for diverse breed types at similar stages of maturity. Some apparent contradictions reported in the literature can be resolved if a curvilinear relationship is assumed between tissue energy and energy intake.

Introduction

Efforts have been made to identify methods that could be used to select for efficient animals, including residual feed intake (RFI). Despite several confounding factors intrinsic to the RFI as frequently applied (Johnson *et al.*, 2003), studies have been conducted to understand the epistemology of RFI. These studies have identified key components that are responsible for the variation in RFI, which include feed intake, digestion of feed, maintenance, growth metabolism, activity, thermoregulation, rate of gain (ADG), body weight (BW), and prolificacy (Johnson *et al.*, 2003; Oddy and Herd, 2001). It is believed that maintenance (i.e. protein turnover, tissue metabolism, physical activity, and stress), feed digestibility and heat increment of fermentation, and body composition account for 47, 19, and 5%; respectively, of the variation in RFI (Herd and Arthur, 2009; Richardson and Herd, 2004). Thus, thorough understanding of energy partitioning is important to accurately identify the most efficient animals.

Current feeding standards (e.g. AFRC, 1993; ARC, 1980; CSIRO, 2007; NRC, 2000) differ in several aspects, including the determination of feed biological values, animal nutrient requirements, and their application (Tedeschi *et al.*, 2005b), but most of them have adopted the metabolisable energy (ME) and (or) net energy (NE) systems to partition energy. Even though the adoption of NE to quantify energy requirements for different purposes (e.g. maintenance, growth, pregnancy, lactation, etc) accounted for several limitations of the dietary energy (IE), digestible energy (DE), and total digestible nutrient (TDN) systems, some limitations of the ME and NE systems still need improvement.

The historical development of animal bioenergetics, and energy partitioning and utilization has been thoroughly summarized by Brody (1945), Kleiber (1961), and Blaxter (1962a), and critically reviewed by various investigators, including Moe (1981), Garrett and Johnson (1983), and Johnson *et al.* (2003). For this reason, the objectives of this paper are to briefly review the concepts and main factors affecting the nutritional energetics of growing ruminants with respect to the partial efficiency of use of ME for growth (k_g), present ideas and suggestions for future research, and to illustrate the use of modelling to apply the current knowledge.

Approaches used to determine efficiency of ME use

The main limitation in using DE as a basis for ruminant nutrition (i.e. overrating high-fibre feeds compared to low-fibre feeds) is partially corrected in the ME and NE systems. The development of techniques to measure energy fluxes was initiated between the late 1920's and early 1950's (Garrett and Johnson, 1983). Since then, the technical apparatus and instrumentation used to determine ME efficiency have evolved but the approaches used to estimate ME efficiency have remained the same to quantify heat production (HE) or the retained energy (RE) by the animal after feeding a diet of known DE content for a given period. It has been shown that the indirect respiration calorimetry technique consistently yields greater RE values than the comparative slaughter technique (Larson and Johnson, 1997), likely because animals at the maintenance level in the comparative slaughter technique are better adapted and retain more energy than expected (Tyrrell, 1987). Preliminary studies by Blaxter (Blaxter, 1962b; Blaxter and Graham, 1955) have led to the understanding of the partial efficiency of use of ME for maintenance (k_m) and productive functions.

The determination of the partial efficiency of use of ME for growth

During the decade of the 1950's, the complications in accurately determining a unique feed energy value to meet animal requirements due to the complexity of factors influencing the partial efficiency of use of ME for various physiological functions surfaced. It has been accepted that at least two NE values and therefore two partial efficiencies (k_m and k_g), are needed. The net availability of ME for growth is determined as the slope of the regression of positive RE (also known as recovered energy) on ME intake (MEI) while the net availability of ME for maintenance is the slope of the regression of negative RE on MEI (Blaxter and Wainman, 1961). In the USA, the work of Lofgreen (1965), using the comparative slaughter technique in beef cattle to determined RE, led to the development of the California NE system (Garrett *et al.*, 1959; Lofgreen and Garrett, 1968) and their revisions (NRC, 1984).

Figure 1A depicts the representation of tissue energy (TE) versus MEI when TE is negative ($TE<0: TE=a_1+b_1 \times MEI$) and TE is positive ($TE>0: TE=a_2+b_2 \times MEI$), meaning that an animal is consuming ME above that required for maintenance. When an animal is fasting (TE<0), MEI is equal to zero, and TE becomes equal to a_1, which is referred to the fasting heat production (H_eE). When an animal is at maintenance (no energy gain or loss: TE=0) and MEI is equal to $-H_eE/b_1$. In this case, b_1 is the k_m and the MEI is the ME required for maintenance (ME_m). When TE=0 and MEI is at ME_m, one can compute a_2 as $-b_2/ME_m$. Substituting a_2 into the equation of TE<0 and rearranging it, yields $TE = b_2 \times (MEI-ME_m) = k_g \times (MEI-ME_m)$.

Even though this two-regression approach (TE<0 and TE>0) solved in part the different efficiencies of use of ME to NE, the main restriction of this approach is that the true relationship between TE and MEI is likely to be curvilinear, showing a diminishing return effect as MEI increases (Chizzotti *et al.*,

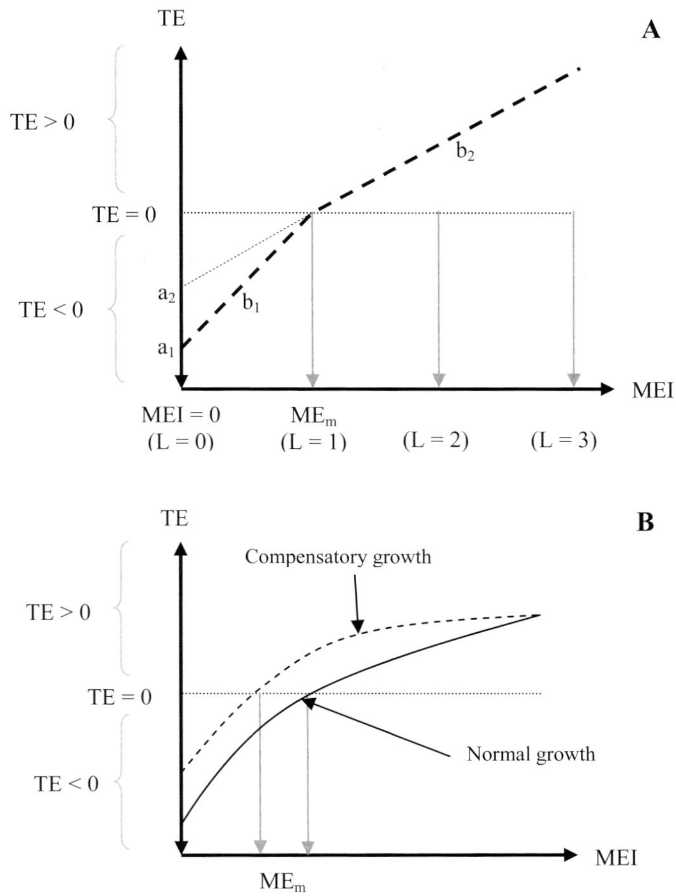

Figure 1. (A) Schematic representation of the relationship between tissue energy (TE, Mcal/kg$^{0.75}$/d) and metabolisable energy intake (MEI, Mcal/kg$^{0.75}$/d) for positive and negative TE conditions and (B) conceptual representation of the curvilinear relationship between TE and MEI for normal and compensatory growths. ME_M is the MEI at maintenance, L is multiple of ME_M intake, and a_1 represents the basal metabolism rate (H_eE, Mcal/kg$^{0.75}$/d).

2008; Ferrell and Jenkins, 1998a,b). We recognize that the assumption that a curvilinear relationship could be approximated by two linear regressions was a matter of convenience as computers were not available in the 50's or 60's.

Altering the partial efficiency of use of ME to NE for growth

Early scientists recognized that the major factors influencing k_g included level of intake, physiological function for which energy is utilized, and the nutrient balance (Meyer and Garrett, 1960); diet composition (ingredients), i.e. dietary fat is converted more efficiently to body fat than glucose from starch or cellulose to body fat (Baldwin, 1995); and environmental conditions, physical activity, sex, gastrointestinal tract architecture, and the nature of absorbed products of ruminal fermentation (Reid *et al.*, 1980).

Dietary effects: nutritional value of the diet

Early studies have indicated that changes in the ruminal volatile fatty acids (**VFA**) profile could impact k_m (Armstrong and Blaxter, 1957a; Armstrong *et al.*, 1957) and k_g (Armstrong and Blaxter, 1957b; Armstrong *et al.*, 1958). For growing animals in positive energy balance (TE > 0), acetic acid was poorly utilized compared to propionic and butyric acids. Later, the analysis of 72 calorimetric experiments in which the percentage of hay and flaked corn varied from 0 to 100% indicated that k_g was negatively correlated with the molar proportion of acetic acid in the rumen fluid (Blaxter and Wainman, 1964). The efficiency of use of acetic, propionic, and butyric acids when infused individually in the rumen of fattening sheep were 32.9, 56.3, and 61.9%; respectively. Other studies with diets containing large amounts of propionate and lactate were used more efficiently for fat deposition compared to those diets with lower proportions of propionate and lactate (Fox and Black, 1984). For mature cows, acetate was utilized for ADG with an efficiency of 69% when fed 30% hay (Tyrrell *et al.*, 1979). Others reported that the effect of acetic acid on efficiency of energy use in growing animals is negligible for practical conditions (Ørskov *et al.*, 1979). Reid *et al.* (1980) indicated that the relation of glucose to acetate utilization may explain better different k_g values than the ratio of acetate to propionate. This observation seems to be supported by those of Ku-Vera *et al.* (1989). These discrepancies are still not fully explained, but their occurrences are assumed to have differing effects at different stages of growth and different combinations of dietary ingredients (Moe, 1981).

The NRC (2000; 2007) compute k_g based on the work of Garrett (1980), who regressed RE on MEI above maintenance to estimate k_g (the b_2 slope of the regression shown in Figure 1A). The equation developed by Garrett (1980) computes k_g using diet ME concentration. Similarly, ARC (1980), CSIRO (2007), and AFRC (1993) utilize metabolisability of the feed in the diet (q_m = ME/IE) to compute k_g. Even though these systems use diet information to compute k_g, the values obtained are remarkably different. In the ARC (1980), CSIRO (2007), and AFRC (1993) systems, as diet digestibility increases, NE for growth (NE_g) increases curvilinearly upward with increasing ME. The ARC (1980), CSIRO (2007), and AFRC (1993) are primarily based on indirect respiration calorimetry data, which as mentioned above tends to overpredict RE (and k_g). In contrast, in the Garrett (1980) system, NE_g increases at an ever-decreasing rate (Johnson *et al.*, 1997), and becomes negative at DE values less than 0.4. This suggests that the NE_g value in the upper range of diet ME concentration is probably being underestimated (Johnson *et al.*, 1997). Both approaches have inherent limitations based primarily upon explicit and implicit assumptions imbedded in the technique used.

Dietary associative effects (both positive and negative) among feeds within diet may also impact the determination of the true nutritional value of the diet if the diet ME is determined at the ingredient level or diet consumption increases (Merchen, 1993). However, the latter one is difficult to determine due to confounding effects of level of intake on digestibility (increased passage rate). Earlier studies have shown depressed animal performance (Peterson *et al.*, 1973) and decreased diet digestibility (Joanning *et al.*, 1981) when processed or unprocessed corn grain and corn silage were fed together.

Composition of the gain: fat versus protein deposition

Theoretical estimates of efficiency of use of ME for energy stored as protein or fat were about 90 and 70% respectively, implying a greater efficiency for protein deposition (Garrett and Johnson, 1983). However, these estimates did not consider the turnover rates of protein. Practical estimates using regression techniques to partition between protein and fat deposition subsequently demonstrated that storing energy as protein in growing animals is energetically less efficient than storing energy as fat (Graham, 1980; Rattray *et al.*, 1974). In fact, fat is deposited from excess of fat intake with an efficiency of 97% (Baldwin, 1995: 4) or synthesized from protein with an efficiency of 67.5% (Baldwin, 1995: 4) and from carbohydrate with an efficiency of 60.1 to 77.8% (Baldwin, 1995: 4).

The lower energetic efficiency of protein deposition is likely influenced by the mix of amino acids available and the energy cost associated with body protein turnover (Baldwin, 1995).

The energy cost of ME for energy stored as protein and fat is similar at approximately 12.5 Mcal/kg if the heats of combustion of 5.686 and 9.367 Mcal/kg (Lofgreen, 1965) and efficiency of deposition of 45 and 75% are assumed for protein and fat, respectively (CSIRO, 2007). However, Rattray *et al.* (1974) reported the energy cost of ME for energy stored as protein is greater than that for energy stored as fat (45.6 Mcal ME/kg or 8.14 Mcal ME/Mcal of protein and 10.2 Mcal ME/kg or 1.1 Mcal ME/Mcal of fat, respectively). Rompala *et al.* (1987) reported the efficiency of use of ME for energy deposited as fat (k_F) and protein (k_P) was 75% and 19%; respectively. Reported values by Reid *et al.* (1980) were 69 and 34%, respectively for k_F and k_P. The variation in estimates may be partially explained by the difficulties associated with partitioning energy intake between fat and protein gain by multiple regression methods due to collinearity between fat, protein, and RE (Roux *et al.*, 1982).

The use of a single partial efficiency to compute NE_g from ME introduces some inaccuracies in estimating weight gain. Feeding standards that employ an average value for k_g will overestimate energy gains when a high proportion of the tissue growth is in the form of protein because its efficiency is lower than fat efficiency. Despite the effect of gain composition on k_g, no efforts have being made to account for this variation in computing NE_g in the past, probably because procedures are too uncertain for practical use and unlikely to improve k_g predictions for a given class of feed (CSIRO, 2007). This effect will be more pronounced when comparing double muscle breeds (e.g. Belgian blue, Marchigiana) with regular cattle (e.g. Angus, Hereford). However, when averaged over the entire feeding period and animals are fed to a constant endpoint, the body composition, composition of gain, and the diluting effect of water gain on cost of lean weight gain may minimize the effects of protein versus fat gain. The ME requirement can be computed as RE_{Fat} and RE_{Prot} divided by their respective efficiencies of deposition (k_F and k_P, respectively) and solved for an overall k_g (Tedeschi *et al.*, 2004).

Protein efficiency is highly variable, depending on physiological stage, feeding level, and feed quality, likely amino acid profile (CSIRO, 2007), which in turn partially dictates the protein turnover. Unlike protein, fat efficiency is thought to be less variable (CSIRO, 2007), but measured values have been contrary to this; values for k_P and k_F have varied from 12 to 22% and 56 to 128% (Rattray *et al.*, 1974; Rattray and Joyce, 1976). When a mean efficiency for protein and fat of 20 and 75%, respectively (Geay, 1984) are used, k_g can be computed as a function of the proportion of protein energy in the RE. Other values of efficiencies for fat (68%) and protein (27%) have been reported to more accurately predict ADG for growing sheep (Cannas *et al.*, 2006; Tedeschi *et al.*, 2010). Chizzotti *et al.* (2008) have also found different values for Nellore and Nellore × *Bos taurus*.

Metabolic modifiers

Anabolic implants are known to shift the composition of gain in cattle by increasing protein deposition and decreasing fat at a particular weight (NRC, 1984; 2000). Implanted animals reach the same body composition at a heavier weight when compared to non-implanted animals (Hutcheson *et al.*, 1997; Perry *et al.*, 1991). Beta agonists added to the feed increase growth rate by reducing fat accretion, reducing protein turnover, and increasing protein accretion (NRC, 1994). Steers fed zilpaterol hydrochloride (Zilmax; Internvet Inc.) had heavier carcass, less rib fat and kidney-pelvic-heart fat, and larger *Longissimus dorsi* area (Vasconcelos *et al.*, 2008) by likely increasing carcass muscle deposition (Leheska *et al.*, 2009). The ADG increased by 36 and 18% for steers and heifers (Montgomery *et al.*, 2009) likely due to the increased deposition of water associated with lean tissue even though the k_g might have decreased.

Plane of nutrition: compensatory growth

Several factors contribute to compensatory growth in ruminant animals including an increase in protein deposition relative to fat, reduced maintenance requirement, increased k_g, increased gut fill, and (or) greater dry matter intake (DMI) (Ryan, 1990). Animals exhibiting compensatory growth have a greater potential to deposit more protein rather than fat specially during the beginning of the re-alimentation period, but more fat than protein during the last part of the re-feeding period (Drew and Reid, 1975; Fox *et al.*, 1972). Fox *et al.* (1972) indicated that compensating steers were leaner at 364 kg, but similar in empty body composition at 450 kg BW compared to non-restricted steers. During nutritional restriction the proportional weights and sizes of liver and digestive tract tissues are decreased and the maintenance requirement is reduced. Upon realimentation, the repletion of the non-carcass tissues is associated with rapid increases in protein deposition (Ryan *et al.*, 1993). Carstens *et al.* (1991) found that accretion rates for carcass protein and fat relative to EBW were similar for compensatory and non-restricted growth steers, but that accretion rates for non-carcass protein were greater during growth compensation. Compensatory growth steers had less empty body fat (EBF; 24.2 vs. 32.4%) and more empty body protein (EBP; 16.6 vs. 14.8%) at similar BW of 500 kg, and consequently had 18% lower NE_g requirements than non-restricted growth steers. Studies based on comparative slaughter technique estimates of RE in sheep (Reid *et al.*, 1980) and cattle (Carstens *et al.*, 1989; Fox *et al.*, 1972) have demonstrated that k_g is elevated 14 to 32% during growth compensation. The estimates of ME available for growth in these studies were based on the assumption that maintenance energy requirements were static. Thus, to the extent that maintenance requirements were reduced during realimentation, the estimates of k_g would be overestimated. Based on serial indirect respiratory calorimetry measurements of HE during realimentation, Carstens *et al.* (1989) found that k_g decreased from 72% on day 3 to 50% on day 30 of realimentation as the proportion of RE_P decreased from 24 to 18% during the first 50 d of realimentation. Elevated rates of EBP deposition during that early phase of compensatory growth are primarily associated with repletion of visceral organ tissues, suggesting the proportional growth of visceral organ tissues affects k_g. These results suggested that because of the change in the composition of the gain, the k_g is likely to be changed and the dietary NE_g value is not independent of the previous plane of nutrition.

Using modelling to integrate and apply knowledge of effects of animal and dietary factors on the efficiency of energy utilization in ruminants

A nutrition model can be defined as an integrated set of mathematical equations and transfer coefficients that describe the various animal physiological functions. The level of aggregation of equations depends on the data available to develop and test them, and the objective of the model (i.e. to understand and describe how the biological system works vs. the development of a model for on farm application). Most nutrition models use a combination of mechanistic and empirical approaches to represent the aggregated response of whole physiological functions to the variables (Tedeschi *et al.*, 2005b). To be useful on farms, models must be able to accurately predict animal nutrient requirements and performance under diverse combinations of animal type, feeds available, and environmental conditions. When done correctly, the effects of each input variable affecting animal responses can be accounted for in accurately formulating diets, predicting costs and returns, and setting priorities for management. To improve its accuracy, prediction equations and coefficients need to be updated with new knowledge and re-evaluated to make sure that predictions are as good as or better than before the changes were made to minimize the risk of use due to offsetting errors.

Accounting for the effect of mature size and breed type

The NRC (2000) developed a size scaling system for their equations used to predict energy and protein retained for differences due to mature size and breed type. This model accounted for 94 and 92% of the predicted energy and protein retained with biases of 2 and -2% respectively. These results indicate the use of the NRC (2000) growth model results in accurate predictions of NE_g requirements across wide variations in breed, mature body size, anabolic implants and nutritional management systems. They concluded that most errors in predicting ADG with this model were likely due to assigning the wrong mature BW, short-term transitory effects of previous nutrition, gut fill, anabolic implants, variation in NE_m requirement, variation in ME value assigned to the feed because of variations in feed composition and extent of ruminal or intestinal digestion, variation in NE and NE_g derived from the ME because of variation in end products of digestion and their metabolisability, and variations in gut fill.

Accounting for the effect of metabolic modifiers

Guiroy *et al.* (2002) used the Cornell Net Carbohydrate and Protein System [CNCPS; (Fox *et al.*, 2004)] to account for animal and dietary effects to evaluate the independent effects of anabolic implants on intake, growth, weight at a target body composition and efficiency of ME use for growth. After accounting for differences in mean BW and composition of gain, implanted steers and heifers had 4.2 and 3.1% higher apparent diet ME values, respectively. Use of the model in this study resulted in determining that anabolic implant response is due to a combination of a reduced proportion of the DMI required for maintenance, reduced energy content of gain, and efficiency of use of absorbed energy. An analysis of data with the CNCPS model collected in a commercial feedlot with steers receiving different combinations of metabolic modifiers indicated over all treatments, the CNCPS accurately predicted net energy requirements, feed net energy values and ADG (D.G. Fox, personal communication). However, the accuracy of ADG prediction varied with level of anabolic compound received. It was concluded that further modifications are needed to more accurately account for the effects of aggressive use of anabolic compounds on efficiency of use of ME.

Predicting ME values of feeds

Tedeschi *et al.* (2005a) used the CNCPS model to evaluate the effect of predicting feed ME values with NRC tabular, an empirical equation that uses actual feed content of carbohydrate fractions (Weiss *et al.*, 1992) or the CNCPS mechanistic rumen model that uses actual feed content of carbohydrate and protein fractions and their digestion rates. The r^2 and bias of the regression of predicted on observed ADG of the growing steers were 0.61 and -11.4%, 0.73 and -2.2%, and 0.80 and 0.4%, respectively. They concluded that accurate prediction of DE values with the mechanistic model indicated the coefficients used for the prediction of ME from DE and NE from ME resulted in accurate prediction of ADG.

Ainslie *et al.* (1993) used the CNCPS to evaluate growth when ME, MP, or essential amino acids limited growth. The ME allowable ADG alone had an r^2 of 0.70 and a bias of 25% across all treatments. However, when ME, MP, or essential amino acid allowable growth was used, the r^2 and bias were 0.82 and 3%, respectively. These results and those in other studies have suggested ME efficiency is reduced when protein, essential amino acids or other nutrients are not adequate to support the ME allowable growth. It is apparent that nutrient deficiencies or excesses impact efficiency of utilization of ME and consequently, k_g can be accurately assessed only if the diets in use are nutritionally balanced.

Conclusions

The development of the partial efficiencies of use of ME to NE for maintenance and growth has dramatically increased our understanding of energy partition for growing cattle in the last century, and has greatly increased our ability to predict growth and to improve animal performance on diets fed. Johnson *et al.* (2003) affirmed that even though tremendous efforts and advancements have been made in the last 100 years, modifications are needed to the determination of requirement for NE_m

and k_g. There are several factors that can affect k_g; the key ones are dietary (energy concentration and end product VFA profile) and composition of the gain. We hypothesize that dietary factors provide the initial base of the energetic efficiency given the proportion of VFA and glucose availability, but animal genetics and stage of maturity are the ones that ultimately controls the proportion of fat and protein being deposited. Therefore, the maximization of the efficiency of use of ME can be achieved if diets are accurately formulated to meet the desired composition of the gain under specific production scenarios. Asynchrony between these two factors is likely to decrease the energy efficiency and be limited by the least efficient one.

A more integrated system of maintenance and growth nutritional energetics is needed. The inability to completely separate maintenance from growth requirements for different levels of energy intake is likely the cause of wrong assignments of energy requirements, causing errors in the prediction of k_g. The increase in energy requirements to maintain greater levels of production is accounted for by increasing the basal metabolic rate of the animal, but at which energy efficiency is it performed? Efforts should be towards developing methods to integrate dietary and animal factors in predicting k_g and specific factors should be determined for diverse breed types. The majority of studies that have examined compensatory growth support the concept that during realimentation animals exhibit reduced maintenance requirement and increased k_g. However, the greater k_g seems contradictory as discussed above because to have the same RE with less ME_m, the slope of the regression between RE and MEI is reduced. These issues can be resolved if a curvilinear relationship between TE and MEI is used as shown in Figure 1B. The pattern shown for compensating animals is temporary and over time (delayed) it shifts to the normal growth.

We believe that future research should focus on methodologies that are capable of quantifying individual animal differences in energetic efficiency, including inter-animal variances in energy requirements for maintenance and growth, and efficiency of use of energy.

References

AFRC, 1993. Energy and Protein Requirements of Ruminants. Agricultural and Food Research Council. CAB International, Wallingford, UK.

Ainslie, S. J., D.G. Fox, T.C. Perry, D.J. Ketchen and M.C. Barry, 1993. Predicting amino acid adequacy of diets fed to Holstein steers. J. Anim. Sci. 71, 1312-1319.

ARC, 1980. The Nutrient Requirements of Ruminant Livestock. Agricultural Research Council. The Gresham Press, London.

Armstrong, D.G. and K.L. Blaxter, 1957a. The heat increment of steam-volatile fatty acids in fasting sheep. Br. J. Nutr. 11, 247-272.

Armstrong, D.G. and K.L. Blaxter, 1957b. The utilization of acetic, propionic and butyric acids by fattening sheep. Br. J. Nutr. 11, 413-425.

Armstrong, D.G., K.L. Blaxter and N.M. Graham, 1957. The heat increments of mixtures of steam-volatile fatty acids in fasting sheep. Br. J. Nutr. 11, 392-408.

Armstrong, D.G., K.L. Blaxter, N.M. Graham and F.W. Wainman, 1958. The utilization of the energy of two mixtures of steam-volatile fatty acids by fattening sheep. Br. J. Nutr. 12, 177-188.

Baldwin, R.L., 1995. Modeling Ruminant Digestion and Metabolism. Chapman & Hall, New York.

Blaxter, K.L., 1962a. The Energy Metabolism of Ruminants. Hutchinson, London, UK.

Blaxter, K.L., 1962b. The utilization of the energy of food by ruminants. In: Proceedings of Energy Metabolism of Farm Animals, 2, Wageningen. Pudoc: 211.

Blaxter, K.L. and N.M. Graham, 1955. Plane of nutrition and starch equivalents. J. Agric. Sci. 46, 292-306.

Blaxter, K.L. and F.W. Wainman, 1961. The utilization of food by sheep and cattle. J. Agric. Sci. 57, 419-425.

Blaxter, K.L. and F.W. Wainman, 1964. The utilization of the energy of different rations by sheep and cattle for maintenance and for fattening. J. Agric. Sci. 63, 113-128.

Brody, S., 1945. Bioenergetics and growth. Reinhold Publishing Corporation, NY.

Cannas, A., L.O. Tedeschi, A.S. Atzori and D.G. Fox. 2006. Prediction of energy requirements for growing sheep with the Cornell Net Carbohydrate and Protein System. In: J. Dijkstra (ed.), Nutrient Digestion and Utilization in Farm Animals. CABI Publishing, Cambridge, MA: 99-113.

Carstens, G.E., D.E. Johnson and M.A. Ellenberger, 1989. Energy metabolism and composition of gain in beef steers exhibiting normal and compensatory growth. In: Proceedings of Energy Metabolism of Farm Animals, 11, Lunteren. Pudoc: 131-134.

Carstens, G.E., D.E. Johnson, M.A. Ellenberger and J.D. Tatum, 1991. Physical and chemical components of the empty body during compensatory growth in beef steers. J. Anim. Sci. 69, 3251-3264.

Chizzotti, M.L., L.O. Tedeschi and S.C. Valadares Filho, 2008. A meta-analysis of energy and protein requirements for maintenance and growth of Nellore cattle. J. Anim. Sci. 86, 1588-1597.

CSIRO, 2007. Nutrient Requirements of Domesticated Ruminants. Commonwealth Scientific and Industrial Research Organization, Collingwood, VIC.

Drew, K.R. and J.T. Reid, 1975. Compensatory growth in immature sheep. II. Some changes in the physical and chemical composition of sheep half-carcass following feed restriction and realimentation. J. Agric. Sci. 85, 205-213.

Ferrell, C.L. and T.G. Jenkins, 1998a. Body composition and energy utilization by steers of diverse genotypes fed a high-concentrate diet during the finishing period: I. Angus, Belgian Blue, Hereford, and Piemontese sires. J. Anim. Sci. 76, 637-646.

Ferrell, C.L. and T.G. Jenkins, 1998b. Body composition and energy utilization by steers of diverse genotypes fed a high-concentrate diet during the finishing period: II. Angus, Boran, Brahman, Hereford, and Tuli sires. J. Anim. Sci. 76, 647-657.

Fox, D.G. and J.R. Black, 1984. A system for predicting body composition and performance of growing cattle. J. Anim. Sci. 58, 725-739.

Fox, D.G., R.R. Johnson, R.L. Preston, T.R. Dockerty and E.W. Klosterman, 1972. Protein and energy utilization during compensatory growth in beef cattle. J. Anim. Sci. 34, 310-318.

Fox, D.G., L.O. Tedeschi, T.P. Tylutki, J.B. Russell, M.E. Van Amburgh, L.E. Chase, A.N. Pell and T.R. Overton, 2004. The Cornell Net Carbohydrate and Protein System model for evaluating herd nutrition and nutrient excretion. Anim. Feed Sci. Technol. 112, 29-78.

Garrett, W.N., 1980. Energy utilization by growing cattle as determined in 72 comparative slaughter experiments. In: Proceedings of Energy Metabolism, 8, Cambridge. Butterworths & Co: 3-7.

Garrett, W.N. and D.E. Johnson, 1983. Nutritional energetics of ruminants. J. Anim. Sci. 57, 478-497.

Garrett, W.N., J.H. Meyer and G.P. Lofgreen, 1959. The comparative energy requirements of sheep and cattle for maintenance and gain. J. Anim. Sci. 18, 528-547.

Geay, Y., 1984. Energy and protein utilization in growing cattle. J. Anim. Sci. 58, 766-778.

Graham, R.M., 1980. Variation in energy and nitrogen utilization by sheep between weaning and maturity. Austr. J. Agric. Res. 31, 335-345.

Guiroy, P.J., L.O. Tedeschi, D.G. Fox and J.P. Hutcheson, 2002. The effects of implant strategy on finished body weight of beef cattle. J. Anim. Sci. 80, 1791-1800.

Herd, R.M. and P.F. Arthur, 2009. Physiological basis for residual feed intake. J. Anim. Sci. 87, E64-71.

Hutcheson, J.P., D.E. Johnson, C.L. Gerken, J.B. Morgan and J.D. Tatum, 1997. Anabolic implant effects on visceral organ mass, chemical body composition, and estimated energetic efficiency in cloned (genetically identical) beef steers. J. Anim. Sci. 75, 2620-2626.

Joanning, S.W., D.E. Johnson and B.P. Barry, 1981. Nutrient digestibility depressions in corn silage-corn grain mixutres fed to steers. J. Anim. Sci. 53, 1095-1103.

Johnson, D.E., C.L. Ferrell and T.G. Jenkins, 2003. The history of energetic efficiency research: Where have we been and where are we going? J. Anim. Sci. 81 (Suppl. 1), E27-E38.

Johnson, D.E., E.M. Larson and M.J. Jarosz, 1997. Extrapolating from ME to NE: unintended consequences. In: Proceedings of Energy Metabolism of Farm Animals, 14, Newcastle. CAB International: 383-386.

Kleiber, M., 1961. The Fire of Life: An Introduction to Animal Energetics. John Wiley & Sons, Inc, New York.

Ku-Vera, K.C., N.A. Macleod and E.R. Orskov, 1989. Energy exchanges of cattle nourished by intragastric infusion of nutrients. In: Proceedings of Energy Metabolism of Farm Animals, 11, Lunteren, the Netherlands. Pudoc: 271-274.

Larson, E.M. and D.E. Johnson, 1997. Predicting net energy of feedstuffs for beef cattle. In: Proceedings of Energy Metabolism of Farm Animals, 14, Newcastle. CAB International: 355-358.

Leheska, J.M., J.L. Montgomery, C.R. Krehbiel, D.A. Yates, J.P. Hutcheson, W.T. Nichols, M. Streeter, J.R. Blanton, Jr. and M.F. Miller, 2009. Dietary zilpaterol hydrochloride. II. Carcass composition and meat palatability of beef cattle. J. Anim. Sci. 87, 1384-1393.

Lofgreen, G.P., 1965. A comparative slaughter technique for determining net energy values with beef cattle. In: Proceedings of Energy Metabolism, 3, Troon. Academic Press: 309-317.

Lofgreen, G.P. and W.N. Garrett, 1968. A system for expressing net energy requirements and feed values for growing and finishing beef cattle. J. Anim. Sci. 27, 793-806.

Merchen, N.R. 1993. Digestion, Absorption and Excretion in Ruminants. In: D.C. Church (ed.), The Ruminant Animal. Waveland Press, Inc., Prospect Heights, Illinois: 172-201.

Meyer, J.H. and W.N. Garrett, 1960. Efficiency of feed utilization. J. Anim. Sci. 26, 638-646.

Moe, P.W., 1981. Energy metabolism of dairy cattle. J. Dairy. Sci. 64, 1120-1139.

Montgomery, J.L., C. R. Krehbiel, J.J. Cranston, D.A. Yates, J.P. Hutcheson, W.T. Nichols, M.N. Streeter, D.T. Bechtol, E. Johnson, T. TerHune and T.H. Montgomery, 2009. Dietary zilpaterol hydrochloride. I. Feedlot performance and carcass traits of steers and heifers. J. Anim. Sci. 87, 1374-1383.

NRC, 1984. Nutrient Requirements of Beef Cattle (6th ed.). National Academy Press, Washington, DC.

NRC, 1994. Metabolic modifiers: Effects on the nutrient requirements of food-producing animals. National Academy Press, Washington, DC.

NRC, 2000. Nutrient requirements of beef cattle (updated 7th ed.). National Academy Press, Washington, DC.

NRC, 2007. Nutrient requirements of small ruminants: sheep, goats, cervids, and new world camelids (6th ed.). National Academy Press, Washington, DC.

Oddy, V.H. and R.M. Herd, 2001. Potential mechanisms for variation in efficiency of feed utilization in ruminants. In: Feed Efficiency in Beef Cattle; Proceedings of the Feed Efficiency Workshop, Armidale, Australia. CRC for Cattle and Beef Quality, C.J. Hawkins Homestead, University of New England: 30-34.

Ørskov, E.R., D.A. Grubb, J.S. Smith, A.J.F. Webster and W. Corrigall, 1979. Efficiency of utilization of volatile fatty acids for maintenance and energy retention by sheep. Br. J. Nutr. 41, 541-551.

Perry, T.C., D. G. Fox and D.H. Beermann, 1991. Effect of an implant of trenbolone acetate and estradiol on growth, feed efficiency, and carcass composition of Holstein and beef steers. J. Anim. Sci. 69, 4696-4702.

Peterson, L.A., E.E. Hatfield and U.S. Garrigus, 1973. Influence of concentration of dietary energy on protein needs of growing-finishing cattle. J. Anim. Sci. 36, 772-781.

Rattray, P.V., W.N. Garrett, N. Hinman and N.E. East, 1974. Energy cost of protein and fat deposition in sheep. J. Anim. Sci. 38, 378-382.

Rattray, P.V. and J.P. Joyce, 1976. Utilization of metabolizable energy for fat and protein deposition in sheep. NZ J Agric. Res. 19, 299-305.

Reid, J.T., O.D. White, R. Anrique and A. Fortin, 1980. Nutritional energetics of livestock: some present boundaries of knowledge and future research needs. J. Anim. Sci. 51, 1393-1415.

Richardson, E.C. and R.M. Herd, 2004. Biological basis for variation in residual feed intake in beef cattle. 2. Synthesis of results following divergent selection. Austr. J. Exp. Agric. 44, 431-440.

Rompala, R.E., S.D.M. Jones, J.G. Buchaman-Smith and H.S. Bayley, 1987. Utilization of dietary energy for protein and lipid deposition in late maturing cattle. In: Proceedings of Energy Metabolism, 10, Virginia. Rowman & Littlefield: 94-97.

Roux, C.Z., H.S. Hofmeyer and E. Jordaan, 1982. The problem of multicollinearity in the estimation of partial efficiencies of protein and fat by regression methods. In: Proceedings of the 9th Symposium on Energy Metabolism of Farm Animals, Lillehammer, Norway. EAAP: 138-140.

Ryan, W.J., 1990. Compensatory growth in cattle and sheep. Nutrition abstracts and reviews. Series B. 60, 653-664.

Ryan, W.J., I.H. Williams and R.J. Moir, 1993. Compensatory growth in sheep and cattle. II. Changes in body composition and tissue weights. Austr. J. Agric. Res. 44, 1623-1633.

Tedeschi, L.O., A. Cannas and D.G. Fox, 2010. A nutrition mathematical model to account for dietary supply and requirements of energy and nutrients for domesticated small ruminants: The development and evaluation of the Small Ruminant Nutrition System. Small Rum. Res. 89, 174-184.

Tedeschi, L.O., D.G. Fox and P.H. Doane, 2005a. Evaluation of the tabular feed energy and protein undegradability values of the National Research Council nutrient requirements of beef cattle. Prof. Anim. Scient. 21, 403-415.

Tedeschi, L.O., D.G. Fox and P.J. Guiroy, 2004. A decision support system to improve individual cattle management. 1. A mechanistic, dynamic model for animal growth. Ag. Syst. 79, 171-204.

Tedeschi, L.O., D.G. Fox, R.D. Sainz, L.G. Barioni, S.R. Medeiros and C. Boin, 2005b. Using mathematical models in ruminant nutrition. Scientia Agricola. 62, 76-91.

Tyrrell, H.F., 1987. Partitioning and utilization of energy by the growing animal. In: Proceedings of Energy Metabolism, 10, Virginia. Rowman & Littlefield: 134-137.

Tyrrell, H.F., P.J. Reynolds and P.W. Moe, 1979. Effect of diet on partial efficiency of acetate use for body tissue synthesis by mature cattle. J. Anim. Sci. 48, 598-606.

Vasconcelos, J.T., R.J. Rathmann, R.R. Reuter, J. Leibovich, J. P. McMeniman, K.E. Hales, T.L. Covey, M.F. Miller, W.T. Nichols and M.L. Galyean, 2008. Effects of duration of zilpaterol hydrochloride feeding and days on the finishing diet on feedlot cattle performance and carcass traits. J. Anim. Sci. 86, 2005-2015.

Weiss, W.P., H.R. Conrad and N.R. St. Pierre, 1992. A theoretically-based model for predicting total digestible nutrient values of forages and concentrates. Anim. Feed Sci. Technol. 39, 95-110.

The energetic cost of maintenance in ruminants: from classical to new concepts and prediction systems

A. Cannas[1], A.S. Atzori[1], I.A.M.A. Teixeira[2], R.D. Sainz[3] and J.W. Oltjen[3]
[1]Dipartimento di Scienze Zootecniche, University of Sassari, 07100 Sassari, Italy,
[2]Departamento de Zootecnia, Universidade Estadual Paulista, 14884-900, Jaboticabal, SP, Brazil
[3]Department of Animal Science, University of California, Davis, CA 95616, USA

Abstract

The cost of maintenance makes up a large part of total energy costs in ruminants. Metabolizable energy (ME) requirement for maintenance (ME_m) is the daily ME intake that exactly balances heat energy (HE). The net energy requirement for maintenance (NE_m) is estimated subtracting ME_m from the HE produced by the processing of the diet. Men cannot be directly measured experimentally and is estimated by measuring basal metabolism in fasted animals or by regression measuring the recovered energy in fed animals. ME_m and NE_m usually, but not always, are expressed in terms of $BW^{0.75}$. However, this scaling factor is substantially empirical and its exponent is often inadequate, especially for growing animals. ME_m estimated by different feeding systems (AFRC, CNCPS, CSIRO, INRA, NRC) were compared by using dairy cattle data. The comparison showed that these systems differ in the approaches used to estimate ME_m and for its quantification. The CSIRO system estimated the highest ME_m, mostly because it includes a correction factor to increase ME as the feeding level increases. Relative to CSIRO estimates, those of NRC, INRA, CNCPS, and AFRC were on average 0.92, 0.86, 0.84, and 0.78, respectively. ME_m is affected by the previous nutritional history of the animals. This phenomenon is best predicted by dynamic models, of which several have been published in the last decades. They are based either on energy flows or on nutrient flows. Some of the different approaches used were described and discussed.

Introduction

The energetic cost of maintenance makes the largest (between 60% and 80%) and most variable part of total energetic costs in ruminant herds (Ferrell and Jenkins, 1985; Cannas *et al.*, 2008). Energy requirements and, in particular, maintenance energy requirements (E_m) have been one of the main objective of research in animal nutrition and started more than 200 years ago with Lavoisier and La Place's experiments. For this reason, this review does not cover the historical development of the research on maintenance requirements, which has been detailed in many publications (e.g. Blaxter, 1989; Baldwin, 1995; Johnson *et al.*, 2003; Ferrell and Oltjen, 2008), but summarizes the information on which there is general consensus and describes current developments in this field. It also makes comparisons among the predictions of some of the most commonly used feeding systems for ruminants, to highlight some of the most significant theoretical and practical differences among them.

Classical definitions of the energetic cost of maintenance

The classic energetic partitioning scheme (NRC, 1981) defines metabolizable energy (ME) as the energy eaten by the animal minus the energy excreted as feces, urine and combustible gases. It can appear only in two forms, heat energy (HE) and energy of products formed, called recovered energy (RE), such as that in meat or milk: ME = HE + RE.

The MErequirement for maintenance (ME_m) is defined as the daily ME intake (MEI) that will exactly balance heat production and produce no loss or gain in body energy reserves (Webster, 1978). This occurs when RE=0 and MEI = HE. ME_m can be subdivided into two main components, defined here following the NRC (1981) nomenclature. The first one is basal metabolism (H_eE), which is the minimal HE needed to support vital processes in a healthy animal in a fasting (postabsorptive, from 48 to 144 hours after the last meal) and resting state and kept in a thermo-neutral environment (NRC, 1981). Basal metabolism is also denominated fasting metabolism, minimal metabolism, postabsorptive metabolism or basal metabolic rate. The second component of ME_m regards HE of animals fed at maintenance in normal living conditions, i.e HE associated with thermal regulation (H_cE), voluntary activity (H_jE), digestion, absorption and assimilation (H_dE), fermentation (H_fE), and waste formation and excretion (H_wE). The HE factors directly related to the processing of the diet by the animal (H_dE, H_fE, H_wE) form the so-called heat increment of feed (H_iE).

Energetic costs can be also defined in terms of net energy (NE), which is: NE = ME - H_iE (Webster, 1978). By analogy, the NE of maintenance (NE_m), i.e. the energy requirements for basal body function, should be defined as $NE_m = H_eE = ME_m - H_iE$ at maintenance (H_iE_m). However, NE_m cannot be directly measured experimentally even reducing H_iE_m to zero, because fasted animals use body energy to sustain life. Thus NE_m is not equivalent to H_eE but is always smaller than it, since $NE_m = k_b \times H_eE$, where k_b is the efficiency of converting energy contained in body reserves to energy useful for basal functions (Birkett and de Lange, 2001). Since k_b probably varies little, the ratio of NE_m and H_eE also varies little. Thus H_eE can be considered a good predictor of NE_m (Birkett and de Lange, 2001). In many publications, however, the two concepts are treated as synonymous (e.g. Webster, 1978; NRC, 1981). In fasting metabolism studies, H_eE is measured directly or indirectly and the efficiency of conversion of ME_m to H_eE (k_m) is estimated as the slope of the regression RE = a + b × MEI, considering two measurements of HE below RE=0. Based on the difference between NE_m and H_eE, as described before, k_m is a composite efficiency, which includes the efficiency of utilizing dietary energy for basal functions ($k_d = NE_m/ME_m$), and k_b, as described above (Blaxter, 1989; Birkett and de Lange, 2001). As said before, k_b probably varies little, whereas k_d, and as a consequence k_m, is markedly influenced by the diet.

Basal metabolism had been related in the past to body surface area, considered proportional to live weight$^{0.67}$ (Blaxter, 1989; Berman, 2003). This relationship between body surface area and body weight (BW) is controversial (Berman, 2003) and is still used only by the AFRC (1995) feeding systems for cattle (but not for sheep). Most current feeding systems express H_eE and ME_m as a function of $BW^{0.75}$ (Table 1). This scaling factor is derived from interspecies comparison of mature animals and might not apply for within species comparisons, especially for growing animals (Thonney *et al.*, 1976). Graham *et al.* (1974) found that in growing lambs H_eE was proportional to $BW^{0.60}$. Similarly, Freetly *et al.* (1995) related HE to $BW^{0.45}$ and $BW^{0.51}$ for growing Suffolk and Texel lambs, respectively, confirming the earlier observations of Thonney *et al.*(1976). On the other hand, Williams and Jenkins (2003) found that the scaling factor BW^1 was the most appropriate for mature ruminants. It appears that the mathematical association between BW and basal metabolism is empirical and thus it is affected by many different factors, such as species, age, body surface, body insulation, mass of visceral organs, and diets used. As more mechanistic and dynamic approaches and models are developed, the importance of scaling HE on BW will be reduced.

The estimation of E_m requires the measurement of HE. Details on the techniques that can be used are reported in many publications (e.g. Webster, 1978; Blaxter, 1989; Baldwin, 1995). In brief, HE can be measured by direct methods, such as direct or indirect calorimetry, or by indirect (which estimate HE as ME-RE) methods, such as comparative slaughter or carbon and nitrogen balance. ME_m can be also estimated as the MEI required to maintain BW stasis. Numerous studies have compared the estimates of maintenance energy requirements obtained with different methods. Very small differences (<0.5%) were observed comparing direct vs. indirect calorimetry measurements (Blaxter, 1989). Comparison of RE estimation by indirect calorimetry or by comparative slaughter also resulted in small differences both in fed monogastrics (Blaxter, 1989) and in fed ruminants (Steen *et al.*, 1997). In general, RE is slightly overestimated with calorimetry and conversely HE is underestimated by using comparative slaughter. Larger differences are observed comparing E_m estimates obtained with calorimetric fasting metabolism studies vs. values obtained in fed animals

Table 1. Basic requirements and corrections applied to estimate NE_m and ME_m requirements of dairy cows by five feeding systems.

	AFRC	CNCPS	CSIRO	INRA	NRC
Basic requirements					
NE: kJ/d [1]	$530 \times (BW/1.08)^{0.67}$	$305.4 \times SBW^{0.75}$	$343 - 304 \times BW^{0.75}$ (2)	$293 \times BW^{0.75}$	$334.7 \times BW^{0.75}$
Efficiency (NE/ME) [3]	$k_m = 0.35\ q_m + 0.503$	$k_m = k_l = 0.64$	$k_m = 0.02\ M/D + 0.5$	$k_l = 0.24\ q_m + 1.94$	$k_m = k_l = 0.644$
Reference unit	ME	ME	ME	UFL (7.11 MJ NEL)	NEL
Corrections					
Breed	no	yes[4]	*B. indicus:* 0.86	No	no
Gender	yes[5]	no	yes [5]	No	no
Age	no	no	1.0-0.84 (0-6 years)	no	no
Activity	yes	yes	yes	yes	yes
Previous nutrition	no[6]	no[6]	no	no [6]	no[6]
Cold stress	no	yes	yes	no	no[7]
temperature	no	yes	yes	no	no
humidity	no	yes	yes	no	no
wind	no	no	no	no	no
rain	no	no	yes	no	no
mud	no	yes	no	no	no
acclimatization	no	yes	yes	no	no
hide depth	no	yes	no	no	no
coat depth	no	yes	yes	no	no
BCS	no	yes	no	no	no
age	no	yes	no	no	no
Heat stress	no	yes [8]	no	no	no
Urea cost	no	yes	no	no	no
Production	no[9]	no[10]	yes[11]	no[9]	no

[1] No activity allowances included; BW = body weight, kg; SBW = shrunk body weight (0.96 full BW, kg). [2] Variation related to the age of the animals, the range reported refers to cows from 2 (highest value) to 6 year-old (the lowest); highest value for new born calves (364 kJ/kg $BW^{0.75}$). [3] q_m = ME/GE at maintenance feeding level; calculated assuming GE= 18.40 MJ; M/D = ME per kg of DM of the diet. [4] Specific multipliers are used for several non-dairy breeds. [5] 15% increase for males. [6] Considered only for heifers (CNCPS and NRC) and beef females (INRA). [7] Considered only for heifers. [8] Based on a current month's effective temperature index. [9] Corrections factors are included in total requirements to account for the effect of the feeding level on diet digestibility. [10] Values are increased for all but dairy breeds. [11] A value equal to 0.09 MEI is added to ME for maintenance.

with calorimetric or indirect methods (Webster, 1978). This probably occurs because fasted animals (always non-productive) decrease their visceral organ activity and their metabolic rates compared to fed animals (Webster, 1978). In this regard, Roux (2008) demonstrated that ME_m estimates obtained by regression of MEI on RE as protein and fat of fed pigs agreed with fasting HE values if in the latter were included the costs of body protein resynthesis.

Factors affecting maintenance and its prediction with current static feeding systems

The E_m is composed of basal metabolism and various other sources of HE. For this reason, many animal, dietary and environmental factors can affect E_m. These variables are discussed in detail in various publications (e.g Blaxter, 1989; CSIRO, 2007). Here they are discussed taking as example the variables used by some of the most recent feeding systems for dairy cattle. Thus, E_m was calculated following the methods proposed by the Agricultural and Food Research Council (AFRC, 1995; AFRC), Fox *et al.* (2004; Cornell Net Carbohydrate and Protein System, version 5.0, CNCPS), Commonwealth Scientific and Industrial Research Organisation (CSIRO, 2007; CSIRO), Institut National de la Recherche Agronomique (INRA, 2007; INRA), and the National Research Council (NRC, 2001; NRC). In particular, their submodels used to calculate ME_m and NE_m were evaluated and compared on a qualitative and quantitative basis. The comparison focused on dairy cows, because of their importance and the vast research carried out to predict their requirements. However, similar comparisons have been published on sheep (Wallach *et al.*, 1984; Cannas, 2004) and goats (Cannas *et al.*, 2008). The requirements were estimated using data collected from 22 published lactation studies with a total of 79 different feeding treatments. The studies used and more details on the methods used for the comparison are by Cannas (2000). The information used included milk yield (28.1±8.3 kg/d, range 16.0 to 41.7 kg/d) and quality, days of lactation, feed and diet composition, dry matter intake (19.5±3.6 kg/d, range 12.42 to 26.7 kg/d), and BW of cows (589±41 kg, range 513 to 717 kg). The animals used were mature Holstein (75 treatments) or Swedish red and white (4 treatments) lactating cows in thermoneutral conditions. For the calculation of E_m, several parameters related to the diet and to the efficiency of its utilization (dietary ME concentration and intake, undigested dry matter, etc.) had to be estimated. This was done by using the CNCPS nutrient supply sub-models (Fox *et al.*, 2004) for all systems and for each of the 79 dietary treatments, using the information found in each publication as inputs. The approaches used to calculate energy requirements differ among systems (Table 1). The NRC and INRA systems use a single NE unit for all functions (maintenance, milk production, pregnancy, etc.): NE for lactation (NEL) for NRC and UFL (forage unit for lactation, which corresponds to 7.113 MJ of NEL) for INRA. In contrast, AFRC, CSIRO and CNCPS calculate NE requirements for each function. These requirements are then converted to ME using the efficiency of conversion of NE to ME specific for each function, and then summed. For this reason, the systems could be compared only using a common unit, ME. The conversion of NE to ME was done using the coefficients of each system (Table 2). The AFRC and INRA systems require the calculation of the ratio of ME/gross energy in the diet (called metabolizability or q_m) and the CSIRO system requires the concentration of ME in the diet, both calculated at a maintenance feeding level. For this reason, ME was calculated at the feeding level required by each feeding system.

Energy requirements for maintenance are calculated by the systems considered by using a basic requirement, which is then adjusted for several factors (Table 1). Except for the AFRC, all systems calculate E_m as a function of $BW^{0.75}$. AFRC uses $(BW/1.08)^{0.67}$ as a basis, where 1.08 converts live weight to fasted BW. Interestingly, this system uses the scaling factor $BW^{0.75}$ for sheep. The basic requirement for cows ranges between 293 (INRA) and 342 (CSIRO) kJ of NE per kg of $BW^{0.75}$ (Table 1). The value for AFRC is, if recalculated on a $BW^{0.75}$ basis, close to that of INRA and CNCPS, being around 303 kJ of NE. The value of NRC is instead closer to that of CSIRO. This pattern is unexpected, given the different origin of these requirements. Values of AFRC and CSIRO originate from fasting metabolism studies conducted on dry cows in several laboratories (ARC, 1980;

Table 2. Maintenance energy requirements and energetic efficiency of conversion of metabolizable energy (ME) to net energy (NE) for maintenance calculated by different systems.

	AFRC	CNCPS	CSIRO	INRA	NRC
Net energy requirements (MJ/d)[1]					
Mean	40.96	39.04	53.22	38.49	43.01
St. dev.	2.01	2.09	4.06	2.01	2.30
Range	37.2-47.3	35.1-45.2	45.6-64.4	34.7-44.8	38.5-49.8
Metabolizable energy requirements (MJ/d)[1]					
Mean	56.32	60.67	72.43	61.97	66.78
St. dev.	3.26	3.26	5.44	3.60	3.60
Range	49.8-64.9	54.4-70.3	63.6-87.4	54.8-71.5	59.8-77.4
Efficiency of conversion of ME to NE for maintenance					
Mean	0.73	0.644	0.73	0.62	0.644
St. dev.	0.01	-	0.02	0.01	-
Range	0.69-0.75	-	0.69-0.76	0.59-0.64	-

[1] Including the minimal physical activity of animals kept in tie-stall barns in thermo neutral conditions.

CSIRO, 2007), those of CNCPS are from the comparative slaughter trials on beef cattle of Lofgreen and Garrett (1968), whereas those of NRC and INRA are from the energy balances carried out in respiration chambers on lactating dairy cattle by Moe *et al.* (1972) and van Es (1978), respectively.

The adjustments used to estimate NE_m and ME_m vary among feeding systems (Table 1). The CNCPS varies ME_m depending on the breed, giving specific values for 34 different breeds in range between 268 and 351 kJ of NE per kg of $BW^{0.75}$ (Fox *et al.*, 2004). The value of basic requirement reported in Table 1 is valid for all dairy breeds and, surprisingly, is intermediate between the lowest and the highest values. The CSIRO distinguishes only between *Bos taurus* and *Bos indicus*. The other systems do not correct for breed. Although not included in the present comparison, the NRC (1996) for beef does correct for breed effect.

A differentiation based on gender is included only in the AFRC and the CSIRO systems, which increase E_m by 15% for males, as done also by the NRC (1996) for beef. This correction was also included in the CNCPS for sheep (Cannas *et al.*, 2004) but later excluded based on evaluations with literature data (Cannas *et al.*, 2006; Tedeschi *et al.*, 2010) and on the findings of Ferrell *et al.* (1979) and Tedeschi *et al.* (2002). Freer *et al.* (1997) stated that experimental data could not support the gender adjustment of the CSIRO (1990) model. Despite this, CSIRO (2007) maintained this adjustment.

The effect of chronological age on E_m of cows is taken into consideration only by CSIRO, which decreases these requirements from 343 to 304 kJ of NE per kg of $BW^{0.75}$ as cows age from 2 to 6 years, based on the fact that the metabolic rate decreases as animals age (CSIRO, 2007). The importance of age on E_m is well known but it is addressed only in terms of additional growth requirements for first and second lactation cows by CNCPS, INRA, and NRC. A limitation of the CSIRO approach is that E_m are more associated with the physiological age (degree of maturity) than with the chronological age (Freetly *et al.*, 1995).

All the systems based their activity costs on the requirements originally proposed by ARC (1980). The AFRC, CNCPS, and CSIRO estimate the cost of physical activity as a function of the hours spent standing, number of position changes, and of the horizontal and vertical component of the distance walked by the cows. CSIRO adds as well the cost of eating and ruminating. NRC includes a fixed cost proportional to BW and the cost of walking in flat or sloped areas. INRA includes the activity of cows kept in tie-stalls in the basic requirements and increases E_m by 10% for cows in free-stalls and by 20% for grazing cows.

Cold stress, in general, results in an increase in E_m and feed intake. No adjustments for cold stress are considered by AFRC and INRA (Table 1). The NRC system adopts a cold stress model for heifers but not for cows. CSIRO and CNCPS adopt complex cold stress models (Table 1). They account for many environmental factors (temperature, wind, rain or mud, cold nights and clear sky, losses from heat radiation), energy intake, animal factors (BW, hair and hide depth, acclimatization). Only the CNCPS model adjusts E_m for heat stress, based on a current month's effective temperature index, which accounts for air temperature, relative humidity, wind speed, hours of sunlight. The CNCPS also includes in E_m the cost of production of urea from ammonia. Being a diet-related correction, this cost cannot be considered properly a E_m component and is not included in the values of Table 2.

The CSIRO systems accounts for the effect of production adding to E_m a factor proportional to total MEI (i.e. 0.09 MEI), (Table 1, note 11). Thus, as the level of production increases, with concomitant increase in MEI, E_m increases as well. The rationale is that an increase in feed intake associated with milk production changes both size and rate of metabolism of visceral organs and tissues (CSIRO, 2007). For similar reasons, earlier versions of the CNCPS (Fox *et al.*, 1992) increased basic E_m by 12% for dairy breeds and by 20% for lactating cows of all breeds but Holsteins. The peculiarity of the Holstein breed was justified as an effect of its high milk production on its hormonal status. The current CNCPS, however, does not apply any type of correction of E_m for dairy cows, either dry or lactating (Fox *et al.*, 2004). The AFRC and INRA systems increase the requirements of lactating animals to account for the reduction of digestibility that occurs at high feeding levels. Thus, AFRC and INRA calculate an artificial increase in the requirement to avoid corrections of the feed value of the diets. For this reason these adjustments were not considered in Table 2.

Maintenance NE and ME requirements were the highest for the CSIRO system and the lowest for the AFRC system (Table 2). Relative to the ME requirements of CSIRO, those of NRC, INRA, CNCPS, and AFRC were on average 0.92, 0.86, 0.84, and 0.78, respectively. CSIRO requirements had also the highest standard deviation (Table 2), due to the combined effect of age and MEI corrections on ME_m. The CSIRO E_m were thus greater than those obtained with the other systems. Indeed, with this model MEI related corrections on E_m averaged 19.29 MJ/d of ME (range 11.55 to 25.90 MJ/d). Without them, the CSIRO requirements would have been lower than those of the other systems (53.18 MJ/d of ME). The effect of production on E_m is still debated. Several experiments showed that increasing the level of production (i.e. the intake of energy), the weight and activity of the visceral organs dramatically increase (Ferrell *et al.*, 1986; Ortigues and Doreau, 1995). Even though these organs comprise a minor portion of the body (11 to 12%, including visceral fat, and 5 to 6% without it), they are responsible for 45% to 50% of total energy expenditures in ruminants at maintenance (Huntington and Reynolds, 1987). For this reason, the CSIRO approach, based on a continuous increase of E_m as MEI increases, seems to be biologically appropriate. However, the extent of the CSIRO correction (on average 27% of the total ME_m) seems to be quite large, probably influenced by the fact that the CSIRO system is the only one among those compared that does not account for the reduction of diet digestibility that occurs at high feeding levels. The low requirements of the AFRC and CSIRO, when the correction factor proportional to MEI is not included, and their high efficiency of conversion of ME to NE (Table 2), with subsequent low ME requirements, are probably due to the fact that this system based its estimates on fasting metabolism studies on non-producing animals, which involve decreased metabolic rates (Webster, 1978). Dawson and Steen (1998) reviewed estimates of ME_m, based on linear regressions of RE (measured by calorimetry) on MEI, and found that they were 34% higher in steers (75 estimates) and 32% higher in lambs (56 estimates) than the values predicted by AFRC. The effect of previous nutrition on E_m is not considered for dairy cows by any of the feeding systems evaluated (Table 1). The CNCPS increases NE_m, but only for growing animals, proportionally to their BCS (5% increase or decrease for each BCS increase, scale 1 to 9, above or below BCS 5), to account for their previous plane of nutrition. Similar corrections (6.8% increase or decrease for each BCS variation, scale 0 to 5, above or below BCS 2.5) are applied to beef cows by INRA. They are used to account for the lower visceral organs mass and activity of underfed animals (low BCS) compared to well-fed animals (high BCS). However, the relationship between body fatness and E_m is controversial.

Reid and Robb (1971) and McNiven (1984) stated that in sheep body fatness did not affect E_m. In contrast, Birnie *et al.* (2000) found that H_eE was significantly higher, on a $BW^{0.75}$ basis, for cows with BCS <2 than for cows with BCS ³4.5. They suggested that in low BCS cows a large percentage of daily energy supply is probably met by catabolism of body proteins, energetically very expensive, causing increases of HE. Wagner *et al.* (1988) found that cows in either a thin (BCS=3) condition or a fat (BCS=7) condition required less (4.4% and 8.9%) E_m than cows in moderate (CS=5) condition. Ferrell *et al.* (1986) observed that in lambs H_eE was altered by previous nutritional treatments. H_eE was not associated with differences in body composition but it was associated with the rate of BW gain, for the strict association of H_eE with visceral organ weights, which in turn were not correlated to BW or its composition (Ferrell *et al.*, 1986). Thus, body fatness and BCS at one point in time cannot be considered reliable indicators of previous nutrition, leading to the need of approaching it with dynamic models (Ferrel and Oltjen, 2008).

Prediction of the cost of maintenance with dynamic models

The interaction between previous plane of nutrition and current nutritional status dynamically affects E_m. This is particularly evident in dairy cows in early lactation, as already discussed in the previous section. Another situation in which previous nutrition clearly affects E_m is when compensatory gain occurs. Compensatory gain, i.e. the more rapid and efficient growth of animals following a period of feed restriction, is a complex phenomenon, including changes in gut fill and feed intake, altered composition of gain, and reduced maintenance energy expenditures. It is now clear that maintenance energy expenditures are reduced, by 10% up to as much as 50%, during restriction and for some time after re-feeding. The duration of reduced maintenance (i.e. compensatory gain) is typically 60 to 90 days. Metabolic basis for changes include altered rates of ion pumping and metabolite cycling (Milligan and Summers, 1986) and altered size and metabolic rate of visceral organs (Sainz *et al.*, 1995). The prediction of variable E_m, such as in the case of compensatory growth, is best achieved with dynamic models, many of which have been published in the last decades. They are based either on energy flows or on nutrient flows, with several cases of integration between the two. Some of the different approaches used in these models will be discussed.

Oltjen *et al.* (2006) developed a model to dynamically estimate E_m, growth, and composition of growth based on visceral protein, muscle protein and fat pools. In this model, the energy driving the growth of muscle and viscera is given by the term NE of gain, considered equal to MEI - HE, where HE was expressed as the sum of HE for maintenance and for gain: $HE = HE_{maint} + HE_{gain}$. In the first version of the model, HE_{maint} was estimated following the approach of the CSIRO (1990) model but with a variable coefficient on empty BW (EBW): $HE_{maint} = a_t\ EBW^{0.75} + 0.09\ MEI$, where the factor 0.09 MEI was the same of CSIRO (1990), previously discussed, and accounted for the effect of MEI on metabolic active tissues and organs. The other factor was $a_t = a_0\ (1 + b\ (MEI_t/MEI_0 - 1)(1-e^{-t/\tau}))$, where t is time (d), b and τ are constants, MEI_0 and α_0 are original values of MEI and of the maintenance coefficient, respectively. The variable a_t results in a lag in change of HE_{maint} after intake changes from MEI_0 to MEI_t. Fit of the Ferrell *et al.* (1986) data showed that the double correction for variable maintenance was not necessary because the coefficient 0.09 on MEI was not different from zero. Therefore, HE_{maint} can be calculated as: $HE_{maint} = \alpha_t\ EBW^{0.75}$. A different approach was used by Williams and Jenkins (2003), who proposed a dynamic model for growing and mature animals. In their model, heat increment of feeding includes components attributable to maintenance, H_iE_m, and production, H_iE_g. The latter is further partitioned in two components. The first is the heat increment to support metabolism (H_iE_v), which represents costs associated with elevation of vital functions necessary for the synthesis of the RE and should be associated to E_m. In classical energy systems H_iE_v is usually charged against the productive processes. The second is the heat increment of RE (H_iE_r), which represents costs directly involved in synthesis of products. Thus in their model MEI = [$ME_m + H_iE_v$] + [H_iE_r + RE], in which the terms in the first parenthesis indicate total maintenance and those in the second the ME for gain. Williams and Jenkins (2003) considered ME_m constant for a particular diet, with average values around 125 kJ/kg of full BW, estimated as the MEI of animals with BW stasis for long periods. The component H_iE_v, also expressed on a full BW basis, was considered linearly proportional to the level of nutrition and was estimated by difference regressing the equation above to experimental data. With this approach Williams and Jenkins (2003) were able to model the reduction of HE during growth experimentally described by Freetly *et al.* (1995). In addition, they suggested that H_iE_v could be used to model the impact of previous level of nutrition on HE, with an equation that described the delayed response of HE with a change in the level of feeding. The integrated model was fully evaluated and implemented in a decision aid software for beef producers.

The two models described used energy flows to predict variable maintenance requirements. However, they did not directly consider the changes in metabolically active organs and tissues caused by different planes of nutrition, which have been associated to altered rates of ion pumping and metabolite cycling, and altered size and metabolic rate of visceral organs (Milligan and Summers, 1986). Thus, Oltjen *et al.* (2006) proposed a model alternative to that previously described to predict HE, in which multiple regression equations including muscle protein (m), visceral protein (v), and the change in muscle (dm/dt) and visceral protein (dv/dt) were used: $HE = b_1\,m + b_2\,v + b_3\,dm/dt + b_4\,dv/dt$ where b_1 and b_2 were 1.023±0.333 and 10.54±3.40 MJ/(d kg), respectively, and b_3 and b_4 were 60.93±13.80 and 282.7±91.6 MJ/kg, respectively. HE per unit protein mass of viscera was about ten times that of muscle. Since viscera respond faster than muscle to changing energy intake, this equation results in a more dynamic HE, in which HE_{maint} and HE_{gain} are aggregated. Thus, this model integrates energy and organ variations to define a flow of energy. An evolution of this approach is the dynamic model for growing beef cattle of Hoch and Agabriel (2004). They focused on anabolism and catabolism of protein and fat rather than on the estimate of maintenance and growth as separate phases. In particular, the synthesis of proteins was based on Michelis-Menten type equations which integrated biochemical flows with classic energy concepts. Maintenance was not explicitly included in the model but could be calculated indirectly, integrating the rate of change of the energetic content of protein and lipids in the body when RE was zero. Garcia *et al.* (2008) compared the Hoch and Agabriel (2004) model with the Davis Growth Model (DGM; Oltjen *et al.*, 1986), using data sets with fluctuating planes of nutrition. Both models predicted body protein well, but the DGM was superior at predicting body fat in compensating animals, possibly because E_m is represented explicitly.

The energy partition schemes used so far are of course arbitrary. Specific physiological functions contribute to the animal's HE costs. The main ones that contribute to H_eE may be grouped as: service functions at the organism level, ion transport at the cellular level, cell repair at the moleculal level. Service functions include work done by the heart, kidneys, and central nervous and respiratory systems, which account for, respectively, about 10%, 8%, 15 to 20% and 7% of H_eE, totaling 40 to 45% of H_eE (Baldwin, 1995). Ion transport, mainly through Na^+, K^+-ATPase (15 to 20%) but including also transport of Ca^{2+} and others (5 to 7%), accounts for another 20 to 30% of H_eE (Baldwin, 1995). Finally, cell repair, which includes not only cell death and replacement but also macromolecule (mainly protein) turnover within cells, accounts for another 25 to 35% of H_eE (Baldwin, 1995). Therefore, these processes may account for 80 to 100% of endogenous HE, which in turn accounts for 70 to 90% of ME_m, thus leaving H_iE a range from 10 to 30% of ME_m. The major component of H_iE, H_dE, is the energy cost associated with the fed state and includes (with approximate % of

ME_m): costs of ingestion (2 to 3%), heat of fermentation (1.5 to 6%), costs of digestive secretions (5%), costs of bond breakage (0 to 0.5%), costs of absorption (0.3 to 1.5%), costs of assimilation (3 to 12.5%), and changes in metabolic activity due to intake (Webster, 1978; Baldwin, 1995; Baldwin and Sainz, 1995). Moreover, the efficiency of ATP resynthesis through oxidative phosphorylation varies due to proton leakage across the mitochondrial membrane (Brand, 2005), introducing potentially large variations in E_m. These details are not easily accounted for in energy flow models, even though remarkable efforts in this direction were carried out by Vetharaniam *et al.* (2001). For this reason, detailed whole animal biochemical models have been developed (e.g. Baldwin, 1995; France *et al.*, 1987; Danfaer *et al.*, 2006). They differ in the level of aggregation considered. Some, e.g. the model of Baldwin (1995), are very detailed and have the objective of increasing the understanding of the biology underlying production processes more than giving accurate predictions. Others (e.g. France *et al.*, 1987, for growing animals; Danfaer *et al.*, 2006, for dairy cows) have an higher degree of aggregation and are more suitable for practical applications. That is why they still retain empirical relationships to estimate fasting metabolism, whereas they introduce biochemically based energy exchanges at tissue and organ level to estimate the effects of their activity and variations on ME_m (e.g. France *et al.*, 1987; Danfaer *et al.*, 2006).

Future developments

Van Milgen and Noblet (2003) suggested that 'maintenance requirements for growing animals remain a fuzzy concept, for which there is no currently a suitable alternative'. As a matter of fact, future research should focus on defining the priorities to improve their estimation or even replacement. Johnson *et al.* (2003) suggested that a priority should be the assessment of individual animal differences in energetic efficiency, and especially E_m, by finding variables that would allow the identification of the specific efficiency of individuals. Possible candidates were considered residual feed intake, HE from heart rate, and, above all, developments in microarray technology to identify key 'genes' controlling maintenance costs (Johnson *et al.*, 2003). Patterns of expressed genes could also reveal individuals particularly efficient for energy utilization. The assessment of individual animal differences can be also based on the utilization of sensor technology, which has increased the amount of information available by continuously monitoring the animals, their environment, and their performances. Systems are already available, and many others will be soon ready, for identifying, tracking and weighing animals, monitoring their body temperature and heart rate, assessing BCS, etc. However, the utilization of this information, to assess individual animal differences and to dynamically predict their requirements and their performances, would require a change in the approach taken when developing nutritional models. Most mechanistic and dynamic models still use the few and mostly static inputs (e.g. intake, BW, milk yield and composition) used by the oldest empirical models. This led to very complex mechanistic models, in which very large numbers of parameters need to be estimated, limiting their flexibility when their predictions are extrapolated to different conditions and inputs (Birkett and de Lange, 2001). If, in future developments, nutritional models will be able to consider and integrate the massive and continuous information that can be obtained with the application electronic and sensor technologies, 'time' could become an explicit variable of the models. This could produce a great reduction in their complexity, because dozens of parameters that are usually estimated could be measured in real time. Detailed biochemical models would still be needed to better understand the underlying metabolic processes and to define the rules which control the input-output responses at whole-animal level.

Acknowledgements

We are grateful to Ana H.D. Francesconi for the technical revision and to Luis O. Tedeschi and Corrado Dimauro for the useful scientific comments.

References

AFRC, 1995. Energy and protein requirements of ruminants. CAB International, Wallingford, Oxon, UK, 159 pp.

ARC, 1980. The nutrient requirements of ruminant livestock. Tech. Rev. Agric. Res. Council Working Party. Commonwealth Agricultural Bureaux, Farnham Royal, UK, 351 pp.

Baldwin, R.L., 1995. Modelling ruminant digestion and metabolism. Chapman & Hall Ltd., London, UK. 578 pp.

Baldwin, R.L. and R.D. Sainz, 1995. Energy partitioning and modeling in animal nutrition. Annu. Rev. Nutr. 15, 191-211.

Berman, A., 2003. Effects of body surface are estimates on predicted energy requirements and heat stress. J. Dairy Sci. 86, 3605-3610.

Birnie J.W., R.E. Agnew and F.J. Gordon, 2000. The influence of body condition on the fasting energy metabolism of nonpregnant, nonlactating dairy cows. J. Dairy Sci. 83, 1217-1223.

Birkett, S. and K. de Lange, 2001. Limitations of conventional models and a conceptual framework for a nutrient flow representation of energy utilization by animals. Br. J. Nutr. 86, 647-659.

Blaxter, K., 1989. Energy metabolism in animals and man. Cambridge University Press, Cambridge, UK, 336 pp.

Brand, M.D., 2005. The efficiency and plasticity of mitochondrial energy transduction. Biochem. Soc. Trans. 33, 897-904.

Cannas, A., 2000. Sheep and cattle nutrient requirement systems, ruminal turnover, and adaptation of the Cornell Net Carbohydrate and Protein System to sheep. Ph.D. dissertation, Cornell University, Ithaca, NY, USA, 337 pp.

Cannas, A., 2004. Energy and protein requirements. In: Pulina G. (ed.), Dairy sheep nutrition. CAB International, Wallingford, Oxon, UK: 31-49.

Cannas, A., A.S. Atzori, F. Boe, and I.A.M.A. Teixeira, 2008. Energy and protein requirements of goats. In: Cannas A. and G. Pulina (eds.), Dairy goats feeding and nutrition. CAB International, Wallingford, UK: 118-146.

Cannas, A., L.O. Tedeschi, A.S. Atzori, and D.G. Fox, 2006. Prediction of energy requirement for growing sheep with the Cornell Net Carbohydrate and Protein System. In: Kebreab, E., Dijkstra, J., Bannink, A., Gerrits, W.J.J. and J. France (eds.), Nutrient digestion and utilization in farm animals: modelling approaches. CAB International, Wallingford, UK: 99-113.

Cannas, A., L.O. Tedeschi, D.G. Fox, A.N. Pell, and P.J. Van Soest, 2004. A mechanistic model to predict nutrient requirements and feed biological values for sheep. J. Anim. Sci. 82, 149-169.

CSIRO, 1990. Feeding standards for Australian livestock. Ruminants. CSIRO Publications, East Melbourne, Australia, 266 pp.

CSIRO, 2007. Nutrient requirements of domesticated ruminants. Freer, M., Dove, H, and J.V. Nolan (editors), CSIRO Publishing, Collingwood, Australia, 270 pp..

Danfaer, A., P. Hutanen, P. Uden, J. Sveinbjornsson, and H. Volden, 2006. The Nordic dairy cow model Karoline – description. In: Kebreab, E., Dijkstra, J., Bannink, A., Gerrits, W.J.J. and J. France (eds.), Nutrient digestion and utilization in farm animals: modelling approaches. CAB International, Wallingford, UK: 383-406.

Dawson, L.E.R., and R.W.J. Steen, 1998. Estimation of maintenance energy requirements of beef cattle and sheep. J. Agric. Sci, Camb., 131, 477-485.

Ferrell, C.L., J.D. Crouse, R.A. Field, and J.L. Chant, 1979. Effects of sex, diet and stage of growth upon energy utilization by lambs. J. Anim. Sci. 49, 790-801.

Ferrell, C.L. and T.G. Jenkins, 1985. Cow type and nutritional environment: nutritional aspects. J. Anim. Sci. 61, 725-741.

Ferrell, C.L., L.J. Koong, and J.A. Nienaber, 1986. Effect of previous nutrition on body composition and maintenance energy costs of growing lambs. Br. J. Nutr. 56, 595-605.

Ferrell, C.L. and J.W. Oltjen, 2008. ASAS centennial paper: net energy systems for beef cattle – Concepts, application, and future development. J. Anim. Sci. 86, 2779-2794.

Fox, D.G., C.J. Sniffen, J.D. O'Connor, J.B. Russell and P.J. Van Soest, 1992. A net carbohydrate and protein system for evaluating cattle diets: III. Cattle requirements and diet adequacy. J. Anim. Sci. 70, 3578-3596.

Fox, D.G., L.O. Tedeschi, T.P. Tylutki, J.B. Russell, M.E. Van Amburgh, L.E. Chase, A.N. Pell, and T.R. Overton, 2004. The Cornell Net Carbohydrate and Protein System model for evaluating herd nutrition and nutrient excretion. Anim. Feed Sci. Technol. 112, 29-78.

France, J., M. Gill, J.H.M. Thornley and P. England, 1987. A model of nutrient utilization and body composition in beef cattle. Anim. Prod. 44, 371-385.

Freer, M., A.D. Moore and J.R., Donnelly, 1997. GRAZPLAN: decision support systems for Australian grazing enterprises. II. The animal biology model for feed intake, production and reproduction and the GrazFeed DSS. Agric. Syst. 54, 77-126.

Freetly, H.C., J.A. Nienaber, K.A. Leymaster and T.G. Jenkins, 1995. Relationships among heat production, body weight, and age in Suffolk and Texel ewes. J. Anim. Sci. 73, 1030-1037.

Garcia, F., R.D. Sainz, J. Agabriel, L.G. Barioni and J.W. Oltjen, 2008. Comparative analysis of two dynamic mechanistic models of beef cattle growth. Anim. Feed Sci. Technol. 143, 220-241.

Graham, N. McC., T.W. Searle, and D.A. Griffiths, 1974. Basal metabolic rate in lambs and young sheep. Aust. J. Agric. Res. 25, 957-971.

Hoch, T. and J. Agabriel, 2004. A mechanistic dynamic model to estimate beef cattle growth and body composition: 2. Model evaluation. Agric. Syst. 81, 17-35.

Huntington, G.B. and C.K. Reynolds, 1987. Oxygen consumption and metabolite flux of bovine portal-drained viscera and liver. J. Nutr. 117, 1167-1173.

INRA, 2007. Alimentation des bovins ovins et caprins. Besoins des animaux – Valeures des aliments. Éditions Quæ, Versailles, France, 307 pp.

Johnson, D.E, C.L. Ferrell, and T.G. Jenkins, 2003. The history of energetic efficiency research: where have been and where are we going? J. Anim. Sci. 81 (suppl. 1), E27-E38.

Lofgreen, G.P. and W.N. Garrett, 1968. A system for expressing net energy requirements and feed values for growing and finishing beef cattle. J. Anim. Sci. 27, 793.

McNiven M., 1984. The effect of body fatness on energetic efficiency and fasting heat production in sheep. Br. J. Nutr. 51, 297-304.

Milligan, L.P. and M. Summers, 1986. The biological basis for maintenance and its relevance to assessing responses to nutrients. Proc. Nutr. Soc. 45:185-193.

Moe, P.W., W.P. Flatt and H.F. Tyrrel, 1972. The net energy value of feeds for lactation. J. Dairy Sci. 55, 945-958.

NRC, 1981. Nutritional energetics of domestic animals and glossary of energy terms. National Academy Press, Washington, DC, 54 pp.

NRC, 1996. Nutrient requirements of beef cattle (7th ed.). National Academy Press, Washington, DC, 242 pp.

NRC, 2001. Nutrient requirements of dairy cattle (7th ed). National Academy Press, Washington DC, 381 pp.

Oltjen, J.W., A.C. Bywater, R.L. Baldwin and W.N. Garrett. 1986. Development of a dynamic model of beef cattle growth and composition. J. Anim. Sci. 62, 86-97.

Oltjen, J.W., R.D. Sainz, A.B. Pleasants, T.K. Soboleva and V.H. Oddy, 2006. Representation of fat and protein gain at low levels of growth and improved prediction of variable maintenance requirement in a ruminant growth and composition model. In: Kebreab, E., Dijkstra, J., Bannink, A., Gerrits, W.J.J. and J. France (eds.), Nutrient digestion and utilization in farm animals: modelling approaches. CAB International, Wallingford, UK: 144-159.

Ortigues, I. and M. Doreau, 1995. Responses of the splanchnic tissues of ruminants to changes in intake: absorption of digestion end products, tissue mass, metabolic activity and implications to whole animal energy metabolism. Ann. Zootech. 44, 321-346.

Reid J.T. and J. Robb, 1971. Relationship of body composition to energy intake and energetic efficiency. J. Dairy Sci. 54, 533-564.

Roux, C.Z., 2008. Use of theoretical efficiencies of protein and fat synthesis to calculate energy requirements for growth in pigs. Br. J. Nutr. 101, 895-901.

Sainz, R.D., F. De la Torre and J. W. Oltjen, 1995. Compensatory growth and carcass quality in growth-restricted and refed beef steers. J. Anim. Sci. 73, 2971-2979.

Steen, R.W.J., L.E.R. Dawson, D.E. Kirkpatrick, and S.D. Johnston, 1997. Estimation of maintenance energy requirements and efficiency of utilization of metabolizable energy for growth in lambs by indirect calorimetry and comparative slaughter. In: McCracken, K., Unsworth, E.F, and A.R.G. Wylie (eds.), Energy metabolsim of farm animals. CAB Inbternationa, Wallingford, UK: 351-354.

Tedeschi, L.O., C. Boin, D.G. Fox, P.R. Leme, G.F. Alleoni and D.P.D. Lanna, 2002. Energy requirement for maintenance and growth of Nellore bulls and steers fed high-forage diets. J. Anim. Sci. 80, 1671-1682.

Tedeschi, L.O., A. Cannas, and D.G. Fox. 2010. A nutrition mathematical model to account for dietary supply and requirements of energy and other nutrients for domesticated small ruminants: the development and evaluation of the Small Ruminant Nutrition System. Small Rumin. Res. 89, 174-184.

Thonney, M.L., R.W. Touchberry, R.D. Goodrich, and J.C. Meiske. 1976. Intraspecies relationship between fasting heat production and body weight: A reevaluation of W^{75}. J Anim. Sci. 43, 692-704.

Van Es, A.J.H. 1978. Feed evaluation for ruminants. I. The system in use from May 1977 onwards in the Netherlands. Livest. Prod. Sci. 5, 331-346.

Van Milgen, J. and J. Noblet, 2003. Partitioning of energy intake to heat, protein, and fat in growing pigs. J. Anim. Sci. 81, E 86-93.

Vetharaniam I., D.G. McCall, P.F. Fennessy and D.J. Garrick, 2001. A model of mammalian energetics and growth: model development. Agric. Syst. 68, 55-68.

Wagner, J.J., K.S. Lusby, J.W. Oltjen, J. Rakestraw, R.P. Wettemann and L.E. Walters, 1988. Carcass composition in mature Hereford cows: Estimation and effect on daily metabolizable energy requirement during winter. J. Anim. Sci. 66, 603-612.

Wallach, D., J.M. Elsen and J.L. Charpenteau, 1984. Maintenance energy requirements of grazing sheep: a detailed comparison between models. Agric. Systems, 15, 1-22.

Webster, A.J.F., 1978. Energy metabolism and requirements. In: Church D.C. (editor), Digestive physiology and nutrition of ruminants. 2nd ed. O&B Books, Corvallis, Oregon, USA: 210-229.

Williams, C.B. and T.G. Jenkins, 2003. A dynamic model of metabolizable energy utilization in growing and mature cattle. I. Metabolizable energy utilization for maintenance and support metabolism. J. Anim. Sci. 81, 1371-1381.

Prediction of partial efficiency of use of metabolizable energy to net energy for gain and maintenance

M.I. Marcondes[1,2], L.O. Tedeschi[1], S. de Campos Valadares Filho[2], M.P. Gionbelli[2] and M.L. Chizzotti[3]
[1]*Texas A&M University, College Station, TX 77843, USA; marcosinaciomarcondes@gmail.com*
[2]*Universidade Federal de Viçosa, Viçosa, MG 36570, Brazil*
[3]*Universidade Federal de Lavras, Brazil*

Introduction

According to CSIRO (2007) the use of a fixed partial efficiency of use of metabolizable energy (ME) to net energy for growth (NEg) will overestimate the energy gain when this gain is formed mostly by protein gain. This happens because the efficiency of energy deposit as protein is lower than the energy deposit as fat (ARC, 1980). Similarly, even though the partial efficiency of use of ME to net energy for maintenance (NEm) was also assumed to be fixed by Garret (1980), Frisch and Vercoe (1977) have shown that maintenance requirements might vary in different rates of gain. This study was conducted to predict the partial efficiency of use of metabolizable energy to NEg (kg) and to NEm (km).

Material and methods

Twenty six comparative slaughter studies were gathered (N=752 animals) and coded by sex (431 bulls, 204 steers, and 117 heifers) and breed (447 Nellore and 305 *Bos indicus* and *B. taurus* crossbreds).The retained energy (RE) was regressed on ME intake (MEI) available for gain within each experiment to obtain the kg using orthogonal regressions. The computed kg was regressed on RE as protein (REp) according to the following equation: a/(b + REp). Sex and breed effects were not tested because of the small number of experiments. The data were also used to develop an equation to predict REp as a function of RE in the empty body gain (REc). A random coefficients model, including sex and breed as fixed effects and studies as random effects, was used to fit the model: $REp = a \times REc^{b}$.

The NEm was estimated as the intercept of the model: $HP = a \times e^{b \times MEI}$ and the ME for maintenance (MEm) was computed when MEI = HP. The km was calculated as NEm/MEm. For each study, one km was estimated within breed or sex. Thus, thirty eight points of km were used. The km was then modeled using a stepwise procedure in a multiple regression, including average daily gain, empty body gain (EBG), empty body weight (EBW), $EBW^{0.75}$, kg, and energy content (Mcal) in the EBW. The selected variables were included in a random coefficients model assuming a random variation to the effect of the study to test breed and sex effects.

Results and discussion

The overall equation to predict kg was: 0.327±0.142/(0.539±0.317 + REp), with an r^2 of 0.963. The meta-analysis of REp indicated no breed effects on the parameters a (P=0.729) and b (P=0.693), and no sex effect on parameters a (P=0.610) and b (P=0.266). The Akaike's Information Criteria was -1670 and the r^2 was 0.975. Therefore, one overall equation was devised to compute REp ($1.1404 \pm 0.033 \times ERc^{-1.137 \pm 0.035}$). Because our database consisted of Nellore and *B. indicus* and *B. taurus* crossbreds, it is necessary to further evaluate differences between *Bos taurus* and *Bos indicus* regarding kg and REp. The principal nutrient requirements systems (ARC, 1980; NRC, 2000; CSIRO, 2007) use diet energy density to predict kg. However this method may not be adequate in few occasions, especially when the diet is not known by those who want to predict the nutrient requirements of the animal. Johnson *et al.* (1997) also reported that in high concentration of metabolizable energy ([ME]), the MEg could be underestimated. The authors suggested yet, that this model might have problems in diets, or feeds, with digestible energy below 0.40. No correlation was found between kg and [ME] (P=0.636) using the database of this study.

The equation to predict km was 0.513±0.024 + 0.173±0.061 × kg + a × EBG, where a is either 0.100(±0.021) or 0.073(±0.021) for *Bos indicus* or crossbreds, respectively. No sex effect was observed in the intercept (P=0.564), kg coefficient (P=0.357) and EBG coefficient (P=0.064), and there were also no breed effects on intercept (P=0.232) and kg coefficient (P=0.054). However, breed affected the EBG (P=0.016), making B. indicus animals more efficient than the crossbreds. Even though the NRC (2000) suggests that *Bos indicus* might have 10% less NEm than *Bos taurus*, our results indicated that this difference may not be exactly in the NEm but in the km in which *Bos indicus* may be more efficient than *Bos taurus*. Taylor *et al.* (1986) and Carstens *et al.* (1989) reported that genetic differences may affect km rather than NEm.

Based on our study, we concluded that REc is a good predictor of REp and that REp may estimate kg satisfactorily. In addition, km can be predicted from kg and EBG, and breed may affect km.

References

ARC, 1980. The nutrient requirements of ruminant livestock. Agricultural Research Council. The Gresham Press, London.

Carstens, G.E., D.E. Johnson, K.A. Johnson, S.K. Hotovy and T.J. Szymanski, 1989. Genetic variation in energy expenditures of monozygous twin beef cattle at 9 and 20 months of age. In: Proceedings of Energy Metabolism of Farm Animals, 11, Lunteren: 312-315.

CSIRO, 2007. Nutrient requirements of domesticated ruminants. Commonwealth Scientific and Industrial Research Organization, Collingwood, VIC.

Frisch, J.E. and J.E. Vercoe, 1977. Food intake, eating rate, weight gains, metabolic rate and efficiency of feed utilization in *Bos taurus* and *Bos indicus* crossbred cattle. Anim. Prod. 25, 343-358.

Garrett, W. N. 1980. Factors influencing energetic efficiency of beef production. J. Anim. Sci. 51, 1434-1440.

Johnson, D.E., E.M. Larson and M.J. Jarosz. 1997. Extrapolating from ME to NE: unintended consequences. In: Proceedings of Energy Metabolism of Farm Animals, 14, Newcastle: 383-386.

NRC, 2000. Nutrient requirements of beef cattle. updated 7th ed. National Academy Press, Washington, DC.

Taylor, S.C.S., R.S. Thiessen and J. Murray, 1986. Inter-breed relationship of maintenance efficiency to milk yield in cattle. Anim. Prod. 43, 37-61.

Analysis of energy balance data from lactating dairy cows

A.B. Strathe[1], J. Dijkstra[2], J. France[3], M.S. Dhanoa[4], S. Lopez[5] and E. Kebreab[1]
[1]Department of Animal Science, University of California, Davis, USA
[2]Animal Nutrition Group, Wageningen University, Wageningen, the Netherlands
[3]Department of Animal and Poultry Science, University of Guelph, Canada
[4]North Wyke Research, Okehampton, Devon, United Kingdom
[5]Instituto de Ganadería de Montaña (CSIC-ULE), Departamento de Producción Animal, Universidad de León, Spain

Introduction

Key parameters of energy evaluation systems for dairy cows are net and metabolizable energy (ME) for maintenance (NE_M, ME_M), the efficiency of utilizing ME intake for milk production (k_l), and for growth (k_g), and the efficiency of utilizing body stores for milk production (k_t). These parameters may be constant or a function of other variables such as metabolizability, i.e. feed quality (AFRC, 1993). Recently, Tolkamp and Kyrizakis (2009) proposed a constant k_l regardless of feed quality, but did not provide data in support of this premise. A meta-analytical approach is best suited to evaluate the hypothesis of a constant k_l as the conclusion reached is valid across studies. Kebreab *et al.* (2003) presented an analysis of a large dataset on energy metabolism in lactating dairy cows and compared alternate functional forms, but did not investigate the effects of metabolizability on efficiency of utilizing ME intake for milk production. The aim of the present study was to investigate the effects of diet quality on efficiency of utilizing ME intake for milk production using Bayesian methods of analysis.

Material and methods

A database containing energy balance observations on 701 individual dairy cows was assembled from 38 calorimetry studies conducted in the UK (Kebreab *et al.*, 2003). This data set was further updated with energy balance data from the Netherlands (Van Knegsel *et al.*, 2007). A full Bayesian hierarchical model was developed for analyzing the data. The model consisted of three levels: within and between study variability, and prior distribution.

$$y_{ij} \sim t(\mu_{ij}, \tau, v)$$

$$\mu_{ij} = \theta_{1i} + (\theta_{2i} + \beta X_{4ij}) X_{1ij} + \theta_{3i} X_{2ij} + \theta_{4i} X_{3ij} \tag{1}$$

where y_{ij} denotes the jth ($1<j<n_i$) observation (milk energy) in the ith study; μ_{ij} is the expected value of the data from the model, the primary covariates are metabolizable energy intake, tissue energy gain, tissue energy loss and metabolizability, which are denoted by X_{1ij}, X_{2ij} X_{3ij} and X_{4ij}, respectively. t refers to a Student's t-distribution with v degrees of freedom and precision τ, i.e. 1/variance. Student's t-distribution was preferred over the normal distribution because it has thicker tail probabilities and thus is more robust towards influential observations. The energy balance Equation 1 has five structural parameters where the first four are specific to the ith study, i.e. θ_{1i} is the intercept (NE_M); θ_{2i} the first slope (k_l); θ_{3i} the second slope (k_l/k_g); θ_{4i} is the third slope (k_t) and the fourth slope β adjusts for the population effect of feed quality on the utilization of the ME intake for milk production. θ_i is a vector of individual study estimates, i.e. $\theta_i=[\theta_{1i}, \theta_{2i}, \theta_{3i}, \theta_{4i}]'$, which are assumed to follow a multivariate normal distribution with a vector of mean population parameters and the between study variance-covariance matrix. The model was developed in the general purpose software for Bayesian modeling: OpenBugs (Thomas *et al.*, 2006; Lunn *et al.*, 2000). Weak priors were used for all parameters (Lunn *et al.*, 2000). Convergence was established using the Gelman-Rubin statistic as the main determinant and running three chains. Convergence was established

after 10,000 samples, i.e. the burn-in period. Inference was based on an additional 100,000 samples from the posterior distribution with a thinning of 20. In the final analysis, an informative prior was introduced for the population parameter ($NE_M \sim N(0.45, 0.04)$). The results of the final analysis are summarized in Table 1, including the Bayesian version of confidence intervals, i.e. credible intervals.

The change in k_l per unit change in metabolizability (β) was estimated to be 0.10 (-0.004, 0.210), indicating that β is not significant different from zero (P=0.059). However, a tendency is implied. The deviance information criteria for model 1 compared with a model with no effect of metabolizability are -2,247 and -2,242, respectively, indicating a better fit to the data. Values of 0.89 and 0.69 are obtained for k_g and k_t, respectively, which were consistent with reported value by Kebreab *et al.* (2003) but different from national research council values of 0.60 to 0.75 for k_g and 0.82 to 0.84 for k_t (Table 1, e.g. AFRC, 1993). In conclusion, metabolizability in the range of 0.42 to 0.76 does not appear to affect milk energy output significantly, however, inclusion of data with wider range than considered in this analysis might lead to a different conclusion.

Table 1. Summary of the posterior distribution of selected parameters presented as posterior mean, standard deviation and 95% credible intervals.

Parameter[1]	Estimate (SD)	Lower	Upper
NE_M, MJ/(kg$^{0.75}$d)	0.34 (0.028)	0.28	0.39
ME_M, MJ/(kg$^{0.75}$d)	0.58 (0.034)	0.52	0.65
k_l	0.58 (0.021)	0.54	0.62
k_g	0.89 (0.056)	0.85	0.95
k_t	0.69 (0.047)	0.60	0.79

[1] NE_M and ME_M net and metabolizable energy for maintenance; k_l and k_g the efficiency of utilizing ME intake for milk production and growth; k_t the efficiency of utilizing body stores for milk production.

References

Agricultural and Food Research Council. 1993. Energy and Protein Requirements of Ruminants. CAB International, Wallingford, U.K.

Kebreab E., J. France, R.E. Agnew, T. Yan, M.S. Dhanoa, J. Dijkstra, D.E. Beever and C.K. Reynolds, 2003. Alternatives to linear analysis of energy balance data from lactating dairy cows. J. Dairy Sci. 86, 2904-2913.

Lunn, D.J., A. Thomas, N. Best and D. Spiegelhalter, 2000. WinBUGS – a Bayesian modelling framework: concepts, structure, and extensibility. Statist. Comp. 10, 325-337.

Thomas A., B. O'Hara, U. Ligges and S. Sturtz, 2006. Making BUGS Open. R News 6:12-17 (available online at http://www.rni.helsinki.fi/~boh/publications/Rnews2006-1.pdf).

Tolkamp B.J. and I. Kyriazakis, 2009. Toward a new practical energy evaluation system for dairy cows Anim. 3, 307-314.

Van Knegsel A.T.M., H. van den Brand, J. Dijkstra, W.M. van Straalen, M.J.W. Heetkamp, S. Tamminga and B. Kemp, 2007. Dietary energy source in dairy cows in early lactation: energy partitioning and milk composition. J. Dairy Sci. 90: 1467-1476.

Energy and protein value of lucerne hay and wheat straw in diets for lambs

A. De Angelis
Dipartimento di Scienze Agronomiche, Agrochimiche e delle Produzioni Animali, Sez. di Scienze delle Produzioni Animali, Università degli Studi di Catania, Via Valdisavoia 5, Catania, Italy

Introduction

Studies of the energy and protein value of forages are increasing due to interest in their use; these may be considered more healthful than concentrates. Previously, there has been a lack of data and of unequivocal and exhaustive nutritive value systems (Ouda and Nsahlai, 2009; France and Kebreab, 2008; Cannas *et al.*, 2008; Kebreab *et al.*, 2008; Cannas, 2004a,b; Cannas and Dattilo, 1991). Lucerne hay is a forage often used in ruminant nutrition and the use of straw is also added to diets due to its low cost and positive effects on the environment. On this basis, the aim of the present work was to contribute to the knowledge on energy and protein value of lucerne hay and wheat straw in diets for lambs.

Material and methods

Data obtained from a national scientific research project entitled 'Energy and protein value of diets for ruminants', were organized and statistically analysed (ANOVA) to point out differences in energy (NE, Mcal/kg dry matter) and protein (NP, NP/GP, net protein/gross protein) efficiencies values of lucerne hay (Lh) and wheat straw (Ws) in diets for lambs.

The experiment, based on comparative slaughter, was carried out, with animals from 50 to 100 days of age, using entire male lambs of Sarda, Appenninica, Bergamasca, and Leccese breeds. Two groups for each breed were fed the two diets, ratio 40/60 (concentrate/forage), containing Lh and Ws, respectively. Ingredients of concentrate were: corn meal, barley meal, soybean meal and mineral-vitamin premix (Galvano *et al.*, 1997, 1995, 1994a,b).

Results and discussion

Statistical analysis of data showed (1) no difference between age of lambs (50-100 days); (2) differences of NE efficiencies only for Appenninica and Leccese breeds; (3) differences of NP efficiencies for all breeds. Its interesting to note that energy efficiency of Ws was higher than Lh for a dairy breed Leccese, well-know to be good users of roughages; this trend was also evident but not significant for the Sarda breed as well.

Moreover, the higher values of Ws than Lh, for NP efficiency, between all the breeds, could be attributed to improved utilization of protein of Ws when associated with the concentrates, promoting complementary and interactive effects within the protein fractions.

In general terms, mean values of NE (Mcal/kg) and NP efficiencies were 1.40 and 0.23 for Lh, 1.43 and 0.37 for Ws, respectively.

Table 1. Comparison between net energy (NE) and protein (NP) efficiencies of lucerne hay (Lh) an wheat straw (Ws).

Breed	NE (Mcal/kg)		NP (net protein/gross protein)	
	Lh	Ws	Lh	Ws
Sarda	1.53[a]	1.61[a]	0.23[a]	0.38[b]
Appenninica	1.42[a]	1.25[b]	0.18[a]	0.24[b]
Bergamasca	1.26[a]	1.19[a]	0.28[a]	0.35[b]
Leccese	1.37[a]	1.65[b]	0.24[a]	0.49[b]

[a,b] $P<0.05$.

References

Cannas, A., 2004a. Energy and protein requirements. In Dairy sheep nutrition 3, CAB International, Wellingford, Oxon, UK: 31.

Cannas, A., 2004b. Feeding of lactating ewes. Dairy sheep nutrition: 31-49.

Cannas, A. and M. Dattilo, 1991. Fibrous agroindustrial by-products: characteristics and improvement. Production and utilization of lignocellulosis: 427-441.

Cannas, A., A.S. Atzorias, F. Boe and I.A.M.A. Teixeira, 2008. Energy and protein requirements of goats. CAB International, Wellingford, Oxon, UK: 31.

France, J. and E. Kebreab, 2008. Mathematical modelling in animal nutrition. CAB International, Wellingford, Oxon, UK.

Galvano, G., V. Scerra, A. De Angelis, F. Galvano, A. Bonomi, A. Sabbioni, P. Superchi and A. Quarantelli, 1994a. ENES. Prontuario di razionamento per agnelloni di razza sarda. 1.3. Fabbisogni degli animali e valore nutritivo degli alimenti. Diete miste a base di fieno di erba medica, paglia di grano e radichette d'orzo. Ed. Un. Catania – I.S.Te.P.A.

Galvano, G., V. Scerra, A. De Angelis, F. Galvano, P. Pollidori, F. Costantini, C. Casoli and S. De Vincenti, 1994b. ENES. Prontuario di razionamento per agnelloni di razza Sarda. 2.1. Fabbisogni degli animali e valore nutritivo degli alimenti. Diete miste a base di fieno di erba medica, paglia di grano e paglia di soia. Ed. Un. Catania – I.S.Te.P.A.

Galvano, G., V. Scerra, A. De Angelis, F. Galvano, F. Malossini, S. Bovolenta, C. Piras and E. Piasenter, 1995. ENES. Prontuario di razionamento per agnelloni di razza Bergamasca. 4.1 Fabbisogni degli animali e valore nutritivo degli alimenti. Diete miste a base di fieno di erba medica, paglia di grano e buccette d'uva. Ed. Un. Catania – I.S.Te.P.A.

Galvano, G., A. De Angelis, F. Galvano, L. Chies, M. Galvano, B. Ciruzzi, A. Vicenti and G. Marsico, 1997. ENES. Prontuario di Razionamento per Agnelloni di Razza Leccese. 3.3 Fabbisogni degli Animali e Valore Nutritivo degli Alimenti. Diete Miste a Base di Paglia di Grano, Fieno di Medica, Frasche d'Ulivo. Ed. Un. Catania – I.S.Te.P.A.

Kebreab, E., J. Dijkstra, A. Bamrink, W.J. Gerrits and J. France, 2008. Nutrient digestion and utilization in farm animals: modelling approachs. CAB International, Wellingford, Oxon, UK.

Ouda, J.O. and I.V. Nsahlai, 2009. Relevance and potential use of *in vitro* gas production measurements to evaluate varying ratios of roughages and protein sources for ruminants. J. Appl. An. Res. 35, 9-16.

Prediction of degradability by near infrared reflectance spectroscopy (NIRS)

A. Foskolos[1], E. Albanell[1], M. Chrenkova[2], S. Calsamiglia[1] and M.R. Weisbjerg[3]
[1]Dept. Ciéncia Animal i dels Aliments, Univ. Autónoma Barcelona, Bellaterra, Spain
[2]Animal Production Research Center Nitra, Slovakia
[3]Faculty of agricultural sciences, University of Aarhus, Denmark

Introduction

Near infrared reflectance spectroscopy (NIRS) can predict chemical values and several parameters of nutritional interest for different feedstuffs. NIRS offers important advantages over traditional chemical methods. It is a physical, non-destructive method, requiring minimal or no sample preparation, no reagents, while no wastes are produced. Degradability parameters are important components of mathematical models of nutritional evaluation systems. Our objective was to evaluate NIRS potential for predicting degradability parameters and effective degradation of original feed samples used in ruminant nutrition.

Materials and methods

In REDNEX funded research, degradability of dry matter (dDM), crude protein (dCP) and neutral detergent fibre (dNDF) of 821 samples (607, 626 and 117 with data on DM, CP and NDF respectively) was determined using the *in situ* technique. Samples were forages (whole crop, straw, hay, silages and tropical forages), concentrates (concentrate, grains and seeds) and by-products (sugar-beet pulp, oil seed by-products and other by-products). Degradability was described in terms of immediately rumen soluble fraction, the degradable but not soluble faction and its rate of degradation. Overall effective degradability of DM, CP (5%/h passage rate) and NDF (2%/h passage rate) were calculated according to the following equation (Ørskov and McDonald, 1979): ED = a + b (c/c + k), where k = fractional outflow rate from the rumen per h.

All samples were recorded from 1,100 to 2,500 nm using a NIRSystems 5000 scanning monochromator (FOSS, Hoganas, Sweden). Reflectance was recorded in 2 nm steps as log 1/Reflectance. Samples were scanned twice in duplicate using ring cup cells and mean spectrum was calculated for each sample. WinISI III (v. 1.6) software was employed for spectra data analysis and development of chemometric models. Prior to calibration, spectra were transformed using standard normal variate, detrend and multiplicative scatter correction transformations and different math treatments (derivate number, subtraction gap and smoothing intervals) were tested. Calibrations were developed by the modified partial least squares (MPLS) regression technique. The precision of the equations obtained was confirmed by external validation on a set of 70, 70 and 13 samples for DM, CP and NDF respectively.

Results and discussion

NIRS equations were selected, taking into account the best values of r^2, SEP and other statistics: ratio of performance to deviation and range error ratio. Table 1 summarizes statistics for calibration and validation of the degradability parameters. Prediction of feeds chemical composition was satisfactory. The effective degradation of DM, CP and NDF was also well predicted by the NIRS equations, as well as fraction a in the case of DM and CP, and fraction b for DM. In the case of c value and the asymptote, R^2 and r^2 coefficients were below acceptable limits. These results are in accordance with Andres *et al.* (2005) and Ohlsson *et al.* (2007), even though current values are higher probably due to the higher variation in the present material.

Table 1. NIRS statistics of degradability parameters of dry matter (dDM), crude protein (dCP) and neutral detergent fibre (dNDF).[1]

Quality parameter	Degradation parameter[2]	Scatter correction[3]	R^2	SEC	r^2	SEP	RPD	RER
CP (%DM)		MSC	0.995	0.958	0.994	1.841	7.86	47.97
NDF (%DM)		DT	0.922	3.483	0.896	4.484	2.85	14.64
dDM	ed	SNV-D	0.935	0.029	0.949	0.042	2.72	19.48
	a	DT	0.906	0.040	0.778	0.078	1.72	9.80
	b	DT	0.880	0.047	0.762	0.084	1.43	8.4
	c	SNV-D	0.813	0.014	0.714	0.036	1.47	6.39
	Asymptote	MSC	0.824	0.025	0.721	0.032	2.70	8.43
dCP	ed	MSC	0.899	0.045	0.926	0.056	2.65	15.03
	a	SNV-D	0.901	0.062	0.833	0.093	2.15	10.26
	b	MSC	0.891	0.071	0.761	0.113	1.92	8.76
	c	MSC	0.625	0.020	0.328	0.024	2.06	7.92
	Asymptote	SNV-D	0.695	0.017	0.598	0.022	3.42	6.36
dNDF	ed	MSC	0.826	0.061	0.766	0.054	2.66	10.90
	b	DT	0.716	0.014	0.676	0.024	1.39	7.33
	c	DT	0.716	0.014	0.676	0.024	1.04	4.16

[1] R^2: determination coefficient of calibration, r^2: determination coefficient of validation, SEC: standard error of calibration, SEP: standard error of prediction, RPD: performance/deviation, RER: range error ratio.
[2] ed: effective degradation, a: immediately rumen soluble fraction, b: degradable but not soluble faction, c: rate of degradation of the b fraction, Asymptote: asymptotic of degradability equations.
[3] SNV: standard normal variate, DT: detrend, MSC: multiplicative scatter correction.

In conclusion, NIRS equations obtained demonstrate the potential but also the limitation of the NIRS technique. Present work on a large number of feeds varying from animal based concentrates to vegetable concentrates, and from forages to straws, shows the potential of NIRS to universally predict chemical composition, solubilities, and effective and potential degradabilities, whereas rate of degradation is not predicted satisfactorily. Future work on dividing feedstuff into groups might improve NIRS equations.

References

Andres S., F.J. Giraldez, J.S. Gonzalez, R. Pelaez, N. Prieto and A. Calleja, 2005. Prediction of aspects of neutral detergent fibre of forages by Chemicals composition and near infrared reflectance spectroscopy. Aust. J. Agric. Res. 56, 187-193.

Ohlsson C., L.P. Houmøller, M.R. Weisbjerg, P. Lund and T. Hvelplund, 2007. Effective rumen degradation of dry matter, crude protein and neutral detergent fibre in forage determined by near infrared reflectance spectroscopy. J. Anim. Phys. Anim. Nut. 91, 498-507.

Ørskov, E.R. and I. McDonald, 1979. The estimation of protein degradability in the rumen from incubation measurements weighted according to rate of passage. J. Agric. Sci. 92, 499-503.

Ruminal redox potential in dairy cows regarding diet composition and live yeast supplementation: A modelling approach.

C. Julien[1], J.P. Marden[2], R. Moncoulon[1], E. Auclair[2] and C. Bayourthe[1]
[1]Université de Toulouse, INRA, UMR 1289 INRA/INPT/ENVT TANDEM, 31326 Castanet-Tolosan, France
[2]Lesaffre Feed Additives, 59520 Marquette-Lez-Lille, France

Introduction

The rumen is an anaerobic and reducing milieu with low redox potential (E_h) values, ranging from -220 to -110 mV, propitious to the development and activity of endogenous anaerobic microflora. References about ruminal E_h values are scarce and factors of variation are not clearly established. One obvious factor could be oxygen entering the rumen during feed intake, but recent research studies (Julien *et al.*, 2009; Marden *et al.*, 2008b) showed that the ruminal microflora alone is capable of re-establishing the reducing condition. Live yeast used as a dietary feed additive for dairy cows is another potent modulator of ruminal E_h (Marden *et al.*, 2008a). Its probiotic effects on animal performance and ruminal parameters seemed to be linked with DMI and diet composition (Desnoyers *et al.*, 2009). However, the impact of DMI and diet composition on ruminal E_h has not yet been studied. Therefore, the objective of this modelling approach was to gain a better insight on the influence of diets on ruminal E_h in the presence or absence of live probiotic yeast.

Material and methods

A database was made up of 10 trials involving live yeast supplementation (10^{10} CFU/g of DM, ACTISAF® Sc 47, Lesaffre Feed Additives, France) tested with 15 diets fed to dry or lactating dairy cows. Average ruminal E_h value (E_h Mean) was measured over a 10-hr period around the morning meal (Marden *et al.*, 2005). Each experimental diet was characterised by selected descriptors of the diet (Table 1). To explore those descriptors in relation to ruminal E_h, a multiple regression analysis was fit using R (http://www.R-project.org/).

Table 1. Dietary descriptors used in the models to predict E_h mean.

Item	Control $R^2 = 0.70$ RSE = 18.0	Yeast $R^2 = 0.49$ RSE = 20.7	Definition
CP		-1.61	g of crude protein / kg DM
CF	0.93	-0.25	g of crude fibre / kg DM
Starch	0.32	-0.33	g of starch / kg DM
UFL	108.6	423.1	Energetic density of the diet (/ kg DM); NE_L in Mcal = 1.7×UFL
PDIA	0.87	1.28	g of protein digested in the intestine from dietary origin / kg DM
StarchType	-64.3 if mix -80.3 if rapid 0 if slow		Type of starch in the energy concentrate: mostly rapidly or slowly degradable starch or mix of both
$DMI^{0.75}$	269.5	309.5	Dry matter intake per kg of metabolic body weight
Intercept	-519.7	-389.4	

Results and discussion

The parameter estimates of two models (with and without live yeast supplementation) for predicting ruminal E_h Mean are presented in Table 1. Despite the lack of data on such ruminal parameter, modelling E_h allowed at a first glance, to observe its strong relationship with diet composition and DMI. Results showed that in absence of live yeast, the level of E_h in the rumen of dairy cows seemed directly linked with the dietary energetic density expressed in terms of carbohydrate content and type of starch. When the diet is supplemented with live yeast, the model explained less variability in ruminal redox level than for control diet.

Enriching our database will allow a gain in accuracy of these models. Further trials involving such physico-chemical measurements with various types of diets will further consolidate these preliminary predicting models of ruminal reducing power. The final aim is to provide field condition tools to assess microbial activity of the rumen which can profitable in dairy nutrition management and to feed additive users.

References

Desnoyers M., S. Giger-Reverdin, G. Bertin, C. Duvaux-Ponter and D. Sauvant, 2009. Meta-analysis of the influence of Saccharomyces cerevisiae supplementation on ruminal parameters and milk production of ruminants. J. Dairy Sci. 92, 1620-1632.

Julien C., A. Troegeler-Meynadier, J.P. Marden, F. Enjalbert and C. Bayourthe, 2009. *In vivo* and *in vitro* measurements of ruminal redox potential: a comparative study. in Joint Annual Meeting, Montreal, Québec, Canada.

Marden J.P., C. Bayourthe, F. Enjalbert and R. Moncoulon, 2005. A new device for measuring kinetics of ruminal pH and redox potential in dairy cows. J. Dairy Sci. 88, 277-281.

Marden J.P., C. Julien, V. Monteils, E. Auclair, R. Moncoulon and C. Bayourthe, 2008a. How does live yeast differ from sodium bicarbonate to stabilize ruminal pH in high-yielding dairy cows? J. Dairy Sci. 91, 3528-3535.

Marden J.P., V. Monteils, E. Auclair, R. Moncoulon, and C. Bayourthe, 2008b. Response of ruminal redox potential (E_h) to an exogenous supply of oxygen in presence of the probiotic yeast Sc 47 in the diet of dairy cows. In 6th INRA-RRI Symposium Gut Microbiom. Clermont-Ferrand (France): 73.

R Development Core Team, 2009. R: A language and environment for statistical computing. R Foundation for Statistical Computing, Vienna, Austria. ISBN 3-900051-07-0, URL http://www.R-project.org.

Integration of the Small Ruminant Nutrition System and of the UC Davis sheep growth model for improved gain predictions

J.W. Oltjen[1], A. Cannas[2], A.S. Atzori[2], L. O. Tedeschi[3], R.D. Sainz[1] and D.G. Fox[4]
[1]Department of Animal Science, University of California Davis, CA 95616, USA
[2]Dipartimento di Scienze Zootecniche, University of Sassari, 07100 Sassari, Italy
[3]Department of Animal Science, Texas A&M University, College Station, TX 77843, USA
[4]Department of Animal Science, Cornell University, Ithaca, NY 14853, USA

Introduction

The Small Ruminant Nutrition System (SRNS) is a mechanistic, static model that uses animal inputs (e.g. body weight, age, body condition score, physical activity, milk production and composition, body reserves, mature weight, pregnancy) and environmental factors (current and previous temperature, wind, rainfall) to predict energy and protein requirements (Cannas *et al.*, 2004; Tedeschi *et al.*, 2010). Live weight gain, empty body gain, and the composition of the gain of growing lambs and kids are predicted based on the energy balance, the degree of maturity and the feeding level of the animals. Feed biological values are predicted based on carbohydrate and protein fractionations and their digestion rates, passage rates of forage, concentrate, and liquid, microbial growth, and physically effective fibre. Thus, specific feed biological values are estimated for specific feeding condition. The SRNS has been able to accurately predict growth of lambs and kids under normal rate of growth (Tedeschi *et al.*, 2010).

The UC Davis Sheep (UCDS) is a mechanistic, dynamic model that predicts post-weaning growth and body composition of lambs (Oltjen *et al.*, 2003). It includes viscera and non-viscera protein pools and a body fat pool and predicts variable energy requirements for maintenance based on the previous nutritional history of the animal, thus accounting for the large effects of changes in viscera size on heat production. Net energy above maintenance is used for viscera and non-viscera tissue growth before it is used for fat accretion. The UCDS has been able to predict gain and composition with good accuracy when remarkable compensatory growth has occurred (Oltjen *et al.*, 2003). However, it lacks a supply submodel, so it unable to predict the dietary energy and protein values, which in this model are inputs.

The SRNS and the UCDS growth models were integrated into a new model to account for the effects of variations in feeding level on nutrient utilization, energy and protein requirements, and composition of the gain of growing sheep.

Methods

The two models were integrated so that the feed values estimated by the SRNS could be used as an input for the UCDS model. The two original models and the newly integrated model were evaluated by inputting the data of Ferrell *et al.* (1986), who studied average daily gain (ADG) in lambs fed the same pelleted diet in two consecutive growth stages (42 d each) with 9 combinations of planes of nutrition, some of which caused compensatory growth in the second period. The assessment of the adequacy of the models was carried out with the Model Evaluation System, which is based on statistical techniques as discussed by Tedeschi (2006).

Results and discussion

The database included a broad range of ADG (from -152 to 412 g/d; mean 176) and dry matter intake (from 343 to 1,820 g/d; mean 1,025). The results of the evaluation showed that the UCDS over predicted, while the SRNS under predicted the ADG during the first period (Table 1), with the SRNS having slightly lower mean bias and RMSEP. The integrated model predicted ADG similarly to the UCDS model but with lower mean bias and root of the mean squared error of prediction (RMSEP). In the second period, during which compensatory growth occurred, the UCDS predicted ADG with much lower mean bias and RMSEP than the SRNS (Table 1), but the precision was similar. This is likely because the SRNS was not designed to account for compensatory growth. The integrated model further improved the evaluation statistics. This bias reduction was mainly due to the fact that the integrated model was able to account for the adjustment of ME due to the variations on feeding levels.

The integration of the UCDS and the SRNS models was able to improve the accuracy of the prediction of the ADG of lambs subjected to different planes of nutrition. Future work should consider a further integration of the two models.

Table 1. Comparison of observed (O) and predicted (P) average daily gains (n = 9).[1]

Variables	P (g/d)	P - O (g/d)	MSEP partition, %			RMSEP (g/d)	r^2	P<	CCC
			U_b	U_s	U_r				
UCDS, days 0-42	213	41	91.5	1.2	7.3	43.2	1.00	0.001	0.98
SRNS, days 0-42	138	-34	85.3	9.0	5.7	37.4	1.00	0.001	0.98
UCDS & SRNS, days 0-42	209	36	86.4	5.3	8.2	39.2	1.00	0.001	0.98
UCDS, days 43-84	190	11	23.1	0.6	76.3	22.2	0.99	NS	1.00
SRNS, days 43-84	144	-35	66.6	9.4	24.0	43.0	0.99	0.007	0.97
UCDS & SRNS, days 43-84	188	9	16.3	9.8	73.9	21.4	0.99	NS	1.00

[1] MSEP = mean squared error of prediction; RMSEP = root of the MSEP; U_b = mean bias; U_s = systematic bias; U_r = random error. r^2 = coefficient of determination of the best fit regression line not forced through the origin; *P* = probability associated to a F test to reject the simultaneous hypothesis that the slope = 1 and the intercept = 0; when NS (*P*>0.1) the hypothesis is not rejected; CCC is concordance correlation coefficient (Tedeschi, 2006).

References

Cannas, A., L.O. Tedeschi, D.G. Fox, A.N. Pell and P.J. Van Soest, 2004. A mechanistic model to predict nutrient requirements and feed biological values for sheep. J. Anim. Sci. 82, 149-169.

Ferrell, C.L., L.I. Koong and J.A. Nienaber, 1986. Effects of previous nutrition on body composition and maintenance energy costs of growing lambs. Brit. J. Nutr. 56, 595-605.

Oltjen, J.W., A.B. Pleasants, T.K. Soboleva and V.H. Oddy, 2003. Ruminant growth and composition model. In: Souffrant, W.B. and C.C. Metges (eds.), Progress in Research on Energy and Protein Metabolism. EAAP Publication 109: 209-212.

Tedeschi, L.O., 2006. Assessment of adequacy of mathematical models. Agric. Syst. 89, 225-247.

Tedeschi, L.O., A. Cannas and D.G. Fox, 2010. A nutrition mathematical model to account for dietary supply and requirements of energy and other nutrients for domesticated small ruminants: The development and evaluation of the Small Ruminant Nutrition System. Small Rum. Res. 89, 174-189.

Note on the calculation of efficiency of feed use for maintenance and gain in feeding systems

J.W. Oltjen[1], R.D. Sainz[1], A. Cannas[2] and L.O. Tedeschi[3]
[1]*Department of Animal Science, University of California, Davis, California, USA*
[2]*Dipartimento di Scienze Zootecniche, University of Sassari, Sassari, Italy*
[3]*Department of Animal Science, Texas A&M University, College Station, Texas, USA*

Introduction

When Lofgreen and Garrett (1968) proposed the California Net Energy System (CNES), they provided methods to estimate net energy for maintenance and gain for different rations fed to growing cattle. These methods have been applied in widely different situations since their introduction, and controversy about adjustments and corrections for various factors has arisen. Application of experimental results to estimate NE_m and NE_g, or k_g and k_m, has often been inconsistent across laboratories, and the objective of this communication is to describe these inconsistencies and show when and how they may introduce error in feed values.

Maintenance

In the original CNES paper, Lofgreen and Garrett (1968) found that the daily net energy for maintenance requirement of growing steers and heifers was 0.077 $BW^{0.75}$ (Mcal). Based on estimation of the retained energy in daily gain (EB) and measurement of daily metabolizeable energy intake (MEI, estimated by multiplying daily feed intake by ration metabolizability), heat production (HP) can be determined by difference: HP = MEI - EB. If the diet to be studied is fed *ad libitum*, then the HP calculated at *ad libitum* may be regressed with the HP at fasting, 0.077 $BW^{0.75}$ (MEI=0). The logarithm of HP to base 10 was used to provide the best estimate according to Lofgreen and Garrett (1968). Then the point on the resulting line where HP = MEI is the metabolizable energy required for maintenance (ME_{maint}, EB=0), or k_m is the ratio of 0.077 to ME_{maint} and NE_m concentration (Mcal/kg DM) of the ration is 0.077/DMI_{maint} where DMI_{maint} is the dry matter intake (kg/$BW^{0.75}$) required to meet maintenance requirements. Since $DMI_{maint} = MEI_{maint}/ME$, then $NE_m = 0.077/(MEI_{maint}/ME)$.

Whether adjusting MEI for discounts in metabolizability with increasing intake (Van Soest, 1982) affects this estimate of ME for maintenance is of concern. Lofgreen and Garrett (1968) do not address this, and it may only be of significance for lower energy diets, since Blaxter (1969) showed that metabolizeability only decreased with intake for diets with metabolizable to gross energy ratios less than 0.62. To determine if errors in ME for maintenance might occur if ME is discounted on low energy diets, we test this possibility with the same 100% forage diet (ME of 2.04 Mcal/kg DM) fed to heifers *ad libitum* that Lofgreen and Garrett (1968) used to illustrate the method to estimate ME for maintenance. If we apply a severe decrease in feed ME of 6% per multiple of maintenance feeding (2.28 here), the original mean MEI for the *ad libitum* fed animals is reduced from 0.2977 to 0.2748 Mcal/$BW^{0.75}$. However, the calculated daily HP (MEI-EB) is also reduced from 0.2563 to 0.2334 Mcal/$BW^{0.75}$. Regressing the log of HP with MEI results in ME for maintenance values (at the point on the line where HP=MEI) of 0.1304 and 0.1303 Mcal/$BW^{0.75}$ for the original and adjusted ME calculations, representing only a 0.11% error for such a large adjustment of ME. It can be concluded that since both MEI and HP are reduced by the discounted ME with intake, little error occurs in calculation of ME_{maint}, NE_m, or k_m.

Growth

Lofgreen and Garrett (1968) state that the efficiency of net energy of a ration for weight gain when fed above maintenance is constant. They proposed that a difference trial, feeding a ration at two levels and measuring the difference in energy balance (ΔEB) brought about by the increase in feed intake between the two levels (ΔDMI), is the method to determine a ration net energy for gain, NE_g. Also, the efficiency of ME use for gain (k_g) in metabolizable energy systems is (ΔEB/ΔMEI). Alternatively, and especially useful when rations have not been fed at two levels, efficiency of gain has been estimated by dividing EB by the feed or ME fed above the maintenance level (ME_{gain}). As above, the question arises as to whether adjusting MEI for discounts in metabolizability with increasing intake affects the estimate of NE_g or k_g. In the case where NE_g is determined by difference there is no concern since the difference in actual energy retained in the animal is divided by observed difference in feed dry matter intake, not ME intake.

The concern is therefore about using imputed MEI values in the denominator of the k_g calculation. It is readily apparent that k_g depends on the discount applied, and is not constant. That is, since k_g is ΔEB/ΔMEI, the ΔMEI depends on the ME at each intake level. If ME is not similar at the different levels, then ΔMEI is not the same as using one ME value multiplied by the difference in feed intakes ME × ($Intake_2$ - $Intake_1$). Hence a ME system is not consistent with the NE system as proposed by Lofgreen and Garrett (1968) since the k_g from the ME system differs from that imputed by the NE system as proposed, that is, for discounted ME values as intake increases, k_g for the ME system is greater than k_g for the NE system.

Also, for NE_g and k_g to be determined at one feeding level, we estimated the efficiency of gain by dividing EB by the feed or MEI fed above the maintenance level, ME_{gain} (equivalent to the difference method above with the maintenance level assumed as the lower level). Both require an estimate of maintenance, and as shown above, one arrives at the same estimate by regressing log HP on MEI whether discounting ME for intake level or not. Then it is apparent that NE_g is EB divided by the dry matter intake above maintenance (DMI - $0.077/NE_m$). The equivalent k_g is EB divided by ME_{gain}. Therefore, k_g is greater if MEI is discounted for level of intake. In the years since Lofgreen and Garrett (1968) published their method, it is has been shown that the cost of maintenance is not constant but variable. This has been incorporated in newer feeding systems, often with adjustments of ME_{gain} and/or k_g. However, the California Net Energy system seems to work because it does not adjust ME for intake levels but rather implicitly accounts for this by calculating NE_g using the actual feed intakes and EB observed. Since EB = $ME_{gain} \times k_g$, it can be shown that both methods provide estimates of EB that differ at all intakes except the experimental one (I_{obs}) on which the calculation was based:

$$EB_D(I) = EB_G(I) \times [(ME_{gainD}(I) \times ME_{gainG}(I_{obs})] / [(ME_{gainG}(I) \times ME_{gainD}(I_{obs})]$$

where the D subscript are for those where ME is discounted for level of intake, I, and the G (Lofgreen and Garrett, 1968) ones are not discounted. So, from a biological standpoint the ME system is more correct than the California NE system. On the other hand, the NE system, although less elegant, produces more accurate estimates.

References

Blaxter, K.L, 1969. The efficiency of energy transformation in ruminants. In: Blaxter, K.L., J. Kielanowsky and G. Thorbek (eds.), Energy metabolism of farm animals. Oriel Press, Newcastle upon Tyne, United Kingdom, EAAP Publication 12: 21-28.

Lofgreen, G.P., and W.N. Garrett, 1968. A system of expressing net energy requirements and feed values for growing and finishing beef cattle. J. Anim. Sci. 27, 793-806.

Van Soest, P.J., 1982. Nutritional Ecology of the Ruminant. O & B Books, Inc., Corvallis, Oregon, USA.

Metabolizable energy requirements and energetic efficiency of Brahman cattle fed various levels of energy under humid tropical condition

A. Chaokaur[1], T. Nishida[2] and K. Sommart[3]
[1]Faculty of Animal Science and Agriculture, Silpakorn University, 76120, Thailand
[2]Japan International Research Center for Agricultural Sciences, Japan
[3]Department of Animal Science, Khon Kaen University, 40002, Thailand

Introduction

NRC (2000) reported that level of feed intake may affect metabolic rate and energy requirement for maintenance (ME_m) of cattle. Previous research indicated that the energy excreted in the feces, urine, gases and energy lost as heat followed the same patterns of increase as intake energy of beef steers (Clark *et al.*, 2007). However, the effect of energy level intake on energy metabolism and utilization has not been elucidated in growing beef cattle under humid tropical conditions in Thailand. The objectives of this study were to determine the metabolizable energy requirements and energetic efficiency of Brahman male cattle fed various levels of energy intake.

Materials and methods

Twenty growing male Brahman cattle (BW 342.7±16.2 kg; age 2 years old) were housed in individual pens. Dietary treatments were assigned to a randomized complete block design with five replications and four treatment levels of energy intake feeding M, 1.4M, 1.8M and *ad libitum*; M was feeding at a maintenance level of 450 kJ/kg $BW^{0.75}$/d (Chaokaur *et al.*, 2007). Analyses (% DM) of the diets: OM and CP were 89.63 and 11.56, respectively. Energy content of diet: GE, DE and ME were 16.42, 11.87 and 10.13 MJ/kg DM, respectively.

During a preliminary 14 day feeding period, heat production measurements (3-day sampling) began using respiration head boxes. Nutrient digestibilities were determined by a 6-day sampling of the complete feces. The animals were fasted for five days for heat production measurements over the last two ddays. Collection samples were rotated for each block following assigned sequence during the experiment. Significance of linear, quadratic and cubic contrasts of dietary energy levels was determined in the randomized block model using GLM procedure (SAS, 1999).

Results and discussion

The results of the balance trials indicated that feeding cattle diets containing higher ($P<0.01$) dietary energy resulted in higher energy retention (Table 1); the regression is shown in Figure 1. Similarly, Freetly *et al.* (2006) also reported that increased recovered energy is associated with a rapid increase in heat production of mature cows after a change in feeding level of realimentation. A significant linear relationship between energy retention (ER) and metabolizable energy intake (MEI) allowed estimation of metabolizable energy requirement for maintenance (ME_m):

$$ER = (0.62 \times MEI) - 265.54\ [r^2 = 0.98;\ n=8;\ P<0.001;\ RSD=6.526] \quad (1)$$

$$ER = (0.58 \times MEI) - 269.54\ [r^2 = 0.94;\ n=12;\ P<0.001;\ RSD=6.093] \quad (2)$$

Therefore, the equations gives a value for net energy requirement for maintenance (NE_m) determined by fasting heat production (FHP) of 266 kJ/kg $BW^{0.75}$/d, and an estimate of ME_m requirement (Equation 1) of 425 kJ/kg $BW^{0.75}$/d. Therefore, the calculated efficiency of utilization of ME for maintenance ($k_m = FHP/ ME_m$) was 0.62. In addition, using Equation 2 the calculated efficiency of utilization of ME for growth (k_g) was 0.58.

Table 1. Energy partition of Brahman cattle fed increasing energy levels (kJ/kg $BW^{0.75}$).

Item	Levels of energy feeding				Fasted	SE	*P*-value[1]		
	M	1.4M	1.8M	*Ad libitum*			L	Q	C
GE intake	738	1,028	1,291	1,516	-	22.464	**	NS	NS
DE intake	554	743	911	1,077	-	19.290	**	NS	NS
ME intake	454	637	789	943	-	18.742	**	NS	NS
Feces excretion	184	284	379	439	108	6.630	**	NS	NS
Urine excretion	15	15	20	24	12	0.380	**	NS	NS
Methane production	85	91	102	111	4	2.456	**	NS	NS
Heat production	436	537	601	665	265	7.880	**	NS	NS
Energy retention	18	99	188	278	-265	11.202	**	NS	NS

[1] ** = $P<0.01$; NS = $P>0.05$; L = linear; Q = quadratic; C = cubic.

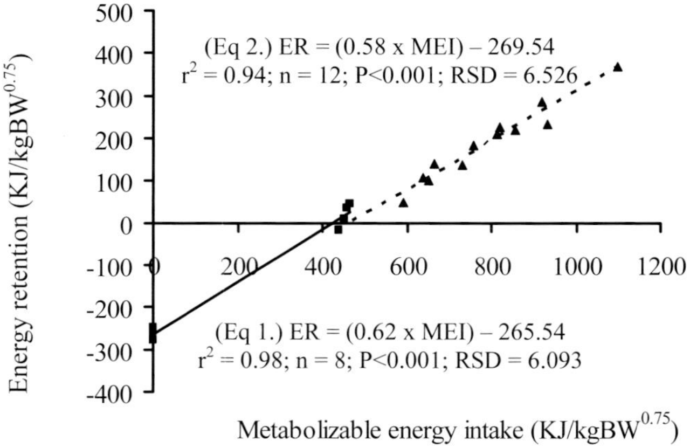

Figure 1. The relationship between metabolizable energy intake and energy retention (the solid line and dash line describe regression Equations 1 and 2, respectively (see text).

References

Chaokaur, A., T. Nishida, I. Phaowphaisal, P. Pholsen, R. Chaithiang, and K. Sommart, 2007. Energy metabolism and energy requirement for maintenance of Brahman steers in tropical conditions. In: I., Ortigues-Marty, N. Miraux and W. Brand-Williams (eds.), Energy and Protein Metabolism and Nutrition. Wageningen Acadademic Publishers, Wageningen, the Netherlands. EAAP No. 124: 505-506.

Clark, J.H., K.C. Olson, T.B. Schmidt, M.L. Linville, D.O. Alkire, D.L. Meyer, G.K. Rentfrow, C.C. Carr, and E.P. Berg, 2007. Effects of dry matter intake restriction on diet digestion, energy partitioning, phosphorus retention, and ruminal fermentation by beef steers. J. Anim. Sci. 85, 3383-3390.

Freetly, H.C., J.A. Nienaber, and T. Brown-Brandl, 2006. Changes in heat production by mature cows after changes in feeding level. J. Anim. Sci. 84, 1429-1438.

NRC, 2000. Nutrient Requirements of Beef Cattle. 7th ed. Natl. Res. Counc., Natl. Acad. Press, Washington, D.C.

SAS, 1999. SAS User's Guide: Statistics, Version 6.12th ed. SAS Institute Inc.Cary, NC.

Estimating maintenance energy using type I and type II regression models

M.S. Dhanoa[1], R. Sanderson[2], S. Lopez[3], J. Dijkstra[4], E. Kebreab[5] and J. France[6]
[1]North Wyke Research, Okehampton, Devon, United Kingdom
[2]Aberystwyth University, Ceredigion, United Kingdom
[3]Universidad de León, Spain
[4]Wageningen University, the Netherlands
[5]University of California, Davis, USA
[6]University of Guelph, Canada

Introduction

Ordinary least squares (OLS) regression (Type I regression model) is commonly used when analysing energy balance data, e.g. whole body energy retention (ER) as a function of metabolizable energy intake (MEI) where the slope is interpreted as the efficiency of MEI utilisation (k_f or k_g) and the estimate of maintenance energy (ME_m) is the level of MEI when ER is = 0. However, MEI is subject to measurement errors which violates the type I model's main assumption that the x-variable is either free of measurement errors or is under control. The effect of errors in the x-variable is that the OLS estimates of efficiency and ME_m are attenuated (biased downwards). In this situation, an 'errors in both variables' regression model (Type II regression model) is more appropriate. Further, the allometric scaling exponent has a major impact on these quantities. The objective of this paper is to evaluate the extent to which estimates of maintenance energy in beef cattle can be affected by the type of regression model and the allometric scaling exponent applied to energy balance data.

Material and methods

Regression models

In OLS, a simple linear model $y_i = \hat{\alpha} + \hat{\beta}x_i + \varepsilon_i$ is fitted to a set of bivariate sample values (x_i, y_i). Parameters $\hat{\beta}$ (slope or Δy per unit of Δx) and $\hat{\alpha}$ (intercept on y-axis) are estimated by minimising $\sum\varepsilon_i^2$ and estimates are given by $\hat{\beta}_{OLS} = \sum(y_i\text{-}\bar{y})\,(x_i\text{-}\bar{x}) / \sum(x_i\text{-}\bar{x})^2$ and $\hat{\alpha}_{OLS} = \bar{y} - \hat{\beta}_{OLS}\,\bar{x}$. There is a choice of type II models depending on whether the ratio of measurement error variances $\lambda = \sigma_\varepsilon^2/\sigma_\delta^2$ is known [σ_ε^2 is the error variance of each value of the y-variable (ER here) and σ_δ^2 the error variance of each value of the x-variable (MEI here)]. If λ is known then the unique maximum likelihood (ML) solution (also called the Deming regression model) is the only choice (Kendal and Stuart, 1966) and slope is estimated by

$$\hat{\beta}_{ML} = \hat{\sigma}_y^2 - \lambda_{ML}\,\hat{\sigma}_x^2 \sqrt{(\hat{\sigma}_y^2 - \lambda_{ML}\,\hat{\sigma}_x^2)^2 + 4\,\lambda_{ML}\,\hat{\sigma}_{xy}^2} \,/\, 2\hat{\sigma}_{xy}$$

where $\lambda_{ML} = \hat{\sigma}_\varepsilon^2/\hat{\sigma}_\delta^2$ with both $\hat{\sigma}_\varepsilon^2$ and $\hat{\sigma}_\delta^2$ assumed constant over the range of the data, $\hat{\sigma}_x^2$ and $\hat{\sigma}_y^2$ being the sample variances of x and y sample values, respectively, and $\hat{\sigma}_{xy}$ the sample covariance. If λ is unknown, other type II models (Dhanoa *et al.*, 2007) are available:

a. Bartlett's 3-group method (Bartlett, 1949) can be used for unknown λ;
b. *Major axis* or *orthogonal* regression (MA) is appropriate when $\lambda = 1$ provided the measurement scales of the x- and y-variables are the same;
c. *Reduced major axis* (RMA), the so-called '*y on x and x on y*' regression, may be applied when measurement scales are non-homogeneous and $\hat{\sigma}_\varepsilon^2 / \hat{\sigma}_y^2 = \hat{\sigma}_\delta^2 / \hat{\sigma}_x^2$, giving $\hat{\lambda} = \hat{\sigma}_y^2 / \hat{\sigma}_x^2$. With this value of λ, the ML solution leads to $\hat{\beta}_{RMA} = \pm\,\hat{\sigma}_y / \hat{\sigma}_x$;
d. Ranged standardisation may be used before calculating the MA slope;
e. The method of *covariate ratio* or *instrument variables* is also an option.

Allometric scaling laws of metabolism in animals

Dodds *et al.* (2001) re-examined the '3/4-law' of metabolism and concluded that present theories for scaling live weight by the exponent 3/4 require assumptions and the exponent 2/3 was also plausible. Demetrius (2006) proposed rules based on a combination of chemiosmotic theory of energy transduction and quantum statistics and gave metabolic rate $P = \gamma C \Delta_p W^{\psi}$, where γ is energy transport efficiency within the organism, C represents proton conductance and Δ_p the electrochemical proton gradient. The scaling exponent, ψ, to which body size (W) is raised is a function of metabolic efficiency (μ) and for animals is given by $\psi = (4\mu - 1)/4\mu$. Prediction is that ψ will have the range $2/3 \leq \psi \leq 3/4$. In animals, equilibrium species, typically large organisms, will have $\psi=3/4$ whereas opportunistic species, typically small organisms, will have $\psi=2/3$.

Results and discussion

Estimates of k_f and ME_m were obtained using OLS applied to respiration calorimetry energy balance data for growing steers (n=31) with correlation, $r_{yx}=0.7932$ and 0.7882 with ψ equivalent to '3/4' and '2/3' respectively. Since λ was not known for this data set RMA was also applied and estimates compared with those derived using OLS (Table 1). It is evident that the choice of regression model in particular and metabolic scaling law has consequences for nutritional concepts applied in practical feed evaluation systems.

Table 1. OLS and RMA estimates of efficiency (k_f) and ME_m for '3/4' and '2/3' metabolic scaling laws in relation to steers of 120 kg LW.

Scaling exponent (ψ)	OLS			RMA		
	Slope (k_f)	ME_m		Slope (k_f)	ME_m	
		MJ/kg LW^{ψ}/d	MJ/d		MJ/kg LW^{ψ}/d	MJ/d
'3/4'	0.3960	0.3784	13.72	0.4993	0.4523	16.40
'2/3'	0.3696	0.5228	12.96	0.4690	0.6553	15.94

References

Bartlett, M.S., 1949. Fitting a straight line when both variables are subject to error. Biometrics 5, 207-212.

Demetrius, L., 2006. The origin of allometric scaling laws in biology. J. Theor. Biol. 243, 455-467.

Dhanoa, M.S., R. Sanderson, S. Lopez, J. Dijkstra, E. Kebreab and J. France, 2007. Alternative regression approaches when modelling energy components. In: Ortigues-Marty, I., N. Miraux, and W. Brand-Williams (eds.), Wageningen Academic Publishers, the Netherlands, EAAP Publication No. 124: 593-594.

Dodds, P.S., D.H. Rothman, and J.S. Weitz, 2001. Re-examination of the '3/4-law' of metabolism. J. Theor. Biol. 209, 9-27.

Kendall, M.G. and A. Stuart, 1966. The Advanced Theory of Statistics, Volume 3. Charles Griffin and Company Limited, London, 552 pp.

Characterization of the starch protein matrix in corn and barley endosperm and quantification of their prolamin content by two different methods

G. Giuberti, A. Gallo, M.R.S. Vilas Boas and F. Masoero
Istituto di Scienze degli Alimenti e della Nutrizione, Università Cattolica del Sacro Cuore, Piacenza, Italy

Introduction

Prolamins are hydrophobic storage protein insoluble in the rumen environment (Lawton, 2002). Recently, Lopes *et al.* (2009) found that vitreous corn types contain higher concentration of prolamin than floury or opaque corn types, contrarily to data obtained by Landry *et al.* (2004). These discrepancies could be due to differences in the fractionation methods used (Philippeau *et al.*, 2000). Several laboratory procedures have been periodically published to quantify prolamins in cereals. However, the lack of a reference method could result in uncertain prolamin quantification. The first objective of this study was to characterize maize and barley endosperm multiple protein fractions by using a conventional fractionation method (LND, Landry *et al.*, 2000). The second was to compare the prolamin extraction values obtained by LND against those achieved from a rapid turbidimetric procedure (L&H) proposed by Larson and Hoffman (2008).

Material and methods

Grain samples of eight commercial barley (*Hordeum vulgare*, L.) and eight commercial corn (*Zea mais*, L.) grown in the 2009 season in the north of Italy were included. Dried samples were assayed in duplicate. In particular, by using different solvent, LND divide the protein matrix into multiple fractions (NPN, albumins + globulins, prolamins and glutelins), whereas L&H is employed only for prolamin quantification. Finally, for L&H, a purified gliadin (Sigma-Aldrich Co, St. Louis, MO, USA) was used as alternative to purified zein as standard reference material for barley samples. Data were analyzed by the GLM procedure (SAS, 2003), while the method comparison was performed in agreement with Bland and Altman (1995).

Results and discussion

Analyses (g/100 g of DM) of corn and barley samples, respectively, are as follow: CP 8.36, 13.51; CL 3.83, 1.46; starch 68.61, 55.92; ash 1.37, 2.74. The multiple protein fraction data performed with LND are shown in Table 1. Barley had the highest albumin + globulin and glutelin contents ($P<0.01$). Whereas, by expressing on a CP basis, the prolamin content was higher for corn than for barley (39.5 vs. 28.6 g/100 g of CP, respectively; $P<0.01$). The prolamin contents were different among grain type ($P<0.01$), while the method x grain type interaction was not significant ($P=0.11$). The coefficient of repeatability (CR) and the relative repeatability standard deviation (RS_r) were lower for LND compared to L&H (0.15 and 2.64% vs. 0.43 and 7.03%; respectively). However, the slope estimate (0.84) and the correlation between L&H and LND ($R^2 = 0.71$; $P<0.01$) indicated a good agreement between methods. Though these methods did not differ ($P=0.56$), L&H seems to be more efficient in the extraction of prolamin in barley rather than in corn, because of a zein underestimation of 20%.

Table 1. Multiple protein fraction (g/100 g DM) in barley and corn samples, prolamin content respectively performed by LND and L&H (g/100 g DM) and comparison between tested methods.

	Protein fraction[1]				Prolamin	
	NPN	Alb+Glo	Glutelin	Recovery	LND	L&H
Barley	0.67	2.05^b	7.61^b	105.1%	3.87	4.14^B
Corn	0.68	0.83^a	2.66^a	101.3%	3.30	2.75^A
S.E.M.[2]	0.051	0.105	0.217	0.234	0.285	0.193
CR[3]					0.15	0.43
RS_r[4], %					2.64	7.03

Comparison between methods[5]

	MD[6]	SR[7]	95% Limits agreement		Slope[8]	SE	*P*[9]	R^2
			Lower	Upper				
L&H vs. LND	-0.08	0.15	-1.51	+1.36	0.84	0.68	0.01	0.71

Factorial analysis

	DF	*P*
Grain type (G)	1	<0.01
Method (M)	1	0.56
G × M	1	0.11
Error of model	28	

^{a,b} P<0.01. [1] Extracted with Landry *et al.* (2000). [2] S.E.M. = standard error of the means. [3] CR = coefficient of repeatability. [4] RS_r = relative repeatability standard deviation. [5] Performed in agreement with Bland and Altman (1995). [6] MD = mean of differences. [7] SR = standard error of differences. [8] Intercept did not differ from 0 (P n.s.). [9] *P* of the model of regression analysis.

References

Bland M.J. and D.G. Altman, 1995. Comparing methods of measurements: Why plotting difference against a standard method is misleading. Lancet 346, 1085-1087.

Landry J., S. Delhaye and C. Damerval, 2000. Improved method for isolating and quantitating a-amino nitrogen as nonprotein, true protein, salt-soluble proteins, zeins, and true glutelins in maize endosperm. Cereal Chem. 77, 620-626.

Landry J., S. Delhaye and C. Damerval, 2004. Protein distribution pattern in floury and vitreous endosperm of maize grain. Cereal Chem. 81, 153-158.

Larson J. and P.C. Hoffman, 2008. Technical Note: A method to quantify prolamin proteins in corn that are negatively related to starch digestibility in ruminants. J. Dairy Sci. 91, 4834-4839.

Lawton J. W., 2002. Zein: A history of processing and use. Cereal Chem. 79, 1-18.

Lopes J.C., R.D. Shaver, P.C. Hoffman, M.S. Akins, S.J. Bertics, H. Gencoglu and J.G. Coors, 2009. Type of corn endosperm influences nutrient digestibility in lactating dairy cows. J. Dairy Sci. 92, 4541-4548.

Philippeau C., J. Landry, and B. Michalet-Doreau, 2000. Influence of the protein distribution of maize endosperm on ruminal starch degradability. J. Sci. Food. Agric. 80, 404-408.

SAS, 2003. Release 9.13. SAS Ins. Inc., Cary, NC, USA.

Cereal cultivar affects protein and starch digestion of grains in different parts of gastrointestinal tract of ruminants

P. Micek and Z.M. Kowalski
Department of Animal Nutrition and Feed Management, University of Agriculture in Krakow, Al. Mickiewicza 24/28, 30-059 Cracow, Poland

Introduction

Genetic progress in cereal breeding results not only in new cultivars but also in variability of nutritive value within grain species. Our aim was to examine the effect of cereal species and cultivar on chemical composition of grain as well as on ruminal degradability and intestinal digestibility of protein (CP) and starch. Additionally, density, porosity, hardness and weight of 1000 grain (WTG) were measured as predictors of ruminal starch and protein digestibility.

Material and methods

The study was carried out on 43 cultivars of wheat, triticale, rye, barley, oats and maize commonly used in Poland in diets for ruminants. Chemical composition of grain was determined using standard methods. Starch content was determined according to Faisant *et al.* (1995). *In situ* ruminal degradability was determined by the method of Michalet-Doreau *et al.* (1987), using 3 dry cows with rumen and duodenum fistulas, fed standard diets. Samples were ground to pass through a 1.5 mm sieve. Effective rumen digestibility (ERD) and digestibility rate constants (a, b, c) were calculated according to Ørskov and McDonald (1979) at a ruminal outflow (k) of 0.06 per h. Intestinal digestibility (ID) of by-pass CP and starch was measured using the mobile nylon bag technique (Peyraud *et al.*, 1988). Density of grain was measured using an electronic device, porosity by means of a gas pycnometer and hardness by modified Vickers microhardness test. The data were subjected to one-way analysis of variance (cereal species as a main experimental factor) using the GLM procedure of SAS.

Results and discussion

There was wide variation between cultivars within cereal species in the chemical composition of grain. The differences among cultivars often exceeded those between the species. As expected, considerable variation was also found in the ERD and ID of protein and starch. ERD of CP ranged from 41.1% for maize to 89.7% for oats. Similarly ERD of starch was highest for oats (99.5%) and lowest for maize (45.3%) with important variation occurring between cultivars within the species (maize type: dent 52.6%, flint 29.8%, hybrids 48.0% in average). At the same time oats grain was characterised by the lowest ID of CP (56.3% vs. 78.9% for maize and 87.7% for wheat). The ERD of CP and starch were significantly (P<0.01) correlated with some physical characteristics of grain, especially WTG (R^2=0.770 and 0.771) or porosity (R^2=0.584 and 0.521, respectively). Ruminal protein degradability could be predicted more accurately by the combination of both these parameters (R^2=0.864) and starch degradability by the combination of WTG and starch contents (R^2=0.838).

To improve the precision of prediction of amino acids or glucose absorbed in the intestines would require data on cultivars or groups of cultivars in the feeding tables instead of one value for each species or to predict using regressions based on chemical composition. Results of this study suggest that physical characteristics of grain can also be used in such equations.

Table 1. Mean chemical composition, density, porosity, hardness, weight of 1000 grain (WTG) and digestibility indicators of cereal grains (means without naked cultivars).

Species	Wheat	Triticale	Rye	Barley	Oats	Maize
No of cultivars	11	6	4	11	6	5
Chemical composition, g/kg d.m.						
Crude protein	159±15[a]	136±12[b]	106±6[c]	119±17[bc]	132±13[b]	118±9[bc]
Starch	666±68	660±45	644±79	639±61	555±77	663±76
NDF	170±31[bc]	207±62[ab]	212±18[ab]	236±18[a]	259±39[a]	150±18[c]
Physical parameters						
Density, $g \cdot cm^{-3}$	1.31±0.02[a]	1.19±0.07[b]	1.24±0.02[b]	1.19±0.0[b]3	1.19±0.07[b]	1.26±0.02[b]
Porosity, %	45.0±2.2[d]	49.4±2.3[c]	47.9±0.5[c]	52.1±2.8[b]	60.6±1.6[a]	40.2±0.5[e]
Hardness, MPa	75.4±11.1[ab]	44.1[bc]±15.7	72.3±34.1[ab]	40.3±20.0[bc]	9.3±2.0[c]	99.9±20.7[a]
WTG, g	45.7±5.9[b]	43.8[b]±8.5	30.2±3.3[b]	45.5±6.9[b]	31.6±3.5[b]	280.9±78.3[a]
Digestibility indicators, %						
Crude protein						
ERD[2]	72.5±4.0[d]	76.3±3.6[cd]	83.6±2.8[b]	80.0±5.8[bc]	89.7±2.2[a]	41.1±8.3[e]
a[1]	37.1±6.4[b]	36.8±4.1[b]	39.7±6.0[b]	37.0±4.7[b]	71.9±7.2[a]	23.3±5.2[c]
b[1]	60.5±6.4[b]	60.1±4.3[b]	54.5±5.1[d]	57.9±5.2[b]	23.5±7.5[c]	73.0±3.7[a]
c[1]/h	0.118±0.026[c]	0.127±0.03[c]	0.273±0.012[a]	0.203±0.078[b]	0.198±0.062[b]	0.022±0.005[d]
Intestinal digestibility	87.7±3.0[a]	83.1±4.3[ab]	70.9±5.9[d]	75.1±8.6[cd]	56.3±6.5[e]	78.9±8.1[bc]
Starch						
ERD[2]	92.5±2.5[b]	95.6±1.5[b]	96.0±1.2[b]	93.8±2.0[b]	99.5±0.4[a]	45.3±9.9[c]
a[1]	43.8±11.5[b]	45.1±8.1[b]	47.5±9.6[b]	39.9±8.5[b]	92.9±5.3[a]	20.4±8.0[c]
b[1]	55.3±11.2[b]	54.1±7.9[b]	51.8±9.6[b]	59.6±8.3[b]	7.0±5.3[c]	79.2±7.9[a]
c[1]/h	0.501±0.199[c]	0.866±0.141[a]	0.842±0.128[a]	0.656±0.200[bc]	0.786±0.271[ab]	0.043±0.012[d]
Intestinal digestibility	100±0	100±0	100±0	100±0	100±0	100±0

[a,b,c,d] Means within a row with different letters are significantly different ($P \leq 0.05$).

[1] Digestibility rate constants.

[2] Effective rumen digestibility.

References

Faisant, N., V. Planchot, F. Kozlowski, M.P. Pacouret, P. Colonna and M. Champ, 1995. Resistant starch determination adapted to products containing high level of resistant starch. Sci. Alim. 15, 83-89.

Ørskov, E.R. and P. McDonald, 1979. The estimation of protein degradability in the rumen from incubation measurements weighted according to rates of passage. J. Agr. Sci. 92, 499-503.

Peyraud, J.L., Ch. Genest-Rulquin and R. Verité, 1988. Mesure de la digestion de l'azote des aliments dans l'intestin des ruminants par la technique des sachets mobiles. 1. Evaluation de la quantité de matières azotées indigestibles en sachet des principaux aliments. Reprod. Nutr. Develop. 28, 129-130.

Enzymatic corn starch degradability in high moisture corn and dried corn grains

M.R.S. Vilas Boas, A. Gallo, G. Giuberti and F. Masoero
Istituto di Scienze degli Alimenti e della Nutrizione, Università Cattolica del Sacro Cuore, Piacenza, Italy

Introduction

Recent studies demonstrated that ruminal maize starch degradability could be influenced by many factors such as particle size (Rémond *et al.*, 2004), vitreousness (Ngonyamo-Majee *et al.*, 2008), stage of maturity of whole plants (Correa *et al.*, 2002) or by the composition of the protein matrix (Philippeau *et al.*, 2000). In particular, Sniffen (2009) reported that as maize becomes more mature, the 7 h ruminal *in vitro* starch degradability (7h-IVSD) was reduced. Prolamins content was also associated to the stage of maturity of corn: the maximum prolamin content in corn grain was measured at the harvest near to the back-layer phase (Hoffman and Shaver, 2009). The aim of this study was to evaluate enzymatic starch degradation of different types of corn and the characteristics that may influence starch degradation.

Material and methods

The set of samples was composed by 8 dry corn grains (CG), 9 high moisture 'shelled' corns (HMS) and 9 high moisture 'ear' corns (HME), collected in farms of the Po Valley. The samples were analysed for DM, starch, CP, albumin/globulin and zein contents. Enzymatic starch degradation was evaluated using a multi-enzymatic procedure (pancreatin and amyloglucosidase) by measuring the amount of glucose released after 8 and 24 h of incubations. The data were analysed by the proc GLM and proc CORR of SAS (SAS, 2003).

Results and discussion

The DM content of samples was higher in CG (890 g/kg) compared to HMC and HME (654 g/kg and 516 g/kg, respectively; $P<0.01$). The three types of samples were similar in CP content, but with different zein and albumin-globulin contents. In particular, zeins increased ($P<0.01$) with the increased of the DM (1.38, 1.93 and 3.70 respectively for HME, HMC and CG; Table 1). On the contrary, the albumin-globulin decreased ($P<0.01$) from 6.13 to 1.98, respectively for HMC and CG. This could be due to a progressive accumulation of starch and zein during kernels ripening and an encapsulation of starch granules by hydrophobic zein (Lasztity, 1984). The starch degradation after 8h of incubation range from 22.1 to 42.4%, with the highest value observed for HME and the lowest for CG (Table 1). The moisture content was related to 8 h and 24 h enzymatic starch degradability (r=0.70 and r=0.84, respectively; $P<0.01$). Moreover, the enzymatic starch degradation was negatively influenced by the zein content at both 8 h and 24 h enzymatic incubation times (r=-0.50 and r=-0.75, respectively; $P<0.01$). On the contrary, albumin-globulin resulted positively related with the enzymatic starch degradation at 24h (r=0.70; $P<0.001$). Philippeau *et al.* (2000) argued about the negative influence of zein on the ruminal *in situ* starch degradation, even if the authors expressed zein as a percentage of starch. In conclusion, more studies need to be done in order to better understand the effect of factors influencing the starch degradation in ruminants.

Table 1. Chemical composition and enzymatic starch degradation at 8 and 24 h of incubation of CG, HMC and HME samples.[1]

Variable	High moisture corn			High moisture ear corn			Corn grain		
	Mean±SD	Min	Max	Mean±SD	Min	Max	Mean±SD	Min	Max
DM	65.4±5.14	56.0	72.0	51.6±5.5	39.8	58.2	98.7±0.87	88.4	90.5
Starch	66.7±2.7	60.3	69.2	51.1±5.51	42.6	57.6	68.7±3.32	61.9	71.9
CP	8.75±0.97	7.34	10.4	8.5±1.45	6.05	11.2	9.70±1.24	8.27	12.4
Zein[1]	1.93±0.72	0.84	2.85	1.38±0.53	0.74	2.19	3.70±1.04	1.43	4.72
Alb/Glob[1]	6.13±1.32	4.44	8.18	5.42±1.23	3.5	7.9	1.98±0.54	0.98	2.8
Deg. 8 h[2]	25.4±4.25	18.8	33.4	42.4±12.4	26.4	61.7	22.1±4.1	16.2	27.0
Deg. 24 h[2]	81.1±9.1	67.3	98.5	90.2±9.7	77.4	104.6	53.5±8.3	35.2	63.5

[1] Albumin/globulin (Alb/Glob) measured by Landry *et al.* (2000); Zein measured by Larson and Hoffman (2008).
[2] Deg. 8 h and Deg. 24 h estimated with a multi-enzymatic method and determined basing on the percentage of incubated starch.

References

Correa, C.E.S., R.D. Shaver, M.N. Pereira, J.G. Lauer and K. Kohn, 2002. Relationship between corn vitreousness and ruminal *in situ* starch degradability. J. Dairy Sci. 85, 3008-3012.

Dorshorst, A. E., P.C. Hoffman, N. M. Esser, M. G. Bertram and K. Seeger, 2009. Factors effecting corn silage starch hydrolysis potential. J. Anim. Sci. 86 (Suppl.), 29.

Landry, J., S. Delhaye and C. Damerival, 2000. Improved method for isolating and quantitating α-amino nitrogen as nonprotein, true protein, slat soluble protein, zeins and true glutelins in maize endosperm. Cereal Chem. 74, 620-626.

Larson, J. and P.C. Hoffman, 2008. Technical note: a method to quantify prolamin proteins in corn that are negatively related to starch digestibility in ruminants. J. Dairy Sci. 91, 4834-4839.

Lasztity, R., 1984. The chemistry of cereal proteins. CRC Press, Inc. Boca Raton, FL.

Hoffman, P.C. and R.D. Shaver, 2009. Corn biochemistry: factors related to starch digestibility in ruminants. In Proceedings of the Cornell Nutrition Conference for Feed Manufactures, 71st Meeting. October 20-22, 2009. East Syracuse, New York.

McAllister, T.A., R.C. Phillippe, L.M. Rode, and K.J. Cheng, 1993. Effect of the protein matrix on the digestion of cereal grains by ruminal microorganisms. J. Anim. Sci. 71, 205-212.

Newbold, J.R., E.A. Lewis, J. Lavrijssen, H.J. Brand, H. Vedder and J. Bakker, 2006. Effect of storage time on ruminal starch degradability in corn silage. J. Dairy Sci. 89 (Suppl. 1), T94.

Philippeau, C., J. Landry and B. Michalet-Doreau, 2000. Influence of the protein distribution on maize endosperm on ruminal starch degradability. J. Sci. Food and Agri. 80, 404-408.

SAS, 2003. SAS/SAT Guide for Personal Computers 9.13. SAS Institute Inc. Cary, NC, USA.

Xylose treatment reduced the *in vivo* rumen degradation of starch and protein in cereals and legumes

M. Larsen, P. Lund, M.R. Weisbjerg and T. Hvelplund
Faculty of Agricultural Sciences, Aarhus University, 8830 Tjele, Denmark

Introduction

Feeding diets that partly shifts the site of starch digestion from the rumen to the small intestine is, in theory, an attractive strategy to overcome some of the nutritional shortcomings associated with meeting the nutrient needs of transition cows. Using the *in situ* technique, Südekum *et al.* (2004) found that xylose treatment reduced the ruminal degradability of both starch and protein in wheat. Therefore, we investigated the *in vivo* effect of xylose treatment of barley, wheat and peas on site of starch digestion and ruminal protein degradability. The starch digestion data has previously been presented by Larsen *et al.* (2009).

Material and methods

The effect of xylose treatment was studied using rolled/cracked barley, wheat and peas either untreated (-Xy, n=4) or treated with xylose (+Xy, n=2) in simple cross-over experiments with 14 d periods using lactating Holstein cows cannulated in the rumen, distal duodenum and terminal ileum. Rumen fluid, duodenal and ileal digesta, and faeces were sampled 12 times during the last 5 d of each 14 d period. Chromic oxide was used as intestinal flow marker. Rumen fluid was sampled medially in rumen contents. Roller mills with either conventional or special rollers were used to roll or crack starch sources. Cracked feeds were treated with xylose in aqueous Ca-Mg lignosulphonate solution under elevated temperature (Südekum *et al.*, 2004). Diets were fed *ad libitum* as TMR and composed of the respective starch source and grass-clover silage balanced with soy bean meal and a mineral premix. The *in vivo* duodenal flow of ruminally undegraded N from starch sources was calculated by correcting the total duodenal N flow for microbial N (DAPA and purines as markers), endogenous N (37.3% of total N; Larsen *et al.*, 2000) and ruminal undegraded N from soy bean meal and grass-clover silage. The non-additive feed evaluation system NorFor was used to calculate rumen undegraded N from soy bean meal and grass-clover silage. Data was analysed according to an incomplete 3×2 factorial design with cow as random factor. The model included the fixed effects of starch source (So), xylose (Xy) and the interaction (So×Xy).

Results and discussion

Xylose treatment reduced the *in vivo* ruminal degradability of both starch and protein ($P<0.01$ and $P=0.02$, respectively; Table 1) in all tested starch sources. With the *in situ* technique, Südekum *et al.* (2004) found xylose treatment of wheat to reduce the ruminal degradability of protein and starch with 30 and 18 percentage units, respectively. The present reduction in ruminal protein degradability was similar, but the reduction in starch degradation was less, e.g. 9 percentage units. Assuming the small intestinal protein digestibility to be unaffected by xylose treatment, the metabolisable protein value of xylose treated feeds might be enhanced.

The percentage of starch intake digested in the small intestine was unaffected by starch source ($P=0.47$; Table 1) and xylose treatment ($P=0.25$). Even so, xylose treatment of wheat numerically increased the small intestinal digestibility from 6 to 13% of ingested starch. However, similar small intestinal digestibilities of starch from rolled wheat and xylose treated wheat was observed (12±3% and 10±5%, respectively), when assessed as the recovery of glucose equivalents in hepatic portal blood using multicatheterised lactating dairy cows (N.B. Kristensen and M. Larsen, unpublished results). Overall, it seemed not possible to increase the absolute amount of starch digested in the small intestine by application of xylose treatment to the starch sources investigated. In contrast, the total tract digestibility of starch was reduced with xylose treatment of barley and wheat, but not with peas (interaction, $P<0.01$).

The reduction in ruminal starch degradation with xylose treatment was accompanied by increased ($P=0.01$; Table 1) total tract NDF degradation. This is a likely reason for the microbial protein synthesis to be unaffected by xylose treatment ($P=0.29$; not shown). Consistent with reduced ruminal starch degradation with xylose treatment, the minimum ruminal pH was higher ($P=0.01$) and the proportion of propionate in ruminal VFA was lower ($P<0.01$).

In conclusion, xylose treatment generally reduced ruminal protein degradability and might increase the metabolisable protein value of barley, wheat and peas. Xylose treatment reduced the ruminal starch degradability, but this was not associated with increased starch digestion in the small intestine. In contrast, the total tract digestibility of starch was reduced for cereals.

Table 1. Effects of xylose (Xy) treatment of barley, wheat and peas (So), means ± SEM.

Item	Barley		Wheat		Peas		SEM	*P*-value		
	-Xy	+Xy	-Xy	+Xy	-Xy	+Xy		So	Xy	So×Xy
Ruminal protein degradability, %										
Total diet protein	76	71	76	68	77	75	4	0.66	0.06	0.60
Starch source	78	62	79	50	81	70	11	0.65	0.02	0.50
Starch digestibility, % of intake										
Ruminal	89	79	90	81	71	66	2	<0.01	<0.01	0.42
Small intestinal	7	5	6	13	8	9	2	0.47	0.25	0.18
Total tract	98	88	99	96	91	92	1	<0.01	<0.01	<0.01
NDF digestibility, % of intake										
Ruminal	58	56	57	68	66	69	2	0.02	0.07	0.10
Total tract	55	58	60	67	68	71	1	<0.01	0.01	0.28
Ruminal environment										
Mean pH	5.7	5.7	5.8	6.0	5.8	6.2	0.1	0.25	0.06	0.17
Minimum pH	5.4	5.4	5.3	5.8	5.6	5.9	0.1	0.06	0.01	0.08
Total VFA, mM	149	149	145	127	151	125	6	0.19	<0.01	0.08
Acetate, mol%	58	60	56	62	61	65	1	<0.01	<0.01	0.11
Propionate, mol%	27	24	28	22	21	19	1	<0.01	<0.01	0.18
Butyrate, mol%	11	13	12	13	13	13	1	0.12	0.28	0.22

References

Larsen, M., T.G. Madsen, M.R. Weisbjerg, T. Hvelplund and J. Madsen, 2000. Endogenous amino acid flow in the duodenum of dairy cows. Acta Agric. Scand. Anim. Sci. 50, 161-173.

Larsen, M., P. Lund, M.R. Weisbjerg and T. Hvelplund, 2009. Digestion site of starch from cereals and legumes in lactating dairy cows. Anim. Feed Sci. Technol. 153, 236-248.

Südekum, K.H., M. Klein, M. Paschke-Beese and O. Schade, 2004. Ruminal nutrient degradation of untreated and chemically treated wheat grain. J. Dairy Sci. 87 (Suppl. 1), 438-439.

Effect of N source (soybean meal vs. whole lupin) and of yeast addition on digestion and ruminal N metabolism in sheep

M. Doreau[1], S. Laverroux[1], F. Chaucheyras-Durand[2] and C. Poncet[1]
[1]INRA, Unité de Recherches sur les Herbivores, 63122 Saint-Genès Champanelle, France
[2]Lallemand Animal Nutrition, 31702 Blagnac Cedex, France

Introduction

Protein from grain legumes are highly degradable in the rumen; however, coarse grinding may reduce rumen protein degradation compared to fine grinding when determined *in situ* (Moss *et al.*, 1997). Live yeasts enhance ruminal protein synthesis under specific conditions (Chaucheyras-Durand *et al.*, 2008) and may decrease ruminal degradability by reducing bacterial proteolysis and peptidolysis (Chaucheyras-Durand *et al.*, 2005). This experiment compared whole lupin seeds to soybean meal and the addition of yeast on ruminal N metabolism in sheep.

Material and methods

Four sheep fitted with rumen and duodenal cannulas were used in a Latin square design. They received diets of 60% hay, 23% maize, 1% mineral-vitamin premix and 16% protein feed, either soybean meal or coarsely cracked white lupin seeds, in the absence (diets S and L) or presence (diets SY and LY) of live yeast (4 x 10^9 cfu/day of *Saccharomyces cerevisiae* CNCM I-1077) distributed twice daily through rumen cannula. Diets S and L contained 16.6 and 14.7% crude protein, of which 49 and 43% were from protein sources, respectively. Sheep were fed at 08:00 and 20:00 h. Each period lasted 5 weeks. Total tract digestibility and N balance were determined by total faeces and urine collection. Ruminal digestibility was determined by the double marker technique using ^{103}Ru-phenanthroline and ^{51}Cr-EDTA. Microbial protein synthesis was measured using $^{15}(NH_4)_2SO_4$. Rumen liquid samples were taken between 08:00 and 18:00 h, at 8 times for pH and NH3 determination and at 4 times for non-ammonia N determination. Statistical analysis was performed using GLM procedure of SAS; protein source (S vs. L), yeast and protein source x yeast interaction were included in the model. Ruminal kinetics were analysed using the repeated option of GLM.

Results and discussion

Ammonia concentration in rumen liquid was higher with soybean meal than with lupin ($P<0.01$) and was decreased by yeast addition ($P<0.05$). The yeast effect was greatest between 1 and 6 h after feeding. Non-ammonia N concentration in rumen liquid did not vary with protein source, but decreased with yeast ($P=0.05$), especially 2 h after feeding ($P<0.01$).

Organic matter (OM) ruminal digestibility tended to be lower for L than for S, but OM total tract digestibility did not differ between diets (Table 1). No difference among diets was observed for microbial, nonmicrobial and total N duodenal flows, but efficiency of microbial N synthesis tended to be higher for lupin diets. Lupin diets resulted in more faecal N and less urinary N than soybean meal diets. Yeast and the interaction between yeast and protein source were not significant for any parameter.

This experiment shows that N duodenal flow of coarsely cracked lupin is similar to that of soybean meal, both from microbial and nonmicrobial flows. As Froidmont *et al.* (2008) did not find any effect of lupin particle size on *in vivo* nonmicrobial N flow, the higher ruminal degradability which is widely accepted for lupin compared to soybean meal may be an artifact due to the *in situ* method. Indeed lupin ruminal degradation results in insoluble protein and peptides which contribute to N duodenal flow (Rémond *et al.*, 2003). As OM digested in the rumen tends to be lower for lupin

Table 1. Effect of diets containing soybean meal or lupin, with or without live yeast.

	S	L	SY	LY	SEM	S vs. L[1]
Dry matter intake (g/d)	1,103	1,110	1,113	1,102	5.2	NS
N intake (g/d)	28.8	25.3	29.0	25.2	0.13	0.01
OM total tract digestibility (%)	74.1	73.5	74.4	73.8	0.54	NS
OM ruminal digestibility (%)	47.5	44.7	48.0	45.5	1.12	0.06
NAN duodenal flow (g/d)	27.3	26.6	27.2	26.7	0.79	NS
NAN duodenal flow (% N intake)	94.8	105.3	93.7	106.2	2.98	0.01
Nonmicrobial NAN[2] duodenal flow (g/d)	8.1	6.5	7.8	6.4	0.37	NS
Microbial N duodenal flow (g/d)	19.3	20.2	19.3	20.3	0.63	NS
Microbial N (g/kg OMDR[2])	40.2	44.3	39.5	44.5	2.03	0.07
N intestinal digestibility (% duodenal N)	77.0	75.9	76.0	76.2	0.55	NS
Faecal N (% N intake)	24.3	27.9	24.6	27.6	0.65	0.01
Urinary N (% N intake)	62.5	56.9	62.4	57.5	1.35	0.01

[1] Yeast and protein source x yeast effects are not significant.
[2] NAN: non ammonia N; OMDR: organic matter digested in the rumen.

than for soybean meal, microbial N efficiency tends to be higher for lupin. The high particle size of lupin did not impair intestinal digestibility, as shown by Froidmont *et al.* (2008). The higher urinary N with soybean meal may be due to the higher N intake, because retained N does not vary with protein source. The effect of yeast on ruminal soluble N does not result in changes in microbial and nonmicrobial N flows. This discrepancy may be due either to differences in rumen pools or passage rate, as shown by Doreau and Jouany (1998), or to changes in microbial communities involved in production and utilisation of soluble N.

References

Chaucheyras-Durand, F., S. Masséglia and G. Fonty, 2005. Effect of the microbial additive *Saccharomyces cerevisiae* CNCM I-1077 on protein and peptide degrading activities of rumen bacteria grown *in vitro*. Curr. Microbiol. 50, 96-101.

Chaucheyras-Durand, F., N.D. Walker and A. Bach, 2008. Effects of active dry yeasts on the rumen microbial ecosystem: Past, present and future. Anim. Feed Sci. Technol. 145, 5-26.

Doreau, M. and J.P. Jouany, 1998. Effect of a *Saccharomyces cerevisiae* culture on nutrient digestion in lactating dairy cows. J. Dairy Sci. 81, 3214-3221.

Froidmont, E., M. Bonnet, R. Oger, V. Decruyenaere, J.M. Romnée, Y. Beckers and N. Bartiaux-Thill, 2008. Influence of the grinding level and extrusion on the nutritional value of lupin seed *Lupinus albus*) for cattle in the context of the Dutch protein evaluation system. Anim. Feed Sci. Technol. 142, 59-73.

Moss, A.R., D.I. Givens, H.F. Grundy and K.P.A. Wheeler, 1997. The nutritive value for ruminants of lupin seeds from determinate plants and their replacement of soya bean meal in diets for young growing cattle. Anim. Feed Sci. Technol. 68, 11-23.

Rémond, D., M.P. Le Guen and C. Poncet, 2003. Degradation in the rumen and nutritional value of lupin (*Lupinus albus* L.) seed proteins: effect of extrusion. Anim. Feed Sci. Technol. 105, 55-70.

Part 10. Evaluation and modelling of feed value and requirements: monogastrics

Metabolic utilization of energy in monogastric animals and its implementation in net energy systems

J. Noblet[1], S. Dubois[1], E. Labussière[1], B. Carré[2] and J. Van Milgen[1]
[1]INRA, UMR1079 SENAH, F-35590 Saint-Gilles, France
[2]INRA, UR83 Recherches Avicoles; 37380 Nouzilly, France

Abstract

The evaluation of the energy content of feeds for monogastric animals has been most commonly based on their DE or ME contents. However, the closest estimate of the 'true' energy value of a feed should be its NE content, which takes into account differences in metabolic utilization of ME of nutrients. This review considers first some methodological aspects of NE determination with emphasis on the impact of feeding level on fasting heat production and subsequent calculation of NE value. Experimental data indicate that the NE/ME ratio varies greatly with the chemical composition of diets and nutrients (fat>starch>protein=fibre) and NE systems are then better in predicting the performance of monogastric animals. However, literature data also suggest that the impact of moving to a NE system would be more important and therefore more justified for pigs and veal calves than for poultries. In any case, the accuracy of the NE value is highly dependent on the accuracy of measuring DE or ME values or the content of digestible nutrients.

Introduction

The cost of feed is the most important cost of meat production from monogastric animals (~60%) and energy represents the greatest proportion of this cost. Therefore, it is important to estimate precisely the energy value of feeds, either for least-cost formulation or for adapting feed supply to the energy requirements of animals. Evaluation of the energy content of feed for monogastric animals has been most commonly based on the DE or ME contents. However, the closest estimate of the 'true' energy value of a feed should be the NE content, which accounts for differences in metabolic utilization of ME of nutrients for maintenance and production. In addition, NE is the only system in which energy requirements and diet energy values are expressed on a basis that is theoretically independent of the feed characteristics. In several parts of the world, NE systems have been implemented, especially for pigs. However, NE systems are used very little for poultry or veal calves. The objective of this review paper is to consider recent contributions regarding the efficiency with which ME is used in pigs, poultry and veal calves while focussing on growing animals. More complete information can be obtained in recent reviews (Pirgozliev and Rose, 1999; Noblet and van Milgen, 2004; Noblet, 2006). It should be kept in mind that the ultimate energy value of a feed for any monogastric animal depends on the chemical composition of the feed, animal factors such as body weight, physiological stage or species, and technological factors such as particle size, pelletting, extrusion or the addition of enzymes that affect primarily the digestion of nutrients and energy. This latter aspect of feed energy evaluation will not be considered in the current review (Noblet and Le Goff, 2001; Noblet and van Milgen, 2004).

Methodological aspects

Not all gross energy (GE) of a feed is available for meeting the requirements of animals since variable proportions of GE are lost in faeces, in urine, as fermentation gases (i.e. methane, hydrogen) and as heat increment (HI). The DE content of a feed corresponds to its GE content minus faecal energy losses after digestion in the digestive tract. Even though related to digestion, energy losses as gas and heat originating from hindgut fermentation are not considered in the calculation of DE. The ME content of a feed corresponds to the difference between the DE content and energy losses in urine and gases. Most of the energy lost in gases is due to methane production, which is negligible in poultry and milk-fed veal calves, very small in growing pigs (<0.5% of GE) and significant (1-3% of GE intake) in adult sows (Le Goff *et al.*, 2002) or veal calves receiving solid feed in addition to milk (3-9% of GE from solid feed, Labussière *et al.*, 2009a). Most ME values reported in the literature and in feeding tables for monogastric animals ignore energy losses as methane.

Net energy is defined as the ME content minus HI associated with feed utilization (i.e. the energy cost of ingestion, digestion, and metabolic utilization of ME) and the energy cost corresponding to a 'normal' level of physical activity (Figure 1). The NE/ME ratio (or k) corresponds to the efficiency of ME utilization for NE. The NE/ME ratio also corresponds to 1-(HI/ME). However, the HI/ME ratio of a given feed is not constant over a large range of ME intakes for a given animal and depends on several physiological factors. For instance, the HI is lower for ME supplied below the maintenance energy requirement than for ME supplied above maintenance (Noblet *et al.*, 1993; 1994a,b; Birkett and de Lange, 2001). The HI is also lower when ME is used for fat deposition compared with protein deposition (Noblet *et al.*, 1999). As the proportion of fat deposition typically increases more rapidly than the protein deposition with increasing ME intake, the HI/ME should, at least theoretically, be lower at higher levels of ME intake. Therefore, to compare different feeds for HI or the efficiency of ME utilization, it is necessary to calculate these values under similar conditions, at protein and amino acid supplies meeting the requirement and/or a constant composition of the gain and/or at a given physiological stage.

For a growing animal, NE intake is calculated as the sum of retained energy (RE) at a given production feeding level and the fasting heat production at zero activity (FHP) (Noblet *et al.*, 1994a). This NE value and the corresponding k value then correspond to a combined utilization of energy for meeting requirements for maintenance and growth. The RE is either measured by the comparative slaughter technique or, more frequently, calculated as the difference between ME intake and HP estimated by calorimetry. The FHP is either measured directly in fasting animals or obtained from literature data. It can also be calculated by extrapolating HP measured at different feeding levels to zero ME intake (Figure 2; FHPr). However, the latter method, even though it has been widely used in the past, has important limitations. First, it consists of extrapolating HP measured at feed intake levels typically

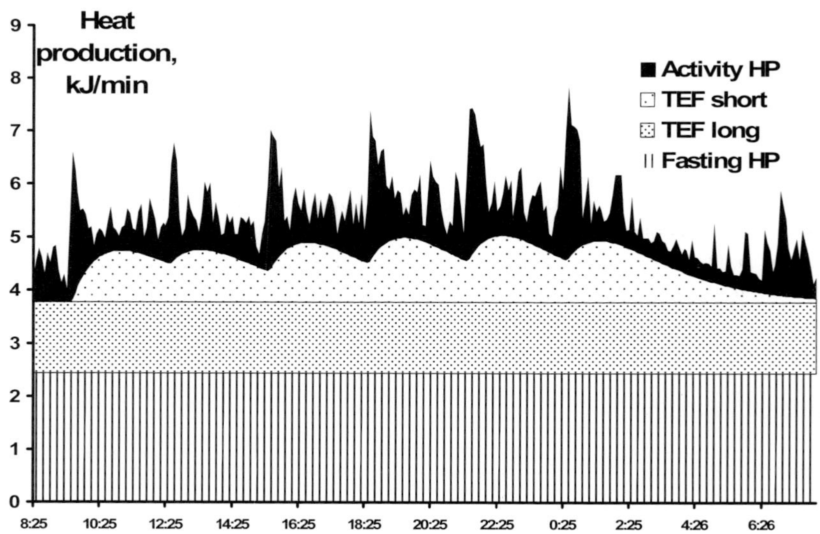

Figure 1. Dynamics of components of heat production in a group of broilers (1 kg BW) fed six meals per day. Measurements were obtained by indirect calorimetry both in the fed and fasting state (INRA data; HP: heat production; TEF: thermic effect of feed).

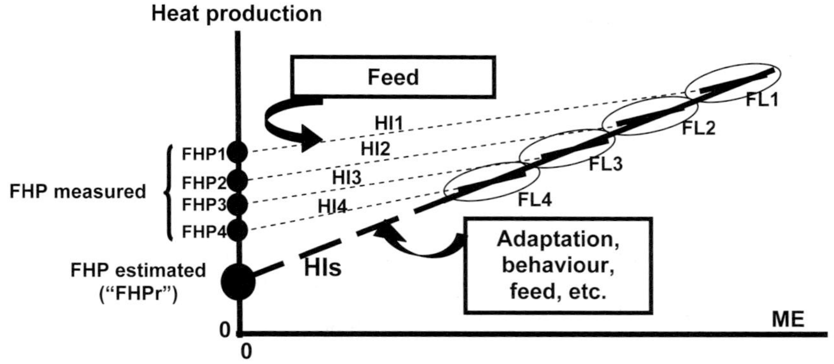

Figure 2. Schematic representation of the effect of feeding level (FLi) on heat production and fasting heat production (FHP) in monogastric animals; each FHPi corresponds to the FHP measured on animals receiving the FLi during the immediate previous period; FHPr (s for statistical) is obtained from the regression between HP and ME and calculated as HP at zero ME intake and the slope is the 'statistical' heat increment (HIs); the slope between FHPi and HPi corresponds to measured heat increment (HIm) (adapted from Koong et al., *1982, De Lange* et al., *2007 and Labussière* et al., *2009b).*

ranging between 60 to 100% of *ad libitum* to HP at zero feed intake, with subsequent inaccuracies in the slope and intercept. Second and more importantly, the measured FHP is not constant and is affected by the feeding level prior to fasting, especially in growing animals (Koong *et al.*, 1982; De Lange *et al.*, 2006; Labussière *et al.*, 2008a, 2010). Apparently, the animal adapts its basal energy expenditure to the level of feed intake and/or growth. The latter authors observed that FHPr was markedly lower than measured FHP with subsequent lower values for NE and k value, and a higher HI (Figure 2). They also observed that HI, calculated as HP minus the measured FHP and expressed per unit of ME, is constant for different feeding levels. Furthermore, the degree of adaptation of FHP and HP to feeding level also depends on animal characteristics such as the genotype (Renaudeau *et al.*, 2007). These observations question the use of FHPr as an estimate of FHP for calculating NE values. The direct measurement of FHP according to indirect calorimetry methods immediately after a fed period is highly preferable. If it is not possible to obtain these measurements, literature values of FHP can be used as an alternative. The HP also depends on climatic factors with an increased HP and reduced RE if the animals are kept below thermoneutrality. It is therefore recommended to keep the animals above thermoneutrality to avoid bias in estimating NE and k.

From a practical point of view and to avoid bias in the calculation of NE for different feeds, it is necessary to carry out energy balance measurements in similar animals (i.e. same sex, same breed and in the same body-weight range), to keep these animals within their thermoneutral zone, to minimize variation in behaviour, and to feed the animals at about the same feed intake level with balanced diets so that the animals can express their growth potential. Under these circumstances, an erroneous estimate of FHP will affect the absolute NE value, but not the ranking between feeds. This also means that NE should not be measured in animals fed ingredients for which the chemical characteristics are very different from those of a complete balanced diet.

While measurements of DE and, to a lesser extent, of ME are relatively easy and can be undertaken on a large number of feeds at a reasonable cost, the actual measurement of NE is far more complex and expensive. The best alternative is then to use reliable NE prediction equations established from measurements carried out under similar and standardized conditions. In our laboratory and for pigs, we proposed prediction equations to estimate the NE value of ingredients and complete diets based on DE or ME content, combined with information on chemical characteristics (Noblet *et al.*, 1994a).

Different predictors (i.e. independent variables) can be used originating from measured chemical composition, existing feeding tables or digestibility trials.

Heat production can be measured directly through direct calorimetry, estimated from gas exchanges through indirect calorimetry or calculated as the difference between ME intake and energy gain obtained by the comparative slaughter technique. The latter technique can easily be used in small animals such as poultry, but is much more difficult to perform in large animals. The most commonly used method is indirect calorimetry, which consists of measuring oxygen consumption, and carbon dioxide and methane production. These measurements, combined with the urinary energy production, are then used to calculate HP (Brouwer, 1965). This method also allows measurements over a short period of time (i.e. a few days) with possibilities of combination of measurements at different feeding levels (including fasting) on the same animal without adaptation. Modelling methods can be implemented to partition the total HP between different components, which can be used in the further interpretation of energy balance data (Van Milgen *et al.*, 1997; Figure 1).

In conclusion, the NE value of a feed and the corresponding k value should be evaluated according to standardized and adequate methods. The values are dependent on assumptions (FHP), conditions of measurement (e.g. climate, activity.) and the composition of the energy gain. This means that data on NE and k available in the literature should be interpreted with caution and may not be directly comparable.

Utilization of ME in pigs

Over the last 50 years, several experiments have been carried out by different laboratories to quantify the effect of diet and animal factors on HI or k in pigs (see the review of Noblet, 2006). The most recent and complete study was carried out by INRA with measurements on 61 diets (Noblet *et al.*, 1994a; Noblet, 2006). From their trials and other results, these authors showed that the maintenance energy requirement and FHP in 30-100 kg growing pigs are proportional to $BW^{0.60}$, and not to the commonly used metabolic BW ($BW^{0.75}$) (Noblet *et al.*, 1999). The FHP at thermoneutrality and zero activity averaged 750 kJ/kg $BW^{0.60}$/d and this value was confirmed in later studies under similar conditions (Van Milgen *et al.*, 2001a; Le Bellego *et al.*, 2001; De Lange *et al.*, 2006). On this basis, the efficiency of utilizing ME for NE in growing pigs (k_g) averaged 74% for the 61 diets, but it varied with the chemical composition of the diet (g/kg DM) according to the following equation (Noblet *et al.*, 1994a):

$$k_g = 74.7 + 0.036 \times EE + 0.009 \times Starch - 0.023 \times CP - 0.026 \times ADF \ (RSD = 1.2)^1.$$

A similar equation was proposed for adult sows fed at maintenance energy level (Noblet *et al.*, 1993). The variation in k_g is due to differences in efficiencies of ME utilization between nutrients, with the highest values for fat (~90%) and starch (~82%), and the lowest values (~60%) for dietary fibre (DF) and CP (Noblet *et al.*, 1994a). These values were confirmed experimentally later in our laboratory (Van Milgen *et al.*, 2001a) and are similar to those provided by the Rostock group in fattening pigs (Schiemann *et al.*, 1972). Measurements conducted in pigs having different BW and composition of BW gain suggested that the efficiency of ME for NE was little affected by the composition of BW gain, at least under practical conditions (Noblet *et al.*, 1994b). Similarly, the ranking between nutrients for their efficiencies was similar in adult sows fed at maintenance level and in growing pigs. Finally, the heat increment of using dietary protein for either protein retention or lipid retention appeared to be similar (Van Milgen *et al.*, 2001a), which means that the NE value of dietary CP does not depend on its final utilization.

[1] EE: ether extract, CP: crude protein; ADF: Acid Detergent Fiber.

Utilization of ME in poultry

As for pigs, several trials or theoretical assumptions over the last 70 years have been carried out to quantify the utilization of ME (or digestible nutrients) for NE in poultry (see the review of Pirgozliev and Rose, 1999). Some studies were carried out with unbalanced diets with makes the interpretation of results more difficult. The most comprehensive series of measurements were conducted more than 60 years ago in USA by Fraps (1946) and later by the Rostock group (Schiemann *et al.*, 1972), mainly focussing on feed ingredients. These studies mostly focused on starch, DF and CP with little variation in fat content of diets or feedstuffs. A recent study of Carré *et al.* (2002) was conducted on complete feeds (n=28) fed to 3-5 wk old broilers while varying the nutrient composition of the diets. The comparative slaughter technique was used to quantify energy retention and NE was calculated using a FHP value (500 kJ/ kg $BW^{0.60}$/d) measured in respiration chambers (Van Milgen *et al.*, 2001b). Later studies carried out at INRA suggested that FHP in 0.5 to 3.0 kg broilers was proportional to $BW^{0.70}$ and FHP was estimated at 445 kJ/ kg $BW^{0.70}$/d (Warpechowski *et al.*, unpublished data).

In several literature studies, efficiencies of DE or ME for NE for maintenance and growth (or fattening) in poultry have been quantified, and some of these values are listed in Table 1. As for pigs, the lowest efficiency is observed for CP and the highest for fat. However, the difference between the most extreme values (i.e. CP and EE) appears to be somewhat lower in poultry (65 to 85%) than in pigs (60 to 90%). Another major difference between poultry and pig concerns dietary fibre, which is not digested in poultry (Carré *et al.*, 1990). Consequently, no efficiency value for fibre exists (nor is needed) for poultry. The efficiency values reported by Schiemann *et al.* (1972) were obtained in adult fattening birds which deposited only fat with a subsequent higher average efficiency. From that point of view, the study of Carré *et al.* (2002) is probably more representative of modern broilers production.

An alternative approach to study the effect of diet on the efficiency of ME utilization for NE was tested in recent trials conducted at INRA (Noblet *et al.*, 2007; 2009). The approach consisted in preparing diets focussing in each trial on one specific nutrient (in exchange for starch). The effects of CP, EE and DF contents were evaluated and measurements were made in respiration chambers in group-housed, growing broilers between 3 and 7 weeks of age offered feed *ad libitum*. The summary of these results is presented in Table 2. The most surprising result of these trials is the absence of an effect of dietary CP on HP, HI and NE/ME in broilers. A study conducted simultaneously in pigs and broilers confirmed this major difference between these species (Noblet *et al.*, 2003; Table 4). Table 2 also indicates that the replacement of starch by fat did not result in a significant increase in the NE/ME ratio of the diet, which contrasts with results reported in Table 1. Finally, the presence of high levels of (undigestible) NDF in broilers diets did not change the HP and the NE/ME ratio of the diet. The latter was even numerically higher with the high DF diet. Overall and unlike pigs, these studies suggest that changes in diet composition have little effect on the efficiency of using ME for NE in broilers. The extrapolation of this conclusion obtained on compound feeds to ingredients that differ widely from the composition of a standard diet would need further studies.

Table 1. Efficiencies of ME from digestible nutrients for NE in poultry (%).

Reference	Production[1]	Crude protein	Ether extract	Starch + sugars	Diet
Schiemann *et al.*, 1972	M + fattening	61	84	75	73[2]
De Groote, 1974	M + growth	60	90	75	74[2]
Carré *et al.*, 2002	M + growth	68	84	77	76

[1] M: maintenance.

[2] Assuming that 25, 20 and 55% of ME is provided as CP, EE and carbohydrates.

Table 2. Effect of replacing starch by protein, fat or fibre on heat production and efficiency of using ME for NE in broilers: compilation of INRA data.[1]

Trial	Diets	kJ/kg $BW^{0.70}$/d			NE/ME %
		ME	HP	AHP	
1	18.0% CP	1,609	853	146	75.1
	22.7% CP	1,609	846	153	74.8
2	2.8% EE	1,873	904	141	75.0
	9.7% EE	1,877	901	152	75.7
3	9.5% NDF	1,503	912	170	71.3
	17.7% NDF	1,521	923	175	72.3

[1] Measurements carried out in groups of broilers weighing on average 1.4 kg; the indirect calorimetry method in respiration chambers was used; AHP: Activity heat production; complementary details by Noblet *et al.* (2007) for trial 1 and Noblet *et al.* (2009) for trial 2; trial 3: unpublished data. In trials 1 and 2, the variation in CP or EE content was created by replacement for starch; in trial 3, the increased NDF level resulted from dilution by dietary fibre provided by wheat bran, maize bran and soybean hulls. In trial 1, data have been adjusted for a similar ME intake while observed values are given for trials 2 and 3. None of the differences between treatments within each trial were significant (P>0.05).

Utilization of ME in veal calves

Veal calves commonly behave as monogastric animals with 90% or more of their daily GE provided by milk replacer and not fermented in the rumen. Although some trials on energy metabolism have been conducted in the 70s to 90s, these were not designed specifically to study energy utilization. In two recent experiments (Labussière *et al.*, 2008b; 2009b) conducted with 60-260 kg BW calves receiving milk replacers differing in CP content and at different feeding levels, measured values of FHP (Van Milgen *et al.*, 1997; Figure 1) were proportional to $BW^{0.85}$ and FHP increased with increasing feed intake (+0.28 kJ/kJ additional ME intake, Labussière *et al.*, 2008a). The k value of ME from milk replacer averaged 85% and was not affected by BW (or age) of the animals, composition of BW gain or level of energy intake. The greater k value observed in veal calves compared with those observed in pigs and poultry may be due to the direct deposition of dietary fat as body fat (Van den Borne *et al.*, 2007) and the high digestibility of nutrients, resulting in lower relative weight of the digestive tract and a lower energy cost of digesta transit. In a third trial conducted with calves receiving milk replacer and solid feeds, the efficiency of using ME from solid feed (i.e. a mixture of starch, dietary fibre and protein) varied between 68 to 76% (Labussière *et al.*, 2009a). These values are close to those obtained in pigs and reflect the combined utilization of intestinally digested and ruminally fermented energy in these feeds. Therefore, the use of a NE system to evaluate feedstuffs in veal calves is important when using liquid milk replacers and solid feeds simultaneously.

Comparative utilization of ME in monogastric animals

Except for veal calves that are fed almost exclusively with milk replacers, monogastric animals are usually fed with diets containing cereals, and protein-rich and fat-rich ingredients. These diets have a similar 'profile', even though differences may exist in the protein, fat, starch and dietary fibre content. To compare the efficiency of ME utilization between species, results obtained from our laboratory were compiled and average values are given in Table 3. It is remarkable that the HI (expressed as a % of ME) of 'standard' diets is very similar for growing pigs, broilers and turkeys. As mentioned previously and probably due to differences in diet composition (high fat) and diet structure (liquid and highly digestible), the HI was markedly lower in milk-fed veal calves. It is also remarkable

Table 3. Heat increment (HI) and activity heat production (AHP) in monogastric animals offered feed close to ad libitum *intake.*[1]

Animal category	Feed	HI % ME	AHP % ME	Reference
Growing pig	Standard	25	8-10	Noblet *et al.*, 1994a; Van Milgen *et al.*, 2001a
Growing broiler	Standard	25	8-10	Noblet *et al.*, 2009
Growing turkey	Standard	24	8-14	Rivera-Torres *et al.*, 2010
Veal calf	Milk	16	8-10	Labussière *et al.*, 2009b
	Milk + solid feed	20	8-10	Labussière *et al.*, 2009a

[1] According to the same method described in Figure 1 and for animals kept in cages with moderate physical activity; AHP is a component of HI and has been evaluated according to the information provided by force sensors placed below the metabolism cage (Van Milgen *et al.*, 1997).

that the average and 'minimal' cost of physical activity was similar between species represented about 10% of the ME intake. The estimated cost of activity is minimal because the animals were housed in metabolism cages and the cost of physical activity in a production setting may depend on housing conditions, physiological stage and the feeding level. This result also means that the NE value of feeds for pigs or poultry corresponds to that obtained under conditions of 'minimal' level of activity. An increase in physical activity must be considered as an additional energy requirement.

As the HI of standard compound feeds was similar between pigs and poultry, one may be tempted to use results obtained in one species for another species. Very few comparisons between species of animals fed the same diet have been conducted. However, the example given in Table 4 for growing pigs and broilers fed diets differing in protein content discourages the use of efficiency values across species. In other words, it is not possible to calculate NE values of feeds for different species according to common equations, at least not for pigs and poultry. If a common system can be used for different birds remains to be confirmed.

Table 4. Efficiency of using energy from crude protein in growing pigs and broilers (adapted from Noblet et al.*, 2003).*[1]

Species	Pigs		Broilers	
Dietary protein level	Normal	Low	Normal	Low
Body weight, kg	57.6	57.2	1.47	1.46
Energy balance, kJ/kg $BW^{0.60}$/d				
ME intake	2,564	2,566	1,626	1,642
Heat production	1,402[a]	1,346[b]	862	861
Fasting heat production	735	731	446	456
Heat increment[2]	667[a]	614[b]	417	404
NE/ME (×100)	73.9[a]	75.9[b]	74.8	75.0

[1] The reduction in dietary CP content consisted of replacing soybean protein concentrate by maize starch with supplementation of amino acids to meet the requirements.
[2] Sum of TEF and AHP (see Figure 1).

Net energy systems

An energy system corresponds to a method of predicting the energy value of compound feeds and ingredients for a given type of animals. It combines a mode of expression with a calculation method. Most NE systems are based on the utilization of ME for maintenance and for growth or fattening and are based on prediction equations taking into account either digestible nutrients or DE (or ME) and some chemical characteristics (see reviews of Pirgozliev and Rose, 1999 for poultry and Noblet, 2006 for pigs). Some systems have been established from measurements on animals (Schiemann *et al.*, 1972; Noblet *et al.*, 1994a; Carré *et al.*, 2002) while others have been proposed based on literature data, biochemical information, and/or theoretical assumptions (Emmans, 1994; Boisen and Verstegen, 1998). An adequate system to express energy values and requirement is important as it typically serves as a reference for express requirements and values for other nutrients (e.g. protein, amino acids, minerals, etc.).

The system proposed for pigs by Noblet *et al.* (1994a) and further developed and validated in subsequent studies (Noblet and van Milgen, 2004, Noblet, 2006) proposes NE prediction equations that are applicable to ingredients and compound feeds and different stages of pig production (Noblet, 2006). As mentioned previously, the efficiency of ME for NE differs widely between nutrients in pigs. Under the assumption that NE represents the best estimate of the 'true' energy value of feeds, the energy values of protein-rich or fibrous feeds will be overestimated when expressed on a DE or ME basis, while fat or starch sources will be underestimated in DE and ME systems (Table 4). Unfortunately, there is no general agreement on prediction equations or systems that can be implemented in poultry nutrition. According to the NE/ME ratios provided by Carré *et al.* (2002) for broilers, the result of moving from ME to NE for estimating the energy value of feeds should be similar to what is observed in pigs (Table 5), but to a smaller extent. However, our experimental results (Table 2) failed to confirm this. There is currently not enough information available to propose a NE system for other monogastric animals, such as veal calves.

It is important to point out that specific and accurate DE (or ME) values or digestible nutrient contents should be used when calculating NE values. For instance, energy digestibility differs between growing pigs and adult sows, with subsequent different NE values of feeds for both stages (Le Goff and Noblet, 2001). In fact, reliable information on digestibility of energy or of nutrients is the most limiting factor for predicting energy values of feeds for pigs or poultry. The lack of comprehensive information on the effects of technology (e.g. pelletting, extrusion, enzyme addition) or about differences in digestion between poultry strains or species, or between physiological stages (e.g. growing vs. adult.) is a major limiting factor for getting accurate estimates of energy values for monogastric animals, irrespective of the energy system used.

Table 5. Relative DE, ME and NE values of ingredients for growing pigs.[1]

	DE	ME	NE	NE/ME, %
Animal fat	243	252	300	90
Corn	103	105	112	80
Wheat	101	102	106	78
Reference diet	100	100	100	75
Pea	101	100	98	73
Wheat bran	68	67	63	71
Soybean meal	107	102	82	60

[1] From Sauvant *et al.* (2004). Within each system, values are expressed as percentages of the energy value of a diet containing 68% wheat, 16% soybean meal, 2.5% fat, 5% wheat bran, 5% peas and 4% minerals and vitamins.

Finally, moving from a DE or ME system to a NE system will affect the ranking of ingredients. This will have technical and economical consequences if the choice of feed ingredients is important and variable over time. This variety in ingredients available for monogastric animals feeds is likely to increase in the near future (e.g. more by-products) due to the competition for feed resources between human nutrition, industry, biofuels and animal nutrition. The use of a NE system would then be even more justified. Furthermore, it has been demonstrated that, at least in pigs, NE systems are better capable to predict performance of (growing) animals than are DE or ME systems (Noblet, 2006). Again, this advantage of NE systems is likely to increase when the chemical composition of diets becomes more variable (e.g. low protein, high dietary fibre).

Conclusion

This review indicates that NE is a better predictor than DE or ME of the 'true' energy value of poultry or pig feeds. Available information for pigs indicates an obvious interest of formulating on a NE basis, and NE systems should be implemented in pig production for getting a reliable prediction of performance. For poultry, the conclusions are less clear and convincing with no demonstrated advantage of a NE system over a ME system for predicting performance. Further investigations are necessary to evaluate the potential interest of a NE system for poultry. The use of two different types of feeds in veal calves (i.e. milk replacer and solid feed) should encourage the development of a NE system. Finally, even though NE is the final objective in energy evaluation of feeds, attention should be paid to the accurate estimation of DE or ME values, which are the most important factors of variation of the energy value of feeds for monogastric animals.

References

Birkett, S.H. and C.F.M. de Lange, 2001. Limitations of conventional models and a conceptual framework for a nutrient flow representation of energy utilization by animals. Br. J. Nutr. 86, 647-659.

Boisen, S. and M.W.A. Verstegen, 1998. Evaluation of feedstuffs and pig diets. Energy or nutrient-based evaluation systems? II. Proposal for a new nutrient-based evaluation system. Acta Agric Scand A, Anim Sci 48, 95-102.

Brouwer, E., 1965. Report of sub-committee on constants and factors. In: Blaxter, K.L. (ed.), Energy Metabolism. Academic Press, London, UK, EAAP Publication N° 11: 441-443.

Carré, B, L. Derouet and B. Leclercq, 1990. The digestibility of cell-wall polysaccharides from wheat (bran or whole grain), soybean meal, and white lupin meal in cockerels, muscovy ducks, and rats. Poultry Sci. 69, 623-633.

Carré, B., M. Lessire and H. Juin, 2002. Development of the net energy system for broilers. Proceedings, Eastern Nutrition Conference, Guelph (Canada): 140-149.

De Groote, G. 1974. A comparison of a new net energy system with the metabolizable energy system in broiler diet formulation, performance and profitability. Br. Poultry Sci. 15, 75-95.

De Lange, K., J. van Milgen, J. Noblet, S. Dubois and S. Birkett, 2006. Previous feeding level influences plateau heat production following a 24 h fast in growing pigs. Br. J. Nutr. 95, 1082-1087.

Emmans, G., 1994. Effective energy: a concept of energy utilization applied across species. Br. J. Nutr. 71, 801-821.

Fraps, G.S., 1946. Composition and productive energy of poultry feeds and rations. Texas Agricultural Experiment Station Bulletin n° 678.

Koong, L.J., J.A. Nienaber, J.C. Pekas and J.T. Yen, 1982. Effects of plane of nutrition on organ size and fasting heat production in pigs. J. Nutr. 112, 1638-1642.

Labussière, E., S. Dubois, J. van Milgen, G. Bertrand and J. Noblet, 2008a. Fasting heat production and energy cost of standing activity in veal calves. Br. J. Nutr. 100, 1315-1324.

Labussière, E., S. Dubois, J. van Milgen, G. Bertrand and J. Noblet, 2008b. Effect of dietary crude protein on protein and fat deposition in milk fed veal calves. J. Dairy Sci. 91, 4741-4754.

Labussière, E., S. Dubois, J. van Milgen, G. Bertrand and J. Noblet J, 2009a. Effect of solid feed on energy and protein utilization in milk fed veal calves. J. Anim. Sci. 87, 1106-1119.

Labussière, E, G. Maxin, S. Dubois, J. van Milgen, G. Bertrand and J. Noblet, 2009b. Effects of feed intake on protein and fat deposition in milk fed veal calves. Animal 3, 557-567.

Labussière, E., J. van Milgen, S. Dubois and J. Noblet, 2010. The maintenance energy expenditure is affected by feeding level in growing animals. International Symposium on Energy and Protein Metabolism (in press).

Le Bellego, L., J. van Milgen and J. Noblet, 2001. Energy utilization of low protein diets in growing pigs. J. Anim. Sci. 79: 1259-1271.

Le Goff, G. and J. Noblet, 2001. Comparative digestibility of dietary energy and nutrients in growing pigs and adult sows. J. Anim. Sci 79, 2418-2427.

Le Goff, G, L. Le Groumellec, J. van Milgen and J. Noblet, 2002. Digestive and metabolic utilization of dietary energy in adult sows: influence of level and origin of dietary fibre. Br. J. Nutr. 87, 325-335.

Noblet, J, X.S. Shi and S. Dubois, 1993. Metabolic utilization of dietary energy and nutrients for maintenance energy requirements in pigs: basis for a net energy system. Br. J. Nutr. 70, 407-419.

Noblet, J, H. Fortune, X.S. Shi and S. Dubois, 1994a. Prediction of net energy value of feeds for growing pigs.J. Anim. Sci 72, 344-354.

Noblet, J, X.S. Shi and S. Dubois, 1994b. Effect of body weight on net energy value of feeds for growing pigs. J. Anim. Sci 72, 648-657.

Noblet, J, C. Karege, S. Dubois and J. van Milgen, 1999. Metabolic utilization of energy and maintenance requirements in growing pigs: effect of sex and genotype. J. Anim. Sci 77, 1208-1216.

Noblet, J and G. Le Goff, 2001. Effect of dietary fibre on the energy value of feeds for pigs. Anim. Feed Sci. Technol. 90, 35-52.

Noblet, J., J. van Milgen, B. Carré, P. Dimon, S. Dubois, M. Rademacher and S. van Cauwenberghe, 2003. Effect of body weight and reduction of dietary crude protein on energy utilisation in growing pigs and broilers. In: Souffrant, W.B. and C.C. Metges (editors), Progress in research on energy and protein metabolism. Wageningen Academic Publishers, Wageningen, the Netherlands, EAAP Series n° 109: 205-208,

Noblet, J. and J. van Milgen, 2004. Energy value of pig feeds: Effect of pig body weight and energy evaluation system. J. Anim. Sci. 82 Number 13, E. Suppl., E229-E238. (http://www.asas.org/symposia/04esupp/E229.pdf).

Noblet, J, 2006. Recent advances in energy evaluation of feeds for pigs. In: Garnsworthy, P.C. and J. Wiseman (eds.), Recent advances in Animal Nutrition 2005. Nottingham University Press, Nottingham, UK: 1-26.

Noblet J., S. Dubois, J. van Milgen, M. Warpechowski and B Carré, 2007. Heat production in broilers is not affected by dietary crude protein. In: Ortigues-Marty I. (ed.), Energy and Protein Metabolism and Nutrition. Wageningen Academic Publishers, Wageningen, the Netherlands, EAAP publication n° 124: 479-480.

Noblet, J., M. Warpechowski, S. Dubois, J. van Milgen and B. Carré, 2009. Influence de la teneur en matières grasses de l'aliment sur l'utilisation métabolique de l'énergie chez le poulet en croissance. Proceedings 8èmes Journées de la Recherche Avicole: 177-181.

Pirgozliev, V. and S.P. Rose, 1999. Net energy systems for poultry feeds: a quantitative review. World's Poultry Sci. J. 55, 23-36.

Renaudeau, D., B. Bocage and J. Noblet, 2006. Influence of energy intake on protein and lipid deposition in Creole and Large White growing pigs. Anim. Sci. 82, 937-945.

Rivera-Torres, V., J. Noblet, S. Dubois and J. van Milgen, 2010. Energy partitioning in male growing turkeys. Poultry Sci. 89, 530-538.

Sauvant, D, J.M. Perez and G. Tran, 2004. Tables of composition and nutritional value of feed materials: pigs, poultry, cattle, sheep, goats, rabbits, horses, fish. Wageningen Academic Publishers, Wageningen, the Netherlands, 304 pp.

Schiemann, R, K. Nehring, L. Hoffmann, W. Jentsch and A. Chudy, 1972. Energetische Futterbevertung und Energienormen. VEB Deutscher Landwirtschatsverlag, Berlin, Germany, 344 pp.

Van den Borne, J., G.E. Lobley, M.W.A. Verstegen, J.M. Mujilaert, S.J.J. Alferink and W.J.J. Gerrits, 2007. Body fat deposition does not originate from carbohydrates in milk-fed calves. J. Nutr. 137, 2234-2241.

Van Milgen, J., J. Noblet, S. Dubois and J.F. Bernier, 1997. Dynamic aspects of oxygen consumption and carbon dioxide production in swine. Br. J. Nutr. 78, 397-410.

Van Milgen, J., J. Noblet and S. Dubois, 2001a. Energetic efficiency of starch, protein and lipid utilization in growing pigs. J. Nutr. 131, 1309-1318.

Van Milgen, J., J. Noblet, S. Dubois, B. Carré and H. Juin, 2001b. Utilization of metabolizable energy in broilers. Proceedings International Animal Agriculture and Food Science Conference. Poultry Sci. 80 (Supplement 1), 170.

Future aspects of feed evaluation

P. Bikker[1] and M. Verstegen[2]
[1]*Wageningen UR, Livestock Research, P.O. Box 65, 8200 AB Lelystad, the Netherlands; paul.bikker@wur.nl*
[2]*Wageningen UR, Animal Nutrition Group, P.O. Box 338, 6700 AH Wageningen, the Netherlands*

Abstract

This paper deals with non-classical aspects of feed value and their inclusion in feed evaluation. Common feed ingredients not only contribute to the energy and amino acids supply of farm animals, but have a number of additional effects. The use of these properties of feed ingredients would allow optimising diets that contribute to animal health and well-being, product quality, reduced environmental costs, etc. This requires both the description and quantification of such properties of feed ingredients and the requirements or dose related effects in the target animals. With this information, feed optimisation may contribute to meeting future conditions of animal production with regard to optimal use of resources, respect for animal health and well-being and minimising environmental costs.

Introduction

Production of animal feed requires a combination of feed ingredients to optimally meet the requirements of the target animal. Therefore the feeding value of the individual ingredients should be properly known. Feed evaluation systems are used to estimate the contribution of each feed ingredient to the requirements of the target animal for a number of different nutrients, e.g. energy, amino acids and minerals. Traditionally attention is focused on the energy and protein value since these are the major nutrients required for maintenance processes, body gain and production of milk and eggs. Improvements have been made regarding the contribution of underlying nutrients (e.g. individual amino acids and different carbohydrates), absorption from the digestive tract (digestibility, endogenous losses), and utilization in the animal. These aspects are discussed in other contributions at this symposium. In our paper we will focus on non-classical aspects of feed value and feed evaluation, i.e. characteristics of feed which are generally not accounted for in feed evaluation. These may deviate from classical feed values because of their effect on specific body functions or because their effect depends on specific conditions and may not be generally applicable. We will first address some general characteristics and limitations of feed evaluation and subsequently move to some new developments and possibilities. We will discuss these in the context of future (intensive) animal production in which increasingly conditions are applied with regard to the demands of animal products, efficient use of limiting resources, attention for animal health and well being and consequences for the environment.

Some characteristics and limitations of feed evaluation

Although the concept of feed evaluation is relatively simple, the allocation of feed values to different ingredients, its practical application is hampered by a number of difficulties and limitations. Feed evaluation systems generally assume that values for different feed ingredients are constant and additive and linearly contribute to the nutrient supply of the animal, irrespective of its physiological state. These assumptions are a simplified representation of nutrient utilisation in the animal as illustrated below.

The requirements of animals can be experimentally determined or calculated using a (factorial) model. The feed value indicates how much a diet or a feed ingredient contributes to meet these requirements. This may suggest that feed value and requirements are independent, but in fact they are closely interrelated and the distinction between the two is a fundamental aspect of feed evaluation. An example is the evaluation of amino acids based on apparent, standardised or true ileal digestibility. Using the apparent digestibility, all endogenous losses, representing metabolic costs of digestion are allocated to the respective feed ingredients. This results in relatively high calculated metabolic costs in low protein feed ingredients and consequently low apparent digestibility values. Using the true digestibility, endogenous losses are not allocated to the feed ingredients and need to be accounted for in the requirements of the animal. However, since these metabolic costs depend on the composition of the feed ingredients (e.g. contents of fibre and ANF), this implies an interaction between diet composition and the animal's requirement. To overcome these problems, the use of standardised ileal digestibility was introduced, in which the apparent digestibility coefficients are corrected for basal endogenous losses. These basal endogenous losses are related to intake of a common diet and are included in the requirements of the animal, irrespective of diet composition. The remaining specific endogenous losses are regarded as a property of the respective feed ingredients (Stein *et al.*, 2007).

The general principle of this example is that metabolic costs of feed intake and nutrient utilisation in general may best be included in the maintenance requirements of the animal. However, metabolic costs or other effects caused by specific feed ingredients should be assigned to those ingredients to realise a correct evaluation. Although this has been implemented in the use of standardised ileal amino acid digestibility coefficients, the interaction between diet composition and metabolic costs of utilisation is still implicated. This is illustrated by the work of Grala *et al.* (1997) who reported that feedstuffs that increased endogenous nitrogen losses, caused increased urinary nitrogen excretion and reduced utilisation of ileally digested nitrogen for nitrogen retention in pigs. This effect was more prominent for ingredients with a high content of antinutritional factors (ANF) than for ingredients with a high fibre content (Grala *et al.*, 1998). Zhu *et al.* (2005) reported a decrease in protein deposition and threonine utilisation when diets with a high pectin content were fed (Figure 1A). Their suggestion that microbiota in the gut may play a role is supported by our study in growing pigs in which the withdrawal of antimicrobial growth promoter showed an increased response to dietary threonine supply (Bikker *et al.*, 2007; Figure 1B).

These studies indicate that the costs of specific endogenous amino acid production for replacing the endogenous proteins which are lost during digestion of feed ingredients in the gastrointestinal tract (GIT) vary between ingredients. This represents an interaction between feed composition and nutrient requirements. This interaction is not adequately accounted for in the present amino acid evaluation systems. On the other hand, feed ingredients that reduce the metabolic costs of homeostasis in the animal (e.g. health, behaviour) may have a higher nutrient value. Other examples of interaction between feed value and nutrient requirements will be discussed later.

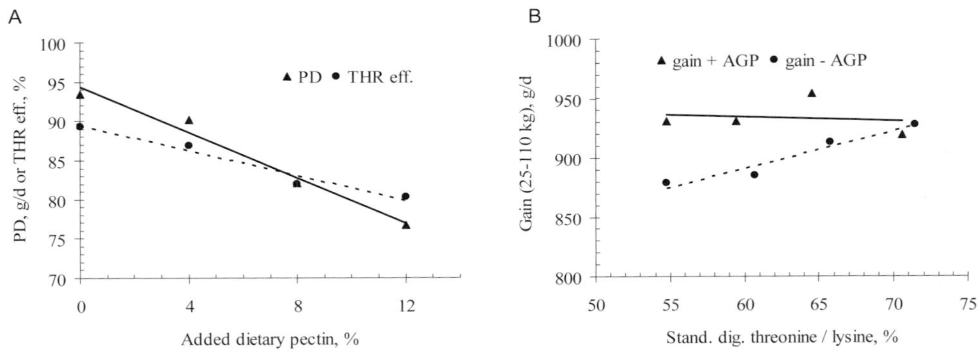

Figure 1. (A) Influence of dietary pectin on protein deposition (PD) and efficiency of Thr utilisation (Zhu et al.*, 2005) and (B) Influence of dietary Thr on body gain in growing pigs fed diets with or without antimicrobial growth promoter (Bikker* et al.*, 2007).*

An important prerequisite in feed formulation is that feeding values of different feed ingredients are additive irrespective of the level and combination of ingredients. In a number of situations this may not be a correct representation of the animal's physiology. For example, Bakker (1996) illustrated this may not be the case for nutrient digestibility when high levels of fat and non-starch polysaccharides (NSP) are combined in the diet. Also, the use of a constant net energy value of dietary protein and fat is a simplification since their energy value is higher when amino acids or fatty acids are incorporated in body tissue and lower when these nutrients are catabolised. Similar limitations arise when non-classical feed values are used as will be discussed later. Further improvements can be made by inclusion of these characteristics in metabolic models in which the fate of nutrients in relation to the physiological state of the animals and the feed intake level is accounted for.

After addressing some of the limitations of feed evaluation in the remaining part of this paper, we will focus on some non-classical aspects of feed value and their inclusion in feed evaluation and diet preparation. To include new characteristics in feed evaluation, it is necessary to establish their effect on relevant response traits and develop a system to describe the response and quantify the effect of different feed ingredients. We will briefly address some potential developments in the following areas: animal health, animal well-being, feed intake, product quality and environmental consequences.

Feed value and animal health

Poor animal health is of major concern because of impaired efficiency of production, negative effects on well-being of the animal and on public acceptance of animal production. Feed composition can have an important impact on animal health, e.g. in relation to nutrient deficits, optimal function of the digestive tract and nutrient requirements for specific body functions, e.g. immune response.

Risk of mineral deficiencies

It is not likely that nutrient deficits these days play an immediate role for animal health, but adequate nutrient supply is a prerequisite for good animal health, especially in the case of dietary minerals. Because of accumulation of minerals (P, Cu, Zn) in soil and surface water, and limited global reserves of phosphate, there is public pressure to reduce the gross level of these minerals in animal diets. This seems quite possible since inclusion levels in feed often exceed the requirements of the animals. However, especially for Cu and Zn, precise requirements are not well established. Therefore it is important that precise information of optimal supply will become available (e.g. Revy *et al.*, 2006). Unlike other nutrients, Cu and Zn levels are generally evaluated on a total (gross) basis because of the number of complex interactions with other dietary ingredients which can influence their absorption. Moreover, the absorption coefficient is largely determined by the supply in relation to the requirements (e.g. Weigand and Kirchgessner, 1980). Nevertheless, more quantitative insight in mineral availability is required to reduce unnecessary high dietary inclusions levels while avoiding the risk of shortage for the animal. This evaluation would include interactions between specific minerals (e.g. Cu and Zn), nutrients (e.g. dietary fibre) and the influence of phytate and phytase. The influence of phytase on Zn availability has been shown in several studies (e.g. Revy *et al.*, 2006) to determine a dose-response relationship. In addition, more emphasis on absorbability of dietary minerals presumably would also allow a better evaluation of organically bound minerals as compared to common inorganic salts.

Health of the digestive tract

The major interaction between diet composition and animal health occurs in the digestive tract and is related to digestive disturbances and infectious diseases. The use of antibiotics in intensive animal production systems to combat these diseases is of major public concern. Since the EU-ban on antimicrobial growth promoters, the reduction in antibiotic use in feed was to some extend compensated by an increase in the therapeutic use, e.g. in the Netherlands (Fidin, 2009). In addition, the use of pharmaceutical levels of Zn in diets for weaned piglets result in high levels of faecal Zn excretion, which are unlikely to be acceptable in the long term for environmental reasons (e.g. Jondreville *et al.*, 2003). This emphasises the importance to further study the interaction between diet composition and animal health and include these in feed optimisation. In the past decade a large number of vegetable components with prebiotic or phytogenic properties, which can influence physiological characteristics in the GIT have been studied with regard to their potential as alternative for antimicrobial growth promoters. However, relatively little attention has been paid to the contribution of common feed ingredients to digestive health of the animal. Also, the development of new feed characteristics to account for these effects did not receive much attention yet.

Fermentation of nutrients in the intestinal tract

Despite the potential benefits of specific purified prebiotic substances (e.g. inulin, fructo-oligosaccharides, lactose and lactulose), common feed ingredients are the most important substrate for microbiota in the digestive tract. Most feed ingredients contain substantial amounts of NSP that are resistant to digestive secretions and potentially available for bacterial fermentation. Also resistant starch, e.g. native potato starch, is fermented to a large extent. In addition, other nutrients, especially undigested dietary protein can be used as substrate by microbiota, thus promoting growth of proteolytic bacteria. It is generally accepted that diet composition has a major impact on intestinal health of monogastrics. However, it is still a matter of debate which feed (fibre) characteristics are most suited to describe and clarify the effects in the GIT and what levels may be required for optimal health and performance.

With regard to dietary protein, evidence has accumulated in the last few years that low protein diets are beneficial for intestinal health as identified through reduced numbers of enterotoxigenic *Escherichia coli* after experimental infection (Wellock *et al.*, 2008b; Opapeju *et al.*, 2009), reduced postweaning diarrhoea (Heo *et al.*, 2009), firmer faeces (Wellock *et al.*, 2006, Wellock *et al.*, 2008b), and reduced intestinal (Bikker *et al.*, 2006; Nyachoti *et al.*, 2006) and faecal ammonia concentrations (Heo *et al.*, 2009). In general, no consistent effects of dietary protein content on intestinal morphology were found, but increased villus height to crypt depth ratios were incidentally reported (e.g. Opapeju *et al.*, 2009). Effects on the intestinal microbiota population are less clear. Often culture based methods were used to enumerate effects on lactobacilli, clostridia or coliform bacteria. Some studies indicated that a reduction of dietary crude protein may reduce both lactobacilli and coliform bacteria, presumably because of reduced substrate availability. Obviously, culture based methods only address a small fraction of the intestinal microbiota and may not be adequate to assess effects on intestinal health. For example, using molecular techniques Opapeju *et al.* (2009) reported reduced richness and diversity of microbiota in the colon of piglets fed low protein diets. It is expected that these techniques will add to our understanding and definition of a healthy microbiota population in the digestive tract. In addition, to these benefits of low protein diets, lower (relative) weights of the small intestine, liver and kidney have been reported, suggesting a lower metabolic burden of these diets (Wellock *et al.*, 2006; Nyachoti *et al.*, 2006). On the other hand, several authors reported a decreased growth performance using low protein diets with (e.g. Nyachoti *et al.*, 2006) or without amino acid supplementation (e.g. Wellock *et al.*, 2006). Thus, it would be desirable to include both the effect of protein content on growth performance and animal health in feed evaluation. For this aim it should be considered to also specify rate of protein degradation and protein entering the large intestine to further optimise the ratio between energy from protein and energy from carbohydrates as substrate for the microbiota to minimise growth of pathogenic bacteria (e.g. Kim *et al.*, 2008).

The effect of dietary fibre on enteric health is both evident and controversial. Some studies indicate that highly digestible diets with a low fibre content are optimal for maximising growth performance e.g. in weaned piglets (e.g. Bikker *et al.*, 2006; Mateos *et al.*, 2006). In addition, Hopwood *et al.*

(2004) reported an increase in intestinal colonisation with *E. coli* and incidence of diarrhoea when pearl barley rich in soluble NSP was added to a diet based on cooked rice. Nevertheless, a large number of recent studies have reported beneficial effects of various sources of dietary fibre on characteristics related to intestinal health. These include increased production of short-chain fatty acids, including lactic acid and butyric acid, in the small and large intestine with soluble NSP (Wellock *et al.*, 2008a), sugar beet pulp (SBP) (Jeaurond *et al.*, 2008), and SBP and wheat bran (Hermes *et al.*, 2009). Furthermore, reduced ammonia concentrations in the intestinal tract have been reported, for example for wheat bran, SBP, and native starch (Bikker *et al.*, 2006; Jeaurond *et al.*, 2008). Increased numbers of lactobacilli and reduced numbers of coliform bacteria or clostridia have been reported using wheat bran, SBP, native starch (Bikker *et al.*, 2006; Jeaurond *et al.*, 2008; Hermes *et al.*, 2009), high soluble non-viscous NSP (inulin), but not when using insoluble NSP (Wellock *et al.*, 2008a). These results indicate potential benefits of fibre inclusion on intestinal health of piglets. However, it seems essential to further specify dietary fibre properties and their interactions with other feed characteristics which influence animal performance and health. For example, fibre inclusion may be more beneficial in high protein diets providing more substrate for proteolytic bacteria (Bikker *et al.*, 2006; Hermes *et al.*, 2009). Kim *et al.* (2008) reported a decrease in natural post-weaning diarrhoea when oat hulls were added to a diet based on extruded rice but not a diet based on wheat, diet indicating the importance of the basal diet composition. Wellock *et al.* (2008a) reported better effects of soluble NSP versus insoluble NSP and suggested that the soluble NSP should have a low viscosity whereas Kim *et al.* (2008) suggested to use mostly insoluble NSP. In addition, studies from our group, e.g. Awati (2005) suggest to use rate of fermentability as a criterion for further characterising dietary fibre. Indeed this could be relevant, especially for optimising the balance of protein and carbohydrate as substrate for microbial fermentation in the digestive tract.

Viscosity is an important feed characteristic in relation to dietary fibre. Effects of viscosity have been studied in recent years using model substances (Lalles *et al.*, 2006; Fledderus *et al.*, 2007) or soluble and non-soluble NSP. Fledderus *et al.* (2007) reported an increase in gastric protein hydrolysis and protein digestibility in piglets with inclusion of 1% carboxy methyl cellulose (CMC) in the weaning diet and hypothesised reduced gastric emptying rate as potential mechanism. Lalles *et al.* (2006) reported an increase in sodium-dependent glucose absorption with CMC inclusion in weaning diets. Sugita-Konishi *et al.* (2004) showed that CMC drastically inhibited the adhesion of *E. coli* K88 (isolated from swine intestine) to Caco-2 cells *in vitro*. The chelating properties of CMC may be involved in inhibiting adhesion of bacteria to the intestinal cells. In contrast, others reported an increase in enteropathogenic β-haemolytic *E. coli* in pigs with CMC inclusion in the diet (Hopwood *et al.*, 2002). The addition of guar gum or cellulose to a standard diet was shown to increase ileal *Bifidobacteria* and *Enterobacteria* populations in growing pigs (Owusu-Asiedu *et al.*, 2006).These results illustrate the importance of the physico-chemical characteristics of diets for digestive physiology and health.

At this stage we suggest to review and analyse the large amount of available information to identify the most promising fibre characteristics and validate and quantify the effects on intestinal health and growth performance. This may allow to optimise animal diets with regards to the effect of common feed ingredients on intestinal health.

Diet and immune response

Besides the effect on intestinal health, there are many more interactions between feed and health of poultry and pigs. These interactions represent possible additional values for specific nutrient or ingredients. These include both the influence of (suboptimal) conditions of animals on nutrient requirements as well as specific effects of diet components on animal health. For example, several studies have shown that withdrawal of antibiotics from the diet may influence the optimal amino acid to energy ratio (Bikker *et al.*, 2003) and optimal amino acid pattern, especially with regard to threonine (Bikker *et al.*, 2007) and possibly tryptophan (Le Floc'h and Sève, 2007). Tryptophan requirements may be increased when the immune system is stimulated and increased tryptophan supply may limit the consequences of immune system activation on animal performance and stress (Le Floc'h and Sève, 2007; Koopmans *et al.*, 2009). Other amino acids with specific effects on animal health include arginine, glutamine and cysteine (Lalles *et al.*, 2009). For a number of other components or ingredients (e.g. β-glucans, fatty acids, spray dried animal plasma) immunomodulatory effects have been suggested but apart from blood plasma, results are often inconsistent and require further research to confirm their role in maintaining animal health (e.g. Gallois *et al.*, 2009). Nevertheless, these examples illustrate that there is an array of components with a feed value beyond the classical energy and protein value. This list can be further extended with effects on insulin resistance and glucose tolerance in sows, viability in neonatal piglets etc. Moreover, until now, effects of nutrients in breeding animals on metabolic programming of health and performance in offspring only has received limited attention and potential effects are scarcely exploited in animal diets. It is beyond the purpose of this paper to review these studies. Nevertheless, the increased interest in this direction with new genetic tools will give new insights in short and long term effects of nutrients and potential influence on animal performance and health.

Animal behaviour and feed intake

Adequate attention for well being of animals is required in future intensive animal production systems. Feed supply and diet composition have a major influence on satiety in animals that need to be (drastically) restricted in nutrient intake for reasons of health, product quality and costs of production, e.g. broiler breeders and pregnant sows. Insight in feed intake regulation and the role of diet composition is also required to stimulate or restrict feed intake for optimal production performance.

Taking the sow as an example, much research has been conducted how to induce satiety in restrictedly fed sows during gestation. Brouns *et al.* (1995) showed that among different fibrous feed ingredients included in a high percentage in sow diets, only SBP significantly reduced voluntary feed intake. Feed intake was inversely related to the inclusion level of SBP. In many studies restricted feeding of high fibre diets to sows reduced abnormal behaviour as indicated by non-feeding oral activities and high water intake (e.g. Robert *et al.*, 1997; D'Eath *et al.*, 2009). Van der Peet *et al.* (2003a,b) showed that diets high in SBP supplied at equal energy intake reduced oral behaviour and improved litter size in first and second parity sows. De Leeuw *et al.* (2005a) showed that stabilised post-prandial blood glucose and insulin levels presumably play an important role in the increased satiety and reduced activity in sows fed high fibre diets. These effects may be largely mediated by increased hindgut fermentation. In many studies, SBP was used as high fibre ingredient, whereas insight in the effect of relevant nutrients would allow using other ingredients as well. Therefore, De Leeuw *et al.* (2005b) compared effects of diets high in fermentable NSP from SBP to a mixture of other fibrous ingredients and concluded that this mixture was almost equally effective in reducing oral behaviour and activity and stabilising indicators of glucose metabolism (Figure 2). Fermentable NSP is an important characteristic of fibrous feed ingredients with regard to their effect on satiety in pregnant sows.

The reduced activity of sows fed high fibre diets might imply that these diets reduce maintenance requirements thus allowing more nutrients for productive purposes. Indeed, Rijnen *et al.* (2003a) reported a reduced activity of group-housed sows fed a diet high in fermentable NSP from SBP as compared to a high starch diet. The reduced energy expenditure was 2.3-3.7 kJ/g fermentable NSP, thus increasing the net energy content by approximately 20%. Although it is unlikely that this interaction between diet composition and nutrient requirements is specifically related to fermentable NSP from SBP, Rijnen *et al.* (2003b) did not find a similar reduction in energy expenditure in pigs

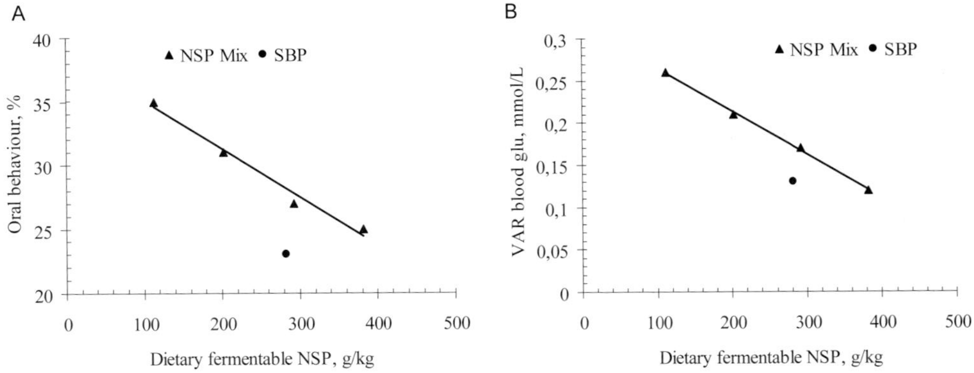

Figure 2. (A) Influence of dietary fermentable non-starch polysaccharides (NSP) on oral behaviour and (B) on variation in eight consecutive blood samples in hourly intervals from 4-12 h. after the meal in sows (De Leeuw et al.*, 2005b).*

fed diets high in fermentable NSP from soybean hulls or coconut meal. At present, only for SBP this energy saving effect has been reported.

Apart from (fermentable) NSP in pig diets, a number of other dietary factors may influence feed intake of pigs and poultry. Many authors studied the effect of specific feedstuffs or feed characteristics on feed intake, but no systematic evaluation system to optimise diets for feed intake is available. Nonetheless, Sola-Oriol *et al.* (2009a,b) used preference tests to determine the effect of feed ingredients and texture of the diet on voluntary feed intake. This may be a useful pragmatic start to rank individual feed stuffs. Additional insight in (physico-chemical) factors regulating feed intake is required to clarify differences between feedstuffs and also take into account the effects of variation within feed stuffs.

Nutrition and product quality

Pork and to a lesser extent poultry and eggs have long been regarded as bulk product and limited attention is paid to the effect of diet on product quality, apart from specific niche markets. e.g. products high in omega fatty acids. Nevertheless, there is gradually increasing interest in quality of animal products both related to the production system (ethical quality) and eating quality e.g. intra-en intermuscular lipid content. In pig production, presumably the biggest challenge is the future production of boars while avoiding off-odour meat due to boar taint. The role of diet composition has been shown, especially with regard to the contribution of skatol. At present, the most promising effects have been shown for high levels of inulin or native potato starch (Zamaratskaia and Squires, 2009). Effects are promising and should be extended to responsible diet characteristics (e.g. fermentable NSP) and dose-response relationships to include this information in feed optimisation. The same approach can be used for other examples of dietary effects on product quality, e.g. the enrichment of meat or eggs with high levels of omega fatty acids.

Impact of livestock on the environment

At the global level, livestock contributes to 15% of total food energy, 25% of the food protein and to the supply of a number of essential micronutrients. Consumption of meat and eggs per capita increased by 37% and 64% over the last 25 years, largely because of economic growth and urbanization. For the next decade an increase in meat consumption of 19% is expected (FAO, 2009). At the same time, livestock produces 18% of global greenhouse gases, and uses 33% of agricultural cropland for feed production (Steinfeld, 2006). Some of the resources previously available to animal production at low cost become increasingly costly, either because of growing competition for the resources from other economic sectors (production of food and biofuels) or because of the greater value society puts on consequences for the environment (e.g. air and water quality). For many years, the discussion about the environmental impact of (industrialised) animal production focussed on the effects of minerals excreted in the manure, mainly nitrogen and phosphorus. Because of the separation of industrialised livestock production from the land used for feed production, the large concentration of waste products is often regarded as an environmental burden rather than a valuable source of nutrients for crop production. Today, there is a growing concern about the (environmental) costs of feed production as well. Continued growth of livestock production will increase pressure on the environment and natural resources and requires approaches that allow increased production with lower environmental consequences. Inclusion of the demand for natural resources and the environmental costs for production and use of animal feeds as a value of individual feed ingredients will give insight in environmental costs of animal diets and may allow optimising diets with regard to environmental impact. This is illustrated below.

Feed characteristics and ammonia emission

Intensive animal production is a large contributor to ammonia emissions. Ammonia emission from animal houses and manure storage largely depends on urea concentration of the urine and pH of the manure. Several studies showed that lowering dietary crude protein content will reduce the nitrogen excretion and ammonia emission (e.g. Canh, 1998; Figure 3A). Inclusion of fermentable NSP in the diet stimulates microbial protein synthesis and excretion of nitrogen in the faeces rather than excretion as urea in the urine (e.g. Kreuzer and Machmüller, 1993; Figure 3B). The use of acidifying salts as calcium sulphate, chloride or benzoate to reduce slurry pH and may reduce ammonia emission by as much as 50% (Canh, 1998).

In a nitrogen balance experiment with 26 diets followed by *in vitro* emission studies, Bakker *et al.* (pers. comm.) showed that the effect of the level of crude protein, fermentable NSP and acidifying salts in the diet had a linear and additive effect on the ammonia emission. Cumulative ammonia emission during a 7 day measuring period was (nutrients in g/kg):

NH_3 (g/7d) = -5.35 + 0.056 × CP - 0.05 × acidifying salt - 0.010 × fNSP, R^2=93.2.

The *in vitro* prediction correlated very well with large scale *in vivo* measurements of the ammonia emission on a selection of these diets. This type of information allows optimising diets for a specific environmental aim, e.g. reduced ammonia emission.

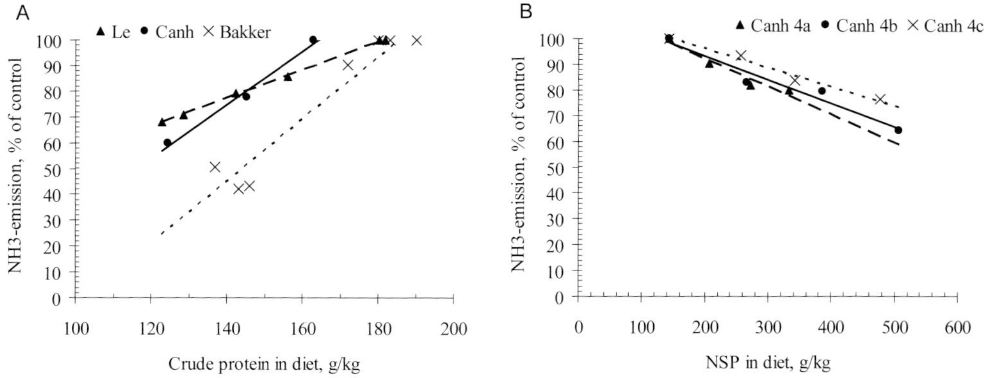

Figure 3. (A) Influence of crude protein and (B) non-starch polysaccharides (NSP) content in the diet on relative ammonia emission from pig manure in studies at our institute.

Use of live cycle assessment of feed and animal production

In recent years, increased attention has been paid to the environmental impact of animal production, beyond mineral excretion in manure. Live Cycle Assessment (LCA) is used as a method to evaluate the environmental impact of livestock products including pork, chicken and eggs (e.g. De Vries and De Boer, 2010). Two types of environmental impact are considered during the life cycle of a product: use of resources such as land and fossil energy and emission of pollutants. In general, LCA regarding animal production include global warming potential or climate change due to CO_2, N_2O and CH_4 (in kg of CO_2 equivalents), eutrophication due to nitrogen and phosphorus, acidification due to NH_3, NO_2, NO_x and SO_2 (in SO_2 equivalents), fossil energy use (MJ) and land use (e.g. De Vries and De Boer, 2010). In some studies this is further extended to other criteria including terrestrial ecotoxicity (Cu, Zn, Cd, Ni and Pb) and pesticide use.

Until now, LCA has been used to compare different systems of livestock production, e.g. conventional and free-range production of chickens and eggs. However, only a limited number of studies used LCA for evaluation of the effect of alternative diet compositions whereas feed is the largest input in intensive animal production systems. Van der Werf *et al.* (2005) showed the effect of by-product based diets compared to cereal based diets in finishing pigs on total energy use of the diet. Also Lammers *et al.* (2010) reported a higher energy use but similar global warming potential (CO_2) for a by-product based diet with dried distillers grain with solubles and glycerol compared to a traditional corn-soy bean meal diet used for swine production in Iowa. Eriksson *et al.* (2005) showed that in Sweden pea-rapeseed diets reduced the energy use and global warming potential compared to a cereal-soybean meal reference diet, whereas inclusion of synthetic amino acids also reduced the eutrophication and acidification. Beneficial effects of the use of enzymes (e.g. phytase, xylanase) and pure amino acids on acidification and eutrophication, energy use and global warming potential also have been reported (Nielsen and Wenzel, 2007; Nielsen *et al.*, 2008; Lammers *et al.*, 2010). Although the calculated effects largely depend on the assumed improvement in nutrient utilisation by the enzyme inclusion, these results indicate that LCA is a powerful method to add new characteristics to feed ingredients that represent the contribution of the feedstuffs on the environmental effects of livestock production. Inclusion of these characteristics when composing animal feeds would first increase awareness of the environmental cost of ingredients and diets and second allow feed optimisation with respect to these values. In the long term, this may influence feed composition and availability of feed ingredients, e.g. stimulate an increased use of locally produced feedstuffs to reduce transport costs, a reduction of energy input in drying of wet by-products, etc. (Vellinga *et al.*, 2009). Thus, use of these feedstuff properties would allow diet production at lower environmental costs.

Summary and conclusions

Common feed ingredients not only contribute to the energy and amino acids supply of farm animals, but have a number of additional effects, e.g. on metabolic costs of nutrient utilisation, feed intake, satiety and behaviour, animal health and well-being, quality of animal products, and environmental costs of feed and animal production. These non-classical values of feed are generally not taken into account in feed evaluation and only to a limited extend in feed optimisation. Nevertheless, the use of these properties of feed ingredients would allow optimising diets that contribute to animal health and well-being, product quality, lower environmental costs etc. This requires both the description and quantification of such properties of feed ingredients and the requirements or dose related effects in the target animals. Allocating these properties to both commonly used and specific ingredients or additives would enable to account for additive or conflicting characteristics of ingredients and allow flexibility in feed optimisation to find the economically optimal solutions for a required purpose. Hence, feed production using non-classical feed values may contribute to meeting future conditions of animal production with regard to optimal use of resources, respect for animal health and well-being and minimising environmental costs.

References

Awati, A., 2005. Prebiotics in piglet nutrition? Fermentation kinetics along the GI tract. PhD Thesis Wageningen University, Wageningen.

Bakker, G.C.M., 1996. Interaction between carbohydrates and fat in pigs: impact on energy evaluation of feeds. PhD Thesis Wageningen University, Wageningen.

Bikker, P., A. Dirkzwager, J. Fledderus, P. Trevisi, I. Le Huërou-Luron, J.P. Lalles and A. Awati, 2006. The effect of dietary protein and fermentable carbohydrates levels on growth performance and intestinal characteristics in newly weaned piglets. J. Anim. Sci. 84, 3337-3345.

Bikker, P. and A. Dirkzwager, 2003. Withdrawal of anti microbial growth promoters from pig diets increases the amino acid requirements for growth performance. In: Souffrant, W.B. and C.C. Metges (editors), Progress in research on energy and protein metabolism. EAAP pub. no. 109: 593-596.

Bikker, P., J. Fledderus, L. le Bellego and M. Rovers. 2007. Growth response of pigs to dietary threonine: lysine ratio is affected by the withdrawal of anti microbial growth promoters. In: Ortigues-Marty, I. (ed.). Energy and protein metabolism and nutrition. EAAP pub. no. 124. Wageningen Academic Publishers, Wageningen, the Netherlands.

Brouns, F., S.A. Edwards and P.R. English, 1995. Influence of fibrous feed ingredients on voluntary intake of dry sows. Anim. Feed Sci. Technol. 54, 301-313.

Canh, T.T., 1998. Ammonia emission from excreta of growing-finishing pigs as affected by dietary composition. PhD Thesis Wageningen University, Wageningen.

D'Eath, R.B., B.J. Tolkamp, I. Kyriazakisc and A.B. Lawrence, 2009. 'Freedom from hunger' and preventing obesity: the animal welfare implications of reducing food quantity or quality. Anim. Behav. 77, 275-288.

De Leeuw, J.A., A.W. Jongbloed, H.A.M. Spoolder and M.W.A. Verstegen, 2005a. Effects of hindgut fermentation of non-starch polysaccharides on the stability of blood glucose and insulin levels and physical activity in empty sows. Livest. Prod. Sci. 96, 165-174.

De Leeuw, J.A., J.J. Zonderland, H. Altena, H.A.M. Spoolder, A.W. Jongbloed and M.W.A. Verstegen, 2005b. Effects of levels and sources of dietary fermentable non-starch polysaccharides on blood glucose stability and behaviour of grouphoused pregnant gilts. Appl. Anim. Behav. Sci. 94, 15-29.

De Vries, M. and I.J.M. De Boer, 2010. Comparing environmental impacts for livestock products: A review of life cycle assessments. Livest. Sci. 128, 1-11.

Fidin, 2009. Antibioticumrapportage 2008. Fidin werkgroep antibioticumbeleid, Den Haag, the Netherlands, 5 p.

Fledderus, J., P. Bikker and J.W. Kluess, 2007. Increasing diet viscosity using carboxymethylcellulose in weaned piglets stimulates protein digestibility. Livest. Sci. 109, 89-92.

Gallois, M., H.J. Rothkötter, M. Bailey, C.R. Stokes and I.P. Oswald, 2009. Natural alternatives to in-feed antibiotics in pig production: can immunomodulators play a role? Animal 3, 1644-1661.

Grala W., M.W.A. Verstegen, P. Van Leeuwen, J. Huisman, A.J.M. Jansman and S. Tamminga, 1997. Nitrogen balance of pigs as affected by feedstuffs causing different endogenous nitrogen flow at the terminal ileum. Livest. Prod. Sci. 48, 143-155.

Grala W., M.W.A. Verstegen, A.J.M. Jansman, J. Huisman and J. Wasilewko, 1998. Nitrogen utilization in pigs fed diets with soybean and rapeseed products leading to different ileal endogenous nitrogen losses. J Anim. Sci. 76, 569-577.

Heo, J.M., J.C. Kim, C.F. Hansen, B.P. Mullan, D.J. Hampson and J.R. Pluske, 2009 Feeding a diet with decreased protein content reduces indices of protein fermentation and the incidence of postweaning diarrhea in weaned pigs challenged with an enterotoxigenic strain of *Escherichia coli*. J. Anim. Sci. 87, 2833-2843.

Hermes, R.G., F. Molist, M. Ywazaki, M. Nofrarías, A. Gomez de Segura, J. Gasa and J.F. Pérez, 2009. Effect of dietary level of protein and fiber on the productive performance and health status of piglets. J. Anim. Sci. 87, 3569-3577.

Hopwood, D.E., D.W. Pethick, J.R. Pluske and D.J. Hampson, 2004. Addition of pearl barley to a rice-based diet for newly weaned piglets increases the viscosity of the intestinal contents, reduces starch digestibility and exacerbates post-weaning colibacillosis. Br. J. Nutr. 92, 419-427.

Jeaurond, E.A., M. Rademacher, J.R. Pluske, C.H. Zhu and C.F.M. de Lange, 2008. Impact of feeding fermentable proteins and carbohydrates on growth performance, gut health and gastrointestinal function of newly weaned pigs. Can. J. Anim. Sci. 88, 271-281.

Jondreville C., P.S. Revy and J.Y. Dourmad, 2003. Dietary means to better control the environmental impact of copper and zinc by pigs from weaning to slaughter. Livest. Prod. Sci. 84, 147-156.

Kim, J.C., B.P. Mullan, D.J. Hampson and J.R. Pluske, 2008. Addition of oat hulls to an extruded rice-based diet for weaner pigs ameliorates the incidence of diarrhoea and reduces indices of protein fermentation in the gastrointestinal tract. Br. J. Nutr. 99, 1217-1225.

Koopmans, S.J., M. Ruis, R. Dekker and M. Korte, 2009. Surplus dietary tryptophan inhibits stress hormone kinetics and induces insulin resistance in pigs. Physiol. Behav. 98, 402-410.

Kreuzer, M. and A. Machmüller, 1993. Reduction of gaseous nitrogen emission from pig manure by increasing the level of bacterially fermentable substrates in the ration. In: Verstegen, M.W.A., L.A. den Hartog, G.J.M.van Kempen and J.H.M. Metz (eds.), Nitrogen flow in pig production and environmental consequences. Pudoc Scientific Publishers, Wageningen, the Netherlands: 151-156.

Lallès, J.P., G. Boudry, C. Favier and B. Sève, 2006. High-viscosity carboxymethylcellulose reduces carbachol-stimulated intestinal chloride secretion in weaned piglets fed a diet based on skimmed milk powder and maltodextrin. Br. J. Nutr. 95, 488-495.

Lalles, J.P., P. Bosi, P. Janczyk, S. J. Koopmans and D. Torrallardona, 2009. Impact of bioactive substances on the gastrointestinal tract and performance of weaned piglets: a review. Animal 3, 1625-1643.

Lammers, P.J., M.D. Kenealy, J.B. Kliebenstein, J.D. Harmon, M.J. Helmers and M. S. Honeyman, 2010. Nonsolar energy use and one-hundred-year global warming potential of Iowa swine feedstuffs and feeding strategies J. Anim. Sci. 88, 1204-1212.

Le Floc'h, N. and B. Sève, 2007. Biological roles of tryptophan and its metabolism: Potential implications for pig feeding. Livest. Sci. 112, 23-32.

Mateos, G.G., M. Martin, M.A. Latorre and R. Lazaro, 2006. Inclusion of oat hull in diets for young pigs based on cooked maize or cooked rice. Anim. Sci. 82, 57-63.

Nielsen, P.H. and H. Wenzel, 2007. Environmental assessment of Ronozyme® P5000 CT phytase as an alternative to inorganic phosphate supplementation to pig feed used in intensive pig production. Int. J. of Life Cycle Assessment 12, 514-520.

Nielsen, P.H., R. Dalgaard, A. Korsbak and D. Pettersson, 2008. Environmental assessment of digestibility improvement factors applied in animal production. Use of Ronozyme® WX CT xylanase in Danish pig production. Int. J. of Life Cycle Assessment 13, 49-56.

Nyachoti, C.M., F.O. Omogbenigun, M. Rademacher and G. Blank, 2006. Performance responses and indicators of gastrointestinal health in early-weaned pigs fed low-protein amino-acid supplemented diets. J. Anim. Sci. 84, 125-134.

Opapeju, F.O., D.O. Krause, R.L. Payne, M. Rademacher and C.M. Nyachoti, 2009. Effect of dietary protein level on growth performance, indicators of enteric health, and gastrointestinal microbial ecology of weaned pigs induced with postweaning colibacillosis. J. Anim. Sci. 87, 2635-2643.

Owusu-Asiedu, A.J.F. J. Patience, B. Laarveld, A.G. Van Kessel, P.H. Simmins and R.T. Zijlstra, 2006. Effects of guar gum and cellulose on digesta passage rate, ileal microbial populations, energy and protein digestibility, and performance of grower pigs J. Anim. Sci. 84, 843-852.

Revy, P.S., C. Jondreville, J.Y. Dourmad and Y. Nys, 2006. Assessment of dietary zinc requirement of weaned piglets fed diets with or without microbial phytase. J. Anim. Phys. Anim. Nutr. 90, 50-59.

Rijnen, M.M.J.A., M.W.A. Verstegen, M.J.W. Heetkamp, J. Haaksma and J.W. Schrama, 2003a. Effects of dietary fermentable carbohydrates on behavior and heat production in group-housed sows. J. Anim. Sci. 81, 182-190.

Rijnen, M.M.J.A., M.W.A. Verstegen, M.J.W. Heetkamp, J.W. Schrama, 2003b. Effects of two different dietary fermentable carbohydrates on activity and heat production in group-housed growing pigs. Anim. Sci. 81, 1210-1219.

Robert, S., J. Rushen and C. Farmer, 1997. Both energy content and bulk of food affect stereotypic behaviour, heart rate and feeding motivation of female pigs. Appl. Anim. Behav. Sci. 54, 161-171.

Solà-Oriol, D., E. Roura and D. Torrallardona, 2009a. Feed preference in pigs: Effect of cereal sources at different inclusion rates. J. Anim. Sci. 87, 562-570.

Solà-Oriol, D., E. Roura and D. Torrallardona, 2009b. Feed preference in pigs: Relationship with feed particle size and texture. J. Anim. Sci. 87, 571-582.

Stein, H.H., B. Sève, M.F. Fuller, P.J. Moughan and C.F.M. de Lange, 2007. Amino acid bioavailability and digestibility in pig feed ingredients: Terminology and application. J. Anim. Sci. 85, 172-180.

Steinfeld H., P. Gerber, T. Wassenaar, V. Castel, M. Rosales and C. de Haan, 2006. Livestock's long shadow: environmental issues and options. Rome, Italy: FAO, 392 pp.

Sugita-Konishi Y., K. Kobayashi, S. Sakanaka, L.R. Juneja and F. Amano, 2004. Preventive effect of sialylglycopeptide-nondigestive polysaccharide conjugates on Salmonella infection. J. Food Agr. Chem. 52, 5443-5448.

Van der Peet-Schwering, C.M.C., 2004. Long-term performance and behavior of sows fed high levels of non-starch polysaccharides. PhD Thesis Wageningen University, Wageningen.

Van der Werf, H.M.G., J. Petit and J. Sanders, 2005. The environmental impacts of the production of concentrated feed: the case of pig feed in Bretagne. Agric. Syst. 83, 153-177.

Vellinga, T., H. van Laar, M.A. Thomassen, I.J.M. de Boer, P. Berkhout and H. Aiking, 2009. Milieueffecten van diervoeders. Report 205, Animal Sciences Group, Wageningen UR, Lelystad, the Netherlands, 36 pp.

Wellock, I.J., P.D. Fortomaris, J.G.M. Houdijk and I. Kyriazakis, 2006. The effect of dietary protein supply on the performance and risk of post-weaning enteric disorders in newly weaned pigs. Anim. Sci. 82, 327-335.

Wellock, I.J., P.D. Fortomaris, J.G.M. Houdijk and I. Kyriazakis, 2008a. The interaction between dietary protein supply, weaning age and enterotoxigenic Escherichia coli infection on the performance and health of newly weaned pigs: Performance. Animal 2, 834-842.

Wellock, I.J., P.D. Fortomaris, J.G.M. Houdijk and I. Kyriazakis, 2008b. The interaction between dietary protein supply, weaning age and enterotoxigenic Escherichia coli infection on the performance and health of newly weaned pigs: Health. Animal 2, 825-833.

Weigand, E. and M. Kirchgessner, 1980. Total true efficiency of zinc utilization: determination and homeostatic dependence upon the zinc supply status in young rats. J. Nutr. 110, 469-480.

Zamaratskaia, G. and E.J. Squires, 2009. Biochemical, nutritional and genetic effects on boar taint in entire male pigs. Animal 3, 1508-1521.

Zhu, C.L., M. Rademacher and C.F.M. de Lange, 2005. Increasing dietary pectin level reduces utilization of digestible threonine intake, but not lysine intake, for body protein deposition in growing pigs. J. Anim. Sci., 83, 1044-1053.

Maintenance energy and nitrogen requirements of the Icelandic horse

S. Ragnarsson[1] and J.E. Lindberg[2]
[1]Hólar University Collage, 551 Saudárkrókur, Iceland
[2]Department of Animal Nutrition and Management, Swedish University of Agricultural Sciences, P.O. Box 7024, 75007 Uppsala, Sweden

Introduction

The Icelandic horse is a popular riding horse and is used for breeding, competition and leisure in more than 20 countries, counting almost 200,000 horses. The largest part of the population is found in Iceland, Germany and Scandinavia (Worldfengur, 2009). They stand an average of 134 to 143 cm and generally weigh 330 to 430 kg. This breed has lived isolated on Iceland for over 1000 years and through the centuries, survived natural disasters and harsh condition (Sveinsson, 2003). There is a scarcity of information about the feed utilization of the Icelandic horse and it is common belief that Icelandic horses require low amounts of feed to maintain condition (Þórhallsdóttir and Eiríksson, 1996). Therefore, the aim of this study was to test the hypothesis that the energy and nitrogen (N) maintenance requirements of mature Icelandic horses are lower than in other horse breeds.

Material and methods

For estimating the maintenance energy requirements of stabled Icelandic horses, results from 3 experiments were used (Ragnarsson, 2009). Experiment I and experiment II were total collection digestibility trials using different types of haylages, both designed as 4×4 Latin-square. Experiment III was a pilot-study using one haylage, fed in different amounts to 9 horses in 3 equal groups. In all experiments, gross energy in food and faeces was determined using an adiabatic bomb calorimeter and digestible energy (DE) calculated by multiplying the gross energy with the energy digestibility. Nitrogen (N) was determined by the Kjeldahl method and the value multiplied by 6.25 for estimating crude protein (CP) content. In experiments I and II urine was also collected for N balance calculations. Each of the experiments lasted for 3 months. In all experiments the horses body weight (BW) was recorded weekly by using an electronic livestock scale and their body condition scored.

Results and discussion

In all studies all horses were in good health and moderate to moderately fleshy body condition (Henneke *et al.*, 1983). By combining the data from all experiments (Table 1), 17 estimates from 3 month periods were obtained and the maintenance requirements (zero BW change) of mature sedentary and stabled Icelandic horses was estimated to be 0.54 MJ DE/kg $BW^{0.75}$/d, using linear regression between BW change and DE intake ($Y=100.21\times-54.058$, $r^2=0.69$). When applied to a 370 kg horse (average weight of the experimental horses, Table I) the calculated maintenance

Table 1. Average daily DE intake, digestible N intake and BW response in mature sedentary Icelandic horses fed different amounts for 3 months.

Experiment	MJ DE/kg $BW^{0.75}$	DCP g/kg $BW^{0.75}$	n	BW response	Initial average BW
I	0.49	3.9	4	-5	348
II	0.51	4.8	4	-9.5	358
III	0.34	2.0	3	-18	385
III	0.52	3.1	3	0	370
III	0.61	3.9	3	+11	391

requirements were 45.6 MJ DE/d. The minimum maintenance requirements recommended by NRC (2007) correspond to 47 MJ/day and the German feeding standards (Coenen, 2001) using the standard value (0.6 MJ DE/kg $BW^{0.75}$/d) corresponds to 50.6 MJ DE/d. However, according to Coenen (2001) the maintenance requirements can range from 0.48 to 0.62 MJ DE/kg $BW^{0.75}$/d and the current estimate (0.54 MJ DE/kg $BW^{0.75}$/d) falls within this range.

In experiments I and II the calculated N balance was numerically positive for all forages. From the results it is possible to estimate the N requirements by regressing N retention and apparent digested N. The estimated value was 400 mg digestible N/kg $BW^{0.75}$/d or 2.5 g DCP/ kg $BW^{0.75}$/d for zero N retention ($Y=0.1146\times-45.854$, $r^2=0.72$). This corresponds to 0.58 DCP/kg BW/d which is similar to NRC (1989). However, because the N balance can underestimate true nitrogen losses from the body due to dermal and sweat losses, in addition to measurement errors, some allowances for nitrogen retention greater then zero should be made when determining maintenance requirements. Therefore N requirements of sedentary Icelandic horses could be comparable to the German (Coenen, 2002) recommendations, 3 g DCP/kg $BW^{0.75}$ and to the 1.08 g CP/kg BW/day suggested as minimum requirements by NRC (2007).

The hypothesis was rejected as the obtained energy and N maintenance requirements were comparable with other published estimates on minimum requirements for mature horses. The current study suggests that the Icelandic horse has similar energy and N requirements as other horse breeds.

References

Coenen, M., 2001. German feeding standards. In: Pagan, J.D. and J. Geor, (eds.), Advances in equine nutrition II. Nottingham University Press: 365-378.

Coenen, M., 2002. Evaluating the protein requirements of horses: The German system for digestible protein. In: Proceedings of the 1st European workshop on equine nutrition. Dijon, France: 153-158.

Henneke, D.R., G.D. Potter, J.L. Kreider and B.F. Yeater, 1983. Relationship between conditions score, physical measurement and body fat percentage in mares. Equine Veterinary Journal 15, 371-372.

NRC, 1989. Nutrient requirements of horses. 5th revised edition. National Research Council. National Academy Press. Washington, DC.

NRC, 2007. Nutrient requirements of horses. 6th revised edition. National Research Council. National Academy Press. Washington, DC.

Ragnarsson, S., 2009. Digestibility and metabolism in Icelandic horses fed forage-only diets. Swedish University of Agricultural Sciences, Acta Universitatis Agriculturae Sueciae, Faculty of Veterinary Medicine and Animal Science, Doctoral thesis no. 2009: 92.

Sveinsson, I., 2003. The Icelandic horse, its history, development and use. Seminar for genressursutvalgene for husdyr i Norden: 1-13.

Worldfengur, (2009). The studbook of origin for the Icelandic horse. www.worldfengur.com [assessed 10th November 2009].

Þórhallsdóttir, A.G. and T. Eiríksson, (1996). Samanburður á meltanleika gróffóðurs hjá hrossum og sauðum [Comparison of forage digestibility between horses and sheep]. Ráðunautafundur:189-193.

The maintenance energy expenditure is affected by feeding level in growing animals

E. Labussière, J. van Milgen, S. Dubois and J. Noblet
INRA, UMR SENAH,35590 Saint-Gilles, France

Introduction

The factorial approach used for calculating energy requirements of animals considers these as the sum of requirements for maintenance, for production (e.g. growth, milk, eggs) and possible requirements for thermoregulation, additional physical activity or work. Maintenance requirements are usually estimated from the regression of heat production (HP) or energy gain on metabolizable energy (ME) intake in animals fed at variable levels of ME intake, and its extrapolation to zero ME intake (fasting HP, FHP) or to zero energy gain (ME_m). This approach considers that FHP or ME_m do not depend on ME intake, despite indications that they do (Koong *et al.*, 1982). The objective of our study is to re-examine data obtained in recent experiments conducted at INRA in pigs and veal calves to propose an alternative method for calculating ME_m that accounts for the effect of feed intake on FHP.

Materials and methods

In experiment 1 (van Milgen *et al.*, 2001), 5 groups of 5 to 6 Piétrain × (Landrace × Large White) barrows (60 kg average BW) received a basal diet (1.7 MJ ME/kg $BW^{0.60}$/d) or a mix of the basal diet and a dietary supplement (+0.9 MJ ME/kg $BW^{0.60}$/d) that provided ME from wheat starch, vegetable oil, balanced protein (casein) or unbalanced protein (corn gluten meal). In experiment 2 (de Lange *et al.*, 2006), 6 Piétrain × (Landrace × Large White) barrows (71 kg average BW) were each fed at 60, 80 or 100% of the *ad libitum* level (2.6 MJ ME/kg $BW^{0.60}$/d) according to a double latin-square design. In experiment 3 (Labussière *et al.*, 2009), 16 male Prim'Holstein calves received at 3 stages of growth (average BW during each stage: 73, 152 and 237 kg) one of 4 feeding levels calculated as 79, 87, 95 and 103% of a reference level (711, 674 or 609 kJ ME/kg $BW^{0.85}$/d at stages 1, 2 and 3, respectively). In each experiment, animals were housed in an open-circuit respiration chamber for 7 days (6 days in the fed state followed by a fasting day) for measuring the kinetics of HP. Total HP was partitioned through a modelling procedure (Van Milgen *et al.*, 1997) into FHP, heat production due to physical activity (AHP) and feed intake (thermic effect of feeding, TEF). Efficiency of utilizing ME for maintenance and growth (k_{mg}) was calculated as (FHP + retained energy)/ME. The ME_m was calculated as FHP/k_{mg}. Data were analyzed by ANOVA for the effect of diet in experiment 1, for the effects of feeding level, experimental period and group in experiment 2, and for the effects of feeding level and stage of fattening in experiment 3. Linear regressions between HP and ME intake in the fed state were calculated in experiments 2 and 3 (SAS, 2004).

Results and discussion

In experiment 1, an increase of 50% in ME intake resulted in an 11% increase in FHP (780 vs. 701 kJ/kg $BW^{0.60}$/d), irrespective of the source of additional ME supply. The k_{mg} for each diet was affected by dietary supplement (P<0.01) and averaged 79.4% for diets with starch and vegetable oil, and 75.5% for diets with corn gluten meal and casein, irrespective of their amino acid profile. In experiment 2 (Table 1), the increase in ME intake resulted in an increase in FHP ranging from 632 to 740 kJ/kg $BW^{0.60}$/d, while it did not affect k_{mg} (76.5%). In experiment 3 (Table 2), FHP in veal calves increased by 12% (310 vs. 276 kJ/kg $BW^{0.85}$/d, on average for the 3 stages) between extreme feeding levels; these values are higher than FHP calculated from the regression between HP and ME (167 kJ/kg $BW^{0.85}$/d). The k_{mg} remained constant (83.9%) at the different stages and feeding levels. Values for ME_m in pigs (from 830 to 1,059 kJ/kg $BW^{0.60}$/d) and calves (from 327 to 371 kJ/kg $BW^{0.85}$/d) calculated from measured FHP and k_{mg} values were higher than those estimated from the regression between HP and ME (777

Table 1. Effect of feeding level on heat production and energy requirements in pigs (experiment 2).

		Feeding level (% of *ad libitum*)			rsd	Probability	
		60	80	100		Feeding level	Period
ME intake	kJ/kg $BW^{0.60}$/d	1,157[a]	2,059[b]	2,554[c]	32	<0.01	<0.01
FHP	kJ/kg $BW^{0.60}$/d	632[a]	666[a]	740[b]	59	0.02	0.38
TEF	% ME	15.2	17.9	18.7	3.6	0.25	0.17
AHP	% ME	12.1[b]	8.9[a]	7.6[a]	1.8	<0.01	0.34
k_{mg}	%	76.0	76.4	77.0	4.2	0.92	0.52
ME_m	kJ/kg $BW^{0.60}$/d	832[a]	871[a]	962[b]	54	<0.01	0.07

[a,b,c,d] $P<0.01$.

Table 2. Effect of feeding level on heat production and energy requirements in calves (experiment 3).

		Feeding level (% of a reference)				rsd	Probability	
		79	87	95	103		Feeding level	Stage of fattening
ME intake	kJ/kg $BW^{0.85}$/d	531[a]	581[b]	625[c]	681[d]	19	<0.01	<0.01
FHP	kJ/kg $BW^{0.85}$/d	276[a]	282[a]	298[b]	310[b]	16	<0.01	<0.01
TEF	% ME	6.5	8.3	8.0	8.4	2.3	0.16	0.54
AHP	% ME	8.9	8.7	8.0	7.9	1.5	0.24	0.16
k_{mg}	%	84.7	83.0	84.1	83.8	2.4	0.39	0.15
ME_m	kJ/kg $BW^{0.85}$/d	327[a]	341[b]	356[c]	371[d]	15	<0.01	<0.01

[a,b,c,d] $P<0.01$.

kJ/kg $BW^{0.60}$/d in experiment 2 and 265 kJ/ kg $BW^{0.85}$/d in experiment 3); corresponding efficiencies of ME utilization were also lower for the regression method (64.4 vs. 76.5% for experiment 2 and 63.3 vs. 83.9% for experiment 3). This alternative approach for calculating ME_m and k_{mg} modifies the partitioning of energy intake between maintenance and production and takes into account the metabolic adaptation of the animal to nutrients supply and the effect of feed intake on ME_m. Although this effect has been identified before, it has not yet been integrated in energy requirement systems.

References

De Lange, K., J. Van Milgen, J. Noblet, S. Dubois, S. Birkett, 2006. Previous feeding level influences plateau heat production following a 24 h fast in growing pigs. Br. J. Nutr. 95, 1082-1087.

Koong, L.J., J.A. Nienaber, J.C. Pekas and J.T. Yen, 1982. Effects of plane of nutrition on organ size and fasting heat production in pigs. J. Nutr. 112, 1638-1642.

Labussière, E.,G. Maxin, S. Dubois, J. van Milgen, G. Bertrand and J. Noblet, 2009. Effects of feed intake on protein and fat deposition in milk fed veal calves. Animal 3, 557-567.

SAS, 2004. SAS.STAT® 9.1 User's Guide. SAS Institute Inc., Cary, NC.

Van Milgen, J., J. Noblet, S. Dubois and J.F. Bernier, 1997. Dynamic aspects of oxygen consumption and carbon dioxide production in swine. Br. J. Nutr. 78, 397-410.

Van Milgen, J., J. Noblet and S. Dubois, 2001. Energetic efficiency of starch, protein and lipid utilization in growing pigs. J. Nutr. 131, 1309-1318.

Changes in fatty acid profile of the subcutaneous fat in Iberian and Landrace × Large White pigs: possible effects on energy utilization

R. Barea[1], B. Isabel[2], R. Nieto[1], C. López-Bote[2] and J.F. Aguilera[1]
[1]Instituto de Nutrición Animal (IFNA), Estación Experimental del Zaidín (CSIC), Camino del Jueves s/n, 18100 Armilla, Granada, Spain
[2]Departamento de Producción Animal, Facultad de Veterinaria, Universidad Complutense, Avda. Puerta de Hierro, s/n, 28040 Madrid, Spain

Introduction

The fatty acid profile of pigs is known to be related to genotype. The Iberian pig is an obese Mediterranean breed with higher whole-body fat content and increased energy cost of protein and fat deposition than lean-type pigs. It has been reported that indigenous pig breeds present higher contents of saturated and monounsaturated fatty acids (Serra *et al.*, 1998). This fact may be consequence of variation in the novo lipid synthesis, which may have effects on the overall efficiency of energy utilization. The aim of this experiment was to study the possible effect of pig genotype on the fatty acid profile of the deposited fat during the growing and fattening stages, and how this fact might influence the overall efficiency of energy utilization.

Materials and methods

Sixteen Iberian (IB) and sixteen Landrace x Large White (LD) pigs were fed one of two diets differing in crude protein (CP) content (12 or 17% of dry matter; diets A and B, respectively) using a pair-feeding procedure (8 pigs per dietary treatment and breed). Diets were offered at 0.8 × *ad libitum* of the IB pigs, which have higher intake capacity. Subcutaneous fat biopsies (at the dorso-lumbar region, 2 cm depth) were taken from both pig genotypes at approximately 40, 50, 65, 90 and 120 kg bodyweight (BW; n=16 pigs/breed for samples taken at 40 kg BW, and n=8 for the rest of sampling points). Biopsies included the inner and outer layers of the subcutaneous backfat. They were analyzed separately and expressed as % of total fatty acid methyl esters (FAMEs). The effects of genotype, CP content, backfat layer and their interactions were analyzed by an analysis of variance using the GLM procedure of SAS. The effect of moment (or BW) of biopsying was evaluated by a three-way repeated-measures analysis using the MIXED procedure of SAS. Experimental protocol was approved by the Bioethical Committee of the CSIC.

Results and discussion

No effects of dietary CP content were observed in the FAMEs composition of subcutaneous backfat of both breeds. The effect of genotype on the FAMEs profile was more relevant in the inner than in the outer layer, although monounsaturated fatty acids (MUFA) were significantly higher in IB pigs only in the outer layer. In addition, the outer layer was more unsaturated than the inner, which is in agreement with results reported by Daza *et al.* (2005). Due to the importance of these factors, we decided to present the FAMEs profile of the inner layer in the present paper. The evolution of level of total saturated (SFA), unsaturated (UFA) and polyunsaturated (PUFA) fatty acids in the inner layer as pigs aged is shown in Figure 1.

The SFA were 4% higher (and the UFA 3% lower) in IB than in LD pigs (44.5 vs. 42.6% for SFA and 57.4 vs. 55.5%, for UFA, respectively). The PUFA were higher in LD than in IB pigs (11.3 vs. 8.7%, respectively). These results are in close agreement with those reported by Serra *et al.* (1998) and Morales *et al.* (2003). The higher SFA and lower PUFA contents in obese pigs have been related not only to genetic variation in energy partitioning between lipid and protein deposition but also to a higher feed intake capacity of those pigs as compared with lean-type pigs (Morales *et al.*, 2003).

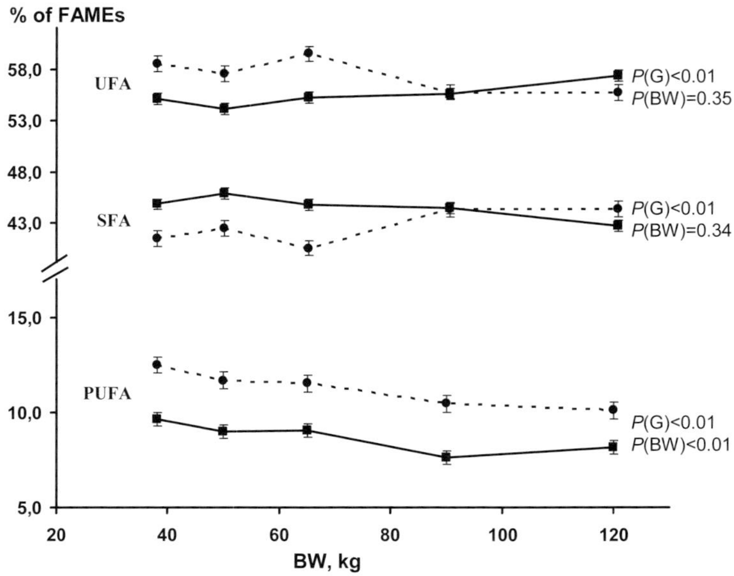

Figure 1. Effects of pig genotype (G) and bodyweight (BW) on total unsaturated fatty acids (UFA), saturated fatty acids (SFA) and polyunsaturated fatty acids (PUFA) in the inner subcutaneous backfat layer of Iberian (continuous line) and Landrace (dashed line) pigs.

However, in the current experiment, the differences in fatty acids profile were similar to those reported in the literature even if both breeds were fed at 0.8 × *ad libitum* of the IB pigs. The PUFA, MUFA, C18:2 (n-6) (linoleic acid) and C18:1 (oleic acid) were also affected by the BW in both genotypes. We conclude that both genetic variation and pattern of fatty acid unsaturation during growth might contribute to explain the differences in FAMEs composition of adipose tissue of IB and LD pigs.

References

Daza, A., A. Mateos, A.I. Rey, and C.J. López-Bote, 2005. Feeding level in the period previous to the late fattening phase influences fat composition at slaughter in freeranged Iberian pigs. Arch. Anim. Nutr. 59, 227-236.

Morales, J., M.D. Baucells, J.F. Pérez, J. Mourot and J. Gasa, 2003. Body fat content, composition and distribution in Landrace and Iberian finishing pigs given *ad libitum* maize- and a corn -sorghum-maize-based diets. Anim. Sci. 77, 215-224.

Serra, X., F. Gil, M. Pérez-Enciso, M.A. Oliver, J.M. Vázquez, M. Gispert, I. Díaz, F. Moreno, R. Latorre, and J.L. Noguera, 1998. A comparison of carcass, meat quality and histochemical characteristics of Iberian (Guadyerbas line) and Landrace pigs. Liv. Prod. Sci. 56, 215-223.

Modelling protein deposition: defining pig genotypes and deriving their nutritional requirements

N. Duijvesteijn, B.A.N. Silva and E.F. Knol
IPG, Institute for Pig Genetics, P.O. Box 43, 6640 AA Beuningen, the Netherlands

Introduction

To meet the current and future demand for animal protein, famers need to improve the efficiency of lean meat production. Cost-effective management, where the expressed genetic potential of the pig is optimized, is essential. To produce the most efficient pig, several criteria have to be met such as an optimal ratio between protein and lipid deposition and an acceptable feed conversion and growth rate.

Several models for estimating protein deposition (Pd) have been proposed of which De Greef (1992) assumes a linear increase of protein deposition with an increase in energy intake until a maximum is reached (referred to as Pdmax). In this study we modelled the Pd for different genotypes and sexes during the finishing period of pigs:

Pd = f (Weight, Feed Intake | genotype, sex), where Pd is a function of weight and feed intake depending on genotype and sex. The final objective is estimating the genetic potential of different pig genotypes for Pd to derive the ideal lysine and energy requirements.

Material and methods

TOPIGS collects data on regular bases of purebred and crossbred animals to determine accurate breeding values for traits related to the breeding goal of which weight, backfat and feed intake are among those. Data was available of two different types of farms. On the experimental farm the pigs were weighed 3 times (T-start, T-med, T-end) and back fat was measured twice (T-med and T-end) with individual feed registration (n=3,144) using crossbred animals. On the nucleus farms, pigs are also weighed 3 times, twice with backfat measurements and individual feed registration was only between T-med and T-end (n=12,590). On the nucleus farms, only boars were tested and on the experimental farm, mainly gilts and castrates were tested. The different weight measurements are divided into 2 phases; phase 1 (25-75 kg live weight) and phase 2 (75-115 kg live weight). Equations from the TMV model (1994) were adjusted to calculate Pd using back fat and live weight as inputs.

Pd and feed intake per phase were inputs for the linear plateau model described by De Greef (1992). The linear-plateau model is a non linear regression model (NLIN) which is programmed in SAS (version 9.1.3). The model estimates the Pdmax, break point (Xh, kg/day) and the steepness of the slope (rate constant referred to as r.c.) for every genotype and sex. Equations derived from Van Milgen *et al.* (2008) were used to calculate lysine and energy requirements based on protein and lipid deposition.

Results and discussion

Estimates for Pd for phase 1 are based on crossbred data and are higher than phase 2 (on average 162 g/day vs. 129 g/day for castrates and gilts and 170 g/day vs. 164 g/day for boars respectively). Differences between boars, castrates and gilts in Pd for the second phase (weight between 75-115 kg's) are presented Table 1. Corrected for farm, damline and sireline the results suggest a higher Pdmax in boars than in castrates and gilts, with a r.c. for boars being also higher, suggesting a more efficient utilization of feed (energy) to deposit protein. Differences between sire- and damlines exist and are significant for feed intake (energy intake) to achieve Pdmax (Table 2). Also differences between damlines and sirelines for the Pdmax, Xh and r.c. were found, which indicates more selection

Table 1. Differences between sexes for Pd in phase 2.

Sexe	r.c.	Xh (kg/day)	Pdmax (g/day)
Boars	40.7^{b}	2.63	164.2^{b}
Castrates	18.2^{a}	2.92	121.9^{a}
Gilts	26.6^{ab}	2.67	136.4^{a}

a,b within a column, means with different subscripts differ (P<0.05).

Table 2. Differences between purebred lines (damlines and sirelines) for Pd in phase 2.

Line	r.c.	Xh (kg/day)	Pdmax (g/day)
Dam1	19.8	2.93^{ab}	130.2
Dam2	21.3	2.84^{abc}	133.8
Sire1	28.2	3.05^{b}	148.3
Sire2	40.1	2.50^{a}	146.8
Sire3	24.7	2.64^{ac}	147.8

a,b within a column, means with different subscripts differ (P<0.05).

pressure on growth in sirelines than damlines. For pigs below 45 kg body weight, Pd is determined primarily by energy intake and thus reflects the relationship between Pd and energy intake, rather than Pdmax, but the genetic potential is not lower in this phase. The Pdmax can be achieved between 45-75 kg (Möhn and De Lange, 1998) if pigs are fed optimal; therefore our model is assuming a constant Pdmax between 20 and 75 kg of live weight and a decreasing Pdmax after that because of an increase of lipid deposition.

Conclusions

The output of information from our model can be used to determine lysine and energy requirements, making it possible to differentiate requirements between genotypes and sex to achieve the genetic potential for Pdmax (when all other factors are optimal. i.e. management and health status). Validation studies have been performed and confirm the differences estimated between genotypes and sexes. Kept in mind that the data collected might not be in an optimal environment, the process of estimating genetic potential for Pdmax is ongoing.

References

De Greef, K.H., 1992. Prediction of production; Nutrition induced tissue partitioning in growing pigs. Ph.D. Thesis. Wageningen, Agric. Univ., Wageningen, the Netherlands.

Mohn, S. and C.F. de Lange, 1998. The effect of body weight on the upper limit to protein deposition in a defined population of growing gilts. J. Anim. Sci. 76, 124-133.

Technisch Model Varkensvoeding (TMV). 1994. Proefstation voor de varkens houderij, Rosmalen, the Netherlands.

Van Milgen, J., A. Valancogne, S. Dubois, J-Y. Dourmad, B. Sève and J. Noblet, 2008. InraPorc, A model and decision support tool for the nutrition of growing pigs. J. Anim. Feed Sci. and Tech. 143, 387-405.

Dietary lysine requirement for sows at maintenance

R.S. Samuel[1], S. Moehn[1], P.B. Pencharz[2] and R.O. Ball[1,2]
[1]Swine Research and Technology Centre, 4-10 Agriculture/Forestry Centre, University of Alberta, T6G 2P5, Edmonton, Alberta, Canada; ron.ball@ualberta.ca
[2]Research Institute, Hospital for Sick Children, M5G 1X8, Toronto, Ontario, Canada

Introduction

The recommended daily lysine requirement of non-pregnant sows at maintenance, 36 mg/kg$^{0.75}$ (NRC, 1998), was estimated from research in the 1960's with growing pigs and sows (Pettigrew, 1993). Ball *et al.* (2008) suggested that, due to selection for positive production traits, including larger and faster growing litters and larger sows with increased body protein mass, the current estimate may be too low for modern sows. Therefore, it is necessary to determine the lysine maintenance requirement of modern, high productivity sows.

Materials and methods

Animals, diets, housing, feeding, and indigestible marker

Non-pregnant Hypor Hybrid (Hypor Inc.) sows (n=4, parity=2), from the University of Alberta's Swine Research and Technology Centre, were adapted to 2.2 kg of a semi-synthetic diet, based on corn, premix and amino acids, containing 14.0 MJ ME/kg and 1.09 g/kg lysine (equivalent to 41.4 mg/kg$^{0.75}$, or 115% of NRC (1998) requirement). Sows were housed individually and fed one-half of their daily feed allowance twice daily. Unlabelled phenylalanine, equivalent to the isotope dose, was mixed into individual batches and fed during adaptation periods between study days. An indigestible marker was mixed into the diets (10 g/kg) and analyzed as acid insoluble ash for determination of energy and nitrogen digestibility (McCarthy *et al.*, 1977). Each sow received 6 experimental diets, in random order, providing lysine intakes of 19.8, 25.2, 30.6, 41.4, 46.8 and 52.2 mg/kg$^{0.75}$. The lowest lysine intake was equivalent to 55% of NRC (1998) requirement. Sows were adapted to each diet for 3 meals or 1.5 d.

Indirect calorimetry, isotope infusion, and sample collection

On study days, sows received 11 meals equivalent to 1/26th of their daily ration every 30 minutes. Comparability of this protocol to once daily feeding and adequacy of the adaptation period was previously verified (Moehn *et al.*, 2004). Sows were individually housed in respiration chambers for the measurement of indicator amino acid oxidation (IAAO) during a primed, constant infusion of L-[1-^{13}C]phenylalanine and, simultaneously, heat production (HP). The study period consisted of, at minimum, 0.5 h of equilibration before 1.5 h of background measurements and 4 h of collection period. The gas exchange was recorded for O_2, CO_2, and CH_4 in 1 min intervals. Expired CO_2 and blood plasma were collected in 30 min intervals for determination of ^{13}C enrichment.

Statistics

Statistical analysis was performed using mixed procedure and breakpoint analysis was performed using the non-linear mixed procedure in SAS (2002). The classification variable was lysine intake and individual animals were treated as random variables. Model statements were tested using the Kenward-Roger degrees of freedom method. Least square means were compared using the 'pdiff' option. Data are presented as means ± SEM. Values were considered significant at $P<0.05$.

Results and discussion

Total tract digestibility of nitrogen (P=0.83) and carbon (P=0.15) were not different among diets at 73.7±1.0% and 93.7±0.2%, respectively. Phenylalanine flux was not affected by dietary lysine intake (P=0.39). Plateaus in oxidation were achieved within 1.5 h from the start of infusion. Oxidation of the indicator amino acid was lowest (P<0.05) when lysine intake was greater than 46.8 mg/kg$^{0.75}$ and was confirmed by results from breakpoint analysis which determined the requirement as 49±11 mg/kg$^{0.75}$ (R^2=0.65). Heat production per 30 minute period and the mean RQ were lowest (P<0.05) and there was a trend (P<0.10) for lower CO_2 production when sows were fed 46.8 mg/kg$^{0.75}$ lysine. These results were supported by breakpoint analysis of calorimetry data; a breakpoint was calculated at 46.8±0.02 mg/kg$^{0.75}$ for minimum heat production per 30 min period (R^2=0.67), RQ (R^2=0.63), and CO_2 production (R^2=0.40). The dietary lysine requirement for this population of sows was determined to be 49 mg/kg$^{0.75}$ (or 2.6 g/d for a 200 kg sow); this exceeds the NRC (1998) recommendation by 30% but is similar to the recommendation of Pettigrew (1993) and GfE (2008). The combined results of lower HP, RQ, and CO_2 production when sows received adequate intake of dietary lysine demonstrate that energy metabolism was also most efficient at this intake.

References

Ball, R.O., R.S. Samuel and S. Moehn, 2008. Nutrient requirements of prolific sows. Advances in Pork Production. 19, 223-236.

Brouwer, E., 1965. Report of sub-committee of the European Association for Animal Production on constants and factors. In: K.L. Blaxter (editor), Energy metabolism. Academic Press, New York, NY: 441-443.

GfE, 2008. Recommendations for the supply of energy and nutrients to pigs. Frankfurt, Germany, DLG Verlags GmbH.

McCarthy, J.F., J.P. Bowland and F.X. Aherne, 1977. Influence of method upon the determination of apparent digestibility in the pig. Can. J. Anim. Sci. 57, 131-135.

Moehn, S., R.F.P. Bertolo, P.B. Pencharz and R.O. Ball, 2004. Pattern of carbon dioxide production and retention is similar in adult pigs when fen hourly, but not when fed a single meal. BMC Physl. 4, 11-18.

National Research Council, 1998. Nutrient requirements of swine. 10th ed. Washington, D.C., National Academy Press.

Pettigrew, J.E., 1993. Amino acid nutrition of gestating and lactating sows. Biokyowa Technical Review-5. Nutri-Quest, Inc., St. Louis, MO.

SAS (Statistical Analysis System), 2002. Release 9.1 Edition. SAS Institute, Inc., Cary, NC.

Effect of dietary tryptophan to lysine ratio on performance of growing pigs fed wheat-barley or corn-soybean meal based diets

M. Naatjes[1]*, J.K. Htoo*[2]*, K.H. Tölle*[3] *and A. Susenbeth*[1]
[1]*Institute of Animal Nutrition and Physiology, Christian-Albrechts-University, 24098 Kiel, Germany*
[2]*Evonik Degussa GmbH, Rodenbacher Chaussee 4, 63457 Hanau, Germany*
[3]*Training and Research Center Futterkamp, 24327 Blekendorf, Germany*

Introduction

Tryptophan (Trp) is usually the fourth limiting amino acid (AA) in typical swine diets. To date, there is considerable variation in optimum dietary ratio of Trp relative to Lys among published data. For example, Guzik *et al.* (2005) estimated an optimal standardized ileal digestible (SID) Trp:Lys ratio to be greater than 19.5% for starter pigs while Susenbeth (2006), by analysis of literature data, suggested a total Trp:Lys ratio of 17% for growing pigs. Thus, the objective of this study was to evaluate the optimum SID Trp:Lys ratio in wheat-barley and corn-soybean meal (SBM) based diets fed to 15- to 35-kg pigs.

Material and methods

A 28-d growth assay was conducted with 880 mixed-sex pigs [Pietrain × dbNaima; initial body weight (BW) of 14.4±0.93 kg] with 11 pigs per pen and 5 pen replicates per treatment. Two Trp-deficient basal diets (0.14% SID Trp, 13.8 MJ ME/kg), based on wheat-barley or corn-SBM, were formulated, using analyzed ingredient AA contents and published SID AA values (AminoDat® 3.0) to meet requirements of AA other than Trp and Lys. The dietary Lys level (1.05% SID Lys) was set to be marginally limiting for the pigs used in this study (GfE, 2008). L-Trp was added to both basal diets at the expense of wheat or corn to create 7 SID Trp:Lys ratios and a positive control diet (diet 8; 1.15% SID Lys) was also formulated for both diet types. Analyzed AA data revealed that corn-SBM diets (on average) contained a higher content of crude protein (CP; 19.6 vs. 16.7%) and AA (e.g. 1.27 vs. 1.16% for Lys; 0.23 vs. 0.20% for average Trp) than wheat-barley based diets. The analyzed total Trp:Lys ratios were 14.3, 15.3, 16.3, 17.3, 18.3, 19.3 and 20.3% in diets 1 to 7 of wheat-barley based diets, and 15.6, 16.5, 17.4, 18.3, 19.3, 20.2 and 21.1% in diets 1 to 7 of corn-SBM based diets, respectively. The corresponding analyzed SID Trp:Lys ratios were given in Table 1.

Data were analyzed by ANOVA using the GLM procedure of SAS with barn (block), initial BW and dietary treatments included in the model. Exponential [at 95% of maximum (max.) response] and curvilinear-plateau [$y = L + U + (R - x)^2$, where $(R - x)$ is zero at values of $x > R$] regression analyses were conducted to estimate the optimum Trp:Lys ratios.

Results and discussion

Performance results are given in Table 1. The average daily gain (ADG) and feed:gain (F:G) of pigs were improved (P<0.03, linear) with increasing Trp:Lys ratios for both wheat-barley and corn-SBM based diets. Feed intake was increased (P<0.01) in both wheat-barley and tended to increase (P<0.10) for corn-SBM based diets by graded level of Trp:Lys ratio.

The optimal SID Trp:Lys ratios estimated by the exponential regression (95% of max. response) exceeded the highest tested SID Trp:Lys ratio (i.e. >19.7%) based on ADG and F:G response data (Table 2). Curvilinear-plateau regression estimated an optimal SID Trp:Lys of 20 and 18.5% based on ADG and F:G response data, respectively. The ADG and F:G response data of pigs fed corn-SBM diets did not reach the plateau (Table 1) and thus, data were not analyzed by regressions.

Table 1. Effect of dietary SID Trp:Lys ratio on performance of growing pigs.[1]

Diets	1	2	3	4	5	6	7	8[2]	SEM	*P*, linear (1 to 7)
Wheat-barley-soybean meal based diets										
SID Trp:Lys, %	13.1	14.2	15.3	16.4	17.5	18.6	19.7	18.3		
ADG, g/d	582	610	645	651	658	660	682	702	0.025	0.003
ADFI, kg/d	1.09	1.16	1.21	1.16	1.19	1.22	1.23	1.20	0.032	0.004
F:G, g/g	1.91	1.91	1.86	1.80	1.78	1.83	1.79	1.69	0.046	0.013
Corn-soybean meal based diets										
SID Trp:Lys, %	14.3	15.4	16.5	17.6	18.7	19.8	20.9	19.3		
ADG, g/d	639	658	674	708	693	713	729	751	0.025	0.004
ADFI, kg/d	1.13	1.16	1.16	1.18	1.24	1.18	1.19	1.20	0.031	0.072
F:G, g/g	1.77	1.76	1.72	1.67	1.68	1.64	1.63	1.59	0.046	0.006

[1] Data are least square means.

[2] Positive control diet (adequate level of Lys and other amino acids).

Table 2. Estimated optimum Trp:Lys ratios in wheat-barley-soybean meal based diets.

Item	Exponential (95% of max. response)	SID Trp:Lys	Curvilinear-plateau	SID Trp:Lys
ADG, g/d	$y = 582 + 102.2\ [1 - e^{-0.347\ (x - 13.1)}]$; $r^2 = 0.963$	21.7%[1]	$y = 674 - 1.86\ (20.0 - x)^2$; $r^2 = 0.949$	20.0%[1]
F:G, g/g	$y = 1.93 - 0.161\ [1 - e^{-0.287\ (x - 13.1)}]$; $r^2 = 0.791$	23.5%[1]	$y = 1.80 + 0.005\ (18.5 - x)^2$; $r^2 = 0.815$	18.5%

[1] Estimated SID Trp:Lys exceeded the highest tested SID Trp:Lys ratio of 19.7%.

Performance data indicate that the optimal SID Trp:Lys ratio in corn-SBM diets was greater than the highest tested SID Trp:Lys ratio of 20.9%.

A greater performance response ($P<0.01$) was observed for pigs fed corn-SBM diets compared with that of wheat-barley diets. However, it should be mentioned that the CP and AA content of corn-SBM diets were considerably higher than that of wheat-barley diets and thus, direct comparison of the optimal SID Trp ratios between the two types of diet is difficult. The highest performance was achieved with the highest tested SID Trp:Lys for both diet types suggesting that dietary optimal Trp:Lys does not seem to be dependent on ingredient composition.

References

AminoDat® 3.0. Platinum version, 2005. Evonik Degussa GmbH, Hanau-Wolfgang, Germany.

GfE, 2008. Recommendations for the supply of energy and nutrients to pigs. DLG-Verlag, Frankfurt am Main: 62-68.

Guzik, A.C., M.J. Pettit, E. Beltranena, L.L. Southern and B.J. Kerr, 2005. Threonine and tryptophan ratios fed to nursery pigs. J. Anim. Physiol. Anim. Nutr. 89, 297-302.

Susenbeth, A., 2006. Optimum tryptophan:Lysine ratio in diets for growing pigs: Analysis of literature data. Livestock Sci. 101, 32-45.

Tryptophan requirements of growing pigs: a dose response study

E. Vinyeta[1]*, P. Bikker*[1]*, E. Corrent*[2] *and M. Rovers*[3]
[1]*Schothorst Feed Research, P.O. Box 533, 8200 AM Lelystad, the Netherlands*
[2]*Ajinomoto Eurolysine S.A.S., 153 rue de Courcelles, 75817 Paris Cedex 17, France*
[3]*Orffa Additives BV, Vierlinghstraat 51 4251 LC Werkendam, the Netherlands*

Introduction

An optimal ratio between ileal digestible amino acids in the diet is a prerequisite for a high body gain, efficient feed utilisation and high carcass lean meat content in growing-finishing pigs. The recommended ratio of apparent ileal digestible (AID) tryptophan (Trp) to lysine (Lys) for growing pigs varies between 0.17 and 0.19 according to NRC (1998) and CVB (2007) respectively. However, recent studies in weaned piglets suggested beneficial effects of an increase in standardised ileal digestible (SID) Trp/Lys ratio between 0.18 and 0.22 (Corrent *et al.*, 2009), whereas recent information on Trp requirements of growing pigs is limited. Therefore, the aim of this project is to determine the dose response relationship between the dietary AID Trp/Lys ratio and body weight gain and feed conversion ratio in growing pigs (from 23 to 50 kg body weight).

Material and methods

The experiment comprised four dietary treatments with increasing Trp/Lys ratios (AID Trp/Lys 0.14, 0.17, 0.20 and 0.23; SID Trp/Lys 0.15, 0.18, 0.21 and 0.24) during a six week period with growing female pigs, from 23 kg to 50 kg body weight. Lys was the second limiting nutrient after Trp and was supplied 10-15% below estimated requirements for a maximum protein deposition. During the experimental period, dietary net energy (NE) was maintained constant at 9.7 MJ/kg and AID Lys/NE ratio was reduced every two weeks being 0.86, 0.79 and 0.72 g/MJ during week 1-2, week 3-4 and week 5-6 respectively to assure that Lys remained a limiting nutrient. The ratio of other essential amino acids relative to Lys was 10% above the ideal ratio (CVB, 2007) and was kept constant during the experiment.

In total 144 pigs (GY × (GY × SL)) with initial body weight of 23.1 kg were allocated to the treatments. Each treatment comprised six replicates of six animals. Body weight (BW), and feed intake (FI) were recorded and average daily gain (ADG) and feed conversion ratio (FCR) calculated. Data were analyzed by ANOVA followed by a least significant difference-test using GenStat 12 software. In addition, Trp requirements were determined using a linear-plateau (broken-stick) and a quadratic model with non-linear regression procedures with analysed AID or SID Trp/Lys ratios as independent variable and ADG, FI and FCR as dependent response variables.

Results and discussion

The results of the effects of dietary Trp level on growth performance of the pigs are described in Table 1. Increasing dietary AID Trp/Lys ratio from 0.14 to 0.20 significantly stimulated FI from 1.2 to 1.6 kg/d ($P=0.001$), ADG from 509 to 752 g/d ($P=0.001$) and BW at the end of the experimental period (after 42 d) from 44.2 to 54.6 kg ($P=0.001$). FCR was reduced from 2.49 to 2.15 ($P=0.009$). Regression analyses showed linear and quadratic effects of the AID Trp/Lys ratio.

The AID Trp/Lys requirements for a maximum ADG were estimated at 0.19 and 0.22 with the linear-plateau (broken-stick) and quadratic model respectively. Concomitant values for SID Lys/Trp were 0.20 and 0.23. Both in linear-plateau and quadratic models, the optimum Trp/Lys ratios for FCR were similar to that for growth rate.

Table 1. Influence of Trp level in the diet of growing pigs, on performance in the 6-week grower period (23-50 kg BW).

Treatments	1	2	3	4	SEM	Effects		
AID Trp/Lys	0.14	0.17	0.20	0.23		Trp	Lin	Quad
BW at start, kg	23.0	23.1	23.0	23.1	0.04	0.16	0.205	0.313
ADG, g/d	509[a]	645[b]	752[c]	753[c]	31.1	0.001	0.001	0.041
FI, g/d	1,186[a]	1,429[b]	1,590[c]	1,582[bc]	53.0	0.001	0.001	0.027
FCR	2.49[b]	2.29[ab]	2.15[a]	2.15[a]	0.07	0.009	0.001	0.140
BW at end, kg	44.2[a]	50.1[b]	54.5[c]	54.6[c]	1.30	0.001	0.001	0.027

[a,b,c,d] Values within a row with different letters are significantly different ($P<0.01$).

The AID Trp/Lys requirements in the present trial were higher than those reported by Schutte *et al.* (1995). In that study, optimal AID Trp/Lys was determined at 0.18 for maximum ADG (741 g/d) in growing pigs (20-40 kg BW) although experimental diets in that study were not limiting in AID Lys. The Trp requirements expressed as g true ileal digestible (TID) Trp/kg diet estimated by Guzik *et al.* (2005) using a broken-line regression model were 1.67 and 1.34 for pigs weighing 30.9 and 51.3 kg respectively equivalent to a TID Trp/Lys ratio of 0.19 at both body weights. In that study it was not quite clear whether Lys or energy was the second-limiting nutrient. Eder *et al.* (2003) determined SID Trp requirements according to an exponential model. In the 25-50 kg BW range, 95% of the maximum FI, ADG and N-retention were achieved at concentrations between 1.96 and 2.00 g of SID Trp/kg diet, corresponding to an average of SID Trp/Lys ratio of 0.23. It appears that the statistical model is a major factor of variation in the determination of a nutrient requirement and determined requirement must be associated with the model interpretation. For instance, if a linear plateau model depicts individual response, quadratic or curvilinear functions may represent the response of the population (Pomar *et al.*, 2003).

In conclusion, Trp requirements for maximum ADG in fast growing pigs were determined at 0.19 AID or 0.20 SID Trp/Lys with the linear plateau model, and 0.22 AID or 0.23 SID Trp/Lys with the quadratic model. It should be emphasized that Trp requirements were estimated while Lys was the second limiting reference nutrient.

References

Corrent, E., A. Simongiovanni, J. van Milgen and N. Le Floc'h, 2009. Amino acids requirements in piglets with special emphasis on tryptophan and valine. Recent Advances in Animal Nutrition, Univ. of Notttingham, UK.

CVB, 2007. Tabellenboek Veevoeding. Voedernormen landbouwhuisdieren en voederwaarden veevoeders, Lelystad, the Netherlands.

Eder, K., H. Nonn, H. Kluge, and S. Peganova, 2003. Tryptophan requirement of growing pigs at various body weights. J. Anim. Phys. Anim. Nutr. 87, 336-346.

Guzik, A.C., J.L. Shelton, L.L. Southern, B.J. Kerr and T.D. Bidner, 2005. The tryptophan requirement of growing and finishing barrows. J. Anim. Sci. 83, 1303-1311.

NRC, 1988. Nutrient requirements of swine. 10th. revised Edition. National Academy Press, Washington, USA.

Pomar, C., I. Kyriazakis, G.C. Emmans, and P.W. Knap, 2003. Modeling stochasticity: Dealing with populations rather than individual pigs. J. Anim. Sci. 2003. 81, E178-186.

Schutte, J.B., A.J.M.A. Verstraeten, N.P. Lenis, J. de Jong and J.Th.M. Van Diepen, 1995. Requirement of young pigs for apparent ileal digestible tryptophan. Neth. J. Agr. Sci. 43, 287-296.

An excessive supply of leucine aggravates the effect of a valine deficiency in post-weaned piglets

M. Gloaguen[1,2], N. Le Floc'h[1], L. Brossard[1], Y. Primot[2], E. Corrent[2] and J. Van Milgen[1]
[1]*Institut National de la Recherche Agronomique (INRA), Unité Mixte de Recherche (UMR) 1079, 35000 Rennes, France*
[2]*Ajinomoto Eurolysine S.A.S., 75817 Paris Cedex 17, France*

Introduction

The actual NRC (1998) recommendation of valine (Val) for pigs, expressed on a standardized ileal digestible (SID) basis and relative to Lys, is 68%. The three branched-chain amino acids (BCAA) Val, isoleucine (Ile) and leucine (Leu) partially share a common route of catabolism and an excessive supply of Leu and Val (for example due to the use of blood cells as a protein source) appears to increase the Ile requirement in piglets (Dean *et al.*, 2005). Relatively little information is available about the Val requirement in pigs and the possible interaction with other BCAA. The objective of this study was therefore to study how an excessive dietary Leu supply affects the response to Val in post-weaned piglets.

Material and methods

In 2 experiments, 6-wk old barrows and female piglets (Pietrain×(Large White×Landrace), 12 kg initial BW) were used in a randomized-block design. Blocks of piglets were created based on piglets of similar BW, sex and genetic origin. Piglets were housed individually and had free access to feed and water. Each piglet within a block received a different experimental diet that provided 1.0% SID Lys. Average daily feed intake (ADFI), gain (ADG) and gain-to-feed ratio (G:F) during a 3-wk period were used as response criteria.

In experiment 1, 16 blocks of 4 animals were used to study the interaction between Val and Leu in a 2×2 factorial design (60 and 70% SID Val:Lys, and 111 and 165% SID Leu:Lys). Diets were based on corn (45%), barley (15%), wheat (15%) using soybean meal or corn gluten meal as main protein sources. In experiment 2, 15 blocks of 5 animals were used in dose-response study to determine Val requirement, while providing an excessive supply of Leu. Five diets were formulated to contain 60, 65, 70, 75 and 80% SID Val:Lys with 165% SID Leu:Lys. Diets were based on corn (78%), corn gluten meal (9%) and soybean meal (7%).

Data was subjected to ANOVA with diet as fixed effect, and litter as a random effect. Valine requirement was determined by a curvilinear-plateau model (Robbins *et al.*, 2006).

Results and discussion

Results of experiment 1 are given in Table 1. A deficient Val supply (60% SID Val:Lys) reduced ADFI, resulting in a reduction in ADG and G:F. An excessive supply of Leu (165% SID Leu:Lys) had no effect on performance when the SID Val:Lys was adequate (70%). However, the reduction in performance due to a deficient Val supply appeared to be greater when this was accompanied by an excessive supply of Leu, compared with a moderate supply of Leu (111% SID Leu:Lys). The ADFI, ADG and G:F were respectively 13, 34 and 16% lower when the Val-deficiency was combined with an excessive supply of Leu compared with those that received a moderate supply of Leu. Leucine can exert growth-depressing effects due to the antagonism between Ile and Val (Langer *et al.*, 2000). This effect is generally observed when Leu is supplied in excess whereas Val or Ile are supplied below the requirement. The addition of Val and Ile can suppress the growth-depressing effect (Rogers *et al.*, 1967).

Table 1. Interaction between the standardized ileal digestible (SID) valine and leucine supply on piglet performance (experiment 1).

Val:Lys SID %	60		70		RSD	*P*-value		
Leu:Lys SID %	111	165	111	165		Val	Leu	Val×Leu
ADFI (g/d)	634[b]	563[a]	736[c]	715[c]	85	<0.001	0.037	0.24
ADG (g/d)	325[b]	242[a]	465[c]	420[c]	69	<0.001	<0.001	0.28
G:F	0.51[b]	0.44[a]	0.63[c]	0.59[c]	0.09	<0.001	0.008	0.39

[a-c] Within a row, means without a common superscript differ (P<0.05); RSD=Residual standard deviation.

Results of experiment 2 are given in Table 2. For the 3 response criteria, the curvilinear-plateau model estimated that 72% SID Val:Lys was required to reach the plateau. Providing 60% SID Val:Lys reduced ADFI by 28%, ADG by 38% and G:F by 13%. Previous findings from our laboratory estimated a Val requirement at 70% SID Val:Lys (Barea *et al.*, 2009) and this value was not different from that estimated in the current experiment (P=0.65). Although we anticipated that an excessive supply of Leu would increase the catabolism of Val, thereby lowering the Val availability and increasing the Val requirement, this did not occur. However, at 60% SID Val:Lys in experiment 2 compared with results of Barea *et al.* (2009), the reduction of ADFI and ADG due to a Val deficiency was greater by 15% and 11% respectively. The mechanisms of the response of the animal to an increasing supply of Val in the presence or absence of an excess supply of Leu remain unclear.

In conclusion, a 70% SID Val:Lys ratio allow to obtain the highest piglet performance even in conditions of excess Leu supply.

Table 2. Effect of an increasing Val supply on the performance in piglets (experiment 2).

Val:Lys SID %	60	65	70	75	80	RSD	*P*-value
ADFI (g/d)	613[a]	689[ab]	784[b]	763[b]	784[b]	104	<0.01
ADG (g/d)	330[a]	391[b]	478[c]	450[bc]	461[bc]	72	<0.01
G:F	0.54[a]	0.57[b]	0.61[c]	0.59[bc]	0.59[bc]	0.05	<0.01

[a,c] Within a row, means without a common superscript differ (P<0.05); RSD=Residual standard deviation.

References

Barea, R., L. Brossard, N. Le Floc'h, Y. Primot, D. Melchior and J. van Milgen, 2009. The standardized ileal digestible valine-to-lysine requirement ratio is at least seventy percent in postweaned piglets. J. Anim. Sci. 87, 935-947.

Dean, D. W., L.L. Southern, B.J. Kerr and T.D. Bidner, 2005. Isoleucine requirement of 80- to 120-kilogram barrows fed corn-soybean meal or corn-blood cell diets. J. Anim. Sci. 83, 2543-2553.

Langer, S., P.W. Scislowski, D.S. Brown, P. Dewey and M.F. Fuller, 2000. Interactions among the branched-chain amino acids and their effects on methionine utilization in growing pigs: Effects on plasma amino- and keto-acid concentrations and branched-chain keto-acid dehydrogenase activity. Br. J. Nutr. 83, 49-58.

NRC. 1998. Nutrient requirement of swine.10th ed. Natl. Acad. Press, washington, DC, USA. 110-112.

Robbins, K.R., A.M. Saxton and L.L. Southern, 2006. Estimation of nutrient requirements using broken-line regression analysis. J. Anim. Sci. 84 Suppl, E155-165.

Rogers, Q.R., R.I. Tannous and A.E. Harper. 1967. Effects of excess leucine on growth and food selection. J. Nutr. 91, 561-572.

Optimum isoleucine to lysine ratio in wheat and barley based diets fed to growing pigs

J.K. Htoo[1], A.D. Quant[2] J.H. Cho[2], B.J. Kerr[3], G.L. Cromwell[2] and M.D. Lindemann[2]
[1]Evonik Degussa GmbH, Rodenbacher Chaussee 4, 63457 Hanau, Germany
[2]University of Kentucky, 607 W. P. Garrigus Bldg., Lexington, KY 40546-0215, USA
[3]USDA, Swine Odor and Manure Management Research Unit, Ames, IA 50011-3310, USA

Introduction

Reducing the level of dietary crude protein (CP) is an effective means to reduce nitrogen excretion but requires reliable information about the pigs' requirements for potentially limiting amino acids (AA). Isoleucine (Ile) may be limiting in low CP pig diets that are supplemented with lysine (Lys), threonine, methionine and tryptophan (Liu *et al.*, 1999). The objective of this study was to estimate the optimum dietary standardized ileal digestible (SID) Ile:Lys ratio in 25- to 40-kg pigs that were fed wheat-barley based diets.

Material and methods

A 27-d growth assay was conducted with 80 high lean (PIC 327 × PIC C23) pigs [initial body weight (BW) = 24.00±2.2 kg], which were blocked by BW and allotted to 5 dietary treatments with 4 pigs (2 barrows and 2 gilts) per pen and 4 pen replicates per treatment. A wheat-barley based basal diet was formulated to meet requirements of AA other than Ile (0.35% SID basis) and Lys. The SID Lys level was set at 0.91% (87% of recommended Lys level for 20 to 40 kg PIC pigs) in all diets to avoid underestimation of the Ile:Lys ratio. Other essential AA were balanced to meet the ideal AA ratio (Rademacher *et al.*, 2001) relative to the adequate Lys level (1.05% SID Lys). The Ile-adequate diet (Diet 5) was obtained by supplementing L-Ile at the expense of wheat to the basal diet to contain 0.62% SID Ile. Diets 2, 3 and 4 were produced by blending diet 1 and 5 in varying proportions to create 5 SID Ile:Lys ratios: 39, 46, 53, 61 and 68%, respectively. The BW of individual pigs and per pen feed disappearance were recorded weekly. Blood samples were taken from all pigs on d 27 for determining plasma urea nitrogen (PUN).

Data were analyzed by ANOVA using the GLM procedure of SAS with initial BW and dietary treatment included in the model. Using ADG and ADFI as response criteria exponential [at 95% of maximum (max.) response; Robins *et al.*, 1979] and curvilinear-plateau model [$y = L + U + (R - x)^2$, where $(R - x)$ is zero at values of $x>R$; Robins *et al.*, 2006] regression analysis were conducted to estimate the optimum Ile ratios. Additionally the PUN response data were subjected to a one slope linear broken-line regression (Robins *et al.*, 2006) to derive the optimum Ile ratio.

Results and discussion

Performance results are given in Table 1. During the 27-d period, the ADG and final BW increased linearly ($P<0.01$) and quadratically ($P<0.05$) as the Ile:Lys ratios increased and seems to maximize at the SID Ile:Lys ratio of 53.3% (Table 1). Average daily feed intake (ADFI) also increased ($P<0.05$; linear) by graded levels of Ile:Lys ratio. The gain:feed was not affected by the dietary treatments. The PUN level decreased ($P<0.05$; linear) by increasing levels of Ile:Lys ratio.

The SID Ile:Lys ratios to optimize ADG were 50.1 and 51.4%, respectively based on exponential regression (95% of max. response) and curvilinear-plateau regression analyses, respectively. The exponential (95% of max. response) and curvilinear-plateau regressions estimated optimum SID Ile:Lys ratios of 52.6% and 53.6%, respectively based ADFI response (Table 2). Linear broken-line regression analysis estimated a SID Ile:Lys of 60.7% to minimize the PUN level. Overall, the

Table 1. Effect of dietary SID Ile:Lys ratio on performance and PUN of starter pigs.[1]

	SID Ile:Lys, (%)					SEM	*P*-values	
Parameter	38.7	46.0	53.3	60.7	68.0		Linear	Quadratic
Initial BW, kg	24.01	23.98	24.01	23.98	24.00	1.250	1.000	0.991
BW at d 27, kg	39.33	42.98	44.22	43.63	43.49	0.898	0.007	0.014
ADG, kg/d	0.569	0.703	0.749	0.727	0.722	0.033	0.007	0.015
ADFI, kg/d	1.238	1.425	1.537	1.457	1.495	0.067	0.023	0.072
Feed:gain	2.176	2.032	2.050	2.004	2.095	0.113	0.601	0.351
PUN d 27, mg/dl	12.51	11.78	11.35	9.86	10.62	0.510	0.003	0.326

[1] Data are least square mean, representing 4 pens (a total of 8 barrows, 8 gilts) per treatment.

Table 2. Optimal Ile:Lys ratios estimated by regression analyses.

Response criteria	Regression equation	r^2	SID Ile:Lys (%)
ADG (kg/d)			
Exponential	$y = 0.589 + 0.164\,[1 - e^{-0.263\,(x - 38.7)}]$	0.97	50.1
Curvilinear-plateau	$y = 0.733 - 0.001\,(51.4 - x)^2$	0.98	51.4
ADFI (kg/d)			
Exponential	$y = 1.236 + 0.259\,[1 - e^{-0.215\,(x - 38.7)}]$	0.92	52.6
Curvilinear-plateau	$y = 1.496 - 0.001\,(53.6 - x)^2$	0.94	53.6
Plasma urea nitrogen (mg/dL)			
Linear broken-line	$y = 10.32 + 0.102\,(60.7 - x)$	0.91	60.7

average of the ADFI, ADG and PUN optima estimated an optimum SID Ile:Lys ratio of 54% for 24- to 35-kg pigs fed a wheat-barley based diet containing 1.17% SID leucine (Leu) and 0.76% SID valine (Val). This result is in agreement with Wiltafsky *et al.* (2009) who also reported an optimal SID Ile:Lys of 54% based on the ADG and ADFI in 8- to 25-kg pigs fed wheat-barley-corn based diets that contained 1.05% SID Leu and 0.67% SID Val.

References

Liu, H., G.L. Allee, E.P. Berg, K.J. Touchette, J.D. Spencer and J.W. Frank, 2000. Amino acid fortified corn diets for late finishing barrows. J. Anim. Sci. 78 (Suppl. 2), 66.

Rademacher, M., W.S. Sauer and A.J.M. Jansman, 2001. Standardized ileal digestibility of amino acids in pigs. Feed Additives, Degussa AG, Hanau-Wolfgang, Germany, 11 pp.

Robbins, K.R., H.W. Norton and D.H. Baker, 1979. Estimation of nutrient Requirements from growth data. J. Nutr. 109, 1710-1714.

Robbins, K.R., A.M. Saxton and L.L. Southern, 2006. Estimation of nutrient requirements using broken-line regression analysis. J. Anim. Sci. 84 (E. Suppl.), E155-E165.

Wiltafsky, M.K., J. Bartelt, C. Relandeau and F.X. Roth, 2009. Estimation of the optimum ratio of standardized ileal digestible isoleucine to lysine for eight- to twenty-five-kilogram pigs in diets containing spray-dried blood cells or corn gluten feed as a protein source. J. Anim Sci. 87, 2554-2564.

Meta-analysis of the response of growing pigs to different levels of isoleucine in the diet

J. Van Milgen[1,2], M. Gloaguen[1,2,3], N. Le Floc'h[1,2], L. Brossard[1,2], Y. Primot[3] and E. Corrent[3]
[1]INRA, UMR1079 SENAH, 35590 Saint-Gilles, France
[2]Agrocampus Ouest, UMR1079 SENAH, 35000 Rennes, France
[3]Ajinomoto Eurolysine s.a.s., 75817 Paris, France

Introduction

Isoleucine has often been considered to be among the next-limiting amino acids for growth in pigs, after Lys, Met, Met+Cys, Thr and Trp. However, there are conflicting reports concerning the requirement for Ile. Differences in requirement estimates may be due to differences in experimental conditions. The initial two steps of the catabolism of the branched-chain amino acids (BCAA; i.e. Ile, Val, and Leu) are catalyzed by enzyme-complexes shared by the three BCAA. An excessive supply of one BCAA may therefore stimulate the catabolism of the other BCAA. The objective of this study is to perform a meta-analysis of literature data and to identify the sources of variation in the reported Ile requirement.

Data collection and methods of analyses

Forty-three dose-response experiments were collected, originating from 20 publications. Because of their very low Ile content, blood cells were used in 33 experiments, while the other 10 experiments used other protein sources. Within an experiment, diets only differed in the supply of L-Ile with a minimum of 4 levels of Ile. The standardized ileal digestible (SID) Ile content was reported in the experiments or recalculated from feed ingredients using table values (Sauvant *et al.*, 2004). The SID Ile content was expressed relative to the NRC (1998) requirement (in g/kg diet), thereby initially ignoring the fact that after Ile other nutrients (e.g. Lys) could be second-limiting. A quadratic regression analysis was carried out initially for each experiment to assess if the increase in Ile content resulted in an increase in average daily gain (ADG). Experiments for which there was no indication (P>0.25) of a response to Ile were not considered further in the statistical analysis.

The Ile requirement (in g/kg diet) declines as the pig matures. For example, the NRC assumes that the requirement declines by 15% between 12 and 25 kg BW (from 5.9 to 5.0 g SID Ile/kg diet). This means that although a given diet may be Ile-deficient at the start of the experiment, the same diet may be Ile-sufficient at the end of the experiment. In certain studies, the anticipated Ile requirement at the end of the experimental period was 25% lower compared with the requirement at the start of the experiment. To account for this, we modified the traditionally used 'broken-stick' model into a 'bent-stick' model to analyze the response curves. The transition phase (that links the linear increase to the plateau) was determined by the anticipated decline in Ile-requirement during the experimental period. For a given response curve, the model was parameterized to include the slope of the initial linear increase, the plateau value, and the mid-point of the transition phase expressed relative to the NRC requirement for Ile (R*). We assumed that the slope of the initial linear increase was common for all experiments, whereas the other 2 parameters varied between experiments.

Results and discussion

In 32 of the 33 experiments where blood cells were used, there was a trend (P<0.25) for at least a linear response in ADG to the increasing Ile supply. However, when other protein sources were used, this trend was observed in only 2 of the 10 experiments (Figure 1). The reduction in ADG (if observed) due to a deficient supply of Ile was mainly due to a reduction in feed intake. A 10% reduction in Ile content (below the requirement) resulted in a 20% reduction in ADG.

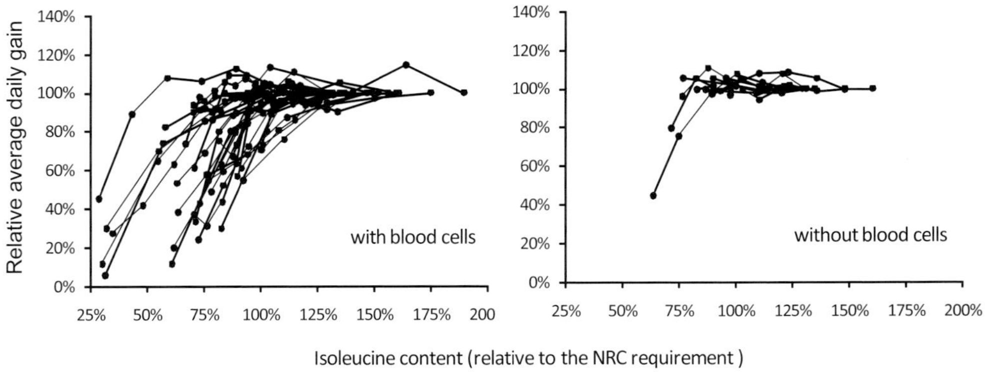

Figure 1. Relative average daily gain (expressed relative to that observed at the highest level of Ile supplementation) as a function of the Ile content in the diet in the presence (left panel) or absence (right panel) of blood cells as a protein source.

For experiments where a response to the Ile supply was observed ($P<0.25$), the average R* was 0.99, but values ranged from 0.59 to 1.29. This indicates that, on average, these experiments confirmed the current NRC requirement for Ile. However, R* is the midpoint of the transition phase, and pigs receiving an Ile-supply corresponding to R* would receive less Ile than required at the beginning of the experiment. The R* will also be underestimated if Lys (or another nutrient) would be second-limiting after Ile. For the 13 experiments where the authors declared Lys to be second-limiting, R* averaged 1.01. Accounting for the fact that Lys was second-limiting in these experiments resulted in an average adjusted R* value of 1.10 (range 0.85 to 1.28). In the 19 experiments where Lys was not declared to be second-limiting, re-calculation of the amino acid composition of the diets indicated that in 12 experiments, Lys or another amino acid may have been limiting, resulting in an underestimation of the Ile-requirement.

In conclusion, there is no strong evidence of a response to Ile supplementation when diets do not contain blood cells. The Ile requirement may be higher than the current NRC requirement estimate when blood cells are used as a protein source. This response may be due to a large supply of the other BCAA (Leu and Val), or of other large neutral amino acids (His, Phe and Tyr).

References

NRC, 1998. Nutrient requirements of swine. National Academy Press, 189 pp.

Sauvant, D., J.-M. Perez and G. Tran, 2004. Tables of composition and nutritional value of feed materials. INRA Editions, 304 pp.

Using near infrared spectroscopy to predict nutritional value of soybean meals

C. Gady[1], Y.G. Liu[2], Pierre Dalibard[1] and P.A. Geraert[1]
[1]*Adisseo France SAS, 10 place du Général De Gaulle, 92160, Antony, France*
[2]*Adisseo Asia Pacific Pte Ltd,1 Loyang Way 4, 179803, Singapore*

Introduction

Numerous works have shown substantial variation in nutritional value of soybean meals (SBM) due to several factors such as genotype, crushing, antinutritional factors, implying a need for better characterization (De Coca-Sinova *et al.*, 2008; Karr-Lilienthal *et al.*, 2004). During the past two decades, ADISSEO has been conducting *in vivo* studies to evaluate digestibility of amino acids (Green *et al.*, 1987) and energy (Bourdillon *et al.*, 1990) of SBM. Using these data, models were then developed with the Near Infrared Spectroscopy (NIRS). The aim of the present study was to assess whether NIRS is a practical and accurate approach of nutritional assessment of SBM.

Materials and methods

In the present study, 4 samples of SBM were obtained from 4 major producing countries (non dehulled SBM from India, dehulled SBM from Argentina, Malaysia and USA), and predicted by NIRS equations for proximate, total and digestible amino acids and apparent metabolizable energy (AME).

A performance trial was carried out at Bangkok Animal Research Center (BARC, Thailand) to evaluate the influence of those 4 sources of SBM when formulating corn-cassava-soybean meals-based diets. Each SBM source was used in 2 treatments, one with standard specifications (those of the US SBM), and another with individual specifications (Table 1). 768 newly hatched male broiler chicks were allocated to 8 treatments in a randomized complete block experiment, with 6 replicates per treatment. The diets were formulated to have similar energy (12.4 and 12.6 MJ/kg ME) and protein content (21 and 19% respectively for starter diets 0-18 days and grower diets 19-40 days) and met the nutritional recommendations for broilers. Body weight and feed intake were then measured for growth and feed conversion ratio (FCR) was calculated at 18 and 40 days. Traits were subjected to analysis of variance as a 4×2 factorial arrangement in randomized complete block design.

Table 1. Nutrient composition of soybean meals used for formulating diets (individual specifications issued from NIRS).

SBM source	India	Argentina	USA	Malaysia
Moisture (%)	10.44	10.61	10.66	11.13
Fat (%)	1.36	2.12	2.63	2.55
Crude fiber (%)	6.05	3.75	3.94	2.28
Ash (%)	8.04	6.81	6.22	5.54
ME (MJ per kg)	9.05	9.78	9.95	10.68
Protein (%)	46.0	46.5	47.5	47.5
Digest. Lys (%)	2.36	2.43	2.46	2.69
Digest. Met (%)	0.55	0.57	0.56	0.59
Digest.M+C (%)	1.03	1.12	1.13	1.18
Digest. Trp (%)	0.56	0.59	0.60	0.64
Digest. Thr (%)	1.56	1.64	1.61	1.75
Digest. Arg (%)	3.06	3.16	3.10	3.39

Results and discussion

Results are presented in Table 2 and illustrate a global good performance with a precision found at 2% variation for most performance traits. When diet formulations were based on standard US SBM specifications, performance classified SBM from low quality to high quality in the following order: India, Argentina, USA and Malaysia origins. The Indian SBM showed the poorest quality with 0.14 points difference in FCR and 100 g live weight as compared to the highest quality SBM of Malaysia origin. When formulating with individual specifications, the differences between SBM qualities were smoothed, with birds achieving similar live weight and FCR. The Indian SBM showed the poorest quality with 0.06 points difference in FCR and 51 g live weight as compared to the highest quality SBM of US origin. This study demonstrated that there is significant variations between SBM sources that have to be defined for their contribution in both digestible amino acids and energy. Considering these variations, formulating diets with accurate and individual SBM specifications issued from NIRS allowed ensuring similar level of performance as compared to feed formulation derivating from standard specification. Also, our results suggest that using individual SBM specifications issued from NIRS lead to better flock homogeneity, not dependant of SBM source.

Table 2. Performance of male broilers (0-40 days) fed soybean meals of different sources basing on individual and standard (US) specifications.

SBM source	SBM specification	Final weight (g)	Weight gain (g)	Feed intake (g)	FCR	Livability %
India	Standard (US)	2812^{b}	2771^{b}	4762	1.765^{d}	92.71
Argentina	Standard (US)	2837^{b}	2797^{b}	4675	1.672^{b}	98.96
USA	Standard (US)	2869^{ab}	2830^{ab}	4765	1.693^{b}	96.88
Malaysia	Standard (US)	2908^{a}	2869^{a}	4656	1.625^{a}	98.96
India	Individual (Ind.)	2814^{b}	2775^{b}	4771	1.739^{cd}	96.88
Argentina	Individual (Arg.)	2815^{b}	2776^{b}	4710	1.711^{bc}	96.88
USA	Individual (US)	2865^{ab}	2825^{ab}	4736	1.676^{b}	100.00
Malaysia	Individual (Mal.)	2831^{b}	2792^{b}	4686	1.690^{b}	98.96
Pooled SEM		20.54	20.55	30.73	0.016	1.82

^{a,b,c,d} Means in the same column lacking a common superscript letter are different ($P<0.05$).

References

Bourdillon, A., B. Carré, L. Conan, J.J. Duperray, G. Huyghebaert, B. Leclercq, M. Lessire, J. McNab and J. Wiseman, 1990. European reference method for the *in vivo* determination of metabolisable energy with adult cockerels: reproducibility, effect of food intake and comparison with individual laboratory methods. Br. Poult. Sci., 31, 557-565.

De Coca-Sinova, A., D.G. Valencia, E. Jiménez-Moreno, R. Lázaro and G.G. Mateos, 2008. Apparent ileal digestibility of energy, nitrogen, and amino acids of soybean meals of different origin in broilers. Poult. Sci. 87, 2613-2623.

Green, S., S.L. Bertrand, M.J.C. Duron and R. Maillard, 1987. Digestibilities of amino acids in maize, wheat and barley meals determined with intact and caecectomised cockerels. Br. Poult. Sci. 28, 631-641.

Karr-Lilienthal, L.K., N.R. Merchen, C.M. Grieshop, M.A. Flahaven, D.C. Mahan, N.D. Fastinger, M. Watts and G.C. Fahey, Jr., 2004. Ileal amino acid digestibilities by pigs fed soybean meals from five major soybean-producing countries. J Anim Sci. 82, 3198-3209.

Nutrient utilization and growth performance in growing pigs fed a cereal-based diet with chicory forage (*Cichorium intybus* L) inclusion

E. Ivarsson and J.E. Lindberg
Department of Animal Nutrition and Management, P.O. Box 7024, Swedish University of Agricultural Science, 750 07 Uppsala, Sweden

Introduction

There is a great interest in dietary fibre in pig nutrition due to its possibility to improve gut health and well-being (Wenk, 2001). However, inclusion of fibre-rich feedstuffs in pig feeds decreases the energy content in the diet by decrease the enzymatic digestion in the foregut and increase the microbial digestion in the hindgut (Anguita *et al.,* 2006). The digestibility coefficient of dietary fibre ranges from about zero in highly lignified and water-insoluble fibres to 0.80-0.90 in pectin and water-soluble fibres. Consequently, the energy and nutrient utilization of dietary fibre is highly dependent on the fibre source (Noblet and Le Goff, 2001).

Chicory is a perennial herb that can be complementary to traditional forage crops and is used in organic production. Nutritionally, chicory forage has a high content (150 g/kg DM) of soluble uronic acid, the building block in pectin (Voragen *et al.,* 2001). The aim of this study was to evaluate the effect of feeding growing pigs with different inclusion levels of chicory forage on growth performance and the coefficient of total tract apparent digestibility (CTTAD) of dietary components and energy.

Materials and methods

Eighteen seven weeks old pigs (castrated males and females) with an initial weight of 17.2±0.8 kg from six different litters were used in an 18-days randomized block experiment. The study was approved by ethical committee of the Uppsala region.

Three diets were formulated, one nutritionally balanced cereal-based basal diet (B) and two diets where a part of the cereal was substituted to 80 or 160 g/kg air-dried chicory forage (CH80 and CH160). The CTTAD of dietary components and energy was calculated using the marker-ratio technique with TiO_2 (2.5 g/kg) as indigestible marker. The cereals (wheat and barley) and the air-dried chicory forage were grounded through a 3.5-mm screen before mixed with the other ingredients. The diets were given as a pellet and the pigs had *ad libitum* access to feed and water throughout the experiment.

Feed intake was registered daily and the pigs were weighed weekly, faeces samples for digestibility determination were collected once a day during days 15-18 and stored at -20 °C. Feed and faeces samples were freeze dried and milled through a 1-mm sieve before analysis.

Dry matter, ash, crude protein, TiO_2, total non-starch polysaccharides (NSP), the constituent sugars and lignin were analyzed as described by Högberg and Lindberg (2006). Gross energy was measured with a bomb calorimeter (Parr 6300 Oxygen Bomb Calorimeter, Illinois, USA). Dietary fibre was calculated as NSP+lignin and CTTAD was calculated using the indicator technique.

Statistical analysis was performed with procedure Mixed in SAS (SAS institute, Cary, NC, USA, version 9.1). The model included diet (C, CH80, CH160) as fixed factor and litter as random factor.

Results and discussion

Inclusion of chicory forage did not affect the daily weight gain, daily feed intake or feed conversion ratio (*P>0.05*). This is in agreement with previous studies where chicory has been included in diets for weaned piglets (Ivarsson *et al.*, 2010) but is in contrast to studies were grass meal has been included in the diet for growing pigs (Vestergaard *et al.*, 1996). The CTTAD of organic matter tended to be lower (*P=0.06*) whereas crude protein and energy was lower in diet CH160 than in diets B and CH80 (*P<0.05*; Table 1). The difference in CTTAD of energy between diet CH160 and diet B was smaller than reported with inclusion of other forage crops (Raj *et al.*, 2006). The CTTAD of dietary fibre, NSP and uronic acid was higher in the chicory diets than in diet B, and diet CH160 had a higher CTTAD of dietary fibre and NSP than diet CH80 (*P<0.05*), this is agreement with previous studies (Ivarsson *et al.*, 2010). This study confirms that chicory forage is a highly digestible fibre source for pigs and should have potential to be an ingredient in pig diets with only minor impairment on the energy and nutrient utilization of the diet.

Table 1. Effect of diet on coefficient of total tract apparent digestibility of dietary components and energy. Least square means ± s.e.

	B	CH80	CH160	s.e.	*P*-value
n	6	6	6		
Organic matter	0.83[a]	0.82[ab]	0.81[b]	0.052	0.06
Crude protein	0.78[a]	0.77[ab]	0.74[b]	0.011	0.03
Dietary fibre	0.32[c]	0.45[b]	0.52[a]	0.014	<0.01
NSP	0.42[c]	0.56[b]	0.62[a]	0.014	<0.01
Uronic acid	0.36[b]	0.85[a]	0.89[a]	0.014	<0.01
Energy	0.80[a]	0.79[ab]	0.77[b]	0.070	0.05

References

Anguita M., N. Canibe, J.F. Pérez and B.B. Jensen, 2006. Influence on the amount of dietary fiber on the available energy from hindgut fermentation in growing pigs: Use of cannulated pigs and *in vitro* fermentation. J. Anim. Sci. 84, 2766-2778.

Högberg, A. and J.E. Lindberg, 2006.The effect of level and type of cereal non-starch polysaccharides on the performance, nutrient utilization and gut environment of pigs around weaning. Anim. Feed Sci. Technol. 127, 200-219.

Ivarsson, E., B.E. Frankow-Lindberg, K. Andersson and J.E. Lindberg, 2010. Growth performance, digestibility and faecal coliform bacteria in weaned piglets fed a cereal-based diet including either chicory (*Cichorium intybus* L) or ribwort (*Plantago lanceolata* L) forage. Anim. (In press).

Noblet, J. and G. Le Goff, 2001. Effect of dietary fibre on the energy value of feeds for pigs. Anim. Feed Sci. Technol. 90, 35-52.

Raj, St., G. Skiba, D. Wermeko and H. Fandrejewski, 2006. Digestibility of energy and nutrients in pigs previously fed a high-fibre diet. J. Anim. Feed. Sci. 15, 591-598.

Vestergaard, EM., V. Danielsen, A. Eklund Larsen and C. Bejerholm, 1996. Dried grass meal for finishing pigs and pregnant sows. Report Nr. 50 National Institute of Animal Science, Research Centre Foulum, Denmark, 36 pp.

Voragen, F., G. Beldman and H. Schols, 2001. Chemistry and enzymology of pectins. In: McCleary, B.V. and L. Prosky (eds.), Advanced dietary fibre technology. Blackwell Science Ltd, Oxford, UK: 379-398.

Wenk, C., 2001.The role of dietary fiber in the digestive physiology of the pig. Anim. Feed Sci. Technol. 90, 21-33.

Nutrient utilization and growth performance in broiler chickens fed a cereal-based diet with chicory forage (*Cichorium intybus* L.) inclusion

H.Y. Liu and J.E. Lindberg
Department of Animal Nutrition and Management, P.O. Box 7024, Swedish University of Agricultural Science, 750 07 Uppsala, Sweden

Introduction

Chicory (*Cichorium intybus* L.) is a perennial herb with a high content of soluble fibre in the form of uronic acid (building block in pectin), with potential as forage crop in organic production. Despite a high fibre content, chicory forage is well utilized by weaned pigs (Ivarsson *et al.*, 2010), and could potentially also be a useful feed resource for broiler chickens. However, a more limited capacity to digest plant fibre sources in chickens (Jørgensen *et al.,* 1996) may impose constraints in terms of nutrient and energy utilization, and performance.

Material and methods

One-hundred and sixty 1-day old broilers (Ross 308) were fed five nutritionally balanced diets (4 replicates per diet, 8 chickens per replicate) in a 30 d growth trial (27 d growth performance + 3 d excreta collection). The basal diet (B) was cereal-based (g/kg: 550 wheat, 187 barley, 160 soybean meal) and in the four experimental diets part of the cereal was substituted with 60 or 120 g/kg air-dried chicory forage from either the June (CH_1) or September (CH_2) harvest. Each diet was offered *ad libitum*.

Dry matter, ash, crude protein, total non-starch polysaccharides (NSP) and the constituent sugars were analyzed, as well as TiO_2, which was included in all diets (5 g/kg) as a indigestible marker (Högberg and Lindberg, 2006). Gross energy was measured with a bomb calorimeter (Parr 6300 Oxygen Bomb Calorimeter, Illinois, USA).

The coefficients of total tract apparent digestibility (CTTAD) and ileal apparent digestibility (IAD) were calculated with the marker-ratio technique. The apparent metabolizable energy (AME) was calculated as: AME = gross $\text{energy}_{\text{diet}} \times$ CTTAD of gross $\text{energy}_{\text{excreta}}$.

Statistical analysis was performed with procedure GLM in SAS (SAS Institute, Cary, NC, U.S.A., version 9.1) to determine dietary effects by one way ANOVA. Linear contrasts were used to assess the effect of fibre level and time of harvest. The significance was considered when $P<0.05$.

Results and discussion

The CTTAD of all dietary components was affected by fibre level (Table1). The CTTAD of organic matter was higher ($P<0.05$) for diet B than for diet CH60, as well as for diet CH120. Diet CH60 had a higher ($P<0.05$) CTTAD than diet CH120 for all dietary components, except for organic matter. The AME showed a progressive decline with increasing fibre level ($P<0.05$), which was in agreement with Jørgensen *et al.* (1996) when including pea fibre, wheat bran or oat bran in broiler diets. In the current study, uronic acid was the NSP residue with highest digestibility values. Inclusion of 60 g/kg chicory forage increased the uronic acid content without depressing the digestibility. Moreover, the higher CTTAD of uronic acid and xylose in CH_2 indicated an effect of time of harvest, reflecting in different chemical composition between CH_1 and CH_2. Chickens fed diet CH60 had higher average daily gain (28.3 g/d) during day 0-13 ($P<0.05$) than those fed diets B (25.5 g/d) and CH120 (24.8 g/d). The feed conversion ratio (FCR; g feed intake/g average daily gain) during day 0-13 was higher in chickens fed diet CH120 (1.7) than in those fed diet CH60 (1.4) ($P<0.05$). During day 0-27, the

Table 1. Effects of chicory forage inclusion on the CTTAD of dietary components and energy, the IAD of crude protein and AME (MJ/kg dry matter). Least square means ± s.e.

Diet					s.e.	*P*-value			
B	$CH_1$60	$CH_2$60	$CH_1$120	$CH_2$120		B vs. CH60	B vs. CH120	CH60 vs. CH120	CH_1 vs. CH_2
Organic matter									
0.73	0.70	0.72	0.69	0.69	0.01	0.04	0.002	0.08	0.3
NSP									
0.31	0.29	0.31	0.21	0.27	0.02	0.6	0.006	0.004	0.06
Arabinose									
0.31	0.37	0.31	0.12	0.24	0.02	0.2	0.0001	<0.0001	0.2
Xylose									
0.29	0.21	0.26	0.13	0.20	0.02	0.04	<0.0001	0.0008	0.004
Uronic acid									
0.48	0.38	0.49	0.30	0.36	0.04	0.3	0.005	0.01	0.03
IAD crude protein									
0.79	0.75	0.73	0.76	0.74	0.02	0.04	0.1	0.4	0.3
Energy									
0.74	0.71	0.73	0.69	0.70	0.01	0.07	0.0008	0.02	0.1
AME									
13.2	12.7	12.9	12.2	12.3	0.14	0.02	<0.0001	0.002	0.3

FCR was higher in diets with high fibre inclusion ($P<0.05$). Average daily gain was higher and FCR was lower during day 0-13 in chickens fed chicory forage harvested in September than in June.

In conclusion, chicory forage appears to be a good quality fibre source for broiler chickens and has a minor negative impact on digestibility and growth performance.

References

Högberg, A. and J.E. Lindberg, 2006.The effect of level and type of cereal non-starch polysaccharides on the performance, nutrient utilization and gut environment of pigs around weaning. Anim. Feed Sci. Technol. 127, 200-219.

Ivarsson, E., B.E. Frankow-Lindberg, K. Andersson and J.E. Lindberg, 2010. Growth performance, digestibility and faecal coliform bacteria in weaned piglets fed a cereal-based diet including either chicory (*Cichorium intybus* L) or ribwort (*Plantago lanceolata* L) forage. Animal 'In press'.

Jørgensen, H., X.Q. Zhao, K.E. Bach Knudsen and B. Eggum, 1996. The influence of dietary fibre source and level on the development of the gastrointestinal tract digestibility and energy metabolism in broiler chickens. Brit. J. Nutr. 75, 379-395.

SAS, 2009. Release 9.1 SAS Inst. Inc., Cary, NC, USA.

Nutritional value of byproducts from beet biofuel production evaluated in digestibility trials with sows and sheep

J.V. Nørgaard, T. Hvelplund, J.A. Fernández and H.D. Poulsen
Dept. of Animal Health and Bioscience, Aarhus University, P.O. Box 50, 8830 Tjele, Denmark

Introduction

Sugar and fodder beet have a potential yield much higher than that of maize, but when only used as a feedstuff for ruminants, the profit is diminished by high handling costs. A scenario where the beets are fractionated into compounds with a collective higher value can be accomplished by using the sugar containing juice fraction for e.g. biogas or bioethanol production and the remaining fibrous fraction as feedstuff for ruminants or sows. Dietary fibre for pigs is interesting as a source of energy and because of welfare aspects (De Leeuw *et al.*, 2008). *Ad libitum* feeding of roughage such as beet pulp might alleviate behavioural problems, since the sows will feel satiety. The objective was to evaluate the nutritional value of pressed top and root fractions from sugar and fodder beet as roughage for sows and ruminants.

Materials and methods

A total of 30 sows were housed in cages for 12 days and were fed one of 5 diets. Urine and faeces were collected during the last 7 days. The sows were fed 1 kg basal diet and either 4 kg root pulp or 3 kg top pulp from sugar and fodder beets. Basal diet nutrient digestibility was estimated by feeding 6 sows 2.5 kg basal diet as the only feed. In the experiment with sheep, 25 Leicester wethers were fed one of 5 diets for 21 days. The wethers were housed individually and faeces were collected for 7 days. The wethers were fed either 3500 g root or top pulp combined with 325 g of hay, 8 g urea and 2.9 g of sodium sulphate. Hay nutrient digestibility was estimated with 5 sheep fed 1080 g of hay, 8 g urea and 2.9 g of sodium sulphate.

The diets consisted of 4 types of beet pulp: Root or top fractions from Angus sugar beet (A) or Colosse fodder beet (C). The pulp was prepared by a cold mechanical pressing of the cut root or top fractions. Daily rations were immediately packed and stored at -18 °C.

The chemical composition of pulp did only vary little between the two beet varieties. However, the top fraction contained significantly more ash (150 vs. 34 g/kg DM), crude protein (175 vs. 53 g/kg DM), and crude fibre (174 vs. 63 g/kg DM), crude fat (24 vs. 2 g/kg DM), starch (34 vs. 15 g/kg DM) and less total sugars (105 vs. 721 g/kg DM) compared with the root fractions. Data were analyzed using generalized linear models including pulp fraction, variety and their interaction term.

Results and discussion

There were no major differences in digestibility between sugar and fodder beet pulp fractions in neither sows (Table 1) nor sheep (Table 2). The apparent digestibilities of pulp in root fractions were significantly higher than those found in the top fractions.

The apparent digestibility of dry matter and organic matter was high in the sow compared to the sheep. Full grown sows can ferment and utilize large amounts of fibres (Fernandez *et al.*, 1986; Serena *et al.*, 2008). Digestibility of sugars may have been complete in the upper part of the sow's intestine thereby leaving a smaller amount of dry matter for microbial fermentation in the hind gut. Also, taking the larger body weight of the sows and thus the relatively smaller feed allowance into account, the capacity for fermentation may therefore be high compared to sheep.

Table 1. Apparent digestibility (%) of sugar (A) and fodder (C) beet pulp fractions in sows.

	Root A	Root C	Top A	Top C	SEM	*P*-value		
						Fraction	Strain	F x S
Dry matter	97.2	95.6	56.9	67.9	2.4	<0.001	0.05	0.10
Organic matter	98.0	96.5	62.6	71.1	2.2	<0.001	0.10	0.03
Ash	69.3	75.5	27.7	47.6	7.6	<0.001	0.08	0.35
Energy	96.3	94.2	56.9	65.9	2.9	<0.001	0.18	0.04
Crude protein	79.5	74.4	46.6	56.1	5.4	<0.001	0.66	0.16
Retained prot., g/d	-86	-171	-36	-238	89	0.92	0.10	0.49

Table 2. Apparent digestibility (%) of sugar (A) and fodder (C) beet pulp fractions in sheep.

	Root A	Root C	Top A	Top C	SEM	*P*-value		
						Fraction	Strain	F x S
Dry matter	88.0	88.0	58.7	62.4	1.5	<0.001	0.19	0.18
Organic matter	89.9	89.3	65.4	66.7	1.5	<0.001	0.78	0.49
Ash	30.0	66.0	26.9	37.6	3.2	<0.001	<0.001	<0.001
Energy	87.2	86.3	60.9	62.7	1.6	<0.001	0.76	0.35
Crude protein	23.7	40.0	53.9	57.0	2.8	<0.001	0.002	0.02
Crude fat	-386	-509	15	22	30	<0.001	0.05	0.03
NDF	71.5	73.1	66.2	67.0	4.3	0.17	0.75	0.92
ADF	66.0	66.8	58.0	62.5	5.5	0.25	0.61	0.72

The apparent digestibility in sheep of crude protein and fat was low for the root fraction. This may be due to the low crude protein and fat content of the root fractions making the contribution of endogenous and undigested microbial protein and fat in faeces high and consequently the apparent digestibility low. The NDF apparent digestibility was also relatively low in sheep. Due to the high concentration of sugars the rumen pH may have been lower than required for optimal growth conditions of the NDF degrading microflora.

Conclusions

Both root and top pulp can be used as feeds for sows and ruminants, but as the top fraction contains a higher proportion of fibre than the root fraction their nutritional value are significantly lower.

References

Fernandez, J.A., H. Jørgensen and A. Just, 1986. Comparative digestibility experiments with growing-pigs and adult sows. Anim. Prod. 43, 127-132.

De Leeuw, J.A., J.E. Bolhuis, G. Bosch and W.J.J. Gerrits, 2008. Effects of dietary fibre on behaviour and satiety in pigs. Proc. Nutr. Soc. 67, 334-342.

Serena, A., H. Jørgensen and K.E. Bach Knudsen, 2008. Digestion of carbohydrates and utilization of energy in sows fed diets with contrasting levels and physicochemical properties of dietary fiber. J. Anim Sci. 86, 2208-2216.

The effect of potato protein concentrate on short chain fatty acid profile of caecal digesta in broiler chickens

A. Tuśnio, B. Pastuszewska, M. Taciak, A. Mieczkowska and S. Smulikowska
The Kielanowski Institute of Animal Physiology and Nutrition, Polish Academy of Sciences, Instytucka 3, 05-110 Jabłonna, Poland

Introduction

Carbohydrates not digested in the small intestine of monogastric animals are partly fermented to short chain fatty acids (SCFA). SCFA can be also a product of bacterial degradation of resistant proteins, escaping enzymatic digestion (Geypens *et al.*, 1997). Proteolytic fermentation is generally considered as detrimental for animal health as it can lead to the formation of potentially toxic substances (Cone *et al.*, 2005). It was, however, found that fermentation of resistant protein modifies also SCFA profile resulting in a greater proportion of butyrate (Morita *et al.*, 1998) which is an important energy source for colonocytes. Part of potato protein, often used in piglets and chicken diets, is resistant to microbial degradation (Morita *et al.*, 2004). The aim of the present study was to determine the effect of dietary levels of potato protein concentrate (PPC) on the SCFA profile in the caecum of broiler chickens.

Material and methods

Four groups of 24 Ross female broilers were fed from the first day of life on wheat- based diets containing 0, 5, 10 and 15% of PPC added at the expense of isoprotein amounts of soybean oil meal. The diets contained 220 g protein and 12.7 MJ ME/kg. From the age of 8 days the birds were kept in individual cages. At the age of 35 days the birds were sacrificed, full and empty caeca were weighed and caecal digesta was sampled. pH was measured and adjusted to 8.2. The samples were stored at -18 °C. The SCFA content was measured using gas chromatography method with isocaproic acid as an internal standard. The data were submitted to one-way analysis of variance according to ANOVA procedure using the Statgraphics package (version 5.1).

Results and discussion

The concentration of total SCFA and of acetate, propionate and butyrate, pH and weight of caecal digesta and tissue are given in Table 1. Contrary to expectations, neither the concentrations of all three acids (sum of SCFA) nor of any single acid were affected by the presence and level of PPC. The concentration of isobutyric and isovaleric acids, which are indicative for fermentation of branch-

Table 1 Effect of potato protein concentrate (PPC) level on pH, concentration of short chain fatty acids (SCFA) in caecal digesta (µM/g digesta) and caecal digesta and tissue weights (g/100 g b.w.)

Item	Sum of SCFA	Acetate	Propionate	Butyrate	pH	Caecal digesta	Caecal tissue
Dietary treatment							
0% PPC	50.61	35.48	6.36	8.77	6.46	0.42	0.35[a1]
5% PPC	54.96	36.82	8.42	9.73	6.65	0.44	0.36[a]
10% PPC	52.07	36.69	8.15	7.23	6.60	0.42	0.41[a]
15% PPC	51.30	35.92	6.99	8.39	6.27	0.36	0.51[b]
SEM	3.04	2.11	0.63	0.79	0.12	0.05	0.03
P-value	0.76	0.97	0.09	0.18	0.13	0.79	<0.01

[1] Means in the same column with different superscripts are significantly different ($P \leq 0.05$).

chained amino acids, did not differ (not shown). The average proportions of acetate, propionate and butyrate in caecal digesta of birds fed on control wheat-soybean meal diet were about 70:13:17, respectively, which is in agreement with data observed by Rehman *et al.* (2008) who obtained the same proportions (70:12:18) of SCFA in digesta of broilers fed diets with inulin.

Positive effect of resistant protein on fermentation was found in rats when it was compared with protein almost fully digested in the small intestine (Morita *et al.*, 2004). Our results indicate that in chickens PPC does not modify the fermentation when added to a wheat-soybean diet. This suggest that hindgut substrates did not differ enough to alter the fermentative activity of microflora (Biggs and Parsons, 2009).

The relative caecal tissue weight was the only parameter significantly affected by the PPC level and was the greatest on the diet containing 15% of PPC. A tendency to the increase of caecal tissue mass with the raise of PPC content was observed. The trophic effect of a diet containing prebiotic compounds on the intestinal tissue is generally considered as positive but it is difficult to explain in view of the absence of any differences in SCFA concentration.

References

Biggs P. and C.M. Parsons, 2009. The effects of whole grains on nutrient digestibilities, growth performance, and cecal short-chain fatty acid concentrations in young chicks fed ground corn-soybean meal diets. Poultry Sci. 88, 1893-1905.

Cone J.W., A.W. Jongbloed, A.H. Van Gelder and L. de Lange, 2005. Estimation of protein fermentation in the large intestine of pigs using a gas production technique. Anim. Feed Sci. Technol. 123-124, 463-472.

Geypens B., D. Claus, P. Evenepoel, M. Hiele, B. Maes, M. Peeters, P. Rutgeerts and Y. Ghoos, 1997. Influence of dietary protein supplements on the formation of bacterial metabolites in the colon. Gut 41, 70-76.

Morita T., S. Kasaoka, A. Oh-hashi, M. Ikai, Y. Numasaki and S. Kiriyama, 1998. Resistant proteins alter cecal short-chain fatty acid profiles in rats fed high amylose cornstarch. J. Nutr. 128, 1156-1164.

Morita T., S. Kasaoka, S. Kiriyama, 2004. Physiological functions of resistant proteins: proteins and peptides regulating large bowel fermentation of indigestible polysaccharide. J. AOAC Int. 87, 92-796.

Rehman H., P. Hellweg, D. Taras, J. Zentek, 2008. Effects of dietary inulin on the intestinal short chain fatty acids and microbial ecology in broiler chickens as revealed by denaturing gradient gel electrophoresis. Poultry Sci. 87, 783-789.

Modelling methionine requirements in growing chicken by using the dietary methionine efficiency

F. Liebert, J. Farke and C. Wecke
Department of Animal Sciences, Division Animal Nutrition Physiology, Goettingen University, Kellnerweg 6, 37077 Goettingen, Germany

Introduction

Amino acid (AA) requirement studies in growing animals mostly make use of the factorial approach or dose-response-studies with supplementation of the limiting AA. However, the statistical procedures for approximation of dose-response effects in supplementation studies are differently adapted to physiological conditions and measure an average response due to a varying mixture of native and supplemented limiting AA in crystalline or liquid form. Modelling the AA requirements may provide advantages, when the applied model is appropriate. Applications of an exponential N utilization model for statistical evaluation of observed N-balance data have been described earlier (Samadi and Liebert, 2008; Wecke and Liebert, 2010). The principle of the approach is to obtain model parameter of dietary N- and AA utilization from N-balance studies for feed protein evaluation and growth-dependent AA supply. The objective of this study was to apply this approach to provide Met requirement estimates at different dietary Met:Cys ratios, at different ages and rates of protein deposition.

Material and methods

N-balance studies with 88 male, fast-growing chickens (Ross 308) were conducted (starter, grower) using a low-Met basal diet based on corn (47.5%), soybean meal (22.5%), field beans (11%), peas (10%), and crystalline AA other than Met. The basal diet was supplemented with two levels of DL-Met in the absence or presence of additional L-Cys×HCl×H_2O (Table 1).

For both age periods, N-balance data were recorded and statistically evaluated based on the basal exponential function:

$$ND = NR_{max}T\ (1 - e^{-b \times NI}) - NMR$$

where ND is the daily N-balance (mg/$BW_{kg}^{0.67}$), $NR_{max}T$ is the theoretical maximum for daily N-retention (=ND+NMR; mg/$BW_{kg}^{0.67}$), NI is the daily N-intake (mg/$BW_{kg}^{0.67}$), NMR is the daily N-maintenance requirement (mg/$BW_{kg}^{0.67}$); b is the slope of the N-retention curve (protein quality parameter, independent of NI), and e is the basic number of the natural logarithm (ln). Dietary protein quality (b) and Met-efficiency (bc^{-1}) data were derived according to earlier applications of the procedure (Samadi and Liebert, 2008; Wecke and Liebert, 2010). Met-requirements depending on aimed growth response and realized dietary Met-efficiency were derived according to the equation:

$$LAAI = [\ln NR_{max}T - \ln(NR_{max}T - NR)] : 16 \times bc^{-1}$$

Table 1. Methionine content and Met:Cys ratio of the experimental diets.

Diet	1 Control	2 +0.1% DL-Met	3 +0.2% DL-Met	4 +0.1% DL-Met; +0.1% L-Cys	5 +0.2% DL-Met; +0.2% L-Cys
Methionine (g/16 g N)	1.36	1.90	2.44	1.90	2.41
Dietary Met:Cys ratio	42:58	51:49	57:43	40:60	40:60

where LAAI is the daily intake of the limiting AA (mg/$BW_{kg}^{0.67}$) necessary to achieve the aimed level of NR (ND + NMR) with given dietary AA-efficiency (bc^{-1}).

Results and discussion

Table 2 provides the observed relative changes of model parameters for dietary protein quality (b) and AA-efficiency (bc^{-1}), when setting the results with control diet as a standard (=100).

Increased protein quality in both age periods shows that Met was the limiting AA. Accordingly, Met efficiency remained high when DL-Met was supplemented in starter diets. In grower diets, Met efficiency declined due to higher metabolic need for Cys, provided from Met-degradation. The Met efficiency improved when both DL-Met and L-Cys were added (diets 4 and 5). As an example for chicken starter period (Table 3), the observed Met efficiency was utilized to derive Met requirement estimates for both graded daily protein deposition and dietary Met efficiency (assumed -10% or +10% of the observed average).

In conclusion, in addition to the aimed protein deposition, the assumed variation of Met-efficiency is an important dietary factor influencing the recommended Met-supply. A Met:Cys ratio of 40:60 improved the efficiency of Met utilization in older chicken and needs more attention in future studies. For robust recommendations of the optimal Met content in the diet, it is important to establish reliable feed intake patterns that will depend on genetic and environmental factors.

Table 2. Relative changes of protein quality (b) and Met-efficiency (bc^{-1}) due to the diets under study during starter or grower period (observed parameters for diet 1=100).

Diet	1	2	3	4	5
Protein quality, starter/grower	100/100	146/127	174/161	141/190	151/191
Met-efficiency, starter/grower	100/100	105/88	97/86	102/134	86/104

Table 3. Example for modelling the Met requirement in growing chicken depending on daily protein deposition, dietary Met efficiency, and assumed daily feed intake.

Protein deposition	6 g/d			8 g/d			10 g/d		
Met-efficiency	-10%	Mean	+10%	-10%	Mean	+10%	-10%	Mean	+10%
	Daily Met requirement								
mg/$BW_{kg}^{0.67}$	263	237	215	383	344	313	548	493	448
mg[1]	165	149	135	241	216	197	344	310	282
Feed intake[1] (g/d)	Required Met content in feed (%)								
50	0.33	0.30	0.27	0.48	0.43	0.39	0.69	0.62	0.56
60	0.28	0.25	0.23	0.40	0.36	0.33	0.57	0.52	0.47
70	0.24	0.21	0.19	0.34	0.31	0.28	0.49	0.44	0.40

[1] BW=500 g.

References

Samadi and F. Liebert, 2008. Modelling the optimal lysine to threonine ratio in growing chickens depending on age and efficiency of dietary amino acid utilisation. Br. Poult. Sci. 49, 45-54.

Wecke, C. and F. Liebert, 2010. Optimal dietary lysine to threonine ratio in pigs (30-110 kg BW) derived from observed amino acid efficiency. J. Anim. Physiol. Anim. Nutr. (in press).

Maintenance requirements for methionine & cystine for poultry

M.A. Bonato, N.K. Sakomura, E.B. Malheiros, I.A.M.A. Teixeira and R M. Gous
Departamento de Zootecnia, UNESP Jaboticabal, SP, Brazil

Introduction

Maintenance can be defined as the state in which the intake of nitrogen is exactly equal to the sum of the losses, to keep the body nitrogen content constant (Owens and Pettigrew, 1989). Methionine+cystine (M&C) are essential for body maintenance, as part of metabolic processes and also for cell synthesis and renewal (Baker, 1991). This study aimed to estimate the digestible M&C requirement for maintenance using adult roosters of different body weights and composition.

Material and methods

The methodology of Nonis and Gous (2008) was applied, using 36 Bovans White® (mean body weight 2.02±0.2 kg), and 36 Cobb® adult roosters (5.54±0.6 kg). Six treatments, comprising a range of daily amino acid intakes, were replicated among six birds of each strain. First the birds were subjected to 48 h of fasting and subsequently received the experimental diets for 72 h (N-balance period). Protein-containing diets were fed by intubation each day and excreta were collected every 24-h after feeding. A nitrogen (N)-free diet was offered *ad libitum* throughout the balance period. Body protein content (BP) of the birds was determined at the end of the trial by slaughter.

Two basal feeds were used, one being N-free and containing energy, minerals and vitamins, and the other containing protein with the same levels of ME, minerals and vitamins as in the N-free feed (concentrated). The amino acid composition was according to Rostagno *et al.* (2005) but with all the amino acids to 40% of the requirement for Bovans White and 60% for Cobb, and the M&C reduced by 20% of others. Daily intakes of M&C (methionine in parenthesis) was 0 (0), 8.62 (5.18), 17.2 (10.4), 34.5 (20.7), 69 (41.5) and 138 (82.9), and 0 (0), 10.7 (5.77), 24 (13), 37.4 (20.2), 50.7 (27.4) and 64 (34.6) mg/kg body weight for the Bovans White and Cobb strains. For this range in intake of M&C, were weighed different amounts of concentrated diet in the proportions 0, 2.5, 5, 10, 20 and 40 g, and 0, 10, 12.5, 25, 37.5 and 50 g, being completed for 40 and 50 g with a N-free diet, for the Bovans White and Cobb strains, respectively.

Nitrogen balance (NB) was calculated as the difference between N intake and N excreted. A linear regression expressing NB in terms of M&C intake (mg/$BP_m^{0.73}$) (Emmans and Fisher, 1986) was fitted and the daily requirement for maintenance (NB = 0) determined. The linear regressions of the two strains were compared (Kaps and Lamberson, 2004) using the GLM procedure of SAS 9.0 (2002).

Results and discussion

For the Bovans White® strain, the regression equation expressing NB in terms of M&C intake, based on $BP_m^{0.73}$, was -303 (±70.7) + 3.24 (±0.2) (r^2=0.86). The equivalent equation for the Cobb® males was -247 (±76.2) + 3.1 (±0.4) (r^2=0.66). As there were no significant differences in intercept or slope (P<0.05) between the two regressions a common regression (Figure 1) could be used for both strains. The daily M&C intake at zero N-balance for the two strains, and hence the intake required for maintenance, was 87.2 (±20.5) mg/$BP_m^{0.73}$.

The results were higher than those found by Edwars and Baker (1999) (34.99 mg/$BPm^{0.73}$) and lower than Leveille *et al.* (1960) (548.46 mg/$BPm^{0.73}$). As the authors did not analyzed body protein, these values were calculated based on 16% CP. These differences could be because of different methodologies used and in the studies.

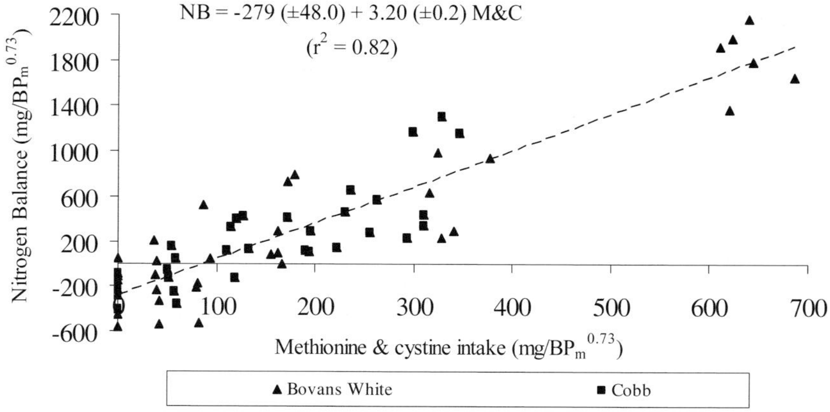

Figure 1. Regression of nitrogen balance on daily methionine & cystine intake (mg/BP$_m^{0.73}$) to determine the M&C for maintenance.

Conclusion

The daily M&C requirement of poultry for maintenance was the same for birds with different body weight and composition, estimated at 87.2 (±20.5) mg/BP$_m^{0.73}$.

Acknowledgments

Fundação de Apoio à Pesquisa do Estado de São Paulo (FAPESP) for supporting the research and scholarship.

References

Baker, D.H., 1991. Partitioning of nutrients for growth and other metabolic functions: efficiency and priority considerations. Poult. Sci. 70, 1797-1805.

Edwards, H.M. and D.H. Baker, 1999. Maintenance Sulfur Amino Acid Requirements of Young Chicks and Efficiency of Their Use for Accretion of Whole-Body Sulfur Amino Acids and Protein. Poult. Sci. 78, 1418-1423.

Kaps, M. and W.R. Lamberson, 2004. Biostatistics for animal science. CABI Publishing, Wallingford, USA, 445pp.

Leveille, G.A., R. Shapiro and H. Fisher, 1960. Amino acid requirements for maintenance in the adult rooster IV. The requirements for methionine, cystine, phenylalanine, tyrosine and tryptophan; the adequacy of the determined requirements. J. Nutr. 72, 8-15.

Nonis, M.K. and R.M. Gous, 2008. Threonine and lysine requirements for maintenance in chickens. S. Afr. J. Anim. Sci. 38, 75-82.

Owens, F.N. and J.E. Pettigrew, 1989. Subdividing amino acid requirements into portions for maintenance and growth. In: Friedman, M. (ed.), Absorption and utilization of amino acids. Boca Raton, USA: CRC Press: 15-30.

Rostagno, H.S., L.F.T. Albino, J.L. Donzele, P.C. Gomes, R.F.M. Oliveira, D.C. Lopes, A.S. Ferreira and S.L.T. Barreto, 2005. Tabelas brasileiras para aves e suínos: composição de alimentos e exigências nutricionais. 2a. ed. Viçosa, Brasil: Departamento de Zootecnia, Universidade Federal de Viçosa.

SAS, 2002. SAS user's guide: statistics. Cary, USA: Statistical Analysis System Institute.

Maintenance requirements of threonine for poultry

M.A. Bonato, N.K. Sakomura, I.A.M.A. Teixeira, E.B. Malheiros and J.B.K. Fernandes
Departamento de Zootecnia, UNESP Jaboticabal, SP, Brazil

Introduction

It is essential to feed poultry according to amino acids requirements to avoid excess and environmental pollution. Never before has the economic success of feed manufacturers been so heavily dependent on optimal and accurate amino acid composition as today (D'Mello, 2003). The threonine (Thr) is important not only for protein deposition, but also for mucin production and digestive system (Ball *et al.*, 1999). This study aimed to estimate the digestible Thr requirement for maintenance using adult roosters of different body weights and composition.

Material and methods

The methodology of Nonis and Gous (2008) was applied, using 36 Bovans White® (mean body weight 2.11±0.2 kg), and 36 Cobb® adult roosters (5.78±0.7 kg). Six treatments, comprising a range of daily amino acid intakes, were replicated among six birds of each strain. First the birds were subjected to 48 h of fasting and subsequently received the experimental diets for 72 h (N-balance period). Protein-containing diets were fed by intubation each day and excreta were collected every 24-h after feeding. A nitrogen (N)-free diet was offered *ad libitum* throughout the balance period. Body protein content (BP) of the birds was determined at the end of the trial by slaughter.

Two basal feeds were used, one being N-free and containing energy, minerals and vitamins, and the other containing protein with the same levels of ME, minerals and vitamins as in the N-free feed. The amino acid composition was according to Rostagno *et al.* (2005) but with all the amino acids to 40% of the requirement for Bovans White and 70% for Cobb, and the Thr reduced by 20% of others. Daily intakes of Thr was 0, 7.96, 15.9, 31.8, 63.6 and 127, and 0, 10.6, 42.5, 63.8, 85.1 and 106 mg/kg body weight for the Bovans White and Cobb strains. For this range in intake of Thr, were weighed different amounts of concentrated diet in the proportions 0, 2.5, 5, 10, 20 and 40 g, and 0, 5, 20, 30, 40 and 50 g, being completed for 40 and 60g with a N-free diet, for the Bovans White and Cobb strains, respectively.

Nitrogen balance (NB) was calculated as the difference between N intake and N excreted. A linear regression expressing NB in terms of Thr intake (mg/$BP_m^{0.73}$) (Emmans and Fisher, 1986) was fitted and the daily requirement for maintenance (NB = 0) determined. The linear regressions of the two strains were compared (Kaps and Lamberson, 2004) using the GLM procedure of SAS 9.0 (2002).

Results and discussion

For the Bovans White® strain, the regression equation expressing NB in terms of Thr intake, based on $BP_m^{0.73}$, was -130 (±57.8) + 2.91 (±0.2) (r^2=0.74). The equivalent equation for the Cobb® males was -369 (±111) + 2.19 (±0.4) (r^2=0.55). As there were no significant differences in intercept or slope (P<0.05) between the two regressions a common regression (Figure 1) could be used for both strains. The daily Thr intake at zero N-balance for the two strains, and hence the intake required for maintenance, was 75.5 (±29.8) mg/$BP_m^{0.73}$.

The results were expressed on the basis of body protein weight at maturity because there is no demand for amino acids for the maintenance of lipid, which can vary considerably, even among individuals with similar body weight (Emmans and Fisher, 1986). It is therefore the preferred scale

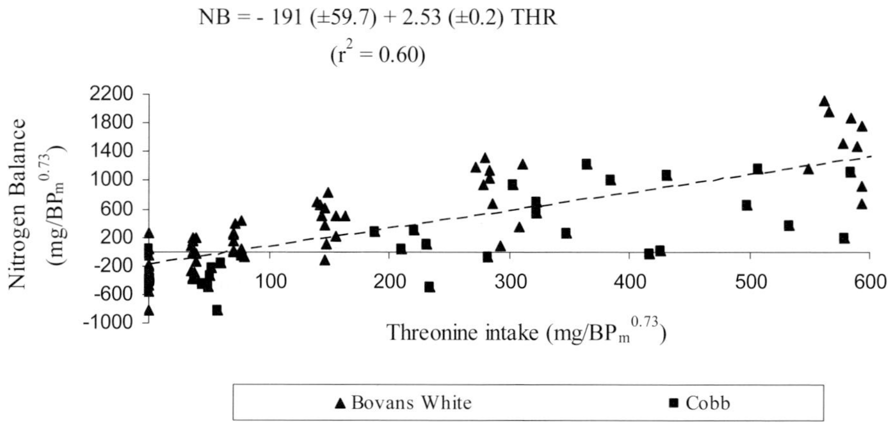

Figure 1. Regression of nitrogen balance on daily threonine intake (mg/BP$_m^{0.73}$), to determine the Thr for maintenance.

when comparing the maintenance requirements of birds of different body weight, mature body weight and degree of fatness.

Conclusion

The Thr daily requirement of poultry for maintenance was the same for birds with different body weight and composition, estimated at 75.5 (±29.8) mg/BP$_m^{0.73}$.

Acknowledgments

Fundação de Apoio à Pesquisa do Estado de São Paulo (FAPESP) for supporting the research and scholarship.

References

Ball, R.O., G. Law, R.F.P. Bertolo and P.B. Pencharz, 1999. Adequate oral threonine is critical for mucin production and mucosal growth by neonatal piglet gut. In: International Symposium on Protein Metabolism and Nutrition, 7. Proceedings 96, 31.

D'Mello, J.F.P., 2003. Amino acids as multifunctional molecules. In: D'Mello, J.F.P. (ed.), Absorption and utilization of amino acids. Wallingford, USA: CABI Publishing: 1-14.

Emmans, G.C. and C. Fisher, 1986. Problems in nutritional theory. In: Fisher, C.; Boorman, K. N., eds. Nutrient requirements of poultry and nutritional research. London, England: Butterworths:, 9-39.

Kaps, M. and W.R. Lamberson, 2004. Biostatistics for animal science. Wallingford, USA: CABI Publishing, 445pp.

Nonis, M.K. and R.M. Gous, 2008. Threonine and lysine requirements for maintenance in chickens. S. Afr. J. Anim. Sci. 38, 75-82.

Rostagno, H.S., L.F.T. Albino, J.L. Donzele, P.C. Gomes, R.F.M. Oliveira, D.C. Lopes, A.S. Ferreira and S.L.T. Barreto, 2005. Tabelas brasileiras para aves e suínos: composição de alimentos e exigências nutricionais. 2a. ed. Viçosa, Brasil: Departamento de Zootecnia, Universidade Federal de Viçosa.

SAS, 2002. SAS user's guide: statistics. Cary, USA: Statistical Analysis System Institute.

Valine requirement of weaned piglets fed low-protein diet

J. Trautwein[1], G. Dusel[1] and J. Bartelt[2]
[1]University of Applied Sciences Bingen, Berlinstr. 109, 55411 Bingen, Germany
[2]Lohmann Animal Health GmbH & Co. KG, Heinz-Lohmann-Str. 4, 27472 Cuxhaven, Germany

Introduction

For environmental and economic reasons as well as for protective purposes, the piglets are meant to be saved from digestive disorders by means of a reduction of the CP content in their diets. Therefore, a requirement-based supplementation of crystalline essential AA is a way of avoiding excessive CP levels in piglet diets. According to some authors, Val is, after Lys, Met/Cys, Thr and Trp, next limiting amino acid (Mavromichalis *et al.*, 1998; Figueroa *et al.*, 2003). Due to the admission of L-valine as feed supplement (EC Regulation No. 403/2009 of the Commission), crystalline valine may be added now. Consequently, the question of valine requirement in relation to age/weight and performance intensity arises.

The objective of the test was the investigation of the effects of different valine to lysine ratios for early weaned piglets with reduced crude protein content in a dose-response-study. The diets were supplemented with L-Lysine HCL, DL-Methionine, L-Threonine and L-Tryptophan to fulfil the corresponding requirements at high performance conditions.

Material and methods

In the pretest (152 piglets, 9-24 kg LW), a basal diet, deficient in valine at sub-optimal lysine content (lysine second-limiting) was tested. Afterwards, this diet was used as basic feed for the dose-response-study.

In the dose-response-study 160 weaned piglets (German-Landrace x Piétrain, ♂/♀ 1:1, initial body weight 9 kg, final body weight 22 kg) were divided into four treatment-groups (A-D, 9 double-pens and 2 large pens with 11 piglets per treatment) and fed with different valine:lysine ratio diets. The feed composition and treatment design is described in Table 1. The test was carried out for four weeks at the University of Applied Sciences Bingen/Germany. The weekly feed intake per pen and individual live weight were measured.

Results and discussion

Significant treatment effects were found for feed intake and feed conversion. The daily weight gain was improved by 9% in group C compared to the valine deficient diet (group A). However, the effect was not significant. Additionally, the optimum of the daily valine intake was calculated by using the broken line model with 5.94 g standardized ileal digestible valine per animal for a body weight gain of 512 g/d. Based on the average daily intake of 8.75 g standardized ileal digestible lysine in the trial, a Val:Lys ratio (stand. ileal digestible) of 0.68 could be calculated. Wiltafsky *et al.* (2009) determined an optimal valine:lysine ratio of 0.65 to 0.67 (stand. ileal digestible). Fully corresponding with Chung and Baker (1992), a valine ratio of 0.68 should be in the ideal protein of piglets. A higher ratio was published by Barea *et al.* (2009) who derived a valine:lysine ratio of at least 0.7 (standardised ileal digestible).

Conclusion

In accordance to newly published data, the optimum dietary standardized ileal digestible valine to lysine ratio should range from 0.67 to 0.70 or from 0.70 to 0.73 (gross). A study of the AA contents

Table 1. Basic-feed composition and treatment design.

Basic-feed (all treatments)

AA-content (%)	Total basis	Stand. ileal digestibl.[1]
Lys	1.09	1.00
Met / Cys	0.67	0.60
Thr	0.73	0.65
Trp	0.25	0.22
Ile	0.61	0.52
Leu	1.18	1.02
CP (%)	16.2	
MJ ME / kg	13.8	

Treatment				
Stand. ileal digestible Val(g) / kg	A	B	C	D
feed	5.7	6.3	6.9	7.5
Val : Lys ratio (s. dig.)	0.57:1	0.63:1	0.69:1	0.75:1

Composition (%) basic diet without AA supplementation	
Wheat	25.8
Corn	17.2
Soya bean meal (44)	12.7
Barley	10.7
Corn (processed)	10.1
Oat flakes (rolled)	5.1
Oat groats	5.1
Wheat bran	5.0
Whey powder	2.0
Whey fat concentrate	2.0
Vegetable fat/oil	1.6
Lignocelluloses	1.0
Vitamine-premix	1.7

[1] Calculated with INRA-tables (2004).

Table 2. Influence of L-valine on performance parameters of piglets (± SD).

Group	A	B	C	D	SEM	*P*-value[1]
stand. ileal digestible Val (g/kg)	5.7	6.3	6.9	7.5		
Initial body weight	8.7±1.8	8.6±1.7	8.6±1.7	8.8±1.7	0.136	0.982
Final body weight	20.9±4.0	21.4±3.0	22.0±3.6	21.7±3.3	0.278	0.557
Daily weight gain(g)	454±95	473±74	496±84	480±77	6.680	0.163
Daily feed intake (g)	929±131^{a}	829±81^{b}	882±99^{b}	861±110^{b}	8.940	<0.01
Feed convers. ratio	1.93±0.2^{a}	1.76±0.1^{b}	1.78±0.2^{b}	1.80±0.2^{b}	0.015	<0.001

[1] ANOVA, post hoc test – Tukey-HSD test ($\alpha<0.05$); STATISTICA for Windows, StatSoft.

in piglet feed carried out throughout Europe by Ajinomoto Eurolysine S.A.S in 2006, revealed that two thirds of all piglet feeds in Europe do not contain adequate valine supplements (Trautwein *et al.,* 2009).

References

Barea, R., L. Brossard, N. Le Floc´h, Y. Primot, D. Melchio and J. van Milgen, 2009. The standardized ileal digestible valine to lysine requirement ratio is at least 70% in post-weaned piglets. J. Anim. Sci. 87, 935-947.

Chung, T.K. and D.H. Baker, 1992. Ideal amino acid pattern for 10-kilogram pigs. J. Anim. Sci. 70, 3102-3111.

Figueroa, J.L., A.J. Lewis, P.S. Miller, R.L. Fischer and R.M. Diedrichsen, 2003. Growth, carcass traits, and plasma amino acid concentration of gilts fed low-protein diets supplemented with amino acids including histidin, isoleucine and valine. J. Anim. Sci. 81, 1529-1537.

Mavromichalis, I., D.M. Webel,. J.L. Emmert, R.L. Moser and D.H. Baker, 1998. Limiting order of amino acids in a low-protein corn-soybean meal-whey-based diet for nursery pigs. J. Anim. Sci. 76, 2833-2837.

Trautwein, J.T., G. Dusel and J. Bartelt, 2009. Releasing potentials in piglet feed-L-Valine-supplement achieves ideal protein. Feed Magazine, 11-12/09, 24-28.

Wiltafsky, M.K., B. Schmidtlein and F.X. Roth, 2009. Estimation of the optimum dietary ratio standardized ileal digestible valine to lysine for eight to twenty five kilograms of body weight pigs. J. Anim. Sci. 87, 2544-2553.

Digestible lysine levels for gilts and barrows with high genetic potential for lean meat gain from 30 to 95 kg

P.F. Campos[1], B.A.N. Silva[2], R.J. Cogo[3], N. Duijvesteijn[2]
[1]UFV, Universidade Federal de Viçosa, 36571-000 Viçosa, Minas Gerais, Brazil
[2]IPG, Institute for Pig Genetics, P.O. Box 43, 6640 AA Beuningen, the Netherlands
[3]TOPIGS do Brasil, 13070-056 Campinas, São Paulo, Brazil

Introduction

In the past years, the pork industry in general has become more industrialized and more specialized leading to focused breeding and selection of leaner and more efficient pigs. This breeding selection allowed raising the slaughter weight without compromising carcass quality. As the nutritional requirements of pigs varies according to growth stage, the nutrient levels needed to feed these animals must optimize the potential of the pigs during each stage of its development in order to obtain the maximum of efficiency and carcass quality (Abreu *et al.*, 2007). Average amino acid and particularly lysine requirements for fattening pigs are well known. Nevertheless, some difficulties remain to derive practical recommendations for adaptating the amino acid supplies to different genetic lines, growing stage or, more generally, to different growth potentials (Fontes *et al.*, 2005). The aim of this study was to determine the ileal digestible Lys levels that optimize performance of TOPIGS Tempo sired barrows and gilts with diets formulated to meet all essential amino acid requirements on a digestible ideal basis.

Materials and methods

Seven hundred and sixty eight crossbred (TOPIGS Tempo × TOPIGS 40) pigs (384 gilts and 384 barrows) were used to determine the optimal ileal digestible Lys levels during starter (SP = 30 to 55 kg), grower (GP = 55 to 75 kg) and finisher (FP = 75 to 95 kg) periods. This study was conducted at a commercial farm in Brazil during 2009 summer. Each phase consisted of four different ileal digestible Lys levels, which were: SP = 0.90, 1.00, 1.10, and 1.20%; GP = 0.80, 0.90, 1.00, and 1.10%; and FP = 0.70, 0.80, 0.90, and 1.00%. Treatments consisted of iso-ME diets in each phase. To reach the higher Lys levels, synthetic amino acids were added to the corn-soybean basal diets, always maintaining the same ideal ratio between Lys and the essential amino acids for all diets in each period. Pigs had *ad libitum* access to feed and water. Equations for predicting the initial and final tissue composition of the experimental animals were fitted using live weight and backfat thickness measurements as predictors. For each period, the experimental data were submitted to GLM procedures (SAS, 2008), using 12 animals per pen as the experimental unit and with the main effects of sex and dietary treatment.

Results and discussion

General results are shown in Table 1. The average minimum and maximum ambient temperatures registered during the study were 19.9 and 29.6 °C, respectively. Lys levels affected the average daily feed intake (ADFI) ($P<0.05$) of pigs during the SP. During the same period the levels also influenced the average daily gain (ADG) ($P<0.01$), where the best gain was obtained at the level of 1.10% (938 g/d). Fontes *et al.* (2005), observed a similar effect of the levels of Lys on ADG from 30 to 60 kg. Protein deposition (PD) increased ($P<0.01$) until the level of 1.10% (148 g/d). Lipid deposition (LD) was lowest at the level of 1.20% (96 g/d; $P<0.01$). The best feed conversion (FC) was observed at the level of 1.00% (2.24 kg/kg; $P<0.01$). During the GP, ADFI increased until the level of 1.00% ($P<0.01$). Similar effect on ADFI was also observed by Fontes *et al.* (2000). Also PD was affected ($P<0.05$), where the best deposition rate (202 g/d) already occurred at 0.80% of digestible Lys. Lipid deposition was also lowest at the level of 0.80% (190 g/d; $P<0.05$). The best feed conversion (FC) was observed at the level of 0.80% (2.58 kg/kg; *P<0.05*). For the FP, the levels of Lys affected the ADFI (*P<0.05*). ADG was not affected (*P>0.05*). The best FC was observed at the level of 0.70% (3.17 kg/kg; *P<0.01*). The treatments did not influence PD in this period, indicating that the level of 0.70% is sufficient to reach maximum PD. LD was influenced (*P<0.01*), whereas the lowest value was observed at 1.00% of digestible Lys. It was concluded that for the starter phase, the levels of digestible Lys that allowed the best performance were 1.00% for gilts and 0.95% for barrows. Lys needs for a maximal growth rate were 0.82 and 0.75% in growing phase and 0.72 and 0.65% in finishing phase, in gilts and barrows, respectively.

Table 1. General performance results according to phase and levels of ileal digestible lysine (least square means corrected for sex).

Variables	Levels of ileal digestible lysine %				Significance
Starter phase	0.90	1.00	1.10	1.20	
Average daily gain, g/d	901[a]	918[a]	938[b]	936[b]	*P*<0.05
Backfat thickness (mm)	8.05[c]	7.66[b]	7.74[b]	7.48[a]	*P*<0.01
Av. protein deposition, g/d	141[a]	144[a]	148[b]	149[b]	*P*<0.01
Av. daily feed intake, kg/d	2.10[b]	2.06[a]	2.09[b]	2.07[a]	*P*<0.05
Grower phase	0.80	0.90	1.00	1.10	
Average daily gain, g/d	1,058	1,051	1,066	1,034	ns
Backfat thickness (mm)	9.38[a]	9.44[a]	9.38[a]	9.57[b]	*P*<0.01
Av. protein deposition, g/d	202[b]	198[bc]	203[b]	194[ac]	*P*<0.05
Av. daily feed intake, kg/d	2.73[a]	2.76[ac]	2.82[b]	2.78[bc]	*P*<0.001
Finisher phase	0.70	0.80	0.90	1.00	
Average daily gain, g/d	894	892	886	872	ns
Backfat thickness (mm)	12.33	12.18	12.19	11.80	ns
Av. protein deposition, g/d	98	104	100	108	ns
Av. daily feed intake, kg/d	2.84[a]	2.85[a]	2.98[b]	2.82[a]	*P*<0.001

[a,b,c] Within a line, different letters means a significant difference.

References

Abreu, M.L.T., J.L. Donzele, R.F.M. Oliveira, A.L.S. Oliveira, D. Haese and A.A. Pereira, 2007. Dietary digestible lysine requirements, based on the ideal protein concept, for barrows with high genetic potential from 30 to 60 kg. R. Bras. Zootec. 36, 62-67.

Fontes, D.O., J.L. Donzele, R.F.M. Oliveira, D.C. Lopes, A.S. Ferreira and F.C.O. Silva, 2005. Levels of Lysine for gilts with high genetic potential for lean gain from 30 to 60 kg. R. Bras. Zootec. 34, 81-89.

Fontes, D.O., J.L. Donzele, A.S. Ferreira, R.F.M. Oliveira and C.A.G.G. Junior, 2000. Lysine levels for gilts with high genetic potential for lean gain from 60 to 95 kg. R. Bras. Zootec. 29, 784-793.

Energy requirement of broiler breeder hens with different body weights

C. Salas, R.D. Ekmay, J. England, S. Cerrate and C.N. Coon
University of Arkansas, USA

Introduction

Energy represents a challenge for formulating diets for broiler breeder hens because of their regimens of feed restriction. Overweight broiler breeders have been shown to produce less hatching eggs with reduced fertility, reduced hatchability and poor egg shell quality. Ongoing research shows that breeder hens may need less than the approximate 2.0 MJ ME/day that the industry is presently feeding.

Materials and methods

The feed allocation during the growing period of Cobb 500 broiler breeder pullets was weekly adjusted to produce 3 different growth curves; a control group (SBW) reared according to Cobb guidelines, a group reared 20% heavier (HBW) and a group reared 20% lighter (LBW). At 21 wks, 324 pullets of each group were transferred to breeder cages and fed 1 of 6 diets (each diet had 12.0 MJ ME and 20.8-14.3% CP). Feed allocation was adjusted to provide 24 g of ideal protein/d to all the birds and energy intakes of 1.38, 1.51, 1.63, 1.76, 1.88 and 2.0 MJ ME/hen/d at peak production. Production performance, body weight, body composition, egg fertility, hatchability and chick weight were monitored through a 40 wk production period.

Results and discussion

Table 1 shows production results for the flock. Differences were found in age of maturity and numbers of eggs. The LBW birds reached sexual maturity 7 d later than the SBW. This is in disagreement with Renema *et al.* (2007), who found that hen body weight did not affect age of sexual maturity when they were photo-stimulated at 22 wks. At 65 wks, birds fed 1.63 and 1.38 MJ ME/day produced the highest and the lowest number of eggs per hen housed (182.8 and 170). Pearson and Herron (1981) also reported maximum egg production for breeders fed 1.73 kcal ME/day when compared to allowances of 1.52 and 1.88 kcal ME/day (floor pens). This slight difference with our results can be attributed to a higher activity in birds housed in floor pens.

Table 1. Production parameters of breeder hens reared on 3 different growth curves and fed 6 different energy intakes (40 wk production period).

		Age maturity (wks)	Eggs/hen housed	% Peak prod. (30 wks)	Final Body weight (g)	Average Egg weight (g)
Growth curve	HBW	25.9^{b}	181.0^{a}	83.6%	$3,724^{a}$	66.8^{a}
	SBW	26.0^{b}	178.6^{a}	86.4%	$3,600^{ab}$	65.7^{b}
	LBW	26.9^{a}	173.3^{b}	86.4%	$3,563^{b}$	64.7^{c}
Energy intake (kcal ME/day)	330	25.9	170.2^{d}	82.4%	$3,724^{d}$	66.8^{ab}
	360	25.9	175.2^{cd}	82.6%	$3,978^{c}$	67.2^{a}
	390	25.7	182.8^{a}	86.9%	$4,309^{ab}$	67.1^{a}
	420	25.9	180.1^{abc}	84.3%	$4,224^{b}$	66.0^{c}
	450	25.7	181.0^{ab}	88.0%	$4,281^{ab}$	65.6^{c}
	480	25.6	176.4^{bc}	88.9%	$4,451^{a}$	66.2^{bc}

[a,b] Means with different letter in same column differ ($P<0.05$).

All the birds gained weight during the 40 wk production period. The hens fed 1.38 MJ ME/day were always under the breed target weight and lost weight at 30-33 wk and 45-50 wk periods. In agreement with the present study, Pearson and Herron (1981) also reported an increase in breeder body weight with an increase in energy allowance (1.52, 1.73 and 1.88 kcal ME/day), with gain differences of more than 600 g between the lower and the higher energy intake.

The body composition of 8 birds per treatment was determined every 10 wks from 30 to 65 wks by DXA technology. In general, all groups gained fat mass until wk 45; afterwards all groups decreased fat deposition until the end of production period at wk 65. Lean mass deposition presented a different tendency; all groups gained lean mass until wk 35, declined lean deposition from 35 to 45 wks and increased it until wk 65. This indicates that breeders shift lean mass allocation after peak of production, with a higher accumulation of protein tissue instead of fat tissue at the end of the production period.

The breeders were artificially inseminated every 5 weeks to monitor egg fertility, hatchability and chick weight. Egg fertility and hatchability were higher than 90% and 85%, respectively, for all breeder groups. The day old chicks from the LBW hens were always lighter than the progeny from the HBW and the SBW hens ($P<0.05$). Overall, the chicks that hatched from LBW and SBW hens that were fed the 1.38 and 1.51 MJ ME/day were the smallest chicks. The day-old chicks from breeders fed 1.63, 1.76, 1.88, and 2.0 MJ ME/day (independent of the growth curve) weighed ≥42 g in all the hatchability studies. Spratt and Leeson (1987) who fed breeders 1.36, 1.61 and 1.88 MJ/d found that birds fed higher energy levels produced heavier day-old chicks between 32 to 35 wks but no effect before or after that period.

The present study shows that breeders of different body weights kept in cages perform well with an energy intake of 1.63 MJ ME/day/hen throughout the 40 wk production period, approximately 18.5% less energy that the current allowance utilized by the industry.

References

Pearson, R.A. and K.M. Herron, 1981. Effects of energy and protein allowances during lay on the reproductive performance of broiler breeder hens. Br. Poultry Sci. 22, 227-239.

Renema, R.A., F.E. Robinson, and M.J. Zuidhof, 2007. Reproductive efficiency and metabolism of female broiler breeders as affected by genotype, feed allocation, and age of photostimulation. 2. Sexual maturation. Poult. Sci. 86, 2267-2277.

Spratt, R.S. and S. Leeson, 1987. Broiler breeder performance in response to diet protein and energy. Poult. Sci. 66, 683-693.

Prediction of energy digestibility and energy content in forage fed to horses

J.E. Lindberg[1] *and S. Ragnarsson*[2]
[1]*Department of Animal Nutrition and Management, Swedish University of Agricultural Sciences, P.O. Box 7024, 75007 Uppsala, Sweden*
[2]*Holar University Collage, 551 Saudárkrókur, Iceland*

Introduction

Forages can make up the whole, or a large part, of the diet for horses. Therefore, a correct prediction of the energy value in this feed source is crucial for valid estimates of the daily energy intake.

According to Fonnesbeck (1968), the major difference among forages in organic matter (OM) digestibility and energy digestibility is due to different carbohydrate composition of the forages. This can be related to very distinct differences in the digestibility of the cell wall constituents as compared to the cell content (Fonnesbeck, 1969). In contrast, Smolders *et al.* (1990) considered the *in vitro* digestibility of OM to be a better predictor of the energy digestibility in horse feeds than its chemical composition alone.

In the present study the possibility to predict the coefficient of energy digestibility (dE) and the content (MJ/kg DM) of digestible energy (DE) in forages fed to horses from the dietary content of neutral detergent fiber (NDF), acid detergent fiber (ADF) and dietary fiber (DF), and the coefficient of *in vitro* digestibility of organic matter (IVDOM) was evaluated.

Material and methods

The *in vivo* digestibility data (average of four observations) from eight batches of preserved forage, produced in the north of Iceland and fed at maintenance level of energy intake to mature Icelandic horses in balance trials, were used (Ragnarsson, 2009). All forages were plastic wrapped and wilted, and baled in large (250-450 kg) round or square bales and fed straight out of the bales in long form. The forage-only diets were fed for a total of 20 days, comprising 14 days of adaptation and 6 days for total collection of faeces, to determine the total tract apparent digestibility of energy. During adaptation periods horses had access to salt lick stones and were fed minerals during non-collection periods.

NDF was analyzed according to Chai and Udén (1998) using undiluted ND solution, sodium sulphate and amylase. ADF was analyzed according to Goering and van Soest (1970) and DF as described by Högberg and Lindberg (2006). The IVDOM was analyzed according to Lindgren (1979) using rumen fluid as inoculum. Gross energy was measured with a bomb calorimeter (Parr 1241 Oxygen Bomb Calorimeter, Illinois, USA).

The effect of fiber content and IVDOM on the dE and DE in preserved forage was subjected to linear regression analysis.

Results and discussion

The range in content (g/kg DM) of dietary components in the forages were: crude protein 93-200, NDF 503-639, ADF 270-411 and DF 534-645. The range in dE was 0.47-0.73, and the range in GE and DE (MJ/kg DM) was 19.2-20.1 and 9.0-14.1, respectively.

The dE and DE of the forages were linearly related ($P<0.05$) to the content of NDF, ADF and DF, and the IVDOM (Table 1). In accordance, Smolders *et al.* (1990) showed a linear decline in the digestibility of organic matter (dOM) with increasing dietary fiber content in different feedstuffs. However, Smolders *et al.* (1990) did not consider the predictive accuracy from chemical analysis alone to be satisfactory. In the current study, the best prediction of dE and DE in the forages was found for the IVDOM (r^2=0.95). This was in accordance with the predictions of dOM in roughages reported by Smolders *et al.* (1990). As shown in Table 1, neither of the fiber measures used (NDF, ADF or DF) were able to predict the variation in dE (r^2=0.47-0.60) and DE (r^2=0.53-0.63) in the forages with a high accuracy.

In conclusion, neither of the common fiber measures used to evaluate the nutritional value of feedstuffs, such as NDF and ADF, nor the more advanced DF measure, was able to predict the variation in dE and DE in preserved forage for horses with an acceptable accuracy as reflected in the low R^2 values.

Table 1. The relationship between the coefficient of energy digestibility (dE) and the content (MJ/kg DM) of digestible energy (DE), and the dietary content (g/kg DM) of NDF, ADF and DF, and the coefficient of in vitro *digestibility of organic matter (IVDOM) in preserved forage.*[1]

	dE				DE			
	NDF	ADF	DF	IVDOM	NDF	ADF	DF	IVDOM
Intercept	1.296	0.9430	1.4391	-0.2638	25.94	18.97	28.92	-5.65
SE	0.248	0.150	0.280	0.080	4.844	2.849	5.406	1.545
Slope	-0.0013	-0.0010	-0.0015	0.0116	-0.0256	-0.0222	-0.0299	0.2335
SE	0.0004	0.0004	0.0005	0.0011	0.0086	0.0085	0.0093	0.0208
R^2	0.57	0.48	0.60	0.95	0.59	0.53	0.63	0.95
RSD	0.0036	0.0043	0.0033	0.0004	1.36	1.57	1.23	0.15
P-value	0.030	0.058	0.023	0.0001	0.025	0.040	0.018	0.0001

[#] SE=standard error; RSD=residual standard deviation.

References

Chai, W. and P. Udén, 1998. An alternative oven method combined with different detergent strengths in the analysis of neutral detergent fiber. Anim. Feed Sci. Technol., 74, 281-288.

Fonnesbeck, P.V., 1968. Digestion of soluble and fibrous carbohydrate of forage by horses. J. Anim. Sci., 27, 1336-1344.

Fonnesbeck, P.V., 1969. Partitioning the nutrients of forage for horses. J. Anim. Sci., 28, 624-633.

Goering, H.K. and P.J. van Soest, 1970. Forage fiber analysis. USDA Agricultural Research Service, Handbook number 379. U.S. Department of Agriculture, Washington DC, USA.

Högberg, A. and J.E. Lindberg, 2006.The effect of level and type of cereal non-starch polysaccharides on the performance, nutrient utilization and gut environment of pigs around weaning. Anim. Feed Sci. Technol. 127, 200-219.

Lindgren, E., 1979. The nutritional value of roughages determined *in vivo* and by laboratory methods. Department of Animal Nutrition and Management, Swedish University of Agricultural Sciences, Report 45, 63 pp.

Ragnarsson, S., 2009. Digestibility and metabolism in Icelandic horses fed forage-only diets. Swedish University of Agricultural Sciences, Acta Universitatis Agriculturae Sueciae, Faculty of Veterinary Medicine and Animal Science, Doctoral thesis no. 2009: 92.

Smolders, E.A.A., A. Steg and V.A. Hindle, 1990. Organic matter digestibility in horses and its prediction. Neth. J. Agric. Sci. 38, 435-447.

Part 11. Influence of energy/protein metabolism and nutrition on product quality

Impact of maternal and early postnatal nutrition on muscle growth and carcass quality in the pig

C. Rehfeldt, C. Kalbe and D. Lösel
Research Unit Muscle Biology & Growth, Leibniz Institute for Farm Animal Biology (FBN), Wilhelm-tahl-Allee 2, 18196 Dummerstorf, Germany

Abstract

Imbalanced maternal nutrient supply may affect foetal growth. Pigs with intrauterine growth retardation (IUGR) and a lower number of muscle fibres due to natural variation of foetal growth in this litter-bearing species exhibit less potential to develop high carcass and meat quality. Related to myogenesis, both maternal and early postnatal nutrition may have permanent effects on lean growth and body composition. More precisely, restriction of energy or protein supply to sows during gestation has long-term consequences for growth performance and carcass quality of the progeny, whereas overfeeding with energy or protein seems to have only marginal effects. Maternal energy restriction reduces the overall growth of the progeny, whereas protein restriction selectively impairs skeletal muscle growth, which results in lower relative lean and higher relative fat accretion and consequently suboptimal carcass composition. Supplementation of the maternal diet with arginine may provide a novel approach to prevent IUGR by enhancing the transfer of nutrients from mother to foetus or to improve postnatal growth performance and skeletal muscle growth. Carnitine supplementation to gestating sows or to piglets of low birth weight probably provides a possibility to compensate for impaired foetal growth. Results from research, such as those described in this review, will enable a more complete understanding of the interaction between maternal/early postnatal nutrition, skeletal muscle growth and carcass quality in pigs.

Introduction

Inadequate nutrition *in utero* may have serious consequences for foetal development and result in intrauterine growth retardation (IUGR). This phenomenon is observed in the pig as a litter bearing species. Genetic selection for greater litter size has increased the number of piglets born per litter during the last 10-20 years associated with increased variation in birth weight and a higher number of light piglets (Nissen and Oksbjerg, 2009). IUGR in turn is known to exert long-term effects on the adult organism, which is also termed 'foetal programming' in that postnatal growth, in particular lean growth, and ultimate carcass quality are adversely affected (Bee, 2004; Gondret *et al.*, 2006; Rehfeldt *et al.*, 2008c), which is of great economical importance. To develop strategies to manage prenatal environment to enhance pig production, it is important to understand the factors and mechanisms that affect foetal growth and muscle development and their consequences for ultimate product quality.

Foetal development and myogenesis

Prenatal growth in the pig, as in other mammals, is determined by the genotype of the conceptus, but largely depends on the maternal uterine milieu of hormones, nutrients, and growth factors (Wu *et al.*, 2006). The supply of nutrients to the embryo/foetus and its capacity to utilize the available substrates is of major importance for foetal growth. Nutrient partitioning and utilization in the foeto-maternal unit are under the control of hormones and growth factors although, conversely, nutrition may also influence the hormonal status (Brameld *et al.* 1998; Robinson *et al.*, 1999). Maternal characteristics, as determined by genetic (breed, genotype) and environmental (nutrition, housing, etc.) factors, further modify these interactions. The maternal diet controls foetal growth directly by providing glucose, amino acids, and other essential nutrients and metabolites for the conceptus (Robinson *et al.*, 1999; Wu *et al.*, 2006). These are transferred across the placenta by passive and active transport mechanisms. The major substrates reaching the *porcine* foetus are glucose, placentally-derived lactate and fructose, and amino acids, while transfer of fatty acids by the epitheliochorial placenta is very low in the pig (Pere, 2003). As the placenta is central to extrinsic regulation of foetal growth, all maternal factors that influence placental function are of major importance for foetal growth (Ashworth *et al.*, 2001).

Prenatal skeletal muscle development (myogenesis) has an irreversible impact on postnatal growth and muscle accretion, which are main constituents of carcass and meat quality in the pig (Figure 1; Rehfeldt *et al.*, 2004b; Stickland *et al.*, 2004). Prenatal myogenesis is highly significant for postnatal growth, carcass and meat quality, because it determines the number of skeletal muscle fibres. The postnatal increase in skeletal muscle mass results mainly from an increase in muscle fibre size, which in turn is limited by genetic and physiological factors. Muscle structural and functional properties affect an animal's growth performance and influence meat quality traits *post mortem* (Larzul *et al.*, 1997; Rehfeldt *et al.*, 2000; Lefaucheur *et al.*, 2001).

Skeletal myogenesis and its control has been the subject of several comprehensive reviews (Florini *et al.*, 1996; Brameld *et al.*, 1998; Buckingham, 2001; Maltin *et al.*, 2001; Lefaucheur *et al.*, 2001; Wigmore and Evans, 2002; Picard *et al.*, 2002). The elementary events during myogenesis are stem cell commitment, proliferation and apoptosis of myoblasts, differentiation and fusion of myoblasts to myotubes and, finally, their maturation into muscle fibres. Post-myogenic muscle growth is characterised by an increase in length and diameter of myofibres, particularly during postnatal life. Once formed, the multinucleated myofibres commence production of myofibrillar proteins. The myofibre types are distinguished by different isoforms of myofibrillar myosin associated with different contractile and metabolic properties. Throughout life, the myofibres remain in a state of dynamic adaptation in response to hormones, mechanical activity and innervation, which modulate differential gene expression during functional acclimatisation.

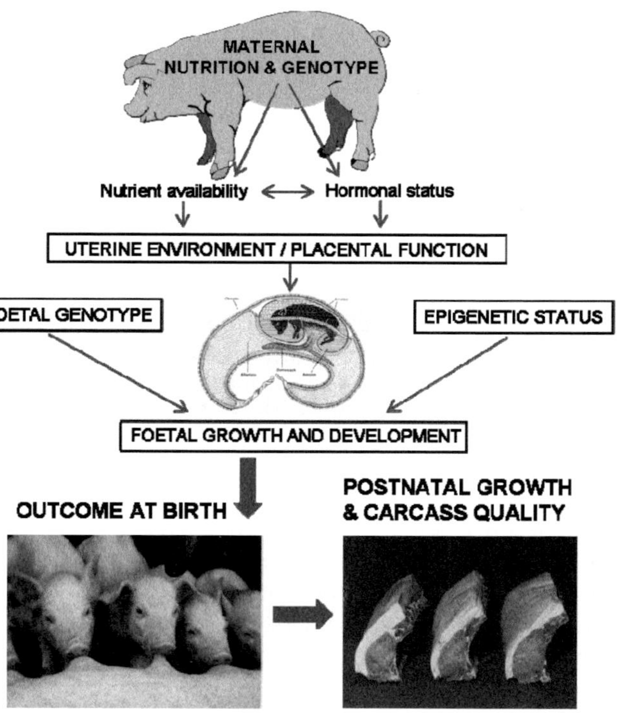

Figure 1. Regulation of foetal growth and development and consequences on the postnatal phenotype in the pig. The arrows indicate the direction of effects. This schematic does neither include potential feedback by foetus to the dam, nor interactions among foetuses.

Porcine primary myofibres form between about days 35 to 60, and secondary myofibres between about days 55 to 90 days of gestation (Wigmore and Stickland, 1983a; Lefaucheur *et al.*, 1995). A third generation of very small diameter fibres forms shortly after birth (Lefaucheur *et al.*, 1995), which may explain increases in total fibre number observed in porcine muscle between birth and 5 weeks of age (Rehfeldt *et al.*, 2000). About 20 secondary myofibres form around each primary fibre in the pig. Primary myofibres comprise slow myosin heavy chain and exhibit slow myosin ATPase activity from about day 60 of gestation. During late gestation, some secondary fibres located adjacent to single primary fibres, beside some exceptions, convert to the slow type and thereby generate clusters of slow myofibres, which is typical for porcine muscle (Handel and Stickland, 1987a,b; Lefaucheur *et al.*, 1995).

Factors that impact foetal and early postnatal development and consequences for growth and carcass quality

Uterine conditions/capacity

In the pig, foetal and birth weights have been shown to be inversely correlated with litter size (Milligan *et al.*, 2002; Quiniou *et al.*, 2002; Berard *et al.*, 2008) indicating competition for nutrients among littermates *in utero*. The primary reason why foetal growth is compromised is the lack of essential nutrients reaching the foetus. This proposition is further supported by the relationship between foetal size and placental size (Michael *et al.*, 1983; Ashworth *et al.*, 1999) or placental blood flow (Wootton *et al.*, 1977; Pere and Etienne, 2000). Pigs of low birth weight exhibit reduced myofibre and myonuclei numbers and lower DNA content in their muscles (Wigmore and Stickland, 1983a,b; Handel and Stickland, 1987a,b; Dwyer and Stickland, 1991; Rehfeldt and Kuhn, 2006). In addition, low birth weight pigs show reduced growth and accretion of lean tissue, but higher levels of fat deposition and tend to develop poorer meat quality at market weight (Powell and Aberle, 1981; Bee, 2004; Gondret *et al.*, 2006; Rehfeldt and Kuhn, 2006; Rehfeldt *et al.*, 2008c; Fix *et al.*, 2010). Reduced meat quality in terms of tenderness scores, drip loss, pH value 45 min *post mortem*, and impedance values have been recently reported, whereas intramuscular fat content was highest in low birth weight pigs in consistency with their higher degree of fatness (Gondret *et al.*, 2006; Rehfeldt and Kuhn, 2006; Rehfeldt *et al.*, 2008c). Conclusively, pigs with compromised foetal growth with a low number of myofibres exhibit less potential to develop a high carcass and meat quality.

Imbalanced maternal nutrition (energy and protein)

Nutrient supply to the foetus depends on maternal nutrient supply. Therefore, imbalanced nutrition of the gestating sow by energy and/or protein intake may result in IUGR.

Progeny of gilts or sows receiving a diet deficient in energy (Robinson, 1969; Buitrago, 1974; Pond *et al.*, 1986; Pond and Mersmann, 1988) during gestation had reduced foetal and birth weights. Low energy diets (to ~50% or ~28% of the standard) were associated with a reduction in muscle weights, skeletal myofibre and myonuclear numbers and muscle protein accretion in newborn and young pigs (Robinson, 1969; Buitrago *et al.*, 1974), which indicates the possibility of a permanent effect on skeletal muscle growth. Indeed, slower growth, back fat and lean tissue accretion in the progeny demonstrated the long-term effects of severe undernutrition (down to ~35-42%) during pregnancy with the lean percentage not being affected (Pond and Mersmann, 1988). Liver, kidneys and stomach were lighter, whereas heart and brain cortex tended to be heavier in this progeny. Mild maternal feed restriction to 60% of the standard during early gestation had only marginal impact on offspring growth (Bee, 2004). The progeny of the low energy fed sows exhibited even faster early postnatal growth and higher weaning weight, but there were no effects on muscle weights and carcass quality at market weight.

As reviewed by Rehfeldt *et al.* (2004a) and Nissen and Oksbjerg *et al.* (2009) an increase in sow feed intake above standard requirements during various periods of gestation did not result in any beneficial effects on foetal growth. The effects on postnatal growth, feed efficiency and myofibre number in the offspring are inconsistent and reveal positive, negative or no effects, which does not allow clear conclusion.

When gilts or sows received diets deficient (0; 0.5; 0.7; 6.5; or 8.5%) in protein (Pond *et al.* 1969; 1986; 1987; 1991; 1992; Mahan, 1977; Schoknecht *et al.*, 1993; 1994; Kalbe *et al.*, 2010) foetal and birth weights of the progeny were reduced. Studies on restricted protein feeding have revealed decreases in placental growth (Pond *et al.*, 1991), amino acid concentration in allantoic fluid and foetal plasma (Wu *et al.*, 1998b) and in placental and endometrial synthesis of nitric oxide and polyamines (Wu *et al.*, 1998a). These results suggest that protein deficiency may impair placental transport of amino acids from the maternal to the foetal blood. In response to maternal protein restriction to 0.7% foetuses of day 63 of gestation exhibited lower weights of liver, kidneys, gastrointestinal tract, cerebrum and cerebrellum as well as a higher DNA concentration in longissimus muscle indicative of lower protein synthetic activity (Pond *et al.*, 1991). With the exception of cerebrellum organ weights were also reduced in newborn piglets. When gilts were fed 6.5% instead of 12.1% crude protein throughout gestation birth weight and its constituents were reduced in newborn piglets at day 1 of age (Rehfeldt *et al.*, 2008a; Kalbe *et al.*, 2010). Absolute and relative weights of the kidneys were mostly affected, but also throat, tongue, diaphragm, colon and stomach were lighter. The relative weight of internal organs, however, tended to be increased. When adjusted to equal birth weight there were only little differences suggesting that all components of body growth proportionally decreased. Along with birth weight, the semitendinosus (ST) muscle was lighter, which was associated with a lower total number of myofibres, total DNA, protein and creatine kinase activity. The expression of the embryonic MyHC isoform in longissimus muscle and the RNA/protein ratio as an average of three muscles tended to be greater. These results suggest that prenatal myogenesis was negatively affected resulting in less mature skeletal muscle at birth. Interestingly, the number of cells in terms of total DNA was reduced in loin subcutaneous (SC) fat.

Decreased postnatal growth performance was reported in response to severe maternal protein restriction (Pond *et al.*, 1969; 1987; Schoknecht *et al.*, 1993), and pig weight gain during suckling was also affected at less severe restriction (Kusina *et al.*, 1999). However, this may be largely related to a decrease in the milk yield of the sow. When the piglets from gilts fed 6.5% instead of 12.1% crude protein during gestation were cross-fostered to adequately fed gilts, postnatal growth of the progeny was only numerically reduced (Rehfeldt *et al.*, 2008b). *Vice versa*, piglets of adequately fed sows and then cross-fostered to protein restricted (0.5%) sows tended to weigh less at weaning (Schoknecht *et al.*, 1993).

Carcass quality in response to protein restriction has been examined in a few studies. At market weight pigs from sows severely protein restricted (< 0.5%) throughout gestation had reduced indices of carcass lean, with carcass fat not affected, which was not the case in offspring of sows that had been protein-restricted during specific periods of gestation only (Pond *et al.*, 1987; Schoknecht *et al.*, 1993). Both loin area and backfat thickness, and weight of some internal organs (heart, liver, kidneys, stomach and small intestine) were reduced in mature female offspring (119 weeks) from sows that had been severely restricted throughout gestation (Pond *et al.*, 1990). It was concluded that protein restriction throughout pregnancy resulted in permanent whole body weight stunting of progeny. Carcass weight tended to be reduced and significantly lower carcass lean and higher carcass fat percentages were observed in pigs at market weight born to gilts fed at a low (6.5%) vs. an adequate (12.1%) protein level (Rehfeldt *et al.*, 2008b) suggesting that the lower potential for skeletal muscle growth observed in newborn piglets was accompanied by increased fat deposition as it is known from piglets of low birth weight by natural variation (see 3.1). ST muscle tended to be lighter, contained less DNA, and numerically both ST and longissimus muscle contained still less fibres. Importantly, the absolute weights of the heart and lungs were reduced and that of the kidneys tended to be smaller.

We have recently studied the effects of excess dietary protein (30% vs. 12.1%) in the maternal sow diet (Rehfeldt *et al.*, 2008a; Kalbe *et al.*, 2010) as we know from studies on rats that this may also impair foetal growth (Daenzer *et al.*, 2002). Indeed, birth weight was reduced to a similar degree as found in response to the low (6.5%) protein diet. Surprisingly, as early as the first day of life no significant difference in body weight and any constituents remained compared with control piglets. Markers of prenatal myogenesis were not negatively affected. Interestingly, total DNA in loin SC fat tended to be decreased suggesting a suboptimal adipose tissue development *in utero*. At market weight these pigs did not differ from the controls except of lower absolute and relative weights of the lungs and relative liver weight. Structural and biochemical analyses of muscle tissues revealed no effects except a tendency for lower density of capillaries.

It is to be expected that also early postnatal nutrition has permanent effects on lean tissue accretion as there is an enormous increase in myofibre number during the 2 to 3 first weeks of age in the pig. Pigs given a low (14%) protein diet from 1.8 to 15 kg live weight grew at a slower rate and contained more fat and less protein than pigs given a high (25%) protein diet, which was associated with reduced muscle weight and muscular DNA content and less DNA and greater fat cell size in SC adipose tissue (Campbell and Dunkin, 1983a,b). Subsequent growth rate was diminished in that these pigs needed longer time to achieve 75 kg live weight both with high (+26 d) and low (+14 d) energy feeding, whereas final body composition did not differ.

Maternal and early postnatal supplementation with specific nutrients

Arginine

Besides being a proteinogenic amino acid, l-arginine is a precursor for the formation of creatine, proline, glutamate, polyamines, and nitric oxide (NO), and thus plays an important role in physiological function and homeostasis in animals (for review see Wu *et al.*, 2007; Wu *et al.*, 2009).

Arginine supplementation has been shown to improve postnatal growth performance and skeletal muscle growth in pigs (He *et al.*, 2009; Tan *et al.*, 2009). In addition, there is increasing evidence that arginine-dependent metabolic pathways play also an important role in pig foetal development. This has been derived from the highly enrichment of arginine, ornithine and glutamine in the allantoic fluid at day 40 of gestation which was associated with the greatest rates of polyamine and NO synthesis in the first half of gestation stimulating placental angiogenesis and growth (Wu *et al.*, 2007). Indeed, dietary supplementation of gilts with 1% l-arginine-HCl from d 30-114 of gestation increased the number of piglets born alive by 2 per litter, without affecting individual piglet birth weight (Mateo *et al.*, 2007). Thus, a negative correlation between litter size and birth weight was no longer observed. Similarly, Ramaekers *et al.* (2006) found that the addition of 25 g/d of arginine to the basal diet of sows from d 14 to 28 of gestation increased litter size by 0.8 to 1.08 piglets. Supplementing sows with 26 g/d of arginine from d 14 to 28 of gestation increased the number of viable foetuses by 4 on day 75 of gestation without affecting the weight of individual foetuses (Berard and Bee, 2010). In addition, the number of primary myofibres in ST muscle was increased in foetuses of arginine-supplemented sows, whereas the number of secondary fibres and total fibre number were not influenced. However, as the number of secondary fibres is not fixed at day 75 of gestation and as the secondary fibres form around primary fibres, the authors suppose that the foetuses of arginine-supplemented sows may still increase total fibre number until birth. Signs of increased myogenic proliferation and delayed differentiation (by DNA, protein, creatine kinase activity, MYF5 mRNA expression) have been observed in these foetuses (Kalbe *et al.*, 2009). These studies on the supplementation of gestating sows and gilts implicate a possibility to improve reproductive performance of sows (in terms of litter size) without affecting the growth performance of the offspring negatively. If the total myofibre number would indeed be increased in the progeny, arginine-supplementation of sows may even have the potential to improve postnatal growth, muscle development, carcass and meat quality of the offspring.

In suckling piglets, plasma concentrations of arginine and its precursors, ornithine and citrulline, decrease from days 3 to 14 of life. At the same time increased ammonia concentrations and reduced nitrate and nitrite concentrations in plasma indicated a reduced ammonia detoxification and systemic NO synthesis (Flynn *et al.*, 2000). This obvious arginine deficiency is due to a decline in synthesis of arginine and citrulline from glutamine and proline in enterocytes (Wu, 1997). In addition, sow's milk does not meet the arginine requirement of suckling piglets (Wu *et al.*, 2000). These observations made arginine an essential amino acid for maximal growth of suckling piglets. In fact, in piglets weaned at d 7 of life, the addition of 0.2% and 0.4% arginine to milk replacer powder increased weight gain from d 7 to 21 after birth (Kim *et al.*, 2004). In addition, increased plasma concentrations of insulin and growth hormone in the 0.4% group as well as reduced urea concentration indicated that arginine improved the efficiency of nutrient utilization for enhancing tissue protein synthesis and growth performance. Milk-replacer supplemented with 0.6% of arginine fed to piglets from d 7 to 14 of life, increased body weight gain, plasma insulin concentration, and fractional protein synthesis rate in longissimus muscle associated with a stimulation of mTOR signalling pathways (Yao *et al.*, 2008).

To study whether arginine is effective in stimulating the early postnatal increase in myofibre number, we supplemented suckling piglets of low birth weight from days 7 to 28 of age with 0.48 g/kg body weight/d arginine-HCl (Lösel *et al.*, 2010). Preliminary results reveal no increases in body weight and myofibre number, but a trend for higher total DNA and slightly advanced maturity of skeletal muscle by creatine kinase and LDH activities. In summary, with regard to the importance of adequate development during the suckling period for later growth performance, arginine supplementation to suckling piglets may be beneficial in enhancing the efficiency of feed utilization for protein accretion.

Carnitine

l-carnitine is essential for the transport of long-chain fatty acids into the mitochondrial matrix, where β-oxidation of activated fatty acids occurs. Besides this catalytic function, carnitine regulates the mitochondrial acetyl-CoA:CoA-SH ratio by buffering excess acetyl groups and thereby maintaining a pool of free CoA for the reactions in the citric acid cycle. Thus, carnitine plays a central role in fat and carbohydrate metabolism (for review see Stephens *et al.*, 2007). Carnitine is provided by absorption from the diet and by endogenous biosynthesis in the liver and the kidneys (Evans and Fornasini, 2003).

Carnitine supplementation to weaning and growing-finishing pigs has been shown to improve growth performance and to alter body composition in terms of reduced fat accretion and increased protein accretion (e.g. Owen *et al.*, 1996; 2001). Supplementation of sows or gilts with carnitine during gestation and/or lactation revealed effects on their reproductive performance (for review see Eder, 2009). Some studies report an increased number of piglets born alive (Ramanau *et al.*, 2004; 2008), whereas this effect of maternal carnitine-supplementation during gestation was not observed in other studies (Doberenz *et al.*, 2006; Musser *et al.*, 1999; 2007). Inconsistent results were also obtained with respect to birth weight that was either increased (Musser *et al.*, 1999; Eder *et al.*, 2001; Birkenfeld *et al.*, 2006b; Ramanau *et al.*, 2008) or unaffected (Doberenz *et al.*, 2006). In one study, birth weight of piglets born to carnitine-supplemented sows was even decreased, but these piglets displayed compensatory growth during the suckling period and were heavier at weaning than control piglets (Ramanau *et al.*, 2004). Interestingly, the total muscle fibre number in ST muscle of newborn piglets born to sows supplemented with 50 ppm carnitine during gestation was increased which is indicative of a greater maturity at birth (Musser *et al.*, 2001). Total myofibre number was not determined in the study of Ramanau *et al.* (2006), and carnitine supplementation during gestation and lactation did not affect fibre size and fibre type distribution in longissimus and ST muscles of weaning piglets on day 28 of life.

There is also evidence that carnitine-supplementation to sows improves early postnatal performance of their offspring. Carnitine-supplementation during gestation and lactation increased weight gain and individual piglet weaning weight (Musser *et al.*, 1999; Eder *et al.*, 2001; Ramanau *et al.*, 2004; 2008). This effect was attributed to the increased milk production of sows supplemented with carnitine during gestation (Ramanau *et al.*, 2004) resulting from an intensified suckling activity of the piglets (Birkenfeld *et al.*, 2006a). The authors suggested that the increased suckling persistence indicated a greater vitality resulting from the improved energy balance during the early postnatal phase or from an enhanced intrauterine development. Changes in the maternal and foetal IGF system (Musser *et al.*, 1999; Doberenz *et al.*, 2006; Brown *et al.*, 2008) and increased placental capacity for glucose transport (Doberenz *et al.*, 2006) are likely responsible for the positive effects of maternal carnitine-supplementation. It was also shown that carnitine has direct effects on the metabolism of pig foetuses (Xi *et al.*, 2008). After weaning, no effect of maternal carnitine-supplementation on growth performance was observed (Birkenfeld *et al.*, 2005; Ramanau *et al.*, 2006). In addition, body composition at slaughter was not affected (Ramanau *et al.*, 2006). In contrast, Musser *et al.* (2007) found an increased loin depth and percentage lean at day 179 of life in offspring of carnitine-supplemented sows.

Whereas the effects of maternal carnitine supplementation are well described, the number of studies on the influence of additional carnitine given to neonatal piglets is limited. Carnitine supplementation to sows during lactation (50-200 ppm) increased the carnitine concentration in milk (Ramanau *et al.*, 2004; Birkenfeld *et al.*, 2006a,b). In fact, this indirect carnitine supplementation to piglets did not increase litter weight at weaning on day 21. Similarly, piglets born to control sows, but suckled by sows that received carnitine during gestation and lactation, did not differ in body weight gain from piglets suckled by control sows (Birkenfeld *et al.*, 2006a). Even in piglets that were fed a soybean-based (i.e. free of milk-derived compounds) replacement diet with low or high fat content from d 3 of life, the addition of 800 ppm carnitine did not affect growth performance until d 21 of life (Hoffman *et al.*, 1993). Similar to the above studies, oral administration of 400 mg/d carnitine from day 7 to 27 of life to suckling piglets of low and medium birth weight did not increase weaning weight (Lösel *et al.*, 2009). However, similar to growing and finishing pigs, fat deposition was reduced in response to treatment. In addition, in piglets of low birth weight, the total myofibre number in ST muscle was increased by 13% compared with the control group thereby reaching the unchanged level of medium birth weight piglets. This effect was attributed to an improved energy balance through intensified fatty acid oxidation which may have stimulated myogenic proliferation and thus enabled compensatory postnatal myofiber formation in disadvantaged piglets. Thus, carnitine supplementation to piglets of low birth weight probably provides a means to attenuate the negative consequences of low birth weight on carcass and meat quality at market weight.

References

Ashworth, C.J., A.M. Finch, K.R. Page, M.O. Nwagwu and H.J. McArdle, 2001. Causes and consequences of fetal growth retardation in pigs. Reprod. Suppl. 58, 233-246.

Ashworth, C.J. and H.J. McArdle, 1999. Both placental amino acid uptake and fetal amino acid concentrations differ between small and normally grown porcine fetuses. Early Hum. Dev. 54, 90-91.

Bee, G., 2004. Effect of early gestation feeding, birth weight, and gender of progeny on muscle fiber characteristics of pigs at slaughter. J. Anim. Sci. 82,826-836.

Berard, J. and G. Bee, 2010. Effects of dietary l-arginine supplementation to gilts during early gestation on foetal survival, growth and myofiber formation. Animal, in press.

Berard, J., M. Kreuzer, and G. Bee, 2008. Effect of litter size and birth weight on growth, carcass and pork quality, and their relationship to *post mortem* proteolysis. J Anim Sci. 86:2357-2368.

Birkenfeld, C., H. Kluge and K. Eder, 2006a. L-carnitine supplementation of sows during pregnancy improves the suckling behaviour of their offspring. Br. J. Nutr. 96, 334-342.

Birkenfeld, C., J. Doberenz, H. Kluge and K. Eder, 2006b. Effect of l-carnitine supplementation of sows on l-carnitine status, body composition and concentrations of lipids in liver and plasma of their piglets at birth and during the suckling period. Anim. Feed Sci. Technol. 129, 23-38.

Birkenfeld, C., A. Ramanau, H. Kluge, J. Spilke and K. Eder, 2005. Effect of dietary L-carnitine supplementation on growth performance of piglets from control sows or sows treated with L-carnitine during pregnancy and lactation. J. Anim. Physiol. Anim. Nutr. 89, 277-283.

Brameld, J.M., P.J. Buttery, J.M. Dawson and J.M.M. Harper, 1998. Nutritional and hormonal control of skeletal-muscle cell growth and differentiation. Proc. Nutr. Soc. 57, 207-217.

Brown K.R., R.D. Goodband, M.D. Tokach, S.S. Dritz, J.L. Nelssen, J.E. Minton, J.J. Higgins, X. Lin, J. Odle, J.C. Woodworth, and B.J. Johnson, 2008. Effects of feeding L-carnitine to gilts through day 70 of gestation on litter traits and the expression of insulin-like growth factor system components and L-carnitine concentration in foetal tissues. J. Anim. Physiol. Anim. Nutr. 92, 660-667.

Buckingham, M., 2001. Skeletal muscle formation in vertebrates. Curr. Opin. Genet. Dev. 11, 440-448.

Buitrago, J.A., E.F. Walker Jr., W.I. Snyder and W.G. Pond, 1974. Blood and tissue traits in pigs at birth and at 3 weeks from gilts fed low or high energy diets during gestation. J. Anim. Sci. 38, 766-771.

Campbell, R.G. and A.C. Dunkin, 1983a. Influence of feeding level and of protein level in early life on the cellularity of adipose tissue and body fat content of growing pigs. Br. J. Nutr. 49, 109-118.

Campbell, R.G. and A.C. Dunkin, 1983b. The influence of protein nutrition in early life on growth and development of the pig. 2. Effects on the cellularity of muscle and subcutaneous adipose tissue. Br. J. Nutr. 50, 619-626.

Daenzer, M., S. Ortmann, S. Klaus and C.C. Metges, 2002. Prenatal high protein exposure decreases energy expenditure and increases adiposity in young rats. J. Nutr. 132, 142-144.

Doberenz, J., C. Birkenfeld, H. Kluge and K. Eder, 2006. Effects of l-carnitine supplementation in pregnant sows on plasma concentrations of insulin-like growth factors, various hormones and metabolites and chorion characteristics. J. Anim. Physiol. Anim. Nutr. 90, 487-499.

Dwyer, C.M. and N.C. Stickland, 1991. Sources of variation in myofiber number within and between litters of pigs. Anim. Prod. 52, 527-533.

Eder, K., 2009. Influence of l-carnitine on metabolism and performance of sows. Br. J. Nutr. 102, 645-654.

Eder, K., A. Ramanau and H. Kluge, 2001. Effect of L-carnitine supplementation on performance parameters in gilts and sows. J. Anim. Physiol. Anim. Nutr. 85, 73-80.

Evans, A.M. and G. Fornasini, 2003. Pharmacokinetics of L-carnitine. Clin. Pharmacokinet. 42, 941-967.

Fix, J.S., J.P. Cassady, W.O. Herring, J.W. Holl, M.S. Culbertson and M.T. See. 2010. Effect of piglet birth weight on body weight, growth, backfat, and longissimus muscle area of commercial market swine. Livest. Sci. 127, 51-59.

Florini, J.R., D.Z. Ewton and S.A. Coolican, 1996. Growth hormone and the insulin-like growth factor system in myogenesis. Endocr. Rev. 17, 481-517.

Flynn, N.E., D.A. Knabe, B.K. Mallick and G. Wu, 2000. Postnatal changes of plasma amino acids in suckling pigs. J. Anim. Sci. 78, 2369-2375.

Gondret, F., L. Lefaucheur, H. Juin, I. Louveau and B. Lebret, 2006. Low birth weight is associated with enlarged muscle fiber area and impaired meat tenderness of the longissimus muscle in pigs. J. Anim. Sci. 84, 93-103.

Handel, S.E. and N.C. Stickland, 1987a. Muscle cellularity and birthweight. Anim. Prod. 44, 311-317.

Handel, S.E. and N.C. Stickland, 1987b. The growth and differentiation of porcine skeletal muscle fibre types and the influence of birthweight. J. Anat. 152, 107-119.

He, Q., X. Kong, G. Wu, P. Ren, H. Tang, F. Hao, R. Huang, T. Li, B. Tan, P. Li, Z. Tang, Y. Yin and Y. Wu, 2009. Metabolomic analysis of the response of growing pigs to dietary l-arginine supplementation. Amino Acids 37, 199-208.

Hoffman, L.A., D.J. Ivers, M.R. Ellersieck and T.L. Veum, 1993. The effect of L-carnitine and soybean oil on performance and nitrogen and energy utilization by neonatal and young pigs. J. Anim. Sci. 71, 132-138.

Kalbe, C., J. Block, L. Lefaucheur, B. Stabenow, W. Otten, C.C. Metges and C. Rehfeldt, 2010. Effects of maternal protein intake during pregnancy on body composition and skeletal muscle properties of newborn piglets. Proc. Soc. Nutr. Physiol. 19, 22.

Kalbe, C., M. Porm, J. Bérard, G. Bee and C. Rehfeldt, 2009. Effects of arginine supplementation to gilts during early gestation on fetal myogenesis. J. Anim. Sci. 87, E-Suppl 2/J. Dairy Sci. 92, E-Suppl. 1, 235.

Kim, S.W., R.L. McPherson and G. Wu, 2004. Dietary arginine supplementation enhances the growth of milk-fed young pigs. J. Nutr. 134, 625-630.

Kusina, J., J.E. Pettigrew, A.F. Sower, M.E. White, B.A. Crooker and M.R. Hathaway, 1999. Effect of protein intake during gestation and lactation on the lactational performance of primiparous sows. J. Anim. Sci. 77, 931-941.

Larzul, C., L. Lefaucheur, P. Ecolan, J. Gogue, A. Talmant, P. Sellier, P. Le Roy and G. Monin, 1997. Phenotypic and genetic parameters for longissimus muscle fiber characteristics in relation to growth, carcass, and meat quality traits in large white pigs. J. Anim. Sci. 75, 3126-3137.

Lefaucheur, L., 2001. Myofiber typing and meat production. Slov. Vet. Res. 38, 5-28.

Lefaucheur, L., F. Edom, P. Ecolan and G.S. Butler-Browne, 1995. Pattern of muscle-fiber type formation in the pig. Dev. Dyn. 203, 27-41.

Lösel, D., C. Kalbe and C. Rehfeldt, 2009. L-carnitine supplementation during suckling intensifies the early postnatal skeletal myofiber formation in piglets of low birth weight. J. Anim. Sci., 87, 2216-2226.

Lösel, D., S. Görs and C. Rehfeldt, 2010. Effects of l-arginine supplementation to suckling piglets on plasma metabolites and skeletal muscle properties at weaning. J. Anim. Sci. (Abstract), accepted.

Mahan, D.C., 1977. Effect of feeding various gestation and lactation dietary protein sequences on long-term reproductive performance in swine. J. Anim. Sci. 45, 1061-1072.

Maltin, A., M.I. Delday, K.D. Sinclair, J. Steven and A.A. Sneddon, 2001. Impact of manipulations of myogenesis in utero on the performance of adult skeletal muscle. Reproduction 122, 359-374.

Mateo, R.D., G. Wu, F.W. Bazer, J.C. Park, I. Shinzato and S.W. Kim, 2007. Dietary l-arginine supplementation enhances the reproductive performance of gilts. J. Nutr. 137, 652-656.

Michael, B.S. Ward and W.M. Moore, 1983. Relationship of fetal to placental size, the pig model. Eur. J. Obstet. Gynecol. Reprod. Biol. 16, 53-62.

Milligan, N., D. Fraser and D.L. Kramer, 2002. Within-litter birth weight variation in the domestic pig and its relation to pre-weaning survival, weight gain, and variation in weaning weights. Livest. Prod. Sci. 76, 181-191.

Musser, R.E., S.S. Dritz, D.L. Davis, M.D. Tokach, J.L. Nelssen, R.D. Goodband and K.Q. Owen, 2007. Effects of L-carnitine in the gestating sow diet on fetal muscle development and carcass characteristics of the offspring. J. Appl. Anim. Res. 31, 105-111.

Musser, R.E., R.D. Goodband, M.D. Tokach, K.Q. Owen, J.L. Nelssen, S.A. Blum, S.S. Dritz and C.A. Civis, 1999. Effects of L-carnitine fed during gestation and lactation on sow and litter performance. J. Anim. Sci. 77, 3289-3295.

Musser, R.E., R.D. Goodband, K.Q. Owen, D.L. Davis, M.D. Tokach, S.S. Dritz and J.L. Nelssen, 2001. Determining the effect of increasing L-carnitine additions on sow performance and muscle fiber development of the offspring. J. Anim. Sci. 79 (Suppl. 2), 157.

Nissen, P.M. and N. Oksbjerg, 2009. Quantification and prenatal effects on productivity in pigs. In: Managing Prenatal Development to Enhance Livestock Productivity. Pages 203-241. Editors P.L. Greenwood, A.W. Bell, P.E. Vercoe and G.J. Viljoen. Springer Dordrecht, Heidelberg, London, New York., 37-69.

Owen, K.Q., H. Jit, C.V. Maxwell, J.L. Nelssen, R.D. Goodband, M.D. Tokach, G.C. Tremblay and S.I. Koo, 2001b. Dietary L-carnitine suppresses mitochondrial branched-chain keto acid dehydrogenase activity and enhances protein accretion and carcass characteristics of swine. J. Anim. Sci. 79, 3104-3112.

Owen, K.Q., J.L. Nelssen, R.D. Goodband, T.L. Weeden and S.A. Blum, 1996. Effect of L-carnitine and soybean oil on growth performance and body composition of early-weaned pigs. J. Anim. Sci. 74, 1612-1619.

Pere, M.C., 2003. Materno-foetal exchanges and utilisation of nutrients by the foetus, comparison between species. Reprod. Nutr. Dev. 43, 1-15.

Pere, M. C. and M. Etienne. 2000. Uterine blood flow in sows: effects of pregnancy stage and litter size. Reprod. Nutr. Dev. 40, 369-382.

Picard, B., L. Lefaucheur, C. Berri and M.J. Duclos, 2002. Muscle fibre ontogenesis in farm animal species. Reprod. Nutr. Dev. 42, 415-431.

Pond, W.G., J.T. Yen and L.H. Yen, 1986. Response of nonpregnant versus pregnant gilts and their fetuses to severe feed restriction. J. Anim. Sci. 63, 472-483.
Pond, W.G., J.T. Yen and H.J. Mersmann, 1987. Effect of severe dietary protein, nonprotein calories or feed restriction during gestation on postnatal growth of progeny in swine. Growth 51, 355-371.
Pond, W.G., H.J. Mersmann and J.T. Yen, 1988. Maternal tissue repartitioning in pregnant primiparous swine in response to restriction of calories or feed. Proc. Soc. Exp. Biol. Med. 188, 128-135.
Pond, W.G., R.R. Maurer and J. Klindt, 1991. Fetal organ response to maternal protein deprivation during pregnancy in swine. J. Nutr. 121, 504-509.
Pond, W.G., R.R. Maurer, H.J. Mersmann and S. Cummins, 1992. Response of fetal and newborn piglets to maternal protein restriction during early or late pregnancy. Growth Dev. Aging 56, 115-127.
Pond, W.G., D.N. Strachan, Y.N. Sinha, E.F. Walker, Jr., J.A. Dunn and R.H. Barnes, 1969. Effect of protein deprivation of swine during all or part of gestation on birth weight, postnatal growth rate and nucleic acid content of brain and muscle of progeny. J. Nutr. 99, 61-67.
Pond, W.G., J.T. Yen, H.J. Mersmann and R.R. Maurer, 1990. Reduced mature size in progeny of swine severely restricted in protein-Intake during pregnancy. Growth Dev. Aging 54, 77-84.
Pond, W.G., J.T. Yen and L.H. Yen, 1986. Response of nonpregnant versus pregnant gilts and their fetuses to severe feed restriction. J. Anim. Sci. 63, 472-483.
Powell, S.E. and E.D. Aberle, 1981. Skeletal muscle and adipose tissue cellularity in runt and normal birth weight swine. J. Anim. Sci. 52, 748-756.
Quiniou, N., J. Dagorn and D. Gaudre, 2002. Variation of piglets birth weight and consequences on subsequent performance. Livest. Prod. Sci. 78, 63-70.
Ramaekers, P., B. Kemp and T. van der Lende, 2006. Progenos in sows increases number of piglets born. J. Anim. Sci. 84 (Suppl.1), 394.
Ramanau, A., H. Kluge, J. Spilke and K. Eder, 2004. Supplementation of sows with L-carnitine during pregnancy and lactation improves growth of the piglets during the suckling period through increased milk production. J. Nutr. 134, 86-92.
Ramanau, A., H. Kluge, J. Spilke and K. Eder, 2008. Effects of dietary supplementation of L-carnitine on the reproductive performance of sows in production stocks. Livest. Sci. 113, 34-42.
Ramanau, A., R. Schmidt, H. Kluge and K. Eder, 2006. Body composition, muscle fibre characteristics and postnatal growth capacity of pigs born from sows supplemented with L-carnitine. Arch. Anim. Nutr. 60, 110-118.
Rehfeldt, C., I. Fiedler, G. Dietl and K. Ender, 2000. Myogenesis and postnatal skeletal muscle cell growth as influenced by selection. Livest. Prod. Sci. 66, 177-188.
Rehfeldt, C., I. Fiedler and N.C. Stickland, 2004b. Number and size of muscle fibres in relation to meat production. p. 1-37. In M.F.W. Te Pas, M.E. Everts, and H.P. Haagsman, (eds.), Muscle Development of Livestock Animals, Physiology, Genetics, and Meat Quality, CAB Int., Wallingford, Oxon, UK.
Rehfeldt, C., C. Kalbe, J. Block, G. Nuernberg, B. Stabenow, D. Loesel, C.C. Metges, 2008a. Effects of maternal low and high protein diets on body composition and skeletal muscle properties of newborn piglets. J. Anim. Sci. 86 (E-Suppl.2)/J. Dairy Sci. 91 (E-Suppl. 1), 422.
Rehfeldt, C., C. Kalbe, J. Block, G. Nürnberg, B. Stabenow and C.C. Metges, 2008b. Long-term effects of low and high protein feeding to pregnant sows on offspring at market weight. Proc. Int. Congress Meat Sci.& Technol. 10-15 August 2008, Cape Town, South Africa, 7B.15 (CD).
Rehfeldt, C. and G. Kuhn. 2006. Consequences of birth weight for postnatal growth performance and carcass quality in pigs as related to myogenesis. J. Anim. Sci. 84 (E-Suppl), E113-E123.
Rehfeldt, C., P.M. Nissen, G. Kuhn, M. Vestergaard, K. Ender and N. Oksbjerg, 2004a. Effects of maternal nutrition and porcine growth hormone (pGH) treatment during gestation on endocrine and metabolic factors in sows, fetuses and pigs, skeletal muscle development, and postnatal growth. Domest. Anim. Endocrinol. 27, 267-285.
Rehfeldt, C., A. Tuchscherer, M. Hartung and G. Kuhn, 2008c. A second look at the influence of birth weight on carcass and meat quality in pigs. Meat Sci. 78, 170-175.
Robinson, D.W., 1969. The cellular response of porcine skeletal muscle to prenatal and neonatal nutritional stress. Growth 33, 231-240.
Robinson, J.J., K.D. Sinclair and T.G. McEvoy, 1999. Nutritional effects on foetal growth. Anim. Sci. 68, 315-331.

Schoknecht, P.A., G.R. Newton, D.E. Weise and W.G. Pond, 1994. Protein restriction in early-pregnancy alters fetal and placental growth and allantoic fluid proteins in swine. Theriogenology 42, 217-226.

Schoknecht, P.A., W.G. Pond, H.J. Mersmann and R.R. Maurer, 1993. Protein restriction during pregnancy affects postnatal growth in swine progeny. J. Nutr. 123, 1818-1825.

Stephens, F.B., D. Constantin-Teodosiu and P.L. Greenhaff, 2007. New insights concerning the role of carnitine in the regulation of fuel metabolism in skeletal muscle. J. Physiol. 581, 431-444.

Stickland, N.C., S. Bayol, C. Ashton and C. Rehfeldt, 2004. Manipulation of muscle fibre number during prenatal development. In M.F.W. Te Pas, M.E. Everts, and H.P. Haagsman, (eds.), Muscle Development of Livestock Animals, Physiology, Genetics, and Meat Quality, CAB Int., Wallingford, Oxon, UK: 69-82.

Tan, B., Y. Yin, Z. Liu, X. Li, H. Xu, X. Kong, R. Huang, W. Tang, I. Shinzato, S.B. Smith and G. Wu, 2009. Dietary l-arginine supplementation increases muscle gain and reduces body fat mass in growing-finishing pigs. Amino Acids 37, 169-175.

Wigmore, P.M. and D.J. Evans, 2002. Molecular and cellular mechanisms involved in the generation of fiber diversity during myogenesis. Int. Rev. Cytol. 216, 175-232.

Wigmore, P.M. and N.C. Stickland, 1983a. Muscle development in large and small pig fetuses. J. Anat. 137, 235-245.

Wigmore, P.M. and N.C. Stickland, 1983b. DNA, RNA and protein in skeletal muscle of large and small pig fetuses. Growth 47, 67-76.

Wootton, R., I.R. McFayden and J.E. Cooper, 1977. Measurement of placental blood flow in the pig and its relation to placental and fetal weight. Biol. Neonate 31, 333-339.

Wu, G., 1997. Synthesis of citrulline and arginine from proline in enterocytes of postnatal pigs. Am. J. Physiol. 272, G1382-1390.

Wu, G., F.W. Bazer, T.A. Davis, L.A. Jaeger, G.A. Johnson, S.W. Kim, D.A. Knabe, C.J. Meininger, T.E. Spencer and Y.-L. Yin, 2007. Important roles for the arginine family of amino acids in swine nutrition and production. Livest. Sci. 122, 8-22.

Wu, G., F.W. Bazer, T.A. Davis, S.W. Kim, P. Li, J.M. Rhoads, M.C. Satterfield, S.B. Smith, T.E. Spencer and Y. Yin, 2009. Arginine metabolism and nutrition in growth, health and disease. Amino Acids 37, 153-168.

Wu, G., F.W. Bazer, J.M. Wallace and T.E. Spencer, 2006. Intrauterine growth retardation: Implications for the animal sciences. J. Anim. Sci. 84, 2316-2337.

Wu, G., C.J. Meininger, D.A. Knabe, F.W. Bazer and J.M. Rhoads, 2000. Arginine nutrition in development, health and disease. Curr. Opin. Clin. Nutr. Metab. Care 3, 59-66.

Wu, G.Y., W.G. Pond, S.P. Flynn, T.L. Ott and F.W. Bazer, 1998. Maternal dietary protein deficiency decreases nitric oxide synthase and ornithine decarboxylase activities in placenta and endometrium of pigs during early gestation. J. Nutr. 128, 2395-2402.

Wu, G.Y., W.G. Pond, T.L. Ott and F.W. Bazer, 1998. Maternal dietary protein deficiency decreases amino acid concentrations in fetal plasma and allantoic fluid of pigs. J. Nutr. 128, 894-902.

Xi, L., K. Brown, J. Woodworth, K. Shim, B. Johnson and J. Odle, 2008. Maternal dietary L-carnitine supplementation influences fetal carnitine status and stimulates carnitine palmitoyltransferase and pyruvate dehydrogenase complex activities in swine. J. Nutr. 138, 2356-2362.

Yao K., Y.L. Yin, W. Chu, Z. Liu, D. Deng, T. Li, R. Huang, J. Zhang, B. Tan, W. Wang and G. Wu, 2008. Dietary arginine supplementation increases mTOR signalling activity in skeletal muscle of neonatal pigs. J. Nutr. 138, 867-872.

Lipid supplementation in small ruminant nutrition and dairy products quality: implications for human nutrition

M. Mele[1] and S. Banni[2]
[1]*Dipartimento di Agronomia e Gestione dell'Agroecosistema, Università di Pisa, Via S. Michele degli Scalzi, 2, Pisa, Italy*
[2]*Dipartimento di Biologia Sperimentale, Università di Cagliari, Monserrato, Cagliari, Italy*

Abstract

This review deals with: (1) the effects of dietary lipid supplementation on milk fatty acid composition in small ruminants, taking into consideration also the different responses observed among ruminant species, and (2) the implications of milk fat modification on human health. In the last years, an increasing number of studies have focused on lipid supplementation in dairy goats and ewes. Some studies have focused on vegetable lipid sources (i.e. linseed, sunflower, rapeseed, soybean, olive), either as whole seed (raw and processed) or as oil (protected and unprotected), whereas few of them have dealt with the effects of marine oil in combination with vegetable fat on milk composition in small ruminants. Responses to lipid supplementation can vary significantly among ruminant species, as a consequence of differences in eating behaviour, rumination, buffering of rumen pH, rumen digestion kinetics and rumen passage rate. Recently, the metabolic mechanisms regulating milk fat synthesis have been studied across ruminant species, by using the milk fat depression model. At present, effective nutritional strategies based on dietary lipid supplementation of small ruminants are available, to obtain dairy products enriched in conjugated linoleic acid (CLA), vaccenic acid (VA) and alpha-linolenic acid (LNA) Specific clinical studies are needed to provide the scientific evidences that enriched dairy products may be beneficial to human health.

Introduction

Lipid metabolism plays a central role in the energy metabolism of dairy ruminants. In fact, milk fat represents the highest (i.e. more than 50%) part of the energetic cost of milk synthesis (Bauman *et al.*, 1985), mainly due to energy requirements for fatty acids (FA) neosynthesis and desaturation. The interest in including lipid sources in the diet of dairy ruminants is not recent. Lipids are often used in ruminant diets mainly with the aim to meet the high energy requirements of the animals during the first phase of lactation. Indeed, the high caloric value of lipids can be useful to overcome limitations in energy supplies in high yielding ruminants (Chilliard, 1993). During the last decade, a lot of research in animal science has dealt with the use of dietary lipid supplements, as a consequence of the increasing interest in the improvement of milk FA profile for human health purposes. Since the discovery of the putative positive biological effects of several FA (i.e. conjugated linoleic acid (CLA), vaccenic acid (VA)) of ruminant milk on human health, several studies have aimed at elucidating the metabolic mechanisms that regulate mammary lipid synthesis and elaborating optimal nutrition strategies to enrich milk fat with beneficial FA. Together with pasture regimen, dietary lipid supplementation with vegetable or marine fat is considered the main dietary strategy, in order to improve milk FA composition in dairy ruminants (Mele, 2009). In the literature, the use of dietary vegetable and marine fat sources have been tested mostly on dairy cows (Chilliard *et al.*; 2007; Glasser *et al.*, 2008) and, to a smaller extent, on small ruminants, with very few studies on dairy ewes (Chilliard *et al.*, 2003; Pulina *et al.*, 2006). The present paper is a review of the role of lipid supplementation in dairy ewes and goats nutrition, as a tool to improve the healthfulness of dairy products, taking into consideration also the different responses to dietary fat supplementation among ruminants. In addition, the implications of milk fat modification on human health are discussed.

Effect of dietary lipid supplementation on milk fatty acids with putative beneficial effects on human health.

In the last years, an increasing number of studies on lipid supplementation in the diet of dairy goats and ewes have been published. Some studies have tested vegetable lipid sources (i.e. linseed, sunflower, rapeseed, soybean, olive), either as whole seed (raw and processed) or as oil (protected and unprotected) (Antongiovanni *et al.*, 2002; Chilliard *et al.*, 2003, 2007; Bernard *et al.*, 2005, 2009; Luna *et al.*, 2005; Mele *et al.*, 2006, 2007, 2008; Zhang *et al.*, 2006; Bouattour *et al.*, 2008; Gomez-Cortes *et al.*, 2008a,b, 2009; Hervas *et al.*, 2008), whereas few of them included the use of dietary lipid supplementation in combination with different forage concentrate ratios or with different forage basis (Chilliard *et al.*, 2003, 2007; Mele *et al.*, 2006, 2008; Bernard *et al.*, 2009). Fewer studies regarded the use of marine oils in the diet of dairy goats and ewes (Kitessa *et al.*, 2001, 2003; Sanz-Sampelayo *et al.*, 2002; Cattaneo *et al.*, 2006; Toral *et al.*, 2010a,b).

The addition of adequate lipid sources (more than 3% of dry matter intake) to ruminant feeds can change the composition of milk FA, whose profile reflects that of the diet, being affected by rumen biohydrogenation and activity of rumen microorganisms (Chilliard *et al.*, 2007). In general, lipid sources rich in polyunsaturated FA (PUFA) cause a decrease of medium chain FA (MCFA) and branched chain FA content and an increase of C18 FA content in milk. The consequent decrease in the saturated:unsaturated FA ratio and increase in the concentration of unsaturated C18 FA improves milk fat quality. In contrast, butyric acid percentage in milk fat is rarely affected by lipid supplements, probably because this FA is partly synthesized by metabolic pathways different to that of acetyl-CoA carboxylase (Chillard *et al.*, 2007). Linseed, soybean, sunflower and olive are the main sources of unsaturated plant lipids (as seeds or as protected or unprotected oil) tested in the diet of lactating small ruminants (Chilliard *et al.*, 2003; Mele, 2009). When comparing data from several studies (Figures 1 and 2), dietary inclusion of linseed (as oil or seed), as a source of alpha-linolenic acid (LNA), caused an increase of CLA, VA and LNA. Among plant oils rich in 18:2 n-6, soybean and sunflower oils are the most frequently studied sources. These lipid sources markedly increase milk CLA, VA and LNA content, especially when included in the diet as unprotected oil (Mele *et al.*, 2006, 2008; Chilliard *et al.*, 2007; Hervas *et al.*, 2008) (Figures 1 and 2).

The inclusion of the same amounts of unsaturated plant oils in the diet allows to reach higher levels of CLA and VA in dairy goats milk than in dairy sheep milk. For example, the same concentration of CLA and VA in milk fat was obtained by the use of soybean oil or sunflower oil at 4% or 5.5% of dry matter intake in dairy goats or sunflower oil at 6.0% of dry matter intake (150 g/d of sunflower oil) in dairy ewes (Figures 1 and 2).

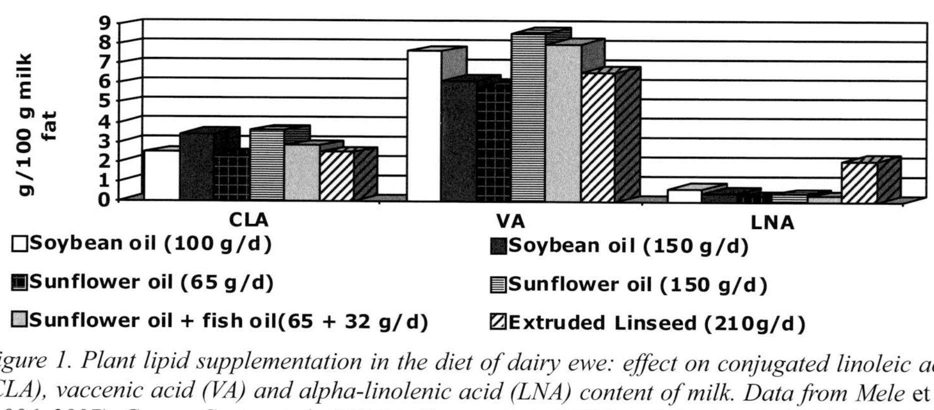

Figure 1. Plant lipid supplementation in the diet of dairy ewe: effect on conjugated linoleic acid (CLA), vaccenic acid (VA) and alpha-linolenic acid (LNA) content of milk. Data from Mele et al. *(2006, 2007), Gomez-Cortes* et al. *(2008b), Hervas* et al. *(2008) and Toral* et al. *(2010a). Each bar corresponds to a mean treatment value observed in a single study.*

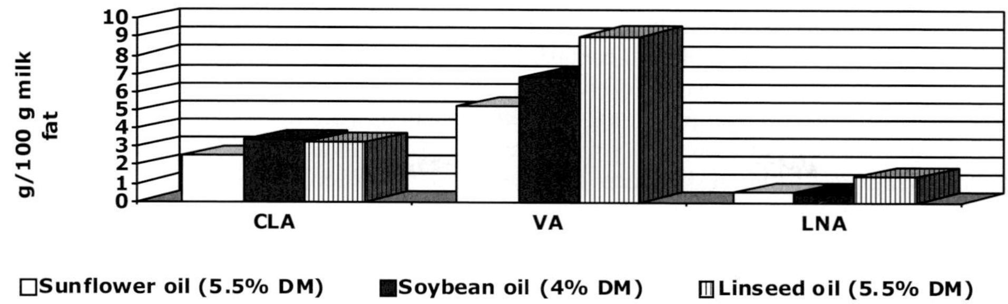

Figure 2. Plant lipid supplementation in the diet of dairy goat: effect on conjugated linoleic acid (CLA), vaccenic acid (VA) and alpha-linolenic acid (LNA) content of milk. Data from Chilliard et al. *(2007), Mele* et al. *(2008) and Bernard* et al. *(2009). Each bar corresponds to a mean treatment value observed in a single study.*

When regressing the milk CLA content against the dietary lipid (soybean or linseed oil) intake data from several studies on dairy ewes (Luna *et al.*, 2005, Mele *et al.*, 2006, 2007; Zhang *et al.*, 2006; Gomez-Cortes *et al.*, 2008b), the levels of CLA in milk linearly increased as the amounts of vegetable oil in the diet increased (Figure 3). However, responses varied depending on the nature of the basal diet (i.e. hay, silage, grasses, legumes) and on the forage:concentrate ratio in the diet. In general, when the amount of concentrate did not exceed 50% of total dry matter intake, the inclusion of unsaturated vegetable oils in the diet favoured the accumulation of VA in the rumen, with an increasing secretion of CLA in milk (Mele *et al.*, 2006, 2008; Chilliard *et al.*, 2007).

A higher intake of concentrate, especially when associated with maize silage, induces a shift in rumen biohydrogenation towards the synthesis of *trans*-10, *cis*-12 C18:2 and *trans*-10 C18:1 (Chilliard *et al.*, 2007). Among animal sources, fish oil is more effective than plant oils in enhancing CLA, VA and omega-3 FA in milk fat, especially when fish oil is fed in combination with oil supplements rich in linoleic acid (Figure 1).

This phenomenon was first demonstrated in dairy cows (Shingfield *et al.*, 2005) and, more recently, in dairy sheep (Toral *et al.*, 2010a). Also the use of marine algae in association with sunflower oil enhanced CLA, VA and long-chain PUFA n-3 in milk from dairy ewes, but in this case a large increase of *trans*-10 C18:1 was also observed (Toral *et al.*, 2010b). In dairy goats, adding 1.1% of fish oil in the diet doubled milk omega-3 FA (Cattaneo *et al.*, 2006). Similar results were obtained also in dairy ewes (Kitessa *et al.*, 2003).

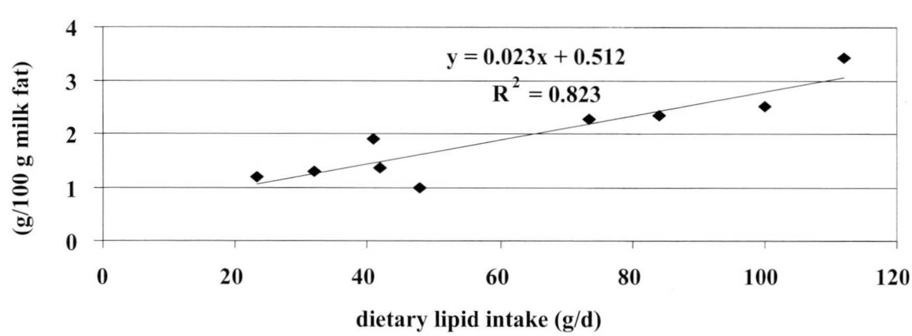

Figure 3. Relationship between daily lipid (soybean or linseed oil) intake and milk CLA content in dairy sheep. Data from Luna et al. *(2005), Mele* et al. *(2006, 2007), Zhang* et al. *(2006) and Gomez-Cortes* et al. *(2008b, 2009). Each point corresponds to a mean treatment value observed in a single study.*

A common effect observed after the inclusion of an unsaturated lipid source in the diet of small ruminants is the increase of *trans* FA other than that of VA in milk fat. The highest increases of *trans* FA were observed when the diet of dairy ewes was supplemented with soybean oil at 6% of dry matter intake or with soybean oil plus marine algal oil, which caused an increase of the content of *trans*-10 C18:1 in milk higher than 200% and 400%, respectively (Reynolds *et al.*, 2006; Gomez-Cortes *et al.*, 2008b; Toral *et al.*, 2010b). A significant increase of *trans* C18:1 content in milk fat also occurred, even if to a minor extent, in response to linseed (as oil or extruded seed) supplementation (Mele *et al.*, 2007, Gomez-Cortes *et al.*, 2009). Interestingly, in sheep milk high levels of *trans*-10 C18:1 are usually accompanied by similar or even higher contents of VA, whereas in cow milk these two FA are inversely related, so that the *trans*-10 C18:1 content may largely exceed that of VA (Shingfield and Griinari, 2007).

In conclusion, at present, effective nutritional strategies based on dietary lipid supplementation of small ruminants allow to obtain levels of enrichment of CLA, VA and LNA higher than 2.5%, 5% and 1.5% of milk fat, respectively, without any negative influence on milk yield and composition. However, in several studies a concomitant increase of *trans* C18:1 isomer other than VA was observed. Because of the negative effect of *trans* FA on serum LDL cholesterol the total content of these FA in the enriched cheese should be carefully considered in the overall evaluation of the nutritional properties of this dairy product. Finally, specific clinical studies are needed in order to provide the scientific evidences that enriched dairy products may be beneficial to human health.

Lipid supplementation and milk fat response: inter-species differences.

In small ruminants the inclusion of plant lipids in the diet enhanced milk fat secretion in several studies (Chilliard *et al.*, 2003), whereas in dairy cows high concentrate diets and plant oils decreased milk fat content (Shingfield and Griinari, 2007). Moreover, dietary lipids positively affected milk yield in dairy ewes, but did not influence consistently milk yield in dairy goats (Chilliard *et al.*, 2007). Similarly to dairy ewes, lipid supplementation in the diet of dairy cows increased milk production (Chilliard *et al.*, 2007). These differences in the responses to lipid supplementation demonstrate inter-species differences in metabolic mechanisms, which have not been fully elucidated yet. The main difference between dairy cows and small ruminants in the response to lipid supplementation, is that under a certain dietary regimen dairy cows seem more prone to the milk fat depression (MFD) syndrome than goats and ewes. Several studies reported that high concentrate diets in association with plant lipid supplementation led to milk fat depression in dairy cows (Bauman and Griinari, 2001). In contrast, when dairy ewes and goats were fed a high concentrate diet supplemented with plant oils or marine oils, milk fat secretion either significantly increased or remained unchanged (Chilliard and Ferlay, 2004; Mele *et al.*, 2005; Pulina *et al.*, 2006).

Bauman and Griinari (2001) proposed the biohydrogenation (BH) theory in order to explain the MFD observed in dairy cows fed high concentrate diet supplemented with unsaturated lipid sources or marine oils. This theory is based on the synthesis of some unique FA during the BH of PUFA in the rumen. During the last ten years, several FA have been tested as putative inhibitors of milk fat synthesis (Shingfield and Griinari, 2007). At present, *trans*-10, *cis*-12 CLA is the only biohydrogenation intermediate shown unequivocally to have depressive effects on milk fat synthesis in cows and small ruminants, either when infused in the abomasums or when administrated as calcium soap in the diet. However, several differences in the responses to *trans*-10, *cis*-12 CLA treatment were observed among species. Recently, Shingfield *et al.* (2009) demonstrated that calcium salts of CLA containing the *trans*-10, *cis*-12 isomer inhibited milk fat synthesis in goats, but the response varied with the supply of dietary FA, putting in evidence that dairy goats are less sensitive to the anti-lipogenic effects of *trans*-10, *cis*-12 CLA. On the other hand, previous studies reported that dairy sheep and cows responded similarly to *trans*-10, *cis*-12 CLA supplementation when doses were compared on a metabolic body weight basis (Lock *et al.*, 2006; Sinclair *et al.*, 2007). In dairy cows, dietary fish oil supplementation increased *trans*-10, *cis*-12 CLA content in milk and was associated with MFD (Shingfield *et al.*, 2010). Recent studies reported that dietary supplementation of ewes with fish oil significantly reduced milk fat content, but *trans*-10, *cis*-12 CLA in milk was detected in very low concentration (Toral *et al.*, 2010a). The authors suggested that the induction of MFD by fish oil could be because other CLA or C18:1 isomers may be putative inhibitors of milk fat synthesis or because of a reduction in milk fat fluidity (consequent of the reduced availability of oleic acid). Therefore, although the MFD response to fish oil supplementation is similar between dairy cows and sheep, the metabolic cause involved could differ. Recently, an exhaustive review has been published on the differences in the nutritional regulation of mammary lipogenesis among ruminants (Shingfield *et al.*, 2010). The authors examined the responses of dairy cows and small ruminants to the FA involved in the MFD and concluded that the sensitivity to the anti-lipogenic effects of *trans*-10, *cis*-12 CLA is several-fold lower in goats than in cows. Moreover, diets causing MFD in cows typically increase milk fat secretion in dairy goats and sheep. Differences between species related to eating behaviour, rumination, buffering of rumen pH, rumen digestion kinetics and transit rates are involved in the formation of specific BH intermediates which cause MFD (Shingfield *et al.*, 2010). Moreover, several authors proposed that differences in the regulation of cellular processes in the mammary gland and the relative importance of key enzymes in the synthesis of milk FA may reflect inter-species differences between dairy cows and goats (Shingfield *et al.*, 2009).

Interestingly, regardless of the ruminant species considered and the extent of the phenomenon, when MFD occurs the energy surplus is not partitioned towards milk yield. In dairy cows, it has been suggested that the energy surplus derived from the decreased milk fat positively affected the energy balance of lactating animals, with beneficial effects on reproductive performance of animals in the first part of the lactation (De Veth *et al.*, 2009). On the basis of these results, dietary supplementation with *trans*-10, *cis*-12 CLA-rich products may be an useful tool to improve the energy balance of animals in early lactation. However, because the metabolic response of small ruminants often differ significantly from that of dairy cows, further research is needed to demonstrate the effects of *trans*-10, *cis*-12 CLA on the energy balance of dairy goats and ewes.

Nutritional impact of dairy fat on human health

Dairy fat is characterized by a high concentration of saturated FA, palmitic acid (16:0) being the most abbundant reaching up to 30% of total FA in milk from cow. Other saturated FA are also present in relatively high concentrations. Dietary saturated FA have been associated with cardiovascular diseases, because some of them increase cholesterolemia in a dose-response manner, increasing both LDL and HDL cholesterol. Dairy fat also contains *trans* FA which increase LDL cholesterol and decrease HDL cholesterol.

Therefore, theoretically, dairy fat should have strong detrimental effects on cholesterolemia and thereby on cardiovascular diseases. Surprisingly, most of the studies show an inverse correlation between dairy fat consumption and cardiovascular disease incidence.This apparent discrepancy may reveal that some dairy fat components may have a strong protective effect on cardiovascular diseases. One of the candidate is CLA which had several beneficial activities in different experimental models and in humans. Most probably, also VA possesses similar activities as precursor of CLA. In fact, it has been shown that VA is readily converted to CLA in experimental animals (Banni *et al.*, 2001) and humans (Turpeinen *et al.*, 2002), and as such it may exert similar effects.

Conjugated linoleic acid

The presence of CLA and its seasonal variation in milk is known since the 30's, when Booth *et al.* (1935) reported that milk fat from grazing cows showed a great increase in the absorption in the ultra-violet region at 230 nm.

However, it was not until the 80's that the biological activities of CLA were discovered. Since the first report of Pariza's research group on CLA antimutagenic activity (Ha *et al.*, 1987), research on CLA has increased, with the further discovery of several other biological activities in different experimental models, including protection against cancer and atherosclerosis, stimulation of certain immune functions, reduction of body fat and normalization of impaired glucose tolerance in type 2 diabetes (see the book by Yurawecz *et al.*, 2007 for a thorough review on CLA biological activities). Most of the early data were obtained using a mixture of CLA isomers. This happened because in order to have sufficient material to perform the experiments, CLA was synthesised by alkali isomerisation of linoleic acid from sunflower oil. The isomeric composition consisted of at least 12 isomers, as a combination of four positional isomers (8,10; 9,11; 10,12; 11,13) and three geometrical (t,t; c,t or t,c; c,c) isomers. Because among them the most concentrated isomers were *cis*-9, *trans*-11 CLA and *trans*-10, *cis*-12 CLA, studies have focused mainly on these two isomers.

As a consequence, the peculiar biological activities attributed to CLA were a combined result of the effects of the single isomers. Only after single isomers were commercially available in sufficient amounts to carry out studies in experimental animals and humans, it was possible to distinguish the effects of single isomers. More than 90% of CLA in dairy products is *cis*-9, *trans*-11 CLA and, among the most important effects, the anticarcinogenic and antiatherogenic properties in several experimental models, and the ability to decrease LDL cholesterol in humans have been described for this isomer.

Among different experimental models where CLA had anticarcinogenic activity, the mammary gland is the target organ which has been more extensively studied. One of the most employed experimental model is the time frame of CLA feeding during the period of pubescent mammary gland development in the rat and the switch to control diet after administration of the carcinogen (either DMBA or MNU) to yield a reduction of mammary tumour of about 50% (Banni *et al.*, 1999). This model has been extensively employed to demonstrate a dose-response activity of CLA on different parameters correlated to its anticarcinogenic activity. Similar results were obtained by feeding naturally occurring CLA; in fact, as mentioned above, CLA preparation used in previous studies consisted of a mixture of isomers synthetically produced from vegetable oils. The use of an enriched CLA butter became necessary to reach a level of CLA in the diet close to 1%, an amount shown to be chemo-preventive in models using synthetic CLA. The results showed that naturally occurring CLA was able to inhibit mammary tumour yield by 53% (Ip *et al.,* 1999). Therefore, it became clear that either naturally occurring or synthetic CLA when present in the diet are able to exert similar anticarcinogenic activities.

Another peculiar effect of *cis*-9, *trans*-11 CLA isomer is its antiatherogenic effects in different experimental models. Probably the most striking is its effect on rabbits, the best experimental model for atherogenesis. In this model, *cis*-9, *trans*-11 CLA in the diet decreased atherogenesis by 50% when fed as a component of a semi-.purified diet containing 0.2% cholesterol and reduced established lesions by 26% when fed as part of a cholesterol-free diet (Kritchevsky *et al.*, 2004). Cancer and atherosclerosis have some common features among them such as sustained cell proliferation and inflammation. It is likely that *cis*-9, *trans*-11 CLA affects both by increasing apoptosis (Ip *et al.*, 2000) and by interfering with eicosanoid formation (Urquhart *et al.*, 2002). In fact, it has been demonstrated that CLA can be metabolized (Banni, 2002) and modifies both FA availability and metabolism by competing with other PUFAs and in particular with n-6 PUFAs and/or by activating nuclear receptors such as PPARs thereby modifying phospholipid composition, bioactive lipid molecule formation, and FA beta-oxidation repartition between mitochondria and peroxisomes. These effects may diverge in different tissues according to their lipid composition, metabolism, gene regulation of lipid metabolism and energy repartition.

However, data on the anticarcinogenic and antiatherogenic effects of CLA on humans are difficult to obtain, because cancer and atherosclerosis take years to develop. Indeed, studies in humans have failed so far to unravel this issue.

Therefore, to establish whether CLA may also be active in humans, studies should focus on CLA's mechanism(s) of action, thus clarifying at which level CLA acts and if it interferes in those pathways shared by both humans and experimental animals, such as PUFA metabolism. Several studies on humans, (Riserus *et al.*, 2002; Smedman *et al.*, 2004, 2005; De Roos *et al.*, 2005; Jaudszus *et al.*, 2005; Ramakers *et al.*, 2005; Zulet *et al.*, 2005) and experimental animals (Turnock *et al.*, 2001; Hontecillas *et al.*, 2002; Yu *et al.*, 2002; Ogborn *et al.*, 2003; Sher *et al.*, 2003; Coen *et al.*, 2004; Lai *et al.*, 2005; de Roos *et al.*, 2005; Zhao *et al.*, 2005) have shown conflicting results on the activity of CLA isomers in the metabolic syndrome. None of these studies however took into consideration that modulation of metabolic syndrome parameters could be the direct results of lipid metabolism perturbation in different tissues, exerted by CLA isomers. Several experimental models have been employed to understand the mechanism of action of CLA. One rodent model of obesity, the zucker rat, has been widely employed because it mimics the consequences of obesity found in humans, including visceral obesity, hyperglycemia, insulin resistance, hypertension, fatty liver, dyslipidemia, and pro-inflammatory state, named metabolic syndrome. CLA improved significantly several of these parameters in different studies. The effect of CLA on fat mass is variable (Nagao *et al.*, 2005). When pure isomers were tested it seemed that a mixture of isomers as well as *cis*-9, *trans*-11 CLA alone was effective in improving insulin resistance (Ryder *et al.*, 2001), in another study only *trans*-10, *cis*-12 CLA was effective in improving muscle glucose transport (Henriksen *et al.*, 2003). All these studies used short-term feeding and a very high dose, 1.5% of CLA in the diet, which corresponds, in terms of plasma concentrations, to about 17-18 g /d in humans, based on our previous data. Furthermore, no further investigations were conducted to find a correlation between FA incorporation and metabolism in different tissues and metabolic changes related to visceral obesity, hyperglycemia, insulin resistance, hypertension, fatty liver, dyslipidemia, and pro-inflammatory state.

Obesity is characterized by an overflow of FA in extra-adipose tissues, increasing FA beta oxidation both in mitochondria and peroxisomes at the expense of a lower glucose disposal. An increased availability of FA for peroxisomal beta oxidation impairs peroxisomal beta oxidation of eicosanoids, sustaining inflammation. CL, is preferentially beta oxidized in peroxisomes in experimental animals and humans, and is able to activate PPAR alpha which is a transcription factor of the enzymes of the peroxisomal beta oxidation. As a result, CLA should improve eicosanoid catabolism and act as an anti-inflammatory agent, by increasing peroxisomal beta oxidation (Iannone *et al.*, 2009, Banni *et al.*, 2010).

An overall perturbation of lipid metabolism by CLA may also involve cholesterol metabolism. In fact, some interesting studies showed that c,9t11 isomer was able to significantly decrease LDL cholesterol (Tricon *et al.*, 2004).

Furthermore, a significant reduction in inflammatory parameters such as interleukin-6, interleukin-8 and tumour necrosis factor-alpha, and in the extent of platelet aggregation was found in volunteers with a daily intake of 200 g/week of cheese naturally rich in c9,t11 CLA (Sofi *et al.*, 2009).

Data in the literature clearly show that CLA seems to positively interfere with lipid metabolism, in particular with n-6 PUFAs for eicosanoid formation, mitochondrial and peroxisomal beta oxidation, energy expenditure and eicosanoid catabolism, and with cholesterol metabolism. Most of these data are confined to experimental models, but human studies, even if scarce (especially long-term trials), suggest that CLA enriched products may have a positive impact on human health, extending the dietary inclusion of dairy products to those patients affected by metabolic syndrome.

Acknowledgements

We are grateful to Ana H.D. Francesconi, from the University of Sassari, for the technical and linguistic review of the manuscript.

References

Antongiovanni, M., P. Secchiari, M. Mele, A. Buccioni, A. Serra, G. Ferruzzi, S. Rapaccini and A. Pistoia, 2002. Olive oil calcium soaps and rumen protected methionine in the diet of lactating ewes: effect on milk quality. Ital. J. Anim. Sci. 1, 55-63.

Banni S., R. Montisci, R. Sanfilippo, G. Finco, D. Sanna, E. Giordano, E. Murru, L. Cordeddu, G. Carta, D. Banni and A. Marchi, 2010. Physiological response to lipid peroxidation in ischemia and reperfusion during carotid endarterectomy. Lipids Health Dis. 9, 41; doi:10.1186/1476-511X-9-41 (In Press).

Banni, S., 2002. Conjugated linoleic acid metabolism [Review]. Curr Opin Lipidol 13, 261-266.

Banni, S., Angioni, E., Casu, V., Melis, M.P., Carta, G., Corongiu, F.P., Thompson, H. and C. Ip, 1999. Decrease in linoleic acid metabolites as a potential mechanism in cancer risk reduction by conjugated linoleic acid. Carcinogenesis 20, 1019-1024.

Banni, S., Angioni, E., Murru, E., Carta, G., Melis, M.P., Bauman, D., Dong, Y. and C. Ip, 2001. Vaccenic acid feeding increases tissue levels of conjugated linoleic acid and suppresses development of premalignant lesions in rat mammary gland. Nutr. Cancer. 41, 91-97.

Bauman, D.E. and J.M. Griinari, 2001. Regulation and nutritional manipulation of milk fat: low fat milk syndrome. Livest. Prod. Sci. 70, 15-29.

Bauman, D.E., S.N. McCuthceon, W.D. Steinhour, P.J. Eppard and S.J. Sechen, 1985. Sources of variation and prospects for improvement of productive efficiency in the dairy cow: a review. J. Anim. Sci. 60, 583-592.

Bernard, L., J. Rouel, C. Leroux, A. Ferlay, Y. Faulconnier, P. Legrand and Y. Chilliard, 2005. Mammary lipid metabolism and milk fatty acid secretion in Alpine goats fed vegetable lipids. J. Dairy Sci. 88, 1478-1489.

Bernard, L., K.J. Shingfield, J., Rouel, A. Feraly and Y. Chilliard, 2009. Effect of plant oils in the diet on performance and milk fatty acid composition in goats fed diets based on grass hay or maize silage. Br. J. Nutr. 101, 213-224.

Booth, R.G., Kon, S.K., Dann, W.J. and Moore, T., 1935. A study of the seasonal variation in butter fat. II. A seasonal spectroscopic variation in the fatty acid fraction. 29, 133-137.

Bouattour, M.A., R. Casals, E. Albanell, X. Such and G. Caja., 2008. Feeding soybean oil to dairy goats increases conjugated linoleic acid in milk. J. Dairy Sci. 91, 2399-2407.

Cattaneo, D, V. Dell'Orto, G. Varisco, A. Agazzi, G. Savoini, 2006. Enrichment in *n*–3 fatty acids of goat's colostrum and milk by maternal fish oil supplementation. Small Rum. Res. 64, 22-29.

Chilliard, Y., 1993. Dietary fat and adipose tissue metabolism in ruminants, pigs, and rodents: a review. J. Dairy Sci. 76, 3897-3931.

Chilliard, Y., A. Ferlay, J. Rouel and G. Lamberet, 2003. A review of nutritional and physiological factors affecting goat milk lipid synthesis and lipolysis. J. Dairy Sci. 86, 1751-1770.

Chilliard, Y. and A. Ferlay, 2004. Dietary lipids and forages interactions on cow and goat milk fatty acid composition and sensory properties. Reprod. Nutr. Dev. 44, 467-492.

Chilliard, Y., F. Glasser, A. Ferlay, L. Bernard, J. Rouel and M. Doreau, 2007. Diet, rumen biohydrogenation and nutritional quality of cow and goat milk fat. Eur. J. Lipid Sci. Technol. 109, 828-855.

Coen, P.M., P.M. Cummins, Y.A. Birney, R. Devery and P.A. Cahill, 2004. 'Modulation of nitric oxide and 6-keto-prostaglandin F (1alpha) production in bovine aortic endothelial cells by conjugated linoleic acid.' Endothelium: J. Endothel. Cell Res. 11, 211-220.

De Roos, B., G. Rucklidge, M. Reid, K. Ross, G. Duncan, M.A. Navarro, J.M. Arbones-Mainar, M.A. Guzman-Garcia, J. Osada, J. Browne, C.E. Loscher and H.M. Roche, 2005. Divergent mechanisms of *cis*9, *trans*11-and *trans*10, *cis*12-conjugated linoleic acid affecting insulin resistance and inflammation in apolipoprotein E knockout mice: a proteomics approach. FASEB Journal 19, 1746-1748.

De Veth, M.J., D.E. Bauman, W. Koch, G.E. Mann, A.M. Pfeiffer and W.R. Butler, 2009. Efficacy of conjugated linoleic acid for improving reproduction: A multi-study analysis in early-lactation dairy cows. J. Dairy. Sci. 92, 2662-2669.

Glasser, F., A. Ferlay and Y. Chilliard, 2008. Oilseed supplements and fatty acid composition of cow milk: a meta-analysis. J. Dairy Sci. 91, 4687-4703.

Gomez-Cortes, P., A. Bach, P. Luna, M. Juarez and M.A. De la Fuente, 2009. Effects of extruded linseed supplementation on n-3 fatty acids and conjugated linoleic acid in milk and cheese from ewes. J. Dairy Sci. 92, 4122-4134.

Gomez-Cortes, P., P. Frutos, A.R. Mantecon, M. Juarez, M.A. De la Fuente and G. Hervas, 2008a. Addition of olive oil to dairy ewes diets: effects on milk fatty acid profile and animal performance. J. Dairy Sci. 91, 3119-3127.

Gomez-Cortes, P., P. Frutos, A.R. Mantecon, M. Juarez, M.A. De la Fuente and G. Hervas, 2008b. Milk production, conjugated linoleic acid content and *in vitro* ruminal fermentation in response to high levels of soybean oil in dairy ewe diet. J. Dairy Sci. 91, 1560-1569.

Ha Y.L., Grimm N.K. and M.W. Pariza, 1987. Anticarcinogens from fried ground beef: heat-altered derivatives of linoleic acid. Carcinogenesis 8, 1881-1887.

Henriksen, E.J., M.K. Teachey, Z.C. Taylor, S. Jacob, A. Ptock, K. Kramer and O. Hasselwander, 2003. Isomer-specific actions of conjugated linoleic acid on muscle glucose *trans*port in the obese Zucker rat. Am. J. Physiol. - Endocrinology & Metabolism 285, E98-E105.

Hervas, G., P. Luna, A.R. Mantecon, N. Castanares, M.A. de la Fuente, M. Juarez and P. Frutos, 2008. Effect of diet supplementation with sunflower oil on milk production, fatty acid profile and ruminal fermentation in lactating dairy ewes. J. Dairy Res. 75, 399-405.

Hontecillas, R., M.J. Wannemeulher, D.R. Zimmerman, D.L. Hutto, J.H. Wilson, D.U. Ahn and J. Bassaganya-Riera, 2002. Nutritional regulation of porcine bacterial-induced colitis by conjugated linoleic acid. J. Nutr. 132, 2019-2027.

Iannone A., A. Petroni, E. Murru, L. Cordeddu, G. Carta, M. P. Melis, S. Bergamini, L. Della Casa, L. Cappiello, R. Carissimi, M. O'Shea, D. Bell, E. De Santis and S. Banni, 2009. Impairment of 8-Iso-PGF2ALPHA Isoprostane Metabolism by Dietary Conjugated Linoleic Acid (CLA). Prost. Lukotr. Ess. Fatty acids, 80, 279-287.

Ip, C., M.M. Ip, T. Loftus, S. Shoemaker and W. Shea-Eaton, 2000. Induction of apoptosis by conjugated linoleic acid in cultured mammary tumor cells and premalignant lesions of the rat mammary gland. Cancer Epidemiol. Biomarkers Prev. 9, 689-696.

Ip, C., S. Banni, E. Angioni, G. Carta, J. McGinley, H.J. Thompson, D. Barbano and D. Bauman, 1999. Conjugated linoleic acid-enriched butter fat alters mammary gland morphogenesis and reduces cancer risk in rats. J. Nutr. 129, 2135-2142.

Jaudszus, A., M. Foerster, C. Kroegel, I. Wolf and G. Jahreis, 2005. *Cis*-9,*trans*-11-CLA exerts anti-inflammatory effects in human bronchial epithelial cells and eosinophils: comparison to *trans*-10,*cis*-12-CLA and to linoleic acid. Biochim. Biophys. Acta 1737, 111-118.

Kitessa, S.M., D. Peake, R. Bencini and A.J. Williams, 2003. Fish oil metabolism in ruminants III. Transfer of n-3 polyunsaturated fatty acids (PUFA) from tuna oil into sheep's milk. Anim. Feed Sci. Tech. 108, 1-14.

Kitessa, S.M., S.K.Gulati, J.R. Ashes, E. Fleck, T.W. Scott and P.D. Nichols, 2001. Utilisation of fish oil in ruminants. II. Transfer of fisdh oil fatty acids into goat's milk. Anim. Feed Sci. Tech. 89, 201-208.

Kritchevsky, D., S.A. Tepper, S. Wright, S.K. Czarnecki, T.A. Wilson and R.J. Nicolosi, 2004. Conjugated linoleic acid isomer effects in atherosclerosis: Growth and regression of lesions. Lipids 39, 611-616.

Lai C., J. Yin, D. Li, L. Zhao, S. Qiao and J. Xing, 2005. Conjugated linoleic acid attenuates the production and gene expression of proinflammatory cytokines in weaned pigs challenged with lipopolysaccharide. J Nutr 135, 239-244.

Lock, A.L., B.M. Teles, J.W. Perfield and D.E. Bauman, 2006. A conjugated linoleic acid supplement containing *trans*-10, *cis*-12 reduces milk fat synthesis in lactating sheep. J. Dairy Sci. 89, 1525-1532.

Luna, P., J. Fontecha, M. Juarez and M.A. De la Fuente, 2005. Changes in the milk and cheese fat composition of ewes fed commercial supplements containing linseed with special reference to the CLA content and isomer composition. Lipids. 40, 445-454.

Mele, M., 2009. Designing milk fat to improve healthfulness and functional properties of dairy products: from feeding strategies to a genetic approach. It. J. Anim. Sci. in press.

Mele, M., A. Buccioni, F. Petacchi, A. Serra, S. Banni, M. Antongiovanni and P. Secchiari, 2006. Effect of forage/concentrate ratio and soybean oil supplementation on milk yield, and composition from Sarda ewes. Anim. Res. 55, 273-285.

Mele, M., A. Buccioni and A. Serra, 2005. Lipid requirements in nutrition of dairy ewe. Ital. J. Anim. Sci. 4 (Suppl. 2), 53-62.

Mele, M., A. Serra, A. Buccioni, G. Conte, A. Pollicardo and P. Secchiari, 2008. Effect of soybean oil supplementation on milk fatty acid composition from Saanen goats fed diets with different forage:concentrate ratios. Ital. J. Anim. Sci. 7, 297-311.

Mele, M., A. Serra, G. Conte, A. Pollicardo, M. Del Viva and P. Secchiari, 2007. Whole extruded linseed in the diet of dairy ewes during early lactation: effect on the fatty acid composition of milk and cheese. It. J. Anim. Sci. 6 (Suppl. 1), 560-562.

Nagao, K., N. Inoue, Y.M. Wang, B. Shirouchi and T. Yanagita, 2005. Dietary conjugated linoleic acid alleviates nonalcoholic fatty liver disease in Zucker (fa/fa) rats. J. Nutr. 135, 9-13.

Ogborn, M. R., E. Nitschmann, N. Bankovic-Calic, H.A. Weiler, S. Fitzpatrick-Wong and H.M. Aukema, 2003. Dietary conjugated linoleic acid reduces PGE2 release and interstitial injury in rat polycystic kidney disease. Kidney Int. 64, 1214-1221.

Pulina, G., A. Nudda, G. Battacone and A. Cannnas, 2006. Effects of nutrition on the contents of fat, protein, somatic cells, aromatic compounds and undesirable substances in sheep milk. Anim. Feed Sci. Tech. 131, 255-291.

Ramakers, J.D., J. Plat, J.L. Sebedio and R.P. Mensink, 2005. Effects of the individual isomers *cis*-9,*trans*-11 vs. *trans*-10,*cis*-12 of conjugated linoleic acid (CLA) on inflammation parameters in moderately overweight subjects with LDL-phenotype B. Lipids 40, 909-918.

Riserus, U., S. Basu, S. Jovinge, G.N. Fredrikson, J. Arnlov and B. Vessby, 2002. Supplementation with conjugated linoleic acid causes isomer-dependent oxidative stress and elevated C-reactive protein: a potential link to fatty acid-induced insulin resistance. Circulation 106, 1925-1929.

Ryder JW, C.P. Portocarrero, X.M. Song, L. Cui, M. Yu, T. Combatsiaris, D. Galuska, D.E. Bauman, D.M. Barbano, M.J. Charron, J.R. Zierath and K.L. Houseknecht, 2001. Isomer-specific antidiabetic properties of conjugated linoleic acid. Improved glucose tolerance, skeletal muscle insulin action and UCP2 gene expression. Diabetes. 50, 1149-1157.

Sanz-Sampelayo, M.R., L. Perez, J.J. Martin Alonso, L. Amigo and J. Boza, 2002. Effect of concentrates with different contents of protected fat rich in PUFAs on the performance lactating Granadina goats part II. Milk production and composition. Small Rum Res. 43, 141-148.

Sher, J., A. Pronczuk, T. Hajri and K.C. Hayes. 2003. Dietary conjugated linoleic acid lowers plasma cholesterol during cholesterol supplementation, but accentuates the atherogenic lipid profile during the acute phase response in hamsters. J. Nutr. 133, 456-460.

Shingfield, K.J, C.K. Reynolds, B. Lupoli, V. Toivonen, M.P. Yurawecz, P. Delmonte, J.M. Griinari, A.S. Grandison and D.E. Beever, 2005. Effect of forage type and proportion of concentrate in the diet on milk fatty acid composition in cows given sunflower or fish oil. Anim. Sci. 80, 225-238.

Shingfield, K.J. and J.M. Griinari, 2007. Role of biohydrogenation intermediates in milk fat depression. Eur. J. Lipid Sci. Technol. 109, 799-816.

Shingfield, K.J., J. Rouel and Y. Chilliard, 2009. Effect of calcium salts of a mixture of conjugated linoleic acids containing *trans*-10, *cis*-12 in the diet on milk fat synthesis in goats. Br. J. Nutr. 101, 1006-1019.

Shingfield, K.J., L.Bernard, C. Leroux and Y. Chilliard, 2010. Role of *trans* fatty acids in the nutritional regulation of mammary lipogenesis in ruminants. Animal. (in press).

Sinclair, L.A., A.L. Lock, R. Early and D.E. Bauman, 2007. Effects of *trans*-10, *cis*-12 conjugated linoleic acid on ovine milk fat synthesis and cheese properties. J. Dairy Sci. 90, 3326-3335.

Smedman, A., B. Vessby and S. Basu, 2004. Isomer-specific effects of conjugated linoleic acid on lipid peroxidation in humans: regulation by alpha-tocopherol and cyclo-oxygenase-2 inhibitor. Clin. Sci. 106, 67-73.

Smedman, A., S. Basu, S. Jovinge, G.N. Fredrikson and B. Vessby, 2005. Conjugated linoleic acid increased C-reactive protein in human subjects. Brit. J. Nutr. 94(5), 791-5.

Sofi, F., A. Buccioni, F. Cesari, A.M. Gori, S. Minieri, L. Mannini, A. Casini, G.F. Gensini, R. Abbate and M. Antongiovanni, 2009. Effects of a dairy product (pecorino cheese) naturally rich in *cis*-9, *trans*-11 conjugated linoleic acid on lipid, inflammatory and haemorheological variables: a dietary intervention study. Nutr Metab Cardiovasc Dis 20, 117-124.

Toral, P.G., G. Hervás, P. Gómez-Cortés, P. Frutos, M. Juárez and M.A. De la Fuente, 2010b. Milk fatty acid profile and dairy sheep performance in response to diet supplementation with sunflower oil plus incremental levels of marine algae. J. Dairy Sci. 93, 1655-1667.

Toral, P.G., P. Frutos, G. Hervás, P. Gómez-Cortés, M. Juárez and M.A. De la Fuente, 2010a. Changes in milk fatty acid profile and animal performance in response to fish oil supplementation, alone or in combination with sunflower oil, in dairy ewes. J. Dairi Sci. 93, 1604-1615.

Tricon, S., G.C. Burdge, S. Kew, T. Banerjee, J.J. Russell, E.L. Jones, R.F. Grimble, C.M. Williams, P. Yaqoob and P.C. Calder, 2004. Opposing effects of *cis*-9,*trans*-11 and *trans*-10,*cis*-12 conjugated linoleic acid on blood lipids in healthy humans. Am. J. Clin. Nutr. 80, 614-620.

Turnock, L., M. Cook, H. Steinberg and C. Czuprynski, 2001. Dietary supplementation with conjugated linoleic acid does not alter the resistance of mice to *Listeria monocytogenes* infection. Lipids 36, 135-138.

Turpeinen, A.M., M. Mutanen, A. Aro, I. Salminen, S. Basu, D.L. Palmquist and J.M. Griinari, 2002. Bioconversion of vaccenic acid to conjugated linoleic acid in humans. Am. J. Clin. Nutr. 76, 504-510.

Urquhart, P., S.M. Parkin, J.S. Rogers, J.A. Bosley and A. Nicolaou, 2002. The effect of conjugated linoleic acid on arachidonic acid metabolism and eicosanoid production in human saphenous vein endothelial cells. Biochim. Biophys. Acta 1580, 150-160.

Yu, Y., P. H. Correll and J.P. Vanden Heuvel, 2002. Conjugated linoleic acid decreases production of pro-inflammatory products in macrophages: evidence for a PPAR gamma-dependent mechanism. Biochim. Biophys. Acta 1581, 89-99.

Yurawecz, M.P., J.K.G. Kramer, S. Banni, O. Gudmundsen and M.W. Pariza (eds.), 2007. Advances in conjugated linoleic acid research, Volume 3. AOCS Press Champaign, IL, USA.

Zhang, R., A.F. Mustafa and X. Zhao, 2006. Effects of flaxseed supplementation to lactating ewes on milk composition, cheese yield, and fatty acid composition of milk and cheese. Small Rum. Res. 63, 233-241.

Zhao, L., J. Yin, D. Li, C. Lai, X. Chen and D. Ma, 2005. Conjugated linoleic acid can prevent tumor necrosis factor gene expression by inhibiting nuclear factor binding activity in peripheral blood mononuclear cells from weaned pigs challenged with lipopolysaccharide. Arch. Anim. Nutr. 59, 429-438.

Zulet, M.A., A. Marti, M.D. Parra and J.A. Martinez, 2005. Inflammation and conjugated linoleic acid: mechanisms of action and implications for human health. J. Physiol. Biochem. 61, 483-494.

More muscular, younger cattle have a lower incidence of dark cutting

P. McGilchrist[1,2], K.L. Thomson[1,3], D.W. Pethick[1,2], R.H. Jacobs[1,3] and G.E. Gardner[1,2]
[1]Australian Cooperative Research Centre for Beef Genetic Technologies, Australia
[2]School of Veterinary & Biomedical Science, Murdoch University, Murdoch, WA 6150, Australia
[3]Department of Agriculture & Food of Western Australia, South Perth, WA, 6151, Australia

Introduction

Beef cattle in Australia are commonly selected for muscularity and growth to increase efficiency and profitability. However, it is unknown if selection for greater muscling and growth will impact on the incidence of dark cutting. Dark cutting in beef carcasses is one of the largest problems affecting meat quality within the beef industry world-wide, costing million of dollars per annum in discounted beef. Dark cutting or pH>5.7 is also the largest cause of carcasses failing to grade under the Meat Standards Australia carcass grading system. Dark cutting is caused by low muscle glycogen levels at slaughter. Muscle glycogen at slaughter is a function of basal muscle glycogen minus the quantity of glycogen utilised during the stressful pre-slaughter period. Controlled experiments in cattle and sheep have shown that the high muscled and younger animals have lower muscle responsiveness to adrenaline (Martin *et al.*, 2010; McGilchrist *et al.*, 2009) as well as increased glycogen storage (McGilchrist *et al.*, 2009). This is likely to result in higher levels of muscle glycogen at slaughter and reduced incidence of dark cutting. The aim of this study was to assess whether a decrease in adrenaline responsiveness would impact at a broader industry level. Thus, we hypothesised that younger cattle with greater phenotypic muscling would have a lower incidence of dark cutting when slaughtered at a commercial processing plant.

Materials and methods

Meat Standards Australia (MSA) beef carcass grading data was obtained from a Western Australian (W.A.) meat processor between February 2002 and December 2008 containing 204,072 carcass records. The data was standardised for carcass quartering site and carcasses with missing measurements were omitted. The effects of muscling and age on ultimate pH (pH_u) were analysed using a linear mixed effects model. Muscling was indicated by eye muscle area (EMA) adjusted for hot standard carcass weight (HSCW), and age was indicated by ossification score (Oss). The model included fixed effects for year, month, gender, finishing system (either grass fed or sourced from an accredited Australian Lot Feeders Association feedlot) and region of production in W.A., covariates for HSCW, EMA, Oss, lot-size and tropical breed content were tested. All relevant first order interactions between fixed effects, interactions between covariates with HSCW and squared terms of each covariate were tested and removed if non-significant ($P>0.05$). MSA grader, producer and lot number were used as random terms.

Results and discussion

The linear mixed effects model including the random terms explained 16.12% of variation in the pH_u measurements. The most marked effects of EMA and Oss were seen at low HSCW (i.e. ~150 kg) with an increase in EMA of 30 cm^2, or a decrease in Oss of 60 points both reducing pH_u by ~0.05 pH units ($P<0.001$, Figure 1). This decrease in pH_u equates to a 12% reduction in the proportion of dark cutters. For both covariates this effect diminished beyond about 350 kg HSCW. Higher growth rate carcasses (i.e. HSCW corrected for Oss) had lower pH_u ($P<0.001$). HSCW also had an effect with heavier carcasses having a lower pH_u ($P<0.001$).

In support of our initial hypothesis, carcasses with greater muscularity (i.e. higher EMA at the same HSCW), had reduced pH_u. Furthermore, younger (i.e. lower Oss) and faster growing cattle had a

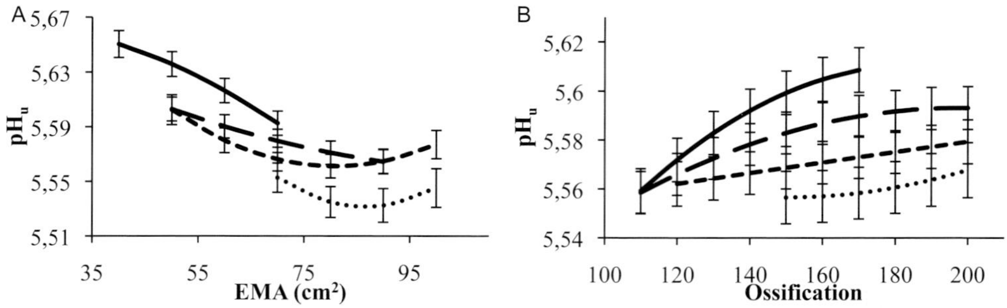

Figure 1. Effect of (A) EMA and (B) Ossification on pH_u ± s.e.m. at different hot standard carcass weights (HSCW). Straight line = 150 HSCW; large dashed line = 250 HSCW; small dashed line = 350 HSCW; dotted line = 450 HSCW.

reduced pH_u. From these results, it can be assumed that more muscular and younger cattle have higher storage of glycogen at slaughter, subsequently reducing the incidence of dark cutting in these types of animals. A shift in muscle fibre type proportions may explain this result. The more glycolytic type IIX myofibres have displayed increased GLUT4 density in cattle (Hocquette *et al.*, 1995), and decreased β-adrenergic receptor density in rats (Martin *et al.*, 1989) compared with the more oxidative type I and IIA myofibres. Therefore more muscular and younger cattle, which have a higher proportion of type IIX myofibres (Brandstetter *et al.*, 1998; Wegner *et al.*, 2000), will have greater storage and reduced mobilisation of muscle glycogen. In conclusion, this data demonstrated that slaughtering more muscular cattle at younger ages will help alleviate the dark cutting beef syndrome.

References

Brandstetter, A.M., B. Picard and Y. Geay, 1998. Muscle fibre characteristics in four muscles of growing bulls I. Postnatal differentiation. Livest. Prod. Sci. 53, 15-23.

Hocquette, J.F., F. Bornes, M. Balage, P. Ferre, J. Grizard and M. Vermorel, 1995. Glucose-transporter (GLUT4) protein content in oxidative and glycolytic skeletal muscles from calf and goat. Biochem. J. 305, 465-470.

Lacourt, A. and P.V. Tarrant, 1985. Glycogen depletion patterns in myofibres of cattle during stress. Meat Sci. 15, 85-100.

Martin, K.M., P. McGilchrist, J.M. Thompson and G.E. Gardner, 2010. Progeny of high muscling sires have reduced muscle response to adrenaline in sheep. Animal. In Press.

Martin, W., S. Murphree and J. Saffitz, 1989. Beta-adrenergic receptor distribution among muscle fibre types and resistance arterioles of white, red, and intermediate skeletal muscle. Circul. Res. 64, 1096-1105.

McGilchrist, P., P.L. Greenwood, D.W. Pethick and G.E. Gardner, 2009. Selection for muscling increases glycogen and reduces response to adrenaline in muscle. In: Y. Chilliard, F. Glasser, Y. Faulconnier, F. Bocquier, I. Veissier and M. Doreau (eds.) XIth Int. Symp. Rumin. Physiol. Wageningen Academic Publishers, Wageningen, the Netherlands: 598-599.

Wegner, J., E. Albrecht, I. Fiedler, F. Teuscher, H.J. Papstein and K. Ender, 2000. Growth- and breed-related changes of muscle fiber characteristics in cattle. J. Anim. Sci. 78, 1485-1496.

Dietary fats in transition dairy goats: effects on milk FA composition

D. Cattaneo, M. Ferroni, F. Caprino, V. Moretti, A. Agazzi, G. Invernizzi and G. Savoini
Dipartimento di Scienze e Tecnologie Veterinarie per la Sicurezza Alimentare, Università degli Studi di Milano, Via Trentacoste 2, 20134 Milano, Italy

Introduction

Dietary fats not only provide energy for dairy ruminants in the transition period, but also play a role in modifying milk fatty acid (FA) composition. In particular, rumen inert or partially inert fats (calcium salts of FA, encapsulated FA etc.) may reduce susceptibility of polyunsaturated FA (PUFA) to rumen bio-hydrogenation and enhance their flow to the small intestine for absorption. Healthfulness of milk could be improved by increasing the incorporation of FA with putative beneficial actions on human health, such as long-chain n-3 PUFA of marine origin eicosapentaenoic acid (EPA C20:5 n-3) and docosahexaenoic acid (DHA C22:6 n-3) into milk fat (Savoini *et al.*, 2010).

Aim of the study was to evaluate the effects, on milk fatty acid composition, of sources of rumen inert fats fed to transition dairy goats: hydrogenated palm oil (HPO), rich in saturated FA or protected fish oil (PFO), rich in long-chain n-3 PUFA.

Material and methods

Eighteen second parity Saanen dairy goats were selected for this study and divided into three homogenous groups of six animals each. Starting from wk 2 before the expected kidding date to wk 3 of lactation the goats were fed one of the following experimental diets: C, basal transition diet (36.0% alfalfa hay, 14.8% beetpulp dry, 13.6% triticale silage, 10.0% corn, 8.0% wheat bran and cereal by-products, 5.7% barley, 4.6% field bean, 3.8% sunflower meal, 2.0% alfalfa meal, 0.9% molasses, 0.6% mineral premix); C + HPO, basal transition diet plus 47 g/d HPO (hydrogenated palm oil) as source of saturated FA; C + PFO, basal transition diet plus 47 g/d PFO (protected fish oil; microencapsulated) as source of long-chain n-3 PUFA. Analysis (DM basis) of the basal diet was: CP 15.0%, ether extract 3.9%, NDF 38.4%. Fatty acid profile (% total FA) of the basal diet (C) and of HPO and PFO supplements were respectively: myristic 0.40, 1.50, 6.90; palmitic 17.75, 46.30, 19.10; palmitoleic 0.20, <0.10, 7.60; stearic 3.35, 40.90, 4.10; oleic 30.75, 4.90, 16.50; linoleic 36.05, 0.20, 2.40; linolenic 8.20, <0.10, 3.40; EPA (C20:5 n-3) <0.10, <0.10, 17.60; DHA (C22:6 n-3) <0.10, <0.10, 10.20. Dry matter intake (DMI), milk yield and milk characteristics were recorded and analyzed on day 4, 7, 15, and 21 of lactation. Separate aliquots of individual milk samples were obtained at 21 d of lactation, stored at -20 °C and subsequently analyzed by gas chromatography to determine fatty acid composition of milk, following the procedure described in Cattaneo *et al.*, 2006. The data were analyzed as repeated measures by a MIXED procedure of SAS, including the effects of treatment, day of treatment and interaction (SAS, 2006).

Results and discussion

Dietary fat supplements had no significant influence on DMI. Milk yield increased during the first three weeks of lactation in C+HPO compared to C goats (3.65 vs. 2.42 kg/d, $P<0.01$) while no significant difference in milk yield was shown for C+PFO (3.26 kg/d) compared to C goats ($P=0.33$). Milk protein and fat contents were not affected by dietary treatments and values were respectively 3.90 and 5.45% (C), 4.00 and 6.03% (C+HPO) and 4.74 and 5.92% (C+PFO). Effect of dietary fats on milk FA profile is presented in Table 1. Percentages of C8:0 and C12:0 were significantly increased by PFO supplementation, whereas content of C18:0 was decreased ($P<0.01$) by fat supplements and was significantly lower with PFO than HPO (5.36 vs. 10.03 g/100 g total FA, $P<0.01$). Feeding 47 g/d of fish oil (2.5% of TMR as fed) increased the concentrations of EPA and

DHA in milk fat from undetectable levels, in C goats, to respectively 0.40 and 0.27 g/100 g total FA ($P<0.01$) and more than doubled the content of beneficial n-3 PUFA (2.52 vs. 1.21 g/ 100 g total FA, $P<0.01$). Apparent transfer efficiency of EPA and DHA in milk fat were respectively 7 and 8.1%, similarly to previous results in dairy goats supplemented with unprotected fish oil (Cattaneo *et al.,* 2006). Feeding PFO increased milk *trans* C18:1 fatty acids, intermediate products in the ruminal metabolism of PUFA. This suggests that part of encapsulated fish oil was released as free FA and altered rumen biohydrogenation pathways of C18:2 and C18:3 of the basal diet, inhibiting the final reduction of *trans* C18:1 to stearic acid. In conclusion, feeding encapsulated fish oil was an effective means of enriching milk fat with beneficial n-3 fatty acids, without affecting DMI or milk yield and gross composition.

Table 1. Milk fatty acid composition (g/100 g total FA) at 21 d of lactation in dairy goats fed a basal diet (C), a basal diet plus 47g/d of hydrogenated palm oil (C + HPO), or a basal diet plus 47g/d of protected fish oil (C + PFO).

Fatty acids	C	C + HPO	C + PFO	SEM	P-value
6:0	3.21	2.86	4.38	0.52	0.60
8:0	2.86^{b}	3.07^{b}	3.72^{a}	0.24	0.03
10:0	8.62	9.18	11.24	0.94	0.09
12:0	3.29^{B}	3.52^{b}	4.90^{Aa}	0.41	0.01
14:0	9.02	9.11	9.94	0.48	0.26
14:1	0.07	0.09	0.13	0.05	0.07
16:0	25.22	26.38	24.57	0.08	0.29
16:1 n7	0.82^{bB}	1.07^{aB}	1.69^{A}	0.08	0.01
18:0	13.80^{C}	10.03^{B}	5.36^{A}	2.26	0.01
cis 18:1	22.71	21.27	17.84	1.06	0.24
trans 18:1	2.60^{B}	4.41^{b}	7.02^{Aa}	1.09	0.01
18:2 n6	3.90^{A}	4.04^{A}	3.37^{B}	0.12	0.01
18:3n3	0.36	0.39	0.35	0.09	0.57
18:3 n6	0.12	0.04	0.04	0.02	0.07
18:4 n3	0.75	1.18	1.06	0.02	0.46
20:4 n6	0.22^{Aa}	0.27^{b}	0.29^{B}	0.02	0.01
20:5n3 EPA	0.01^{B}	0.02^{B}	0.40^{A}	0.09	0.01
22:5n3	0.08^{B}	0.09^{B}	0.43^{A}	0.06	0.01
22:6n3 DHA	0.02^{B}	0.02^{B}	0.27^{A}	0.06	0.01
PUFAs	5.47	6.07	6.31	0.06	0.16
n-3 PUFAs	1.21^{B}	1.70^{B}	2.52^{A}	0.06	0.01
≥20C n3 PUFAs	0.10^{B}	0.13^{B}	1.10^{A}	0.07	0.01

^{a,b} $P<0.05$ and ^{A,B} $P<0.01$.

References

Cattaneo, D., V. Dell'Orto, G. Varisco, A. Agazzi and G. Savoini. 2006. Enrichment in n-3 fatty acids of goat's colostrums and milk by maternal fish oil supplementation. Small Rum Res. 64, 22-29.

SAS, 2006. SAS Institute User's Guide Version 9.1: Statistics. SAS Institute, Cary, NC, USA.

Savoini, G., A. Agazzi, G. Invernizzi, D. Cattaneo, L. Pinotti and A. Baldi. 2010. Polyunsaturated fatty acids and choline in dairy goats nutrition: production and health benefits. Small Rum Res. 88, 135-144.

Metabolic imprinting effect in beef production: influence of nutrition manipulation during an early growth stage on carcass characteristics and intramuscular fat content of longissimus muscle in Wagyu (Japanese Black)

T. Gotoh[1], K. Etoh[1], K. Saitoh[2], K. Metoki[2], S. Kaneda[2], T. Abe[2], T. Etoh[1], Y. Shiotsuka[1], R. Fujino[1], K. Matsuda[1], H. Suzuki[2], H. Hasebe[2], F. Ebara[1], J. Wegner[3] and S. Tabata[1]
[1]*Kuju Agricultural Research Center, Faculty of Agriculture, Kyushu University 8780201, Japan*
[2]*Kuju, Japan, National Livestock Breeding Center (NLBC), 961-8581, Fukushima, Japan*
[3]*Research Institute for Biology of Farm Animals, 18196 Dummerstorf, Germany*

Introduction

In Japan, Wagyu (Japanese Black cattle) are known for their excellent marbled beef which is achieved by feeding grain with a considerable amount of concentrate (4,000 to 5,000 kg altogether, until slaughter at 28-30 months of age). More than 90% of the concentrate is imported from foreign countries. Cattle have an important ecological niche that capitalises on the symbiotic relationship between fibre fermenting ruminal microbes and mammalian demand for usable nutrients as a ruminant. Our main goal is to produce a high-quality safe beef product while maximising the use of domestic grass resources. We would like to challenge this project by applying Wagyu's potential, which is a high ability to accumulate intramusucular fat. It has been shown that alterations in foetal and early *postnatal* nutrition and endocrine status may result in developmental adaptations that permanently change the structure, physiology, and metabolism in the adult life of rats, mice, domestic species and humans (Levin *et al.*, 2000). This observed by nutritional treatment during not only foetal but also the neonatal early growth period. This phenomenon is referred to as 'metabolic imprinting or metabolic programming' based on medical research regarding 'the developmental origins of health and disease (DOHaD)'. However, there are very few reports of metabolic imprinting in cattle. We have already presented effects of nutrition during the early growth period on meat-quality related gene expression, myofiber type composition, etc. till 20 months of age (Gotoh *et al.*, 2007). In this study, we investigated how the metabolic imprinting effect of differences in feeding during an early growth stage influences on the carcass characteristics and the intramuscular fat content in Wagyu at slaughter age.

Material and methods

Japanese Black steers were randomly allocated to 2 groups. The high energy group (HE: n=12) was treated by intensified nursing (maximum intake of 1.8 kg per day) till 3 months of age and was fed a high-concentrate diet for 4 to 10 months of age. On the contrary, the Roughage group (R: n=11) was treated by normal nursing (maximum intake of 0.6 kg per day) and was fed only roughage (orchardgrass hay) *ad libitum* from 4 to 10 months of age. Furthermore, nursing was used for each quality milk replacer in every group. After feeding at 10 months of age, both groups were fed only roughage (orchardgrass hay) *ad libitum* from 10 to 14 months of age. Subsequently all animals were put out to the same pasture and grazed until 20 months of age. Consequently, after that both groups were fed only roughage (orchardgrass hay) *ad libitum* from 21 to 31 months of age and then slaughtered at 31 months of age. Samples of tissues from the *longissimus* muscles (LM) in all animals were collected (Figure 1). After slaughter, we selected 5 carcasses from each group and measured weights of muscle, fat, bone and others in both groups. We also investigated intramuscular fat content by Soxhlet methods in LM in both groups.

Results and discussion

The average live weight at 30 months of age was always significantly higher in the HE group (576±40 kg) than in the R group (527±36 kg) ($P<0.05$). The weights of carcass and total fat in carcass were significant larger in group HE than in group R ($P<0.05$) (Figure 2). In the muscle weight in carcass,

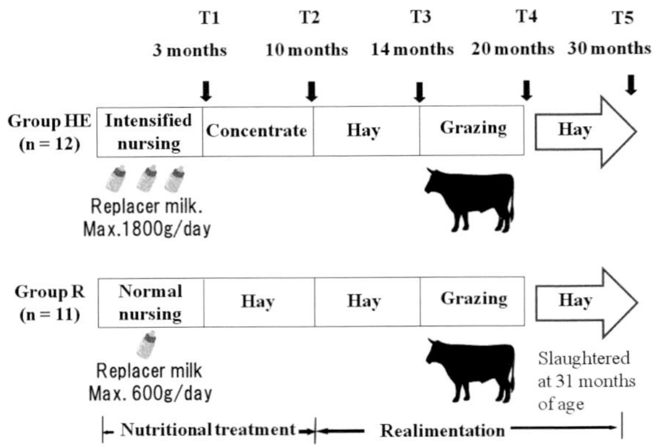

Figure 1. Feeding system used in this study.

the larger tendency observed in group HE than in group R (P=0.06) (Figure 2). By investigating the main 18 muscles, the following muscles were found significantly larger in group HE than in group R: *M. infraspinatus*, *M. serratus ventralis cervicis*, *M. serratus ventralis thoracis*, *M. supraspinatus*, and *M. semimembranosus* (P<0.05). Regarding fat accumulation, the weights of subcutaneous and perirenal fat were significantly larger in group HE than in group R (P<0.05). On the other hand, the weight of fat in visceral cavity was significantly smaller in group HE than in group R (P<0.05). Meanwhile, there were no significant differences in the weights of intermuscular fat and fat in thoracic cavity between groups HE and R. The intramuscular fat content in LM was significantly larger in group HE (13.2±4.7%) than in group R(9.4±3.6%) (P<0.05). These results indicate that the high energy treatment during the early growth phase influenced carcass characteristics and intramuscular fat content of LM in Wagyu fattened by only grass at slaughter age (31 months).

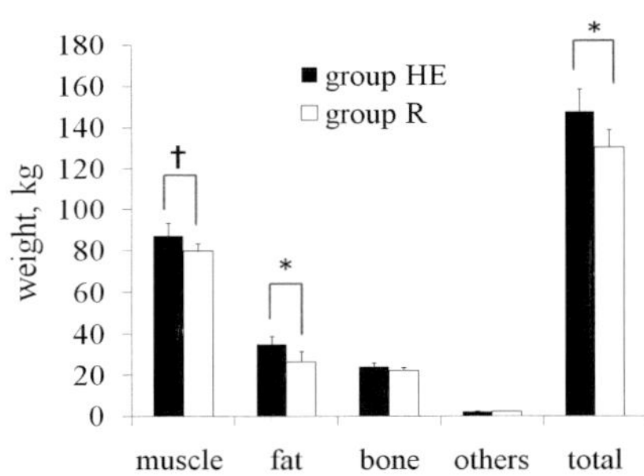

*Figure 2. Weights of muscle, fat, bone, and others in carcasses of groups HE and R. Group HE: n=5. Group R: n=5. * Significantly different between groups HE and R (*P<0.05*); *†* P=0.06.*

References

Gotoh, T., T. Fumita, T. Etoh, K. Hayashi, Y. Nakamura, J. Wegner and H. Iwamoto, 2007. Influence of metabolic imprinting during early growth period on muscle development and intramuscular adipogenesis in Japanese Black steers. In: Ortigues-Marty, I. (ed.), Energy and protein metabolism and nutrition, EAAP No 124, Wageningen Academic Publishers, the Netherlands: 197-198.

Levin, B.E., 2000. The obesity epidemic: metabolic imprinting on genetically susceptible neural circuits. Obes. Res. 8: 342-347.

Genetic and production factors that influence the content of iron and zinc in the meat of prime lambs

L. Pannier[1], G.E. Gardner[1], A.J. Ball[2], D.L. Hopkins[2], R.H. Jacob[2], S.I. Mortimer[2], K.L. Pearce[1] and D.W. Pethick[1]
[1]School of Veterinary & Biomedical Science, Murdoch University, Murdoch, WA 6150, Australia
[2]Australian Cooperative Research Centre for Sheep Industry Innovation, Australia

Introduction

Red meats, especially beef and lamb, have always been assumed to contain substantial concentrations of iron and zinc, and accordingly marketing campaigns often use these attributes to differentiate against other meats. However there are limited studies on large data sets that properly benchmark these mineral levels, and what factors contribute to this nutritive value. Nutrient levels need to account for 10% or 25% of the recommended daily intake to achieve a 'source' or 'good source' claim from one serving of food. For iron and zinc this equates to 0.88 and 2.2 mg/100g of fresh meat, assuming a serve size of 135g (average for men and women; Food Standards Australia and New Zealand, 2004). There is little evidence to suggest that animals fed high dietary iron or zinc levels will induce higher iron or zinc levels in muscle, and therefore research is needed to understand the biology of the phenotypes and genotypes contributing to this nutritive value. The objectives of this study were to estimate the levels of iron and zinc in lamb from diverse genetic backgrounds and production systems, and to quantify the genetic and non-genetic factors that impact on iron and zinc concentrations.

Material and methods

Details of the design of the Sheep CRC's Information Nucleus Flock were presented by Fogarty *et al.* (2007). Briefly, approximately 2000 lambs were produced in 2007 from Merino and crossbred ewes located at 7 research sites across Australia (Katanning WA, Cowra NSW, Kirby NSW, Struan SA, Turretfield SA, Hamilton VIC, and Rutherglen VIC) which represented a broad cross section of Australian production systems. These ewes were artificially inseminated with semen from 93 key industry sires representing major production types in the Australian sheep industry. Progeny lambs were measured and sampled for carcass, meat and growth traits, and were slaughtered at a target average carcass weight of 21.5 kg. At 24 h *post mortem* the *longissimus thoracis lumborum* muscle was excised and samples (40 g) were taken for mineral analysis which was carried out according to the USEPA method 200.3 (USEPA, 1991). A linear mixed effects model was used to analyse the data and included fixed effects (and their relevant interactions) for site, sex, birth type-rear type, dam breed within sire type, sire type and kill group within site. Sire and dam identification were included as random terms and age at slaughter was used as a covariate in a second model to demonstrate the impact of age on iron and zinc levels. Sire estimates for iron and zinc concentration were analysed for their association with sire Australian Sheep Breeding Values (ASBVs) for post weaning weight (PWWT), C-site fat depth (PFAT), and eye muscle depth (PEMD) using a general linear model. This included sire type as a fixed effect, sire ASBV for PWWT, PFAT and PEMD as covariates, and the first order interaction between sire type and ASBVs.

Results and discussion

The phenotypic linear mixed effects model used 1965 observations. Site, sex and kill group within site were significant ($P<0.01$) for the iron analysis. The average level of iron in lamb muscle was 2.05 mg/100 g (Std Dev = 0.4). All sites (range 1.84-2.38 mg/100 g) had greater iron levels than that required to claim lamb as a 'source' of iron and had on average 95% of that required to achieve a 'good source' claim, with lambs at Cowra having the lowest and lambs at Kirby having the highest levels. Females had about 4% higher iron concentration than males and lambs from groups killed at a later age generally had more iron (Figure 1A). For the zinc analysis, site, sire type and kill group within site were significant (*P*<0.01). The average level of zinc in lamb muscle was 2.31 mg/100 g (Std Dev = 0.4). All sites, except for Cowra (2.15 mg/100 g), had greater zinc levels than that required to claim lamb as a 'good source' of zinc. Maternal sired lambs had 5% more iron than Merino and terminal sired lambs, and lambs of groups killed at a later age generally had a higher zinc concentration (Figure 1B). The effect of age at slaughter was significant (*P*<0.01; Figure 1) increasing iron and zinc concentration by almost 50% and 24% across the age range, indicating that age at slaughter is a key driver of iron levels, and to a lesser extent for zinc. Given that older animals express more oxidative fibres (Brandstetter *et al.*, 1998) it is expected that muscle type is an important determinant of mineral levels, however this aspect was not part of this study. Sire effects (independent of breed) on iron and zinc levels were significant (*P*<0.01) and within terminal sires the sire estimates for zinc reduced by 0.05mg/100g for every unit increase in PEMD ASBV (*P*<0.05). This trait may require attention given the current industry focus on selecting sires with high PEMD. No sire estimates correlated with the sire ASBVs for iron. In conclusion, both genetic and non-genetic factors had a significant impact on iron and zinc levels, and in the current data lamb achieved the levels required to underpin a claim of 'source' for iron, and 'good source' for zinc.

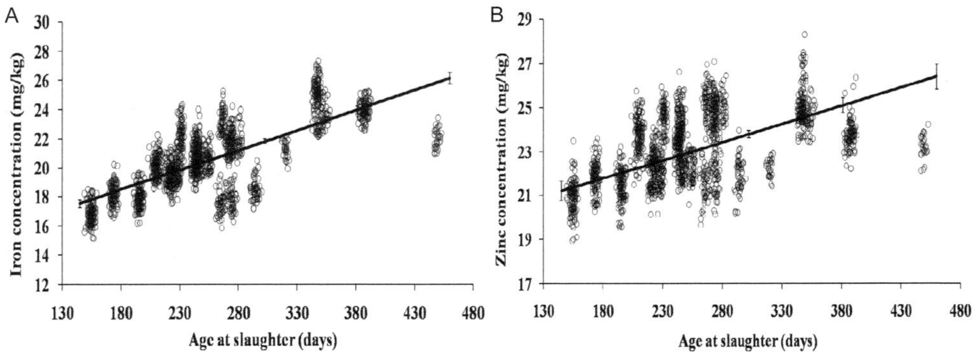

Figure 1. Effect of slaughter age on (A) iron and (B) zinc concentration (mg/kg wet tissue). Each cluster of points represents raw data from individual slaughter groups at the 7 sites.

References

Brandstetter, A.M., B. Picard and Y. Geay, 1998. Muscle fibre characteristics in four muscles of 5 growing bulls. I. Postnatal differentiation. Livest Prod Sci. 53, 15-23.

Fogarty, N.M., R.G. Banks, J.H.J. Van der Werf, A.J. Ball and J.P. Gibson, 2007. The information nucleus – a new concept to enhance sheep industry genetic improvement. Proc Ass Adv Anim Breed Genet. 17, 29-32.

Food Standards Australia and New Zealand, 2004. Initial Assessment Report of Proposal 293 Nutrition, Health and Related Claims, Canberra, Australia.

US Environmental Protection Agency (USEPA), 1991. Methods for the determination of metals in environmental samples. EPA/600/4-91-010.

Genetic and production factors that influence the content of intramuscular fat in the meat of prime lambs

D.W. Pethick[1], L. Pannier[1], G.E. Gardner[1], G.H. Geesink[2], A.J. Ball[2], D.L. Hopkins[2], R.H. Jacob[2], S.I. Mortimer[2] and K.L. Pearce[1]
[1]School of Veterinary & Biomedical Science, Murdoch University, Murdoch, WA 6150. Australia
[2]Australian Cooperative Research Centre for Sheep Industry Innovation, Australia

Introduction

Consumers are showing an increasing demand for juicy, flavoursome and tender meat, but in parallel are seeking lower fat, healthier food options. Intramuscular fat (IMF) has an important impact on palatability due to its specific contribution to juiciness and flavour and has been shown to contribute 10-15% of the variance in palatability (Dikeman, 1987). This can be compromised if the fat content is less than 3%, but is enhanced as IMF levels increase from 3-6% (Savell and Cross, 1988). Lamb contains about 4-5% IMF (Pethick *et al.*, 2005), and studies have demonstrated that this is the level required to achieve consumer satisfaction (Hopkins *et al.*, 2006). In recent years, increasing selection for lean growth has been linked to a decline in IMF level with subsequent detrimental effects on meat eating quality (Pethick *et al.*, 2005). Thus maintaining sufficient levels in lamb meat is important to ensure the sensory appeal for consumers. The objectives of this study were to estimate the levels of IMF of slaughter progeny representing diverse genetic backgrounds and production systems, and to test the effect of genetic and non-genetic factors that affect IMF levels in prime lambs.

Material and methods

Details of the design of the Sheep CRC's Information Nucleus Flock were presented by Fogarty *et al.* (2007). Briefly, about 2,000 lambs were produced in 2007 from artificially inseminated matings to Merino and crossbred ewes located at 7 research sites across Australia (Katanning WA, Cowra NSW, Kirby NSW, Struan SA, Turretfield SA, Hamilton VIC, and Rutherglen VIC) which represented a broad cross section of Australian production systems. The lambs were progeny of 93 key industry sires representing major production types in the Australian sheep industry and were measured and sampled for carcass, meat and growth traits. Lambs were slaughtered at a target average carcass weight of 21.5 kg. At 24 h *post mortem* the *longissimus thoracis lumborum* muscle was excised and samples (40 g) were taken for IMF analysis using a near infrared procedure in a Technicon Infralyser 450 (Perry *et al.*, 2001). IMF data was analysed using a linear mixed effects model including fixed effects (and their relevant interactions) for site, sex, birth type-rear type, dam breed within sire type, sire type and kill group within site. Sire and dam identification were included as random terms and hot carcass weight (HCWT) was used as a covariate. Sire estimates were analysed for their association with Australian Sheep Breeding Values (ASBVs) for post weaning weight (PWWT), C-site fat depth (PFAT), and eye muscle depth (PEMD) using a general linear model including sire type as fixed effect, sire ASBV for PWWT, PFAT and PEMD as covariates, and the first order interaction between sire type and ASBVs.

Results and discussion

The phenotypic linear mixed effects model used 1,881 observations. There were significant ($P<0.01$) effects for site, sex, sire type, kill group within site and HCWT. On average, lambs at Katanning (4.83%) had proportionately more IMF% compared to lambs at the other sites (average of 4.52 IMF% for other sites), and lambs at Rutherglen (3.62%) and Hamilton (4.00%) had proportionately less IMF% than lambs at the other sites. Females had about 0.20 IMF% units more than males, and Merino sired lambs had 0.42 IMF% units more than terminal and maternal sired lambs. Within each site there were marked differences between kill groups which varied by as much as 1 IMF% unit, however these differences showed no obvious trend across sites. HCWT had a positive relationship with IMF, with levels increasing by 1 IMF% unit across the 20kg range in HCWT (Figure 1A). Previous studies have described this relationship in cattle which demonstrated a linear increase in IMF by 0.47% for every 10 kg of HCWT between a range of 200-400 kg carcass weight (Duckett *et al.*, 1993; Pugh *et al.*, 2002). Sire effects (independent of breed) on IMF levels were significant (P<0.01) and sire estimates demonstrated a positive association with PFAT ASBV (P<0.05) (Figure 1B), implying the current selection objectives targeting sires with low PFAT ASBVs are likely to be reducing IMF content. In conclusion, both genetic and non-genetic factors had significant impacts on IMF and in the current data the average level of IMF (4.25%, Std.Dev =1.00) reached the Australian target to maintain consumer satisfaction.

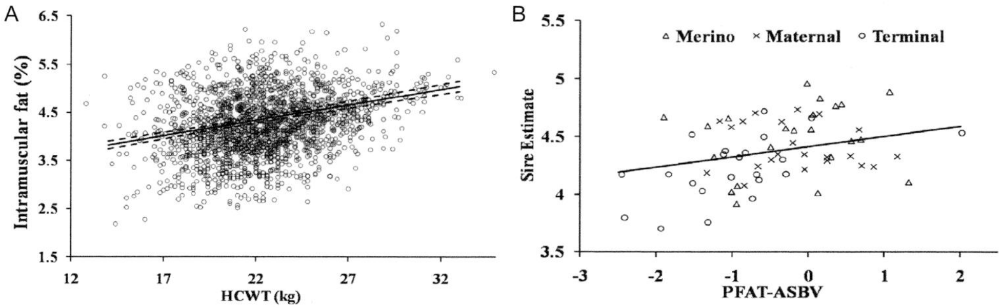

Figure 1.(A) Effect of hot carcass weight on IMF concentration (%);(B) Effect of PFAT ASBV on sire estimates of IMF for Merino, Maternal and Terminal sires.

References

Dikeman, M.E., 1987. Fat reduction in animals and the effect on palatability and consumer acceptance of meat products. Proceedings of 40th Annual RMC: 93-103.

Duckett, S.K., D.G. Wagner, L.D. Yates, H.G. Dolezal and S.G. May, 1993. Effects of time of feed in beef nutrient composition. J. Anim. Sci. 71, 2079-2088.

Fogarty, N.M., R.G. Banks, J.H.J. van der Werf, A.J. Ball and J.P. Gibson, 2007. The information nucleus – a new concept to enhance sheep industry genetic improvement. Proc Ass Adv Anim Breed Genet. 17, 29-32.

Hopkins, D.L., R.S. Hegarty, P.J. Walker and D.W. Pethick, 2006. Relationship between animal age, intramuscular fat, cooking loss, pH, shear force and eating quality of aged meat from young sheep. Austr. J. Exp. Agr. 46, 978-884.

Perry, D., W.R. Shorthose, D.M. Ferguson and J.M. Thompson, 2001. Methods used in the CRC program for the determination of carcass yield and beef quality. Austr. J. Exp. Agr. 41, 953-957.

Pethick, D.W., D'Souza, D.N., Anderson, C.A. and Muir, L.L., 2005. Eating quality of Australian lamb and sheep meats. Austr. J. Exp. Agr. 45, 465-601.

Pugh, A.K., D.W. Pethick, G.E. Tudor and B.L. McIntyre, 2002. The development of marbling in steers and heifers. Final report for the CICF of WA, Murdoch Uni., Perth.

Savell JW, Cross HR. 1988. The role of fat in the palatability of beef, pork, and lamb. In: Farrell, D.J. (ed.), Designing foods: animal product options in the marketplace. National Academy Press, 345.

The effect of linseed, Nutex® and Valomega® on production parameters and milk fatty acid pattern in dairy cows

W.M. Van Straalen and B.M. Tas
Schothorst Feed Research B.V., P.O. Box 533, 8200 AM Lelystad, the Netherlands

Introduction

The demand for milk with a higher content of unsaturated fatty acids and adapted W-3/W-6 ratio is increasing. A way to increase the unsaturated fatty acids in milk, is to increase the intake of linseed (Chilliard *et al.*, 2009). A possible disadvantage of the use of (untreated) linseed is that free α-linolenic acid will be hydrogenated in the rumen. To reduce the hydrogenation of α-linolenic acid, several extruded linseed products are available (Nutex® Dumoulin, Belgium and Valomega®, Valorex, France). The aim of this study was to test the effect of different linseed products on feed intake, milk production and –composition and fatty acid pattern of milk fat.

Material and methods

An experiment with 30 dairy cows in mid lactation with three treatments (linseed, Nutex® and Valomega®) was carried out. In the pre-period of three weeks all animals received the same diet. The basal diet was made up with (on DM basis) 42.5% grass silage, 42.5% corn silage, 5% dried alfalfa and 10% soybean meal and was given *ad libitum*. The concentrate allowance was based on production level. During the main period of 6 weeks the treatments were implemented. Linseed products were fed based on the linolenic acid content (1.4 kg/d linseed; 2.0 kg/d Nutex® and 1.6 kg/d Valomega® for the 2nd calvers and older animals and 75% from this amount for heifers). Feed intake and milk production and composition were averaged per week. Data from the pre-period were used as covariates in the statistical analysis and treatment effects were tested using ANOVA using Genstat software (VSN International Ltd., UK).

Results and discussion

There were no significant differences between treatment in feed intake and milk production parameters (Table 1). The supply of linseed products resulted in a numerical lower milk protein and fat content, and an increase of the proportion of unsaturated fatty acids in the milk (pre-period compared to main

Table 1. Average milk production and composition during the pre-period (week 1 to 3) and effect of linseed products on these parameters during the main period (week 4 to 9).

	Pre-period	Treatment			Lsd[1] (P<0.05)
		Linseed	Nutex®	Valomega®	
Basal diet (kg DM/d)	17.28	17.27	17.35	17.22	0.48
Concentrate[2] (kg DM/d)	5.49	5.50	5.45	5.53	0.22
Total (kg DM/d)	22.77	22.74	22.80	22.77	0.48
Milk (kg/d)	33.46	32.60	33.15	33.38	1.28
Fat (%)	4.04	3.92	3.86	3.92	0.20
Protein (%)	3.55	3.46	3.49	3.47	0.07

[1] Lsd = least significant difference.

[2] The amount of concentrate during the main period includes the linseed products.

period). Effects were less pronounced as found by Chilliard *et al.* (2009), which can be explained by the lower dose of linseed products.

There was no significant difference between treatments in fatty acid pattern with the exception of the higher linolenic acid content on the treatment with Nutex® compared to Valomega®. However, the supply of Valomega® resulted in a significant higher content of CLA in milk compared to the linseed treatment (Table 2). The effect on milk production parameters with Nutex® were later in time compared to linseed or Valomega®, which might be an indication that linolenic acid in Nutex® is less hydrogenated in the rumen than in linseed or Valomega®.

It can be concluded that linseed, Nutex® and Valomega® showed comparable effects on feed intake, milk production parameters and milk fatty acid composition.

Table 2. Fatty acid profile (g/100g fat) of milk during the pre-period (week 1 to 3) and the effect of linseed products on these parameters during the main period (week 5, 7 and 9).

Chemical structure	Pre-period	Treatment[1]			Lsd[2] (P<0.05)
		Linseed	Nutex®	Valomega®	
C6:0	2.24	2.14	2.20	2.14	0.120
C8:0	1.46	1.35	1.39	1.33	0.067
C10:0	3.59	3.19	3.27	3.15	0.184
C12:0	4.35	3.79	3.84	3.73	0.190
C14:0	12.48	11.42	11.5	11.66	0.313
C16:0	30.60	26.17	26.75	26.18	1.226
C16:1	1.87	1.56	1.65	1.64	0.123
C18:0	8.82	10.79	10.06	10.41	0.988
C18:1	16.67	18.95	17.83	18.28	1.483
C18:2	1.86	1.79	1.89	1.82	0.125
CLA, 9,11	0.44	0.64	0.69	0.71	0.115
C18:3	0.68	0.92^{ab}	0.98^{b}	0.91^{a}	0.066
Saturated fatty acids	70.60	65.60	65.72	65.18	1.405
MUFA	22.68	25.94	25.29	25.79	0.927
PUFA n-3	0.85	1.11^{ab}	1.14^{b}	1.08^{a}	0.046
Total CLA	0.45	0.63^{a}	0.70^{ab}	0.74^{b}	0.094
Trans-fatty acids	2.28	3.48	3.80	3.92	0.470

[1] Figures within the same row with different superscript differ significantly (P<0.05).
[2] Lsd = least significant difference.

References

Chilliard Y., C. Martin, J. Rouel and M. Doreau, 2009. Milk fatty acids in dairy cows fed whole crude linseed, extruded linseed, or linseed oil, and their relationship with methane output. J. Dairy Sci. 92, 5199-5211.

Effect of energy and protein restriction during nursing period on intramuscular fat accumulation of *longissimus dorsi* muscle of pigs during growing-finishing period

A. Ashihara[1], G. Yoshioka[2] and S. Ohtani[3]
[1]National Institute of Livestock and Grassland Science, Tsukuba, 305-0901, Japan
[2]Swine Research Department, Gifu Prefectural Livestock Research Institute, Minokamo, 505-0037, Japan
[3]Faculty of Applied Biological Sciences, Gifu University, Gifu, 501-1193, Japan

Introduction

We observed previously that high intramuscular fat (IMF) accumulation in *longissimus dorsi* muscle (LDM) in finishing pigs (n=151) related to be low growth rate during the nursing period (Ashihara *et al.*, 2008). From this observation, we hypothesized that the restriction of nutrition during the nursing period might increase accumulation of IMF in LDM. To test this hypothesis, we conducted an experiment to determine the effect of energy and protein restriction during nursing period (from 3 to 28 days of age) on accumulation of IMF in pigs during growing-finishing period.

Material and methods

Female piglets (n=8) were assigned to one of two groups on two days of age. During 3 to 28 days of nursing period, one group was fed a milk substitute to meet nutrient requirements of nursing pigs (Control). The other were given a feed amount of which was controlled as they had half of the energy and protein as the Control pigs had (Low). Vitamins and minerals were adequate in the feed of both groups. After the nursing period, all pigs were fed *ad libitum* until their body weight reached 115 kg and then slaughtered. The samples of LDM were taken from between the 1st and the last ribs of the right side of the carcasses. LDM weight and volume percentages of IMF were measured. Size and the numbers of adipose cells were measured from serial transverse LDM sections which were stained with Sudan III. The contents of type I collagen in LDM were determined by an ELISA method.

Results and discussion

The feed intake of the Low group was not different from that of the Control group during growing-finishing period. Although the body weight gains of the Low group were halves of the Control group during nursing period, the body weight gains during growing-finishing period of the two groups were not different (Table1). The weights of LDM and its myofiber area in the Low group were lower than those of the Control group (P<0.05). The contents of IMF and type I collagen in LDM in the Low group were higher than those of the Control group when the pigs reached 115 kg of body weight (P<0.05) (Table2). An important finding of this experiment was that higher accumulation of IMF in LDM of the Low group was associated with suppressed development of LDM and its higher contents of type I collagen and adipose cell number.

The restriction of energy and protein during the nursing period in pigs resulted in suppression of muscle development and this suppression did not recover throughout their growth. The suppression of muscle development might enlarge the spaces between each myofiber. The higher collagen I content might be a resultant of promoted development of connective tissue filling the spaces between each myofiber. We conclude that the suppression of muscle development and the promotion of development of connective tissue induced accumulation of IMF in LDM.

Table 1. Body weight gain and feed conversion ratio of pigs in Control group and in Low group.[1]

		Control (n=4)	Low (n=4)	*P-value*
Body weight gain				
from 3 to 28 days of age	kg/day	0.13±0.05	0.06±0.03	0.044
from 29 days of age to the slaughter days of age	kg/day	0.69±0.02	0.66±0.02	0.522
Feed conversion ratio				
from 3 to 28 days of age	kg/kg	1.2±0.3	1.2±0.4	0.931
from 29 days of age to the slaughter days of age	kg/kg	2.8±0.1	2.9±0.1	0.447

[1] Each data represents the means±SD. Control, pigs were given the milk substitute to meet the nutrient requirements from 3 to 28 days of age; Low, the half of Control; The *P*-value shows the level of significance for both groups.

Table 2. Area of muscle fiber, muscle weight and type 1 collagen content, intramuscular fat content, adipose cell size and number in Longissimus dorsi *muscles at the body weight of 115 kg in Control group and in Low group.*[1]

		Control (n=4)	Low (n=4)	P-value
Area of muscle fiber	10^2 μm^2	90.2±2.9	75.7±9.4	0.016
Muscle weight	kg	1.5±0.1	1.1±0.1	0.044
Type I collagen contents	mg/g	0.7±0.1	0.9±0.1	0.011
Intramuscular fat content	%	3.3±0.5	5.6±0.6	0.045
Adipose cell size	10^4 μm^3	28.6±3.4	35.7±6.0	0.106
Adipose cell number	per mm^2	8.1±3.2	13.5±2.0	0.050

[1] Each data represents the means±SD. Control, pigs were given the milk substitute to meet the nutrient requirements from 3 to 28 days of age; Low, the half of Control; The *P*-value shows the level of significance for both groups.

References

Ashihara, A., G Yoshioka, N Imaeda, M. Yayota and S Ohtani, 2008. Investigation of factors affecting intramuscular fat accumulation in pigs -growth performance, carcass traits, lipid fractions in serum and adipose cell size and number. Nihon Chikusan Gakkaiho (Japanese) 79, 227-234.

Effect of feeding dairy cows diets with different fermentable energy on milk cheese-making features

L. Calamari, F. Piccioli Cappelli, P. Bani, M.G. Maianti and G. Bertoni
Istituto di Zootecnica, Facoltà di Agraria, Via Emilia Parmense, 84, 29122 Piacenza, Italy

Introduction

The cheese-making aptitude of milk depends on chemical composition, renneting properties, bacteriological and fermentability traits. These traits are influenced by genetic, nutritional, environmental and management factors, Among the nutritional factors, the energy of the diet (i.e. carbohydrates composition and their fermentability), modifying the fermentations and the microbial syntheses in the rumen, influences fat and protein content of milk, titratable acidity and clotting properties (Bertoni and Piva, 1997; Piccioli *et al.*, 2009). A variable milk fermentability, also called acidification rate (AR), has been observed (Montel *et al.*, 2003; Coulon *et al.*, 2003). Furthermore, the lactation phase modifies the AR and milk from small farms generally performs better (Calamari *et al.*, 2003), but there is a lack of knowledge which brake any attempt to improve it. The trial aimed to evaluate the effect of changes in the dietary carbohydrates composition on milk traits, with particular regard to the AR.

Material and methods

The trial was carried out in an experimental barn, on 3 Italian Friesian cows in mid lactation and fed diets based on corn silage, alfalfa hay, ryegrass hay and concentrate. According to a Latin square experimental design with periods of 21 d each, three different isoproteic diets (CP was 15.6% on DM basis) were compared:, diet B with high starch (24%) and 50:50 forage:concentrate ratio (F:C) on DM basis; diet LS with low starch (19%) and 50:50 F:C; diet HF with high forage (F:C = 57:43) and high starch (24%).

During the trial, feed samples were taken weekly and analyzed for chemical-nutritional features. Individual DMI and milk yield were measured daily, and representative milk samples from morning milking were collected every 3-4 d and analyzed for: fat, protein, lactose, urea and titratable acidity (FT120, Foss Electric, Denmark); SCC (Fossomatic, Foss Electric); renneting properties (Formawin, Foss Electric); AR measured either on raw milk either on milk after pasteurization (80 °C × 10') (Calamari *et al.*, 2003); some milk enzymes involved in the antimicrobial activity of milk or related to the udder health (GGT, GOT, LDH, lactoperoxidase and xanthine oxidase) (Calamari *et al.*, 2003). Fecal samples were also collected at the end of each period and analyzed for pH and DM.

The results obtained during the last 7 d of each period were processed (Mixed procedure of SAS) considering the effect of diet, period, time (d from start of each period) and cow.

Results and discussion

The NDF content of the diets was 34.6±0.31% of DM (NDF from forage was 23.8±0.64% of DM), 37.8±1.09% of DM (24.2±1.85%) and 34.5±0.84% of DM (26.6±0.94%) for the B, LS, and HF diets respectively. The NDF contents of all diets were greater than the minimum suggested by the NRC (2001).

The treatments did not affect DMI (24.4±1.84 kg/head/d) and milk yield (31.1±3.14 kg/head/d); only a slight lower milk yield was observed with HF diet (29.6 kg/head/d). The different carbohydrates composition of the diets, and their different fermentability, did not affect the pH (on average

6.44±0.21) and the DM content of the feces. This suggests the digestive processes and the fermentation activity in the hindgut were not altered by the diets.

Despite the diets were isoproteic, a significant effect of the F:C ratio of the diets on milk urea was observed, with lower values (*P*<0.05) in milk for HF diet (21.8 mg/100 ml) compared with those of the other two diets (26.1 and 27.9 mg/100 ml in B and LS respectively). This confirms a diet effect on N metabolism. The F:C of the diets affected only slightly the milk fat content (4.46% in HF vs. 4.23 and 4.35% in B and LS respectively, n.s.).

A significant effect of dietary starch content on milk protein content was observed, with slightly lower values (*P*<0.05) in milk from LS diet (3.73%) compared with those of the other two diets (3.77 and 3.79% in B and HF respectively). These differences did not affect the titratable acidity (on average 3.81±0.28 °SH/50 ml) and the renneting properties (clotting time: 16.56±2.42 min; curd firming rate: 4.83±1.60 min; curd firmness: 33.62±6.51 mm).

The low SCC (62.2±32.7 n/µl) and LDH activity (167±57 U/l) of milk indicate a good udder health of the cows involved, without any effect of the diets. Also the activities of enzymes involved in the antimicrobial activity of milk were not affected by the diets. Furthermore, a slow milk AR was observed, without diet effect.

In conclusion our results confirm that high fermentable diets, if well balanced, do not alter digestive functions and animal health. Moreover they slightly affect milk fat and protein contents; furthermore, neither the milk acidification rate (although low), nor the renneting properties (always optimal) were modified by the different content of fermentable energy of the diets used in this trial.

References

Bertoni, G. and G. Piva, 1997. The evaluation of the dairy cows diet efficacy thorough milk quality. In: Enne, G. and G.F. Greppi (eds.), Animal production: advances in technology, accuracy and management. Elsevier, Amsterdam: 33-48.

Calamari, L, P. Bani, M.G. Maianti and G. Bertoni, 2003. New researches on the factors affecting milk acidification rate. Proc. 38 Simp. Int. Zootecnia 'Milk & research': 137-144.

Coulon, J.B., E. Rock and Y. Noël, 2003. Caractéristiques nutritionnelle des produits laitiers et variations selon leur origine. INRA Prod. Anim. 16, 275-278.

Montel, M.C., E. Beuvier and A. Hauwuy, 2003. Pratiques d'élévage, microflore du lait et qualités des produits laitiers. INRA Prod. Anim. 16, 279-282.

NRC, 2001. Carbohydrates, In: Nutrient Requirements of Dairy Cattle. Seventh Revised Edition. National Academic Press, Washington, D.C., 34-42.

Piccioli Cappelli, F., M.G. Maianti, S. Ferrandi, L. Calamari and P. Bani, 2009. Effects of diets with different content of starch and protein fed to dairy cows in early lactation on milk yield and traits. Italian J. of Anim. Sci., 8, Suppl. 2, 319-321.

Effects of dietary lysine levels on expression of adipogenesis related genes in muscle of growing pigs

M. Katsumata, A. Ishida, T. Kyoya and K. Nakashima
National Institute of Livestock and Grassland Science, Tsukuba 305-0901, Japan

Introduction

Feeding a low lysine diet to fattening pigs enhanced intramuscular fat (IMF) content in l*ongissimus dorsi (l. dorsi)* muscle (Katsumata *et al.*, 2005). However, the underlying mechanisms how dietary low lysine enhances IMF content in porcine muscle is still unknown. One possibility is that dietary low lysine promotes differentiation of intramuscular preadipocytes to adipocytes. If so, dietary low lysine may up-regulate expression of adipogenesis related genes, in particular transcription factors relevant to differentiation of preadipocytes such as PPARγ and C/EBPα (Mandrup and Lane, 1997), in porcine muscle. From these backgrounds, we conducted this study to elucidate effects of dietary lysine levels on expression of adipogenesis related genes in muscle of growing pigs.

Material and methods

We formulated two diets, low lysine (LL) and control diets. The diets were isoenergetic and isoprotein; providing 14.3 MJ/kg digestible energy and 16.1% crude protein. The control diet contained all essential amino acids in the recommended amounts including 1.16% lysine. The LL diet was similar but contained only 0.73% lysine. All the contents of those nutrients are expressed as on fed basis. Barrows aged 6 weeks were fed these diets for either one week (Experiment 1; n=6) or three weeks (Experiment 2; n=6). The average initial body weight of the barrows was 13.5 kg. *L. dorsi* and *rhomboideus* muscles were taken at the end of the experiments. Abundances of mRNA of adipogenesis related genes in the muscles were determined. Statistical significance between the two treatment groups was assessed by analysis of variance for a randomized-block design, where litter was block and dietary lysine level was main effect, using the GLM procedure of SAS (version 9.1, 2003).

Results and discussion

Experiment 1: Abundances of mRNA of SREBP1 both in *l. dorsi* and *rhomboideus* muscles were higher in the pigs fed the LL diet (Figure 1A; $P<0.05$). Further, abundances of mRNA of fatty acid synthase (FAS) in *l. dorsi* muscle and those of PPARγ and adiponectin in rhomboideus muscle tended to be higher in the LL group (Figure 1A; $P<0.10$). However, abundances of mRNA of C/EBPα were not affected by dietary lysine levels (Figure 1A). Experiment 2: Abundances of mRNA of FAS in *l. dorsi* muscle and those of PPARγ and adiponectin in *rhomboideus* muscle were higher in the LL group (Figure 1B; $P<0.05$). Abundances of PPARγ and adiponectin in *l. dorsi* muscle tended to be higher in the LL group (Figure 1B; $P<0.10$). However, those of C/EBPα were not affected by dietary lysine levels (Figure 1B). Thus, dietary low lysine was incapable to up-regulate expression of C/EBPα, an important transcription factor regulating adipocyte differentiation, in porcine muscle. This suggests that differentiation of intramuscular preadipocytes was not involved in underlying mechanisms of enhanced accumulation of IMF due to dietary low lysine. Although PPARγ is a transcription factor well known as a master regulator of adipocyte differentiation (Mandrup and Lane, 1997), it is also recognized to promote lipogenesis (Schadinger *et al.*, 2005). Further, an important function of another transcription factor SREBP1 also is to promote lipogensis (Horton, 2002). Thus, we infer that enhanced lipogenesis in already existed adipocytes rather than newly differentiation of preadipocytes to adipocytes plays a pivotal role in accumulation of IMF due to dietary low lysine in porcine muscle.

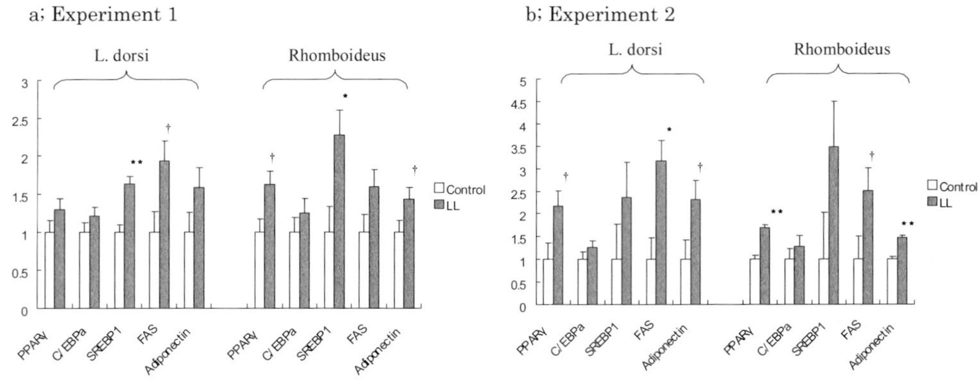

Figure 1. Effects of dietary lysine levels on abundances of m RNA of adipogenesis related genes in l. dorsi *and* rhomboideus *muscles of growing pigs. Bars are means ± pooled SE (n=6). (A) Experiment 1; (B) Experiment 2. Means from the control group are expressed as † (*P<*0.10), * (*P<*0.05) and ** (*P<*0.01).*

References

Horton, J.D, 2002. Sterol regulatory element-binding proteins: transcription activators of lipid synthesis. Biochem. Soc. Trans. 30, 1091-1095.

Katsumata, M., S. Kobayashi, M. Matsumoto, E. Tsuneishi and Y. Kaji, 2005. Reduced intake of dietary lysine promotes accumulation of intramuscular fat in the Longissimus dorsi muscle of finishing gilts. Anim. Sci. J. 76, 237-244.

Mandrup, S. and M.D. Lane, 1997. Regulating adipogenesis. J. Biol. Chem. 272, 5367-5370.

Schadinger, S.E., N.L.R. Bucher, B.M. Schreiber and S.R. Farmer, 2005. PPARγ2 regulates lipogenesis and lipid accumulation in steatotic hepatocytes. Am. J. Physiol. 288, E1195-E1205.

The magnitudes of promoted accumulation of intramuscular fat in responding to dietary low lysine in five distinct skeletal muscles of fattening pigs

T. Kyoya[1,2], A. Ishida[2], K. Nakashima[2], A. Toyoda[3], Y. Nakamura[3] and M. Katsumata[2]
[1]Department of Biological Production, Tokyo University of Agriculture and Technology, Fuchu-shi, Tokyo, 183-8509, Japan
[2]National Institute of Livestock and Grassland Science, Tsukuba, Ibaraki 305-0901, Japan
[3]College of Agriculture, Ibaraki University, Ami, Ibaraki 300-0393, Japan

Introduction

Some studies have shown that intramuscular fat (IMF) content in *longissimus dorsi* (*l. dorsi*) muscle is increased by feeding of low lysine diet in fattening pigs (Witte *et al.*, 2000; Katsumata *et al.*, 2005). However, the underlying mechanisms how dietary low lysine enhances IMF content in porcine muscle is still unknown. Higher IMF content are associated with higher proportion of oxidative or type I myofibres (Fernandez *et al.* 1995). Thus, magnitude of promoted accumulation of IMF in responding to dietary low lysine seems to be higher in muscles with higher proportion of oxidative or type I myofibre. We conducted this experiment to elucidate the relationships between magnitudes of promoted accumulation of IMF in responding to dietary low lysine and myofibre proportions in five distinct skeletal muscles. Five distinct skeletal muscles, *l. dorsi*, *rhomboideus*, *biceps fermoris* (*b. fermoris*), *gastrocnemius* and *soleus* muscles were selected in order to cover a wide range of myofibre proportion. According to our results, the order of proportions of oxidative and type I myofibre of these muscles are as follows:

oxidative fibres: *soleus* > *rhomboideus* ≥ *b. fermoris* ≥ *gastrocnemius* > *l. dorsi*.

type I fibres: *soleus* > *rhomboideus* > *b. fermoris* = *gastrocnemius* = *l. dorsi*.

Materials and methods

We formulated three diets, low lysine 1 (LL1) (3.43Mcal/kg digestible energy (DE), 14.1% crude protein (CP) and 0.60% lysine), LL2 (3.44Mcal/kg DE, 13.0% CP and 0.50% lysine) and control (3.43 Mcal/kg DE, 14.4% CP and 1.00% lysine) diets. All the contents are expressed as-fed basis. Thirty barrows aged 10 week, average body weight of which were 36 kg, were used in this experiment. In order to assess the initial IMF contents, the muscle samples were taken from six barrows at the beginning of the experiment. The rest, 24 barrows, were assigned to the control or low lysine group. Control group was given control diet from 10 to 18 week of age. Low lysine group was given LL1 diet from 10 to 14 week of age and LL2 diet from 14 to 18 week of age. The muscle samples were taken at the end of each period. Statistical significance between the two treatment groups was assessed using ANOVA for randomized block design, where litter was the block and diet was the main effect.

Results and discussion

Initial IMF contents varied among the five different skeletal muscles studied. *Rhomboideus*, a muscle with relatively high proportion of oxidative and type I fibres, had the highest content (3.3%), whereas, *soleus*, a muscle with extremely high proportion of oxidative and type I fibres, had the lowest content (1.3%). Although the proportions of oxidative and type I myofibre are higher in *rhomboideus* than in *l. dorsi*, the magnitude of promoted accumulation of IMF in responding to dietary low lysine were higher in *l. dorsi* (4.6 times higher than the initial contents) than in *rhomboideus* (2.3 times higher than the initial contents). In addition, although the proportions of oxidative and type I myofibres in *soleus* were the highest, the IMF contents in the *soleus* of pigs fed the low lysine diets for eight weeks was as low as 2.4%. In conclusion, although the magnitude of promoted accumulation of IMF in responding to dietary low lysine vary among five distinct skeletal muscles, it is unlikely to be associated with the proportion of oxidative or type I myofibre.

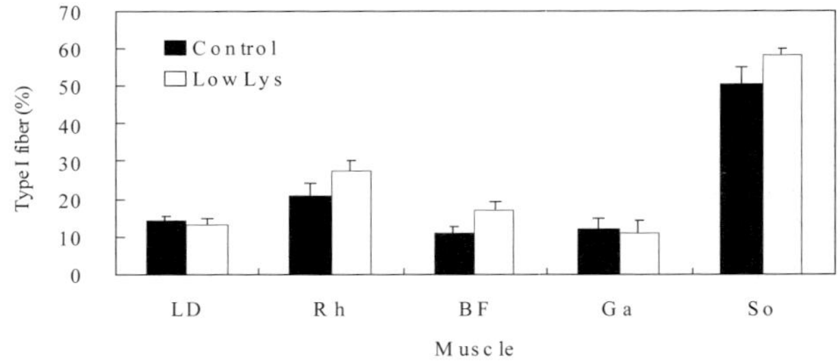

Figure 1. Effect of dietary low lysine for 8 weeks on percentage of type I myofibres in l. dorsi *(LD),* rhomboideus *(Rh),* b. fermoris *(BF),* gastrocnemius *(Gs), and* soleus *(So) muscles of pigs. Means±S.E. (n=6).*

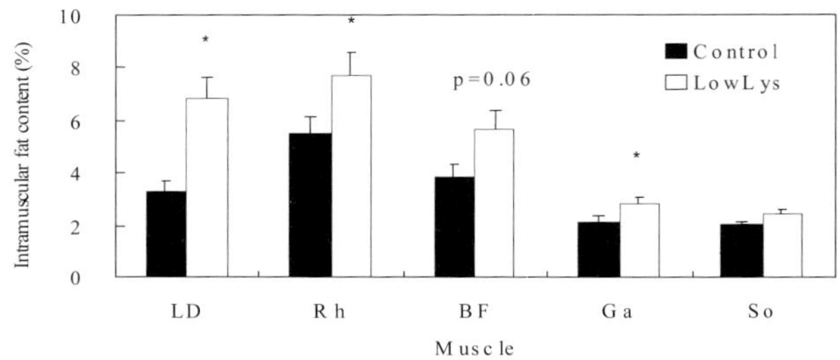

Figure 2. Effect of dietary low lysine for 8 weeks on intramuscular fat content in l. dorsi *(LD),* rhomboideus *(Rh),* b. fermoris *(BF),* gastrocnemius *(Gs), and* soleus *(So) muscles of pigs. Means±S.E. (n=6);* * P*<0.05.*

References

Fernandez, X., J. Mourot, A. Mounier and P. Ecolan, 1995. Effect of muscle type and food deprivation for 24 hours on the composition of the lipid fraction in muscles of large white pigs. Meat Sci. 41, 335-343.

Katsumata, M., S. Kobayashi, M. Matsumoto, E. Tsuneishi and Y. Kaji, 2005. Reduced intake of dietary lysine promotes accumulation of intramuscular fat in the longissimus dorsi muscles of finishing gilts. Ani. Sci. J. 76, 237-244.

Witte, D.P., M. Ellis, F.K. McKeith and E.R. Wilson, 2000. Effect of dietary lysine level and environmental temperature during the finishing phase on the intramuscular fat content of pork. J. Ani. Sci. 78, 1272-1276.

Influence of metabolizable energy intake on carcass traits and beef quality

P.V.R. Paulino, I.M. De Oliveira, S.C. Valadares Filho, M.I. Marcondes, C.A. Neves and S.F. Dos Reis
Departmento de Zootecnia, Universidade Federal de Viçosa, Av. P.H. Rolfs, sn. 36570-000. Viçosa-MG, Brazil

Introduction

Energy level in the diet and thus energy intake are important factors that affect carcass composition and meat palatability (Maiorano *et al.*, 2007). Declining energy intake limits animal growth and has a negative effect on the collagen matrix, resulting in decreased meat tenderness (Thompson, 2002). The Brazilian beef cattle systems are mainly based on tropical forage, where the animals graze along the year, being susceptible to fluctuations that occur on forage amount and quality. Thus, it is not uncommon to observe fluctuations on energy intake, which impact growth, muscle metabolism and probably carcass and meat quality traits (Lehnert *et al.*, 2006). So, the aim of this study was to asses the influence of energy intake on carcass traits and beef quality of beef steers.

Material and methods

Forty eight beef steers of three genetic groups (Nellore, F1 Angus × Nellore and F1 Simmental × Nellore) with initial body weight (BW) of 265.6±6.4 kg; 325.3±4.7 kg and 324.6±6.0 kg, respectively, were used.

The animals were randomly assigned into 3 different feeding regimes: maintenance (M - 12 animals – 4 from each genetic group) or *ad libitum* (AL) with two concentrate allowance levels: 1 (AL1) or 2% (AL2) of body weight on concentrate (36 animals – 6 on each concentrate level, within each genetic group). Corn silage was used as roughage source. The experimental treatments were designed in order to allow increasing metabolizable energy intake (MEI). After 112 days on feed all animal were slaughtered. A *Longissimus dorsi* (LD) sample was taken from each animal at the 12th rib, and immediately snap-frozen using liquid for myofiber classification. Another sample was taken for measurement of LD shear force (SF), fat (EE) and collagen content. After 24 h of chilling, rib eye are (REA) and subcutaneous fat thickness (SFT) were measured.

Data were analyzed by ANOVA and treatments means were compared by orthogonal contrasts using SAS.

Results and discussion

MEI of the animals fed at M (123.03 kcal/kg$^{0.75}$) was much lower (P<0.05) than the values observed for AL1 and AL2 animals (257.02 and 267.12 kcal/kg$^{0.75}$, respectively). As a result, carcass weight (CW), carcass fat content and dressing were greater on AL than on M fed animals (Table 1). REA and SFT did not differ (P<0.05) between AL fed animals but was greater (P>0.05) than that of the M fed animals (Table 1), demonstrating lower muscle development and fat accretion due to energy restriction.

Muscle fiber frequency was not altered (P>0.05) by MEI (Table 1) and glycolytic fibers were the most predominant in LD muscle (58.90%). Probably the time on feed, mostly for the M fed animals, was not sufficient to elicit changes in muscle fibers frequency. Lehnert *et al.* (2006) did not find any difference on the proportions of muscle fibers from restricted fed animals during 114 days when compared to the AL ones. LD shear force did not differ (P>0.05) between the animals fed *ad libitum*, but was lower than that observed on M fed animals (Table 1). This result may suggest a more intense *cold shortening* on the carcass of M animals as they were much leaner and practically

Table 1. Carcass traits, meat quality and LD myofiber frequency.

Variables	Feeding regime			*P*-value		CV (%)
	AL1	AL2	M	M vs. AL	AL1 vs. AL2	
CW (Kg)	264.39	281.22	200.17	<0.0001	0.0059	6.8
Dressing (%)	57.57	58.88	57.03	0.0139	0.0074	2.4
REA (cm^2)	75.84	75.35	58.96	<0.0001	0.8122	8.5
SFT (mm)	6.71	6.40	1.68	<0.0001	0.6820	41.0
Carcass EE (%)	21.54	21.43	15.38	<0.0001	0.8839	11.0
SF (kgf)	3.59	3.68	4.04	0.0975	0.3975	16.8
Glycolytic (%)	61.42	58.05	57.22	0.4280	0.2902	16.0
Intermediate (%)	29.75	31.08	32.52	0.2638	0.4775	18.0
Oxidative (%)	8.83	10.87	10.24	0.8731	0.4128	74.3

[1]CV (%) = coefficient of variation.

devoided of subcutaneous fat, which protects the carcass during cooling (Fiems *et al.*, 2000). LD myofibrilar fragmentation index (MFI) and collagen content were also evaluated, but no difference among treatments were detected ($P>0.05$), with mean values of 51.67% and 6.36 mg/g of muscle, respectively.

Acknowledgements

This research was finanacially supported by FAPEMIG.

References

Fiems, L.O., S. De Campeneere, S. De Smet, G. Van de Voorde, J.M. Vanacker and Ch.V. Boucqué, 2000. Relationship between fat depots in carcasses of beef bulls and effect on meat colour and tenderness. Meat Sci. 56, 41-47.

Lehnert, S.A., K.A. Byrne, A. Reverter, G.S. Nattrass, P.L. Greenwood, Y.H. Wang, N.J. Hudson and G.S. Harper, 2006. Gene expression profiling of bovine skeletal muscle in response to and during recovery from chronic and severe undernutrition. J. An. Sci. 84, 3239-3250.

Maiorano, G., C. Cavone, R.J. McCormick, A. Ciarlariello, M. Gambacorta and A. Manchisi, 2007. The effect of dietary energy and vitamin E administration on performance and intramuscular collagen properties of lambs. Meat. Sci. 76, 182-188.

Thompson, J., 2002. Managing meat tenderness. Meat. Sci. 62, 295-308.

Dietary β-alanine modifies dipeptide levels in the brain and breast muscle of chickens

S. Tomonaga[1], Y. Kaji[2], M. Matsumoto[2], H. Murakami[2], K. Kaneko[3], Y. Kido[4] and M. Furuse[1]
[1]Faculty of Agriculture, Kyushu University, Higashi-ku, Fukuoka-shi, Japan
[2]National Agricultural Research Center for Kyushu Okinawa Region, Koshi-shi, Japan
[3]Fukuoka Prefectural Yame Agricultural High School, Yame-shi, Japan
[4]Faculty of Human Environment, Kyoto Prefectural University, Sakyo-ku, Kyoto-shi, Japan

Introduction

Carnosine (β-alanyl-L-histidine) and anserine (β-alanyl-1-methyl-L-histidine) are present in high levels in the muscle and brain of mammalian and avian species (Tamaki *et al.*, 1976; Biffo *et al.*, 1990). They are recognized as antioxidants (Kohen *et al.*, 1988). Carnosine is synthesized from β-alanine (β-Ala) and histidine, and anserine is mainly formed from carnosine by a methylation reaction (Boldyrev and Severin 1990). In the present study, whether dietary β-Ala could increase these dipeptide levels of the brain and breast muscle was investigated in chickens.

Materials and methods

In experiment 1, male layer chicks (Julia) were orally administered either distilled water or β-ala (22 mmol/kg) in a volume of 1 ml twice a day for five consecutive days (from 2 to 6 days old) using a syringe. Twelve hours after the final administration, birds were killed by an overdose of sodium pentobarbital, and the brain and breast muscle were removed and stored at -85 °C prior to analysis. In experiment 2, male broiler chicks (Chanky, 4 weeks old) were fed a commercial diet (control) or the diet containing 0.5, 1 or 2% of β-Ala for 4 weeks. At 8 weeks of age, all birds were sacrificed after fasting for 12 hours. The brain and breast muscle were collected and stored at -30 °C prior to analysis. In experiment 3, male broiler chicks (Chanky) were used. Other materials and methods were the same as experiment 1. Dipeptides and amino acids in all samples were analyzed by HPLC as described elsewhere (Tomonaga *et al.*, 2005). Data in experiments 1and 3 were statistically analyzed by t-test. In experiment 2, one way analysis of variance was used. When significant effects were detected ($P<0.05$), comparisons between means were carried out using the Tukey–Kramer test. Statistical analysis was conducted using a commercially available package StatView (SAS, 1998).

Results and discussion

In experiment 1, oral administration of β-Ala increased both dipeptide levels in the brain. The treatment also increased carnosine, but decreased anserine levels in the breast muscle. In experiment 2, dietary β-Ala dose-dependently increased both dipeptide levels in the brain. However, both dipeptides in the breast muscle were not influenced. In experiment 3, oral administration of β-Ala increased both dipeptide levels in the brain. The treatment also increased both dipeptides in the breast muscle (Figure 1). In amino acids, taurine levels in the brain and breast muscle were significantly decreased in all experiments. It was previously reported that depletion of taurine altered functions of the muscle (De Luca *et al.*, 1996) in rats. Dietary β-Ala might alter physiological functions of birds due to the depletion of tissue taurine. These results suggest that dietary β-ala could modify dipeptide levels in the brain and breast muscle of chickens and the effects would be altered by age and/or type of chickens.

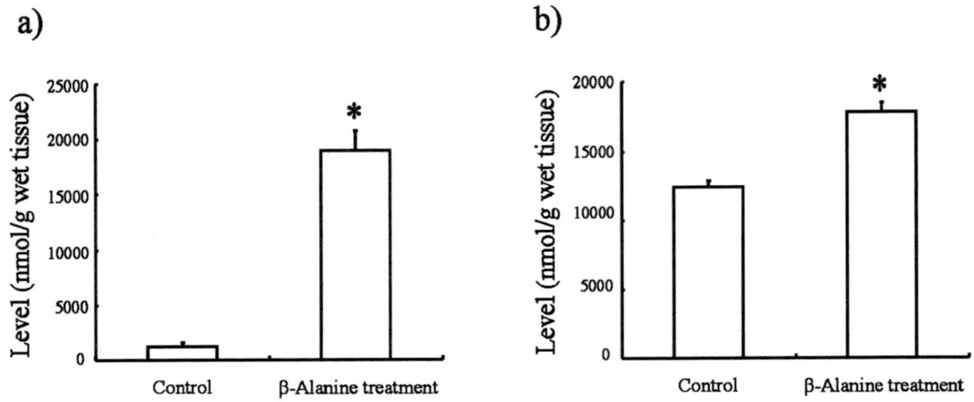

*Figure 1. Effects of β-alanine on (A) carnosine and (B) anserine levels in the breast muscles of broiler chicks. The numbers of birds used were seven in all groups. Data are expressed as mean ± standard error of the mean. *Significantly different from the control group (*P*<0.0001).*

References

Biffo, S., M. Grillo and F.L. Margolis, 1990. Cellular localization of carnosine-like and anserine-like immunoreactivities in rodent and avian central nervous system. Neuroscience 35, 637-651.

Boldyrev, A.A. and S.E. Severin, 1990. The histidine-containing dipeptides, carnosine and anserine: distribution, properties and biological significance. Adv. Enzyme Regul. 30, 175-694.

De Luca, A., S. Pierno, D.C. Camerino, 1996. Effect of taurine depletion on excitation-contraction coupling and Cl^- conductance of rat skeletal muscle. Eur. J. Pharmacol. 296, 215-222.

Kohen, R., Y. Yamamoto, K.C. Cundy and B.N. Ames, 1988. Antioxidant activity of carnosine, homocarnosine, and anserine present in muscle and brain. Proc. Natl. Acad. Sci. USA 85, 3175-3179.

Pegova, A., H. Abe and A. Boldyrev, 2000. Hydrolysis of carnosine and related compounds by mammalian carnosinases. Comp. Biochem. Physiol. B 127, 443-446.

Statistical Analysis Systems (SAS). 1998. Stat View, Version 5. SAS Institute, Cary, NC.

Tamaki, N., H. Iizumi, N. Masumitu, A. Kubota and T. Hama, 1976. Species specificity on the contents of anserine and carnosine. Yakugaku Zasshi 96, 1481-1486 (in Japanese).

Tomonaga, S., Y. Kaji, T. Tachibana, D.M. Denbow and M. Furuse, 2005. Oral administration of β-alanine modifies carnosine concentrations in the muscles and brains of chickens. Anim. Sci. J. 76, 249-254.

Part 12. *In vitro* techniques and feed nutritive value

In vitro methods to determine rate and extent of ruminal protein degradation[1]

G.A. Broderick[1] *and S. Colombini*[2]
[1]*USDA Dairy Forage Research Center, Madison, WI, USA*
[2]*Università degli Studi, Dipartimento di Scienze Animali, Italy*

Abstract

Rates of ruminal protein degradation are required in most of the current nutritional models; however, protein degradation in the rumen is very complex and difficult to measure reliably using simple *in vitro* and *in situ* methods. *In vitro* assays based on using mixed ruminal organisms in which microbial N incorporation is quantified or in which microbial growth is inhibited have proven to be the most robust for estimating rate and extent of protein degradation. Use of gas production to estimate N incorporation or release has become widely adopted. Recent developments have allowed study of the time course of protein breakdown so degradation rates can be determined. The inhibitor *in vitro* procedure has been useful in a number of applications but is limited to short term incubations of 4 to 6 h. Also, this assay has not been widely adapted beyond the laboratory where it was developed, possibly because it requires large numbers of assays for ammonia and total AA. Under some conditions, extents of protein hydrolysis obtained using several different cell-free proteases have been correlated to extents of *in situ* degradation but, thus far, cell-free proteases have not proven completely reliable for predicting rates of ruminal protein degradation. Future research might be directed toward identifying how protein degradation activities of cell-free proteases differ from that of the proteases elaborated by mixed ruminal organisms.

Introduction

Ruminants obtain the amino acids (AA) required for tissue protein synthesis and metabolism from protein produced by the microbial population in the rumen plus dietary protein that escapes microbial degradation in the rumen. Microbial protein formation is largely a function of fermentability of dietary carbohydrates; the AA pattern of this source is of better quality than most of the dietary ingredients commonly fed to domestic animals (Schwab, 1996). However, microbial protein will not meet all of the AA requirements of ruminants at normal to high levels of productivity. There is a large body of literature documenting production responses to increased supply of rumen-undegraded protein (RUP) in dairy cattle (e.g. Faldet *et al.*, 1991; Broderick, 1992; Brito and Broderick, 2007), although effects have often been surprisingly small (Santos *et al.*, 1998). Similar responses have been reported in other ruminant species. Protein feeding systems now require information on the amount of dietary protein escaping the rumen. Most of these systems predict RUP from simple models using single rates of ruminal degradation (kd) and passage (kp) for all of the protein in a specific feedstuff (e.g. NRC, 2001). Protein escape is computed from the ratio: kp / (kd + kp) (Ørskov and McDonald, 1979). The Cornell model (Fox *et al.*, 2004) applies different degradation and passage rates to each of several crude protein (CP) fractions within a feed. However, Grabber and Coblentz (2009) showed inconsistencies between the CP fractions based on the Cornell model and those determined using biological methods (*in situ* and protease assays); these differences resulted in conflicting estimates of ruminal protein degradability and highlighted the need to complement data on pool sizes of CP fractions with estimates of degradation rates made using ruminal microbes. Lanzas *et al.* (2007) concluded that better assays are necessary for determining the degradation rates used for the various protein fractions in the Cornell model.

[1] Mention of any trademark or proprietary product in this paper does not constitute a guarantee or warranty of the product by the USDA or the Agricultural Research Service and does not imply its approval to the exclusion of other products that also may be suitable.

If one can match the metabolizable protein (MP) and metabolizable AA supplied from ruminal microbes with MP from RUP of complementary AA pattern, then productivity could be maintained at lower N intakes. Generally, unreliable information on protein degradation causes producers to over-feed protein. Nutrition advisors are even more prone to recommending excessive dietary protein as a risk management practice. The N in this extra protein, whether it is degraded in the rumen or contributes MP that goes unutilized, is excreted completely in the urine. Routine assays that accurately and rapidly characterize degradabilty of feedstuff proteins will greatly improve the economic and environmental sustainability of ruminant production.

Although *in vivo* methods might be considered the 'gold standard', they are not appropriate for routine measurement of rate and extent of protein degradation in the rumen. These techniques suffer from considerable variation due to differences among animals and many other factors; this variation necessitates use of a lot experimental replication to obtain reliable results. The cost of adequate replication, when added to the expense of maintaining and sampling large animals, often makes *in vivo* studies prohibitively expensive. Moreover, an elaborate and costly *in vivo* study will reflect protein degradation determined under a limited set of conditions, restricted as to dietary composition and intake and level of production. *In situ* methods have been most widely applied for quantifying kinetic parameters of protein degradation. While assumed to reflect actual behavior of feedstuffs within the *in vivo* rumen, the *in situ* approach is subject to number of errors including: (1) non-digestion loss of soluble and particle-associated nitrogenous materials from the bag, (2) restricted access from microbial attack of the feeds encapsulated within the bag, and (3) the need to quantify microbial contamination of the feed residues remaining within the bag (Broderick and Cochran, 2000). The NRC (2001) model's RUP values, which derive from *in situ* results, were more than 25% greater than those obtained by omasal sampling; this difference may have resulted from the NRC model using ruminal passage rates that are too rapid (Broderick *et al.*, 2010). Furthermore, Huhtanen *et al.* (2008) and Rinne *et al.* (2009) both reported that, rather than using RUP values from *in situ* incubations, applying constant ruminal degradation coefficients for forage protein improved RUP predictions. A recent meta-analysis (Huhtanen and Hristov, 2009) showed that ruminal protein degradability predicted by NRC (2001) was not a significant factor for predicting milk protein yield and milk N efficiency in dairy cows.

Ruminal protein degradation is a very complex process. Comminution of large feed particles through rumination and other processes increases surface area available for microbial colonization. Dissolution of soluble dietary proteins also increases exposure to proteolytic attack. Numerous organisms elaborate an array of protein, peptide and AA degrading activities (Jouany, 1996; Wallace, 1996; Wallace and Cotta, 1988; Paster *et al.*, 1993; Attwood and Reilly, 1995; Atasoglu *et al.*, 2004). Soluble and insoluble proteins from the diet (plus smaller amounts of endogenous protein) are attacked by these enzymes. Although soluble N appears to be more rapidly degraded in the rumen (Krishnamoorthy *et al.*, 1983; Volden *et al.*, 2002), N solubility did not give reliable estimates of *in vivo* protein degradation of feedstuffs (Nocek *et al.*, 1979, Stern and Satter, 1984). Ahvenjarvi *et al.* (2009) recommended that RUP be considered the sum of intact protein, peptides and free AA of feed origin that pass out of the rumen undegraded. This logical definition implies that the commonly used term, 'bypass protein', is in fact a misnomer. Dietary protein 'escapes' the rumen and escape is the net resultant of competing processes – the balance between rates of microbial attack on protein, peptides and free AA and the passage of these fractions. Thus, RUP values are not constants but alter with rates of passage, which in turn vary with proportion of forage, particle size and specific gravity of the diet (Welch, 1982; Hristov *et al.*, 2003). Moreover, degradation rates have been observed to vary substantially from day to day in the same animals consuming similar amounts of the same diets (Ahvenjarvi *et al.*, 2009). Reports of variation in microbial populations among animals fed the same diet and, following diet alteration, of return of the microbial population to one characteristic of the animal (Palmonari *et al.*, 2010), also suggest that microbial degradation activities are not fixed entities.

A successful assay for protein degradation then must somehow mimic or predict the behavior of feed proteins under ruminal microbial attack, ideally yielding a rate quantitatively similar to some standard mean value and reflecting normal variation under *in vivo* conditions. However, feed protein characteristics that are only correlated with *in vivo* degradation rate could also be useful. Examples of the latter would be extent of hydrolysis with commercial proteases or a chemical property, such as near infrared reflectance, which describes aspects of protein structure that are related with, and can be calibrated to, rates of microbial protein degradation. This paper will concentrate on laboratory procedures for quantifying rates of microbial protein degradation in the rumen. Continuous culture methods will not be discussed.

Ruminal *in vitro* methods to estimate of protein degradability

Measuring formation of protein breakdown products using *in vitro* incubations with mixed ruminal organisms would seem a reliable way to quantify degradation rate. However, this approach is complicated by the fact that the microbes utilize as well as produce the nitrogenous compounds formed from protein degradation; moreover, microbial and residual feed proteins present in ruminal inocula are also degraded. Microbial catabolism of background inoculum proteins can be accounted for, at least theoretically, by using blanks that contain all components of the incubation except the test proteins. However, microbial incorporation of the breakdown products from test proteins cannot be easily 'blanked out' and degradation will not be a simple function of the net accumulation of nitrogenous end-products. It is necessary either to distinguish newly formed microbial protein from feed protein remaining undegraded or somehow to account for this underestimation of end-product formation. An example of this potential flaw is illustrated by an early report on use of *in vitro* ammonia production to assess protein degradation. Results showing an inverse relationship between starch content of feed and apparent ammonia release led to speculation that starch inhibited AA deamination (Warner, 1956). Different approaches to account for microbial utilization of protein breakdown products are discussed below.

Quantifying microbial growth

An approach for accounting for microbial uptake of degradation end-products was developed by Raab *et al.* (1983) by using a ruminal *in vitro* system to which starch was added in graded amounts to a series of incubations also containing the test protein. Net (blank corrected) gas production (Menke *et al.*, 1979) and ammonia concentration were measured for up to 24 h. A negative slope from regressing ammonia concentration on gas production resulted from greater microbial uptake of ammonia for growth at greater additions of starch. Regressions of ammonia concentration on gas production were extrapolated to zero gas production; that intercept, ammonia concentration at zero gas production, was assumed to represent the ammonia concentration that would have resulted when no N was incorporated into microbial protein. In this way, correction was made for microbial uptake of N in ammonia (and presumably free AA). Later, these workers found virtually all of the N end-product present as ammonia, with very little released N remaining as free AA and peptides, after 6 or 24 of incubation (Krishnamoorthy *et al.*, 1990). In both reports, extent of protein degradation most often was determined after 24 h incubations; however, proportionately less protein was degraded in the time period between 6 and 24 h. This suggested that the system may not exhibit linear microbial growth and protein breakdown at times later than 6 h. Computations of degradation rates from extents of degradation at 6 and 24 h suggested that results were more satisfactory at 6 h. Karlsson *et al.* (2009) modified this gas production approach, making multiple measurements of ammonia and gas production on the same incubation vessels, to estimate degradation rate and extent for protein concentrates. Thus, it was unnecessary to assume a linear relationship between microbial N uptake and gas production over all incubation times. An alternative method, such as incorporation of ^{15}N-ammonia, might be used to confirm that fermentation and gas production do not become uncoupled from microbial growth (Blümmel *et al.*, 1997).

Cone and colleagues have published a number of reports on using gas production to quantify microbial catabolism of feed nutrients in the rumen. Their most recent system for estimating protein degradation rate is based on pre-incubating ruminal inoculum with readily fermented carbohydrates to depress concentrations of nitrogenous compounds, making the medium N-limiting for microbial growth; gas production in subsequent incubations of test proteins in this medium was presumed to reflect release of N from protein degradation (Cone *et al.*, 2009). Results obtained with 19 feedstuffs (ranging from 7 to 56% CP) yielded a strong, negative correlation ($R^2 = 0.85$) between gas production over the first 20 h and ruminal protein escape, estimated using *in situ* incubations. The authors applied this calibration curve to estimate RUP from gas production of test proteins. However, because theoretical concerns about *in situ* methodology, it might be preferable to incubate standard proteins for which *in vivo* degradation rates are known to develop the calibration curve to determine degradation rates for the unknown proteins. The authors noted that, unlike the *in situ* method, release of N from the soluble ('washout') CP fraction and insoluble CP both contribute to degradation rates estimated from gas production in the N-limiting medium. This fraction includes small particles that have degradation rates similar to the large particles that do not wash out of *in situ* bags (Gierus *et al.*, 2006). The wide availability of systems for measuring gas production makes application of this and similar methodologies very timely.

An *in vitro* procedure was developed to estimate rate and extent of protein degradation in uninhibited ruminal inoculum by using ^{15}N-ammonia to quantify microbial uptake of protein breakdown products (Hristov and Broderick, 1994). Incubations were conducted for 6 h using buffered ruminal inoculum containing soluble carbohydrates and ^{15}N-ammonium sulfate. Degradation rate was computed from net (i.e. blank-corrected) release of ammonia plus net synthesis of microbial CP (estimated from ^{15}N-enrichment of isolated microbial cells and total solids). The degradation rates and ruminal escapes (computed assuming passage rate = 0.06/h) obtained for seven test proteins indicated that this approach yielded apparently reliable results. The procedure may prove useful with problematic protein sources such as legume and grass silages, which contain large amounts of background nonprotein N.

Inhibition of microbial growth

Borchers (1967) attempted to circumvent microbial uptake of protein degradation products by adding toluene, an inhibitor of AA deamination, to *in vitro* inocula and then quantifying degradation from AA release. Broderick (1978) added 1.0 mM hydrazine to ruminal *in vitro* incubations containing limited amounts of feed protein; protein degradation was quantified from N accumulation as AA plus ammonia, after correction for blanks without added protein. Subsequent work indicated that hydrazine was more effective than toluene for inhibiting AA catabolism but that direct utilization of AA was not prevented totally (Broderick and Baltrhop, 1979). This led to the inclusion of the antibiotic chloramphenicol with hydrazine in the inhibitor *in vitro* (IIV) system to completely shut off microbial uptake of protein degradation products and the computation of degradation rate from net release of N as ammonia and free AA (Broderick, 1987). Neither compound depressed proteolytic activity in short term incubations of up to 6 h.

The IIV method was used to estimate protein degradation rate by incubating limited substrate concentrations (i.e. under first-order conditions) with the inhibited ruminal inoculum; extent of degradation was computed using the observed rate and an assumed ruminal passage (typically 0.06/h). The IIV procedure successfully predicted relative differences in lactation performance of cows fed solvent and expeller soybean meal (SBM) (Broderick *et al.*, 1990), identified the optimal extent of heating required for protecting protein in soybeans (Faldet *et al.*, 1991), predicted RUP content of alfalfa hays (Tremblay *et al.*, 2000), and served as the basis of a solubility test (Hsu and Satter, 1995) and a near infrared spectrometric calibration (Tremblay *et al.*, 1996) to estimate protein degradability in roasted soybeans. However, the IIV procedure has several significant limitations: (1) accumulation of ammonia, AA and small peptides may make the system subject to end-product inhibition with rapidly degraded proteins; (2) degradation rates determined for feeds containing high levels of ammonia and free AA (e.g. grass and legume forage silages) are less accurate because breakdown of more slowly degraded residual protein must be quantified from appearance of additional ammonia and AA in the presence of high backgrounds; and (3) estimation of degradation rates from the gentle slopes obtained with slowly degraded proteins appears to be less accurate. These latter two problems also plague *in situ* methodology. Veresegyházy *et al.* (1993) reported that more than 19 m*M* of added ammonia or 12 m*M* of added free AA inhibited the rate of casein degradation in ruminal inoculum *in vitro*.

Regardless of these limitations, degradation rates determined with the IIV method compared favorably with those determined *in vitro* using uninhibited ruminal inoculum in which degradation rates were estimated using ^{15}N to account for microbial N uptake (Hristov and Broderick, 1994). Recent work showed that preincubation of ruminal inocula prior to addition of inhibitors increased degradation rates and reduced variation (Broderick *et al.*, 2004b) and that casein degradation followed simple first-order kinetics while degradation of solvent and expeller SBM were best described by bi-exponential models (Broderick *et al.*, 2004a). Additionally, a number of factors that influenced degradation rates measured for the rapidly degraded protein casein were not related to rates observed for solvent and expeller SBM (Broderick *et al.*, 2004a), suggesting that casein may not be a useful standard protein to include in incubations. Presumably, compounds other than hydrazine and chloramphenicol that inhibit microbial utilization of protein breakdown products, without altering proteolysis, would also be effective in this system. However, increasing chloramphenicol concentrations in inocula not containing hydrazine did not give quantitative recovery of ammonia and free AA (Broderick *et al.*, 2004b).

A Michaelis-Menten, enzyme kinetic approach based on the IIV system was developed partly to correct for the underestimation of degradation rates for rapidly degraded proteins (Broderick and Clayton, 1992). Rather than following the time-course of degradation in incubations using a single, limited amount of protein substrate (about 2-3 mg CP/ml of ruminal fluid equivalent in the inoculum), several protein levels, from very small to very high (2 to 45 mg CP/ml of ruminal fluid equivalent), are incubated for a single, relatively short time period in the Michaelis-Menten approach. Fractional degradation rate is estimated as the tangent through the origin of the velocity versus substrate concentration curve; the slope of this line is the ratio of the maximum velocity to the Michaelis constant (i.e. kd = Vmax / Km). Nonlinear regression analysis using the integrated Michaelis-Menten equation with data from 2-h incubations yielded direct estimates of this ratio (Broderick and Clayton, 1992).

Degradation rates estimated by the Michaelis-Menten IIV method for certain proteins were more rapid and more consistent with *in vivo* estimates of ruminal protein escape than were those obtained using the limited substrate IIV approach. For example, mean estimates of rate and escape for protein in casein and alfalfa hay were, respectively, about 0.30/h and 15% and 0.07/h and 40% by the limited substrate method, and 0.75/h and 8% and 0.20/h and 25% by the Michaelis-Menten procedure. Literature values of ruminal escape for casein (McDonald and Hall, 1954; Broderick, 1978) and alfalfa hay (NRC, 2001) were more similar to the Michaelis-Menten results. Degradation rates and ruminal escapes also were determined with the limited substrate and Michaelis-Menten IIV procedures for several legume forages: rates were similar for non-tannin containing forages but, for tannin-containing legumes, rates averaged less than 0.01/h in the limited substrate system and averaged about 0.05/h with the Michaelis-Menten approach (Broderick and Albrecht, 1997). It should be noted that determination of degradation rates with either IIV method requires methodology for rapid ammonia and total AA analyses.

Another limitation of the IIV method has been that it does not account for formation of small peptides produced during protein degradation. Broderick and Wallace (1988) quantified ruminal concentrations of peptides, amino acids and ammonia in sheep supplemented with different N sources and concluded that rapidly degraded proteins, such as casein, yielded substantial levels of peptides during degradation. Ruminal accumulation of peptides containing hydrophobic amino acids (Chen *et al.*, 1987) and proline (Yang and Russell, 1992) indicated that structure and the amino acid composition of the peptides influenced peptide utilization by the microorganisms in the rumen. Therefore, specific peptides may persist in the rumen depending on the composition of the protein fed to the animal. A widely used reagent applied for detection of amino acids is *o*-phthaldialdeide (OPA). In the presence of a sulfhydryl compound, OPA reacts with primary amines to form a complex that absorbs strongly at 340 nm and fluoresces at 450 nm (Roth, 1971). The OPA fluorescent reaction was adapted for estimating protein degradability in the IIV system (Broderick *et al.*, 2004b). However, fluorescent intensity of peptide OPA-reaction products diminishes dramatically with increasing peptide chain length; thus, the OPA fluorescence method largely fails to detect peptides due to quenching of fluorescence of derivatized peptides. Church *et al.* (1985) developed a spectrophotometric OPA assay, based on the absorbance at 340 nm; this method was a rapid, convenient, and sensitive procedure for quantifying proteolysis of milk proteins. Our preliminary results showed mean relative recoveries of 104 and 16% for dipeptides in, respectively, the OPA colorimetric (OPA-C) and OPA fluorimetric (OPA-F) assay. Similarly the average relative responses for tri- and oligopetides were 102 and 11% by OPA-C and OPA-F. The OPA-C and OPA-F assays were used to measure degradation rates of different concentrates using the Michaelis-Menten approach (Colombini and Broderick, 2010). In this experiment, a higher passage rate was applied for the soluble-N than for the insoluble N; a blended digestion rate, determined using the IIV assay, was applied to both N fractions to predict RUP. The OPA-C assay resulted in higher degradation rates for all the feeds analyzed due to the contribution of peptides to the fraction degraded. Using this assay, estimated RUP for solvent SBM (30%) and expeller SBM (42%) were similar to *in vivo* values reported by Reynal and Broderick (2003). The method also predicted degradation rates of intensively grazed tropical grasses, showing differences among species (Reis *et al.*, 2010), and quantified the effects of condensed tannins and maceration on degradation of protein in legume forages (Broderick and Grabber, 2009).

Neutze *et al.* (1993) applied the limited substrate IIV system described above, except they measured protein degradation from the net release of trichloroacetic acid-soluble N. The advantage of their method is that extent of degradation can be measured by simple Kjeldahl or Dumas N assays rather than by a dual ammonia plus total AA analyses. That they obtained more complete recovery of degraded CP present as small peptides was indicated by the fast rates observed for casein. Degradation rates reported for a number of SBM samples (Neutze *et al.*, 1993) were of similar magnitude to rates determined from ammonia and total AA release for several other samples of SBM (Broderick, 1987; Broderick and Clayton, 1992).

Other techniques used *in vitro* have included the application of dyes (Mahadevan *et al.*, 1979) or radioisotopes (Wallace, 1983) to label the test proteins to quantify their breakdown from release of these labels in incubations with mixed ruminal organisms or proteases. These methods worked well for soluble proteins, but proved unreliable with insoluble proteins due to non-homogeneous labeling (Broderick *et al.*, 1988).

Use of proteolytic enzymes to estimate of ruminal protein degradability

Use of cell-free 'commercial' proteases to quantify ruminal protein degradation followed closely on the wide spread application of N solubility as an index of protein degradability and there is an extensive literature on this topic. Ideally, these proteases would yield estimates of protein degradation rates that are similar to those obtained using mixed ruminal organisms. However, it seems clear that these enzymes have substantially different specificities from the proteases elaborated by the organisms residing in the rumen. Rather than determining rate, estimates of extent of ruminal degradation made with proteases most often have been correlated to *in situ* estimates of extent of degradation (Michalet-Doreau and Ould-Bah, 1992; Vanzant *et al.*, 1996; Cone *et al.*, 2004). A common approach has been to try to find incubation times that will yield extents of degradation comparable to those found *in situ*. Theoretical concerns about using the *in situ* method as a standard were discussed earlier. Finding appropriate assays with which to calibrate enzyme digestion has been a major limitation to progress in identifying the best *in vitro* techniques to adopt and researchers using these enzymes cannot always be faulted for relying heavily on methods such as calibrating proteases to *in situ* results. But reverting to prediction of *in situ* extents of degradation, rather than attempting to quantify rates of ruminal protein degradation, has limited progress. In the original version of the Cornell model (Sniffen *et al.*, 1992) a curve-peeling technique was applied with an *in vitro* procedure using Streptomyces griseus protease to quantify degradation rates and amounts of each protein fraction.

A review from several years ago compared results from regressing extent of proteolysis (measured as total N solubilized by enzyme action) obtained with *Streptomyces griseus* protease, ficin, bromelain, papain, pancreatin, and neutral proteases from *Aspergillus oryzae* and *Bacillus subtilis* on *in situ* extents of degradation (Broderick and Cochran, 2000). Conditions used in these incubations, including incubation times and enzyme activities, varied greatly and compromised conclusions. Poos-Floyd *et al.* (1985) compared five proteases and found that, *S. griseus*, the protease with the lowest correlation to *in vivo* results, actually solubilized the most protein N. Assoumani *et al.* (1992) reported that *S. griseus* protease (at pH 8.0) digested proportionately lower amounts of more slowly degraded feeds and showed poor correlation with RDP values determined *in situ*. Subsequently, other studies have been conducted to test the effects of incubation pH, enzyme concentrations, incubation times and different feeds in the *S. griseus* procedure (Licitra *et al.*, 1998; Cone *et al.*, 2002; Chaudhry, 2005). Aufrere *et al.* (1991), using a version of the assay that has been widely adopted, reported $r^2 = 0.88$ from regressing N solubilized by *S. griseus* protease at pH 8.0 on *in situ* degradation for a large number of feeds. Also, many of the degradation rates for protein fraction B3 in the database of the Cornell model were determined using this method (Lanzas *et al.*, 2008). Licitra *et al.* (1998) modified the *S. griseus* method by using a constant ratio of enzyme activity to substrate true protein and found improved predictions of *in situ* degradability. Chaudhry (2007), also using *S. griseus*, obtained rates of degradation for casein, wheat gluten and maize gluten that approximated those reported for the IIV system; *S. griseus* proved more effective than papain in these experiments. Coblentz *et al.* (1997) had good results estimating *in situ* degradation using an *in vitro* assay based on ficin but not on *S. griseus*. Antoniewicz and Kosmala (1997) had poor results with alfalfa forages using the *S. griseus* method of Aufrere. Moreover, Gosselink *et al.* (2004) found that neither the *S. griseus* assay nor *in situ* incubations yielded degradabilities that were correlated to *in vivo* determinations of forage protein escape from the rumen. These contrasting results indicate the need of further experimentation.

Including enzymes to remove the carbohydrates that are normally digested *in vivo*, but which occlude protein attack *in vitro*, shows promise for *in vitro* assays of protein degradation. Added carbohydrases improved the predictions of *in situ* protein degradation for cereal grains in systems using *S. griseus* and a neutral protease (Assoumani *et al.*, 1992), ficin (Kosmala *et al.*, 1996), and bromelain (Tomankova and Kopecny, 1995). For example, Tomankova and Kopecny (1995) observed an $r^2 = 0.07$ without amylase but an $r^2 = 0.65$ with amylase addition to bromelain incubations with 19 grains. Although Cone *et al.* (1996) found a slight reduction in correlations to *in situ* results when they added two carbohydrases to *S. griseus* incubations with a number of concentrates and forages, Abdelgadir *et al.* (1997) observed that addition of either cellulase or a broad-spectrum carbohydrase to *S. griseus* protease resulted in improved estimates of *in vivo* protein degradability for alfalfa and prairie hays. Antoniewicz and Kosmala (1997) observed high correlations between extent of protein digestion using ficin or pancreatin and extent of *in situ* digestion of 24 alfalfa forages.

Ideal performance of a cell-free protease system *in vitro* would be successful prediction of both rate and extent of protein degradation. In their early work, Krishnamoorthy *et al.* (1983) used incubations with *S. griseus* protease at high activity (6.6 units/ml) to estimate kinetically the sizes of pools corresponding to protein fractions B2, B3 and C in the Cornell model; they also used *S. griseus* protease at low activities (0.066 and 0.022 unit/ml) to estimate degradation rates of SBM. Although Krishnamoorthy *et al.* (1983) observed reproducible degradation patterns and rates for SBM at 0.022 unit/ml of *S. griseus* protease, no *in situ* or *in vivo* degradation rates were reported for the same protein source. Roe *et al.* (1991) were unable to predict *in situ* degradation rates of four SBM and two distillers grains using *S. griseus*, ficin, or neutral fungal protease plus amylase. Inspection of the degradation patterns for all six proteins revealed considerable deviation from that obtained using the *in situ* assay (Roe *et al.*, 1991). Mahadevan *et al.* (1987) observed substantially different degradation rates using proteases extracted from mixed ruminal organisms and those from *S. griseus* or Aspergillus oryzae. For example, *S. griseus* protease yielded digestion rates for blood meal and fish meal, two proteins known to be very resistant to ruminal degradation, that were more rapid than the rate obtained for solvent SBM, while proteases from mixed ruminal microbes yielded digestion rates for blood meal and fish meal that were only about half as rapid as that for SBM. Kohn and Allen (1995a) developed an assay for protein degradation based on proteases extracted from mixed ruminal microbes using butanol and acetone. These proteases were used to assess degradability of several forage proteins (Kohn and Allen, 1995b). Although largely successful, the difficulty of isolating even small amounts of protease from mixed ruminal microbes decreases applicability of this approach.

Because ruminal microbial proteases have specificities that differ substantially from those of commercial proteases, simple use of individual proteolytic enzymes likely will not yield direct estimates of ruminal degradation rate. Luchini *et al.* (1996) characterized the proteolytic activity of mixed ruminal microbes using 13 artificial substrates, and then elaborated blends of commercial proteases to try to match this proteolytic activity. Although three different blends had hydrolytic activities toward the artificial substrates that were similar to mixed ruminal microbes, none of the blends degraded standard feed proteins at similar rates. For example, degradation rates obtained for 15 proteins using one of these protease blends ranged only from 0.01 to 0.10/h, while degradation rates obtained with ruminal inoculum ranged from 0.01 to 0.23/h; rates deviated most with the more rapidly degraded proteins (Luchini *et al.*, 1996). All of the protease blends were inaccurate for assessing degradability for heat-treated soybean products. Carbohydrate digesting enzymes were not tested in this system.

Despite some success in using cell-free proteases to predict extent of *in situ* degradability, clearly these enzymes, as applied thus far, have not captured many of the important characteristics of the activities of ruminal organisms that determine rates and extents of ruminal protein degradation. Mohamed and Chaudhry (2008) have published a thorough review that discusses many of the approaches using proteases to estimate ruminal protein degradability.

References

Abdelgadir, I.E.O., R.C. Cochran, E.C. Titgemeyer and E.S. Vanzant, 1997. *In vitro* determination of ruminal protein degradability of alfalfa and prairie hay via a commercial protease in the presence or absence of cellulase or driselase. J. Anim. Sci. 75, 700-713.

Antoniewicz, A.M. and I. Kosmala, 1997. Estimation of ruminal effective degradability of dried lucerne forage based on *in vitro* digestion by protease from *Streptomyces griseus*, ficin or pancreatin. Proceedings of the International Symposium on *In vitro* Techniques for Measuring Nutrient Supply to Ruminants. British Society of Animal Science, Reading, England.

Antoniewicz, A.M., I. Kosmala, T. Hvelplund and A. van Vuuren, 1997. Use of pepsin-pancreatin system to estimate intestinal digestibility of protein in ruminant feeds and undegraded residues. Proceedings of the International Symposium on *In vitro* Techniques for Measuring Nutrient Supply to Ruminants. British Society of Animal Science, Reading, England.

Assoumani, M.B., F. Vedeau, L. Jacquot, and C.J. Sniffen, 1992. Refinement of an enzymatic method for estimating the theoretical degradability of proteins in feedstuffs for ruminants. Anim. Feed. Sci. Technol. 39, 357-368.

Atasoglu, C., A.Y. Guliye and R.J. Wallace, 2004. Use of stable isotopes to measure de novo synthesis and turnover of amino acid-C and -N in mixed micro-organisms from the sheep rumen *in vitro*. Brit. J. Nutr. 91, 253-261.

Attwood, G.T. and K. Reilly, 1995. Identification of proteolytic rumen bacteria isolated from New Zealand cattle. J. Appl. Bacteriol. 79, 22-29.

Aufrere, J., D. Graviou, C. Demarquilly, R. Verite, B. Michalet-Doreau and P. Chapoutot, 1991. Predicting *in situ* degradability of feed proteins in the rumen by two laboratory methods (solubility and enzymatic degradation). Anim. Feed. Sci. Technol. 33, 97-116.

Ahvenjarvi, S., T. Stefanski and P. Huhtanen, 2009. *In vitro* method for determining the ruminal degradation rate of rapeseed meal protein using 15N isotope labelled ammonia nitrogen. Anim. Feed. Sci. Technol. 153, 88-100.

Blümmel, M., H.P.S. Makkar and K. Becker, 1997. *In vitro* gas production: a technique revisited. J. Anim. Physiol. Anim. Nutr. 77, 24-34.

Borchers, R, 1967. Proteolytic activity of rumen fluid *in vitro*. J. Anim. Sci. 24, 1033-1038.

Brito, A.F. and G.A. Broderick, 2007. Effects of different protein supplements on milk production and nutrient utilization in lactating dairy cows. J. Dairy Sci. 90, 1816-1827.

Broderick, G.A, 1978. *In vitro* procedures for estimating rates of ruminal protein degradation and proportions of protein escaping the rumen undegraded. J. Nutr. 108, 181-190.

Broderick, G.A, 1987. Determination of protein degradation rates using a rumen *in vitro* system containing inhibitors of microbial nitrogen metabolism. Brit. J. Nutr. 58, 463-476.

Broderick, G.A, 1992. Relative value of fish meal versus solvent soybean meal for lactating dairy cows fed alfalfa silage as sole forage. J. Dairy Sci. 75, 174-183

Broderick, G.A. and K.A. Albrecht, 1997. Ruminal *In vitro* degradation of protein in tannin-free and tannin- containing forage legume species. Crop Sci. 37, 1884-1891.

Broderick, G.A. and J.E. Balthrop, Jr, 1979. Chemical inhibition of amino acid deamination by ruminal microbes *in vitro*. J. Anim. Sci. 49, 1101-1111.

Broderick, G.A. and M.K. Clayton, 1992. Rumen protein degradation rates estimated by nonlinear regression analysis of Michaelis-Menten *in vitro* data. Brit. J. Nutr. 67, 27-42.

Broderick, G.A., and R.C. Cochran, 2000. *In vitro* and *in situ* methods for estimating digestibility. Pages 53-85 in Feeding Systems and Feed Evaluation Models. M. K. Theodorou and J. France, ed. CABI Publishing, CAB International, Wallingford, England, UK.

Broderick, G.A. and J.H. Grabber, 2009. Effect of condensed tannins and maceration on *in vitro* ruminal degradation of protein in legume hay. J. Dairy Sci. 92(E-suppl. 1), 467 (abstract).

Broderick, G.A., P. Huhtanen, S. Ahvenjärvi, S. M. Reynal and K.J. Shingfield, 2010. Quantifying ruminal nitrogen metabolism using the omasal sampling technique in cattle – a meta-analysis. J. Dairy Sci. 93, in press.

Broderick, G.A., M.L. Murphy, and P. Udén, 2004a. Effect of inhibitor concentration and end-product accumulation on estimates of ruminal *in vitro* protein degradation. J. Dairy Sci. 87, 1360-1371.

Broderick, G.A., D.B. Ricker and L.S. Driver, 1990. Expeller soybean meal and corn by-products versus solvent soybean meal for lactating dairy cows. J. Dairy Sci. 73, 453-462.

Broderick, G.A., P. Udén, M.L. Murphy and A. Lapins, 2004b. Sources of variation in rates of *in vitro* ruminal protein degradation. J. Dairy Sci. 87, 1345-1359.

Broderick, G.A. and R.J. Wallace, 1988. Effects of dietary nitrogen source on concentrations of ammonia, free amino acids and fluorescamine-reactive peptides in the sheep rumen. J. Anim Sci. 66, 2233-2238.

Broderick, G.A., R.J. Wallace, E.R. Ørskov and L. Hansen, 1988. Comparison of estimates of ruminal protein degradation by *in vitro* and *in situ* methods. J. Anim. Sci. 66, 1739-1745.

Chaudhry, A.S., 2005. Comparing commercial enzymes to estimate *in vitro* proteolysis of purified and semi purified proteins. J. Anim. Physiol. Anim. Nutr. 89, 403-412.

Chaudhry, A.S., 2007. Enzymic and in sacco methods to estimate rumen degradation of food protein in cattle. J. Sci. Food Agric. 87, 2617-2624.

Chen, G., H.J. Strobel, J.B. Russell and C.J. Sniffen, 1987. Effect of hydrophobicity on utilization of peptides by ruminal bacteria *in vitro*. Appl. Environ. Microbiol. 53, 2021-2025.

Church, F.C., D.H. Porter, G.L. Catignani and H.E. Swaisgood, 1985. An o-phthalaldehyde spectrophotometric assay for proteinases. Anal. Biochem. 146, 343-348.

Coblentz, W.K., J.O. Fritz, R.C. Cochran, W.L. Rooney and K.K. Bolsen, 1997. Protein degradation in response to spontaneous heating in alfalfa hay by *in situ* and ficin methods. J. Dairy Sci. 80, 700-713.

Colombini, S. and G.A. Broderick, 2010. Application of an O-phthalaldehyde assay that detects oligopeptides for estimating rumen protein degradation in vitro. In: Crovetto, G.M. (ed.) Energy and protein metabolism and nutrition. Wageningen Academic Publishers, Wageningen, the Netherlands, p. 713-714.

Cone, J.W., A.A. Kamman, A.H. Gelder and V.A. Van Hindle, 2002. Rumen escape protein in concentrate ingredients determined with the nylon bag and enzymatic technique. Anim. Feed Sci. Technol. 97, 247-254.

Cone, J.W., M.A.M. Rodrigues, C.M. Guedes and M.C. Blok, 2009. Comparison of protein fermentation characteristics in rumen fluid determined with the gas production technique and the nylon bag technique. Anim. Feed. Sci. Technol. 153, 28-38.

Cone, J.W., A.H. Van Gelder, A.A. Mathijssen-Kamman, and V.A. Van Hindle, 2004. Rumen escape protein in grass and grass silage determined with a nylon bag and an enzymatic technique. Anim. Feed Sci. Technol 114, 195-204.

Cone, J.W., A.H. Van Gelder, A. Steg and A.M. Van Vuuren, 1996. Prediction of *in situ* rumen escape protein from *in vitro* incubation with protease from Streptomyces griseus. J. Sci. Food Agric. 72, 120-126.

Faldet, M.A., V.L. Voss, G.A. Broderick and L.D. Satter, 1991. Chemical, *in vitro* and *in situ* evaluation of heat treated soybean proteins. J. Dairy Sci. 74, 2548-2554.

Fox, D.G., L.O. Tedeschi, T.P. Tylutki, J.B. Russell, M.E. Van Amburgh, L.E. Chase, A.N. Pell and T.R. Overton, 2004. The Cornell Net Carbohydrate and Protein System model for evaluating herd nutrition and nutrient excretion. Anim. Feed. Sci. Technol. 112, 29-78.

Gierus, M., L. de Jonge and G.A.L. Meijer, 2006. Physico-chemical characteristics and degradation rate of soluble protein obtained from the washout fraction of feeds. Livest. Prod. Sci. 97, 219-229.

Gosselink, J.M.J., J.P. Dulphy, C. Poncet, J. Aufrère, S. Tamminga and J.W. Cone, 2004. Rumen escape nitrogen from forages in sheep: comparison of *in situ* and *in vitro* techniques using *in vivo* data. Anim. Feed Sci. Technol. 116, 35-51.

Grabber, J.H. and W.K. Coblentz, 2009. Polyphenol, Conditioning, and Conservation Effects on Protein Fractions and Degradability in Forage Legumes. Crop Sci. 49, 1511-1522.

Hristov, A.N., S. Ahvenjarvi, T.A. McAllister, and P. Huhtanen, 2003. Composition and digestive tract retention time of ruminal particles with functional specific gravity greater or less than 1.02. J. Anim. Sci. 81, 2639-2648.

Hristov, A. and G.A. Broderick, 1994. *In vitro* determination of ruminal protein degradability using [15N]-ammonia to correct for microbial nitrogen uptake. J. Anim. Sci. 72, 1344-1353.

Hsu, J.T. and L.D. Satter, 1995. Procedures for measuring the quality of heat-treated soybeans. J. Dairy Sci. 78, 1353-1361.

Huhtanen, P. and A. N. Hristov, 2009. A meta-analysis of the effects of dietary protein concentration and degradability on milk protein yield and milk N efficiency in dairy cows. J. Dairy Sci. 92, 3222-3232.

Huhtanen, P., J.I. Nousiainen, M. Rinne, K. Kytölä and H. Khalili, 2008. Utilization and partitioning of dietary nitrogen in dairy cows fed grass silage based diets. J. Dairy Sci. 91, 3589-3599.

Jouany, J.P., 1996. Effect of rumen protozoa on nitrogen utilization by ruminants. J. Nutr. 126, 1335S-1346S.

Karlsson, L., M. Hetta, P. Udén and K. Martinsson, 2009. New methodology for estimating rumen protein degradation using the *in vitro* gas production technique. Anim. Feed. Sci. Technol. 153, 193-202.

Kohn, R.A. and M.S. Allen, 1995a. *In vitro* protein degradation of feeds using concentrated enzymes extracted from rumen contents. Anim. Feed. Sci. Technol. 52, 15-20.

Kohn, R.A. and M.S. Allen, 1995b. Prediction of protein degradation of forages from solubility fractions. J. Dairy Sci. 78, 1774-1788.

Kosmala, I., A. Antoniewicz, J. De Boever, T. Hvelplund and J. Kowalczyk, 1996. Use of enzymatic solubility with ficin (EC 3.4.22.3) to predict *in situ* feed protein degradability. Anim. Feed. Sci. Technol. 59, 245-254.

Krishnamoorthy, U., C.J. Sniffen, M.D. Stern and P.J. Van Soest, 1983. Evaluation of a mathematical model of rumen digestion and an *in vitro* simulation of rumen proteolysis to estimate the rumen-undegraded nitrogen content of feedstuffs. Brit. J. Nutr. 50, 555-568.

Krishnamoorthy, U., H. Steingass and K.H. Menke, 1990. The contribution of ammonia, amino acids and short peptides to estimates of protein degradability *in vitro*. J. Anim. Physiol. Anim. Nutr. 63, 135-141.

Lanzas, C., L.O. Tedeschi, S. Seo and D.G. Fox, 2007. Evaluation of protein fractionation systems used in formulating rations for dairy cattle. J. Dairy Sci. 90, 507-521.

Licitra, G., F. Lauria, S. Carpino, I. Schadt, C.J. Sniffen and P.J. Van Soest, 1998. Improvement of the *Streptomyces griseus* method for degradable protein in ruminant feeds. Anim. Feed. Sci. Technol. 72, 1-10.

Luchini, N.D., G.A. Broderick and D.K. Combs, 1996. Characterization of the proteolytic activity of commercial proteases and strained ruminal fluid. J. Anim. Sci. 74, 685-692.

Mahadevan, S., J.D. Erfle and F.D. Sauer, 1979. A colorimetric method for the determination of proteolytic degradation of feed proteins by rumen microorganisms. J. Anim. Sci. 48, 947-953.

Mahadevan, S., J.D. Erfle and F.D. Sauer, 1987. Preparation of protease from mixed rumen microorganisms and its use for the *in vitro* determination of the degradability of true protein in feedstuffs. Can. J. Anim. Sci. 67, 55-64.

McDonald, I.W. and R.J. Hall, 1954. The conversion of casein into microbial proteins in the rumen. Biochem. J. 67, 400-405.

Menke, K.H., L. Raab, A. Salewski, H. Steingass, D. Fritz and W. Schneider, 1979. The estimation of the digestibility and metabolizable energy content of ruminant feedingstuffs from the gas production when they are incubated with rumen liquor *in vitro*. J. Agric. Sci. (Camb.) 93, 217-222.

Michalet-Doreau, B. and M.Y. Ould-Bah, 1992. *In vitro* and in sacco methods for the estimation of dietary nitrogen degradability in the rumen: a review. Anim. Feed. Sci. Technol. 40, 57-86.

Mohamed, R. and A.S. Chaudhry, 2008. Methods to study degradation of ruminant feeds. Nutr. Res. Rev. 21, 68-81.

NRC, 2001. Nutrient Requirements of Dairy Cattle, Seventh Revised Edition. National Research Council, National Academy Press, Washington, DC.

Neutze, S.A., R.L. Smith and W.A. Forbes, 1993. Application of an inhibitor *in vitro* method for estimating rumen degradation of feed protein. Anim. Feed. Sci. Technol. 40, 251-265.

Nocek, J.E., K.A. Cummins and C.E. Polan, 1979. Ruminal disappearance of crude protein and dry matter in feeds and combined effects in formulated rations. J. Dairy Sci. 62, 1587-1598.

Ørskov, E.R. and I. McDonald, 1979. The estimation of protein degradability in the rumen from incubation measurements weighted according to rate of passage. J. Agric. Sci. (Camb.) 92, 499-503.

Palmonari, A., D.M. Stevenson, D.R. Mertens, C.W. Cruywagen and P.J. Weimer, 2010. pH dynamics and bacterial community composition in the rumen of lactating dairy cows. J. Dairy Sci. 93, 279-287.

Paster, B.J., J.B. Russell, C.M.J. Yang, J.M. Chow, C.R. Woese and R. Tanner, 1993. Phylogeny of the ammonia-producing ruminal bacterial *Peptostreptococcus anaerobius*, *Clostridium sticklandii*, and *Clostridium aminophilum* spp. nov. International J. Systematic Bacteriol. 43, 107-110.

Poos-Floyd, M., T. Klopfenstein and R.A. Britton, 1985. Evaluation of laboratory techniques for predicting ruminal protein degradation. J. Dairy Sci. 68, 829-839.

Raab, L., B. Cafantaris, T. Jilg and K.H. Menke, 1983. Rumen protein degradation and biosynthesis. 1. A new method for determination of protein degradation in rumen fluid *in vitro*. Brit. J. Nutr. 50, 569-582.

Reis, R.B., S. Colombini, A. Miller, D.K. Combs and G.A. Broderick, 2010. Protein fractions and rates of degradation of tropical forages from intensively grazed pastures. In: Crovetto, G.M. (eds.) Energy and protein metabolism and nutrition. Wageningen Academic Publishers, Wageningen, the Netherlands, p. 717-718.

Reynal, S.M. and G.A. Broderick, 2003. Effects of feeding dairy cows protein supplements of varying ruminal degradability. J. Dairy Sci. 86, 835-843.

Rinne, M., J. Nousiainen, and P. Huhtanen, 2009. Effects of silage protein degradability and fermentation acids on metabolizable protein concentration: A meta-analysis of dairy cow production experiments. J. Dairy Sci. 92, 1633-1642.

Roe, M.B., L.E. Chase and C.J. Sniffen, 1991. Comparison of *in vitro* techniques to the *in situ* technique for estimation of ruminal degradation of protein. J. Dairy Sci. 74, 1632-1640.

Roth, M, 1971. Fluorescence reaction of amino acids. Anal. Chem. 43, 880-882.

Santos, F.A.P. J.E.P. Santos, C.B. Theurer and J.T. Huber, 1998. *In vivo* and *in situ* measurements of forage protein degradation in beef cattle. J. Dairy Sci. 81, 3182-3213.

Schwab, C.G, 1996. Rumen-protected amino acids for dairy cattle: Progress towards determining lysine and methionine requirements. Anim. Feed Sci. Technol. 59, 87-101.

Sniffen. C.J., J.D. O'Connor, P.J. Van Soest, D.G. Fox and J.B. Russell, 1992. A net carbohydrate and protein system for evaluating cattle diets: I. Carbohydrate and protein availability. J. Anim Sci. 70, 3562-3577.

Stern, M.D. and L.D. Satter, 1984. Evaluation of Nitrogen Solubility and the Dacron Bag Technique as Methods for Estimating Protein Degradation in the Rumen. J. Anim Sci. 58, 714-724.

Tomankova, O. and J. Kopecny, 1995. Prediction of feed protein degradation in the rumen with bromelain. Anim. Feed. Sci. Technol. 53, 71-80.

Tremblay, G.F., G.A. Broderick and S.M. Abrams, 1996. Estimating ruminal protein degradation of roasted soybeans using near infrared reflectance spectroscopy. J. Dairy Sci. 79, 276-282.

Tremblay, G.F., R. Michaud, G. Bélanger, K.B. McRae and H.V. Petit, 2000. *In vitro* ruminal undegradable proteins of alfalfa cultivars. Can. J. Plant. Sci. 80, 315-325.

Vanzant, E.S., R.C. Cochran, E.C. Titgemeyer, S.D. Stafford, K.C. Olson, D.E. Johnson, and G. St. Jean, 1996. *In vivo* and *in situ* measurements of forage protein degradation in beef cattle. J. Anim. Sci. 74, 2773-2784.

Veresegyházy, T., S. Fekete and M. Hegedûs, 1993. Factors influencing ruminal bacterial activity. I. Effect of endproduct concentration on proteolytic activity. J. Anim. Physiol. Anim. Nutr. 70, 1-5.

Volden, H., L.T. Mydland and V. Olaisen, 2002. Apparent ruminal degradation and rumen escape of soluble nitrogen fractions in grass and grass silage administered intraruminally to lactating dairy cows. J. Anim Sci. 80, 2704-2716.

Wallace, R.J, 1983. Hydrolysis of 14C-labelled proteins by rumen microorganisms and by proteolytic enzymes prepared from rumen bacteria. Brit. J. Nutr. 50, 345-355.

Wallace, R.J, 1996. Ruminal microbial metabolism of peptides and amino acids. J. Nutr. 126, 1326S-1334S.

Wallace, R.J. and M.A. Cotta, 1988. Metabolism of nitrogen-containing compounds. In: Hobson, P.N. (ed.), The Rumen Microbial Ecosystem. Elsevier Applied Science, London, UK. 217-249.

Warner, A.C.I, 1956. Proteolysis by rumen micro-organisms. J. Gen. Microbiol. 14, 749-762.

Welch, J.G, 1982. Rumination, particle size and passage from the rumen. J. Anim. Sci. 54, 885-894.

Yang, C.M.J. and J.B. Russell, 1992. Resistance of proline-containing peptides to ruminal degradation *in vitro*. Appl. Environ. Microbiol. 58, 3954-3958.

In situ and *in vitro* methods to determine intestinal digestion of protein and amino acids in ruminants

S. Calsamiglia[1], S. Boucher[2], S. Gargallo[1], C. Schwab[3] and A. Ferret[1]
[1]*Universitat Autonoma de Barcelona, Spain*
[2]*William H. Miner Agricultural Research Institute Institute, NY, USA*
[3]*University of New Hampshire, NH, USA*

Abstract

Most feeding systems recognize that intestinal digestibility of rumen undegradable protein varies among feeds and different samples of the same feed (mainly in heat processed feeds). In spite of the evidence that intestinal digestibility of individual AA varies within feed, the difficulty in its determination results in the assumption that digestibility of each individual AA is constant and equal to the digestibility of total protein. Measuring intestinal digestion of proteins and AA *in vivo* is difficult, expensive and time consuming. Therefore, alternative methods have been developed for routine analysis of intestinal digestibility of protein and AA. The mobile bag technique was developed to estimate intestinal digestion of protein and AA in ruminants. While it has proven useful in identifying differences in intestinal digestibility of protein and AA between and within feeds, the collection of bags in the feces should be challenged, and a full independent validation against *in vivo* models is needed. A modified three-step *in vitro* procedure has been recently proposed and validated against an *in vivo* animal model to be used in the determination of intestinal digestion of protein and AA in ruminants. While results on few samples are promising, validation with a larger set of samples is required. The availability of an easy, cheap and reliable method to determine intestinal digestion of protein and AA in ruminants may help in improving diet formulation and the efficiency of N utilization in ruminants.

Introduction

Current cattle nutritional models (e.g. NRC, 2001; INRA, 2007) are generally based on metabolizable protein requirements/supply and, in some cases, they also provide recommendations for the supply of some or all essential amino acids (AA). Calculations of the supply of metabolizable protein and/or AA require the determination of their rumen degradability and intestinal digestibility. Average digestibility of feeds is provided in tables, although large variations within samples occur. Furthermore, it is assumed that the digestibility of each individual AA in rumen undegradable protein (RUP) is the same as the digestibility of total RUP, primarily due to a lack data. However, variation in digestibility of individual AA has been recognized in poultry (NRC, 1994), swine (NRC, 1998) and cattle (Prestløkken and Rise, 2003). Measuring intestinal digestibility of protein and AA in individual feeds in ruminants is difficult, time consuming and expensive, with limited *in vivo* values available (Hvelplund and Madsen, 1990). However, the availability of simple, reliable, and inexpensive *in vitro* methods that can be used to routinely analyze feeds for these parameters needs to be identified and validated *in vivo* (Stern *et al.*, 1997), and should allow for a more accurate ration formulation.

The objective of this review is to summarize current status of *in situ* and *in vitro* methodologies available to determine intestinal digestibility of protein and AA in ruminants.

The *in situ* mobile bag technique

The *in situ* mobile bag technique (MBT) is commonly used to estimate intestinal digestibility of RUP and AA in ruminants (NRC, 2001; INRA, 2007). It was originally introduced to measure protein digestibility in pigs (Sauer *et al.*, 1983) but was modified to study postruminal digestion of feedstuffs by ruminants (Hvelplund, 1985). The MBT in ruminants starts with an *in situ* ruminal incubation of feeds. This initial step is justified by the changes in digestibility of the protein remaining after ruminal incubation compared with the intact original protein, although it appears to be only required in some feeds (Erasmus *et al.*, 1994; Prestløkken and Rise, 2003; Hvelplund *et al.*, 1992; Volden and Harstad, 1995; Vanhatalo *et al.*, 1996). The second step is the incubation of bags in a pepsin-HCl solution to mimic the conditions of the abomasum. This second incubation does not appear to be necessary for all feeds, but should be recommended in a standardized protocol (Voigt *et al.*, 1985; Vanhatalo *et al.*, 1995). After this incubation, bags are introduced through a duodenal cannula and collected from the terminal ileum via an ileal cannula, or from the feces. Estimates obtained from fecal collection of bags assume that bags and feed residues are not contaminated with microbial protein from the large intestinal fermentation and that protein leaving the ileum is not further digested by microbes in the large intestine, both of which extremes are questioned. Because remaining N after digestion is often very small, the contribution of potentially contaminating microbial protein is important, mainly in low N and high fiber feeds (Vanhatalo *et al.*, 1995; Vanhatalo and Varvikko, 1995). However, Hvelplund (1985) and Kohn and Allen (1992) suggested that microbial contamination was small enough to be ignored. Once collected, the bags are rinsed, and the CP and AA content of the undigested residue that remains in the bags is determined. Again, the CP and AA that disappeared from the bags are assumed to have been absorbed by the animal. Therefore, total tract measurements using this technique can be considered as estimates of true, rather than apparent, digestibility.

The original development of the method in ruminants indicated that good correlation was only found between intestinal digestion of protein *in vivo* and ileal collection of bags, with the collection of bags in the feces poorly correlated (Hvelpund, 1985). Although other researchers have found better correlations with fecal collection of bags, these were feed dependent and many of the correlations were not conducted against *in vivo* ileal digestibility, but against table values or other *in vitro* methods (Voigt *et al.*, 1985; Cone *et al.*, 2006; Prestløkken and Rise, 2003; De Boever *et al.*, 1995; Masoero *et al.*, 1994). While it is reasonable to justify the fecal collection of bags from feces based on the difficulty in maintaining ileally cannulated ruminant animals and in retrieving bags through the ileal cannula, this should not preclude criticisms on the fundamental and scientific validation of the method. Based on the discrepancies on the need to preincubate samples in the rumen and/or in the HCl-pepsin solution, as well as the discrepancies on site of bag collection, a standardize procedure suitable for all type of samples should include these three steps and the collection of samples should be in the ileum. Moreover, a stronger validation of the method for protein and AA against *in vivo* data is also required.

Despite the potential sources of variation that have been observed, several studies have shown that the MBT may be useful in predicting intestinal digestibility of protein. Hvelplund (1985) regressed *in vivo* intestinal protein digestion values for seven feeds measured in intestinally cannulated cows on estimates of the same feeds obtained using the MBT with ileal collection of bags and found a correlation of 0.81. Others have attempted to provide evidence of the suitability of the technique by correlating MBT values of intestinal digestibility of protein against table values (De Boever *et al.*, 1995; Masoero *et al.*, 1994), but the validity of the approach should be questioned.

Although the MBT provides an easy and fast way to determine intestinal digestion of proteins and AA, standardization and validation of the technique is still necessary. The seach for alternative techniques that are easier, cheaper and more reliable is also warranted.

Lysine availability tests

During heat-processing of feeds, Maillard reaction products have been measured. These have been used as indicators of nutritive damage of food protein, which may be indicators of changes in intestinal digestibility of protein. There are two different approaches: (1) measure the products derived from lysine (Lys) damage or (2) measure available Lys.

Furosine test

During heating, ε-*N*-deoxyfructosyl-L-lysine is formed during the early Maillard reaction. Upon acid hydrolysis, ε-*N*-deoxyfructosyl-L-lysine is released in a constant ratio of 50% lysine, 30% furosine, 20% pyridosine, and 10% other compounds (Mauron, 1990). Therefore, the amount of furosine measured after acid hydrolysis is used to calculate the percentage blocked Lys (Mauron, 1990). The furosine assay is attractive because furosine and AA content of feeds can all be determined in the same procedure. However, furosine analysis only indicates damage due to the early phases of the Maillard reaction.

Carboxymethyllysine test

If feeds are subjected to more severe heat treatments, many of the Amadori compounds react and the content of furosine decreases (Erbersdobler and Somoza *et al.*, 2007). The limitations of the furosine assay can be partially overcome by analyzing more severely damaged feeds for carboxymethyllysine, a product of oxidative degradation of ε-*N*-deoxyfructosyl-L-lysine. Analyzing foods for carboxymethllysine is less common than furosine because it is a more difficult procedure and normal heat processing conditions are generally not severe enough to induce the advanced Maillard reaction (Erbersdobler and Somoza *et al.*, 2007). However, when feeds are stored for prolonged periods of time and/or processing conditions are more severe, carboxymethllysine analysis may be a useful tool for assessing thermal damage of feeds. A negative relationship between furosine and carboxymethllysine content of foods and digestibility of protein and AA is well documented (Hurrell and Carpenter, 1981). However, the Maillard reaction is not the only reaction that can occur in feeds that results in decreased protein and AA digestibility.

Racemization test

Amino acid racemization and the resultant formation of protein cross-links can also depress digestibility of protein and AA in heated foods. Therefore, all of these reactions may need to be considered when assessing the effect of heat processing conditions on protein quality. With the exception of glycine, all AA have at least one asymmetric carbon atom that can exist in two mirror images, D and L (Friedman, 1999). Amino acids found in nature occur in the L configuration. However, heat processing and alkali treatments of feeds can induce AA racemization resulting in the isomerization of L-AA to the D configuration (Friedman, 1999). The nutritional consequence of racemization is the loss of AA (Friedman, 1999).

Lysinoalanine test

Racemization reactions also lead to protein cross-linking reactions. These reactions occur because of the formation of cross-linking compounds such as lysinoalanine (LAL). The formation of LAL not only renders the Lys that participates in the reaction unavailable to the animal, but it also decreases the digestibility of total protein and other AA because LAL impairs the approach of proteolytic enzymes to the peptide chain (Boschin *et al.*, 2003). However, because LAL formation occurs more readily at a higher pH, this reaction is more prevalent in alkali treated proteins compared with proteins simply exposed to heat treatments (Bunjapamai *et al.*, 1982). Lysinoalanine content of a feed can be determined when analyzing feeds for AA content.

Lysine reactivity tests

An alternative to measuring Lys-derived products, several methods have also been developed to estimate available Lys. These methods are advantageous because they quantify the total amount of reactive Lys in a feed irrespective of the reaction that resulted in Lys damage. Of the methods that quantify reactive Lys, the most widely utilized and studied are the 1-fluoro-2,4-dinitrobenzene (FDNB) and the guanidination methods (Hurrell and Carpenter, 1981). The FDNB and guanidiantion methods work by attaching a chemical group to the reactive ε-amino group of Lys and measuring the resultant product. In both reactions, if the ε-amino group of Lys is bound to another compound, then it is not available to react and the resultant products are not formed; therefore, the amount of reactive Lys can be calculated based on the amount of product formed as a result of the respective reactions. Research has shown that the guanidination procedure is more accurate in estimating reactive Lys especially in feeds containing starch and sugars (Hurrell and Carpenter, 1981). Faldet *et al.* (1992) evaluated the relationship between heat treatment of soybeans on reactive Lys content as measured using the FDNB method, and observed that as temperature increased the length of time needed to affect the reactive Lys concentration of the soybeans decreased. The authors reported a high correlation between reactive Lys consumed and body weight gain in rats ($r=0.99$; $P<0.05$). This indicates that the FDNB method provides accurate estimates of reactive Lys in soybeans. However, intestinal digestion of protein may be affected by chemical reactions other than bound Lys.

Lysine availability tests are useful to determine the degree of heat damage in proteins, but they do not necessarily determine potential decreases in intestinal protein digestion or information on individual AA digestibility. Although the technique may be appropriate for comparing different samples within the same feeds, it is not recommended for use among various feeds.

Enzymatic techniques

Most enzymatic techniques follow a similar protocol; feed samples are incubated at optimal temperature and pH with an enzyme or combination of enzymes. After the incubation period, protein digestion is measured as the amount of amino acids and small peptides released from the protein divided by the amount of protein in the original sample. Determination of protein digestion requires that the method used will estimate amounts of these amino acids and peptides that are released. Various techniques differ in the enzyme or combination of enzymes used and the method selected for determining the digested protein.

In vitro single enzyme techniques

Single enzyme systems have been used with variable success to estimate intestinal digestion of proteins (Buchanan, 1969; Saunders *et al.*, 1973; Rhinehart, 1975). Pepsin-insoluble N (PIN) was strongly correlated to total tract unavailable N in forages (Goering *et al.*, 1972; Shelford *et al.*, 1980) and other protein supplements (Loerch *et al.*, 1983). In contrast, Zinn and Owens (1982) and Britton *et al.* (1986) indicated that PIN was a poor predictor of total tract CP digestibility when tested across feeds. Other single-enzyme techniques have been proposed with variable results and limited accuracy, probably due to the limited spectrum of specificity for any single enzyme. Because most enzymes have specificity for individual peptide bonds, protein digestion may be more dependent on the number of bonds sensitive to that enzyme in the protein than to the actual intestinal digestibility. Therefore, multienzyme methods are more likely to work across feeds.

Multienzyme techniques

A combination of enzymes with different activity and specificity and similar to those found in the small intestine is more likely to work. Hsu *et al.* (1977) speculated that AA released from enzymatic digestion in a non-buffered solution will result in pH changes of the solution. Samples were incubated in a unbuffered pH 8 solution that contained a mixture of purified enzymes (trypsin-chymotrypsin-pepsin), and correlated the changes in pH after 10 min against *in vivo* rat digestibility results obtained in rats. The correlation was very high ($r=0.90$). This result was later confirmed by others (Marshall *et al.*, 1979; Bodwell *et al.*, 1980; Satterlee *et al.*, 1981). We attempted to apply the method to samples for ruminants (Calsamiglia, unpublshed data), but were unable to achieve a reasonable correlation with intestinal digestion estimates obtained using the three-step procedure of Calsamiglia and Stern (1995). It is likely that the technique is very sensitive to the buffering capacity of samples, and rumen preincubated samples may not be appropriate to be used with this techinque. Other limitations of the technique are that the digestibility of individual AA cannot be determined and the digestibility of structurally stable proteins will be underestimated using this technique due to the short incubation period (Porter *et al.*, 1984). So although relatively simple and economical, the multienzyme technique does not appear to be adequate for estimating digestibility of total protein and individual AA in a wide variety of feedstuffs for ruminants. Therefore, other *in vitro* methods for estimating protein and AA digestibility need to be considered.

Immobilized digestive enzyme assay (IDEA)

Porter *et al.* (1984) developed the IDEA to rapidly estimate protein digestibility of human foodstuffs. Protein substrate solutions were made by dissolving the protein in HCl (pH 2.0) and continuously circulating the mixture via a peristaltic pump through a jacketed column containing glass beads with immobilized pepsin. After 18 h, the protein solution was collected in a reservoir bottle and continuously circulated through a similar column containing the immobilized trypsin, chymotrypsin, and intestinal peptidase glass beads. After 18 h, the protein solution was collected in a reservoir bottle, and protein digestibility of the samples was determined using the reaction of orthopthaldialdehyde (OPA) and 2-mercaptoethanol with α-amino groups. The digestibility of the protein sources compared well with published values of digestibility, but the procedure was not validated with *in vivo* samples. Schasteen *et al.*, (2007) modified the IDEA to provide a more rapid and accurate prediction of protein digestibility, and developed an IDEA™ kit specifically designed to estimate protein digestibility of SBM (Novus International Inc., St. Charles, MO). The authors used the calculated IDEA™ value for 17 SBM samples to determine if the IDEA™ kits could be used to accurately estimate digestibility of individual AA in poultry. The relationship between the IDEA™ value of the 17 samples and *in vivo* standardized AA digestibility estimates of the same samples determined in cecectomized roosters was observed for all AA, and the R^2 values ranged from 0.73 to 0.91. The IDEA™ values were also highly correlated with *in vivo* standardized Lys and Met digestibility values (R^2 = 0.86 and 0.88, respectively). The authors then validated the prediction equations that resulted from the regression analysis of the IDEA™ values and *in vivo* standardized AA digestibility using a separate set of 5 SBM samples not included in the original analysis. The authors reported only minor differences between digestibility coefficients measured *in vivo* and those predicted from the IDEA™ test.

Boucher *et al.* (2009a) evaluated the use of the IDEA™ kits to estimate intestinal digestibility of RUP AA using 3 soybean meal (SBM), 3 SoyPlus® (SP; West Central, Ralston, IA), 5 distillers dried grains with solubles (DDGS), and 5 fish meal (FM) samples obtained from the Feed Analysis Consortium, Inc. (Champaign. IL, USA). One of the SP, SBM, and DDGS samples were heated to decrease intestinal digestibility of RUP. In order evaluate the kits for use in ruminant nutrition, samples were ruminally incubated *in situ* for 16 h prior to analysis with the IDEA™ kits. The IDEA values of the rumen undegraded residues were regressed on *in vivo* digestibility of RUP AA estimates obtained using the precision-fed cecectomized rooster assay (Boucher *et al.*, 2009b,c). The IDEA values of the rumen residues were highly correlated with *in vivo* digestibility of RUP AA of the soy-product and DDGS samples (R^2 for total RUP-AA = 0.83, Y = 24.3 + 152.2X for DDGS; R^2 for total RUP-AA = 0.95, Y = 62.8 + 38.8X for soy products; $P<0.05$), but for the FM samples, correlation of the IDEA values was much lower (R^2 for total RUP-AA = 0.47; Y = 72.2 + 132.0X). However, because the IDEA values are used to predict, not measure, AA digestibility, the relationship between the IDEA value of the intact feed and digestibility of RUP AA measured *in vivo*, was examined to determine if the IDEA value of the feed could be used to predict RUP-AA digestibility. Based on this analysis, the IDEA value of the intact feed can be used to predict RUP-AA digestibility of rumen residue samples (R^2 = 0.90 for all feed types; Y = 36.0 + 54.0X for

DDGS; Y = 54.6 + 41.2X for soy products; Y = 54.3 + 122.7X for FM; $P<0.05$). Therefore, for a more efficient, cost-effective analysis, it is recommended that for future IDEA analyses, the IDEA values of the intact samples be determined, and that the IDEA value of the intact feed be compared to *in vivo* RUP-AA digestibility to develop accurate prediction equations.

The advantage of the IDEA kits is that it takes only 1 day to complete, and AA analysis is not needed on the final undigested product. This saves time and money when obtaining AA digestibility estimates. However, because the kits are specific to a particular feed type, the use of the kits is limited. In addition, in the experiments described here, a small sample size was used for the evaluation of each kit; therefore, further analysis is needed of the IDEA kits to develop more accurate prediction equations for estimating the digestibility of RUP AA of feed ingredients.

Pepsin-pancreatin *in vitro* assays

Calsamiglia and Stern (1995) developed a three-step procedure (TSP) that simulates physiological conditions of ruminants, is rapid, reliable, and inexpensive, applicable to a wide variety of protein supplements, and reflects differences in protein digestion. The system was designed to simulate ruminal degradation (*in situ* incubation), and abomasal (HCL-pepsin) and intestinal (pancreatic extract) digestion. Protein is precipitated with trichloracetic acid and soluble protein is considered digested. Preincubation of samples in the rumen affected intestinal digestion of some but not all feeds, and the HCl-pepsin digestion before pancreatin digestion increased CP digestion of all proteins tested by a mean of 23 percentage units. Results were highly correlated ($R^2 = 0.91$, n=34, $P<0.001$) with *in vivo* intestinal digestion. The TSP provides a more physiological environment for digestion and contains proteolytic enzymes with specificity and activity similar to those found in the small intestine. Similar approaches, although validated against intestinal digestion determined with the MBT technique, were used by Antoniewicz *et al.* (1992) and Van Straalen *et al.* (1993), resulting in good correlations. Since the development of the TSP, the procedure has been used in a variety of studies to estimate intestinal digestibility of RUP and results incorporated in the feed library of NRC (2001).

Gargallo *et al.* (2006) modified the TSP so that digestibility of RUP-AA could be determined. In the modified TSP, the pepsin and pancreatin digestion steps are conducted in nylon bags (pore size 50 μm) incubated in a Daisy^{II} incubator bottles (Ankom Technology, Macedon, NY). A final undegraded, undigested residue can be collected at the end of the pancreatin incubation and analyzed for AA and CP content. Digestibility of RUP-AA can be calculated based on AA disappearance from the bags. However, while the authors reported a high correlation between RUP digestibility estimates obtained using the original and modified TSP ($R^2 = 0.84$, $P<0.001$), they did not validate the use of this modified *in vitro* technique to estimate RUP-AA digestibility with *in vivo* data. Also, as with the MBT described above, the assumption with the modified TSP that what disappears from the bag is equal to what would be absorbed by the animal may not always be accurate. However, if validated with *in vivo* data, the modified TSP can potentially be used to estimate digestibility of total RUP and RUP-AA on a routine basis for a wide variety of feedstuffs. Therefore, further evaluation of the modified TSP to estimate RUP-AA digestibility is warranted.

Borucki-Castro *et al.* (2007) used the TSP to determine intestinal RUP AA digestibility of various soy products by analyzing the rumen undegraded feed residue and the final supernatant obtained from the TSP for AA content via GC-MS. The authors compared the intestinal RUP-AA digestibility coefficients obtained *in vitro* with RUP-AA digestibility estimates obtained using the MBT in lactating cows with collection of the bags in the feces. However, the authors reported no agreement between *in vitro* determined RUP-AA digestibility coefficients and those obtained using the MBT. The RUP-AA digestibility values obtained using the MBT were higher compared with the values obtained with the *in vitro* method; therefore, the authors concluded that the TSP cannot be used as a reliable replacement for *in situ* measurements of RUP-AA digestibility. However, the discrepancy between the *in vitro* and *in vivo* RUP-AA digestibility measurements may have been due to the fact that the authors collected the bags from the feces. The effect of microbial degradation of undigested feed protein would be reflected in the RUP-AA digestibility coefficients obtained using the MBT but not in the *in vitro* RUP-AA digestibility estimates.

Boucher *et al.* (2009a) evaluated digestibility of AA and RUP AA of the same samples described above for the IDEA™ kits using the modified TSP of Gargallo *et al.* (2006). Samples were analyzed via the modified TSP and fed to cecectomized roosters in both the intact form and after the16 h ruminal *in situ* incubation. Amino acid digestibility estimates (feed samples) and RUP-AA digestibility estimates (rumen residue samples) obtained via the modified TSP are presented in Table 1. *In vitro* AA and RUP-AA digestibility estimates obtained with the modified TSP generally agreed well with *in vivo* estimates for both the intact feeds and rumen residues. *In vitro* digestibility estimates (X) for all AA and RUP AA were highly correlated to *in vivo* estimates (Y; $R^2 = 0.93$ for total RUP-AA; Y = 31.8 + 0.7X; $R^2 = 0.75$ for total AA in feed; Y = 45.2 + 0.5X; $P<0.05$). In addition, the relationship between *in vitro* AA digestibility of feed samples and *in vivo* RUP AA digestibility (using the rumen residue samples) was examined. *In vitro* AA digestibility was highly correlated to *in vivo* RUP-AA digestibility for all AA (R^2 for total AA = 0.76; Y = 44.0 + 0.5X; $P<0.05$). The strength of the correlation was equal for the correlation between *in vivo* RUP-AA digestibility and *in vitro* AA digestibility of the intact feeds and *in vivo* RUP-AA digestibility and *in vitro* RUP-AA digestibility (data not shown). Therefore, further investigation into estimating RUP-AA digestibility from *in vitro* digestibility of AA in the intact feedstuff is warranted because it could potentially eliminate the need to use live animals for any step in determining RUP-AA digestibility. The advantage of

Table 1. Average in vitro *AA and RUP-AA digestibility estimated using the modified three-step procedure of protein concentrates commonly fed to dairy cows.*[1,2]

	Feed samples				Rumen residue samples			
Digestibility, %	SP	SBM	DDGS	FM	SP	SBM	DDGS	FM
Arginine	98.0	98.5	89.2	81.4	93.3	99.2	92.9	86.6
Histidine	96.5	97.2	86.4	81.0	91.3	97.4	92.1	91.3
Isoleucine	95.6	96.8	85.0	75.8	92.1	98.6	93.8	93.4
Leucine	95.3	96.2	86.7	78.3	92.0	98.5	95.5	93.5
Lysine	96.4	97.5	82.9	82.5	93.4	98.3	86.8	94.0
Methionine	96.0	96.5	87.0	76.1	86.5	98.6	94.9	89.6
Phenylalanine	96.1	96.5	86.9	74.0	92.5	98.7	94.6	90.9
Threonine	96.2	96.5	84.5	78.4	90.6	98.2	89.7	91.6
Tryptophan	98.8	100.0	100.0	76.2	94.5	100.0	100.0	97.7
Valine	95.3	96.3	83.6	77.1	91.3	98.2	92.5	91.9
Essential AA	96.3	97.1	86.1	78.7	92.1	98.5	93.6	91.7
Nonessential AA	96.7	97.4	87.3	78.4	91.9	98.5	94.6	88.6
Total AA	96.5	97.2	86.8	78.6	89.9	98.1	93.2	86.9
CP modified TSP	95.6	96.2	84.2	78.0	91.8	97.8	91.8	88.9
CP original TSP	-	-	-	-	78.6	73.0	76.7	73.0
Minimum	95.3	96.2	82.9	74.0	86.5	97.4	86.8	86.6
Maximum	98.8	100.0	100.0	82.5	94.5	100.0	100.0	97.7
Standard deviation	1.2	1.2	4.9	2.8	2.2	0.7	3.5	2.9

[1] Samples heated in the lab were not included in arriving at the average values reported in this table.

[2] Heated samples are not included in the averages presented, therefore, n=2 for SP, n=2 for SBM, n=4 for DDGS, n=5 for FM.

the modified TSP is that it can be performed by any lab equipped with a Daisy^{II} incubator, and can be used to estimate RUP-AA digestibility of any feed ingredient. Validation of the procedure with different feed types, including forages, is needed.

References

Antoniewicz, A.M., A.M. van Vuuren, C.J. Van der Koelen and I. Kosmala, 1992. Intestinal digestibility of rumen undegraded protein of formaldehyde treated feedstuffs measured by mobile bag and *in vitro* technique. Anim. Feed. Sci. Technol. 39, 111-124.

Bodwell, C.E., L.D. Satterlee and R.L. Hackler, 1980. Protein digestibility of the same protein preparations by human and rat assays and by *in vitro* enzymic digestion methods. Am. J. Clin. Nutr. 33, 677-686.

Borucki-Castro, S.I., L.E. Phillip, H. Lapierre, P.W. Jardon and R. Berthiaume, 2007. Ruminal degradability and intestinal digestibility of protein and amino acid in treated soybean meal product. J. Dairy Sci. 90, 810-812.

Boschin, G., A. D'Agostina, A. Rinaldi and A. Arnoldi, 2003. Lysinoalanine content of formulas for enteral nutrition. J. Dairy Sci. 86, 2283-2287.

Boucher, S.E., S. Calsamiglia, C.M. Parsons, H.H. Stein, M.D. Stern, P.S. Erickson, P.L. Utterback and C.G. Schwab, 2009. Intestinal digestibility of amino acids in rumen undegradable protein estimated using a precission-fed cecectomized rooster bioassay: Soybean meal and SoyPlus. J. Dairy Sci. 92, 4489-4498.

Boucher, S.E., S. Calsamiglia, C.M. Parsons, M.D. Stern, M Ruiz Moreno, M. Vázquez-Añón and C.G. Schwab, 2009. *In vitro* digestibility of individual amino acids in rumen-undegraded protein: The modified three-step procedure and the immobilized digestive enzyme assay. J. Dairy Sci. 92, 3939-3950.

Boucher, S.E., S. Calsamiglia, C.M. Parsons, H.H. Stein, M.D. Stern, P.S. Erickson, P.L. Utterback and C.G. Schwab, 2009. Intestinal digestibility of amino acids in rumen undegradable protein estimated using a precission-fed cecectomized rooster bioassay: Soybean meal and SoyPlus. II. Distillers dried grains with solubles and fish meal. J. Dairy Sci. 92, 6056-6067.

Britton, R.A., T.J. Klopfenstein, R. Cleale, F. Goedeken and V. Wilkerson, 1986. Methods of estimating heat damage in protein sources. Proc. Dist. Grain Conf. Cincinnati, OH. 41, 67.

Buchanan, R.A, 1969. *In vivo* and *in vitro* methods of measuring nutritive value of leaf-protein preparations. Br. J. Nutr. 23, 533-544.

Bunjapamai, S., R.R. Mahoney and I.S. Fagerson, 1982. Determination of D-amino acids in some processed foods and effect of racemization on *in vitro* digestibility of casein. J. Food Sci. 47, 1229-1234.

Calsamiglia, S. and M.D. Stern, 1995. A three-step *in vitro* procedure for estimating intestinal digestion of protein in ruminants. J. Anim. Sci. 73, 1459-1465.

Cone, J.W., A.H. Van Gelder, A.A. Mathijssen-Kamman and V.A. Hindle, 2006. Post-ruminal digestibility of crude protein from grass and grass silage in cows. Anim. Feed Sci. Technol. 128, 42-52.

De Boever, J.L., J.M. Vanacker, D.F. Bogaerts and C.V. Boucque, 1995. Protein evaluation of cattle compound feeds: comparison of in sacco measurements and tabular values. Neth. J. Agric. Res. 43, 297-311.

Erasmus, L.J., P.M. Botha and C.W. Cruywagen, 1994. Amino acid profile and intestinal digestibility in dairy cows of rumen-undegradable protein from various feedstuffs. J. Dairy Sci. 77, 541-551.

Erbersdobler, H.F. and V. Somoza, 2007. Forty years of furosine – Forty years of using Maillard reaction products as indicators of the nutritional quality of foods. Mol. Nutr. Food Res. 51, 423-430.

Faldet, M.A., L. D.Satter and G.A. Broderick, 1992. Determining optimal heat treatment of soybeans by measuring available lysine chemically and biologically with rats to maximize protein utilization by ruminants. J. Nutr. 122, 151-160.

Friedman, M., 1999. Chemsitry, nutrition and microbiology of D-amino acids. J. Agric. Food Chem. 47, 3457-3479.

Gargallo, S., S. Calsamiglia and A. Ferret, 2006. Technical note: A modified three-step *in vitro* procedure to determine intestinal digestion of proteins. J. Anim. Sci. 84, 2163-2167.

Goering, H.K., C.H. Gordon, R.W. Hemken, D.R. Waldo, P.J. Soest and L.W. Smith, 1972. Analytical estimates of nitrogen digestibility in heat damaged forages. J. Dairy Sci. 55, 1275-1280.

Hsu, H.W., D.L. Vavak, L.D. Satterlee and G.A. Miller, 1977. A multienzyme technique for estimating protein digestibility. J. Food Sci. 42, 1269-1273.

Hurrell, R.F. and K.J. Carpenter, 1981. The estimation of available lysine in foodstuffs after Maillard reactions. Prog. Food Nutr. Sci. 5, 159-176.

Hvelplund, T, 1985. Digestibility of rumen microbial protein and undegraded dietary protein estimated in the small intestine of sheep and by in sacco procedure. Acta. Agric. Scand. Suppl. 25, 132-144.

Hvelplund, T. and J. Madsen, 1990. The digestibility of amino acids in the small intestine. In: A study of the quantitative nitrogen metabolism in the gastro-intestinal tract, and the resultant new protein evaluation system for ruminants. The AAT-PBV System. T. Hvelplund and J. Madsen (eds.) Institute of Animal Science, Copenhagen, Denmark: 18-21.

Hvelplund, T., M.R. Weisbjerg and L.S. Andersen, 1992. Estimation of the true digestibility of rumen undegraded dietary protein in the small intestine of ruminants by the mobile bag technique. Acta Agric. Scand. Sect. A, Animal Sci. 42, 34-39.

INRA, 2007. Alimentation des bovines, ovins et caprins. Besoins des animaux. Valeur des aliments. INRA, Vesaukkes Cedex.

Kleinschmit, D.H., J.L. Anderson, D.J. Schingoethe, K.F. Kalscheur and A.R. Hippen, 2007. Ruminal and intestinal degradability of distillers grains plus solubles varies by source. J. Dairy Sci. 90, 2909-2918.

Kohn, R.A. and M.S. Allen, 1992. Storage of fresh and ensiled forages by freezing affects fibre and crude protein fractions. J. Sci. Food Agric. 58, 215-220.

Loerch, S.C., L.L. Berger, S.D. Plegge and G.C. Fahey, Jr, 1983. Digestibility and rumen escape of soybean meal, blood meal, meat and bone meal and dehydrated alfalfa nitrogen. J. Anim. Sci. 57, 1037-1047.

Marshall, H.F., G.W. Wallace and L.D. Satterlee, 1979. Prediction of protein digestibility by *in vitro* procedure using human, porcine and rat pancreatin preparations. Nutr. Rep. Int, 19, 901-913.

Masoero, F., L. Fiorentini, F. Rossi and A. Piva, 1994. Determination of nitrogen intestinal digestibility in ruminants. Anim. Feed Sci. Technol. 48, 253-263.

Mauron, J., 1990. Influence of processing on protein quality. J. Nutr. Sci. Vitaminol. 36, S57-69.

National Research Council, 1994. Nutrient Requirements of Poultry. 9th rev. ed. Natl. Acad. Sci. Washington, D.C.

National Research Council, 1998. Nutrient Requirements of Swine. 10th rev. ed. Natl. Acad. Sci. Washington, D.C.

National Research Council, 2001. Nutrient Requirements of Dairy Cattle. 7th rev. ed. Natl. Acad. Sci. Washington D.C.

Novus International, 2008. IDEA. Poultry. Available at: http://www.novusint.com/species/poultry/products.aspx?product=idea. Accessed April 1, 2008.

Porter, D.H., H.E. Swaisgood and G.L. Catignani, 1984. Characterization of an immobilized digestive enzyme system for determination of protein digestibility. J. Agric. Food Chem. 32, 334-339.

Prestløkken, E. and O. Rise, 2003. Protein and amino acid digestibility in dairy cows measured with mobile nylon bags recovered in ileum or in faeces. Acta Agric. Scand. Sect. A, Animal Sci. 53, 11-20.

Rhinehart, D., 1975. A nutritional characterization of the distiller's grain protein concentrates. M. S. Thesis. University of Nebraska, Lincoln.

Satterlee, L.D., J.G. Kendrick, D.K. Jewell and W.D. Brown, 1981. Estimating apparent protein digestibility from *in vitro* assays. In: C.E. Bodwell, J.S. Adkins and D.T. Hopkins (eds.), Protein quality in humans: assessment and *in vitro* estimation. AVI Publishing Co., Westport, CT: 316-339.

Sauer, W.C., H. Jorgensen and R. Berzins, 1983. A modified nylon bag technique for determining apparent digestibilities of protein in feedstuffs for pigs. Can. J. Anim. Sci. 63, 233-237.

Saunders, R.M., M.A. Connor, A.N. Booth, E.M. Bickoff and C.O. Kohler, 1973. Measurement of digestibility of alfalfa protein concentrates by *in vivo* and *in vitro* methods. J. Nutr. 103, 530-535.

Schasteen, C.S., J. Wu, M.G. Schulz and C.M. Parsons, 2007. Correlation of an immobilized digestive enzyme assay with poultry true amino acid digestibility for soybean meal. Poult. Sci. 86, 343-348.

Shelford, J.A., E.B. Cathcart, R.T. France and J.L. Dobbs, 1980. A survey of composition and heat damage in dehydrated alfalfa products over the growing season from two B.C. processing plants. Can. J. Anim. Sci. 60, 197-200.

Stern, M.D., A. Bach and S. Calsamiglia, 1997. Alternative techniques for measuring nutrient digestion in ruminants. J. Anim. Sci. 75, 2256-2276.

Van Straalen, W.M., F.M.H. Dooper, A.M. Antoniewicz, I. Kosmala and A.M. Van Vuuren, 1993. Intestinal digestibility in dairy cows of protein from grass and clover measured with mobile nylon bag and other methods. J. Dairy Sci. 76, 2970-2981.

Vanhatalo, A. and T. Varvikko, 1995. Effect of rumen degradation on intestinal digestion of nitrogen of 15N-labelled rapeseed meal and straw measured by the mobile-bag technique in cows. J. Agric. Sci. 125, 253-261.

Vanhatalo, A., I. Aronen and T. Varvikko, 1995. Intestinal nitrogen digestibility of heat-moisture treated rapeseed meals as assessed by the mobile-bag method in cows. Anim. Feed Sci. Technol. 55, 139-152.

Vanhatalo, A., P. Dakowski and P. Huhtanen, 1996. Effects of stage of growth and duration of rumen incubation time on intestinal digestibility of rumen-undegradable nitrogen of grass by mobile- bag method in cows. Acta Agric. Scand. 46, 1-10.

Vanhatalo, A., P. Dakowski and P. Huhtanen, 1996. Effects of stage of growth and duration of rumen incubation time on intestinal digestibility of rumen-undegradable nitrogen of grass by mobile-bag method in cows. Acta. Agric. Scand. Sect. A, Animal Sci. 46, 1-10.

Varvikko, T. and A. Vanhatalo, 1990. The effect of differing types of cloth and of contamination by non-feed nitrogen on intestinal digestion estimates using porous synthetic-fibre bags in a cow. Br. J. Nutr. 63, 221-229.

Voigt, J., B. Piatkowski, H. Englemann and E. Rudolph, 1985. Measurement of the postruminal digestibility of crude protein by the bag technique in cows. Arch. Tierernaehr. 8, 555-562.

Volden, H. and O.M. Harstad, 1995. Effect of rumen incubation on the true indigestibility of feed protein in the digestive tract determined by nylon bag techniques. Acta Agric. Scand. Sect. A, Animal Sci. 45, 106-115.

Application of an *O*-phthalaldehyde assay that detects oligopeptides for estimating rumen protein degradation *in vitro*

S. Colombini[1] and G.A. Broderick[2]
[1]Università degli Studi di Milano, Dipartimento di Scienze Animali, 20133 Milano, Italy
[2]Agricultural Research Service, USDA US Dairy Forage Research Center, 1925 Linden Drive West, Madison 53706, USA

Introduction

O-phthalaldehyde (OPA) reacts with primary amines to form a complex that absorbs at 340 nm and fluoresces at 450 nm (Roth, 1971). The disadvantage of using OPA fluorescence is that there is little contribution from small peptides because fluorescent intensity of N-terminal peptide products diminishes dramatically with increasing peptide chain length (Randall *et al.*, 1994). Church *et al.* (1983) developed an OPA colorimetric assay, based on absorption at 340 nm, to quantify proteolysis of milk protein by measuring the contribution of small peptides. The aim of the present work is to evaluate the OPA colorimetric (OPA-C) assay vs. the OPA fluorimetric (OPA-F) assay to determine the rate of rumen protein degradability of common feedstuffs by applying the inhibitor *in vitro* system with Michaelis-Menten saturation kinetics (Broderick and Clayton, 1992) and to determine rumen undegraded protein (RUP) by applying different passage rates to the soluble and insoluble N.

Material and methods

Three protein sources (1 expeller-extracted soybean meal, ESBM; 2 solvent-extracted soybean meals, SSBM1 and SSBM2; and 1 casein) were incubated in a rumen *in vitro* system treated with hydrazine and chloramphenicol to inhibit microbial uptake of protein degradation products. Samples were weighed to give a range of amounts of N (from 0.2 to 3 mg N/ml inoculum) and incubated for 2 hours with 10 ml of rumen fluid and 5 ml of buffer. At the end of incubations, fermentation was stopped by the addition of TCA. Samples were analyzed for N soluble in McDougall's buffer (SN) and for N not precipitable by TCA, which was considered the fraction already degraded at t=0 (FD0). TCA supernatants were analyzed for NH3 (phenol-hypochlorite assay), total amino acids (TAA) (OPA-F) and free aminoacids plus oligopeptides (OPA-C) by a flow injection analyzer. Velocity of protein degradation was calculated as the sum of (1) Ammonia-N plus N in free TAA or (2) Ammonia-N plus N in free TAA and peptides. Rate of degradation was estimated with the following equation:
$S_t = S_0 - t \times k_d \times K_m + K_m \times \ln (S_0/S_t)$;
where S_0 (mg/N ml SRF) is the amount of N added at the t=0, t is the duration of incubation (2 h) and K_m (mg/N ml SRF) is the Michaelis-Menten constant. The parameters K_m and k_d were estimated by non linear regression (NLIN) in SAS.
The k_d was adjusted (Adj k_d) for the FD0 fraction as follows:
Adj $k_d = [(k_d \times t) - \text{FD0}]/t$;
The rumen undegraded protein (%) (RUP) was calculated by applying 2 different rates of passage for insoluble (k_{pi}) and soluble (k_{ps}) N:
$\text{RUP} = [(\text{Total N} - \text{SN}) \times k_{pi}) / (k_{pi} + \text{Adj } k_d)] + [((\text{SN} - \text{FD0}) \times k_{ps}) / (k_{ps} + \text{Adj } k_d)]$
where k_{pi}= 0.06/h and k_{ps}=0.16/h.

Results and discussion

Nitrogen content and solubility of the samples are reported in Table 1. The OPA-C assay resulted in higher degradation rates for all the feeds analyzed due to the contribution of peptides to the fraction degraded (Table 2). Degradation rates determined by the OPA-C and OPA-F assays ranked proteins similarly; however, the OPA-F rate observed for ESBM was not different from that for SSBM2. The RUP contents were significantly lower using the OPA-C assay. Both assays ranked the 2 SSBM samples differently for RUP content. SSBM RUP content reported by NRC (2001) is

34.6% and similar to SSBM2 (determined by OPA-C) whilst the value for ESBM is higher (69%) than that found in the present study. The *in vivo* RUP content (45%) of ESBM reported by Reynal and Broderick (2003) is similar to the value obtained with both *in vitro* assays. Ruminal degradation of dietary protein as predicted by NRC (2001) did not appear to be a significant factor in predicting milk protein yield and milk N efficiency in dairy cows (Huhtanen and Hristov, 2009).

Table 1. Nitrogen, soluble N, free amino acids and oligopeptides content of the samples.

Sample	N (% DM)	Soluble N (% total N)	Free AA (% total N)	Oligopeptides (% total N)
Casein	14.58	99.9	0.27	0.34
ESBM	7.33	4.5	0.53	1.38
SSMB1	7.22	17.3	0.76	1.28
SSBM2	7.36	37.4	1.04	1.20

Table 2. Nitrogen degradation rates and RUP of the samples with the two different assays.

Sample		adj. rate (/h)	RUP (% CP)
Method OPA-C	Overall	0.272	30.5
	ESBM	0.086[c]	41.8[a]
	SSBM1	0.231[b]	24.4[c]
	SSBM2	0.181[b]	34.6[b]
	Casein	0.590[a]	21.3[c]
Method OPA-F	Overall	0.215	35.2
	ESBM	0.072[c]	47.5[a]
	SSBM1	0.188[b]	28.6[c]
	SSBM2	0.157[bc]	38.0[b]
	Casein	0.444[a]	26.7[c]
	SEM	0.022	1.57
Probability	Sample	<0.001	<0.001
	Method	0.015	0.002
	Sample×method	0.082	0.91

[a,b,c] LS-Means in column for each method with different superscripts differ ($P<0.05$).

References

Broderick, G.A. and M.K. Clayton, 1992. Rumen protein degradation rates estimated by non-linear regression analysis of Michaelis Menten *in vitro* data. Brit. J. Nutr. 67, 27-42.

Church, F.C., H.E. Swaisgood, D.H. Porter and G.L. Catignani, 1983. Spectrophotometric assay using o-phthaldialdehyde for determination of proteolysis in milk and isolated milk proteins. J. Dairy Sci. 66, 1219-1227.

Huhtanen, P. and A.N. Hristov, 2009. A meta-analysis of the effects of dietary protein concentration and degradability on milk protein yield and milk N efficiency in dairy cows. J. Dairy Sci. 92, 3222-3232.

National Research Council, 2001. Nutrient Requirements of Dairy Cattle. 7th rev. ed. Natl. Acad. Sci., Washington, DC.

Randall, C.S., T.R. Malefyt and L.A. Sternson, 1994. Approaches to the analysis of peptides. In: Lee, V.H.L. (ed.), Peptide and protein drug delivery. Marcel Dekker Inc., New York, USA: 203-246.

Reynal, S.M. and G.A. Broderick, 2003. Effects of feeding dairy cows protein supplements of varying ruminal degradability. J. Dairy Sci. 86, 835-843.

Roth, M., 1971. Fluorescence reaction for amino acids. Anal. Chem. 43, 880-882.

SAS Institute, 2000. Release 8.1, USA. Cary, NC: SAS Institute Inc.

Chemical composition, tannins content and *in vitro* fermentability of narrow-leaf lupin (*Lupinus angustifolius* L.) seeds

B. Ronchi[1], P.P. Danieli[1], R. Primi[1], U. Bernabucci[1] and P. Bani[2]
[1]*Dipartimento di Produzioni Animali, Università della Tuscia, Via S. C. de Lellis, 01100, Viterbo, Italy*
[2]*Istituto di Zootecnica, Università Cattolica del Sacro Cuore, Via E. Parmense, 84, 29100 Piacenza, Italy*

Introduction

Increasing grain legume production in Europe to be used in animal feeding should be a priority to provide a suitable alternative to the imported protein sources, *in primis* soybean products (Nemecek *et al.*, 2008). *Lupinus sp.* is an ancient crop widely cultivated in the Mediterranean Region and South American Andes, with chemical and nutritional properties of proteins similar to those of soybean (Pastor-Cavada *et al.*, 2009). Narrow-leaf lupin (*Lupinus angustifolius* L.) is one of the species best suited for growing conditions in the Mediterranean area (Annicchiarico, 2008) and it can be profitably used in animal diets (Petterson, 2000). In addition, the moderate content of phenolic antioxidant compounds (tannins) showing positive effects on the animal immune system (Nakagami *et al.*, 1995) and on the protein digestibility (El Gharras, 2009), may be interesting for ruminant feeding. The aim of the present study was to assess the nutritional value, total and condensed tannin content and *in vitro* ruminal fermentability of seeds of narrow-leaf lupin grown under specific agricultural conditions.

Material and methods

Seeds belonging to three different genetic lines of narrow-leaf lupin (26A, 42A and 48A) obtained from a two year (2007-2008) agronomic trial including two times of sowing (Mid November and Mid February) were analyzed for crude protein (CP), crude fiber (CFom), fiber fractions (NDF, ADF, ADL), crude fat (EE), ash (AOAC, 2006), total tannins (TT) (Riedl *et al.*, 2007), condensed tannins (CT) (Broadhurst and Jones, 1978) and amino acidic profile (Cohen and Michaud, 1993). Non-fibrous carbohydrates (NFCs) were calculated as [100-(NDF+CP+EE+ASH)]. *In vitro* gas production (IVGP) and DM degradation were performed incubating 0.5 g substrate, 40 ml of Menke and Steingass (1988) medium and 10 ml of strained rumen fluid for 48 h in 125 ml glass bottles. *In vitro* gas production (IVGP48) was measured by a pressure transducer and at the end pH was measured. After filtration and drying of fermented substrate at 105 °C, DM disappearance was determined. All data were analyzed by ANOVA and means were compared with LSD using Statistica 7 (StatSoft. Inc., USA); significance was declared at $P<0.05$).

Results and discussion

Nutritional characterization of lupin seeds (Table 1) showed a good CP content (29.1-35.5% on DM basis) with small but significant ($P<0.05$) differences also of CFom and ash between genetic lines. The effect ($P<0.01$) of sowing time was only significant for EE and ADF. Amino acidic profile of protein showed a lower content of essential (100.47±7.42 mg/g DM), sulfur (2.90±0.25) and aromatic (23.51±3.41) amino acids than the values reported in literature for soybean meal (ILSI, 2010). Condensed tannins were always found below the method detection limit (0.05 mg/g Catechin Equivalents on DM basis) while moderate levels of total tannins (0.20±0.01 mg/g Catechin Equivalents on DM basis) were found, with significant differences ($P<0.001$) due to the sowing time. *In vitro* gas production data pointed out a rapid fermentability of lupin with almost 50% of IVGP48 produced after the first 7 h of incubation. There was a significant difference ($P<0.001$) between years in terms of total IVGP48 and gas production rate (327.4 vs. 316.7 ml/g OM; 7.51 vs. 6.87%/h for year 2007 and 2008, respectively). No significant effects of variety were found on IVDMD although a significant interaction variety x year was found. IVDMD was 94.25 and 93.63%

Table 1. Nutritional characterization of narrow-leaf lupin seeds (mean ± SD are expressed as % on DM basis).

Line	CP	EE	CFom	NDF	ADF	ADL	Ash	NFCs
26A	30.5±1.7^{b}	4.9±0.7	18.3±0.3^{b}	30.5±0.3	22.1±1.5	5.5±0.5	4.1±0.1^{a}	30.0±3.0
42A	32.8±3.1^{b}	4.3±1.1	19.9±0.1^{a}	29.3±0.4	22.1±0.5	5.7±2.3	3.9±0.1	29.6±3.8
48A	34.0±1.3^{a}	4.9±1.1	18.7±0.7^{b}	33.1±3.5	22.3±1.5	4.6±2.0	3.8±0.1^{b}	29.7±3.8

^{a,b} P<0.05; ^{A,B} P<0.01.

for year 2007 and 2008 respectively, i.e. in agreement with IVGP48 data. Final incubation pH was lower for 2007 than 2008 lupin seeds (6.91 vs. 6.94, P<0.01) and differences were also observed among varieties (6.91, 6.92 and 6.96 respectively for 48A, 26A and 42A variety) ranking them in an inverse order with respect to IVGP48.

The findings of present study suggest that narrow-leaf lupin could potentially be a good feedstuff contributing at the same time to the energy and protein balance of animal ration. Agricultural practices, such as sowing date, may have a relevant influence on the accumulation of bioactive chemicals over the nutritional characteristic of narrow-leafed lupin. The level of tannins observed may be considered low if compared to other grains (e.g. sorghum, some varieties) and forage legumes (e.g. sanfoin, bird's foot trefoil). All genetic lines showed a good rumen potential digestibility, but *in vivo* studies are needed to evaluate optimal feeding schemes utilizing narrow-leaf lupin as a combined energy-protein source for livestock farming.

References

Annicchiarico, P., 2008. Adaptation of cool-season grain legume species across climatically-contrasting environments of Southern Europe. Agron. J. 100, 1647-1654.

AOAC, 2006. Official Methods of Analysis, Association of Official Analytical Chemists. 18th Ed. Gaithersburg, USA: AOAC Press.

Broadhurst, R.B. and W.T. Jones, 1978. Analysis of condensed tannins using acidified vanillin. J. Sci. Food Agr. 29, 788-794.

Cohen, S. and D. Michaud, 1993. Synthesis of a fluorescent derivatizing reagent, 6-Aminoquinolyl-N-Hydroxysuccinimidyl Carbamate, and its application for the analysis of hydrolysate Amino Acids via High-Performance Liquid Chromatography. Anal. Biochem. 211, 279-287.

El Gharras, H., 2009. Polyphenols: food sources, properties and applications – a review. Int. J. Food Sci. Tech. 44, 2512-2518.

ILSI, 2010. International Life Sience Insitute, Crop Composition Database, version 3,0; http://www.cropcomposition.org, search criteria: soybean seed, all locations, all years.

Menke, K.H. and H. Steingass, 1988. Estimation of the energetic feed value obtained from chemical analysis and *in vitro* gas production using rumen fluid. Anim. Res. Develop. 28, 7-55.

Nakagami, T., N. Nanaumi-Tamura, K. Toyomura, T. Nakamura and T. Shgeehisa, 1995. Dietary flavonoids as potential natural biological response modifiers affecting the autoimmune system. J. Food Sci. 60, 653-656.

Nemecek, T., J-S. von Richthofen, G. Dubois, P. Casta, R. Charles and H. Pahl, 2008. Environmental impacts of introducing grain legumes into European crop rotations. Europ. J. Agronomy 28, 380-393.

Pastor-Cavada, E., R. Juan, J.E. Pastor, M. Alaiz and J. Vioque, 2009. Analytical nutritional characteristics of seed proteins in six wild *Lupinus* species from Southern Spain. Food Chem. 117, 466-469.

Petterson, D.S., 2000. The use of lupins in feeding systems – review. Asian-Aust. J. Anim. Sci. 13, 861-882.

Riedl, K.M. J.H. Lee, M. Renita, S.K.S. Martin, S.J. Schwartz and Y. Vodovotz, 2007. Isoflavone profiles, phenol content, and antioxidant activity of soybean seeds as influenced by cultivar and growind location in Ohio. J. Sci. Food Agr. 87, 1197-1206.

Theodorou, M.K., B.A. Williams, M.S. Dhanoa, A.B. McAllan and J. France, 1994. A simple gas production method using a pressure transducer to determine the fermentation kinetics of ruminant feeds. Anim. Feed Sci. Tech., 48: 185-197.

Protein fractions and rates of degradation of tropical forages from intensively grazed pastures

R.B. Reis[1], *S. Colombini*[2], *A. Miller*[3], *D.K. Combs*[3] *and G.A. Broderick*[4]
[1]*Universidade Federal de Minas Gerais, CNPq, Brazil*
[2]*Università degli Studi, Dipartimento di Scienze Animali, Italy*
[3]*University of Wisconsin-Madison, Madison, WI, USA*
[4]*USDA-ARS, US Dairy Forage Research Center, Madison, WI, USA*

Introduction

Tropical forages are known for low levels of crude protein (CP) and high levels of fiber. Forage quality and utilization are most affected by forage species and maturity. Intensive rotational grazing systems offer better forage quality for the animals, but knowledge of composition and degradation patterns of CP of these forages is limited. Compared with temperate forages, tropical forages have greater annual dry matter (DM) yield; however, increased yield is usually associated with decreased forage quality and feeding value. Brown and Pitman (1991), using the *in situ* technique, found that concentrations of soluble and degradable N in Limpograss were relatively small. They observed a long lagtime (11 h) before N degradation began and, using *in vitro* procedures, they also showed there was little or no ammonia released during the first 8 h of incubation. Lopes *et al.* (2003) found CP degradation rate and effective degradability of 0.033/h and 60.3%, respectively, for *Pennisetum purpureum* using the *in situ* technique. The objective of this research was to quantify the CP fractions and rates of degradation of intensively managed tropical forages.

Material and methods

Tropical grasses from different research plots managed intensively in Brazil were collected using 95% of light interception as the criterion (Da Silva, 2002). The grasses were *Brachiaria brizanta cv Braquiarão*, cv *Marandu, cv MG-5*; *Brachiaria hibrida cv Mulato, Cynodon dactylon cv Tifton 85; Panicum maximum cv colonião, cv mombaça, cv tanzânia* and *Pennisetum purpureum cv cameron, cv napier.* The NPN (NH_3, TAA and oligopeptides) (fraction A) was determined using a semi-automated method (Colombini and Broderick, 2010) modified to include automated data collection. Soluble crude protein N (fraction B_1) was determined using a modification of the method of Licitra *et al.* (1996). Fraction C was the nitrogen found in acid detergent fiber (ADIN). Fraction B_3 was calculated as the difference between the CP recovered in neutral detergent fiber (NDIN) and ADIN. Fraction B_2 was the total CP minus the sum of fractions A, B_1, B_3 and C. Fractional degradation rates (k_d) were estimated according to Broderick and Clayton (1992) using non-linear regression analysis of the integrated Michaelis-Menten equation ($k_d = V_{max}:K_m$). The statistical analysis was performed using GLM Procedure of SAS (SAS Inst. Inc., Cary, NC).

Results and discussion

Average N, neutral detergent fiber (NDF) and acid detergent fiber (ADF) content of the forages was 2.78; 62.0 and 34.0% of DM, respectively. Although there was a significant amount of N bound to the NDF (37.8% of total N), only 5% remained in the ADF. This means that the average B_3 fraction was 32.5% of total N (Table 1). Fraction C was 5.2% of total N, which was considerably lower if compared with tropical forage not intensively managed (Lopes *et al.*, 2003). Having a large proportion of total CP bound to the fiber fraction can depress N utilization; however, much of this CP was present as N fraction B_3 in these forages. These forages had 40.2% of the total N as soluble N, which implied that readily fermentable carbohydrates would need to be fed for improving N efficiency. The high soluble N content contributed to a total N degradation rate of 0.207/h. Mean RUP content of the forages was 40% of total N partially due to the high passage rate (k_p) of the soluble N fraction (Table 1).

Table 1. Nitrogen, N bound to the fiber fractions, soluble N, rate of degradability and rumen-undergraded protein (RUP) of tropical forages intensively managed.

Forages	N (%, DM)	NDFN (%, total N)	ADIN (%, total N)	Soluble N (%, total N)	Adj. Rate (/h)	RUP (% CP)
B. brizantha	2.23[c]	27.5[c]	4.91[a]	47.04[a]	0.145[a]	28.3[b]
B. hibrida	2.51[c]	29.0[c]	6.21[a]	48.10[a]	0.118[b]	30.7[b]
C. dactylon	3.21[a]	45.3[a]	4.75[a]	45.53[a]	0.079[c]	40.4[a]
P. maximum	2.80[bc]	43.7[a]	5.30[a]	36.91[b]	0.071[c]	40.9[a]
P. purpureum	2.92[b]	36.5[b]	5.39[a]	33.28[b]	0.081[c]	42.0[a]
SEM	0.19	3.00	0.62	3.54	0.008	2.11

[a,b,c] LS-Means within column with different superscripts differ ($P<0.05$).

Cynodon dactylon and *Panicum maximum* contained about 45% of total N as NDIN. However, *Cynodon dactylon* had less than 5% of total N as ADIN suggesting that approximately 90% of the N bound to the NDF is potentially available. The *Brachiaria brizantha* and the *Brachiaria hibrida* showed the highest soluble N fraction (47 and 48%) and values of k_d (0.118 and 0.145/h). The *Brachiaria brizantha* has 11.6% of the total N as oligopeptides. Estimates of RUP ranged from 28% (*Brachiaria brizantha*) to 42% (*Pennisetum purpureum*) of total N. The RUP values for *Cynodon* determined by Vendramini *et al.* (2008) ranged from 38.9 to 43.3, depending on N fertilization, and were similar to the average value obtained in the present study. Most of the information in the available literature derives from *in situ* studies and forages that were not intensively managed. In the NRC (2001) library there is no information about intensively managed C4 pasture grasses. When compared with cool season grasses under same management these tropical forages have higher NDIN, lower ADIN and comparable rates of N degradation. Intensively managed tropical forages have high contents of soluble N with high rates of degradation. The high proportion of plant N bound to NDF may be an advantage for feeding these forages because NDIN is thought to be available to the animal.

References

Broderick, G.A. and M.K. Clayton, 1992. Rumen protein-degradation rates estimated by nonlinear regression analysis of Michaelis-Menten *in vitro* data. Br. J. Nutrition 67, 27-42.

Brown, R.H. and W.D. Pitman, 1991. Concentration and degradation of nitrogen and fiber fractions in selected tropical grasses and legumes. Trop. Grassl. 25, 305-312.

Colombini, S. and G.A. Broderick, 2010. Application of an O-phthalaldehyde assay that detects oligopeptides for estimating rumen protein degradation in vitro. In: Crovetto, G.M. (ed.) Energy and protein metabolism and nutrition. Wageningen Academic Publishers, Wageningen, the Netherlands, p. 713-714.

Da Silva, S.C., 2002. Características morfo-fisiológicas e respostas funcionais de plantas forrageiras e animais submetidos a pastejo. Tese (Livre Docência) – Escola Superior de Agricultura 'Luiz de Queiroz', Piracicaba, 476 p.

Lopes, F.C.F., L.J.M. Aroeira, N.M. Rodrigues, F. Deresz, I.B.M. Sampaio, I. Borges, H. Maldonado-Vasquez and A. Vittor, 2003. Rumen degradability of elephantgrass (*Penissetum purpureum*, Schumack) grazed by lactating crossbred Holstein x Zebu dairy cows. Arq. Bras. Med, Vet. Zoot. 55, 694-701.

Licitra, G., T.M. Hernandez and P.J. Van Soest, 1996. Standardization of procedures for nitrogen fractioning of ruminants. Anim. Feed Sci. Techn. 57, 347-358.

National Research Council, 2001. Nutrient requirements of dairy cattle. 7th rev. ed. Natl. Acad. Sci., Washington, DC.

Vendramini, J.M.B., L.E. Sollenberger, A.T. Adesogan, J.C.B. Dubeux Jr., S.M. Interrante, R.L. Stewart Jr. and J.D. Arthington, 2008. Protein fractions of Tifton 85 and Rye-Ryegrass due to sward management practices. Agron. J. 100, 463-469.

Evaluating the protein value of forages using a modified gas test

B. Edmunds[1,2]*, K.-H. Südekum*[1]*, F.J. Schwarz*[3]*, M. Schuster*[2] *and H. Spiekers*[2]
[1]*Institute of Animal Science, University of Bonn, Endenicher Allee 15, 53115 Bonn, Germany*
[2]*Bavarian State Research Centre for Agriculture (Grub), Prof. Dürrwaechter-Platz 3, 85586 Poing, Germany*
[3]*Animal Nutrition Weihenstephan, TU Munich, Weihenstephaner Berg 3, 85350 Freising, Germany*

Introduction

Utilisable crude protein (uCP) is the sum of microbial CP and undegraded dietary CP (UDP) reaching the duodenum. Previously used *in vivo* methods for estimating uCP are time consuming, costly and uncomfortable for animals. The current method of estimating uCP in Germany is a calculation derived from *in vivo* data and takes into account metabolisable energy (ME) and UDP (Lebzien *et al.*, 1996; Lebzien and Voigt, 1999). Unfortunately, the calculation does not take into account degradation characteristics and more intimate nutritional properties of individual feedstuffs, and pre-recorded UDP estimates are crude and outdated. In addition, values are singular and information on different levels of feeding is not known. A slightly modified version of the Hohenheim gas test (Menke *et al., 1979*) has been used to calculate uCP (Steingass *et al.*, 2001). This method was employed in the present study in an attempt to standardise the procedure whilst improving and extending the current data base of uCP values from products of various grass conservation methods.

Materials and methods

Samples comprising 88 silages, 24 hays, 49 artificially dried pellets and their fresh counter parts, and 67 freshly cut grasses were collected from different regions throughout Bavaria over a two-year period. Samples were dried at 50 °C for 36 hours and analysed for proximate parameters. The gas test followed the basic method of Menke and Steingass (1988) whereby 200 mg dried feedstuff was incubated in a rumen fluid/buffer solution for a maximum of 24 hours. In the present study the method was modified by the addition of 2 g/l NH_4HCO_3 and a reduction of 2 g/l $NaHCO_3$ in the buffer solution. A blank containing only the rumen fluid/buffer solution and three standard samples with long running averages were included with every run to correct for gas and uCP values between each different batch of rumen fluid. Four replicates divided in two incubators were incubated for 6 and 24 hours to allow calculation of uCP at different rates of passage. At each incubation time point, gas readings were taken and the syringe contents analysed for ammonia nitrogen. Utilisable crude protein was calculated as:

$$\text{uCP g/kg DM} = \frac{NH_3\text{-}N_{blank} + N_{sample} - NH_3\text{-}N_{sample}}{\text{weight (mg)} \times DM} \times 10000$$

where $NH3\text{-}N_{blank}$ is mg nitrogen coming from ammonia in the blank, N_{sample} represents the total nitrogen (mg) content of the 200 mg dried sample, and $NH3\text{-}N_{sample}$ is ammonia nitrogen (mg) from the syringe incubating the sample. The uCP values were calculated at passage rates of 2, 4 and 6%/h (PR6, PR4, PR2). Data was analysed using the GLM procedure of SAS (version 9.2) and means were deemed significant at ($P<0.05$).

Results and discussion

Table 1 shows CP, ME and uCP means for different grass products at three assumed rates of passage. Pellets gave higher uCP values (PR6 and 4) than other products due probably to the increase in Maillard products from the high heat treatment. It is as of yet unknown how much of this additional uCP in the pellets is truly digestible intestinally. The lowest uCP results came, as expected, from silage and hay. The lack of significant difference between pellets and hay at PR2 can be attributed

Table 1. Product means for crude protein (CP; g/kg DM) metabolisable energy (ME; MJ/kg DM) and utilisable crude protein (uCP; g/kg DM) at passage rates (PR) of 6, 4 and 2%/h.

Grassland product	n	CP	ME	PR6	PR4	PR2
Pellets	49	181±25	10.2±0.7	211±20[a]	196±25[a]	170±39[a]
Fresh	67	168±41	10.3±0.6	189±27[b]	179±28[b]	161±34
Hay	24	135±40	9.5±0.9	166±22[c]	159±23[c]	148±37
Silage	88	168±21	10.1±0.7	165±17[c]	161±21[c]	154±32[b]

to the high standard deviation and low number of hay samples. The difference between pellets and silage at PR2 was only weakly significant (P=0.043).

The higher standard deviations at the slower passage rate are caused by higher variability in ammonia values at 24 hours. It is still uncertain whether this variability is due to the experimental design or biological fluctuations in the rumen fluid. It is important that such questions are answered before the method is widely employed.

The modHFT was correlated against uCP estimated from the current equation using *in situ* estimated UDP values (r^2=0.679, P<0.0001) (Nibbe *et al.*, 2001). The uCP values estimated from the modified gas test in the present study are slightly higher than those calculated using the current method (Lebzien and Voigt, 1999). This is in part due to the method of correction using the protein standard. A linear relationship has been shown to exist when uCP is presented on a log scale (H. Steingass, unpublished) allowing the 6 hour uCP mean for the protein standard to be estimated from the 8 and 24 hour means provided by University of Hohenheim. The resulting 6 hour estimation is 37 g/kg higher than the actual recordings, hence uCP values for faster rates of passage are overestimated. Nevertheless a method of correction using standardised means is necessary for intra- and inter-laboratory comparison. Additionally, the Lebzien and Voigt calculation gives no indication of feeding level so a stronger correlation cannot be expected. Some *in vivo* trials would be beneficial. Although the method requires further standardisation and validation, it can be concluded that it holds promise for analysing the uCP values of individual grass products with simplicity and labour efficiency being further advantages.

References

Lebzien, P., J. Voigt, M. Gabel and D. Gaedeken, 1996. Zur Schätzung der Menge an nutzbarem Rohprotein am Duodenum von Milchkühen. J. Anim. Physiol. Anim. Nutr. 76, 218-223.

Lebzien, P. and J. Voigt, 1999. Calculation of utilizable crude protein at the duodenum by two different approaches. Arch. Anim. Nutr. 52, 363-369.

Menke, K.H., L. Raab, A. Salewski, H. Steingass, D. Fritz and W. Schneider, 1979. The estimation of the digestibility and metabolisable energy content of ruminant feedingstuffs from the gas production when they are incubated with rumen liquor *in vitro*. J. Agric. Sci. Camb. 93, 217-222.

Menke, K.H. and H. Steingass, 1988. Estimation of energetic feed value obtained from chemical analysis and *in vitro* gas production using ruminal fluid. Anim. Res. Dev. 28, 7-12.

Nibbe, D., P. Lebzien, H. Spiekers, H. Steingass and K.-H. Südekum, 2001. Vergleich verschiedener *in vitro*- und *in situ*-Verfahren zur Beurteilung des Protein-wertes von Raps- und Sojaextractionsschrot. 113. VDLUFA-Kongress, Berlin pg 111.

Steingass, H., D. Nibbe, K.-H. Südekum, P. Lebzien and H. Spiekers, 2001. Schätzung des nXP-Gehaltes mit Hilfe des modifizierten Hohenheimer Futterwerttests und dessen Anwendung zur Bewertung von Raps- und Sojaextraktionsschroten. 113. VDLUFA-Kongress, Berlin: 114 (Abstr.).

Estimation of indigestible NDF (iNDF) in forages

M.R. Weisbjerg[1], P. Lund[1], M. Chrenkova[2] and M.K. Larsen[1]
[1]Faculty of Agricultural Science, Aarhus University, P.O. Box 50, 8830 Tjele, Denmark
[2]Animal Production Research Centre Nitra, Hlohovecká 2, 951 41 Lužianky, Slovakia

Introduction

Indigestible NDF (iNDF) is an important parameter in models for prediction of energy and protein value of feeds. Therefore, rapid laboratory methods for the prediction of iNDF are needed. Lignin measured as acid detergent lignin (ADL) is used as predictor for iNDF in the Cornell Net Energy and Protein evaluation system, where iNDF is predicted as 2.4 times ADL, and it is also known that iNDF concentration is highly correlated with organic matter digestibility (Huhtanen *et al.*, 2006), supporting the idea of simple prediction equations based on these well known feed characteristics. However, several observations in our lab have shown, that both forage type and growth season influence these relationships, which makes simple universal prediction models problematic. The aim of the present experiment was to examine potential laboratory methods for iNDF estimation on grasses and legumes.

Material and methods

Twenty one different forage species representing 13 perennial ryegrasses, 2 festuloliums, 2 hybrid ryegrasses, 1 cocksfoot, 1 lucerne, 1 white clover, and 1 red clover sown in plots were harvested at 3 harvesting times in the first (primary) growth, and once in second, third and fourth growth in 2008, resulting in a total of 126 samples. Samples were dried at 60°C. Chemical analyses were performed according to conventional methods after milling on a 1 mm screen in a hammer mill. *In vitro* OM (IVOM) digestibility was determined according to the Tilley and Terry method. *In situ* iNDF was determined after milling in a cutter mill with 1.5 mm screen according to the NorFor standard after 288 h rumen incubation in Dacron bags with 12 μm pore size.

Results and discussion

Samples varied highly in both chemical composition and availability (Table 1). NDF concentration in dry matter (DM) varied from 18 to 60%, ADL in DM from 0.8 to 6.2%, and *in vitro* digestibility of organic matter varied from 62 to 83%. Mean concentrations of iNDF in percentance of DM across harvesting times and varieties were: Perennial ryegrass 5.4, festulolium 7.2, hybrid ryegrass 6.3, cocksfoot 7.2, lucerne 10.3, white clover 5.7 and red clover 8.1. Mean iNDF/ADL ratio was 2.79, which is considerable higher than the 2.4 used in the Cornell system, however closer for legumes (2.39) than for grasses (2.85). The ratio of indigestible OM (indigestible OM (indig) = 100-IVOM) over iNDF (Indig/iNDF) was 5.39, however much higher for grasses (5.63) than for legumes (3.94).

Simple correlations between iNDF concentration and other measures are given in Table 1. iNDF in percentance of DM was most correlated (r=0.81) with ADL (%DM), and correction for ash content or use of ADL in NDF did not increase correlations (results not shown). Correlation to IVOM was less (r=-0.73). iNDF concentration in NDF was highly correlated (r=0.89) with ADL (%DM) whereas, as expected, the correlation to IVOM was much less (r=-0.51). Plots and analysis of residuals (not shown) reveal that there was systematic bias. For iNDF vs. ADL concentrations simple regressions underestimated iNDF for legumes, especially white clover, and for growths later than first growth. For iNDF vs. IVOM, iNDF was overestimated for lucerne and underestimated for growths later than first. Due to differences between forage types in NDF concentration and lignification, it was anticipated that inclusion of more parameters in multiple regression prediction equations might account for these systematic deviations. From Stepwise multiple regression analysis, IVOM, ADL

Table 1. Variation in main parameters and correlations to iNDF concentration in DM or NDF for all (N=126) and ratios for all, grasses (N=108) and legumes (N=18).

		Mean	Std. dev.	Min.	Max.	Correlation to iNDF	
						(% DM)	(% NDF)
NDF (%DM)	All	45.3	7.7	17.9	59.8	0.28	-0.17
ADL (%DM)		2.17	0.93	0.79	6.21	0.81	0.89
ADL (%NDF)		5.07	3.00	1.64	2.13	0.51	0.80
IVOM (% OM)		72.7	4.9	62.3	83.3	-0.73	-0.51
iNDF (%DM)		6.12	3.35	2.35	16.12	1	0.88
iNDF (%NDF)		13.7	7.6	5.1	42.7	0.88	1
iNDF/ADL ratio		2.79	0.75	1.22	4.57		
Indig[1]/iNDF ratio		5.39	2.08	2.17	11.31		
iNDF/ADL ratio	Grasses	2.85	0.75	1.27	4.57		
Indig[1]/iNDF ratio		5.63	2.09	2.26	11.31		
iNDF/ADL ratio	Legumes	2.39	0.70	1.22	3.59		
Indig[1]/iNDF ratio		3.94	1.34	2.17	6.97		

[1] Indig = indigestibility of organic matter (%OM) = 100-IVOM.

and NDF concentration, and ADL/NDF ratio were found to be the four most important predictors. Compared with simple regressions, inclusion of more variables increased R^2 of the prediction model, however systematic effects of forage type and harvest time on residuals were still obvious (not shown).

ADL in DM correlated well with iNDF in NDF, indicating that a combination of prediction of iNDF/NDF ratio with an analysis of NDF in DM might be an attractive solution for prediction of iNDF in DM.

In conclusion, multiple instead of simple regressions reduced systematic bias, but results were not yet fully satisfactory for universal prediction equations for iNDF across feed types.

Acknowledgements

The work has been carried out with financial support from Ministry for Food, Agriculture and Fisheries, 3304-FVFP-07-766-01, and from the Commission of the European Communities, FP7, KBB-2007-1.

References

Huhtanen, P., S. Ahvenjärvi, M.R. Weisbjerg and P. Nørgaard, 2006. Digestion and passage of fibre in ruminants. In: Sejrsen, K., T. Hvelplund and M.O. Nielsen (eds.), Ruminant physiology. Digestion, metabolism and impact of nutrition on gene expression, immunology and stress. Proceedings Xth ISRP. Wageningen Academic Publishers, Wageningen, the Netherlands: 87-135.

Effect of lignin linkages with other plant cell wall components on *in vitro* and *in vivo* NDF digestibility of forages and potential energy yield

E. Raffrenato[1], R. Fievisohn[2], K.W. Cotanch[2], R.J. Grant[2], L.E. Chase[1] and M.E. Van Amburgh[1]
[1]*Cornell University, Ithaca, NY, USA*
[2]*W.H. Miner Agricultural Research Institute, Chazy, NY, USA*

Introduction

The association of phenolic components with carbohydrates is generally accepted as the greatest barrier to carbohydrate utilization (Besle *et al.*, 1994). In general lignin is associated with carbohydrate through covalent bonds, hydroxycinnamic (mainly *p*-coumaric (CA) and ferulic acids (FA)) acids that attach to lignin and hemicellulose through ester and ether bonds as bridges forming lignin/phenolic-carbohydrate complexes. To our knowledge, there is no work that compares conventional and bmr corn silages and effects of maturity in grasses and alfalfas on NDF digestibility and energy content. Our objective was, therefore, to evaluate the effect of ester and ether linked phenolic acids on NDF digestibility *in vitro* and *in vivo*.

Material and methods

Thirty forages, conventional and bmr corn silages, alfalfas, and immature and mature grasses, were analyzed for NDF, ADF, Klason lignin (KL) and ADL. Forages were incubated in-vitro for 24 h and 96 h NDFD (Goering and Van Soest, 1970) and estimates of rate of digestion (k_d) were made. Intact samples, NDF, ADF, and KL residues were analyzed for ester- and ether-linked PCA (EsPCA, EtPCA) and ester- and ether-linked FA (EsFA, EtFA). A multiple regression with stepwise selection of variables ($P<0.10$) was conducted within each forage group using the independent variables lignin type (ADL or KL) and their difference and phenolic acid content with their specific linkages, on an NDF basis. For the *in vivo* comparison, three corn silages, a conventional low digestibility (CLD), a conventional high digestibility (CHD) and a bmr corn silage (BMR) were fed to 6 fistulated cows for 3 weeks in three iso-NDF diets. After 10 days on diet, samples of TMR, feces, urine, plasma, and rumen samples were taken for 3 days every 3 h. Phenolic acids in forages, TMRs, rumen and feces samples were analyzed using a modified procedures by Iiyama *et al.* (1990) and Jung and Shalita-Jones (1990). Extraction was by 2N and 4N NaOH treatment, followed by chromatography separation by HPLC with a diode array detector using a reverse-phase column.

Results and discussion

Most of the significant correlations among digestion parameters and phenolic acids were obtained when pooling all forages, however, values were not consistent among groups. The stepwise selection procedure reached significance with an R^2 that varied between 0.56 and 0.99 for all forages except alfalfa (Table 1). This data demonstrated that factors affecting rates and extents of digestion of NDF vary by forage type and maturity. The data from the in-vivo study supported the in-vitro results, where the highest total tract NDFD (70%) was from the corn silage with the lowest phenolic acid and EsPCA content in the ADF fraction. Further, the data support the existence of an acid-soluble lignin fraction liberated during fermentation that can inhibit digestion and helps explain discrepancies in energy availability from plants with high digestibility.

Table 1. Independent variables selected and their solution for each forage group and dependent variable.

Dependent variable	Forage group							
	Grasses		Imm. grasses		Conv. corn		Bmr corn	
k_d	ADL:	-0.72	EtFA:	0.87	ADL:	-0.74	KL:	-0.45
	EtFA:	-0.68			EsFA:	-1.07	EsPCA:	1.04
R^2		0.97		0.75		0.79		0.84
24 h ivNDFD	ADL:		ADL:	-1.12	EtPCA:	-0.75	KL:	-0.62
	EtFA:		EsPCA:	0.36			EsPCA:	1.09
R^2		0.98		0.97		0.56		0.92
96 h ivNDFD	ADL:	-0.90	KL:	0.16			(KL-ADL):	-0.57
	EtFA:	-0.40	ADL:	-0.59			EtFA:	0.70
			EtFA:	0.48			EsPCA:	0.34
R^2		0.97		0.99				0.84

Significance level = 0.10; k_d = rate of digestion, %/h; ivNDFD = *in vitro* neutral detergent fiber digestibility; ADL = acid detergent lignin; KL = klason lignin; EtFA = etherified ferulic acid; EsFA = esterified ferulic acid; EtPCA = etherified *p*-coumaric acid; EsPCA = esterified *p*-coumaric acid.

References

Besle, J.M., A. Cornu and J.P. Jouany, 1994. Roles of structural phenylpropanoids in forage cell wall digestion. J. Sci. Food Agric. 64, 171-190.

Goering, H.K. and P.J. Van Soest, 1970. Forage Fiber Analyses (Apparatus, Reagents, Procedures, and some Applications). US Agricultural Research Service, USA.

Iiyama, K., T.B.T. Lam and B.A. Stone, 1990. Phenolic acid bridges between polysaccharides and lignin in wheat internodes. Phytochemistry. 29, 733-737.

Jung, H.G. and S.C. Shalita-Jones, 1990. Variation in the Extractability of Esterified *p*-Coumaric and Ferulic Acids from Forage Cell Walls. J. Agric. Food Chem. 38, 397-402.

Degradation characteristics of protein in feeds for ruminants

M. Chrenková[1], Z. Čerešňáková[1] and M.R. Weisbjerg[2]
[1]Animal Production Research Centre Nitra, Hlohovecká 2, 951 41 Lužianky, Slovakia
[2]Faculty of Agricultural Sciences, Aarhus University, Denmark

Introduction

Protein degradation in the rumen is important for the protein value of feeds and thereby the utilization of feed nitrogen by ruminants, and can be estimated using different methods. The *in situ* method estimates the protein degradation profile in the rumen environment. The method used in the Cornell Net Carbohydrate Protein System (CNCPS) estimate soluble and insoluble nitrogen (N) fractions in feeds with different degradation characteristics. In CNCPS, crude protein (N×6.25) is partitioned into 5 fractions, A, B_1, B_2, B_3 and C with different degradation characteristics.

The aim of the present experiment was to compare protein degradation characteristics estimated *in situ* with estimates using buffer soluble N and CNCPS estimates for soluble nonprotein N (fraction A) and soluble true protein (B_1) in selected concentrate feeds and silages.

Material and methods

Thirty one feed samples (4 grains, 5 legume seeds, 3 oil seeds, 8 oilseed by-products, 5 dry distiller grains with solubles (DDGS) and 6 silages (3 maize and 3 lucerne) were used. Silages were freeze dried. Samples were milled on hammer mill with a 3 mm screen for *in situ* and 1 mm screen for chemical and solubility analysis. Chemical analyses were according to conventional methods. Soluble N (SN) was measured in borate-phosphate buffer. The insoluble N fraction (IP) was defined as the buffer insoluble N fraction. The soluble true protein (TP) was separated from the NPN by precipitation with tungstic acid. NPN (fraction A) was calculated as the difference between total feed nitrogen and the N content of the residue after filtration according Licitra *et al.* (1996). The fraction B_1 (soluble true protein) was calculated as the difference between TP and IP.

Effective protein degradation (EPD) and degradation parameters (a, b, c) were determined by the *in situ* method. Three rumen fistulated cows (fed twice a day, diet 70% forage and 30% concentrate on dry matter basis) were used for 3, 6, 12, 16, 24, 48, 72 and 96 h of incubation time of samples (with a minimum of three bags per animal, incubation and feed). Pore size of bag material was 47μm. Protein degradation parameters and EPD were calculated using the Neway programme. EPD was calculated using an outflow rate of 0.06 per·h.

Results and discussion

Content of total crude protein (CP), SN and fraction A in feed samples varied considerable (Table 1). Soluble N proportion of total N was lowest in maize grain and distillers grain, followed by palm kernel cake, rapeseed meal and sunflower cake (20-28%), and highest in blue lupine, pea and field beans (60-73%). Soluble N was about 50% of total CP in oil seeds. Very high concentration of buffer soluble N and fraction A was found in lucerne silage as a result of intensive protein hydrolysis during ensiling.

The fraction B_1 (soluble true protein) varied highly in the tested feeds, and was very low in maize grain, distillers grain and DDGS (<3%) and in silages. It was high (59-64%) in legume grains (pea, field beans and blue lupine, respectively). Feeds with high proportion of SN of total N also had high crude protein degradability (EPD) *in situ*, except for rapeseed meal, sunflower cake, soybean meal and triticale.

Table 1. Crude protein solubility and degradation characteristics for selected concentrate feeds and silages (mean values).

Feed	n	CP (g/kgDM)	% of CP					
			Solubility			*In situ*		
			SN	A	B_1	0 h inc.	'a' fract.	EPD
Maize	2	100.0	15.9	13.6	2.4	27.3	28.8	47.0
Triticale	2	112.8	37.1	11.0	26.6	54.8	28.5	87.3
Pea	2	237.1	70.5	11.3	59.2	64.5	55.4	85.3
Field bean	2	300.0	59.8	12.9	46.5	68.8	53.3	87.5
Lupine (blue)	1	325.9	73.2	9.5	63.7	55.7	52.1	83.1
Linseed	1	235.6	55.8	8.1	41,7	41.5	25.5	73.9
Rapeseed	2	194.1	45.5	10.6	34.9	39.2	23.9	75.0
Linseed cake	2	375.6	48.2	8.2	36.6	24.3	4.7	62.9
Soybean meal	1	499.3	6.7	2.6	20.7	13.5	15.3	58.2
Sunflow. cake	1	317.4	28.4	8.8	19.2	33.8	21.3	69.0
Palmker. cake	1	176.4	20.4	10.4	67.0	15.5	12.8	30.4
Linseed meal	1	176.4	59.2	9.5	49.7	18.7	23.3	72.0
Rapeseed meal	1	366.6	27.0	9.9	17.1	25.5	27.5	67.1
Sunflow. meal	1	375.8	41.2	10.3	51.8	30.6	31.8	80.4
Distillers grain	2	317.5	15.6	16.5	-	45.9	54.8	52.9
DDGS	3	293.1	5.7	9.1	0.9	28.4	33.3	48.2
Maize silage	3	62.4	34.5	18.1	2.52	68.9	66.3	69.0
Lucerne silage	3	186.4	64.5	57.4	5.68	73.9	76.4	78.4

SN: buffer soluble N; A: soluble nonprotein N; B_1: soluble true protein; 0 h inc.: washing looses; 'a' fract.: immediately rumen soluble fraction; EPD: effective crude protein degradation.

Correlations between fraction A and *in situ* 'a' was high for silages (r=0.698 for all samples and for silages r=0.928). The correlation between SN and EPD was modest across all tested feeds (r=0. 630).

Conclusion

Buffer solubility only explains some of the variation in EPD. Fractionation of protein according to solubility properties as done in the CNCPS system improves the description of the soluble fraction compared with the *in situ* method.

Acknowledgements

The work has been carried out with financial support from the Commission of the European Communities, FP7, KBB-2007-1 and Ministry of Agriculture.

References

Licitra, G., T.M. Hernandez and P.J. Van Soest, 1996: Standardization of procedures for nitrogen fractionation of ruminant feeds. Anim. Feed Sci. Technol. 57, 347-358.

Rumen protein degradability in tropical grasses: Comparison of results obtained using fluorimetric and colorimetric O-phthalaldehyde assays for degradation products

S. Colombini[1], R.B. Reis[2], G.A. Broderick[3] and D.K. Combs[4]
[1]Department of Animal Science, University of Milan, 20133 Milan, Italy
[2]Universidade Federal de Minas Gerais, CNPq, Brazil
[3]Agricultural Research Service, USDA, US Dairy Forage Research Center, Madison 53706, USA
[4]Department of Dairy Science, University of Wisconsin, Madison 53706, USA

Introduction

The official NRC (2001) method to determine the rates of rumen protein degradation is based on the *in situ* technique. However, the potential contamination with microbial constituents is an inherent obstacle and source of variation associated with estimating true nutrient digestibility of feeds by the *in situ* technique, especially for forage samples (Nocek, 1988). An inhibitor *in vitro* method to determine the rate of rumen protein degradability based on the release of total amino acids (TAA) and oligopeptides has been developed and validated on concentrate feeds with satisfactory results (Colombini and Broderick, 2010). The aim of this study was to compare rates of rumen protein degradation obtained for tropical grasses in the inhibitor *in vitro* system using two different *O*-phthalaldehyde (OPA) methods: OPA-fluorimetric assay (OPA-F), which detects only free amino acids, or OPA-colorimetric assay (OPA-C), which also responds to oligopeptides.

Material and methods

Fresh forage was collected from six tropical grass species (*Brachiaria brizantha, Brachiaria hybrida, Cynodon dactylon, Panicum maximum, Pennisetum purpureum)* in various grazing systems to yield a total of 109 forage samples. Forages were incubated in a rumen *in vitro* system that had been treated with hydrazine and chloramphenicol. Samples were weighed to give a range of amounts of N (from 0.4 to 2.0 mg N/ml inoculum) and incubated for 2 hours in 50 ml centrifuge tubes with 10 ml of rumen fluid and 5 ml of buffer. Duplicate incubations were conducted with each forage. All the incubations and the mathematical computations were done as described in Colombini and Broderick (2010). Data were analyzed by mixed procedures in SAS.

Results and discussion

The average N (% DM) of the samples was 2.79% and was significantly different among species ($P<0.001$) with the lowest mean for *Brachiaria brizantha* (2.23%) and the highest for *Cynodon dactylon* (3.21%). Degradation rate was significantly higher for the OPA-C method and, consequently, the RUP was lower (Table 1). *Brachiaria* ssp. had the highest degradation rate with both the assays; however, the OPA-C method ranked differently *Brachiaria brizantha* (adj. rate=0.145/h) and *Brachiaria hybrida* (adj. rate=0.118/h) species. The rate reported by NRC (2001) for *Cynodon* hay is 0.081/h and close to the rate obtained with the OPA-C assay; however, the NRC (2001) RUP value is lower (29.8%) than that obtained in the present study (40.4). The RUP values for *Cynodon* determined by Vendramini *et al.* (2008) ranged from 38.9 to 43.3, depending on N fertilization, and were similar to the average value obtained in the present study. Regressing the adj. OPA-C rates on the adj. OPA-F rates yielded a regression coefficient of 0.81 (Figure 1) and a slope of 1.22; the more rapid rate by OPA-C is due to detection of oligopeptides formed during protein degradation in the rumen. The contribution from oligopeptides to mean degradation rate, calculated as the difference between the adj. rate determined with the OPA-C and the OPA-F assays, was 0.016/h.

Table 1. Degradation rates and rumen undegradable protein (RUP) of the samples.

Method	Sample	Rate (/h)	Adj. Rate (/h)	RUP (% CP)
OPA-C	Overall	0.207	0.099	36.5
	Brachiaria brizantha	0.271^{a}	0.145^{a}	28.3^{b}
	Brachiaria hybrida	0.240^{b}	0.118^{b}	30.7^{b}
	Cynodon dactylon	0.178^{c}	0.080^{c}	40.4^{a}
	Panicum maximum	0.181^{c}	0.071^{c}	40.9^{a}
	Pennisetum purpureum	0.164^{d}	0.082^{c}	42.0^{a}
OPA-F	Overall	0.147	0.083	39.5
	Brachiaria brizantha	0.183^{a}	0.116^{a}	31.9^{b}
	Brachiaria hybrida	0.176^{a}	0.103^{a}	33.3^{b}
	Cynodon dactylon	0.133^{b}	0.066^{b}	43.8^{a}
	Panicum maximum	0.126^{b}	0.061^{b}	43.8^{a}
	Pennisetum purpureum	0.115^{b}	0.069^{b}	44.7^{a}
SEM		0.0074	0.0052	1.32
P				
Sample		<0.001	<0.001	<0.0001
Method		<0.001	<0.001	<0.0001
Sample*Method		0.0014	0.149	0.981

[a,b,c,d] LS-Means within column for each method with different superscripts differ ($P<0.05$).

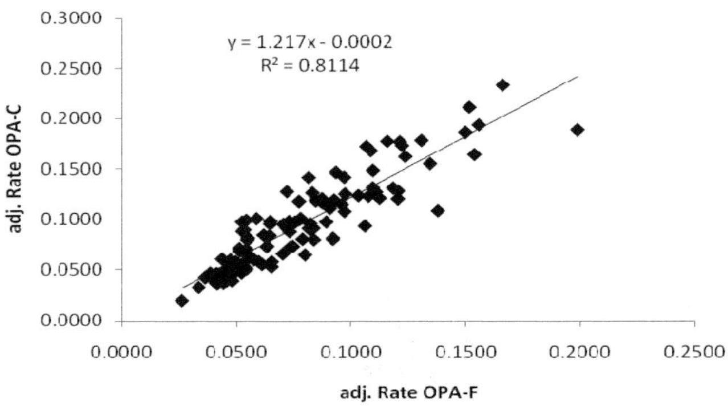

Figure 1. Relationship between the adj. degradation rates determined by colorimetric assay (OPA-C) with those determined by fluorimetric assay (OPA-F).

References

Colombini, S. and G.A. Broderick, 2010. Application of an O-phthalaldehyde assay that detects oligopeptides for estimating rumen protein degradation in vitro. In: Crovetto, G.M. (ed.) Energy and protein metabolism and nutrition. Wageningen Academic Publishers, Wageningen, the Netherlands, p. 713-714.

National Research Council, 2001. Nutrient requirements of dairy cattle. 7th rev. ed. Natl. Acad. Sci., Washington, DC.

Nocek, J.E., 1988. *In situ* and other methods to estimate ruminal protein and energy digestibility: A review. J. Dairy Sci. 71, 2051-2069.

SAS Institute, 2000. Release 8.1, USA. Cary, NC: SAS Institute Inc.

Vendramini, J.M.B., L.E. Sollenberger, A.T. Adesogan, J.C.B. Dubeux, S.M. Interrante Jr., R.L. Stewart Jr. and J.D. Arthington, 2008. Protein fractions of Tifton 85 and Rye-Ryegrass due to sward management practices. Agron. J. 100, 463-469.

Author index